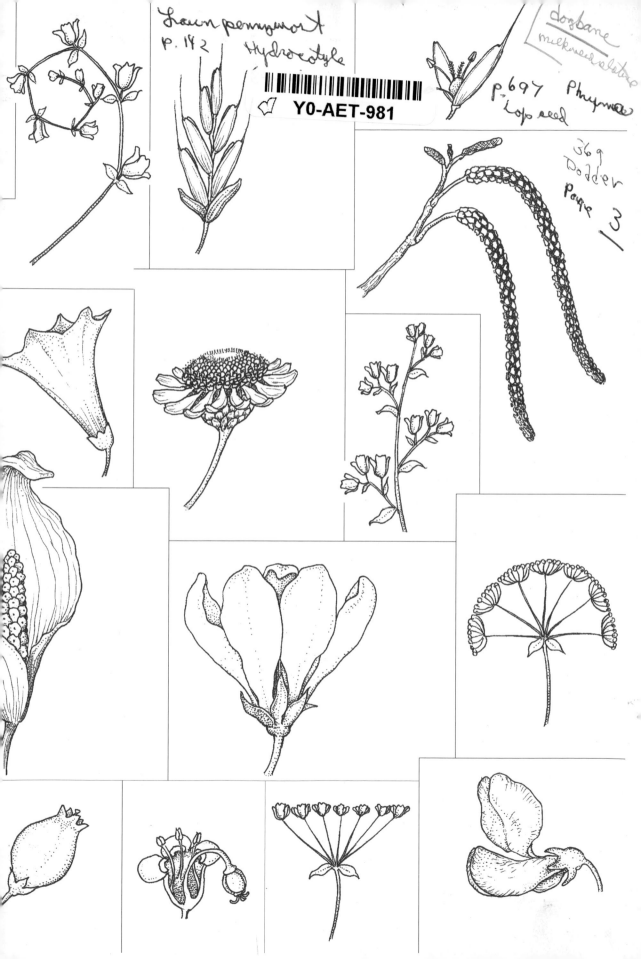

Lawn pennywort
p. 142   Hydrocotyle

Y0-AET-981

dogbane
milkweed relations

p. 697   Phymae
Lopseed

369
Dodder

Page 3

The Plants of Pennsylvania

for Lester Breininger

Best wishes

Ann Fowler Rhoads

Jan. 2, 2004

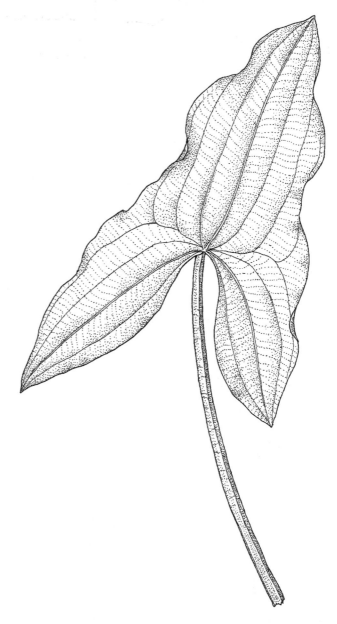

University of Pennsylvania Press   Philadelphia

THE PLANTS OF PENNSYLVANIA

An Illustrated Manual   Ann Fowler Rhoads and Timothy A. Block

Illustrations by Anna Aniśko
Morris Arboretum of the University of Pennsylvania

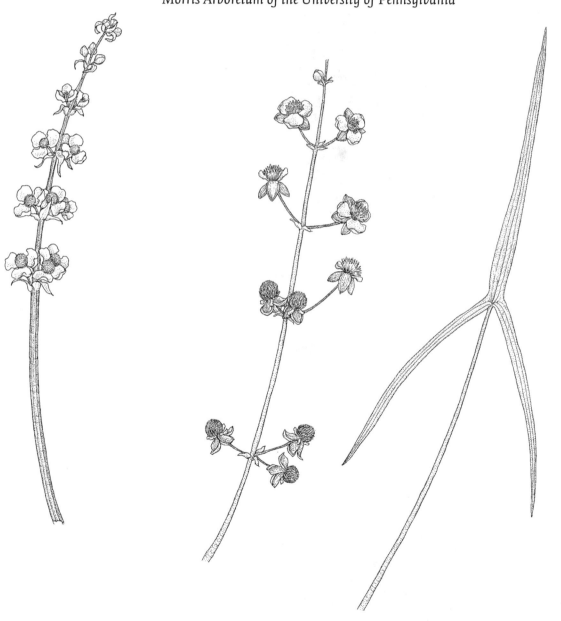

Publication of this volume was supported in part by a grant from the Pennsylvania Department of Community and Economic Development.

10 9 8 7 6 5 4 3 2 1

Published by
University of Pennsylvania Press
Philadelphia, Pennsylvania 19104-4011

Library of Congress Cataloging-in-Publication Data
Rhoads, Ann Fowler.
    The plants of Pennsylvania : an illustrated manual /
Ann Fowler Rhoads and Timothy A. Block ; illustrations by
Anna Aniśko.
        p.   cm.
    Includes bibliographic references.
    ISBN 0-8122-3535-5
        1. Botany—Pennsylvania—Classification. 2. Plants—
Identification. I. Block, Timothy A. II. Title
QK183.R56 2000
581.9748—dc21   99-089946

Title page illustration: leaf and inflorescence variation in
Sagittaria latifolia.

CONTENTS

ACKNOWLEDGMENTS

An editorial advisory board consisting of Jerry G. Chmielewski, Carl S. Keener, John Kunsman, Paul W. Meyer, James C. Parks, and Alfred E. Schuyler assisted in defining the scope of the book and setting standards for the treatments. Most advisory board members also contributed keys and descriptions for specific sections; they are acknowledged within. Many people assisted by reviewing sections of the book, testing keys, and comparing descriptions and illustrations for consistency. Especially noted are the efforts of Elizabeth Zacharias and Jim Von Culin.

Preparation of the manual was supported in part by grants from the William Penn Foundation, the McLean Contributionship, and the Wild Resources Conservation Fund.

INTRODUCTION

This book is a tool for the identification of native and naturalized plants growing in Pennsylvania. It includes all vascular plants, that is, ferns and fern allies, gymnosperms, and flowering plants. All species known to grow in the state are included, with the exception of some species of very limited occurrence that have apparently not spread beyond the early ships' ballast or garden refuse heaps on which they were initially collected. An effort was made to keep the language simple and direct to make the book useful to amateurs as well as more experienced botanists. Keys and descriptions are based on specimens collected in Pennsylvania as were flowering and fruiting dates. Illustrations were drawn from live material or herbarium specimens collected in Pennsylvania. Lack of space prohibited illustrating every plant; instead, an effort was made to show features that assist in identification. All illustrations are at ½ life size unless otherwise noted.

FORMAT

Plants are grouped in four categories: ferns and fern allies, gymnosperms, dicots, and monocots. In each category, families are arranged alphabetically, as are the genera and species within them. The only exception is the genus *Carex*, in which species are grouped in sections within each of the two subgenera. The sections and the species within them are arranged alphabetically. Family definitions follow Cronquist (1981) with minor exceptions. Family descriptions reflect the family as it is represented in Pennsylvania, as do genus descriptions. No attempt was made to include complete morphological descriptions of species; rather, emphasis has been placed on characteristics useful in confirming an identification. For the most part, species descriptions do not repeat characteristics common to the whole family or genus.

The keys use characteristics that can be seen using, at most, a 10× hand lens. They are artificial in the sense that there is no attempt to depict taxonomic relationships. The keys to woody plants are based mainly on vegetative characteristics; both summer and winter keys are provided where appropriate. Keys to herbaceous plants rely more heavily on characteristics of the flowers and fruits. More specific information about the distribution of each plant described in this volume may be obtained by consulting *The Vascular Flora of Pennsylvania: Annotated Checklist and Atlas* (Rhoads and Klein 1993). Unless otherwise specified, plants are native to Pennsylvania; geographic origin is noted for those not indigenous to the state.

The majority of keys and descriptions are the work of the authors, who take full responsibility for editorial decisions regarding the handling of problematic taxa. The contributors of other sections are noted as appropriate.

HOW TO USE THE MANUAL

The keys in this book are designed to guide the user through a series of decisions leading to a proper identification. At each step of the key, the user must examine a characteristic or characteristics of the plant and decide which of a pair of corresponding statements best fits the specimen in question. Each statement and its corresponding member are preceded by the same capital letter(s). There are a few simple rules that, if followed carefully, should ease the task of arriving at a correct identification.

1. Read each statement of a pair carefully and thoroughly before deciding which statement best fits the plant you are looking at.
2. Be sure you understand any botanical terminology used in the statements before moving on to the next step. Consult the illustrated glossary if you are not certain.
3. Look at more than one example of the characteristic being asked about. For example, if the key asks you to determine the number of petals present in a flower, examine more than one flower.
4. Don't dwell on the exceptions! Most plant species are quite variable in their characteristics. Throughout the keys and descriptions you will see qualifiers such as "often," "usually," "mostly" to account for this variability.
5. After reaching a tentative identification, read the family, genus, and species descriptions and look at the illustrations (if appropriate) to confirm that you have come to the proper conclusion.
6. Make your observations carefully and thoroughly. Take your time, especially if you are new to "keying out" plants.

The initial key(s) will lead you to a family (or a genus within a family). Under each family heading is a brief description of that family followed by a key to the genera within. Each genus heading is followed by a description and a key to the species within. The "keying out" process isn't finished until you reach a species identification. Both scientific and common names of species are given along with a brief description that provides information useful for confirming an identification.

You will find a small metric ruler and a 10× magnifier indispensable. A pocket-size notebook is also handy for recording your observations. As with any skill, the accuracy of plant identification improves with practice. While every effort has been made to reduce the amount of technical "jargon" in this book, there simply are no substitutes for many specific botanical terms. You will find, however, that you quickly develop a comfortable familiarity with most of these terms with regular use.

SYMBOLS AND ABBREVIATIONS

U.S. Fish and Wildlife Service wetland codes

code		probability of occurring in wetlands under natural conditions
OBL	obligate wetland species	99%
FACW	facultative wetland species	67–99%
FAC	facultative species	34–66%
FACU	facultative upland species	1–33%
UPL	upland species	1%

Endangered and threatened species

🌿 plants listed as endangered or threatened under the Federal Endangered Species Act or listed as endangered, threatened, rare, vulnerable or undetermined by the Pennsylvania Natural Diversity Inventory (DCNR 1987).

Other abbreviations and symbols

flr. flowering time
frt. fruiting time
syn: indicates a synonymous name

Geographic regions

Figure 1 shows the six geographic regions used in most of the species descriptions to indicate where each plant grows in the state. In addition to these six regions, the descriptions occasionally include the abbreviations W, E, N, S, and C to refer to the western, eastern, northern, southern, or central third or half of the state. For plants of very limited distribution, individual counties are sometimes listed (see Figure 2).

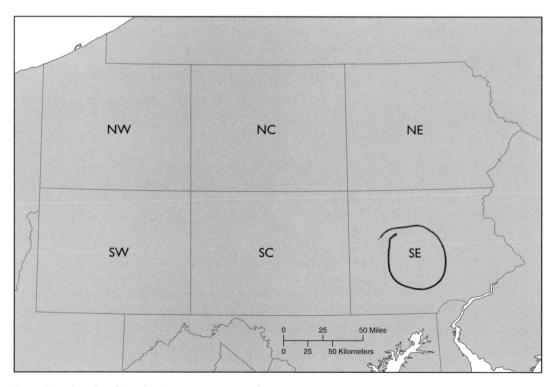

Figure 1. Geographic regions of Pennsylvania.

Figure 2. Counties of Pennsylvania.

COLLECTING AND PRESERVING PLANT SPECIMENS

Our knowledge of the plants of Pennsylvania is derived from the accumulated record of botanical exploration built up in museum collections of herbarium (dried plant) specimens. These collections reflect early activity (in the 1800s) and continue to be added to to document changes in the flora or new discoveries. Major herbarium collections in Pennsylvania are located at the Academy of Natural Sciences of Philadelphia, the Carnegie Museum of Natural History in Pittsburgh, Pennsylvania State University, and the State Museum in Harrisburg. Numerous smaller, more regional collections exist at colleges and universities throughout the state.

It is important to continue to record changes in the flora. When species new to the state or to a county are found, specimens should be deposited in a recognized herbarium, unless the plant is so rare that collecting a specimen would endanger its continued existence in the wild. Specimens should be pressed flat between sheets of newspaper or other absorbent paper until they are thoroughly dry. A plant press or other weights may be used to flatten the specimens. In cases of extreme rarity, a portion of the plant or even a photograph may be substituted for a more complete specimen. In any case, full details of date, location, collector, and habitat where the plant was growing should be included on the accompanying label. Standard herbarium mounts are on acid-free paper, 11½ by 16½ inches, with a label affixed in the lower right corner. Of course, plant collections should be made only with the permission of the landowner or a permit from the appropriate managing agency in the case of public land.

BRIEF FLORISTIC DESCRIPTION OF PENNSYLVANIA

Pennsylvania lies within the eastern deciduous forest region as described by Braun (1950). Northern hardwood forests dominate in the northern third of the state and extend south along the Allegheny Front. The southern two-thirds of the state is primarily Appalachian oak forest. A recently completed classification of plant communities of Pennsylvania (Fike, 1999) lists 105 distinct types ranging from terrestrial forests to palustrine community types.

Floristic diversity is greatly increased by assemblages of rare species associated with specialized geological substrates such as serpentinite, diabase, limestone, glacial till, and peat (Figures 3 and 4). A diversity of physiographic provinces is represented also (Figure 5). Along Lake Erie, the Central Lowlands Province contains elements of the Great Lakes flora. In the southeastern corner of the state, the Atlantic Coastal Plain is represented by a narrow sliver along the Delaware Estuary. Many species more abundant to the south and east are represented in this part of the state, and many formerly found here have been lost due to the concentration of human activity.

Within the Piedmont Province, the lower Susquehanna River valley contains a significant number of southern species at their northern limit of range. Similarly, the Appalachian Plateaus Province, especially the glaciated portions in the northeast and northwest, is home to many northern plants at their southern limit of range.

A total of approximately 3400 different kinds of plants have been found growing spontaneously in Pennsylvania. Two-thirds of them are believed to be native.

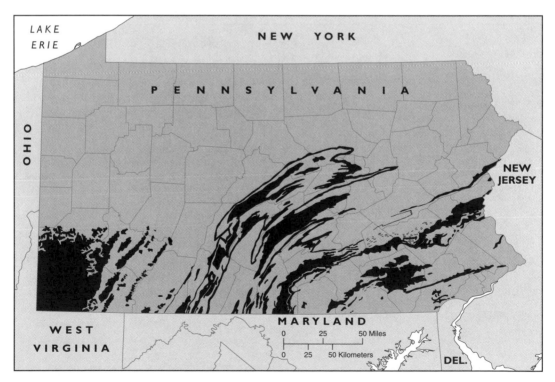

Figure 3. Limestone and dolomite areas. Adapted from Map 15, Limestone and Dolomite Distribution in Pennsylvania. Commonwealth of Pennsylvania Department of Environmental Resources, Bureau of Topographic and Geologic Survey.

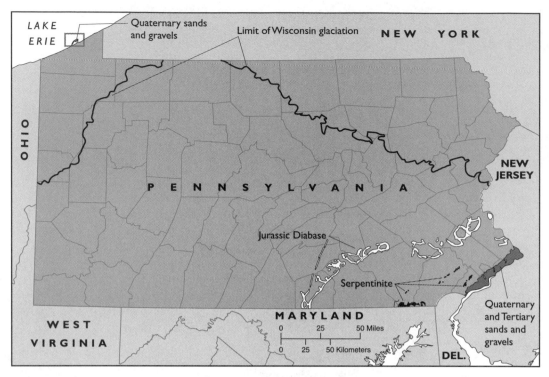

Figure 4. Selected geologic features. Adapted from Map 7, Geologic Map of Pennsylvania, and Map 59, Glacial deposits of Pennsylvania. Commonwealth of Pennsylvania Department of Environmental Resources, Bureau of Topographic and Geologic Survey.

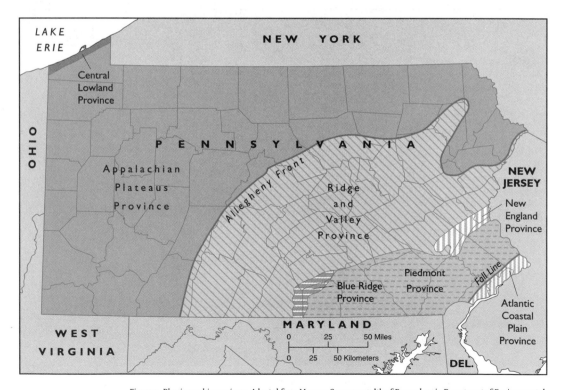

Figure 5. Physiographic provinces. Adapted from Map 13, Commonwealth of Pennsylvania Department of Environmental Resources, Bureau of Topographic and Geologic Survey.

The remainder have arrived since the period of earliest European settlement. A total of 116 species that once grew here naturally have apparently been extirpated from the state.

ENDANGERED AND THREATENED SPECIES PROGRAMS

The designation of endangered, threatened, and rare plants of Pennsylvania and their protection began with the passage of P.L. 597, No. 170 (32 P.S. §§ 5301–5314), the Wild Resource Conservation Act of 1982. The regulations implementing the act have since been amended to update the lists of classified species.

Plants listed as endangered or threatened receive official protection; those classified as rare, vulnerable, or undetermined do not, with the exception of ginseng (*Panax quinquefolius*), which is covered by CITES (Conference on International Trade in Endangered Species). The program is administered by the Bureau of Forestry of the Pennsylvania Department of Conservation and Natural Resources in cooperation with The Nature Conservancy and the Western Pennsylvania Conservancy.

Plant lists are reviewed annually by the Rare Plant Forum and the Vascular Plants Technical Committee of the Pennsylvania Biological Survey and recommendations are made for updates as needed. Any discoveries of classified species should be reported to the Pennsylvania Natural Diversity Inventory, Bureau of Forestry, Department of Conservation and Natural Resources, PO Box 8552, Harrisburg, PA 17105-8552.

GENERAL KEYS TO FAMILIES

MASTER KEY

Q. leaves parallel-veined; flower parts mostly in 3s .... Key 15, p. 39

Q. leaves with reticulate venation; flower parts numerous or in 2s, 4s, or 5s

    R. plants with at least some unisexual flowers .... Key 16, p. 41

    R. flowers perfect

        S. flowers with 0–1 perianth whorls .... Key 17, p. 44

        S. flowers with both sepals and petals

            T. plants with >1 ovary per flower .... Key 18, p. 47

            T. plants with only 1 ovary per flower (a single carpel or 2 or more fused or partly fused carpels)

                U. ovary inferior .... Key 19, p. 48

                U. ovary superior

                    V. stamens more numerous than the petals or corolla lobes .... Key 20, p. 49

                    V. stamens same number as or fewer than the petals or corolla lobes

                        W. petals distinct .... Key 21, p. 53

                        W. petals united

                            X. corolla regular

                              Y. stamens as many as the corolla lobes ....Key 22, p. 54

                              Y. stamens fewer than the corolla lobes ....Key 23, p. 56

                          X. corolla irregular ....Key 23, p. 56

SECTIONAL KEYS

Key 1. Herbaceous plants that lack green color and/or well-developed stems and/or leaves at flowering time

A. plants lacking green color

    B. plants not rooted in the ground, parasitic and attached to the stems of other plants at maturity

        C. stems slender, elongate, bright orange to yellowish, sprawling or twining .... Cuscutaceae, p. 369

        C. stems 5–15 mm, stiff, peg-like protrusions from the twigs of *Picea* .... *Arceuthobium* in Viscaceae, p. 710

    B. plants rooted in the ground

        D. flowers in a dense fleshy spike; leaves reduced to scales or appearing after flowering

            E. spike surrounded by a hood-like spathe; scale-like leaves not present below the inflorescences; all parts of the plant with a skunky odor.... .... *Symplocarpus* in Araceae, p. 724

            E. spike not enclosed by a spathe; scale-like leaves present below the inflorescence; plants lacking a skunky odor .... *Conopholis* in Orobanchaceae, p. 526

        D. flowers not in a dense fleshy spike; leaves preceding flowering, reduced to scales, or absent

            F. flowers in an umbel on a naked stem; plants with an onion odor..... .... *Allium* in Liliaceae, p. 858

            F. flowers solitary or in a spike, raceme or cluster, but not in an umbel; plants not smelling of onion

G. flowers regular
   H. stems 4–6 mm thick, translucent white, pink, or reddish, not at all green .... Monotropaceae, p. 506
   H. stems slender, wiry, greenish
      I. flowers yellow, 5-merous.....*Hypericum* in Clusiaceae, p. 353
      I. flowers greenish-white, 4-merous .... *Bartonia* in Gentianaceae, p. 436
G. flowers slightly to strongly irregular
   J. flowers 3-merous; ovary inferior .... Orchidaceae, p. 875
   J. flowers 4- or 5-merous; ovary superior .... Orobanchaceae, p. 526

A. plants green
  K. floating or submergent aquatic plants
    L. plants floating, tiny, without distinct leaves or stems .... Lemnaceae, p. 854
    L. plants submergent, rooted in the substrate or attached to rocks
      M. plants branching, strap-like; attached to rocks .... Podostemaceae, p. 537
      M. plants erect, rooted in the bottom, leaves reduced to tiny scales .... *Myriophyllum* in Haloragaceae, p. 448
  K. terrestrial plants
    N. stems spiny and flattened; leaves not present .... Cactaceae, p. 299
    N. stems not flattened or spiny; leaves scale-like
      O. flowers in a dense terminal head surrounded by an involucre .... *Tussilago* in Asteraceae, p. 250
      O. flowers not in a dense head
         P. flowers yellow, 5-merous; leaves opposite .... *Hypericum* in Clusiaceae, p. 353
         P. flowers greenish-white, 4-merous, leaves opposite or alternate.... .... *Bartonia* in Gentianaceae, p. 436

Key 2. Submersed or floating-leaf aquatic plants

A. plants without distinct stems or leaves
  B. plants tiny, free-floating, solitary or in small colonies
    C. individual plants tiny, elliptical, solitary, in small clusters, or attached by strap-like stalks .... Lemnaceae, p. 854
    C. forming small dichotomously branching colonies .... *Riccia* (an aquatic liverwort, non-vascular)
  B. plants attached to rocks in fast-flowing water; branching, thallus-like...... .... Podostemaceae, p. 537
A. plants with distinct stems and/or leaves, usually rooted in the bottom (floating fragments may also be present)
  D. leaves not apparent, very similar to the stems in texture, or reduced to minute bracts or scales
    E. stems with the appearance of whorled branches
      F. vascular tissues lacking; spiral oogonia present in the axils .... *Chara* and *Nitella* (non-vascular green algae)
      F. vascular tissues present; spiral oogonia not present
         G. whorled appearance due to a pair of narrow opposite leaves with expanded bases and narrowed tips, and leafy axillary shoots .... Najadaceae, p. 874

G. whorled appearance due to leaves with dichotomously divided fili-
form segments ....Ceratophyllaceae, p. 343
E. stems simple or branching, but not whorled
H. stems simple, erect, arising singly from a slender rhizome and forming
a loose underwater turf; leaves merely tiny scales..............................
.... *Myriophyllum* in Haloragaceae, p. 448
H. stems flexuous, branching; leaves present but so slender they may ap-
pear as stems
I. leaves alternate; flowers and fruits in a terminal spike or raceme.....
.... Potamogetonaceae, p. 988
I. leaves opposite; flowers and fruits in small axillary clusters ....
....Zannichelliaceae, p. 1002
D. obvious leaves present
J. floating leaves present
K. leaves compound, shaped like a 4-leaf clover .... *Marsilea*, p. 93
K. leaves simple
L. floating leaf blades peltate or with a deep basal notch or sinus
M. leaves peltate (petiole attached centrally)
N. leaf blade elliptic, 4–12 cm long and half as wide ....
.... *Brasenia* in Cabombaceae, p. 298
N. leaf blade more or less circular
O. leaf blade 3–7 dm wide; flowers solitary ....
.... Nelumbonaceae, p. 510
O. leaf blade 3–6 cm in diameter; flowers in an umbel
P. umbel axillary; flowers small, whitish ....
.... *Hydrocotyle* in Apiaceae, p. 142
P. umbel subtended by a pair of opposite leaves; flowers
showy, yellow ..... *Nymphoides* in Menyanthaceae, p. 505
M. leaves not peltate, petiole attached at the summit of a deep notch
or sinus
Q. leaves cordate
R. leaf blade <5 cm wide, margins crenate ....
.... *Hydrocotyle* in Apiaceae, p. 142
R. leaf blade >5 cm wide, margins entire
S. flowers in an umbel subtended by one or more leaves...
.... *Nymphoides* in Menyanthaceae, p. 505
S. flowers solitary on long, naked peduncles ....
.... *Nuphar* in Nymphaeaceae, p. 511
Q. leaves not cordate
T. leaves with divergent basal lobes (sagittate) ....
.... *Sagittaria* in Alismataceae, p. 718
T. leaves with a narrow basal sinus, but not sagittate or cor-
date
U. leaves nearly round with a deep, narrow basal sinus;
flowers white or pink ....
.... *Nymphaea* in Nymphaeaceae, p. 512
U. leaves longer than wide with a more open sinus; flowers
yellow....................... *Nuphar* in Nymphaeaceae, p. 511
L. floating leaves unlobed or at the most subcordate at the base, nei-
ther peltate nor deeply notched

V. floating leaves crowded in a terminal rosette at the tips of stems; submersed leaves opposite or alternate

    W. leaf blades <1 cm long; petioles not swollen .... Callitrichaceae, p. 302

    W. leaf blades >1 cm long; petioles swollen in the middle .... Trapaceae, p. 689

V. floating leaves not in a rosette; submersed leaves alternate or basal

    X. leaves narrow and ribbon-like, >20 times as long as wide, without a distinct petiole

        Y. leaves mostly basal, with the ends floating, rounded at the tip

            Z. leaves without a midvein; flowers (and fruits) in 1 or more dense, spherical heads ..... Sparganiaceae. p. 999

            Z. leaves with a distinct midvein; flowers in whorls of 3 ... . *Sagittaria* in Alismataceae, p. 718

        Y. leaves alternate; tips of the blades acute .... *Glyceria* and *Leersia* in Poaceae, p. 933, 939

    X. at least the floating leaves with an expanded elliptical blade <20 times as long as wide; petiole present

        AA. veins of the leaf parallel; leaves all cauline with submersed and floating leaves differing in shape or all leaves basal

            BB. leaves all cauline; inflorescence a spike or raceme of greenish or reddish flowers .... Potamogetonaceae, p. 988

            BB. leaves all basal, sometimes partially emergent; inflorescence a dense, elongate spike of yellow flowers .... *Orontium* in Araceae, p. 722

        AA. leaf venation netted; all leaves cauline; submersed and floating leaves similar

            CC. flowers pink, in an erect, terminal spike or raceme .... *Polygonum* in Polygonaceae, p. 544

            CC. flowers yellow, solitary in the leaf axils .... *Ludwigia* in Onagraceae, p. 520

J. floating leaves absent, plants entirely submersed or the inflorescence and a few terminal leaves extending above the water surface

  DD. leaves with filiform segments bearing tiny stalked bladders .... Lentibulariaceae, p. 488

DD. leaves various, lacking bladders

    EE. leaves all basal

      FF. leaves flat

        GG. leaves <20 times as long as wide, firm

           HH. leaves 1–2 mm wide and 2–5 cm long, in pairs or clusters along slender stolons; plants mat-forming; flowers solitary on pedicels arising from the nodes of the stolons ...... Scrophulariaceae, p. 661

           HH. leaves wider and/or longer, in a short, stiff, basal rosette; plants not mat-forming; flowers in a terminal head or raceme

II.   leaves rounded at the tip; inflorescence a few-flow-ered raceme .... *Lobelia* in Campanulaceae, p. 307

II.   leaves acute at the tip; flowers in a dense white head . ... Eriocaulaceae, p. 837

GG. leaves >20 times as long as wide, flexuous, ribbon-like, with the ends often floating

    JJ.   midvein no more prominent than the other veins of the leaf, tiny cross veins giving a uniformly checkered appear-ance throughout; flowers in 1 or more dense, spherical heads .... Sparganiaceae, p. 999

    JJ.   midvein more prominent than the other veins of the leaf; flowers not in dense spherical heads

        KK. the central third of the leaves more densely reticulate than the edges; plants without milky sap; flowers soli-tary on long, slender pedicels .... *Vallisneria* in Hydrocharitaceae, p. 839

        KK. leaves uniform in the intensity of reticulate venation; plants often with milky sap; flowers in a whorled in-florescence .... *Sagittaria* in Alismataceae, p. 718

FF.   leaves (or sterile stems) filiform or terete, not strongly flattened, elon-gate and limp to short and quill-like or greatly reduced

    LL.   leaves (or sterile stems) stiff or firm, holding their shape out of water

        MM. leaves (or sterile stems) tufted, forming a basal rosette

            NN. leaf bases expanded and containing sporangia .... Isoetaceae, p. 62

            NN. leaf (or stem) bases not bearing sporangia .... *Eleocharis* in Cyperaceae, p. 816

        MM. leaves (or stems) neither tufted nor forming a basal rosette

            OO. flowering stems with a single, terminal spikelet and a bladeless basal sheath .... *Eleocharis* in Cyperaceae, p. 816

            OO. flowering stems with a diffuse panicle >half the height of the entire plant....*Juncus* in Juncaceae, p. 843

    LL.   leaves (or sterile stems) limp, not firm or rigid out of water (flowering stems may be more rigid and shortly emergent)

        PP.   sterile stems terete or filamentous, each with a bladeless, tubular sheath at the base; flowering stems with a single terminal spikelet .... *Eleocharis* in Cyperaceae, p. 816

        PP.   basal leaves filamentous; flowering stem lacking a basal sheath and bearing a single lateral spikelet .... *Schoenoplectus* in Cyperaceae, p. 827

EE. leaves all or mostly cauline

    QQ. leaves compound or deeply dissected (simpler emergent leaves may also be present)

        RR. leaves bearing tiny bladders .... Lentibulariaceae, p. 488

        RR. leaves without bladders

            SS. leaves opposite or whorled

                TT. leaves pinnately divided .... *Myriophyllum* in Haloragaceae, p. 448

                TT. leaves fan-like, the filamentous segments repeatedly forking

UU. leaves remaining rigid out of water, crowded toward the stem tips ....Ceratophyllaceae, p. 343

UU. leaves limp out of water, not crowded toward the stem tips

VV. emergent leaves not present; submersed leaves opposite, with a distinct petiole; flowers white, solitary ....
.... *Cabomba* in Cabombaceae, p. 299

VV. emergent leaves simple, coarsely toothed; submersed leaves sessile; flowers in a composite head with conspicuous yellow rays ....
.... *Megalodonta* in Asteraceae, p. 230

SS. leaves alternate or scattered

WW. leaves pinnately or bipinnately divided

XX. flowers in whorls on inflated terminal peduncles ......
.... *Hottonia* in Primulaceae, p. 560

XX. flowers axillary or in umbels; peduncles not inflated

YY. submersed leaves once pinnate; emersed leaves simple; base of the petiole not expanded or sheathing....*Proserpinaca* in Haloragaceae, p. 450

YY. submersed leaves pinnate to bipinnate; emersed leaves once pinnate; petiole with an expanded, sheathing base .... *Sium* in Apiaceae, p.146

WW. leaves fan-like, the filamentous segments repeatedly forking .... *Ranunculus* in Ranunculaceae, p. 577

QQ. leaves simple, unlobed, but sometimes toothed

ZZ. leaves reduced to tiny, scattered scale-like bracts ....
.... *Myriophyllum* in Haloragaceae, p.448

ZZ. leaves not scale-like

AAA. leaves alternate, with or without stipules

BBB. leaves kidney-shaped, the largest 4–5 cm wide ....
.... *Heteranthera* in Pontederiaceae, p. 987

BBB. leaves elliptic, lanceolate or linear, the leaf bases sometimes cordate or auriculate

CCC. leaves with a basal sheath or sheathing stipules variously fused to the leaf base or free; flowers in a terminal spike .... Potamogetonaceae, p. 988

CCC. sheath or stipules absent; flowers axillary

DDD. leaves linear, entire, 2–6 mm wide and to 15 cm long....*Zosterella* in Pontederiaceae, p. 988

DDD. leaves lanceolate, shallowly dentate, 1.5–3 cm long ....*Lobelia* in Campanulaceae, p. 307

AAA. leaves opposite or whorled, without stipules, but sometimes with a broader, sheathing base

EEE. leaves not over 0.5 mm wide ....Zannichelliaceae, p. 1002

EEE. leaves >0.5 mm wide

FFF. leaves whorled

GGG. leaves linear or tapered to the tip; stem terete; plants completely submergent unless stranded due to receding water levels ....
.... Hydrocharitaceae, p. 837

GGG. leaves linear to narrowly oblanceolate; stem 4-angled; plants rarely completely submergent .... *Galium* in Rubiaceae, p. 631

FFF. leaves opposite

HHH. leaves with a narrow blade and a broader, sheathing base; flowers axillary .... Najadaceae, p. 874

HHH. leaves tapering or truncate at the base, broadest in the blade; flowers various

III. leaves 1 cm or less long; flowers and fruits sessile or on short stalks in the leaf axils

JJJ. leaves linear, spatulate or obovate, weakly 3-nerved, bidentate at the apex .... Callitrichaceae, p. 302

JJJ. leaves oblong to narrowly obovate, 0–1-nerved, not bidentate at the apex

KKK. ovary formed of 2–3 fused carpels; fruit a capsule .... Elatinaceae, p. 375

KKK. carpels distinct or nearly so; fruit a cluster of follicles .... *Crassula* in Crassulaceae, p. 365

III. leaves >1 cm long; flowers in axillary or terminal inflorescences

LLL. flowers irregular .... *Gratiola* in Scrophulariaceae, p. 668

LLL. flowers regular or nearly so

MMM. ovary inferior ... *Ludwigia* in Onagraceae, p. 520

MMM. ovary superior

NNN. flowers blue or white .... *Veronica* in Scrophulariaceae, p. 676

NNN. flowers yellow

OOO. stems strongly erect; leaves with an evident midvein; leaves and stem with red or black dots ....*Lysimachia* in Primulaceae, p. 560

OOO. stem lax; leaves 3-nerved, without dark dots .... *Hypericum* in Clusiaceae, p. 353

Key 3. Woody angiosperms with flowers appearing before the leaves

A. leaf scars opposite or whorled

B. both calyx and corolla present

C. petals separate

D. ovary superior, 2-lobed .... Aceraceae, p. 118

D. ovary inferior, not lobed .... *Viburnum* in Caprifoliaceae, p. 318

C. petals fused, at least near the base

E. ovary inferior .... *Lonicera* in Caprifoliaceae, p. 312

E. ovary superior

F. trees; flowers purple.............. *Paulownia* in Bignoniaceae, p. 262

F. shrubs; flowers yellow .... *Forsythia* in Oleaceae, p. 513

B. one or both perianth whorls absent

G. flowers staminate or perfect

H. stamens 2 or 4

I. stamens 2 (rarely 4)... Oleaceae, p. 512

I. stamens 4 .... *Cornus* in Cornaceae, p. 362

H. stamens 5 or more

J. stamens 5–10, usually 8
  K. shrubs; sepals or calyx lobes 4, spreading ....
  .... *Shepherdia* in Elaeagnaceae, p. 374
  K. trees; sepals or calyx lobes usually 5, erect ....
  .... Aceraceae, p. 118
J. stamens 15–20 .... Cercidiphyllaceae, p. 344
G. flowers pistillate
  L. sepals borne on a well developed hypanthium ....
  .... *Shepherdia* in Elaeagnaceae, p. 374
  L. sepals not on a hypanthium
    M. styles 2 and distinct, or 1 with 2 stigma lobes
      N. ovary 2-lobed .... Aceraceae, p. 118
      N. ovary not lobed .... Oleaceae, p. 512
    M. styles 3–5 .... Cercidiphyllaceae, p. 344
A. leaf scars alternate
  O. flowers perfect
    P. both sepals and petals present
      Q. plant flowering in the fall ....
      .... *Hamamelis virginiana* in Hamamelidaceae, p. 451
      Q. plants flowering in the spring
        R. flowers tightly clustered, subtended by showy white to pinkish
           bracts .... *Cornus florida* in Cornaceae, p. 364
        R. flowers not tightly clustered, not subtended by showy bracts
          S. petals not fused
            T. stamens >20
              U. flowers white to pinkish or purplish ....
              .... *Magnolia* in Magnoliaceae, p. 498
              U. flowers purplish-brown .... Annonaceae, p. 133
            T. stamens usually 20 or fewer
              V. petals shorter than the sepals
                W. trees .... *Ulmus* in Ulmaceae, p. 690
                W. shrubs
                  X. stamens 5 (rarely 4) ....
                  .... *Ribes* in Grossulariaceae, p. 445
                  X. stamens 6 .... *Berberis* in Berberidaceae, p. 254
              V. petals longer than the sepals; stamens usually about 20
                Y. style 1
                  Z. bark bearing prominent horizontal lenticels ....
                  .... *Prunus* in Rosaceae, p. 613
                  Z. bark not bearing prominent horizontal lenticels...
                  .... *Pyrus* in Rosaceae, p. 617
                Y. styles 5
                  AA. flowers <3.5 cm across ....
                  .... *Amelanchier* in Rosaceae, p. 596
                  AA. flowers 3.5–5 cm across ....
                  .... *Chaenomeles* in Rosaceae, p. 600
          S. petals fused at least toward the base
            BB. corolla irregular
              CC. shrubs or small trees
                DD. low shrub .... *Cytisus* in Fabaceae, p. 403
                DD. small tree .... *Cercis* in Caesalpiniaceae, p. 299

CC. vines .... *Wisteria* in Fabaceae, p. 423

BB. corolla regular, or nearly so .... *Rhododendron* in Ericaceae, p. 384

P. one or both perianth whorls absent

EE. sepals and petals both absent .... *Ulmus* in Ulmaceae, p. 690

EE. sepals present (sometimes very small)

FF. flowers with a narrow funnelform hyphanthium ....
.... Thymelaeaceae, p. 687

FF. flowers without a hyphanthium ....
.... *Xanthorhiza* in Ranunculaceae, p. 588

O. flowers unisexual

GG. plants monoecious

HH. staminate flowers in catkins

II. pistillate flowers in cone-like catkins .... Betulaceae, p. 256

II. pistillate flowers solitary or in small clusters

JJ. pistillate flowers terminal on the twigs ....
.... Juglandaceae, p. 460

JJ. pistillate flowers axillary on the twigs...Fagaceae, p. 424

HH. staminate flowers not in catkins

KK. staminate flowers in solitary spherical heads ....
.... Platanaceae, p. 536

KK. staminate flowers in racemes of ovoid heads ....
.... *Liquidambar* in Hamamelidaceae, p. 452

GG. plants dioecious

LL. staminate and/or pistillate flowers in catkins

MM. twigs aromatic when crushed

NN. staminate catkins mostly >1.5 cm long; pistillate catkins
subglobose, bur-like in fruit ....
.... *Comptonia* in Myricaceae, p. 509

NN. staminate catkins mostly >1.5 cm long; pistillate catkins
cylindric, 5–10 mm long, not bur-like in fruit ....
.... *Myrica* in Myricaceae, p. 509

MM. twigs not aromatic when crushed

OO. inner bark of twigs bright orange ....
.... *Maclura* in Moraceae, p. 508

OO. inner bark of twigs not bright orange ....
.... Salicaceae, p. 641

LL. flowers not in catkins

PP. twigs aromatic when crushed .... Lauraceae, p. 487

PP. twigs not aromatic when crushed ....
.... *Maclura* in Moraceae, p. 508

Key 4. Woody gymnosperms with leaves needle- or scale-like, or fan-shaped

A. leaves needle- or scale-like

B. evergreen shrubs with flattened needle-like leaves, green (not white-
striped) on the underside; seeds borne singly, surrounded by a fleshy red
aril when mature .... Taxaceae, p. 116

B. evergreen (or deciduous) trees or shrubs with needle-like or scale-like
leaves; seeds borne in a woody or fleshy cone, without an aril

C. evergreen trees or shrubs; leaves opposite or in whorls of three, appressed and scalelike or subulate and spreading; cone woody or fleshy,
<2 cm long .... Cupressaceae, p. 107

C. evergreen or deciduous trees; needle-like leaves alternate, in fascicles or crowded on short lateral spurs; cone woody, mostly >2 cm long ....
.... Pinaceae, p. 110
A. leaves fan-shaped .... Ginkgoaceae, p. 109

Key 5. Woody angiosperms with opposite or whorled, compound leaves

A. leaves palmately compound
B. shrubs; leaflets short-stalked, leaflet margins entire ....
.... *Vitex* in Verbenaceae, p. 700
B. shrubs to large trees; leaflets sessile, leaflet margins variously toothed .....
.... Hippocastanaceae, p. 452
A. leaves pinnately compound
C. leaves trifoliate (occasionally with some leaves having 5 leaflets)
D. vines .... *Clematis* in Ranunculaceae, p. 573
D. trees or shrubs
E. shrub; fruit a bladdery capsule; leaflet margins finely toothed; twigs and young branches brown .... Staphyleaceae, p. 686
E. large shrub or tree; fruit a double samara; leaflet margins coarsely toothed or lobed; twigs and young branches green ....
.... *Acer negundo* in Aceraceae, p. 120
C. leaves with 5 or more leaflets
F. vines
G. leaflets toothed .... *Campsis radicans* in Bignoniaceae, p. 262
G. leaflets entire to somewhat undulate ....
.... *Clematis* in Ranunculaceae, p. 573
F. shrubs or trees
H. branches remaining green for 3– 4 years; often with some leaves trifoliate .... *Acer negundo* in Aceraceae, p. 120
H. 3- to 4-year-old branches gray or brown; all leaves with 5 or more leaflets
I. shrubs .... *Sambucus* in Caprifoliaceae, p. 316
I. trees
J. terminal bud lacking; lateral buds hidden by leaf base ....
.... *Phellodendron* in Rutaceae, p. 639
J. terminal bud present; lateral buds not hidden by leaf base
K. crushed leaves with a foul odor; fruit a dehiscent capsule....
.... *Euodia* in Rutaceae, p. 638
K. crushed leaves without a foul odor; fruit a single samara ...
.... *Fraxinus* in Oleaceae, p. 514

Key 6. Woody angiosperms with opposite or whorled, simple leaves

A. leaves entire
B. leaves 2–4 dm long, often whorled .... Bignoniaceae, p. 261
B. leaves smaller, rarely >2 dm
C. at least some leaves whorled; small shrubs to 1 m ....
.... *Kalmia* in Ericaceae, p. 380
C. leaves opposite (rarely whorled); trees, shrubs, or vines
D. plants obligately parasitic, growing attached to and not separable from the branches of a host plant .... Viscaceae, p. 710
D. plants not obviously parasitic, not attached to another plant

E. leaves densely silvery-downy and scurfy with rusty scales below ....
.... *Shepherdia* in Elaeagnaceae, p. 374
E. leaves glabrous or pubescent but not scurfy with rusty scales
   F. leaves with punctate dots apparent when held up to light ....
   .... *Hypericum* in Clusiaceae, p. 353
   F. leaves without punctate dots
      G. bases of the petioles meeting or joined by stipules, or stipular scars completely encircling twigs
         H. vines .... *Lonicera* in Caprifoliaceae, p. 312
         H. trees or shrubs
            I. stipules present; leaves often 3 at a node ....
            .... *Cephalanthus* in Rubiaceae, p. 631
            I. stipules absent or early deciduous but stipular scars present; leaves all opposite
               J. terminal buds with 2 valvate scales; fruit a drupe
                  K. young leaves, petioles, young twigs and terminal buds covered with brown scurfy scales ....
                  .... *Viburnum* in Caprifoliaceae, p. 318
                  K. young leaves, petioles, young twigs and terminal buds glabrous or glaucous to sparsely pubescent ....
                  .... Cornaceae, p. 362
               J. terminal bud scales imbricate (or the number of scales difficult to interpret); fruit a berry or capsule...................
               .... Caprifoliaceae, p. 310
      G. bases of the petioles not joined in any way
         L. leaves and twigs spicy-aromatic when crushed ....
         .... Calycanthaceae, p. 303
         L. leaves and twigs not aromatic .... Oleaceae, p. 512
A. leaves toothed or lobed
   M. leaves 3–5-lobed (sometimes also toothed)
      N. shrubs; buds naked or with 2 valvate scales; leaves often dark-dotted below; fruit a berry-like drupe .... *Viburnum* in Caprifoliaceae, p. 318
      N. shrubs or trees; buds imbricate; leaves never dark-dotted below; fruit a double samara .... Aceraceae, p. 118
   M. leaves toothed but not lobed
      O. low shrub; leaves evergreen, often some entire ....
      .... *Paxistima* in Celastraceae, p. 343
      O. upright shrubs or trees; leaves deciduous, all toothed
         P. leaves double-toothed .... *Rhodotypos* in Rosaceae, p. 617
         P. leaves single-toothed
            Q. lateral buds hidden (or nearly so) under petiole base ....
           .... *Philadelphus* in Hydrangeaceae, p. 456
           Q. lateral buds evident
              R. twigs 4-angled
                S. twigs remaining green, sometimes with corky wings or ridges ....*Euonymus* in Celastraceae, p. 341
                S. twigs brown, without corky wings or ridges
                   T. pith continuous .... Loganiaceae, p. 494
                   T. pith chambered or hollow....*Forsythia* in Oleaceae, p. 513
             R. twigs rounded

U. bark exfoliating in thin strips, even on younger branches ....
.... *Deutzia* in Hydrangeaceae, p. 455

U. bark exfoliating only on older branches or not at all
  V. pith hollow .... *Paulownia* in Bignoniaceae, p. 262
  V. pith solid
    W. leaves palmately veined, often some subopposite or alternate
    .... Cercidiphyllaceae, p. 344
    W. leaves pinnately veined, opposite or occasionally subopposite
      X. leaf margins serrate
        Y. lateral veins strongly curved toward the apex; branches
        frequently spine-tipped ....
        .... *Rhamnus* in Rhamnaceae, p. 590
        Y. lateral veins not strongly curved toward the apex;
        branches never spine-tipped .... Caprifoliaceae, p. 310
      X. leaf margins coarsely serrate-dentate or crenate to nearly
      entire
        Z. buds with imbricate scales; stems easily broken; fruit a
        2-celled capsule .... *Hydrangea* in Hydrangeaceae, p. 455
        Z. bud scales valvate; stems not easily broken; fruit a 1-
        seeded drupe .... *Viburnum* in Caprifoliaceae, p. 318

Key 7. Woody angiosperms with alternate, compound leaves

A. leaves palmately compound or trifoliate
  B. leaves trifoliate
    C. stems unarmed .... *Cytisus* in Fabaceae, p. 403
    C. stems armed
      D. stems bearing prickles between the nodes ....
      .... *Rubus* in Rosaceae, p. 621
      D. stems bearing spines only at the nodes ....
      .... *Poncirus* in Rutaceae, p. 640
  B. leaves palmately compound
    E. upright shrubs to small trees; stipular spines present ....
    .... *Acanthopanax* in Araliaceae, p. 152
    E. vines or sprawling shrubs; unarmed, bristly, or armed with internodal
    prickles
      F. stems bristly or armed with prickles .... *Rubus* in Rosaceae, p. 621
      F. stems unarmed
        G. plants climbing by means of tendrils often with pad-like endings
        .... *Parthenocissus* in Vitaceae, p. 712
        G. plants climbing by twining, without tendrils ....
        .... Lardizabalaceae, p. 486
A. leaves pinnately compound
  H. plants armed with thorns, spines or prickles
    I. leaves bi-pinnately compound
      J. leaflets rounded at the apex; leaves even-pinnate; pinnately and bi-
      pinnately compound leaves often present on the same tree; thorns
      present at the nodes .... *Gleditsia* in Caesalpiniaceae, p. 300

J. leaflets pointed at the apex; leaves odd-pinnate; all leaves bi-pinnate; spines or prickles found at the nodes and on the internodes .... *Aralia* in Araliaceae, p. 152

I. leaves once compound

K. stems with internodal prickles; upright to sprawling shrubs or woody vines .... Rosaceae, p. 592

K. stems armed at the nodes and sometimes also between the nodes; trees or shrubs

L. leaves showing pellucid dots when held up to the light .... Rutaceae, p. 638

L. leaves not showing pellucid dots .... *Robinia* in Fabaceae, p. 417

H. plants unarmed

M. leaves bi-pinnately compound

N. leaflets rounded at the apex; often with once-pinnate leaves on the same tree .... *Gleditsia* in Caesalpiniaceae, p. 300

N. leaflets more or less pointed at the apex; all leaves bi-pinnate

O. leaflets coarsely toothed or lobed; fruit a papery capsule about 5 cm long .... *Koelreuteria* in Sapindaceae, p. 656

O. leaflets entire; fruit a legume

P. main veins of the leaflets near the margins; twigs smooth with prominent lenticels .... Mimosaceae, p. 505

P. main veins of the leaflets near the centers; twigs coarse without prominent lenticels .... *Gymnocladus* in Caesalpiniaceae, p. 300

M. leaves once compound

Q. leaflets toothed or lobed

R. leaflets coarsely few-toothed near the base, the teeth gland-tipped .... Simaroubaceae, p. 681

R. leaflets toothed or lobed along most or all of the margin

S. leaflets coarsely and irregularly toothed or lobed; fruit a papery capsule about 5 cm long .... *Koelreuteria* in Sapindaceae, p. 656

S. leaflets regularly toothed; fruit variable but not a papery capsule

T. pith chambered .... *Juglans* in Juglandaceae, p. 462

T. pith continuous

U. leaf base partly or entirely surrounding the lateral bud

V. pith occupying <½ the diameter of the stem .... *Gleditsia* in Caesalpiniaceae, p. 300

V. pith occupying >½ the diameter of the stem .... *Rhus* in Anacardiaceae, p. 130

U. lateral buds not hidden by the leaf base

W. leaflets double toothed

X. leaflets 3–5, sometimes deeply cleft and appearing lobed; wood of the stems and roots bright yellow .... *Xanthorhiza* in Ranunculaceae, p. 588

X. leaflets 13–17, regularly toothed; wood of the stems light brown .... *Sorbaria* in Rosaceae, p. 627

W. leaflets single toothed

Y. stipules present; fruit a pome; shrubs or small trees .... *Sorbus* in Rosaceae, p. 627

Y. stipules absent; fruit a nut, generally with a 4-parted husk; large trees .... *Carya* in Juglandaceae, p. 460

Q. leaflets entire (or occasionally some lobed)
    Z. leaves with 3 leaflets
        AA.   leaflets showing pellucid dots when held up to light.......................
               .... *Ptelea* in Rutaceae, p. 640
        AA.   leaflets not showing pellucid dots
           BB. climbing, twining, or sprawling vines; sometimes with some leaflets toothed or shallowly lobed
               CC. some simple, entire, or lobed leaves present on the same plant .... *Solanum dulcamara* in Solanaceae, p. 685
               CC. all leaves compound
                   DD. stipules present; flowers violet-purple; fruit a narrow legume .... *Pueraria* in Fabaceae, p. 417
                   DD. stipules absent; flowers greenish-white; fruit a yellowish-white berry-like drupe ....
                        .... *Toxicodendron* in Anacardiaceae, p. 132
           BB. upright shrubs; all leaflets entire.................. Fabaceae, p. 397
    Z. leaves with 5 or more leaflets
        EE.   vines, or twining or sprawling vine-like shrubs
           FF.   some simple, entire or lobed leaves present on the same plant
                .... *Solanum dulcamara* in Solanaceae, p. 685
           FF.   all leaves compound .... *Wisteria* in Fabaceae, p. 423
        EE.   upright shrubs or trees
           GG. leaf rachis winged .... *Rhus copallina* in Anacardiaceae, p. 131
           GG. leaf rachis not winged
               HH. shrubs, 1–4 m tall
                   II.   low, dense shrubs, 1.5 m tall or less; leaflets silky pubescent .... *Potentilla* in Rosaceae, p. 610
                   II.   taller shrubs, generally 2–4 m; leaflets glabrous to sparsely pubescent below ....
                        .... *Amorpha* in Fabaceae, p. 399
               HH. trees or large tree-like shrubs
                   JJ.   pinnate and twice pinnate leaves present on the same tree .... *Gleditsia* in Caesalpiniaceae, p. 300
                   JJ.   all leaves once pinnate
                     KK. lower leaflets with few gland-tipped teeth near the base .... Simaroubaceae, p. 681
                     KK. leaflets strictly entire
                         LL. base of the petiole not conspicuously expanded ....
                         Toxicodendron vernix in Anacardiaceae, p. 133
                       LL. base of the petiole expanded into a prominent pulvinus ....
                         .... *Sophora* in Fabaceae, p. 418

Key 8. Woody angiosperms with alternate, simple leaves

A. leaves evergreen
    B. plants armed .... *Ilex opaca* in Aquifoliaceae, p. 151
    B. plants unarmed

C. vines .... *Hedera* in Araliaceae, p. 154
C. sub-shrubs or shrubs
    D. plants generally only semi-woody; primarily prostrate or creeping (sometimes with a few erect stems) .... Ericaceae, p. 375
    D. plants definitely woody; upright
        E. small, dark, triangular stipules or stipular scars present .... *Ilex* in Aquifoliaceae, p. 150
        E. neither stipules nor stipular scars present .... Ericaceae, p. 375
A. leaves deciduous
  F. leaves lobed
    G. leaves palmately lobed
      H. vines
        I. leaves lobed but otherwise entire .... Menispermaceae, p. 504
        I. leaves lobed and also variously toothed .... Vitaceae, p. 711
      H. trees or shrubs
        J. plants armed with spines and/or prickles .... *Ribes* in Grossulariaceae, p. 445
        J. plants unarmed
            K. leaves and stems aromatic when crushed
                L. leaf apex obtuse to rounded....*Sassafras* in Lauraceae, p. 487
                L. leaf apex truncate to indented .... *Liriodendron* in Magnoliaceae, p. 497
            K. leaves and stems not aromatic when crushed
                M. leaves strongly whitened and densely tomentose below .... *Populus alba* in Salicaceae, p. 643
                M. leaves not strongly whitened below
                    N. young stems and petioles densely villous with stiff, glandular hairs .... *Rubus odoratus* in Rosaceae, p. 625
                    N. young stems and petioles not villous
                        O. lateral buds hidden by the base of the petiole .... Platanaceae, p. 536
                        O. lateral buds not hidden
                            P. shrubs
                              Q. bark of older stems shredding lengthwise; fruit an inflated follicle .... *Physocarpus* in Rosaceae, p. 609
                              Q. bark of older stems not shredding; fruit a capsule or berry-like
                                  R. shrub to 6 m tall; stamens numerous; fruit a 5-celled capsule surrounded by linear bracts and calyx lobes .... *Hibiscus syriacus* in Malvaceae, p. 500
                                R. shrub to about 2 m tall; stamens 4 or 5; fruit berry-like .... *Ribes* in Grossulariaceae, p. 445
                        P. trees
                            S. broken petioles showing milky juice .... Moraceae, p. 507
                          S. broken petioles not showing milky juice

T. leaf apex truncate or indented ....
.... *Liriodendron* in Magnoliaceae, p. 497

T. leaf apex pointed ...... *Liquidambar* in Hamamelidaceae, p. 452

G. leaves pinnately lobed

  U. leaves with shallow, tooth-like lobes

    V. leaf blades distinctly asymmetrical at the base ....
.... *Hamamelis* in Hamamelidaceae, p. 451

    V. leaf blades generally symmetrical at the base

      W. plants armed with thorns .... *Crataegus* in Rosaceae, p. 600

      W. plants unarmed

        X. leaves not aromatic when crushed; pith stellate in cross sec-
tion .... *Quercus* in Fagaceae, p. 425

        X. leaves aromatic when crushed; pith round in cross section ....
.... *Comptonia* in Myricaceae, p. 509

  U. leaves more deeply lobed

    Y. leaf apex truncate or indented, all leaves lobed ....
.... *Liriodendron* in Magnoliaceae, p. 497

    Y. leaf apex rounded, or acute to acuminate

      Z. leaves not aromatic when crushed; pith stellate in cross section;
buds clustered at the tips of the twigs ....
.... *Quercus* in Fagaceae, p. 425

      Z. leaves aromatic when crushed; pith not stellate in cross section;
buds solitary at the tips of the twigs ....
.... *Sassafras* in Lauraceae, p. 487

F. leaves toothed or entire

  AA. leaves toothed

    BB. plants armed

      CC. shrubs

        DD. leaves remotely toothed to nearly entire; spines usually
3-branched... *Berberis canadensis* in Berberidaceae, p. 254

        DD. leaves regularly toothed; thorns unbranched ....
.... *Chaenomeles* in Rosaceae, p. 600

      CC. small trees

        EE. spines or thorns occurring only on the branch tips ......
.... *Prunus* in Rosaceae, p. 613

        EE. thorns abundant on the branches ....
.... *Crataegus* in Rosaceae, p. 600

    BB. plants unarmed

      FF. vines

        GG. tendrils present .... *Vitis* in Vitaceae, p. 712

        GG. tendrils absent .... *Celastrus* in Celastraceae, p. 340

      FF. trees or shrubs

        HH. leaf blades asymmetrical at base

          II. teeth rounded and irregular; buds naked ....
.... *Hamamelis* in Hamamelidaceae, p. 451

          II. teeth pointed; buds covered by scales

            JJ. lateral veins extending into the teeth

              KK. leaf blades about as wide as long and
heart shaped .... Tiliaceae, p. 688

    KK. leaf blades rounded or cordate at the base but the leaf not heart shaped .... Ulmaceae, p. 689

   JJ. lateral veins anastomosing near the margin ....
                              .... *Celtis* in Ulmaceae, p. 689

  HH. leaf blades essentially symmetrical at base

   LL. leaves palmately veined

    MM. broken petiole showing milky juice .... Moraceae, p. 507

    MM. broken petiole not showing milky juice ..... Rhamnaceae, p. 590

   LL. leaves pinnately veined

    NN. lateral leaf veins extending into the teeth

     OO. one lateral vein extending into each tooth, the number of lateral veins equaling the number of teeth

      PP. buds clustered at the tips of the twigs, or the buds 1–2.5 cm long .... Fagaceae, p. 424

      PP. buds solitary at the tips of twigs, <1 cm long

       QQ. pith rounded in cross section ....
           .... Ulmaceae, p. 689

       QQ. pith stellate in cross section ....
           .... *Castanea* in Fagaceae, p. 424

     OO. teeth more numerous than the lateral veins

      RR. teeth (and lobes, if present) confined to the upper half of the blade .... *Spiraea* in Rosaceae, p. 628

      RR. blades regularly toothed

       SS. branches green .... *Kerria* in Rosaceae, p. 608

       SS. branches gray or brown

        TT. slender, little-branched shrubs ....
           .... *Spiraea* in Rosaceae, p. 628

        TT. trees or much-branched shrubs

         UU. leaves heart-shaped, deeply cordate at the base .... Tiliaceae, p. 688

         UU. leaves ovate to oblong or deltoid, truncate or rounded to shallowly cordate at the base

          VV. leaves averaging >5 cm long ....
           .... Betulaceae, p. 256

          VV. leaves averaging <5 cm long ....
           .... Ulmaceae, p. 689

    NN. lateral leaf veins anastomosing before reaching the teeth

     WW. lateral veins curving strongly toward the apex and rarely branching

      XX. leaves with small dark glands or scales along the upper surface of the midrib.....*Aronia* in Rosaceae, p. 599

      XX. leaves without glands or scales on the midrib ....
        .... Rhamnaceae, p. 590

     WW. lateral veins not curving strongly toward the apex and branching

      YY. leaves with numerous orange resin dots below ....
        .... Myricaceae, p. 509

      YY. resin dots absent

ZZ.   petiole distinctly flattened toward the base of the blade .... *Populus* in Salicaceae, p. 641

ZZ.   petiole rounded or canaliculate
   AAA.   buds covered by a single scale .... *Salix* in Salicaceae, p. 645
   AAA.   buds covered by multiple scales, or naked
      BBB.   lenticels prominent and horizontally oriented; petioles usually bearing 1–3 glands .... *Prunus* in Rosaceae, p. 613
      BBB.   lenticels (if present) not horizontally oriented; petioles not glandular
         CCC.   leaves with stellate pubescence below .... Styracaceae, p. 687
         CCC.   leaves not pubescent, or pubescence (if present) not stellate
            DDD.   stipules or stipular scars present (though sometimes very small)
               EEE.   broken petiole showing milky juice .... Moraceae, p. 507
               EEE.   broken petiole not showing milky juice
                  FFF.   leaf scars with 1 vein scar .... Aquifoliaceae, p. 149
                  FFF.   leaf scars with >1 vein scar .... Rosaceae, p. 592
            DDD.   neither stipules nor stipular scars present
               GGG. buds naked .... Clethraceae, p. 353
               GGG. buds covered by multiple scales
                  HHH.   pith chambered .... *Itea* in Grossulariaceae, p. 445
                  HHH.   pith not chambered
                     III.   leaves acute to rounded at the base.... .... Ericaceae, p. 375
                     III.   leaves cuneate to attenuate at the base
                        JJJ.   large shrub or small tree to 6 m tall; fruit a 1-seeded drupe .... Symplocaceae, p. 687
                        JJJ.   much-branched shrub to about 3 m tall; fruit an achene bearing a pappus .... *Baccharis* in Asteraceae, p. 188

AA.   leaves entire
   KKK.   plants armed
      LLL.   vines with tendrils; stems bristly or armed with spines .... Smilacaceae, p. 996
      LLL.   trees or shrubs
         MMM. trees armed with thorns .... *Maclura* in Moraceae, p. 508
         MMM. shrubs armed with spines
            NNN. shrubs with long spreading, arching or prostrate branches .... *Lycium barbarum* in Solanaceae, p. 682
            NNN. shrubs with short erect branches .... *Berberis thunbergii* in Berberidaceae, p. 254
   KKK.   plants unarmed
      OOO. vines

PPP. leaves 15–40 cm long, reniform to round-cordate ....
.... *Aristolochia* in Aristolochiaceae, p. 156

PPP. leaves smaller, ovate, some leaves lobed or compound

    QQQ. leaves palmately veined, peltate near the margin; some leaves palmately 3–7-lobed .... Menispermaceae, p. 504

    QQQ. leaves pinnately veined, not peltate; some leaves pinnately 3–5-lobed or compound .... *Solanum dulcamara* in Solanaceae, p. 685

OOO. trees or shrubs

  RRR. leaves averaging >1.5 dm long

    SSS. stipules or stipular scars encircling twigs ... *Magnolia* in Magnoliaceae, p. 498

    SSS. stipules or stipule scars absent .... Annonaceae, p. 133

  RRR. leaves averaging <1.5 dm long

    TTT. leaves palmately veined .... *Cercis* in Caesalpiniaceae, p. 299

    TTT. leaves pinnately veined

      UUU. lower surface of leaves silvery-scurfy or covered with orange resin dots

        VVV. lower leaf surface silvery-scurfy ....
.... *Elaeagnus* in Elaeagnaceae, p. 373

        VVV. lower leaf surface covered with orange resin dots

        WWW. leaves aromatic when crushed ....
.... *Myrica* in Myricaceae, p. 509

        WWW. leaves not aromatic .... *Gaylussacia* in Ericaceae, p. 379

      UUU. leaves not silvery-scurfy or covered with orange resin dots

        XXX. leaves and branches aromatic when crushed

        YYY. stipular scars encircling the twigs .... Magnoliaceae, p. 497

        YYY. stipular scars absent .... Lauraceae, p. 487

        XXX. leaves and branches not aromatic

        ZZZ. pith stellate in cross section; fruit an acorn; terminal buds clustered at the ends of the branches ....
.... *Quercus* in Fagaceae, p. 425

        ZZZ. pith round in cross section; fruit variable but not an acorn; usually with only one terminal bud per branch or terminal bud lacking

        AAAA. base of the petiole hollow and surrounding the lateral bud .... Tymelaeaceae, p. 687

        AAAA. base of the petiole not surrounding the lateral bud

        BBBB. lateral veins strongly curving toward the apex and not branching, or not branching before anastomosing near the margin

        CCCC. carefully torn leaf blade remaining connected by fibrous threads; buds with valvate scales .... *Cornus alternifolia* in Cornaceae, p. 363

        CCCC. torn leaf blade not showing fibrous threads; buds naked .... *Rhamnus frangula* in Rhamnaceae, p. 592

        BBBB. lateral leaf veins not strongly curving toward apex, or, if curving strongly toward the apex then branching before anastomosing

        DDDD. trees

EEEE. broken petiole showing milky juice ....
.... *Maclura* in Moraceae, p. 508

EEEE. broken petiole not showing milky juice
    FFFF. pith diaphragmed; fruit a single-seeded drupe, blue-black when ripe, 5–8 mm across ..... *Nyssa* in Cornaceae, p. 365
    FFFF. pith not diaphragmed; fruit a several-seeded berry, orange-pink when ripe, 2–3 cm across.....Ebenaceae, p. 373

DDDD. shrubs
  GGGG. trailing shrubs with arching or prostrate branches; occasionally armed with spines........... *Lycium barbarum* in Solanaceae, p. 682
  GGGG. upright shrubs; never armed
    HHHH. petioles as long as the blades (or nearly so) ....
................................*Cotinus* in Anacardiaceae, p. 130
    HHHH. petioles distinctly shorter than blades
      IIII. stipules or stipular scars present.........................
.... Rosaceae, p. 592
      IIII. neither stipules nor stipular scars present
        JJJJ. leaves 7–15 cm long; fruit a large yellow pear-shaped drupe ending in persistent calyx lobes .... Santalaceae, p. 654
        JJJJ. leaves 3–8 cm long; fruit a berry, capsule or small drupe
          KKKK. leaves with purple petioles ....
.... *Nemopanthus* in Aquifoliaceae, p. 152
          KKKK. leaves with green petioles ....
.... Ericaceae, p. 375

Key 9. Evergreen and semievergreen trees, shrubs, and woody vines in winter

A. leaves needle- or scale-like
  B. shrubs with flattened needle-like leaves, green (not white-striped) on the underside; seeds borne singly, surrounded by a fleshy red aril when mature .... Taxaceae, p. 116
  B. trees or shrubs with needle-like or scale-like leaves; seeds borne in a woody (or fleshy) cone
    C. trees or shrubs; leaves opposite or in whorls of three, appressed and scale-like or subulate and spreading; cone woody or fleshy, <2 cm long .... Cupressaceae, p. 107
    C. trees; needle-like leaves alternate, in fascicles, or crowded on short lateral spurs; cone woody, mostly >2 cm long .... Pinaceae, p. 110
A. leaves blade-like, more or less normally expanded
  D. leaves opposite or whorled
    E. leaves toothed (some entire leaves may be present)
      F. vine-like shrub, creeping or climbing by aerial rootlets ....
.... *Euonymus fortunei* in Celastraceae, p. 342
      F. low but upright shrub without aerial rootlets ....
.... *Paxistima* in Celastraceae, p. 343
    E. all leaves entire

G. plants obligately parasitic, growing attached to and not separable from the branches of a host plant.....*Arceuthobium* in Viscaceae, p. 710

G. plants not obviously parasitic, not growing attached to a host plant

    H. vines .... *Lonicera japonica* in Caprifoliaceae, p. 314

    H. shrubs

        I. at least some leaves whorled .... Ericaceae, p. 375

        I. all leaves opposite .... *Ligustrum* in Oleaceae, p. 515

D. leaves alternate

  J. plants armed .... *Ilex opaca* in Aquifoliaceae, p. 151

  J. plants unarmed

    K. vines

        L. leaves simple, palmately veined .... *Hedera* in Araliaceae, p. 154

        L. leaves palmately compound .... Lardizabalaceae, p. 486

    K. trees or shrubs

        M. stipules or stipular scars present

            N. small, dark, triangular stipules or stipular scars present (though sometimes very small) .... *Ilex* in Aquifoliaceae, p. 150

            N. stipular scars completely encircling the twigs .... *Magnolia virginiana* in Magnoliaceae, p. 499

        M. neither stipules nor stipular scars present .... Ericaceae, p. 375

Key 10. Winter key to deciduous vines and vine-like shrubs or shrubs with twining, climbing or sprawling branches

A. leaf scars opposite

  B. vein scars 3 or more; leaf scars small, crescent-shaped, on the ends of raised bases .... *Lonicera* in Caprifoliaceae, p. 312

  B. vein scars 1; leaf scars larger, not on raised bases

    C. stems soft, woody only near the base of the plant .... *Clematis* in Ranunculaceae, p. 572

    C. stems entirely woody. .... *Campsis* in Bignoniaceae, p. 262

A. leaf scars alternate

  D. tendrils present

    E. tendrils attached to the persistent petiole bases ..... Smilacaceae, p. 996

    E. tendrils attached to the stem opposite the leaf scars .... Vitaceae, p. 711

  D. tendrils lacking

    F. plants climbing by means of aerial rootlets .... *Toxicodendron radicans* in Anacardiaceae, p. 132

    F. aerial roots lacking

        G. stems bristly or armed with spines or prickles

            H. spines present only at the tips of branches .... *Lycium barbarum* in Solanaceae, p. 682

            H. spines, prickles or bristles present along the stem

                I. leaf scars irregular, on raised persistent petiole bases .... *Rubus* in Rosaceae, p. 621

                I. leaf scars linear, narrow, not raised ... *Rosa* in Rosaceae, p. 618

        G. stems unarmed

            J. vein scars 3 or more

K. buds naked; aerial rootlets often present ....
.... *Toxicodendron radicans* in Anacardiaceae, p. 132
K. buds covered with scales; aerial rootlets lacking
  L. vein scars 3; leaf scars U- or V-shaped ....
  .... *Aristolochia* in Aristolochiaceae, p. 156
  L. vein scars >3; leaf scars approximately circular ....
  .... Menispermaceae, p. 504
J. vein scar 1, or several joined to appear as 1
  M. bud scales ending in a mucronate tip ... *Celastrus* in Celastraceae, p. 340
  M. bud scales acute, obtuse, or difficult to distinguish
    N. wart-like or spine-like projections present at the side of the leaf scar
    .... *Wisteria* in Fabaceae, p. 423
    N. no projections present at the side of the leaf scar ....
    .... Solanaceae, p. 681

Key 11. Winter key to deciduous trees and upright shrubs with opposite or whorled leaf scars

A. vein scar 1, or several in a straight or curved line, crowded or joined to appear as 1
  B. vein scars several in a straight or curved line, crowded or joined to appear as 1 .... Oleaceae, p. 512
  B. vein scar clearly 1
    C. terminal buds absent
      D. leaf scars often 3 at a node; twigs not armed; lateral buds submerged in the bark; fruit a globose head ... *Cephalanthus* in Rubiaceae, p. 631
      D. leaf scars usually 2 at a node; twig tips often ending in a spine; lateral buds not submerged in the bark; fruit a dark, berry-like drupe....
      .... *Rhamnus* in Rhamnaceae, p. 590
    C. terminal buds present
      E. terminal buds naked, often partially expanded into a small leafy shoot .... *Hypericum* in Clusiaceae, p. 353
      E. terminal buds covered by scales
        F. pith hollow or chambered .... *Forsythia* in Oleaceae, p. 513
        F. pith solid
          G. twigs green, or reddish (where exposed to sun), sometimes with corky wings or ridges ... *Euonymus* in Celastraceae, p. 341
          G. twigs gray or brown, without corky wings or ridges
            H. leaf scars connected by stipular scars ....
            .... *Symphoricarpos* in Caprifoliaceae, p. 316
            H. leaf scars not connected by stipular scars ....
            .... *Ligustrum* in Oleaceae, p. 515
A. vein scars 3 or more and distinct
  I. terminal buds absent
    J. leaf scars fan-shaped, extending around the twigs and meeting ....
    .... *Sambucus* in Caprifoliaceae, p. 316
    J. leaf scars not meeting (although they may be joined by stipular scars)
      K. stipular scars present

L. leaf scars connected by stipular scars ....
.... *Philadelphus* in Hydrangeaceae, p. 456
L. stipular scars present but not connecting leaf scars ....
.... Staphyleaceae, p. 686
  K. stipular scars absent
    M. crushed twigs with a spicy-aromatic smell .... Calycanthaceae, p. 306
    M. crushed twigs not spicy-aromatic
      N. leaf scars nearly or completely surrounding lateral buds ....
.... *Phellodendron* in Rutaceae, p. 639
      N. leaf scars clearly below the lateral buds
        O. vein scars 3 .... Cercidiphyllaceae, p. 344
        O. vein scars 7 or more .... *Catalpa* in Bignoniaceae, p. 262
I. terminal buds present
  P. terminal buds naked or distinct bud scales impossible to distinguish
    Q. buds stalked, obviously naked, covered with rusty or golden pubes-
cence; lateral veins of the outer pair of bud leaves apparent; buds not
strongly flattened .... *Viburnum* in Caprifoliaceae, p. 318
    Q. buds sessile, not obviously naked (but no scales distinguishable), cov-
ered with short light brown hairs; buds strongly flattened ....
.... *Euodia* in Rutaceae, p. 638
  P. terminal buds clearly covered by scales
    R. bud scales valvate
      S. buds and twigs densely covered with brown or silvery peltate scales
.... *Shepherdia* in Elaeagnaceae, p. 374
      S. buds and twigs not densely scaly
        T. leaf scars raised on persistent petiole bases (petiole base late
deciduous); twigs often brightly colored (red, reddish-brown,
green, purple or yellow) .... *Cornus* in Cornaceae, p. 362
        T. leaf scars not raised on petiole bases; twigs grayish or brownish
.... *Viburnum* in Caprifoliaceae, p. 318
    R. bud scales imbricate
      U. vein scars close and nearly joined in a C- or U-shaped line ....
.... Oleaceae, p. 512
      U. vein scars separate and distinct
        V. leaf scars raised on persistent petiole bases; fruit a berry............
.... *Lonicera* in Caprifoliaceae, p. 312
        V. leaf scars not raised on petiole bases; fruit variable but not a berry
          W. vein scars 5 or more (usually in 3 groups); terminal buds usu-
ally 1.5 cm or more long .... Hippocastanaceae, p. 452
          W. vein scars 3 (rarely 5); terminal buds usually <1.5 cm long
            X. leaf scars narrow, V- or U-shaped .... Aceraceae, p. 118
            X. leaf scars not V- or U-shaped
              Y. twigs coarse, easily broken; fruit small capsules in ter-
minal panicles or corymbs ....
.... *Hydrangea* in Hydrangeaceae, p. 454
              Y. twigs slender; fruit a slender beaked capsule about 1 cm
long, in few-fruited cymes, late dehiscent...................
.... *Diervilla* in Caprifoliaceae, p. 312

Key 12. Winter key to deciduous trees and upright shrubs with alternate leaf scars

A. stems armed with thorns, spines, or prickles, or some branches with thorn-like endings
  B. some branches with thorn-like endings
    C. terminal buds present; stipule scars absent .... Rosaceae, p. 592
    C. terminal buds absent; stipule scars present
      D. crushed twigs with a distinct aroma
        E. branches brownish or grayish; crushed twigs with a bitter almond aroma .... *Prunus* in Rosaceae, p. 613
        E. branches remaining green; crushed twigs with a spicy-citrus aroma .... *Poncirus* in Rutaceae, p. 640
      D. crushed twigs without a particular aroma ....
        .... *Chaenomeles* in Rosaceae, p. 600
  B. stems with thorns, spines or prickles (branches not with thorn-like endings)
    F. stems with regular spines definitely associated with the nodes (internodal prickles sometimes also present)
      G. stems usually with 2 spines at each node, spines never branched
        H. terminal buds present .... *Zanthoxylum* in Rutaceae, p. 641
        H. terminal buds absent .... *Robinia* in Fabaceae, p. 417
      G. stems usually with 1 spine at each node or spines absent from some nodes, spines sometimes branched
        I. spines not branched
          J. shrubs
            K. much-branched shrubs, rarely to 2 m tall
              L. spines beside or occupying the position of the leaf scar .... *Berberis* in Berberidaceae, p. 254
              L. spines below the leaf scars .... *Ribes* in Grossulariaceae, p. 445
            K. shrub or shrub-like tree, generally 3 m or more tall .... *Acanthopanax* in Araliaceae, p. 152
          J. small to large trees
            M. terminal buds absent .... *Maclura* in Moraceae, p. 508
            M. terminal buds present .... *Crataegus* in Rosaceae, p. 600
        I. at least some spines branched
          N. shrubs .... *Berberis* in Berberidaceae, p. 254
          N. small to large trees .... *Gleditsia* in Caesalpiniaceae, p. 300
    F. stems armed with scattered irregular prickles ....
    .... *Aralia* in Araliaceae, p. 152
A. stems unarmed
  O. leaf or stipule scars encircling or nearly encircling the twigs
    P. stipule scars completely encircling the twigs; bud scales 1 or 2
      Q. leaf scars encircling the buds .... Platanaceae, p. 536
      Q. leaf scars not encircling the buds
        R. terminal buds not flattened, enclosed in single scales ....
        .... *Magnolia* in Magnoliaceae, p. 498
        R. terminal buds more or less flattened, enclosed in valvate scales....
        .... *Liriodendron* in Magnoliaceae, p. 497

P. leaf or stipule scars not quite completely encircling the twigs; bud scales several or difficult to distinguish

S. trees; buds 1–2.5 cm long, sharp-pointed ..... *Fagus* in Fagaceae, p. 425

S. low shrubs; buds <1.5 cm long, not sharp-pointed ....
.... *Xanthorhiza* in Ranunculaceae, p. 588

O. leaf or stipule scars absent or not nearly encircling the twigs

T. leaf scars 2-ranked (occurring alternately on opposite sides of the twigs)

U. buds clearly naked

V. stipule scars absent; buds not stalked .... Annonaceae, p. 133

V. stipule scars present; buds stalked ....
.... *Hamamelis* in Hamamelidaceae, p. 451

U. buds covered by scales or the nature of their covering difficult to distinguish

W. leaf scars completely encircling the lateral buds ....
.... Thymelaeaceae, p. 687

W. leaf scars not encircling the lateral buds

X. vein scar 1

Y. vein scar line-like .... Ebenaceae, p. 373

Y. vein scar elliptic .... *Zelkova* in Ulmaceae, p. 692

X. vein scars 3 or more

Z. pith triangular or stellate in cross section

AA.   pith triangular in cross section ....
.... *Betula* in Betulaceae, p. 258

AA.   pith stellate in cross section

BB.   bark gray, smooth; buds with several scales ....
.... *Carpinus* in Betulaceae, p. 260

BB.   bark brownish; buds with 2 or 3 scales ....
.... *Castanea* in Fagaceae, p. 424

Z. pith rounded in cross section

CC.   stipule scars absent ..... *Cercis* in Caesalpiniaceae, p. 299

CC.   stipule scars present

DD.   buds mostly with 2 scales .... Tiliaceae, p. 688

DD.   buds with 3 or more scales

EE.   bud scales clearly in 2 ranks

FF.   vein scars 3 ....
.... *Ulmus* in Ulmaceae, p. 690

FF.   vein scars >3 ....
.... *Morus* in Moraceae, p. 508

EE.   bud scales not in 2 ranks

GG.   buds closely appressed to the stem; pith often chambered or diaphragmed ....
.... *Celtis* in Ulmaceae, p. 689

GG.   buds not closely appressed to the stem; pith solid

HH. shrubs..
.... *Corylus* in Betulaceae, p. 260

HH. trees

II.   axillary buds directed to one side of the twig .... *Ulmus parvifolia* in Ulmaceae, p. 691

II.   axillary buds not directed to one side of the twig ....
      .... *Ostrya* in Betulaceae, p. 261

T. leaf scars not 2-ranked (occurring at any location around the twigs)

  JJ.   drooping clusters of white or whitish berries persistent through winter; generally confined to swamps, fens, or marshes ....
      .... *Toxicodendron vernix* in Anacardiaceae, p. 133

  JJ.   fruit not drooping clusters of white or whitish berries or not persisting through winter; habitat variable

    KK. crushed twigs aromatic

      LL.   trees

        MM. abundant, peg-like short lateral shoots present ....
        .... *Larix* in Pinaceae, p. 110

        MM. short lateral shoots absent ....
        ....*Sassafras* in Lauraceae, p. 487

      LL.   shrubs

        NN. seldom-branching shrubs; aroma fruity ....
        ....*Cotinus* in Anacardiaceae, p. 130

        NN. much-branched shrubs; aroma spicy

          OO. twigs covered with orange resin dots

            PP.   stipule scars present ....
            ....*Comptonia* in Myricaceae, p. 509

            PP.   stipule scars absent....................................
            .... *Myrica* in Myricaceae, p. 509

          OO. twigs not covered with orange resin dots ....
          ....*Lindera* in Lauraceae, p. 487

    KK. crushed twigs not aromatic

      QQ. most (if not all) leaf scars on abundant, short lateral shoots
      .... Ginkgoaceae, p. 109

      QQ. leaf scars spaced more or less normally along branches; short shoots may be present but not as abundant

        RR. pith stellate in cross section

          SS.   vein scars 3 per leaf scar ....
          .... *Quercus* in Fagaceae, p. 425

          SS.   vein scars >3 per leaf scar ....
          .... *Liquidambar* in Hamamelidaceae, p. 451

        RR. pith more or less rounded or somewhat angled in cross section, or the pith difficult to distinguish

          TT.   twigs covered with silvery-scurfy scales ....
          .... *Elaeagnus* in Elaeagnaceae, p. 373

          TT.   twigs not covered with silvery-scurfy scales

            UU. pith diaphragmed or chambered

              VV.   vein scar 1 ....
              .... *Halesia* in Styracaceae, p. 687

              VV.   vein scars 3

                WW. pith hollow between diaphragms (chambered) ....
                ... *Juglans* in Juglandaceae, p. 462

                WW. pith solid between diaphragms ...
                .... *Nyssa* in Cornaceae, p. 365

            UU. pith solid (not diaphragmed), or the pith difficult to distinguish

XX. lateral buds partially or totally buried in the leaf scar and rupturing the leaf scar when expanding

    YY. bark on older branches light gray, smooth, with prominent horizontal lenticels .... Mimosaceae, p. 505

    YY. bark on older branches dark gray, rough, lenticels not evident .... *Gleditsia* in Caesalpiniaceae, p. 300

XX. lateral buds not buried in the leaf scar

    ZZ. first (or only) scale of the lateral bud directly above the leaf scar

        AAA. buds with only 1 scale .... *Salix* in Salicaceae, p. 645

        AAA. buds with >1 scale... *Populus* in Salicaceae, p. 641

    ZZ. first scale of the lateral bud not directly above the leaf scar

        BBB. pith salmon-pink .... *Gymnocladus* in Caesalpiniaceae, p. 300

        BBB. pith whitish, greenish, brownish, or difficult to distinguish

            CCC. terminal flower buds much larger than the lateral buds; lateral buds often crowded toward tips of twigs, the higher ones larger than the lower .... *Rhododendron* in Ericaceae, p. 384

            CCC. terminal and lateral buds approximately the same size or terminal buds only somewhat larger; lateral buds not crowded toward tips of twigs or if crowded toward the tips, then without distinct size difference

            DDD. leaf scars surrounding or nearly surrounding the lateral buds

                EEE. pith occupying ½ or more of the diameter of the stem .... *Rhus* in Anacardiaceae, p. 130

                EEE. pith occupying <½ the diameter of the stem .... *Ptelea* in Rutaceae, p. 640

            DDD. leaf scars not nearly surrounding the lateral buds

                FFF. vein scar 1

                    GGG. vein scar line-like .... Ebenaceae, p. 373

                    GGG. vein scar not line-like

                        HHH. cup-like bases of capsules persistent in axillary or terminal clusters .... *Ceanothus* in Rhamnaceae, p. 590

                        HHH. without persistent cup-like capsule bases

                            III. slender little-branched shrubs .... *Spiraea* in Rosaceae, p. 628

                            III. much-branched shrubs or trees

                              JJJ. trees

                                KKK. stipule scars absent .... *Oxydendrum* in Ericaceae, p. 384

                                KKK. stipule scars present .... *Rhamnus* in Rhamnaceae, p. 590

                            JJJ. shrubs

                                LLL. buds and sometimes twigs orange glandular-dotted .... *Gaylussacia* in Ericaceae, p. 379

                                LLL. buds and twigs not orange glandular-dotted

                                    MMM. twigs generally brown or gray; leaf scars triangular

NNN. buds with 2 exposed scales .... *Nemopanthus* in Aquifoliaceae, p. 152

NNN. buds with 4–6 exposed scales .... *Ilex* in Aquifoliaceae, p. 150

MMM. twigs generally reddish or green; leaf scars narrow

OOO. buds with 2 scales visible .... *Lyonia* in Ericaceae, p. 382

OOO. buds with >2 scales visible .... *Vaccinium* in Ericaceae, p. 386

FFF. vein scars 3 or more

PPP. buds apparently naked, sulfur-yellow or brown .... *Carya* in Juglandaceae, p. 460

PPP. buds covered by scales

QQQ. shrubs with bark on older stems shredding

RRR. stipule scars present .... *Physocarpus* in Rosaceae, p. 609

RRR. stipules scars absent .... *Ribes* in Grossulariaceae, p. 445

QQQ. trees, or shrubs without bark shredding on older branches

SSS. buds on distinctive long stalks .... *Alnus* in Betulaceae, p. 256

SSS. buds not on long stalks

TTT. lateral buds with 2–3 scales

UUU. pith occupying about ½ or more of the diameter of the stem .... Simaroubaceae, p. 681

UUU. pith occupying <½ the diameter of the stem

VVV. twigs striped gray and green; fruit a legume .... *Amorpha* in Fabaceae, p. 399

VVV. twigs not striped; fruit an inflated papery capsule ... . *Koelreuteria* in Sapindaceae, p. 656

TTT. lateral buds with 4 or more scales

WWW. true terminal bud lacking .... *Maclura* in Moraceae, p. 508

WWW. true terminal bud present

XXX. leaf scars large, triangular or 3-lobed .... *Carya* in Juglandaceae, p. 460

XXX. leaf scars smaller, not triangular or 3-lobed

YYY. lateral buds usually covered by the persistent bases of fallen leaves or inflorescences

ZZZ. very low shrub, usually not over 1 m tall .... *Potentilla* in Rosaceae, p. 610

ZZZ. tall shrub to 6 m .... *Hibiscus syriacus* in Malvaceae, p. 500

YYY. lateral buds not covered by leaf or inflorescence bases

AAAA. terminal buds generally >8 mm long and 3–4 times longer than broad

BBBB. pith whitish; vein scars mostly 3 .... *Amelanchier* in Rosaceae, p. 596

BBBB. pith brownish; vein scars mostly 5 .... *Sorbus* inRosaceae, p. 627

AAAA. terminal buds generally <8 mm long and <3–4 times as long as wide

CCCC. twigs (and often branches) with prominent, horizontally elongated lenticels; crushed twigs often with a bitter almond aroma .... *Prunus* in Rosaceae, p. 613
CCCC. lenticels (if present) not prominent or horizontally elongated; crushed twigs not aromatic
 DDDD. leaf scars mostly crowded near ends of current year's growth; twigs glossy reddish- or copper-brown; terminal buds much larger than lateral buds ..... *Cornus alternifolia* in Cornaceae, p. 363
 DDDD. leaf scars more evenly spaced; twigs brownish or grayish; terminal buds and lateral buds of about equal size
  EEEE. fruit a persistent legume ..... *Amorpha* in Fabaceae, p. 399
  EEEE. fruit (if persistent) berry- or pome-like .... Rosaceae, p. 592

Key 13. Ferns and fern allies; nonflowering, spore-bearing plants

A. leaves >2 cm long with a distinct stipe (petiole) and a broad, well-veined blade; mostly basal, arising from a rhizome .... Ferns, p. 71
A. leaves 2 cm or less long; scale-like, needle-like, opposite, subopposite, or whorled, or >2 cm long and grass-like from a basal corm
 B. stems erect, conspicuously jointed, green, ribbed, leaves small and scale-like, whorled, fused at the base, non-green .... Equisetaceae, p. 59
 B. stems not as above
  C. leaves grass-like with sporangia in the base, 2 cm or more long, arising from a corm; plants of wet places .... Isoetaceae, p. 62
  C. leaves scale-like, <2 cm long, spreading or appressed, arranged along erect, spreading, or horizontal stems; plants of dry to moist habitats
   D. plants <2 cm tall, creeping-spreading or erect and moss-like; spores of 2 distinct sizes .... Selaginellaceae, p. 70
   D. plants >3 cm tall, spreading to erect; spores all alike .... Lycopodiaceae, p. 64

Key 14. Plants with flowers in a head surrounded by an involucre

A. flowers crowded on a thick fleshy axis (spadix) surrounded by a single sheathing bract (spathe) .... Araceae, p. 721
A. flowers not in a spathe and spadix type inflorescence
 B. leaves all basal, <5 mm wide, parallel-veined
  C. bracts of the involucre glabrous; flowers pink or yellow
   D. flowers yellow, almost hidden by the glabrous involucral bracts ...... Xyridaceae, p. 1001
   D. flowers pink, not hidden by the bracts .... *Armeria* in Plumbaginaceae, p. 537
  C. involucral bracts densely white-woolly; flowers white .... Eriocaulaceae, p. 837
 B. cauline leaves present (although sometimes very reduced), principal leaves >5 mm wide, mostly net-veined
  E. calyx absent, or reduced to a pappus consisting of hairs or scales

F. ovary inferior; anthers forming a ring around the style ....
.... Asteraceae, p. 162
F. ovary superior; anthers not forming a ring around the style
G. sap milky... Euphorbiaceae, p. 388
G. sap not milky .... Nyctaginaceae, p. 510
E. calyx present; pappus not present
H. ovary inferior
I. leaves opposite or whorled
J. leaves opposite, leaves and stem prickly.....Dipsacaceae, p. 371
J. leaves whorled, leaves and stem not prickly ....
.... *Sherardia* in Rubiaceae, p. 638
I. leaves alternate .... *Eryngium* in Apiaceae, p. 141
H. ovary superior
K. sap milky .... *Euphorbia* in Euphorbiaceae, p. 392
K. sap not milky
L. leaves alternate, trifoliate .... *Trifolium* in Fabaceae, p. 419
L. leaves opposite, simple
M. plants with a mint odor; ovary deeply 4-lobed; style 1 ....
.... Lamiaceae, p. 463
M. plants not aromatic; ovary not lobed; styles 2 ....
.... *Dianthus* in Caryophyllaceae, p. 327

Key 15. Plants with parallel-veined leaves or 3-merous flowers or both

A. woody or herbaceous, climbing or twining vines; leaves ovate, cordate-ovate to hastate, net-veined between the parallel, curved-convergent main veins
B. tendrils present; inflorescence an umbel; fruit a berry ....
.... Smilacaceae, p. 996
B. tendrils not present, inflorescence a spike or panicle; fruit a 3-winged capsule .... Dioscoreaceae, p. 836
A. herbs, not climbing or twining; leaves parallel-veined or net-veined (or reduced to bladeless sheaths)
C. flowers in a dense, fleshy spike (spadix) subtended and sometimes enclosed by a sheathing bract (spathe) .... Araceae, p. 721
C. flowers not in a spadix and spathe arrangement
D. perianth inconspicuous and scarious, reduced to a few bristles, or absent
E. each flower subtended by 1 or 2 bracts or scales in addition to those that may be present at the base of the inflorescence
F. flowers arranged in spikelets; fruit an achene or caryopsis
G. each floret subtended by a single bract or scale; leaf sheaths tubular; stems mostly triangular in cross section ....
.... Cyperaceae, p. 727
G. each floret subtended by a pair of opposing bracts; leaf sheaths with overlapping edges; stems mostly round in cross section.. Poaceae, p. 891
F. flowers not arranged in spikelets; fruit a capsule

H. flowers in a single dense, white-woolly terminal head ....
.... Eriocaulaceae, p. 837
H. flowers not in a white-woolly head
  I. inflorescence a spike; flowers 4-merous..........................
....Plantaginaceae, p. 534
  I. inflorescence a raceme or branched and umbel-like; flowers 3-merous
    J. flowers in a raceme; ovaries 3, only slightly fused at the base .... Scheuchzeriaceae, p. 996
    J. flowers solitary or in small clusters in a branched or umbel-like inflorescence; ovary 1 ...... Juncaceae, p. 843
E. individual flowers not subtended by bracts
  K. flowers unisexual, pistillate and staminate flowers in separate parts of the inflorescence
    L. flowers in a single dense elongate spike; leaves flat ....
.... Typhaceae, p. 1000
    L. flowers in 2–several spherical heads; leaves strongly keeled ....
.... Sparganiaceae, p. 999
  K. flowers perfect
    M. inflorescence branched or umbel-like, flowers solitary or in small clusters .... Juncaceae, p. 843
    M. inflorescence a spike or raceme
      N. inflorescence lateral .... Acoraceae, p. 717
      N. inflorescence terminal
        O. inflorescence a fleshy spike; flowers bright yellow.................
.... *Orontium* in Araceae, p. 722
        O. spike (or raceme) not fleshy, flowers brownish or greenish
          P. flowers 4-merous, sessile; fruit a circumsessile capsule ....
....Plantaginaceae, p. 534
          P. flowers 3-merous, each with a slender, erect pedicel; fruit a cluster of 3 follicles separating at the base but remaining attached at the tip ....Juncaginaceae, p. 853
D. perianth of sepals and petals, or tepals readily evident
Q. flowers irregular
  R. ovary inferior; fertile stamens 1 or 2, fused to the style ....
.... Orchidaceae, p. 875
  R. ovary superior; fertile stamens 3 or 6, distinct
    S. petals and sepals similarly colored .... Pontederiaceae, p. 987
    S. petals and sepals unlike in color
      T. upper petals blue, the lower white; flowers in small cymes subtended by a folded bract........................ Commelinaceae, p. 725
      T. petals all yellow; flowers in a dense terminal head ....
.... Xyridaceae, p. 1001
Q. flowers (at least the corolla) regular
  U. sepals green or scarious; petals variously colored, not green, distinct
    V. leaves opposite or whorled
      W. leaves in a single whorl of 3; sepals and petals each 3 ....
....*Trillium* in Liliaceae, p. 870

  W. leaves opposite; sepals 2; petals 5 ....
    .... *Claytonia* in Portulacaceae, p. 558
 V. leaves alternate or all basal
  X. flowers subtended by overlapping bracts in a dense terminal head; petals yellow .... Xyridaceae, p. 1001
  X. flowers in a more open or larger inflorescence; petals blue, pink, or white
   Y. ovaries several to many, distinct or nearly so; fruit an achene
    Z. ovaries numerous, each with its own stigma; leaves all basal, linear or with a simple, expanded blade ....
     .... Alismataceae, p. 718
    Z. ovaries 2–3, style 1 with 2 or 3 branches at the tip; leaves cauline, deeply divided into 3–7 linear lobes ....
     .... Limnanthaceae, p. 491
   Y. ovary 1, compound; fruit a capsule or achene
    AA. leaves linear with a sheathing base ....
     .... Commelinaceae, p. 725
    AA. leaves deeply divided into 3–7 linear lobes ....
     .... Limnanthaceae, p. 491
 U. sepals and petals (tepals) similar in color and/or texture, or the perianth united
  BB. ovary superior
   CC. stamens 9; ovaries 6, united only at the base; fruit a cluster of follicles .... Butomaceae, p. 724
   CC. stamens 3–6; ovary 1, the carpels fused except perhaps at the top; fruit a capsule or berry
    DD. stamens 3; tepals 6; plants creeping on wet shores or submergent .... Pontederiaceae, p. 987
    DD. stamens and tepals (or tepal lobes) 4 or 6; plants erect
     EE. leaves stiff and spine-tipped .... Agavaceae, p. 717
     EE. leaves not stiff or spine-tipped ... Liliaceae, p. 856
  BB. ovary inferior
   FF. stamens 3; leaves folded and fused along the margins except at the base .... Iridaceae, p. 840
   FF. stamens 6; leaves not folded and fused .... Liliaceae, p. 856

Key 16. Herbaceous dicots with at least some unisexual flowers

A. leaves simple
 B. leaves all basal
  C. flowers in dense spikes .... Plantaginaceae, p. 534
  C. flowers in panicles .... *Rumex* in Polygonaceae, p. 554
 B. cauline leaves present
  D. leaves densely silvery-scaly beneath ....
   .... *Croton* and *Crotonopsis* in Euphorbiaceae, p. 391, 392
  D. leaves variously pubescent or glabrous, but not silvery-scaly
   E. sap milky .... Euphorbiaceae, p. 388
   E. sap not milky

F. leaves opposite or whorled
    G. flowers solitary in the axils; creeping plants of muddy habitats............
    .... Callitrichaceae, p. 302
    G. flowers in axillary or terminal inflorescences; upright plants (or vines)
       of drier habitats
       H. inflorescences axillary
          I. vines; principal leaves lobed ..... *Humulus* in Cannabaceae, p. 309
          I. plants erect or decumbent, but not vines; leaves entire or toothed
             J. leaves linear, entire ....Plantaginaceae, p. 534
             J. leaves ovate or ovate-lanceolate, distinctly toothed ....
              .... Urticaceae, p. 693
       H. inflorescences terminal
          K. styles 1; stamens 3 ....Valerianaceae, p. 695
          K. styles 5; stamens 10 .... Caryophyllaceae. p. 323
F. leaves alternate, at least above
    L. calyx and corolla both present (although sometimes very small)
       M. vines
          N. plants twining, non-tendril-bearing; woody at the base..............
          .... Menispermaceae, p. 504
          N. plants tendril-bearing; completely herbaceous ....
          .... Cucurbitaceae, p. 368
       M. erect or creeping plants
          O. leaves fleshy, sessile; ovaries 4 or 5; stamens 8 or 10 ....
          .... *Sedum* in Crassulaceae, p. 366
          O. leaves not fleshy; ovaries and stamens more numerous ....
          .... *Dalibarda* in Rosaceae, p. 604
    L. perianth absent, or a single whorl of tepals present
       P. vines
          Q. twining vines without tendrils; inflorescence not an umbel; ovary
             inferior; fruit a papery, winged capsule .... Dioscoreaceae, p. 836
          Q. tendril-bearing vines; inflorescence an umbel; ovary superior;
             fruit a berry .... Smilacaceae, p. 996
       P. erect or spreading plants
          R. flowers very small, in few-flowered axillary clusters
             S. style 1 .... *Parietaria* in Urticaceae, p. 693
             S. styles 2 or 3
                T. styles branched ....
               ..... *Acalypha* and *Phyllanthus* in Euphorbiaceae, p. 389, 396
               T. styles unbranched
                   U. pistillate flowers enclosed by a pair of partially fused,
                     triangular bracts; stamen 5 ....
                   .... *Atriplex* in Chenopodiaceae, p. 344
                 U. pistillate flowers not so enclosed; stamens 3 ....
                 .... Amaranthaceae, p. 124
          R. flowers in spikes, racemes or panicles
             V. tepals petal-like, white or pink ....Phytolaccaceae, p. 533
            V. tepals not petal-like
               W. stipules forming a sheath (ocrea) at each node ....
              .... *Rumex* in Polygonaceae, p. 554
             W. stipules, sheathing or otherwise, not present

X. individual flowers distinct along the axis of a spike ....
.... *Pachysandra* in Buxaceae, p. 297
X. individual flowers in small clusters which in turn form a spike or
    panicle
    Y. tepals acute, scarious, intermingled with similar bracts ....
    .... Amaranthaceae, p. 124
    Y. tepals and bracts not acute or not scarious, tepals not mingled
      with bracts .... Chenopodiaceae, p. 344
A. leaves compound
  Z. leaves palmately compound or ternate
    AA. flowers in umbels
      BB. leaves alternate or basal................. *Sanicula* in Apiaceae, p. 144
      BB. leaves whorled .... *Panax* in Araliaceae, p. 154
    AA. flowers in spikes, racemes or panicles
      CC. flowers on a dense fleshy axis (spadix) surrounded by a sheath-
        ing bract (spathe) .... *Arisaema* in Araceae, p. 722
      CC. flowers not in a spadix and spathe type inflorescence
        DD. vine; leaves all opposite; perianth conspicuous ....
        .... *Clematis* in Ranunculaceae, p. 572
        DD. erect plant; leaves alternate above; perianth minute or
          absent
          EE. leaves palmately lobed or compound ....
          .... Cannabaceae, p. 309
          EE. leaves 2 or 3 times ternate ....
          ....*Thalictrum* in Ranunculaceae, p. 585
  Z. leaves 1–3 times pinnate
    FF. climbing or sprawling vines
      GG. plants climbing by axillary tendrils; fruit an inflated papery
        capsule .... *Cardiospermum* in Sapindaceae, p. 655
      GG. plants climbing by twining petioles and leaf rachises; fruit an
        achene with a persistent feathery style ....
        .... *Clematis* in Ranunculaceae, p. 572
    FF. erect plants
    HH. flowers in umbels
      II. styles 2; fruit splitting into 2 dry, achene-like parts ....
      .... Apiaceae, p. 134
      II. styles 5; fruit a fleshy berry .... Araliaceae, p. 152
    HH. flowers not in umbels
      JJ. flowers in globose heads....*Sanguisorba* in Rosaceae, p. 626
      JJ. flowers in panicles
        KK. cauline leaves opposite ....
        ....*Valeriana* in Valerianaceae, p. 695
        KK. cauline leaves alternate
          LL. flowers with only 1 perianth whorl or none;
            fruit an achene ....
            ....*Thalictrum* in Ranunculaceae, p. 585
          LL. flowers with both calyx and corolla; fruit a
            follicle................. *Aruncus* in Rosaceae, p. 600

Key 17. Herbaceous dicots with perfect flowers lacking a perianth or with only a single perianth whorl

A. flowers with perianth lacking
  B. aquatic plants
    C. stamens 4–8 .... Haloragaceae, p. 447
    C. stamens 1 .... Podostemaceae, p. 537
  B. emergent plants of shallow water, or terrestrial plants
    D. leaves cordate, entire .... Saururaceae, p. 657
    D. leaves deeply lobed or compound .... Ranunculaceae, p. 566
A. flowers with a single perianth whorl present
  E. ovary inferior
    F. stamens more numerous than the perianth lobes
      G. perianth lobes 1–3; stamens 6 or 12 .... Aristolochiaceae, p. 155
      G. perianth lobes 4; stamens 8–many
        H. leaves pinnately compound; stamens many ....
        .... *Sanguisorba* in Rosaceae, p. 626
        H. leaves simple; stamens 8 ....
        ....*Chrysosplenium* in Saxifragaceae, p. 658
    F. stamens the same number as the perianth lobes or fewer
      I. leaves opposite or whorled
        J. flowers surrounded by 4 conspicuous white bracts ....
        .... *Cornus* in Cornaceae, p. 362
        J. flowers not not surrounded by 4 white bracts
          K. flowers in dense terminal heads or clusters
            L. stems prickly; flowers in dense heads with prickly involucral bracts at the base .... Dipsacaceae, p. 371
            L. stems not prickly; flowers in dense clusters without involucres .... Valerianaceae, p. 695
          K. flowers not in dense heads or clusters
            M. leaves whorled
              N. flowers in an umbel; leaves in a single whorl ....
              .... Araliaceae, p. 152
              N. flowers not in umbels; leaves in several to many whorls
              .... *Galium* in Rubiaceae, p. 631
            M. leaves opposite
              O. leaves linear; plants low, matted ....
              .... *Scleranthus* in Caryophyllaceae, p. 333
              O. leaves broader; plants erect or prostrate
                P. style 1 .... *Ludwigia* in Onagraceae, p. 520
                P. styles 2 .... *Chrysosplenium* in Saxifragaceae, p. 658
      I. leaves alternate or basal
        Q. perianth lobes and stamens each 5
          R. leaves compound or lobed; flowers in heads or umbels
            S. styles 2; fruit splitting into 2 dry, achene-like parts ....
            .... Apiaceae, p. 134
            S. styles 5; fruit a fleshy berry ...... *Aralia* in Araliaceae, p. 152
          R. leaves simple, not lobed; flowers solitary in the axils or in few-flowered terminal or axillary cymes ....
          .... *Comandra* in Santalaceae, p. 654
        Q. perianth lobes and stamens each <5

T. leaves pinnately compound
   U. stipules present .... *Sanguisorba* in Rosaceae, p. 626
   U. stipules absent .... *Proserpinaca* in Haloragaceae, p. 450
T. leaves simple, although sometimes pinnately lobed
   V. perianth lobes and stamens each 3 ....
   .... *Proserpinaca* in Haloragaceae, p. 450
   V. perianth lobes and stamens each 4
      W. styles 2; leaves oval to round ....
      ....*Chrysosplenium* in Saxifragaceae, p. 658
      W. style 1; leaves lanceolate or linear ....
      .... *Ludwigia* in Onagraceae, p. 520
E. ovary superior
   X. ovaries >1 per flower, distinct at least in the upper half
      Y. stipules conspicuous; leaves pinnately compound ....
      .... *Sanguisorba* in Rosaceae, p. 626
      Y. stipules none, or leaves simple
         Z. ovaries distinct .... Ranunculaceae, p. 566
         Z. ovaries fused at the base
            AA. leaves entire; flowers in a raceme .... Phytolaccaceae, p. 533
            AA. leaves serrate; flowers in a 2–4-branched cyme ....
            .... *Penthorum* in Saxifragaceae, p. 660
   X. ovary 1 per flower (although it may consist of several fused carpels and have several styles)
      BB. stamens >2 times as many as the perianth lobes
         CC. leaves tubular, pitcher-like .... Sarraceniaceae, p. 657
         CC. leaves not pitcher-like
            DD. perianth inconspicuous, shorter than the stamens
               EE. herbs; flowers white, in terminal racemes ....
               .... Ranunculaceae, p. 566
               EE. tree; flowers pink, in dense clusters at the ends of long peduncles .... Mimosaceae, p. 506
            DD. perianth well-developed, showy
               FF. leaves entire
                  GG. sap milky.................... Euphorbiaceae, p. 388
                  GG. sap not milky
                     HH. leaves linear, terete, succulent; plants of dry sites ....
                     .... *Talinum* in Portulacaceae, p. 559
                     HH. leaves broad, cordate or deeply notched at the base; aquatic or wetland plants ....
                     .... Nymphaeaceae, p. 511
               FF. leaves compound, dissected, or lobed
                  II. perianth 5-parted; juice not milky or colored....
                  .... Ranunculaceae, p. 566
                  II. perianth 4- or 8-parted; juice milky or colored
                  .... Papaveraceae, p. 529
      BB. stamens twice as many as the lobes of the perianth or fewer
         JJ. stipules forming a sheath (ocrea) at each node ....
         ....Polygonaceae, p. 543
         JJ. stipules not forming a sheath

KK. style 1 (or none, in the case of sessile stigmas)
    LL. stamens more numerous than the perianth parts
        MM. plants white or reddish; leaves reduced to scales ....
        .... Monotropaceae, p. 506
        MM. plants green; leaves not reduced to scales
            NN. corolla irregular .... Fumariaceae, p. 433
            NN. corolla regular
                OO. leaves opposite; flowers mostly axillary
                    PP. plants with 2 deeply lobed leaves and a single axillary flower between them ....
                    .... *Podophyllum* in Berberidaceae, p. 256
                    PP. plants with several to many pairs of entire leaves; flowers in dense axillary clusters ....
                    .... *Ammannia* in Lythraceae, p. 495
                OO. leaves alternate or basal; flowers mostly terminal .....
                .... Brassicaceae, p. 271
    LL. stamens the same number or fewer than the perianth parts
        QQ. corolla irregular .... Fumariaceae, p. 433
        QQ. corolla regular
            RR. leaves alternate or basal
                SS. perianth parts and stamens 6, 8, or 9 ....
                .... Berberidaceae, p. 254
                SS. perianth parts 4 or 5; stamens 1–5
                    TT. leaves lobed or compound ..... Rosaceae, p. 592
                    TT. leaves simple
                        UU. leaves lobed, toothed, or entire; flowers basal ....
                        cleistogamous flowers of Violaceae, p. 700
                        UU. leaves entire; flowers not basal
                            VV. flowers solitary or in small clusters...
                            .... *Parietaria* in Urticaceae, p. 693
                          VV. flowers in dense spikes ....
                            ...... *Celosia* in Amaranthaceae, p. 128
            RR. leaves opposite or rarely whorled
                WW. flowers axillary, sessile or nearly so; perianth 4-lobed
                .... Lythraceae, p. 494
                WW. flowers in a terminal inflorescence
                    XX. flowers in dense clusters subtended by a 5-lobed involucre .... Nyctaginaceae, p. 510
                    XX. flowers not surrounded by an involucre
                      YY. flowers in a loose open cyme or panicle; calyx and bracts herbaceous ....
                      .... Caryophyllaceae, p. 323
                    YY. flowers in dense spikes or heads; calyx and bracts scarious ....
                      .... *Froelichia* in Amaranthaceae, p. 128
KK. styles >1
    ZZ. leaves opposite or whorled

AAA. leaf margins crenate; stamens normally 8 ....
....*Chrysosplenium* in Saxifragaceae, p. 658
AAA. leaf margins entire
    BBB. leaves opposite; styles 2 .... Caryophyllaceae, p. 323
    BBB. leaves whorled; styles 3 .... Molluginaceae, p. 506
ZZ. leaves alternate
    CCC. stamens and styles each 10; flowers in a raceme ....
....Phytolaccaceae, p. 533
    CCC. stamens and styles <10; flowers in small clusters or glomerules, not in racemes .... Chenopodiaceae, p. 344

Key 18. Herbaceous dicots with petals and sepals both present and >1 ovary per flower

A. style 1, although it may be branched and have several stigmas
  B. ovaries 2 or 3; corolla regular; sap milky in most species
    C. petals and sepals each 3; ovaries 2 or 3, joined at the base ....
....Limnanthaceae, p. 491
    C. petals and sepals each 5; ovaries 2, joined at the tip
      D. flowers solitary or in cymes; styles partly united; anthers not fused to the stigma .... Apocynaceae, p. 148
      D. flowers in umbels or small cymes; styles distinct; filaments fused in a tube around the ovary; anthers fused to the stigma ....
....Asclepiadaceae, p. 158
  B. ovaries 4 or more; corolla regular or bilateral; sap not milky
    E. petals distinct; ovaries 5 or more; stamens numerous, their filaments fused to form a tube around the style ....Malvaceae, p. 499
    E. petals united; ovaries 4; stamens 2, 4, or 5, their filaments not fused to form a tube
      F. leaves alternate; stamens 5; corolla regular; foliage not aromatic ....
.... Boraginaceae, p. 263
      F. leaves opposite; stamens 2 or 4; corolla slightly or very irregular; foliage often aromatic .... Lamiaceae, p. 463
A. styles as many as the ovaries, or not well-developed
  G. flowers irregular
    H. leaves deeply lobed or cleft; flowers spurred; stamens numerous; fruit a follicle.. Ranunculaceae, p. 566
    H. leaves toothed or shallowly lobed; flowers not spurred; stamens 5; fruit a capsule .... *Heuchera* in Saxifragaceae, p. 658
  G. flowers regular
    I. sepals (or sepal-like bracts) 3
      J. plants aquatic; leaves peltate and floating or deeply divided and submersed .... Cabombaceae, p. 298
      J. plants growing on mud or terrestrial; leaves various
        K. leaves opposite; plants of tidal mudflats ....
.... *Crassula* in Crassulaceae, p. 365
        K. leaves alternate, basal, or whorled; plants of various habitats
          L. ovary deeply 2- or 3-lobed; stamens 3 or 6; leaves deeply pinnately lobed ....*Floerkea* in Limnanthaceae, p. 491
          L. ovaries and stamens numerous; leaves simple, palmately lobed, or compound .... Ranunculaceae, p. 566

I. sepals 4 or more
   M. petals united, at least part way .... Apocynaceae, p. 148
   M. petals separate
      N. leaves peltate, floating; flowers >10 cm broad, their ovaries embedded in a broad receptacle .... Nelumbonaceae, p. 510
      N. leaves not peltate; flowers smaller, not embedded in a receptacle
         O. leaves succulent...Crassulaceae, p. 365
         O. leaves not succulent
            P. sepals easily detached, generally falling before the petals .... Ranunculaceae, p. 566
            P. sepals persistent, firmly attached to an hypanthium or an hypanthium-like base
               Q. carpels fewer than the petals
                  R. leaves simple or shallowly lobed ...... Saxifragaceae, p. 657
                  R. leaves compound
                     S. flowers yellow; leaves trifoliate or once-pinnate .... *Agrimonia* or *Waldsteinia* in Rosaceae, p. 594, 629
                     S. flowers white; leaves ternately twice-compound........... .... *Aruncus* in Rosaceae, p. 600
               Q. carpels as many as or more than the petals....Rosaceae, p. 592

Key 19. Dicots with perfect flowers, sepals and petals both present, and a single inferior ovary

A. stamens more numerous than the petals
   B. stamens >2 times as many as the petals; stems flattened, spiny; leaves lacking .... Cactaceae, p. 299
   B. stamens twice as many as the petals; stems and leaves normally developed
      C. style 1
         D. terrestrial plants ....Onagraceae, p. 517
         D. plants of aquatic or semi-aquatic habitats
            E. petals yellow, 1 cm or more long .... *Ludwigia* in Onagraceae, p. 520
            E. petals whitish or greenish, minute .... *Myriophyllum* in Haloragaceae, p. 448
      C. styles 2 or more
         F. sepals 2; styles 3 .... Portulacaceae, p. 557
         F. sepals 4 or 5; styles 2 .... Saxifragaceae, p. 657
A. stamens the same number as the petals or corolla lobes, or fewer
   G. petals separate
      H. petals and stamens each 2 .... *Circaea* in Onagraceae, p. 517
      H. petals and stamens each 4 or 5
         I. petals 4
            J. flowers in dense heads surrounded by 4 showy white bracts .... *Cornus* in Cornaceae, p. 362
            J. flowers not in heads, not surrounded by white bracts
               K. submersed leaves finely dissected
                  L. floating leaves triangular or rhombic with swollen areas in the petioles.. Trapaceae, p. 689
                  L. floating and emersed leaves pinnatifid or reduced to small scales .... *Myriophyllum* in Haloragaceae, p. 448

  K. submersed leaves with entire blades, or absent

    M. submersed leaves with a flat, expanded blade ....

    ....*Ludwigia* in Onagraceae, p. 520

    M. submersed leaves reduced to minute scales ....

    .... *Myriophyllum* in Haloragaceae, p. 448

 I. petals 5

  N. leaves simple

    O. flowers solitary or in panicles or cymes ...... Saxifragaceae, p. 657

    O. flowers in umbels .... Apiaceae, p. 134

  N. leaves compound or dissected

    P. flowers in spike-like racemes .... *Agrimonia* in Rosaceae, p. 594

    P. flowers in umbels

      Q. styles 5; fruit a fleshy berry .... Araliaceae, p. 152

      Q. styles 2 or 3; fruit various

        R. leaves alternate or basal; fruit splitting into 2 dry achene-like parts .... Apiaceae, p. 134

        R. leaves in a single whorl; fruit berry-like ....

        .... *Panax* in Araliaceae, p. 154

G. petals united, at least at the base

 S. cauline leaves alternate

  T. corolla irregular, the tube split on the upper side ....

  ....*Lobelia* in Campanulaceae, p. 307

  T. corolla regular

    U. corolla <3 mm wide; flowers in an open panicle ....

    .... *Samolus* in Primulaceae, p. 564

    U. corolla larger; inflorescence variable, but not a panicle ....

    .... Campanulaceae, p. 304

 S. cauline leaves opposite or whorled, or the leaves all basal

  V. leaves in whorls of 4–8 .... Rubiaceae, p. 630

  V. leaves opposite or basal

    W. stipules present .... Rubiaceae, p. 630

    W. stipules absent, or very tiny and inconspicuous

      X. stamens 3 .... Valerianaceae, p. 695

      X. stamens 4 or 5

        Y. flowers in dense peduncled heads .... Dipsacaceae, p. 371

        Y. flowers solitary from the axils, paired on a slender pedicel, or in terminal cymes

          Z. flowers 4-merous, axillary or in terminal cymes ....

          .... *Houstonia* in Rubiaceae, p. 636

          Z. flowers 5-merous, axillary or paired at the end of a slender peduncle .... Caprifoliaceae, p. 310

Key 20. Herbaceous dicots with perfect flowers, both sepals and petals present, ovary 1, superior, stamens more numerous than the petals

A. flowers regular

 B. leaves modified into tubular pitchers or reduced to scales

  C. leaves modified into tubular pitchers .... Sarraceniaceae, p. 657

  C. leaves reduced to scales

    D. plants lacking chlorophyll; scale leaves alternate or irregularly arranged .... Monotropaceae, p. 506

D. plants green; scale leaves opposite; flowers yellow ....
.... *Hypericum* in Clusiaceae, p. 353
B. leaves not modified into pitchers or reduced to scales
  E. sepals 2
    F. leaves entire, thick and succulent; sap watery ..... Portulacaceae, p. 557
    F. leaves toothed, lobed or compound, not succulent; sap milky or colored
    .... Papaveraceae, p. 529
  E. sepals 3 or more
    G. stamens >2 times as many as the petals
      H. leaves compound
        I. plants sticky-glandular pubescent; leaves palmately compound
        with 3 entire leaflets .... *Polanisia* in Capparaceae, p. 310
        I. plants glabrous or with short, non-glandular hairs; leaflets
        toothed .... Ranunculaceae, p. 566
      H. leaves simple
        J. aquatic plants with large basal leaves .... Nymphaeaceae, p. 511
        J. terrestrial plants with at least some cauline leaves (basal leaves
        may also be present)
          K. style 1
            L. plants with numerous simple leaves and several to many
            flowers .... Cistaceae, p. 350
            L. plants with 2 deeply lobed leaves and a single flower in the
            axil between them .... *Podophyllum* in Berberidaceae, p. 256
          K. styles 2 or more
            M. leaves opposite, with pellucid dots; flowers yellow ....
            .... Clusiaceae, p. 353
            M. leaves alternate, lacking pellucid dots; flower color various
            .... Malvaceae, p. 499
    G. stamens 2 times as many as the petals or fewer
      N. stamens more numerous than the petals but <2 times as many
        O. styles 2–5; leaves opposite or whorled
          P. flowers yellow .... Clusiaceae, p. 353
          P. flowers not yellow
            Q. stamens 9, in fascicles of 3 each alternating with glands ...
            .... Clusiaceae, p. 353
            Q. stamens not fascicled .... Caryophyllaceae, p. 323
        O. style 1; leaves alternate, whorled, or basal
          R. sepals 4; petals 4
            S. stamens 6, 4 long and 2 short; flowers in racemes without
            bracts .... Brassicaceae, p. 271
            S. stamens 6, all the same length, or >6; flowers in a raceme,
            each subtended by a bract .... Capparaceae, p. 310
          R. sepals 5 or 6; petals 3 or 6
            T. leaves simple
              U. sepals 6; petals 6; stamens 11 ....
              .... *Cuphea* in Onagraceae, p. 495
              U. sepals 5; petals 3; stamens 6–9 ....
              .... *Lechea* in Cistaceae, p. 351
            T. leaves pinnately compound .... Caesalpiniaceae, p. 299

N. stamens 2 times as many as the petals
    V. petals 3
        W. leaves compound or deeply lobed, alternate ...... Limnanthaceae, p. 491
        W. leaves simple, opposite or whorled
            X. leaves in a single whorl; flower large, solitary, terminal ....
            .... *Trillium* in Liliaceae*, p. 870
            X. leaves opposite; flowers minute in the axils ....
            .... *Crassula* in Crassulaceae, p. 365
    V. sepals and petals each 4 or more
        Y. sepals and petals each 6 or more
            Z. leaves 2, opposite, deeply lobed, peltate, with a single flower between them .... *Podophyllum* in Berberidaceae, p. 256
            Z. leaves >2, neither peltate nor lobed; flowers several to many
                AA. style 1; sepals and petals borne at the rim of a hypanthium; leaves thin .... *Lythrum* in Lythraceae, p. 495
                AA. styles as many as the petals; hypanthium not well-developed; leaves fleshy .... *Crassula* in Crassulaceae, p. 365
        Y. sepals and petals each 4 or 5
            BB. leaves compound, or at least divided nearly to the base
                CC. leaves opposite
                    DD. leaves pinnately compound; flowers yellow .... Zygophyllaceae, p. 715
                    DD. leaves palmately divided or compound; flowers pink, purple or red .... Geraniaceae, p. 442
                CC. leaves alternate
                    EE. styles 5; leaves with 3 obcordate leaflets .... Oxalidaceae, p. 527
                    EE. style 1; leaves various; but if trifoliate then the leaflets not obcordate
                        FF. principal leaves 2 times pinnatifid but not truly compound .... *Ruta* in Rutaceae, p. 641
                        FF. principal leaves compound
                            GG. leaves palmately compound .... Capparaceae, p. 310
                            GG. leaves pinnately compound .... Caesalpiniaceae, p. 299
            BB. leaves simple; margins entire, toothed or shallowly lobed
                HH. style 1
                    II. hypanthium present with petals and sepals borne on the rim
                        JJ. anthers opening by longitudinal slits .... Lythraceae, p. 494
                        JJ. anthers opening by terminal pores .... Melastomaceae, p. 503
                    II. hypanthium not present; petals and sepals attached to the receptacle
                      KK. sepals all nearly the same size and shape .... Pyrolaceae, p. 565

---

* not a Dicot, but included here because of its net-veined leaves.

        KK. sepals of 2 types, the outer 2 much narrower and often shorter than the inner 3 .... Cistaceae, p. 350

HH. styles 2 or more
    LL. ovary lobed with a style on each lobe
        MM. lobes of the ovary 2; leaves not succulent, mostly basal .... Saxifragaceae, p. 657
        MM. lobes of the ovary 4 or 5; leaves succulent, mostly cauline .... Crassulaceae, p. 365
    LL. ovary not lobed, the styles arising together from its summit
        NN. leaves toothed; flowers minute, sessile in the axils .... Elatinaceae, p. 375
        NN. leaves entire; flowers not as above
            OO. flowers yellow; leaves with pellucid dots .... Clusiaceae, p. 353
            OO. flowers white, pink or red; leaves without pellucid dots .... Caryophyllaceae, p. 323

A. flowers irregular
    PP. sepals petal-like or with a spur
        QQ. none of the sepals with a spur; stamens 6, 7, or 8; leaves entire .... Polygalaceae, p. 541
        QQ. one of the sepals prolonged into a spur or sac; stamens <6 or >8; leaves not entire
            RR. leaves crenate or toothed; stamens 5 .... Balsaminaceae, p. 253
            RR. leaves palmately divided; stamens numerous .... Ranunculaceae, p. 566
    PP. sepals not petal-like, usually green
        SS. sepals 2, distinct .... Fumariaceae, p. 433
        SS. sepals 4 or more, usually more-or-less fused
            TT. lower 2 petals forming a keel which encloses the (usually 10) stamens .... Fabaceae, p. 397
            TT. lower petals neither forming a keel nor enclosing the stamens
                UU. leaves compound
                    VV. flowers 4-merous
                        WW. plants climbing by tendrils; leaflets >3 .... *Cardiospermum* in Sapindaceae, p. 655
                        WW. plants erect; leaflets 3 .... *Polanisia* in Capparaceae, p. 310
                    VV. flowers 5-merous
                        XX. leaves pinnately or bipinnately compound ... . Caesalpiniaceae, p. 299
                        XX. leaves palmately compound .... Geraniaceae, p. 442
                UU. leaves simple, deeply lobed to entire
                    YY. styles 3; upper petals larger than the lower .... Resedaceae, p. 589
                    YY. style(s) 1 or 2; lower petals larger than the upper
                      ZZ. sepals and petals each 4; stamens 6 .... *Iberis* in Brassicaceae, p. 290
                      ZZ. sepals and petals each 4–7; stamens 8–12

AAA. styles 2 .... *Saxifraga* in Saxifragaceae, p. 660
AAA. style 1
    BBB. leaves entire or nearly so ...*Cuphea* and *Lythrum* in Lythraceae, p. 495
    BBB. leaves palmately lobed .... Geraniaceae, p. 442

Key 21. Herbaceous dicots with perfect flowers, both sepals and petals present, ovary 1, superior, stamens as many as the petals or fewer, corolla regular or irregular, petals distinct

A. leaves compound or dissected
    B. flowers solitary on leafless stems
        C. corolla irregular, with a spur; petals 4 or 5
            D. petals 5; sepals 5 .... *Viola* in Violaceae, p. 701
            D. petals 4; sepals 2 .... Fumariaceae, p. 433
        C. corolla regular; petals 8 .... *Jeffersonia* in Berberidaceae, p. 255
    B. flowers produced on leafy stems
        E. flowers in loose, open peduncled clusters
            F. petals 6; stamens 6; leaves ternately 3 times compound.................
            .... *Caulophyllum* in Berberidaceae, p. 255
            F. petals 5; stamens 5; leaves deeply lobed or dissected ....
            ....Geraniaceae, p. 442
        E. flowers in other types of inflorescences
            G. leaves once compound; flowers sessile in the leaf axils ....
            .... Caesalpiniaceae, p. 299
            G. leaves twice compound; flowers in dense, peduncled heads ....
            .... Mimosaceae, p. 505
A. leaves entire to deeply lobed, but not compound
    H. leaves opposite
        I. sepals 2 or 3
            J. sepals 2; petals 3 or 5 .... Portulacaceae, p. 557
            J. sepals and petals each 2 or 3 .... Elatinaceae, p. 375
        I. sepals and petals each 4–6 (or more)
            K. leaves deeply palmately lobed .... Geraniaceae, p. 442
            K. leaves entire to serrate
                 L. style 1
                    M. hypanthium well-developed, the sepals and petals borne at its
                        rim .... Lythraceae, p. 494
                    M. hypanthium not present
                        N. stamens alternate with the corolla lobes ....
                        .... Gentianaceae, p. 436
                        N. stamens opposite the corolla lobes ..... Primulaceae, p. 559
                 L. styles 2–5
                    O. ovary and capsule 4- or 5-locular .... Linaceae, p. 491
                    O. ovary and capsule with a single locule
                        P. petals separate to the base .... Caryophyllaceae, p. 323
                        P. petals united at the base ... *Sabatia* in Gentianaceae, p. 440
    H. leaves alternate or basal
        Q. leaves shallowly to deeply palmately lobed
            R. flowers irregular, one petal spurred .... *Viola* in Violaceae, p. 701
            R. flowers regular, not spurred

S. vines; flowers with a conspicuous fringed corona in addition to the calyx and corolla .... *Passiflora* in Passifloraceae, p. 532

S. herbs; flowers without a corona .... Saxifragaceae, p. 657

Q. leaves entire, serrate, crenate, or pinnately lobed

 T. styles 2 or more

  U. leaves all basal or nearly so

   V. leaves linear, glabrous, lacking stalked glands .... *Armeria* in Plumbaginaceae, p. 537

   V. leaves not linear, covered with conspicuous stalked glands .... *Drosera* in Droseraceae, p. 372

  U. leaves cauline .... Linaceae, p. 491

 T. style 1 or none

  W. flowers with a tubular hypanthium bearing the sepals, petals, and stamens at its rim .... *Lythrum* in Lythraceae, p. 495

  W. flowers with a very short hypanthium, or none

   X. flowers irregular, spurred, or with petals of different sizes .... Violaceae, p. 700

   X. flowers regular, spur not present

    Y. petals and sepals each 4 .... Brassicaceae, p. 271

    Y. petals and sepals each 5

     Z. leaves pinnately lobed ... *Erodium* in Geraniaceae, p. 442

     Z. leaves entire or serrate

      AA. flowers in a terminal inflorescence or solitary and terminal

       BB. flowers solitary and terminal .... *Parnassia* in Saxifragaceae, p. 659

       BB. flowers in a terminal umbel or raceme .... Primulaceae, p. 559

      AA. flowers axillary ...... *Hybanthus* in Violaceae, p. 700

Key 22. Herbaceous dicots with perfect flowers, both sepals and petals present, petals united, corolla regular, ovary 1, superior, stamens as many as the petals or fewer

A. leaves all basal

 B. leaves covered with stalked glands .... *Drosera* in Droseraceae, p. 372

 B. leaves not glandular

  C. flowers 4-merous, corolla scarious....*Plantago* in Plantaginaceae, p. 534

  C. flowers 5-merous, corolla petaloid, not scarious

   D. style 1; flowers in umbels .... *Dodecatheon* in Primulaceae, p. 560

   D. styles 3 or 5; flowers in heads ..... *Armeria* in Plumbaginaceae, p. 537

A. cauline leaves present

 E. ovary deeply lobed, appearing as 2 or 4 separate ovaries

  F. leaves opposite .... Lamiaceae, p. 463

  F. leaves alternate

   G. styles 2; ovary 2-lobed .... Convolvulaceae, p. 358

   G. style 1; ovary 4-lobed .... Boraginaceae, p. 263

 E. ovary 1, not lobed or divided

  H. leaves opposite or whorled

   I. flowers in dense short spikes or heads; corolla 4- or 5-lobed

J. corolla 4-lobed
    K. corolla scarious; leaves linear ..... *Plantago* in Plantaginaceae, p. 534
    K. corolla petaloid; leaves broader .... *Phyla* in Verbenaceae, p. 697
J. corolla 5-lobed .... *Froelichia* in Amaranthaceae, p. 128
I. flowers in loose spikes or short, crowded racemes; corolla 4–12-lobed or divided
    L. stamens opposite the corolla lobes .... Primulaceae, p. 559
    L. stamens alternate with the corolla lobes, or inserted so low that its hard to tell if they are alternate or opposite
        M. leaves connected by a stipular line .... Loganiaceae, p. 494
        M. stipules lacking, although leaves may be slightly connate
            N. corolla 5-lobed
                O. stigmas 3 .... Polemoniaceae, p. 538
                O. stigma 1, although it may be 2-lobed
                    P. leaves entire; ovary unilocular .... Gentianaceae, p. 436
                    P. leaves dentate or angled; ovary 2- or 4-locular ....
                        .... Solanaceae, p. 681
            N. corolla lobes 4 or 6–12 .... Gentianaceae, p. 436
                Q. leaves whorled .... *Veronicastrum* in Scrophulariaceae, p. 680
                Q. leaves opposite .... Gentianaceae, p. 436
H. leaves alternate
    R. leaves deeply lobed or compound
        S. twining or trailing vines
            T. corolla funnelform .... Convolvulaceae, p. 358
            T. corolla rotate .... *Solanum* in Solanaceae, p. 684
        S. plants erect or prostrate but not twining
            U. corolla rotate or saucer-shaped
                V. anthers fused around the style; petals not fringed ....
                .... *Solanum* in Solanaceae, p. 684
                V. anthers separate; petals fringed ....
                .... *Phacelia* in Hydrophyllaceae, p. 459
            U. corolla campanulate to funnelform, tubular, or salverform
                W. leaves trifoliate .... *Menyanthes* in Menyanthaceae, p. 505
                W. leaves simple or compound but not trifoliate
                    X. ovary 3-locular, style divided into 3 branches at the tip ....
                    .... Polemoniaceae, p. 538
                    X. ovary unilocular, style bifid .... Hydrophyllaceae, p. 456
    R. leaves entire, toothed, or shallowly lobed
        Y. leaves reduced to small scales; flowers 4-merous ....
        .... *Bartonia* in Gentianaceae, p. 436
        Y. leaves not scale-like; flowers mostly 5-merous
            Z. flowers axillary or in the forks of the branches
                AA.   corolla salverform .... Convolvulaceae, p. 358
                AA.   corolla rotate
                    BB.   corolla lobes 5; stamens 5, fused around the style ....
                    .... Solanaceae, p. 681
                    BB.   corolla lobes 4; stamens 2, separate ....
                    .... Scrophulariaceae, p. 661
            Z. flowers terminal
                CC.   corolla rotate or saucer-shaped

      DD. corolla lobes 4; stamens 2, separate .... Scrophulariaceae, p. 661

      DD. corolla lobes 5; stamens 5

          EE. anthers united; fruit a berry .... *Solanum* in Solanaceae, p. 684

          EE. anthers not united; fruit a capsule ....
              .... *Verbascum* in Scrophulariaceae, p. 674

CC. corolla campanulate, funnelform, or salverform

      FF. inflorescence a panicle .... *Collomia* in Polemoniaceae, p. 538

      FF. inflorescence a cyme or raceme

          GG. flowers about 2 mm wide, in an open raceme ....
              .... *Samolus* in Primulaceae, p. 564

          GG. flowers 3–6 mm wide, in a helicoid cyme ....
              .... *Heliotropium* in Boraginaceae, p. 266

Key 23. Herbaceous dicots with perfect flowers, both sepals and petals present, petals united, ovary 1, superior, corolla irregular or stamens fewer than the corolla lobes

A. stamens with well-developed anthers 5

    B. ovary deeply 4-lobed with a central style .... Boraginaceae, p. 263

    B. ovary not deeply lobed

        C. flowers nearly regular, not spurred ....
            .... *Verbascum* in Scrophulariaceae, p. 674

        C. flowers strongly bilabiate and spurred ....
            ....*Impatiens* in Balsaminaceae, p. 253

A. stamens with 2–4 well-developed anthers

    D. corolla spurred or saccate at the base

        E. calyx deeply 5-lobed .... Scrophulariaceae, p. 661

        E. calyx 2-parted

           F. aquatic plants with finely dissected leaf segments bearing traps ......
              .... Lentibulariaceae, p. 488

           F. terrestrial plants; leaves various but without traps ....
              .... Fumariaceae, p. 433

    D. corolla not spurred or saccate

        G. leaves alternate or basal

           H. stamens 2

              I. flowers borne on a leafy stem ....
                 .... *Veronica* in Scrophulariaceae, p. 676

              I. flowers borne on a leafless scape .... Plantaginaceae, p. 534

           H. stamens 4

              J. flowers solitary and terminal or solitary in the axils

                 K. flowers solitary and terminal on leafless scapes or basal pe-
                    duncles .... Scrophulariaceae, p. 661

                 K. flowers axillary .... Pedaliaceae, p. 532

              J. flowers in terminal spikes or racemes

                 L. calyx split to the base on one side; corolla 3–5 cm long and
                    wide .... Pedaliaceae, p. 532

                 L. calyx 4-lobed, or divided into lateral halves, or equally 5-lobed;
                    corolla <3 cm long .... Scrophulariaceae, p. 661

        G. leaves opposite or whorled

           M. ovary deeply 4-lobed with a central style; plants often with a square
              stem and aromatic when crushed .... Lamiaceae, p. 463

M. ovary not 4-lobed; plants seldom square-stemmed or aromatic
   N. stamens 2
      O. corolla scarious ....Plantaginaceae, p. 534
      O. corolla petaloid
         P. flowers in terminal racemes or spikes or solitary or paired in the axils .... Scrophulariaceae, p. 661
         P. flowers in axillary racemes or spikes
            Q. corolla nearly regular, 4-lobed .... Scrophulariaceae, p. 661
            Q. corolla distinctly 2-lipped .... *Justicia* in Acanthaceae, p. 117
   N. stamens 4
      R. corolla distinctly 2-lipped or strongly irregular
         S. flowers in dense axillary heads or spikes .... *Phyla* in Verbenaceae, p. 697
         S. flowers solitary in the axils, in racemes, panicles, or in terminal spikes
            T. upper lip of the corolla very small or lacking .... Lamiaceae, p. 463
            T. upper lip of the corolla well-developed
               U. upper lip of the corolla composed of 4 lobes, the lower of 1 lobe .... *Trichostema* in Lamiaceae, p. 486
               U. upper lip composed of 2 lobes, the lower lip 3-lobed
                  V. calyx very irregular, the upper 3 lobes subulate the lower broadly triangular .... *Phryma* in Verbenaceae, p. 697
                  V. calyx regular or irregular, 2-, 4-, or 5-lobed, but the upper lobes never subulate .... Scrophulariaceae, p. 661
      R. corolla nearly regular and about equally lobed, the upper lobe sometimes slightly smaller and/or the lower lobe extending farther than the upper
      W. corolla salverform, with a slender tube of almost uniform diameter
         X. stamens inserted near the middle of the corolla tube .... *Verbena* in Verbenaceae, p. 698
         X. stamens near the base of the corolla tube .... *Buchnera* in Scrophulariaceae, p. 666
      W. corolla funnelform, salverform, or campanulate, with a flaring tube
         Y. calyx 5-lobed, split to the base on one side .... Pedaliaceae, p. 532
         Y. calyx 4-lobed, or about equally 5-lobed but not split to the base
            Z. calyx with 5 narrow lobes much longer than the tube; flowers pink, lavender, or blue
               AA. the upper leaves often with 2 small lobes at the base; corolla campanulate .... *Agalinis* in Scrophulariaceae, p. 663
               AA. none of the leaves lobed; corolla funnelform .... *Ruellia* in Acanthaceae, p. 117
            Z. calyx with 4 lobes or with short, broad lobes; flowers yellow .... Scrophulariaceae, p. 661

FERNS AND FERN ALLIES

The ferns and fern allies (or vascular spore bearers) are like seed plants vegetatively in that they have stems, roots, and leaves. Reproductively, the vascular spore bearers are different in fundamental and significant ways. These plants produce not seeds but single-celled spores, usually in vast numbers. Spread by wind, these spores grow into separate, independent organisms called gametophytes. The field botanist may find fern gametophytes on shady, moist soil banks or moldering logs where they appear as tiny, green, heart shaped ribbons, barely visible to the naked eye. Gametophytes of most fern allies are subterranean and unlikely to be seen. Though tiny, the gametophytes are where the sexual phase of the life cycle occurs. The familiar spore-producing plant, the sporophyte, develops from the gametophyte.

EQUISETACEAE Horsetail Family
James D. Montgomery

Terrestrial or emergent aquatic plants with creeping rhizomes; stems erect, ribbed, hollow, and jointed; leaves tiny, whorled, fused into a sheath around the stem above the joint and projecting as teeth on the sheath; branches, if any, whorled at the joints, alternating with the leaves; sporangia borne in terminal strobili; spores green.

Equisetum L.

Characteristics of the family.

A. stems annual, usually with whorls of branches; strobili rounded at the tip
  B. stems regularly branched; the first branch internode mostly longer than the main stem sheath
    C. stem sheath teeth reddish-brown, lacking white margins; branches branched again .... E. sylvaticum
    C. stem sheath teeth dull brown with white margins; branches unbranched .... E. arvense
  B. stems unbranched, irregularly branched, or with regular short branches; first branch internode shorter than to barely equaling the main stem sheath
    D. ridges on the main stem 12–24; sheaths tightly appressed to the stem; spores normal .... E. fluviatile
    D. ridges on the main stem 10–14; sheaths flaring outward from the main stem; spores abortive .... E. x littorale
A. stems perennial, evergreen, not generally branched (damaged plants may branch); strobili pointed at the tip
  E. teeth of the stem sheaths persistent, prominent, with black centers and white margins; stems 0.5–5 mm in diameter; sheaths green with a black rim .... E. variegatum

E. teeth of the stem sheaths mostly falling off; stems 3–17 mm in diameter
 F. upper stem sheaths white, with a black band near the middle; spores normal .... *E. hyemale*
 F. upper stem sheaths green, lacking a dark band; spores abortive ....
  .... *E. x ferrissii*

Equisetum arvense L.   Field horsetail
Stems annual, the vegetative ones regularly branched with ascending to horizontal branches, 1–10 dm tall, 0.8–4.5 mm in diameter, with 4–14 ridges; the first branch internode longer than the subtending sheath; sheaths green, teeth narrow with white margins that are narrower than the dark central stripe; fertile stems appearing in early spring, pale brown or pinkish, unbranched, succulent, withering soon after shedding spores; strobili rounded at the tip; common in moist meadows, roadside ditches, and open fields; throughout; FAC.

Equisetum x ferrissii Clute   Intermediate scouring-rush
Stems evergreen, unbranched, 2.5–18 dm tall, 3–11 mm in diameter, with 14–32 ridges; sheaths on lower part of the stem with dark bands and white above, on the upper part of the stem all green, 7–17 mm long, usually longer than broad; sheath teeth deciduous or partly persistent, leaving a dark rim on the tip of the sheath; stems all alike; strobili pointed; spores abortive; uncommon in moist gravelly or sandy circumneutral soils of shores and fields; scattered. Specimens from Pennsylvania formerly labeled as *E. laevigatum* are this hybrid. [syn: *E. hyemale* var. *affine* x *laevigatum*]

Equisetum fluviatile L.   Water horsetail
Stems annual, unbranched, irregularly branched, or with whorls of short branches from the middle nodes, 3.5–12 dm tall, 2.5–9 mm in diameter, with 12–24 shallow ridges; first branch internode shorter than the main stem sheath; sheath appressed to the stem, with narrow dark teeth; fertile stems like the vegetative; strobili rounded at the tip; occasional in shallow water of marsh borders, wet meadows, and edges of ponds and rivers; scattered throughout; OBL.

Equisetum hyemale L.   Scouring-rush
Stems evergreen, normally unbranched, 2–20 dm tall, 3–17.5 mm in diameter, with 14–50 ridges; sheath with a dark band across the middle and tan to ashy-gray above, about as long as wide, appressed to the stem; sheath teeth deciduous; stems all alike; strobili pointed; frequent in moist sandy or gravelly slopes, including stream banks, railroad embankments, and roadsides; scattered throughout, except at the highest elevations; FACW. Ours is var. *affine* (Engelm.) A.A. Eaton

Equisetum x littorale Kühlew.   Shore horsetail
Stems annual, unbranched or irregularly branched, 2–10 dm tall, 1.2–4 mm in diameter, with 10–14 ridges; first branch internode (if present) shorter than to nearly equaling the stem sheath; sheaths somewhat flaring at the top, with black teeth exhibiting very narrow, white margins; stems all alike, but fertile ones rare; spores abortive; very rare along river and lake shores; scattered, usually with the parents; OBL. [syn: *E. arvense* x *fluviatile*]

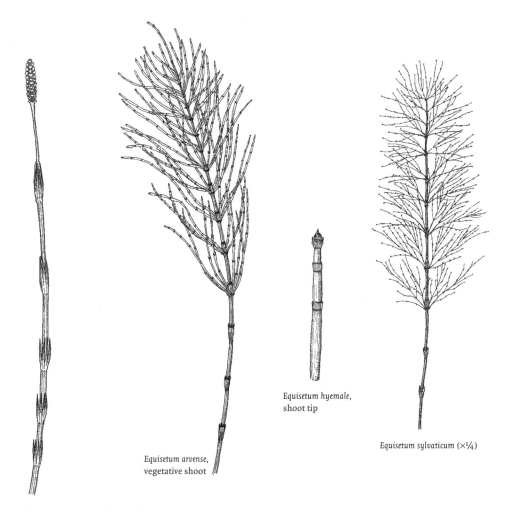

Equisetum hyemale,
shoot tip

Equisetum sylvaticum (×¼)

Equisetum arvense,
vegetative shoot

Equisetum arvense, fertile shoot

Equisetum sylvaticum L.   Woodland horsetail
Stems annual, branched in regular whorls with the branches again branched,
2.5–7 dm tall, 1.5–3 mm in diameter, with 8–18 ridges; first branch internode
longer than the main stem sheath; sheaths reddish-brown, papery, with brown
teeth usually cohering in 3 or 4 groups; fertile stems at first non-green, becoming green and branched after spores are shed; strobili rounded at the tip; occasional in moist open woods and wet meadows in subacid soil; mostly E and N;
FACW.

Equisetum variegatum Schleich.   Variegated scouring-rush
Stems evergreen, unbranched, 0.6–4.5 dm tall, 0.5–3 mm in diameter, with 3–12
ridges; sheaths green with black rims, somewhat flaring; sheath teeth persistent, obtuse, with broad white margins and dark central stripes, narrow tip deciduous; stems all alike; strobili pointed; very rare in damp soil of stream banks
and sandy flats, in circumneutral to alkaline soil; NW; FACW; ✿.

ISOËTACEAE Quillwort Family

W. Carl Taylor and Daniel F. Brunton

Plants tufted, grass-like, submersed or emergent aquatics in lakes, ponds, streams, and river shores; rootstock brown, 2-lobed below, leaves bright to dark green, linear, several to many, erect, spreading, or recurved, each containing a vascular strand and expanded at the base to contain a single micro- or megasporangium; sporangia partly covered by a thin veil of tissue (velum) extending downward over a portion of the sporangium, each sporangium containing tens to hundreds of megaspores or thousands of microspores; megaspores white, globose, about 0.5 mm in diameter, each with a pronounced equatorial ridge and 3 converging ridges, textured with spines, short flattened ridges (cristate), or a network of longer ridges (reticulate); microspores dust-like, white or grayish in mass.

Isoëtes L.

Characteristics of the family. For identification to species it is necessary to determine the texture and average size of mature megaspores. Mature spores may be found around the rootstock much of the year, but plants gathered in late summer are easiest to identify because well-developed spores are still contained within the sporangia. A 10× to 20× hand lens is needed to resolve megaspore texture. A microscope fitted with an ocular micrometer should be used to determine spore size. At least 10 spores should be measured to determine the average size. Megaspore textures and diameters are determined from dry spores. Interspecific hybrids of *Isoëtes* are occasionally found where 2 or more species occur together. Hybrid plants are sometimes recognizable by their sporadic occurrence and robust growth, but the presence of megaspores that vary in size, shape, and texture is the best indication of a hybrid.

A. megaspores globose, uniform in size, shape, and texture (fertile species)
   B. megaspores spiny ....*I. echinospora*
   B. megaspores reticulate or cristate
      C. megaspores averaging <0.5 mm in diameter
         D. megaspores regularly reticulate with even ridges; velum covering 10–20% of the sporangium ....*I. engelmannii*
         D. megaspores cristate to irregularly reticulate with uneven ridges; velum covering 50–70% of the sporangium ....*I. valida*
      C. megaspores averaging >0.5 mm in diameter
         E. megaspores irregularly reticulate with more or less uneven ridges...
         ....*I. appalachiana*
         E. megaspores cristate with short branching ridges ....*I. riparia*
A. megaspores mostly subglobose to flattened, irregular in size, shape, and texture, only a few medium-sized and globose (sterile hybrids)
   F. medium-sized, globose megaspores averaging >0.5 mm in diameter, cristate-reticulate .... *I. x brittonii*
   F. medium-sized, globose megaspores averaging <0.5 mm in diameter, cristate-echinate ....*I. x dodgei*

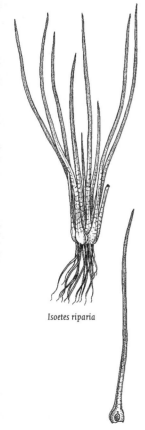

Isoetes riparia

Isoetes riparia, leaf

Isoëtes appalachiana D.F.Brunt. and D.M.Britton   Appalachian quillwort
Emergent aquatic; leaves to about 30 cm long; velum covering 20–40% of the
sporangium; sporangium wall brown-spotted; megaspores 0.45–0.65 mm in diameter, irregularly-reticulate with more or less uneven ridges; shallow water of
lakes, ponds, and river shores; mostly C and W. [syn: *I. engelmannii* A.Br. var.
georgiana Engelm.]

Isoëtes x brittonii D.F.Brunt. & W.C.Taylor   Britton's quillwort
Emergent aquatic; leaves about 20 cm long; velum covering 20–40% of the
sporangium; sporangium wall brown-spotted; megaspores irregular in size
and form; well-formed megaspores averaging >0.5 mm in diameter, cristate-
reticulate; very rare in shallow water, and shores of reservoirs and slow-moving
streams. [syn: *I. appalachiana* x *riparia*; *I. engelmannii* x *riparia* of Rhoads and
Klein, 1993]

Isoëtes x dodgei A.A.Eaton   Dodge's quillwort
Aquatic; leaves about 15 cm long; velum covering 20–40% of the sporangium;
sporangium wall brown- spotted; megaspores irregular in size and form; well-
formed megaspores averaging <0.5 mm in diameter, cristate-spiny; shallow water of lakes; known from a single site in Wyoming Co. [syn: *I. echinospora* x
riparia]

Isoëtes echinospora Durieu   Spiny-spored quillwort
Aquatic; leaves to about 15 cm long; velum covering 20–40% of the sporangium;
sporangium wall more or less brown-spotted; megaspores 0.4–0.5 mm in diameter, covered with thin spines; occasional in shallow water of cold, slightly acidic
lakes, ponds, and slow-moving streams; NE and NW; OBL.

Isoëtes echinospora,
spore (×15)

Isoëtes engelmannii A.Braun   Engelmann's quillwort
Emergent aquatic; leaves to about 30 cm long; velum covering 10–20% of the
sporangium; sporangium wall usually unspotted; megaspores 0.4–0.5 mm in
diameter, reticulate with even ridges; frequent in shallow water and shores of
lakes, ponds, and slow-moving rivers and streams; E and SC, mostly east of the
Allegheny Front; OBL. Suspected hybrids with *I. appalachiana* and *I. riparia* have
been found.

Isoëtes engelmannii,
spore (×15)

Isoëtes riparia A.Braun   Shore quillwort
Aquatic; leaves to about 25 cm long; velum covering 20–40% of the sporangium;
sporangium wall brown-spotted to completely brown; megaspores 0.45–0.65
mm in diameter, cristate with short branching ridges; infrequent in shallow water and shores of slow-moving rivers and streams and intertidal mud flats; OBL.

Isoëtes riparia,
spore (×15)

Isoëtes valida (Engelm.) Clute   Carolina quillwort
Emergent aquatic or terrestrial; leaves to about 30 cm long; velum covering 50–
70% of the sporangium; sporangium wall unspotted; megaspores 0.4–0.5 mm
in diameter, cristate to irregularly reticulate with uneven ridges; rare in woodland seeps, shallow water and shores of slow-moving streams and ponds; S; 🌿.
[syn: *I. caroliniana* (A.A.Eaton) Luebke]

LYCOPODIACEAE Clubmoss Family
James C. Parks

Terrestrial plants, usually with long, creeping horizontal stems (absent in *Huperzia*) and erect, fertile shoots that are dichotomously branched or unbranched; gemmae (modified lateral shoots that serve as asexual propagules) present in some species; leaves small with only 1 unbranched central vein, green; leaf arrangement spiral and spreading, opposite, fused basally with only tips distinct, or sometimes dimorphic and/or twisted giving branchlets a flattened appearance; sporangia borne in the axils of alternating zones of slightly smaller leaves, or more often aggregated into distinct terminal strobili that are sessile or stalked.

A. strobili absent, sporangia borne in the axils of slightly smaller leaves, zones of which alternate with vegetative leaves; gemmae often present; horizontal stems absent .... *Huperzia*
A. strobili present; horizontal stems present
    B. leaves paired, closely appressed to the branchlets with only the tips free and divergent, every other pair dorsoventrally compressed; stems branching in a fan-like manner, branchlets more or less flattened; strobili grouped on an elongated peduncle .... *Diphasiastrum*
    B. leaves spirally arranged, divergent; branchlets appearing flattened in some species; strobili stalked or sessile
        C. horizontal stem slightly to deeply buried; strobili sessile or stalked from the branch tips .... *Lycopodium*
        C. horizontal stem at the soil surface, leafy; strobilus occupying the terminal portion of an erect, leafy peduncle arising from the rhizome ....
        .... *Lycopodiella*

Diphasiastrum Holub

Plants trailing; horizontal stems at the soil surface and covered only with leaf litter, or subterranean; aerial shoots erect and branched 3–5 times in a dichotomous-digitate manner to form a flat, fan-like structure; lateral branches more or less flattened; leaves on erect shoots mostly opposite, usually scale-like and overlapping with only the tips free, in 4 ranks; the lateral rows larger and more spreading than the dorsal and ventral rows; strobili several on long peduncles, with or without a constricted sterile tip. *D. sabinifolium* (Willd.) Holub. of Rhoads and Klein, 1993 is now considered to be a hybrid between *D. tristachyum* and *D. sitchense* (Ruprecht) Holub.

A. horizontal stems buried; branchlets ascending, rounded with evident annual constrictions; leaves bluish-green; peduncles usually 1; strobili lacking sterile tips .... *D. tristachyum*
A. horizontal stems superficial or just beneath the leaf litter; branchlets spreading, flattened, lacking annual constrictions; peduncles mostly 2 from the base, strobili usually with sterile tips .... *D. digitatum*

Diphasiastrum digitatum (Dill. ex A.Braun) Holub   Running-pine
Horizontal stems at the soil surface and covered with leaf litter or shallowly subterranean; erect shoots 1.5–3 dm tall, with 3–4 spreading to recurved, fan-

shaped branches; peduncles 4–11 cm long, sparsely leafy; strobili 2–4 per peduncle, on short, opposite to subopposite branches; strobili with contracted sterile tips to 11 mm long; common in dry, sandy barrens, abandoned fields, scrubby woods, and open forests; throughout. Hybrids with D. tristachyum have been reported at a few sites. [syn: *Lycopodium digitatum* Dill.)

Diphasiastrum tristachyum (Pursh) Holub   Deep-rooted running-pine
Horizontal stems deeply buried, sometimes relatively stout; erect shoots 1.5–3 dm tall, branched; the main branchlets ascending-spreading, dividing 4–6 times to become fan-shaped; branchlets slightly flattened to squarish in cross section, 1–2.2 mm wide with conspicuous annual constrictions; strobili 2–4, 1–2.8 cm long, on sparsely leafy, 3–11 cm long, branching peduncles; occasional in open coniferous forests, sandy barrens, and clearings; throughout. [syn: *Lycopodium tristachyum* Pursh]

Huperzia Bernh.

Plants terrestrial, often on rocky substrates, horizontal stems absent; aerial shoots erect, sometimes sprawling with age, forming clumps; leaves spirally arranged, appressed-ascending to reflexed, irregularly toothed to nearly entire; gemmae-bearing branches in a whorl near the stem tips, flattened, more or less heart-shaped; sporangia kidney-shaped, borne individually in the axils of slightly smaller leaves, zones of which alternate with zones of larger, vegetative leaves. A previous report of *H. selago* (L.) C.Mart. & Schrank from Pennsylvania was based on a misidentified specimen.

A. leaves spreading-reflexed, narrowly obovate-spatulate with several irregular teeth present; gemmae wider than long .... *H. lucidula*
A. leaves ascending to spreading, lance-linear to triangular, entire; gemmae longer than wide .... *H. porophila*

Diphasiastrum digitatum

Huperzia lucidula (Michx.) Trevis.   Shining firmoss
Shoots erect, 1.4–2 dm tall, becoming decumbent with the long, trailing portion covered by leaf litter; leaves dark, lustrous green, margins with 1–8 irregular teeth, mostly toward the tip; sterile leaves 7–11 mm long; fertile leaves 3–6 mm long, bearing yellowish sporangia in the axils; common in moist, shaded, mixed conifer-hardwood forests; throughout; FACW–. [syn: *Lycopodium lucidulum* Michx.]

Huperzia porophila (F.E.Lloyd & Underw.) Holub   Sandstone-loving firmoss
Shoots erect, 12–15 cm tall, clustered and often decumbent; leaves mostly spreading, reflexed basally, ascending and a bit smaller apically, mostly dark green, linear to lanceolate-attenuate or narrowly triangular, sometimes denticulate toward the tip; vegetative leaves 5–8 mm long; fertile leaves 3–6 mm long; very rare on shaded sandstone rocks and cliffs, often within the spray zone of waterfalls; NE; FACU–; ❧. [syn: *Lycopodium porophilum* Lloyd & Underw.]

Lycopodiella Holub

Plants with horizontal leafy stems, flat on the soil surface or arching; upright shoots unbranched, with a single terminal strobilus; leaves spreading or appressed-ascending, not in rows, linear-lanceolate, margins usually with a few teeth.

A. horizontal stems strongly arching; leaves spreading to slightly ascending, margins usually toothed .... *L. alopecuroides*

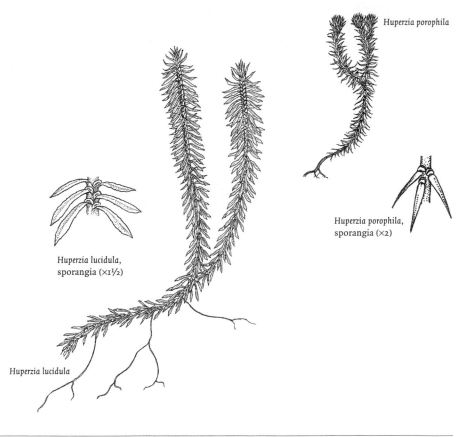

Huperzia porophila

Huperzia porophila, sporangia (×2)

Huperzia lucidula, sporangia (×1½)

Huperzia lucidula

A. horizontal stems flat on the substrate; leaves spreading to appressed-ascending, margins toothed or entire
  B. fertile shoots 3.5–6 cm tall; leaves entire .... L. inundata
  B. fertile shoots 13–40 cm tall; leaves usually with a few teeth
    C. strobili 0–2 mm thicker than the stem and ⅙–⅓ the total height of the fertile shoot .... L. appressa
    C. strobili 3–6 mm thicker than the stem and ⅓–½ the total height of the fertile shoot .... L. margueritae

Lycopodiella alopecuroides (L.) Cranfill   Foxtail bog clubmoss
Horizontal stems strongly arching and rooting at the tip, 2–4 mm in diameter (excluding leaves); leaves spreading, with 1–7 marginal teeth per side; fertile shoots 6–30 cm tall with the strobili occupying the terminal ⅐–⅓; very rare in moist openings; SE; FACW+; ✿. A hybrid with *L. appressa* has been collected occasionally. [syn: *Lycopodium alopecuroides* L.]

Lycopodiella appressa (Chapm.) Cranfill   Appressed bog clubmoss
Horizontal stems 1.5–2 mm in diameter (excluding leaves); leaves appressed, often twisted and curved upward, with 1–7 teeth per side; fertile shoots 13–40 cm tall with the strobili occupying the terminal ⅙–⅓; rare in moist, sandy-peaty openings; SE; FACW+; ✿. [syn: *Lycopodium appressum* (Chapman) F. E. Lloyd & Underw.]

Lycopodiella inundata (L.) Holub   Northern bog clubmoss
Horizontal stems to 0.9 mm in diameter (excluding leaves); leaves spreading, upcurved, with entire margins; fertile shoots 3.5–6 cm tall with the only slightly thicker strobili occupying the terminal ⅓–½; occasional in bogs and moist, acidic soils; scattered; OBL. [syn: *Lycopodium inundatum* L.]

Lycopodiella margueritae J.G.Bruce, W.H.Wagner & Beitel   Marguerite's clubmoss
Similar to *L. inundata* but more robust; leaves with a few marginal teeth, those of the erect, fertile shoots spreading-ascending; strobili 2–4 mm thicker than the lower portion of the fertile shoot; very rare; N; ✿.

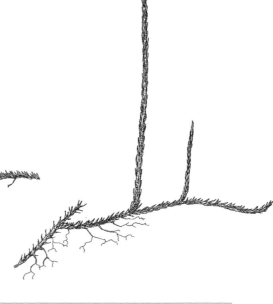

Lycopodiella inundata

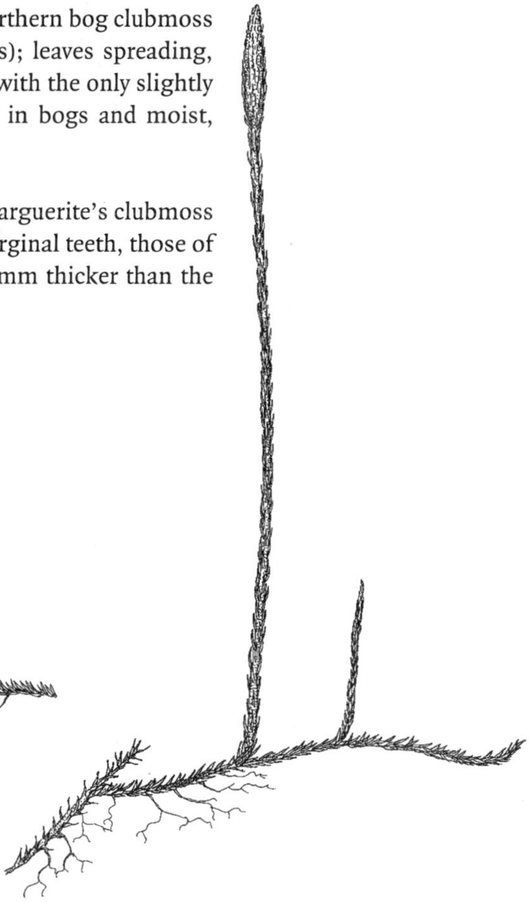

Lycopodiella appressa

Lycopodium L.

Plants trailing; horizontal stems covered only with leaf litter, or subterranean; upright shoots ascending and sparsely dichotomously branched or more erect and treelike; leaves on horizontal stems scattered, appressed; leaves on erect stems spirally arranged, spreading-reflexed to appressed-ascending, linear to linear-lanceolate, entire or sometimes dentate; strobili present, single and sessile or several on a long scaly-bracted peduncle.

A. strobili several, stalked; leaves with a fine hair tip .... L. clavatum
A. strobili solitary, sessile; leaves lacking a hair tip
    B. erect shoots unbranched or branched 1 or 2 times from the base; horizontal stems superficial .... L. annotinum
    B. erect shoots highly branched forming a treelike appearance; horizontal stems subterranean
        C. lateral branchlets flattened in cross section; leaves unequal in size, lateral ones often twisted; leaves on the main axis tightly appressed ....
        .... L. obscurum
        C. lateral branches round in cross section; leaves equal in size, in 6 rows, not twisted; leaves on the main axis tightly appressed or spreading
            D. leaves on the main axis often spreading; leaves on the lateral branches in 1 row each above and below and 4 lateral rows ....
            .... L.hickeyi
            D. leaves on the main axis tightly appressed; leaves on the lateral branches in 2 rows each above and below, and 2 lateral rows ....
            .... L. dendroideum

Lycopodium annotinum L.   Bristly clubmoss
Horizontal stems at the soil surface; erect shoots 0.6–2.5 dm tall, sparingly dichotomously branched from near the base; leaves spreading to reflexed, dark green, linear-lanceolate, 5–8 mm by 0.6–1.2 mm, margins shallowly dentate toward the tip, apex narrowly acute but lacking a hair tip; strobili solitary, sessile on the main shoot apex; occasional in cool, shaded, often moist forests, often on rocky sites; N and at higher elevations along the Allegheny Front; FAC.

Lycopodium clavatum L.   Common clubmoss or running-pine
Horizontal stems at the soil surface, lightly covered with leaf litter or exposed and similar to the erect shoots; erect stems with 3–6 spreading branches; leaves spirally arranged, spreading to ascending, pale green, linear with the tip extended into a hair 2.5–4 mm long; strobili on branched peduncles; common in open woods, bog margins, or rocky barrens; throughout; FAC.

Lycopodium dendroideum Michx.   Round-branched ground-pine
Horizontal stems deeply subterranean; erect stems 1.2–3 dm tall, branched and tree-like; leaves linear, acute, spreading on the main stem below the branches; on side branches the leaves in 6 rows such that when a branch is viewed in cross section 2 rows are directed up, 1 on each side laterally, and 2 down; strobili sessile, in clusters of 2–6 at the tips of the upright shoots; common in bogs and barrens in moist, acidic soils; throughout; FACU.

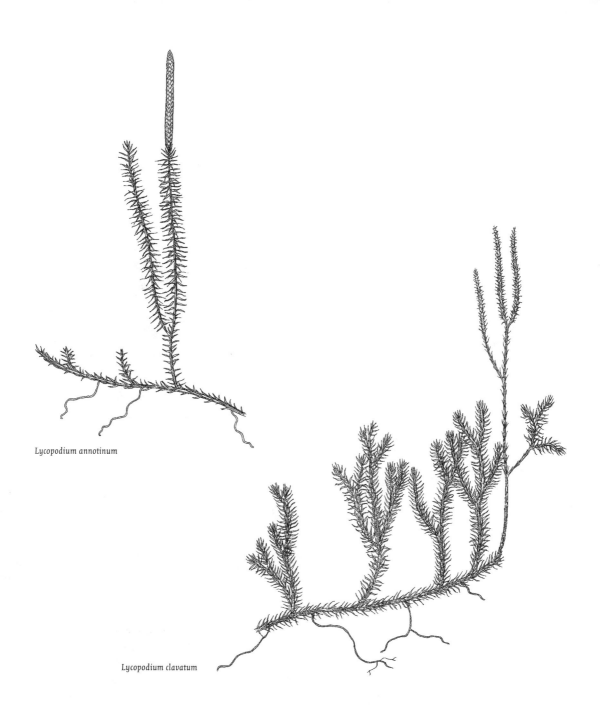

Lycopodium annotinum

Lycopodium clavatum

Lycopodium hickeyi W.H.Wagner, Beitel & Moran   Hickey's ground-pine
Very similar to *L. obscurum* in size and aspect, differing in the ascending branches
and the round cross section of the branches resulting from 6 rows of leaves that
are equal in size and radiate equally, 1 row up, 1 down, and 2 on each side; leaves
not twisted; additional field study is required to determine the distribution of
this species in Pennsylvania.

Lycopodium obscurum

Lycopodium obscurum L.   Flat-branched ground-pine
Similar to L. *dendroideum* and L. *hickeyii*, differing by the tightly appressed leaves
on the main stem axis below the branches and the row of distinctly smaller
leaves directed downward on the branches producing a flat branch; common in
shrubby, successional areas and hardwood forests; throughout; FACU.

SELAGINELLACEAE Spikemoss Family
James D. Montgomery

Plants small, moss-like; stems creeping; leaves small, simple having a midrib
only, either spirally arranged or in 4 ranks; sporangia in the axils of more or less
modified leaves (sporophylls), grouped into distinct 4-sided, terminal strobili;
spores of two types, microspores and megaspores.

Selaginella P. Beauv.

Characteristics of the family.

A. leaves in 4 rows, the lateral ones spreading, ovate; the top and bottom leaves appressed, acuminate .... *S. apoda*
A. leaves spirally arranged, all alike, linear .... *S. rupestris*

Selaginella apoda (L.) Spring   Meadow spikemoss
Stems creeping, forming patches; leaves thin, 1.5–2 mm long, the 2 top rows smaller and appressed, elliptic, 0.8–1.5 mm long; common in moist to wet meadows, borders of marshes, springy slopes, and damp places in lawns; S; FACW.

Selaginella rupestris (L.) Spring   Rock spikemoss
Stems erect, forming mats; leaves thick-textured, 1–1.5 mm long, with bristle tips; uncommon on ledges and rocky slopes, usually in full or partial sun; E and SC.

FERNS
James C. Parks and James D. Montgomery

Herbaceous, nonflowering, spore-bearing plants mainly of moist, shaded habitats; often rhizomatous; leaves (fronds) various from simple to pinnately divided into pinnae that may be divided again into pinnules; the axis of the frond is referred to as the rachis and the axis of a pinna as a rachilla; the stalk at the base of a frond is the stipe; spores are produced in sporangia located on the underside of the fronds in monomorphic species or in separate highly modified fertile fronds in dimorphic species; sporangia are usually clustered in distinct sori that may or may not each have a covering (indusium); juvenile fronds develop from coiled fiddleheads in most genera. The familiar leafy fern plant is the sporophyte phase.

The taxonomy of fern families has undergone considerable change in recent years. For a detailed discussion, the interested reader should consult *Flora of North America*, vol. 2 (Flora of North America Editorial Committee, 1993). Because there is little agreement regarding the classification of ferns at the family level, we have chosen not to present fern families in this manual, but rather to go directly to the genus level. In addition to the genera treated in this key, two fern genera are present in Pennsylvania only as gametophytes. *Vittaria appalachiana* Farrar and Mickel is a thallus and *Trichomanes intricatum* Farrar is a branched filamentous plant resembling a green alga. Found in the deep shade of non-calcareous rock houses, both spread to form small colonies by asexual production of gemmae. Both are relictual and do not produce sporophytes.

A. plants emergent aquatics; leaves clover-like, with 4 wedge-shaped leaflets ....
.... *Marsilea*
A. plants terrestrial; leaves not as above
    B. plants vine-like; pinnae palmately lobed .... *Lygodium*
    B. plants not vine-like, frond simple or pinnately divided
        C. frond single, with a simple to highly divided vegetative blade and, when fertile, an erect branch bearing brownish sporangia; juvenile fronds not developing from curled fiddleheads

D. sterile portion of the frond undivided, ovate-elliptic; fertile portion consisting of an erect axis with two rows of fused sporangia .... *Ophioglossum*

D. sterile portion of the frond variously pinnate to bipinnate, triangular in overall shape in larger species; sporangia several to numerous, distinct .... *Botrychium*

C. fronds several to numerous, arising in clusters or scattered along rhizomes; fronds monomorphic or less commonly dimorphic; individual sporangia very small (barely visible at 10×), usually borne in rust-colored clusters (sori) of various shapes; juvenile fronds developing from coiled fiddleheads

  E. vegetative and fertile fronds (or pinnae) distinctly different in appearance; the fertile portion a green or non-green, highly reduced frond, or separate and highly modified pinnae midway or at the end of the otherwise vegetative frond

    F. fertile frond separate, non-green or highly reduced and green

      G. fertile fronds produced in the early spring, nonpersistent, withering soon after spores are shed, narrow, erect, cinnamon-brown; spores green; vegetative fronds to 1.5 m tall, with cinnamon-brown hairs on stipe and rachis .... *Osmunda*

      G. fertile fronds developing later than the vegetative fronds, pinnate but much reduced, green when young, brown when mature

        H. vegetative fronds pinnate-pinnatifid, arising in whorls from a prominent root crown; fertile fronds inside .... *Matteuccia*

        H. vegetative fronds with a broad, deeply pinnately lobed blade; fertile fronds scattered

          I. vegetative fronds with lobes opposite; lateral veins anastomosing throughout; fertile fronds developing after the vegetative ones, appearing as dark mahogany-colored clusters of spheres, persisting into the winter.... .... *Onoclea*

          I. vegetative fronds with lobes alternate; lateral veins in the lobes anastomosing only near the midvein; fertile fronds much reduced; sori long, aligned end-to-end near midveins .... *Woodwardia*

    F. fertile pinnae occurring in the middle or at the end of 1 or more of the vegetative fronds

      J. fertile portion consisting of 2 or 3 pairs of pinnae in the middle of otherwise vegetative fronds .... *Osmunda claytoniana*

      J. fertile portion consisting of the terminal 1/3–1/2 of otherwise vegetative fronds

        K. fronds to 7 dm tall, pinnate; pinnae with a single basal lobe on the upper side .... *Polystichum acrostichoides*

        K. frond usually >7 dm tall, bipinnate, ultimate segments mostly not lobed .... *Osmunda regalis*

  E. vegetative and fertile fronds weakly to not at all dimorphic, although the fertile fronds may be more slender; sori on the undersurface of the fronds

L. sori near the edge of the blade; indusia, if present, never kidney-shaped

  M. sori with cup-shaped indusia barely visible at vein endings in the notches of the blades; fronds bipinnate-pinnatifid, yellow-green, with abundant hairs and glands .... *Dennstaedtia*

  M. sporangia covered or partly covered by the downward rolled margin of the blade; fertile fronds slightly more slender than the vegetative ones

    N. fronds broadly triangular, to 1.5 m tall, often numerous and forming large clones; basal pinnae as large as the rest of the blade .... *Pteridium*

    N. fronds smaller; basal pinnae not as large as the rest of the blade

      O. fronds wider than long, with a dark maroon stipe to 0.6 m; pinnae forming a spreading-digitate frond .... *Adiantum*

      O. fronds longer than wide, mostly lanceolate or elliptic, pinnate or pinnate-pinnatifid

        P. vegetative fronds about 1/3 shorter than the fertile fronds and with rounded segments; fertile fronds with narrow, elongate segments .... *Cryptogramma*

        P. vegetative and fertile fronds similar in shape, or fertile pinnae slightly more slender

          Q. fronds bipinnate or pinnate-pinnatifid; pinnules oblong to narrowly lanceolate-linear with entire margins and obtuse tips; veins obscure; stipe lustrous black to brown .... *Pellaea*

          Q. fronds bipinnate-pinnatifid, ultimate segments rounded; blade and rachis with copious long hairs; stipe dark brown .... *Cheilanthes*

L. sori usually near the veins of the pinnae; indusia various

  R. blade deeply lobed, leathery; sori lacking indusia, containing both sporangia and sterile, branched hairs (sporangiasters).... .... *Polypodium*

  R. blade once or more pinnately compound, or if merely pinnatifid, then indusia present

    S. sori about as broad as long; indusia various (lacking in *Phegopteris* and *Gymnocarpium*)

      T. stipe and rachis with small, silvery-white, transparent hairs; basal portion of the stipe with sparse or no scales

        U. fronds broadly triangular; sori lacking indusia .... *Phegopteris*

        U. fronds lanceolate-elliptic to narrowly triangular; sori with kidney-shaped indusia .... *Thelypteris*

      T. stipe lacking silvery-white, transparent hairs, however other hairs and/or scales sometimes present; scales of the stipe sparse to abundant

        V. plants >0.4 m tall; blades relatively thick and leathery; scales of the lower stipe very abundant, tan to dark brown; indusia kidney-shaped, or circular with a central attachment (peltate)

W. indusia peltate; ultimate segments of the blade spine-tipped .... *Polystichum braunii*

W. indusia kidney-shaped; ultimate segments crenate to serrate, not spine-tipped .... *Dryopteris*

V. plants <4 dm tall; blades relatively thin; scales of the basal portion of the stipe few, scattered, pale tan; indusia lateral, basal, or none

X. blades broadly triangular; stipe slender, longer than the blade; sori small, indusia absent .... *Gymnocarpium*

X. blades narrower, mostly lance-elliptic, rarely narrowly triangular; stipe equal to or shorter than the blade; indusia present

Y. veinlets in the blade not going to the margins; sori completely enclosed by the bag-like indusia when young; indusium splitting into basal, radiating scales, soon shriveling .... *Woodsia*

Y. veinlets in the blade going to the margins; sori with hood-like indusia enclosing the sporangia when young, and opening to one side then soon shriveling ....

.... *Cystopteris*

S. sori much longer than broad; indusium a flap attached along one side

Z. plants <4 dm tall (mostly <2 dm), growing on shaded rock outcrops or in rocky soil .... *Asplenium*

Z. plants >4 dm tall, growing in soil in rich woods or swamps

AA. sori near the pinnae midveins and parallel to them, linked end-to-end; indusium opening on the side adjacent to the midvein; veinlets anastomosing near the midvein, free toward edge of the blade .... *Woodwardia*

AA. sori parallel to the lateral veinlets of the pinnae; blades lacking anastomosing veins

BB. fronds pinnate; pinnae narrowly lance-oblong............

.... *Diplazium*

BB. fronds pinnate-pinnatifid or bipinnate-pinnatifid

CC. frond pinnate-pinnatifid; groove in the upper surface of the midveins of the pinnae not continuous with that of the rachis .... *Deparia*

CC. frond bipinnate to bipinnate-pinnatifid; groove in the upper surface of the midveins of the pinnae continuous with that of rachis .... *Athyrium*

Adiantum L.

Plants terrestrial or on rocks; rhizomes creeping; stipes shiny black, divided into branches and forming spreading-digitate fronds, from which arise pinnately divided pinnae; sori marginal, oblong, borne beneath the rolled under edges of the pinnules which form a false indusium.

A. segments near the middle of the pinnae generally >3.2 times as long as broad; tips of segments with sharply toothed lobes; segment stalks usually <0.6 mm long .... *A. aleuticum*

A. segments near the middle of the pinnae generally <3.2 times as long as broad; segment tips with rounded or rounded-toothed lobes; segment stalks 0.5–1.5 mm long .... *A. pedatum*

Adiantum aleuticum (Rupr.) Paris   Aleutian maidenhair

Fronds to 4.5 dm tall; segment stalks 0.2–0.9 mm long, usually <0.6 mm, giving the blade a compact appearance; segments near the middle of the pinnae 3–4 times as long as broad, with tips sharply toothed or lobed, the lobes separated by sinuses 0.6–4 mm deep; very rare on serpentine barrens; SE; 🌱.

Adiantum pedatum L.   Northern maidenhair

Fronds up to 7.5 dm tall; segment stalks 0.5–1.5 mm long, giving the blade an open appearance; segments near the middle of the pinnae about 3 times as long as broad, crenulate or round-lobed, the lobes separated by sinuses 0.1–2 mm deep; frequent in moist, rich woods in subacid to circumneutral soils; throughout; FAC–.

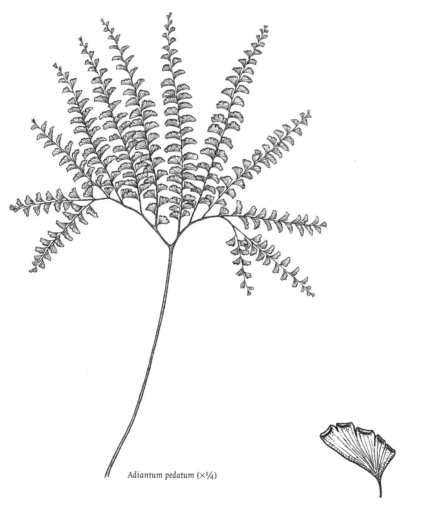

Adiantum pedatum (×¼)

Adiantum pedatum, pinna (×1)

Asplenium L.

Plants terrestrial or frequently on rock faces; rhizomes short, erect, scaly; fronds mostly of 1 type; stipe and rachis green, maroon, or black; blades extremely diverse, from undivided to bipinnate-pinnatifid; veins mostly free, anastomosing in some species; sori linear or crescent-shaped; indusium opening from 1 side. Polyploidy and hybridization are common.

A. frond undivided; margin entire to deeply lobed
- B. blade entire, long, and often rooting at the tip; stipe short, green............
  .... *A. rhizophyllum*
- B. blade deeply lobed or dissected, not rooting at the tip; stipe and basal portion of the rachis maroon
  - C. frond triangular to lanceolate; lobes variable in number and size, very narrowly triangular to straight-sided with the sinuses going nearly to the rachis; spores not well developed (abortive) .... *A. x ebenoides*
  - C. frond narrowly triangular with an elongate tip; lobes broad, rounded, progressively less deeply divided toward the tip; spores well-developed (viable) .... *A. pinnatifidum*

A. frond variously divided into pinnae
- D. pinnae not deeply subdivided; frond oblong-elliptic; stipe and rachis lustrous dark maroon to black nearly to the tip
  - E. pinnae broadly ovate to round; fronds spreading, horizontal, or ascending at the tip .... *A. trichomanes*
  - E. pinnae oblong, 3 or more times as long as wide; most fronds erect
    - F. pinnae with basal auricles that overlap the rachis; fronds dimorphic, some small, reflexed, most larger, erect; plants mostly growing in soil .... *A. platyneuron*
    - F. pinnae lacking basal auricles, not overlapping the rachis; fronds all erect; plants growing on calcareous rocks .... *A. resiliens*
- D. pinnae deeply divided, pinnatifid to bipinnate-pinnatifid; frond triangular to lanceolate, widest near the base; stipe green or maroon
  - G. stipe green or maroon at the base; rachis green; fronds triangular
    - H. pinnae bipinnate-pinnatifid with 4–7 pairs of pinnae, the ultimate segments not lacerate toward the tip .... *A. montanum*
    - H. pinnae bipinnate or bipinnate-pinnatifid with 2–5 pairs of pinnae, the ultimate segments lacerate toward the tip .... *A. ruta-muraria*
  - G. stipe maroon; rachis maroon for ½ its length; fronds commonly lanceolate .... *A. bradleyi*

Asplenium bradleyi D.C.Eaton   Bradley's spleenwort
Rhizome short-creeping, occasionally branched; blade narrowly lanceolate-oblong, pinnate-pinnatifid to commonly bipinnate, 2–17 cm long by 1–6 cm wide, sparsely pubescent; pinnae distinct, 5–15 pairs, variable in shape but mostly lanceolate-ovate, margins dentate-denticulate; sori 3 or more per pinna; rare and local in crevices of dry, shaded, acid rock outcrops; SE; ✵. Hybridizes with *A. pinnatifidum* to form *A. x gravesii* (Graves's spleenwort).

Asplenium x ebenoides R.R.Scott   Scott's spleenwort
Rhizome ascending to erect; sterile fronds smaller and horizontal; the spore-bearing fronds larger and erect; stipe reddish-purple; blade variable and often irregular in shape and margin, commonly narrowly triangular to lanceolate, pin-

natifid, sometimes pinnate toward the base, to 20 cm long; rachis reddish- to purplish-brown at the base, fading to green toward the tip; blade segments irregular in size, shape, and number, often narrowly triangular; rare on moss-covered limestone, always with the parental species nearby. [syn: *A. platyneuron* x *rhizophyllum*]

Asplenium montanum Willd.　　　　　　　　　　　　　Mountain spleenwort
Rhizome creeping, sometimes arched up at the tip, unbranched, producing clusters of stems; stipe maroon-brown at the base, green toward the tip; blade triangular, to 11 cm long by 1–10 cm wide, delicate bluish-green, pinnate to bipinnate at the base, pinnatifid toward the tip; pinnae triangular, coarsely incised to bipinnate; sori few per ultimate segment, often confluent; infrequent on shaded cliffs of noncalcareous rock; scattered. Hybridizes with *A. pinnatifidum* to form *A.* x *trudellii* Wherry.

Asplenium pinnatifidum Nutt.　　　　　　　　　　　　Cliff spleenwort
Rhizome short, creeping to erect; stipe maroon at the base, green toward the tip; blade narrowly triangular, sometimes irregular in outline, to 1.7 dm long, base cordate-auriculate to truncate, margin pinnatifid, progressively less so toward the tip, sometimes with 1 pair of distinct pinnae at the base; pinnae and lobes usually rounded, broadly ovate but sometimes narrower and irregular, especially toward the tip; sori elongate, several per lobe, often confluent when mature; rare in crevices of dry, lightly shaded cliffs of noncalcareous rocks; SE and SW; ❧.

Asplenium platyneuron (L.) Britton, Stearns & Poggenb.　　　Ebony spleenwort
Rhizome short, creeping, covered with dark brown to black scales; vegetative fronds smaller and spreading, fertile ones larger and erect; stipe lustrous reddish-brown, scaly at the base; blade linear to oblanceolate, once pinnate, to 5 dm long by 2–5 cm wide, glabrous to sparsely pubescent; rachis lustrous reddish-brown; pinnae 15–45 pairs, mostly alternate, oblong to quadrangular; margins crenate to serrate, sometimes deeply so; sori in 1–12 pairs on both sides of the midvein; common in dry to moist forests, often at the base of rocks; throughout, but less frequent in the northernmost counties; FACU.

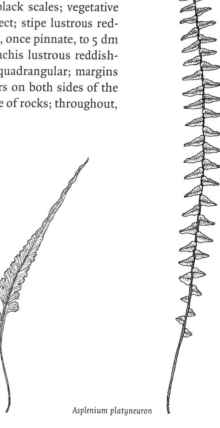

Asplenium pinnatifidum　　　　　Asplenium platyneuron

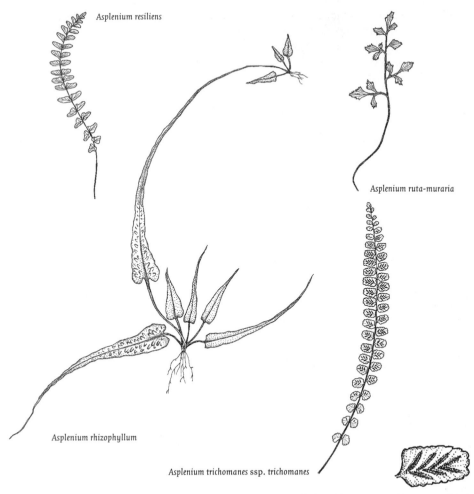

Asplenium resiliens

Asplenium ruta-muraria

Asplenium rhizophyllum

Asplenium trichomanes ssp. trichomanes

Asplenium trichomanes ssp. trichomanes, sori (×2)

Asplenium resiliens Kunze   Black-stemmed spleenwort
Rhizome short, erect, with black scales; fronds all erect; stipe lustrous black;
blade linear to narrowly oblanceolate, once pinnate throughout, 9–20 cm long
by 1–2 cm wide, somewhat leathery; rachis lustrous black throughout; pinnae in
20–40 pairs, mostly opposite, oblong; margins entire to crenate; sori 2–5 pairs
per pinna; rare on limestone rocks; SC; 🌿.

Asplenium rhizophyllum L.   Walking fern
Rhizome short, erect, with dark brown scales; stipe relatively short, reddish-
brown at the base and green toward the tip; blade simple, narrowly triangular to
linear-lanceolate, to 3 dm long, leathery with a cordate-auriculate base, margin
entire to sinuate; apex rounded to long-attenuate and rooting at the tip; sori numerous; frequent on shaded, usually moss-covered rock faces and boulders
where it forms clonal patches, mostly on limestone; scattered.

Aspleium ruta-muraria L.   Wall rue spleenwort
Rhizome short-creeping, often branched; stipe dull reddish-brown at the base,

green above, blade triangular-ovate, once pinnate to pinnate-pinnatifid, 2–6 cm long by 1–4 cm wide; pinnae triangular, deeply cleft into 3–6 ovate-lanceolate ultimate segments with lacerate margins at the ends; veins free, not reaching the margin; rare on shaded, moist limestone cliffs; mostly SE.

Asplenium trichomanes L.   Maidenhair spleenwort
Rhizome short-creeping, sometimes branched; fronds mostly appressed to the rock face or arched-ascending toward the tips; stipe and rachis lustrous reddish-brown to black; blade linear, to 2 dm long and 1.5 cm broad, pinnate with 15–35 pairs of pinnae; pinnae mostly broadly ovate to oblong, base broadly tapered, margin shallowly crenate-serrate with an obtuse tip; veins free; frequent in shaded rock crevices throughout, except NW; 2 varieties:

A. fronds arching upward and outward, away from the rock face; pinnae mostly alternate, spaced more distantly, suborbicular, the lower margin often auriculate; mostly on noncalcareous rock .... ssp. *trichomanes*
A. fronds closely appressed to the rock face, spreading sinuously; pinnae mostly opposite, more crowded, rarely auriculate; only on calcareous rocks ....
.... ssp. *quadrivalens* D.E.Mey.

Athyrium Roth

Athyrium felix-femina (L.) Roth ex Mert.   Lady fern
Plants terrestrial; rhizome short-creeping to ascending; fronds to 1 m tall; stipe straw-colored (or reddish), darker basally, swollen at the base, scales present, scattered, light to dark brown; blade broadly ovate, elliptic, or lanceolate, bipinnate to bipinnate-pinnatifid; pinnae sessile due to a decurrent wing running onto the rachis, or short-stalked, oblong, lanceolate with acuminate tips; pinnules oblong-lanceolate to narrowly triangular, pinnatifid with serrate margins, base often asymmetric; stipe and midribs grooved above, the grooves largely continuous; sori elongate and curved or hooked at the end, indusia attached laterally with ciliate and sometimes glandular margins; rich, moist woods; FAC; 2 varieties:

Athyrium felix-femina
var. *angustism*,
sori (×2)

A. stipe scales brown to dark brown; blade elliptic to broadly lanceolate, widest near the middle; pinnae usually sessile with decurrent wings; pinnules oblong; indusia glandless .... var. *angustum* (Willd.) G.Lawson
Northern lady fern; common throughout.
A. stipe scales light brown; blade ovate to broadly ovate, widest near base; pinnae usually stalked; pinnules narrowly triangular; indusia frequently glandular .... var. *asplenoides* (Michx.) Farw.
Southern lady fern; frequent; mostly S.

Botrychium Sw.

Plants terrestrial; rhizomes very short, erect, with fleshy roots; plants producing a single frond each year (a second frond may be produced if the first is damaged); fronds with an erect fertile portion and a spreading vegetative blade; the vegetative blade few-lobed to 4 times pinnate; fertile portion erect, branched, panicle-like; sporangia large and globose, opening by a slit.

A. vegetative blades triangular, mostly 5–50 cm long and 2–4 times pinnate; frond mostly >2.5 dm tall; fertile segments sometimes absent

B. vegetative blade thin, 3–4 times pinnate; frond appearing in spring, shedding spores in early summer, and deciduous in autumn .... B. *virginianum*

B. vegetative blades leathery, 2–3 times pinnate; frond appearing in late summer, shedding spores in autumn, and remaining green through the winter

    C. terminal segments acute at the tip; margins with pointed teeth ....
    .... B. *dissectum*

    C. terminal segments rounded at the tip; margins with rounded teeth

        D. segments of the vegetative blade all similar in shape, terminal segments not elongate; plants of fields and open woods....B. *multifidum*

        D. vegetative blade with elongate terminal segments; plants of alluvial woods .... B. *oneidense*

A. vegetative blades oblong to narrowly triangular, mostly 2–10 cm long and pinnatifid to bipinnate-pinnatifid; fronds mostly <1 dm tall; fertile segments always present

E. vegetative blade triangular, sessile; segments acute .... B. *lanceolatum*

E. vegetative blades elliptic or ovate, short-stalked; segments obtuse or rounded

    F. vegetative blade pinnate to bipinnate, segments lanceolate to ovate, with a midrib .... B. *matricariifolium*

    F. vegetative blade almost entire to pinnatifid, segments spoon-shaped to fan-shaped, without a midrib .... B. *simplex*

Botrychium dissectum Spreng.   Cut-leaved grape fern
Frond produced in late summer, remaining green, or turning bronze in winter, and lasting until the following spring; vegetative blade to 3 dm long, triangular, at least tripinnate, segments sharply and irregularly serrate with acute apices, terminal segments longer than broad, and much more elongate than lateral segments of the pinnae; spores shed in autumn; common in moist open woods, thickets, and abandoned fields; throughout; FAC. Occurs in several forms varying in the degree of dissection of the pinnae.

Botrychium lanceolatum (J.G.Gmel.) Ångstr.   Triangle moonwort
Frond produced in late spring, usually withering soon after spores are shed in early summer; vegetative blade pinnate-pinnatifid to bipinnate-pinnatifid, 1–6 cm long, triangular, bright shiny green, sessile; ultimate segments lanceolate with acute apices, midrib present; fertile segment erect, slightly longer than the vegetative blade; occasional in moist to wet woods and on hummocks; scattered throughout, but more common E; FACW. Ours is ssp. *angustisegmentum* (Pease & Moore) Clausen.

Botrychium matricariifolium (Döll) A.Br. ex Koch   Daisy-leaved moonwort
Frond produced in late spring, withering soon after spores are shed in early summer; vegetative blades 1–10 cm long, pinnate-pinnatifid to bipinnate-pinnatifid, lanceolate to ovate, light, dull green, short-stalked; ultimate segments ovate with obtuse apices, midrib present; fertile segment overtopping the vegetative portion; frequent in moist open woods, thickets, and abandoned fields; FACU.

Botrychium multifidum (J.G.Gmel.) Rupr.   Leathery grape fern
Frond produced in midsummer, remaining green or turning bronze in winter, and lasting to the following summer; spores shed in late summer; vegetative

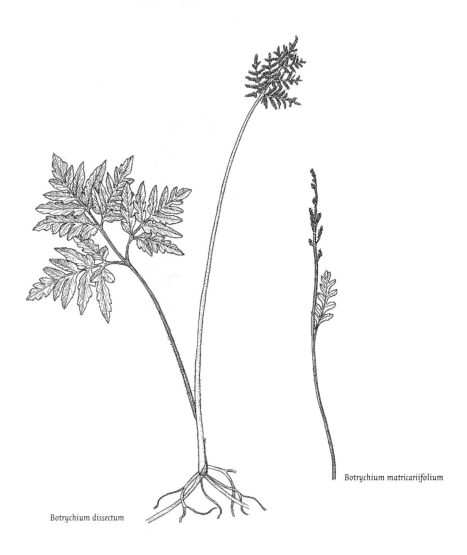

Botrychium matricariifolium

Botrychium dissectum

blade to 3 dm long, triangular, at least tripinnate; segments rounded-serrate to entire, with obtuse apices; terminal segments nearly as broad as long and similar to the lateral segments in shape; rare in abandoned fields, thickets, and barrens; mostly E, with a few widely scattered sites W; FACU.

Botrychium oneidense (Gilbert) House   Blunt-lobed grape fern
Frond produced in late summer, remaining green (not turning bronze) in winter, and lasting into the following summer; spores shed in late summer; vegetative blades to 2 dm long, triangular, 2–3 times pinnate; segments rounded-serrate, with obtuse apices; terminal segments longer than broad, and more elongate than the lateral segments; uncommon in alluvial soil in moist rich woods and floodplain forests; throughout. B. oneidense is often confused with B. dissectum. The difference in habitat between this and other grape ferns is pronounced; there are no records of this species from abandoned fields where other grape ferns are often found.

Botrychium virginianum

Botrychium simplex Hitchc. Least moonwort
Frond produced in late spring, withering soon after spores are shed in early summer; vegetative blades nearly simple, lobed, pinnatifid, or pinnate, 1–6 cm long, linear to ovate-lanceolate, dull light to medium green, short- to long-stalked; segments, if any, fan shaped, blunt, lacking a midrib; fertile portion shorter than to longer than the vegetative blade; rare in moist woods, meadows, and barrens; E and NW; FACU.

Botrychium virginianum (L.) Sw. Rattlesnake fern
Frond produced in spring, withering in autumn; spores shed in late spring; vegetative blades 3–4 times pinnate, 1.5–5 dm long, triangular, sessile; fertile portion, if present, overtopping the vegetative blade; common in rich loamy woods and moist wooded slopes; throughout; FACU.

Cheilanthes Sw.

Cheilanthes lanosa (Michx.) D.C.Eaton Hairy lip fern
Small to medium-sized plant with short-creeping rhizomes; fronds all alike,

lance-oblong, evergreen, bipinnate-pinnatifid, 1.5–3 dm tall; rachis chestnut brown; blades sparsely hairy on both sides; ultimate segments oblong, lobed; sori nearly round, marginal, completely or partially covered by the false indusia formed by rolled under margins of the pinnae; occasional on dry cliffs and rock outcrops, in subacid to circumneutral soils; SE and SC.

Cryptogramma R.Br.

Cryptogramma stelleri (Gmel.) Prantl   Slender rockbrake
Small plants of rock crevices; rhizomes creeping; fronds deciduous, bipinnate, somewhat dimorphic; the vegetative fronds lanceolate-ovate, to 1 dm long with rounded pinnae; the fertile fronds to 2.5 dm, narrower and longer than the vegetative; sori continuous along the margins of the fertile segments with a false indusium formed by the rolled under margins of the pinnae; rare on moist calcareous cliffs and ravines; NC; FACU–; 🌿.

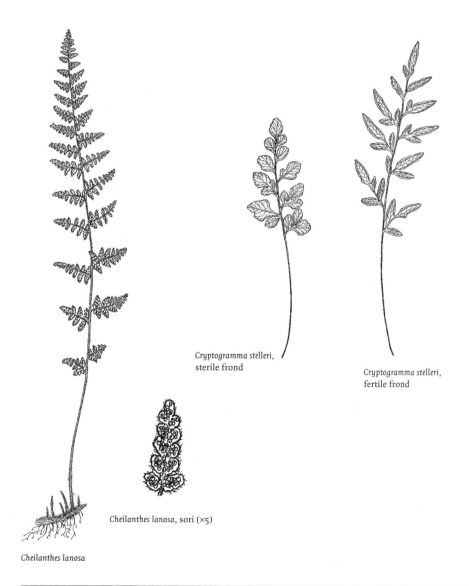

Cryptogramma stelleri, sterile frond

Cryptogramma stelleri, fertile frond

Cheilanthes lanosa, sori (×5)

Cheilanthes lanosa

Cystopteris Bernh.

Plants terrestrial or on rocks; rhizome short- to long-creeping; stipe slender, sparsely scaly basally, grooved above, glabrous; blade ovate-lanceolate to narrowly triangular, thinly herbaceous, bipinnate-pinnatifid, glabrous or with glandular hairs on the blades and midribs; margins of the ultimate segments crenulate, dentate, or serrate; veins free, branching, and ending at the margin in teeth or notches; sori between the midrib and pinnule margins, round; indusium hood-like, arching over the sorus, often obscure at maturity. Many of the species are very similar morphologically and field identification is further complicated by the presence of hybrids.

A. fronds narrowly triangular (widest at the base), glands present; bulblets commonly present on the rachis
    B. fronds densely glandular; apex long-attenuate; bulblets 1 cm or more in diameter, abundant .... *C. bulbifera*
    B. fronds with scattered glands; apex acuminate; bulblets <1cm, not abundant .... *C. tennesseensis*
A. fronds lanceolate to narrowly elliptic (widest above the base), glabrous or with a few glands; bulblets rarely present
    C. rhizome extending beyond the current season's fronds, covered with golden-tan hairs; first basal pinnule of the basal pinna stalked, deeply dissected, base cuneate; stipe straw-colored at the base .... *C. protrusa*
    C. rhizome not extending beyond the current season's fronds; first basal pinnule of the basal pinna mostly sessile; stipe base usually dark-purple
        D. fronds with widely scattered glands on the rachis and midribs; basal pinnae not more distant from the second pair than are the subsequent pairs from each other .... *C. laurentiana*
        D. fronds glandless; first pair of pinnae somewhat to considerably more distant from second pair than are the subsequent ones from each other
            E. pinnae diverging from the rachis at an acute angle and often curving toward the tip of the blade; blade ovate, 2–2.5 times as long as wide; pinnules ovate with crenulate teeth and an obtuse or rounded, sessile or short-stalked base .... *C. tenuis*
            E. pinnae diverging from the rachis at 90%, not curving; blade lanceolate, 3–4 times as long as wide; pinnules ovate to lanceolate with serrate-dentate margins and a truncate to broadly obtuse, sessile base .... *C. fragilis*

Cystopteris bulbifera (L.) Bernh.   Bulblet bladder fern
Rhizome short-creeping, covered with old stipe bases; fronds to 7.5 dm; stipe relatively short, usually green to straw-colored; blade triangular to narrowly so with long-attenuate tip, bipinnate-pinnatifid; pinnae closely spaced; rachis, midrib, and blade with copious, small glandular hairs; spherical, bilobed bulblets 1 cm or more in diameter common on the upper side of the rachis and sometimes on the pinnae; occasional on shaded, calcareous cliffs; scattered throughout; FAC.

Cystopteris fragilis (L.) Bernh.   Fragile fern
Rhizome short-creeping, hairs absent; fronds several, clustered at the tip of the rhizome, to 4 dm long, even small fronds usually fertile; stipe purplish and with scattered scales basally, usually fading to green in the rachis; blade narrowly

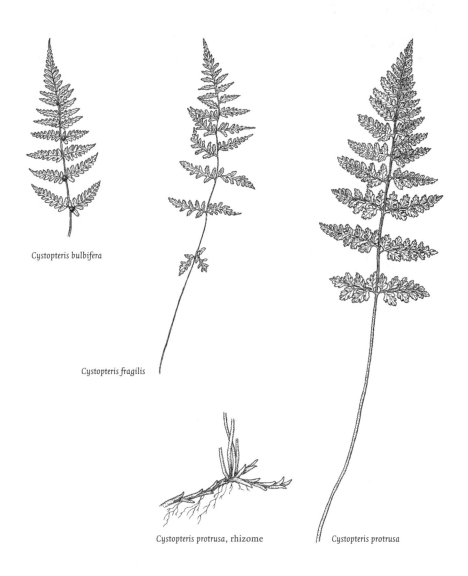

Cystopteris bulbifera

Cystopteris fragilis

Cystopteris protrusa, rhizome

Cystopteris protrusa

lanceolate-elliptic, bipinnate-pinnatifid; basal pair of pinnae noticeably more separated from the next distal pair than are the subsequent pairs; lower basal pinnules sessile, base truncate to broadly obtuse, margins serrate-dentate; occasional on alkaline rock; throughout; FACU.

Cystopteris laurentiana (Weath.) Blasdell   Laurentian bladder fern
Rhizome short-creeping, covered with old stipe bases; fronds several, clustered, to 4.5 dm long; stipe dark-purplish with scattered scales at the base, pale above; blade ovate to narrowly ovate, widest above the base, bipinnate-pinnatifid; small (<5 mm) misshapen bulblets sometimes present; very rare and scattered on alkaline rocks; N; ❦. [*C. bulbifera* x *fragilis* of Rhoads & Klein, 1993]

Cystopteris protrusa (Weath.) Blasdell   Protruding bladder fern
Rhizome creeping, covered with golden-tan hairs that are most noticeable near the tip, apex extending >1 cm beyond the current season's fronds; stipe with a few scales basally, straw-colored to green; blade bipinnate-pinnatifid, lowest pair of pinnae not noticeably more separated than the more distal pairs; lower

basal pinnules short-stalked, ovate-triangular, deeply and narrowly dissected with a broadly obtuse base and mostly serrate-dentate margins; fronds seasonally dimorphic, early fronds smaller, pinnules less deeply divided, margins rounder, often sterile; later fronds usually fertile, to 4.5 dm long, segments more deeply divided; common in moist woods, usually terrestrial; S.

Cystopteris tennesseensis Shaver   Tennessee bladder fern
Rhizome short-creeping, covered with old stipe bases, hairs absent; fronds to 4.5 dm tall; stipe mostly straw-colored, darker and with scattered scales basally; blade triangular with short-attenuate tip, bipinnate-pinnatifid; pinnae closely spaced, not as deeply dissected as in the otherwise similar *C. bulbifera*; bulblets sometimes present, <1 cm in diameter; rare on shaded alkaline cliffs and abutments; S; ❧.

Cystopteris tenuis (Michx.) Desv.   Fragile fern, MacKay's brittle fern
Rhizome short-creeping, hairs absent; fronds several, clustered at the tip of the rhizome, to 4 dm tall, even small fronds usually fertile; stipe purplish, with a few

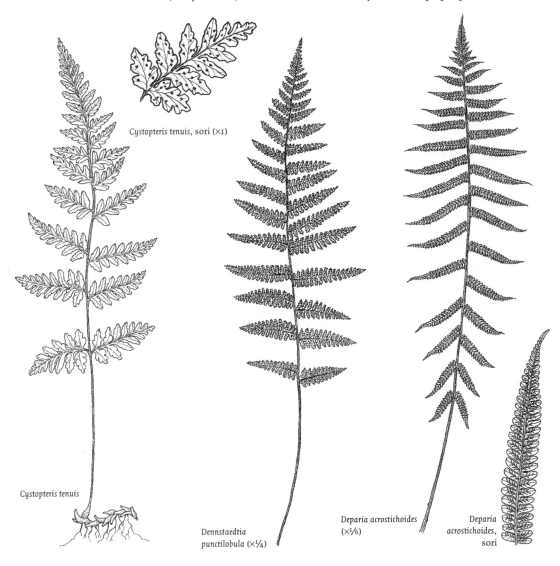

Cystopteris tenuis, sori (×1)

Cystopteris tenuis

Dennstaedtia
punctilobula (×¼)

Deparia acrostichoides
(×⅙)

*Deparia
acrostichoides*,
sori

scattered scales basally, glabrous and paler distally; blade lanceolate to narrowly elliptic, usually widest just below middle, twice pinnate-pinnatifid; the lowest pair of pinnae more distant from the next pair than are the subsequent pairs; pinnae lanceolate; lower basal pinnules sessile, ovate, not usually deeply dissected, with crenulate margins and a broadly obtuse to tapered base; common on alkaline rocks; throughout.

Dennstaedtia Bernh.

Dennstaedtia punctilobula (Michx.) T.Moore Hay-scented fern
Plant terrestrial; clone-forming; rhizome slender, long-creeping, dark reddish-brown and densely hairy; fronds clustered, 0.4–1.3 m tall; stipe dull brown basally, lighter above; stipe, rachis, midribs, and blade covered with numerous hairs, some glandular; blade dull yellowish-green, ovate-lanceolate to narrowly triangular, bipinnate-pinnatifid; pinnae narrowly triangular, pinnules numerous, truncate-sessile, pinnatifid with basal segments opposite, distal ones alternate, margins irregularly crenate; veins of the ultimate segments pinnately branched, not going to the margins; sori small, globose, numerous, produced near the blade margins on short, lateral pinnule lobes; common in mesic woods and borders; throughout; UPL. In areas overbrowsed by deer vast populations of this unpalatable fern dominate the forest floor.

Deparia Hook. & Grev.

Deparia acrostichoides (Sw.) M.Kato Silvery glade fern
Plant terrestrial; rhizome short-creeping; fronds to 1.2 m tall; stipe deeply grooved above with the groove continuing into the rachis, darker in color at the base, straw-colored above; blade oblong, lanceolate, or oblanceolate, pinnate-pinnatifid, acuminate; stipe, rachis and midribs covered with pale brown multicellular hairs and sometimes also narrow scales; pinnae oblong, long-acuminate, sessile or with a short stalk, base truncate, margins entire or slightly crenate, lateral veins mostly unbranched and ending just short of the edge in a swelling, basal pair of pinnae often angled toward the stipe; sporulating in mid-summer, fertile fronds usually slightly smaller; sori elongate, parallel to the lateral veinlets of the pinnae, mostly straight; indusia flap-like, opening on the upper edge; common in rich, moist woods and shady slopes; throughout; FAC. [syn: *Athyrium thelypteroides* (Michx.) Desv.]

Diplazium Sw.

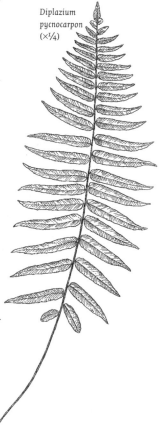

*Diplazium
pycnocarpon*
(×¼)

Diplazium pycnocarpon (Spreng.) Broun Narrow-leaved glade fern
Plant terrestrial with a creeping rhizome; fronds to 1.1 m tall; stipe with a few tan scales at the base, deeply grooved above; blade lanceolate, pinnate; pinnae linear to narrowly lanceolate, tip acuminate, base often slightly asymmetric, truncate, margin crenulate with a row of clear cells, veinlets branching 1 or 2 times and ending just short of the edge in a swelling; pinnae dark green, the grooves of midribs continuous with that of the rachis; sporulating in mid summer, fertile fronds slightly smaller; sori elongate, straight, along lateral veinlets of the pinnae; indusia flap-like, opening on the upper edge; scattered in moist, wooded glades and alluvial thickets; throughout, but missing from the Valley and Ridge Physiographic Province; FAC. [syn: *Athyrium pycnocarpon* (Spreng.) Tidestr.]

Dryopteris Adans.

Plants terrestrial; rhizomes short-creeping or erect; fronds mostly large, evergreen or deciduous, pinnate-pinnatifid to tripinnate; stipes scaly at least at the base; sori round with kidney-shaped indusia. In the field the species are fairly easy to distinguish; however, the numerous hybrids, which can be recognized by their aborted spores, pose a challenge. The key below includes the 2 most frequently encountered hybrids, others can be inferred by their abortive spores and intermediate characteristics.

A. spores all or nearly all well-developed, normal in size and shape (fertile species)
  B. blades pinnate-pinnatifid throughout or bipinnate at the base only
    C. sori at or near the margin of the segments; blades evergreen, leathery, blue-green .... *D. marginalis*
    C. sori midway between the margin and midrib or closer to the midribs of the segments; blades deciduous, or with smaller evergreen vegetative ones, herbaceous in texture and green
      D. lowest pinnae ovate (broadest above the base); fertile fronds ovate to ovate-lanceolate, sides not parallel; scales at the stipe base dark brown or with a dark brown stripe
        E. blade ovate, tapering abruptly to the tip; sori nearer the midveins than the margins .... *D. goldiana*
        E. blades gradually narrowed to the tip, ovate-lanceolate; sori about midway between the midveins and margins .... *D. celsa*
      D. lowest pinnae triangular (broadest at the base); fertile fronds lanceolate and parallel-sided; scales at the stipe base tan
        F. basal pinnae 1.5–2 times longer than wide; pinnae of fertile fronds nearly in the plane of the blade .... *D. clintoniana*
        F. basal pinnae about as wide as long, pinnae of fertile fronds turned at an angle to the plane of the blade .... *D. cristata*
  B. blades bipinnate to tripinnate throughout
    G. indusia and axes of segments glandular; fronds evergreen; first basal pinnule of the basal pinna often shorter than the second....*D. intermedia*
    G. indusia and axes of segments not glandular; fronds deciduous; first basal pinnule of the lowest pinna often longer than the second
      H. first basal pinnule not much wider than the one above; blades ovate-lanceolate; fronds more or less erect ....*D. carthusiana*
      H. first basal pinnule about as wide as the next two above; blades ovate-triangular; fronds spreading.................................. *D. campyloptera*
A. spores few or misshapen (hybrids)
  I. blades tripinnate, lowest pinnae nearly as long as middle ones ....
  .... *D. x triploidea*
  I. blades bipinnate, lowest pinnae much shorter than middle ones ....
  .... *D. x boottii*

Dryopteris x boottii (Tuckerm.) Underw.   Boott's hybrid wood fern
Fronds bipinnate at least at the base, evergreen, 3–8 cm long, lanceolate-ovate; basal pinnae triangular; scales light tan to brown; indusia and axes of the pinnules glandular; frequent in swamps, wet thickets, and moist woods; scattered throughout; FACW. [syn: *D. cristata x intermedia*]

Dryopteris campyloptera (Kunze) Clarkson   Mountain wood fern
Fronds tripinnate-pinnatifid, deciduous, 2.5–9 dm long, ovate-triangular; basal
pinnae with the first lower pinnule much longer than second and twice as wide
as the first pinnule above; scales light brown, sometimes with a darker patch at
the base; indusia and axes not glandular; rare in cool moist woods; N and SC; 🌿.

Dryopteris carthusiana (Vill.) H.P.Fuchs   Spinulose wood fern
Fronds bipinnate to tripinnate-pinnatifid, deciduous, 1.5–7 dm long, ovate;
basal pinnae with first lower pinnule often longer than the second and slightly
wider than the pinnule above; scales light brown; indusia and axes glabrous,
without glands; common in swampy woods, moist wooded slopes, stream
banks, around old houses, foundations, and in conifer plantations; throughout;
FAC+. *D. carthusiana* is often confused with *D. intermedia*. The easiest way to distinguish these common species is by the hatpin-like glands of *D. intermedia*,
which are lacking in *D. carthusiana*. Hybridizes with *D. clintoniana* [*D. x benedictii*
Wherry], *D. cristata* [*D. x uliginosa* (A. Br.)Druce], and *D. marginalis* [*D. x pittsfor-
densis* Slosson] in addition to *D. intermedia* as described below.

Dryopteris celsa (W.Palmer) Small   Log fern
Fronds pinnate-pinnatifid, deciduous, 6.5–12 dm long, ovate, gradually tapering
to the tip; pinnae ovate, widest above the base; scales dark brown at least at the
base, or tan with a dark stripe; sori about midway between the margin and
midvein of the segments; indusia glabrous; rare on seepage slopes, hummocks,
and logs in swamps; mostly SE; OBL; 🌿. Hybridizes with *D. cristata*, *D. goldiana*,
and *D. marginalis* [*D. x leedsii* Wherry].

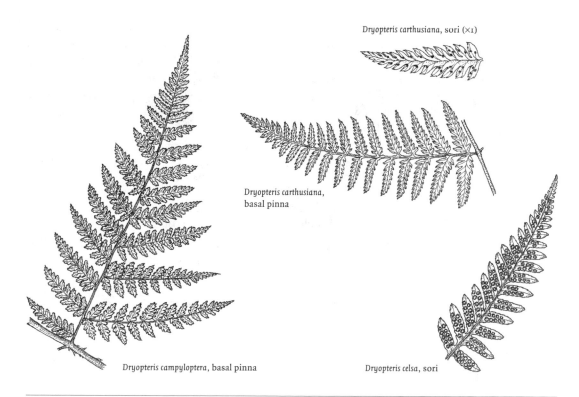

Dryopteris carthusiana, sori (×1)

Dryopteris carthusiana,
basal pinna

Dryopteris campyloptera, basal pinna

Dryopteris celsa, sori

Dryopteris clintoniana (D.C.Eaton) Dowell   Clinton's wood fern
Fronds pinnate-pinnatifid, 4.5–7(10) dm long, lanceolate with nearly parallel
sides; fertile fronds deciduous and erect, with 1–several smaller, more or less
prostrate, evergreen vegetative fronds; pinnae elongate-triangular, widest at the
base, about twice as long as wide, in the plane of the blades or only slightly
twisted; scales light brown, sometimes with a dark brown center; sori about
midway between the margin and midvein of the segments; indusia glabrous;
rare in swampy woods, especially red maple swamps; N; FACW+; ✿. Hybridizes
with D. intermedia [D. x dowellii (Farw.) Wherry] and D. marginalis [D. x burgessii].

Dryopteris cristata (L.) A.Gray   Crested wood fern
Fronds somewhat dimorphic, the larger fertile fronds erect, deciduous, pinnate-
pinnatifid, 3.5–7 dm long, narrow-lanceolate, with parallel sides; several
smaller evergreen vegetative fronds forming a rosette; pinnae triangular, widest
at the base, often twisted at right angles to the plane of the blade; basal pinnae
about as wide as long; scales light brown; sori about midway between the margin and midrib of the segments; indusia glabrous; common in wooded swamps,
or open shrubby wetlands; throughout; FACW+. Hybridizes with D. marginalis
[D. x slossonae Wherry] in addition to those listed above.

Dryopteris goldiana (Hook.) A.Gray   Goldie's wood fern
Fronds pinnate-pinnatifid, sometimes bipinnate at the base, deciduous, green
and often mottled at the tip, 3.5–12 dm long, ovate, abruptly tapering to the tip;
pinnae ovate, widest above the base; scales glossy dark brown to almost black;
sori nearer the midvein than the margin; indusia glabrous; scattered in rich
moist woods, ravines, or at the edges of swamps; mostly SE and SW; FAC+. Hybridizes with D. marginalis [D. x neowherryi W.H.Wagner] in addition to those
mentioned above.

Dryopteris intermedia (Muhl.) A.Gray   Evergreen wood fern, fancy fern
Fronds tripinnate-pinnatifid, evergreen, 3–9 dm long, ovate; basal pinna with
first lower pinnule usually shorter than the second and not much, if any, wider
than the one above; scales light brown; indusia, and usually also the axes of the
segments, with minute glandular hairs; common in moist rocky woods, conifer
plantations, and swampy woods; throughout; FACU. Hybridizes with D. marginalis in addition to those mentioned above.

Dryopteris marginalis (L.) A.Gray   Marginal wood fern
Fronds pinnate-pinnatifid, sometimes bipinnate at the base, evergreen, blue-
green in color and leathery in texture, 3–10 cm long; blades ovate-lanceolate;
stipe scales bright tan, in a dense tuft at the base of the stipe; sori near the margins of the segments, indusia glabrous; common on rocky wooded slopes and
ravines, edges of woods, stream and road banks, and rock walls throughout;
FACU–.

Dryopteris x triploidea Wherry   Triploid hybrid wood fern
Fronds tripinnate, sub-evergreen (green well into winter, but eventually turning
brown) 3–9 dm long, ovate; scales light tan; indusia and axes of the pinnules
glandular; frequent in swamps, damp woods, thickets, roadside banks, and
stone walls; throughout where both parental species are found; FAC. [syn: D.
carthusiana x intermedia]

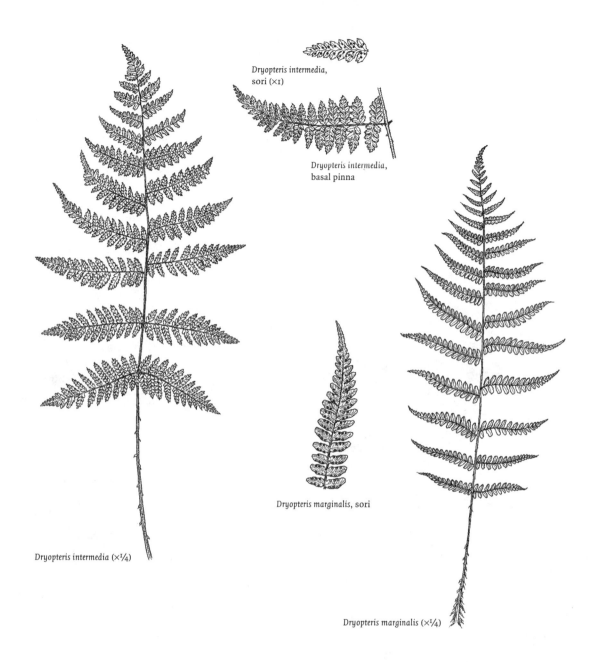

Dryopteris intermedia, sori (×1)

Dryopteris intermedia, basal pinna

Dryopteris marginalis, sori

Dryopteris intermedia (×¼)

Dryopteris marginalis (×¼)

Gymnocarpium Newman

Plants terrestrial from slender, long-creeping rhizomes; fronds slender, lax and delicate; stipe 1.5–3 times as long as the blade; blade 3–14 cm long, broadly triangular, ternate, 2–3 times pinnate-pinnatifid, glabrous, or glandular on the lower surface, especially on the midribs; pinnae crenulate; the lowest pair of pinnae the longest, inequilateral with lower edge pinnules longer than upper edge members of a pair; sori small, round, in one row on each side of the midrib, indusia absent; spores brownish.

A. first basal pinnules of the lowest pinnae sessile .... *G. dryopteris*
A. first basal pinnules of the lowest pinnae stalked .... *G. appalachianum*

Gymnocarpium appalachianum K.M.Pryer & Haufler   Appalachian oak fern
Very similar to *G. dryopteris*, differing in the characters in the key; rare and local in talus slopes and maple-birch-hemlock forests, usually on or near cool, shaded, moist rocks; SC; 🌿.

Gymnocarpium dryopteris (L.) Newman   Common oak fern
Lowest pinnae 1–12 cm long, more or less perpendicular to the rachis and approximately the same size as the remainder of the frond, creating a ternate appearance to the frond; basal pinnae with the first lower pinnules sessile, their ultimate segments equal to or longer than the next lower pairs, the members of the pairs of pinnules on either side of the midrib about equal in size; the second pinnae usually sessile and resembling the lowest basal pinnules of the first pinnae; occasional in cool, coniferous or mixed forests in ravines, often on talus in pockets of humus; mostly N; UPL. A hybrid with *G. appalachianum* (*G.* x *heterosporum* W.H.Wagner) was known from single site in Bedford Co. which has since been destroyed.

Lygodium Sw.

Lygodium palmatum (Bernh.) Sw.   Climbing or Hartford fern
Plant terrestrial; rhizome long-creeping; fronds climbing by a twining rachis, 2–3 m long; vegetative portion below, with each pinna divided into two palmately lobed pinnules; fertile portion above, dichotomously forked into contracted,

Lygodium palmatum

Gymnocarpium dryopteris

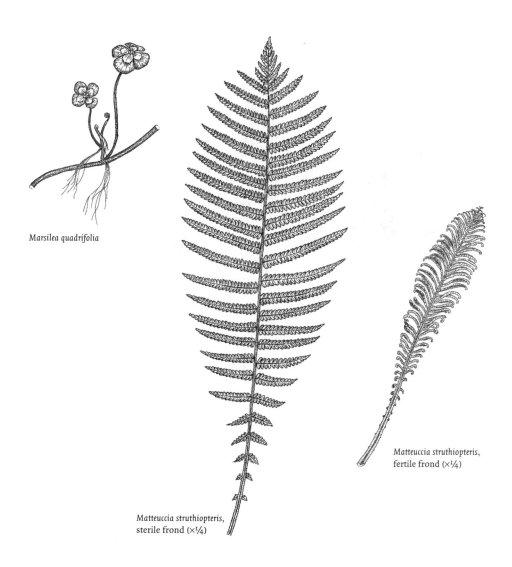

Marsilea quadrifolia

Matteuccia struthiopteris,
fertile frond (×¼)

Matteuccia struthiopteris,
sterile frond (×¼)

palmately lobed segments; rare, but sometimes locally abundant in moist thickets, barrens, and edges of swampy, open woods, in acidic, peaty soil; NE and a few scattered sites elsewhere; FACW; 🌿.

Marsilea L.

Marsilea quadrifolia L.   European water-clover
Plant an emergent aquatic, forming colonies in shallow water; horizontal stems cord-like, bearing roots at nodes and internodes; sterile fronds divided into 4 wedge-shaped pinnae and resembling a 4-leaf clover; stipe 5.4–16.5 cm; sporangia borne in 2 bean-shaped sporocarps on a short, branched stalk near the base of the stipe of the sterile fronds; occasionally naturalized in shallow water of ponds and streams; E; native to Europe; OBL.

Matteuccia Tod.

Matteuccia struthiopteris (L.) Tod.   Ostrich fern
Plant terrestrial; fronds strongly dimorphic; vegetative fronds to 1.3 m tall; stipe

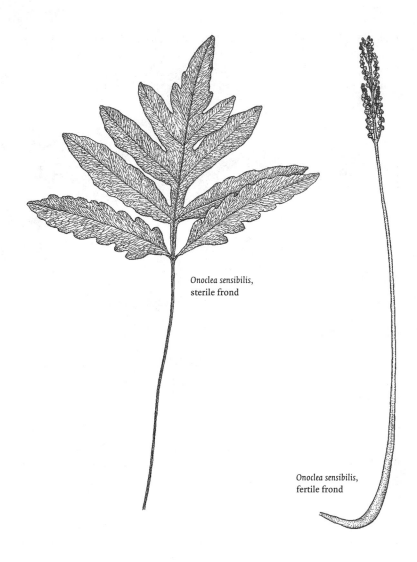

Onoclea sensibilis,
sterile frond

Onoclea sensibilis,
fertile frond

black and swollen at the base with light brown scales; blade elliptic to oblanceolate, pinnate-pinnatifid; pinnae linear, longest near or above the middle of the frond, tapering to the base and tip, veins free; fertile fronds appearing in mid to late summer and persisting all winter, to 6.5 dm tall, blade oblanceolate, pinnate, green when young but soon turning brown, edges of the pinnae strongly recurved, covering the single row of sori; frequent in moist alluvial flats, floodplains, swamps, and rich woods; scattered throughout; FACW.

Onoclea L.

Onoclea sensibilis L.   Sensitive fern

Plant terrestrial; fronds widely spaced, strongly dimorphic with vegetative fronds produced in late spring and dying in autumn, fertile fronds produced in midsummer and persisting through the winter and spring; vegetative fronds with many tan scales on the unopened fiddleheads but glabrous at maturity; stipe swollen and darker in color at the base, lighter above; blade broadly triangular, deeply pinnatifid with 5–11 opposite, linear lobes on each side of rachis, the lowest lobes almost distinct, margins entire or commonly irregularly coarsely sinuate-laciniate; fertile fronds with contracted pinnae that are rolled up to form bead-like spheres, green when young but soon turning dark brown; common in sunny or shaded swamps, marshes, moist meadows, and roadside banks; throughout; FACW.

Ophioglossum L.

Plants small, terrestrial, producing a single fleshy frond divided into a simple, lateral, vegetative blade that is 2–10 cm long by 2–5 cm wide and an erect fertile portion with two rows of sporangia embedded in a fleshy axis.

A. veins of the blade with areoles enclosing a second set of areoles; apex pointed ....*O. engelmannii*
A. veins of the blade with areoles enclosing free secondary veinlets; apex rounded
    B. vegetative blade pale green, broadest near the middle, with no persistent basal sheath .... *O. pusillum*
    B. vegetative blade broadest near the base, with a persistent basal sheath .....
    .... *O. vulgatum*

Ophioglossum engelmannii Prantl   Limestone adder's-tongue
Vegetative blade ovate to ovate-lanceolate, pale green, often folded, apex pointed, veins with a second network inside the first set of areoles; very rare in limestone glades and pastures; SC; FACU; 🌿.

Ophioglossum pusillum Raf.   Northern adder's-tongue
Vegetative blade elliptic, broadest near the middle and narrowing to the base, dull green, apex rounded, veins with free included veinlets inside the primary areoles; sheath absent at the base of the stalk; uncommon in wet meadows and moist open woods; throughout.

Ophioglossum vulgatum L.   Southern adder's-tongue
Vegetative blade ovate, broadest near the base, bright green, rounded at the apex, veins with free veinlets within the primary areoles; sheath persistent at the base of the stipe; rare in moist to dry open woods and floodplains; SE; believed to be extirpated; FACW; 🌿. Ours is var. *pycnostichum* Fernald. [syn: *O. pycnostichum* (Fernald) Löve & Löve]

Ophioglossum pusillum

Osmunda L.

Plants terrestrial; rhizomes stout, creeping, covered with old stipe bases and fibrous roots; fronds to 1.5 m, pinnate-pinnatifid to bipinnate, with reddish or pale hairs but lacking scales; fertile and vegetative fronds fully dimorphic or with distinct fertile sections within the otherwise vegetative fronds; sporangia not in sori, when mature appearing like opened "clam shells"; fertile segments green when young, cinnamon-red when mature.

A. vegetative fronds bipinnately compound; fertile pinnae at the ends of the fronds .... *O. regalis*

A. vegetative fronds pinnate or pinnatifid; fertile pinnae borne in the middle of the fronds or on separate, fully fertile fronds

    B. stipe and rachis with dense reddish hairs, especially at the stipe-rachis joints; ultimate segments acute, green; lowest pinna segments not overlapping the rachis; fertile fronds separate .... *O. cinnamomea*

    B. stipe and rachis glabrescent, with only scattered hairs; ultimate segments of the pinnae obtuse- rounded, bluish-green; lowest pinna segments often overlapping the rachis; fertile pinnae in the middle of the frond .... *O. claytoniana*

Osmunda cinnamomea L.   Cinnamon fern

Fronds dimorphic; stipe and rachis with rusty brown hairs, especially in tufts at pinna-rachis joints; vegetative fronds to 1.5 m tall; vegetative blade broadly lanceolate-ovate, pinnate-pinnatifid; ultimate segments broadly acute apically, margin entire to shallowly crenulate; veins branching dichotomously; first basal pinnule of the basal pinna not overlapping the rachis; fertile fronds appearing in early spring, to 0.5 m tall, initially greenish, soon turning cinnamon-red and soon afterward withering; common in swamps, vernal ponds, seeps, and stream margins in acid soils throughout; FACW.

Osmunda claytoniana L.   Interrupted fern

Fronds with a section of fertile pinnae midway on the rachis; stipe light brown and hairy when young, glabrescent when mature and lacking tufts of hairs at the pinna-rachis joints; blade broadly lanceolate-ovate to elliptical, bluish-green, pinnate-pinnatifid; vegetative pinnae oblong, deeply pinnatifid, segments closely spaced, obtuse-rounded apically, margins entire, veins branching dichotomously; first basal segment of the basal pinna commonly overlapping the rachis; fertile pinnae greatly reduced in size, composed of clusters of sporangia that form in late spring and later wither leaving a gap; common in rich woods, seeps, swamps, bog edges, and hummocks; throughout; FACU–.

Osmunda regalis L.   Royal fern

Fronds clustered, to 1.5 m tall; blade ovate-elliptic, glabrescent, coarsely bipinnate; vegetative pinnae widely spaced, oblong-ascending; pinnules ternate, narrowly lanceolate-triangular, margins entire; fertile pinnae at the ends of the fronds, bipinnate, consisting of clusters of sporangia, at first green, turning brown-black and persisting; frequent in swamps, bogs, and moist acidic soils; throughout; OBL. Ours is var. *spectabilis* (Willd.) A.Gray.

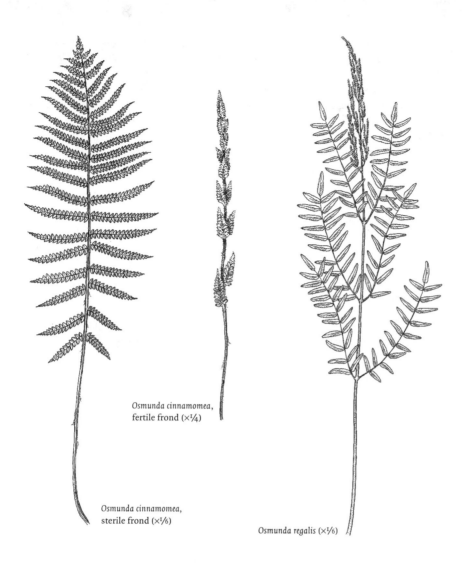

Osmunda cinnamomea,
fertile frond (×¼)

Osmunda cinnamomea,
sterile frond (×⅙)

Osmunda regalis (×⅙)

Pellaea Link

Small to medium-sized plants of rock crevices or masonry walls; fronds bipinnate, pinnate-pinnatifid, or pinnate with glabrous ultimate segments; sori marginal in a continuous line, with a poorly defined false indusium formed by the
rolled under edge, in addition the segments often rolling lengthwise during dry
periods.

A. stipes dull brown or purple, rough hairy; fertile fronds larger than vegetative
 fronds .... *P. atropurpurea*
A. stipes shining brown, glabrous except for a few scales at the base; fertile and
 vegetative fronds alike .... *P. glabella*

Pellaea atropurpurea (L.) Link   Purple cliffbrake
Rhizome short-creeping; fronds evergreen, blue-green, somewhat dimorphic,
the fertile ones taller than the vegetative, pinnate above to bipinnate toward the
base, ultimate segments oblong to lance-linear; stipe dull brown or purple, with

short hairs; occasional in calcareous cliffs and ledges or occasionally in old masonry walls; mostly E of the Allegheny Front.

Pellaea glabella Mett. ex Kuhn   Smooth cliffbrake
Rhizome short-creeping; fronds all alike, evergreen, up to 20 cm long, pinnate to pinnate-pinnatifid toward the base; blade blue-green, segments oblong; stipe shining brown or purple, glabrous; occasional in exposed calcareous cliffs and ledges, also in old masonry walls; scattered, mostly SE. Ours is var. *glabella*.

Phegopteris (C.Presl) Fée

Plants terrestrial; rhizome cord-like, often with tan scales; blades triangular, tapering to the apex, pinnate-pinnatifid; lowest pinnae deeply pinnatifid, pinna pairs sometimes connected with a leafy wing on the rachis, ultimate segments with crenate margins; veinlets in ultimate segments unbranched or branched; surface with white hairs and dark tan scales beneath; sori round, small, lacking indusia; sometimes included in *Thelypteris*.

A. lowest pinna pair connected to the next above by a wing along the rachis; scales on the midrib beneath narrowly triangular, pale tan ..... *P. hexagonoptera*

A. lowest pinna pair not connected to the next above; scales on the midrib beneath broadly triangular, dark tan .... *P. connectilis*

Pellaea atropurpurea

Pellaea glabella

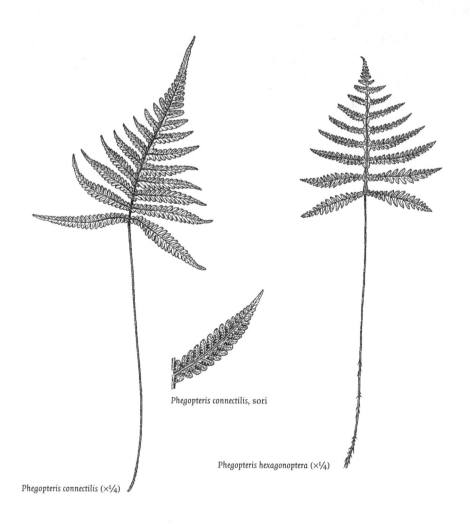

Phegopteris connectilis, sori

Phegopteris hexagonoptera (×¼)

Phegopteris connectilis (×¼)

Phegopteris connectilis (Michx.) D.Watt   Long beech fern
Blade triangular, mostly longer than broad; (6)12–25 cm long; basal pinna pair linear, sessile, margins of the ultimate segments mostly entire, veinlets mostly not forked; common in cool, shaded woods; mostly N. [*Thelypteris phegopteris* (L.) Sloss. of Rhoads and Klein, 1993]

Phegopteris hexagonoptera (Michx.) Fée   Broad beech fern
Blades broadly triangular, (8)15–33 cm long, the lowest pinnae the longest, ovate, margin of the ultimate segments irregularly crenate, veinlets usually forked; common in mesic woods; throughout. [*Thelypteris hexagonoptera* (Michx.) Weath. of Rhoads and Klein, 1993]

Polypodium L.

Small to medium-sized plants of rocky ledges; rhizomes long-creeping; fronds leathery, evergreen, pinnatifid; sori round to oval, without an indusium, but often including modified sporangia (sporangiasters) among the true sporangia.

A. blades triangular (broadest near the base); segments mostly acute at the tip; sporangiasters abundant (>40 per sorus) .... *P. appalachianum*

A. blades narrowly elliptic (broadest near the middle or straight-sided); segments rounded at the tip; sporangiasters fewer (<40 per sorus) ....
.... *P. virginianum*

Polypodium appalachianum Haufler & Windham   Appalachian polypody
Rhizome long-creeping, tip with golden-brown scales, sometimes with a faint darker central stripe; blades elongate-triangular, usually broadest at or near the base, pinnatifid; segments linear to oblong, acute to narrowly rounded at the tip, entire to denticulate; sori round, with >40 glandular sporangiasters; common on rocks, boulders, ledges, cliffs, and rocky woods; throughout. A hybrid with *P. virginianum*, that is best recognized by its aborted spores, occurs at scattered locations.

Polypodium virginianum L.   Common polypody, rock cap
Rhizome long-creeping, tip with golden-brown scales, often with a darker central stripe; blades oblong to narrowly lanceolate, usually broadest near the middle and tapering to the base, pinnatifid; segments oblong, rounded to broadly acute at the tip, entire or crenulate; sori round, with <40 glandular sporangiasters; common on rocks, boulders, ledges, cliffs, and rocky woods; throughout, except on the central Allegheny Plateau.

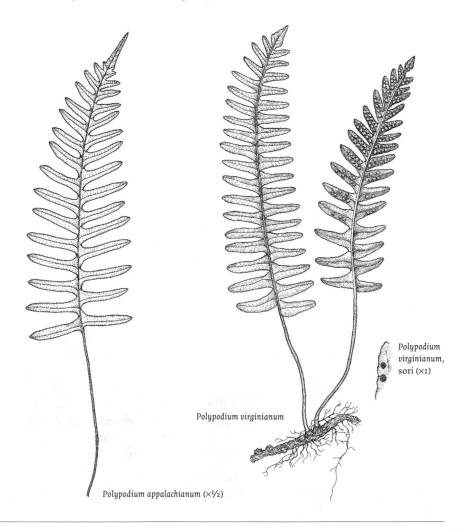

Polypodium virginianum, sori (×1)

Polypodium virginianum

Polypodium appalachianum (×½)

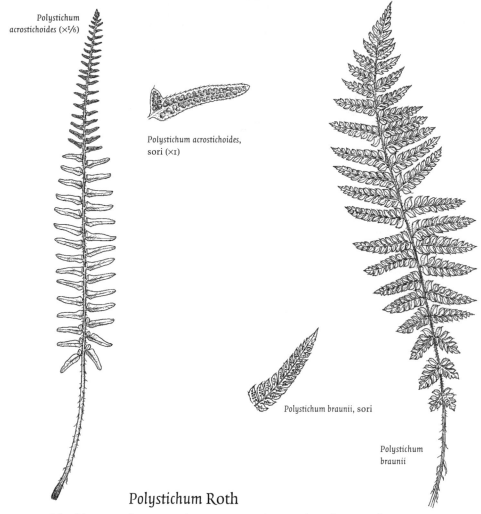

Polystichum
acrostichoides (×⅙)

Polystichum acrostichoides,
sori (×1)

Polystichum braunii, sori

Polystichum
braunii

Polystichum Roth

Plants terrestrial; rhizomes short-creeping to erect; stipes scaly at least at the base; fronds evergreen, pinnate or bipinnate; segments spinulose-toothed; sori round, with peltate indusia.

A. fronds once pinnate, pinnae serrulate; fertile pinnae at the end of the frond, sharply reduced in size, sori covering the lower surface .... *P. acrostichoides*

A. fronds bipinnate; fertile pinnae not restricted to the apex of the frond, not reduced in size, sori distinct ....*P. braunii*

Polystichum acrostichoides (Michx.) Schott   Christmas fern
Fronds, lanceolate, once pinnate; pinnae linear-lanceolate, sharply serrulate, with a conspicuous basal lobe on the upper side; fertile pinnae only on the upper half of the blade, abruptly reduced in size compared to the vegetative pinnae below; sori round, but appearing to cover the lower surface of the fertile pinnae; common in rich moist woods, stream banks, and shaded roadsides; throughout; FACU–. Hybridizes with *P. braunii* to form *P.* x *potteri* Barrington, found occasionally where both parental species occur.

Polystichum braunii (Spenn.) Fée   Braun's holly fern
Fronds, broadly elliptic-lanceolate, bipinnate; pinnae oblong-lanceolate, re-

Pteridium aquilinum var. *latiusculum* (×¹⁄₆)

duced at the base of the blade, with somewhat enlarged upper basal pinnules; pinnules lobed or sharply serrate; fertile pinnae similar to the vegetative ones; sori round, distinct; rare in cool, rocky, shaded ravines; NE; 🌿.

Pteridium Gled. ex Scop.

Pteridium aquilinum (L.) Kuhn   Northern bracken fern
Plants terrestrial, frequently spreading to form extensive clones; fronds to 1.5 m tall, arising singly, widely spaced; stipe dark maroon basally, green above, frequently with low, conical nectaries at the junction of the stipe and the lowest pinnae (best seen on fiddleheads and young fronds); stipe and rachis grooved on the upper surface; blade broadly triangular, bipinnate-pinnatifid on the basal pinnae, progressively less divided toward the tip; basal pinnae broadly triangular, often equal to the entire terminal part of blade; terminal segments of the pinnae oblong, 2–4 times as long as wide, with entire margins; fertile fronds,

which are rarely present, with noticeably thinner segments due to inrolled margins which form a false indusium covering the sori; common in dry, acidic forests, barrens, and clearings; throughout; FACU. Ours is var. *latiusculum* (Desv.) Underw. ex A.Heller.

Thelypteris Schmidel

Plants terrestrial; rhizomes cord-like, long-creeping; stipe with a few scales at the base; blades delicate, ovate-elliptic to lanceolate, pinnate-pinnatifid, midribs grooved on the upper side; surfaces covered with distinctive needle-like, white hairs; sori small, medial to submedial, indusia round-reniform, sometimes glandular.

A. frond elliptical overall, widest in the middle and tapering to a point at both ends .... *T. noveboracensis*
A. base of the frond only slightly narrower than the middle
    B. veinlets of the pinnae lobes unbranched; veins and blade with small, sessile glands beneath .... *T. simulata*
    B. veinlets of the pinnae lobes forked; glands lacking .... *T. palustris*

Thelypteris noveboracensis (×¼)

Thelypteris palustris, pinnule (×2½)

Thelypteris noveboracensis (L.) Nieuwl.   New York fern
Fronds elliptic, to 6 dm tall; pinnae numerous, deeply pinnatifid, margins entire to crenulate; pinnae size decreasing from the middle to a very small pair at the base; stipe < 20% of the length of the blade; veinlets in the pinnae simple; common in moist woods, swamps, wooded ravines, or seeps; throughout; FAC.

Thelypteris palustris Schott   Marsh fern
Fronds to 9 dm tall; pinnae deeply pinnatifid with entire margins, the basal pair only slightly shorter than the middle ones; veinlets in pinnae lobes usually forked; blades not glandular beneath; fertile fronds produced late in the season, their margins strongly revolute; common in swamps and marshes; throughout; FACW+. Ours is var. *pubescens* (G.Lawson) Fernald.

Thelypteris simulata (Davenp.) Nieuwl.   Massachusetts fern
Fronds to 8 dm tall; pinnae deeply pinnatifid, the basal pair only slightly shorter than those in the middle of the frond, margins entire; veinlets in pinnae lobes unbranched; midribs and lower surface of blade bearing small, sessile, glands as well as hairs; margins of the fertile fronds not strongly revolute; occasional in bogs and swamps; mostly NE; FACW.

Woodsia R. Br.

Small plants of rocky cliffs; rhizome short-creeping, ascending to erect, scaly and often with clusters of old stipe bases; fronds lanceolate to ovate, pinnate-pinnatifid or bipinnate; stipe and rachis scaly; pinnae lance-ovate, deeply pinnatifid, ultimate segments with irregular to dentate margins; veinlets ending before reaching the margins, often in a swollen tip; sori in a single row between the midrib and the margins of the ultimate segments; indusium cup-like, arising from beneath the sorus and initially enclosing it, splitting into radiating scales or hairs which soon wither and become obscure.

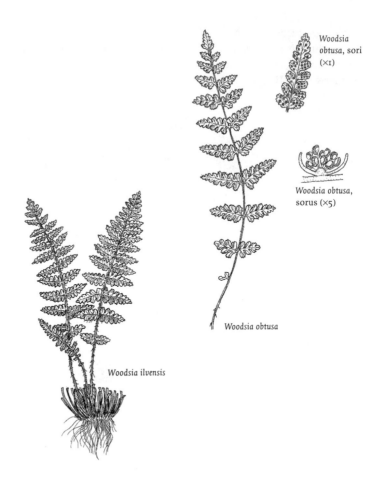

Woodsia obtusa, sori (×1)

Woodsia obtusa, sorus (×5)

Woodsia obtusa

Woodsia ilvensis

A. stipe breaking at a distinct joint leaving persistent bases all of the same length; scales of the lower stipe commonly a uniform tan color; frond lanceolate; blade not glandular beneath .... *W. ilvensis*

A. stipe lacking joint, old stipe bases of varied lengths; scales of the lower stipe commonly tan with a dark central stripe; frond ovate; blade glandular beneath .... *W. obtusa*

Woodsia ilvensis (L.) R.Br.   Rusty woodsia
Fronds narrowly lanceolate, to 25 cm long; pinnae lanceolate-triangular, deeply pinnatifid, lower surface with both hairs and scales; indusium of hair-like segments; occasional on sunny, dry cliffs; E and SC.

Woodsia obtusa (Spreng.) Torr.   Blunt-lobed woodsia
Fronds ovate, 0.8–6 dm tall, pinnae ovate-triangular with 5 or more pairs of pinnules; pinnule margins obtusely lobed to dentate, bearing stalked glands; indusium of 4–5 broad, scale-like lobes; common on shaded cliffs; mostly S. Very similar to *Cystopteris tenuis* from which it most easily distinguished by veins that end prior to meeting the edge of the blade and abundant scales on the rachis and stipe.

Woodwardia Sm.

Plants terrestrial; rhizomes slender, long-creeping; fronds arising singly; sterile fronds pinnatifid or pinnate-pinnatifid, with conspicuous anastomosing veins forming a band along the midrib; fertile fronds contracted or similar to the sterile ones; sori elongate, in a single row along each side of the midribs of the pinnae and with flap-like indusia.

A. sterile fronds pinnately lobed, lobes alternate along the rachis; veins anastomosing 2 or more times, free only at the blade margins .... *W. areolata*

A. sterile fronds pinnate-pinnatifid, veins of the pinna lobes anastomosing once near the midvein, otherwise free to the margins .... *W. virginica*

Woodwardia areolata (L.) T.Moore   Netted chain fern
Sterile fronds few, widely separated, 5–8 dm tall, stipe reddish-brown at the base, with scattered brown scales, not swollen; blade bright green, ovate, pin-

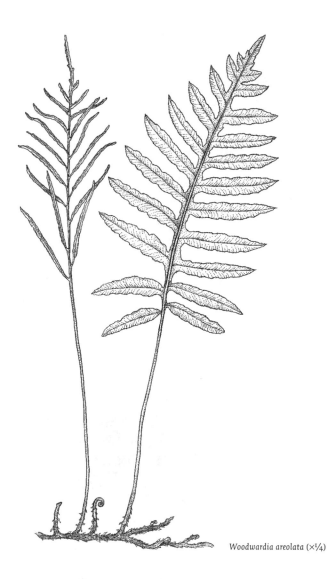

Woodwardia areolata (×¼)

nately lobed with 7–12 alternate lobes on each side, margin entire; fertile frond also alternately pinnately lobed but highly revolute; sori elongate, end-to-end near the midveins of the pinnae; indusium opening on the side adjacent to the midvein; rare in moist or wet woods and acidic bogs; mostly SE; FACW+; 🌿. Sterile specimens are best distinguished from *Onoclea sensibilis*, which they resemble, by their alternate pinnae.

Woodwardia virginica (L.) Sm.   Virginia chain fern
Fronds to 1 m tall; stipe dark purple to black at the swollen base; blade dark green, broadly lanceolate, pinnate-pinnatifid; pinnae mostly alternate, linear to narrowly lanceolate, tip acuminate, deeply pinnatifid with segment margins finely serrulate; veinlets ending in a swelling near the margin; sori elongate, end-to-end along the rachis and the midveins of the pinnules; indusium flap-like, opening toward the midveins; occasional in moist, acidic swamps, bogs, marshes and ditches; mostly N; OBL. Sterile fronds are similar to *Osmunda cinnamomea*, from which they are best distinguished by the presence of anastomosing veins and the lack of reddish, fuzzy hairs on the stipe and pinna-rachis joints.

GYMNOSPERMS

CUPRESSACEAE Cypress Family

Evergreen or deciduous trees or shrubs with appressed scale-like or subulate leaves; dioecious or monoecious; seed cones <2 cm long, fleshy and berry-like, or woody with peltate or basally attached scales; pollen cones terminal or axillary, mostly solitary but in drooping panicles in *Taxodium*; opening early spring.

A. deciduous trees with alternate, linear, needle-like leaves .... *Taxodium*
A. evergreen trees or shrubs with opposite or whorled, scale-like or subulate leaves
    B. trees or shrubs; leaves appressed and scale-like or subulate, opposite in 4 equal ranks, or in whorls of 3; cone fleshy, blue or black, berry-like, indehiscent .... *Juniperus*
    B. trees; leaves subulate or more commonly appressed and scale-like, opposite and 4-ranked; cones woody, dehiscent
        C. branchlets nearly terete in cross section; prominent white bands of stomata visible where the tiny scale-like leaves overlap; cones spherical with peltate scales .... *Chamaecyparis*
        C. branchlets strongly flattened; stomatal bands not apparent; cones ellipsoid; cone scales basally attached .... *Thuja*

Chamaecyparis Spach

Chamaecyparis thyoides (L.) Britton, Stearns & Poggenb.   Atlantic white-cedar Narrow, upright evergreen tree with appressed, scale-like leaves in the mature phase; juvenile foliage subulate; cones woody, spherical with peltate scales; seeds winged; our only native stands were apparently extirpated very early from coastal plain bogs in Bucks and Philadelphia counties, a naturalized population is established at one Westmoreland County site; also cultivated; OBL; ❦.

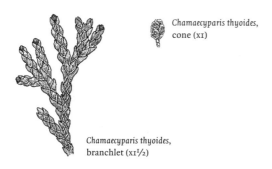

Chamaecyparis thyoides, cone (x1)

Chamaecyparis thyoides, branchlet (x1½)

Juniperus L.

Evergreen trees or shrubs with subulate and/or appressed scale-like leaves in pairs or whorls of 3; cones fleshy, blue or black, glaucous, individual cone scales barely distinguishable; seeds not winged.

A. narrow, upright trees; leaves opposite, 4-ranked or occasionally in whorls of 3; juvenile leaves subulate; adult leaves appressed and scale-like; cones terminal .... *J. virginiana*
A. low, spreading shrubs; all leaves subulate, in whorls of 3; cones axillary .... *J. communis*

Juniperus communis L.   Common juniper
Low growing, spreading evergreen shrub with subulate needles in whorls of 3 and glaucous, berry-like cones 6–9 mm long; infrequent in dry, open woods, slopes, or pastures, mostly SE; native, but declining due to habitat loss and severe deer browsing. Ours is var. *depressa* Pursh.

Juniperus virginiana L   Eastern red-cedar
More or less narrow, upright evergreen tree with tightly appressed scale-like leaves in the mature phase; juvenile foliage subulate; cones 4–6 mm long, fleshy, glaucous-blue, berry-like; common in old fields, early successional woods, serpentine barrens, limestone and diabase glades, and other moist to dry open sites; mostly S; FACU.

Taxodium Richard

Taxodium distichum (L.) Richard   Bald cypress
Deciduous tree to 40 m tall; bark reddish, peeling in long strips; needles linear, alternate on slender terminal twigs that are shed with the leaves; cones drooping at the ends of branches, subglobose, with peltate scales; seeds angled but not winged; planted and rarely naturalized; native from southern DE and MD south; OBL.

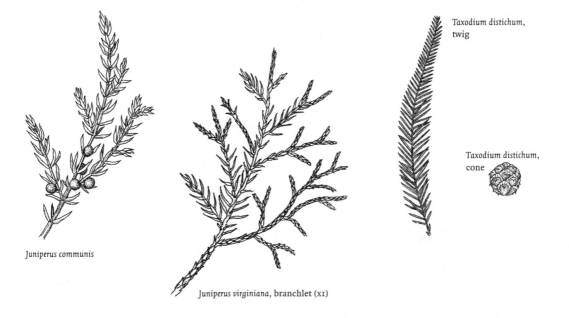

Taxodium distichum, twig

Taxodium distichum, cone

Juniperus communis

Juniperus virginiana, branchlet (x1)

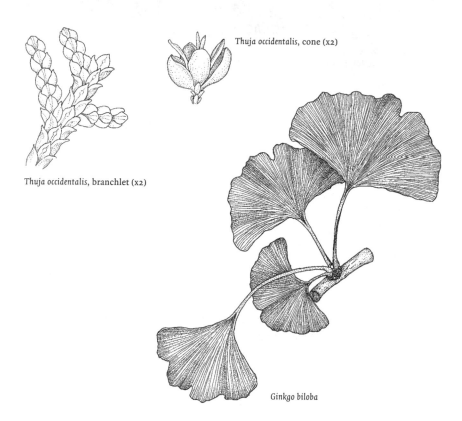

Thuja occidentalis, branchlet (x2)

Thuja occidentalis, cone (x2)

Ginkgo biloba

Thuja L.

Thuja occidentalis L.   Arbor-vitae, northern white-cedar
Evergreen tree with appressed, scale-like leaves and strongly flattened branch-lets; cones woody, <2 cm long with basally attached, paired cone scales; seeds narrowly winged; there are no known native stands in Pennsylvania although the species occurs naturally in adjacent states to the north and south; planted and/or naturalized at a few scattered locations; FACW.

GINKGOACEAE Ginkgo Family

Deciduous trees, branches with terminal long shoots and lateral short shoots; leaves fan-shaped with parallel-dichotomous venation; dioecious; pollen cones catkin-like, opening before the leaves; seeds naked except for a fleshy outer layer.

Ginkgo L.

Ginkgo biloba L.   Maidenhair tree
Trees to 30 m tall , seeds obovoid to elliptic, about 2.5 cm long, the fleshy outer layer malodorous; widely planted and rarely naturalized; native to China.

PINACEAE Pine Family

Monoecious, evergreen or deciduous trees with needle-like leaves; seed cones woody, maturing in 1 or 2 years; cone scales spirally arranged and bearing 2 seeds each on the upper surface; in some species the cone scales are subtended by bracts that extend beyond the scales, in others the bracts are shorter and inconspicuous; pollen cones borne singly or clustered, opening in the early spring, then dropping; seeds winged.

A. deciduous trees; leaves needle-like, alternate, and clustered on short lateral shoots .... *Larix*

A. evergreen trees; leaves needle-like, borne singly or in clusters (fascicles) of 2–5
  B. leaves in clusters (fascicles) of 2–5 with a scaly sheath at the base at least when young; cone scales mostly with a thickened portion at the tip (umbo) and often bearing a spine .... *Pinus*
  B. leaves borne singly, cone scales without a thickened portion at the tip
    C. twigs roughened by persistent raised, peg-like leaf bases (sterigmata)
      D. leaves sessile, square in cross section, tapering at the tip, spirally arranged on the twig.. *Picea*
      D. leaves narrowed to a short petiole, flattened, with a minute notch at the tip, 2-ranked, giving the twig a flattened appearance .... *Tsuga*
    C. twigs smooth or nearly so
      E. leaf scar circular and flush with the twig surface; bracts of the cone scales shorter than the scales .... *Abies*
      E. leaf scar slightly raised on one side; bracts of the cone scales conspicuously longer than the scales .... *Pseudotsuga*

Abies Mill.

Abies balsamea (L.) Mill.   Balsam fir
Evergreen tree to 20 m tall; distinguished by its smooth to somewhat scaly bark; smooth twigs with circular leaf scars and large, upright seed cones that disintegrate in place; infrequent in cool swamps or bogs in peaty soils; N; FAC.

Larix Mill.

Deciduous trees with needle-like leaves that are alternate on long shoots and clustered on short spur shoots; seed cones axillary, bearing few to many thin scales.

A. cone scales 15–20, glabrous and shining on outside; needles light green to blue-green, with inconspicuous white bands beneath .... *L. laricina*

A. cone scales >40, glabrous or pubescent on the outside; needles blue-green, with faint to conspicuous white bands beneath
  B. cone scales reflexed at the apex, glabrous on the outside .... *L. kaempferi*
  B. cone scales straight or slightly incurved, pubescent on the outside ....
  .... *L. decidua*

Larix decidua Mill.   European larch
Tree to 35 m tall with slender conic crown; needles blue-green, flat above, keeled beneath with faint white stomatal bands; cones ovate, 2.5–4 cm; cone scales

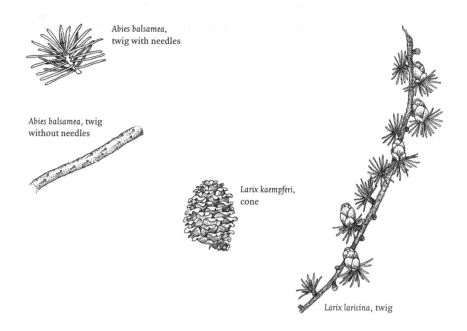

Abies balsamea,
twig with needles

Abies balsamea, twig
without needles

Larix kaempferi,
cone

Larix laricina, twig

straight or slightly incurved, not reflexed, short, pubescent on the outside; forest plantations; scattered; native to Europe; UPL.

Larix kaempferi (Lamb.) Carr.   Japanese larch
Tree to 30 m tall with conic crown; leaves blue-green, flat above, keeled beneath with conspicuous white stomatal bands; cones 2–3 cm; cone scales reflexed at the margin; forest plantations; scattered; native to Japan.

Larix laricina (Du Roi) K.Koch   American larch, tamarack
Tree to 20 m tall with narrowly conic crown when young; needles light yellow-green to blue-green, keeled beneath with faint white stomatal bands; cones 10–15 mm; cone scales glabrous and shining on the outside, slightly incurved at the margin; infrequent in sphagnum bogs and peatlands; N; FACW.

Picea A.Dietr.

Evergreen trees with needle-like leaves borne singly in a spiral arrangement on the twig; needles square in cross section and borne on a persistent woody base (sterigma); seed cones woody; cone scales not thickened at the tip; seeds winged.

A. twigs pubescent; cone scales fan-shaped, broadest near the margin; cones 1.5–4.5 cm long
    B. needles 0.8–2.5 cm, mostly sharp-pointed, yellow-green to dark green; cones 2.3–4.5 cm long .... *P. rubens*
    B. needles 0.6–1.5 cm, mostly blunt-tipped, blue-green; cones 1.5–2.5 cm long .... *P. mariana*
A. twigs glabrous or only slightly hairy; cones 2.5–16 cm long
    C. needles dark green; branchlets strongly drooping on mature trees; cones 12–16 cm long .... *P. abies*

C. needles blue-green, glaucous; branchlets not drooping; cones 2.5–11 cm long
  D. cone scales fan-shaped, margin entire .... *P. glauca*
  D. cone scales diamond-shaped, margin toothed or ragged .... *P. pungens*

Picea abies (L.) H.Karst.   Norway spruce
Evergreen tree to 30 m tall with narrow, conical crown; secondary branches drooping in mature specimens; needles dark green; cones 12–16 cm long, cylindrical; cone scales widest near the middle, margins toothed; forest plantations and other cultivated sites throughout; native to Europe.

Picea glauca (Moench) Voss   White spruce
Evergreen tree to 30 m tall with broadly conic crown; needles blue-green; cones 2.5–6 cm; cone scales fan-shaped, margin entire; forest plantations; scattered; native farther north; FACU.

Picea mariana (Mill.) Britton, Stearns & Poggenb.   Black or bog spruce
Evergreen tree to 20 m tall with narrow conic crown but often stunted and shrub-like; needles dark bluish-green, mostly blunt-tipped; cones 1.5–2.5 cm; cone scales fan-shaped with irregularly toothed margin; rare in sphagnum bogs; NE and NC; FACW–.

Picea pungens Sarg.   Colorado blue spruce
Evergreen tree to 30 m tall with narrowly conical crown; needles strongly glaucous, blue-green, sharp-pointed; cones 6–11 cm, cone scales elliptic with a ragged margin; forest plantations and other cultivated sites; scattered; native farther west.

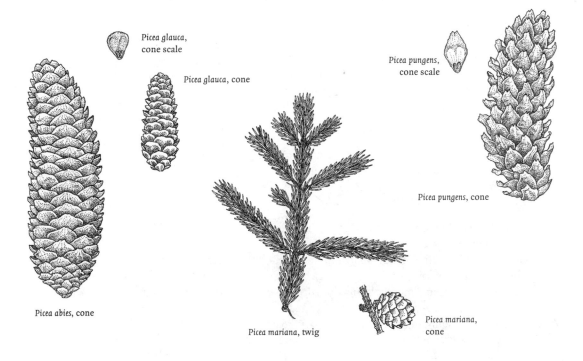

Picea glauca,
cone scale

Picea glauca, cone

Picea pungens,
cone scale

Picea pungens, cone

Picea abies, cone

Picea mariana, twig

Picea mariana,
cone

Picea rubens Sarg.   Red spruce
Evergreen tree to 30 m tall with narrowly conic crown; needles yellow-green to
dark green, sharp-pointed; cones 2.3–4.5 cm; cone scales fan-shaped with an
entire to irregularly toothed margin; infrequent in cool, moist woodlands and
margins of bogs and swamps; mostly N; FACU.

Pinus L.

Evergreen trees with needle-like leaves borne in fascicles of 2–5; seed cones
woody with spirally arranged scales, maturing in 2 years; the exposed tips of the
cone scales form a thickened apophysis with an umbo (scar), which may or may
not bear a sharp spine or prickle; seeds with a broad papery wing.

A. leaves 5 per fascicle; scaly bracts at the base of the fascicles not persistent;
    cones cylindrical, usually curved .... *P. strobus*
A. leaves 2–3 per fascicle; scaly bracts at base of fascicle persistent; cones
    broader at the base
    B. needles <9 cm long, mostly in 2s
        C. cone scales spineless
            D. bark of upper trunk and larger branches orange; needles blue-green;
                cones symmetrical, persisting on the branches after opening..........
                .... *P. sylvestris*
            D. bark of upper trunk and branches brown; needles yellow-green;
                cones asymmetrical, persisting on the branches mostly in a closed
                condition .... *P. banksiana*
        C. cone scales with a definite spine
            E. cones 6–9 cm with very stout, spreading or upwardly curving spines
                .... *P. pungens*
            E. cones 3–7 cm with slender, straight spines
                F. rounded or flat-topped tree to 18 m; needles strongly twisted .....
                    .... *P. virginiana*
                F. erect tree to 40 m; needles slightly twisted .... *P. echinata*
    B. needles 7–18 cm long, in 2s or 3s
        G. needles in 3s
            H. needles twisted; cones persistent, armed with definite spines ....
                .... *P. rigida*
            H. needles mostly in 2s (rarely 3s), not twisted; cones deciduous, bear-
                ing short spines .... *P. echinata*
        G. needles in 2s
            I. fresh needles breaking cleanly if bent; cone scales without spines ...
                .... *P. resinosa*
            I. fresh needles not breaking cleanly if bent; cone scales with a short
            spine
                J. twigs glaucous; needles 7–12 cm; cones dull brown .... *P. echinata*
                J. twigs not glaucous; needles to 17 cm; cones glossy yellow-brown
                .... *P. nigra*

Pinus banksiana Lamb.   Jack pine
Evergreen tree to 27 m, crown spreading in age; needles 2 per fascicle, 2–5 cm,
twisted, yellow-green; cones 3–5.5 cm long, asymmetrical, unarmed, often remaining closed on the branches for many years; forest plantations; scattered;
native farther north; FACU.

Pinus echinata Mill.   Short-leaf pine
Evergreen tree to 40 m tall with slender, often glaucous branchlets; needles in 2s (or rarely 3s), straight or only slightly twisted, 5–12 cm; cones slenderly ovoid, 4–6 cm long, with short, sharp spines; rare on wooded slopes or ridges, in low nutrient soils; mostly SC; 🌿.

Pinus nigra Arnold   Austrian pine
Evergreen tree to 30 m tall (or more) with an upright crown; needles in 2s, dark green and stiff, 10–17 cm; cones symmetrical, ovoid, 5–7 cm long, nonpersistent; cone scales with a small spine or nearly spineless; forest plantations and other cultivated sites; native to Eurasia.

Pinus pungens Lamb.   Table-mountain pine
Evergreen tree to 12 m tall with a broad, open crown; needles in 2s, twisted, 3–7 cm; cones broadly ovoid, 6–10 cm long, often asymmetrical at the base; cone scales thickened with a stout, spreading or upwardly curved spine; occasional on dry gravelly or sandy slopes and ridge tops; SE and SC.

Pinus resinosa Aiton   Norway pine, Red pine
Evergreen tree to 37 m tall with narrowly rounded crown; needles in 2s, dark green, breaking cleanly when bent (if fresh), 11–15 cm; cones broadly ovoid to globose, 3–5 cm long, nonpersistent; cone scales without spines; dry slopes and mountain tops; C and NC; native, but also frequently planted; FACU.

Pinus rigida Mill.   Pitch pine
Evergreen tree to 30 m, the trunk often with leafy shoots; needles 3 per fascicle, 5–10(15) cm, straight or twisted; cones broadly ovoid with a nearly flat base

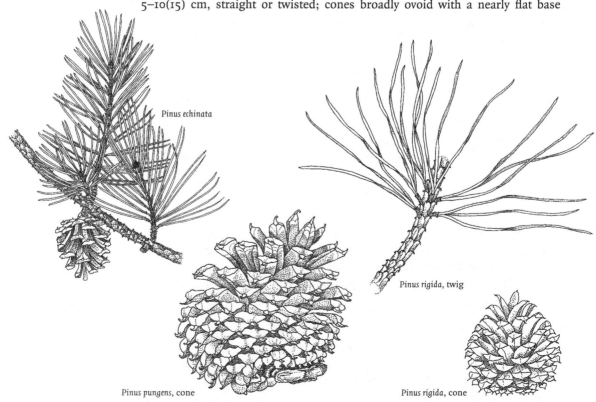

Pinus echinata

Pinus rigida, twig

Pinus pungens, cone

Pinus rigida, cone

Pseudotsuga menziesii, cone

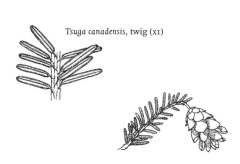

Tsuga canadensis, twig (x1)

Tsuga canadensis, cone

when open, 4.5–8 cm, with a slender, downcurved spine, long-persistent on the branches; frequent in barrens and other moist to dry, sterile, acidic soils throughout, highly fire tolerant; FACU.

Pinus strobus L.   Eastern white pine

Evergreen tree to 40 m, symmetrical when young, flat-topped and irregular in age; needles in 5s, slender and flexible; cones cylindrical, often slightly curved; common in forests throughout; native, but also frequently cultivated; FACU.

Pinus sylvestris L.   Scots pine

Evergreen tree to 20 m tall with an upright to somewhat flattened crown; upper trunk and larger branches of mature trees with a distinct orange color; needles in 2s, twisted, 3–6 cm; foliage bluish; cones broadly ovoid, 3–6 cm, symmetrical, unarmed; frequent in forestry plantations, hedgerows, and old fields; native to Eurasia.

Pinus virginiana Mill.   Virginia pine

Evergreen tree to 15 m tall with irregularly rounded or flattened crown; needles in 2s, strongly twisted, 2–8 cm; cones ovoid, 3–7 cm long, with slender stiff spines; frequent in barrens, slopes, and ridge tops in dry sandy or shaly soils; mostly S.

Pseudotsuga Carrière

Pseudotsuga menziesii (Mirb.) Franco   Douglas-fir

Evergreen tree to 40 m; needles borne singly, 1.5–3 cm, bluish-green; cones 4–10 cm long, ellipsoid or ovoid, with conspicuous 3-lobed bracts extending beyond the cone scales; forest plantations and other cultivated sites; native farther west.

Tsuga Carrière

Tsuga canadensis (L.) Carrière   Canada hemlock

Evergreen tree to 30 m; needles with a short petiole, arranged in two ranks giving branchlets a flattened appearance; cones small, borne at the branch tips; common in cool, moist woods and shaded slopes throughout; FACU. The state tree of Pennsylvania.

TAXACEAE Yew Family

Evergreen shrubs; leaves needle-like, flat, green beneath, 2-ranked; branchlets flattened or forming a V- shaped trough; monoecious or dioecious; seeds naked, surrounded only by a fleshy, red, cup-shaped aril; pollen cones globose, clustered, with peltate sporophylls, opening in the early spring, then dropping.

Taxus L.

Characteristics of the family.

A. leaves abruptly pointed; scales of the winter buds keeled
  B. arrangement of the leaves flatly 2-ranked; branchlets green ... T. canadensis
  B. arrangement of the leaves irregularly 2-ranked, forming a V-shaped trough; branchlets brown .... T. cuspidata
A. leaves gradually acuminate; scales of the winter buds not keeled; branchlets green .... T. baccata

Taxus baccata L.   English yew
Similar to T. canadensis but more erect, dioecious; leaves gradually acuminate; numerous cultivated forms, occasionally naturalized in disturbed woods; SE; native to Eurasia.

Taxus canadensis Marshall   Canadian yew
Low spreading evergreen shrub to 1.5 m tall, monoecious; cool, moist rocky slopes or ravines under mixed or coniferous (rarely deciduous) forest canopy; formerly found throughout most of the state, but declining due to habitat loss and severe deer browsing; FAC.

Taxus cuspidata Siebold & Zucc.   Japanese yew
Similar to T. canadensis, but more erect, dioecious; twigs brown, leaves 2-ranked but forming a V- shaped trough; numerous cultivated forms, occasionally naturalized in disturbed woods; SE; native to Asia.

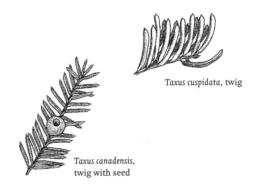

Taxus cuspidata, twig

Taxus canadensis,
twig with seed

ANGIOSPERMS, DICOTS

ACANTHACEAE Acanthus Family

Herbs with simple, opposite leaves lacking stipules; flowers perfect; calyx 4–5-lobed; corolla sympetalous, regular or irregular, 5-lobed; stamens 2 or 4, epipetalous; ovary superior, bilocular with a terminal style; placentation axile, ovules 2 per locule, each with a hooked funiculus that ejects the mature seed; fruit an explosively dehiscent capsule.

A. stamens 2; corolla strongly 2-lipped .... Justicia
A. stamens 4; corolla with 5 nearly equal lobes .... Ruellia

Justicia L.

Justicia americana (L.) Vahl   Water-willow
Upright emergent aquatic; stems 5–10 dm tall, simple, glabrous; leaves linear to lanceolate, long cuneate at the base; flowers short spikes on axillary peduncles; calyx lobes narrowly lanceolate; corolla 2-lipped, pale violet to nearly white with purple markings on the lower lip, 8–13 mm, upper lip concave; stamens 2; colonial by rhizomes and forming large colonies; frequent in shallow water along the shorelines of lakes and rivers throughout; flr. late Jun–early Sep; OBL.

Justicia americana,
flower (×1)

Ruellia L.

Perennial herbs with showy, lavender-blue flowers in axillary clusters or terminating lateral branches; calyx regular, 5-lobed; corolla funnelform, with 5 nearly equal lobes; stamens 4, in 2 pairs; fruit club-shaped or obovate, seeds 3–8 per locule.

A. calyx lobes lanceolate, 2 mm or more wide .... R. strepens
A. calyx lobes linear-subulate, 1 mm or less wide
    B. leaves sessile or subsessile, petioles if present <3 mm .... R. humilis
    B. leaves narrowed to a petiole-like base 3 mm or more .... R. caroliniensis

Ruellia caroliniensis (Walter ex J.F.Gmel.) Steud.   Carolina petunia
Stems 3–8 dm tall, with spreading and/or minutely curved hairs; leaves lanceolate to ovate, 4–12 cm; corolla 2.5–5 cm; fruit glabrous or hairy; very rare, collected once on a riverbank; SE; flr. early Jul–early Aug; believed to be extirpated; ⚘.

Ruellia humilis Nutt.   Fringe-leaved petunia
Stems 2–6 dm tall, leaves sessile or subsessile, 3–8 cm; flowers in sessile to subsessile crowded clusters from the axils of several upper leaves; calyx lobes narrowly linear; corolla 3–7 cm; fruit glabrous; very rare on limestone barrens and quarry waste; SC; flr. late Jun–late Jul; UPL; ⚘.

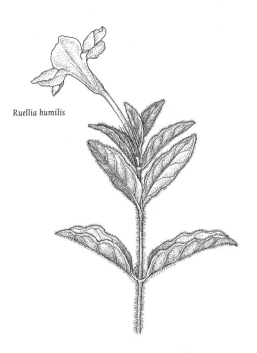

Ruellia humilis

Ruellia strepens L.   Limestone petunia
Stems 3–10 dm tall, puberulent in 2 opposite strips; leaf blades ovate to lance-ovate, petioles 5–20 mm; flowers from a few nodes near the middle of the stem, 1–3 on a common peduncle or terminal on a leafy axillary branch; calyx lobes lanceolate; corolla 3.5–5 cm; fruit glabrous; rare on rich wooded slopes, bluffs, and roadsides on limestone; S; flr. late Jun–late Jul; FAC; ❦.

ACERACEAE Maple Family

Trees or shrubs with opposite (sometimes whorled on vigorous growth), simple, or occasionally compound leaves; flowers small, structurally or functionally uni-sexual, usually 5-merous, with an annular or lobed disk, petals small or lacking; inflorescences variable; fruit a pair of 1-seeded samaras united at the base but eventually separating.

Acer L.

Characteristics of the family.

Summer key

A. leaves tri-foliate or pinnately compound .... *A. negundo*
A. leaves simple
    B. broken petioles clearly showing milky juice
        C. leaves 1–2 dm across, glabrous below, teeth and lobes rounded ....
        .... *A. platanoides*
        C. leaves 5–10 cm across, pubescent below, teeth and lobes pointed ....
        .... *A. campestre*

B. broken petioles not showing milky juice
  D. leaves 3-lobed and finely double-toothed; bark of young trunk and branches striped .... *A. pensylvanicum*
  D. leaves 3–5-lobed and coarsely toothed; bark of young trunk and branches not striped
    E. sinuses between the lobes extending about ²⁄₃ of the way (or more) to the midvein
      F. large trees; leaves distinctly whitened below .... *A. saccharinum*
      F. small trees or shrubs; leaves green below (or sometimes reddish)
        G. shrubs; leaves 3-lobed (or occasionally with some leaves 5-lobed) .... *A. ginnala*
        G. small trees; leaves 5–9-lobed .... *A. palmatum*
    E. sinuses between the lobes extending halfway (or less) to the midvein
      H. bases of the sinuses between the principal lobes rounded
        I. stipules present (often stalked and resembling tiny leaves); sides of leaf blades turned under giving the leaves a claw-like appearance .... *A. nigrum*
        I. stipules absent; leaf blades essentially flat .... *A. saccharum*
      H. bases of the sinuses between the principal lobes forming a sharp angle
        J. leaves essentially glabrous below .... *A. pseudoplatanus*
        J. leaves glaucous or pubescent below
          K. shrubs or small trees; leaves generally pubescent below .... *A. spicatum*
          K. large trees; leaves glaucous below, pubescence, if present, confined to the principal veins .... *A. rubrum*

Flowering key

A. inflorescence a panicle or a raceme
  B. inflorescence a slender drooping raceme
    C. bark of young trunk and branches striped .... *A. pensylvanicum*
    C. bark of young branches and twigs green but not striped .... *A. negundo*
  B. inflorescence a panicle
    D. flowers yellowish-white, fragrant .... *A. ginnala*
    D. flowers greenish-yellow, not fragrant
      E. panicles drooping .... *A. pseudoplatanus*
      E. panicles erect or ascending .... *A. spicatum*
A. inflorescence neither a panicle nor a raceme
  F. flowers appearing with the leaves
    G. broken petioles showing milky juice
      H. inflorescences glabrous .... *A. platanoides*
      H. inflorescences pubescent .... *A. campestre*
    G. broken petioles not showing milky juice
      I. flowers greenish-yellow .... see couplet I in summer key
      I. flowers purple .... *A. palmatum*
  F. flowers appearing before the leaves
    J. ovaries and young fruits pubescent .... *A. saccharinum*
    J. ovaries and young fruits glabrous
      K. flowers red, sessile or short-stalked .... *A. rubrum*
      K. flowers yellowish-green, on long pedicels .... *A. negundo*

Winter key

A. buds stalked, covered by 2 exposed valvate scales
    B. twigs and buds glabrous; bark white striped .... A. *pensylvanicum*
    B. twigs and buds pubescent; bark not white striped .... A. *spicatum*
A. buds sessile, covered by >2 exposed imbricate scales
    C. buds white tomentose .... A. *negundo*
    C. buds not white tomentose
        D. terminal buds usually 7–15 mm long
            E. buds reddish; opposite leaf scars meeting .... A. *platanoides*
            E. buds green; opposite leaf scars not meeting .... A. *pseudoplatanus*
        D. terminal buds usually <7 mm long
            F. terminal buds rounded or blunt pointed; 4 or fewer pairs of bud scales visible
                G. bud scales glabrous
                    H. terminal buds 4–6 mm long
                        I. twigs foul smelling when broken; bark flaking on older trunks .... A. *saccharinum*
                        I. twigs not foul smelling when broken; bark rough but not usually flaking on older trunks .... A. *rubrum*
                    H. terminal buds 2–3 mm long .... A. *ginnala*
                G. bud scales pubescent, at least toward the tips .... A. *campestre*
            F. terminal buds acutely pointed
                J. 4–8 pairs of bud scales visible
                    K. buds brown; hairs at upper edge of leaf scar brown .... A. *saccharum*
                    K. buds dark brown to almost black; hairs at upper edge of leaf scar pale .... A. *nigrum*
                J. fewer than 4 pairs of bud scales visible .... A. *palmatum*

Acer campestre L.   Hedge maple
Shrub or small tree with a rounded crown; leaves 3–5-lobed; flowers yellowish-green; inflorescence a few-flowered, erect, pubescent corymb; samaras 2.5–3.5 cm long, spreading at nearly 90°; cultivated and occasionally spreading to moist, rocky, disturbed woods; flr. May; native to Europe.

Acer ginnala Maxim.   Amur maple
Shrub or small tree; leaves 3-lobed, the middle lobe much longer than the lateral ones; flowers yellowish-white; inflorescence a long-peduncled panicle; samaras 2–3 cm long, the wings nearly parallel; cultivated and occasionally escaped; flr. late May–Jun; native to Asia.

Acer negundo L.   Box-elder
Tree; young twigs green; leaves pinnately compound with 3–5 (9 on vigorously growing shoots) leaflets; flowers yellowish-green, petals absent; inflorescence few-flowered, drooping, fascicled; samaras 3–4 cm long; common in low moist areas and along stream banks; throughout; flr. Apr–May; FAC+.

Acer nigrum Michx. f.   Black maple
Large tree; leaves similar to A. *saccharum* but with drooping edges and usually pubescent below; flowers and fruits similar to A. *saccharum*; rich woods, ravines,

Acer negundo

Acer nigrum

and river banks; mostly W; flr. late Apr–May; FACU. Considered a subspecies or variety of *A. saccharum* by some authors.

Acer palmatum Thunb.   Japanese maple
Shrub or small tree; leaves 5–9-lobed; flowers purple; inflorescence a small, glabrous corymb; samaras 2–3 cm long, spreading at an obtuse angle; cultivated and occasionally escaped to disturbed woods; late Apr–May; native to Asia.

Acer pensylvanicum L.   Moosewood, striped maple
Small tree; leaves cordate or truncate at base; flowers yellow, petals present; inflorescence a slender, drooping raceme; samaras 2.5–3 cm long; cool, moist, rocky woods; throughout except extreme SE; flr. late Apr–May; FACU.

Acer platanoides L.   Norway maple
Large tree with broadly spreading crown; leaves with 5–7 acuminate lobes and a few large teeth; flowers yellow, petals present; inflorescence a many-flowered, erect, glabrous corymb; samaras 3.5–4.5 cm long, spreading at nearly 180°; cultivated and frequently escaped throughout; flr. Apr–early May; native to Europe; UPL.

Acer pseudoplatanus L.   Sycamore maple
Large tree; leaves 5-lobed; flowers yellow-green, petals present; inflorescence a slender, drooping panicle; samaras 3–4 cm long, diverging at an acute angle; cultivated and occasionally spreading to waste ground and urban woods; flr. late Apr–Jun; native to Europe.

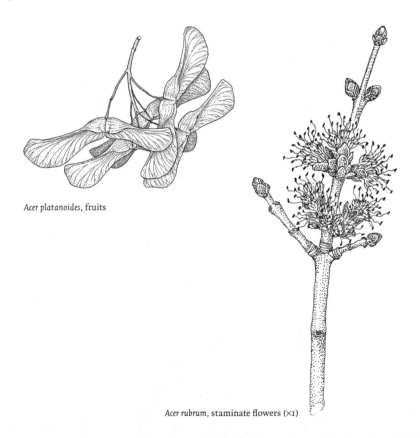

Acer platanoides, fruits

Acer rubrum, staminate flowers (×1)

Acer rubrum L.   Red maple
Large tree; leaves coarsely toothed, 3–5-lobed, petioles often red; flowers red, petals absent; inflorescence a rounded cluster; samaras 1.5–2.5 cm long, reddish; wooded slopes, swamps, bogs, and moist areas; throughout; flr. late Mar–early May; FACW.

Acer saccharinum L.   Silver maple
Large tree; leaves deeply 5-lobed and strongly whitened below; flowers greenish- or yellowish-red, petals absent; inflorescence a rounded cluster; samaras 4–8 cm long; moist woods, stream banks, and alluvial soils; throughout; flr. Apr–May; FACW.

Acer saccharum Marshall   Sugar maple, rock maple
Large tree; leaves 3–5-lobed, flat, with rounded sinuses, often glaucous below; flowers greenish-yellow, petals absent; inflorescence a drooping corymb; samaras 2.5–4 cm long; moist woods, wooded slopes, ravines, and alluvial areas; throughout; flr. Apr–May; FACU.

Acer spicatum Lam.   Mountain maple
Shrub to small trees; leaves 3-lobed to obscurely 5-lobed, coarsely serrate; flowers greenish-yellow, petals absent; inflorescence a pubescent, erect panicle; samaras 1.5–2.5 cm long; moist, rocky woods; throughout; flr. May–Jun; FACU–.

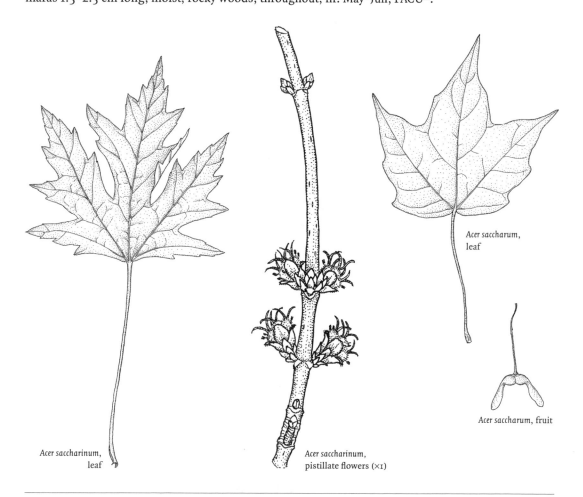

Acer saccharinum,
leaf

Acer saccharinum,
pistillate flowers (×1)

Acer saccharum,
leaf

Acer saccharum, fruit

AMARANTHACEAE Amaranth Family

Erect, spreading, or prostrate annual or perennial herbs, monoecious or dioecious, or flowers perfect; leaves alternate or opposite, simple, usually entire, stipules absent; flowers small, petals absent, sepals usually present, stamens usually as many as the sepals; inflorescences simple or compound spikes or axillary clusters; fruit a utricle, dehiscent or indehiscent.

A. leaves alternate
    B. flowers unisexual ....Amaranthus
    B. flowers perfect ....Celosia
A. leaves opposite .... Froelichia

Amaranthus L.

Leaves alternate; monoecious or dioecious; mature anthers with 2 locules; style very short or absent; fruits 1-seeded.

A. plants dioecious
    B. flowers pistillate
        C. sepals absent (rarely 1 or 2 and less than 1 mm long)
            D. mature fruits 1.5–2 mm long; most leaves linear-lanceolate ....
                .... A. cannabinus
            D. mature fruits 2.5–4 mm long; most leaves broadly lanceolate to ovate
                .... A. tuberculatus
        C. sepals present, 1 mm or more long
            E. sepals 1 or 2 .... A. rudis
            E. sepals 5
                F. bracts about as long as or somewhat exceeding the sepals ....
                .... A. arenicola
                F. bracts 2–3 times as long as the sepals .... A. palmeri
    B. flowers staminate
        G. bracts 4–6 mm and longer than the sepals .... A. palmeri
        G. bracts 1–2.5 mm and shorter than or about as long as the sepals
            H. outer sepals with a heavy midvein, usually longer than the inner sepals
                I. apex of outer sepals acuminate, the midvein extended as a rigid spine .... A. rudis
                I. apex of outer sepals obtuse, the midvein not extended as a rigid spine .... A. arenicola
            H. outer sepals with a slender midvein, not longer than the inner sepals
                J. most leaves linear-lanceolate .... A. cannabinus
                J. most leaves broadly lanceolate to ovate .... A. tuberculatus
A. plants monoecious
    K. stems with paired spines at the bases of the leaves .... A. spinosus
    K. stems without spines
        L. flowers mostly in axillary clusters, a small terminal panicle or cluster often present
        M. stems erect, 3–12 dm tall .... A. albus
        M. stems prostrate or spreading, rarely erect to 3 dm

    N. plants fleshy, succulent; sepals about ½ as long as the mature fruits .... *A. pumilus*

    N. plants not succulent; sepals about as long as the mature fruits.... .... *A. blitoides*

  L. flowers mostly in long terminal spikes, axillary clusters sometimes present

    O. sepals 5 (rarely 4)

      P. bracts 3.5–7 mm long, much longer than the mature fruits

        Q. sepals obtuse or emarginate at apex with a short mucronate tip .... *A. retroflexus*

        Q. sepals acute to long acuminate at apex

          R. inflorescence stiff, unbranched or with a few long branches .... *A. powellii*

          R. inflorescence drooping with many short lateral branches.... .... *A. hybridus*

      P. bracts 1.2–2 mm long, shorter to slightly longer than the mature fruits

        S. sepals curved outward at apex .... *A. caudatus*

        S. sepals straight at apex .... *A. cruentus*

    O. sepals 3 (rarely 2)

      T. leaf apices mostly emarginate or truncate .... *A. blitum*

      T. leaf apices acute .... *A. deflexus*

Amaranthus albus L.   Tumbleweed

Erect, much branched plant to 1 m tall and wide; stems glabrous or sparsely pubescent, whitened; leaves glabrous, whitened along the veins beneath, elliptic to ovate or obovate to spatulate, obtuse and mucronate at apex, cuneate at base; fruit dehiscent; seeds dark reddish-brown, flattened and round in outline; common in disturbed ground and alluvium; mostly S; flr. late Jul–Oct; native to central North America; FACU.

Amaranthus arenicola I.M.Johnston   Amaranth

Erect plant to 2 m tall; leaves long-petioled, oblong to lanceolate; male flower sepals 5, of nearly equal lengths; female flower sepals 5, the outer ones 2–2.5

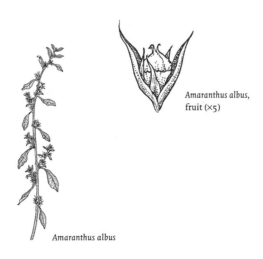

Amaranthus albus,
fruit (×5)

Amaranthus albus

mm long and the inner ones 1.5–2 mm long; fruit dehiscent; seeds dark reddish-brown; rare on waste ground, dumps, and other disturbed sites; flr. summer; native to western U.S.; UPL.

Amaranthus blitoides S.Watson   Prostrate pigweed, tumbleweed
Densely branched, spreading annual; stems to 6 dm long, glabrous to sparsely pubescent; leaves obovate to spatulate or elliptic, rounded to acute at apex, cuneate to attenuate at base; fruit dehiscent; seeds black; occasional in fields, on roadsides, railroad tracks, and other waste areas; flr. late Jul–Oct; native to western North America.

Amaranthus blitum L.   Amaranth
Spreading prostrate annual forming mats; stems to 15 dm long, green or with a reddish tinge, glabrous, succulent; leaves broadly- to rhombic-ovate, deeply emarginate at apex, cuneate to rounded at base; fruit indehiscent; seeds dark reddish-brown to black; an occasional weed of waste ground; mostly E; flr. Sep–Oct; native to tropical America.

Amaranthus cannabinus (L.) Sauer   Saltmarsh water-hemp,
                                                       water-hemp ragweed
Erect, somewhat succulent annual to 3 m tall; stems glabrous, green, often enlarged at the base, branches ascending; leaves narrowly lanceolate to linear, acuminate to long-attenuate at apex, acute to attenuate at base; fruit indehiscent; seeds dark reddish-brown, ovoid in outline; rare in the uppermost zone of freshwater intertidal marshes; SE; flr. Jul–Sep; OBL; ✺.

Amaranthus caudatus L.   Love-lies-bleeding
Erect annual to about 1 m tall (much taller in cultivation); stems glabrous or sparsely villous, often tinged with red; leaves ovate to lanceolate, acute at apex, acute at base; fruit dehiscent; seeds usually yellowish-white (sometimes reddish or black); cultivated and rarely escaped; mostly SE; flr. Jun–Aug; native to South America.

Amaranthus cruentus L.   Blood amaranth, purple amaranth
Erect annual to about 2 m tall; stems usually pubescent, stout, green or reddish; leaves elliptic or ovate to lanceolate, acute or attenuate at apex, acute to attenuate at base; fruit dehiscent; seeds black or dark reddish-brown; an occasional weed of waste places; flr. Jul–Aug; native to Central America.

Amaranthus deflexus L.   Low amaranth
Prostrate to ascending annual to about 5 dm tall; leaves lanceolate to broadly ovate, petiole about the same length as the blade; fruit indehiscent; seeds shining, obovate; rare on ballast; mostly S; flr. Jul–early Oct; native to tropical America.

Amaranthus hybridus L.   Pigweed
Erect annual to about 2 m tall; stems roughly hairy below, glabrous to villous above, stout, green and tinged with red; leaves lanceolate to ovate, acute or rounded at apex, cuneate or rounded at base, usually pubescent below; fruit dehiscent; seeds black or dark reddish-brown, round in outline; a common weed

of cultivated fields, gardens, and waste ground throughout; flr. Aug–Oct; native
to tropical America.

Amaranthus palmeri S.Watson   Palmer's amaranth
Erect annual to about 1 m tall (much taller in cultivation); leaves long-petioled,
rhombic-ovate to rhombic lanceolate; male flower sepals 5, of unequal lengths;
female flower sepals 5, recurved; midvein of outer sepals of both sexes prolonged into a stiff spine; fruit dehiscent; seeds dark reddish-brown; rare on
waste ground and ballast; SE; flr. Sep–Oct; native to western North America;
FACU.

Amaranthus powellii S.Watson   Amaranth
Erect annual to about 2 m tall; stems stout, green or whitish, glabrous below,
villous above; leaves rhombic-ovate to lanceolate or elliptic, acute to rounded
(sometimes emarginate) at apex, cuneate to rounded at base; fruit dehiscent;
seeds black to reddish-brown; an occasional weed of waste ground throughout;
flr. Jul–Aug; native to western U.S.

Amaranthus pumilus Raf.   Seabeach amaranth
Prostrate to somewhat ascending annual; stems fleshy, to about 3 dm long,
densely branched; leaves clustered at the ends of the branches, obovate to suborbicular, rounded or emarginate at apex, rounded to attenuate or decurrent at
base, fleshy, veins often red-purple; fruit indehiscent, fleshy; seeds dark reddish-
brown to black; represented by a single collection from Philadelphia in 1865;
NC; flr. Jul–Aug; native in coastal areas, adventive in PA; 🍂.

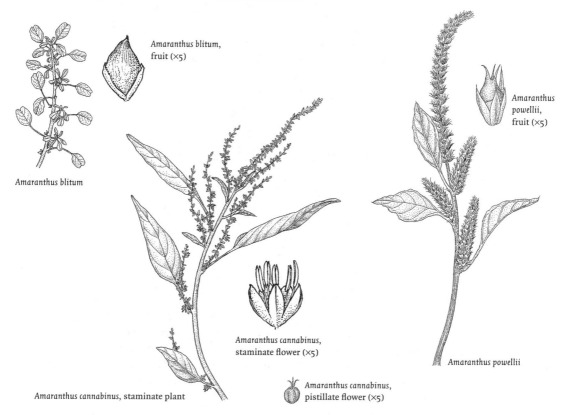

Amaranthus blitum,
fruit (×5)

Amaranthus blitum

Amaranthus powellii,
fruit (×5)

Amaranthus cannabinus,
staminate flower (×5)

Amaranthus cannabinus, staminate plant

Amaranthus cannabinus,
pistillate flower (×5)

Amaranthus powellii

Amaranthus retroflexus L.   Green amaranth
Erect annual to about 3 m tall; stems stout, obtusely angled, green or whitish, villous above; leaves lanceolate to ovate or obovate, acute, obtuse, or emarginate at apex, rounded or acute at base, glabrous above, often villous below; fruit dehiscent; seeds black; a common weed of cultivated fields, gardens, and waste ground; throughout; flr. Jul–Aug; native to tropical America; FACU.

Amaranthus rudis Sauer   Western water-hemp
Erect annual to about 2 m tall; stems stout, often angled, green and sometimes glaucous, unbranched or with a few ascending branches; leaves oblong to lanceolate-oblong, sometimes ovate, rounded or obtuse at apex, acute to attenuate at base; fruit dehiscent; seeds dark reddish-brown to black; occasional on waste ground and ballast; SE; flr. late summer; native to central North America; FACW–.

Amaranthus spinosus L.   Spiny amaranth
Spiny, erect annual to about 1 m tall; stems stout, succulent, often reddish, glabrous below and pubescent above; leaves ovate to lanceolate or rhombic-ovate, rounded, obtuse or emarginate at apex, broadly cuneate at base; fruit dehiscent (rarely indehiscent); seeds black to dark reddish-brown; an occasional weed of cultivated fields, waste ground, and alluvium; SE; flr. Jul–Oct; native to tropical America; FACU.

Amaranthus tuberculatus (Moq.) Sauer   Water-hemp
Erect annual to about 3 m tall or prostrate; stems succulent, stout, green often tinged with red; leaves broadly ovate to linear-lanceolate, apex variable from obtuse to acuminate, often emarginate, base obtuse to cuneate; fruit indehiscent or bursting irregularly; seeds dark reddish-brown to black; rare on river banks and low, disturbed ground; scattered; flr. late summer; native to central North America; FACW.

Celosia L.

Celosia argentea L.   Celosia
Erect annual to 1 m tall; leaves alternate, lanceolate to linear, 8–15 cm long; flowers perfect, mature anthers with 2 locules, style exserted at maturity, sepals pink, yellow, red, or orange; inflorescence a dense spike, terminal and/or axillary, 2–15 cm long (or fan-shaped in cockscomb cultivars); fruit dehiscent; seeds 2–several, disk-like; cultivated and occasionally escaped; mostly SE; flr. late Jul–Oct; native to tropical America.

Froelichia Moench.

Froelichia gracilis (Hook.) Moq.   Cottonweed
Erect to prostrate annual to 7 dm tall, usually branching from near the base; leaves opposite, linear to narrowly lanceolate, to 8 cm long, mostly below the middle of the stem; flowers perfect, mature anthers with 1 locule, style to about 1 mm, sepals united into a conic tube; fruit indehiscent, 1-seeded; occasional in dry, open, sandy soil, often around railroad tracks; SE; flr. Jul–Sep; native to southwestern U.S.

Amaranthus retroflexus

Amaranthus retroflexus,
fruit (×5)

Froelichia gracilis,
flower

Froelichia gracilis

ANACARDIACEAE Cashew Family

Small trees, shrubs, or woody vines with alternate, simple, or compound leaves
lacking stipules; flowers small, regular, perfect or unisexual, 5-merous; sepals
fused at the base; petals distinct; stamens alternate with petals; ovary of 3 fused
carpels, superior; fruit a drupe.

A. leaves compound; pedicels not elongating or becoming plumose
    B. inflorescence dense, terminal (on short lateral shoots in *R. aromatica*), erect
       or ascending; fruits red, conspicuously glandular-hairy. .... *Rhus*
    B. inflorescence loose, axillary, frequently drooping in fruit; fruits whitish,
       glabrous or inconspicuously hairy .... *Toxicodendron*
A. leaves simple; sterile pedicels elongating and developing long, feathery hairs
    .... *Cotinus*

Cotinus Adans.

Cotinus coggygria Scop.   Smoketree
Shrub or small tree with gray-green leaves and feathery purplish inflorescences; leaves simple, oval or obovate, 2.5–7 cm, glabrous; inflorescence to 2 dm due to the elongate, hairy pedicels of sterile flowers; fertile flowers small, yellowish-green; cultivated and occasionally persisting at old garden sites; flr. Jun; native to Eurasia.

Rhus L.

Shrubs or small trees with dense, terminal clusters of red, glandular-hairy fruits; leaves pinnately compound or trifoliate; flowers with 5 hairy petals; ovary with a short, 3-lobed style.

A. small trees or shrubs with few branches; leaves pinnately compound, leaflets 7–29; inflorescences terminating upper branches
   B. twigs, petioles, and rachises hairy
      C. twigs etc. densely velvety-villous; rachis of the leaf not winged; leaflets serrate .... *R. typhina*
      C. twigs etc. short, hairy or puberulent; rachis of the leaf winged; leaflets entire .... *R. copallina*
   B. twigs, petioles, and rachises glabrous .... *R. glabra*
A. bushy shrub; leaves trifoliate; inflorescences terminating short lateral shoots
   .... *R. aromatica*

Rhus aromatica Aiton   Fragrant sumac, squawbush
Much-branched shrub to 2 m tall, often forming thickets, aromatic when bruised; leaflets 3, sessile, with a few teeth; flowers yellow, opening before or with the leaves from scaly, catkin-like buds formed the summer before; fruits 4–5 mm, red, densely hairy; occasional in dry, open woods, limestone prairies, and shale barrens; SC; flr. May, frt. Aug–Sep. Ours is var. *aromatica*.

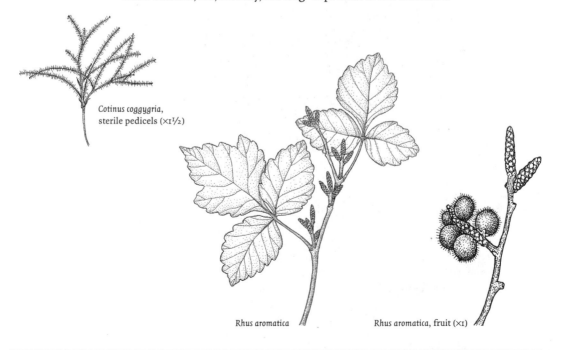

Cotinus coggygria,
sterile pedicels (×1½)

Rhus aromatica   *Rhus aromatica,* fruit (×1)

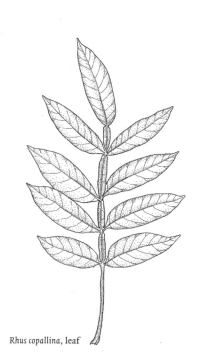

Rhus copallina, leaf

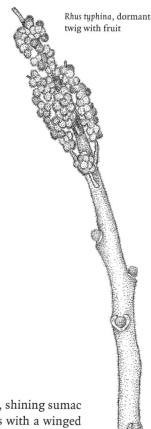

Rhus typhina, dormant twig with fruit

Rhus copallina L.   Winged sumac, shining sumac
Shrub or small tree to 6 m tall with pinnately compound leaves with a winged
rachis; young twigs and petioles short hairy or puberulent; leaflets 5–23, oblong
to lanceolate, 3–8 cm, entire or with few teeth, lustrous above; inflorescence to
15 cm; fruits 4–5 mm, flattened, red, hairy; frequent in dry, open woods, thickets
and old fields throughout; flr. Jul–early Sep, frt. Oct–Nov; 2 varieties:

A. leaflets 11–23, lanceolate or linear-oblong, 1–2 cm wide .... var. *copallina*
A. leaflets 5–13, broadly oblong to narrowly ovate, 1.5–4 cm wide ....
.... var. *latifolia* Engl.

Rhus glabra L.   Smooth sumac
Sparingly branched colonial shrub or small tree to 5 m tall with pinnately compound leaves; young branches and petioles glabrous and somewhat glaucous;
leaflets 11–31, lanceolate to narrowly oblong, 5–10 cm, serrate; inflorescences
often unisexual, staminate more open and larger, to 4.5 dm, the pistillate
smaller and denser; fruits red, 4–5 mm, flattened, densely covered with 2-mm-
long red hairs; frequent in shale barrens, old fields, and dry, open slopes;
throughout; flr. Jun–early Aug, frt. late Aug–Oct.

Rhus typhina L.   Staghorn sumac
Tall shrub or small tree to 10 m tall with pinnately compound leaves; young
twigs, petioles and leaf rachis densely hairy; leaflets 11–31, oblong-lanceolate, 5–
12 cm, acuminate, serrate; inflorescence pyramidal to ovoid, 0.5–2 dm; fruits
red, 1–2 mm, somewhat flattened, densely covered with long, spreading hairs,
often persisting through the winter; common in dry, open soil of old fields,
roadsides, and woods edges; throughout; flr. Jun, frt. late Jul–winter.

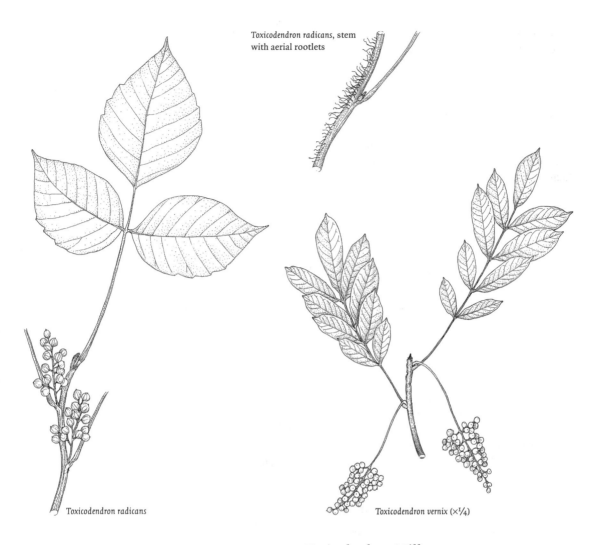

Toxicodendron radicans, stem with aerial rootlets

Toxicodendron radicans

Toxicodendron vernix (×¹/₄)

Toxicodendron Mill.

Small trees, shrubs, or vines containing a poisonous oil (urushiol) that causes an itchy rash in many people; leaves compound; inflorescences axillary; fruits white, glabrous or slightly hairy but not glandular.

A. vines or erect shrubs; leaves trifoliate
    B. aerial rootlets usually present; leaflets not folded .... *T. radicans*
    B. aerial rootlets not present; leaflets somewhat folded along the midrib ......
    .... *T. rydbergii*
A. small trees or tall shrubs; leaves pinnately compound .... *T. vernix*

Toxicodendron radicans (L.) Kuntze   Poison-ivy
Climbing or scrambling vine with abundant aerial rootlets, especially on old stems, or erect shrubs to 1 m; leaflets 3, ovate to elliptic, 5–15 cm, irregularly coarsely toothed or shallowly lobed; terminal leaflet long stalked; lateral leaflets short-stalked or sessile; inflorescence to 1 dm, axillary, drooping in fruit; fruits white, 3–5 mm; common in open woods, roadside thickets, fencerows, floodplains, and edges; throughout; flr. late May, frt. late Jul–Oct; FAC. [syn: *Rhus radicans* L.]

Toxicodendron rydbergii (Small ex Rydb.) Greene Giant poison-ivy
Very similar to *T. radicans*, differing in the absence of aerial rootlets; leaflets somewhat folded along the midrib; rare in dry, rocky woods; N; FAC−; 🍂.

Toxicodendron vernix (L.) Kuntze Poison sumac
Upright shrub or small tree to 5 m tall with pinnately compound leaves; leaflets 7–13, oblong to obovate or elliptic, 4–5 cm, entire, glabrous; leaf rachis often red; inflorescence to 2 dm, axillary, drooping; fruit grayish-white; occasional in swamps, bogs, fens, and marshes; throughout; flr. Jun, frt. Aug–Nov; OBL. [syn: *Rhus vernix* L.]

ANNONACEAE Custard-apple Family

Woody plants with alternate, simple, entire leaves lacking stipules; flowers regular, perfect; sepals 3; petals 6; stamens numerous; pistils 3–5, distinct; ovaries superior; fruit a cluster of 1–several fleshy berries.

Asimina Adans.

Characteristics of the family.

Asimina triloba (L.) Dunal Pawpaw
Deciduous, colony-forming shrub or understory tree to 10 m tall; leaf blade oblong-obovate, 15–35 cm, gradually tapering to the petiole; flowers appearing with the leaves; petals purple, the outer 1.5–2.5 cm; fruits solitary or a few together, irregularly cylindrical, to 15 cm long and 3–4 cm thick, green with edible, yellow, custard-like flesh containing several large flattened seeds; occasional in moist, rich woods; S; flr. early May, frt. Aug–Sep; FACU+.

Asimina triloba

APIACEAE Parsley Family

Herbaceous plants, often aromatic, occasionally toxic, generally with hollow stems; leaves alternate and/or basal, usually compound but occasionally simple, the petiole forming a sheath around the stem; flowers generally perfect, rarely with some staminate, regular, ovary inferior; inflorescence an umbel, usually compound and divided into a number of umbellets; primary branches of the umbel are referred to as rays; the umbel is often subtended by bracts and the umbellets by bractlets; fruit from a single ovary but splitting into two sections at maturity.

A. inflorescence composed of densely flowered heads .... *Eryngium*
A. inflorescence an umbel
    B. leaves evidently simple (not deeply divided)
        C. leaves perfoliate .... *Bupleurum*
        C. leaves petiolate .... *Hydrocotyle*
    B. leaves compound or divided deeply
        D. fruits and ovaries bristly
            E. fruits and ovaries bristly only on the ribs
                F. fruits 3–4 mm long; umbels with 10 or more rays .... *Daucus*
                F. fruits 15–22 mm long; umbels mostly with 4–6 rays .... *Osmorhiza*
            E. fruits and ovaries uniformly bristly
                G. leaves palmately compound .... *Sanicula*
                G. leaves pinnately compound .... *Torilis*
        D. fruits and ovaries glabrous
            H. leaves finely and often irregularly divided or dissected (giving a somewhat feathery appearance)
            I. leaves ultimately divided into fine, thread-like segments
                J. fruits conspicuously winged; flowers yellow .... *Anethum*
                J. fruits not winged (although conspicuously ribbed); flowers white .... *Ptilimnium*
            I. leaves divided into fine but flattened segments, segments not thread-like
                K. bractlets linear to narrowly lanceolate, or absent
                    L. fruits conspicuously winged; umbellets densely many-flowered .... *Conioselinum*
                    L. fruits not winged; umbellets few-flowered
                        M. fruits distinctly flattened; bractlets minute or absent .... *Carum*
                        M. fruits only slightly or not at all flattened; bractlets 2–5, all on one side of umbellet .... *Aethusa*
                K. bractlets broadly lanceolate to ovate or obovate
                  N. umbellets 1–3 (rarely 4–5)
                    O. fruits 5.5–10 mm long; plants 2–6 dm tall, often decumbent .... *Chaerophyllum*
                    O. fruits <5 mm long; plants 0.5–1.5 dm tall, usually erect .... *Erigenia*
                N. umbellets 6 or more
                  P. stems purple-spotted; tall plant to 3 m; bracts generally present .... *Conium*

P. stems not purple-spotted; plant to 1 m; bracts generally absent....
.... *Anthriscus*

H. leaves divided into distinct, regular leaflets, not finely divided

  Q. leaves ternately compound (with ultimate leaflets in groups of 3)

    R. leaflets entire .... *Taenidia*

    R. leaflets toothed or lobed

      S. leaves mostly with 3 leaflets (once ternate)

        T. plants in fruit

          U. fruits pointed on both ends .... *Cryptotaenia*

          U. fruits rounded or notched at the ends

            V. fruits conspicuously winged

              W. fruits only slightly or not at all flattened .... *Thaspium*

              W. fruits conspicuously flattened

                X. fruits 4–5 mm long .... *Peucedanum*

                X. fruits 7–12 mm long .... *Heracleum*

            V. fruits not winged .... *Zizia*

        T. plants in flower

          Y. flowers purple .... *Thaspium*

          Y. flowers white or yellow

            Z. flowers yellow .... *Zizia*

            Z. flowers white

               AA. umbels with 2–7 rays .... *Cryptotaenia*

               AA. umbels with 15 or more rays

                  BB. leaves pubescent .... *Heracleum*

                  BB. leaves glabrous .... *Peucedanum*

      S. leaves mostly with 9 or more leaflets (twice or more ternate)

        CC. leaflets entire .... *Taenidia*

        CC. leaflets toothed or lobed

          DD. rachis of the leaf winged below the leaflets .... *Falcaria*

          DD. rachis of the leaf not winged

             EE. plants in fruit

               FF. fruits conspicuously winged .... *Peucedanum*

               FF. fruits not winged

                  GG. lower leaves mostly 3–4-times trifoliate (having 27–54 ultimate leaflets) .... *Ligusticum*

                  GG. lower leaves mostly twice trifoliate (having 9 ultimate leaflets)

                    HH. fruits prominently ribbed .... *Zizia*

                    HH. fruits not prominently ribbed ....
                                 .... *Aegopodium*

            EE. plants in flower

              II. flowers yellow .... *Zizia*

              II. flowers white

                JJ. sepals present, triangular .... *Ligusticum*

                JJ. sepals absent

                  KK. sheaths of upper leaves >1 cm wide when flattened .... *Peucedanum*

                  KK. sheaths of upper leaves <1 cm wide when flattened .... *Aegopodium*

    Q. leaves pinnately compound
       LL. leaflets entire
          MM. leaves once pinnate; fruits conspicuously winged .... *Oxypolis*
          MM. leaves twice or thrice pinnate; fruits not winged
             NN. bractlets present; flowers white .... *Cicuta*
             NN. bractlets absent; flowers yellow .... *Taenidia*
       LL. leaflets toothed or lobed
          OO. plants in fruit
             PP. fruits conspicuously winged
                QQ. fruits 3–5-winged
                    RR. bracts present.......................................*Thaspium*
                    RR. bracts absent .... *Sium*
                QQ. fruits 2-winged
                    SS. plants 3–6 dm tall .... *Pimpinella*
                    SS. plants generally 1 m or more tall
                        TT. upper leaves greatly reduced, the uppermost
                            often without blades .... *Angelica*
                        TT. upper leaves somewhat smaller than the lower
                            but still pinnately compound .... *Pastinaca*
             PP. fruits not winged (though they may be conspicuously ribbed)
                UU. lateral veins of the leaflets consistently extending to si-
                    nuses between the teeth .... *Cicuta*
                UU. lateral veins of the leaflets not apparently related to si-
                    nuses between the teeth .... *Sium*
          OO. plants in flower
             VV. flowers yellow or purple
                WW. sepals present ....*Thaspium*
                WW. sepals absent .... *Pastinaca*
             VV. flowers white
                XX. leaves mostly once pinnate
                    YY. bracts present... *Sium*
                    YY. bracts absent.................................... *Pimpinella*
                XX. leaves mostly twice or more pinnate
                    ZZ. sepals apparent, triangular .... *Cicuta*
                    ZZ. sepals minute or absent .... *Angelica*

Aegopodium L.

Aegopodium podagraria L.   Goutweed

Branched herb, 4–9 dm tall; basal and lower cauline leaves mostly twice ternate with petioles longer than blades, upper cauline leaves often once ternate or irregular to simple and lobed with petioles shorter than blades; flowers white, sepals absent; umbel with 15–25 rays of nearly equal length; fruits ovoid, compressed, 3–4 mm long, with prominent styles about 1.5 mm long; cultivated and frequently naturalized in fields, thickets, disturbed woods, and roadsides; throughout; flr. late May–late Jun; native to Eurasia; FACU.

Aethusa L.

Aethusa cynapium L.   Fool's parsley
Freely branched herb, 2–7 dm tall; leaves shining, generally deltoid in outline; flowers white, sepals absent; umbels with 7–20 rays; fruits ovoid, about 3 mm long by 2 mm wide; dry roadsides, alluvial terraces, waste ground, and ballast; mostly SE; flr. Jun–Sep; native to Eurasia.

Aethusa cynapium,
umbellet (×1)

Anethum L.

Anethum graveolens L.   Dill
Glabrous and glaucous somewhat branched annual herb to 1.5 m tall, strongly scented; ultimate leaf segments thread-like, 5–20 mm long; flowers yellow, sepals absent; umbels up to 15 cm across with 15–40 rays; fruits 3–5 mm long by 1.5–2.5 mm wide; cultivated and occasionally escaped in fields and rubbish dumps; throughout; flr. Jul–Aug; native to Europe.

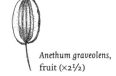

Anethum graveolens,
fruit (×2½)

Angelica L.

Stout, robust herbs to 2 m tall, glabrous except in the inflorescence of some species; leaves long-petioled near the base and progressively reduced upward, the uppermost often reduced to only a sheath and petiole; fruits flattened and winged; flowers white or greenish-white.

A. apex of leaflets obtuse; upper stem pubescent .... *A. venenosa*
A. apex of leaflets acute or acuminate; upper stem glabrous or nearly so
    B. leaflets acuminate, the margins minutely ciliate .... *A. triquinata*
    B. leaflets acute, the margins glabrous .... *A. atropurpurea*

Angelica atropurpurea L.   Purple-stemmed angelica
Leaflets ovate to lanceolate, 4–10 cm long, sharply serrate; umbels 1–2 dm across with 20–45 rays; fruits oblong-elliptic, rounded at the base, 4–6.5 mm long; swamps, moist meadows, stream banks, and wet woods; throughout; flr. Jun–Aug; OBL.

Angelica triquinata Michx.   Angelica
Leaflets lanceolate to oblong, 3–8 cm long, coarsely toothed; umbels 6–15 cm across with 13–25 rays; fruits oblong-elliptic, cordate at the base, 5–7 mm long; banks of mountain streams, wet woods, and floodplains; mostly SW; flr. Jul–Sep; UPL.

Angelica venenosa (Greenway) Fernald   Deadly angelica, hairy angelica
Leaflets oblong to elliptic, 2–4 cm long, finely serrate; umbels 5–15 cm across with 20–35 rays; fruits oblong-elliptic, cordate at the base, 4–7 mm long; dry, open woods, roadside banks, serpentine barrens, and old fields; throughout; flr. Jul–Aug.

Angelica venenosa,
fruit (×2½)

Anthriscus Hoffm.

Anthriscus sylvestris (L.) Hoffm.   Chervil
Freely branching plant to 1 m tall; ultimate leaf segments 1.5–5 cm long; flowers white; umbels large, with 6–10 rays; rays up to 4 cm long; fruits lanceolate, to 6

mm long with a beak about 1 mm long; gravelly banks, roadsides, and alluvial areas; mostly S; flr. May–Jul; native to Europe.

Bupleurum L.

Bupleurum rotundifolium L.   Hare's ear, thoroughwax
Glabrous and glaucous, sometimes purple-tinged annual herb, 3–6 dm tall; leaves elliptic-ovate to suborbicular, the lower leaves often with a mucronate tip; flowers yellow to greenish-yellow, subsessile, sepals absent; umbels with 4–10 rays, bractlets broadly ovate to elliptic, 8–12 mm long; fruits elliptic-oblong, blackish-brown, 2.5–3 mm long; fallow areas, railroad tracks, wharves, and rubbish dumps; S; flr. May–Jun; native to Eurasia.

Carum L.

Carum carvi L.   Caraway
Glabrous biennial to 1 m tall; ultimate leaf segments 5–15 mm long; flowers white to rarely pink; umbels with 7–14 rays, rays 2–4 cm long; fruits elliptic to oblong, 3–4 mm long by 1.5–2 mm wide; cultivated and occasionally naturalized in fields, meadows, and roadsides; throughout; flr. Jun–Aug; native to Eurasia.

Chaerophyllum L.

Chaerophyllum procumbens (L.) Crantz   Slender chervil, spreading chervil
Low herb, 2–6 dm tall, spreading, usually branched from the base, glabrous to only slightly hairy; ultimate leaf segments oblong to ovate; flowers white; fruits

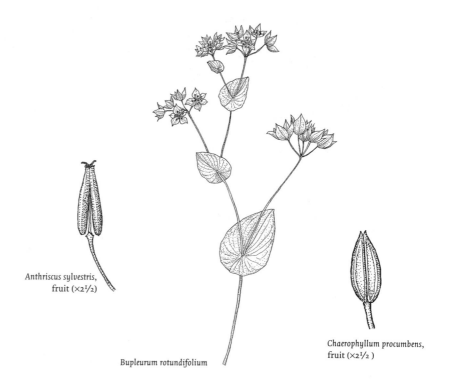

Anthriscus sylvestris, fruit (×2½)

Chaerophyllum procumbens, fruit (×2½)

Bupleurum rotundifolium

elliptic to oblong, 5.5–10 mm long by 2–3 mm wide; rich woods, wooded slopes, and bottomland; mostly S; flr. Apr–May; FACW.

Cicuta L.

Glabrous herbs of wet places, all parts of the plant highly toxic; leaves mostly pinnately compound (rarely ternate); fruits flattened, ribs prominent and corky; flowers white.

A. axils of the upper leaves bearing bulblets .... *C. bulbifera*
A. bulblets absent .... *C. maculata*

Cicuta bulbifera L.   Water-hemlock
Slender herb, 3–10 dm tall; leaflets linear to linear-lanceolate, about 5 mm wide, sparsely toothed to entire; umbels <5 cm across; fruits orbicular, 1.5–2 mm long and wide, but rarely maturing; marshes, swampy meadows, swales, and openings in wet bottomland woods; throughout; flr. Jul–Sep; OBL.

Cicuta maculata L.   Beaver-poison, musquash-root
Stout much branched plant to 2 m tall; leaflets ovate-lanceolate to linear, 3–10 cm long, sharply serrate or dentate to nearly entire; umbels 5–12 cm across; fruits ovoid to orbicular, 2–4 mm long, the ribs pale brown; swamps, marshes, wet meadows, stream banks, and ditches; throughout; flr. Jun–Aug; OBL. Ours is var. *maculata*.

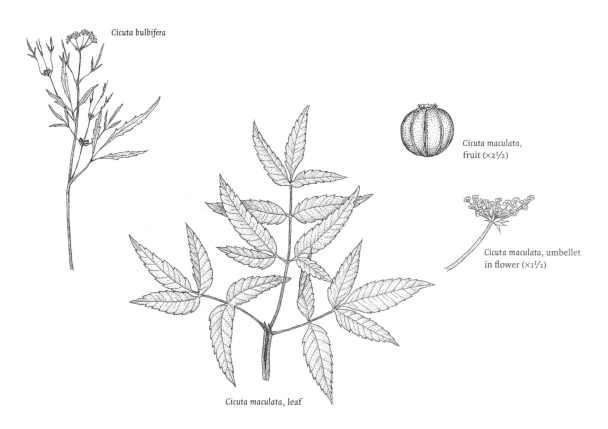

Cicuta bulbifera

Cicuta maculata, fruit (×2½)

Cicuta maculata, umbellet in flower (×1½)

Cicuta maculata, leaf

Conioselinum Hoffm.

Conioselinum chinense (L.) Britton, Stearns & Poggenb.   Hemlock-parsley
Slender to stout herb, 4–15 dm tall; petioles of upper leaves short and broadly winged, leaflets 1.5–4 cm long; flowers white, sepals absent; umbels 3–12 cm across; fruits elliptic to oblong, 4–5.5 mm long by 2–4 mm wide; moist rich woods and stream banks; widely scattered N; flr. Aug–Sep; FACW; ✿.

Conium L.

Conium maculatum L.   Poison-hemlock
Glabrous much branched plant to 3 m tall, stems purple-spotted, all parts of the plant extremely toxic; lower leaves up to 50 cm long, entirely glabrous; flowers white; umbels 4–6 cm across, rays 10–20 or more; fruits broadly ovoid, 2.5–3.5 mm long; roadside ditches, floodplains, and moist woods; throughout; flr. late May–Jul; native to Europe; FACW.

Cryptotaenia DC.

Cryptotaenia canadensis (L.) DC.   Honewort, wild-chervil
Branching, glabrous, perennial herb, 3–8 dm tall; leaflets lanceolate to obovate, 4–15 cm long, sharply and irregularly toothed and sometimes lobed; flowers white, sepals minute or absent; umbels with 2–7 rays, the rays 1–5 cm long; fruits 5–8 mm long, dark; moist woods, wooded stream banks, and seeps; throughout; flr. late May–early Jul; FAC.

Daucus L.

Daucus carota L.   Queen Anne's-lace, wild carrot
Biennial, 4–10 dm tall, glabrous to roughly hairy; leaves finely divided; flowers white (the central flower of each umbellet often purple); umbels 4–12 cm across, usually with 20 or more rays; roadsides, gardens, old fields, and waste ground; throughout; flr. Jun–Sep; native to Eurasia.

Erigenia Nutt.

Erigenia bulbosa (Michx.) Nutt.   Harbinger-of-spring, pepper-and-salt
Delicate perennial, 5–15 cm tall at flowering, taller at maturity; leaves 10–20 cm long; flowers white, sepals absent; umbels most commonly with 3 rays, the rays

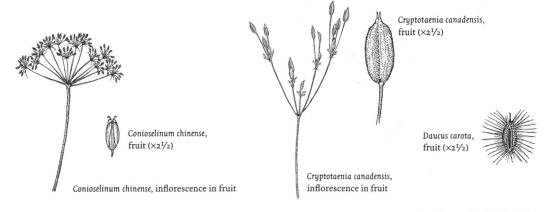

Conioselinum chinense,
fruit (×2½)

Conioselinum chinense, inflorescence in fruit

Cryptotaenia canadensis,
fruit (×2½)

Cryptotaenia canadensis,
inflorescence in fruit

Daucus carota,
fruit (×2½)

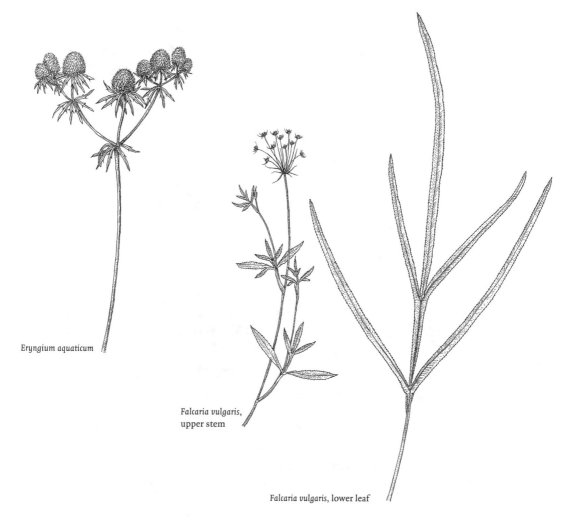

Eryngium aquaticum

Falcaria vulgaris,
upper stem

Falcaria vulgaris, lower leaf

1–2 cm long; seeps and spring heads on wooded slopes; W and lower Susquehanna Valley; flr. Mar–Apr; 🌿.

Eryngium L.

Eryngium aquaticum L.   Marsh eryngo, rattlesnake-master
Erect, somewhat branched perennial to 1 m tall; leaves elongate, lower leaves long-petioled, 1–3 dm long, upper leaves shorter and short-petioled to sessile; flowers white to purple, sepals large; inflorescence a dense head 10–15 mm long; fruits globose to obovoid, not flattened; river swamps, pond banks, and gravelly shores; mostly SE; believed to be extirpated; flr. late Jul–Sep; OBL; 🌿.

Falcaria Host

Falcaria vulgaris Bernh.   Sickleweed
Much branched glaucous herb to 1 m tall, stems filled with pith; lower leaves with petioles, upper leaves sessile at the apex of the sheath, leaflets linear-lanceolate to linear; flowers white, sepals triangular; umbels with 10–18 rays; fruits oblong, 3–4 mm long; open woods and meadows; mostly E; flr. Jul–Aug; native to Eurasia.

Heracleum L.

Large, robust perennials or biennials, somewhat pubescent to tomentose; leaves mostly trifoliate; fruits large, winged; flowers white to rarely pinkish.

A. umbels 1–2 dm across with 15–30 rays .... H. lanatum
A. umbels 3–5 dm across with 50 or more rays .... H. mantegazzianum

Heracleum lanatum Michx.   Cow-parsnip
Perennial, 1–3 m tall, pubescent to tomentose throughout; leaflets broadly ovate to rotund, 1–4 dm long and wide, coarsely serrate and variously lobed, cordate at the base; fruits about 10 mm long and wide; rich woods, wooded roadside banks, marshy flats, stream banks, and ditches; throughout; flr. early May–Jun; FACU–.

Heracleum mategazzianum, fruit (×2½)

Heracleum mantegazzianum Sommier & Levier   Giant hogweed
Perennial or biennial, 2–5 m tall, stems up to 10 cm in diameter; leaflets up to 12 dm long, pinnately lobed, shortly pubescent below; fruits 9–11 mm long by 6–8 mm wide; cultivated and reported as escaped along roadsides, ditches, waste areas and woodlots; Erie and McKean Cos., probably elsewhere; flr. summer; native to Eurasia; designated as a federal noxious weed.

Hydrocotyle L.

Low or creeping plants, stems slender, rooting at the nodes; leaves simple, broadly ovate to orbicular; fruits orbicular and strongly flattened, 1–3 mm wide; flowers white, in few-flowered generally simple umbels; growing in or near water.

A. petiole attached near middle of blade .... H. umbellata
A. petiole attached at margin of blade
    B. leaves mostly 1 cm or less in diameter .... H. sibthorpioides
    B. leaves >1 cm in diameter
        C. leaves lobed to near the middle .... H. ranunculoides
        C. leaves only very shallowly lobed .... H. americana

Hydrocotyle americana L.   Marsh pennywort, navelwort
Leaves 2–5 cm wide, shallowly 6–10-lobed; umbels 2–7-flowered, sessile or nearly so; fruits about 1.5 mm wide; swampy thickets, boggy fields, wet woods, and lake margins; throughout; flr. late Jun–Sep; OBL.

Hydrocotyle ranunculoides L.f.   Floating pennywort
Leaves up to 7 cm wide, 5–6-lobed to nearly the middle; umbels 5–10-flowered, peduncles shorter than the subtending leaves; fruits 2–3 mm wide; shallow water, moist shores, and wet meadows; mostly SC and SE; flr. Jun–Aug; OBL.

Hydrocotyle sibthorpioides Lam.   Lawn pennywort
Leaves rarely >1 cm wide, shallowly 7-lobed, petioles 1–2 cm long; umbels 3–10-flowered, peduncles longer than the subtending leaves; fruits 1–1.5 mm wide; an occasional weed of shady lawns; mostly SE; flr. May–Aug; native to Asia.

Hydrocotyle umbellata L.   Water pennywort
Leaves 2–5 (rarely up to 7) cm wide, shallowly 5–7-lobed; umbels 10- or more

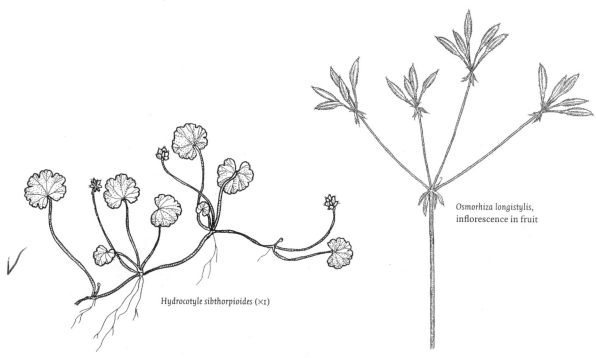

Osmorhiza longistylis,
inflorescence in fruit

Hydrocotyle sibthorpioides (×1)

flowered, peduncles equal to or longer than the subtending leaves; fruits 2–3 mm wide; shallow water and muddy shores; mostly SE; believed extirpated; flr. Jul–Oct; OBL; 🌿.

Ligusticum L.

Ligusticum canadense (L.) Britt.   Lovage
Stout, much-branched herb to 1.5 m tall; lower leaves 3–4-times compound, leaves less compound upward, the uppermost sometimes simple; flowers white, sepals triangular; primary rays 1–2 dm long, terminal umbels sometimes doubly compound; fruits elliptic, 5–7 mm long by 2.5–5 cm wide; mountain woods, stream banks and wooded roadsides; rare SW and SC; flr. Jun; FAC; 🌿.

Osmorhiza Raf.

Herbs, 4–8 dm tall, somewhat pubescent (at least when young); leaves twice ternately compound; fruits linear to clavate, flattened, the base prolonged into a bristly tail; flowers white to greenish-white; flr. May–Jun.

A. plants anise-scented .... O. longistylis
A. plants not anise-scented .... O. claytonii

Osmorhiza claytonii (Michx.) C.B.Clarke   Sweet-cicely
Plant villous-pubescent; leaflets 5–7.5 cm long, coarsely toothed and lobed; fruits 15–22 mm long and about 2 mm wide; rich woods, wooded stream banks, and wet meadows; throughout; FACU–.

Osmorhiza longistylis (Torr.) DC.   Aniseroot
Plant glabrous to only slightly pubescent (pubescent when young), all parts anise-scented; otherwise very similar to O. claytonii; rich woods, moist wooded slopes, and thickets; throughout; FACU.

Osmorhiza longistylis,
fruit (×1)

Oxypolis Raf.

Oxipolis rigidior,
fruit (×2½)

Oxypolis rigidior (L.) Raf.   Cowbane, water-dropwort
Slender or stout plant to 1.5 m tall with few leaves or branches; leaves once pinnate, sessile or nearly so, leaflets linear to oblanceolate to elliptic, 6–14 cm long
by 5–40 mm wide, sparsely but coarsely toothed to entire; flowers white, sepals
absent; umbels up to 15 cm across, loose; fruits rounded at both ends, 4.5–6 mm
long; swamps, bogs, sedge meadows, sandy shores, and abandoned railroad
beds; mostly S; flr. Aug–Sep; OBL.

Pastinaca L.

Pastinaca sativa,
fruit (×2½)

Pastinaca sativa L.   Wild parsnip
Somewhat pubescent biennial up to 1.5 m tall; leaflets oblong to ovate, 5–10 cm
long, variously toothed and lobed; flowers yellow, sepals absent; umbels 1–2 dm
across with 15–25 rays; fruits broadly elliptic to obovate, 5–7 mm long; roadsides, woods edges, fields, meadows, and waste ground; throughout; flr. Jun–
late Jul; native to Eurasia.

Peucedanum L.

Peucedanum ostruthium (L.) W.D.J.Koch   Masterwort
Glabrous perennial, 5–15 dm tall; leaflets sharply toothed, usually 2–3-lobed;
flowers white, sepals absent; umbels up to 15 cm across with 30–60 rays; fruits
suborbicular, 4–5 mm long; cultivated fields and waste ground; scattered
throughout; flr. Jun–Jul; native to Europe.

Pimpinella L.

Pimpinella saxifraga,
fruit (×2½)

Pimpinella saxifraga L.   Burnet-saxifrage
Perennial herb, 3–6 dm tall, stem filled with pith; lower leaves once pinnate, upper leaves much reduced, sometimes only the sheath remaining; flowers white;
fruits ovoid, 2–2.5 mm long; thickets, roadsides, and waste ground; mostly E;
flr. Jun–Sep; native to Eurasia.

Ptilimnium Raf.

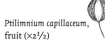

Ptilimnium capillaceum,
fruit (×2½)

Ptilimnium capillaceum (Michx.) Raf.   Mock bishop's weed
Annual herb, 2–8 dm tall; leaves 4–10 cm long, the ultimate leaf segments 5–25
mm long and thread-like; flowers white; fruits broadly ovoid, 2–3 mm long, the
ribs pale brown; swamps and marshes; Philadelphia and Bucks Cos.; flr. Jun–
Aug; OBL; ✿.

Sanicula L.

Glabrous herbs to 14 dm tall; leaves palmately compound, both basal and
cauline leaves present, petioles of basal leaves exceeding the blades in length,
petioles of cauline leaves progressively shorter upward; fruits bristly, ovoid to
oblong or subglobose, only slightly flattened; flowers perfect or staminate, in
dense umbellets, pedicels of the staminate flowers generally exceeding the perfect in length; flr. Jun–Aug.

A. leaves mostly with 3 leaflets (the lateral leaflets often deeply lobed)
    B. pedicels of staminate flowers 3–6 mm long; calyx lobes on fruit forming a beak exceeding the bristles in length .... S. trifoliata
    B. pedicels of staminate flowers <2 mm long; calyx lobes on fruit inconspicuous .... S. canadensis
A. leaves mostly with 5 or more leaflets
    C. sepals about 0.5 mm long; petals yellowish-green, much longer than the sepals .... S. odorata
    C. sepals 1–2 mm long; petals white or whitish-green, equal to or only slightly longer than the sepals. .... S. marilandica

Sanicula canadensis L.   Canadian sanicle, snakeroot
Leaves with 3 (occasionally 5) leaflets; flowers white, sepals generally longer than the petals, anthers white; fruits subglobose, 3–5 mm long by 3–7 cm wide, on pedicels 1–1.5 mm long; rich woods, rocky wooded slopes and roadsides; throughout; UPL; 2 varieties:

A. styles not exceeding the calyx lobes in length .... var. *canadensis*
A. styles about 1.5 times the length of the calyx lobes .... var. *grandis* Fernald

Sanicula marilandica L.   Black snakeroot, black sanicle
Leaves with 5 (occasionally 7) leaflets; flowers greenish-white, anthers greenish-white; fruits ovoid, 4–6 mm long by 4–7 mm wide, sessile or nearly so; moist woods, wooded limestone slopes, bogs, and barrens; throughout; UPL.

Sanicula odorata (Raf.) K.M.Pryer & L.R.Phillippe   Yellow-flowered sanicle,
                                                      fragrant snake-root
Leaves with 5 (sometimes 3) leaflets; flowers greenish-yellow, petals much longer than the sepals, anthers bright yellow; fruits subglobose, 2–3.5 mm long by 3–6 mm wide, on pedicels 0.5–1 mm long; moist rich woods; throughout; FACU.

Sanicula trifoliata E.P.Bicknell   Large-fruited sanicle
Leaves with 3 leaflets; flowers white, petals equal to or shorter than the sepals; fruits ovoid to oblong, 4–7 mm long by 5–8 mm wide, sessile; rich woods, wooded slopes, and stream banks; throughout.

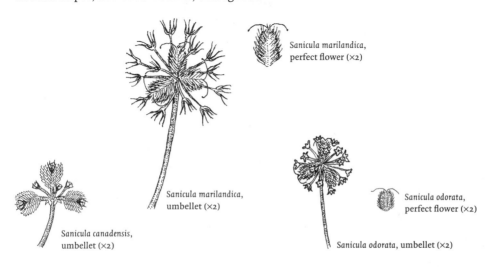

Sanicula marilandica,
perfect flower (×2)

Sanicula marilandica,
umbellet (×2)

Sanicula canadensis,
umbellet (×2)

Sanicula odorata,
perfect flower (×2)

Sanicula odorata, umbellet (×2)

Sium L.

Sium suave, fruit (×2½)

Sium suave Walter   Water-parsnip
Stout plant to 2 m tall; leaflets linear to ovate-lanceolate, 7–17 on the largest leaves, leaves progressively reduced upward; flowers white; umbels 3–12 cm across; fruits oval, 2–3 mm long; swamps, bogs, wet meadows, and pond margins; throughout; flr. Jul–Sep; OBL.

Taenidia Drude

Glabrous herbs, 4–8 dm tall; leaves twice to thrice ternate or pinnate, leaflet margins entire; fruits winged or wingless; flowers yellow; umbels loose and spreading, with 6–14 rays, the rays to 9 cm long.

A. fruits 3–4 mm long and 2–3 mm wide, wings absent .... *T. integerrima*
A. fruits 5–7 mm long and 3–5 mm wide, narrowly winged .... *T. montana*

Taenidia integerrima (L.) Drude   Yellow pimpernel
Leaflets ovate to oblong or elliptic; rocky woods and slopes; throughout; flr. May–Jul.

Taenidia montana (Mack.) Cronquist   Mountain pimpernel
Similar in overall appearance to *T. integerrima*; shale barrens and roadside banks; Bedford Co.; flr. May; ❦.

Thaspium Nutt.

Somewhat branched herbs to 1 m tall; leaves simple to ternate to pinnately compound; fruits conspicuously winged; flowers yellow or purple, sepals ovate to obovate.

A. basal leaves simple or trifoliate .... *T. trifoliata*
A. basal leaves mostly twice pinnate .... *T. barbinode*

Thaspium barbinode (Michx.) Nutt.   Meadow-parsnip
Plant up to 1 m tall, always pubescent around the upper nodes; cauline leaves usually with 3 but occasionally 5 leaflets; flowers pale yellow to cream-colored; umbels 3–6 cm wide; fruits ellipsoid, 4–6 mm long; open woods, wooded slopes, and roadside banks; throughout; flr. May–Jun; UPL.

Thaspium barbinode, fruit (×2½)

Zizia aptera, basal leaf

Thaspium trifoliatum (L.) A.Gray   Meadow-parsnip
Plant 3–8 dm tall, glabrous around the upper nodes; cauline leaves mostly trifo-liate; flowers yellow or purple; umbels 3–8 cm wide; fruits ellipsoid, 3.5–5 mm long; flr. May–early Jul; 2 varieties:

A. flowers yellow .... var. *flavum* S.F.Blake
   rich woods, ravines, and roadsides; mostly W.
A. flowers purple .... var. *trifoliatum*
   woods, wooded slopes, and roadsides; mostly SE.

Torilis Adans.

Erect annual herbs to 8 dm tall, usually with stiff pubescence; leaves pinnately compound, leaflets coarsely toothed; fruits with stiff recurved bristles; flowers white.

A. umbels axillary; rays 2–3 .... *T. leptophylla*
A. umbels terminal; rays 5–12 .... *T. japonica*

Torilis japonica (Houtt.) DC.   Japanese hedge-parsley
Much branched plant to 8 dm tall; leaves once to thrice pinnately compound; petals slightly longer than the sepals; fruits 2–4 mm long; fields, thickets, and waste ground; mostly SE; flr. Jun–Jul; native to Eurasia.

Torilis leptophylla (L.) Rchb.f.   Hedge-parsley
Mostly unbranched plant to 4 dm tall; leaves mostly twice pinnately compound; axillary umbels opposed by a leaf; fruits linear-oblong, bristles yellowish to straw-colored; rubbish dump, Bucks Co., possibly elsewhere; flr. Jun–Jul; native to Europe.

Zizia Koch

Branched glabrous perennial herbs, 3–8 dm tall; both basal and cauline leaves present, mostly once to thrice ternately compound (or some simple); fruits ovate to oblong, flattened; flowers bright yellow; umbellets many-flowered with the central flowers sessile; flr. May–Jun.

A. basal leaves mostly simple .... Z. *aptera*
A. basal leaves once to twice compound .... Z. *aurea*

Zizia aptera (A.Gray) Fernald   Golden-alexander
Both basal and sometimes lower cauline leaves simple; fruits 3–4 mm long by 1.5–2 mm wide; woods, wooded slopes, clearings and roadsides; throughout; FAC.

Zizia aptera,
fruit (×2½)

Zizia aurea (L.) W.D.J.Koch   Golden-alexander
All leaves compound; fruits 3–4 mm long by 2.5–3.5 mm wide; wooded bottom-land, stream banks, moist meadows, and floodplains; throughout; FAC.

Zizia aurea,
fruit (×2½)

APOCYNACEAE Dogbane Family

Erect perennial herbs or creeping vines with opposite or alternate, simple leaves and milky sap; flowers perfect, regular, 5-merous; calyx deeply lobed; petals fused to form a bell-shaped, salverform, or tubular corolla, the lobes overlapping in a spiral pattern in bud; stamens attached to the corolla tube and alternate with the lobes; ovaries superior, 2, distinct below but united in a common style and stigma; fruit a pair of slender, many-seeded follicles.

A. plants trailing, evergreen creepers; flowers solitary in the axils .... *Vinca*
A. plants erect, not evergreen; flowers in small terminal clusters
    B. leaves opposite; corolla bell-shaped to tubular .... *Apocynum*
    B. leaves alternate or irregular; corolla salverform .... *Amsonia*

Amsonia Walter

Amsonia tabernaemontana Walter   Blue-star
Stems erect with alternate or irregularly scattered leaves and terminal clusters of blue flowers; calyx deeply 5-lobed; corolla salverform, 1 cm wide; fruit cylindric, erect, 8–12 cm long; seeds lacking hairs; cultivated and occasionally escaped to fields, woods, or waste ground; SE; flr. May–Jun; native farther south; FACW.

Apocynum L.

Stems erect, with fibrous, shredding outer layers; leaves opposite, petiolate, mucronate; inflorescences terminal or from the upper axils; calyx deeply divided; corolla with short lobes and a triangular scale near the base of the tube opposite each lobe; stamens attached at the base of the corolla tube, the anthers convergent and adhering to the stigma; fruit a pair of long, slender follicles; seeds with a tuft of white hairs; frequent hybridization between species complicates identification.

A. corolla pinkish, 5–9 mm long, with spreading or reflexed lobes ....
.... *A. androsaemifolium*
A. corolla white or greenish, 3–6 mm long, lobes erect or slightly divergent ......
.... *A. cannabinum*

Apocynum androsaemifolium L.   Pink or spreading dogbane
Stems erect-ascending, 2–8 dm tall, simple or branched; leaves drooping, oblong-ovate, 3–8 cm, usually hairy beneath; flowers nodding; corolla bell-shaped, pinkish marked with red inside; follicles 7–15 cm long; common in woods, roadsides, dry pastures, thickets, and barrens; throughout; flr. Jun–Aug.

Apocynum cannabinum L.   Indian-hemp
Erect branched plant to 1.5 m tall; leaves ascending or spreading, ovate to oblong-lanceolate or elliptic, 5–11 cm, glabrous or hairy beneath; flowers erect; corolla white or greenish, cylindric or urn-shaped with erect lobes; follicles 10–20 cm long; common in woods, old fields, sandy flats, limestone bluffs; and cindery waste ground; throughout; flr. Jun–Aug; FACU.

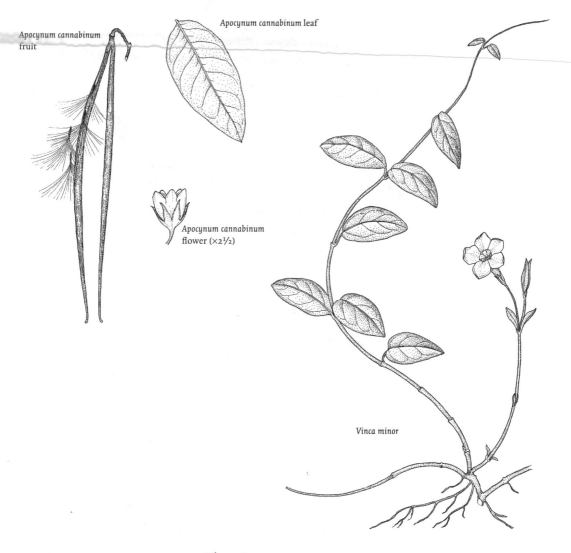

Apocynum cannabinum fruit

Apocynum cannabinum leaf

Apocynum cannabinum flower (×2½)

Vinca minor

Vinca L.

Vinca minor L.   Common periwinkle, creeping myrtle
Evergreen plant with creeping stems rooting at the nodes and erect flowering stems; leaves opposite, entire, petiolate; flowers solitary in the axils; corolla blue (rarely white), salverform; cultivated and occasionally naturalized in woods, fields, or roadsides; throughout; flr. Apr–Jun; native to Europe.

AQUIFOLIACEAE Holly Family

Evergreen or deciduous trees or shrubs with alternate, simple leaves; stipules minute or absent; plants dioecious; flowers unisexual, 4–8-merous, ovary superior, 4–8-locular; fruit a berry-like drupe with 4–8 nutlets.

A. leaves with spines or teeth; petals fused at the base; stamens inserted on the corolla .... *Ilex*
A. leaves entire or nearly so; petals not fused; stamens free .... *Nemopanthus*

Ilex L.

Evergreen or deciduous trees or shrubs with alternate simple leaves; flowers unisexual, white, axillary; fruits black or red, frequently persistent through the winter.

A. leaves thick and leathery, evergreen
    B. trees; leaves spine-tipped; fruit red .... *I. opaca*
    B. shrubs; leaves more or less toothed; fruit black
        C. leaves with only a few teeth above the middle; flowers 5–8-merous ......
        .... *I. glabra*
        C. leaves serrate or crenate; flowers 4-merous .... *I. crenata*
A. leaves thin, deciduous
    D. leaves glabrous beneath except on the midrib and major veins; sepals entire .... *I. laevigata*
    D. leaves hairy or glabrous beneath; sepals ciliate on the margins
        E. short, lateral branch spurs not present; petals entire; nutlets smooth on the back. .... *I. verticillata*
        E. most leaves (and flowers) borne on short, lateral branch spurs; petals ciliate; nutlets grooved on the back
            F. leaves glabrous beneath, except perhaps along the veins; sepals glabrous with ciliate margins .... *I. montana*
            F. leaves pubescent beneath; sepals villous as well as ciliate ... *I. beadlei*

Ilex beadlei Ashe   Mountain holly
Very similar to *I. montana* except the leaves <2 times as long as wide and pubescent beneath; sepals villous as well as ciliate; rare on wooded slopes; NE, and at higher elevations along the Allegheny Front.

Ilex crenata Thunb.   Japanese holly
Upright, evergreen shrub to 3 m tall with shiny, dark green leaves with serrulate or crenate-serrulate margins; fruit 6 mm, black; cultivated and occasionally escaped to urban and suburban woodlands; SE; native to Japan.

Ilex glabra (L.) A.Gray   Inkberry
Stoloniferous, evergreen shrub to 3 m tall; leaves thick and leathery, oblanceolate, 2–5 cm with 1–3 small teeth on each side near the tip; flowers 6–8-merous, staminate flowers in small clusters, pistillate solitary; fruits black, 4–5 mm, persistent; rare in moist, sandy soils; SE, but believed to be extirpated; flr. May, frt. Sep–winter; native, but also frequently cultivated; FAC+; 🔥.

Ilex laevigata (Pursh) A.Gray   Smooth winterberry
Deciduous shrub to 3 m tall; leaves elliptic to oblanceolate or obovate, 4–8 cm, acuminate, serrulate, sparsely villous on the veins beneath; flowers 6–8-merous; sepals entire; staminate flowers on 8–16 mm pedicels; pistillate flowers on 3–8-mm pedicels; fruits 8 mm thick; nutlets smooth on the back; occasional in wooded swamps, wet thickets, and shores; E; flr. May–Jun, frt. Sep; OBL.

Ilex montana (Torr. & A.Gray) A.Gray   Mountain holly
Large deciduous shrub or small tree to 10 m tall with many leaves borne in clusters on short lateral branch spurs; leaves petiolate; blades ovate or lance-ovate, 6–12 cm, sharply serrate, acuminate, glabrous beneath; flowers 6–8-merous;

pistillate flowers and fruits on 3–5-mm pedicels; pedicels of staminate flowers longer; sepals and petals ciliate but otherwise glabrous; fruits red, 6 mm thick; frequent in cool, moist, rocky woods; N and at higher elevations along the Allegheny Front; flr. late Apr–early Jun; frt. Sep.

Ilex opaca Aiton   American holly
Tree to 15 m tall with spiny-margined, evergreen leaves and bright red fruits; leaves 2–5 cm; flowers 4-merous; petals fused at the base; stamens inserted on the corolla tube; rare in moist alluvial woods and wooded slopes; mainly SE; flr. May–early Jun, frt. Oct–winter; native, but also frequently cultivated and escaped; FACU; 🌿.

Ilex verticillata (L.) A.Gray   Winterberry, black-alder
Deciduous shrub to 5 m tall with bright red fruits remaining on the branches into the winter; leaves variable, 4–8 cm, obovate to elliptic or lance-oblong, acuminate, serrate; staminate flowers 4–6-merous; pistillate flowers 5–8-merous; sepals ciliate; fruit 7 mm thick, on pedicels to 3 mm; nutlets smooth on the back; common in swamps, bogs, wet woods, and moist shores; throughout; flr. May–Jun, frt. Sep–winter; FACW+.

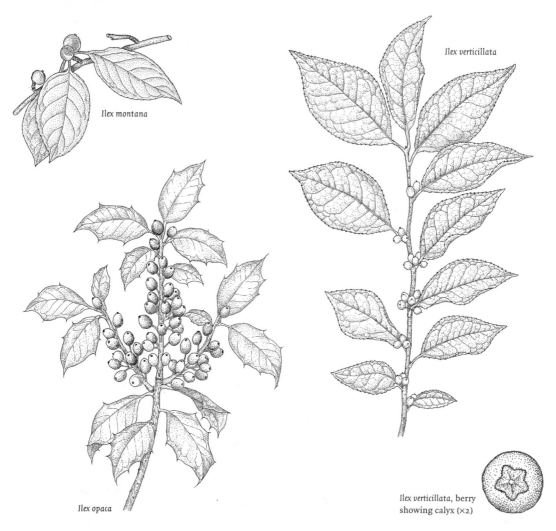

Ilex montana

Ilex verticillata

Ilex opaca

Ilex verticillata, berry showing calyx (×2)

Nemopanthus Raf.

Nemopanthus mucronatus (L.) Trel.   Mountain holly
Deciduous shrub to 3 m tall; leaves alternate or crowded on short lateral spur branches; blades 2–5 cm long, entire or occasionally slightly toothed, elliptic, obtuse to rounded with a mucronate tip; petioles purplish, 5–12 mm; petals and stamens distinct; fruit red, 6 mm thick, solitary on 1–3 cm pedicels; occasional in swamps, bogs, moist woods, and rocky slopes; mostly N; flr. May–Jun, frt. Aug–Sep; OBL.

ARALIACEAE Ginseng Family

Shrubs, small trees, woody vines, or herbs with alternate or whorled, simple or compound leaves with or without stipules; the small, perfect, 5-merous flowers in umbels; ovary inferior, of 2–5 fused carpels; petals distinct; stamens same number as the petals and alternate with them; fruit a berry or drupe.

A. leaves compound, deciduous; plants woody or herbaceous
   B. leaves pinnately compound .... *Aralia*
   B. leaves palmately compound or trifoliate
      C. herbs with a single whorl of leaves and only 1 umbel .... *Panax*
      C. spiny shrubs with many alternate leaves and inflorescences ....
      .... *Acanthopanax*
A. leaves simple, lobed, evergreen; woody climbing vine .... *Hedera*

Acanthopanax Miq.

Acanthopanax sieboldianus Makino   Fiveleaf aralia
Deciduous shrub to 3 m tall with spiny stems and alternate, palmately compound leaves; leaflets 5–7, ovate-oblong, 6–7 cm long, serrate, glabrous; umbel 2.5 cm across, on 5–10-cm peduncles; flowers greenish-white, unisexual; fruit a black, 2–5-seeded berry; plants dioecious; cultivated and occasionally spreading to waste ground or persisting in abandoned garden sites; flr. early Jun; native to Asia.

Aralia L.

Smooth or bristly herbaceous perennials or small prickly trees; leaves 2 or 3 times compound; flowers white or greenish, 5-merous, in 2–many umbels; fruit a berry-like drupe.

A. stems completely herbaceous; plants without bristles or prickles
   B. stemless, the single compound leaf and flowering peduncle arising directly from the rhizome ....*A. nudicaulis*
   B. with a branching stem bearing several alternate leaves and inflorescences
   .... *A. racemosa*
A. stems woody, at least at the base; plants bristly or prickly
   C. bristly, mainly herbaceous plant to 1 m tall with a woody base.... *A. hispida*
   C. prickly shrubs or trees to 10 m tall with very large leaves
      D. inflorescence with a distinct central axis; veins of the leaflets not running to the teeth ....*A. spinosa*

D. central axis of the inflorescence very short or lacking; veins of the leaflets running into the teeth .... *A. elata*

Aralia elata Seem.   Japanese angelica-tree
Shrub or tree to 10 m tall with very large, 2 or 3 times pinnate leaves and spiny
stems; leaflets sessile or nearly so, ovate, 5–12 cm long, acuminate, serrate to
crenate or almost entire; petioles and rachises with or without prickles; inflorescence 3–6 dm, with the central axis very short or lacking and 3–8 spreading secondary branches; flowers white; fruit black; naturalized in disturbed woodlands;
especially SE; flr. late Jul–Aug, frt. Aug–Sep; native to Asia.

Aralia hispida Vent.   Bristly sarsaparilla
Herbaceous perennial to 1 m tall with a bristly, woody base; leaves twice pinnate;
leaflets oblong to ovate, to 10 cm, acute or short-acuminate, doubly serrate; umbels in a loose terminal inflorescence; fruit globose, black; occasional in dry,
rocky woods and barrens; throughout; flr. Jul.

Aralia nudicaulis L.   Wild sarsaparilla
Herbaceous perennial to 4 dm tall lacking an aboveground stem; the single leaf

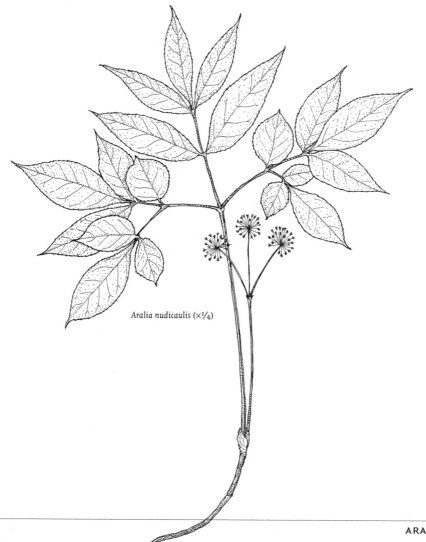

Aralia nudicaulis (×¼)

and leafless peduncle arising from the rhizome; leaf twice compound, the 3 primary segments each divided pinnately into 3–5 leaflets; umbels 1–3; flowers greenish; fruits purplish-black, 6 mm; common in dry to moist woods; throughout; flr. May; FACU.

Aralia racemosa L. Spikenard
Herbaceous perennial to 2 m tall with large, twice compound leaves to 8 dm, each of the 3 primary divisions pinnately compound; leaflets ovate, to 15 cm long, doubly serrate; inflorescence a large terminal panicle of many small umbels; fruit dark purple; frequent in rich woods, wooded slopes and edges; throughout; flr. Jun–Jul.

Aralia spinosa L. Hercules'-club
Shrub or tree to 10 m tall with very large, 2 to 3 times pinnate leaves; leaflets with a definite petiole, ovate, 5–7 cm long, acuminate, serrate, the veins anastomosing before they reach the margins; stem, petioles, and rachises all bearing prickles, those on the leaves short and recurved; umbels in a terminal panicle 10–12 dm long with an elongate central axis; flowers whitish; fruit black; occasional in moist woods, stream banks, and roadsides; mostly W; flr. late Jul–Aug, frt. Aug–Sep; FAC.

Hedera L.

Hedera helix L. English ivy
Evergreen vine climbing by means of aerial rootlets; flowering branches extending horizontally from the climbing stems and lacking rootlets; leaves alternate, simple, suborbicular-cordate, and somewhat palmately 3–5-lobed on vegetative stems, narrowly ovate and unlobed on flowering stems, dark green with lighter veins; inflorescence a raceme of umbels; flowers small, green; fruit black, 2–3-seeded; widely cultivated and occasionally naturalized in disturbed woods; mostly SE; flr. Aug–Sep; native to Eurasia.

Panax L.

Herbaceous perennials with a single, unbranched stem from the fleshy, tuber-like root; leaves in a single whorl of 3, palmately compound or trifoliate; flowers small, greenish or white, in a single umbel.

A. leaflets 5, stalked, acuminate .... *P. quinquefolius*
A. leaflets 3, nearly sessile, obtuse or subacute .... *P. trifolius*

Panax quinquefolius L. Ginseng
Herb to 6 dm tall; leaves mostly 5-foliate; leaflets oblong-obovate, 6–15 cm long, with a long petiolule; flowers greenish, perfect; fruit red, 1 cm; rich, mesic woods; formerly frequent, but declining due to excessive collecting; flr. May; frt. Sep–Oct; ✺.

Panax trifolius L. Dwarf ginseng
Herb to 1–2 dm tall; leaves mostly 3-foliate; leaflets lanceolate to elliptic or oblanceolate, 4–8 cm long, sessile; flowers white, often unisexual; fruit yellow, 5 mm; common in moist woods; throughout; flr. late Apr–early May.

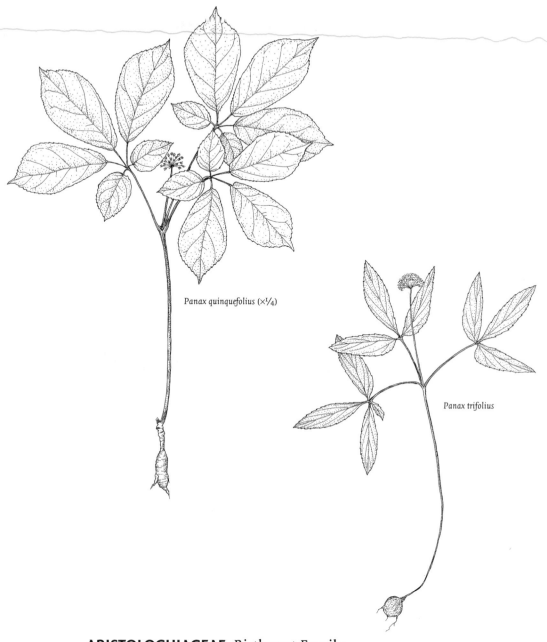

Panax quinquefolius (×¼)

Panax trifolius

ARISTOLOCHIACEAE Birthwort Family

Herbaceous perennials or twining woody vines with aromatic stems and simple, palmately veined, cordate leaves; flowers basal or axillary, perfect, regular or irregular, with a 1- or 3-lobed, tubular, corolla-like calyx; petals absent; stamens 6 or 12, closely appressed or fused to the styles; ovary 1, 6-locular, inferior; fruit a capsule.

A. stemless, colonial herbs from a creeping rhizome; perianth radially symmetrical; stamens 12 .... *Asarum*

A. upright or reclining herbs or woody vines; perianth with bilateral symmetry; stamens 5–6 .... *Aristolochia*

Aristolochia L.

Herbaceous perennials or woody vines with flowers solitary or in small clusters in the axils; calyx tube bent or straight, alternately inflated and constricted, with 1 or 3 lobes at the summit; stamens 6, anthers sessile, adnate to the style; ovary inferior, 6-locular; fruit a lobed or angled capsule.

A. erect or reclining herbs; leaves 5–12 cm across; perianth tube 1–3 cm long, straight or curved
    B. leaf venation pinnate or palmate-pinnate, blades longer than wide; flowers nearly basal on the stem; perianth S-shaped with 3 lobes ....*A. serpentaria*
    B. leaves palmately veined, blade as wide or wider than long; flowers in small axillary clusters; perianth nearly straight with 1 lobe ....*A. clematitis*
A. twining woody vines; leaves 2–4 dm across; perianth tube 4 cm long, strongly curved ....*A. macrophylla*

Aristolochia clematitis L.   Birthwort
Stems erect or reclining, 5–10 dm tall; leaves broadly cordate to reniform, flowers in small axillary clusters; perianth yellow, 3 cm long, straight with a single, large terminal lobe; fruit subglobose, 2.5–3 cm; escaped from cultivation at a few sites; SE; flr. Jun–Aug; native to Europe.

Aristolochia macrophylla Lam.   Dutchman's pipe, pipevine
High-twining, deciduous, woody vine with large, broad-cordate leaves; flowers solitary or a few together in leaf axils; perianth tube purple-brown, 4 cm long, strongly curved and abruptly expanded into a 3-lobed limb; capsule cylindric-ovoid, 5.5–8 cm long by 2.5 cm; rare in rich woods, stream banks, and rocky slopes; W; flr. May–Jun; native, but also occasionally planted.

Aristolochia serpentaria L   Virginia snakeroot
Herbaceous perennial; stem 1–4.5 dm tall from a knotty rhizome; leaves few, ovate to oblong, acuminate, cordate at the base, with slender petioles; flowers very inconspicuous, on short basal branches, sometimes almost subterranean; perianth purple-brown, 1–1.5 cm long, strongly curved with a broad, 3-lobed limb; capsule subglobose, 0.8–1.3 cm; occasional in moist to dry, rich woods; flr. May; S; UPL.

Asarum L.

Asarum canadense L.   Wild ginger
Low-growing, colonial perennial herb with hairy, cordate leaves arising from a creeping rhizome; leaf blades 8–12 cm wide at flowering, enlarging later; flowers borne below on short peduncles arising from the rhizome; perianth purple, tubular at the base with 3 spreading or strongly reflexed lobes; stamens 12; ovary 6-locular, inferior; fruit a fleshy capsule; common in moist, rich woods and wooded floodplains; throughout; flr. late Apr–May; FACU–.

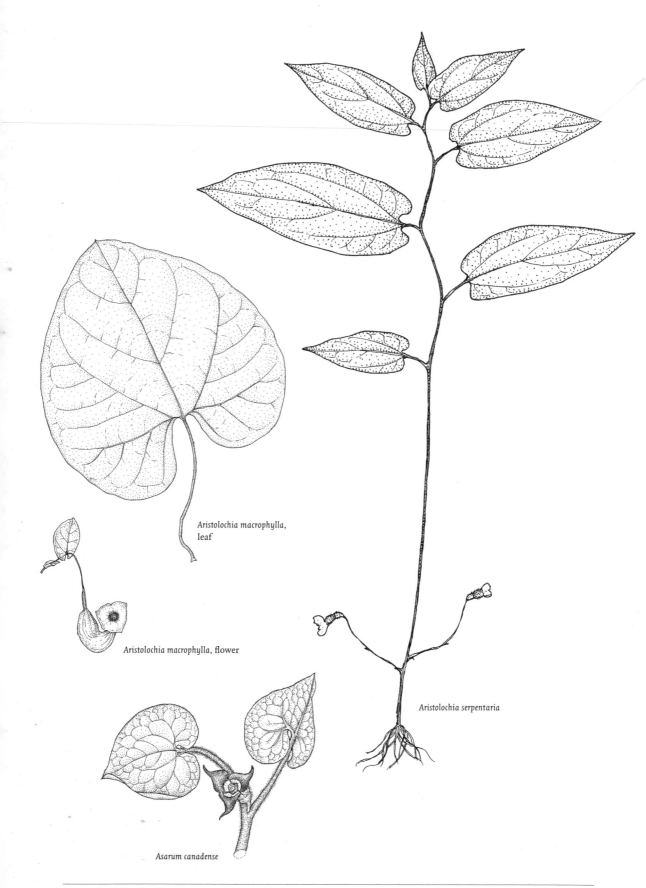

Aristolochia macrophylla,
leaf

Aristolochia macrophylla, flower

Aristolochia serpentaria

Asarum canadense

ASCLEPIADACEAE Milkweed Family

Upright or twining perennials with milky juice (except A. *tuberosa*); leaves simple, entire, opposite, or whorled (alternate in A. *tuberosa*); flowers in umbels or cymose clusters, perfect, regular, 5-merous; calyx present; corolla with a short tube and spreading or reflexed lobes; a well-developed corona, arising from the base of the filaments, occurs between the corolla and the 5 stamens; filaments forming a short sheath around and fused to the style, the anthers also united to each other and to the stigma, the entire structure referred to as the gynostegium; pollen in each of the 2 locules of the anther clumped in a waxy mass known as a pollinium; ovaries 2, superior, free below, united in a single stigma; fruit a pair of follicles; seeds numerous, each with a tuft of long white hairs that serves as a means of dispersal.

A. stems erect; lobes of the corolla strongly reflexed at anthesis .... *Asclepias*
A. stems twining or trailing; lobes of the calyx and corolla spreading or ascending, not reflexed
   B. flowers white or greenish .... *Matelea*
   B. flowers dark purple to purple-brown
      C. corona with 10 lobes .... *Cynanchum*
      C. corona with 5 lobes
         D. corolla lobes with dense long hairs on the upper surface .... *Periploca*
         D. corolla lobes glabrous or with sparse short hairs .... *Vincetoxicum*

Asclepias L.

Erect, perennial herbs with opposite or whorled leaves (alternate in A. *tuberosa*); corolla deeply divided, the lobes reflexed; a corona of 5 prominent scoop-shaped hoods arising from near the top of the filament column, frequently also with 5 horns within or extending from the hoods; anthers with a triangular appendage at the tip; follicles erect.

A. at least some of the leaves in whorls of 3–6
   B. leaves narrowly linear, up to 6 per node .... A. *verticillata*
   B. leaves lance-ovate, 3–4 present at some of the nodes .... A. *quadrifolia*
A. leaves opposite or alternate
   C. leaves alternate below the inflorescence; flowers yellow to orange-red ......
      .... A. *tuberosa*
   C. leaves all opposite
      D. leaves cordate or subcordate at the base, sessile or with the petiole shorter than the basal lobes
         E. horns longer than the hoods .... A. *amplexicaulis*
         E. horns shorter than the hoods .... A. *rubra*
      D. blades of the leaves rounded or tapered to the petiole, not cordate
         F. flowers green, horns not present .... A. *viridiflora*
         F. flowers white, greenish, pink, or purple, horns present
            G. horns conspicuously longer than the hoods
               H. flowers rose-purple; follicles 6–9 cm long .... A. *incarnata*
               H. flowers white, pinkish, or greenish; follicles 9.5–13 cm long...
                  .... A. *exaltata*

G. horns shorter than the hoods
 I. flowers purple or purplish-green; leaves pubescent beneath
  J. corolla lobes pubescent outside; umbels usually 4 or more
   .... *A. syriaca*
  J. corolla lobes glabrous; umbels 1–3 .... *A. purpurascens*
 I. flowers white or pinkish-white; leaves glabrous beneath except perhaps on the veins .... *A. variegata*

Asclepias amplexicaulis Sm.   Blunt-leaved milkweed
Stem simple, erect or decumbent, 3–8 dm tall with 2–5 pairs of sessile, somewhat clasping leaves and a single, long-peduncled, many-flowered, terminal umbel; flowers purple to nearly green; rare in dry pastures, sandy flats, barrens, and roadsides; SE, SC, and NE; flr. Jun–Jul.

Asclepias exaltata L.   Poke or tall milkweed
Stems erect, simple, 8–15 dm tall; leaves opposite, broadly elliptic, thin, tapering at both ends; umbels loose and few-flowered, the pedicels often drooping; flowers white, greenish, or pale purple; frequent in moist upland woods; throughout; flr. Jun–Jul; FACU.

Asclepias incarnata L.   Swamp milkweed
Stems erect, to 1.5 m tall, branched above; leaves opposite, lanceolate to oblong-lanceolate; umbels several, clustered; flowers pink or rose-colored; swamps, floodplains, and wet meadows; throughout; flr. Jun–Aug; OBL; 2 subspecies:

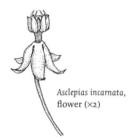
Asclepias incarnata,
flower (×2)

A. stems and leaves glabrous to sparsely pubescent .... ssp. *incarnata* common throughout.
A. stems and leaves densely pubescent .... ssp. *pulchra* (Ehrh. & Willd.) Woods. occasional; SE.

Asclepias purpurascens L.   Purple milkweed
Stems erect to 1 m tall, simple; leaves opposite, elliptic to ovate-oblong; umbels 1 or few, terminal, many-flowered; flowers purple; fruits pubescent; rare in dry to moist thickets, fields and roadsides; mostly SE; flr. Jun–Jul; FACU.

Asclepias purpurascens,
flower (×2)

Asclepias quadrifolia Jacq.   Four-leaved milkweed
Stems erect, simple, 3–8 dm tall; leaves in whorls of 3–4 at the middle nodes, smaller and fewer above and below, lance-ovate, thin; umbels 1–2, terminal, many-flowered; flowers white to pale pink; frequent in dry upland woods; throughout; flr. late May–Jun.

Asclepias rubra L.   Red milkweed
Stems erect, simple, 5–12 dm tall, glabrous; leaves ovate with a broadly rounded to subcordate base, sessile or nearly so; umbels few, small, terminal; flowers purplish-red; very rare in sphagnous wetlands; SE; believed to be extirpated; flr. Jul; OBL; ❦.

Asclepias syriaca L.   Common milkweed
Stems erect, simple, 1–2 m tall, colonial; leaves opposite, oblong-lanceolate to

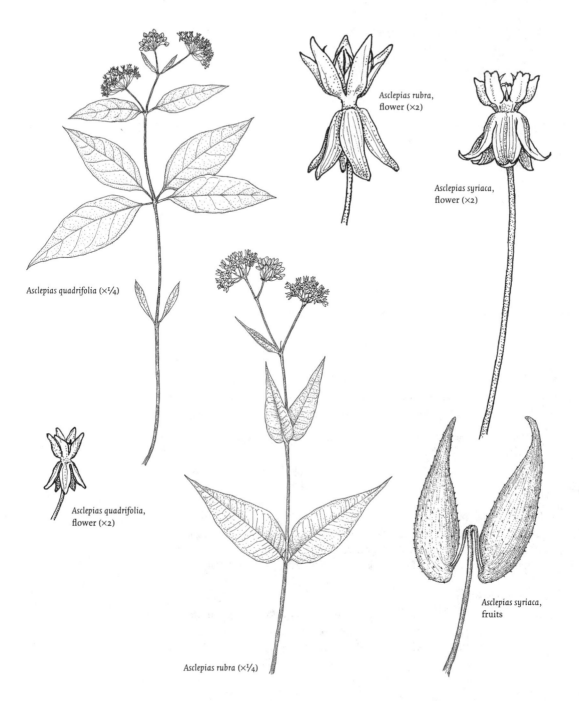

Asclepias quadrifolia (×¼)

Asclepias quadrifolia,
flower (×2)

Asclepias rubra,
flower (×2)

Asclepias syriaca,
flower (×2)

Asclepias syriaca,
fruits

Asclepias rubra (×¼)

oval; umbels several, terminal and in the upper axils, many-flowered; flowers purple to greenish; common in fields, roadsides, and waste ground; throughout; flr. late May–early Aug; FACU–.

Asclepias tuberosa L   Butterfly-weed
Stems erect or ascending, 3–7 dm tall, hairy and frequently branched above; juice not milky; leaves mostly alternate, or opposite on the branches, sessile; umbels axillary on the branches; flowers yellow to orange-red; frequent in dry fields, roadsides, and shale barrens; S; flr. Jul–Aug.

Asclepias variegata L. White milkweed
Stem erect, simple, to 1 m tall, glabrous; leaves opposite, broadly ovate to oblong or obovate, tapering to the petiole; umbels 1–4, many-flowered; flowers white or pink tinged; rare in dry woods; mostly SE; flr. late May–Jul; FACU; ✤.

Asclepias verticillata L. Whorled milkweed
Stems erect, simple, 2–5 dm tall; leaves numerous, whorled, narrowly linear with revolute margins; umbels several, terminal and from the upper nodes; flowers white or greenish; rare in dry, rocky, or sandy soils and barrens; SE and SC; flr. Jul–Aug.

Asclepias viridiflora Raf. Green milkweed
Stems simple, erect to prostrate, 3–8 dm tall; leaves broadly oblong to elliptic, thick and leathery; umbels several, densely flowered, axillary, sessile or nearly so; flowers green; occasional in dry fields and dry rocky slopes including serpentine and limestone; mostly S; flr. late Jun–Aug.

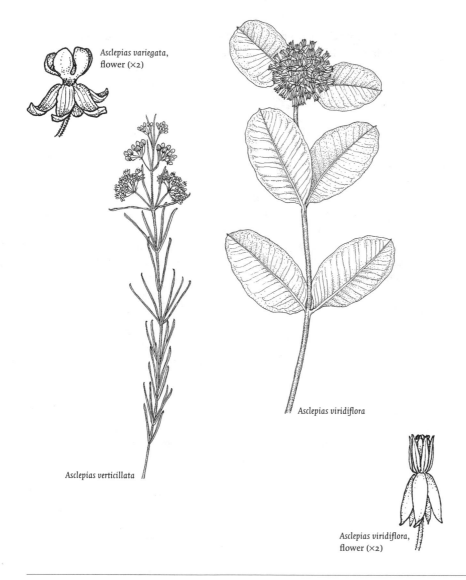

Asclepias variegata,
flower (×2)

Asclepias verticillata

Asclepias viridiflora

Asclepias viridiflora,
flower (×2)

Cynanchum L.

Cynanchum laeve (Michx.) Pers.   Smooth swallow-wort
Glabrous, twining vine 3–4 m tall; leaves deeply cordate-ovate, long-petioled; flowers in axillary umbels or short racemes; corolla creamy or greenish-white; corona erect, white, 10-lobed, about equaling the corolla lobes; very rare on riverbanks; SE; flr. Jul–Aug; FAC; ✤. [syn: *Ampelamus albidus* (Nutt.) Britton.]

Matelea Aubl.

Matelea obliqua (Jacq.) Woodson   Anglepod, oblique milkvine
Twining vine with hairy stems; leaves ovate to rounded, deeply cordate; flowers in peduncled axillary clusters, purplish-brown; corolla deeply divided, with ascending lobes; the corona forming a shallow, 10-lobed cup; rare in mesic woods, wooded edges, and thickets; SE and SC; flr. late Apr–Oct; ✤.

Periploca L.

Periploca graeca L.   Silkvine
Glabrous twining vine to 10 m tall; leaves lance-ovate, 5–10 cm; flowers purple–brown, corolla lobes densely long-hairy on the upper surface, especially toward the margins and the base; lobes of the corona slender and thread-like; fruits 10–15 cm; cultivated and rarely escaped to riverbanks and urban waste ground; flr. Jun; native to Europe.

Vincetoxicum N.Wolf

Vincetoxicum nigrum (L.) Moench   Black swallow-wort
Herbaceous, perennial twining vine; leaves opposite, oblong to ovate, acuminate; flowers dark purple, few, in axillary cymes; corolla small, subrotate with short, broad, pubescent lobes; corona a fleshy, 5-lobed cup nearly the same length as the gynostegium; follicles smooth, slender; roadsides, stream banks, and woods edges; spreading aggressively in some areas; flr. May–Jul; native to Europe. In addition *V. rossicum* (Kleopow) Barbar., which is distinguished by the glabrous corolla lobes and longer peduncles, is known from one Lancaster Co. site.

ASTERACEAE Aster Family

Herbaceous plants (except *Baccharis halimifolia* and *Iva frutescens*, which are shrubs), with alternate, opposite, or whorled leaves; inflorescence a head of several to many florets on a common receptacle surrounded by an involucre of more or less herbaceous bracts; florets of 2 types: disk florets that are tubular and regular, and ray florets that have an irregular, strap-shaped corolla, or sometimes of 1 type only; receptacle chaffy due to the presence of bracts, scales, or bristles between or subtending the florets, or the receptacle smooth or merely somewhat pitted (naked); sepals reduced to a pappus of capillary hairs (bristles), scales, or awns on the top of the achene, or absent; anthers usually fused in a ring surrounding the 2-branched style; ovary inferior, fruit an achene.

Key to groups of genera

A. shrubs
    B. leaves alternate .... *Baccharis*
    B. leaves opposite .... *Iva*
A. herbaceous plants
    C. plants with leaves and/or involucral bracts prickly or spiny (thistles or thistle-like plants) .... Group 1
    C. plants not prickly
        D. heads with only disk florets (discoid) or with very short, inconspicuous rays (<5 mm)
        E. receptacle naked .... Group 2
        E. receptacle chaffy .... Group 3
        D. at least some of the florets with well-developed strap-shaped corollas (rays)
        F. inflorescence entirely of ray florets; juice milky .... Group 4
        F. both disk and ray florets present; juice mostly watery
            G. rays yellow or orange
                H. receptacle chaffy, at least at the margin .... Group 5
                H. receptacle naked .... Group 6
            G. rays white, pink, purple, or blue .... Group 7

Key to the genera of Group 1 (thistles and thistle-like plants)

A. involucre becoming a spiny bur enclosing the seeds
    B. florets perfect, heads all similar .... *Arctium*
    B. florets unisexual, in separate heads, staminate above, pistillate below
        C. involucres of the pistillate heads becoming a bur with hooked bristles... .... *Xanthium*
        C. involucres of the pistillate heads bearing a few straight spines or tubercles .... *Ambrosia*
A. involucre not becoming a spiny bur, although the involucral bracts may be spiny or prickly
    D. leaves not spiny or prickly .... *Centaurea*
    D. leaves spiny or prickly
        E. inflorescence spherical .... *Echinops*
        E. inflorescence not spherical
            F. sap milky; only ray florets present
                G. leaves prickly on the midrib beneath as well as the margin .... *Lactuca*
                G. leaves not prickly on the midrib .... *Sonchus*
            F. sap not milky; only disk florets present
                H. leaves white mottled; filaments united in a tube below .... *Silybum*
                H. leaves not white-mottled; filaments free below
                      I. pappus bristles barbed but not plumose .... *Carduus*
                      I. pappus bristles plumose
                          J. receptacle flat or conical, densely bristly .... *Cirsium*
                          J. receptacle honeycombed, the partitions with short bristles .... *Onopordum*

Key to the genera of Group 2 (heads without rays or with very small and inconspicuous rays; receptacles naked)

A. pappus of scales, awns, or none; plants aromatic
  B. leaves opposite; rays present .... *Dystopia*
  B. leaves alternate; rays absent
    C. inflorescence an elongate spike, raceme, or panicle .... *Artemisia*
    C. inflorescence a flat-topped corymb, or heads solitary
      D. leaves toothed or lobed toward the base .... *Chrysanthemum*
      D. leaves pinnately dissected
        E. receptacle flat .... *Tanacetum*
        E. receptacle conical .... *Matricaria*
A. pappus of capillary bristles; plants not strongly aromatic
  F. plants persistently white-woolly, except perhaps on the upper leaf surfaces
    G. leaves mostly basal, those on the stem greatly reduced
      H. basal leaves cordate or reniform, to 3 dm wide; inflorescence an elongate raceme of heads .... *Petasites*
      H. basal leaves tapering to the petiole, blade not more than 5 cm wide; inflorescence a dense, flat-topped cluster of heads or a solitary head .... *Antennaria*
    G. leaves mostly cauline, alternate, linear
      I. plants dioecious, or pistillate heads with a few central staminate florets .... *Anaphalis*
      I. plants not dioecious, outer florets in each head pistillate, inner few perfect .... *Gnaphalium*
  F. plants not persistently white-woolly
    J. florets yellow or orange .... *Senecio*
    J. florets greenish, white, purple, or pink
      K. twining herbaceous vine .... *Mikania*
      K. upright herbs
        L. heads few-flowered, aggregated in secondary heads .... *Elephantopus*
        L. heads distinct
          M. involucral bracts essentially all the same length and in one row
            N. leaves broadly palmately veined or hastate, petiolate .... *Cacalia*
            N. leaves pinnately veined, the lower tapering to a short petiole, the upper auriculate-clasping .... *Erechtites*
          M. involucral bracts imbricate
            O. leaves opposite or whorled .... *Eupatorium*
            O. leaves alternate
              P. inflorescence an elongate spike, raceme or panicle
                Q. florets purple or pink, no rays present; involucre 7–11 mm long
                  R. flowers all perfect; pappus bristles barbellate or plumose .... *Liatris*
                  R. outer ring of flowers pistillate, inner ones perfect; pappus of capillary bristles .... *Aster brachyactis*
                Q. florets white or pinkish; slender rays present; involucre 3–4 mm long .... *Conyza*
              P. inflorescence highly branched, flat-topped

S. florets purple or pink
    T. flowers purple; involucral bracts not glandular ....
    .... *Vernonia*
    T. flowers pink; involucral bracts with short glandu-
    lar hairs .... *Pluchea*
S. florets white or yellowish .... *Brickellia*

Key to the genera of Group 3 (heads without rays or with very small and inconspicuous rays, receptacles bristly or chaffy)

A. pappus of capillary (and sometimes plumose) bristles
  B. leaves prickly or spiny
    C. leaves white-mottled above .... *Silybum*
    C. leaves not white-mottled
      D. pappus bristles plumose .... *Cirsium*
      D. pappus bristles simple
  B. leaves not prickly or spiny .... *Filago*
A. pappus consisting of scales or awns, or lacking
  E. pappus represented by scales or awns
    F. leaves opposite .... *Bidens*
    F. leaves alternate or basal
      G. leaves prickly or spiny on the margins .... *Echinops*
      G. leaves not prickly or spiny
        H. involucral bracts forming hooked bristles at the tips .... *Arctium*
        H. involucral bracts not forming hooked bristles
          I. involucral bracts with lacerate or fimbriate tips .... *Centaurea*
          I. involucral bracts not lacerate or fimbriate
            J. leaves entire; florets all disk-type and perfect .... *Marshallia*
            J. leaves variously toothed; heads with minute rays
              K. plants erect, to 1 m tall; leaves coarsely serrate or pin-
              natifid .... *Parthenium*
              K. plants prostrate or spreading; leaves remotely serrulate
              .... *Eclipta*
  E. pappus absent
    L. leaves alternate
    M. heads with minute rays .... *Parthenium*
    M. rays not present
      N. involucre becoming a prickly or spiny bur surrounding the
      achenes .... *Xanthium*
      N. involucre not becoming a prickly bur
        O. involucral bracts with a fimbriate or lacerate tip; receptacle
        chaffy throughout .... *Centaurea*
        O. involucral bracts not lacerate or fimbriate
          P. involucral bracts spine-tipped; receptacle bristly through-
          out .... *Centaurea*
          P. involucral bracts not bristle-tipped; receptacle chaffy only
          near the margin .... *Madia*
    L. leaves opposite, at least below
    Q. staminate and pistillate florets in separate heads .... *Ambrosia*
    Q. all heads the same

R. leaves lobed .... *Polymnia*

R. leaves serrate or serrulate, but not lobed

    S. heads with minute rays .... *Eclipta*

    S. rays not present .... *Iva*

Key to the genera of Group 4 (only ray florets present, sap milky)

A. pappus absent .... *Lapsana*

A. pappus present

    B. pappus partly or entirely of thin scales

        C. florets blue; pappus a crown of short scales .... *Cichorium*

        C. florets yellow; pappus of capillary bristles with an outer ring of thin scales .... *Krigia*

    B. pappus of capillary bristles only

        D. at least some of the pappus bristles plumose

            E. leaves mostly basal, flowering stems with only reduced bracts

                F. receptacle lacking chaffy bracts .... *Leontodon*

                F. receptacle with chaffy bracts .... *Hypochoeris*

            E. flowering stems leafy, although the leaves may be reduced upward

                G. plants glabrous; leaves narrow, grass-like, clasping at the base..... *Tragopogon*

                G. plants hairy; leaves lanceolate to oblanceolate, not grass-like ..... *Picris*

        D. pappus bristles not plumose

            H. heads solitary on naked scapes; leaves all basal

                I. plants glabrous

                    J. taprooted plants with sessile leaves .... *Taraxacum*

                    J. creeping plants with petiolate leaves .... *Ixeris*

                I. plants hairy, at least at the base .... *Hieracium*

            H. heads >1 on naked or leafy stems

                K. leaves all or mostly basal, blades entire and unlobed

                  L. achene not beaked .... *Hieracium*

                  L. achene with a long, slender beak .... *Ixeris*

                K. leaves all or mostly cauline at flowering time; or if basal, then toothed

                  M. flowering heads hanging down .... *Prenanthes*

                  M. heads erect at flowering

                      N. achenes terete or angled, barely or not at all flattened

                          O. body of the achene spiny at the apex .... *Chondrilla*

                          O. achene not spiny

                              P. cauline leaves never auriculate-clasping; pappus whitish, tan or brown .... *Hieracium*

                              P. cauline leaves sometimes auriculate-clasping; pappus white .... *Crepis*

                    N. achenes distinctly flattened

                      Q. achenes beaked .... *Lactuca*

                      Q. achenes beakless .... *Sonchus*

Key to the genera of Group 5 (disk and ray florets both present, rays yellow or orange, receptacle chaffy with bracts subtending some or all of the flowers or with long bristles between the flowers)

A. leaves (at least the lower) opposite or whorled
    B. leaves joined at the nodes forming funnel-shaped structures surrounding the stem, or in whorls of 3–4 at each node .... *Silphium*
    B. leaves only slightly or not at all connate, not whorled
        C. involucral bracts of 2 types, the outer herbaceous, the inner membranous and often striate
            D. pappus of 2–3 nonpersistent short awns or teeth, or none
                E. achenes beaked .... *Cosmos*
                E. achenes not beaked .... *Bidens*
            D. pappus of 2–6 well-developed, persistent awns
                F. terrestrial or emergent aquatic; lower leaves not filiform-dissected .... *Bidens*
                F. aquatic with lower, submersed leaves filiform-dissected .... *Megalodonta*
        C. involucral bracts all similar
            G. receptacle chaffy only near the margin, each bract enclosing a ray achene .... *Madia*
            G. receptacle chaffy throughout
                H. aquatic with lower, submersed leaves filiform-dissected .... *Megalodonta*
                H. terrestrial plants lacking filiform-dissected leaves
                    I. rays persistent on the achenes .... *Heliopsis*
                    I. rays not persistent
                        J. erect to ascending plants to 4 dm tall; rays 4–5, yellow .... *Chrysogonum*
                        J. taller, coarser plants to 1 m or more; rays yellow and more numerous or very small and whitish
                            K. leaves coarsely lobed; achenes obovate; pappus absent... .... *Polymnia*
                            K. leaves simple; achenes slender, bearing 2 awns .... *Helianthus*

A. leaves alternate
    L. stems winged .... *Verbesina*
    L. stems not winged
        M. involucre of 2 distinct types of bracts, the inner membranous and erect, the outer herbaceous .... *Coreopsis*
        M. involucral bracts all similar
            N. leaves pinnatifid with a winged rachis and toothed segments, strongly aromatic .... *Anthemis*
            N. leaves variously entire, lobed, or laciniately cleft, not aromatic
                O. receptacle with long hairs or setae among the disk florets .... *Gaillardia*
                O. receptacle with chaffy bracts subtending the disk florets
                    P. rays strongly drooping; stem with appressed hairs .... *Ratibida*
                    P. rays mostly spreading; or if drooping, stem glabrous .... *Rudbeckia*

Key to the genera of Group 6 (disk and ray florets both present, rays yellow or orange, receptacle without bracts or bristles between the flowers although the surface may be pitted)

A. pappus lacking
    B. leaves pinnately dissected .... *Tanacetum*
    B. leaves toothed or pinnatifid, but not dissected .... *Chrysanthemum*
A. pappus present
    C. pappus consisting of awns or scales
        D. leaves, at least the lower ones, opposite .... *Dyssodia*
        D. leaves alternate
            E. rays 3-lobed or 3-cleft at the tip, yellow, but sometimes with a purple blotch near the base
                F. stems not winged; receptacle with long hairs or bristles among the disk florets .... *Gaillardia*
                F. stem usually winged; receptacle lacking hairs or bristles .... *Helenium*
            E. rays not lobed or cleft at the tip, yellow throughout .... *Grindelia*
    C. pappus of fine capillary bristles
        G. leaves opposite (except perhaps the small uppermost ones) .... *Arnica*
        G. leaves alternate or basal
            H. involucral bracts in a single row
                I. flowering stem with only reduced bracts; leaves all basal, expanding after flowering .... *Tussilago*
                I. stem with well-developed basal and cauline leaves at the time of flowering .... *Senecio*
            H. involucral bracts in several overlapping rows
                J. pappus bristles all similar
                    K. heads 3–5 cm wide, few; anthers tailed at the base .... *Inula*
                    K. heads smaller, many; anthers not tailed
                        L. leaves and involucral bracts not glandular-punctate (however, *S. odora* has leaves with translucent dots and anise odor) .... *Solidago*
                        L. leaves and involucral bracts glandular-punctate .... *Euthamia*
                J. pappus distinctly double, inner bristles longer than the short outer ones
                    M. achenes of the ray florets lacking pappus .... *Heterotheca*
                    M. achenes of the ray florets with pappus like those of the disk florets .... *Chrysopsis*

Key to the genera of Group 7 (disk and ray florets both present, rays white, pink, purple, or blue)

A. receptacle chaffy, at least toward the middle
    B. leaves all or mostly opposite
        C. leaves pinnately dissected; achenes beaked .... *Cosmos*
        C. leaves entire, toothed, or lobed; achenes not beaked
            D. leaves linear or linear-lanceolate to lanceolate
                E. rays pink; leaves linear .... *Coreopsis rosea*
                E. rays white (or absent); leaves lanceolate or lance-linear ..... *Eclipta*

  D. leaves broader
    F. plants mostly >1 m tall; leaves up to 3 dm long, somewhat lobed
     .... *Polymnia*
    F. plants rarely >0.5 m tall; leaves ovate to lance-ovate, 2–7 cm long,
     not lobed .... *Galinsoga*
 B. leaves alternate
  G. rays 1–13 mm long, white or pink
   H. leaves pinnately finely divided into filiform segments or if undivided,
    then linear or lance-linear
    I. heads small, many, in a flat-topped inflorescence; rays 1–5 mm
     long .... *Achillea*
    I. heads larger, on long peduncles at the ends of branches; rays 5–
     13 mm long .... *Anthemis*
   H. leaves undivided, broader .... *Parthenium*
  G. rays 3–8 cm long, reddish-purple .... *Echinacea*
A. receptacle naked
 J. pappus of scales, awns or completely lacking
  K. leaves all basal; pappus absent .... *Bellis*
  K. leaves not all basal; pappus present or absent
   L. leaves entire, not aromatic .... *Boltonia*
   L. leaves toothed or dissected, aromatic
    M. leaves finely divided into filiform ultimate segments .... *Matricaria*
    M. leaves dentate to pinnatifid with broad lobes .... *Chrysanthemum*
 J. pappus of long capillary bristles ....
  N. well-formed leaves all basal, only scaly bracts present on the stems .....
   .... *Petasites*
  N. cauline leaves present
   O. rays minute; heads small, involucres no more than 4 mm long ....
    .... *Conyza*
   O. rays conspicuous; heads larger
    P. involucral bracts in a single series (approximately) .... *Erigeron*
    P. involucral bracts of several overlapping lengths
     Q. rays blue, purple, pink, or white .... *Aster*
     Q. rays creamy white .... *Solidago bicolor*

Achillea L.
Jerry Chmielewski

Erect, herbaceous perennials with leafy stems and alternate leaves; heads small, white, radiate, aggregated in a branched, flat-topped inflorescence; involucral bracts in several overlapping series; receptacle chaffy; achenes compressed, glabrous, lacking pappus.

A. leaves finely dissected, rays 4–5 .... *A. millefolium*
A. leaves undivided or coarsely pinnatifid, rays 6–15 .... *A. ptarmica*

Achillea millefolium L.   Common yarrow, milfoil
Stems simple, leafy, 3–9 dm tall; leaves sessile, lanceolate in outline, fern-like, once to twice pinnately parted into linear, toothed segments, about 15 cm long, pubescent or nearly glabrous; heads about 6 mm wide; rays 4–5, white or pink,

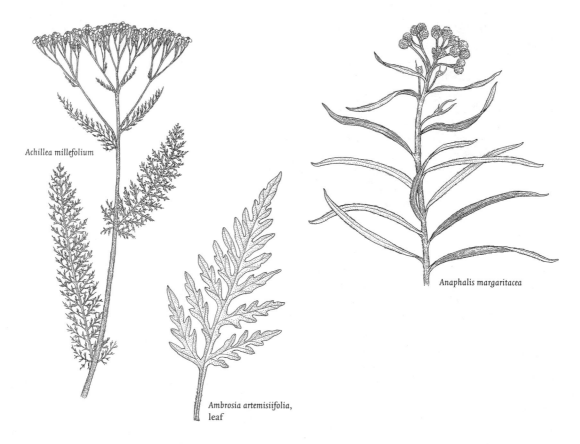

Achillea millefolium

Ambrosia artemisiifolia,
leaf

Anaphalis margaritacea

surrounding smaller central disk florets; common in fields, roadsides, and waste places; throughout; flr. Jun–Sep; native to Europe; FACU.

Achillea ptarmica L.   Sneezeweed
Stems simple or branched above, 3–6 dm tall; leaves linear-lanceolate, 2.5–7.5 cm long, finely toothed, sessile, nearly glabrous; heads 1.2–1.9 cm wide, rays 6–15, white; a rare garden escape of fields and roadsides; flr. Jun–Sep; native to Eurasia.

Ambrosia L.
Jerry Chmielewski

Herbs with opposite (at least the lower), lobed or dissected leaves; heads unisexual, the staminate heads nearly sessile, in a spike or raceme, not subtended by bracts; the pistillate heads below the staminate, in the axils of leaves or bracts, 1-flowered; fruiting involucres becoming closed bur-like structures bearing short stiff spines or tubercules; pistillate florets lacking corollas and pappus; receptacles flat with slender scales.

A. leaves palmately 3–5-lobed or undivided ....*A. trifida*
A. leaves once or twice pinnatifid
    B. taprooted annual; leaves smooth above; fruiting involucres spiny ....
    ....*A. artemisiifolia*
    B. perennial with creeping roots; leaves scabrous above; fruiting involucres
    tuberculate .... *A. psilostachya*

Ambrosia artemisiifolia L. Common ragweed

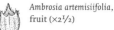

Ambrosia artemisiifolia, fruit (×2½)

Tap-rooted annual to 1 m tall (depauperate plants smaller), hairy to subglabrous; leaves once- or twice-pinnatifid; plants monoecious or dioecious, staminate heads with 15–20 small, yellow-green florets; pistillate heads small, green, sessile, in small clusters in the axils of reduced leaves; fruits bur-like, 4–5 mm long, capped with 6 short spines; common in fields, meadows, cultivated areas, roadsides, and waste ground; throughout; flr. Aug–Oct; FACU.

Ambrosia psilostachya DC. Western ragweed

Similar to *A. artemisiifolia,* but a colonial perennial from creeping roots; leaves once pinnatifid, sessile, rough above; fruiting involucres merely tuberculate; rare on sandy shores or meadows; NW; flr. Aug–Oct.

Ambrosia trifida L. Giant ragweed

Coarse, hairy-stemmed annual to 2 m tall or more; leaves broadly elliptic to ovate, serrate, the larger ones palmately 3–5-lobed; staminate inflorescence 1.5 mm; pistillate involucres 3–5 mm with several sharp spines at the top; common in fields, roadsides, and floodplains; throughout; flr. Aug–Oct; FAC.

Anaphalis DC.

Anaphalis margaritacea (L.) Benth. & Hook Pearly everlasting

Conspicuously white-woolly, rhizomatous, dioecious perennial; leaves numerous, all cauline, linear- lanceolate, the margins often revolute; heads discoid, <1 cm wide, crowded in a short, broad inflorescence; pistillate heads sometimes with a few staminate florets; involucral bracts dry and scarious; pappus of distinct capillary bristles; frequent in dry, sandy, or gravelly soil of fields, woods edges, and roadsides; more common N; flr. Jul–Oct.

Antennaria Gaertn.
Jerry Chmielewski

Stoloniferous perennial herbs; dioecious, forming mat-like colonies of staminate or pistillate plants; leaves often white-tomentose, at least beneath, alternate, simple; heads solitary or clustered at the end of the upright flowering stems; involucre white-tomentose; receptacle flat, naked; corollas tubular; styles of the pistillate florets bifid, those of the staminate florets mostly undivided; pappus of numerous capillary bristles. A difficult genus due to polyploidy and apomixis.

A. basal leaves 0.5–2 cm wide with 1–3 prominent veins
  B. middle and upper cauline leaves terminated by a flat or inrolled scarious tip
    C. young and mature basal leaves bright green and glabrous above or becoming glabrate; phyllary tips white or pink .... *A. howellii*
    C. young basal leaves pubescent; mature basal leaves pubescent or glabrous; phyllary tips variable in color, white, pink, red, brown, dark green, or black ....*A. neglecta*
  B. middle and upper cauline leaves blunt or with subulate or aristate tips

Antennaria neglecta,
pistillate plant

    D. basal leaves 1.5–5.5 cm long by 0.5–2 cm wide; pistillate involucres 6–9 mm high; staminate involucres 3.8–6.5 mm high; pistillate corollas 4.0–5.5 mm long .... *A. howellii*

    D. basal leaves 1–2.5 cm long by 3–8 mm wide; pistillate involucres 4.5–7 mm high; staminate involucres 3.8–5 mm high; pistillate corollas 3.2–4.5 mm long .... *A. virginica*

A. basal leaves 1–6 cm wide with 3–7 prominent veins

    E. inflorescence consisting of a solitary head .... *A. solitaria*

    E. inflorescence a corymb with several heads

        F. pistillate heads 5–7 mm high; pistillate corollas 3–4 mm long; staminate corollas 2–3.5 mm long; basal leaves tomentose above and usually long-petiolate .... *A. plantaginifolia*

        F. pistillate heads 7–10 mm high; pistillate corollas 4–7 mm long; staminate corollas 3.5–5 mm long; basal leaves tomentose or glabrous above, petiole variable in length .... *A. parlinii*

Antennaria howellii Greene   Howell's pussytoe

Typically pistillate, only occasionally staminate; peduncles 15–30 cm; basal leaves 1–3-nerved, oblanceolate to spatulate, 1.5–5.5 cm long by 0.5–2 cm wide, bright green, pubescent or glabrous above, tomentose beneath; cauline leaves linear, 2–4 cm long; heads 5–12; pistillate involucres 6–9 mm long, phyllaries narrow, acute, light brown, white, or pinkish; staminate involucres 3.8–6.5 mm long, phyllaries spreading, white or roseate; pistillate corollas 4.0–5.5 mm long; staminate corollas slightly smaller; fruit 1.2–2 mm, commonly papillate; pappus bristles 6–8 mm long; common in dry fields, pastures, open woods, and rocky barrens; throughout; flr. late Apr–early Jun. [includes *A. neodioica* Greene ssp. *neodioica*, ssp. *canadensis* (Greene) Bayer & Stebbins and ssp. *petaloidea* (Fernald) Bayer & Stebbins of Rhoads and Klein, 1993]

Antennaria neglecta Greene   Overlooked pussytoe

Peduncles of pistillate plants slender, 15–40 cm long, staminate peduncles up to 15 cm long; stolons elongate and flexuous; basal leaves 1–3-nerved, cuneate-oblanceolate to spatulate, 1.5–6.5 cm long by 0.5–1.5 cm wide, young leaves canescent above; cauline leaves linear, 3–8, the upper with a flat or curled scarious tip; heads 2–10; pistillate involucres 6.5–9 mm long, phyllaries brown with obtuse-acute white or scarious tips; staminate involucres 4–6 mm long, phyllaries white, blunt, and petaloid; pistillate corollas 5–6.2 mm long; achenes 1–1.4 mm long; frequent in fields, pastures, and open woods; mostly E; flr. late Apr–early Jun; UPL. [*A. neodioica* Greene in part, of Rhoads and Klein, 1993]

Antennaria neglecta,
staminate plant

Antennaria parlinii Fernald   Parlin's pussytoe
Peduncles initially stout, becoming slender by elongation, those of pistillate
plants 15–50 cm long, staminate peduncles shorter, with or without purple
glands; young stolons mostly decumbent; basal leaves 3–7-nerved, broadly
obovate-spatulate to obovate and petioled, 1–7 cm long by 1–4 cm wide, bright
green and glabrous adaxially or minutely cinereous-tomentose; lower cauline
leaves oblong-lanceolate and crowded; heads 3–10; pistillate involucres 7–11
mm long, with scarious or somewhat petaloid phyllaries; staminate involucres
with lustrous round-spreading white phyllaries; pistillate corollas 4–7 mm long;
staminate corollas 3.5–5 mm long; achenes 1.3–2.2 mm long, glabrous or sparingly to distinctly papillate; pappus bristles 6–8.5 mm long; frequent in clearings, fields, and dry open woods; throughout; flr. Apr–early Jun.

Antennaria plantaginifolia (L.) Hook.   Plantain-leaved pussytoe
Peduncles slender, white-tomentose, those of pistillate plants 5–30 cm long,
staminate peduncles to 20 cm long; young stolons mostly ascending; basal
leaves with 3–7 nerves, obovate to oblong or suborbicular, 1.5–6 cm long by 1.5–
4 cm wide, canescent; upper cauline leaves terminated by a subulate tip; heads
numerous, to 30; pistillate involucres 5–7 mm long, phyllaries often purplish at
the base with narrow white stramineous tips; staminate phyllaries spreading,
broad, white or rarely pink; pistillate corollas 3–4 mm long; staminate corollas
2–3.5 mm; fruit 1–1.5 mm long; pappus bristles 4–5.5 mm; frequent in dry open
woods, fields, pastures, and rocky banks; throughout; flr. Apr–Jun.

Antennaria solitaria Rydb.   Solitary pussytoe
Pistillate and staminate plants similar; peduncles up to 3.5 dm long; stolons
long and creeping, forming extensive prostrate branches; basal leaves with 3–7
prominent nerves, obovate to broadly oblong-spatulate, 2–7 cm long by 1.5–4.5
cm wide, tomentose above; cauline leaves narrow, appressed to the peduncle;
heads solitary, terminal; involucre 8–12 mm long, pistillate phyllaries linear,
acute, brown to purple with thin paler tips; staminate phyllaries broader and
more blunt; fruit papillate; rare in rich woods and clearings; SW; flr. late Apr–
May; 🌿.

Antennaria virginica Stebbins   Virginia or shale-barren pussytoe
Peduncles slender, 6–20 cm long; stolons short, ascending; basal leaves 1–3-
nerved, cuneate-oblanceolate, 1–2.5 cm long, 3–8 mm wide, canescent; cauline
leaves linear, 1–1.4 cm long; heads 2–10; pistillate involucres 4.5–7 mm long,
phyllaries obtuse to acute; staminate involucres 3.8–5 mm long, phyllaries
broad; pistillate corollas 3.2–4.5 mm long; staminate corollas 2.5–3.5 mm long;
fruit 1.1–1.3 mm long; pappus bristles as long as the corolla; rare in dry woods
and openings; SC; flr. late Apr–Jun; 🌿.

Anthemis L.
Jerry Chmielewski

Aromatic herbs with alternate, dissected leaves; heads with elongate white or
yellow rays; involucral bracts nearly equal or in several overlapping rows; receptacle convex, chaffy, at least toward the middle; disk florets numerous; achenes
angled; pappus a short scaly crown or none.

A. rays white; disk 0.5–1.2 cm wide; corolla tube cylindric
   B. plants ill-scented; ray florets sterile, chaffy scales of the receptacle sub-
      tending only the central disk florets .... A. cotula
   B. plants odorless or fragrant; ray florets fertile, chaffy scales of the receptacle
      subtending all disk florets ....A. arvensis
A. rays yellow; disk 1–2 cm wide; corolla tube compressed .... A. tinctoria

Anthemis arvensis L.   Corn chamomile
Annual or biennial; stems bushy, branched, pubescent, 3–6 dm tall; leaves once
or twice pinnate, 2.5–6.5 cm long; heads approaching 2.5 cm wide; ray florets
white, fertile; disk 0.7–1.2 cm wide, chaffy scales of the receptacle subtending all
disk florets; achenes smooth, 10-nerved; frequent in roadsides, woods, fields,
and waste places; mostly E and C; flr. Jun–Oct; native to Europe.

Anthemis cotula L.   Mayweed, stinking chamomile
Annual, stem bushy, branched, 3–6 dm tall; leaves finely tripinnately dissected,
fern-like, 2.5–6.5 cm long; heads approaching 2.5 cm wide; ray florets white,
sterile; disk 0.5–1.2 cm wide; chaffy scales of the receptacle subtending only the
central disk florets; achenes warty; occasional on roadsides, fields, and waste
places; throughout; flr. Jul–Oct; native to Europe; FACU–.

Anthemis tinctoria L.   Yellow chamomile
Herbaceous perennial, stem angular, erect or ascending, 3–9 dm tall, pubescent;
leaves pinnately divided; heads approaching 4 cm wide; ray florets yellow; disk
1–2 cm wide; achenes quadrangular, marked with fine longitudinal lines; rare on
roadsides, fields, and waste places; scattered; flr. Jun–Aug; native to Europe.

Arctium L.

Coarse biennials with large, alternate, mostly cordate leaves and several to many
heads; florets all tubular and perfect; corollas pink or purplish; involucre
subglobose, the bracts narrow, appressed at the base with a spreading, hooked
tip; the receptacle flat, densely bristly; pappus of short bristles that detach separately; the achenes remaining within the spiny involucre which serves as a dispersal unit.

A. heads short-peduncled or subsessile, in a racemiform inflorescence
   B. involucre 1.5–2.5 cm thick .... A. minus
   B. involucre 3–3.5 cm thick .... A. nemorosum
A. heads on long peduncles, inflorescence corymbiform
   C. involucre tomentose, 1.5–2.5 cm thick .... A. tomentosum
   C. involucre not tomentose, 3–4 cm thick ....A. lappa

Arctium lappa L.   Great burdock
Stems 1.5–3 m tall; leaves with solid petioles, the lower broadly ovate and cordate at the base; branches of the inflorescence corymbiform with 3–10-cm-long,
glandular or glandular-hairy peduncles; involucre 3–4 cm wide, equaling or surpassing the florets; occasionally naturalized along floodplains and roadsides;
mostly E, scattered elsewhere; flr. Jul–Sep; native to Eurasia.

Arctium minus (Hill) Bernh.   Common burdock

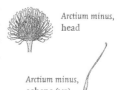

Arctium minus, head

Arctium minus, achene (×1)

Stems to 1.5 m tall; lower leaves with hollow petioles; blade narrowly to broadly ovate with a cordate base; inflorescence racemiform, heads short-pedunculate or subsessile; involucre 1.5–2.5 cm thick, slightly shorter than the florets; widely naturalized along fields, woods, railroad tracks, and waste ground; throughout; flr. Jul–Sep; native to Eurasia; FACU–.

Arctium nemorosum Lej. & Court.   Woodland burdock

Very similar to *A. minus* but with heads larger (involucre 3–3.5 cm thick); infrequent in waste ground; scattered, mostly S; flr. Jun–Oct; native to Europe.

Arctium tomentosum Mill.   Wooly burdock

Similar to *A. lappa* but smaller, rarely >1.3 m tall; lower petioles mostly hollow; involucre 1.5–2.5 cm thick and strongly long-tomentose, involucral bracts only weakly hooked; infrequent on roadsides, floodplains, and waste ground; mostly SE, scattered elsewhere; flr. Jul–Oct; native to Eurasia.

Arnica L.

Arnica acaulis (Walter) Britton, Stearns & Poggenb.   Leopard's-bane

Glandular, hairy perennial to 8 dm tall; leaves mostly toward the base, opposite, sessile, broadly elliptic to ovate; cauline leaves few and much reduced, becoming alternate above; heads few, to 5 cm wide; rays 10–15, yellow; achenes cylindric; pappus white; rare in open woods and thickets on serpentine barrens; SE; flr. May–Jun; FACU; ❧.

Artemisia L.

Aromatic herbs with alternate, entire to finely dissected leaves; heads small, often nodding, lacking rays; outer florets pistillate, the central ones perfect; involucral bracts with scarious margins, often pubescent; receptacle naked or with long hairs; achenes oblong; pappus none.

A. leaves green, glabrous or pubescent, but not white tomentose
    B. leaves pinnate to bipinnate with flat, toothed segments
        C. heads in panicles, drooping, 2 mm wide .... *A. annua*
        C. heads in leafy spikes, erect, 3 mm wide .... *A. biennis*
    B. leaves dissected into linear-filiform, entire segments with revolute margins
        .... *A. campestris*
A. leaves densely white-tomentose, at least beneath
    D. upper surface of leaves white-tomentose
        E. leaves finely divided into linear segments <1 mm wide .... *A. pontica*
        E. leaves coarsely lobed or divided, segments >1.5 mm wide ....
        .... *A. absinthium*
    D. upper surface of leaves green
        F. leaves deeply pinnately dissected .... *A. vulgaris*
        F. leaves linear to obovate, the upper entire, the lower coarsely toothed or
        lobed .... *A. ludoviciana*

Artemisia absinthium L.   Common wormwood

Perennial herb to 1 m tall, often somewhat woody at the base; leaves silvery-hairy to subglabrate above, 2 or 3 times pinnatifid, the segments 1.5–4 mm wide; inflorescence leafy; involucre 2–3 mm long; receptacle with long white hairs between the florets; cultivated and occasionally escaped to waste ground or ballast; mostly S; flr. Aug; native to Europe.

Artemisia annua

Artemisia annua L. — Sweet or annual wormwood
Fragrant, glabrous annual to 3 m tall; leaves 2 or 3 times pinnatifid with narrow, sharply toothed segments; inflorescence a broad, open panicle; heads nodding; cultivated fields, roadsides, and urban waste ground; S; flr. Jun–Oct; native to Eurasia; FACU.

Artemisia biennis Willd. — Biennial wormwood
Coarse, glabrous annual or biennial, nearly odorless; leaves pinnately or bipinnately divided into narrow, sharply toothed lobes; inflorescence dense and spike-like; heads erect, numerous, and crowded; railroad tracks, wharves, and ballast; scattered; flr. Jul–Oct; native to the northwestern U.S.; FACU–.

Artemisia campestris L. — Beach wormwood
Biennial or perennial to 1 m tall, glabrous to slightly hairy; leaves 2 or 3 times pinnatifid into linear, entire segments < 2 mm wide; inflorescence spike-like to diffuse and panicle-like; involucre 2–4.5 mm long, glabrous to villous-tomentose; rare on dry, sandy shores or sand flats; NW and NE; flr. Aug; 🍂. Ours is ssp. *caudata* (Michx.) Hall & Clements. [syn: A. *caudata* Michx.]

Artemisia ludoviciana Nutt. — Western mugwort
Rhizomatous perennial to 1 m tall, unbranched below the inflorescence; leaves densely white-tomentose beneath, lanceolate or lance-elliptic, entire or the lower leaves with irregularly toothed or lobed margins; inflorescence compact and elongate; involucres 2.5–3.5 mm long; increasingly common in vacant lots, roadsides, and waste ground; mostly S; flr. Jul–Sep; native to the western U.S.; UPL.

Artemisia pontica L. — Roman wormwood
Rhizomatous perennial to 1 m tall; leaves densely white-tomentose, 2 or 3 times pinnatifid with the ultimate segments to 1 mm wide; inflorescence narrow and elongate; involucre 2–3 mm long; cultivated and rarely escaped; mostly SE; flr. Aug–Sep; native to Europe.

Artemisia vulgaris

Artemisia vulgaris, leaf

Artemisia vulgaris L. — Common mugwort
Rhizomatous perennial to 2 m tall, stem glabrous or nearly so; leaves densely white-tomentose beneath, pinnately lobed nearly to midrib, segments coarsely toothed; inflorescence with numerous, progressively reduced leaves; involucre 3.5–4.5 mm long; receptacle naked; common in gardens, lawns, roadsides, thickets, and waste ground; throughout; flr. Jun–Oct; native to Eurasia; UPL.

Aster L.

Herbaceous perennials (except A. *brachyactis*); leaves simple, mainly cauline but with basal rosettes in some species; involucral bracts in several overlapping rows, the tips sometimes recurved and always green or greenish; receptacle flat or slightly convex, naked; heads usually with conspicuous blue, purple, pink or white rays (rays absent in *A. brachyactis*); ray flowers pistillate and fertile; disk flowers perfect and fertile, creamy, yellow, reddish or purple; style branches flattened; achenes with several nerves; pappus of capillary bristles, in addition club-shaped bristles and/or an outer whorl of very short bristles sometimes present; mostly late summer and fall blooming. This treatment reflects the traditional broad view of the genus; however, recent evidence suggests that *Aster* should be split into a number of segregate genera. In our flora these would include, in addition to *Aster*, *Doellingeria*, *Ionactis*, *Oclemena*, and *Sericocarpus*.

A. leaves that are both cordate and petiolate present, at least at the base
  B. glands present on the involucral bracts and inflorescence branches ....
  .... *A. macrophyllus*
  B. plants not glandular
    C. leaves entire, except perhaps the lowest ones
      D. some of the cauline leaves sessile or auriculate-clasping ....
      .... *A. undulatus*
      D. none of the leaves auriculate-clasping ....*A. shortii*
    C. leaves mostly toothed
      E. plants glabrous, or slightly puberulent in the inflorescence
        F. inflorescence elongate; separate sterile shoots not present at flowering time
          G. involucral bracts acuminate or acute .... *A. urophyllus*
          G. involucral bracts obtuse or barely acute .... *A. cordifolius*
        F. inflorescence more or less flat-topped; separate sterile shoots present at flowering time .... *A. schreberi*
      E. leaves and/or stem more-or-less hairy
        H. rays white; inflorescence rounded to flat-topped ..... *A. divaricatus*
        H. rays blue or purple
          I. involucral bracts obtuse to acute, their tips short and broad and often purplish .... *A. cordifolius*
          I. involucral bracts acute to acuminate with an elongate green tip
          ....*A. drummondii*
A. plants without leaves that are both cordate and petiolate
  J. an outer whorl of short pappus bristles present in addition to the longer ones
    K. leaves >5 mm wide, with well-developed lateral veins; rays white
      L. involucre 3–5 mm; achenes hairy ....*A. umbellatus*
      L. involucre 4.5–7 mm; achenes glabrous .... *A. infirmus*
    K. leaves 1.2–4 mm wide, 1-nerved; rays violet ....*A. linariifolius*
  J. pappus bristles all the same length
    M. involucre or achenes (and sometimes the leaves and stem) glandular
      N. involucre glandular; achenes glabrous or hairy but not glandular
        O. leaves with a clasping base
          P. leaves sessile and cordate-clasping
            Q. leaves rarely constricted above the clasping base; middle and upper leaves 1.5–3.5 times as long as wide .....*A. patens*
            Q. leaves usually constricted above the enlarged clasping base, the middle and upper leaves 3.5–6 times as long as wide....
            ....*A. phlogifolius*
          P. leaves weakly to strongly auriculate-clasping, not cordate

II. leaves mostly <5 mm wide, entire to subentire .... *A. borealis*

II. leaves >5 mm wide, coarsely toothed .... *A. radula*

GG. heads many; plants 1–2 m tall, growing in moist low ground and clearings

JJ. principal leaves mostly <6 mm wide but occasionally wider, glabrous .... *A. longifolius*

JJ. principal leaves often >1 cm wide, the younger ones somewhat short-hairy above .... *A. praealtus*

CC. rays white

KK. involucral bracts with a spinulose or mucronate tip

LL. involucral bracts inrolled at the tip

MM. plants hairy; leaves not appressed; heads often secund on the branches .... *A. pilosus*

MM. plants glabrous; leaves remaining at flowering time very narrow and appressed; heads not secund .... *A. depauperatus*

LL. involucral bracts flat, not inrolled, but with a mucronate tip ......
.... *A. ericoides*

KK. involucral bracts not spine-tipped

NN. lobes of the disk corollas comprising 45–75% of the expanded limb .... *A. lateriflorus*

NN. lobes of the disk corollas comprising 15–45% of the limb

OO. heads small and numerous; involucre <5 mm long; rays 3–7 mm long

PP. leaves linear to narrowly lanceolate, those of the branches <1.5 cm long .... *A. racemosus*

PP. leaves lanceolate to lance-elliptic, those of the branches >1.5 cm long .... *A. lanceolatus*

OO. heads larger or fewer

QQ. leaves veiny beneath with squarish areolae ....
.... *A. praealtus*

QQ. leaf veins obscure beneath or forming areolae that are distinctly longer than wide

RR. inflorescence short and broad; lobes of the disk corollas <30% of the limb .... *A. borealis*

RR. inflorescence elongate; lobes of the disk corollas 30–45% of the limb .... *A. lanceolatus*

Aster acuminatus Michx.   Wood aster
Rhizomatous, stems 2–8 dm tall, variously hairy to puberulent and slightly sticky; leaves often crowded below the inflorescence forming a pseudo-whorl, elliptic or obovate, acuminate with a few sharp teeth; heads few to many in an open corymbiform inflorescence; rays white; frequent in cool, moist woods; mostly N and at higher elevations along the Allegheny Front; FACU+.

Aster x *blakei* (Porter) House   Aster
A hybrid of *A. acuminatus* and *A. nemoralis*, which is intermediate between them and occurs in several sites where the parental species are both present; NC.

Aster borealis (Torr. & A.Gray) Prov.   Northern bog aster
Stems slender, 1.5–10 dm tall, from long rhizomes, glabrous below, puberulent in lines above; leaves cauline, linear to lance-linear, sessile and slightly auricu-

late-clasping, entire or subentire; heads few (or solitary on small plants) in an open inflorescence; involucre glabrous with strongly imbricate bracts with purple tips or margins; rays 20–50, white, pale blue, or lavender; rare in cold bogs; NE and NW; OBL; ❦.

Aster brachyactis S.F.Blake   Western annual aster
Tap-rooted annual; stems 1–7 dm tall; leaves linear or nearly so, glabrous but with ciliolate margins; heads several or many in an open paniculate inflorescence; rays lacking; achenes appressed-hairy; very rare in gravel along interstate highways, especially where deicing salt is used; NW; native farther west.

Aster cordifolius L.   Blue wood aster
Stems 2–12 dm tall from a branched rootstock, glabrous, or hairy in lines above; leaves sharply toothed, acuminate, the lower ones strongly cordate and petiolate, the upper with shorter petioles, only the very reduced leaves in the inflorescence not cordate; heads numerous, on spreading branches forming a paniculate inflorescence; rays 8–20, pale blue-violet; achenes glabrous; common in woods, meadows, and roadsides; throughout, highly variable. [syn. *A. lowrieanus* Porter in part]

Aster depauperatus (Porter) Fernald   Serpentine aster
Plants from a short rootstock, with an overwintering rosette of basal leaves;

Aster acuminatus

Aster acuminatus,
head (×1)

Aster cordifolius

flowering stems spreading or ascending, slender and wiry, 1–4 dm tall, glabrous; basal leaves oblanceolate, those of the flowering stems narrow, subulate and often appressed; heads very numerous, in an open paniculate inflorescence; rays 9–16, white; disk flowers about the same number as the rays; limited to open areas of serpentine barrens; SE; 🍂.

Aster divaricatus L. White wood aster
Rhizomatous and colonial, stems 2–10 dm tall; leaves cordate and long-petioled below, progressively less so above, glabrous or with some appressed hairs along the main veins beneath; inflorescence corymbiform; rays 5–10, white; common in woods; throughout.

Aster drummondii Lindl. Hairy heart-leaved aster
Stems erect, 4–12 dm tall, from a branching rootstock with short spreading hairs especially above the middle; leaves shallowly toothed, scabrous above, densely short-pubescent beneath; the lowest leaves cordate and long-petiolate, the upper ones progressively less cordate and with shorter, mostly winged petioles; heads numerous in a paniculate inflorescence with spreading or ascending branches; rays 10–20, bright blue; achenes short hairy or glabrous; very rare in stream valleys; SW; 🍂.

Aster dumosus L. Bushy aster
Stems erect, 3–9 dm tall from branching rhizomes, glabrous to densely pubescent; leaves linear-oblanceolate, sessile, entire to sparsely serrate, mostly glabrate; leaves of the branches greatly reduced and very numerous; heads many, in a paniculate inflorescence; involucre glabrous with strongly overlapping bracts; rays 15–33, pale blue or rarely white; disk corollas white to pale yellow becoming purple; rare in serpentine barrens, open woods, moist fields, bogs, and swales; mostly SE; FAC; 🍂.

Aster dumosus

Aster ericoides L. White heath aster
Stems 3–10 dm tall, arising singly from creeping rhizomes, moderately to densely hairy; leaves numerous, linear, sessile, the lower and middle ones shriveling early; leaves of the branches reduced and spine-tipped; heads many, in a paniculate, often pyramidal, inflorescence; peduncles, bracts, and involucres densely pubescent; involucres campanulate, the involucral bracts tipped with tiny white spines; rays 13–20, white; disk corollas yellow, becoming brown; achenes densely hairy; rare on calcareous soils and outcrops; SE; FACU; 🍂. Ours is ssp. *ericoides*.

Aster ericoides, head (×3)

Aster infirmus Michx. Flat-topped white aster
Stems solitary from a fibrous-rooted crown, 4–11 dm tall; leaves nearly all cauline, entire, glabrous or short-hairy along the midrib and larger veins beneath; inflorescence flat-topped, with 5–35(75) heads; involucre 4.5–7 mm high; rays 5–9, white; inner pappus bristles elongate, slightly club-shaped, outer <1 mm; occasional in rocky woods, thickets, and barrens; throughout.

Aster laevis L. Smooth blue aster
Stems 3–10 dm tall from a short rhizome or branched rootstock; plants glabrous throughout and glaucous; basal leaves narrowly spatulate, middle and upper

stem leaves broadly to narrowly lanceolate, sessile with auriculate-clasping bases; heads few to many in a paniculate inflorescence; involucres campanulate, glabrous; involucral bracts with diamond-shaped green spots at the tips; rays pale to dark blue; disk corollas yellow becoming purple; occasional in dry woods, rocky ledges, and roadsides; throughout, but mostly SE and SC; UPL; 2 varieties:

A. larger leaves >2.5 cm wide, strongly auriculate-clasping .... var. *laevis*
A. larger leaves <2.5 cm wide, only slightly or not at all auriculate-clasping ....
.... var. *concinnus* (Willd.) House

Aster lanceolatus Willd.   Panicled aster
Stems erect, 3–15 dm tall from elongate rhizomes, pubescent in lines; leaves all cauline, the lower oblanceolate, sessile with entire margins, upper leaves linear to oblanceolate, glabrate or sparsely pubescent above, margins entire to sharply serrate, leaves of the branches only slightly reduced; heads few to many in a paniculate inflorescence; peduncles pubescent, involucres campanulate, the involucral bracts glabrate, strongly overlapping with appressed tips; rays 16–47, white or bluish; disk flowers yellow becoming purple, the lobes comprising 30–45% of the swollen limb portion; common in old fields, open woods and roadsides. [syn: *A. simplex* Willd.] 3 subspecies:

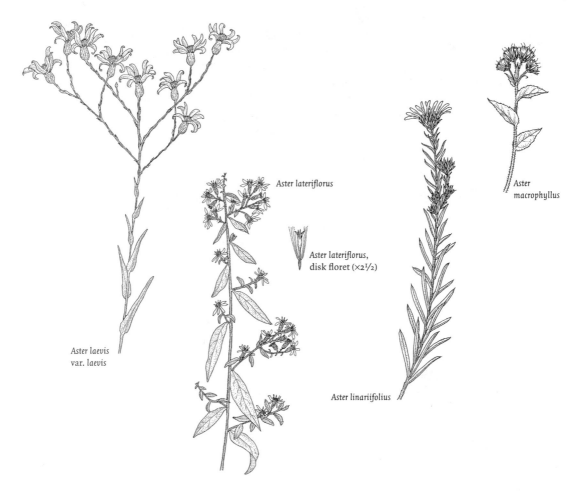

Aster laeviflorus

Aster laeviflorus,
disk floret (×2½)

Aster
macrophyllus

Aster laevis
var. laevis

Aster linariifolius

A. involucre 3–4 mm high .... ssp. *interior* (Wiegand) A.G.Jones
    mostly W.
A. involucre 4–6 mm high
    B. leaves 1–3.5 cm wide, <11 times as long as wide ....
    .... ssp. *simplex* (Willd.) A.G.Jones
      common; throughout.
    B. leaves 3–12 mm wide, 12 or more times a long as wide ...... ssp. *lanceolatus*
      scattered; throughout; FACW.

Aster lateriflorus (L.) Britton.   Calico aster
Stems 3–12 dm tall from a branched rootstock or short rhizome, sparsely to densely pubescent; leaves linear to lance-elliptic, tapering at both ends, some of the larger ones on the branches reflexed so that they point backward, entire or serrate, the main ones 5–15 cm long, those of the branches abruptly reduced in size; heads small, numerous, in a widely branched, paniculate inflorescence (or more simple on smaller plants); rays white; disk corollas pink soon becoming reddish-purple, the lobes making up 50–75% of the swollen limb portion; common in old fields, rocky woods, roadsides, and waste ground; throughout, except in the northernmost counties; FACW–.

Aster linariifolius L.   Stiff-leaved aster
Stems 1–5 dm tall, several from a short rootstock, wiry; leaves numerous, all cauline, linear or nearly so, entire, nerveless except for the midrib; heads several in a corymbiform inflorescence; rays 10–20, violet; disk flowers yellow or purplish; inner pappus bristles elongate, tawny; outer pappus bristles 1 mm or less; occasional in dry rocky woods and edges; mostly E.

Aster longifolius Lam.   Long-leaved blue aster
Similar to *A. praealtus*; leaves generally narrower, lance-linear, mostly <6 mm wide, but occasionally those on the lower stem wider; stem and leaves glabrous; inflorescence narrower and more elongate; rays blue-violet; rare in moist, grassy meadows or openings; NE and C.

Aster macrophyllus L.   Bigleaf aster
Rhizomatous and colonial; flowering stems 2–12 dm tall; accompanied by abundant short, leafy, sterile shoots; the basal and lower leaves broadly cordate with long petioles, the upper sessile and ovate to lanceolate; inflorescence flat-topped, corymbiform; involucre glandular; rays purplish; frequent in moist, often rocky woods; throughout; UPL.

Aster nemoralis Aiton   Leafy bog aster
Rhizomatous, stem 1–8 dm tall; leaves numerous, 40–75 or more, cauline, linear or lance-linear with revolute margins; heads solitary or several in a corymbiform inflorescence; involucral bracts purplish; rays pink or purple; achenes glandular; very rare in sphagnum bogs; NC; FACW+; 🌿.

Aster novae-angliae L.   New England aster
Stems 3–20 dm tall, clustered from a stout rootstock or short rhizome, hairy below and glandular above; leaves mostly cauline, sessile with conspicuous auriculate-clasping bases, softly hairy beneath, scabrous or stiffly hairy above; heads

Aster nemoralis

Aster novae-angliae

Aster novi-belgii

several or many in a leafy inflorescence; involucres and peduncles densely glandular; rays bright purple or pink; achenes densely hairy; common in fields, roadsides, and waste ground; throughout, but more abundant S; FAC.

Aster novi-belgii L.   New York aster
Stems 2–14 dm tall from long creeping rhizomes, puberulent in lines or glabrous; leaves cauline, lanceolate or lance-linear, sessile and more-or-less auriculate-clasping, sharply serrate to entire, glabrous except for the ciliolate margins; heads several to many in an open, leafy-bracted inflorescence; rays blue; very rare in swamps and moist meadows; SE; FACW+; ❦.

Aster oblongifolius Nutt.   Aromatic aster
Stems 1–10 dm tall from a rhizome or short rootstock, branched and glandular above; leaves entire, sessile and weakly to strongly auriculate-clasping, reduced to bracts on the flowering branches, scabrous, short-hairy or glabrous; heads several or many, at the ends of the branches; rays 15–40, blue or purple; achenes finely hairy; infrequent on calcareous hillsides, cliffs, and bluffs; SW and C.

Aster patens Aiton   Late purple aster
Stems slender, 2–15 dm tall from a short rootstock, loosely hairy; leaves scabrous or hairy, at least beneath, sessile and cordate-clasping, entire; heads few to many in a open, branching inflorescence; involucre glandular and/or short hairy; rays bright blue; achenes glabrous; frequent in moist woods and fields; throughout.

Aster paternus Cronquist   White-topped aster

Stems 1.5–6 dm tall, from a branched rootstock, scabrous in the inflorescence; leaves oblanceolate to obovate, toothed, ciliolate; inflorescence flat-topped; involucre glabrous, the bracts with spreading tips; rays 4–8, white; disk flowers white or creamy; achenes densely hairy; pappus with some slightly club-shaped bristles; frequent in dry woods, fields, and barrens; throughout, except in the northernmost counties. [syn: *Sericocarpus asteroides* (L.) Britton, Stearns & Poggenb.]

Aster paternus,
achene (×2½)

Aster phlogifolius Muhl. ex Willd.   Late purple aster

Similar to *A. patens*, but the leaves constricted above the large clasping bases, the middle and upper leaves 3.5–6 times as long as wide; occasional in open woods, old fields, and rocky banks; S.

Aster pilosus Willd.   Heath aster

Stems ascending to erect, 1–15 dm tall, from branched rootstocks or rhizomes; basal rosette leaves oblanceolate to spatulate, rosettes present at flowering and persistent through the winter; lower stem leaves lanceolate to linear, shriveling early; upper leaves linear to lance-elliptic, sessile, entire or slightly toothed, further reduced and becoming subulate in the inflorescence; heads small, in an open, pyramidal inflorescence, often secund on the branches; involucre urn-shaped, glabrous, the bracts with loose, marginally inrolled, green tips; rays white (rarely purplish); disk corollas yellow, becoming purple; common in dry fields, open woods, vacant lots, and roadsides; throughout, except in the northernmost counties; 2 varieties:

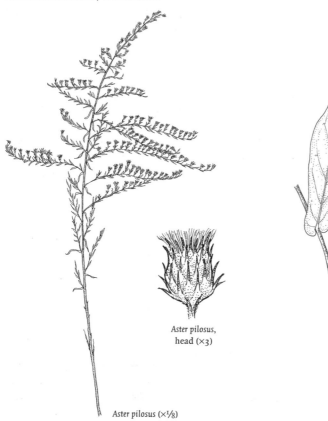

Aster phlogifolius,
leaf

Aster pilosus,
head (×3)

Aster pilosus (×⅛)

A. stems and leaves sparsely to densely hairy .... var. *pilosus* UPL.

A. stems and leaves nearly glabrous .... var. *pringlei* (A.Gray) Britton FACW. [syn: *A. pilosus* var. *dumotus* S.F.Blake, in part]

Aster praealtus Poir.   Veiny-leaved aster
Stems erect, to 2 m or more tall, from a rhizomatous rootstock, pubescent in lines in the inflorescence, glabrous below; lower stem leaves lanceolate, to 1.5 cm wide (or less); upper leaves linear to elliptic-lanceolate with ciliolate margins; heads in a leafy, corymbose-paniculate inflorescence; involucres campanulate; rays 20–35, pale blue-violet; disk corollas yellow, becoming purple, shallowly lobed; woods, fields, thickets, and roadsides; W; FACW; 🍂.

Aster prenanthoides Muhl. ex Willd.   Zig-zag aster
Stems 2–10 dm tall, often somewhat zigzag; pubescent in lines to glabrate below; leaves cauline, scabrous to glabrous above and glabrous or loosely hairy along the midrib beneath, abruptly narrowed to a winged petiole with an auriculate-clasping base; heads several or many in an open, paniculate inflorescence; involucral bracts acute or obtuse with spreading tips; rays 20–35, blue or pale

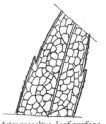

Aster praealtus, leaf surface with areolae (×2)

Aster praealtus, ray floret (×2½)

Aster prenanthoides, leaf

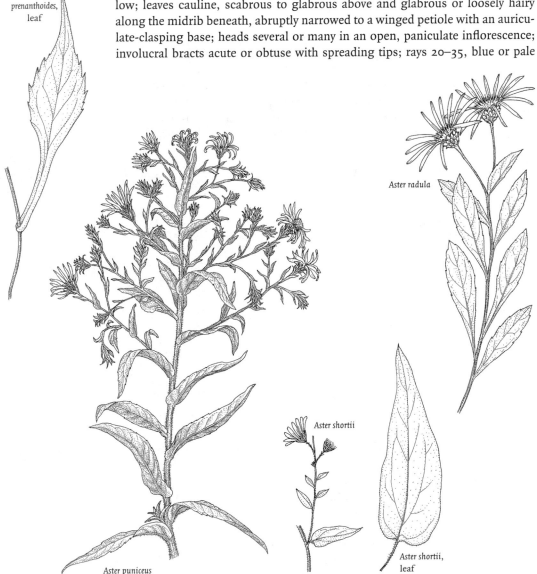

Aster radula

Aster puniceus

Aster shortii

Aster shortii, leaf

purple; achenes sparsely hairy; pappus about as long as the disk corollas; common in swamps and low woods; throughout; FAC.

Aster puniceus L.   Purple-stemmed aster
Stems from a stout rhizome or rootstock, 0.5–2 m tall, usually branched above, hairy, at least in the inflorescence; leaves cauline, sessile, auriculate clasping, irregularly serrate to entire, scabrous or nearly glabrous above and glabrous or hairy along the midrib beneath; heads few to many in a leafy inflorescence; involucral bracts narrow, loose, and spreading; rays blue; common in wet meadows, stream banks, and moist ditches; throughout; OBL; 2 subspecies:

A. stem and branches uniformly spreading-hairy .... ssp. *puniceus*
A. stem and branches puberulent in lines above, glabrous below the inflorescence .... ssp. *firmus* (Nees) A.G.Jones
🌿. [syn: *A. firmus* Nees]

Aster racemosus Elliot   Small white aster
Stems 4–15 dm tall, colonial by long rhizomes, glabrous or puberulent in lines; leaves all cauline, glabrous or slightly scabrous above, linear to narrowly lanceolate, entire or slightly toothed, tapering to a sessile base; heads numerous in a much-branched inflorescence with mostly secund branches; involucre 2.5–3.5 mm, glabrous with overlapping bracts with elongate green tips that are not inrolled; rays 15–30, white, 3–6 mm long; lobes of the disk corollas comprising 40% of the limb; infrequent in low meadows, floodplains, and swamps; scattered; 🌿. [syn: *A. fragilis* Willd., *A. vimineus sensu auct.*, non Lam. var. *subdumosus* Wieg.]

Aster
solidagineus,
head (×1)

Aster radula Aiton   Rough or swamp aster
Rhizomatous, stems 1–12 dm tall, glabrous; leaves cauline, veiny, somewhat serrate, sessile or nearly so; heads few in a sparsely leafy-bracted corymbiform inflorescence or occasionally solitary; rays 15–40, purple; achenes glabrous; rare in wet woods, swamps, seeps, and bogs, in acid soils; S; OBL; 🌿.

Aster schreberi Nees   Schreber's aster
Similar to *A. macrophyllus* but the inflorescence not glandular; glabrous throughout; frequent in woods, moist slopes, and stream banks; throughout.

Aster shortii Lindl. in Hook.   Short's aster
Stems 3–12 dm tall from a branched caudex, glabrous below to hairy above; leaves entire or with only a few teeth, hairy beneath, especially along the veins; lower stem leaves cordate, upper stem leaves similar but reduced in size; heads many in a paniculate inflorescence the lowest branches of which are subtended by petiolate, short-pubescent leaves; rays 10–20, blue; achenes glabrous; infrequent in moist wooded slopes and ravines; SW.

Aster solidagineus Michx.   Narrow-leaved white-topped aster
Stems 2–6 dm tall from a short rootstock, glabrous; leaves entire, linear or narrowly oblong; inflorescence flat-topped; rays 3–6, white; disk flowers white or creamy; achenes densely hairy; pappus with some slightly club-shaped bristles; rare in dry woods, serpentine barrens, and sandy roadsides; SE; 🌿. [syn: *Sericocarpus linifolius* (L.) Britton, Stearns & Poggenb.]

Aster solidagineus

Aster spectabilis Aiton   Showy aster

Rhizomatous, stems 1–9 dm tall, densely glandular or with spreading hairs; leaves clustered toward the base, entire to shallowly and irregularly toothed, the lower ones tapering to a well-developed petiole, the upper much reduced, well-spaced, and sessile; heads few, in an open, corymbiform inflorescence; involucre glandular, the involucral bracts with loose, spreading tips; rays 15–35, violet-purple; very rare on rock outcrops and in sandy woods; SE; 🍃.

Aster tataricus L. f.   Tartarian aster

Stems 5–20 dm tall from a stout rootstock; lower leaves long-petiolate with a large, coarsely toothed blade; the middle and upper leaves smaller, becoming sessile and entire; inflorescence flat-topped; involucre 7–10 mm; rays 15–20, purple or blue, 1–2 cm long; cultivated and rarely escaped to waste ground; SE; native to Siberia.

Aster umbellatus, disk floret (×1½)

Aster undulatus (×⅙)

Aster umbellatus Mill.   Flat-topped white aster

Rhizomatous, stems 1–2 m tall, glabrous below the inflorescence; leaves all cauline, entire, elliptic, sessile or nearly so; heads numerous in a flat-topped, corymbiform inflorescence; rays white; disk flowers creamy; pappus with an inner whorl of long, club-shaped bristles and an outer whorl about 1 mm long; frequent in moist woods, floodplains, and wet fields; throughout; FACW.

Aster undulatus L.   Clasping heart-leaved aster

Stems 3–12 dm tall from a branching rootstock; plants pubescent throughout; leaves cordate or subcordate at the base, narrowed to a winged petiole with an expanded auriculate-clasping base; heads many, in a paniculate inflorescence; rays blue-violet; achenes sparsely hairy; frequent in dry woods, fields, and shaly slopes; throughout.

Aster urophyllus Lindl. in DC.   Aster

Stems from a branched caudex, 4–12 dm tall, glabrous or slightly puberulent above; lower leaves cordate, petiolate, toothed or rarely entire; midstem leaves with a winged, petiole-like base, upper leaves progressively reduced and sessile; heads numerous in a paniculate inflorescence of raceme-like branches; rays white; disk corollas white or pale yellow becoming purple; achenes glabrous; pappus bristles slightly shorter than the disk corolla; occasional in grassy roadside slopes, fields, and woods edges; mostly SE and W. [syn: A. *sagittifolius* Willd.]

Baccharis L.

Baccharis halimifolia L.   Groundsel-tree

Glabrous, glutinous shrub; leaves alternate, somewhat fleshy, coarsely toothed toward the tips; dioecious; pistillate heads with numerous filiform-tubular florets that are shorter than the conspicuous white pappus; staminate heads with shorter pappus; occasional in marshes; SE; flr. Aug–Oct; native to tidal marshes but also adventive along roadsides where deicing salt is used; FACW; 🍃.

Baccharis halimifolia

Bellis L.

Bellis perennis L.   English daisy

Low-growing, scapose perennial with solitary heads approximately 1.5–2 cm wide; leaves hairy, elliptic to ovate or orbicular and narrowed to a winged petiole; rays numerous, white or pink; achenes compressed, 2-nerved, lacking pappus; occasionally naturalized in lawns and floodplains; scattered; flr. Mar–Jul; native to Europe.

Bidens L.

Terrestrial, semi-aquatic or aquatic herbs with opposite, simple or dissected leaves; heads with ray and disk florets or lacking rays; involucre with large, firm outer bracts and a series of smaller membranous inner bracts that grade into those subtending each floret; receptacle flat or convex; disk florets perfect; achenes topped with 0–4 barbed awns.

A. leaves simple
  B. leaves sessile; ray florets present or absent
    C. outer involucral bracts spreading or reflexed; heads mostly nodding in age; anthers partially exerted
      D. outer involucral bracts seldom surpassing the disk, not leafy; rays 1.5–3 cm .... *B. laevis*
      D. outer involucral bracts surpassing the disk, leafy; rays rarely exceeding 1.5 cm, or lacking .... *B. cernua*
    C. outer involucral bracts ascending or erect; heads erect; anthers included .... *B. comosa*
  B. leaves with distinct petioles 1–4 cm long; ray florets lacking
    E. heads narrow with 7–30 florets; awns of the achenes antrorsely barbed .... *B. bidentoides*
    E. heads wider with 30–150 florets; awns of the achenes retrorsely barbed

       F. achenes 3-awned; disk corollas mostly 4-lobed, yellow ..... B. comosa

       F. achenes 4-awned; disk corollas mostly 5-lobed, orange-yellow ....

       .... B. connata

A. leaves compound

    G. rays lacking or inconspicuous (<5 mm long)

       H. leaves 2–3 times pinnate .... B. bipinnata

       H. leaves once pinnate to pinnate-pinnatifid

          I. outer involucral bracts 3–5, not ciliate .... B. discoidea

          I. outer involucral bracts 5 or more, ciliate at least near the base

             J. outer involucral bracts 5–10 .... B. frondosa

             J. outer involucral bracts 10–16 .... B. vulgata

    G. well-developed rays present (>1 cm long)

       K. achenes narrowly cuneate-oblong, 2.5–4 times as long as wide ....

       .... B. coronata

       K. achenes broader, 1.5–2 times as long as wide

          L. outer involucral bracts 10–25, longer than the inner .... B. polylepis

          L. outer involucral bracts 8–10, not longer than the inner ..... B. aristosa

Bidens aristosa (Michx.) Britton   Tickseed-sunflower

Glabrous to slightly hairy annual or biennial; leaves 1–2 times pinnate; rays about 8, 1–2.5 cm long; outer involucral bracts 8, linear, shorter than the inner; achenes flat, narrow and somewhat winged; pappus of 2–4 barbed awns or lacking; rarely naturalized on stream banks; native farther west; flr. Sep–Oct; FACW.

Bidens cernua

Bidens bidentoides

Bidens frondosa

Bidens bidentoides (Nutt.) Britton   Swamp beggar-ticks
Glabrous annual with simple, coarsely toothed leaves; heads erect, nearly cylindrical, discoid; achenes narrowly linear, usually with 2 antrorsely barbed awns;
rare on tidal shores and mudflats; SE; flr. Sep–Oct; FACW+; ✤.

Bidens bidentoides,
achene (×1)

Bidens bipinnata L.   Spanish needles
Glabrous to slightly hairy annual to 1.7 m tall; leaves 2–3 times pinnate; heads
narrow, discoid, the rays not surpassing the disk; outer involucral bracts linear,
acute, shorter than the inner; achenes 10–13 mm long, narrowed above with 3–4
awns; frequent in dry, rocky woods, shale barrens, and roadside banks; S; flr.
Aug–Oct.

Bidens cernua L.   Bur-marigold, stick-tights
Annual to 1 m tall, nearly glabrous; leaves sessile and sometimes even connate,
simple, lance-ovate to linear-ovate and more or less toothed; heads with about 8
rays, usually nodding with age; receptacular bracts blunt, yellowish; common in
swamps, wet shores, and ditches; throughout; flr. Jul–Oct; OBL.

Bidens comosa (A.Gray) Wieg.   Beggar-ticks
Similar to *B. connata* but differing in the larger, leafy outer involucral bracts,
mostly 4-lobed disk corollas, included anthers, and flat achenes with inconspicuous medial nerves and 3 awns; frequent in stream banks, pond edges, and
ditches; throughout; flr. Aug–Oct.

Bidens comosa,
achene with disk
corolla (×3)

Bidens connata Muhl.   Beggar-ticks, stick-tights
Annual to 2 m tall, mostly glabrous; leaves simple, serrate to sometimes more
deeply cleft, with distinct, but sometimes winged petioles; heads erect with 4–9
outer involucral bracts; rays inconspicuous or none, disk corollas mostly 5-
lobed; anthers slightly exerted, achenes cuneate or cuneate-obovate, com-
pressed-quadrangular with evident midribs and 4 retrorsely barbed awns; frequent in swamps, bogs, moist meadows, and along streamlets; throughout,
except at the highest elevations; flr. Aug–Oct; FACW+.

Bidens connata,
achene with disk
corolla (×3)

Bidens coronata (L.) Britton   Tickseed-sunflower
Glabrous annual or biennial to 1.5 m tall; leaves pinnately divided into 3–7
lance-linear to linear segments; heads with about 8 rays, each 1–2.5 cm long;
outer involucral bracts 6–10, linear or linear spatulate; achenes flat, 5–9 mm
long, narrowly cuneate-oblong; pappus of 2 short, strong awns; rare in bogs,
swamps, and wet ditches; SE and NW; flr. Aug–Sep; OBL.

Bidens discoidea (Torr. & A.Gray) Britton   Small beggar-ticks
Glabrous annual to 8 dm tall; leaves trifoliate; leaflets lanceolate to lance-ovate,
serrate; heads numerous, narrow, discoid; outer involucral bracts 3–5, leafy;
achenes narrowly cuneate, 3–6 mm; pappus of 2 short antrorsely hairy awns;
rare in swamps, vernal pools, and swampy ground; E and SC; flr. Aug–Oct;
FACW; ✤.

Bidens frondosa L.   Beggar-ticks, stick-tights
Mostly glabrous annual to 1.2 m tall; leaves pinnately compound; leaflets 3–5,
lanceolate, acuminate, petiolate; heads discoid, campanulate to hemispheric;

Bidens frondosa,
achene (×1)

outer involucral bracts 5–10, leafy; florets orange, stamens exerted; achenes flat, narrowly cuneate, strongly 1-nerved on each face; pappus of 2 retrorsely barbed awns; common in fields, roadsides, and moist open ground; throughout; flr. Aug–Oct; FACW.

Bidens laevis (L.) Britton, Stearns & Poggenb.   Showy bur-marigold
Glabrous annual (or perennial) to 1 m tall; similar to B. *cernua* but heads with prominent rays to 3 cm long, only occasionally nodding; receptacular bracts red-tinged at the tip; pappus awns retrorsely barbed; rare in wet meadows, shores, stream, or pond edges; SE and NW; flr. Aug–Oct; OBL; 🍂.

Bidens polylepis S.F.Blake   Tickseed-sunflower
Similar to B. *aristosa* but with 10–25 outer involucral bracts that are 12–25 mm long, twisted and curled, hispid-ciliate, and surpassing the inner; pappus short or lacking; locally abundant in moist fields, vacant lots, and roadsides; mainly SE; flr. Aug–Oct; native farther west, adventive here.

Bidens vulgata Greene   Beggar-ticks, stick-tights
Similar to B. *frondosa* but with 10–16 outer involucral bracts; florets yellow; anthers mostly included; achenes to 12 mm long; common in moist woods, wet fields, stream banks, and roadsides; throughout; flr. Aug–Oct.

Boltonia L'Hér

Boltonia asteroides (L.) L'Hér.   Aster-like boltonia
Glabrous perennial to 1.5 m tall; leaves broadly linear to lanceolate or lance-elliptic, reduced upward; heads 1.5–2.5 cm wide; rays white or pinkish; achenes winged; pappus of 2 well-developed awns; rare on rocky shores and exposed rocky river beds; SE; flr. Jul–Oct; native, but also cultivated; FACW; 🍂.

Brickellia Elliot

Brickellia eupatorioides (L.) Shinners   False boneset
Stems 3–13 dm tall; leaves alternate to subopposite, lanceolate, gland-dotted beneath; involucral bracts striate, imbricate, the outer bracts much shorter; receptacle flat, naked; heads discoid; florets creamy white; anthers separating at anthesis; achenes 10-ribbed; pappus a single series of plumose bristles; rare on shale barrens, dry slopes, and limestone barrens or prairies; S; flr. late Jul–early Sep. [syn: *Kuhnia eupatorioides* L.]

Cacalia L.

Glabrous perennial herbs with large, alternate, petiolate leaves; juice sometimes milky; heads discoid, cylindrical; florets all perfect; corollas deeply 5-lobed, white or yellowish; receptacle naked; involucre with a single row of bracts; achenes cylindrical, several-nerved; pappus of numerous capillary bristles.

A. heads 20–40-flowered; involucral bracts 10–15; leaves pinnately veined ....
....*C. suaveolens*
A. heads 5-flowered; involucral bracts 5; leaves palmately veined

Brickellia eupatorioides

Brickellia eupatorioides, achene (×1)

Boltonia asteroides

B. leaves green on both sides; stem grooved, but not glaucous ....
.... *C. muhlenbergii*
B. leaves pale and glaucous beneath; stem round or striate, glaucous ....
.... *C. atriplicifolia*

Cacalia atriplicifolia L.   Pale Indian-plantain
Similar to *C. muhlenbergii* but with the stem terete and stem and lower leaf surfaces glaucous; leaves with fewer and larger teeth or shallowly lobed; involucre
6–8 mm high; occasional in open woods, fields and moist banks; throughout;
flr. Jun–Oct.

Cacalia muhlenbergii (Sch.Bip.) Fernald   Great Indian-plantain
Stout perennial to 3 m tall; stem grooved and angled; leaves palmately veined,
lobed and toothed, green on both sides, the lowest to 8 dm wide; heads numerous in a flat-topped inflorescence, 5-flowered, narrowly cylindric; involucral
bracts 5, with a few short outer bracteoles; receptacle with a short projection in
the center; rare in woods and floodplains; S; flr. Jun–Sep; 🦋.

Cacalia suaveolens L.   Sweet-scented Indian-plantain
Stems 0.5–1.5 m tall, ribbed or grooved; the middle and lower leaves triangular
hastate, doubly serrate, and conspicuously petioled, the upper less hastate and
with winged petioles; heads in a flat-topped inflorescence; involucre 8–12 mm
high with a few loose, subulate outer bracts; receptacle flat, deeply pitted; rare in
stream banks, shaly slopes, and meadows; scattered; flr. Jul–Sep.

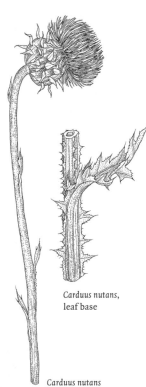

Carduus nutans,
leaf base

Carduus nutans

Carduus L.

Biennials or winter annuals, similar to *Cirsium* but with the pappus consisting of capillary (but not plumose) bristles; stems spiny-winged; florets pink to purple or rarely white; achenes quadrangular to flattened with 5–10 nerves or none.

A. heads 1.5–4.5 cm in diameter, nodding; peduncles naked .... *C. nutans*
A. heads 1.2–1.6 cm in diameter, erect; peduncles spiny-winged ....
.... *C. acanthoides*

Carduus acanthoides L.   Thistle
Plant to 1.5 m tall, very strongly spiny; leaves deeply lobed or pinnatifid, loosely villous beneath; heads clustered or solitary at the ends of the branches; peduncles winged; involucre 1.4–2 cm high, outer and middle bracts spine-tipped; occasional on ballast, roadsides, stream banks, and urban waste ground; mostly E; flr. Jun–Oct; native to Europe.

Carduus nutans L.   Nodding or musk thistle
Plant to 2 m tall; leaves deeply lobed, glabrous or villous along the midvein beneath; heads solitary and nodding on naked peduncles; common in pastures, roadsides, waste ground and ballast; mostly SE; flr. May–Aug; native to Europe; designated as a noxious weed in Pennsylvania.

Centaurea L.

Grayish-puberulent or tomentose herbs with entire or pinnatifid, alternate, or basal leaves; florets all disk type, perfect, or with the peripheral ones sterile and enlarged but still essentially regular; involucral bracts imbricate, spine-tipped or with an enlarged lacerate or pectinate tip; receptacle nearly flat, densely bristly; anthers with an apical appendage; style with a thickened hairy ring; achenes obliquely or laterally attached to the receptacle; pappus much reduced or absent.

A. tips of the involucral bracts lacerate or fringed, not spiny
  B. leaves pinnatifid with narrow lobes .... *C. maculosa*
  B. leaves entire or toothed, or with a few lobes on some of the larger ones
    C. leaves linear, <1 cm wide .... *C. cyanus*
    C. leaves broader, at least some of the lower ones >1 cm wide
      D. tips of the involucral bracts blackish, the middle and outer ones pectinate but not bifid
        E. involucral tips 4–6 mm long; marginal florets not enlarged ....
        .... *C. nigra*
        E. involucral tips 1–3 mm long; marginal florets enlarged and ray-like .... *C. nigrescens*
      D. tips of the involucral bracts tan to dark brown, the middle and outer ones irregularly lacerate and the inner ones distinctly bifid....*C. jacea*
A. involucral bracts spine-tipped
  F. stems angled but not winged .... *C. calcitrapa*
  F. stems winged by the decurrent leaf bases .... *C. solstitialis*

Centaurea calcitrapa L.   Purple star-thistle
Branching biennial to 8 dm tall; leaves pinnatifid with narrow lobes or the upper entire; heads numerous; involucre 12–18 mm, the bracts tipped with 1–3-cm-

long spines; florets few, purple; pappus none; ballast and waste ground; SE; flr. May–Aug; native to Europe.

Centaurea cyanus L.   Bachelor's buttons, cornflower
Annual or winter annual to 1.2 dm tall; leaves linear, entire or with a few teeth or narrow lobes; heads terminal, involucre 11–16 mm the bracts striate with dark, pectinate, or lacerate tips; florets blue (pink, purple, or white), the marginal ones enlarged; cultivated and occasionally spreading to roadsides and fields; throughout, also a prominent component of seed mixtures for meadow establishment; flr. May–Sep; native to Europe.

Centaurea jacea L.   Brown knapweed
Similar to *C. nigra* in habit; tips of the involucral bracts tan to dark brown, the outer ones irregularly lacerate, the inner bifid; marginal florets enlarged; pappus none; naturalized in roadsides, fields, and woods; scattered; flr. Jun–Sep; native to Europe.

Centaurea jacea, involucral bract (×1½)

Centaurea maculosa Lam.   Bushy or spotted knapweed
Biennial or short-lived perennial to 1.5 m tall; leaves pinnatifid with narrow lobes or entire in the inflorescence; head terminal, involucre 10–13 mm high, constricted upward, the bracts with short, dark, pectinate tips; florets pink-purple; commonly naturalized in dry woods, fields, roadsides, and shale barrens; throughout; flr. Jun–Oct; native to Europe.

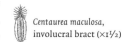

Centaurea maculosa, involucral bract (×1½)

Centaurea nigra L.   Black knapweed, Spanish buttons
Perennial; leaves entire or toothed, the basal ones broadly oblanceolate or elliptic, cauline leaves reduced upward; involucre 12–19 mm, broader than high; tips of the involucral bracts conspicuously blackish, 4–6 mm long, the middle and outer ones regularly pectinate; florets pink-purple, marginal ones not enlarged; occasional in dry, open fields, railroad banks, roadsides, and waste ground; throughout; flr. Jun–Oct; native to Europe.

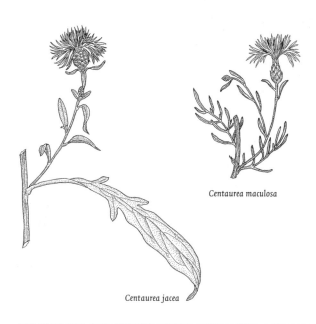

Centaurea maculosa

Centaurea jacea

Centaurea
nigrescens,
involucral bract
(×1½)

Centaurea nigrescens Willd.   Knapweed
Similar to *C. nigra* but the involucre higher than broad; tips of the involucral bracts 1–3 mm; marginal florets enlarged and ray-like; infrequent in fields and roadsides; mostly S; flr. Jul–Oct; native to Europe.

Centaurea solstitialis L   Barnaby's or yellow star-thistle
Annual or biennial to 8 dm with the stem winged by the decurrent leaf bases; basal leaves lyrate or pinnatifid, the upper becoming linear and entire; involucral bracts spine-tipped; florets yellow; rarely naturalized in cultivated fields, fencerows, and gardens; SE; flr. Jun–Oct; believed to have arrived in Delaware Co. in 1849 in seed of *Medicago sativa* imported from Europe.

Chondrilla L.

Chondrilla juncea,
head (×1)

Chondrilla juncea,
achene (×2)

Chondrilla juncea L.   Skeleton-weed; gum-succory
Biennial or perennial, stem to 1.5 m tall with stiff hairs below; basal leaves pinnately divided, stem leaves linear, entire; heads small, scattered along the branches; florets yellow, about 11 per head; involucral bracts white-woolly to glabrous; achenes spiny toward the tip; rarely naturalized in fields, shale barrens, and mine waste; SE and SC; flr. Jul–Sep; native to Europe.

Chrysanthemum L.

Herbs with alternate, pinnatifid leaves; heads radiate; involucral bracts with dry margins or tips; receptacle flat, naked.

A. heads few, receptacle 1–2.5 cm wide .... *C. leucanthemum*
A. heads many, receptacles 4–9 mm wide .... *C. parthenium*

Chrysanthemum leucanthemum L.   Ox-eye daisy
Rhizomatous perennial, stems 3–8 dm tall, simple and glabrous for the most part; leaves alternate, crenate and often lobed or cleft, reduced upward; heads large, solitary at the ends of long, naked peduncles; rays 15–35, white; disk florets yellow; achenes terete, 10-ribbed; common in fields, woods, meadows, and roadsides; throughout; flr. May–Oct; native to Europe.

Chrysanthemum parthenium (L.) Bernh.   Feverfew
Perennial, stems 3–8 dm tall; leaves mainly cauline, pinnatifid and often incised again, hairy beneath; heads several or many in a corymbiform inflorescence; rays 10–20, white; cultivated and occasionally escaped to roadsides and waste ground; mostly S; flr. Jun–Sep; native to Europe.

Chrysogonum L.

Chrysogonum virginianum L.   Green-and-gold
Fibrous-rooted, hairy perennial with glandular, semi-erect stems to 5 dm tall; leaves opposite, ovate to suborbicular, long-petiolate, crenate, 2.5–10 cm by 1.5–6 cm; involucral bracts of 2 series the inner subtending the rays and fusing with adjacent receptacular bracts to partially enclose each achene; heads solitary or few on terminal or axillary peduncles; rays about 5, yellow, 7–15 mm; rare in open woods on limestone; SC; flr. May–Aug; ❦.

Chrysopsis (Nutt.) Elliott

Chrysopsis mariana (L.) Elliot   Golden aster

Fibrous-rooted perennial from a short woody rhizome; stems 3–8 dm tall with long flexuous hairs when young, later becoming glabrate; basal leaves oblanceolate to ovate, petiolate, the upper smaller and sessile; heads in a crowded inflorescence; involucre and peduncles stipitate-glandular, the involucre 7–10 mm; rays 13–21, yellow, about 1 cm long; achenes obovate, 3–5-nerved; pappus double, with inner capillary bristles and short, coarse outer bristles or scales; rare in dry sandy woods, clearings, roadside banks, and serpentine barrens; SE; flr. Jul–Oct; UPL.

Chrysopsis mariana, achene (×1½)

Cichorium L.

Cichorium intybus L.   Blue chicory; blue-sailors

Herbaceous perennial 6–9 dm tall, glabrous or stiff-hairy; leaves pinnately lobed or divided, reduced upward; heads sessile or terminal on stiff, spreading branches, 3.8 cm wide; florets blue (or occasionally white); common in fields, roadsides, and waste ground; throughout; flr. Jun-Sep; native to Europe, designated as a noxious weed in Pennsylvania.

Cichorium intybus, achene (×5)

Chrysogonum virginianum

Cichorium intybus

Chrysopsis mariana

Cirsium Mill.

Heads discoid; involucral bracts imbricate, at least the outer spine-tipped (except *C. muticum*); receptacle flat to subconic, densely bristly; florets perfect or unisexual (in *C. arvense*); corolla purple to yellowish or white; anthers with a narrow apical appendage; style with a thickened hairy ring; achenes glabrous, nerveless, thick-compressed; pappus of numerous plumose bristles.

A. involucre 1–2 cm high; heads numerous, in an open inflorescence; plants colonial ....*C. arvense*
A. involucre >2.5 cm high; heads fewer; plants not colonial
    B. none of the involucral bracts spine-tipped ....*C. muticum*
    B. at least the outer involucral bracts spine-tipped
        C. stem winged by virtue of decurrent leaf bases ....*C. vulgare*
        C. stem not winged
            D. heads surrounded by narrow, spiny-toothed leaves; corollas yellow...
            .... *C. horridulum*
            D. heads not surrounded by spiny-toothed leaves although they may have 1–2 reduced subtending leaves; florets purple
                E. leaves persistently white tomentose beneath
                    F. leaves toothed or shallowly lobed ....*C. altissimum*
                    F. leaves deeply pinnatifid .... *C. discolor*
                E. leaves eventually glabrate although they may be somewhat tomentose early .... *C. pumilum*

Cirsium altissimum (L.) Spreng.   Tall thistle
Perennial to 3 m tall, stem spreading hirsute to glabrate; leaves coarsely spiny-toothed or shallowly lobed, densely white tomentose beneath; heads several or numerous; involucre 2.5–3.5 cm high, outer bracts spiny-tipped; florets pink-purple; frequent in woods, riverbanks, fields, and roadsides; throughout; flr. Jul–Sep.

Cirsium arvense (L.) Scop.   Canada thistle
Colonial perennial; leaves glabrous or more or less white tomentose beneath; heads numerous in an open, branched inflorescence, unisexual or nearly so; the plants polygamo-dioecious; involucre 1–2 cm high; corollas pink-purple, longer than the pappus in staminate heads, shorter than the pappus in pistillate heads; common in fields, pastures, roadsides, and waste ground; throughout; flr. Jun–Sep; native to Eurasia; designated as a noxious weed in Pennsylvania; FACU.

Cirsium discolor,
achene

Cirsium discolor (Muhl.) Spreng.   Field thistle
Similar to *C. altissimum* but plants shorter (to 2 m tall) and leaves deeply pinnatifid and very spiny; an occasional native of abandoned fields, open hillsides, and roadside banks; scattered but especially SE and SC; flr. Aug–Oct; UPL.

Cirsium horridulum Michx.   Yellow or horrible thistle
Stout biennial to 1.5 m tall; leaves pinnatifid, strongly spiny, thinly hairy to glabrate; heads several or solitary, each surrounded by spiny reduced leaves; involucre 3–5 cm high, the outer bracts spiny; corollas pale yellow, white, or lavender; rare in moist, sandy or peaty meadows; SE; flr. May–Jul; FACU–; 🌿.

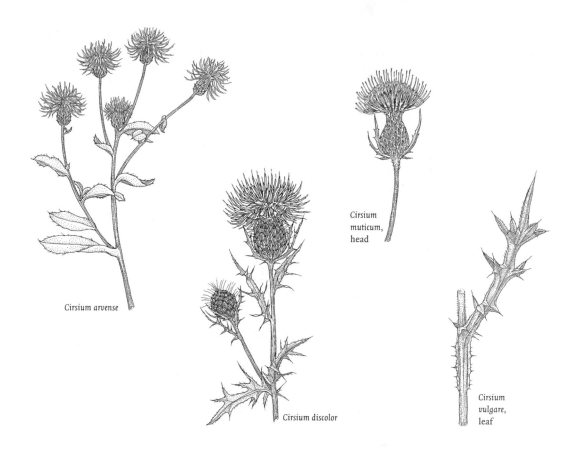

Cirsium arvense

Cirsium muticum, head

Cirsium discolor

Cirsium vulgare, leaf

Cirsium muticum Michx.   Swamp thistle
Coarse biennial to 2 m tall; stem glabrous or with spreading hairs; leaves deeply pinnatifid, only weakly spiny, thinly tomentose to subglabrous beneath; heads few to many; involucre 2–3.5 cm high, bracts glutinous, lacking spines; corolla pink to purple; frequent in swamps, bogs, stream banks, and wet meadows; throughout, except at the highest elevations; flr. Jul–Sep; OBL.

Cirsium pumilum (Nutt.) Spreng.   Pasture thistle
Stout biennial to 8 dm tall; leaves lobed or pinnatifid with numerous spines; leaves green, coarsely villous or short-hirsute; heads few or solitary; involucre 3.5–5 cm high, the outer bracts spine-tipped; corollas purple (or occasionally white); frequent in dry fields, shaly hillsides, sandy floodplains, woods, and roadsides; throughout; flr. Jun–Sep.

Cirsium vulgare (Savi) Ten.   Bull thistle
Coarse biennial to 1.5 m tall; stem conspicuously spiny-winged from the decurrent leaf bases; leaves pinnatifid, the large ones with the lobes coarsely toothed or lobed again, very spiny, thinly white-tomentose to green beneath; heads several; involucre 2.5–4 cm high, the bracts all spine-tipped; florets purple; common in pastures, meadows, and roadsides; throughout; flr. Jul–Oct; native to Eurasia; designated as a noxious weed in Pennsylvania; FACU–.

Conyza canadensis (×⅛)

Conyza Less.

Conyza canadensis (L.) Cronquist   Horseweed

Erect pubescent annual to 1.5 m tall, branching only in the inflorescence (or simple), with numerous alternate leaves and many small heads; involucre 3–4 mm high, the bracts strongly imbricate, brownish; rays white, 0.5–1.0 mm; common in fields, roadsides, railroad tracks, and waste ground; throughout; flr. Jul–Oct; UPL. Most of our material is var. *canadensis*; var. *pusilla* (Nutt.) Cronquist, which is smaller and has a glabrous stem, occurs in the extreme SE.

Coreopsis L.

Herbs with opposite, entire to highly dissected leaves; rays approximately 8, conspicuous, yellow or pink; involucral bracts of 2 types all joined at the base, the outer narrow and herbaceous, the inner more membranous; receptacle flat or slightly convex, chaffy; style branches flattened, with a hairy appendage; achenes flattened parallel to the subtending bracts, usually winged; pappus of 2 awns or teeth, or lacking.

A. rays pink .... *C. rosea*
A. rays yellow
    B. rays with a reddish blotch at the base; achenes not winged .... *C. tinctoria*
    B. rays without a reddish blotch; achenes narrowly winged
        C. leaves simple or pinnatifid .... *C. lanceolata*
        C. leaves ternately or palmately lobed or compound .... *C. tripteris*

Coreopsis lanceolata, achene (×2½)

Coreopsis lanceolata L.   Longstalk tickseed

Glabrous or villous perennial, 2–6 dm tall; leaves mostly on the lower stem, lance-linear to spatulate, simple or with a few lateral lobes; heads few on long naked peduncles; involucre with 8–10 outer bracts; disk yellow, 1–2 cm wide; rays 1.5–3 cm; achenes 2–3 mm with thin, flat wings; cultivated and frequently escaped, also included in meadow seed mixtures; mostly S; flr. May–Sep; native farther west; FACU.

Coreopsis rosea Nutt.   Pink tickseed

Glabrous, rhizomatous perennial to 6 dm tall; leaves linear or occasionally slightly lobed; heads short-pedunculate, disk florets yellow; rays pink, 1 cm; very rare in moist, open sandy soil; SE; believed to be extirpated; flr. Aug; FACW; .

Coreopsis tinctoria, achene (×5)

Coreopsis tinctoria Nutt.   Plains tickseed

Glabrous annual, 4–12 dm tall; leaves subsessile or short petiolate, once or twice pinnatifid into linear or lance-linear segments; heads numerous, disk florets red-purple; rays yellow with a reddish blotch near the base; widely cultivated and occasionally escaped to yards and stream banks; S; flr. Jun–Sep; native to the Great Plains; FAC–.

Coreopsis tripteris, achene (×2½)

Coreopsis tripteris L.   Tall tickseed

Erect, simple perennial to 3 m tall; stems glabrous and glaucous; leaves mainly cauline, mostly trifoliate (the terminal leaflet sometimes again divided), petiolate; heads several to many; disk florets red (yellow at the base); rays yellow, 1–2.5 cm; achenes obovate, 4–7 mm, flattened and narrowly winged; pappus of a few erect bristles, or none; occasional in old fields, thickets, woods edges, and roadsides; mostly W; flr. May–Sep; FAC.

Cosmos Cav.

Herbs with opposite, highly dissected leaves; involucral bracts of 2 types, the outer herbaceous, the inner membranous or almost hyaline; receptacle flat, chaffy; rays white, pink, or orange; achenes quadrangular, linear, beaked; pappus of 2–8 nonpersistent, barbed awns.

A. rays white or pink; ultimate leaf segments 1 mm wide or less ..... *C. bipinnatus*
A. rays orange-yellow or orange-red; ultimate leaf segments 1 mm wide............
.... *C. sulphureus*

Cosmos bipinnatus Cav. Cosmos
Glabrous annual to 2 m with leaves finely dissected into filiform segments; heads numerous; receptacle 1–1.5 mm wide; rays about 8, 1.5–4 cm long, pink or white; achenes 7–16 mm, beaked; pappus of 2–3 short awns or none; cultivated and occasionally escaped to fields and roadsides, also frequently a component of wildflower seed mixes; mostly SE; flr. Jul–Nov; native to Mexico; FACU–.

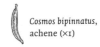

Cosmos bipinnatus, achene (×1)

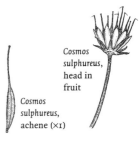

Cosmos sulphureus, head in fruit

Cosmos sulphureus, achene (×1)

Cosmos sulphureus Cav. Orange cosmos
Similar to *C. bipinnatus* but with broader leaf segments and orange-yellow or orange-red rays; cultivated and occasionally escaped; SE; flr. Jul–Sep; native to tropical America.

Coreopsis lanceolata, underside of head

Coreopsis tinctoria

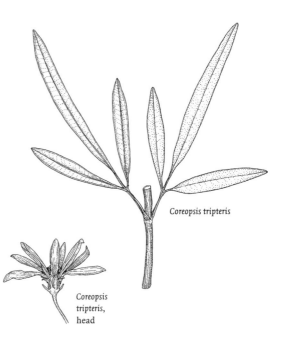

Coreopsis tripteris

Coreopsis tripteris, head

Crepis L.

Tap-rooted annuals or biennials; leaves basal and alternate, the basal ones variously denticulate to pinnatifid, pinnate, or lyrate; cauline leaves often becoming lanceolate or linear, entire, and sessile or even auriculate-clasping; heads several to many in a simple or compound corymb; only ray florets present; the corollas yellow, sometimes reddish or purplish on the lower surface; receptacles glabrous or with ciliate pits; achenes fusiform, narrowed at the top or beaked; pappus of whitish, capillary bristles.

A. involucre and upper stems with stiff, glandular hairs .... *C. capillaris*
A. involucre and upper stem variously hairy or glabrous, but not glandular
    B. involucre and upper stems with pale, stiff, nonglandular hairs with swollen bases .... *C. setosa*
    B. involucre variously pubescent or glabrous, but lacking stiff, spreading hairs
        C. achenes narrowed at the top but not beaked .... *C. tectorum*
        C. achenes distinctly beaked .... *C. vesicaria*

Crepis capillaris (L.) Wallr.   Hawk's-beard
Annual or biennial, 2–10 dm tall, with several to many stems; leaves glabrous or with a few nonglandular hairs, the upper narrowed to the base; heads numerous; the involucres and peduncles usually with stiff, blackish, glandular hairs; receptacles with ciliate pits; achenes not distinctly beaked; occasionally naturalized in fields, woods, lawns, roadsides, ballast, and waste ground; scattered; flr. Jun–Oct; native to Europe.

Crepis setosa Haller f.   Hawk's-beard
Annual, stems 1–8 dm tall, with pale, stiff, nonglandular hairs; cauline leaves auriculate-clasping; heads several; achenes narrowed to a slender beak; fields and waste ground at a few scattered sites; S; flr. Jul; native to Europe.

Crepis tectorum L.   Hawk's-beard
Annual, 1–10 dm tall; leaves glabrous or puberulent, the cauline ones merely sessile; heads many; achenes narrowed above but not distinctly beaked; ballast and waste ground, SE; flr. Jun–Jul; native to Eurasia.

Crepis vesicaria L.   Hawk's-beard
Annual, biennial, or short-lived perennial; stems to 1.5 m tall, much branched; leaves glabrous or pubescent, the upper auriculate-clasping and becoming

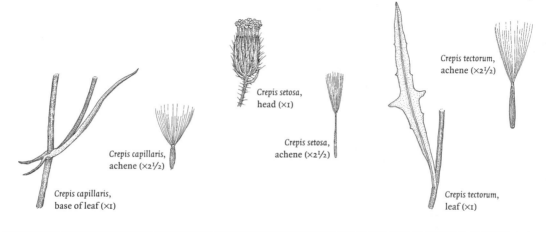

Crepis setosa, head (×1)

Crepis capillaris, achene (×2½)

Crepis setosa, achene (×2½)

Crepis tectorum, achene (×2½)

Crepis capillaris, base of leaf (×1)

Crepis tectorum, leaf (×1)

bract-like; heads many; the inner achenes always distinctly beaked; roadsides; NW; flr. Jun–Jul; native to Europe.

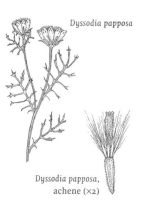

Dyssodia papposa

Dyssodia papposa, achene (×2)

Dyssodia Cav.

Dyssodia papposa (Vent.) A.Hitchc.   Stinking-marigold
Branched annual to 4 dm tall; strongly and unpleasantly aromatic; leaves opposite, pinnatifid or pinnate, glandular punctate; heads numerous, involucral bracts with conspicuous oval glands; ray florets few, inconspicuous; achenes narrow, striate; pappus of 10–20 scales each divided into numerous long, slender bristles; roadsides in a few scattered sites; flr. Jul–Oct; native to western North America.

Echinacea L.

Perennial herbs with simple, alternate, triple-nerved leaves; heads solitary or few, on long peduncles; involucral bracts overlapping, with reflexed green tips; receptacle conical; rays approximately 13, purple; achenes 4-sided, glabrous or slightly hairy on the angles; pappus a short toothed crown.

A. leaves pubescent on both sides .... E. purpurea
A. leaves glabrous on both sides or somewhat short-hairy above ...... E. laevigata

Echinacea laevigata (F.E.Boynton & Beadle) Blake   Smooth purple coneflower
Very similar to E. purpurea, but the stem simple and leaves mostly glabrous and more or less glaucous; open woods and fields, known from a single site in the SE; believed to be extirpated; flr. Jun–Jul; ✿.

Echinacea purpurea (L.) Moench   Purple coneflower
Perennial, stems to 1.8 m tall, simple or few-branched; leaves hairy above and below; cultivated and rarely escaped to fields and waste ground; SE; flr. Jul–Oct; native to the midwestern U.S.

Echinops L.

Echinops sphaerocephalus L.   Globe-thistle
Coarse, branching perennial to 2.5 m tall; leaves pinnatifid, sessile and clasping but not decurrent, densely white-tomentose beneath and with gland-tipped hairs; heads spherical; florets pale bluish; cultivated and occasionally escaped to crevices in limestone rock; scattered; flr. Jul–Aug; native to Eurasia.

Eclipta L.

Eclipta prostrata (L.) L.   Yerba-de-tajo
Weak or spreading annual often rooting at the nodes; leaves opposite, lance-linear to lance-elliptic, 2–10 cm long, serrate; heads in small terminal or axillary

Eclipta prostrata, achene (×2½)

Eclipta prostrata

clusters, minute white rays present; receptacle bearing slender, bristle-like bracts between the florets; achenes 2–2.5 mm, 4-sided, rugose or warty, bearing a small crown of pappus scales; occasional on wet shores and riverbanks; SE and SW; flr. Aug–Oct; FAC.

Elephantopus L.

Elephantopus carolinianus Willd.   Elephant's foot

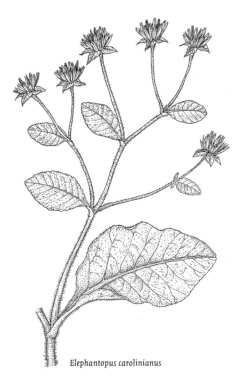

Elephantopus carolinianus, achene (×2)

Perennial herb, 3–10 dm tall, glandular; leaves mostly cauline, broadly elliptic to obovate, 9–25 cm long and 3–10 cm wide; heads with disk florets only, 1–5-flowered, in small glomerules subtended by several leafy bracts; involucre consisting of 4 pairs of bracts, the outer shorter; florets all perfect, purple or whitish; achene pubescent, ribbed; pappus bristles tapering to a triangular base; rare in open woods and serpentine barrens; SE; flr. Aug–Oct; FACU; ❦.

Erechtites Raf.

Erechtites hieracifolia (L.) Raf. ex DC.   Fireweed, pilewort

Erect, unbranched annual to 2.5 m tall; stems grooved or ribbed, glabrous to sparsely hairy; leaves alternate, tapering to a short petiole, the middle and upper often auriculate-clasping; leaf margins sharply serrate to pinnately dissected, the teeth with white callus tips; heads discoid, cylindric to ovoid; involucre 1–1.5 cm high, a single row of equal bracts and a few tiny bracteoles at the base; receptacle flat, naked; florets barely longer than the closed involucre, whitish; pappus white, conspicuous when exposed by the open involucre; achenes 5-angled, with appressed hairs; common in fields, woods, clearings, and disturbed ground; throughout; flr. Aug–Oct; FACU.

Elephantopus carolinianus

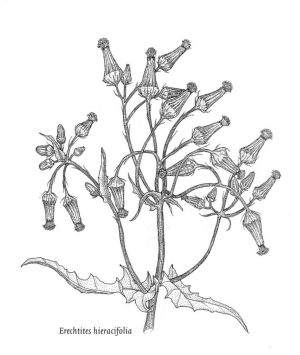

Erechtites hieracifolia

Erigeron L.

Herbs with alternate leaves and solitary to numerous heads; receptacle flat, naked; involucral bracts narrow and evidently imbricate; disk florets numerous, yellow, perfect and fertile; ray florets 50–150, rays not more than 1 mm wide, white or pinkish; achenes 2–4-nerved with a pappus of capillary bristles.

A. stem leaves tapered at the base, not clasping; pappus of the ray florets reduced
    B. leaves numerous, broad, coarsely toothed or serrate ....E. annuus
    B. leaves few, narrow, entire or with a few low teeth .... E. strigosus
A. stem leaves rounded or somewhat clasping at the base; all florets with a well-developed pappus of long bristles
    C. rays >100, 0.5 mm wide or less ....E. philadelphicus
    C. rays about 50, at least 1 mm wide .... E. pulchellus

Erigeron annuus (L.) Pers.   Daisy fleabane
Annual or biennial, stems 6–15 dm tall with coarse, spreading hairs; leaves numerous, large, and coarsely toothed or cleft; heads 10–14 mm wide, few to many in corymbiform clusters; ray florets 40–100, white (or pale lavender); pappus of the disk florets of slender bristles and shorter scales, that of the ray florets scales only; common in fields, roadsides, and waste ground; throughout; flr. May–Nov; FACU.

Erigeron philadelphicus L.   Daisy fleabane
Short-lived perennial with short basal rhizomes; stems 2–10 dm tall; basal leaves oblong to narrowly obovate, upper stem leaves sessile and slightly clasping; heads few to many, nodding in bud but erect in flower, 1.5–2.5 cm wide; ray florets 100–150, white to pale lavender; pappus of simple bristles; common in woods edges, fields, roadsides, and lawns; throughout; flr. May–Aug; FACU.

Erigeron philadelphicus, head

Erigeron pulchellus Michx.   Robin's-plantain
Rhizomatous perennial to 4 dm tall; leaves ovate to lanceolate with somewhat clasping bases, decreasing greatly upward; heads 2.5–4 cm wide; ray florets about 50, bluish, pinkish, or white; frequent in meadows, wooded slopes, woodland edges, and roadsides; throughout; flr. May–Sep; FACU.

Erigeron pulchellus, head

Erigeron strigosus Muhl. ex Willd.   Daisy fleabane, whitetop
Annual or biennial, stems 3–10 dm tall, with fine appressed hairs; leaves few, mostly oblanceolate to elliptic, entire or with a few low teeth; heads few to many, with 50–100 white (or bluish or pinkish) rays; pappus of fine hairs or shorter scales; common in fields, fencerows, and dry grasslands; throughout; flr. May–Oct; FACU+.

Erigeron strigosus, leaf

Erigeron philadelphicus, leaf

Eupatorium L.

Perennial herbs with opposite or whorled leaves; most species more or less glandular throughout; heads small in a corymbiform inflorescence, discoid; florets all perfect; receptacle naked; achenes 5–8-angled, glabrous or slightly hairy along the veins; pappus a single series of capillary bristles.

A. leaves in whorls of 3–7, >2 cm wide; florets pale to dark pink or purplish (Joe-pye-weeds)
    B. leaves triple-nerved due to an elongate pair of basal lateral veins; blade abruptly narrowed to the petiole .... E. dubium
    B. leaves not triple-nerved; blade gradually narrowed to the petiole
        C. inflorescence, or its divisions, flat-topped; florets 9–22 per head ....
        .... E. maculatum
        C. inflorescence rounded or strongly convex, florets 4–8 per head
            D. stems purple throughout, hollow with a large central cavity, glaucous .... E. fistulosum
            D. stems purple only at the nodes, solid or with only a slender cavity, not glaucous .... E. purpureum
A. leaves opposite, alternate above, or in whorls of 3–4, but if so then <1 cm wide; florets white or blue
    E. florets blue.. E. coelestinum
    E. florets white
        F. florets >9 per head
            G. leaves sessile, connate-perfoliate .... E. perfoliatum
            G. leaves petiolate, not connate
                H. involucral bracts not strongly imbricate, subequal, although there may be a few small outer ones
                    I. leaves thick and firm, 3–7 cm long, crenate to crenate-serrate .... E. aromaticum
                    I. leaves thin, 6–18 cm long, sharply serrate .... E. rugosum
                H. involucral bracts imbricate, in 3 series, the outer less than half as long as the inner .... E. serotinum
        F. florets 3–7 per head
            J. leaves mostly <1 cm wide
                K. leaves in whorls of 3–4; involucral bracts broadly rounded to acute .... E. hyssopifolium
                K. leaves opposite; involucral bracts long-acuminate, mucronate ... . E. leucolepis
        J. leaves >1 cm wide
            L. involucral bracts acuminate to attenuate .... E. album
            L. involucral bracts rounded to acute
                M. leaves tapering to a narrow base .... E. altissimum
                M. leaves rounded, subcordate, subtruncate or broadly cuneate at the base
                    N. plants glabrous below the inflorescence; leaves acuminate or narrowly acute .... E. sessilifolium
                    N. plants hairy throughout
                        O. leaves strictly pinnately veined .... E. godfreyanum
                        O. lowest lateral veins prominent giving the leaf a triple-veined appearance

P. leaves broadly ovate, evenly toothed; upper leaves and main branches of the inflorescence opposite .... E. rotundifolium

P. leaves lanceolate, lance-ovate or elliptic-ovate, coarsely and irregularly toothed; upper leaves and main inflorescence branches mostly alternate .... E. pilosum

Eupatorium album L. White-bracted eupatorium
Stems 4–8 dm tall, pubescent; leaves opposite, petiolate to subsessile; blades elliptic to lanceolate, the lowest lateral veins elongate and becoming parallel to the midvein; involucral bracts long-acuminate with white tips and margins; florets white, 5 per head; rare in sandy, open woods, dry slopes, and serpentine barrens; SE; believed to be extirpated; flr. Aug–Oct; ❧.

Eupatorium altissimum L. Tall eupatorium
Stems to 2 m tall, with soft spreading hairs or glabrate below; leaves opposite, sessile or short-petioled, blade with a tapering base and acuminate tip; florets white, 5 per head; occasional on dry, rocky slopes, bluffs, fields, and roadsides, usually on calcareous soils; S; flr. Aug–Nov.

Eupatorium aromaticum L. Small-leaved white-snakeroot
Stems to 8 dm tall, pubescent throughout; leaf blades thick and firm, 3–7 cm long, crenate or crenate-serrate; florets white, 10–19 per head; rare in dry woods and sandy, open areas; mostly SE; flr. Aug–Oct; ❧.

Eupatorium coelestinum L. Mistflower, wild ageratum
Stems 3–9 dm tall; rhizomatous, glandular throughout; leaves opposite, deltoid-ovate, petiolate; florets blue, 35–70 per head; rare in old fields, meadows, and stream banks; S; native, but also cultivated and occasionally escaped; flr. Aug–Oct; FAC; ❧.

Eupatorium dubium Willd. ex Poir. Joe-pye-weed
Stems 4–10 dm tall, purple spotted; leaves in 3s or 4s, strongly triple-veined; blade thick and firm, rugose, abruptly narrowed to a short petiole; inflorescence slightly to strongly convex; florets purple, 5–8 per head; rare in moist, usually sandy, acidic soil of swamps, bogs, marshes, and swales; SE; flr. late Jul–early Oct; FACW; ❧.

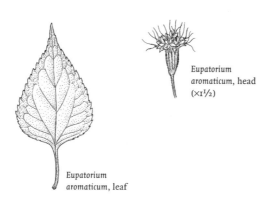

Eupatorium
aromaticum, head
(×1½)

Eupatorium
aromaticum, leaf

Eupatorium fistulosum Barratt   Joe-pye-weed

Stems to 3 m tall, purple throughout and strongly glaucous, hollow below the inflorescence with a large central cavity; leaves in whorls of 4–7; inflorescence strongly convex; florets pink-purple, 5–7 per head; common in floodplains, meadows, moist thickets, and roadsides; throughout; flr. Jul–Oct; FACW.

Eupatorium godfreyanum Cronquist   Eupatorium

Similar to E. *sessilifolium* but stems conspicuously hairy; leaves puberulent, blades lance-ovate, 7–11 cm long, acute to short-acuminate; rare in stream banks, wooded roadsides, and shaly slopes along the lower Susquehanna River; flr. Aug–Oct; ✹. [E. *vaseyi* Porter of Rhoads and Klein, 1993]

Eupatorium hyssopifolium L.   Hyssop-leaved eupatorium

Stems 3–12 dm tall, the entire plant grayish-pubescent; leaves linear-lanceolate, <1 cm wide, in whorls of 3–4 or occasionally opposite or even alternate above, usually with small leafy shoots in the axils; florets white, 5 per head; occasional in dry, sandy or gravelly fields, roadsides, and railroad rights-of-way; SE; flr. Jul–Oct.

Eupatorium leucolepis (DC.) Torr. & A.Gray   White-bracted throughwort

Stems 4–10 dm tall; leaves opposite, linear-oblong to oblanceolate, no more than 1 cm wide, sessile; florets white, 5 per head; very rare in moist gravel pits; SE; believed to be extirpated; flr. Aug; FACW+; ✹.

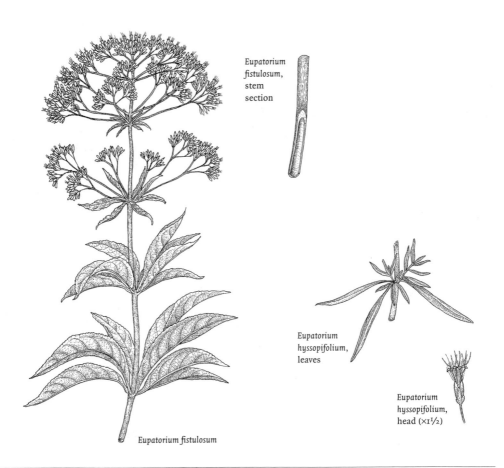

Eupatorium
fistulosum,
stem
section

Eupatorium
hyssopifolium,
leaves

Eupatorium
hyssopifolium,
head (×1½)

Eupatorium fistulosum

Eupatorium maculatum L.   Spotted joe-pye-weed
Stems to 2 m tall, purple spotted to purple throughout, rarely glaucous, solid or
occasionally hollow near the base in age; leaves in whorls of 4–5; inflorescence,
or its divisions, flat-topped; florets purplish, 9–20 per head; common in floodplains, swamps, and alluvial thickets; mostly N; flr. Jul–Sep; FACW.

Eupatorium perfoliatum L.   Boneset
Stems 4–15 dm tall, hairy throughout; leaves opposite, or occasionally 3 per
node, conspicuously connate-perfoliate; inflorescence flat-topped; florets white,
9–23 per head; common on floodplains, swamps, bogs, stream banks, and wet
meadows; throughout; flr. Jul–Oct; FACW+.

Eupatorium pilosum Walt.   Ragged eupatorium
Very similar to *E. rotundifolium* but the leaves narrower and more coarsely
toothed; lower branches of the inflorescence alternate; occasional in moist
woods, sphagnum bogs, or sandy, peaty openings; SE; flr. Jul–Oct; FACW. [syn:
E. rotundifolium L. var *saundersii* (Porter) Cronquist]

Eupatorium purpureum L.   Joe-pye-weed
Stems to 2 m tall, dark purple at the nodes otherwise green, solid or with a slender central cavity toward the base; leaves in whorls of 3–4; inflorescence convex;
florets pale pink or purplish; common in open woods, fields, and floodplains,
occurs in drier, more shaded habitats than our other joe-pye-weeds; mostly S;
flr. Jul–Oct; FAC.

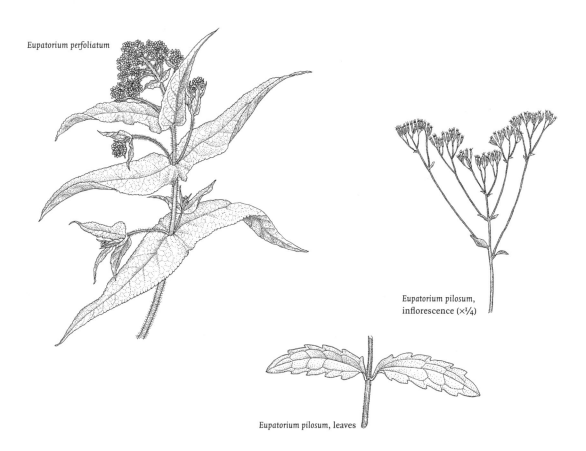

Eupatorium perfoliatum

Eupatorium pilosum,
inflorescence (×¼)

Eupatorium pilosum, leaves

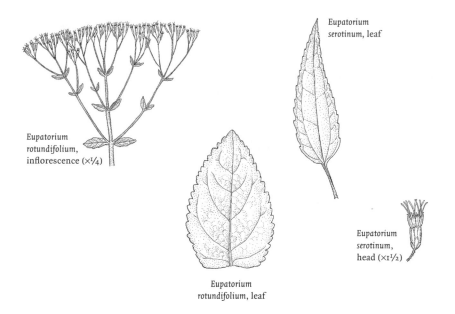

Eupatorium
serotinum, leaf

Eupatorium
rotundifolium,
inflorescence (×¼)

Eupatorium
serotinum,
head (×1½)

Eupatorium
rotundifolium, leaf

Eupatorium rotundifolium L.   Round-leaved eupatorium
Stems 4–12 dm tall, pubescent; leaves sessile, ovate, with obtuse or acute tips
and rounded or truncate base, the 2 lowest lateral veins prominent; lowest
branches of the inflorescence opposite; involucral bracts obtuse to acute with
white tips and margins; florets white, 5 per head; rare in sandy or clayey fields
and open thickets; mostly SE; flr. late Jun–Oct; FAC–; ❧. This species intergrades
with E. pilosum and may not always be separable, polyploidy and apomixis farther
complicate the picture; 2 varieties:

A. principal lateral veins diverging from the base of the midrib; leaves obtuse and
   crenate .... var. *rotundifolium*
A. principal lateral veins diverging above the base of the midrib; leaves acute and
   serrate .... var. *ovatum* (Bigelow) Torr.

Eupatorium rugosum Houtt.   White-snakeroot
Stems 3–15 dm tall, often in a cluster of 2–3, glabrous below the inflorescence;
leaves opposite, blades ovate to subcordate, 6–18 cm long, sharply serrate; florets white, 12–25 per head; common in woods, meadows, and roadsides;
throughout; flr. Jul–Oct; poisonous.

Eupatorium serotinum Michx.   Late eupatorium
Stems to 2 m tall, puberulent throughout; leaves opposite, petioled; blades lanceolate, acuminate, coarsely serrate; florets white, 9–14 per head; increasingly
common in sandy fields, moist thickets, ditches, roadsides, alluvial areas, and
ballast; flr. Aug–Oct; native farther west, apparently adventive here; FAC–.

Eupatorium sessilifolium L.   Upland eupatorium
Stems 6–15 dm tall, glabrous below the inflorescence; leaves opposite, sessile;
blades lanceolate, serrate, acuminate; venation strictly pinnate; florets white, 5–
6 per head; occasional on dry wooded slopes, rocky banks, and roadsides;
throughout, except at the highest elevations; flr. Jul–Oct.

Euthamia Nutt.

Rhizomatous perennial herbs; leaves numerous, alternate, resinous-punctate, narrow, entire; heads small, forming a large flat-topped inflorescence; involucre glutinous; disk florets usually fewer than the ray florets, both yellow; rays minute; achenes several-nerved, short-hairy; pappus of numerous, white, capillary bristles.

A. leaves 3-nerved; heads with 20–35 florets .... E. *graminifolia*
A. leaves 1-nerved; heads with 10–21 florets .... E. *tenuifolia*

Euthamia graminifolia (L.) Nutt.   Grass-leaved goldenrod
Stems to 1.5 m tall; leaves 4–13 cm by 3–12 mm, clearly 3-nerved; heads with 15–25 ray florets and 5–10 disk florets; the rays minute and often scarcely spreading; common in moist fields, roadsides, ditches, and shores; throughout; flr. Jul–Nov; FAC.

Euthamia tenuifolia (Pursh) Greene   Grass-leaved goldenrod
Stems to 1 m tall; leaves linear, 1–3 mm wide and 20–50 mm long, 1-nerved or sometimes with a pair of weak lateral nerves; heads pedunculate, 10–21-flowered; very rare in moist sandy or clayey fields; SE, believed to be extirpated; flr. Jul–Oct; FACU; 🌿.

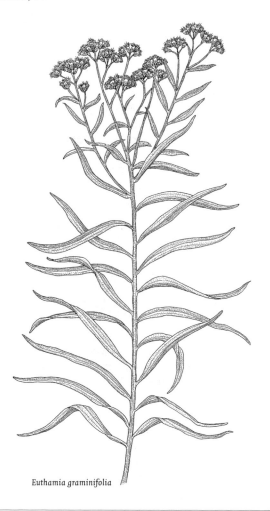

Euthamia graminifolia

Euthamia tenuifolia, leaf

Filago L.

Filago germanica (L.) Huds.   Cotton-rose, herba-impia
Low, white-woolly annuals with alternate, entire leaves; stem and branches terminating with clusters of whitish heads; involucral bracts few; receptacle bristly or chaffy; heads discoid; the outermost florets partly enclosed in a boat-shaped bract, the others bractless; rare on stream banks and fields; SE; flr. Jul–Aug; native to Europe.

Gaillardia Foug.

Gaillardia x grandiflora Van Houtte   Blanket-flower
Hairy perennial to 7 dm tall, tap-rooted but also with slender, creeping roots; leaves linear-oblong to lance-ovate, entire to pinnatifid; heads solitary, long-pedunculate, to 6 cm wide; rays 6–16, yellow often with a purplish blotch at the base, 3-lobed at the tip; the receptacle with numerous soft hairs between, but not subtending the disk florets; disk corollas densely hairy on the lobes; cultivated and occasionally naturalized in old fields or roadsides, also included in roadside wildflower seed mixtures; flr. Jun–Sep; native to western North America.

Galinsoga Ruíz & Pav.

Hairy annuals; leaves opposite, petiolate, weakly to strongly toothed; heads in an open, leafy cyme, small, with a few small, 3-toothed, white rays and yellow disk florets; receptacle conical, chaffy; achenes 4-angled; pappus consisting of scales.

A. ray achenes lacking pappus; teeth of the leaf margins obscure, broadly rounded .... G. parviflora
A. ray achenes with pappus; teeth of the leaf margins distinct, acute ....
.... G. quadriradiata

Galinsoga parviflora Cav.   Small-flowered quickweed
Stems 10–60 cm long, pubescence wanting or minute and appressed; leaves lanceolate-ovate, with small blunt teeth; heads 2–3 mm long; ray florets white, without pappus; disk florets awnless; achenes glabrous or pilose at the summit; occasional in waste places, sidewalks, and streets; scattered; flr. Jun–Nov; native to Central and South America.

Galinsoga
quadriradiata, ray
floret (×2½)

Galinsoga quadriradiata Ruíz & Pav.   Quickweed
Stems bushy, highly branched, 1–1.6 dm tall; leaves petiolate, ovate, coarsely serrate, 2–8 cm long; peduncles coarsely spreading-hispid, often glandular; heads 2–3 mm long, phyllaries ovate, ciliate; ray florets white, 1–2 mm long; achenes densely hispid with an awned pappus; a common weed of waste places, sidewalks, streets, and gardens; throughout; flr. Jun–Nov; native to Central and South America.

Galinsoga
quadriradiata

Gnaphalium L.

White, woolly herbs with alternate, entire leaves; heads discoid, yellow or whitish; the outer florets pistillate with filiform corollas about as long as the pappus;

the few inner ones coarser and perfect; involucre ovoid or campanulate with numerous papery bracts; achenes small, terete or compressed; pappus of capillary bristles falling separately or connate at the base.

A. inflorescence narrow, spike-like; pappus bristles united at the base
    B. involucral bracts rounded to obtuse; achenes sparingly strigose ....
    .... *G. sylvaticum*
    B. involucral bracts acute to acuminate; achenes papillate .... *G. purpureum*
A. inflorescence not spike-like; pappus bristles falling separately
    C. inflorescence of numerous small axillary and terminal clusters overtopped by their subtending leaves; plants <25 cm tall .... *G. uliginosum*
    C. inflorescence panicle-like; plants >25 cm tall
        D. leaves not decurrent at the base; stem woolly but not glandular except perhaps at the base .... *G. obtusifolium*
        D. leaves decurrent at the base; stem glandular-hairy, sometimes also woolly at the base .... *G. macounii*

Gnaphalium macounii Greene   Fragrant cudweed
Similar to *G. obtusifolium* but with the stem glandular-hairy becoming woolly in the inflorescence and the leaves distinctly decurrent; occasional in old fields, cut-over woods, clearings; and roadsides; throughout, except in the SE; flr. Aug–Oct.

Gnaphalium obtusifolium L.   Fragrant cudweed
Erect annual or winter annual 3–10 dm tall; stem thinly white-woolly; leaves numerous, all cauline, lance-linear, white-woolly beneath and glabrous to slightly woolly above; involucre grayish-white, woolly at the base; florets 75–125, only a few perfect; achenes glabrous; pappus bristles distinct; common in dry pastures, old fields, shale barrens, and roadsides; throughout; flr. Aug–Nov.

Gnaphalium purpureum L.   Purple cudweed
Thinly woolly annual or biennial to 4 dm tall with a basal rosette of spatulate or oblanceolate leaves; heads numerous in a terminal, spike-like, leafy bracted inflorescence; involucral bracts acute to acuminate, light brown or tinged with purple; pappus bristles united at the base; occasional in dry, sandy or rocky fields; mostly S; flr. May–Sep. Ours is var. *purpureum*.

Gnaphalium sylvaticum L.   Woodland cudweed
Erect, thinly woolly stems to 6 dm tall; leaves linear or oblanceolate, subglabrate above; inflorescence narrow, spike-like, somewhat leafy-bracted with 10–many heads; involucral bracts rounded or obtuse, light straw-colored or greenish at the base with a conspicuous V-shaped, dark brown spot above the middle; pappus bristles united at the base; very rare, collected once on a dry wooded hillside in Tioga Co.; flr. Sep; ✤.

Gnaphalium uliginosum L.   Low cudweed
Branching annual to 2.5 dm tall; stem densely and loosely white woolly; leaves numerous, mostly cauline, only sparsely woolly; heads in axillary clusters overtopped by their subtending leaves; involucral bracts greenish or brown, acute, not strongly imbricate; moist fields, stream banks, woods, and ditches; throughout; flr. Jun–Nov; native to Europe; FAC.

Gnaphalium uliginosum

Grindelia Willd.

Grindelia squarrosa (Pursh) Dunal   Gum-weed, rosinweed
Tap-rooted, biennial to 1 m tall; leaves alternate, resinous-punctate; heads several to many, about 3 cm wide; rays yellow, 25–40, or rarely absent; achenes 2–3 mm long, pappus of 2–8 finely serrulate awns; ballast, sidewalks, and lawns; flr. Aug–Sep; native to western North America; FACU.

Helenium L.

Herbs with alternate, sessile, and often decurrent leaves producing winged stems; heads radiate; receptacle hemispheric to subglobose; rays yellow, 3-lobed at the tip, usually somewhat deflexed; involucral bracts becoming deflexed; achenes 4–5-angled; pappus of 5–10 scarious, awn-tipped scales.

A. leaves <2 mm wide, not decurrent .... *H. amarum*
A. leaves 5 mm wide or more, their bases decurrent forming wings on the stem
    B. disk florets yellow.. *H. autumnale*
    B. disk florets red-brown or purple-brown .... *H. flexuosum*

Helenium amarum (Raf.) H.Rock   Sneezeweed
Glabrous, glandular annual; 2–5 dm tall; leaves linear or linear-filiform, rarely >2 mm wide; heads on short, naked peduncles extending above the leaves; disk and ray florets yellow; rare along railroad tracks and riverbanks; SE and NW; flr. Sep–Oct; native farther west, but apparently adventive here; FACU–.

Helenium autumnale L.   Common sneezeweed
Fibrous-rooted perennial with winged stems to 1.5 dm tall; leaves numerous, cauline, lance-linear; heads in a leafy inflorescence; disk florets and rays yellow; common in swamps, moist riverbanks, alluvial thickets and wet fields; throughout; flr. Jul–Nov; FACW+.

Helenium flexuosum Raf.   Southern sneezeweed
Fibrous-rooted perennial to 1 m tall; leaves decurrent forming wings on the stem; disk florets red-brown or purple; rays yellow, sometimes with a purple blotch near the base; occasional in moist fields, pastures, shores, and waste ground; throughout; native farther west; flr. Jun–Nov; FAC–.

Grindelia squarrosa

Helenium autumnale

Helenium autumnale, head

Helianthus L.

Coarse annual or perennial herbs with simple, opposite leaves (at least on the lower part of the stem); heads with prominent yellow rays; involucral bracts nearly equal to clearly imbricate, green; receptacle flat to low conical, the chaffy bracts clasping the achenes; achenes somewhat compressed, pappus of 2 deciduous awns.

A. leaves 1 cm wide or less; disk florets red-purple .... H. angustifolius
A. at least some of the leaves >1 cm wide
  B. disk florets red-purple (only occasionally yellow); receptacle nearly flat; leaves, except the lowest, mostly alternate; annuals
    C. involucral bracts ovate or ovate-oblong; central receptacular bracts inconspicuously short-hairy .... H. annuus
    C. involucral bracts narrower, tapering; receptacular bracts white-bearded at the tip .... H. petiolaris
  B. disk florets mostly brown or yellow; receptacle convex to low conical; leaves, except the upper, mostly opposite; perennials
    D. involucral bracts overlapping, broad, firm, appressed with rounded to acute tips .... H. laetiflorus
    D. involucral bracts loosely spreading with long tapering tips
      E. leaves concentrated toward the base, only 1–2 pairs of reduced leaves or bracts on the upper half of the stem .... H. occidentalis
      E. leaves well-distributed along the stem, >8 pairs below the inflorescence
        F. principal midcauline leaves sessile or with petioles <4 mm
          G. leaves clasping; leaves, stems, and involucral bracts densely covered with silky, grayish hairs .... H. mollis
          G. leaf blades tapered or rounded at the base but not clasping
            H. widest part of the leaf near or slightly below the middle
              I. leaves scabrous above and beneath, sometimes folded; stems with appressed hairs .... H. maximilianii
              I. leaves less scabrous beneath, narrowly elliptic to ovate-lanceolate, flat; stems with spreading hairs .... H. giganteus
            H. leaf widest at the base
              J. stems sparsely hairy; leaves triple-veined from just above the base of the blade .... H. hirsutus
              J. stems glabrous and glaucous; lowest pair of lateral veins joining the midrib at the base of the blade .... H. divaricatus
        F. principal midcauline leaves with petioles >5 mm long (and often winged)
          K. stems scabrous or with spreading hairs
            L. leaf blades ovate-elliptic, the largest about 3.5 cm wide, on winged petioles >1.5 cm long .... H. tuberosus
            L. leaf blades lance-elliptic, mostly <3 cm wide, on petioles <1.5 cm long .... H. giganteus
          K. stems glabrous or nearly so, and often glaucous
            M. heads 3 cm wide or less; rays 0.8–1.6 cm long .... H. microcephalus
            M. heads at least 4 cm wide; rays 1.8–4.3 cm long

N. leaves only slightly scabrous, mostly <3 cm wide ....
.... H. *grosseserratus*

N. leaves very scabrous above, usually >3 cm wide

    O. leaf blades coarsely toothed, glabrate or sparsely pubescent beneath .... H. *decapetalus*

    O. leaf blades shallowly toothed to entire, densely pubescent beneath .... H. *strumosus*

Helianthus angustifolius L.   Swamp sunflower

Fibrous-rooted perennial lacking rhizomes; stem to 1.5 m tall, hairy; leaves numerous, linear, sessile, alternate except near the base; disk florets red-purple; rays 10–15; involucral bracts with loose, narrow tips, seldom surpassing the disk; very rare in swamps and moist, sandy ground; SE; believed to be extirpated; flr. Sep–Oct; FACW; ◈.

Helianthus annuus L.   Common sunflower

Coarse, hairy, branched annual, 1–3 m tall; leaves mainly alternate, long-petiolate, broadly ovate; disk red-purple, >3 cm wide; involucral bracts ovate, abruptly narrowed to an acuminate tip; cultivated and frequently escaped to vacant lots, roadsides, and rubbish dumps; mostly S; native to western North America; flr. May–Nov; FAC–.

Helianthus decapetalus L.   Thin-leaved sunflower

Perennial with slender rhizomes; stems 0.5–1.5 m tall, glabrous below the inflorescence; leaves thin, scabrous to subglabrous, sharply serrate and somewhat decurrent onto the 1.5–5-cm-long petiole; disk yellow; involucral bracts very loose, conspicuously ciliate, attenuate-acuminate and surpassing the disk; frequent in fields, moist bottomlands, stream banks, and roadsides; throughout; flr. Jul–Sep; FACU.

Helianthus divaricatus L.   Rough or woodland sunflower

Colonial, rhizomatous perennial; stems to 1.5 m tall, glabrous, and often glaucous below the inflorescence; leaves sessile, scabrous above and loosely hairy, at least on the veins beneath; disk yellow, 1–1.5 cm wide; involucral bracts lance-acuminate or attenuate, rather loose; occasional in dry open woods, wooded slopes, shale barrens, and roadsides; throughout; flr. Jul–Sep.

Helianthus giganteus L.   Swamp sunflower

Perennial with short rhizomes and sometimes short, thickened roots; stem 1–3 m tall, with spreading hairs or subglabrous; leaves strongly scabrous above, triple-nerved from the base, tapering to a short petiole, the upper leaves alternate; disk yellow, 1.5–2.5 cm wide; involucral bracts narrow, often surpassing the disk; occasional in swamps, ditches, and wet fields; scattered; flr. Jul–Oct; FACW.

Helianthus grosseserratus G.Martens   Sawtooth sunflower

Coarse, fibrous-rooted perennial; stems 1–4 m tall, glabrous and often glaucous below the inflorescence; leaves lanceolate, sharply toothed, tapering to a winged, 1–4-cm-long petiole; disk yellow, 1.5–2.5 cm wide; involucral bracts lance-linear, surpassing the disk; cultivated and rarely escaped to fields, thickets and waste ground; flr. Aug–Oct; native farther west; FACW.

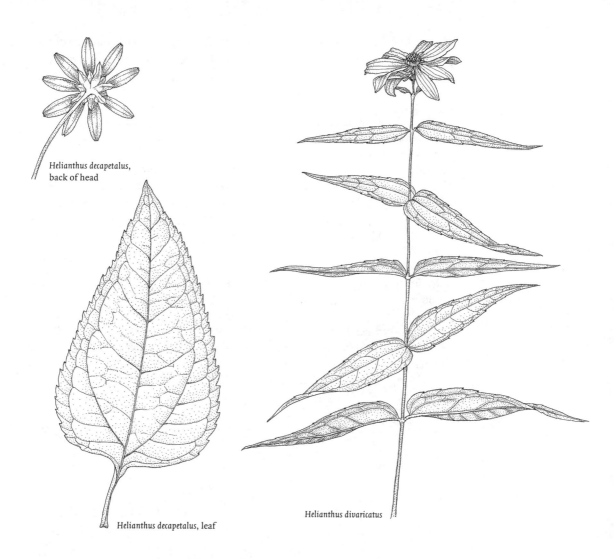

Helianthus decapetalus,
back of head

Helianthus decapetalus, leaf

Helianthus divaricatus

Helianthus hirsutus Raf. Sunflower

Perennial with long rhizomes; stems 0.6–2 m tall, hairy; leaves almost all opposite, hairy, triple-nerved from the broadly rounded to subcordate base, with a 5–15-mm-long petiole; disk yellow, 1.5–2 cm wide; involucral bracts slender, long-pointed with a reflexed tip; rare on shaly slopes, upland meadows, and dry roadside banks; SW; flr. Jul–Sep; ✿.

Helianthus laetiflorus Pers. Showy sunflower

Perennial with long rhizomes; stems 1–2.5 m tall, pubescent or somewhat glabrous above; leaves pubescent and scabrous above and beneath, serrate; petiole to 5 cm long; disk yellow; involucral bracts lance-acuminate; escaped from cultivation to old fields and stream banks; scattered; flr. Aug–Oct; native farther west.

Helianthus maximilianii Schrad. Maximilian's sunflower

Perennial with short rhizomes and thickened, fleshy roots; stems 0.5–3 m tall, pubescent; leaves strongly scabrous above and beneath, often folded along the midrib, gradually narrowed to a short winged petiole, the upper leaves alternate;

heads in an elongate inflorescence; disk yellow, 1.5–2.5 cm wide; involucral bracts narrow, loose, exceeding the disk; cultivated and occasionally escaped to old fields, railroad tracks, or urban waste ground; SE and NW; flr. Jul–Aug; native farther west; UPL.

Helianthus microcephalus Torr. & A.Gray   Small wood sunflower
Fibrous-rooted perennial; stems 1–2 m tall, glabrous and glaucous; leaves scabrous above, resin-dotted and short-hairy beneath, abruptly narrowed to a 1–3-cm-long petiole; disk yellow, 0.5–1 cm wide; involucral bracts few, acuminate or attenuate; occasional in upland woods, rocky slopes, and dry roadside banks; W; flr. Aug–Oct; 🌿.

Helianthus mollis Lam.   Ashy sunflower
Colonial perennial with stout rhizomes; stems 0.5–1 m tall, densely silky-hairy; leaves sessile; disk yellow, 2–3 cm wide; involucral bracts white-hairy and usually finely glandular; rare in clayey fields, shores, ballast, and waste ground; scattered; flr. Jul–Aug; native farther west.

Helianthus occidentalis Riddell   Sunflower
Colonial, rhizomatous perennial 0.5–1.5 m tall; leaves larger near the base, the upper much reduced; disk yellow, 1–1.5 cm wide; involucral bracts with loose, slender tips; very rare, collected once in Warren Co.; UPL; flr. Jul–Sep; 🌿.

Helianthus tuberosus   *Heterotheca subaxillaris*

Helianthus petiolaris Nutt. Sunflower
Similar to *H. annuus*, but stems 0.5–1 m tall; leaves broadly truncate at the base on 1–12-cm petioles; involucral bracts lanceolate; rare on ballast, railroad tracks, and vacant lots; SE and NW; flr. Jun–Oct; native farther west.

Helianthus strumosus L. Rough-leaved sunflower
Rhizomatous perennial; stem 1–2 m tall, glabrous and often glaucous below the inflorescence; leaves mostly opposite or the uppermost alternate, thick, scabrous above, short-hairy or glaucous beneath, decurrent onto the 0.5–3-cm-long petiole; disk yellow; involucral bracts with long acuminate tips somewhat surpassing the disk; occasional in fields, woods, stream banks, and roadsides; throughout; flr. Aug–Sep.

Helianthus tuberosus L. Jerusalem artichoke
Perennial with tuber-bearing rhizomes; stems 1–3 m tall, hairy; leaves numerous, alternate above, scabrous on the upper surface, short-hairy beneath, abruptly narrowed or tapering to a 2–8-cm winged petiole; disk yellow, 1.5–2.5 cm wide; involucral bracts narrow, loose above the middle; cultivated and frequently escaped to fields, woods, railroad tracks, and roadsides; throughout; flr. Sep–Oct; native farther west; FAC.

Heliopsis Pers.

Heliopsis helianthoides (L.) Sweet Ox-eye
Fibrous-rooted perennial to 1.5 m tall; leaves opposite, ovate to lance-ovate, petiolate; heads to 4 cm wide on naked peduncles; outer involucral bracts enlarged and leafy; rays pale yellow; achenes glabrous; pappus absent; common in fields, woods, floodplains, and stream banks; throughout; flr. Jun–Sep.

Heterotheca Cass.

Heterotheca subaxillaris (Lam.) Britton & Rusby Camphorweed
Tap-rooted, glandular, aromatic annual or biennial 2–20 dm tall; leaves ovate or oblong, the lower petiolate, the middle and upper leaves sessile and clasping; heads radiate, yellow; occasional on ballast, waste ground, and sandy dredge spoil; SE; flr. Sep–Oct; native farther south, adventive here; UPL.

Hieracium L.

Fibrous-rooted perennials, many with long or short rhizomes and stolons, sap milky; leaves basal or alternate, variously pubescent or hirsute, mostly with at least some stellate hairs; heads solitary or clustered, involucre cylindric to hemispheric, with imbricate bracts; only ray florets present; corollas yellow or orange; achenes terete or angled, narrowed toward the base, strongly ribbed or grooved; pappus of tan or brown capillary bristles; a large and difficult genus due to hybridization and apomixis.

A. basal leaves well-developed at flowering time; heads single or in clusters of 2–many
    B. heads 1–3; leaves with stellate hairs beneath, at least when young
        C. heads mostly solitary, rarely 2 or 3 .... *H. pilosella*
        C. heads 2–6 .... *H. flagellare*

B. heads several to many; leaves with sparse stellate hairs beneath or none
    D. florets orange-red .... H. *aurantiacum*
    D. florets yellow
        E. basal leaves conspicuously purple-veined .... H. *venosum*
        E. basal leaves not purple-veined
            F. leaves coarsely toothed
                G. basal leaves broadly rounded, truncate or cordate at the base...
                    .... H. *murorum*
                G. basal leaves tapering to the petiole .... H. *lachenalii*
            F. leaf margins entire or with a few tiny denticulate teeth
                H. inflorescence a compact corymbose cluster of heads
                    I. stems and leaves glabrous or nearly so above, hairy only on the midrib beneath .... H. *piloselloides*
                    I. leaves hairy on both sides; stem hairy .... H. *caespitosum*
                H. inflorescence open, paniculate
                    J. inflorescence elongate, cylindrical, with a distinct central axis ....H. *gronovii*
                    J. inflorescence lacking a distinct central axis
                        K. leaves densely long-hairy beneath and to a lesser extent above .... H. *traillii*
                      K. leaves with a few long hairs near the base, but mostly glabrous .... H. *piloselloides*
A. basal leaves not present at flowering time, leaves all cauline
    L. longest hairs on the stem 3–10 mm
        M. leaves glaucous beneath; heads with 8–30 florets .... H. *paniculatum*
        M. leaves not glaucous beneath
            N. heads with >40 florets.. H. *scabrum*
            N. heads with <40 florets..H. *gronovii*
    L. longest hairs on the stems <3 mm
        O. lower stem and lower leaf surfaces with long, bulbous-based hairs and little if any stellate pubescence .... H. *sabaudum*
        O. leaves finely stellate, lacking bulbous-based hairs................. H. *kalmii*

Hieracium aurantiacum L.   Orange hawkweed
Very similar to H. *caespitosum* but plant 1–6 dm and florets orange-red; frequently naturalized in abandoned fields, meadows, and pastures; mostly N; flr. Jun–Sep; native to Europe.

Hieracium caespitosum Dumort.   King-devil
Stems 1–several, 2.5–9 dm tall, long-hairy throughout and with shorter, blackish gland-tipped hairs above, naked or with only a few greatly reduced leaves; basal leaves oblanceolate or narrowly elliptic, hairy; heads 5–30 in a compact cluster; florets bright yellow; common in woods, fields, and roadsides; throughout; flr. May–Nov; native to Europe. [syn: H. *pratense* Tausch]

Hieracium flagellare Willd.   Hawkweed
Leafy shoots spreading-ascending at first, becoming stoloniferous; young stems long-hairy; basal leaves oblanceolate, sparsely setose to nearly glabrous above, thinly stellate with scattered longer hairs beneath; peduncles 1.5–4 dm tall, nearly naked; heads 2–3(6); naturalized in fields, lawns, and roadsides; SE; flr. May–Nov; native to Europe.

Hieracium
caespitosum

Hieracium gronovii L. Hawkweed
Stems solitary, 3–15 dm tall, spreading-hairy toward the base and becoming puberulent to subglabrous upward; basal and lower cauline leaves numerous, broadly oblanceolate to obovate or elliptic, the upper reduced and becoming sessile and clasping; inflorescence elongate and openly cylindric, the peduncles puberulent and long-stipitate glandular; occasional in dry, open woods or thickets, usually in sandy soil; SE and W; flr. Jul–Sep; UPL. [syn: *H. floridanum* Britton]

Hieracium kalmii L. Canada hawkweed
Stems 1.5–15 dm tall, often spreading-hairy below and stellate-puberulent above; leaves numerous, sessile, and broadly rounded to clasping at the base, nearly alike in size, the lower withering early; inflorescence loosely corymbiform to umbel-like; peduncles stellate-puberulent; rare in thickets, clearings, and roadside banks; mostly NE; flr. Jul–Sep; ✥. [syn: *H. canadense sensu* Torr. & A.Gray non Michx.]

Hieracium lachenalii C.C.Gmel European hawkweed
Stems 1.5–10 dm tall, sparsely setose and stellate to subglabrous, leaf blades narrowly to broadly elliptic, coarsely toothed to occasionally nearly entire, tapering to the short petiole; cauline leaves gradually or abruptly reduced upward; inflorescence open, corymbiform; peduncles with glandular and stellate hairs; rare in dry woods, grassy slopes, roadsides, and lawns; mostly SE; flr. Jun–Sep; native to Europe. [syn: *H. vulgatum* Fr.]

Hieracium murorum L. Wall hawkweed, golden lungwort
Stems 1.5–6 dm tall, naked or with only 1 or 2 leaves, hairy at the base; basal leaves with densely long-hairy petioles; leaf blades lance-ovate, coarsely toothed, broadly rounded or subtruncate to cordate at the base; branches of the inflorescence ascending; involucre 8–10 mm; rare in rich, dry woods and shaded roadside banks; SE; flr. May–Jun; native to Europe.

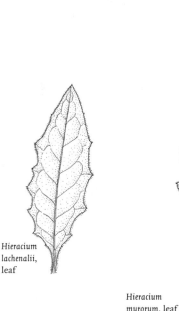

Hieracium
lachenalii,
leaf

Hieracium
murorum, leaf

Hieracium gronovii

Hieracium paniculatum L.   Hawkweed

Stems solitary, 3–15 dm tall, long-hairy below but otherwise glabrous, leafy to the inflorescence; leaves elliptical, minutely denticulate, glaucous beneath; inflorescence open-paniculate, with numerous heads on long, slender peduncles; involucre and peduncles glabrous or with only a few gland-tipped hairs; common in dry, rocky, or sandy woods or slopes; throughout; flr. Jul–Oct.

Hieracium pilosella L.   Mouse-ear hawkweed

Stoloniferous with long rhizomes, peduncles 3–25 cm long, naked except for a single much-reduced leaf and usually bearing a single head of yellow florets; basal leaves oblanceolate, with stellate hairs beneath and scattered longer hairs on both sides; peduncle and involucre usually with some blackish, gland-tipped hairs; occasional in dry fields, pastures, roadsides, and lawns; scattered; flr. May–Jun; native to Europe.

Hieracium piloselloides Vill.   King-devil

Leaves mostly basal, oblanceolate, with scattered long hairs or subglabrous, glaucous; peduncles 2–10 dm long, naked or with 1 or 2 small leaves, glaucous, glabrous except for a few glandular hairs at the top; inflorescence compact to fairly open, of 6–12 heads; involucre 6–8 mm high, with black glandular hairs; occasional in dry fields, meadows, roadsides, and lawns; mostly SE; flr. May–Oct; native to Europe.

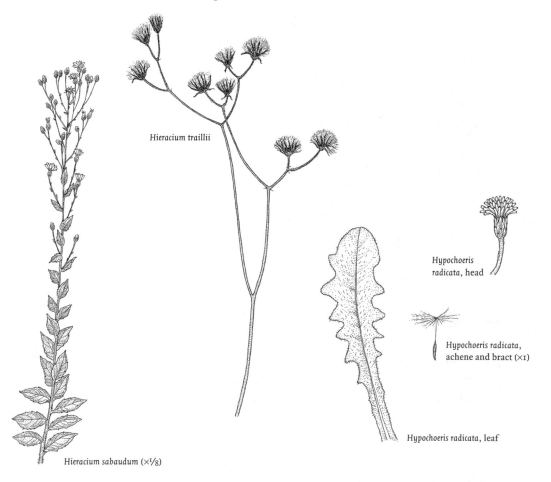

Hieracium traillii

Hieracium sabaudum (×1/8)

Hypochoeris radicata, head

Hypochoeris radicata, achene and bract (×1)

Hypochoeris radicata, leaf

Hieracium sabaudum L.   Hawkweed
Leaves all cauline, elliptical to ovate, with coarse teeth, upper surface sparsely hairy or glabrous, lower leaf surface and stem covered with bulbous-based hairs but not stellate; inflorescence elongate; lower branches and peduncles ascending; involucre 8–10 mm high; rare on roadsides, alluvium, and waste ground; SE; flr. Aug–Oct; native to Europe.

Hieracium scabrum Michx.   Hawkweed
Stems 2–15 dm tall, solitary, densely pubescent with long hairs below and glandular hairs above, especially in the inflorescence; leaves obovate to elliptic, greatly reduced upward with long hairs on the petioles and midribs beneath; inflorescence open-corymbiform to cylindric; peduncles and involucre with abundant blackish, glandular hairs; involucre 7–8 mm high; frequent in open fields, clearings, and woods edges; throughout; flr. Jul–Oct.

Hieracium traillii Greene   Greene's or Maryland hawkweed
Stems 2–7 dm tall, glabrous to very finely pubescent, naked except for 1 or 2 reduced leaves near the base; leaves mostly basal, spatulate to ovate, with numerous long, simple hairs; peduncles and involucres with blackish glandular hairs; very rare on dry wooded slopes, bluffs, and shale barrens; SC; FACU; flr. May–Aug; ◐.

Hieracium venosum L.   Rattlesnake-weed
Stems to 8 dm from a basal rosette of purple-veined leaves, nearly glabrous; inflorescence open; peduncles elongate, slender; involucre 7–10 mm high, glabrous or with a few glandular hairs; frequent in dry, upland woods, wooded slopes, and edges; throughout, except in the northernmost counties; flr. May–Sep.

Hypochoeris L.

Hypochoeris radicata L.   Cat's-ear
Perennial with basal leaves and lightly branched flowering stems to 6 dm tall; leaves toothed or pinnately lobed, covered with stiff, spreading hairs; receptacle with long, chaffy bracts; florets yellow; achenes beaked, with plumose bristles; occasional in lawns, roadsides, and urban waste ground; throughout; flr. Jun–Oct; native to Eurasia.

Inula L.

Inula helenium L.   Elecampane
Coarse perennial herb with a hairy stem to 2 m tall; leaves densely woolly beneath, shallowly toothed, the lower long-petiolate, upper leaves becoming sessile and cordate-clasping; heads few, 6–10 cm wide; rays yellow; anthers sagittate-tailed; cultivated and occasionally naturalized in pastures, roadsides, and waste ground; throughout; flr. Jun–Aug; native to Europe; FACU.

Iva L.

Annual herbs (or shrubs) with opposite leaves and small discoid heads of greenish-white florets; involucre of a few imbricate bracts that grade into the receptacular bracts; corolla of the marginal pistillate florets filiform or nearly obsolete; staminate florets with undivided style; in addition to the species treated below, both of which are native farther west and are known to have been cultivated by native American Indians; the salt marsh shrub, *Iva frutescens* L., was collected early on ballast in Philadelphia.

A. heads subtended by bracts .... *I. annua*
A. heads not subtended by bracts .... *I. xanthifolia*

Iva annua L.   Rough marsh-elder
Annual to 2 m tall; leaves chiefly opposite, lanceolate to broadly ovate, serrate, acuminate; inflorescence of spike-like branches, heads sessile in the axils of reduced leaves; involucre of 3–5 bracts subtending the achenes; pistillate florets with 1–1.5-mm tubular corolla; achenes resin dotted; very rare in moist open areas; C; flr. Sep–Oct; native farther west; FAC.

Iva xanthifolia, head (×2)

Iva xanthifolia Nutt.   Marsh-elder
Coarse annual to 2 m tall with a simple stem; leaves chiefly opposite, long-petiolate, ovate, triple-nerved, coarsely and doubly serrate; inflorescence large with numerous subsessile heads; involucre 1.5–3 mm high with 5 subherbaceous bracts and 5 inner receptacular bracts; corolla of the pistillate florets 0.5 mm high or obsolete; rare on railroad embankments, vacant lots, and fill; flr. Jun–Sep; native farther west; FAC.

Iva xanthifolia, (×¼)

Ixeris (Cass.) Cass.

Ixeris stolonifera A.Gray   Creeping lettuce
Glabrous, creeping perennial; leaves broadly elliptic to orbicular, petiolate; heads solitary or paired on 1-dm-long peduncles; florets yellow; achenes reddish-brown, somewhat compressed with a slender beak as long as the body; rare in lawns, gardens, and nursery beds; SE; flr. May–Sep; native to Japan.

Krigia Schreb.

Small herbs with milky sap; leaves mostly in a basal rosette, entire, denticulate, or pinnately lobed; heads terminal on long peduncles that are often glandular-hairy above; florets yellow or orange; involucral bracts strongly reflexed in fruit; pappus of capillary bristles surrounded by an outer ring of papery scales.

A. achenes 5-angled; pappus with 5–8 bristles .... *K. biflora*
A. achenes 15–20-ribbed; pappus with 10–15 bristles .... *K. virginica*

Krigia biflora (Walt.) S.F.Blake   Dwarf dandelion, two-flowered cynthia
More or less glaucous perennial with stems 1–6 dm tall; leaves oblanceolate to elliptic, mostly basal but with a few subtending branches of the inflorescence, entire to sparsely denticulate; heads 2–6, involucre 7–14 mm high; florets yellow-orange; achenes with 15–20 ribs; frequent in fields, meadows, woods, and sandy banks; E and W; flr. May–Oct; FACU.

Krigia virginica (L.) Willd.   Dwarf dandelion
Annual or winter annual, more or less glaucous; flowering stems 0.5–4 dm tall, usually branched; leaves few, basal, entire to pinnatifid; head solitary; involucre 4–7 mm high; florets yellow; occasional on dry rocky slopes, sandy soils, and shale barrens; SE and SC; flr. May–Aug; UPL.

Lactuca L.

Erect, leafy-stemmed annual or biennial herbs with milky sap; stems mostly unbranched below the inflorescence; leaves alternate, variously entire to pinnatifid; heads numerous in a panicle-like inflorescence; florets all perfect; involucre cylindric or broadened at the base in fruit; corollas yellow or blue; achenes compressed with prominent marginal nerves, beaked or beakless but expanded at the top where the pappus is attached (mature achenes are needed to fully evaluate the beak characteristics); pappus 2 rows of capillary bristles.

A. corollas blue or white; achene beakless or with a short, stout beak <½ as long as the body of the achene
  B. pappus light brown or grayish .... *L. biennis*
  B. pappus white .... *L. floridana*
A. corollas yellow; beak slender, >½ as long as the body of the achene
  C. leaves with prickly margins and midribs, blade often twisted so it is perpendicular to the ground .... *L. serriola*
  C. leaves not prickly on the midrib, blade not twisted
    D. mature achenes with only 1 prominent nerve on each face
      E. involucres <15 mm high; achenes 4.5–6 mm long including the beak
        .... *L. canadensis*

E. involucres 15–22 mm high; achenes 7–10 mm long .... *L. hirsuta*
D. mature achenes with several nerves on each face .... *L. saligna*

Lactuca biennis,
achene (×2)

Lactuca biennis (Moench) Fernald   Blue lettuce

Stems to 2 m tall; leaves variously pinnatifid or toothed, sagittate, glabrous or hairy on the veins beneath; florets blue or rarely white; achenes with a short, stout beak; pappus light brown to grayish; frequent in moist, open soil of woods, stream banks, roadsides, and vacant lots; throughout; flr. Jul–Sep; FACU.

Lactuca canadensis L.   Wild lettuce

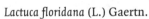

Lactuca canadensis,
achene (×2)

Stems to 2.5 m tall; leaves entire, toothed, or shallowly to deeply pinnately lobed, glabrous to hirsute; heads numerous; florets yellow; achenes strongly flattened, dark, with a single nerve on each face and a slender, elongate beak; pappus white; frequent in meadows, fields, rocky hillsides, and roadside banks; throughout; flr. Jun–Sep; FACU. Several varieties are often recognized based on leaf shape that is extremely variable.

Lactuca floridana (L.) Gaertn.   Woodland lettuce

Stems to 2 m tall; leaves petiolate, not sagittate at the base, elliptic and toothed to distinctly pinnatifid, hairy along the main veins beneath; heads numerous in a paniculate inflorescence; florets blue or rarely white; achenes narrowed upward, with several nerves on each side; beak stout, to $1/3$ the length of the body; pappus white; occasional on rich wooded slopes, meadows, and roadsides; S; flr. Aug–Oct; FACU–. Plants with merely toothed leaves are sometimes recognized as var. *villosa* (Jacq.) Cronquist.

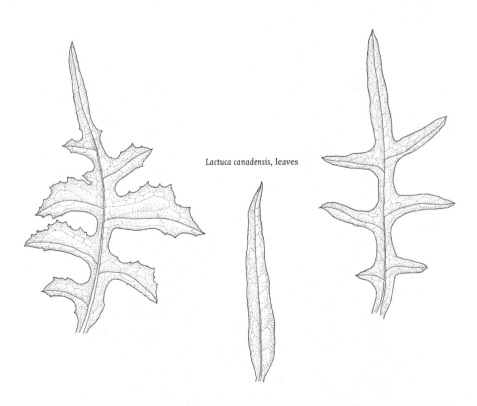

Lactuca canadensis, leaves

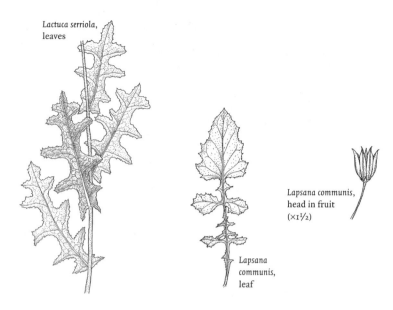

Lactuca serriola, leaves

Lapsana communis, leaf

Lapsana communis, head in fruit (×1½)

Lactuca hirsuta Muhl.   Downy lettuce
Similar to L. *canadensis* but more hairy, leaves more basally disposed, and the involucres and achenes larger; flr. Aug–Sep; 🐝; 2 varieties:

A. plants densely hairy .... var. *hirsuta*
    rare in limestone woods, clearings, and alluvial bottomlands; scattered.
A. plants less hirsute .... var. *sanguinea* (Bigel.) Fernald
    dry, open woods, thickets, and rocky ledges; scattered.

Lactuca saligna L.   Willow-leaf lettuce
Stems to 1 m tall; stem and lower leaf surfaces glaucous; leaves sagittate, linear; heads numerous; florets light yellow; achenes with a long slender beak twice as long as the body, pappus white; occasional in woods, fields, strip mines, and limestone quarries; flr. late Jul–Nov; native to Europe; UPL.

Lactuca saligna, achene (×2)

Lactuca serriola L.   Prickly lettuce
Stems glabrous, glaucous, to 1.5 m tall; leaves prickly on the midribs and margins, pinnately lobed to entire; the blades usually twisted at the base so they are perpendicular to the ground; heads numerous in an open, paniculate inflorescence, drooping in bud but erect in flower and fruit; corollas yellow; achenes hairy, at least on the margins; with a slender beak as long as the body; pappus white; common in fields, woods, strip mines, roadsides, and waste ground; throughout; flr. Jul–Sep; native to Europe; FAC–.

Lactuca serriola, achene (×2)

Lapsana L.

Lapsana communis L.   Nipplewort
Annual to 1 m tall, glabrous or hirsute; leaves alternate, toothed to pinnatifid; involucre cylindrical with 8 inner bracts and a few shorter outer ones; florets yellow; achenes lacking pappus; occasional in moist woods, roadsides, waste ground and ballast; throughout; flr. Jun–Sep; native to Europe.

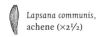

Lapsana communis, achene (×2½)

Marshallia
grandiflora

Marshallia
grandiflora,
leaf

Leontodon taraxacoides

Liatris spicata
(×¼)

Leontodon L.

Fibrous-rooted perennial herbs with a basal rosette of pinnately lobed or toothed leaves; heads few, terminal on simple or branched scapes; florets yellow; at least the inner achenes with plumose pappus bristles.

A. heads solitary; pappus of the outer achenes reduced to a few scales ....
.... *L. taraxacoides*
A. heads usually several; all achenes with a well-developed pappus of plumose bristles .... *L. autumnalis*

Leontodon autumnalis L.   Fall-dandelion
Scapes 1–8 dm tall; basal leaves glabrous to hirsute, lobed to entire; heads terminating the scaly-bracted branches; involucre 7–13 mm high with imbricate bracts; achenes fusiform-columnar, not beaked; pappus wholly of plumose bristles; rare and scattered in lawns, roadsides, waste ground, and ballast; flr. Jul–Sep; native to Europe.

Leontodon
taraxacoides,
head

Leontodon taraxacoides (Vill.) Mérat   Hawkbit
Scapes to 3.5 dm tall, mostly simple; leaves shallowly lobed, hispid-hirsute; heads solitary; involucre 6–11 mm high; achenes with a slight beak; pappus of the inner florets with well-developed plumose bristles and short outer scales, that of the outer florets reduced to a short ragged crown; rare in lawns, grassy banks, and ballast; SE and W; flr. Jun–Aug; native to Europe; FACU.

Liatris Schreb.

Unbranched perennial herbs with an elongate, terminal raceme or spike of discoid heads; involucral bracts imbricate; receptacle naked; florets purple, perfect, with conspicuous purple style branches; achenes 10-ribbed; pappus 1–2 rows of stout, barbed bristles.

A. heads broad, containing >20 florets; larger leaves 1–4 cm wide ...... L. scariosa
A. heads narrower, containing 3–14 florets; leaves all <1 cm wide .... L. spicata

Liatris scariosa (L.) Willd.   Northern blazing-star
Stems 3–8 dm tall with the lower leaves long-petiolate, blades elliptic to broadly oblanceolate; heads approximately 20, subsessile or with spreading peduncles to 5 cm long; florets 25–80 per head, corolla tube hairy toward the base within; rare in dry woods, shaly slopes, and barrens; mostly SC; flr. Aug–Sep; UPL; ✿. Several varieties are described based on the shape and number of cauline leaves, but farther study is needed to determine which are actually present in Pennsylvania.

Liatris spicata (L.) Willd.   Blazing-star
Stems to 2 m tall, leaves linear or linear-lanceolate, numerous, gradually reduced upward; heads crowded in a dense, elongate spike; florets 6–10 per head; occasional in moist fields, meadows, and roadsides on limestone, diabase, and serpentine; mostly SE and W; flr. Jul–early Sep; FAC+.

Liatris spicata, head (×1)

Madia Molina

Madia capitata Nutt.   Tarweed
Coarse, stipitate-glandular, heavy-scented annual to 1 m tall; leaves alternate, linear-oblanceolate, entire; heads clustered along the upper part of the main stem or the ends of the branches, yellow, radiate; receptacle chaffy only near the margin; pappus none; spreading along highways and in urban areas; flr. Jun; native to western North America and Chile.

Marshallia Schreb.

Marshallia grandiflora Beadle & F.E.Boynton   Barbara's buttons
Stems solitary or clustered, to 9 dm tall; leaves mostly toward the base, 3-nerved, broadly oblanceolate to elliptic; heads solitary on 1–3-dm-long peduncles; florets violet or white, all tubular and perfect; receptacle convex and chaffy; receptacular and involucral bracts acute; rare on sandy and rocky stream banks; SW; flr. Jun–Aug; FAC; ✿.

Matricaria L.

Branched, glabrous annuals, weakly to strongly aromatic; leaves bipinnatifid, with linear to filiform ultimate segments; involucral bracts dry with scarious margins; receptacles naked, domed or conical; disk corollas yellow.

A. ray florets lacking, disk florets 4-lobed .... M. matricarioides
A. ray florets present, white; disk florets 5-lobed
    B. receptacle domed; pappus a short crown; plants odorless .... M. perforata
    B. receptacle conical; pappus none; plants aromatic .... M. chamomilla

Matricaria chamomilla L.   Wild chamomile

Plants 2–8 dm tall; aromatic; heads with white ray florets and 5-lobed, yellow disk florets; achenes with 5 raised but not winged ribs; pappus lacking; ballast and waste ground; scattered; flr. May–Oct; native to Eurasia.

Matricaria matricarioides (Less.) Porter   Pineapple-weed

Plants 0.5–4 dm tall, pineapple-scented; heads entirely discoid, domed; corollas 4-lobed, yellow; achenes with a short crown of pappus; common in yards, railroad tracks, roadsides, and waste ground; throughout; flr. Jun–Sep; native to western North America.

Matricaria perforata Mérat   Wild chamomile

Plants 1–7 dm tall, nearly scentless; heads with white ray florets and 5-lobed, yellow disk florets; achenes with 3 strongly thickened, wing-like ribs; pappus a short crown; railroad tracks, streets, and urban waste ground; scattered; flr. Jun; native to Europe; UPL.

Megalodonta Greene

Megalodonta beckii, submersed leaves

Megalodonta beckii (Torr. & Spreng.) Greene   Beck's water-marigold

Aquatic perennial with opposite leaves, the submersed ones filiform-dissected; emersed leaves simple, lanceolate to ovate, sessile, serrate; heads above the water surface, terminal and solitary, 2–2.5 cm wide; rays yellow; achenes subterete, with a pappus of 3–6 awns; rare in calcareous lakes and swamps; NW and NE; flr. Jul–Oct; OBL; ❧.

Mikania Willd.

Mikania scandens (L.) Willd.   Climbing hempweed

Twining herbaceous vine to 5 m tall, puberulent and glandular; leaves opposite, petiolate; blades cordate, palmately veined, entire or with a few teeth; inflores-

Mikania scandens

cences small, axillary; heads discoid; florets all perfect, pink or whitish, 4 per head; frequent in swamps and moist thickets; mostly SE; flr. Jun–early Oct; FACW+.

Onopordum L.

Onopordum acanthium L.   Scotch thistle
Coarse, spiny, branching biennial to 2 m tall with winged stems; leaves toothed or lobed, sessile and decurrent, more or less tomentose; heads 2.5–5 cm wide, involucral bracts spine-tipped; florets purple; pappus of barbed capillary bristles; rare in urban waste ground; mostly SE and SW; flr. Jun–Jul; native to Eurasia.

Parthenium L.

Aromatic herbs with alternate leaves and small, dense, white or yellowish heads; rays few, small and inconspicuous; involucre of dry, scarcely herbaceous bracts; only the ray florets producing achenes, these black, obovate, and partially enclosed by the adjacent bracts; pappus inconspicuous, of 2–3 short awns or scales.

A. leaves merely toothed .... *P. integrifolium*
A. leaves pinnatifid or bipinnatifid .... *P. hysterophorus*

Parthenium integrifolium, leaf

Parthenium hysterophorus L.   Santa-Maria
Annual to 1 m tall, highly branched and hairy and glandular above; leaves pinnatifid or often bipinnatifid; heads small, numerous, in an open, leafy inflorescence; achenes obovate, black; rare in waste ground and barnyards; SE; flr. Aug–Oct; native to tropical America.

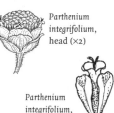

Parthenium integrifolium, head (×2)

Parthenium integrifolium L.   American feverfew
Perennial to 1 m tall; stem simple or branched, hairy above but glabrous below; leaves lance-elliptic to broadly ovate, crenate-serrate, the largest to 2 dm long; heads numerous in a flat-topped inflorescence; the rays scarcely 2 mm long; achenes obovate, black; rare on dry slopes and roadsides; scattered; flr. Jul; 🌿.

Parthenium integrifolium, ray flower (×4)

Petasites Mill.

Petasites hybridus (L.) Gaertn., Meyer & Scherb.   Butterfly-dock
Perennial with large cordate or kidney-shaped basal leaves with long petioles; stem 1–4 dm or taller in fruit; heads numerous, crowded in a narrow raceme, lacking rays; florets purple; cultivated and occasionally naturalized in moist sites and along creeks; mostly SE; flr. Apr–May; native to Europe.

Picris L.

Coarse plants with alternate leaves, the lower petioled, the upper sessile and clasping; heads terminating the loosely branched stems; only ray florets present; corollas yellow; achenes transversely wrinkled, beaked or beakless; pappus of plumose bristles.

A. involucral bracts narrow, imbricate; achene barely if at all beaked ....
.... *P. hieracioides*
A. involucral bracts in 2 series, the outer larger and foliaceous; achene with a
long, slender beak .... *P. echioides*

Picris echioides L.   Bristly ox-tongue
Rough-bristly annual, 3–8 dm tall; leaves toothed to entire; heads several; involucral bracts in 2 series, the outer ovate and spine-tipped; achenes with a long,
slender beak; naturalized in gardens and waste ground; scattered; flr. Jul–Oct;
native to Europe; UPL.

Picris hieracioides,
achene (×2¹⁄₂)

Picris hieracioides L.   Ox-tongue
Biennial to 1 m tall, hispid to subglabrous; leaves lanceolate or oblong; achenes
not beaked, the plumose pappus separating as a unit; occasionally naturalized
on roadsides, fields, and vacant lots; mostly S; flr. Jul–Oct; native to Eurasia.

Pluchea Cass.

Pluchea odorata (L.) Cass.   Marsh fleabane
Annual, stems to 1 m tall, finely glandular-puberulent; inflorescence a dense,
flat-topped cluster of heads; involucral bracts purplish, imbricate in several series, glandular; receptacle flat, naked; florets pinkish-purple, the outer florets
pistillate with filiform corolla shorter than the style, the central florets perfect
but mostly sterile with undivided style; pappus a single series of capillary
bristles, achenes very small; rare in tidal mudflats, wet ditches, and railroad ballast; SE, also locally where salt hay was used as mulch in nursery beds; flr. Aug–
Sep; 🍂.

Polymnia L.

Coarse perennial herbs with large, opposite leaves; heads with or without rays;
involucre a single series of bracts; receptacle flat, chaffy; achenes thick, pappus
none.

A. rays none or whitish, 1–1.5 cm long; achenes 3-angled but not striate ....
.... *P. canadensis*
A. rays yellow, 1–2 cm long; achenes impressed striate .... *P. uvedalia*

Polymnia canadensis L.   Leaf-cup
Stems to 2 m tall; leaves large, to 3 dm with a few pinnate lobes; heads in
crowded cymes; disk florets pale yellow; rays short, whitish; achenes 3-ribbed
and angled but not striate; rare on moist, rocky, wooded hillsides, floodplains,
and roadsides; scattered; flr. Jul–Sep.

Polymnia uvedalia L.   Bear's-foot, leaf-cup
Stem to 3 m tall; leaves to 3 dm or more, deltoid-ovate or elliptic, palmately
lobed; heads in open, leafy cymes; rays bright yellow, 1–2 cm long, rarely reduced; achenes 6 mm long, impressed striate; rare in ravines, thickets, and river
or stream banks; mostly S; flr. Jul–Sep.

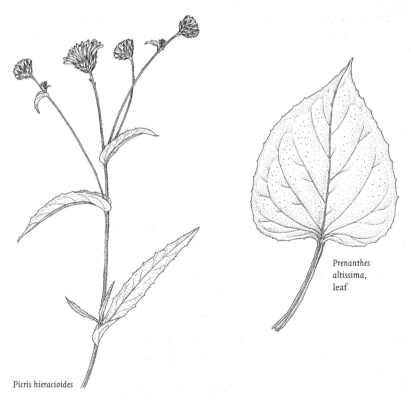

Prenanthes
altissima,
leaf

Picris hieracioides

Prenanthes L.

Tall, leafy perennial herbs with milky juice; leaves alternate, highly variable, frequently lobed, reduced upward; heads often nodding, 5–18 flowered; florets all ray type, perfect; achenes cylindric or tapering, ribbed, with numerous simple pappus bristles.

A. involucre glabrous
    B. florets 5–6 per head; principal involucral bracts 5 .... *P. altissima*
    B. florets 8–18 per head; principal involucral bracts 8
        C. pappus dark red-brown when mature .... *P. alba*
        C. pappus tan or pale brown .... *P. trifoliata*
A. involucre with long coarse hairs
    D. heads loosely ascending to erect .... *P. racemosa*
    D. heads nodding
        E. florets 19–27; principal involucral bracts 12–13 .... *P. crepidinea*
        E. florets 10–11; principal involucral bracts 8 .... *P. serpentaria*

Prenanthes alba L.   Rattlesnake-root
Stems stout to 1.7 dm tall; leaves more or less glaucous, the lower long-petioled and few-lobed to sagittate or hastate-reniform, the upper becoming nearly sessile; heads nodding; involucre with 8 principal bracts that are glabrous and purplish; florets white, pinkish, or lavender; occasional in rocky woods, barrens, and roadsides; throughout; flr. Jul–Oct; FACU.

Prenanthes altissima L.   Rattlesnake-root
Stems to 2 m tall, glabrous or spreading hirsute toward the base, the lower leaves long-petioled, deltoid to sagittate or cordate or sometimes deeply few-

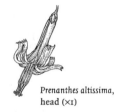

Prenanthes altissima,
head (×1)

lobed, the upper reduced; heads nodding, involucre glabrous with 4–6 principal bracts; florets cream-colored or greenish; common in woods; throughout; flr. Aug–Oct; FACU–.

Prenanthes crepidinea Michx.   Rattlesnake-root
Stems 1–2.5 m tall, glabrous below to puberulent or glandular-puberulent in the inflorescence; leaves petiolate, glabrous or scabrous above, elliptic to deltoid, cordate, or hastate, only slightly reduced upward, basal leaves shriveled by flowering time; heads nodding; involucre coarsely hairy, with 12–13 principal bracts; florets creamy; rare in moist rocky woods and thickets; SW; flr. Aug–Nov; FACU; 🌿.

Prenanthes racemosa Michx.   Glaucous rattlesnake-root
Plant to 1.7 m tall, long-hairy in the inflorescence, otherwise glabrous and glaucous; lower leaves broadly oblanceolate to obovate or elliptic, the upper becoming sessile and clasping; heads loosely ascending to nodding; involucre sparsely to densely long-hairy; florets pink or purplish; very rare in moist, boggy ground; W; believed to be extirpated; flr. Aug–Sep; FACW–; 🌿.

Prenanthes serpentaria Pursh   Lion's-foot
Stems to 1.5 m tall, glabrous or loose-hairy in the inflorescence; leaves with a pinnately few-lobed blade and wing-margined petiole, the basal leaves sometimes trifoliate; heads nodding; involucre with a few long, coarse hairs and 8 principal bracts; florets 10–11, creamy or greenish; occasional in dry woods, clearings, and gravelly roadsides; SE; flr. Aug–early Oct; 🌿.

Prenanthes trifoliata (Cass.) Fernald   Gall-of-the-earth
Stems stout, to 1.2 m tall, glabrous; leaves glabrous above, somewhat hairy beneath, deeply pinnately or palmately few-lobed to hastate or merely toothed; heads nodding; involucre of 8 principal bracts, glabrous; florets 10–12, white to creamy; frequent in sandy or rocky, open woods and shale barrens; throughout; flr. Aug–Oct.

Ratibida columnifera, leaf

Ratibida columnifera, head

Ratibida pinnata, head

Prenanthes trifoliata, leaf

Prenanthes trifoliata

Ratibida Raf.

Short-hairy, perennial herbs to 1.2 m tall with alternate, pinnatifid leaves; heads few to many on naked peduncles, with 3–13 strongly drooping rays; involucre a single series of green, linear or lance-linear bracts; receptacle globose or columnar, with pubescent bracts subtending the disk florets, but not the rays, and clasping the achenes to some extent; achenes compressed at right angles to the bracts, pappus of awn-like teeth or none.

A. receptacle globose or short-cylindrical, shorter than the rays .... R. pinnata
A. receptacle cylindrical, longer than the rays ....R. columnifera

Ratibida columnifera (Nutt.) Wooten & Standl.   Coneflower
Tap-rooted; stems leafy to above the middle; leaves pinnatifid or even bipinnatifid; heads several to many; the disk 2–4.5 times as long as thick; rays yellow or partly purple-brown, spreading or reflexed; pappus an awn-like tooth on the inner angle of the achene; escaped from cultivation to fields and waste ground; flr. Jun–Aug; native farther west, adventive here.

Ratibida pinnata (Vent.) Barnhart   Prairie coneflower
Fibrous-rooted; lower leaves long-petioled, the upper nearly sessile; heads several; disk 1–1.6 times as long as thick, shorter than the rays; rays pale yellow, spreading or reflexed; rare in dry fields, limestone uplands, and open roadsides; SW; flr. Jun–Aug; 🍂.

Rudbeckia L.

Upright herbs with alternate leaves; heads solitary on long peduncles or in an open, branched inflorescence; involucral bracts green and herbaceous, spreading; disk florets numerous, dark purple-brown (or greenish in *R. laciniata*) perfect, on a conical or cylindrical receptacle; ray florets yellow or orange; receptacle chaffy with blunt or awn-tipped bracts partly enclosing the 4-angled achenes, pappus absent or consisting of a minute crown.

A. leaves entire or merely toothed
　　B. receptacular bracts with an acuminate, awn-like tip, glabrous throughout
　　.. R. triloba
　　B. receptacular bracts acute to rounded, pubescent, ciliate or glabrous at the tip
　　　　C. stem and leaves densely covered with stiff hairs; achenes without pappus .... R. hirta
　　　　C. stem and leaves with scattered soft hairs; pappus an inconspicuous low crown .... R. fulgida
A. leaves (at least some of them) deeply pinnatifid or trilobed
　　D. largest leaves 5–7-lobed; stem glabrous and usually glaucous; disk florets greenish .... R. laciniata
　　D. largest leaves 3-lobed; stem pubescent, never glaucous; disk florets purple-brown .... R. triloba

Rudbeckia fulgida Aiton   Eastern coneflower
Rhizomatous perennial with pubescent to nearly glabrous stems to 1 m tall; leaves entire to coarsely toothed, hairy on both sides; heads 2.5–4 cm wide,

Rudbeckia fulgida
var. *fulgida* (×¼)

Rudbeckia
laciniata,
head

Rudbeckia laciniata,
leaf

Rudbeckia
triloba
(×¼)

mostly solitary on long peduncles; disk florets dark purple-brown; ray florets 8–20, yellow or yellow-orange; receptacular bracts glabrous with ciliate margins; pappus a low crown; occasional in moist fields and meadows; SE; flr. Aug–Oct; FAC; ✤; 2 varieties:

A. rays 2.5–4 cm; leaves sharply toothed .... var. *speciosa* (Wender.) Perdue
A. rays 1–2.5 cm long; cauline leaves entire to merely denticulate ...... var. *fulgida*

Rudbeckia hirta L.   Black-eyed-susan
Biennial or short-lived perennial to 1 m tall, hairy throughout; leaves lanceolate to oblong, 3–5 nerved, mostly sessile; heads solitary on long peduncles, 5–8 cm wide; disk florets numerous on a conical receptacle, dark purple-brown; rays 8–21, orange to orange-yellow; pappus none; common in fields, meadows, and roadsides; throughout; native, but also a major component of wildflower seed mixtures; flr. Jun–Oct; FACU–; 2 varieties:

A. leaves ovate to oval, coarsely toothed .... var. *hirta*
    mostly in undisturbed habitats.
A. leaves oblanceolate, entire or only slightly toothed .... var. *pulcherrima* Farw.
    widespread in disturbed habitats.

Rudbeckia laciniata L.   Cutleaf coneflower
Woody-based perennial to 2(3) m tall with erect, glabrous and usually glaucous stems and pinnately lobed or divided leaves; heads 6–10 cm wide, several to many in a branching inflorescence; disk florets greenish, the receptacle becoming cylindrical in fruit; rays yellow and drooping; bracts of the receptacle blunt; pappus a short, toothed crown; common on floodplains, stream banks, and wet fields; throughout; flr. Jul–Sep; FACW. A double-flowered form, var. *hortensis* Bailey, is sometimes found as a garden escape.

Rudbeckia triloba L.   Three-lobed coneflower
Biennial or short-lived perennial 0.5–1.5 m tall; basal leaves broadly ovate or

subcordate, petioled, at least some of them 3-lobed; upper leaves unlobed, narrower, and becoming sessile; heads several on short peduncles in an open, branched inflorescence; rays 6–13, yellow or orange, 1–2 cm long; disk dark purple or brown, 8–15 mm wide; bracts of the receptacle abruptly narrowed to an acuminate awned tip; occasional in moist to dry pastures, old fields, rocky slopes, and edges, frequently on limestone or diabase; scattered; throughout; flr. Aug–Oct; FACU.

Senecio L.

Herbaceous annuals or perennials typically with toothed basal leaves and reduced, more or less pinnatifid, alternate cauline leaves; perennial species have sterile leafy shoots at the base separate from the flowering stem; heads radiate (except *S. vulgaris*), solitary or in corymbose clusters; ray florets yellow, pistillate, and fertile; disk florets perfect; receptacle flat, not chaffy; involucre often with smaller bractlets present at the base; achenes cylindrical, pappus of white, capillary bristles.

A. plants densely glandular-hairy .... *S. viscosus*
A. plants not glandular-hairy
    B. ray florets not present; bracteoles black-tipped .... *S. vulgaris*
    B. ray florets present; bracteoles not black-tipped
        C. plants permanently tomentose at least on the stem, lower leaf surfaces, and involucres
            D. tomentum fine and close; basal leaf blades <5 cm ....
            .... *S. antennariifolius*
            D. tomentum loose; basal leaf blades >5 cm .... *S. plattensis*
        C. plants not persistently tomentose except perhaps at the base or in the leaf axils
            E. undivided basal leaves with the blade tapered and decurrent along the petiole; blades obovate to somewhat rounded .... *S. obovatus*
            E. undivided basal leaves with slender petioles, the blade not decurrent; blades oblanceolate, spatulate, oblong, or ovate
                F. basal leaves cordate .... *S. aureus*
                F. basal leaves tapering to the petiole
                    G. heads rarely >20; stem glabrous at the base .... *S. pauperculus*
                    G. heads 20–100 or more; stem densely woolly at the base ....
                  .... *S. anonymus*

Senecio anonymus A.W.Wood   Appalachian groundsel
Perennial with erect stems to 8 dm tall, persistently tomentose at the base but otherwise glabrate; basal leaves oblanceolate to elliptic, tapering to the petiole, crenate or serrate, cauline leaves reduced, becoming sessile, deeply pinnatifid; heads numerous, 20–100; rare in dry fields, open woods, and serpentine barrens; SE; flr. May–Sep; UPL; 🍂.

Senecio antennariifolius Britton   Shale-barren ragwort; cat's-paw ragwort
Perennial to 4 dm tall from a branching base; basal leaves numerous, elliptic to obovate, glabrate above and white-tomentose beneath, stem leaves reduced upward, somewhat pinnatifid; heads 3–12, involucre densely white tomentose; achenes finely hairy; rare on shale barrens; SC; flr. Apr–Jun; 🍂.

Senecio aureus L. Golden ragwort

Perennial with erect stems 3–8 dm from a branched or creeping rhizome, leafy basal offshoots present; basal leaves with long petioles and cordate bases; cauline leaves reduced and more or less pinnatifid; heads many, with 6–13 rays; achenes glabrous; common in moist fields, woods, floodplains, and roadsides; throughout; flr. May–Jul; FACW.

Senecio obovatus Muhl. Ragwort

Perennial with erect stems to 7 dm tall and slender, superficial rhizomes bearing tufts of leaves, lightly tomentose when young but becoming glabrate; petioles of the basal leaves narrowly winged to the base, blades obovate to orbicular; cauline leaves reduced, becoming sessile upward, pinnatifid; heads usually many; involucral bracts often purple-tipped; achenes glabrous; moist fields, woods, and calcareous slopes; scattered; throughout; flr. May–Jul; FACU–.

Senecio obovatus, head (×1)

Senecio pauperculus Michx. Balsam ragwort

Perennial with erect stems to 5 dm tall, sometimes with short, slender stolons or rhizomes, lightly tomentose when young but soon becoming glabrate; basal leaves oblanceolate to elliptic, tapering to the petiole, crenate, serrate or nearly entire; cauline leaves reduced upward and becoming sessile, pinnatifid; heads few, rarely more than 20; fields, meadows, peaty thickets, stream banks, and roadsides; mostly SE; flr. May–Jun; FAC.

Senecio plattensis Nutt. Prairie ragwort

Biennial or short-lived perennial with a single stiffly erect stem to 7 dm tall, the stem, lower leaf surfaces and involucres persistently floccose-tomentose at least until flowering time; heads several or many; very rare in dry woods and fields; SC; flr. May–Jul; UPL; .

Senecio viscosus L Sticky groundsel

Aromatic, leafy-stemmed annual to 6 dm tall, covered throughout with glandular hairs; leaves pinnatifid and toothed; ray florets present but inconspicuous; achenes glabrous; roadsides, floodplains, railroad tracks, and waste ground; scattered; flr. Jul–Sep; native to Europe.

Senecio obovatus

Senecio obovatus, basal rosette

Senecio aureus, basal leaves

Senecio vulgaris

Senecio vulgaris L. Common groundsel

Leafy-stemmed annual to 4 dm tall, leaves pinnatifid or coarsely and irregularly toothed, the upper sessile and clasping; heads several or many, only disk florets present; involucre subtended by black-tipped bracteoles; achenes with short hairs on the angles; a frequent weed of streets, gardens, roadsides, and vacant lots, where it can often be found blooming throughout the winter months; flr. Feb–Dec; native to Eurasia; FACU.

Senecio vulgaris, head after most achenes were shed (×1)

Silphium L.

Coarse perennial herbs to 2 m or more with opposite or whorled leaves; heads with yellow rays; involucral bracts in 2–several series; receptacle flat, chaffy; achenes flattened, glabrous.

A. leaves opposite, fused at the base and perfoliate; stem square ... S. *perfoliatum*
A. leaves whorled or opposite, tapering to a petiolar base, not fused; stem not square .... S. *trifoliatum*

Silphium perfoliatum L. Cup-plant

Stems to 2.5 m tall, square; leaves opposite and fused at the base forming a funnel-shaped structure that completely surrounds the stem; heads in an open inflorescence, rays yellow; occasional in floodplains, abandoned fields, and moist meadows; mostly S; native, but also cultivated and escaped; flr. Jul–Oct; FACU.

Silphium trifoliatum L. Whorled rosinweed

Stems 1–2 m tall, terete; leaves whorled or opposite, tapering to a petiolar base, not fused; heads several to numerous in an open inflorescence; rays 8–13, disk 1–2 cm wide; occasional on roadsides, dry fields, and meadows; SC and W; flr. Jul–Sep. Ours is var. *trifoliatum*.

Silybum Adans.

Silybum marianum (L.) Gaertn. Milk thistle

Spiny winter annual or biennial with alternate leaves and globose heads consisting entirely of disk florets; rare in waste ground; widely scattered; flr. May–Jul; native to the Mediterranean region.

Solidago L.

Upright perennial herbs 0.5–1(2) m tall from a short, stout caudex or more elongate rhizome; leaves basal and cauline, in some species the basal leaves are large and persistent and the cauline leaves much reduced, in others the leaves are mainly cauline as the basal leaves wither early; inflorescence terminal and pyramidal with recurved-secund branches, or slender, elongate and nonsecund, or of axillary clusters, or a flat-topped corymb; individual heads small; ray and disk florets both present, yellow (white in S. *bicolor*); involucral bracts imbricate, usually green-tipped; receptacle small, flat or slightly convex, naked; achenes subterete or angled, several nerved; pappus of numerous white capillary bristles.

A. inflorescence flat-topped .... S. *rigida*
A. inflorescence various but not flat-topped

B. inflorescence wand- or club-shaped, or entirely or mainly of distinct axillary clusters

  C. inflorescence mainly of axillary clusters

    D. stems upright or arched, not zigzag; leaves lance-ovate

      E. stems terete, purple-glaucous .... *S. caesia*

      E. stems striate-angled, not glaucous .... *S. curtisii*

    D. stems erect, zigzag, green, never glaucous; leaves broadly ovate .... *S. flexicaulis*

  C. inflorescence terminal, club or wand-shaped

      E. heads creamy white .... *S. bicolor*

      E. heads yellow

        F. involucral bracts with conspicuous recurved (squarrose) tips .... *S. squarrosa*

        F. involucral bracts not recurved at the tip

          G. plants glabrous or nearly so; petioles of the lowest leaves with sheathing bases .... *S. uliginosa*

          G. plants pubescent at least in part; petioles of the basal leaves not sheathing the stem

            H. involucral bracts narrowed to a slender point

              I. rays 9–16; leaves and stem with fine, stiffly spreading hairs throughout .... *S. puberula*

              I. rays 6–9; leaves and stem glabrous below the inflorescence or with the stem irregularly short-hairy .... *S. roanensis*

            H. involucral bracts obtuse to merely acute

              J. achenes persistently short-hairy .... *S. simplex*

              J. achenes glabrous when mature

                K. leaves and stem strongly hirsute .... *S. hispida*

                K. leaves and stem essentially glabrous below the inflorescence .... *S. speciosa*

B. inflorescence pyramidal with nodding and/or recurved-secund branches

  L. cauline leaves triple-nerved by virtue of a pair of strong lateral veins that arise near the base of the midrib

    M. leaves entire, somewhat succulent .... *S. sempervirens*

    M. leaves more or less toothed, not succulent

      N. stems glabrous below the inflorescence

        O. basal and lower cauline leaves long-petioled, persistent, middle and upper cauline leaves relatively few and remote .... *S. juncea*

        O. basal leaves and lower cauline leaves shriveled by flowering time .... *S. gigantea*

      N. stems pubescent above, at least to the midpoint

        P. stems pubescent only to the midpoint, glabrate below; involucres 2–3 mm high .... *S. canadensis*

        P. stems pubescent to the base; involucres 3–4.6 mm high .... *S. altissima*

  L. cauline leaves with a distinct midrib, not triple-veined although pinnate secondary veins definitely present

    Q. stems pubescent, at least in the upper half of the plant

      R. leaves glabrous, translucent punctate-dotted, fragrant when bruised .... *S. odora*

R. leaves variously pubescent
    S. cauline leaves entire or obscurely crenate; leaves and stems puberulent throughout .... *S. nemoralis*
    S. cauline leaves sharply toothed, stems with spreading hairs; leaves hairy, at least on the main veins beneath .... *S. rugosa*
Q. stems glabrous below the inflorescence
    T. lowest cauline leaves with clasping bases .... *S. uliginosa*
    T. lowest cauline leaves not clasping the stem
        U. stems strongly ridged or angled, upper leaf surface very scabrous .... *S. patula*
        U. stems terete; upper leaf surface smooth or only slightly scabrous
            V. basal and lower cauline leaves much larger than the middle and upper cauline leaves; leaves glabrous or with a few appressed hairs
            W. basal leaves gradually tapering to the petiole ... *S. juncea*
            W. basal leaves abruptly narrowed to the petiole ... *S. arguta*
            V. basal and lower cauline leaves mostly shriveled at the time of flowering or, if present, not much larger than the mid-cauline leaves; leaves hairy at least on the midrib and main veins beneath .... *S. ulmifolia*

Solidago altissima L.   Late goldenrod
Stems to 2 m tall, solitary or clustered, from long rhizomes, short pubescent throughout; leaves tapering to a sessile base, entire or serrate toward the tips, triple-nerved, lower surface finely pubescent, upper surface scabrous; inflorescence with curved, secund branches, pubescent; common in woods, fields, riverbanks, and roadsides; throughout; flr. Jul–Nov. [*S. canadensis* L. var. *scabra* Torr. & A.Gray of Rhoads and Klein, 1993]

Solidago arguta Aiton   Forest goldenrod
Stems 5–15 dm tall, glabrous except in the inflorescence; leaves mostly basal, glabrous or slightly scaberulous above, the blade broadly elliptic and abruptly narrowed to a long petiole; inflorescence with recurved, secund branches, broad and open to occasionally elongate and narrow; flr. Jul–Oct; 2 varieties:

A. achenes glabrous .... var. *arguta*
   frequent in rocky woods, dry thickets, and roadsides; throughout.
A. achenes hairy .... var. *harrisii* (E.S.Steele) Cronquist
   very rare on shale barrens; SC; 🌿.

Solidago bicolor L.   Silver-rod, white goldenrod
Stems 1–10 dm tall, covered with stiff, spreading hairs throughout; the larger leaves oblanceolate to obovate, reduced upward and becoming bracts; inflorescence narrow and elongate, leafy-bracted below; rays creamy white; common in dry woods, wooded banks, and shale barrens; throughout; flr. Aug–Oct.

Solidago caesia L.   Bluestem or wreath goldenrod
Stems erect or arching, terete, slender, glabrous, purplish-waxy, simple or with a few branches; leaves lance-ovate, sessile, serrate, only slightly reduced upward; heads in axillary clusters; common in rich woods; throughout; flr. Aug–late Oct; FACU.

Solidago canadensis L.   Canada goldenrod
Stems to 2 m tall, puberulent; leaves mainly cauline, numerous, triple-nerved, lance-linear to lance-elliptic, tapering at both ends; inflorescence pyramidal, with strongly recurved-secund branches; fields and roadsides; flr. Jul–Oct; FACU; 3 varieties:

A. pubescence relatively sparse and restricted, lower half of the stem nearly glabrous; leaves thin, sharply serrate to entire
  B. involucre 2–3 mm high; disk florets 3–7 .... var. *canadensis*
     common; throughout.
  B. involucre 3–4 mm high; disk florets 8–13 ....
     .... var. *salebrosa* (Piper) M.E.Jones
     rare, known from only a single Erie Co. site.
A. pubescence dense, covering most of the stem; leaves firm, shallowly few-toothed to entire .... var. *hargeri* Fernald
   frequent; throughout.

Solidago altissima (×¼)

Solidago bicolor (×¼)

Solidago caesia (×¼)

Solidago juncea (×¼)

Solidago gigantea, leaf (×¼)

Solidago flexicaulis (×¼)

Solidago juncea, leaf (×¼)

S. curtisii Torr. & A.Gray   Curtis's goldenrod
Very similar to S. *caesia* but with the stem striate-angled and grooved; rare in floodplain forests; SW; flr. Aug–Oct; 🍂. [S. *caesia* L. var. *curtisii* (Torr. & A.Gray) A.W.Wood of Rhoads and Klein, 1993]

Solidago flexicaulis L.   Zigzag goldenrod
Stems erect from creeping rhizomes, angled at the nodes in a zigzag pattern; leaves ovate, acuminate, narrowing to a winged petiole, sharply dentate, the upper ones reduced; heads in short axillary clusters; frequent in moist woods and rocky, wooded slopes; throughout; flr. Aug–Nov; FACU.

Solidago gigantea Aiton   Smooth goldenrod
Very similar to S. *canadensis* but stems pubescent only in the inflorescence, otherwise glabrous and glaucous; leaves glabrous or with a few hairs along the 3 main veins beneath; lower stem leaves mostly shriveled by flowering time; common in moist fields, meadows, banks, and ditches; throughout; flr. Jul–Oct; FACW; var. *serotina* (Aiton) Cronquist has been distinguished from the typical variety on the basis of wider leaves.

Solidago hispida Muhl.   Hairy goldenrod
Very similar to S. *bicolor* but the rays yellow; dry rocky slopes and wooded roadside banks; mostly SC; flr. Aug–Oct.

Solidago juncea Aiton   Early goldenrod
Plant to 1.2 m tall, often with a reddish stem; basal leaves prominent, blades

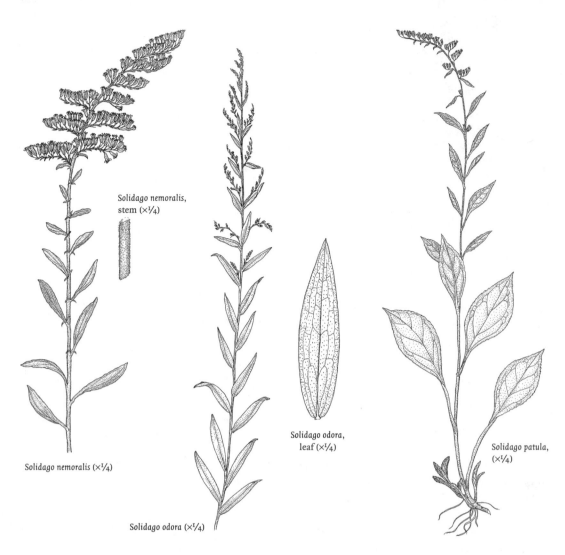

Solidago nemoralis, stem (×¼)

Solidago nemoralis (×¼)

Solidago odora (×¼)

Solidago odora, leaf (×¼)

Solidago patula, (×¼)

narrowly elliptic, more or less serrate, acuminate; cauline leaves few, reduced; leaves and inflorescence glabrous or short hirsute; inflorescence as broad as long with dense, recurved-secund branches; common in fields, meadows, rocky slopes, and roadsides; throughout; flr. early Jul–Oct.

Solidago nemoralis Aiton   Gray goldenrod
Plant to 1 m tall, densely and finely puberulent throughout; basal leaves prominent, weakly triple-nerved, oblanceolate, toothed; inflorescence varying from long and narrow with a nodding tip to wider with recurved-secund branches; common in fields, woods, and roadsides in dry, sterile soils; throughout; flr. Jun–Oct.

Solidago odora Aiton   Sweet goldenrod
Stems to 1.6 m tall, rough-puberulent in the inflorescence and in lines decurrent from the upper leaf bases; leaves cauline, sessile, entire, glabrous, finely translucent-punctate, and smelling of anise when bruised; inflorescence with spreading, recurved-secund branches; occasional in dry open woods and barrens; E; flr. Jul–early Oct.

Solidago patula Muhl. ex Willd.   Spreading goldenrod
Stems to 2 m tall, glabrous below the inflorescence and angular, at least below;
basal leaves prominent, glabrous beneath and strongly scabrous above due to
distinctive bulbous-based hairs, the lowest with elliptic-ovate to elliptic-obovate,
sharply-toothed blades and somewhat sheathing petioles; inflorescence pyramidal with widely spreading recurved-secund branches, or narrower and denser
but still secund in smaller specimens; frequent in swamps, floodplains, and
moist woods; scattered throughout; flr. Aug–early Oct; OBL.

Solidago puberula Nutt.   Downy goldenrod
Stems 2–10 dm tall, entire plant minutely puberulent; basal leaves oblanceolate,
tapering to a long petiole, lower stem leaves lanceolate, becoming reduced upward; inflorescence elongate, with short erect branches; heads with 9–16 rays;
involucral bracts tapering to a narrow tip; occasional in rocky woods, barrens,
and roadsides; E and SC; flr. Aug–early Oct; FACU–.

Solidago rigida L.   Stiff goldenrod
Plants to 1.5 m tall, densely pubescent throughout with stiff, spreading hairs;
leaves firm, elliptic to elliptic-oblong, slightly toothed to entire; inflorescence a
dense, flat-topped corymb; rare in moist fields or thickets on calcareous soils;
scattered E and C; flr. Aug–Oct; UPL; 🌱.

Solidago rigida (×¼)

Solidago puberula (×¼)

Solidago roanensis Porter   Mountain goldenrod
Very similar to S. *puberula* but leaves and stem glabrous below the inflorescence
or with stem irregularly short-hairy; rays 6–9; rare on rocky banks, roadsides,
cut-over woods, and woods edges; SW; flr. Aug–Sep; 🌿.

Solidago rugosa Mill.   Wrinkle-leaf goldenrod
Strongly colonial to 1.5 m tall; stems spreading-hirsute; leaves chiefly cauline,
numerous, rugose-veiny and glabrous or scabrous above, hairy beneath at least
on the veins; inflorescence with recurved-secund branches; common in fields,
woods, floodplains, roadsides, and waste ground; throughout; flr. Aug–Nov;
FAC; 3 varieties (we have chosen not to recognize var. *sphagnophila* Graves):

A. leaves relatively thin, sharp-toothed and acuminate, not strongly rugose
    B. inflorescence with long branches exceeding the subtending leaves ....
    .... var. *rugosa*
        common; throughout.
    B. inflorescence narrow, leafy .... var. *villosa* (Pursh) Fernald
        occasional; throughout.
A. leaves thick and firm, strongly rugose-veiny, blunt-toothed to subentire, acute
    .... var. *aspera* (Aiton) Fernald
    occasional, except at the highest elevations.

Solidago sempervirens L.   Seaside goldenrod
Plant to 2 m tall, somewhat succulent, glabrous except perhaps in the inflorescence; petiole of the lower leaves with a clasping base; cauline leaves numerous, reduced upward; inflorescence pyramidal with spreading-recurved lower
branches; heads secund; occasional on waste ground and roadsides where deicing salts are used; flr. late Jul–Oct; native in coastal areas, but adventive in
Pennsylvania; FACW.

Solidago simplex Kunth   Sticky goldenrod
Plant 1–9 dm tall, glabrous except for some puberulence in the inflorescence;
leaves mostly basal, oblanceolate to obovate, subentire; cauline leaves reduced,
lance-elliptic; inflorescence varying from a dense wand to more open and
branched, but never secund; rare in rock crevices and shores of the Lower Susquehanna River; flr. Aug–Sep; 🌿. Ours is ssp. *randii* (Porter) Ringius var. *racemosa*
(Greene) Ringius. [syn: S. *spathulata* DC. ssp. *randii* (Porter) Cronquist var.
racemosa (Greene) Cronquist]

Solidago speciosa Nutt.
Plant to 1.5 dm tall, coarsely puberulent in the inflorescence but otherwise glabrous, or a bit scabrous; leaves thick and firm, the lower slightly toothed, the
persistent lowest ones abruptly petiolate; inflorescence erect and simple or with
ascending branches, not at all secund; flr. late Aug–Oct; 2 varieties:

A. inflorescence narrow, spike-like, with barely developed branches, often inter-
    rupted; leaves 0.5–2 cm wide .... var. *erecta* MacMill.—slender goldenrod
    rare on dry, acidic, shaly banks; S; 🌿.
A. inflorescence broader with densely flowered ascending branches; middle
    cauline leaves often >2 cm wide .... var. *speciosa*—showy goldenrod
    rare in moist meadows, rocky woods, thickets, and roadsides on diabase and
    limestone; mostly S; 🌿.

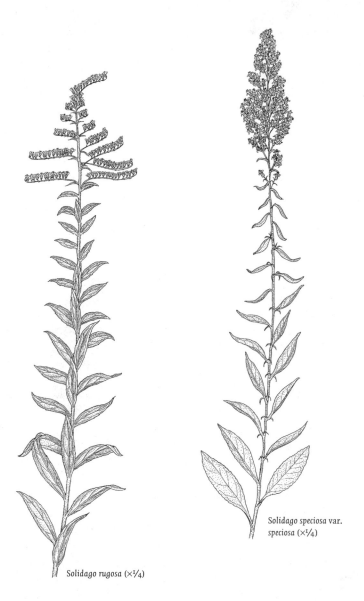

Solidago rugosa (×¼)

Solidago speciosa var.
speciosa (×¼)

Solidago squarrosa Muhl.   Ragged or stout goldenrod
Stems to 1.5 dm tall, glabrous below the inflorescence; leaves glabrous or scabrous above, the lower ones broadly oblanceolate to obovate or elliptic, sharply
serrate, tapering to a long petiole; inflorescence narrow and elongate, leafy-
bracted below; outer involucral bracts with prominent recurved green tips; frequent in mountain woods, rocky roadside banks, and thickets; throughout; flr.
Jul–Oct.

Solidago uliginosa Nutt.   Bog goldenrod
Plants to 1.5 m tall, glabrous except in the inflorescence; basal leaves prominent,
tapering to a long petiole with a sheathing base; blades oblanceolate to narrowly
elliptic, serrate to subentire; cauline leaves reduced; inflorescence longer than
broad, branches straight, not secund, or with short recurved-secund branches
and a recurved tip; infrequent in bogs, swamps, sedge meadows, and fens;
throughout; flr. Aug–Oct; OBL; .

Solidago ulmifolia Muhl. ex Willd. Elm-leaved goldenrod
Stems to 1.2 m tall, glabrous below the inflorescence; leaves sharply serrate, hairy, at least on the midrib and main veins beneath; basal and lower cauline leaves shriveling early; inflorescence relatively broad with a few elongate and divergent recurved-secund branches; frequent on wooded slopes, roadside banks, and shale barrens; mostly S; flr. Jul–early Nov.

Sonchus L.

Tall herbs to 2 m with milky sap; leaves alternate or basal, entire to coarsely toothed or pinnately-lobed, usually auriculate-clasping and often prickly-margined; heads several to many in a panicle-like inflorescence; heads discoid; florets all perfect; corollas yellow; achenes flattened, beakless, pappus of numerous capillary bristles that fall as a unit plus a few outer bristles that fall separately.

A. flowering heads 3–5 cm wide; fruiting involucre 14–22 mm high .... *S. arvensis*
A. flowering heads <2.5 cm wide; fruiting involucre 9–13 mm high
    B. mature achenes with a transversely roughened surface between the nerves; leaf bases mostly with acute lobes as well as prickly teeth .... *S. oleraceus*
    B. mature achenes smooth between the prominent longitudinal nerves; leaf bases generally with rounded (and prickly-toothed) lobes .... *S. asper*

Sonchus arvensis L. Field sow-thistle
Perennial with creeping roots; corollas deep yellow to orange; achenes roughened between the 5 ribs on each side; roadsides, abandoned fields, and waste ground; scattered; throughout; flr. Jul–Aug; native to Europe; UPL; 2 subspecies:

A. involucre and pedicels with numerous stiff hairs with dark glandular tips .....
....ssp. *arvensis*
A. involucre and pedicels lacking glandular hairs ....
.... ssp. *uliginosus* (M.Bieb.) Nyman

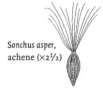

Sonchus asper, achene (×2½)

Sonchus asper (L.) Hill Spiny-leaved sow-thistle
Annual with prickly leaves with conspicuously coiled, clasping, rounded but prickly-toothed basal lobes; achenes very flat, with conspicuous ribs but not otherwise roughened; frequent in abandoned fields, roadsides, and waste ground; throughout; flr. May–Oct; native to Europe; FAC.

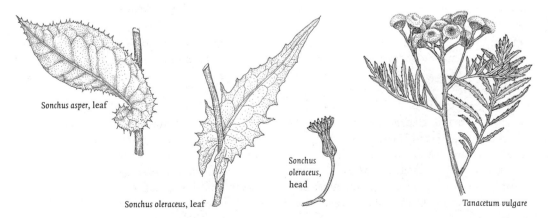

Sonchus asper, leaf

Sonchus oleraceus, leaf

Sonchus oleraceus, head

Tanacetum vulgare

Sonchus oleraceus L.   Common sow-thistle

Sonchus oleraceus,
achene ($\times 2\frac{1}{2}$)

Mostly glabrous annual from a short taproot; leaves pinnatifid, reduced upward; basal lobes generally with an acute tip and prickly teeth; corollas pale yellow; frequent on roadsides, shores, and urban waste ground; mostly S; flr. Jun–Oct; native to Europe; UPL.

Tanacetum L.

Tanacetum vulgare L.   Common tansy

Glabrous, aromatic perennial to 1.5 m tall; leaves alternate, pinnately finely divided; heads 5–10 mm wide, arranged in a flat-topped inflorescence; heads compact, yellow, discoid, occasionally with ray-like extensions developing in the marginal florets; infrequent along roadsides, fencerows, fields, and pastures; throughout; flr. Jul–Oct; native to Europe.

Taraxacum F.H.Wigg.

Biennials with pinnately lobed or toothed leaves in a basal rosette; heads borne singly on naked, hollow scapes; involucral bracts in 2 series; florets all ray type; corollas yellow; mature fruiting heads nearly spherical with strongly recurved involucral bracts and radiating achenes with conspicuous stalked pappus. A taxonomically difficult complex of polyploids and apomictic forms.

A. involucral bracts hooded at the tip; achenes red or reddish at maturity ....
.... *T. laevigatum*
A. involucral bracts not hooded at the tips; achenes gray-brown or tan at maturity .... *T. officinale*

Taraxacum laevigatum (Willd.) DC.   Red-seeded dandelion

Very similar to *T. officinale*, but leaves generally more deeply lobed, heads somewhat smaller, and achenes reddish or purplish; occasional in woods, rocky slopes, and waste ground; throughout; flr. Apr–Sep; native to Eurasia.

Taraxacum officinale Weber   Common dandelion

Characteristics of the genus; a common weed of fields, roadsides, lawns, and waste ground; throughout; flr. Mar–Dec; native to Eurasia; FACU–.

Tragopogon L.

Glabrous biennial herbs; leaves alternate, to 3 dm long, entire, linear with clasping bases and long slender tips; heads solitary at the ends of the branches; involucral bracts in 1 series, elongating in fruit; florets all ray type and perfect; achenes terete or angled, 5–10-nerved with a slender beak topped with numerous plumose bristles.

A. peduncle swollen just below the solitary head
    B. rays pink to purple .... *T. porrifolius*
    B. rays yellow .... *T. dubius*
A. peduncle not swollen .... *T. pratensis*

Tragopogon dubius Scop.   Yellow goatsbeard

Peduncle 3–9 dm long, swollen just below the solitary head; heads to 6 cm wide; rays yellow; achenes 2–4 cm long; frequent in fields, roadsides, and waste places; throughout; flr. May–Aug; native to Europe.

Tragopogon porrifolius L.   Oyster-plant, salsify

Peduncle 6–12 dm long, swollen and hollow just below the solitary head; heads 5–10 cm wide; rays pink to purple; achenes and pappus 5–6 cm long; cultivated and occasionally naturalized in fields and roadsides; mostly S; flr. May–Jul; native to Europe.

Tragopogon pratensis L.   Meadow salsify

Peduncle 3–9 dm long, slenderly cylindric below the solitary head; heads to 6 cm wide; rays yellow; achenes 1.5–2.5 cm long; occasional in fields and waste places; throughout; flr. May–Aug; native to Europe.

Tussilago L.

Tussilago farfara L.   Coltsfoot

Herbaceous perennial; flowering stems arising from a horizontal creeping rootstock, 8–45 cm tall and bearing several 1-cm-long bracts; head solitary, 2.5 cm

Verbesina alternifolia

Tragopogon dubius (×¼)

wide; ray florets yellow, pistillate, fertile, surrounding the central sterile disk florets; leaves appearing after flowering has ceased, basal, heart-shaped, 5–18 cm long, slightly toothed, upright, whitish beneath; frequent on roadsides, railroad rights-of-way, stream banks, and waste places; throughout; flr. Mar–Jun; native to Eurasia; FACU.

Verbesina L.

Verbesina alternifolia (L.) Britton   Wingstem

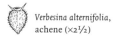

Verbesina alternifolia,
achene (×2½)

Perennial herb to 2.5 m tall with a conspicuously winged stem; leaves alternate, lance-elliptic to ovate, numerous; heads 10–100 in an open inflorescence, yellow, 3–4 cm wide; involucral bracts few, glabrous or subglabrous, small, narrow, and soon deflexed; disk florets and achenes forming a globose head; pappus of 2 awns; frequent in moist, wooded slopes, riverbanks, and shaded lowlands; S; flr. Aug–Oct; FAC. [syn: *Actinomeris alternifolia* (L.) DC.]

Vernonia Schreb.

Tall, late summer-blooming, perennial herbs, unbranched below the inflorescence; leaves alternate, sessile or nearly so, numerous; heads discoid; involucre of numerous appressed, overlapping bracts narrowed at the top; receptacle flat, naked; florets purple, all tubular and perfect; achenes ribbed and hairy; pappus a double row of capillary bristles, the inner longer than the outer.

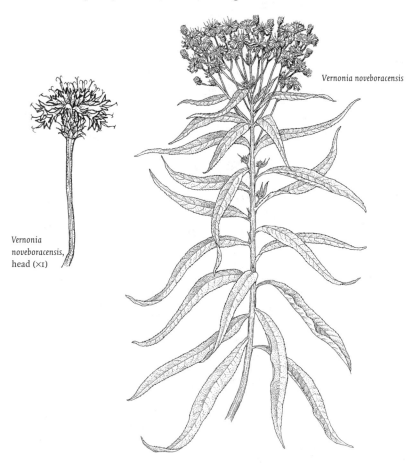

Vernonia noveboracensis

Vernonia noveboracensis, head (×1)

A. principal involucral bracts with a long filiform tip
  B. pappus brownish purple or dark tawny ....V. noveboracensis
  B. pappus bright straw-colored or pale tawny to nearly white........... V. glauca
A. principal involucral bracts obtuse to acuminate .... V. gigantea

Vernonia gigantea (Walter) Trel.   Ironweed
Stems to 3 m tall, mostly glabrous; leaves lanceolate to lance-ovate, gradually
narrowed to the base; involucral bracts obtuse or rounded, appressed, regularly
imbricate; pappus purple, brown, or tan; frequent in moist fields, meadows, and
floodplains; W; flr. Aug–Oct; FAC.

Vernonia glauca,
head (×1)

Vernonia glauca (L.) Willd.   Appalachian or tawny ironweed
Stems 7–15 dm tall, glabrous to somewhat puberulent above; leaves narrowed to
a petiolate base; blades lance-ovate to ovate, shortly acuminate; pappus straw-
colored to nearly white; rare in dry fields, upland wooded slopes, or clearings;
flr. Jul–Oct; SE; ❧.

Vernonia noveboracensis,
achene (×1)

Vernonia noveboracensis (L.) Michx   New York ironweed
Stems to 2 m tall; leaves lanceolate to lance-linear, long attenuate; involucral
bracts abruptly narrowed to a filiform tip; pappus purple or brownish-purple;
common on stream banks, wet fields, pastures, and meadows; throughout, except in the northernmost counties; flr. Jul–Sep; FACW+.

Xanthium L.

Coarse annuals with alternate leaves; inflorescences axillary or terminal; heads
small, unisexual; the staminate heads above, many-flowered with a very reduced
involucre; pistillate heads 2-flowered, lacking a corolla, completely enclosed, except for the styles, by a bur that is covered with hooked prickles.

A. leaves broad, cordate or deltoid at the base; not spiny ....X. strumarium
A. leaves lanceolate, tapering to the base; spiny at the axils ....X. spinosum

Xanthium spinosum L.   Spiny cocklebur
Plants to 1.2 m tall, stems strigose or puberulent; leaves lanceolate, tapering to
each end, short-petiolate, densely silvery-hairy beneath and bearing branched
yellow spines, 1–2 cm long, in the axils; burs mostly solitary or few in the axils;
rare in waste ground, cinders, and ballast; SE and W; flr. Jul–Oct; native to Europe; FACU.

*Xanthium
strumarium,*
head (×1)

Xanthium strumarium L.   Common cocklebur
Plants to 2 m tall, appressed hairy to subglabrous; leaves alternate, with long
petioles, broadly ovate to suborbicular, shallowly 3–5-lobed; staminate heads in
a terminal cluster; pistillate florets in short axillary clusters; burs cylindric to
ovoid, 1–3.5 cm long; a common weed of fields and edges; throughout; native to
Europe; flr. Jul–Sep; FAC; 2 varieties:

A. bases of the prickles of the burs spreading-hairy and stipitate-glandular ....
  .... var. glabratum (DC.) Britton
A. bases of the prickles glandular-puberulent or subglabrous ....
  ....var. canadense (Mill.) Torr. & A.Gray

BALSAMINACEAE Balsam Family

Succulent annuals with watery sap; leaves alternate, coarsely toothed, lacking stipules; flowers irregular, bilabiate, sepals 3, the outer one petaloid and forming an open, sac-like tube with a recurved spur; stamens 5; ovary superior, 5-locular; fruit a capsule that dehisces explosively.

Impatiens L.

Characteristics of the family.

A. flowers in axillary racemes; spur and fruit glabrous
    B. flowers orange with reddish-brown spots within, spur gradually bent parallel with the sac and ⅓–½ its length .... *I. capensis*
    B. flowers pale yellow with darker spots, spur bent at right angles, ⅕–¼ as long as the sac .... *I. pallida*
A. flowers solitary or paired in the axils; spur and fruit pubescent .... *I. balsamina*

Impatiens balsamina L. Garden balsam
Pubescent, to 8 dm tall; leaves oblanceolate, serrate; flowers purple to rose or white, single or paired in the leaf axils; cultivated and occasionally seeding itself in waste places or fields; flr. summer; native to Asia.

Impatiens capensis Meerb. Jewelweed, touch-me-not
Glabrous and more or less glaucous, to 1.5 m tall; leaves ovate, crenate-serrate, long-petioled; flowers on long, drooping pedicels, orange with red-brown spots and a slender, curved spur; common in moist ground of meadows, swamps, stream banks, and open woods; throughout; flr. May–Oct; FACW. Color forms include yellow or whitish flowers with red or pink spots.

Impatiens capensis

Impatiens pallida Nutt. Pale jewelweed, touch-me-not
Very similar to *I. capensis*, larger and more glaucous giving the leaves a bluish-green appearance; flowers pale yellow with a short, abruptly bent spur; frequent in swamps, moist woods, and stream banks; throughout; flr. May–Oct; FACW. A creamy white color form is seen occasionally.

Impatiens pallida

BERBERIDACEAE Barberry Family

Herbaceous perennials or spiny shrubs with yellow wood; flowers regular, perfect, with distinct parts; sepals 4 or 6, in 2 whorls, often dropping early or sometimes petal-like; petals 6–9, in several whorls, or absent; stamens 6 or 12; ovary 1, superior, with a single locule; fruit a berry or capsule with 1 to many seeds.

A. glabrous perennial herbs
    B. flowers white, >1 cm wide, solitary; leaves deeply lobed but not compound
        C. the single flower borne in the axil of 2 peltate, deeply lobed leaves .... *Podophyllum*
        C. flowers on long pedicels arising from the rhizome .... *Jeffersonia*
    B. flowers yellow or purplish, <1 cm wide, in a small, paniculate cyme; leaves 2 or 3 times compound .... *Caulophyllum*
A. spiny shrubs with yellow wood... *Berberis*

Berberis L.

Spiny, deciduous shrubs with yellow wood and small, clustered simple leaves with entire or spiny- toothed margins; flowers solitary, in small umbel-like clusters, or in racemes; the 6 sepals yellow and petal-like in appearance and subtended by several small outer bracts; stamens 6; ovary 1, with 1–several ovules; fruit a red berry.

A. leaf margins entire; spines all simple; flowers borne singly or in small, umbel-like clusters .... *B. thunbergii*
A. leaf margins spiny-toothed; spines on the stems mostly 3-pronged from the base; flowers in a raceme
    B. leaves finely spinulose-serrulate; racemes with 10–20 flowers; petals entire .... *B. vulgaris*
    B. leaves coarsely spinulose-toothed; racemes with 5–10 flowers; petals notched at the tip .... *B. canadensis*

Berberis canadensis Mill.   Allegheny barberry
Erect, rather sparsely branched shrub to 1–2 m tall with spines mostly 3-pronged; leaves obovate to spatulate, 2–6 cm long, with coarsely spiny-toothed margins; second-year twigs brown, purplish, or red; racemes 2–4 cm long with 5–10 flowers; petals notched at the tip; fruit to 1 cm; rare in rocky woods, known from only a single site in Huntingdon Co., but believed to be extirpated; ✴.

Berberis thunbergii DC.   Japanese barberry
Densely branched shrub to 2 m tall with simple spines; leaves 1–2 cm, obovate to spatulate with entire margins; flowers solitary or in small umbel-like clusters of 2–4; fruit to 1 cm, often persisting through the winter; cultivated and frequently naturalized in woods, old fields, roadsides, and hedgerows throughout; flr. Apr, frt. Aug–winter; native to Japan.

Berberis vulgaris L.   European barberry
Shrub to 3 m tall with gray twigs and mostly 3-pronged spines; leaves 2–5 cm long, obovate with finely spinulose margins; racemes 3–6 cm with 10–20 flowers; petals entire; fruit 1 cm; cultivated and escaped to fields, pastures, and dis-

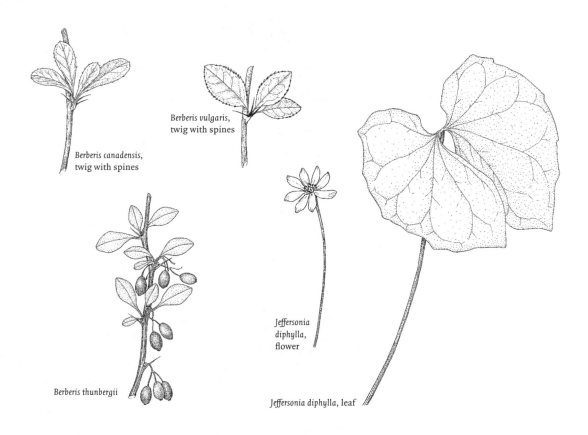

Berberis canadensis, twig with spines

Berberis vulgaris, twig with spines

Berberis thunbergii

Jeffersonia diphylla, flower

Jeffersonia diphylla, leaf

turbed woods; scattered throughout; flr. May–early Jun; frt. Aug–Nov; native to Europe.

Caulophyllum Michx.

Caulophyllum thalictroides (L.) Michx.   Blue cohosh
Erect glaucous perennial, 3–8 dm tall; leaves ternately compound (actually 1 tri-ternately compound leaf), leaflets dark bluish-green, obovate-oblong, shallowly lobed above the middle, 5–8 cm; flowers in a terminal panicle-like cyme, small, greenish-yellow or greenish-purple; sepals 6, petaloid, subtended by 3–4 sepal-like bracts; stamens 6; ovary asymmetrical, splitting early to expose the ripen-ing, fleshy-coated seeds that appear as dark blue drupes; frequent in moist, rich woods throughout; flr. Apr.

Jeffersonia Barton

Jeffersonia diphylla (L.) Pers.   Twinleaf
Glabrous perennial herb; leaves basal, long-petioled, deeply cleft into 2 large symmetrical lobes; flowers solitary on naked scapes, 1–2 dm tall at anthesis; flowers white, 1–3 cm wide, with 8 distinct petals and 4 sepals that soon drop; fruit a 2–3-cm-long leathery capsule opening by the separation of the upper third to form a hinged lid; seeds many; occasional in moist woods, usually on calcareous soils; SW and SC; flr. Apr.

Podophyllum L.

Podophyllum peltatum L.   Mayapple, mandrake
Glabrous perennial herb spreading by stout rhizomes; leaves with circular or semicircular, deeply-lobed, peltate blades 2–3 dm wide, present in terminal pairs on flowering stems or solitary on nonflowering stems; the single waxy white, 3-cm-wide flower is borne below the leaves in the angle of the petioles on a short, nodding pedicel; fruit an ovoid yellow berry, 4–5 cm long; common in mesic woods throughout, where it often forms large colonial patches; flr. May.

BETULACEAE Birch Family

Deciduous trees or shrubs with alternate, simple, sharply serrate leaves with pinnate venation; plants monoecious; staminate flowers in drooping catkins; pistillate flowers in capitate clusters, loose pendulous catkins, or short, erect, somewhat woody catkins; each bract of the catkins bearing 1–3 flowers, bracts enlarging and enclosing the fruit in some species; perianth greatly reduced or absent; fruit a nut, nutlet, or samara.

A. shrubs or occasionally small trees
    B. pith 3-angled; leaves obovate, ovate, or subrotund with retuse, rounded, or obtuse tip and rounded or tapered base; pistillate catkins becoming woody and cone-like and persisting after the small, winged fruits are dispersed... .... *Alnus*
    B. pith round; leaves ovate with acute to acuminate tip and rounded, cordate base; fruit a nut enclosed in a leafy involucre .... *Corylus*
A. small to medium trees, sometimes with clumped trunks
    C. bark brownish, rough and flaky; each nutlet surrounded by an inflated, sac-like, closed bract .... *Ostrya*
    C. bark smooth or becoming rough and platy with age; bracts subtending, but not surrounding, the fruits
        D. bark dark reddish-brown, white, whitish-tan, or yellowish, often papery and exfoliating with prominent horizontal lenticels .... *Betula*
        D. bark smooth and gray, not exfoliating, lenticels not obvious; trunk with a muscular configuration .... *Carpinus*

Alnus Mill.

Tall, deciduous, colonial shrubs or small upright trees with alternate, simple, serrate leaves with pinnate venation; staminate catkins visible all winter, elongating before or with the leaves; pistillate flowers in shorter catkins which become woody in fruit and persist after dispersal of the winged samaras.

A. winter buds stalked; flowers opening before the leaves; fruits with a narrow margin
    B. tall, multistemmed colonial shrubs; leaves with 8–14 principal veins on each side, acute to short acuminate at apex; pistillate catkins 1–1.5 cm
        C. leaves doubly serrate; teeth irregular in size; pistillate catkins bent downward .... *A. incana*
        C. leaves serrulate with fine regular teeth; pistillate catkins erect .... *A. serrulata*

B. upright small trees; leaves with 5–8 principal veins on each side, rounded or notched at apex; pistillate catkins 1.5–3 cm .... *A. glutinosa*
A. winter buds sessile; flowers opening with the leaves; fruits broadly winged ...
....*A. viridis*

Alnus glutinosa (L.) Gaertn.   Black or European alder
Tree to 10 m tall; leaves broadly rounded or slightly notched at the end, finely serrate, with 5–8 principal veins on each side; young leaves, stems, and catkins glutinous; fruits thin-margined; planted and occasionally naturalized on roadsides and old fields; flr. Apr, before the leaves; native to Eurasia; FACW–.

Alnus incana (L.) Moench   Speckled alder
Colonial shrub to 6 m tall with stalked winter buds; leaves doubly serrate with 8–14 principal veins on each side, obtuse to short acuminate at the tip; mature pistillate catkins more or less bent downwards; fruits thin-margined; frequent in bogs and swamps, especially N; flr. Mar–Apr, before the leaves; FACW. Ours is ssp. *rugosa* (Du Roi) Clausen.

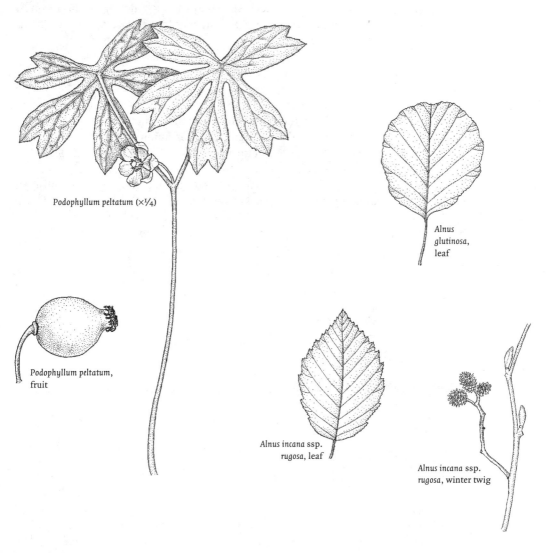

Podophyllum peltatum (×¼)

Podophyllum peltatum,
fruit

*Alnus
glutinosa,*
leaf

Alnus incana ssp.
rugosa, leaf

Alnus incana ssp.
rugosa, winter twig

Alnus serrulata (Drand. ex Aiton) Willd. Smooth alder

Colonial shrub to 6 m tall, with stalked winter buds; leaves with 8–14 principal veins, finely and evenly serrate, obtuse to rounded at the tip; mature pistillate catkins more or less erect; fruits thin-margined; common in low, wet woods and swamps; mostly S; flr. Mar–Apr, before the leaves; OBL.

Alnus viridis (Vill.) DC. Mountain alder

Colonial shrub to 5 m tall with sessile winter buds; leaves with 6–9 principal veins on each side, finely and sharply serrate, glutinous beneath when young; fruit surrounded by a broad, membranous wing; very rare on cool, rocky, wooded slopes; scattered; flr. Jun; FAC; ❧. Ours is ssp. *crispa* (Aiton) Turrill.

Betula L.

Deciduous trees to 30 m tall, most with papery, horizontally exfoliating bark and prominent horizontal lenticels; leaves pinnately veined, serrate; staminate flowers in elongate drooping catkins; pistillate catkins shorter, upright, disintegrating at time of seed dispersal, each scale subtending 3 flowers; fruit a samara with 2 lateral wings and tipped by 2 persistent styles; flr. late Apr–early May, as the leaves are emerging.

Summer key

A. mature bark white or creamy
    B. leaves triangular or rhombic with an acuminate tip
        C. branches not pendulous; leaf tip long-acuminate .... *B. populifolia*
        C. branches pendulous; leaf tip acuminate .... *B. pendula*
    B. leaves ovate with an acute tip
        D. leaves 5–8 cm long; branches erect to slightly drooping ..... *B. papyrifera*
        D. leaves 3–5 cm long; branches stiffly erect .... *B. pubescens*

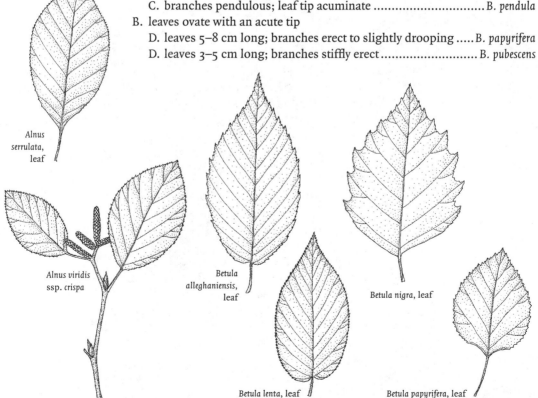

Alnus serrulata, leaf

Alnus viridis ssp. *crispa*

Betula alleghaniensis, leaf

Betula lenta, leaf

Betula nigra, leaf

Betula papyrifera, leaf

A. mature bark light reddish-brown, yellowish-gray, or dark blackish-brown, but not white
 E. twigs slightly to strongly aromatic when scraped or crushed; bark dark and tight or yellowish-gray and exfoliating; leaves rounded to cordate at the base
  F. bark blackish-brown, not exfoliating but becoming rough and platy in age; twigs strongly aromatic .... *B. lenta*
  F. bark yellowish-gray, exfoliating in papery layers; twigs weakly aromatic .... *B. alleghaniensis*
 E. twigs not aromatic; bark light reddish-brown, exfoliating in irregular papery layers; leaves tapered at the base .... *B. nigra*

Winter key

A. mature bark white or creamy
 B. buds <5 mm long; bark close, not easily separable .... *B. populifolia*
 B. buds 5–10 mm long; bark separating into papery layers
  C. branches erect to slightly drooping
   D. bark spontaneously peeling into papery layers, often yellowish or pinkish .... *B. papyrifera*
   D. bark usually not spontaneously peeling, white .... *B. pubescens*
  C. branches pendulous .... *B. pendula*
A. mature bark light reddish-brown, yellowish-gray, or dark blackish-brown, but not white .... see second A in summer key

Betula alleghaniensis Britton   Yellow birch
Tree to 30 m tall with yellowish-gray bark exfoliating in irregular, papery layers; twigs with a slight wintergreen smell when crushed; leaves with 9–12 pairs of lateral veins, rounded to subcordate at the base, pubescent on the veins beneath; common in cool, moist woods; N, and at higher elevations; FAC. [syn.: *B. lutea* Michx.]

Betula lenta L.   Black or sweet birch
Tree to 25 m tall with smooth, dark, nonexfoliating bark that becomes rough and platy with age; twigs with a strong wintergreen smell when crushed; leaves with 9–12 pairs of lateral veins that are impressed above and villous beneath; leaf base rounded to subcordate; common in woods and streambanks throughout; FACU.

Betula nigra L.   River birch
Tree to 30 m tall with tan to reddish-brown bark exfoliating in irregular papery layers; leaves with 6–10 pairs of lateral veins, softly villous on the veins beneath; petioles tomentose; common on floodplains, riverbanks, wet woods, and swamps; mostly SE; FACW.

Betula nigra, samara (×2½)

Betula nigra, pistillate scale (×2½)

Betula papyrifera Marshall   Paper or canoe birch
Tree to 30 m with silvery-white bark bearing black marks at the branches; outer layers of bark readily peeling away to expose a salmon-pink surface beneath; leaves ovate, 5–8 cm, tapered to rounded at the base, pubescent on veins beneath; fruit broadly winged; occasional in cool woods and slopes; NE, and at high elevations along the Allegheny Front; FACU.

Betula pendula Roth   European white birch

Tree to 25 m with white, exfoliating bark with triangular black patches; branches pendulous; leaves triangular-ovate, acuminate, glabrous at maturity; fruit broadly winged; cultivated and occasionally naturalized in roadside thickets and waste ground; scattered; native to Eurasia.

Betula populifolia Marshall   Gray birch

Tree to 10 m, often growing in clumps; bark white with prominent black markings at the branches, exfoliating; twigs strongly glandular; leaves widest near the base, tapering to a long slender point, glabrous beneath; scales of pistillate catkins densely hairy on both sides; fruits broadly winged; common in old fields, open woods, strip mines, and other disturbed sites, especially on dry, sterile soils; mostly E; FAC.

*Betula populifolia,
leaf*

Betula pubescens Ehrh.   European white birch

Tree to 25 m tall with white bark and stiffly erect branches; leaves ovate, 3–5 cm; buds strongly resinous; cultivated and occasionally escaped at abandoned home or nursery sites; scattered; native to Europe. [syn: *B. alba* L. var. *pubescens* (Ehrh.) Spach]

Carpinus L.

Carpinus caroliniana Walter   Hornbeam, ironwood

Deciduous tree to 10 m tall with smooth, gray, sinuous bark; leaves 5–12 cm, oblong-ovate, pinnately veined, and sharply doubly serrate; fruiting catkins 2–5 cm, nutlets subtended by an enlarged, 2- or 3-lobed leafy bract; common in moist woods and stream banks throughout; flr. late Apr–early May, as the leaves are emerging; FAC.

Corylus L.

Stoloniferous, deciduous shrubs to 3 m tall; staminate catkins visible all winter, expanding and drooping in anthesis; pistillate catkins very compact, appearing as a bud with only the pinkish stigmas visible; fruit a hard-shelled, edible nut closely surrounded by a leafy involucre; flr. Mar–early Apr, before the leaves, frt. Aug–Sep.

A. twigs and petioles with stalked glands; mature involucre 1.5–3 cm, with spreading, laciniate lobes ....*C. americana*
A. twigs and petioles lacking glands; mature involucre 4–7 cm, narrowed beyond the nut into a slender, tubular beak .... *C. cornuta*

Corylus americana Walter   American filbert or hazelnut

Colonial shrub to 3 m tall; young twigs and petioles pubescent and also bearing stalked glands; leaves doubly serrate, pubescent beneath; nuts surrounded by a broad, laciniate involucre at maturity; common in rich woods and edges throughout, however, less abundant in the northern tier counties; FACU–.

Corylus cornuta Marshall   Beaked hazelnut

Colonial shrub to 3 m tall; young twigs villous at first, but lacking glands; leaves coarsely doubly serrate, somewhat pubescent beneath, especially on the veins

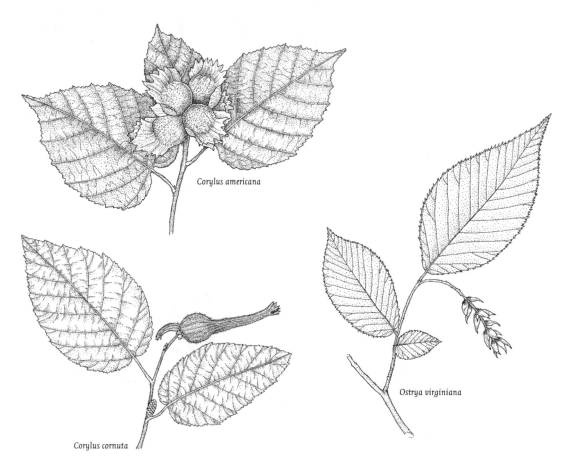

Corylus americana

Corylus cornuta

Ostrya virginiana

and vein axils; involucre narrowed beyond the nut into a long, slender tubular beak; frequent in dry, rocky woods and thickets, especially east of the Allegheny Front; FACU–.

Ostrya Scop.

Ostrya virginiana (Mill.) K.Koch   Hop-hornbeam
Deciduous tree to 20 m tall with rough, brownish, flaky bark; leaves doubly serrate; nutlets surrounded by inflated sac-like involucres that form dangling, cone-like catkins at maturity; frequent on dry, wooded slopes throughout, often on calcareous soils; flr. late Apr–early May, as the leaves are emerging; FACU–.

BIGNONIACEAE Trumpet-creeper Family

Deciduous trees or woody vines with opposite or whorled leaves lacking stipules and showy, perfect, irregular flowers; calyx 5-lobed, regular or irregular; corolla weakly to strongly bilabiate, 4–5-lobed; stamens 2–4; fruit a 2-locular capsule; seeds flattened and winged.

A. woody vine with pinnately compound leaves .... *Campsis*
A. trees with large simple leaves

B. leaves often somewhat lobed, pubescent above and below, margin entire or irregularly toothed; flowers lavender; capsule short, ovoid.... *Paulownia*
B. leaves ovate or cordate-ovate, glabrous above, entire; flowers white (or yellow); capsule elongate .... *Catalpa*

Campsis Lour.

Campsis radicans (L.) Seem. ex Bureau   Trumpet-vine, trumpet-creeper
Deciduous woody vine sprawling or climbing by means of aerial rootlets; leaves pinnately compound; leaflets 5–13, lanceolate to ovate, 4–8 cm long, acuminate, coarsely serrate; flowers 6–8 cm; corolla orange-red, with 5 spreading lobes, only slightly bilabiate; fruit 10–15 cm, flattened, slightly club-shaped; with a broad, thin partition; seeds 1.5 cm, flat, broadly winged on each side; occasional in moist woods, stream banks, thickets, and fencerows; flr. late Jul–Aug; native, but also escaped from cultivation; FAC.

Catalpa Scop.

Deciduous trees 10 to 30 m tall; leaves whorled or opposite, simple, petiolate; blades ovate, cordate-ovate, or rotund, to 3 dm; margins entire; flowers in large terminal panicles; corolla white with purple and yellow spots, campanulate with 5 spreading lobes, bilabiate; stamens mostly 2; fruit an elongate, cylindrical capsule to 5 dm; seeds flat, winged. In addition to the taxa treated below, *C. ovata* G.Don, which has smaller, yellow flowers and leaves glabrous beneath is a rare escape.

A. apex of leaf abruptly acuminate; flowers 3–4 cm wide, conspicuously purple-spotted ....*C. bignonioides*
A. apex of leaf long-acuminate; flowers 5–6 cm wide, faintly purple-spotted .....
.... *C. speciosa*

Catalpa bignonioides Walter   Catalpa, Indian-bean
Tree to 15 m tall; leaves only short-acuminate; corolla with conspicuous purple spots, 3–4 cm across; capsule 6–10 mm thick; cultivated and frequently naturalized in disturbed woods, floodplains, fields, and waste ground; mostly S; native farther south; flr. late Jun–early Jul; UPL.

Catalpa speciosa (Warder ex Barney) Warder ex Engelm.   Catalpa, cigar-tree
Tree to 30 m tall; leaves distinctly acuminate; corolla only faintly purple-spotted, 5–6 cm across; fruit 1–1.5 cm thick; seeds 2.5 cm with rounded wings with a flat fringe of hairs; cultivated and occasionally escaped to low woods, roadsides, and spoil banks; mostly SW and SC; flr. Jul; native farther south and west; FAC.

Paulownia Siebert & Zucc.

Paulownia tomentosa (Thunb.) Steud.   Empress-tree, princess-tree
Deciduous tree to 20 m tall with densely pubescent reddish-brown flower buds conspicuous during the winter; leaves opposite or whorled, simple, long-petioled; blades broadly cordate-ovate to somewhat lobed, 1.5–8 dm, entire or coarsely toothed, velvety pubescent on both sides; flowers in conspicuous upright panicles; corolla purple, bilabiate, 5–7 cm; fruit an ovoid, dehiscent, 2-

Paulownia tomentosa (×¼)

Catalpa speciosa,
fruit (×¼)

locular capsule 3–4 cm long; seeds winged, numerous; cultivated and frequently escaped to roadsides, railroad rights-of-way, urban waste ground, and disturbed woods; mostly SE; flr. May, before the leaves; native to China.

BORAGINACEAE Borage Family

Hairy (except *Mertensia*) herbs with alternate (sometimes opposite at the base), simple, entire leaves lacking stipules; inflorescence typically a helicoid cyme that appears spike-like or raceme-like when fully expanded; flowers perfect, 5-merous; calyx lobes united at the base or free; corolla regular to slightly irregular, sympetalous, the tube often with hairy appendages (fornices) within opposite the lobes; stamens 5, inserted on the corolla tube alternate with the lobes; ovary superior, deeply 4-lobed (except *Heliotropium*); the style arising from between the lobes of the ovary; fruit 1–4, 1-seeded nutlets.

A. plants glabrous throughout .... *Mertensia*
A. plants coarsely to finely hairy
    B. ovary shallowly lobed, style terminal .... *Heliotropium*
    B. ovary deeply 4-lobed, style arising from between the lobes
        C. nutlets covered with prickles
            D. each flower subtended by a bract .... *Lappula*
            D. at least the upper flowers without subtending bracts (there may be small bracts between the flowers)
                E. inflorescence branches secund; corollas 1.5–2 mm wide; nutlets erect .... *Hackelia*
                E. inflorescence not secund; corollas 5–12 mm wide; nutlets spreading widely in fruit .... *Cynoglossum*
        C. nutlets without prickles
            F. corolla irregular, the 5 lobes oblique and unequal .... *Echium*
            F. corolla regular, lobes equal
                G. flowers axillary, solitary or in small clusters
                    H. corolla white, with 5 rows of glandular hairs in the throat; nutlets grayish-tan, wrinkled .... *Buglossoides*
                    H. corolla pale yellow to yellow-orange, the throat lacking hairs or other appendages; nutlets smooth and shining, white .... *Lithospermum*
                G. flowers in terminal inflorescences
                  I. corolla tubular or tubular-campanulate, the lobes erect or barely spreading
                    J. flowers without bracts; scales (fornices) present in throat of corolla tube .... *Symphytum*
                    J. flowers subtended by leafy bracts; corolla tube without scales at the throat .... *Onosmodium*
                I. corolla broadly funnelform, salverform or rotate
                  K. individual flowers not subtended by bracts .... *Myosotis*
                  K. at least the lowest flowers subtended by bracts
                    L. corolla pale yellow to yellow-orange .... *Lithospermum*
                    L. corolla blue or white
                      M. bracts present throughout the inflorescence; nutlets rough
                      N. corolla blue, 6–20 mm wide .... *Anchusa*
                      N. corolla white, 1–2 mm wide .... *Plagiobothrys*
                    M. the upper flowers lacking bracts; corolla blue or white, often with a yellow eye, 1–9 mm wide; nutlets smooth and shining .... *Myosotis*

Anchusa L.

Anchusa azurea Mill.   Showy bugloss, alkanet
Single stemmed, hairy perennial with petioled basal leaves, the upper reduced and clasping; flowers blue, in a bracteate, helicoid false raceme; nutlets erect, coarsely reticulate-ridged; cultivated and occasionally escaped to waste ground; scattered; flr. May–Jul; native to the Mediterranean region.

Buglossoides Moench

Buglossoides arvense (L.) I.M.Johnst.   Bastard alkanet, corn gromwell
Branched annual to 8 dm tall; leaves linear to narrowly oblanceolate, 1-nerved; flowers solitary in the upper axils, 2–4 mm wide, whitish; nutlets deeply wrinkled and pitted; frequently naturalized in dry pastures, limestone slopes, roadsides, and railroads; SE and SC; flr. Apr–May; native to Eurasia. [syn: *Lithospermum arvense* L.]

Cynoglossum L.

Erect, leafy herbs with large leaves; flowers in axillary or terminal helicoid cymes that appear raceme-like when fully expanded; corolla tube nearly closed by hairy appendages at the top; nutlets prickly, divergent at the base, attached near the top.

A. inflorescence terminal with a common, naked peduncle; calyx inconspicuous in fruit
  B. blades of basal leaves to 3 dm; corolla 8–12 mm wide; nutlets 5–6 mm .... *C. virginianum*
  B. blades of basal leaves 1–1.5 dm; corolla 5–7 mm wide; nutlets 3–4 mm .... *C. boreale*
A. inflorescence of numerous axillary (false) racemes; calyx conspicuous in fruit .... *C. officinale*

Cynoglossum boreale Fernald   Northern hound's-tongue, wild comfrey
Very similar to *C. virginianum* but smaller overall; rare in open woods and roadsides; NE and NC; flr. May–Jun; 🌿. [syn: *C. virginianum* L. var. *boreale* (Fernald) Cooperr.]

Cynoglossum officinale L.   Hound's-tongue
Softly pubescent biennial with erect stem to 1 m tall, branched above; basal leaves to 3 dm, tapering to a narrowed base, upper leaves progressively smaller and becoming sessile; corolla dull red or red-purple, 8 mm wide; the calyx and nutlets spreading widely in fruit; occasional in limestone banks, pastures, railroad beds, and waste ground; throughout; flr. May–early Jul; native to Eurasia.

Cynoglossum virginianum L.   Wild comfrey
Erect, unbranched perennial to 8 dm tall; basal leaves oblong, blade 1–3 dm, tapering to a winged petiole; cauline leaves gradually reduced and expanded or clasping at the base; flowers in 1–4 racemes branching from a long terminal peduncle; corolla blue; occasional in rich open woods and wooded slopes; W and S; flr. May–Jun.

Echium L.

Echium vulgare L.   Viper's bugloss, blueweed
Erect, rough-hairy biennial to 8 dm tall; basal leaves oblanceolate, upper gradually reduced; flowers in numerous, short helicoid cymes that are alternately arranged along the terminal portion of the stem axis, each flower subtended by a

Echium vulgare, flower (×1)

Hackelia virginiana

Echium vulgare

Cynoglossum
virginianum

bract; corolla blue, style and 4 filaments longer than the corolla tube; nutlets more or less roughened; frequent in dry fields, roadsides, railroad banks, and waste ground; mostly S; flr. late Jun–Sep; native to Europe.

Hackelia Opiz

Hackelia virginiana, flower (×3)

Hackelia virginiana, fruit (×3)

Hackelia virginiana (L.) I.M.Johnst.   Beggar's-lice, stickseed
Erect widely branched biennial to 1.5 m tall; leaves oblong elliptic, the lower petioled, upper sessile and gradually passing into floral bracts; flowers in spreading, 1-sided racemes, the terminal flowers often lacking subtending bracts or with very tiny ones; corolla white or pale blue; nutlets erect, 3–4 mm, with several rows of hooked prickles; common in dry to moist woods, wooded slopes, and roadsides; throughout; flr. Jul–Aug; FACU.

Heliotropium L.

Heliotropium europaeum L.   European heliotrope
Erect hairy annual 2–5 dm tall; leaves with long petioles; flowers in terminal helicoid cymes that are clustered in groups of 3–5 and appear spike-like when fully expanded; axis and sepals densely white-hairy; corolla white; fruit globose or depressed-ovoid, splitting into 4 nutlets; an infrequent garden escape of waste ground, cinders, and ballast; mostly S; flr. Jul–Oct; native to Europe.

Lappula Moench

Lappula squarrosa (Retz.) Dumort.   Beggar's-lice, stickseed
Erect, rough-hairy, branching annual to 8 dm tall; leaves linear to linear-lanceolate; flowers blue, 1.5 mm wide, in numerous terminal, helicoid cymes, each flower subtended by a bract; nutlets erect, 3–4 mm, with several rows of hooked prickles; infrequent on railroad ballast and cinders; scattered; flr. Jun–Jul; native to Eurasia.

Lithospermum L.

Erect or ascending perennials with hairy stems and leaves; flowers orange, yellow, or yellowish-white in leafy-bracted cymes or solitary in the upper axils; corolla funnelform or salverform; nutlets smooth or pitted, basally attached.

A. flowers in a leafy-bracted cyme; corolla yellow or orange
    B. corolla lobes 3–6 mm long; foliage canescent .... *L. canescens*
    B. corolla lobes 8–11 mm long; foliage roughly hirsute .... *L. caroliniense*
A. flowers axillary; corolla yellowish-white .... *L. latifolium*

Lithospermum canescens (Michx.) Lehm.   Hoary puccoon, Indian-paint
Several-stemmed perennial covered with short white hairs, 1–4 dm tall from a stout taproot; leaves lanceolate to narrowly oblong; corolla orange to yellow, 1–1.5 cm wide; nutlets yellowish-white, smooth, and shining; rare on river bluffs, dry rocky hillsides, and barrens, on limestone; mainly SC; flr. late Apr–May; 🌿.

Lithospermum caroliniense (J.F.Gmel.) MacMill. Golden puccoon, hispid gromwell
Erect stems, 3–6 dm tall, clustered from a woody root crown; leaves numerous, hairy, linear to lanceolate; corolla yellow-orange, 1.5–2.5 cm wide; nutlets creamy white, smooth, and shining; very rare on open sandy barrens and roadsides; NW; flr. Jun–Jul; 🌿.

Lithospermum latifolium Michx.   American gromwell
Erect stems, 4–8 dm tall; leaves lanceolate to lance-ovate with 2 or 3 prominent lateral veins on each side; flowers solitary in the upper axils; corolla yellowish-white, 5–7 mm; nutlets ovoid, white, and shining, 3.5–5 mm, smooth or slightly pitted; rare on rich, wooded limestone slopes and hilltop woods; SW; flr. late May–Jun; 🌿.

Lithospermum caroliniense

Mertensia Roth

Mertensia virginica (L.) Pers. ex Link   Virginia bluebell
Erect or ascending, glabrous, leafy perennial 3–7 dm tall; upper leaves nearly sessile, obovate to oblanceolate, the lower long-tapering at the base; flowers in a bractless, terminal cyme, blue (occasionally white or pink and often pink in bud); corolla tube elongate, expanded at the end but barely lobed; frequent on rich wooded slopes and forested floodplains; throughout, except at the highest elevations; flr. late Apr–early May; FACW.

Myosotis L.

More or less strigose herbs with small flowers in terminal bractless, helicoid cymes that appear raceme-like when fully expanded; basal leaves oblanceolate to spatulate; the upper smaller, elliptic to oblong; calyx regular or irregular; corolla regular; blue (or yellowish or white), often with a yellow eye, the 5 lobes spreading and the tube partly closed by 5 appendages; nutlets smooth and shining.

A. hairs of the calyx not hooked
    B. corolla 5–10 mm wide; nutlets not longer than the style .... M. scorpioides
    B. corolla 2–5 mm wide; nutlets surpassing the style .... M. laxa
A. calyx with at least some hooked hairs
    C. corolla 5–8 mm wide .... M. sylvatica
    C. corolla 1–4 mm wide
        D. calyx irregular, 3 lobes larger than the other 2; corolla white
            E. fruiting pedicels erect; nutlets 1.2–1.5 mm .... M. verna
            E. fruiting pedicels divergent; nutlets 1.4–2.2 mm .... M. macrosperma
        D. calyx regular, the lobes all equal; corolla blue (occasionally white, or yellow turning blue)
            F. fruiting pedicels as long or longer than the calyx .... M. arvensis
            F. fruiting pedicels shorter than the calyx
                G. flowers present nearly to the base of the plant, the lowest among leaves; nutlets longer than the style .... M. stricta
                G. flowers present only in the upper half of the plant, well above the leaves; style surpassing the nutlets .... M. discolor

Myosotis arvensis (L.) Hill   Forget-me-not
Annual or biennial, 1–4 dm tall, branched above; the inflorescence not more than half the height of the plant; corolla blue or occasionally white, 2–4 mm wide; nutlets surpassing the style; cultivated and occasionally escaped to fields and waste ground; scattered; flr. May–Oct; native to Eurasia; UPL.

Myosotis discolor Pers.   Yellow and blue scorpion-grass
Slender annual, 1–5 dm tall; racemes about half the height of the plant, naked except for 1 or 2 basal bracts; flowers yellowish changing to blue, 1–2 mm wide; nutlets short to sometimes longer than the style; a very rare introduction, moist rocks; SE; flr. May–Jul; native to Europe; UPL.

Myosotis laxa Lehm.   Wild forget-me-not
Short-lived perennial or annual with slender stems 1–4 dm tall and decumbent at the base; inflorescence terminal; corolla blue, 2–5 mm wide; nutlets surpassing the style; frequent in wet open ground and swamps throughout except at the highest elevations; flr. late May–Sep; OBL.

Myosotis macrosperma, calyx (×2)

Myosotis macrosperma Engelm.   Big-seed scorpion-grass
Very similar to M. verna, but larger; stems to 6 dm tall; corolla 1–2 mm wide; locally abundant in low, moist woods; SC and SW; flr. May.

Myosotis scorpioides L.   Forget-me-not, water scorpion-grass
Stoloniferous, creeping perennial 2–6 dm tall; inflorescence terminal; corolla blue, 5–10 mm wide; style equaling or surpassing the nutlets; cultivated and frequently escaped to ditches, streambanks and floodplains; throughout; flr. Jun–Sep; native to Europe; OBL.

Myosotis stricta Link ex Roem. & Schult.   Forget-me-not
Annual or winter annual with branched stems to 2 dm tall and flowers nearly to the base; corolla blue, 1–2 mm wide; nutlets longer than the style; rare in pastures, dry banks, and lawns; scattered; flr. May; native to Eurasia. [syn: M. micrantha auct., non Pallas ex Lehm.]

Myosotis sylvatica Hoffm.   Garden forget-me-not
Short-lived perennial or annual; stems to 5 dm tall, decumbent at the base; racemes naked, corolla blue (or white), 5–8 mm wide; nutlets 1.5–2 mm; cultivated and rarely naturalized; SE; flr. May; native to Eurasia; UPL.

Myosotis verna Nutt.   Spring forget-me-not, early scorpion-grass
Annual or winter annual to 4 dm tall; lowest leaves oblanceolate, upper oblong and sessile; the inflorescence occupying the upper half of the plant; corolla white, 1–2 mm wide; occasional in dry, open woods, rocky ledges, and roadside banks; mostly S; flr. May–Jun; FAC–.

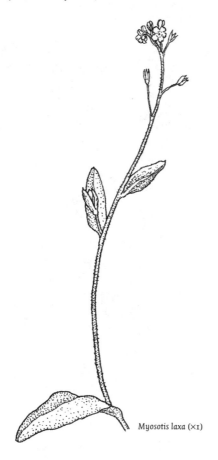

Myosotis laxa (×1)

Myosotis scorpioides,
calyx (×2)

Onosmodium Michx.

Coarse, hairy perennials with leafy stems; flowers in leafy-bracted, helicoid cymes; corolla hairy on the outside, nearly tubular with erect, pointed lobes; nutlets smooth to somewhat pitted, usually only 1 or 2 maturing.

A. stems to 6 dm tall; corolla yellow, lobes acuminate .... *O. virginianum*
A. stems to 1.2 m tall; corolla white or greenish, lobes acute .... *O. molle*

Onosmodium molle Michx.   False gromwell, marble-seed

Multistemmed with a woody base; leaves numerous, lanceolate to narrowly ovate, the lower smaller and dropping early; corolla 8–16 mm long, dull white or greenish, lobes 1.5–2 times as long as wide; nutlets dull; very rare on dry, calcareous hillsides and old pastures; SC and SW; flr. late Jun–early Jul; ✿. Ours is var. *hispidissimum* (Mack.) I.M.Johnst.

Onosmodium virginianum (L.) A.DC.   Virginia false gromwell

Leaves oblanceolate, narrowed to the base; corolla 7–10 mm long, yellow, the lobes 2–3 times as long as wide; nutlet whitish, more or less pitted; very rare on sandy banks and limestone bluffs; SE; believed to be extirpated; flr. late Jun–early Jul; ✿.

Onosmodium molle
var. *hispidissimum*

Plagiobothrys Fisch. & C.A.Mey.

Plagiobothrys scouleri (Hook. & Arn.) I.M.Johnst.　　　　　Meadow plagiobothrys
Slender annual with clustered, prostrate or ascending stems to 2 dm tall; leaves
linear, the lower opposite; flowers 1–2 mm wide, white, in a loosely flowered,
terminal helicoid cyme that appears spike-like when fully expanded; nutlets 1.5–
2.2 mm, roughened; rare in moist depressions in open meadows; flr. Jun; native
farther west.

Symphytum L.

Symphytum officinale L.　　　　　　　　　　　　　　　　　Comfrey
Hispid-hirsute perennial to 1.2 m tall with large leaves, the basal petiolate and
the upper gradually reduced and sessile but conspicuously decurrent; flowers
blue, purplish, or yellowish-white, 12–18 mm, in compact cymes that are terminal or in the upper axils; nutlets 5–6 mm, black and shining; cultivated and occasionally naturalized in fields, roadsides, and waste ground; scattered; flr. Jun–
Aug; native to Eurasia.

BRASSICACEAE Mustard Family

Annual, biennial, or perennial herbs with pungent watery juice; plant surface
glabrous or often with uni-cellular, simple or branched, but mostly nonglandular hairs; leaves alternate or basal, simple to pinnately compound or lobed, lacking stipules; basal and stem leaves similar or different in shape; inflorescences
mostly terminal, racemose or corymbose-racemose, often elongating in fruit;
flowers perfect; sepals 4, distinct, some or all often saccate at the base; petals 4,
distinct, often more or less clawed, alternating with the sepals, sometimes very
small or lacking; stamens 6 (usually 4 long and 2 short) occasionally reduced to
4 or 2; ovary superior, composed of 2 fused carpels; fruit a silique (a 2-valved,
mostly dehiscent capsule with a persistent internal septum), varying in shape
from broad and flat to long and slender, occasionally segmented, and often
tipped by an indehiscent beak. Well-formed fruits are necessary for identification of most taxa.

A. flowers pink, purple, white, whitish-green, or creamy
　　B. flowers pink to purple
　　　　C. petals distinctly unequal, 2 larger and 2 smaller .... *Iberis*
　　　　C. petals all the same size
　　　　　　D. stem leaves pinnately or palmately divided .... *Cardamine*
　　　　　　D. stem leaves simple, lobed or unlobed
　　　　　　　　E. tubers present.. *Cardamine*
　　　　　　　　E. tubers not present
　　　　　　　　　　F. stem and leaves conspicuously stipitate-glandular....*Chorispora*
　　　　　　　　　　F. stem and leaves not conspicuously stipitate-glandular
　　　　　　　　　　　　G. stem and leaves glabrous and succulent .... *Cakile*
　　　　　　　　　　　　G. stem and leaves variously hairy
　　　　　　　　　　　　　　H. ovary >3 times as long as wide
　　　　　　　　　　　　　　　　I. basal leaves lanceolate to lance-ovate

J. leaves pubescent with at least some branched hairs present .... *Hesperis*

J. leaves glabrous or with a few simple hairs .... *Iodanthus*

  I. basal leaves lyrate-pinnatifid

K. leaves scabrous or hispid .... *Raphanus*

K. leaves glabrous or with a few simple hairs .... *Iodanthus*

H. ovary <3 times as long as wide .... *Lunaria*

B. flowers white, whitish-green, or creamy

L. petals with conspicuous purple veins .... *Eruca*

L. veins of the petals not differently colored or petals absent

M. petals deeply 2-lobed or notched .... *Berteroa*

M. petals not notched or lobed or petals absent

N. leaves all basal, no cauline leaves present ....*Draba*

N. cauline leaves present, although sometimes much reduced

O. cauline leaves all simple, variously toothed, but not deeply lobed

P. leaf blades broadly reniform to ovate or ovate-cordate, venation palmate to palmate-pinnate ....*Alliaria*

P. leaf blades narrower, pinnately veined or only the midvein visible

Q. cauline leaves perfoliate or with clasping auriculate bases

R. fruits >3 times as long as wide, flattened parallel to the plane of the septum or not flattened

S. petals <10 mm long; fruits flat or subterete, often curved .... *Arabis*

S. petals 10–12 mm long; fruits 4-angled....*Conringia*

R. fruits <3 times as long as wide, flattened perpendicular to the plane of the septum

T. ovules (and seeds) 1 per locule

U. ovary (and fruit) rounded at the base and notched at the apex .... *Lepidium*

U. ovary (and fruit) subcordate at the base and acute at the apex .... *Cardaria*

T. ovules (and seeds) >1 per locule

V. ovary (and fruit) tapering at the base, cordate at the apex .... *Capsella*

V. ovary (and fruit) circular, ellipsoid or obovoid, notched at the apex .... *Thlaspi*

Q. cauline leaves not perfoliate or clasping

W. ovaries (and fruits) nearly to quite as wide as long..... .... *Lepidium*

W. ovaries (and fruits) >3 times longer than wide

X. cauline leaves few, widely spaced on the stem

Y. flowers <5 mm wide; stems hairy at the base... ....*Arabidopsis*

Y. flowers >5 mm wide; stems glabrous at the base

Z. basal leaves round or cordate, with long petioles; cauline leaves sessile, narrowly oblong to ovate ....*Cardamine*

        Z. basal leaves similar in shape to cauline leaves, short-petiolate or nearly sessile .... *Arabis*

    X. cauline leaves numerous, often overlapping on the stem

        AA. plants variously hirsute or pubescent, with at least some of the hairs branched

            BB. hairs spreading

                CC. petals 2–2.5 cm long; fruits straight .... *Hesperis*

                CC. petals 3–5 mm long; fruits often curved .... *Arabis*

            BB. hairs appressed

                DD. hairs centrally attached with >3 radiating branches .... *Alyssum*

                DD. hairs 2-branched .... *Lobularia*

        AA. plants glabrous or with simple hairs only .... *Armoracia*

  O. at least some cauline leaves deeply lobed to compound

    EE. leaves trifoliate or palmately lobed or divided .... *Cardamine*

    EE. leaves deeply pinnately lobed or divided

        FF. inflorescences all axillary .... *Coronopus*

        FF. inflorescences terminal

            GG. ovary (and fruit) <3 times as long as wide, flattened, rounded at the base and notched at the summit .... *Lepidium*

            GG. ovary (and fruit) >3 times as long as wide, cylindrical or nearly so

                HH. valves of the fruit lacking a distinct midvein, opening explosively and coiling .... *Cardamine*

                HH. valves of the fruit with a distinct midvein, not opening explosively, not coiling .... *Nasturtium*

A. flowers yellow

  II. cauline leaves simple, entire or variously toothed but not deeply lobed

    JJ. cauline leaves perfoliate or auriculate clasping at the base

        KK. ovary <3 times as long as wide

            LL. ovary (and fruit) flattened perpendicular to the septum .... *Lepidium perfoliatum*

            LL. ovary (and fruit) not flattened, or flattened parallel to the septum

                MM. valves of the fruit keeled .... *Camelina*

                MM. valves of the fruit not keeled .... *Rorippa austriaca*

        KK. ovary >3 times as long as wide

            NN. leaves strictly entire; fruits strongly 4-angled, not beaked... .... *Conringia*

            NN. lower leaves dentate to pinnatifid; fruits terete

                OO. foliage dull, often glaucous .... *Brassica*

                OO. foliage glossy dark green .... *Barbarea*

    JJ. cauline leaves not perfoliate or clasping

        PP. ovary (and fruit) orbicular; leaves linear with entire margins .... *Alyssum*

        PP. ovary (and fruit) linear; leaves broader with toothed margins

QQ. hairs branched; fruits not beaked .... *Erysimum*
QQ. hairs simple; fruits strongly beaked .... *Sinapis*
II. at least some of the cauline leaves pinnately lobed or divided
RR. petals pale yellow to whitish with purple veins .... *Eruca*
RR. petals light to deep yellow without purple veins
SS. ovary (and fruit) <3 times as long as wide .... *Bunias*
SS. ovary (and fruit) >3 times as long as wide
TT. cauline leaves all lobed or compound, although the upper ones sometimes much reduced
UU. lower leaves 2–3 times pinnate-pinnatifid .... *Descurainia*
UU. leaves once pinnate or pinnatifid
VV. plants with appressed retrorse hairs; bracts present in much of the inflorescence .... *Erucastrum*
VV. plants glabrous or with spreading hairs; flowers lacking bracts except perhaps the lowest
WW. cauline leaves auriculate clasping at the base
XX. petals 6–8 mm long .... *Barbarea*
XX. petals 2–5 mm long .... *Rorippa*
WW. cauline leaves not auriculate clasping
YY. petals >10 mm
ZZ. petals light yellow with darker yellow veins; seeds in 1 row per locule ....
.... *Coincya*
ZZ. veins of petals not darker; seeds in 2 rows per locule .... *Diplotaxis*
YY. petals 2–8 mm
AAA. plants pubescent; valves 3-nerved, ovules (or seeds) in 1 row per locule...
.... *Sisymbrium*
AAA. plants glabrous or with a few scattered hairs; valves with 1 nerve, ovules in 2 rows per locule ... *Rorippa*
TT. uppermost cauline leaves not lobed, or only slightly lobed or toothed
BBB. leaves dull or even glaucous .... *Brassica*
BBB. leaves glossy dark green .... *Barbarea*

Alliaria Scop.

Alliaria petiolata (M.Bieb.) Cavara & Grande   Garlic-mustard
Biennial with simple or branched stem to 1 m tall, smelling strongly of garlic; basal leaves long-petioled; blade palmately veined, ovate-cordate; upper leaves similar but with progressively shorter petioles and venation becoming palmate-pinnate; sepals erect, not saccate; petals white; fruiting pedicels nearly as thick as the divaricately ascending, 2–4-cm-long, terete fruits; an invasive weed of shady, moist areas including woods, floodplains, and waste ground throughout; flr. late Apr–Jun; native to Europe; FACU–.

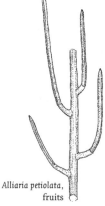

Alliaria petiolata,
fruits

Alliaria petiolata,
upper leaf

Alyssum L.

Alyssum alyssoides (L.) L.   Alyssum
Grayish-green, 1–2-dm-tall annual or biennial covered with appressed stellate hairs; stems erect, decumbent, or ascending; cauline leaves linear-oblanceolate, entire, to 4 cm long; petals pale yellow or white, 3–4 mm, notched at the tip, pubescent on the back; inflorescences elongating in fruit; fruits round, 3–4 mm in diameter, strongly flattened at the margins but bulging in the center over the seeds; seeds 2 per locule, narrowly winged; occasional in dry sandy or gravelly waste ground; SE and SC; flr. late Apr–Jun; native to Europe.

Arabidopsis Heynh. in Holl & Heynh.

Arabidopsis thaliana (L.) Heynh. in Holl & Heynh.   Mouse-ear cress
Freely branched annual, 1–4 dm tall; stems pubescent near the base, glabrous above; leaves mostly basal, 1–4 cm long, narrowly oblanceolate, with a slender petiole, densely covered with simple and branched hairs; cauline leaves few, nearly sessile; racemes elongating in fruit; petals white, approximately 0.3 mm; fruits linear, 1–1.5 cm long, <1 mm wide; seeds in 1 row per locule; common in cultivated fields and waste ground; mostly S; flr. Mar–Jun; native to Eurasia.

Arabis L.

Mostly erect, with simple or branched stems; hairy to glabrous; basal leaves entire to pinnate-lyrate; cauline leaves reduced, mostly sessile; inflorescence racemose, elongating in fruit; petals white, greenish, or cream; fruits elongate, erect to spreading or recurved; seeds mostly winged.

A. mature fruiting pedicels recurved .... A. *canadensis*
A. mature fruiting pedicels erect to ascending
  B. fruits terete .... A. *glabra*
  B. fruits flattened
    C. fruits strictly erect, appressed to the stem axis .... A. *hirsuta*
    C. fruits ascending to widely spreading
      D. cauline leaves not auricled or sagittate at the base
        E. fruits 2–4.5 cm, ascending... A. *lyrata*
        E. fruits 5–11 cm, widely spreading to recurved .... A. *laevigata*
      D. cauline leaves auricled or sagittate at the base
        F. pedicels <3 mm long, spreading at right angles ....A. *shortii*
        F. pedicels >3 mm long, ascending
          G. leaves and stem glabrous or rarely sparsely pubescent with only simple hairs; fruits curved
            H. lower cauline leaves entire to serrate-dentate; basal leaves entire to serrate; petals barely longer than the sepals .... A. *laevigata*
            H. lower cauline leaves sharply dentate to laciniate; basal leaves lyrately pinnatifid; petals half again as long as the sepals .... A. *missouriensis*
          G. leaves and stem pubescent; fruits straight ....A. *patens*

Arabis canadensis L.   Sicklepod
Biennial with a thick taproot; stem simple, rarely branched; sparsely hairy at the base, glabrous above; hairs mostly simple, a few 2-branched; basal leaves not persistent, cauline leaves short-petioled to sessile, 2.5–12 cm long, acuminate, denticulate to subentire; flowers in long, loose racemes; petals white to creamy, 3–5 mm; pedicels recurved to strongly reflexed at maturity; fruits 7–10 cm, curved, pendulous; common on rocky, wooded slopes, in rich soil; throughout; flr. May–Jul.

Arabis glabra (L.) Bernh.   Towercress, tower-mustard
Biennial from a taproot; stems 4–12 dm tall, glabrous and glaucous above, coarsely hairy below with branched hairs; basal leaves pubescent, petiolate, dentate to deeply divided; cauline leaves becoming sessile and entire, auriculate-clasping; petals yellowish-white, 5–7 mm; fruiting pedicels and fruits erect; fruits glabrous, 4–10 cm, slightly flattened; occasional in dry, rocky soil in open fields and edges; scattered; flr. May–Jun.

Arabis hirsuta (L.) Scop.   Hairy rockcress
Biennial with simple or branched stem 2–7 dm tall, hairy throughout with simple and branched hairs; basal leaves oblanceolate to broadly spatulate, entire or dentate, 2–8 cm long; cauline leaves oblong to spatulate, sessile, auriculate; petals white, 3–9 mm; pedicels and fruits erect or ascending; fruits 3–6 cm long, glabrous; flr. late May–Jun; FACU; �*/; 2 varieties:

A. stem pubescence appressed, forked ...... var. *adpressipilis* (M.Hopkins) Rollins
   rare in woods, banks, or rocky ledges, usually on limestone; scattered.
A. stem pubescence spreading, mostly simple ....
   .... var. *pycnocarpa* (M.Hopkins) Rollins
   rare on dry cliffs and ledges of calcareous shale or limestone; S.

Arabis laevigata (Muhl. ex Willd.) Poir.   Smooth rockcress
Biennial with a simple or branched stem 3–9 dm tall, glabrous and glaucous throughout, hairs simple; basal leaves not persistent; cauline leaves overlapping, 3–20 cm long, sessile, and usually sagittate; petals white, barely exceeding the sepals; pedicels widely spreading; fruits recurved-spreading at maturity, 5–10 cm long, curved to nearly straight; dry woods and hillsides; flr. Apr–Jun; 2 varieties:

A. cauline leaves auriculate, often dentate .... var. *laevigata*
   common throughout.
A. cauline leaves not auriculate, usually entire .... var. *burkii* Porter
   occasional, mostly S.

Arabis lyrata L.   Lyre-leaved rockcress
Biennial or perennial with slender, erect stem 0.7–3.5 dm tall, sparsely hairy with mostly simple hairs near the base; leaves mostly basal, lyrate-pinnatifid to subentire, 2–4 cm long; cauline leaves few, becoming linear, 1–4 cm long; petals white, 6–8 mm long; pedicels and fruits loosely ascending or spreading; fruits slender, 2–4 cm long, straight or slightly curved; frequent on serpentine barrens and dry, rocky slopes and rock outcrops; throughout; flr. Apr–Aug; FACU.

Arabis missouriensis Greene   Missouri rockcress
Biennial from a taproot; stems 2–5 dm tall, usually branched, glabrous through-

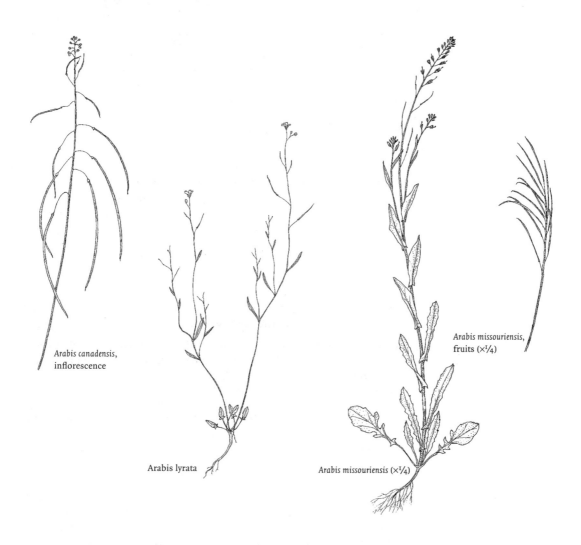

Arabis canadensis,
inflorescence

Arabis lyrata

Arabis missouriensis (×¼)

Arabis missouriensis,
fruits (×¼)

out; basal leaves lanceolate to spatulate, 2–9 cm long, petioled; cauline leaves numerous, 1–8 cm long, more or less appressed, the uppermost sessile with a sagittate base; petals creamy-white to yellowish, 5–7 mm; fruiting pedicels erect or ascending; fruits at first erect, becoming recurved, falcate, 6–9 cm long; rare on dry slopes; E; flr. Apr–early Jun; 🍂.

Arabis patens Sull. Spreading rockcress
Biennial with simple, or more rarely forked, spreading hairs throughout; stems erect, 3–6 dm tall, branched or simple; basal leaves ovate to oblanceolate, 1.5–6 cm long; cauline leaves few, sessile with auriculate-clasping bases; petals white, 5–7 mm; fruiting pedicels ascending or divergent; fruits suberect or ascending, 2.5–4.5 cm, straight or slightly curved; rare in moist, rocky woods; SE and SC; flr. Apr–Jul; 🍂.

Arabis shortii (Fernald) Gleason Toothed rockcress
Biennial, stem simple or branched, slender, leafy, 2–6 dm tall; basal leaves broadly obovate to oblanceolate, 4–16 cm long, with simple hairs above and 3- or 4-branched hairs beneath; cauline leaves numerous, oblong or obovate, auriculate-clasping, 1–6 cm long; petals white, 2–3 mm; fruiting pedicels widely

spreading; fruits widely spreading to ascending, 1.5–3 cm long, usually pubescent; infrequent in rich, moist woods; scattered; flr. Apr–May. Plants with glabrous fruits have sometimes been recognized as var. *phalacrocarpa* (M.Hopkins) Steyerm.

Armoracia Gaertn.

Armoracia rusticana (Lam.) Gaertn., B.Mey. & Scherb.   Horseradish
Glabrous perennial with stout branched stems to 1 m and fleshy roots; basal leaves 3–5 dm, long-petioled, crenate-serrate to irregularly dentate; cauline leaves reduced upward; flowers white; petals 5–7 mm; sepals not saccate; inflorescences elongating in fruit; fruits orbicular to ovoid, 4–6 mm long, somewhat compressed perpendicular to the plane of the septum; style 0.3 mm long with broad persistent stigma; seeds 4–6 per cell; cultivated for its pungent roots, which are grated and used as a condiment; occasionally escaped to roadsides and old fields; flr. Apr–May; native to Eurasia.

Barbarea R.Br.

Erect, mostly glabrous biennials, 2–8 dm tall with stems branched above; basal leaves lyrate-pinnatifid; cauline leaves somewhat auriculate; flowers in racemes; petals yellow; siliques linear, tardily dehiscent, slightly compressed to almost terete; seeds in 1 row in each locule, not winged.

A. basal leaves with <4 pairs of lateral lobes; pedicels <1 mm thick; fruits <3.5 cm long .... *B. vulgaris*
A. basal leaves with 4–10 pairs of lateral lobes; pedicels >1 mm thick; fruits >4 cm long .... *B. verna*

Barbarea vulgaris,
lower leaf

Barbarea verna (Mill.) Asch   Early wintercress
Glabrous biennial; stems, 1–few, 3–8 dm tall, branched above; leaves all pinnatifid, basal ones with 4–10 pairs of lateral lobes; cauline leaves with 3–8 pairs of oblong or linear lobes; petals bright yellow, 6–8 mm long; fruits 4.5–8 cm long on short pedicels of nearly equal diameter; common in fields, roadsides and waste ground; SE and SC; flr. Apr–Jul; native to Eurasia.

Barbarea vulgaris (L.) R.Br.   Wintercress, yellow rocket
Erect, mostly glabrous biennial; stem 2–8 dm tall, branched above; basal leaves petiolate, lyrate-pinnatifid with 1–4 pairs of lateral lobes; upper cauline leaves sessile, entire to dentate; petals bright yellow, 6–8 mm long; pedicels slender, 3–5 mm long; fruits 1–3 cm tipped by a 2–3 mm style; common in moist fields and roadsides; throughout; flr. Apr–Jun; native to Eurasia; FACU.

Barbarea vulgaris,
fruits (×1)

Berteroa DC.

Berteroa incana (L.) DC.   Hoary alyssum
Erect, densely pubescent annual or biennial, 3–11 dm tall; hairs mostly stellate and appressed; stems sparingly branched at the base and more freely so above; basal leaves oblanceolate, entire, 3–5 cm long with a slender petiole; cauline leaves numerous, gradually reduced upward, appressed ascending, sessile; petals white, 4–6 mm; fruits oblong-elliptic, 5–7 mm long, moderately inflated,

only tardily dehiscent; seeds 3–7 per locule, wing-margined; occasional in waste ground; scattered; flr. May–Aug; native to Europe.

Brassica L.

Annuals, glabrous or with simple hairs; stems erect, usually branched above; basal leaves petiolate; cauline leaves short petiolate or sessile, sometimes auriculate-clasping; flowers in racemes; sepals erect or ascending, inner pair saccate at the base; petals yellow, clawed; stamens 6, 4 long and 2 short; fruit narrowly linear, terete to somewhat 4-angled, erect to spreading; seeds in 1 row in each locule.

A. upper cauline leaves sessile, auriculate-clasping .... B. rapa
A. upper cauline leaves petiolate to sessile, not auriculate-clasping
    B. mature fruits and pedicels erect, appressed to the rachis; fruit more or less 4-angled .... B. nigra
    B. mature fruits and pedicels ascending to nearly erect but not appressed; fruit nearly terete .... B. juncea

Brassica juncea (L.) Czern.   Brown mustard, Chinese mustard
Glabrous, usually glaucous annual; stems 4–10 dm tall, branched above; lower leaves petiolate, lyrate-pinnatifid, up to 2.5 dm long; cauline leaves reduced upward, the uppermost sessile but not auriculate; petals pale yellow, >6 mm long; fruits subterete, 2–4 cm, divaricately ascending to nearly erect but not appressed; occasional in fields and waste ground; throughout; flr. May–early Oct; native to Eurasia.

Brassica nigra (L.) W.D.J.Koch   Black mustard
Annual, sparsely to densely hairy near the base; stems 4–15 dm tall, branching above; basal leaves petiolate, lyrate-pinnatifid to lobed and coarsely serrate; cauline leaves similar but reduced upward, becoming sessile but not auriculate; petals yellow, with a slender claw, 7–9 mm long; fruits ascending-appressed, 1–2 cm, somewhat 4-angled; frequent in fields, low ground, and floodplains; throughout; flr. Jun–Oct; native to Eurasia.

Brassica rapa L.   Field mustard
Glabrous or sparingly hirsute, usually glaucous, taprooted annual, 2–10 dm tall; basal leaves petiolate, lyrate-pinnatifid with 2–4 lateral lobes; cauline leaves be-

Brassica rapa, fruits

Brassica rapa, lower leaf

coming sessile, auriculate-clasping, lanceolate, and subentire; sepals greenish-yellow, somewhat spreading; petals yellow, 6–10 mm long; pedicels stout, ascending, 1–2.5 cm; fruits 3–7 cm long with an 8–15-mm beak that narrows to a slender style; frequent in fields and waste ground; throughout; flr. Apr–Oct; native to Europe.

Bunias L.

Bunias orientalis L.   Turkish rocket

Hairy, glandular, odoriferous perennial with simple to much-branched stems, 4–10 dm ; basal leaves petiolate, lanceolate or pinnatifid, often with 1 pair of divergent triangular basal lobes; cauline leaves becoming sessile; inflorescence elongating in fruit; sepals and petals yellow; petals 6–8 mm; fruits ovoid, asymmetrical, often with only 1 locule developed, glabrous but warty, 1–2 seeded, indehiscent; occasional in moist meadows and waste ground; scattered; flr. May–Jun; native to Europe.

Bunias orientalis,
fruit (×1)

Cakile Mill.

Cakile edentula (Bigelow) Hook.   American sea-rocket

Succulent annual, erect to 8 dm tall, much branched; early leaves very succulent, lobed but never pinnatifid; later leaves smaller, less lobed; petals white to pale lavender; fruits slightly constricted between segments, 1.6–2.6 cm, lower segment cylindrical, upper segment 4-angled, with a long, flattened beak; rare on dunes and sand plains along Lake Erie; NW; flr. Jun–Aug; FACU; 🍂.

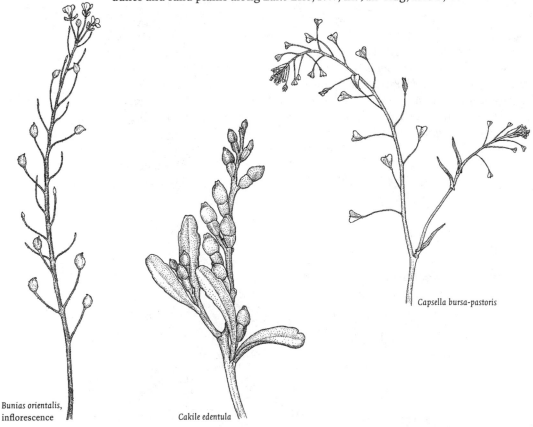

Capsella bursa-pastoris

Bunias orientalis,
inflorescence

Cakile edentula

Camelina Crantz

Annuals or biennials with erect stems; leaves mostly cauline with prominent auricles; inflorescence an elongate raceme; petals yellow to whitish; stamens 6, of 3 different lengths; fruits obovoid or pear-shaped, flattened parallel to the septum.

A. fruits 7–9 mm long, truncate above; styles 1.5–2 mm; seeds deeply grooved....
.... *C. sativa*
A. fruits 5–7 mm long, obtuse at apex; styles 2–2.5 mm; seeds finely alveolate....
.... *C. microcarpa*

Camelina microcarpa Andrz. ex DC.   Small-fruited false flax
Annual, conspicuously hirsute with simple or occasionally forked hairs; stem 3–10 dm tall, simple or slightly branched; leaves mostly cauline, 2–8 cm, the upper sagittate-auriculate; inflorescence greatly elongated and often compound; petals pale yellow fading to white, 4–5 mm; fruiting pedicels spreading-ascending; fruits broad, 5–7 mm long, obtuse at the apex; occasional in fields and waste ground in sandy soil; SE and SC; flr. May–Jul; native to Eurasia.

Camelina sativa (L.) Crantz   False flax, gold-of-pleasure
Annual or biennial, 3–6 dm tall; stems single, branched above, mostly glabrous but sometimes with simple and branched hairs; basal leaves linear-oblong, irregularly toothed; cauline leaves with prominent auricles; petals yellow; fruiting pedicels ascending; siliques 7–9 mm long, nearly truncate at the tip; infrequent in open, sandy ground; mostly S; flr. May–Jul; native to Eurasia; UPL.

Capsella Medik.

Capsella bursa-pastoris (L.) Medik.   Shepherd's-purse
Annual, pubescent below with simple and stellate hairs, 1–5 dm tall; basal leaves oblanceolate, subentire to lyrate-pinnatifid; cauline leaves mostly sessile and clasping with acute auricles, remotely serrate-dentate; racemes many-flowered, elongating in fruit; petals white, 2–3 mm; fruiting pedicels slender, widely spreading to loosely ascending; fruits 4–8 mm long, triangular-obcordate, strongly flattened perpendicular to the septum; seeds numerous; a common weed of cultivated fields, roadsides, and waste ground; throughout; flr. Apr–Oct; native to Eurasia; FACU.

Capsella bursa-pastoris,
fruit (×1)

Cardamine L.

Annual, biennial, or perennial herbs; glabrous or with simple hairs; stems erect to decumbent; leaves entire, pinnate, or palmately divided, basal and cauline leaves usually present; flowers in racemes; petals white, pink, or purple; fruits linear, narrow, dehiscent, usually flattened, opening from the base, often explosively; the septum with thickened margins caused by the failure of the valves to extend completely to the edge; seeds in 1 row in each locule, not winged.

A. leaves simple
    B. stems erect from a short tuber-like rhizome; petals 7–16 mm; fruits 2–4 cm long
        C. upper stems glabrous or with minute, appressed hairs; petals white ....
        .... *C. bulbosa*

  C. upper stems with conspicuous, spreading hairs; petals pink ....
  .... *C. douglassii*
 B. stems decumbent, stoloniferous at the base, tuber-like rhizome not
  present; petals 5–7 mm; fruits <1.8 cm long .... *C. rotundifolia*
A. leaves compound
 D. cauline leaves palmately divided into leaflets or deeply incised lobes; some
  leaves arising directly from the rhizome independently of the stem axis
  E. rhizome leaves and cauline leaves similar
   F. cauline leaves usually 2; rhizomes uniform in diameter ... *C. diphylla*
   F. cauline leaves usually 3; rhizomes not uniform in diameter
    G. cauline leaves whorled or nearly so; rhizomes jointed, consisting
     of segments weakly joined together .... *C. concatenata*
    G. cauline leaves distinctly alternate; rhizomes with alternate en-
     larged and slightly constricted regions, not jointed ..... *C. maxima*
  E. leaflets of the cauline leaves distinctly narrower than those arising di-
   rectly from the rhizome .... *C. angustata*
 D. cauline leaves pinnately divided into leaflets or lobes; no separate rhizome
  leaves present
  H. petals 6–13 mm; flowers 6–16 mm wide; perennials .... *C. pratensis*
  H. petals 1.5–3 mm, or none; flowers 3–4 mm wide; annuals or biennials
   I. basal leaves numerous; cauline leaves 2–5; stem bases and petioles
    with spreading hairs .... *C. hirsuta*
   I. basal leaves few or none; cauline leaves 4–10 or more
    J. cauline leaves numerous, >1 dm long, little reduced upward,
     densely overlapping, with auriculate bases .... *C. impatiens*
    J. cauline leaves mostly 4–10, barely overlapping, <1 dm long, not
     auriculate
     K. cauline leaves mostly 4–10 cm long, their lateral leaflets rela-
      tively broad, oval or broadly oblong
      L. leaflets sessile, decurrent along the rachis, stems erect .....
      .... *C. pensylvanica*
      L. leaflets with short petioles; stems flexuose (zigzag) ....
      .... *C. flexuosa*
     K. cauline leaves mostly 2–4 cm long, with narrowly linear
      leaflets .... *C. parviflora*

Cardamine angustata O.E.Schulz   Toothwort
Perennial with fleshy rhizomes constricted at intervals; stems 2–4 dm tall, glabrous or sparsely hairy; cauline leaves usually 2, nearly opposite, palmately
lobed with 3 narrowly lanceolate segments; rhizome leaves trifoliate with
broader leaflets; petals pinkish-lavender; fruits 2–3 cm; occasional in moist
woods, thickets, and stream banks; S; flr. Apr–May; FACU. [syn: *Dentaria
heterophylla* Nutt.]

Cardamine bulbosa (Schreb. ex Muhl.) Britton, Stearns & Poggenb.
                Bittercress, springcress
Perennial with stems mostly simple, erect, glabrous, 2–5 dm tall, arising from
tuber-like rhizomes; leaves simple, cordate-ovate, the basal ones long-petioled,
the upper sessile; petals white, 7–16 mm; fruits linear to lanceolate, 2–3 cm;
common in low, wet ground, shallow water, swamps, or springs; throughout,
except in the northernmost counties; flr. Apr–Jul; OBL.

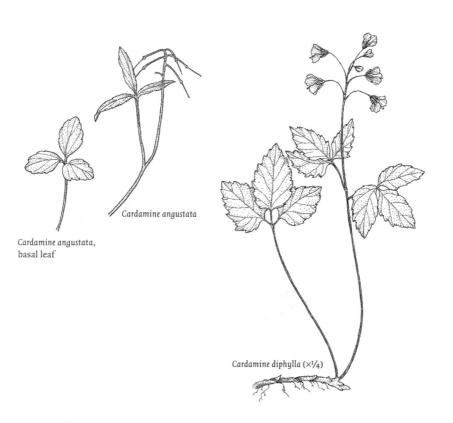

Cardamine angustata

Cardamine angustata,
basal leaf

Cardamine diphylla (×¼)

Cardamine
bulbosa

Cardamine concatenata (Michx.) O.Schwarz — Toothwort

Perennial with rhizomes constricted at intervals; stems 2–4 dm tall, hairy above; cauline leaves 3, whorled or nearly so; rhizome and cauline leaves similar, deeply palmately 3-lobed or 3-foliate, lobes or leaflets highly variable; inflorescences with 3–10 flowers; petals white or purplish, 1–2 cm; fruits 1.5–2.5 cm, ascending; common in deciduous woods throughout, except in the northernmost counties; flr. Mar–early May; FACU. [syn: *Dentaria laciniata* Muhl. ex Willd.]

Cardamine diphylla (Michx.) Wood — Two-leaved toothwort

Perennial with a stout, evenly thickened rhizome; stem glabrous, 1.5–3 dm tall; cauline leaves 2, nearly opposite; cauline and rhizome leaves similar, 3-foliate; leaflets ovate, coarsely toothed with minute marginal hairs; petals white, 1.5–2 cm; fruits ascending, 1.5–3 cm, frequently infertile; common in rich woods and floodplains; N and W; flr. Apr–Jun; FACU. [syn: *Dentaria diphylla* Michx.]

Cardamine douglassii Britton — Purplecress

Perennial with 1–few erect stems arising from a fleshy, tuber-like rhizome; stems with spreading hairs above; basal leaves petiolate, orbicular; cauline leaves few, becoming sessile and oblong; petals pink or purple, 8–12 mm; fruits erect, 2.5–4 cm; infrequent on rich wooded hillsides, calcareous springs, and bottomlands; mostly W; flr. Apr–early May; FACW+.

Cardamine flexuosa With. — Bittercress

Biennial or short-lived perennial with fibrous roots; stems flexuose (zigzag), erect, hairy toward the base; leaves pinnate, the lowest with 7–13 rounded leaflets, the upper similar but larger; petals white, 2–3 mm; fruits nearly erect, 1.5–2 cm long and 1 mm wide; rare in woods and openings; mostly S; native to Europe; OBL.

Cardamine hirsuta, dehiscing fruit (×1)

Cardamine hirsuta L. — Hairy bittercress

Annual with ascending stems 0.5–3 dm tall; leaves mostly basal, pinnate with orbicular to ovate, short-petioled leaflets, leaflets of cauline leaves narrower; stems, petioles, and upper surface of cauline leaves sparsely hairy; petals white, 1.5–2 mm; stamens 4; siliques erect, their valves coiling tightly from the bottom when shed and forcibly expelling the seeds; a common weed of lawns, gardens, and stream margins in moist soil; mostly S, but spreading rapidly; flr. Mar–Apr; native to Europe; FACU.

Cardamine impatiens L. — Bittercress

Annual or biennial; stems erect, glabrous, 2.5–8 dm tall; leaves very numerous, not much reduced upward, pinnate with 13–19 narrow, sharply-toothed, sparsely ciliate leaflets; leaf bases sagittate-auriculate; petals white, 2–3 mm or lacking; fruits 1.5–2 cm long and 1 mm wide, ascending; occasional in moist woods and slopes; scattered and spreading rapidly; flr. May; native to Europe.

Cardamine maxima (Nutt.) A.W.Wood — Large toothwort

Perennial with a long, stout rhizome with alternating enlarged and slightly constricted regions; flowering stems glabrous, simple, to 3 dm tall; cauline leaves 3, alternate; cauline and rhizome leaves 3(5)-foliate with ovate leaflets; inflorescences 4–10-flowered; petals white to pink, 5–7 mm; fruits ascending, acuminate, 2–3.5 cm long and 2.5 mm wide; very rare in low, wet woods; scattered; flr. Apr–May; ✿. [syn: *Dentaria maxima* Nutt.]

Cardamine parviflora L. — Small-flowered bittercress

Glabrous annual with erect stems 1–3 dm tall; leaves pinnate, the basal with obovate to suborbicular leaflets, the cauline with 7–9 linear or narrowly oblong leaflets; petals white, 2–3 mm; fruits erect, 1–2 cm long and 0.8–1 mm wide; occasional on dry, rocky ledges or shaly slopes; S; flr. Apr–Jul; FACU. Ours is var. *arenicola* (Britton) O.E.Schulz.

Cardamine pensylvanica Muhl. ex Willd. — Pennsylvania bittercress

Biennial or short-lived perennial; stems few, 1–7 dm tall, erect or decumbent,

Cardamine pratensis

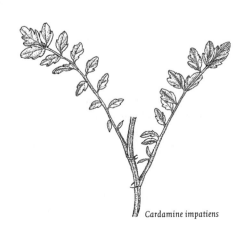

Cardamine impatiens

hirsute toward the base; leaves pinnate, glabrous, lowest leaves with 5–8 rounded leaflets, leaflets of the upper cauline leaves progressively narrowed, with decurrent bases; petals white, 2–4 mm; fruits ascending to erect, slender, 2–3 cm; common in low, wet ground, swamps, springs, and stream margins; throughout; flr. Apr–Oct; OBL.

Cardamine pratensis L.   Cuckoo-flower, lady's-smock
Erect perennial, 2–7 dm tall, glabrous or slightly pubescent; leaves pinnate, the basal ones long-petioled with 3–9 rounded leaflets, petioles of the cauline leaves progressively shorter and leaflets more linear; petals white to pink, 8–15 mm; fruits erect, 2–3 cm long and 1.5 mm wide; rare in swamps, wet meadows, moist shores, and alluvial woods; scattered; flr. May–Jun; OBL. Ours appears to be var. *pratensis* (see Rollins, 1993), a Eurasian introduction.

Cardamine rotundifolia Michx.   Mountain watercress
Glabrous perennial; stems trailing, 2–4 dm tall, occasionally rooting at the nodes; roots fibrous, slender stolons sometimes present; leaves simple, petiolate, reduced upward; petals white, 5–7 mm; fruits slender, 1–1.5 cm, often infertile; frequent in springs, seeps, and stream edges; throughout, except in the northernmost counties; flr. Apr–early Jun; OBL.

Cardaria Desv.

Cardaria draba (L.) Desv.   Hoarycress
Rhizomatous perennials with stout stems, glabrous or with short, simple hairs; leaves simple, dentate with sagittate bases; inflorescences terminal; flowers inconspicuous; petals white; stamens 6; fruits inflated, globose, subglobose to cordate, indehiscent, glabrous or pubescent; seeds 1–2 per locule; scattered on roadsides, fields, and waste ground; flr. May–early Jun; native to Eurasia. This species and *C. pubescens* (C.A.Mey.) Jarm., which differs mainly in its pubescent sepals and fruits, are noxious weeds of alkaline soils in the western states and Canadian provinces.

Chorispora R.Br.

Chorispora tenella (Pallas) DC.   Chorispora
Annual, 1–5 dm tall, with stalked glands and sparse hairs; leaves elliptic-oblong to oblanceolate, the lower petiolate; blades 3–8 cm long, sinuate-dentate; racemes elongated, the lowest flowers with leafy bracts; petals magenta with a narrow claw; blade to 5 mm; fruits 3.5–4.5 cm long, widely spreading to ascending, slightly constricted between the seeds; beaks tapered to a point; rare in field margins, roadsides, and waste ground; SE; native to Eurasia.

Coincya Porta & Rigo ex Rouy

Coincya monensis (L.) Greuter & Burdet   Coincya
Annual or short-lived perennial with a taproot and grayish-green foliage; stems branched, hispid below; basal leaves pinnately cleft, hispid on both surfaces; cauline leaves with progressively fewer and narrower lobes; sepals erect, the inner 2 saccate at the base; petals lemon-yellow, often with darker veins, the narrow claw slightly longer than the sepals; sepals erect, 5–10 mm; fruits ascend-

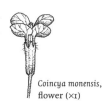

Coincya monensis,
flower (×1)

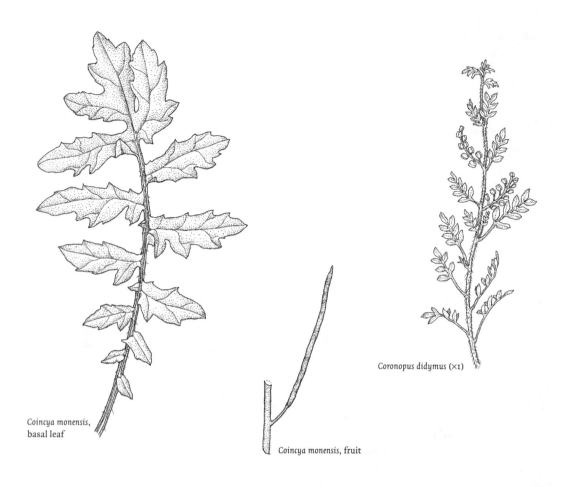

Coincya monensis,
basal leaf

Coincya monensis, fruit

Coronopus didymus (×1)

ing, terete, 3–8 cm, slightly constricted between the seeds, with an 8–12 mm, somewhat flattened beak containing 1–3 seeds; seeds in 1 row in each locule; frequent on roadside banks, fields, and waste ground; SE, spreading rapidly along highways; flr. May–Jul; native to Europe. [*Hutera cheiranthos* (Vill.) Gomez-Campo of Rhoads and Klein, 1993]

Conringia Heister ex Fabr.

Conringia orientalis (L.) Dumort.   Hare's-ear mustard, treacle-mustard
Glabrous, glaucous annual or winter annual; stems mostly simple or sparingly branched, 3–7 dm long; basal leaves entire or subentire, 5–9 cm long, obovate to oblanceolate, narrowed to the base; cauline leaves sessile, cordate-clasping; racemes few-flowered, corymbiform, becoming elongated; petals pale yellow to creamy, 7–12 mm, slenderly clawed; fruits nearly terete to 4-angled, 8–13 cm, tapered to a slender tip, somewhat constricted between the seeds; seeds in 1 row in each locule, wingless; infrequent in waste ground and ballast; mostly S; flr. May–early Jun; native to Eurasia.

Coronopus Zinn.

Small, leafy, densely-branched plants with minute flowers in crowded axillary racemes; glabrous or with simple hairs; petals white or lacking; fertile stamens

2–6; fruit apparently indehiscent, but the single-seeded, inflated valves eventually separating from the septum.

A. stems hairy; fruit constricted between the valves; valves reticulate ....
....*C. didymus*

A. stems glabrous; fruit not constricted between the valves; valves reticulate and
warty .... *C. squamatus*

Coronopus didymus (L.) Sm.   Wartcress, swinecress
Annual or winter annual with spreading, much-branched stems 1–5 dm long, glabrous to somewhat hairy; leaves numerous, ovate-oblong, pinnatifid, dentate or entire; racemes axillary, many-flowered; petals lacking; fertile stamens 2; fruits constricted at the septum; valves with a reticulate surface; occasional in paving joints, waste ground, or ballast; flr. May–Aug; native to South America.

Coronopus didymus,
fruit (×3)

Coronopus squamatus (Forssk.) Asch.   Wartcress, swinecress
Annual or biennial with prostrate to ascending stems 0.5–3 dm long, leafy and much branched; leaves obovate to oblanceolate in shape, twice pinnate; inflorescences axillary, condensed, not elongating in fruit; siliques nearly reniform, valves marked with prominent raised veins and warty protuberances; seeds 2 per locule; rare in waste ground or ballast; native to Europe.

Descurainia Webb & Berth.

Pubescent annuals or biennials, 1–8 dm tall, with minute, branched hairs often mixed with simple ones; leaves very fine-textured, bipinnately divided; flowers yellow, 2–4 mm wide; petals clawed; stamens 6; anthers yellow; fruits linear or club-shaped; seeds in 1 or 2 rows per locule.

A. fruits club-shaped, approximately 1 cm long, seeds in 2 rows .... *D. pinnata*
A. fruits narrowly linear, 2–3 cm long, seeds in 1 row .... *D. sophia*

Descurainia pinnata (Walt.) Britt.   Tansy-mustard
Annual, stems 1–7 dm tall, branched above; leaves dark green, the lower bipinnate-pinnatifid with ultimate segments ovate to oblanceolate or linear; the upper pinnate with mostly linear segments; leaves and stems moderately to densely pubescent and glandular; fruits club-shaped, 5–10 mm; seeds in 2 rows per locule; rare in ballast and waste ground; SE; flr. May; native Canada to NC and TX, but adventive in Pennsylvania. Ours is var. *brachycarpa* (Richardson) Detling.

Descurainia sophia (L.) Webb ex Prantl   Herb-sophia
Annual or biennial; stems 2.5–7.5 dm tall, branched above; leaves bi- to tripinnate, the ultimate lobes usually linear; leaves, stem, and pedicels sparsely to densely pubescent with minute, branched hairs and sometimes also simple hairs; fruits narrowly linear, nearly terete, 10–20 mm long, straight or slightly incurved; seeds in 1 row; rare on roadsides, waste ground, and ballast; SE; flr. May–Aug; native to Eurasia.

Diplotaxis DC.

Erect plants, glabrous or with sparse, simple, hairs near the base; leaves pinnatifid, reduced upward; petals yellow; stamens 6, 4 long and 2 short; fruits linear, with a short beak.

A. leaves mostly cauline; fruits on 1–2-mm stalks .... D. tenuifolia
A. leaves mostly basal; fruits sessile .... D. muralis

Diplotaxis muralis (L.) DC.   Wall- or sand-rocket
Annual or biennial with several ascending stems to 4 dm tall; glabrous or with a
few scattered hairs on the lower stem and petioles; leaves mostly in basal rosettes, lyrate-pinnatifid, petiolate; petals pale yellow fading to brown, 6–7 mm
long; fruits erect-spreading, 2.5–4.5 cm long; occasional on roadsides and
waste ground; SE; flr. May–Oct; native to Europe.

Diplotaxis tenuifolia (L.) DC.   Wall-rocket
Glaucous perennial with erect stems to 7 dm tall, simple to much branched, glabrous or with a few scattered hairs; lower leaves coarsely lobed or pinnatifid, to
1 dm; upper leaves entire or with linear lobes; petals lemon yellow, broadly obovate, approximately 11 mm long; fruits 2–4.5 cm long, on 1–2-mm stalks; occasional on roadsides and waste ground in dry, gravelly soil; mostly E; flr. Jun–Oct;
native to Europe.

Draba L.

Pubescent, low, very early blooming annuals with a basal rosette of leaves and
slender erect stems; petals white; fruits oblong to elliptic, 0.5–2.2 cm long, flattened parallel to the septum. This treatment includes the genus Erophila, following Rollins (1993).

A. petals deeply notched, fruits 4–8 mm long .... D. verna
A. petals not notched; fruits 0.5–2.2 cm long .... D. reptans

Draba reptans (Lam.) Fernald   Whitlow-grass
Annual, 2–15 cm tall, pubescent below with simple and forked hairs; all but a
few leaves in a basal rosette; stems filiform, lightly branched; petals white; fruits
linear to narrowly oblong, notched, 0.5–2.2 cm long, glabrous or pubescent
with simple hairs; rare on dry slopes and ledges; SE; flr. Apr–May; 🌿.

Draba verna L.   Whitlow-grass
Scapose annual with slender stems 5–15 cm, pubescent below with branched
hairs; leaves in a basal rosette, entire to dentate, 1–2 cm long; petals white,
deeply notched; fruits elliptic to oblong or obovate, 4–8 mm long, glabrous, flattened parallel to the septum; common in dry waste ground, lawns, roadsides,
and open woods; S; flr. Mar–May; native to Europe. [Erophila verna (L.) Chev. of
Rhoads and Klein, 1993]

Eruca Miller

Eruca vesicaria (L.) Cav.   Garden-rocket
Annual; stems simple or branched from the base, 2–10 dm tall, more or less
hairy; leaves petioled, pinnatifid, 5–15 cm long, greatly reduced, and becoming
sessile upward; petals white to yellowish with purple veins, 1.5–2 cm long; fruits
erect, 1.5–2 cm long and 3–5 mm thick; a rare weed of grassy fields; SE; flr. May–
Jun; native to Eurasia.

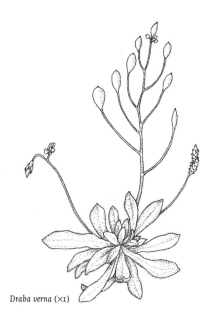

Draba verna (×1)

Erucastrum K.B.Presl

Erucastrum gallicum (Willd.) O.E.Schulz Dog-mustard, French-rocket
Annual or winter annual with appressed, retrorse hairs; stems 1.5–8 dm tall, simple to freely branched; leaves 3–20 cm long, pinnatifid; raceme elongate, bracteate throughout; petals pale yellow, 4–7 mm; fruits linear, somewhat constricted between the seeds, 1–4.5 cm long with a slender, seedless beak; seeds in 1 row in each locule, wingless; rare on roadsides and waste ground; SE; flr. Jun–Sep; native to Eurasia.

Erysimum L.

More or less densely pubescent plants with strongly appressed, 2–5-branched hairs aligned parallel to long axis of the stems and leaves; leaves simple, unlobed; petals yellow, clawed; fruits linear, terete or 4-angled in cross section; seeds in 1 row per locule.

A. pedicels more or less the same diameter as the fruits; fruits more or less at right angles to the rachis; constricted between the seeds .... *E. repandum*
A. pedicels definitely more slender than the fruits; fruits ascending to erect, not constricted
    B. fruits at most 2 cm long; pedicels ⅓ to ½ as long as the fruits ....
    .... *E. cheiranthoides*
    B. fruits >2 cm; pedicels <⅓ as long as the fruits
        C. hairs mostly 2-branched; fruits divaricately ascending.....*E. inconspicuum*
        C. hairs mostly 3- or 4-branched; fruits erect and closely appressed ....
        .... *E. hieracifolium*

Erysimum cheiranthoides L. Treacle-mustard, wormseed-mustard
Sparsely pubescent, leafy plant 4–10 dm tall with scattered, mostly 2-branched

hairs; petals bright yellow, 3.5–5.5 mm long; fruits 1.5–2 cm long, sparsely covered with appressed, 3-branched hairs and somewhat flattened perpendicular to the septum, thus some specimens can appear to have seeds in 2 rows, especially after pressing; occasional on roadsides and waste ground, usually in moist soil; E and S; flr. May–Oct; native to Eurasia; FAC.

Erysimum hieracifolium L.   Tall wormseed-mustard
Erect pubescent perennial, 3–6 dm tall with a few branches above; hairs appressed, mostly 3–4 branched; basal leaves 4–6 cm long and 3–10 mm wide, petioled; cauline leaves narrower, becoming sessile; racemes 20–60-flowered, greatly elongated in fruit; petals yellow, 8–10 mm; fruits linear-terete, 3–6 cm long and 1.5 mm wide, appressed pubescent; valves with a distinct midrib; rare on dry, rocky banks, but apparently spreading; NC; flr. Jun; native to Eurasia.

Erysimum inconspicuum (S.Watson) MacMill.   Treacle-mustard
Erect, grayish-pubescent perennial with simple stems 3–8 dm tall, branched above; hairs appressed, mostly 2-branched; basal leaves entire or sparsely dentate, petiolate, 3–6 cm long and 2–4 mm wide; cauline leaves gradually reduced and sessile; petals pale yellow, 5–7 mm with a narrow claw; fruits terete to quadrangular, 3–5 cm long and 1 mm wide, densely pubescent with 2-branched hairs; infrequent in dry, alkaline soils of roadsides, railroad banks, and waste ground; W; flr. Jun; native farther west.

Erysimum repandum L.   Treacle-mustard
Greenish-pubescent annual; stems erect, 1.5–4 dm tall, freely branched from near the base; leaves 1–6 cm long, oblanceolate to linear, dentate to almost entire, petioled below, sessile above; raceme 15–30 flowered; petals yellow, approximately 7 mm, slender-clawed; fruiting pedicels 3–4 mm, nearly as thick as the fruits; fruits 3–7 cm, just under 1 mm wide, constricted between the seeds, pubescent; valves with a midrib; infrequent on roadsides, waste ground, and ballast; SE and SC; flr. Apr–Aug; native to Eurasia.

Hesperis L.

Hesperis matronalis L.   Dame's-rocket
Erect biennial, 5–13 dm tall, with leafy stems and showy purple to white flowers; both simple and branched hairs present; leaves pubescent, simple, lanceolate to ovate-lanceolate, 5–20 cm long, the lower long-petioled, the upper sessile; flowers fragrant; sepals pubescent, the inner pair strongly saccate at the base; petals 1.8–2.5 cm; fruits terete, 4–15 cm; seeds in 1 row per locule, wingless; common in low woods, floodplains, wet meadows, and roadside ditches; throughout; flr. May–Jun; native to Europe; FACU–.

Iberis L.

Iberis umbellata L.   Candytuft
Annual to 3 dm; leaves alternate, elliptic-lanceolate, entire or with a few teeth; inflorescence compact, as wide as high and remaining so in fruit; flowers purple, 2 petals larger and 2 smaller; fruits flattened perpendicular to the septum, obovate with a deep notch; occasionally naturalized in fields and roadsides; SE; flr. Jun–Aug; native to Europe.

Iodanthus Torr. & A.Gray

Iodanthus pinnatifidus (Michx.) Steud.   Purple-rocket

Perennial; stems single, 3–8 dm tall, branched above, glabrous or with a few un-branched hairs; leaves glabrous, incised or dentate, acute to acuminate; petioles winged and somewhat clasping at the base; upper leaves sessile; inflorescence racemose; petals pinkish-purple to white, 7–14 mm; fruits terete, 2–4 cm, glabrous; seeds in 1 row per locule, wingless; rare in moist alluvial woods and wooded slopes; SW; flr. May–Jun; FACW; ❦.

Lepidium L.

Annual, biennial, or perennial herbs, glabrous or with simple hairs; basal leaves petiolate, entire to lobed or dissected; flowers in corymbose racemes that usually elongate in fruit; petals white or yellow; sepals erect to spreading, not saccate at the base; fruits strongly flattened, dehiscent; seeds 1 per locule.

A. upper cauline leaves perfoliate or sagittate
    B. upper leaves perfoliate; fruiting pedicels glabrous; flowers yellow ....
    .... *L. perfoliatum*
    B. upper leaves sagittate; pedicels pubescent; flowers white .... *L. campestre*
A. upper cauline leaves tapering to the petiole or base
    C. fruits 5–6 mm long; stamens 6 .... *L. sativum*
    C. fruits 1.5–3.5 mm long; stamens 2
        D. petals conspicuous, as long as to twice the length of the sepals ....
        .... *L. virginicum*
        D. petals inconspicuous (shorter than the sepals) or absent
            E. fruits oblong-obovate to suborbicular (widest above the middle); axis of the inflorescence with straight, usually subclavate, minute papillae .... *L. densiflorum*
            E. fruits orbicular to broadly ovate, widest at or below the middle; axis of the inflorescence with curved, subappressed, minute trichomes, or rarely glabrous .... *L. virginicum*

Lepidium campestre (L.) R.Br.   Fieldcress

Annual or biennial, erect, densely pubescent, 2–5 dm tall, branched above; stems leafy to the inflorescence; basal leaves petiolate, lyrately lobed to entire; stem leaves sessile with auriculate, clasping bases; flowers numerous; petals white, 2 mm long; fruiting pedicels slightly flattened; fruits ovate to broadly oblong, 4 mm broad, winged, upper surface concave, covered with scale-like hairs; style shorter than to equaling the wings; common in fields and waste ground; throughout; flr. Apr–Aug; native to Eurasia.

Lepidium campestre,
fruit (×1½)

Lepidium densiflorum Schrader   Wild pepper-grass

Annual with erect stems 2.5–5 dm; pubescent with flat, obtuse hairs; basal leaves serrate to pinnatifid; upper cauline leaves tapering to the base (not auriculate), toothed; flowers white, stamens mostly 2; fruiting pedicels somewhat flattened; fruits 2–3.3 mm, nearly round to slightly obovate, winged; style lacking; a common weed of waste ground and roadsides in dry to moist soil; throughout; flr. May–Sep; geographic origin not determined; FAC.

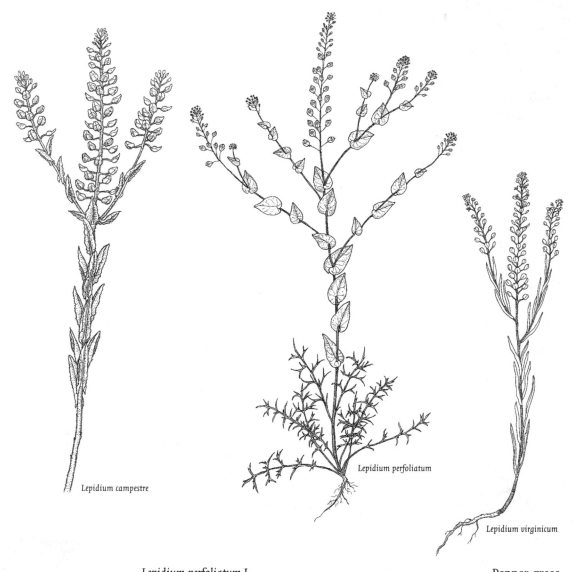

Lepidium campestre

Lepidium perfoliatum

Lepidium virginicum

Lepidium perfoliatum, fruit (×1½)

Lepidium virginicum, fruit (×1½)

Lepidium perfoliatum L. Pepper-grass
Annual with erect branching stems 2–5 dm; lower leaves pinnately dissected, upper broadly ovate to rounded, deeply cordate and auriculate; flowers yellow, stamens 6; fruit elliptic, 4 mm by 3 mm; rare in waste ground; SE; flr. May; native to Europe; UPL.

Lepidium sativum L. Gardencress
Annual, 2–4 dm tall, glabrous and glaucous; leaves pinnately dissected with a few linear segments; flowers white; stamens 6; fruit 5–7 mm long with a deep notch; rarely escaped from cultivation to waste ground and roadsides; SE and SC; flr. Jun–Jul; native to Europe.

Lepidium virginicum L. Poor-man's-pepper, wild pepper-grass
Annual with only sparse pubescence; stems 1.5–6 dm tall, simple below and freely branched above; basal leaves oblanceolate to linear-oblanceolate, incised to nearly entire, lower and middle stem leaves sharply toothed, reduced upward; racemes numerous, many-flowered; sepals glabrous or slightly pilose on the

back; stamens 2; pedicels nearly terete; fruits glabrous, broadly elliptical to nearly orbicular, 2.5–4 mm long, barely winged; style not extending beyond the shallow apical notch; common in dry or moist soil of roadsides, fields, and waste ground; throughout; flr. May–Nov; FACU–.

Lobularia Desv.

Lobularia maritima (L.) Desv.   Sweet alyssum
Perennial with grayish pubescence due to appressed, 2-pronged hairs; stems 1–3 dm tall, branching from the base and spreading; leaves linear to linear-lanceolate, 1–5 cm long by 1–4 mm wide, acute; flowers fragrant; petals white, 3–4 mm long; fruits greenish or purplish, broadly ovate, obovate, or suborbicular, 2–3.5 mm long; seeds 1 per locule; cultivated and rarely escaped to waste ground and alluvium; SE and SW; flr. Apr–Nov; native to the Mediterranean region.

Lunaria L.

Lunaria annua L.   Honesty, moneyplant
Biennial or perennial herb to 1 m, with simple dentate leaves; upper leaves sessile or subsessile; flowers showy, purple; fruits oblong-oval, 3–5 cm long and nearly as wide, strongly flattened parallel to the septum; cultivated and occasionally escaped to roadsides and waste ground; mostly S; flr. Apr–May, frt. Jun; native to Europe. *L. rediviva* L. (perennial honesty) has been collected rarely, it differs by the petiolate upper leaves.

Nasturtium R.Br.

Nasturtium officinale R.Br.   Watercress
Aquatic or semiaquatic perennial; stems hollow, submersed or partly emergent, rooting at the nodes; leaves pinnate; lateral leaflets ovate to broadly oblong, sessile, nearly entire; terminal leaflet larger; petals white, 3–4 mm; fruits narrowly oblong, 1–1.5 cm long by 2–3 mm wide, rounded at both ends; seeds in 2 rows; cultivated and extensively naturalized in springs, seeps, ditches, and quietly flowing water; throughout; native to Europe; flr. Apr–Sep; OBL. *N. microphyllum* Boenn. ex Rchb.f., with smaller fruits and seeds in 1 row, and hybrids are occasionally encountered.

Nasturtium officinale

Raphanus L.

Taprooted annuals or biennials, scabrous or hispid with simple spreading or appressed hairs; basal leaves petiolate, lyrate-pinnatifid; petals yellow or purplish-white; fruits weakly to strongly constricted between the seeds.

A. fruits strongly constricted between the seeds; petals yellow fading to white....
.... *R. raphanistrum*
A. fruits not strongly constricted between the seeds; petals purplish to white ....
.... *R. sativus*

Raphanus raphanistrum L.   Wild radish, white charlock
Annual or biennial with a taproot, sparsely hirsute-hispid, stems 3–8 dm tall;

basal leaves lyrate-pinnatifid; cauline leaves several, petiolate; petals yellow fading to white, often purple-veined, 1.5–2 cm long; fruits 4–8 cm long and 3–6 mm wide, finally breaking between the 1-seeded, prominently grooved segments; occasional in waste ground; scattered; flr. May–Nov; native to the Mediterranean region.

Raphanus sativus L.　　　　　　　　　　　　　　　　　　　Garden radish
Hispid annual or biennial with a fleshy elongated taproot; stems 4–12 dm tall, freely branched; lower leaves pinnately lobed; petals purplish to white with darker venation, 1–2.5 cm; fruits 3–6 cm long and 5–10 mm thick, with a conical beak, slightly depressed between the seeds but usually not breaking transversely; cultivated and occasionally escaped to fields and roadsides; scattered; flr. Jun–Sep; native to the Mediterranean region.

Rorippa Scop.

Annual, biennial, or perennial plants of semiaquatic or terrestrial habitats; stems solid; leaves entire to pinnately lobed, pubescence lacking or composed of simple hairs; racemes terminal and lateral; petals yellow; fruits plump, short to elongate; seed in 1 or 2 rows per locule, wingless.

A. petals longer than the sepals; perennials with creeping roots or rhizomes
    B. cauline leaves deeply pinnatifid or lobed; fruit equaling or longer than the
        pedicel .... R. sylvestris
    B. cauline leaves entire to serrate, never lobed; fruit shorter than the pedicel...
    .... R. austriaca
A. petals about as long as the sepals; annuals or biennials with a taproot ....
.... R. palustris

Rorippa austriaca,
lower leaf

Rorippa austriaca (Crantz) Besser　　　　　　　　　　　Field yellowcress
Nearly glabrous perennial spreading from thick, fleshy horizontal roots; stems 4–10 dm tall, finely pubescent toward the base, otherwise glabrous; basal leaves petiolate; cauline leaves linear-oblong to oblanceolate, 3–10 cm long, entire to coarsely serrate, sessile with clasping auricles; petals yellow, 3–5 mm, exceeding the sepals; fruits spheroid-ovoid, about 3 mm long by 2.4–3 mm wide, usually sterile; rarely established in fields and roadsides; SE; flr. May–Jul; native to Europe; FAC–.

Rorippa palustris (L.) Besser　　　　　　　　　Marsh or yellow watercress
Glabrous to densely hirsute annual, biennial, or short-lived perennial; stems 3–10 dm tall, erect or rarely decumbent; basal and lower cauline leaves short-petiolate to sessile, oblong to oblanceolate, 6–20 cm long, irregularly serrate to deeply incised or pinnately divided, apex broadly acute or attenuate; petals yellow, 1–3.5 mm; fruits subglobose to elongate-cylindrical, straight or curved upward, 3–14 mm long; containing 20–80 seeds; common on wet shores and low open ground; throughout; flr. May–Sep; OBL. Several poorly defined varieties have been described, but the differences cited seem to be mainly a response to water depth and season of growth.

Rorippa austriaca

Rorippa sylvestris (L.) Besser　　　　　　　　　　Creeping yellowcress
Nearly glabrous, creeping, mat-forming perennial; stems 1.5–6 dm tall, decum-

bent to ascending; leaves petiolate, dark green; cauline leaves with 4–6 pairs of lateral segments that are lanceolate to oblong or obovate; terminal segment lance-oblong with rounded apex; racemes numerous, from the stem apex and upper leaf axils, dense and short; petals yellow, 2.8–5.5 mm; fruits 1–2 cm, linear and often curved upward; rarely fertile; propagates vegetatively by creeping roots and root fragments; frequent in wet soil of roadsides, stream banks, and waste ground; throughout; native to Europe; flr. May–Oct; FACW.

Sinapis L.

Hispid or glabrous annuals; stems erect, leafy; lower leaves petiolate, upper becoming sessile; sepals yellowish, widely spreading, not saccate at the base; petals yellow, obovate, with a narrow claw; fruits linear or oblong, strongly beaked, terete or somewhat flattened; seeds in 1 row, wingless.

A. valves of the fruits pubescent; pedicels slender, mostly at right angles to the rachis .... *S. alba*
A. valves of the fruits glabrous; pedicels thick, erect or spreading ....*A. arvensis*

Sinapis alba L.   White-mustard
Mostly hispid annual; stems 2–6 dm tall; leaves petiolate, lyrate-pinnatifid; petals 11 mm long and 5 mm wide; pedicels widely spreading, 5–12 mm long, fruits 2–4.5 cm with a straight or sometimes curved 1.5–3-cm beak; infrequent in waste ground and ballast; S; flr. late May–Sep; native to the Mediterranean region.

Sinapis arvensis L.   Charlock, wild-mustard
Annual, hispid at least below, or sometimes glabrous; stems 2–6 dm tall; lower leaves lyrate, upper sessile, simple, dentate; petals 10 mm long, pedicels ascending, 3–5 mm; fruits 2–3.5 cm long, slightly constricted between the seeds and with a straight, conical beak; frequently naturalized in fields, gardens, and waste ground; throughout; flr. May–Nov; native to the Mediterranean region.

Sisymbrium L.

Annuals or perennials, glabrous or with simple hairs; stems erect, branched above; basal and cauline leaves similar, lyrate-pinnatifid to deeply lobed, green or glaucous; sepals erect to spreading, the outer pair saccate at the base; petals yellow; stamens 6, 4 long and 2 short; fruits mostly terete, linear to tapering, dehiscent, glabrous or pubescent; valves with a prominent midrib; seeds in 1 row, not winged.

A. fruits tapering from base to tip, closely appressed to the rachis .... *S. officinale*
A. fruits linear, terete, spreading or loosely ascending .... *S. altissimum*

Sisymbrium altissimum L.   Tumble-mustard
Erect glaucous annual; stem 3–15 dm tall, with wide-spreading branches above, sparsely to heavily hirsute at the base; lower leaves to 1.5 dm long, pinnatifid; upper leaves pinnately divided into linear segments; sepals 4 mm; petals pale yellow, 6–8 mm; pedicels stout, spreading; fruits 5–10 cm, about the same diameter as the pedicels, spreading; occasionally naturalized in fields, roadsides, and waste ground; throughout; flr. May–Sep; native to Eurasia; FACU–.

Sisymbrium officinale (L.) Scop. Hedge-mustard
Strongly hirsute-hispid annual; stems simple to loosely branched, 3–8 dm tall, stiff; basal leaves lyrate-pinnatifid, to 2 dm long; cauline leaves usually sessile, much reduced with 4–6 linear to narrowly lanceolate lobes; sepals 2 mm long; petals pale yellow, 3–4 mm long; racemes elongate in fruit with fruits erect and closely appressed to the rachis; fruits subulate, 8–15 mm, beaked, sparsely pubescent to glabrous; valves 3-nerved, tardily dehiscent; occasionally naturalized in gardens, fields, and waste ground; flr. May–Aug; native to Europe.

Thlaspi L.

Annual, mostly glabrous herbs with simple, dentate basal leaves and auriculate clasping cauline leaves; flowers in a raceme that elongates greatly in fruit; pedicels widely spreading; petals white; sepals not saccate; stamens 6; fruits flattened at right angles to the septum, or convex on one or both sides, dehiscent, notched at the apex and often winged; valves keeled; seeds 2–8 per cell.

A. fruits 9–15 mm long and 7–12 mm wide, the wing to 4 mm wide ..... *T. arvense*
A. fruits 4–8 mm long and 4–6 mm wide, the wing to 1 mm wide
    B. stems completely glabrous to the base, <3 dm tall; plants not smelling of garlic .... *T. perfoliatum*
    B. stems with long hairs near the base, mostly >3 dm tall; plants with a strong garlic odor .... *T. alliaceum*

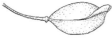

Thlaspi alliaceum, fruit (×2)

Thlaspi alliaceum (L.) Jacq. Garlic pennycress
Annual with long, simple hairs on the lower stem and a strong garlic smell; foliage bright yellowish-green, lower leaves petiolate, middle and upper auriculate-clasping, entire to dentate; fruits obovate 5–7 mm long by 4–6 mm broad, convex on both sides, but slightly asymmetrical; roadsides and farm fields; SE, but spreading rapidly; flr. Mar–Apr; native to Europe.

Thlaspi arvense, fruit (×2)

Thlaspi arvense L. Field pennycress
Glabrous annual or winter annual, 1–5 dm tall; basal leaves few, narrowed to a short petiole; cauline leaves gradually reduced upward becoming sessile and auriculate, dentate to wavy-margined; flowers white, on slender, wide-spreading to upcurved pedicels; fruits strongly compressed with a narrow, perpendicular septum, winged, ascending; a common weed of roadsides, fields, and waste ground; throughout; flr. Apr–Jul; native to Europe; UPL.

Thlaspi perfoliatum, fruit (×2)

Thlaspi perfoliatum L. Pennycress
Glabrous and glaucous annual, 0.5–2 dm tall; basal leaves few; cauline leaves ovate-cordate with rounded auricles, entire to sinuate-denticulate; flowers tiny; sepals with white margins; petals white; stamens shorter than the petals; fruiting pedicels slender, to 8 mm, horizontal; fruits 5–7 mm long, convex beneath, broadly obcordate, conspicuously winged above the middle and notched; stigma sessile in the base of the notch; occasional in fields and roadsides; SE and SC; flr. Apr–May; native to Eurasia.

BUXACEAE Box Family

Low, spreading herbs; leaves simple, coarsely toothed at the end, alternate or clustered at the ends of the upright branches; stipules not present; flowers unisexual, in an axillary or terminal raceme with staminate above and pistillate below, 4-merous; sepals distinct; petals absent; stamens 4, distinct; ovary superior, 3-celled; fruit a capsule.

Pachysandra Michx.

Characteristics of the family; plants 1.5–3 dm tall, spreading by horizontal stolons.

A. inflorescence lateral from near the base; leaves deciduous or semievergreen... .... *P. procumbens*
A. inflorescence terminal; leaves evergreen .... *P. terminalis*

Pachysandra procumbens Michx.   Allegheny-spurge
Leaves deciduous or semipersistent, dull green with darker blotches; flowers pinkish in an axillary raceme arising near the base of the upright stem; cultivated and occasionally naturalized; mostly SE; flr. Apr; native in the southeast as far north as WV, but apparently not occurring naturally in PA.

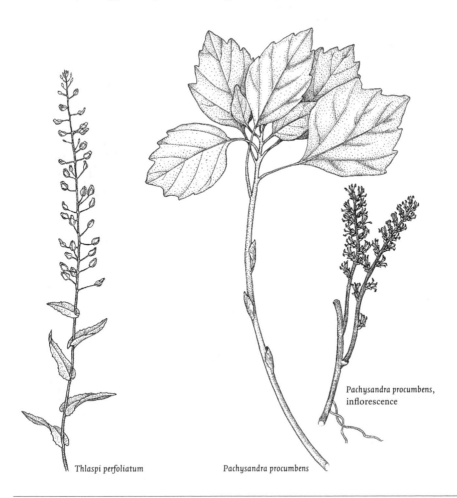

Pachysandra procumbens,
inflorescence

Thlaspi perfoliatum   Pachysandra procumbens

Pachysandra terminalis Siebold and Zucc.　　　　　Japanese pachysandra
Leaves glossy green, evergreen; flowers white, in a terminal raceme; frequently cultivated as a groundcover and occasionally naturalized in urban or suburban woods; mostly SE; flr. late Mar–early May; native to Japan.

CABOMBACEAE Watershield Family

Aquatic perennial plants; leaves floating or submerged, arising from rhizomes; flowers perfect, regular, solitary on long peduncles from the axils of floating leaves, slightly emergent above the water surface; fruit few-seeded, leathery, indehiscent.

A. underwater leaves divided into thread-like segments; floating leaves (if present) entire, blades 1–2 cm long, linear-elliptic .... *Cabomba*
A. all leaves floating; blades elliptic, 4–12 cm long, entire .... *Brasenia*

Brasenia Schreb.

Brasenia schreberi J.F.Gmel.　　　　　Purple wen-dock, watershield
All leaves usually floating, peltate; flowers reddish-purple; sepals and petals similar, corolla 12–20 mm long; stamens 12–18; fruits 1- or 2-seeded; occasional in quiet water of lakes and streams; throughout; flr. Jun–Jul; OBL.

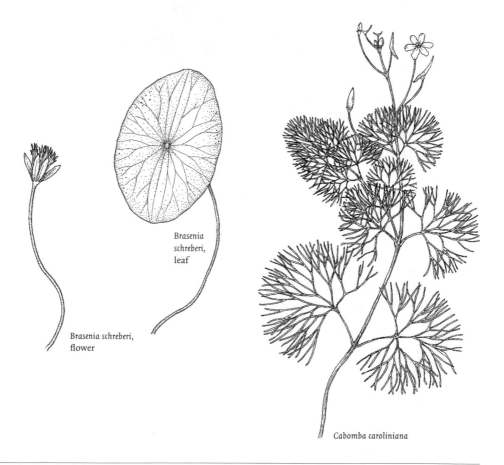

Brasenia schreberi, leaf

Brasenia schreberi, flower

Cabomba caroliniana

Cabomba Aubl.

Cabomba caroliniana A.Gray   Fanwort
Submerged leaves opposite, 3–4 times divided giving a fan-like, feathery appearance; floating leaves alternate, peltate; flowers white to pinkish; sepals 3, similar to the petals; petals 3, yellow at the base, corolla 6–12 mm long; stamens 6; fruits 3-seeded; occasional in lakes and ponds; widely scattered; throughout; flr. Jul–Aug; native to the southeastern U.S.; OBL.

CACTACEAE Cactus Family

Stems prostrate and spreading; branches jointed, fleshy, and flattened, often armed with spines, leafless; flowers bright yellow, 5–8 cm wide, regular, perfect; ovary inferior; sepals and petals numerous; stamens numerous, shorter than the petals; fruit red to red-purple, fleshy, many-seeded, 3–5 cm long.

Opuntia Miller.

Opuntia humifusa (Raf.) Raf.   Eastern prickly-pear cactus
Characteristics of the family; dry, shaly cliffs and barrens; mostly S and E; flr. Jul; 🐝.

CAESALPINIACEAE Caesalpinia Family

Deciduous trees or herbs with alternate, simple or pinnately compound leaves with a swollen base (pulvinus) and stipules; flowers nearly regular to strongly zygomorphic, but not truly papilionaceous, slightly to distinctly perigynous; sepals 5, distinct or the 2 upper fused; petals 5, distinct, nearly equal to very unequal; stamens 5–10; ovary 1, simple; fruit a dehiscent or indehiscent legume.

A. trees or tall shrubs; anthers versatile, opening lengthwise
   B. small trees or large shrubs; leaves simple; flowers magenta-pink, appearing before the leaves .... *Cercis*
   B. large trees; leaves compound; flowers greenish-yellow or whitish, appearing with or after the leaves
      C. stems thorny (except on thornless cultivars); leaflets 1 cm wide or less; fruits 2–4.5 dm .... *Gleditsia*
      C. stems not thorny; leaflets 1.5–4 cm wide; fruits 8–25 cm ... *Gymnocladus*
A. herbs; anthers basifixed, opening by apical pores or short slits
   D. pedicels bearing 2 small bracts near the middle; stamens 5 or 10, all with normal anthers .... *Chamaecrista*
   D. pedicels without bracts; stamens 10, the 3 uppermost sterile .... *Senna*

Cercis L.

Cercis canadensis L.   Redbud, Judas-tree
Deciduous tree to 12 m tall; leaves alternate, simple, broadly cordate, 5–12 cm, entire; flowers magenta-pink, appearing before the leaves in small clusters

Cercis canadensis leaf (×¼)

along the branches, perfect; calyx and corolla irregular, 5-merous, apparently papilionaceous; stamens 10, distinct, anthers versatile, opening longitudinally; occasional to locally abundant in dry to moist, rich woods on limestone or diabase; S; flr. late Apr–May; native, but also cultivated.

Chamaecrista (L.) Moench

Erect or ascending annual herbs with even-pinnate compound leaves with 5–20 pairs of leaflets; petioles bearing 1 or more saucer-shaped glands; pedicels with 2 small bracts near the middle; flowers perfect, 5-merous; petals yellow, unequal; stamens 5 or 10; anthers all well-developed, basifixed, opening by an apical pore or short slit; fruits dehiscent, the valves coiling elastically when they open.

A. petiolar glands stalked; stamens 5, anthers to 3 mm; petals very unequal, the longest 6–8 mm .... *C. nictitans*
A. petiolar glands sessile; stamens 10, anthers to 8 mm; petals nearly equal, 1–2 cm .... *C. fasciculata*

Chamaecrista fasciculata (Michx.) Greene   Partridge-pea, prairie senna
Erect or ascending, pubescent annual to 5 dm tall; leaflets oblong, 1–2 cm, acute or obtuse with a short mucronate tip; flowers in axillary racemes, pedicels 1–2 cm; petals 5, 1–2 cm , 4 of them bearing a red mark at the base; stamens 10, very unequal, all fertile; fruits linear-oblong, flat, 3–6 cm; occasional in stream banks, dry sandy ground, and serpentine barrens; S; flr. late Jul–Aug; FACU.

Chamaecrista nictitans (L.) Moench   Wild sensitive-plant
Erect annual; stem 1–5 dm tall, glabrous to villous; leaflets 6–15 mm; flowers axillary, solitary or in small clusters; petals very unequal; stamens 5; fruits oblong, 2–4 cm long by 3–6 mm wide, straight and flat; frequent in dry, open sandy ground of roadsides, old fields, and railroad embankments; throughout, except in the northernmost counties; flr. late Jul–early Sep.

Gleditsia L.

Gleditsia triacanthos L.   Honey-locust
Deciduous tree to 20 m tall with large, usually branched thorns on the trunk and branches; leaves 1 or 2 times pinnate with small, elliptical leaflets; the once-pinnate leaves and racemes of greenish-yellow flowers borne on short lateral branch spurs; leaves of long shoots bipinnate; flowers perfect or unisexual (plants polygamous), in separate racemes; fruits dark brown, 2–4.5 dm, flattened and often slightly twisted, with sweet pulp between the seeds; frequent on stream banks and floodplains; mostly S; flr. late May–early Jun; native, but also frequently planted; FAC–.

Gymnocladus Lam.

Gymnocladus dioicus (L.) K.Koch   Kentucky coffee-tree
Deciduous tree to 20 m tall; leaves twice-pinnately compound, 6–9 dm; leaflets ovate-acuminate, 3–8 cm; stipules lacking; flowers greenish-white, in long terminal panicles, regular, perfect, or unisexual; sepals and petals each 5; stamens

10, distinct, alternately long and short; anthers versatile, opening longitudinally; fruits flat and thick, 8–25 cm long by 3–5 cm wide containing several large seeds; infrequent in moist woods and bottomlands; native, also occasionally planted.

Senna Mill.

Erect perennial herbs with leaves evenly once-pinnate and 1 or more large glands on the petiole or rachis; pedicels without small bracts (except at the base); flowers somewhat irregular, perfect, with 5 sepals and 5 yellow petals; stamens 10, the 3 uppermost sterile; anthers basifixed, opening by apical pores or short slits; fruits compressed, segmented and only tardily dehiscent, or 4-sided and dehiscent.

A. gland located near the base of the petiole; leaflets oblong, elliptic or ovate; fruits flattened, somewhat segmented
    B. petiolar gland distinctly stalked; ovary villous; fruit segments as long as wide .... S. hebecarpa
    B. petiolar gland not stalked; ovary appressed-hairy; fruit segments twice as wide as long .... S. marilandica
A. petiolar gland located between the lowest pair of leaflets; leaflets obovate; fruits not segmented .... S. obtusifolia

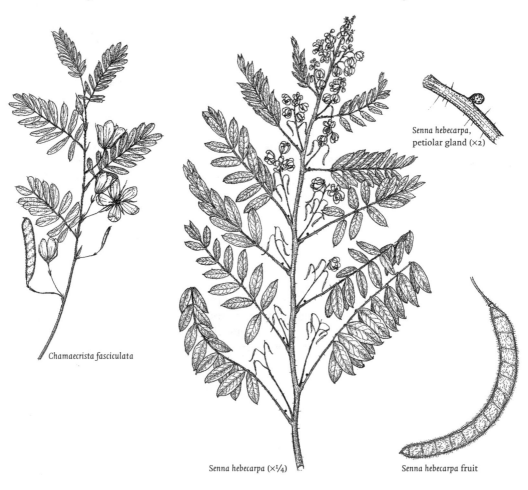

Senna hebecarpa, petiolar gland (×2)

Chamaecrista fasciculata

Senna hebecarpa (×¼)

Senna hebecarpa fruit

Senna hebecarpa (Fernald) H.S.Irwin & Barneby   Northern wild senna
Erect perennial; stems 0.9–2 m tall, mostly unbranched and usually clumped; petioles with a club-shaped gland with a short stipe located near the base; leaflets in 6–10 pairs, oblong or elliptic, 2–5 cm, mucronate; flowers in several dense axillary racemes forming a terminal panicle; fruits 7–12 cm long by 5–9 mm wide, slightly constricted at intervals forming nearly square segments; occasional on stream banks, sandy shores, moist old fields, or wetland edges; throughout, except in the northernmost counties; flr. early Jul–Aug.

Senna marilandica (L.) Link   Southern wild senna
Erect perennial very similar to S. *hebecarpa* except the petiolar glands short-cylindric or dome-shaped, not stalked; leaflets in 4–8 pairs; flowers less numerous, in several axillary racemes; fruits 6–10 cm long by 8–11 mm wide, the segments distinctly wider than long; very rare on dry roadsides and thickets; S; flr. Jul; FAC+; 🌿.

Senna obtusifolia (L.) H.S.Irwin & Barneby   Coffeeweed, sicklepod
Erect malodorous annual; stems to 1 m tall; leaflets in 2–3 pairs, the terminal pair obovate and 4–7 cm long, others smaller; petiolar gland located between the lowest pair of leaflets; flowers solitary or paired in the upper axils; fruits 4-sided, strongly curved, not segmented, dehiscent, 10–18 cm long by 3–6 mm wide; rare in rich soil and waste ground; SE; flr. late Sep–early Oct; native farther south, adventive in Pennsylvania.

CALLITRICHACEAE Water-starwort Family

Small, much branched, annual herbs of aquatic or moist sites; leaves opposite, entire, and lacking stipules; flowers unisexual, axillary, much reduced; pistillate flowers consisting of a bilocular ovary, each locule flattened and 2-lobed, and a cleft style; staminate flowers consisting of 1–3 stamens; fruits dry at maturity and separating into 4 single-seeded nutlets with pit-like markings on the surface.

Callitriche L.

Characteristics of the family; flr. and frt. May–Oct.

A. plants mostly terrestrial; fruit borne on a short pedicel .... *C. terrestris*
A. plants mostly aquatic; fruits sessile
   B. fruit shallowly grooved, margins rounded .... *C. heterophylla*
   B. fruit distinctly grooved, margins winged
      C. fruit obovate, longer than wide, 1–1.4 mm .... *C. palustris*
      C. fruit suborbicular, 1.5–2 mm .... *C. stagnalis*

Callitriche heterophylla

Callitriche heterophylla,
fruit (×10)

Callitriche heterophylla Pursh emend Darby   Water-starwort
Very similar to *C. palustris* but fruits 1 mm or less, wider above the middle and about as long as wide, the margins rounded, only shallowly grooved; common in ponds, slow-moving streams, and muddy shores; throughout; OBL.

Callitriche palustris L.   Water-starwort
Slender, submergent annual of aquatic habitats with only the stem tips floating
or emergent; submersed leaves linear, emergent ones spatulate to obovate, to 3
mm wide, triple-nerved; pistillate and staminate flowers 1 each per axil; fruits
compressed, grooved, and somewhat winged, longer than wide; occasional in
ponds, stream bottoms, swamps, springs, and muddy shores; throughout; OBL.

*Callitriche palustris,
fruit* (×10)

Callitriche stagnalis Scop.   Water-starwort, water-chickweed
Very similar to *C. palustris* except fruits suborbicular, 1.5–2 mm long, and distinctly wing-margined; streams, swamps, and ditches; mainly in the SE, but
spreading; native to Europe; OBL.

Callitriche terrestris Raf. emend Torr.   Water-starwort
Tiny, creeping plant with spatulate or oblanceolate leaves 1–2 mm long; fruits to
1 mm wide, borne on a short pedicel, strongly flattened and grooved forming 4
disk-like segments; occasional, forming small patches on moist, shaded soil of
pond edges and stream banks; S; FACW+.

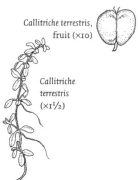

*Callitriche terrestris,
fruit* (×10)

*Callitriche
terrestris*
(×1½)

CALYCANTHACEAE Strawberry-shrub Family

Deciduous, aromatic shrubs; leaves opposite, entire, lacking stipules; flowers
perfect, regular, perigynous with numerous undifferentiated perianth lobes arranged on the outside of the hypanthium; stamens numerous; pistils 5–35, distinct, spirally arranged; fruit consisting of individual achenes enclosed within
the enlarged hypanthium.

Calycanthus L.

Calycanthus floridus L.   Carolina allspice, strawberry-shrub, sweetshrub
Stoloniferous shrub to 3 m tall; leaves opposite, entire, petiolate; tepals linear-

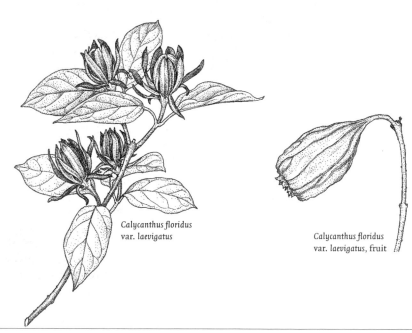

*Calycanthus floridus
var. laevigatus*

*Calycanthus floridus
var. laevigatus, fruit*

oblong, dark purplish-brown; flowers fragrant; flr. early Jun; native mostly VA to FL; 2 varieties:

A. twigs, petioles, and lower leaf surfaces glabrous ....
.... var. *laevigatus* (Willd.) Torr. & A.Gray
cultivated, but perhaps originally native at a few Pennsylvania sites; ❧.
A. twigs, petioles, and lower leaf surfaces hairy .... var. *floridus*
cultivated and occasionally escaped.

CAMPANULACEAE Bellflower Family

Herbs with milky or colored juice and alternate, simple leaves lacking stipules; flowers perfect; calyx 5-lobed; petals united, corolla 5-lobed, regular or irregular; stamens 5, attached to the base of the corolla tube; anthers separate or variously fused; ovary inferior, 2–5-locular; fruit a capsule.

A. corolla highly irregular, essentially 1-sided; carpels 2 .... *Lobelia*
A. corolla regular; carpels 3–5
    B. corolla campanulate to funnelform (or if rotate, the flowers in a long terminal spike) .... *Campanula*
    B. corolla rotate; flowers sessile in the axils of the broadly ovate, clasping leaves .... *Triodanis*

Campanula L.

Herbs with alternate leaves and showy, regular flowers; corolla rotate or campanulate, blue or white, 5-lobed; anthers distinct; ovary 3–5-locular; capsules opening by lateral pores.

A. inflorescence an elongate terminal spike; corolla rotate .... *C. americana*
A. inflorescence various, but not as above; corolla campanulate or funnelform
    B. stems weak and sprawling, 3-angled .... *C. aparinoides*
    B. stems erect or ascending, terete or obscurely angled
        C. flowers solitary on slender pedicels; cauline leaves linear ....
        .... *C. rotundifolia*
        C. flowers on short pedicels, in an erect, slender, one-sided raceme; cauline leaves lanceolate .... *C. rapunculoides*

Campanula americana L.   Tall bellflower
Erect annual or biennial to 2 m tall, often freely branched; leaves lanceolate to ovate-oblong with a winged petiole, gradually reduced upward and merging with floral bracts; flowers blue, solitary or in small clusters; corolla rotate, 2.5 cm wide; fruit opening by round pores near the top; occasional in moist woods, rocky wooded slopes, and stream banks; mostly S; flr. late Jul–Aug; FACU. [syn: *Campanulastrum americanum* (L.) Small]

Campanula aparinoides Pursh   Marsh bellflower
Rhizomatous perennial with weak angled stems with scabrous edges; leaves linear to narrowly lanceolate with scabrous margins and midveins; flowers solitary on long slender pedicels; corolla funnelform, pale blue or white; occasional in

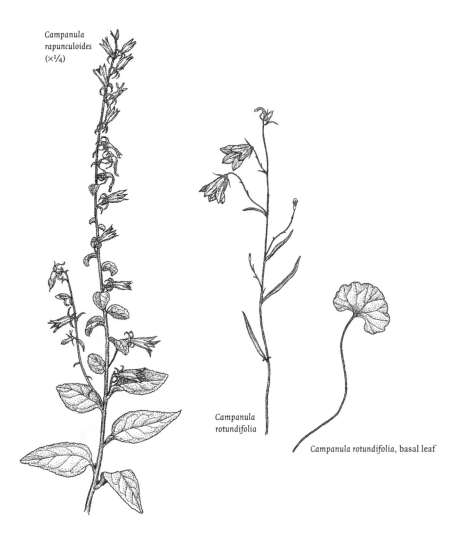

Campanula rapunculoides (×¼)

Campanula rotundifolia

Campanula rotundifolia, basal leaf

moist shores, swamps, and wet open ground; throughout; flr. late Jun–early Sep; OBL.

Campanula rapunculoides L.   Creeping bellflower
Unbranched perennial to 1 m tall; leaves irregularly serrate and hairy beneath; flowers in an unbranched, 1-sided, terminal raceme; corolla blue, 2–3 cm, nodding; occasional on roadsides and woods edges; throughout; flr. Jul–Aug; native to Eurasia.

Campanula rotundifolia L.   Harebell
Glabrous perennial with broadly ovate to cordate, nonpersistent basal leaves; cauline leaves numerous, linear; flowers several to many in a loose panicle or raceme; corolla blue, 1.5–3 cm, campanulate; infrequent on dry, rocky slopes, bluffs, and cliffs; E and C; flr. early Jun–Aug; FACU.

Lobelia L.

Perennial (except L. inflata) herbs with flowers in terminal racemes; corolla strongly irregular with a prominent 3-lobed segment and 2 much reduced (upper) lobes, the corolla tube split nearly to the base on the upper side; filaments and anthers connate, the 2 lower anthers shorter and bearded near the tip; ovary 2-locular, opening near the top.

A. corolla red .... *L. cardinalis*
A. corolla blue, lavender, or white
    B. leaves in an underwater, basal rosette, only the flowers above the water surface .... *L. dortmanna*
    B. terrestrial or emergent aquatic species; leaves all or mostly cauline
        C. low trailing or creeping plants, branching and rooting at the nodes ....*L. chinensis*
        C. erect, unbranched or lightly branched plants
            D. flowers 1.5–4.5 cm long; corolla tube with lateral openings near the base in addition to the slit on the upper side
                E. calyx with prominent auricles 2–5 mm long; bracteoles at or above the middle of the pedicels .... *L. siphilitica*
                E. calyx with small, inconspicuous auricles; pedicels with bracteoles at the base .... *L. puberula*
            D. flowers 7–18 mm long; corolla tube without lateral openings
              F. calyx inflated in fruit .... *L. inflata*
              F. calyx not inflated in fruit
                G. leaves lance-oblong to obovate, at least some >1 cm wide .... *L. spicata*
                G. leaves linear to narrowly lanceolate, <1 cm wide
                    H. bracteoles at or above the middle of each pedicel....*L. kalmii*
                    H. bracteoles at the base of each pedicel .... *L. nuttallii*

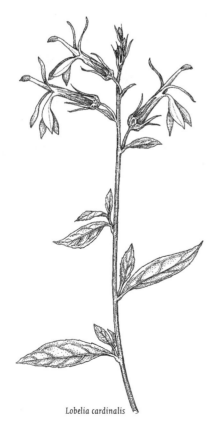

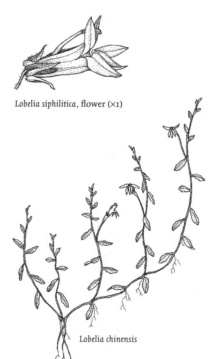

Lobelia siphilitica, flower (×1)

Lobelia chinensis

Lobelia inflata

Lobelia cardinalis

Lobelia cardinalis L.   Cardinal-flower
Erect, usually unbranched perennial 5–15 dm tall; flowers in a terminal raceme, corolla brilliant scarlet, the 2 upper lobes as long as the lower 3, the tube formed by the stamens arching above the spreading corolla lobes; frequent in wet meadows, swamps, riverbanks, and lake shores; throughout; flr. Jul–Sep; FACW+.

Lobelia chinensis Lour.   Chinese lobelia
Low growing, creeping, forming colonies; stems branching and rooting at the nodes; leaves lanceolate, shallowly and irregularly dentate; flowers solitary in the axils; corolla pale lavender, 1.5 cm wide; frequent on tidal river banks; SE, where it was first collected in 1985; flr. Jul–Sep; native to the Asia-Pacific region.

Lobelia dortmanna L.   Water lobelia
Aquatic perennial with a rooted, submergent rosette of hollow, linear leaves and an elongate flowering stem; the raceme of small white or pale blue flowers extending above the water surface; rare in glacial ponds and lakes; NE; flr. Jul–Aug; OBL; 🌿.

Lobelia inflata L.   Indian-tobacco
Erect, loosely hairy annual with branched stems to 1 m tall; leaves sessile, oblong-ovate; racemes terminal, leafy-bracted at the base; corolla blue (or white); 6–8 mm, lower lip bearded at the base within; calyx with linear lobes, strongly inflated in fruit; common in woods, old fields, meadows, and roadsides; throughout; flr. Jul–Sep; FACU.

Lobelia kalmii L.   Brook lobelia
Slender, glabrous perennial to 4 dm tall with spatulate basal leaves and linear cauline leaves; flowers, blue (or white) in terminal racemes; corolla 7–13 mm, lower lip glabrous within; pedicels with bracteoles at or above the middle; very rare in calcareous swamps, moist pastures, and fens; scattered; flr. Aug–early Oct; OBL; 🌿.

Lobelia nuttallii Roem. & Schult.   Nuttall's lobelia
Slender, erect perennial, glabrous above to sparsely pubescent below; lower leaves narrowly lanceolate, upper linear; flowers in terminal racemes; pedicels 3–8 mm, with basal bracteoles; corolla 1 cm, pale blue with a white center and 2 greenish spots; very rare in low woods, moist sandy or peaty thickets, and wet meadows; SE; flr. Jul–Sep; believed to be extirpated; FACW; 🌿.

Lobelia puberula Michx.   Downy lobelia
Erect, short-pubescent perennial with unbranched stems to 1.5 m tall; leaves oblong or oblong-ovate, the lower ones rounded at the apex; flowers blue in a terminal raceme; pedicels 2–5 mm long bearing bracteoles near the base; corolla tube with 2 basal openings, lower lip glabrous within; very rare in moist, sandy old fields, gravel pits, and serpentine barrens; SE; flr. late Aug–Sep; FACW–; 🌿.

Lobelia siphilitica L.   Great blue lobelia
Stout, erect perennial to 1.5 m tall with narrowly oblong or elliptic, sessile leaves and a crowded terminal raceme; flowers blue (white), 2–3 cm, on ascending pedicels 4–10 mm long with a pair of bracteoles at or above the middle; corolla tube with openings near the base on each side; calyx lobes with prominent leafy,

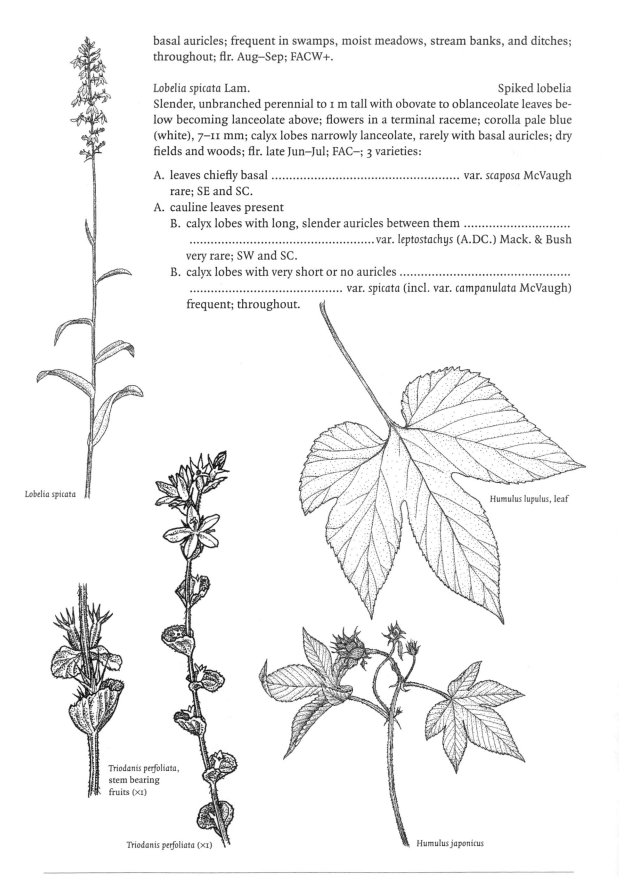

basal auricles; frequent in swamps, moist meadows, stream banks, and ditches; throughout; flr. Aug–Sep; FACW+.

Lobelia spicata Lam.   Spiked lobelia
Slender, unbranched perennial to 1 m tall with obovate to oblanceolate leaves below becoming lanceolate above; flowers in a terminal raceme; corolla pale blue (white), 7–11 mm; calyx lobes narrowly lanceolate, rarely with basal auricles; dry fields and woods; flr. late Jun–Jul; FAC–; 3 varieties:

A. leaves chiefly basal .... var. *scaposa* McVaugh
   rare; SE and SC.
A. cauline leaves present
   B. calyx lobes with long, slender auricles between them ....
      ....var. *leptostachys* (A.DC.) Mack. & Bush
      very rare; SW and SC.
   B. calyx lobes with very short or no auricles ....
      .... var. *spicata* (incl. var. *campanulata* McVaugh)
      frequent; throughout.

Lobelia spicata

Humulus lupulus, leaf

Triodanis perfoliata, stem bearing fruits (×1)

Triodanis perfoliata (×1)

Humulus japonicus

Triodanis Raf.

Triodanis perfoliata (L.) Nieuwl.   Venus' looking-glass
Erect, mostly unbranched annual with sessile, broadly ovate, clasping leaves
bearing 1–3 sessile flowers in the axils, the lowest flowers often cleistogamous;
corolla blue-violet, 1.2 cm wide and rotate; filaments of the stamens bearing
hairs at the base; anthers distinct; capsules opening by pores with upward curling valves located on the sides of each of the 3 locules; roadsides, woods edges,
fields, railroad cinders, and dry waste ground; mostly S; flr. late Jun–early Aug;
FAC. Most of our plants are the typical variety, but var. *biflora* (Ruiz & Pav.)
Greene with the pores of the capsule closer to the top of the fruit and the floral
bracts longer than wide has been collected several times. In addition, the very
similar *Legousia speculum-veneris* (L.) Fisch. ex A.DC., a European plant that differs
by its branched stems, glabrous filaments, and lack of cleistogamous flowers,
has been collected a few times on ballast.

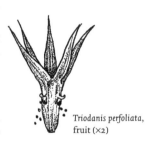
Triodanis perfoliata,
fruit (×2)

CANNABACEAE Hemp Family

Herbaceous plants with generally opposite leaves; flowers small, greenish, unisexual; petals absent; male inflorescences many-flowered, female inflorescences
few-flowered; fruit an achene.

A. plants erect; leaves palmately compound .... *Cannabis*
A. plants vine-like; leaves simple, palmately lobed or unlobed .... *Humulus*

Cannabis L.

Cannabis sativa L.   Hemp, marijuana
Plant usually dioecious; pistillate flowers in small clusters on short leafy
branches, each flower partially enclosed by a bract; fruit closely subtended by the
persistent bract; occasionally planted and persisting or escaping to waste
ground; scattered throughout; flr. late summer; native to Eurasia; FACU; designated as a noxious weed in Pennsylvania.

Cannabis sativa, leaf

Humulus L.

Plants dioecious; pistillate flowers paired on short spikes, each pair of flowers
subtended by a bract; fruit enclosed by the persistent calyx and covered by the
subtending bract; stems usually rough.

A. lower leaves 5–7-lobed (rarely 9-lobed); usually each inflorescence bract enclosing 1 flower or fruit .... *H. japonicus*
A. lower leaves 3-lobed (rarely 5-lobed); usually each inflorescence bract enclosing 2 flowers or fruits .... *H. lupulus*

Humulus japonicus Siebold & Zucc.   Japanese hops
Bases of sinuses between leaf lobes forming an acute angle; inflorescence bracts
weakly or not at all glandular; occasional in meadows, roadsides, and waste
ground; flr. Jul–early Oct; native to Asia; FACU.

Humulus lupulus L.   Brewer's hops, common hops
Bases of sinuses between leaf lobes rounded; inflorescence bracts usually

Humulus lupulus,
inflorescence

strongly glandular; occasional in moist alluvial soil, woods edges, thickets, and waste ground; throughout; flr. Jul–Aug; FACU; 3 varieties:

A. midrib on the underside of the central lobes of the leaves with <20 hairs per cm .... var. *lupulus*
cultivated and escaped; native to Europe.
A. midrib on the underside of the central lobes of the leaves with >20 hairs per cm
   B. hairs present between the veins on the undersides of the leaves ....
   .... var. *pubescens* E.Small
   B. hairs absent between the veins on the undersides of the leaves ....
   .... var. *lupuloides* E.Small

CAPPARACEAE Caper Family

Annual, mostly glandular-pubescent herbs with alternate, trifoliate, or palmately compound leaves; flowers in terminal racemes, regular to strongly irregular, perfect; sepals 4, distinct or slightly fused at the base; petals 4, distinct; stamens 6–many; ovary superior, of 2 carpels but unilocular; fruits oblong to narrowly fusiform, dehiscent by 2 valves.

A. spines present at the base of the petiole; petals entire ....*Cleome*
A. spines not present; petals notched .... *Polanisia*

Cleome L.

Cleome hasslerana Chodat   Spider-flower
Stems to 1 m tall, branched above; leaves with 5–7 oblanceolate leaflets and a pair of short spines at the base of the petiole; flowers white to pink, the petals all on one side; fruits with stipes nearly twice as long as the pedicels; waste ground; mostly S; flr. Jul–Oct; native to tropical America.

Polanisia Raf.

Polanisia dodecandra (L.)   Clammyweed
Stems 1.5–5 dm tall; leaves with 3 oblong leaflets; flowers irregular, with purplish sepals and yellowish- white petals; stamens 8–12 (or more), longer than the petals; fruits sessile or with a short stipe; rare in dry sandy or gravelly, alluvial soils; scattered; flr. Jul–Sep; FACU; 2 subspecies:

A. largest petal 3.5–6.5 mm .... ssp. *dodecandra*
native.
A. largest petal 8–13 mm .... ssp. *trachysperma* (Torr. & A. Gray) Iltis
native farther west, adventive here.

CAPRIFOLIACEAE Honeysuckle Family

Shrubs, woody vines, or herbs with opposite, simple, or compound leaves with or without stipules; flowers mostly 5-merous, sympetalous, regular, or slightly

Cleome hasslerana (×¼)

Polanisia dodecandra

to strongly bilabiate; stamens 4 or 5, inserted on the corolla tube; ovary inferior to half inferior, 3–5-carpellate with axile placentation; fruit a drupe, capsule, or achene.

A. woody shrubs or vines
  B. leaves pinnately compound .... *Sambucus*
  B. leaves simple
    C. leaf margins toothed
      D. flowers white or rarely pinkish; corolla rotate to broadly campuna-late; fruit a fleshy 1-seeded drupe .... *Viburnum*
      D. flowers yellow, becoming reddish with age; corolla funnelform fruit a smooth capsule with a persistent, elongated style and stigma ....
      .... *Diervilla*
    C. leaf margins entire
      E. twining vines .... *Lonicera*
      E. shrubs
        F. flowers weakly to strongly bilabiate, in pairs on axillary pe-duncles, their ovaries more or less united .... *Lonicera*
        F. flowers regular, not paired or if so their ovaries not united
          G. style elongate; fruit white or red, with 2 stones ....
          .... *Symphoricarpos*
          G. stigma essentially sessile; fruit blue-black, with a single flat-tened stone .... *Viburnum*

A. herbs

    H. stems creeping, with a woody base; leaves evergreen .... *Linnaea*

    H. stem upright; perennial, but leaves not persistent .... *Triosteum*

Diervilla Mill.

Diervilla lonicera Mill.   Bush-honeysuckle

Low shrub to 1.2 m tall with arching branches; leaves simple; blades 8–15 cm, acuminate; margins serrulate and finely ciliate; flowers terminal and in the upper axils, in clusters of 3–7; corolla 12–20 mm, yellow becoming reddish with age, funnelform, nearly regular, 5-lobed; stamens 5; fruit a slender capsule with persistent style; frequent in dry woods and rocky slopes; throughout; flr. Jun–Jul.

Linnaea L.

Linnaea borealis L.   Twinflower

Evergreen, trailing subshrub with erect peduncles to 1 dm bearing a pair of flowers; leaves simple, petiolate; blades broadly oval, 1–2 cm, entire; flowers 12–15 mm, nodding; corolla pink to white, shallowly 5-lobed, tube hairy within; stamens 4, inserted near the base of the corolla tube; fruit dry, indehiscent; very rare in cool, moist woods; scattered; FAC; 🌿.

Lonicera L.

Shrubs or twining vines with opposite, simple, mostly entire leaves; flowers in pairs on axillary peduncles, often with fused ovaries, or in terminal whorls sub-

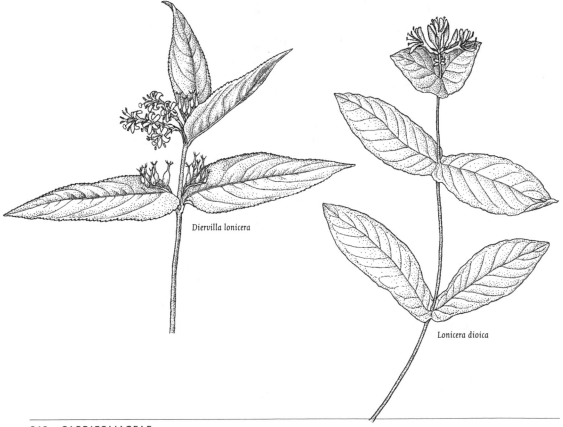

Diervilla lonicera

Lonicera dioica

tended by one or more pairs of fused, disk-like leaves that surround the stem; corolla tubular, 5-lobed, often bilabiate and/or bulging at the base; stamens 5; fruit a few-seeded berry.

A. trailing shrubs or twining vines
    B. flowers axillary; leaves never fused .... L. *japonica*
    B. flowers in terminal whorls; uppermost pairs of leaves below the inflorescence fused
        C. leaves pubescent above and below; young stems hairy and glandular ...
        .... L. *hirsuta*
        C. stems and upper surfaces of leaves glabrous
            D. corolla red (yellowish inside), scarcely bilabiate, the 5 lobes nearly equal, or the lower one slightly larger .... L. *sempervirens*
            D. corolla strongly bilabiate, pale yellow to purplish .... L. *dioica*
A. erect shrubs
    E. pith solid, white; ovaries fused or distinct
        F. corolla regular or nearly so
            G. ovaries evidently separate, divergent; fruit red .... L. *canadensis*
            G. ovaries wholly united, fruit blue .... L. *villosa*
        F. corolla strongly 2-lipped
            H. flowers white or purplish, very early, before the leaves; naturalized species of well-drained, disturbed habitats
                I. branchlets glabrous; leaves broad-ovate or obovate, acute; corolla tube glabrous .... L. *fragrantissima*
                I. branchlets with short reflexed hairs, or nearly glabrous; leaves ovate-oblong to lanceolate, acuminate; corolla tube with long hairs on the outside.. L. *standishii*
            H. corolla yellow; flowers appearing after the leaves; a rare native of northern wetlands .... L. *oblongifolia*
    E. pith hollow, white or tan; ovaries distinct
        J. leaves acuminate; peduncles <5 mm, mostly shorter than the petioles....
        .... L. *maackii*
        J. leaves acute or obtuse; peduncles >5 mm
            K. leaves pubescent, at least beneath; peduncles 0.5–1.5 cm
                L. corolla pubescent outside, white turning yellow, bulging on one side at the base
                    M. bracts and sepals ciliate but not glandular .... L. *morrowii*
                    M. ovaries, bracts, and sepals glandular .... L. *xylosteum*
                L. corolla nearly glabrous outside, pink turning yellow, barely bulging at the base .... L. x *bella*
            K. leaves glabrous, peduncles 1.5–2.5 cm .... L. *tatarica*

Lonicera x *bella* Zabel   Pretty honeysuckle
Deciduous shrub to 6 m tall; leaves slightly hairy beneath; peduncles 5–15 mm, sparsely hairy; corolla pink fading to yellow, glabrous; fruit yellow to red; cultivated and escaped to roadsides and stream banks; scattered; flr. late May, frt. Jul. [L. *morrowii* x *tatarica* of Rhoads and Klein, 1993]

Lonicera canadensis Marshall   Fly-honeysuckle
Deciduous shrub to 2 m tall; leaves triangular-ovate to oblong, 2–3 cm long, glabrous to sparsely hairy beneath, ciliate; flowers on axillary peduncles; corolla

yellowish, spurred at the base; fruits red, distinct and widely divergent; occasional in cool, moist woods; mostly N and W; flr. May, frt. Jun–Jul; FACU.

Lonicera dioica L.   Mountain honeysuckle
Deciduous straggling vine with the uppermost 1–2 pairs of leaves below the inflorescence fused; flowers pale yellow to purplish; corolla bulging on one side near the base and hairy inside; fruit red; flr. late May–early Jun, frt. late Jun–early Jul; FACU; 3 varieties:

A. hypanthium glabrous
    B. leaves glabrous beneath; outside of corolla tube and style glabrous or sparsely hairy .... var. *dioica*
        occasional on moist cliffs, rocky wooded banks, and thickets; mostly E and C.
    B. leaves sparsely to densely villous beneath; outside of corolla tube glandular and villous; style hirsute .... var. *glaucescens* (Rydb.) Butters
        occasional in mossy woods and bogs; W and C.
A. hypanthium densely glandular .... var. *orientalis* Gleason
    very rare on clayey, rocky banks; known from a single site in Butler Co.

Lonicera fragrantissima Lindl. & Paxton   Fragrant honeysuckle
Deciduous shrub to 3 m tall; leaves broadly oval, leathery, strongly apiculate, glabrous except for hairs on the margins and the midribs; flowers very fragrant; corolla tube glabrous; fruit red; cultivated and occasionally escaped to wooded slopes and thickets; flr. Feb–Mar, before the leaves, frt. May; native to China.

Lonicera hirsuta Eaton   Hairy honeysuckle
Woody, deciduous, twining vine; leaves broadly oval, hairy on both sides, the uppermost 1–2 pairs below the inflorescence fused; flowers in spikes of 1–4 crowded whorls, yellow to orange; corolla pubescent and slightly swollen at the base; rare in moist woods, swamps, and rocky thickets; NE and NW; flr. early Jun, frt. early Jul; FAC; ✵.

Lonicera japonica Thunb.   Japanese honeysuckle
Semievergreen trailing or twining vine; leaves ovate to oblong, 4–8 cm, occasionally toothed or lobed; flowers in pairs on axillary peduncles; corolla 3–5 cm, strongly bilabiate, white to occasionally pinkish turning yellow with age; fruit black; commonly established as an invasive weed of disturbed woods, thickets, old fields, banks, and roadsides; flr. Jun, frt. Sep–Oct; mostly S; native to Asia; FAC–.

Lonicera maackii (Rupr.) Maxim.   Amur honeysuckle
Deciduous shrub to 5 m tall; leaves ovate to lance-ovate, 3.5–8.5 cm, acuminate; flowers in pairs on short axillary peduncles; corolla 1.5–2 cm, white turning yellow; fruit dark red; cultivated and established as an aggressive weed of disturbed woods, floodplains, old fields, thickets, and roadsides; mostly S; flr. late May–early Jun, frt. Oct–Nov; native to Asia.

Lonicera morrowii A.Gray   Morrow's honeysuckle
Deciduous shrub to 3 m tall; leaves oblong to narrowly elliptic, softly pubescent beneath; flowers paired on axillary, 5–15-mm peduncles; corolla pubescent out-

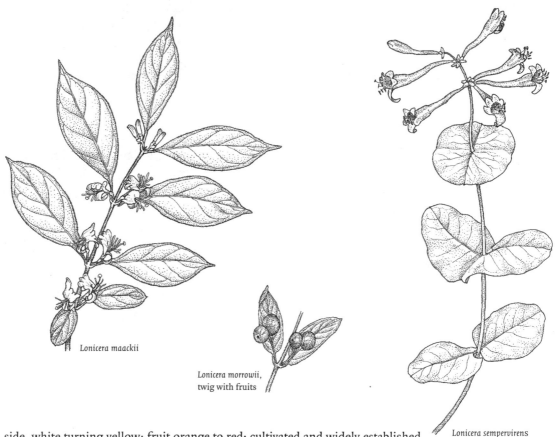

Lonicera maackii

Lonicera morrowii,
twig with fruits

Lonicera sempervirens

side, white turning yellow; fruit orange to red; cultivated and widely established as an invasive weed of disturbed woods, old fields, floodplains, roadsides, and thickets; throughout; flr. May, frt. late Jun–Jul; native to Japan; FACU.

Lonicera oblongifolia (Goldie) Hook.   Swamp fly-honeysuckle
Deciduous shrub to 2 m tall; leaves oblong to oblanceolate, 3–7 cm, short pubescent beneath; peduncles axillary, 2–4 cm; corolla yellow, 10–15 mm, hairy; ovaries glabrous, partly or wholly fused; style hairy; fruit red; rare in bogs and swamps; NE and NW; flr. early Jun, frt. Jul–Aug; OBL; ✇.

Lonicera sempervirens L.   Trumpet honeysuckle
Deciduous, woody twining vine; leaves broadly oval, glabrous (sometimes with a few long hairs beneath), glaucous, the uppermost 1–2 pairs below the inflorescence fused; corolla nearly regular, tubular with 5 shallow lobes, red or yellow outside and yellow inside; fruit red; occasional in fencerows, thickets, and roadsides; mostly SE; flr. late May–Jun, frt. Jul; FACU.

Lonicera standishii Jacques   Honeysuckle
Semievergreen shrub to 3 m tall; branches bearing short reflexed hairs; leaves oblong-ovate to ovate-lanceolate, 4.5–6 cm long; flowers in pairs on short bristly peduncles, creamy white, fragrant; corolla bilabiate with long hairs on the outside of the tube; fruit red, the 2 ovaries united nearly to the top; cultivated and occasionally escaped to wooded slopes and edges; SE; flr. Mar–early Apr, before the leaves, frt. May–Jun; native to China.

Lonicera tatarica L.   Tartarian honeysuckle
Deciduous shrub to 3 m tall; leaves ovate to oblong, 3–6 cm, glabrous beneath; peduncles axillary, 1.5–2.5 cm; ovaries distinct; corolla white to pink, glabrous, nearly regular, the lobes equal to or longer than the tube; style hirsute; fruit red; cultivated and widely established as an invasive weed of disturbed ground, woods edges, and roadsides; throughout; flr. May, frt. Jun–early Jul; native to Eurasia; FACU.

Lonicera villosa (Michx.) Roem. & Schult. Water-berry, mountain fly-honeysuckle
Deciduous shrub to 1 m tall; leaves oval to oblong, 2–8 cm, obtuse or rounded at the tip, hairy; flowers on axillary peduncles 3–10 mm, yellow; corolla 10–15 mm with 5 subequal lobes about the same length as the tube; style glabrous; fruit blue; very rare in bogs, swamps, and wet thickets; N; flr. May, frt. Jul; ✤. [syn: *L. caerulea* L.]

Lonicera xylosteum L.   European fly-honeysuckle
Deciduous shrub to 3 m tall with downy young shoots; leaves ovate to obovate, 2.5–6 cm long, acute or obtuse, hairy above and below; flowers in pairs on peduncles that are longer than the petioles; corolla yellowish-white; fruits dark red, joined only at the base; cultivated and escaped at scattered sites; flr. late Apr–May, frt. Jul; native to Eurasia.

Sambucus L.

Coarse, deciduous shrubs with opposite, pinnate leaves; twigs with a large pith and prominent lenticels; leaflets acuminate, serrate; flowers small, white, in large terminal inflorescences; corolla regular, rotate or saucer-shaped, 5-lobed; stamens 5; style short, 3–5-lobed; fruits red or black, 5-mm berries with 3–5 seeds.

A. pith white; inflorescence 5-rayed from the base, flat; fruits black ....
.... S. canadensis
A. pith brown; inflorescence strongly convex; fruits red .... S. pubens

Sambucus canadensis L.   American elder
Shrub to 3 m tall, spreading and forming thickets; leaflets 5–11, the lower sometimes lobed; inflorescence 5-rayed from the base, flat-topped or only slightly convex, 5–15 cm wide; fruits purple-black; common in woods, fields, stream banks, moist fields, and swamps; throughout; flr. Jun, frt Aug–Sep; FACW–.

Sambucus racemosa L.   Red-berried elder
Shrub to 3 m tall; leaflets 5–7; inflorescence strongly convex or pyramidal, with an elongate central axis; fruit red; frequent in ravines, moist cliffs, and cool, rocky woods; throughout; flr. May, frt. late Jun–Jul; FACU. Ours is var. *pubens* (Michx.) House. [syn: *S. pubens* Michx.]

Symphoricarpos Duhamel

Summer-blooming, deciduous shrubs with opposite, simple, entire leaves with short petioles; flowers nearly regular; corolla campanulate, deeply 5-lobed, densely hairy within; stamens 5; style at least as long as the corolla tube; stigma capitate; ovary inferior, 4-celled, but 2 abortive; fruits red or white berry-like drupes with 2 nutlets, crowded in dense axillary and terminal clusters.

A. fruit red; corolla 2–4 mm .... S. orbiculatus
A. fruit white; corolla 5–8 mm
    B. branches pubescent; style and stamens longer than the corolla tube; style 6–8 mm .... S. occidentalis
    B. branches glabrous; style and stamens shorter than the corolla tube; style 2–3 mm .... S. albus

Symphoricarpos albus (L.) S.F.Blake   Snowberry
Shrub to 1 m tall with slender, upright, mostly glabrous branches; leaves thin, oval to broadly ovate, 2–5 cm, obtuse, glabrous or pubescent beneath; flowers in axillary and terminal clusters; corolla pinkish, 6 mm; style 2–3 mm; fruit subglobose, white, 6–10 mm; flr. early Jun, frt. Sep; FACW–; 2 varieties:

A. stems to 1 m; leaves pubescent beneath .... var. albus
    rocky, wooded, limestone slopes and barrens; native, but also occasionally escaped from cultivation.
A. stems to 2 m or more; leaves and stems glabrous ....
    .... var. laevigatus (Fernald) S.F.Blake
    cultivated and occasionally escaped to roadsides and waste ground; native to the Pacific slope.

Symphoricarpos occidentalis Hook.   Wolfberry
Stoloniferous shrub to 1 m tall with stiff, upright, pubescent branches; leaves elliptic or ovate, 2–7 cm, obtuse, grayish-green and pubescent, becoming leathery; flowers in dense axillary and terminal clusters; corolla pinkish-white, 6 mm, deeply 5-lobed; style 4–7 mm, slightly longer than the corolla; fruit subglobose, white, 1 cm; cultivated and rarely escaped to railroad embankments and alluvium; native to the Midwest.

Symphoricarpos orbiculatus Moench   Coralberry, Indian-currant
Stoloniferous shrub to 2 m tall with slender upright, pubescent branches; leaves oval to suborbicular, 1.5–3.5 cm, obtuse or acute, pubescent, glaucous beneath; flowers sessile, in upper axillary or terminal clusters; corolla yellowish-white, 3–4 mm; style 2 mm; fruits purplish-red, ellipsoid, 3–6 mm long, persistent; occasional in wet woods, thickets and old fields; mostly S; flr. Jun, frt. Sep–Dec; native, but also cultivated and sometimes escaped; UPL.

Triosteum L.

Coarse, hairy, erect perennial herbs with paired leaves that are sessile or fused; flowers axillary, yellowish-green to purplish, 5-merous; sepals long and narrow, leaf-like in texture and persistent in fruit; ovary inferior, 3–5-celled; fruit a dryish, pubescent berry with a few bony seeds.

A. leaves ovate to ovate-oblong; flowers 3–4 per axil, yellowish-green to purplish; sepals puberulent
    B. leaves fused and thus perfoliate; fruit yellow-orange, subglobose ....
    .... T. perfoliatum

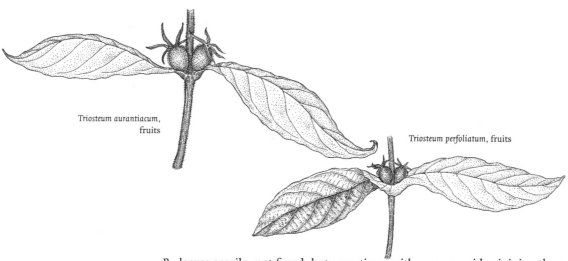

Triosteum aurantiacum, fruits

Triosteum perfoliatum, fruits

    B. leaves sessile, not fused, but sometimes with a narrow ridge joining them across the node; fruit orange-red, ovoid .... *T. aurantiacum*
A. leaves lanceolate; flowers solitary in the axils, yellowish; sepals hispid-ciliate .... *T. angustifolium*

Triosteum angustifolium L.   Horse-gentian, feverfew
Herb to 3–8 dm tall; stems sparsely retrorse hispid with hairs 1.5–3 mm long and also shorter glandular hairs; leaves narrowed to a sessile base; flowers solitary in the axils; sepals hispid-ciliate; corolla yellowish; fruits orange-red; rare in woods and thickets; SE; flr. May–early Jun, frt. Jul–Aug; 🌱.

Triosteum aurantiacum E.P.Bicknell   Wild-coffee
Similar to *T. perfoliatum* but leaves tapering to a narrow, sessile base or forming a narrow ridge across the node; corolla purplish-red; fruit orange-red; occasional on moist, rocky, limestone slopes and rich wooded ravines; throughout; flr May–Jun, frt. late Jun–early Oct.

Triosteum perfoliatum L.   Perfoliate horse-gentian
Coarse herb to 1.3 m tall; densely hairy-glandular with hairs <0.5 mm long; leaves obovate, fused at the base and thus perfoliate; flowers 3–4 per axil; sepals finely and uniformly pubescent on the back and margin; corolla yellowish or purplish; fruit subglobose, orange; occasional in moist woods and thickets on limestone, diabase, or other rich soils; throughout; flr. May–early Jun, frt. late Jun–early Oct.

Viburnum L.

Deciduous shrubs with opposite, simple leaves, linear stipules present near the base of the petiole in a few species, but mostly lacking; winter buds with 1 or 2 pairs of bud scales or naked; flowers small and clustered in terminal cymes, regular, 5-merous, white or rarely pink; stigma 3-lobed, on a short stylopodium atop the ovary; inflorescence sometimes with larger, sterile flowers around the margin; fruit a drupe with a single stone.

A. most of the leaves palmately lobed
    B. leaves without glands on the petioles; flowers all alike; fruit blue-black .... *V. acerifolium*

B. leaves with glands on the petioles; inflorescence with larger, sterile marginal flowers (or flowers all sterile in some cultivated forms); fruit red

  C. petiolar glands stalked, rounded on top .... *V. trilobum*

  C. petiolar glands mostly sessile, concave .... *V. opulus*

A. leaves not lobed, pinnately veined

  D. leaf blades mostly >10 cm long when fully expanded, base cordate .... *V. lantanoides*

  D. leaf blade <10 cm long, base cordate, rounded or cuneate

    E. lateral veins of the leaf blades branching repeatedly and anastomosing before reaching the margin

      F. leaf margins entire or crenulate

        G. leaf margins entire to crenulate, distinctly revolute; blade shiny above, veiny beneath .... *V. nudum*

        G. leaf margins entire to irregularly denticulate, not revolute; blade dull above, veins indistinct .... *V. cassinoides*

      F. leaf margins toothed

        H. leaf margin denticulate to nearly entire; cymes stalked... *V. cassinoides*

        H. leaf margin finely and evenly serrate; cymes sessile (note: if only one branch of the inflorescence remains the determination of sessile vs. stalked can be tricky)

          I. leaf tip sharply acuminate .... *V. lentago*

          I. leaf tip rounded to acute .... *V. prunifolium*

    E. lateral veins of the leaf blades simple or branching several times and ending in a tooth

      J. stipules present; upper leaves subsessile or with petioles <5 mm .... *V. rafinesquianum*

      J. stipules absent; petioles >5 mm

        K. petioles glabrous; leaves glabrous beneath except for tufts of hairs in the larger vein axils .... *V. recognitum*

        K. petioles hairy; leaves hairy beneath at least on the veins

          L. buds naked .... *V. lantana*

          L. bud scales present

            M. leaves oblong-obovate, with an unpleasant odor when crushed; inflorescence with an elongate axis, its lowest branches opposite .... *V. sieboldii*

            M. leaves ovate, ovate-lanceolate, broadly ovate or suborbicular, without an unpleasant odor; branches of the inflorescence all emanating from a single point (umbelliform)

            N. leaves with 8–12 pairs of lateral veins; branches horizontal; inflorescences flat-topped, with larger marginal flowers..... .... *V. plicatum*

            N. leaves with 5–8 pairs of lateral veins; branches upright; inflorescence convex, flowers all the same

              O. fruit blue-black; common native shrub of southeastern counties .... *V. dentatum*

              O. fruit orange or red; escaped ornamental

                P. leaves broadly ovate, acute, pubescent on both sides .... *V. dilatatum*

                P. leaves ovate-lanceolate with an acuminate tip and coarse hairs on the veins beneath but otherwise glabrous .... *V. setigerum*

Viburnum acerifolium L.   Maple-leaved viburnum

Shrub 1–2 m tall; young stems, petioles and lower leaf surfaces stellate pubescent; leaves palmately 3-lobed, 6–12 cm, coarsely toothed; flowers in 3–5-cm-wide cymes; fruit purple-black, 6–8 mm; common in woods; throughout; flr. early Jun, frt. Sep; UPL.

Viburnum cassinoides L.   Witherod

Shrub to 4 m tall; stems smooth or brown-scurfy when young; leaf blades ovate to lanceolate, to 12 cm, with denticulate to nearly entire margins; inflorescence stalked; fruit blue-black, 6–7 mm, stone flattened; frequent in swamps, bogs, moist woods, and barrens; throughout; flr. late May–early Jun, frt. Aug–Sep; FACW.

Viburnum dentatum L   Southern arrow-wood

Shrub to 5 m tall with gray-brown or reddish bark; leaves lance-ovate to rotund, 4–10 cm, sharply toothed; petioles and lower leaf surfaces stellate pubescent; fruits blue-black, 4–7 mm; occasional in swamps and wet woods; SE; flr. late May–early Jun, frt. Sep–Oct; FAC.

Viburnum dilatatum Thunb.   Linden viburnum

Upright shrub to 3 m tall with pubescent branchlets; leaves suborbicular to broadly ovate or even obovate, 6–12 cm long, abruptly short-acuminate, coarsely toothed, pubescent on both sides; cymes 8–12 cm across; fruit red, 8 cm, persistent; cultivated and occasionally escaped; SE; flr. early Jun, frt. Oct; native to Asia.

Viburnum lantana L.   Wayfaring-tree

Upright shrub to 5 m tall with densely stellate-pubescent buds, branchlets, and lower leaf surfaces; buds naked; leaves ovate to oblong-ovate, 5–12 cm long,

Viburnum acerifolium

Viburnum cassinoides

Viburnum lentago

Viburnum lentago, twig with terminal bud

Viburnum lantanoides

acute or obtuse, closely denticulate, sparingly pubescent and wrinkled above; cyme 6–10 cm across; fruit 8 mm, red turning black; cultivated and occasionally escaped to disturbed woods and roadsides; scattered; flr. late May, frt. Sep; native to China.

Viburnum lantanoides Michx.   Hobblebush
Shrub to 2 m tall with naked winter buds; leaves broadly ovate, 10–18 cm long, cordate at the base, serrate; cyme sessile with larger marginal flowers; fruit red, 8–10 mm; occasional in cool, moist woods and ravines; N, and at higher elevations, but declining due to over browsing by deer; flr. late Apr–May, frt. Aug; FACU. [syn: *V. alnifolium* Marshall].

Viburnum lentago L.   Nannyberry, sheepberry
Tall shrub to 10 m, mostly glabrous; petioles wing-margined; leaves ovate to oblong, 5–8 cm, sharply serrulate; cymes sessile, 5–10 cm wide; fruit 10–12 mm, blue-black with a whitish bloom; stone flat, nearly smooth; occasional in moist woods, swamps, and roadside edges; throughout; flr. May, frt. late Jul–Aug; FAC.

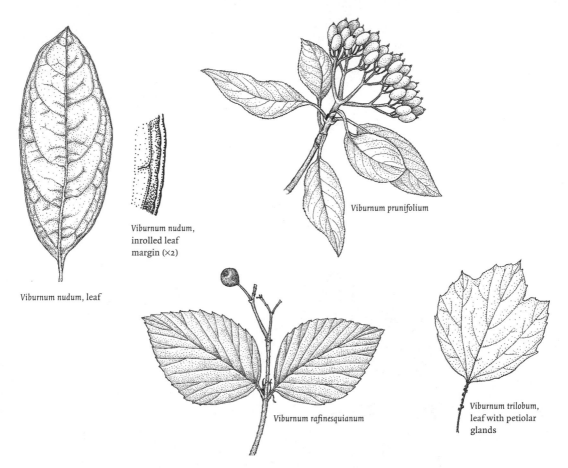

Viburnum nudum, inrolled leaf margin (×2)

Viburnum nudum, leaf

Viburnum prunifolium

Viburnum rafinesquianum

Viburnum trilobum, leaf with petiolar glands

Viburnum nudum L.   Possum-haw
Shrub to 4 m tall, young stems glabrous or scurfy, winter buds reddish-brown to grayish; leaves oblong or oblong-lanceolate, shiny above and veiny beneath; margins entire or wavy, narrowly revolute; fruit 6–12 mm, blue-black; rare in wet woods, swamps, and margins of vernal ponds; SE; flr. Jun, frt. Aug–Nov; OBL; 🌿.

Viburnum opulus L.   Guelder-rose
Very similar to *V. trilobum* and often confused with it; differing in the mostly sessile, concave-topped petiolar glands; cultivated and frequently escaped to woods, fields, and roadsides; throughout; flr. May, frt. Aug–Oct; native to Eurasia.

Viburnum plicatum Thunb.   Doublefile viburnum
Shrub to 2 m tall with showy, flat-topped cymes with enlarged marginal flowers arrayed along the nearly horizontal branches; leaves 6–10 cm; fruit red turning blue-black; cultivated and occasionally escaped; SE and SW; flr. late May–early Jun; native to Asia. [syn: *V. tomentosum* Thunb.]

Viburnum prunifolium L.   Black-haw
Shrub or small tree to 8 m tall with glabrous (except when very young) stems and leaves; leaf blades oblong, elliptic, or obovate, 3–8 cm, serrulate; cymes sessile,

5–10 cm wide; fruit blue-black, 9–15 mm; common in successional woods, thickets, old fields, and roadsides; S; flr. May, frt. late Jul–Sep; FACU.

Viburnum rafinesquianum Schult. — Downy arrow-wood

Shrub to 1.5 m tall with glabrous or sparsely stellate-pubescent young stems; leaves short-petioled to subsessile, blades ovate or ovate-lanceolate, coarsely serrate, petiole and lower leaf surface pubescent, stipules present; fruit blue-black, 6–9 mm, flattened; stone flattened and grooved on both sides; occasional on dry slopes, open woods, and barrens; mostly E and SC; flr. late May, frt. Aug–Sep.

Viburnum recognitum Fernald — Northern arrow-wood

Very similar to *V. dentatum*, differing in the glabrous petioles and nearly glabrous lower leaf surfaces (hairs limited to tufts in the larger vein axils); common in swamps, boggy woods, wet pastures and stream banks; throughout; flr. late May–early Jun, frt. Sep–Oct; FACW–.

Viburnum setigerum Hance — Tea viburnum

Deciduous shrub to 4 m tall, with erect stems and scaly winter buds; leaves ovate-lanceolate, acuminate, glabrous except the veins, which are covered with long coarse hairs beneath; inflorescence peduncled; fruits orange-red; cultivated and occasionally escaped to disturbed woods; SE; flr. May, frt. Sep; native to China.

Viburnum sieboldii Miq. — Siebold viburnum

Small tree or large shrub with scaly winter buds; leaves oblong-obovate, coarsely toothed, rank-smelling; lowest branches of the inflorescence opposite; pedicels and peduncle turning bright red; fruit pink turning blue-black and dropping early; cultivated and occasionally escaped to disturbed woods and stream banks; SE; flr. May, frt. Aug–early Sep; native to Japan.

Viburnum trilobum Marshall — Highbush-cranberry

Shrub to 5 m tall with glabrous stems; petioles with 1–6 stalked glands and slender stipules near the base; leaf blades palmately 3-lobed, coarsely toothed, hairy below; cymes 5–10 cm wide with enlarged marginal flowers; fruit red, 8–12 mm; stone flat; rare in swamps, fens, and wet woods; scattered; flr. late May, frt. Aug–Sep; FACW; .

CARYOPHYLLACEAE Pink Family

Annual or perennial herbs with opposite or whorled, entire leaves and swollen nodes; stipules present or absent; flowers perfect, regular; perianth 4–5-merous; sepals distinct or fused at the base; petals present or absent, distinct, sometimes deeply lobed or bifid; stamens mostly 5 or 10; ovary superior, of 2–5 fused carpels, frequently unilocular with free-central or basal placentation; styles distinct or more or less united from the base; fruit a dehiscent capsule or a 1-seeded utricle.

A. leaves actually or apparently whorled
    B. leaves lance-ovate, mostly in whorls of 4 .... *Silene stellata*
    B. leaves linear, opposite, but appearing whorled due to the presence of leafy axillary tufts
        C. leaves fleshy, 2–5 cm; stipules present .... *Spergula*
        C. leaves not fleshy, linear subulate to bristle-like, 0.8–3 cm; stipules absent
            D. sepals distinct to the base; petals present
                E. styles 3, fewer than the sepals .... *Minuartia michauxii*
                E. styles 4 or 5, same number as the sepals ....*Sagina*
            D. sepals fused at the base to form a hypanthium; petals absent ....
            .... *Scleranthus*
A. leaves distinctly opposite
    F. stipules present
        G. leaves oval to elliptic, thin; petals absent............................. *Paronychia*
        G. leaves linear or linear-subulate, often fleshy; petals present (but shorter than the sepals) .... *Spergularia*
    F. stipules not present
        H. sepals distinct or nearly so
            I. styles (4)5
                J. leaves linear subulate, stamens 4 or 5 ....*Sagina*
                J. Leaves broader, stamens 10
                    K. capsule opening by 10 teeth ....*Cerastium*
                    K. capsule opening by 5 teeth, each one slightly bifid at the tip....
                    .... *Myosoton*
            I. styles (2)3
              L. stamens 3–5
                M. petals deeply notched or absent; inflorescence a cyme ....
                  .... *Stellaria*
                M. petals fringed; inflorescence umbel-like .... *Holosteum*
              L. stamens 10
                N. petals deeply notched or absent .... *Stellaria*
                N. petals entire or at most emarginate
                  O. flowers terminal and axillary; leaves <5 mm wide
                    P. leaves ovate, 2.5–8 m long .... *Arenaria*
                    P. leaves subulate, linear, or linear oblanceolate, 5–30 mm long .... *Minuartia*
                O. flowers from the upper axils; leaves >5 mm wide ....
                .... *Moehringia*
         H. sepals united in a tube or cup
             Q. calyx with additional bracts at the base
                R. sepals each with 5–7 strong ribs, connate without membranous margins ....*Dianthus*
                R. sepals each with 1 main rib, connate by membranous, ribless margins .... *Petrorhagia*
             Q. calyx without bracts at the base
                S. styles 5
                  T. sepals fused for <½ their length, calyx lobes leaf-like, 2–3 cm long ....*Agrostemma*
                  T. sepals fused for >½ their length, calyx lobes not leaf-like
                    U. calyx neither glandular nor inflated .... *Lychnis*

U. calyx glandular-pubescent, often inflated .... *Silene*
S. styles 2 or 3
    V. petals absent; sepals forming a hypanthium; fruit nondehiscent ....
    .... *Scleranthus*
    V. petals present, but sometimes short-lived; sepals not forming a hypan-
    thium; capsule splitting into 3–6 valves
    W. calyx 10-nerved; styles 3 .... *Silene*
    W. calyx 5-nerved; styles 2
        X. calyx not winged on the angles; leaves sessile but not clasping or
        connate .... *Saponaria*
        X. calyx winged on the angles; leaves clasping or even connate ....
        .... *Vaccaria*

Agrostemma L.

Agrostemma githago L.   Corn cockle
Hairy annual to 1 m tall; leaves linear or lanceolate; flowers solitary at the ends of branches; calyx lobes longer than the tube; petals pink, 2–3 cm long; stamens 10; styles 4 or 5; fruit a 14–18-mm capsule; frequent in cultivated fields, road-sides, and waste ground; S; flr. May–Aug; native to Europe.

Arenaria L.

Arenaria serpyllifolia L.   Thyme-leaved sandwort
Scabrous-puberulent annual, branched at the base, erect or ascending, 2–30 cm tall; leaves ovate to ovate-lanceolate, 2.5–8 mm long; flowers in small clusters on pedicels longer than the scarious-margined sepals, 5-merous; petals white; stamens 10; styles (2)3; fruit a dehiscent capsule with as many teeth as styles; a frequent weed of dry sterile soils; throughout; flr. Apr–Jun; native to Europe; FAC; 2 subspecies:

A. capsule pear-shaped, swollen at the base .... ssp. *serpyllifolia*
    common.
A. capsule ovoid-cylindrical, not swollen at the base ....
    .... ssp. *leptoclados* (Rchb.f.) Nyman
    very rare. [*A. leptoclados* (Rchb.f.) Guss. of Rhoads and Klein, 1993]

Cerastium L.

Low annuals or perennials with opposite leaves, frequently viscid-pubescent throughout; flowers in terminal cymes or solitary, 4- or 5-merous; petals white, 2-lobed, rarely absent; stamens (5)10; styles 5; capsule longer than the sepals, opening by 10 teeth.

A. leaves, stems, and sepals densely white tomentose or villous
    B. low, spreading to ascending plant of serpentine barrens ....
    .... *C. arvense* var. *villosissimum*
    B. erect plant of gardens and waste ground .... *C. tomentosum*
A. leaves, stems, and sepals not white tomentose, although straight or irregular hairs may be present
    C. petals 2–3 times as long as the sepals; leaves linear to narrowly oblong or
    lanceolate .... *C. arvense*

*Arenaria
serpyllifolia*
(×1)

C. petals slightly longer than, equal to, or shorter than the sepals; leaves narrowly to broadly ovate to obovate or elliptic
    D. uppermost bracts of the inflorescence entirely herbaceous, lacking scarious tips and margins; pedicels arched
        E. pedicels longer than the calyx; hairs of the sepals not extending beyond the tip .... *C. nutans*
        E. pedicels shorter than the calyx; hairs of the sepals extending beyond the tip .... *C. glomeratum*
    D. uppermost bracts of the inflorescence with scarious tips and margins; pedicels rarely curved
        F. leaves 1–2 cm long .... *C. fontanum*
        F. leaves 0.5–1 cm long
            G. petals as long or slightly longer than the sepals, the notch 1–1.5 mm deep .... *C. pumilum*
            G. petals shorter than the sepals, the notch <1 mm deep ....
            .... *C. semidecandrum*

Cerastium arvense L.   Field chickweed
Glabrous to densely white-villous perennial, branched at the base, ascending or erect to 4 dm tall; leaves linear to narrowly ovate; petals showy, white, bifid, 2–3 times as long as the sepals; UPL; 3 varieties:

A. plant white-villous throughout .... var. *villosissimum* Pennell
    rare on serpentine barrens; flr. May–Aug; ❧.
A. plant glabrous or viscid-pubescent, not white-villous
    B. plant glabrous or nearly so .... var. *arvense*
        occasional on moist to dry, rocky slopes and sandy fields; flr. Apr–Oct.
    B. plant viscid-pubescent .... var. *viscidulum* Gremli
        waste ground at a former zinc mine; flr. May–Jun; native to Eurasia.

Cerastium fontanum,
flower (×2)

Cerastium fontanum,
capsule (×4)

Cerastium fontanum,
inflorescence
bracts (×4)

Cerastium fontanum Baumg.   Common mouse-ear chickweed
Short-lived perennial, viscid-puberulent; leaves ovate or obovate, rounded to acute, 1–2 cm long; inflorescence crowded or becoming more open with age; bracts of the inflorescence scarious-margined; hairs of the sepals not extending beyond the sepal tips; petals equaling or slightly shorter than the sepals, notched to 1 mm or more; a common weed of cultivated ground; throughout; flr. Apr–Oct; native to Eurasia; FACU–. Ours is var. *triviale* (Link) Jalas. [syn: *C. vulgatum* L.]

Cerastium glomeratum Thuill.   Mouse-ear chickweed
Annual or winter annual, glandular-pubescent; leaves ovate to obovate, 1–2.5 cm long; inflorescence becoming open with age; sepals with long, forward pointing hairs that extend beyond the tips; petals slightly shorter than to equaling the sepals, bifid, the notch 1 mm or more deep, or occasionally petals absent; occasional in fields, roadsides, and waste ground; mostly S; flr. Apr–Jun; native to Eurasia; UPL. [syn: *C. viscosum* L.]

Cerastium nutans Raf.   Nodding chickweed
Viscid-pubescent annual, 1–4.5 dm tall; leaves narrowly lance-oblong to oblanceolate; inflorescence open, the bracts completely herbaceous; sepals broad, thin, and blunt with hairs that do not extend beyond the tips; petals slightly

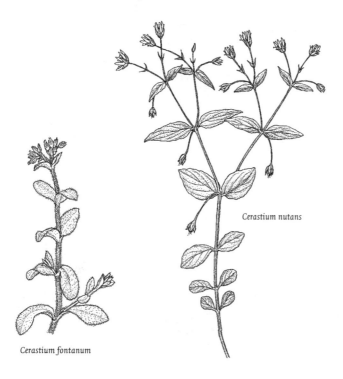

Cerastium nutans

Cerastium fontanum

longer than to shorter than the sepals, notched to 1–2 mm; frequent on rich wooded slopes and alluvium; mostly S; flr. Apr–Aug; FAC.

Cerastium pumilum Curtis   Small mouse-ear chickweed
Annual, very similar to *C. semidecandrum*, but pedicels erect and petals equaling to slightly exceeding the sepals, notched to 1–1.5 mm and seeds minutely papillate; rare in a pasture and cracks in urban paving; SE; flr. Jun; native to Europe.

Cerastium semidecandrum L.   Small mouse-ear chickweed
Viscid-pubescent annual to 2 dm tall; leaves 0.5–1 cm, oblanceolate to spatulate below, broadly elliptic above; inflorescence compact to open, the bracts with scarious tips and margins; sepals stipitate-glandular; petals shorter than the sepals, the notch <1 mm; stamens 5 or 10; rare and scattered in dry, sandy woodland edges, alluvial shores, and beaches; SE and NW; flr. May; native to Eurasia.

Cerastium tomentosum L.   Snow-in-summer
Erect, white-tomentose, rhizomatous perennial to 3 dm tall; leaves linear-lanceolate; flowers in clusters of 7–15; petals much exceeding the sepals; cultivated and sometimes escaped; scattered; flr. Apr–Jul; native to Europe.

Dianthus L.

Annual or perennial herbs with opposite leaves; flowers solitary or clustered in a terminal head-like inflorescence; calyx subtended by 1–3 pairs of bracts, sepals fused for more than half their length to form a ribbed tube; petals 5 with a narrow claw and spreading blade, white, pink, or deep purple-red; stamens 10; styles 2; capsule dehiscent by 4 valves.

A. flowers on long peduncles, solitary .... *D. deltoides*
A. flowers in crowded, terminal clusters
    B. leaves 1–2 cm wide; calyx glabrous .... *D. barbatus*
    B. leaves <1 cm wide; calyx villous-puberulent .... *D. armeria*

Dianthus
armeria,
flower (×1)

Dianthus armeria L.   Deptford pink
Erect annual or biennial, 2–6 dm tall; leaves basal and cauline, linear to lanceolate; flowers in clusters of 3–9, petals pink with white spots, the blade 4–5
mm; fruit as long as the calyx; common in old fields, waste ground, and roadsides; throughout; flr. May–Aug; native to Europe; UPL.

Dianthus barbatus L.   Sweet-William
Stout glabrous perennial, 3–6 dm tall; cauline leaves lanceolate to oblanceolate,
the basal leaves wider; flowers in dense heads; petal white to dark purple-red,
limb 5–10 mm, toothed at the summit; fruit 1 cm; cultivated and sometimes escaped; mostly S; flr. May–Aug; native to Eurasia.

Dianthus deltoides L.   Maiden or meadow pink
Rhizomatous perennial, 1–4 dm tall; basal leaves oblanceolate, cauline leaves
linear-lanceolate; flowers solitary on 1–4-cm pedicels; petals red-purple, violet,
or white, blade 5–10 mm long; fruit equaling the calyx; rare and scattered in dry
fields and alluvial sand and gravel bars; flr. May–Jul; native to Europe.

Holosteum L.

Holosteum umbellatum L.   Jagged chickweed
Glaucous annual with tufted, unbranched stems that are glabrous above and below and stipitate-glandular near the middle; leaves stipitate-glandular around
the margins; flowers in terminal, umbel-like inflorescences; petals slightly
longer than the sepals, white, fringed at the tips; stamens 3–5; styles 3; capsule
longer than the calyx; rare in sandy railroad banks and waste ground; SE; flr.
Apr–Jun; native to Eurasia.

Lychnis L.

Perennials with opposite, entire leaves lacking stipules; flowers perfect; calyx tubular, 10-ribbed but with 5 lobes, glabrous or hairy but never glandular or inflated; petals 5, entire, shallowly notched or 2–4 lobed; stamens 10; styles 5; capsule dehiscent by 5 teeth.

A. plants white-tomentose; calyx lobes twisted .... *L. coronaria*
A. plants green, variously pubescent; calyx lobes not twisted
    B. 10–20 pairs of cauline leaves present; leaves 2–5 cm wide; petals red,
       bifid .... *L. chalcedonica*
    B. 2–5 pairs of cauline leaves present; leaves <1.5 cm wide; petals pink, deeply
       4-lobed .... *L. flos-cuculi*

Lychnis chalcedonica L.   Maltese-cross
Stem erect, hairy, 3–6 dm tall with 10–20 pairs of ovate cauline leaves; inflorescence 10–50-flowered, crowded, terminal; petals crimson red, deeply bilobed;

capsule 1 cm; cultivated and sometimes persisting in old gardens and fields; mostly NE; flr. Jun–Jul; native to Asia.

Lychnis coronaria (L.) Desr.   Rose-campion, mullein-pink
Plants erect, 4–8 dm tall, densely white-tomentose; leaves basal and cauline, ovate to ovate-lanceolate, reduced upward; flowers few, on 5–10-mm pedicels; calyx with twisted, lanceolate lobes; petals reddish-purple, entire to shallowly notched; capsule ovoid, opening by 5 teeth; cultivated and occasionally escaped to fields, pastures, cemeteries, and dumps; scattered; flr. Jun–Aug; native to Europe.

Lychnis flos-cuculi L.   Ragged-robin
Stems erect, 3–8 dm tall, often branched, thinly hairy above; leaves lanceolate to lance-oblong, sessile; inflorescence branched; calyx campanulate; petals pink or rarely white, irregularly 4-lobed; capsule opening by 5 teeth; rare and scattered in fields, pastures, and roadsides; flr. late May–early Jun; native to Europe; FACU.

Minuartia L.

Low, annual herbs with narrowly lanceolate, opposite leaves; stipules not present; flowers mostly on slender pedicels in terminal cymes; sepals 5, distinct; petals 5, entire, white; stamens 10; styles 3; fruit a capsule with 3 valves.

A. sepals acute, the veins prominent
    B. primary leaves with tufts of smaller leaves in the axils creating a whorled
        appearance .... *M. michauxii*
    B. primary leaves lacking axillary tufts of secondary leaves .... *M. patula*
A. sepals obtuse, the veins barely visible .... *M. glabra*

Minuartia glabra (Michx.) Mattf.   Appalachian sandwort
Glabrous annual with linear or linear-oblanceolate, obtuse leaves; flowers few to many in a terminal inflorescence; sepals obtuse or somewhat acute, faintly 1-nerved; petals longer than the sepals; fruit broadly conical, separating to the base; a rare native of exposed sandstone rocks; NE; flr. May–Aug; UPL; ✹. [syn: *Arenaria groenlandica* (Retz.) Spreng. var. *glabra* (Michx.) Fernald]

Minuartia michauxii (Fernald) Farw.   Rock sandwort
Glabrous or sometimes hairy annual or perennial with short nonflowering shoots at the base as well as the 1–4-dm flowering stems; leaves narrow, subulate with tufts of secondary leaves in the axils; inflorescence open, with slender pedicels; sepals broadly lanceolate, acute with scarious margins; petals entire; fruit splitting to the middle or beyond; occasional in dry, open outcrops of limestone or serpentine; E and C; flr. May–Jun. [syn: *Arenaria stricta* Michx.]

Minuartia patula (Michx.) Mattf.   Sandwort
Much branched annual, glabrous or finely glandular-puberulent; leaves 1–2 cm by 0.5–1 mm, lacking axillary tufts; sepals narrowly lanceolate; petals slightly longer than the sepals; fruit splitting to the middle; locally abundant on zinc contaminated slopes of Blue Mountain; E; flr. Jun; native farther west; UPL. [syn: *Arenaria patula* Michx.]

Moehringia L.

Moehringia lateriflora (L.) Fenzl   Blunt-leaved sandwort
Rhizomatous, colonial, puberulent perennial; leaves ovate to elliptic-oblong and obtuse; sepals ovate or obovate, 2–3 mm, 3–5 veined; flowers 5-merous, solitary or in terminal cymes; petals 4–6 mm, white; stamens 10; styles 3; capsule separating to the base into 6 valves; occasional in wet meadows, swamps, swales, and low woods; scattered throughout; flr. May–Jun; FAC. [syn: *Arenaria lateriflora* L.]

Myosoton Moench

Myosoton aquaticum, capsule (×1)

Myosoton aquaticum (L.) Moench   Giant chickweed
Perennial, stems weak, decumbent, rooting at the lower nodes; leaves opposite, ovate to lance-ovate, 2–8 cm by 1–4 cm, sessile or the lower with short petioles; flowers in leafy cymes, 5-merous; sepals distinct, lance-ovate, glandular-hairy; petals white, distinctly longer than the sepals, bifid to the base; stamens 10; styles 5; capsules splitting into 5 valves that are shortly bifid at the tips; occasional in seeps, stream banks, alluvial woods, and moist roadsides; flr. May–Oct; native to Europe; FACW. [syn: *Stellaria aquatica* (L.) Scop., *Alsine aquaticum* (L.) Britton]

Moehringia lateriflora

Myosoton aquaticum

Paronychia Mill.

Annual herbs <25 cm tall with branching stems and opposite, oval, or elliptic leaves and conspicuous, membranous stipules; flowers in small cymes or in the forks of the stems; sepals distinct or united at the base, hooded at the tip; petals absent; stamens 5, inserted at the base of the calyx; styles 2-parted; fruit a utricle.

A. stems glabrous; sepals not conspicuously mucronate .... P. canadensis
A. stems pubescent; sepals with a distinct mucronate tip
    B. plant erect; largest leaves to 2 cm long; united portion of the style much shorter than the ovary .... P. fastigiata
    B. plant matted; largest leaves to 1.2 cm long; united portion of the style as long as the ovary .... P. canadensis

Paronychia canadensis (L.) Wood   Forked chickweed
Stems slender, erect, and glabrous; leaves 5–30 mm long, elliptic to oval, punctate; calyx 1–1.5 mm; styles divided nearly to the base, shorter than the ovary; fruit longer than the sepals; frequent in open woods in dry, rocky, or sandy soil; throughout; flr. May–Oct.

Paronychia fastigiata (Raf.) Fernald   Whitlow-wort
Stems erect or diffusely spreading, pubescent; leaves lance-elliptic, 5–20 mm long, serrulate; sepals lance-linear, 1–3-nerved; styles divided above, the united basal portion shorter than the ovary; fruit barely if at all longer than the sepals; open woods and edges in dry, rocky, or sandy soils; flr. Jul–Sep; 2 varieties:

A. sepals with a stout white awn to 0.2 mm .... var. *nuttallii* (Small) Fernald rare, SC; ✺.
A. sepals lacking an awn .... var. *fastigiata* common; S.

Paronychia montana (Small) Pax & Hoffm.   Forked chickweed
Diffusely spreading, pubescent stems to 25 cm tall; leaves 7–12 mm long, entire; stipules lance-ovate and ciliolate; united portion of the style as long as the ovary; rare in dry woods and shale barrens; SC; flr. Jul–Sep. [syn: *P. fastigiata* (Raf.) Fernald var. *pumila* (A.Wood) Fernald]

Petrorhagia (Seringe) Link

Erect, mostly glabrous annuals or perennials; stems with numerous pairs of narrow, opposite, entire leaves lacking stipules; flowers solitary or clustered in head-like terminal cymes; calyx subtended by 1–3 pairs of overlapping bracts, tubular, 5-nerved, membranous; petals 5, pink, blade obcordate; stamens 10; styles 2; capsule dehiscent by 4 teeth.

A. flowers in dense, head-like clusters; calyx about equaling the bracts ....
.... P. prolifera
A. flowers mostly solitary; calyx longer than the subtending bracts ... P. saxifraga

Petrorhagia prolifera (L.) Ball & Heywood   Childing pink
Glabrous annual, 3–5 dm tall; leaves linear to linear-oblanceolate at the base; flowers mostly in 3–7-flowered dense cymes; calyx 10–13 mm; petal blade 1-

veined; capsule 5–8 mm; rare in dry fields and banks; SE; flr. Jun–Oct; native to Europe.

Petrorhagia saxifraga (L.) Link   Saxifrage pink
Tufted perennial, stems decumbent at the base, 1–4 dm tall; leaves linear, 2-veined, 5–10 mm by 1 mm; flowers mostly solitary and terminal; calyx 3–6 mm; petal blade 3-veined; cultivated and rarely escaped; SE and NW; flr. Jun; native to Eurasia.

Sagina L.

Low growing annual or perennial herbs to 10 cm tall; leaves opposite, linear subulate, connate at the base; stipules not present; flowers solitary, terminal or axillary, 4- or 5-merous; petals white or lacking; stamens and styles each 4 or 5; fruit a capsule.

A. perianth mostly 4-merous; matted, wiry perennials spreading by offshoots.... .... *A. procumbens*
A. perianth 5-merous; ascending or decumbent annuals lacking offshoots
    B. pedicels glabrous; capsules longer than broad .... *S. decumbens*
    B. pedicels glandular pubescent; capsules globose .... *S. japonica*

Sagina decumbens (Elliot) Torr. & A.Gray   Pearlwort
Slender branching annual often with short axillary shoots; pedicels glabrous; flowers 5-merous; petals as long as the sepals or absent; rare in wet or dry sandy soil and cracks between paving stones; SE; flr. May–Oct; FAC.

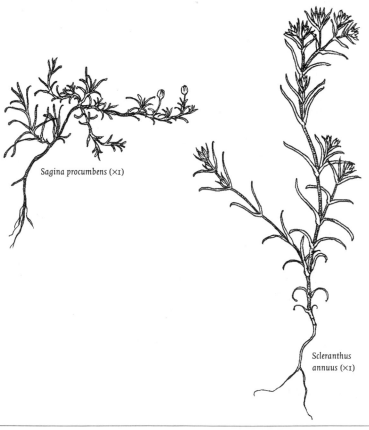

Sagina procumbens (×1)

Scleranthus annuus (×1)

Sagina japonica (Swartz) Ohwi   Japanese pearlwort

Very similar to *S. decumbens* but with glandular-hairy pedicels and more globose capsules; moist shores, lawns, and compacted soil in paving and around buildings; SE and W; flr. Jun–Jul; native to Asia.

Sagina japonica, fruit (×5)

Sagina procumbens L.   Bird's-eye, pearlwort

Glabrous perennial with decumbent to ascending stems, often with basal rosettes or sterile offshoots; leaves mucronate; flowers mostly 4-merous; occasional in moist soil of paving joints, lawns, and dooryards; throughout; flr. May–Oct; FACW–.

Saponaria L.

Saponaria officinalis L.   Bouncing-bet, soapwort

Rhizomatous, glabrous perennial with upright, leafy stems 4–8 dm tall; leaves numerous, elliptic-ovate to lance-elliptic; inflorescence terminal, with leafy bracts; calyx cylindric; petals pinkish, fragrant; stamens 10; styles 2–3; fruit a dehiscent capsule; common on roadsides, railroad banks, and other waste ground; flr. Jun–Oct; native to Europe; FACU–.

Scleranthus L.

Scleranthus annuus L.   Knawel

Low, spreading glabrous or puberulent annuals with forking stems to 15 cm tall; leaves linear-subulate, opposite but often with tufts of additional leaves in the axils; flowers terminal or axillary, sessile or subsessile, lacking petals, and with a cup-shaped hypanthium which is adherent to the 1-seeded, indehiscent fruit; styles 2; frequent in dry open ground, roadsides and shale barrens; mostly S; flr. May–Oct; native to Eurasia; FACU–.

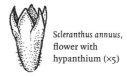

Scleranthus annuus, flower with hypanthium (×5)

Silene L.

Annual or perennial herbs with opposite, entire leaves lacking stipules; flowers perfect or unisexual (plants then dioecious); calyx tube 10–30-veined with 5 short teeth, sometimes inflated; petals 5, usually with auricles at the base of the blade and with an additional pair of appendages on the inner surface; stamens 10; styles usually 3 (rarely 4 or 5), ovary stalked; capsule mostly splitting into twice as many valves as styles.

A. leaves in whorls of 4 at most nodes .... *S. stellata*
A. leaves opposite
    B. calyx glabrous or finely puberulent
        C. stems glutinous below the nodes
            D. calyx 1.5–3 cm long, inflated in fruit .... *S. noctiflora*
            D. calyx 6–16 mm long, not inflated in fruit
                E. calyx 6–9 mm long; stalk of the capsule (within the calyx) 1 mm
                .... *S. antirrhina*
                E. calyx 12–16 mm long; stalk of the capsule 7–8 mm ...... *S. armeria*
        C. stems not glutinous below the nodes
            F. calyx conspicuously inflated, with prominent veinlets connecting the 20 longitudinal veins .... *S. vulgaris*

Silene antirrhina

        F. calyx little if at all inflated, venation not conspicuously reticulate
            G. calyx 7–12 mm long; petals deeply bifid .... S. *cserii*
            G. calyx 15 mm long; petals only shallowly notched .... S *nivea*
    B. calyx pubescent
        H. petals bright red .... S. *virginica*
        H. petals white or pink
            I. styles mostly 4 or 5
                J. petals white; flowers fragrant .... S. *latifolia*
                J. petals pink or red-purple; flowers not fragrant .... S. *dioica*
            I. styles mostly 3
                K. petals deeply 2-lobed .... .S. *dichotoma*
                K. petals shallowly notched to entire
                    L. calyx glandular .... S. *gallica*
                    L. calyx not glandular
                        M. stems <2 dm tall; petals pink .... S. *caroliniana*
                        M. stems 2–3 dm tall; petals white .... S. *nivea*

Silene antirrhina L.   Sleepy catchfly
Slender erect annual or biennial to 8 dm tall; upper internodes glutinous, sticky
to the touch; leaves oblanceolate to linear; inflorescence open, flowers numerous; calyx narrowly ovoid, not becoming inflated in fruit; petals small, pink or
purplish; frequent in dry open woods, fields, and waste ground; throughout; flr.
May–Jul.

Silene armeria L.   Garden catchfly
Glaucous, glabrous, or sparsely puberulent annual to 7 dm tall; stems viscid below each node; leaves ovate-lanceolate, sessile, and somewhat clasping; flowers
in an open or compact inflorescence; calyx 10-nerved, somewhat inflated above;
petals pink to purplish, shallowly notched; capsule long-stalked; formerly cultivated, found in abandoned gardens and waste ground; throughout; flr. May–
Sep; native to Europe.

Silene caroliniana Walter   Wild pink
Taprooted pubescent perennial with tufted, unbranched stems 0.8–2 dm tall;
leaves spatulate to oblanceolate, the basal ones petiolate, the upper sessile; flowers in dense, terminal cymes; calyx cylindric, densely glandular-hairy; petals
pink (or white), entire or slightly notched; occasional in dry open woods, rocky
slopes, roadside banks, and shale barrens, often on calcareous soils; mostly S;
flr. Apr–Jun. Ours is ssp. *pensylvanica* (Michx.) Clausen.

Silene cserii Baumg.   Campion
Similar to S. *vulgaris* but with broadly ovate leaves and a smaller, less inflated calyx with 10 ribs and inconspicuous reticulate venation; rare in alluvium and dry,
open soil along railroad tracks; SE; flr. May–Aug; native to Europe.

Silene dichotoma Ehrh.   Forked catchfly
Hairy annual or biennial, 3–8 dm tall with branching stems; leaves lanceolate
to oblanceolate, the lower ciliate-petiolate, the upper sessile; inflorescence
branched and leafy-bracted; flowers perfect; calyx tubular, with 10 green, hairy
nerves; petals white to reddish; stamens longer than the calyx tube; occasional
in fields and roadsides; S; flr. Jun; native to Eurasia.

Silene dioica (L.) Clairv.   Red campion

Dioecious perennial, very similar to *S. latifolia* but with petals red-purple, not fragrant, opening in the morning; rare in ballast, waste ground, lawns, and roadsides; scattered; flr. May; native to Europe.

Silene gallica L.   Catchfly

Hairy annual 1–4 dm tall, glandular above; leaves oblanceolate to spatulate, the cauline leaves narrower; flowers erect in a leafy-bracted raceme-like inflorescence; calyx 10-nerved, glandular, and hairy; petals white to pink, entire or slightly toothed; cultivated and rarely escaped to waste ground; scattered; flr. Jul–Aug; native to Eurasia.

Silene latifolia Poir.   White campion

Hairy annual or short-lived perennial, dioecious; stems 4–12 dm tall with 10 or more pairs of lanceolate to broadly elliptic cauline leaves; flowers white, fragrant, opening in the evening; calyx of the pistillate flowers 20-nerved, becoming inflated in fruit; petals deeply bilobed; styles usually 5; fruit 10–15 mm, dehiscent by 10 teeth; common in fields, roadsides, and waste ground; throughout; flr. May–Oct; native to Europe. [*S. alba* (Mill.) Krause of Rhoads and Klein, 1993]

Silene latifolia, flower

Silene nivea (Nutt.) Otth.   Snowy campion

Rhizomatous perennial with slender, erect stems 2–3 dm tall, glabrous or puberulent; leaves mainly cauline, sessile or short-petiolate, lanceolate; flowers on slender pedicels in the upper axils; calyx inflated, tubular-campanulate, without prominent veins; petals white, slightly notched; styles 3; fruit opening by 6 teeth; rare in alluvial thickets and woods; mostly SC and SW; flr. Jun–Aug; FAC.

Silene noctiflora L.   Night-flowering catchfly, sticky cockle

Annual, 2–8 dm tall, coarsely hairy below and with glandular hairs above; leaves lance-ovate, petiolate below, becoming sessile above; flowers perfect, calyx inflated, glandular-hairy on the nerves; flowers opening in the evening; blade of the petals 7–10 mm long, pink above, yellowish beneath; cultivated and escaped to upland woods and fields; flr. May–Oct; native to Europe.

Silene stellata (L.) W.T.Aiton   Starry campion

Perennial with erect simple, puberulent stems 3–10 dm tall; leaves in whorls of 4; flowers numerous in a paniculate inflorescence; petals white, deeply fringed; styles 3; capsule dehiscent by 6 teeth; frequent on wooded slopes, roadside banks, and barrens; throughout; flr. Jun–Aug.

Silene virginica L.   Fire pink

Short-lived perennial 2–8 dm tall, the stems glandular and somewhat pubescent; leaves oblanceolate to spatulate; inflorescence open, 7–11-flowered with leafy bracts; calyx broadly tubular; petals crimson-red, bifid; styles 3; capsule dehiscent by 6 teeth; occasional in upland woods, wooded slopes, and stream banks; W; flr. May–early Jul.

Silene vulgaris (Moench) Garcke   Bladder campion

Glabrous and glaucous perennial; stems erect or decumbent, 2–8 dm tall; leaves mostly cauline, lance-ovate to oblanceolate, the cauline leaves often clasping; inflorescence an open cyme with 5–30 flowers; calyx campanulate, inflated,

Silene vulgaris

membranous with 20 nerves and prominent connecting veinlets; petals white, deeply 2-lobed; styles 3; capsule stalked, opening by 6 teeth; common in old fields, roadsides, railroad banks, and waste ground; throughout; flr. May–Sep; native to Europe.

Spergula L.

Annuals with whorled, succulent, linear leaves; stipules present but small; stems simple or branched; inflorescence terminal, branched; petals white, shorter or longer than the sepals; stamens 10 (or 5); styles 5; fruit a 5-valved capsule.

A. leaves channeled beneath; seeds globose and only narrowly wing-margined...
.... *S. arvensis*
A. leaves not channeled beneath; seeds flattened and broadly winged ....
.... *S. morisonii*

Spergula arvensis L.   Corn spurrey
Plant to 4 dm tall; seeds rounded and only narrowly wing-margined; infrequent in fields and dry roadside banks; throughout; flr. May–Sep; native to Eurasia.

Spergula morisonii Boreau   Spurrey
Plant to 2 dm tall; seeds flattened, with a brownish wing about half as wide as the body; sandy waste ground and dredge spoil; SE; flr. May; native to Europe.

Spergularia (Pers.) J.Presl. & C.Presl.

Low, succulent annual or perennial herbs, glabrous or sparsely glandular-pubescent above; leaves opposite, linear, with secondary axillary tufts; stipules pale, scarious; flowers in branched inflorescences; sepals 5, petals 5, shorter than the sepals; stamens 2–10; styles 3; fruit a 3-valved, dehiscent capsule.

Spergula morisonii, node with whorled leaves and stipule (×1½)

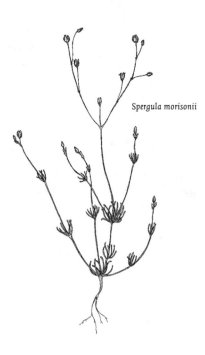

Spergula morisonii

A. sepals at least 4 mm; seed with a thin wing .... *S. media*
A. sepals <4 mm; seeds not winged
    B. petals pink; stamens 10; plants of nonsaline habitats .... *S. rubra*
    B. petals white; stamens 2–3; plants of saline habitats .... *S. marina*

Spergularia marina (L.) Griseb.   Saltmarsh sand-spurrey
Simple or much-branched annual, erect or prostrate; petals white or pink; stamens 2–5; seeds wingless; salted roadsides, very rare; SW; flr. Aug; native to Eurasia.

Spergularia media (L.) C.Presl. ex Griseb.   Sand-spurrey
Erect or prostrate, much-branched annual or perennial to 4 dm tall; petals white or pink; stamens 9 or 10; seeds with marginal wings 0.1–0.4 mm wide; salted roadsides, very rare; W; flr. Aug–Sep; native to Europe.

Spergularia rubra (L.) J.&C.Presl.   Purple sand-spurrey
Simple or much-branched annual or short-lived perennial 5–30 cm tall, often with tufts of leaves in the axils; petals pink; stamens (6)10; seed wingless; infrequent in dry, open, sterile soils; E; flr. May–Jun; native to Europe; FACU.

Stellaria L.

Low annuals or perennials with opposite leaves, lacking stipules; stems mostly weak, ascending or decumbent, often 4-angled; flowers perfect, solitary in the leaf axils or in terminal or axillary cymes; sepals 5, distinct; petals 5, white, 2-lobed, or sometimes absent; a distinct hypanthium present in S. *alsine*; stamens 10 (often 3 or 5 in S. *media*); styles 3; fruit a capsule dehiscent to about the middle by 6 valves.

A. leaves ovate, obovate, or broadly lanceolate, <3 times as long as wide
    B. leaf blade <3 cm long, the middle and lower leaves petiolate
        C. petals present; seeds >0.8 mm .... *S. media*
        C. petals absent; seeds <0.8 mm .... *S. pallida*
    B. leaf blade 2–9 cm long, leaves mostly sessile or with a short, winged petiole
        D. leaves sessile .... *S. pubera*
        D. the middle and lower leaves narrowed to a short winged petiole ....
        .... *S. corei*
A. leaves linear, lanceolate, or narrowly elliptic, >3 times as long as wide
    E. petals longer than the sepals
        F. bracts green, herbaceous; seeds 2–2.5 mm .... *S. holostea*
        F. bracts scarious; seeds 0.7–1.6 mm
            G. flowers few; sepals 3.5–4.5 mm .... *S. longifolia*
            G. flowers many; sepals 4.5–5.5 mm .... *S. graminea*
    E. petals shorter than the sepals or absent
        H. all the flowers subtended by tiny scarious bracts; flower with a distinct hypanthium .... *S. alsine*
        H. bracts of the lower flowers herbaceous, leafy; hypanthium not present
        .... *S. borealis*

Stellaria alsine,
flower (×5)

Stellaria alsine Grimm   Bog chickweed

Slender perennial with decumbent or ascending stems, often rooting at the nodes; leaves elliptic to oblanceolate or linear-oblong, the upper sessile, the lower petiolate; inflorescences axillary with tiny, scarious bracts throughout; pedicels slender, broadened at the summit; flowers with a definite hypanthium; petals white, shorter than the sepals; occasional in springs, swamps, seeps, and stream edges; mostly E and N; flr. May–Aug; native to Eurasia; OBL.

Stellaria borealis Bigelow   Northern stitchwort

Rhizomatous perennial with weak, much-branched stems to 5 dm tall; leaves lanceolate to lance-linear, usually narrowed at the base; flowers 5–50 in a weak-stemmed terminal cyme; lower floral bracts green, the upper reduced and scari-ous; petals white, shorter than the sepals, 2-lobed almost to the base or lacking; rare in springy slopes, sphagnous swamps, and stream banks; scattered; flr. May–Aug; FACW.

Stellaria corei Shinners   Chickweed

Very similar to *S. pubera* but with the middle and lower leaves narrowed to a short, winged petiole and sepals lance-acuminate, 7–10 mm long and glabrous

Stellaria longifolia

Stellaria media (×1)

except for the ciliate margins; rare on rocky hillsides, bluffs, and open woods; SW; flr. Apr–Jun.

Stellaria graminea L.   Lesser stitchwort
Perennial; stems weak, 3–5 dm tall, glabrous or scabrous on the angles; leaves linear to lance-linear; inflorescence terminal, diffuse, with scarious bracts; sepals lanceolate, strongly 3-nerved; petals shorter than to barely longer than the sepals, lobed almost to the base; common in swampy woods, moist meadows, stream banks, or moist slopes; throughout; native to Europe; flr. May–Sep; FACU–.

Stellaria holostea L.   Greater stitchwort
Rhizomatous perennial with glaucous and slightly hispidulous stems; leaves narrowly lanceolate, sessile; inflorescence with green, ciliate bracts throughout; petals longer than the sepals, notched less than halfway to the base; seeds 2–2.5 mm; very rare in rocky woods and roadsides; SW; flr. Apr–Jun; native to Eurasia.

Stellaria longifolia Muhl. ex Willd.   Long-leaved stitchwort
Perennial; stems weak, 1.5–4.5 dm tall, glabrous or sometimes scabrous on the angles; leaves linear; inflorescences terminal or lateral, widely branched, the bracts scarious; sepals lanceolate, weakly 3-nerved; petals slightly longer than the sepals, deeply 2-lobed; common in marshy open ground, swamps, rich woods, and moist roadsides; throughout; flr. May–Sep; FACW.

Stellaria media (L.) Vill.   Common chickweed
Weak-stemmed, bright green annual or winter annual; stems puberulent in 2 broad lines; leaves ovate, to 3 cm long; flowers in terminal leafy cymes; sepals 3–8 mm long; petals shorter, 2-lobed; stamens 3 or 5; fruit ovoid; a common weed of fields and gardens; throughout; flr. Mar–Nov; native to Europe; UPL.

Stellaria media, flower (×2)

Stellaria pallida (Dumort.) Piré   Lesser chickweed
Very similar to *S. media* but with paler, yellowish-green leaves and mainly cleistogamous flowers lacking petals; sepals 3 mm or less; stamens 1–3; mature capsule erect; seeds to 0.8 mm; rare in urban paving cracks; flr. Apr; native to Europe.

Stellaria pubera Michx.   Great or star chickweed
Erect or ascending perennial; stem 1.5–4 dm tall with 1 or 2 lines of pubescence; taller, mostly vegetative stems produced later in the season; leaves elliptic to ovate or lanceolate, sessile; flowers in a leafy, open terminal cyme, the petals longer than the sepals and lobed nearly to the base; occasional in moist woods; S; flr. Apr–Jun.

Vaccaria Wolf

Vaccaria hispanica (Mill.) Raushcert   Cow-cockle
Taprooted, glabrous, and glaucous annual to 6 dm tall, branched above; inflorescence loose and open; calyx strongly 5-ribbed and winged on the angles; petals pink; stamens 10; styles 2–3, fruit a capsule; cultivated and escaped to railroad banks and waste ground; mostly S; flr. Jun–Jul; native to Europe.

CELASTRACEAE Staff-tree Family

Evergreen or deciduous shrubs or woody vines with alternate or opposite, simple leaves; flowers perfect or unisexual, regular, 4–5-merous, small, greenish; ovary superior, 2–5-locular; fruit a capsule opening to expose the fleshy aril that surrounds each seed.

A. twining vines; leaves alternate .... *Celastrus*
A. erect or prostrate shrubs, vines climbing by aerial rootlets, or small trees; leaves opposite
    B. leaves elliptic, ovate, or obovate, evergreen or deciduous; ovary 3–5-locular; aril red or orange .... *Euonymus*
    B. leaves linear-oblong, evergreen; ovary 2 locular; aril whitish .... *Paxistima*

Celastrus L.

Twining woody vines to 4–5 m tall or more; leaves alternate, serrate, suborbicular to oblong or ovate; flowers small, greenish, regular with a prominent staminal disk, perfect or functionally unisexual; plants dioecious or polygamodioecious; fruit a 3-valved capsule that splits open to reveal the bright orange or red aril.

A. flowers in axillary clusters; leaves suborbicular to obovate, less than twice as long as wide .... *C. orbiculatus*
A. flowers in terminal panicles; leaves oblong to oblong-ovate, twice as long as wide .... *C. scandens*

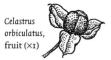

Celastrus orbiculatus, fruit (×1)

Celastrus orbiculatus Thunb.   Oriental bittersweet
Leaves nearly as wide as they are long; flowers in axillary clusters; widely naturalized in disturbed woods, fields, fencerows, and edges; flr. May–Jun; frt. Sep–Nov; native to Japan and China; UPL.

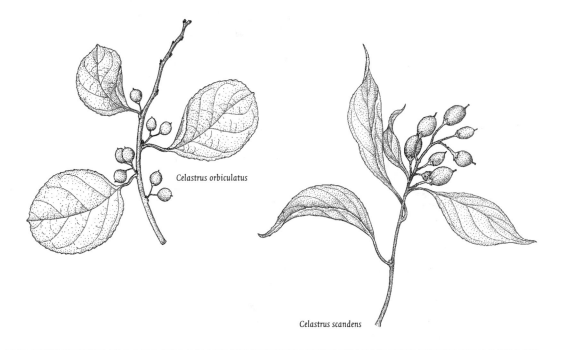

Celastrus orbiculatus

Celastrus scandens

Celastrus scandens L.   American bittersweet

Leaves narrower, longer than wide; flowers in a terminal panicle; occasional in dry fields, rocky ledges, woods, and hedgerows; throughout; appears to be declining due to competition from *C. orbiculatus*; flr. May–Jul, frt. Sep–Nov; FACU–.

Euonymus L.

Erect or prostrate shrubs, vines climbing by aerial rootlets, or small trees; twigs sometimes 4-angled or winged; leaves opposite, simple, serrate; flowers small, greenish, yellowish, or purplish, perfect, 4- or 5-merous; fruit a 3–5-locular capsule splitting to expose the bright red or orange arils that enclose the seeds.

A. stems creeping or climbing by aerial rootlets, at least in the juvenile phase; leaves evergreen .... *E. fortunei*
A. stems upright to prostrate, but not climbing; leaves deciduous
  B. twigs with corky wings; leaves subsessile .... *E. alatus*
  B. twigs not winged; leaves distinctly petiolate
    C. flowers 4-merous; fruits smooth
      D. leaves hairy beneath; flowers brownish-purple; aril red ....
      .... *E. atropurpureus*
      D. leaves glabrous beneath; flowers yellowish or greenish-white; aril orange
        E. leaves 2.5–8 cm long; anthers yellow, seeds white .... *E. europaeus*
        E. leaves 7–14 cm long; anthers purple, seeds red ... *E. hamiltonianus*
    C. flowers 5-merous; fruits strongly tuberculate
      F. erect shrubs with divergent 4-angled branches; upper leaves lanceolate to ovate .... *E. americanus*
      F. low prostrate shrubs with stems rooting at the nodes; upper leaves obovate .... *E. obovatus*

Euonymus alatus (Thunb.) Siebold   Burning-bush, winged euonymus
Deciduous shrub to 2.5 m tall with conspicuous corky wings on the twigs; leaves sessile or with very short petioles, elliptic to obovate, finely serrate, usually turning bright red in the autumn; flowers 4-merous, green; fruit purplish; aril orange; cultivated and frequently naturalized in woods and along stream banks, fencerows, and edges; mostly SE and SW; flr. Apr–Jun, frt. Sep–Oct; native to China and Japan.

Euonymus americanus L.   Hearts-a-bursting, strawberry-bush
Deciduous shrub to 2 m tall with stiffly divergent branches, but frequently reduced to low, nonflowering shoots due to excessive browsing by deer; twigs green, 4-angled; leaves elliptic or ovate, 3–7 cm, acute to acuminate; flowers 5-merous, greenish-purple, 10–12 mm wide; fruits crimson, strongly tuberculate; aril orange; occasional in moist woods, floodplains, and wet thickets; SE and SW; flr. May–Jun, frt. Sep–Oct; FAC.

Euonymus americanus fruit

Euonymus atropurpureus Jacq.   Burning-bush, wahoo
Erect, deciduous shrub to 6 m tall; leaves elliptic to lance-ovate; blade 6–12 cm, acuminate, finely serrate, finely hairy beneath; flowers 4-merous, brownish-purple, 6–8 mm wide; fruits bright pink, smooth, 1.5 cm, deeply 4-lobed; aril red; occasional in moist woods and floodplain thickets on limestone or diabase; mostly S; flr. Jun–Jul, frt. Aug–Sep; FACU.

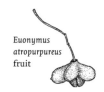
Euonymus atropurpureus fruit

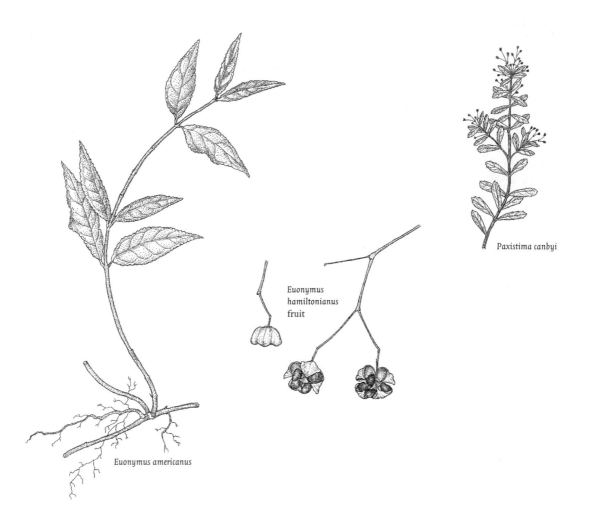

Euonymus hamiltonianus fruit

Paxistima canbyi

Euonymus americanus

Euonymus europaeus L   European spindletree
Deciduous shrub or small tree to 6 m tall; leaves ovate or oblong-lanceolate, acuminate; blades 2.5–8 cm, glabrous beneath; flowers 4-merous, greenish-white, in long-peduncled axillary cymes; fruit smooth, deeply 4-lobed, red or pink; aril orange; cultivated and occasionally escaped; mostly SE; flr. May–Jun, frt. Sep–Oct; native to Eurasia.

Euonymus fortunei (Turcz.) Hand.-Mazz.   Wintercreeper
Evergreen vine, trailing or climbing by aerial rootlets; leaf blades ovate-elliptic, 2–3 cm long on sterile shoots, 7–10 cm long on fertile branches, serrate, dark green or reddish, with lighter veins; flowers 4-merous, greenish, produced on horizontal, non-rootlet-bearing branches; cultivated and occasionally naturalized in floodplain forests; SE; flr. May–Jun, frt. Oct–winter; native to China.

Euonymus hamiltonianus Wall.   Spindletree
Deciduous shrub or small tree to 3 m tall, leaves oblong-lanceolate, acute, or short acuminate, finely toothed; blade 7–14 cm; flowers 4-merous, yellowish

with purple anthers; fruit 12 mm, smooth, deeply 4-lobed, pink; aril orange; cultivated and occasionally escaped to stream banks, hedgerows, and roadsides; scattered; flr. May–Jun, frt. Sep–Nov; native to Asia. [includes *E. yedoensis* Koehne of Rhoads and Klein, 1993]

Euonymus obovatus Nutt.   Running strawberry-bush
Deciduous shrub to 3 dm tall, rooting along the prostrate stems; upper leaves obovate; blades 3–6 cm; lower leaves narrower and smaller; flowers 5-merous, greenish-purple; fruit 3-lobed, strongly tuberculate, crimson; aril scarlet; occasional in wet, deciduous woods and wooded hillsides; flr. May–Jun, frt. Aug–Sep; W.

Paxistima Raf.

Paxistima canbyi A.Gray   Canby's mountain-lover
Low-growing, evergreen shrub to 3 dm tall; leaves leathery, opposite, narrow-oblong, 1–2 cm; flowers axillary, small, perfect; fruit a 2-locular capsule; aril whitish; very rare on calcareous cliffs and slopes; SC; flr. Apr–May, frt. Aug–Sep; 🌿.

CERATOPHYLLACEAE Hornwort Family

Perennial aquatic plants, generally submerged, without roots (free-floating) or rarely rooted; leaves 2–4 times palmately and/or dichotomously divided, the ultimate segments very narrowly linear, giving a feathery appearance; plants monoecious, both male and female flowers solitary or in few-flowered clusters, sessile in the axils of the leaves, very small, sepals and petals absent; fruit an achene with 2–several marginal spines, 4.5–7 mm long, ovoid; flr. early summer, frt. Aug–Sep.

Ceratophyllum L.

A. ultimate leaf segments somewhat flattened, clearly and regularly serrate on one side .... *C. demersum*
A. ultimate leaf segments thread-like, entire or with a few irregularly spaced teeth .... *C. muricatum*

Ceratophyllum demersum L.   Coontail, hornwort
Leaves often divided no more than twice; achenes usually with only 2 prominent spines near the base; occasional in quiet waters of lakes, ponds, and streams; throughout; OBL.

Ceratophyllum muricatum Cham.   Hornwort
Similar to *C. demersum*; leaves often divided 3–4 times; achenes mostly with several spines; occasional in quiet waters of lakes and ponds; throughout; OBL.

Ceratophyllum demersum (×1)

CERCIDIPHYLLACEAE Katsura-tree Family

Deciduous trees with simple, entire, cordate leaves that are opposite on long shoots or clustered on short spur branches; dioecious; pistillate flowers a cluster of 3–5 distinct ovaries with elongated styles; staminate flowers with 15–20 magenta-colored stamens; fruit a cluster of 1–2-cm-long follicles that remain on the tree after splitting to release the many winged seeds.

Cercidiphyllum Siebold & Zucc.

Cercidiphyllum japonicum Siebold & Zucc. ex J.Hoffm. & Schult.   Katsura-tree
Tree to 30 m tall; flowers magenta, solitary in the axils; cultivated and occasionally naturalized in disturbed woods; flr. Apr, before the leaves; SE; native to Japan.

CHENOPODIACEAE Goosefoot Family

Annual or perennial herbs with alternate (rarely opposite) leaves; inflorescence generally of clusters (glomerules) of flowers in the leaf axils or aggregated in spikes or panicles; flowers small, regular, perfect or unisexual, each flower subtended by a bract and a pair of bracteoles; sepals 3–5, or absent in some staminate flowers; petals absent; stamens same number as the sepals and opposite them; ovary superior, of 2–3 fused carpels, 1-locular with a single, spirally twisted or curved ovule; styles 1–3; fruit an utricle or nutlet surrounded by a persistent calyx or bracteoles. Mature pistillate flowers are required for reliable identification.

A. fruit enclosed by a pair of triangular or rhombic bracts ....*Atriplex*
A. fruit surrounded or enclosed by a persistent 3–5-lobed calyx
    B. leaves, especially of the inflorescence, spiny-tipped .... *Salsola*
    B. leaves not spiny-tipped
        C. fruiting calyx winged
            D. a continuous wing encircling the calyx .... *Cycloloma*
            D. each sepal bearing a flat dorsal wing ....*Kochia*
        C. fruiting calyx not winged .... *Chenopodium*

Atriplex L.

Monoecious annuals; leaves alternate, or opposite below, with a whitish-mealy surface, at least when young; flower clusters in the axils or in terminal spikes or panicles; pistillate flowers lacking a perianth and enclosed between 2 broadly triangular or rhombic bracteoles that may have various surface protuberances; seeds of 2 sizes, small and black or larger and brown; flr./frt. Aug–Nov.

A. leaves linear to lanceolate or oblong..*A. littoralis*
A. principal leaves triangular-hastate
    B. leaves tapered at the base, basal lobes, if present, pointing toward the leaf tip .... *A. patula*
    B. leaves cordate or truncate at the base with a pair of basal lobes pointing outward .... *A. prostrata*

Atriplex littoralis L.   Seashore orach

Stems to 1 m tall, branching and becoming somewhat woody at the base; leaves linear to linear-oblong, entire or coarsely toothed on the upper half; inflorescence a terminal spike of glomerules densely packed above and separated below; bracteoles 5–7 mm long, broadly triangular to ovate, more or less spongy-thickened and with margins united at the base; seeds brown and 2–2.8 mm wide or black and 1.5–2 mm wide; occasional in waste ground and roadsides; SE; native throughout the northern hemisphere but mostly adventive in Pennsylvania. [syn: *A. patula* L. var. *littoralis* (L.) A.Gray]

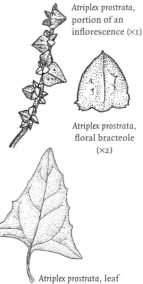

Atriplex littoralis, floral bracteole (×2)

Atriplex littoralis, portion of an inflorescence (×1)

Atriplex littoralis, leaf

Atriplex patula L.   Spreading orach, spearscale

Stems erect or occasionally prostrate, 1.5–15 dm tall, branched from the base; leaves alternate, the upper ones narrowly lanceolate or with basal lobes and irregularly serrate, the lower triangular or triangular hastate with upward-pointing basal lobes, young leaves whitish mealy on the surface becoming glabrate when older; bracteoles of the pistillate flowers 3–7 (or more) mm long, rhombic or triangular-hastate; brown seeds 2.5–3 mm wide, black seeds 1–2 mm wide; brackish or rich soils, generally in waste ground; mostly S; native to Eurasia; FACW.

Atriplex prostrata, portion of an inflorescence (×1)

Atriplex prostrata Boucher ex DC.   Halberd-leaved orach

Stems to 1 m tall, erect, ascending, or procumbent; lower leaves triangular hastate with outward pointing basal lobes; upper leaves triangular-hastate to lanceolate, entire or toothed; inflorescence spike-like; bracteoles with a weakly developed spongy layer and fused at the base, often with 2 tubercles on the dorsal surface; black seeds most abundant, 1–1.5 mm wide; brown seeds 1.5–2.5 mm; occasional in low, moist, brackish or rich soils and roadsides where salt accumulates; throughout; native throughout the northern hemisphere, but probably adventive in Pennsylvania. [syn: *A. patula* L. var. *hastata* auct. non (L.) A. Gray; *A. hastata sensu* Aellen, non L.]

Atriplex prostrata, floral bracteole (×2)

Atriplex prostrata, leaf

Chenopodium L.

Annuals (mostly), leaves, stem and/or calyx of many species with a whitish-mealy surface; leaves simple, entire, toothed or lobed, alternate; flowers small, perfect, or sometimes pistillate, in cymes, or more often clusters (glomerules) that may be axillary or aggregated in terminal spikes or panicles; calyx lobes 3 or 5, incurved, covering the fruit; stamens 1–5; styles 2–5; fruit a 1-seeded utricle with a papery or membranous fruit wall that is easily removed or sometimes adherent to the lens-shaped black seed; flr./frt. late summer–early fall.

A. leaves, stems, or calyx with resinous glands or glandular hairs; plants aromatic
  B. calyx lobes with glandular hairs; stems with spreading glandular hairs
    C. glands on the calyx lobes stalked; leaves irregularly lobed; stems erect. .... *C. botrys*
    C. glands on the calyx lobes sessile; leaves irregularly toothed; stems prostrate to ascending .... *C. pumilio*
  B. calyx lobes not glandular; stems glabrous or with appressed, nonglandular hairs .... *C. ambrosioides*

A. plants not resinous, glandular, or aromatic, although some or all parts may have a whitish-mealy surface
    D. calyx becoming bright red and fleshy in fruit .... *C. capitatum*
    D. calyx green, or whitish-mealy, not becoming fleshy
        E. leaves triangular-hastate, margins entire or merely wavy ....
        ....*C. bonus-henricus*
        E. leaves variously toothed or entire but not triangular-hastate
            F. leaves on the flowering branches mostly elliptic or elliptic-rhombic and entire (lower leaves may be broader and more or less toothed)
            G. leaves oblong to ovate, about 3 times as long as wide ....
            ....*C. standleyanum*
            G. leaves linear, about 5 times as long as wide
                H. leaves entire or with basal lobes, terminal lobe lanceolate and <0.8 cm wide .... *C. pratericola*
                H. leaves lobed, toothed or entire at base, terminal lobe tapering, 1 cm or more wide .... *C. foggii*
            F. leaves all more or less toothed
        E. calyx lobes 3 or 4; seeds mostly vertical .... *C. glaucum*
        E. calyx lobes 5; seeds mostly horizontal
            I. sepals rounded on the back, the midvein flat or only slightly raised
                J. seed and fruit with a rounded edge .... *C. urbicum*
                J. seed and fruit with a sharply angled edge
                      K. sepals covering the mature fruit .... *C. murale*
                      K. sepals not covering the fruit at maturity
                          L. leaves 5–20 cm long, with 1–4 coarse teeth per side; seed 1.5–2.5 mm wide .... *C. simplex*
                          L. leaves smaller, serrate; seed up to 1.5 mm wide ..*C. strictum*
            I. sepals with a prominent keel along the midrib making the calyx appear almost star-shaped
                M. seeds 1–1.5 mm wide; keel as much as half the width of the sepal; inflorescence erect
                    N. sepals prominently keeled but not winged .... *C. album*
                    N. keels of the sepals distinctly winged .... *C. berlandieri*
                M. seeds 1.5–2.3 mm wide; keel not as wide; inflorescence drooping
                .... *C. bushianum*

Chenopodium album var. album

Chenopodium album var. album, fruit (×5)

Chenopodium album L.   Lamb's-quarters
Erect branched annual to 1 m tall; leaves more or less whitish-mealy, rhombic-ovate to lanceolate, toothed; the dense glomerules of flowers forming spikes that are grouped in terminal panicles; calyx 5-lobed, whitish-mealy, covering the fruit; sepals keeled; fruit wall roughened, cellular-reticulate, adhering to the seed; leaves and inflorescences often turning reddish late in the season; a common weed of cultivated ground; throughout; FACU+; 2 varieties:

A. lower leaves 1.5–2 times as long as wide; stems seldom purple at the nodes; inflorescence erect .... var. *album*
   native to Europe
A. lower leaves up to 1.5 times as long as wide; stems consistently purple at the nodes; inflorescence flexuose... var. *missouriense* (Aellen) Bassett & Crompton

Chenopodium
ambrosioides

Chenopodium
botrys, leaf

Chenopodium
bushianum,
leaf

Chenopodium ambrosioides L.   Mexican-tea, wormseed

Erect, aromatic annual to 1 m tall (or more); leaves with sessile yellow resin glands on the lower surface, deeply pinnatifid to merely serrate, the upper reduced and becoming entire; flowers sessile in small glomerules arranged in elongate, bracteate spikes forming terminal panicles; calyx 3–5-lobed; styles 3 or 4; seeds horizontal or erect; common in fields and waste ground; mostly S; native to tropical America; FACU.

Chenopodium berlandieri Moq.   Goosefoot

Very similar to *C. album*, differentiated by the prominently winged sepals and roughened, cellular-reticulate fruit wall; rare in disturbed, open ground; scattered.

Chenopodium berlandieri,
fruit (×5)

Chenopodium bonus-henricus L.   Good-king-henry

Erect or ascending perennial to 7 dm tall; leaves long-petioled, broadly triangular-hastate with entire or wavy margins; flowers in a terminal panicle composed of aggregated small glomerules; sepals 4 or 5; seeds mostly erect; very rare in disturbed roadsides; native to Europe.

Chenopodium botrys L.   Jerusalem-oak; feather-geranium

Aromatic annual 2–6 dm tall; leaves pinnatifid, becoming entire above; flowers in a terminal inflorescence of small lateral cymes; calyx 5-lobed, glandular-hairy; seeds horizontal or erect; waste ground along rivers; mostly S; native to Europe; UPL.

Chenopodium botrys,
fruit (×5)

Chenopodium bushianum Aellen   Pigweed

Annual, 1–2 m tall when well-developed with leaves ovate or rhombic, 6–10 cm; inflorescence large and drooping; sepals 5, whitish-mealy, keeled; seeds 1.5–2.3 mm wide; frequent in cultivated and waste ground; throughout except at the highest elevations. [syn: *C. berlandieri* Moq. var. *bushianum* (Aellen) Cronquist]

Chenopodium
bushianum, fruit
(×5)

Chenopodium capitatum (L.) Asch.   Strawberry-blite, Indian-paint

Erect branched annual 2–6 dm tall; leaves with long petioles, blades triangular or triangular-hastate, coarsely dentate; flowers in globose clusters forming a terminal spike; calyx 3-lobed, becoming enlarged, fleshy and bright red in fruit; rare in woodland clearings and burned areas; scattered; ✿.

Chenopodium foggii Wahl   Goosefoot

Erect branched annual to 8 dm tall; leaves lanceolate or lance-ovate, 2–4 cm by

4–15 mm, entire or the larger ones with a few teeth near the base or even sub-hastate; inflorescence whitish-mealy, consisting of several glomerules of flowers forming short terminal or subterminal spikes; sepals 5, nearly covering the fruit; rare on dry, shaly slopes; scattered; 🍂. [syn: *C. pratericola* Rydb. in part]

Chenopodium glaucum L.   Oak-leaved goosefoot
Annual, stems erect to prostrate, branched from the base, 0.5–2.5 dm; leaves lanceolate to oblong or ovate with obtuse apex and cuneate base, coarsely toothed to entire or undulate, whitish-mealy beneath; inflorescence of glomerules arranged in axillary or terminal, usually interrupted, spikes that are shorter than the subtending leaves; calyx 3- or 4-lobed; occasional in stream banks, roadsides, gardens, and cultivated land; scattered but mostly SE; native to Europe; FACW–.

Chenopodium
murale, fruit (×5)

Chenopodium murale L.   Nettle-leaved goosefoot; sowbane
Malodorous; stem erect, branched, 1–8 dm tall; leaves rhombic-ovate with an acute to obtuse apex and cuneate to subcordate base; leaf margins sharply and coarsely toothed, sometimes slightly whitish-mealy; inflorescences 6–7 cm long, terminal and axillary panicles of closely spaced glomerules or occasionally solitary flowers; calyx 5-lobed, whitish-mealy, the lobes rounded or only slightly keeled, covering the fruit at maturity; rare in waste ground; S; native to Europe.

Chenopodium
pratericola,
fruit (×5)

Chenopodium pratericola Rydb.   Narrow-leaved goosefoot
Erect annual to 8 dm tall; leaves lanceolate or lance-ovate, entire, densely whit-ish-mealy beneath; inflorescences terminal and axillary spikes of subglobose glomerules; calyx 5-lobed, covering the seed at maturity; stigmas 2; seed horizontal; rare in waste ground; E; native to western U.S.

Chenopodium pumilo,
fruit (×10)

Chenopodium pumilio R.Br.   Goosefoot
Annual with spreading or prostrate stems 2–4 dm; leaves with numerous large yellow glands beneath, pinnatifid with 2–4 coarse teeth per side; flowers in small

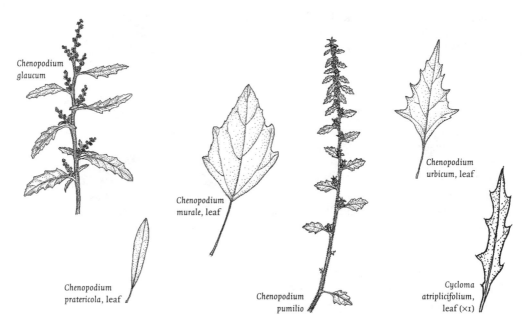

Chenopodium
glaucum

Chenopodium
murale, leaf

Chenopodium
pratericola, leaf

Chenopodium
pumilio

Chenopodium
urbicum, leaf

Cycloma
atriplicifolium,
leaf (×1)

glomerules arranged in short axillary spikes; calyx 5-lobed and covered with large sessile yellow glands; seeds erect; roadsides and waste ground; SE; native to Australia.

Chenopodium simplex (Torr.) Raf.   Maple-leaved goosefoot
Erect, bright green annual to 1.5 m tall; leaves broadly ovate to deltoid, with 1–4 large teeth on each side; inflorescence a terminal panicle of short, interrupted spikes; calyx sparsely or not at all whitish-mealy, although the inflorescence branches are; fruit wall separating easily from the seed; occasional in rocky woods, edges, and waste places; throughout. [*C. gigantospermum* Aellen of Rhoads and Klein, 1993]

Chenopodium standleyanum Aellen   Woodland goosefoot
Erect or arching annual to 1 m tall (or more); leaves green or slightly whitish-mealy, oblong-ovoid to lanceolate with an acute to acuminate apex and a cuneate base; margins mostly entire or with a few teeth toward the base on the lowest leaves; glomerules widely spaced forming flexuose spikes or panicles; calyx 5-lobed; sepals slightly keeled, partially covering the fruit at maturity and only slightly whitish-mealy; occasional in dry, open woods and thickets; SE and SC.

Chenopodium standleyanum, fruit (×5)

Chenopodium strictum Roth   Goosefoot
Erect, branched annual to 1 m tall; leaves oblong-ovate to ovate-lanceolate with a cuneate base and finely serrate margins, the upper leaves becoming entire; inflorescence of glomerules arranged in terminal spikes 1–3 cm long; calyx 5-lobed; sepals slightly keeled, reflexed at maturity; rare in dry waste ground and cinders; scattered. Ours is var. *glaucophyllum* (Aellen) Wahl.

Chenopodium urbicum L.   Upright goosefoot
Stems erect, mostly simple, 3–10 dm tall; leaves triangular with outward pointing lobes toward the base, becoming lanceolate above, slightly whitish-mealy beneath when young; inflorescences erect, terminal and axillary panicles or spikes of glomerules; calyx 5-lobed, only partly covering the fruit at maturity; sepals rounded or only slightly keeled, glabrous; waste ground; mostly S; native to Europe.

Cycloloma Moq.

Cycloloma atriplicifolium (Spreng.) Coult.   Winged pigweed
Densely branched, erect, or spreading annual to 8 dm tall; leaves alternate, coarsely toothed; white tomentose when young, becoming glabrate, falling early; flowers perfect, 5-merous; sepals curved over the ovary and developing a continuous membranous wing that surrounds the fruit; seed borne horizontally, black, with a conspicuous annular embryo; occasional in sandy disturbed soil, scattered; flr. Jul–Nov; FAC.

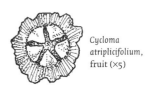

Cycloma atriplicifolium, fruit (×5)

Kochia Roth

Kochia scoparia (L.) Schrader   Summer-cypress
Erect branched annual to 6 dm tall; leaves linear-lanceolate, sessile, hairy; flowers in terminal spikes, star-shaped; the 5 sepals incurved around the fruit

and bearing a horizontal wing on the back; highway edges where salt is used; SW; flr. Aug–Oct; native to Europe.

Salsola L.

Tap-rooted annuals with linear leaves; flowers solitary or in small clusters in the upper axils surrounded by spine-tipped bracts and bracteoles; flowers perfect; calyx 5-lobed, each lobe keeled or somewhat winged, incurved over the ovary and developing fruit, with the tips joined and erect; stamens 5; styles 2; seed horizontal; flr./frt. Aug–Oct.

A. sepals soft with obscure midvein .... *S. tragus*
A. sepals stiff with the midvein extended in a sharp point .... *S. kali*

Salsola kali

Salsola kali L.   Salsola, barilla
Stems 3–6 dm tall, stout and much branched from the base, often red-tinged; leaves thick and succulent, 3–7 cm long; bracts and bracteoles spiny, longer than the flowers, reflexed in fruit; sepals stiff with a distinct midvein that is extended as a spiny tip; fruit wingless or sometimes winged; alluvial meadows, waste ground, and railroad ballast; native to coastal areas, but adventive at scattered sites in Pennsylvania; FACU.

Salsola tragus L.   Russian-thistle
Plants densely branched, breaking off at the ground and becoming tumble-weeds; stems 3–6 dm tall and often red-tinged; leaves narrowly linear 1–3 cm, reduced upward and transforming into spiny bracts and bracteoles that surround the flower and fruit; sepals soft at anthesis, with an obscure midrib; later developing a transverse wing; waste ground, especially in sandy soil; scattered; FACU. [syn: *S. kali* L. ssp. *tragus* (L.) Nyman]

CISTACEAE Rockrose Family

Herbs or subshrubs with alternate (sometimes nearly opposite or even whorled in *Lechea*), simple leaves; flowers in cymes, perfect, regular (except the calyx); sepals 5, the 2 outer ones narrower than the 3 inner; petals 3 or 5 (or lacking in cleistogamous flowers), distinct; stamens 3 to numerous; ovary superior; stigmas sessile or nearly so; fruit a 3-valved capsule separating to the base and surrounded by the persistent calyx.

A. flowers with 5 conspicuous yellow petals .... *Helianthemum*
A. flowers with 3 minute reddish petals .... *Lechea*

Helianthemum Mill.

Hairy perennials or subshrubs with alternate, nearly sessile leaves; early season flowers with 5 conspicuous yellow petals, 8–15 mm long; later blooms cleistogamous and lacking petals; sepals 5, the 2 outer ones narrower and often shorter than the inner 3; stamens 10–50 in the open flowers, 4–6 in cleistogamous ones.

A. early season, petal-bearing flowers solitary (or rarely 2) at the tips of the stems (although later overtopped by branches) .... *H. canadense*
A. early season, petal-bearing flowers in clusters of 3–15
    B. stems clustered from a branching caudex; leaves cuneate at the base .... *H. bicknellii*
    B. stems solitary and scattered from a creeping rhizome, leaves attenuate at the base .... *H. propinquum*

Helianthemum bicknellii

Helianthemum bicknellii Fernald   Bicknell's hoary rockrose
Stems 2–5 dm tall, clustered from a branching caudex and bearing terminal clusters of 6–10 open flowers; branches from the upper axils later bearing terminal and axillary clusters of cleistogamous flowers; leaves linear-oblong to narrowly elliptic, densely stellate-pubescent; rare on dry rocky slopes, open woods, and serpentine barrens; scattered; flr. Jun–early Jul; ✿.

Helianthemum canadense (L.) Michx.   Frostweed
Stems clustered, erect, 1.5–3 dm tall, each with a single open terminal flower, later much branched and to 6 dm with small, terminal and axillary clusters of cleistogamous flowers; leaves elliptic to oblanceolate, densely stellate-tomentose beneath and greener above; the outer sepals half as long as the inner, fused to the inner in the cleistogamous flowers with only a tiny tip free; occasional on dry, sandy or rocky ground, open woods, or barrens; mostly SE; flr. May–Jun.

Helianthemum propinquum E.P.Bicknell   Frostweed
Stems 1–3 dm tall, solitary and erect from a creeping rhizome; leaves linear-spatulate to oblong, attenuate at the base; open flowers in terminal clusters of 2–6, later overtopped by the lateral branches bearing cleistogamous flowers; outer sepals of the cleistogamous flowers shorter than and fused to the inner with the free tip <0.5 mm; very rare in dry, sandy ground or barrens; mostly E; flr. late May–Jun; ✿.

Lechea L.

Perennial herbs with erect flowering stems and overwintering basal shoots; leaves small, alternate, opposite, or whorled; flowers small, numerous, in a terminal panicle; sepals 5, the 2 outer ones linear or lanceolate and the 3 inner ones ovate to obovate and keeled; petals 3, smaller than the sepals and soon falling; stamens 3–15; stigma sessile; fruit 3-valved with 1–6 seeds; flr. late Jun–Sep, frt. late Jul–Nov.

Helianthemum canadense

A. outer sepals only about ½ as long as the inner sepals
    B. fruit (and calyx) twice as long as thick .... *L. racemulosa*
    B. fruit (and calyx) approximately as wide as long
        C. fruiting calyx tapered at the base .... *L. pulchella*
        C. fruiting calyx abruptly rounded at the base .... *L. intermedia*
A. outer sepals as long as or longer than the inner sepals
    D. hairs of the stem spreading; inner sepals glabrous except perhaps on the keel .... *L. villosa*
    D. hairs of the stem appressed; inner sepals hairy

*Lechea
intermedia*

*Lechea
villosa*

Lechea intermedia Legg. ex Britton   Pinweed

Stems 2–6 dm tall, with thin appressed pubescence; leaves oblong-lanceolate to narrowly elliptic, only sparsely hairy on the margins and midribs beneath or glabrous; panicle ⅓–½ the height of the plant and narrowly cylindric; calyx subglobose, obtuse to abruptly rounded at the base; the outer sepals short; occasional in dry sandy fields and dry ridges; mostly E and C.

Lechea minor L.   Thyme-leaved pinweed

Stems 2–5 dm tall with spreading or ascending hairs; leaves opposite or whorled, sparsely hairy on the margins and midribs; flowers in a compact, leafy panicle; outer sepals green, hairy, longer than the inner, the fruit slightly longer than the calyx; rare in dry soil of woods, slopes, and serpentine barrens; mostly SE; ✿.

Lechea pulchella Raf.   Pinweed

Stems 2–8 dm tall with thin appressed pubescence; leaves narrowly lanceolate to oblanceolate, sparsely hairy beneath on the margins and midribs; panicle ½ the total height of the plant; flowers crowded; fruits subglobose to broadly ellipsoid, often longer than the sepals; calyx acute at the base; frequent in dry, open woods, sandy fields, and barrens; SE and SC.

Lechea racemulosa Lam.   Pinweed

Stems 2–4 dm tall with thin appressed hairs; leaves often whorled, glabrous above but hairy on the margins and midribs beneath; panicle ½ the total height of the plant with numerous ascending branches; outer sepals shorter than to about as long as the inner; fruiting calyx with a subcylindric base differing in texture from the softer upper portion; frequent in dry fields and shaly slopes; throughout.

Lechea villosa Elliot   Pinweed

Stems 2–8 dm tall with long, spreading hairs; leaves alternate, or some whorled, hairy beneath and on the margins and midveins above; flowers densely clustered on short, leafy, lateral branches; inner sepals glabrous except for the often sparsely hairy keel; outer sepals about as long as the inner, coarsely hairy; stamens 3; fruit subglobose; occasional in open woods, old fields, and sand plains; E and NW.

Lechea intermedia,
fruit (×5)

Lechea pulchella,
fruit (×5)

Lechea racemulosa,
fruit (×5)

Lechea villosa,
fruit (×5)

CLETHRACEAE Clethra Family

Deciduous shrubs or small trees with alternate, simple leaves lacking stipules; flowers perfect, regular, 5-merous, small, white or pinkish, in terminal racemes, blooming in mid-summer, ovary superior, trilocular; stamens 10, in 2 whorls; fruit a dehiscent capsule with numerous seeds.

Clethra L.

A. leaves glabrous, tip obtuse or subacute; filaments glabrous .... *C. alnifolia*
A. leaves pubescent beneath, tip acuminate; filaments hairy .... *C. acuminata*

Clethra acuminata Michx.   Mountain pepperbush
Shrub or small tree to 6 m tall; leaf blades oblong to elliptic, 10–20 cm, serrulate, acuminate, pubescent beneath at least on the veins; racemes spreading or recurved, 8–15 cm; rachis, pedicels, and calyx tomentose; petals white; filaments hirsute; style glabrous; fruits 3–4 mm, densely villous; very rare on rocky, wooded slopes; SW; flr. Jul–Aug; ✤.

Clethra alnifolia L.   Sweet pepperbush
Shrub to 3 m tall; leaves obovate-oblong, 5–10 cm, serrate above the middle, entire below, obtuse or subacute, glabrous; racemes erect, 5–15 cm; flowers white, fragrant; fruits pubescent, 3 mm; occasional in low, wet woods and swamps; SE; flr. Jul–Aug; FAC+.

Clethra
alnifolia

CLUSIACEAE St. John's-wort Family

Annual or perennial herbs or shrubs; leaves opposite, entire, usually with translucent dots (observable when held up to light) and often black-dotted, without stipules; flowers perfect, regular; ovary superior; sepals often persisting in fruit; fruit a many-seeded capsule.

A. flowers yellow to orange; stamens distinct or fascicled, not alternating with glands .... *Hypericum*
A. flowers pinkish to purplish; stamens in 3 fascicles of 3, alternating with large orange glands .... *Triadenum*

Hypericum L.

Stamens 5–many, sometimes persisting at the base of the fruit; inflorescence terminal; fruit 1- or 3-celled.

A. woody shrubs
    B. flowers 4-merous
        C. leaves linear-oblanceolate to linear-oblong, mostly 2–4 mm wide ....
        .... *H. stragulum*
        C. leaves elliptic-oblong, mostly 7–15 mm wide................. *H. crux-andreae*
    B. flowers 5-merous
        D. petals 8–20 mm long; capsules 7–15 mm long .... *H. prolificum*
        D. petals 5–8 mm long; capsules 3.5–7 mm long .... *H. densiflorum*

A. plants herbaceous
    E. leaves scale-like, 1–3 mm long .... *H. gentianoides*
    E. leaves not scale-like, >3 mm long
        F. petals 2–3 cm long .... *H. pyramidatum*
        F. petals 2–10 mm long
            G. stamens united at base into 3 or 5 fascicles; capsule 3-celled
                H. stems distinctly ridged below the base of each leaf ....
                .... *H. perforatum*
                H. stems rounded below the bases of the leaves
                    I. styles united at base; persisting as a beak on the capsule ....
                    .... *H. adpressum*
                    I. styles not united at base; capsule not beaked ..... *H. punctatum*
            G. stamens not united into fascicles; capsules 1-celled
                J. styles united at base, persisting as a beak on the capsule
                    K. stems arising singly from a rhizome .... *H. ellipticum*
                    K. stems clustered, not rhizomatous .... *H. sphaerocarpum*
                J. styles not united at base; capsule not beaked
                    L. stamens >20 .... *H. denticulatum*
                    L. stamens <20
                        M. inflorescence a raceme .... *H. drummondii*
                        M. inflorescence a cyme
                            N. sepals broadest near the middle; fruit ellipsoid
                            O. ultimate bracts elliptic, leaf-like .... *H. boreale*
                            O. ultimate bracts subulate .... *H. mutilum*
                        N. sepals broadest well below the middle; fruit ovoid or conic
                            P. leaves linear, lanceolate, or narrowly obovate
                              Q. leaves 5–7-veined .... *H. majus*
                              Q. leaves 1–3-veined
                                R. leaves generally 5–10 times as long as wide ....
                                .... *H. canadense*
                                R. leaves generally 2–5 times as long as wide .....
                                .... *H. dissimulatum*
                          P. leaves deltoid-lanceolate to deltoid ovate ....
                          .... *H. gymnanthum*

Hypericum adpressum Raf. ex W.Bartram   Creeping St. John's-wort
Herbaceous perennial, 3–8 dm tall; stems arising from a horizontal rhizome, seldom branching; leaves linear-oblong to narrowly elliptic, 3–6 cm long, acutish at apex, tapering to base; inflorescence many-flowered; sepals lanceolate to ovate, 2–7 mm long; petals 6–8 mm long; capsule ovoid, 4–6 mm long; swamps and damp ground; extreme SE; believed to be extirpated; flr. Jul–Aug; OBL; 🌱.

Hypericum boreale (Britt.) E.P.Bicknell   Dwarf St. John's-wort
Herbaceous perennial, 1–4 dm tall; stems arising from slender rhizomes, much-branched; leaves elliptic to ovoid, obtuse or rounded at both ends, sessile, 1–2 cm long; capsule ellipsoid, 4–5 mm long; rare on open peat of bog edges, also swampy hummocks and wet meadows; mostly NE, scattered elsewhere; flr. Jul–Sep; OBL.

Hypericum canadense L.   Canadian St. John's-wort
Herbaceous annual, 1–4 dm tall, few-branched; leaves linear to narrowly oblanceolate, 1–3 cm long, narrowed to the base; inflorescence few-flowered; sepals
lanceolate, acute to acuminate, 4–6 mm long; petals 2–4 mm long; capsule
ovoid to cylindric, 5–6 mm long; common in sandy stream banks, moist ground
with sphagnum, low thickets, and swales; mostly E, scattered elsewhere; flr. Jul–
Sep; FACW.

*Hypericum
canadense*

Hypericum crux-andreae (L.) Crantz   St. Peter's-wort
Deciduous shrub, 3–8 dm tall, much-branched; leaves elliptic-oblong, 1.5–3.5
cm long, rounded at apex, cordate to somewhat clasping at the base; flowers
mostly solitary; sepals of two sizes, the outer larger and wider; petals showy,
longer than the sepals; swamps in sandy soil; extreme SE; believed to be extirpated; flr. Jul–Aug; FACU; ❧.

Hypericum densiflorum Pursh   Bushy St. John's-wort
Deciduous shrub up to 2 m tall, much-branched above; leaves linear to narrowly
elliptic or oblanceolate, 2–4 cm long, obtuse at the apex, narrowing to the base;
inflorescence a many-flowered cyme; petals 5–8 mm long; capsule with a persistent beak; rare in rocky stream banks, swampy meadows, and sphagnum bogs;
mostly SW, scattered elsewhere; flr. Jul–Aug; FAC+; ❧.

Hypericum denticulatum Walter   Coppery St. John's-wort
Herbaceous perennial, 2–5 dm tall, stems distinctly 4-angled; leaves elliptic to
narrowly oblong, 1–3 cm long; inflorescence a few- to many-flowered cyme; petals 4–5 mm long and persistent in fruit; capsule 3–5 mm long, with persistent
styles; bogs and wet woods; extreme SE; believed to be extirpated; flr. Jul–Sep;
FACW–; ❧.

Hypericum dissimulatum E.P.Bicknell   St. John's-wort
Similar to *H. canadense*; leaves linear-oblong, sessile; rare on moist, sandy, or
peaty soil; mostly S and E; flr. Jul–Sep; FACW.

Hypericum drummondii (Grev. & Hook.) Torr. & A.Gray   Nits-and-lice
Herbaceous annual, 2–6 dm tall, much-branched, stems slender and wiry; leaves
linear, 5–15 mm long; flowers mostly solitary and axillary; sepals lanceolate, 4–6
mm long and longer than the petals; capsule ovoid, 4–5 mm long; very rare in
dry slopes and stony fields; W; flr. Jul–Sep; UPL; ❧.

Hypericum ellipticum Hook.   Pale St. John's-wort
Herbaceous perennial, 2–5 dm tall, stems arising from a horizontal rhizome,
usually unbranched; leaves elliptic, 1–4 cm long, obtuse or rounded at both ends
or sometimes tapering to the base; inflorescence a few- to many-flowered cyme;
sepals up to 6 mm long; petals 5–7 mm long; capsule ovoid, 5–6 mm long; frequent in sandy stream banks, stream borders, marshes, boggy areas, and swales;
throughout; flr. Jul–Aug; OBL.

*Hypericum
ellipticum*

Hypericum gentianoides (L.) Britton, Stearns & Poggenb.   Orange-grass, pineweed
Herbaceous annual, 1–4 dm tall, very much-branched; leaves scale-like, 1–3 mm

Hypericum gentianoides

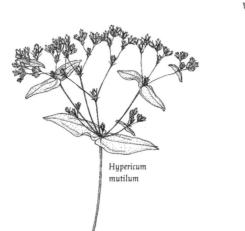

Hypericum mutilum

Hypericum perforatum

long, appressed to the stem; flowers mostly solitary at the nodes, 2–3 mm across; capsule conic, 5–7 mm long; common in abandoned fields, dry hillsides, barrens, and open shale slopes; mostly S, scattered N; flr. Jun–Sep; UPL.

Hypericum gymnanthum Englem. & A.Gray   Clasping-leaved St. John's-wort
Herbaceous annual, 2–6 dm tall, seldom-branched; leaves deltoid-lanceolate to deltoid-ovate, 1.5–3 cm long, rounded to cordate at the base, sessile; inflorescence very loose and open; sepals about twice as long as petals; capsule lanceolate to narrowly ovoid, 3–5 mm long; muddy shores or edges of intermittent ponds or mud holes; Centre and Lehigh Cos.; believed to be extirpated; flr. Jul–Sep; OBL; ⚘.

Hypericum majus (A.Gray) Britton   Canadian St. John's-wort
Herbaceous annual, similar in habit to *H. canadense* but usually somewhat taller; leaves lanceolate, 2–4 cm long, obtuse or acute at the apex, rounded to acute at the base, the margins of a pair of leaves meeting around the stem; flowers 5–10 mm across; capsule narrowly ovoid, 5–8 mm long; very rare on swampy ground and sand plains along Lake Erie; flr. Jul–Sep; FACW; ⚘.

Hypericum mutilum, fruit (×2)

Hypericum mutilum L.   Dwarf St. John's-wort
Herbaceous perennial, 1–8 dm tall, much-branched above; leaves lanceolate to elliptic or ovate, 1–4 cm long, obtuse to acute at the apex, obtuse to rounded at the base; inflorescence many-flowered; flowers 3–4 mm across; capsule ellipsoid, 2–4 mm long; common in stream banks, moist fields, swamps, and ditches; throughout; flr. Jul–Sep; FACW.

Hypericum perforatum, flower (×2)

Hypericum perforatum L.   St. John's-wort
Herbaceous perennial, 4–8 dm tall, much-branched; leaves linear-oblong, 2–4 cm long on the main axis and 1–2 cm long on the branches; inflorescence a many-flowered rounded or flattened compound cyme; petals about twice as long as the sepals and black-dotted near the margins; common in roadsides, fields, and waste places; throughout; flr. Jun–Sep; native to Europe.

Hypericum prolificum L.   Shrubby St. John's-wort
Deciduous shrub, up to 2 m tall, much-branched, twigs 2-edged; leaves linear to

narrowly elliptic or oblong, 3–6 cm long; inflorescence few-flowered, terminal or axillary; flowers to about 2 cm across; capsule 8–14 mm long; occasional in low fields, swamps, and thickets; mostly W and S; flr. Jul–Sep; FACU. [*H. hypericoides* (L.) Crantz of Rhoads and Klein, 1993]

Hypericum punctatum Lam.   Spotted St. John's-wort
Herbaceous perennial, 5–10 dm tall, seldom-branched; leaves oblong to elliptic or ovate, 2–6 cm long; inflorescence many-flowered, crowded; flowers 8–15 mm across; sepals and petals black-dotted; capsule ovoid, 4–6 mm long; common in floodplains, roadsides, moist fields, and thickets; throughout; flr. Jun–Aug; FAC–.

Hypericum punctatum,
flower (×2)

Hypericum pyramidatum Aiton   Great St. John's-wort
Herbaceous perennial, 7–15 dm tall, often much-branched; leaves lanceolate to elliptic, 4–10 cm long, acute to obtuse at apex, sessile to somewhat clasping at base; flowers mostly solitary, terminal, 4–5 cm across; capsule ovoid, 1.5–2 cm long; occasional on alluvial shores, rocky banks, and swamps; throughout; flr. Jun–Aug; FAC.

Hypericum sphaerocarpum Michx.   St. John's-wort
Herbaceous perennial, 3–7 dm tall, often branching near the base; leaves linear-oblong to linear-elliptic, 3–7 cm long, acute to obtuse at the apex, narrowing to the base; inflorescence compact; petals about twice as long as sepals; capsule globose to ovoid, 5–7 mm long; rocky shores; Allegheny Co.; flr. Jun–Aug; native west and southwest of Pennsylvania; FAC.

Hypericum stragulum W.P.Adams & N.Robson   St. Andrew's-cross
Deciduous decumbent shrub with several prostrate stems giving rise to nu-

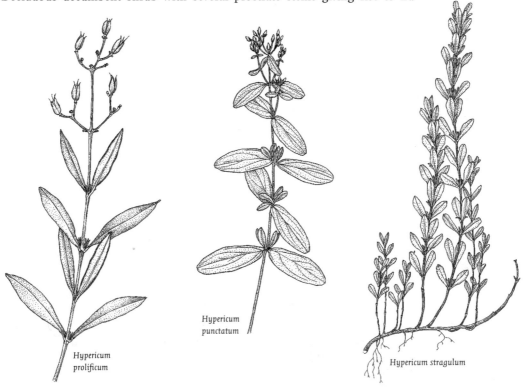
*Hypericum
prolificum*

*Hypericum
punctatum*

Hypericum stragulum

merous erect branches; leaves linear to oblanceolate, 2–3 cm long, obtuse to rounded at the apex, narrowed to the base; petals narrowly oblong-elliptic, 8–10 mm long; occasional in open woods, thickets, dry sandy soil, and serpentine barrens; mostly SE, scattered SC and SW; flr. Jul–Aug; FACU; ❧.

Triadenum Raf.

Flowers pinkish to purplish; stamens 9; inflorescences few-flowered cymes, terminal or axillary; fruit 3-celled, topped by persistent styles.

A. sepals rounded or obtuse at the apex, 3–5 mm long when mature ..... T. fraseri
A. sepals acute or acuminate at the apex, 5–8 mm long when mature ....
.... T. virginicum

Triadenum fraseri (Spach) Gleason   Marsh St. John's-wort
Erect perennial herb, 3–6 dm tall, glabrous; leaves oblong to ovate-oblong or elliptic, 3–6 cm long and 1–3 cm wide, more or less rounded at the apex, cordate to subcordate at the base, sessile, dark-spotted below, translucent dotted; flowers 1–1.5 cm across; capsule ovoid to cylindric, 7–12 mm long, styles 0.6–1.3 mm long; bogs, swamps, moist meadows, and seeps; throughout; flr. Jul–Aug; OBL.

Triadenum virginicum (L.) Raf.   Marsh St. John's-wort
Similar to T. fraseri in habit and leaves; flowers 1.5–2 cm across; capsule cylindric, 8–12 mm long, styles 2–3 mm long; marshes, bogs, swampy woods, stream banks, and ditches; throughout; flr. Jul–Aug; OBL.

CONVOLVULACEAE Morning-glory Family

Herbaceous, usually twining vines (rarely erect); leaves alternate, simple; flowers with a 4- or 5-lobed, funnel-shaped corolla; sepals 4 or 5, distinct; stamens attached to the corolla tube and alternate with the lobes; style 1; stigmas 1–4, sometimes lobed; ovary apparently 1–6-celled, fruit a rounded capsule.

A. calyx subtended and all or mostly covered by 2 large, leafy bracts .... Calystegia
A. calyx not subtended by a pair of large bracts
  B. leaf bases cordate; stigma with 2 or 3 small lobes .... Ipomoea
  B. leaf bases truncate or sagittate; stigma deeply split into 2 linear segments
  .... Convolvulus

Calystegia (R.Br.) A.Gray

Erect or prostrate, twining or trailing perennial herbs; leaves oblong, ovate, or hastate; flowers white or pink, 1–4 on axillary peduncles and subtended by 2 large bracts; sepals 5; corolla funnel-shaped; stigmas 2; ovary 1–4-celled; fruit a rounded capsule.

A. flowers single; corolla 4–7 cm
  B. flowering stems trailing; leaves subtending the flowers with petioles more than half as long as the midvein
    C. peduncles round in cross section

  D. bracts subtending the calyx flat, up to 1.5 cm wide, not overlapping
   .... *C. sepium*
  D. bracts inflated at the base, 1.5–4 cm wide, overlapping .... *C. silvatica*
 C. peduncles winged ....*C. hederacea*
 B. flowering stems erect; petioles of the leaves subtending flowers less than
  half as long as the midvein .... *C. spithamaea*
A. flowers mostly double; corolla 2–3.5 cm
 E. leaves hastate, basal lobes acute ....*C. hederacea*
 E. leaves cordate ....*C. pubescens*

Calystegia hederacea Wall.   Japanese bindweed
Stems climbing or trailing; leaves hastate, 4–8 cm long, with acute basal lobes; flowers single or double, on winged peduncles; corolla pink, 2–3.5 cm long; occasional in hedges, flower beds, fields, and rubbish dumps; SE; flr. May–Aug; native to Japan.

Calystegia pubescens Lindl.   Japanese bindweed
Plants trailing or climbing, densely soft-pubescent throughout; leaves short-petioled, sagittate or hastate, 4–8 cm long with acute basal lobes; corolla pink, double, 2–3 cm long; occasional in fallow fields, vacant lots, roadsides, stream banks, and waste ground; mostly S; flr. Jun–Aug; native to China.

Calystegia sepium (L.) R.Br.   Hedge bindweed, wild morning-glory
Stems twining or trailing to 3 m long, much branched, glabrous to pubescent; leaves hastate, 5–10 cm long; peduncles from many nodes, 1-flowered; calyx hidden by 2 ovate-cordate bracts; corolla white or pink, 4–8 cm long; common in fields, woods edges, and waste ground; throughout; flr. Jun–Sep; FAC–.

Calystegia silvatica (Kit.) Griseb.   Bindweed, morning-glory
Very similar to *C. sepium* but with square leaf sinuses and inflated, obtuse bracts; occasional in fields, stream banks, roadsides, and moist ground; scattered; flr. Jun–Sep. Ours is var. *fraterniflora* (Mack. & Bush) R.Br.

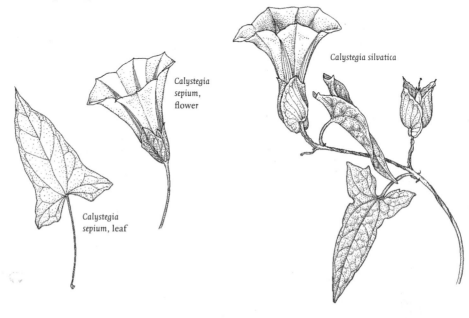

Calystegia sepium, flower

Calystegia sepium, leaf

Calystegia silvatica

Calystegia spithamaea (L.) Pursh   Low bindweed

Stems erect, to 4 dm long, usually not twining; glabrous to pubescent, unbranched or branching near the base; main leaves 3–8 cm long mostly on short petioles; blades oblong, obtuse at both ends or subcordate to sagittate at the base; peduncles present on the lower part of the stem, exceeding the leaves in length; corolla white, 4–7 cm long; frequent in fields, thickets, roadsides, and waste ground; throughout; flr. May–early Aug.

Convolvulus L.

Convolvulus arvensis L.   Field bindweed

Perennial with trailing or twining stems to 1 m long and becoming matted; leaves variable, 2–5 cm long, with long petioles, ovate to oblong, usually hastate or sagittate at the base; peduncles longer than the subtending leaves, 1- or 2-flowered; corolla white or pinkish, 1.5–2.5 cm long, funnel-shaped; style 1, stigmas 2; a weed of fields, roadsides, railroad banks, and waste ground; flr. May–early Oct; native to Europe.

Ipomoea L.

Trailing, creeping, or twining herbs; leaves deeply cordate at the base, entire or 3–5-lobed; bracts lacking below the calyx; the outer sepals larger than the inner; flowers funnel-shaped to nearly bell-shaped, or with a slender tube and abruptly spreading limb; style 1; stigma 2- or 3-lobed; capsule rounded, seeds 4–6.

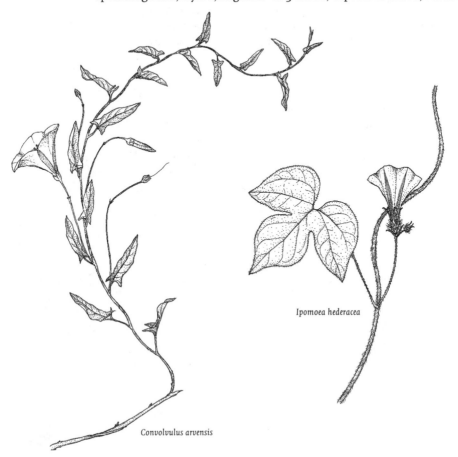

Ipomoea hederacea

Convolvulus arvensis

A. corolla blue, purple, pink, or white, funnel- or bell-shaped
    B. leaves deeply lobed
        C. stigma 3-lobed; ovary and capsule 3-celled ....*I. hederacea*
        C. stigma unlobed or 2-lobed; ovary and capsule 2-celled .... *I. lacunosa*
    B. leaves entire or with a few coarse teeth
        D. stems creeping, but not twining ....*I. pes-cáprae*
        D. stems twining
            E. stigma 3-lobed; ovary 3-celled .... *I. purpurea*
            E. stigma 2-lobed; ovary 2-celled
                F. corolla 5–8 cm; sepals 1.3–2 cm, glabrous .... *I. pandurata*
                F. corolla 1–2 cm; sepals 1–1.5 cm, ciliate .... *I. lacunosa*
A. corolla red with a narrow tube and abruptly flaring limb (salverform)
    G. leaves cordate, entire or shallowly lobed .... *I. coccinea*
    G. leaves deeply pinnately divided .... *I. quamoclit*

Ipomoea coccinea L.   Red morning-glory
Glabrous to finely pubescent annual with stems 1–3 m long; leaves broadly ovate and deeply cordate at the base; flowers few, on long axillary peduncles; corolla red, 2–4 cm long with a slender tube and abruptly spreading limb; stamens and style slightly longer than the corolla tube; capsule 4-celled; rare on roadsides, waste places, fencerows, and old fields; SE; flr. Aug–Oct; native to tropical America.

Ipomoea hederacea Jacq.   Ivy-leaved morning-glory
Hairy annual with slender stems 1–2 m long; leaves deeply 3-lobed, cordate; peduncles 1–3-flowered; flowers funnel-shaped, 3–5 cm long; corolla blue or purplish with a white tube; stigma 3-lobed; ovary 3-celled; cultivated fields, gardens, shale barrens, and ballast; mostly SE and SC; flr. late Jun–Oct; native farther south.

Ipomoea lacunosa L.   White morning-glory
Glabrous or sparsely hairy annual with stems 1–3 m long; leaves ovate, 3–8 cm long, and deeply cordate; peduncles shorter than the subtending petioles, 1–5-flowered; corolla white (or pink or pale purple), 1–2 cm long; ovary 2-celled; rare in fields and along stream banks and railroad tracks; S; flr. Aug–Oct; FACW.

Ipomoea pandurata (L.) G.Mey.   Man-of-the-earth, wild potato-vine
Perennial from a deep tuber-like root; stems glabrous, to 5 m long; leaves ovate and deeply cordate; peduncles usually longer than the petioles, bearing 1–7 flowers in a terminal cluster; corolla white with a red-purple center, 5–8 cm; ovary 2-celled; occasional in calcareous uplands, thickets, and roadsides; S; flr. Jul–early Oct; FACU.

Ipomoea pes-cáprae (L.) Roth   Goat-foot morning-glory, railroad vine
Glabrous and somewhat succulent perennial with creeping stems; leaves suborbicular, notched at the apex and rounded or cordate at the base; corolla purple, 4–5 cm long; rare in wet, open sandy soil and beaches; mostly W; native to the tropics; FAC.

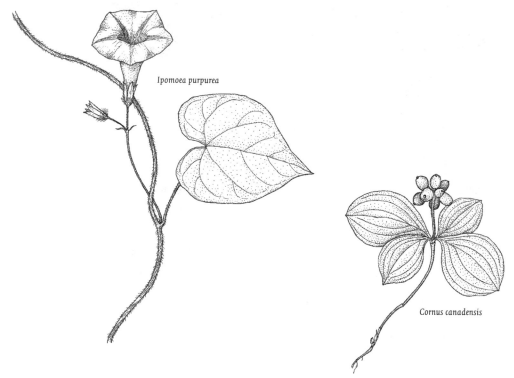

Ipomoea purpurea

Cornus canadensis

Ipomoea purpurea (L.) Roth   Common morning-glory

Hairy annual with stems to 5 m long; leaves broadly ovate to heart-shaped with a deeply cordate base; peduncles as long as the adjacent leaves; flowers blue, purplish, or white, 4–7 cm long, funnel-shaped; stigma 3-lobed; fruit a 3-celled capsule; widely cultivated and occasionally escaped to hillsides, fields, and roadsides; flr. Aug–Oct; native to tropical America; UPL.

Ipomoea quamoclit L.   Cypress-vine

Glabrous annual with slender stems 1–5 m long; leaves pinnately divided nearly to the midrib into narrow lobes; peduncles longer than the subtending petioles; corolla red with narrow tube and abruptly flaring limb; stamens and style slightly longer than the corolla tube; fruit 4-celled; rare in gardens, streets, and waste ground; flr. Jul–Oct; native to tropical America; UPL.

CORNACEAE Dogwood Family

Deciduous trees or shrubs (except *C. canadensis*, which is herbaceous) with opposite, alternate, or whorled simple leaves lacking stipules; flowers perfect or unisexual, 4- or 5-merous; calyx lobes very small; petals distinct; stamens as many as petals; ovary inferior; fruit a drupe.

A. pith solid; leaves mostly opposite (rarely alternate or whorled), entire, the lateral veins curving and paralleling the leaf margin; flowers perfect, 4-merous. ....*Cornus*

A. pith diaphragmed; leaves alternate, occasionally with a few scattered teeth or lobes, lateral veins not curving; flowers unisexual, 5-merous ....*Nyssa*

Cornus L.

Trees, shrubs, or herbs with opposite, whorled, or alternate entire leaves with characteristic pinnate veins that curve to conform with the leaf margin before anastomosing; flowers small, perfect, 4-merous, clustered in compact to open inflorescences, which are sometimes surrounded by 4 showy white bracts; fruits red, blue, white, or black drupes.

A. leaves alternate, but clustered toward the branch tips .... *C. alternifolia*
A. leaves opposite or whorled
    B. flowers in a compact cluster surrounded by 4 showy white bracts; fruit red
        C. small tree .... *C. florida*
        C. herbs from woody rhizomes .... *C. canadensis*
    B. flowers in an open, flat-topped to strongly convex cyme without showy subtending bracts; fruit blue or white
        D. sepals 1–2 mm; pith of 2-year-old stems brown; fruit blue ....
        ....*C. amomum*
        D. sepals <1 mm; pith tan or white; fruit white, black, or blue
           E. twigs dark blood-red; fruit white .... *C. sericea*
           E. twigs gray, yellowish-green, olive, or pinkish-brown; fruit blue or white
               G. leaves with 3–4 pairs of lateral veins; fruit white
                  H. leaves thinly strigose to subglabrous above and below; inflorescence distinctly convex .... *C. racemosa*
                  H. leaves scabrous above, strigose above and below, or loosely short-hairy below; inflorescence flat-topped ..... *C. drummondii*
               G. leaves with 6–8 pairs of lateral veins; fruit blue .... *C. rugosa*

Cornus alternifolia L.f.   Alternate-leaved dogwood, pagoda dogwood
Shrub or small tree to 6 m tall, pith white; leaves alternate, crowded at branch tips; flowers small, white, in hemispherical inflorescences 6–10 cm wide; fruit blue-black, on red pedicels; frequent in low moist woods and shaded ravines; throughout; flr. late May–early Jun; frt Jul–Aug.

Cornus amomum Mill.   Kinnikinik, red-willow
Shrub to 3 m tall with dark maroon-red twigs and dark brown pith; inflorescence a flat cluster of small white flowers; fruit blue, 6–9 mm; swamps, stream banks, moist woods, fields, and thickets; flr. late Jun–early Jul, frt. Sep; FACW; 2 subspecies:

Cornus amomum, inflorescence

A. leaves with 4–6 veins pairs of veins, lower surface greenish and nonpapillose with appressed and loosely spreading whitish or rusty hairs along the veins....
.... ssp. *amomum*
common throughout.
A. leaves with 3–5 pairs of veins, lower surface whitish papillose with appressed whitish hairs along the veins .... ssp. *obliqua* (Raf.) J.S.Wilson
mostly W.

Cornus canadensis L.   Bunchberry, dwarf cornel
Erect herbaceous plant 1–2 dm tall from a woody rhizome; leaves in a whorl of 4–6 at the summit and paired below; flowers in a terminal cluster surrounded by 4 showy, white bracts; fruits red, clustered; occasional in cool, damp woods,

Cornus canadensis, flower

bogs, and swamp edges; N and at high elevations along the Allegheny Front; flr. Jun, frt. late Jul–Aug; FAC–.

Cornus drummondii C.A.Mey.   Roughleaf dogwood
Shrub to 6 m tall; pith white or tan; leaves with 3–4 pairs of lateral veins, scabrous and rough to the touch above, papillose-whitened below with appressed or spreading hairs; inflorescence a flat cluster of small white flowers; fruit blue, 4–8 mm, borne on red pedicels; known from only a single site on disturbed, moist soil in Delaware Co.; flr. early Jun, frt. Aug; native farther west, adventive in Pennsylvania; FAC.

Cornus florida L.   Flowering dogwood
Tree to 10 m tall with widely spreading, horizontal branches; leaves opposite; flowers small, yellowish-green, clustered and surrounded by 4 showy, white bracts; fruit ellipsoid, bright red, 1–1.5 cm; common in mesic woods, edges, and old fields; throughout, except in the northernmost counties; flr. early May, before the leaves, frt. Aug–Sep; FACU–.

Cornus racemosa Lam.   Silky dogwood
Erect, rhizomatous shrub, 1–5 m tall, often forming extensive thickets; twigs

Cornus racemosa

Cornus florida

gray-brown with tan or white pith; leaves opposite, with 3–4 pairs of lateral veins; flowers small, whitish, in strongly convex paniculate cymes; fruit white, 5–8 mm, on red pedicels; common in swampy meadows, moist old fields; and thickets; throughout; flr. Jun–early Jul, frt. Aug–Sep; FAC–.

Cornus rugosa Lam. Round-leaved dogwood
Shrub or small tree, 1–4 m tall; twigs yellowish-green, sometimes with red blotches, pith white; leaves opposite with 6–8 pairs of lateral veins; flowers white in a flat or slightly convex inflorescence; fruit light blue, 6 mm; occasional in well-drained rocky woods and cliffs; E and C; flr. Jun, frt. Oct.

Cornus sericea L. Red-osier dogwood
Shrub, 1–3 m tall, often forming thickets; twigs bright red, pith white; leaves opposite, with 5–7 pairs of veins; flowers small, white in a flat to slightly convex inflorescence; fruit white; infrequent in swamps, moist fields, and thickets; scattered; flr. May, frt. Jul; FACW+.

Nyssa L.

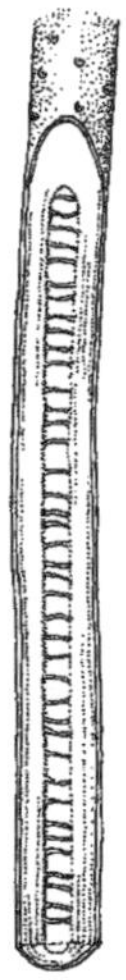
Nyssa sylvatica, diaphragmed pith

Nyssa sylvatica Marshall Sourgum, blackgum, tupelo
Deciduous tree to 30 m tall with strongly horizontal branches, flowers and most leaves clustered on short lateral spurs; pith diaphragmed; leaves obovate to elliptic, 3–10 cm long and 2–6 cm wide; leaf margins entire or with a few coarse, angular teeth; plants dioecious or polygamodioecious; staminate flowers in short, umbel-like racemes on 1–3-cm peduncles and consisting of 5 minute sepals, 5–8 somewhat fleshy, deciduous petals, and 5–12 stamens; pistillate flowers in 2s or 3s, clustered at the end of a peduncle and often including 5–10 functional or sterile anthers; fruit a dark blue drupe 1–1.2 cm long; common in dry to moist woods, rocky slopes, and ridge tops; throughout except in the northernmost counties; flr. May; frt. Jul–Sep; FAC.

CRASSULACEAE Stonecrop Family

Succulent, low-growing herbs; leaves fleshy, simple, opposite, alternate or whorled; flowers regular, 4- or 5-merous; sepals and petals distinct or united at the base; stamens as many or twice as many as the petals; ovary superior, carpels distinct or nearly so; fruit a cluster of follicles.

A. stamens as many as the petals; flowers axillary .... *Crassula*
A. stamens twice as many as the petals; flowers in terminal cymes .... *Sedum*

Crassula L.

Crassula aquatica (L.) Schönland Water-pigmyweed
Upright annual, stems 2–10 cm tall, branched from the base, rooting at the lower nodes; leaves linear, opposite, and connate; flowers axillary, greenish-white, 1 mm wide, 3- or 4-merous; rare on freshwater intertidal mudflats; SE; believed to be extirpated; OBL; ✿.

Sedum L.

Succulent, erect or creeping perennial herbs; leaves alternate, opposite or whorled; leaf margins entire, crenate or toothed; flowers in a terminal cyme; calyx 4- or 5-lobed; petals 4 or 5, distinct or barely united at the base; stamens 8–10; carpels 4 or 5, separate or united at the base; fruit a cluster of follicles.

A. plants creeping, rhizomatous or rooting at the nodes; the flowering branches erect, <2 dm tall
  B. leaves in whorls of 3
    C. leaves flattened, obovate; flowers 4-merous .... *S. ternatum*
    C. leaves terete or only slightly flattened, elliptical; flowers 5-merous ......
    .... *S. sarmentosum*
  B. leaves alternate ....*S. spurium*
    D. leaves linear-oblong to subglobose; flowers white .... *S. album*
    D. leaves ovate; flowers yellow .... *S. acre*
A. stems erect, clustered from a crown, mostly 2–6 dm tall
  E. leaves strongly reduced in size upward ....*S. telephium*
  E. leaves not strongly reduced upward
    F. leaves oblanceolate, broadest above the middle .... *S. kamtschaticum*
    F. leaves broadest at or below the middle
      G. margins of leaves entire
        H. flowers yellow, 4-merous; plants spring blooming...........*S. rosea*
        H. flowers pink, 5-merous; plants fall blooming .... *S. spectabile*
      G. margins of leaves toothed or crenate
        I. stems, branches, and pedicels narrowly winged ..... *S. telephioides*
        I. stems and branches not winged
          J. flowers yellow, 4-merous ....*S. rosea*
          J. flowers pink, 5-merous
            K. stamens distinctly longer than the petals; follicles fused at the base, not stipitate .... *S. spectabile*
            K. stamens not longer than the petals; follicles distinct, each with a short stipe ....*S. alboroseum*

Sedum acre L.   Love-entangle, mossy stonecrop
Evergreen, creeping, mat-forming with erect flowering stems to 5–10 cm tall; leaves yellow-green, alternate, crowded, overlapping, ovate-terete; inflorescence branched; flowers 5-merous; petals yellow; cultivated and frequently naturalized on cliffs, roadsides, and waste ground; SE and SC; flr. Jun; native to Eurasia.

Sedum x alboroseum Baker   Garden orpine
Stems upright, clustered, to 6 dm tall, glaucous; leaves alternate or opposite, elliptic, coarsely dentate toward the apex; inflorescence a dense cyme; petals greenish-white; carpels pink; cultivated and occasionally spreading to waste ground; SE; flr. late Jul–Oct. [syn: S. x erythostictum Miq.]

Sedum album L.   White orpine or stonecrop
Mat-forming from creeping stems with erect flowering stems 1–2 dm tall; leaves alternate, linear-oblong to subglobose, terete or subterete; inflorescence much branched; flowers 5-merous; petals white; occasionally spreading from cultivation; SE; flr. Jun; native to Eurasia.

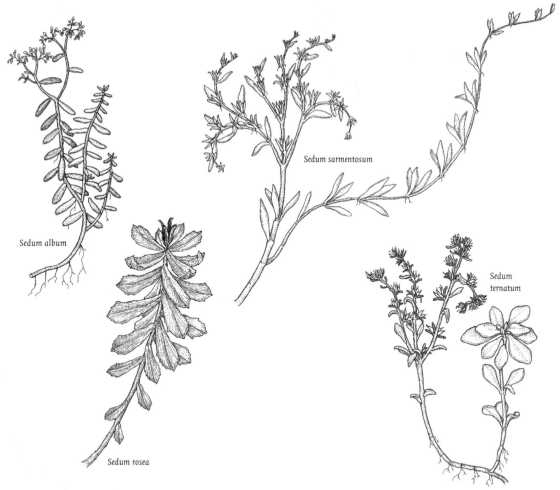

Sedum sarmentosum

Sedum album

Sedum rosea

Sedum
ternatum

Sedum kamtschaticum Fisch. & B.Mey.   Orange stonecrop
Stems to 3 dm tall, decumbent; leaves alternate, spatulate or linear-oblanceolate, toothed toward the tip; inflorescence a dense or lax cyme; flowers yellow; cultivated and occasionally escaped; SE; flr. Jun; native to Asia. Ours is ssp. *ellacombianum* (Praeger) R.T.Clausen.

Sedum rosea (L.) Scop.   Roseroot stonecrop
Stems arising from a thick, scaly rhizome, erect, 1–4 dm tall; leaves sessile, glaucous, oblanceolate or obovate, 2–4 cm, entire or toothed at the tip; dioecious, inflorescence compact; flowers mostly 4-merous; petals yellow; follicles fused except for the short, spreading beaks; very rare on moist cliffs and ledges; E; flr. late May; FACU–; 🌿.

Sedum sarmentosum Bunge   Orpine
Mat-forming with long, creeping stems rooting at the nodes, flowering stems erect to about 1 dm tall; leaves pale green, mostly in whorls of 3, elliptical, thick but somewhat flattened; inflorescence branched; flowers 5-merous; petals yellow, forming a short tube at the base to which the stamens are attached; cultivated and frequently escaped to dry, rocky soils, rock crevices, and roadsides; SE and SC; flr. early Jun; native to China.

Sedum spectabile Boreau
Orpine
Stems clustered, upright, to 5 dm tall, unbranched; leaves alternate, opposite, or in whorls of 3, elliptical, coarsely dentate, glaucous; inflorescence a large dense cyme; flowers pink with prominent exert stamens; cultivated and occasionally persisting in waste ground; scattered; flr. Aug–Oct; native to Japan.

Sedum spurium M.Bieb.
Orpine
Creeping with erect stem tips; leaves opposite, thick and flat with coarsely crenate, glandular margins, obovate or obovate-oblong, tapered at the base; inflorescence compactly branched; flowers mostly 5-merous; petals pink or white; cultivated and occasionally established on roadsides or waste ground; E; flr. Jul; native to Eurasia.

Sedum telephioides Michx.
Allegheny stonecrop
Stems 2–6 dm tall, clustered; leaves alternate or opposite, elliptic, narrowed to the base; 3–8 cm long by 1.5–3.5 cm wide with a few coarse teeth, glaucous; inflorescence much branched, the branchlets narrowly winged; flowers mostly 5-merous; petals white to pale pink; follicles with a short stalk; rare in rocky woods, dry cliffs, ledges, and shale barrens; SC; flr. Aug–early Sep; ✺.

Sedum telephium L.
Garden orpine, live-forever
Stems clustered, erect, to 5 dm tall; leaves alternate, ascending, oblong or ovate-oblong, toothed toward the apex; inflorescence dense; flowers red-purple; stamens not exceeding the petals; cultivated and frequently escaped to woods, old fields, and moist ditches; throughout; native to Eurasia.

Sedum ternatum Michx.
Wild stonecrop
Fibrous-rooted with creeping stems rooting at the nodes; leaves obovate, 1–2 cm long, entire; mostly whorled and forming a rosette at the ends of sterile shoots; inflorescence with 2–4 strongly secund branches; flowers 4-merous; petals white; occasional in rocky banks, cliffs, and woods; S; flr. May.

CUCURBITACEAE Gourd Family

Annual, herbaceous vines climbing by coiling tendrils, monoecious; leaves alternate, simple, usually lobed, and lacking stipules; flowers unisexual, regular; sepals 5, united at the base; petals 5 or 6, united; stamens 5; ovary inferior, of 3 united carpels; fruit a berry, pepo, or capsule.

A. flowers yellow; ovaries and fruits smooth .... *Cucurbita*
A. flowers white or greenish-yellow; ovaries and fruits prickly
    B. staminate corolla 6-lobed; fruits not clustered, each a 4-seeded capsule opening by 2 pores .... *Echinocystis*
    B. staminate corolla 5-lobed; fruits clustered, 1-seeded, indehiscent .... *Sicyos*

Cucurbita L.

Cucurbita pepo L.
Pumpkin, squash
Stems rough, trailing; leaves triangular-ovate; flowers 5–10 cm wide; corolla

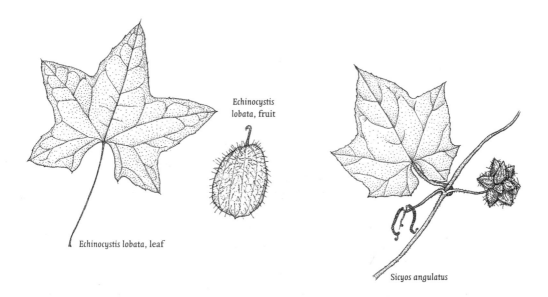

Echinocystis
lobata, fruit

Echinocystis lobata, leaf

Sicyos angulatus

campanulate, yellow, 5-lobed; fruit a pepo; cultivated and occasionally occurring in waste ground; long cultivated from American origins.

Echinocystis Torr. & A.Gray

Echinocystis lobata (Michx.) Torr. & A.Gray   Prickly cucumber, wild balsam-apple
Stem 5–8 m, angular or grooved with branching tendrils; leaves 5-lobed; staminate flowers in elongate, upright racemes, 6-merous; pistillate flowers solitary or in small clusters on a short peduncle; fruit an inflated, 3–5-cm-long membranous capsule opening by 2 pores; common in moist alluvial soil, stream banks, and woods edges; throughout; flr. Jul–Sep; FAC.

Sicyos L.

Sicyos angulatus L.   Bur cucumber
Stem to 8 m, angular and somewhat sticky-pubescent; leaves shallowly 3–5-lobed; staminate flowers in short racemes, 5-merous; pistillate flowers in head-like clusters; fruits in clusters of 3–10, approximately 15 mm long, indehiscent; frequent in moist open soil, stream banks, roadsides, and waste ground; mostly S, but scattered elsewhere; flr. Jul–late Aug; FACU.

CUSCUTACEAE Dodder Family

Parasitic, nonphotosynthetic, annual herbs with twining yellowish or orange stems attached to the host plant by penetrating projections (haustoria); leaves alternate, reduced to tiny bracts; flowers in dense lateral clusters, whitish, small, regular, perfect; calyx 4- or 5-lobed or with sepals distinct, persistent in fruit; corolla 4- or 5-lobed, with small fringed scales at the sinuses just below the 4 or 5 stamens; ovary superior, 2-celled with 2 distinct styles; fruit a small rounded capsule; seeds 1–4.

Cuscuta L.

Characteristics of the family; the numbers of perianth parts not always consistent, several flowers should be examined; the more common species make a conspicuous orange mass of filiform stems sprawled over wetland vegetation in late summer; flr. Jun–Sep, frt. Jul–Oct.

A. stigmas slender and elongate; fruit splitting along a ring near the base ....
.... C. epithymum
A. stigmas capitate; fruit splitting irregularly
    B. each flower subtended by 1 or more bracts; sepals distinct .... C. compacta
    B. individual flowers without bracts; sepals united at the base
        C. flowers mostly 5-merous
            D. corolla lobes triangular, acute
                E. calyx knobby below the sinuses; perianth 1.5–2 mm long ....
                .... C. pentagona
                E. calyx smooth; perianth 2–2.5 mm long .... C. campestris
            D. corolla lobes broadly rounded .... C. gronovii
        C. flowers mostly 4-merous
            F. corolla lobes acute
                G. mature styles <1 mm; tip of corolla lobes erect or ascending; calyx and corolla smooth .... C. polygonorum
                G. mature styles about 1 mm or longer; tip of corolla lobes inflexed; calyx and corolla papillate.. C. corylii
            F. corolla lobes obtuse or rounded............................... C. cephalanthii

Cuscuta campestris Yunck.   Dodder
Very similar to C. pentagona, but with calyx smooth and the perianth 2–2.5 mm long; rare in thickets and waste ground, parasitic on various wild and cultivated hosts including alfalfa and clover; mostly SE; 🌿.

Cuscuta cephalanthii Engelm.   Buttonbush dodder
Flowers mostly 4-merous, sessile or subsessile in compact clusters; calyx shorter than the corolla tube; corolla cylindric-campanulate, the erect or spreading lobes shorter than the tube; fruit 3 mm, depressed-globose; seeds 1 or 2; rare in swamps and moist thickets, parasitic on a variety of shrubs and herbs; S; 🌿.

Cuscuta compacta Juss. ex Choisy   Dodder
Flowers 5-merous, sessile in small, dense clusters, each with 3–5 bracts below the calyx; fruit ovoid to globose, 3–5 mm thick, the withered corolla clinging to the top; styles not persistent; rare in moist thickets and stream banks, parasitic on a variety of shrubs and herbs; SE; 🌿.

Cuscuta corylii Engelm.   Hazel dodder
Flowers mostly 4-merous, in dense or loose clusters, distinctly pediceled; calyx half as long as the corolla and narrowly cylindric, its lobes triangular with acute, inflexed tips; fruit 2–2.5 mm, slightly thickened at the top; rare in dry rocky woods, clearings, and hillsides, parasitic on various shrubs and herbs; scattered; 🌿.

Cuscuta epithymum (L.) L.   Clover dodder
Flowers 5-merous, subsessile in dense clusters; calyx lobes triangular-ovate,

shorter than the corolla; corolla lobes triangular, shorter than the tube, acute and spreading; fruit 1.5 mm, capped by the withered corolla; rare in old fields and woods edges, parasitic on clover and other legumes; native to Europe; designated as a federal noxious weed.

Cuscuta gronovii Willd. ex Schultz   Common dodder

Flowers 5-merous, sessile or subsessile in dense clusters; calyx scarcely reaching the middle of the corolla tube; scales heavily fringed, reaching the sinuses of the corolla; fruit globose to depressed-globose, cupped below by the withered corolla, somewhat glandular-warty toward the summit and with a thickened ring around the base of the styles; seeds 2–4; common in low wet areas throughout, parasitic on a wide range of woody and herbaceous hosts. Var. *latiflora* Engelm., with a longer calyx, is limited to a few sites in the SE.

Cuscuta gronovii,
flower (×5)

Cuscuta pentagona Engelm.   Field dodder

Flowers 5-merous, sessile or with short pedicels; calyx about as long as the corolla tube, its lobes broader than long; corolla lobes 1 mm, lanceolate to triangular, spreading to reflexed; fruit depressed-globose, thin-walled; rare in old fields, thickets, and wet banks, parasitic on many hosts; S; 🌿.

Cuscuta
pentagona,
flower (×5)

Cuscuta polygonorum Engelm.   Smartweed dodder

Flowers mostly 4-merous, in dense or loose clusters; calyx as long as or longer than the corolla tube; corolla lobes longer than the tube, erect or ascending; style <1 mm; fruit depressed-globose, cupped at the base by the persistent corolla; seeds 2–4; rare on moist shores and riverbanks, parasitic on *Polygonum* and other hosts; scattered; 🌿.

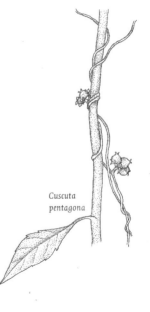

Cuscuta
pentagona

DIPSACACEAE Teasel Family

Perennial herbs with opposite leaves lacking stipules; inflorescence a dense head surrounded by an involucre and with bracts and/or hairs present on the receptacle between the flowers; flowers perfect; calyx with 4 or 5 segments and more numerous teeth or bristles; petals united; corolla 4- or 5-lobed, regular to irregular; stamens 2 or 4; ovary inferior; fruit dry, indehiscent.

A. stem, and usually the leaves and involucre, prickly .... *Dipsacus*
A. stem not prickly
    B. calyx teeth 4 or 5; receptacle bearing bracts but not hairy .... *Succisella*
    B. calyx teeth 8 or more; receptacle hairy but lacking bracts .... *Knautia*

Dipsacus L.

Tall biennials to 2 m tall; prickly on the angles of the stem and leaf midribs; leaves opposite, sessile or even fused at the base; inflorescence ovoid to subcylindric, 3–10 cm long, surrounded by narrow, spreading, or up-curved involucral bracts the longest of which exceed the head; flowers pale lavender to white; 4-merous; flr. Jul–Aug; native to Europe, brought by the earliest settlers for use in the preparation of wool for spinning.

A. stem leaves entire to crenate
    B. bracts of the receptacle ending in a stout, straight awn .... *D. sylvestris*

B. bracts of the receptacle with strongly recurving spine tips .... D. sativus
A. stem leaves more or less deeply laciniate-pinnatifid .... D. laciniatus

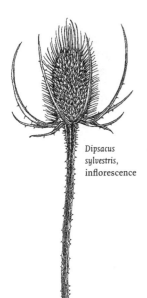

Dipsacus sylvestris, inflorescence

Dipsacus laciniatus L.   Cut-leaved teasel
Similar to *D. sylvestris* but with irregularly laciniate-pinnatifid leaves, rare and scattered on roadsides and waste ground.

Dipsacus sativus (L.) Honck.   Fuller's teasel
Similar to *D. sylvestris*, distinguished by the spiny, recurved tips of the bracts of the receptacle; rare and scattered on roadsides and waste ground. [syn: *D. fullonum* sensu Mill., non L.]

Dipsacus sylvestris Huds.   Common teasel
Bracts on the receptacle straight, awn-tipped; roadsides, fields, and waste ground; common throughout; FACU–. [syn: *D. fullonum* L.]

Knautia L.

Knautia arvensis (L.) Duby   Bluebutton
Hairy perennial with dense, hemispherical heads on long peduncles; leaves coarsely toothed to deeply pinnatifid; flowers lilac-purple, in heads 1.5–4 cm thick; receptacle densely hairy between the flowers; involucral bracts about equaling the head; infrequent in fields and roadsides, mostly E; flr. late Jun; native to Europe.

Succisella Beck

Succisella inflexa (Kluk) Beck   Devil's-bit
Branched, subglabrous perennial 4–10 dm tall; leaves lance-linear, reduced on the branches; inflorescence 1–1.5 cm thick; corolla pale blue; calyx 4-lobed, lacking awns; fruit glabrous, 8-ribbed, longer than the receptacular bracts; rare in moist meadows and pond margins; NW; flr. Jul; native to Europe. [syn: *Scabiosa australis* (Wulfen) Rchb.f.]

DROSERACEAE Sundew Family

Insectivorous, perennial herbs; leaves alternate or basal, petiolate, covered with sticky-tipped, glandular hairs; inflorescence an erect raceme that is nodding at the tip when not yet fully expanded, flowers perfect, 5-merous; petals distinct; stamens same number as the petals; ovary superior with 3 distinct styles, each 2-lobed nearly to the base; fruit a capsule; flr. late Jun–Aug.

Drosera rotundifolia

Drosera L.

A. leaves suborbicular, broader than long .... D. rotundifolia
A. leaves spatulate, longer than broad .... D. intermedia

Drosera intermedia Hayne   Spatulate-leaved sundew

Leaves in a basal rosette or alternate on a somewhat elongated stem; scape 2–25 cm long with 1–20 flowers; petals white; rare on open peat at the edges of bogs and glacial lakes; E; OBL.

Drosera intermedia, leaf

Drosera rotundifolia L.   Round-leaved sundew

Leaves in a basal rosette; scape 5–30 cm long with 2–25 flowers; petals white or pink; occasional in sphagnum bogs and peaty edges; throughout; OBL.

EBENACEAE Ebony Family

Deciduous trees with alternate, simple leaves lacking stipules; dioecious; flowers 4-merous, regular; stamens inserted on the corolla tube, 16 in staminate flowers; pistillate flowers with up to 8 sterile stamens; ovary superior; fruit a fleshy berry with up to 8 large seeds.

Diospyros

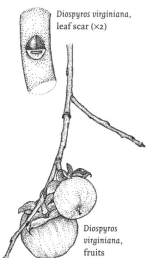

Diospyros virginiana, leaf scar (×2)

Diospyros virginiana L.   Persimmon

Tree to 15 m tall with bark broken in regular small squares; leaves oval or oblong, petiolate; blades entire, 8–15 cm long; flowers greenish-yellow; corolla urn-shaped, petals united at the base; fruit 2–4 cm, orange, edible after frost; occasional in open woods, floodplains, and old fields; S; flr. May, frt. Oct–Nov; FAC–.

Diospyros virginiana, fruits

ELAEAGNACEAE Oleaster Family

Deciduous shrubs covered with silvery and/or brown scales on most surfaces; leaves alternate or opposite, simple, entire, without stipules; flowers perfect or unisexual, perigynous, lacking petals; hypanthium tubular, 4-lobed, and somewhat petaloid; fruit drupe- or berry-like (actually an achene enclosed by the fleshy hypanthium).

A. leaves alternate; flowers perfect or some staminate; stamens 4 .... *Elaeagnus*
A. leaves opposite; flowers unisexual; plants dioecious; stamens 8 .... *Shepherdia*

Elaeagnus L.

Deciduous, frequently spiny shrubs or small trees with alternate, simple leaves; flowers perfect (or staminate) in lateral clusters; fruits drupe-like, red or yellow.

A. leaves covered with silvery scales on both surfaces; fruit 1 cm, yellow, mealy...
.... *E. angustifolia*
A. leaves becoming green and glabrescent above; fruit 6–8 mm, red, juicy ....
.... *E. umbellata*

Elaeagnus angustifolia L.   Russian-olive

Deciduous shrub or small tree to 7 m tall; leaves, twigs, and other parts densely covered with silvery scales; flowers about 1 cm long, yellow within, silvery-scaly on the outside, fragrant; fruit ellipsoid, 1 cm, yellow with silvery scales; introduced and occasionally escaped to waste ground and old fields; mostly S; flr. May, frt. Sep; native to Eurasia; FACU.

Elaeagnus umbellata Thunb.   Autumn-olive

Deciduous shrub to 5 m tall; leaves becoming green and glabrescent above, silvery-scaly beneath; flowers about 1.2 cm long, yellowish-white, scaly on the outside, fragrant; fruit subglobose to ovoid, 6–8 mm, red with scales; extensively naturalized in old fields, abandoned pastures, and other open ground; mostly S, a serious weed in some parts of the state; flr. late May, frt. Sep–Nov; native to Asia.

Shepherdia Nutt.

Shepherdia canadensis (L.) Nutt.   Buffalo-berry, soapberry

Deciduous shrub to 2.5 m tall, unarmed, dioecious; leaves opposite, ovate to elliptic, 2–5 cm long, nearly glabrous above, the lower surface densely covered with silvery scales mixed with scattered brown ones; flowers yellow, 4 mm across; fruit ovoid, 4–6 mm, yellowish-red; rare on wet, shaly banks and slumps along Lake Erie; NW; flr. late Jun; frt. Jul; UPL; 🌿.

Elaeagnus umbellata

Shepherdia canadensis

ELATINACEAE Waterwort Family

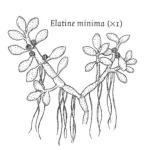

Elatine minima (×1)

Small, submersed aquatic annuals with simple, opposite leaves; flowers perfect, regular; sepals, petals, and stamens each 2 or 3; ovary superior, 2- or 3-locular; fruit a capsule with several elongate seeds the surfaces of which are decorated with ridges enclosing rounded or angular areolae; flr. Jul–Aug.

Elatine L.

A. flowers mostly 2-merous; areolae of the seeds rounded .... E. minima
A. flowers mostly 3-merous; areolae of the seeds angular at the ends ....
.... E. americana

Elatine americana,
seed (×40)

Elatine americana (Pursh) Arn.   American waterwort
Creeping or floating; leaves linear to obovate, 3–8 mm long; flowers mostly 3-merous; seeds marked with angular, 6-sided areolae; rare on muddy tidal shores; SE; believed to be extirpated; OBL; ✺. [syn: E. triandra Schkuhr]

Elatine minima,
seed (×20)

Elatine minima (Nutt.) Fisch. & C.A.Mey.   Small waterwort
Creeping with erect branches to 5 cm tall; leaves oblong to narrowly obovate, not over 4 mm long; flowers 2-merous; seeds with narrow, rounded areolae; rare in shallow water of northern lakes and muddy shores; OBL; ✺.

ERICACEAE Heath Family
John Kunsman

Deciduous or evergreen shrubs or small trees, typically growing in acidic soils; leaves simple, alternate, opposite or whorled; flowers perfect, mostly regular, 4- or 5-merous; corolla of united or separate petals; ovary superior or inferior; fruit a capsule or berry.

A. leaves evergreen
  B. leaves opposite or whorled
    C. leaves often >1 cm long and 8 mm wide; corolla united; fruit 5-celled....
    .... Kalmia
    C. leaves mostly <1 cm long and 8 mm wide; corolla of separate petals;
      fruit 2–3(5)-celled .... Leiophyllum
  B. leaves alternate
    D. leafy part of the stem trailing, creeping, or prostrate
      E. leaves with cordate bases and petioles 5 mm or more; corolla 10 mm
        or more long; fruit a capsule .... Epigaea
      E. leaves with acute, obtuse, or rounded bases and petioles <5 mm; co-
        rolla <10 mm long; fruit a berry
        F. leaves widest at or below the middle, mostly <7 mm wide; flower
          parts in 4s; seeds >10
          G. plant with neutral odor when bruised, stem, leaf margins, and
            lower leaf surfaces glabrous or nearly so; corolla pink; fruit
            red .... Vaccinium

G. plant with wintergreen odor when bruised, stem, leaf margins, and lower leaf surfaces with scattered brownish hairs; corolla and fruit white .... *Gaultheria*

F. leaves widest above the middle, often 7 mm or more in width; flower parts in 5s; seeds 5 or 10 .... *Arctostaphylos*

D. leafy part of the stem erect or ascending

H. lower leaf surfaces with dense scales or woolly hairs, or strongly whitened

I. twigs and lower leaf surfaces with brown scales, or strongly whitened; corolla united; capsule as wide as long or wider

J. twigs and lower leaf surfaces with brown scales; inflorescence a raceme, the base of each pedicel with a leaf-like bract ....
.... *Chamaedaphne*

J. twigs and lower leaf surfaces whitened; inflorescence umbel-like, pedicels without a leaf-like bract .... *Andromeda*

I. twigs and lower leaf surfaces with dense woolly hairs; corolla of separate petals; capsule longer than wide .... *Ledum*

H. lower leaf surfaces without scales or dense woolly hairs (scattered hairs may be present), green or pale green

K. leaves entire

L. leaves >8 mm wide; corolla partly to wholly united; capsule 5-celled

M. corolla saucer-like; leaf blades mostly 4–8 cm long; capsule as wide as long or wider .... *Kalmia*

M. corolla bell-shaped, with well-developed lobes; leaf blades mostly 8 cm or more long; capsule longer than wide ....
.... *Rhododendron*

L. leaves <8 mm wide; corolla of separate petals; capsule 2–3(5)-celled .... *Leiophyllum*

K. leaves toothed

N. plant with wintergreen odor when bruised; mature fruit red; twigs rounded .... *Gaultheria*

N. plant with neutral odor when bruised; mature fruit blue; twigs sharply angled .... *Gaylussacia*

A. leaves deciduous

O. trees; leaves with blades often >8 cm long and 4 cm wide, petioles often 7 mm or more .... *Oxydendrum*

O. shrubs; leaves with blades usually <8 cm long and 4 cm wide, petioles <7 mm

P. lower leaf surfaces glandular-dotted

Q. fruit a berry; corolla 3–6(9) mm long; glandular dots usually yellowish or golden .... *Gaylussacia*

Q. fruit a capsule; corolla 7–13 mm long; glandular dots brown or black
.... *Lyonia*

P. lower leaf surfaces sometimes with glandular hairs, but not glandular-dotted

R. branchlets (under magnification) with a warty or pebbly appearance; fruit a berry; ovary inferior .... *Vaccinium*

R. branchlets with a generally smooth surface (hairs may be present), not warty or pebbly; fruit a capsule or berry; ovary superior or inferior

S. flower parts in 4s; capsule 4-celled; midrib of lower leaf surfaces with scattered chaff-like scales in addition to hairs .... *Menziesia*

S. flower parts in 5s; capsule 5-celled, or fruit a berry; midrib of lower leaf surfaces glabrous, or with hairs

  T. corolla >1.3 cm long; capsule >9 mm long; leaf margins usually ciliate; terminal bud present .... *Rhododendron*

  T. corolla 1.3 cm or less; capsule 9 mm or less; leaf margins glabrous or minutely hairy; axillary bud present at the end of twigs

    U. leaves entire

      V. fruit a berry from an inferior ovary; corolla with open bell shape, the stamens exposed; petioles 2–4 mm; lower leaf surfaces usually hairy .... *Vaccinium*

      V. fruit a capsule from a superior ovary; corolla cylindric, the stamens included; petioles 4–6 mm; lower leaf surfaces usually glabrous except on the veins .... *Lyonia*

    U. leaves toothed (sometimes obscurely so in *Lyonia*)

      W. leaf teeth usually blunt; calyx not subtended by a pair of bracts; buds appressed, scales 2(3); corolla 2–5 mm long .... *Lyonia*

      W. leaf teeth relatively sharp; calyx subtended by a pair of bracts; buds divergent, scales 3 or more; corolla 7–9 mm long .... *Leucothoe*

Andromeda L.

Andromeda polifolia L.   Bog-rosemary

Evergreen shrub to 0.5 m tall; twigs whitened; leaves narrowly oblong, entire, revolute, whitened below; inflorescence umbel-like; corolla white or pink, globe-shaped, 5–6 mm long; capsule globe-shaped, 5 mm in diameter, 5-celled; rare in bogs and peaty wetlands; N; flr. May; OBL; ❦. Ours is var. *glaucophylla* (Link) DC.

Andromeda
polifolia

Arctostaphylos Adans.

Arctostaphylos uva-ursi (L.) Spreng.   Bearberry

Prostrate evergreen shrub; bark peeling; leaves spatulate, entire, green, glabrous in age; inflorescence a raceme; corolla white, urn-shaped, 4–6 mm long; berry

Arctostaphylos
uva-ursi, fruit
(×1)

Arctostaphylos
uva-ursi

Chamaedaphne calyculata

red, rather dry, 5–10-seeded; very rare in dry openings and rocky ledges; NE and NW; believed to be extirpated; flr. early May, frt. Aug–Sep; FAC; 🌿. Ours is ssp. *coactilis* Fernald & J.F.Macbr.

Chamaedaphne Moench

Chamaedaphne calyculata (L.) Moench   Leatherleaf
Evergreen shrub to 1 m tall; twigs with brown scales; leaves irregularly toothed, with brown scales below; inflorescence a raceme, the base of each pedicel with a leaf-like bract; corolla white, urn-shaped, 6–7 mm long; capsule globe-shaped, about 3 mm in diameter, 5-celled; frequent in bogs and acidic wetlands; N, and at high elevations along the Allegheny Front; flr. late Apr–early May; OBL. Ours is var. *angustifolia* (Aiton) Rehder.

Epigaea L.

Epigaea repens L.   Trailing-arbutus
Creeping evergreen shrub; twigs bristly; leaves oblong to oval, ciliate, base often heart-shaped, petiole 5 mm or more; flowers clustered, fragrant; corolla white or pink, tubular with spreading lobes, hairy within, 10–20 mm long; capsule globe-shaped; frequent in dry openings, woods borders, and banks; throughout; flr. Apr–May.

Gaultheria L.

Low evergreen shrubs with wintergreen odor when bruised; leaves toothed or entire; flowers regular, solitary in the leaf axils, 4- or 5-merous; fruit a many-seeded berry.

A. leafy stems erect, the leaves mostly >10 mm long and 6 mm wide, toothed; flower parts in 5s; fruit red .... *G. procumbens*
A. leafy stems trailing, the leaves (or most of them) <10 mm long and 6 mm wide, fringed with hairs; flower parts in 4s; fruits white .... *G. hispidula*

Gaultheria hispidula (L.) Bigelow   Creeping snowberry
Creeping evergreen shrub; twigs bristly; leaves ovoid to elliptic, ciliate with scat-

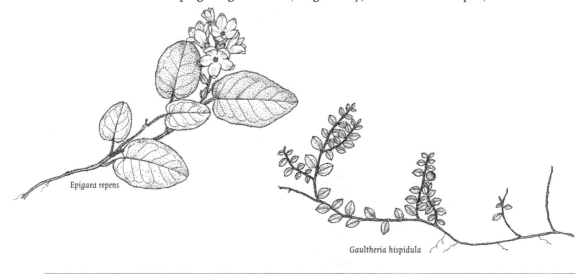

Epigaea repens

Gaultheria hispidula

tered brown hairs below; flower parts in 4s; corolla greenish-white, bell-shaped, 2–3 mm long; berry white; rare in wet woods and bogs; N; flr. Jun, frt. Sep; FACW; 🌿.

Gaultheria procumbens L.   Teaberry
Low evergreen shrub to 20 cm tall; twigs glabrous or minutely pubescent; leaves elliptic to obovate or orbicular, toothed, glabrous or nearly so; flower parts in 5s; corolla white, urn-shaped, 6–12 mm long; berry red; common in dry to wet woods and barrens; throughout; flr. Jul, frt. Sep–spring; FACU.

Gaylussacia Kunth

Deciduous or evergreen shrubs; leaves alternate, entire or toothed, often glandular-dotted; inflorescence a raceme; flower parts in 5s; corolla urn-shaped; ovary inferior; fruit a 10-seeded berry.

A. leaves deciduous, entire or ciliate, the lower surface with glandular dots or globules; twigs rounded or nearly so
  B. upper leaf surfaces of with glandular dots or gland-tipped hairs; mature twigs hairy; fruits black
    C. leaves with rounded tips and hairless or inconspicuously pubescent margins; base of pedicel naked or with a scale .... *G. baccata*
    C. leaves with distinctly cuspidate tips and ciliate margins; base of pedicel with leaf-like bract .... *G. dumosa*
  B. upper leaf surfaces nonglandular and glabrous; mature twigs hairless or nearly so, often somewhat whitened; fruits blue .... *G. frondosa*
A. leaves evergreen, toothed, the lower surface nonglandular; twigs sharply angled .... *G. brachycera*

Gaylussacia baccata (Wangenh.) K.Koch   Black huckleberry
Deciduous shrub to 1 m tall; twigs hairy; leaves elliptic to oblong, entire, glandular-dotted on both surfaces; corolla greenish white to reddish, tubular, 3–6 mm long; berry black; common in dry to wet woods and thickets; throughout; flr. May, frt. Jul–Sep; FACU.

Gaylussacia brachycera (Michx.) A.Gray   Box huckleberry
Colonial, evergreen shrub to 0.5 m tall; twigs angled, glabrous; leaves ovoid to

Gaylussacia brachycera, fruit (×1)

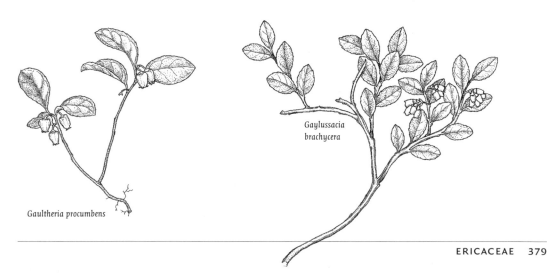
Gaylussacia brachycera

Gaultheria procumbens

Gaylussacia frondosa

Gaylussacia dumosa

elliptic, toothed, glabrous or nearly so; corolla white, tubular, 3–6 mm long; berry blue; rare in dry oak woods; SC; flr. May; frt. Aug; 🌿.

Gaylussacia dumosa (Andr.) Torr. & A.Gray   Dwarf huckleberry
Deciduous shrub to 0.5 m tall; twigs hairy; leaves oblanceolate to obovate, ciliate (even appearing toothed), tip cuspidate, glandular-dotted on both sides; base of pedicel with a leaf-like bract; corolla white, tubular, 5–9 mm long; berry black; very rare in wet woods and thickets; SE; flr. Jun, frt. Jul–Aug; believed to be extirpated; FAC; 🌿.

Gaylussacia frondosa (L.) Torr.   Dangleberry
Deciduous shrub to 2 m; twigs glabrous, often whitened; leaves elliptic to obovate, entire, often somewhat whitened, glandular-dotted below; corolla white, tubular, 3–6 mm long; berries blue, on long, drooping pedicels; frequent in dry to wet woods and thickets; mostly E; flr. May; frt. Jul–Oct; FAC.

Kalmia L.

Evergreen shrubs with alternate, opposite, or whorled, entire leaves; flowers regular, 5-merous, clustered; corolla saucer-shaped, the stamens arching backward into pouches on the corolla; fruit a 5-celled, globose capsule.

A. leaves with well-developed petioles, margins flat and the lower surfaces green or pale; twigs rounded, not whitened
    B. leaves alternate; corolla 15–30 mm wide; capsule 4–7 mm wide .... *K. latifolia*
    B. leaves whorled; corolla 6–14 mm wide; capsule 2.5–4.5 mm wide .... *K. angustifolia*
A. leaves sessile or nearly so, revolute, whitened beneath; twigs 2-edged, whitened .... *K. polifolia*

Kalmia angustifolia L.   Sheep laurel

Evergreen shrub to 1 m tall with rounded twigs; leaves whorled, elliptic to oblong, paler below, margins flat or nearly so; inflorescence axillary; corolla pink to rose-purple, 6–14 mm wide; capsule 2.5–4.5 mm wide; frequent in dry woods, barrens, and bogs; E; flr. Jun–early Jul; FAC.

Kalmia latifolia L.   Mountain laurel

Evergreen shrub to 4 m tall with rounded twigs; leaves alternate, elliptic, green below, margins flat; inflorescence terminal; corolla white to pink, 15–30 mm wide; capsule 4–7 mm wide; common in dry woods and rocky slopes; throughout; flr. May–early Jun; FACU. The state flower of Pennsylvania.

Kalmia polifolia Wangenh.   Bog laurel

Evergreen shrub to 1 m tall with 2-edged, whitened twigs; leaves opposite, narrowly elliptic, strongly whitened below, sessile or nearly so; inflorescence terminal; corolla pink to rose-purple, 12–20 mm wide; capsule 5–7 mm wide; rare in bogs and peaty wetlands; NE; flr. May; OBL.

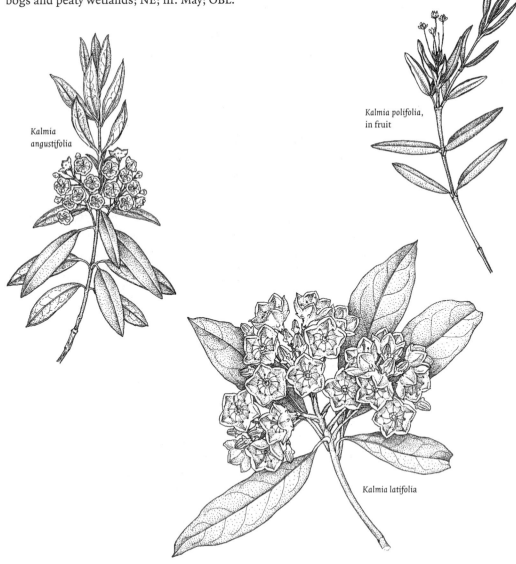

Kalmia angustifolia

Kalmia polifolia, in fruit

Kalmia latifolia

Ledum L.

Ledum groenlandicum, fruit (×2)

Ledum groenlandicum Oeder   Labrador-tea
Evergreen shrub to 1 m tall with densely rusty-woolly twigs; leaves entire, elliptic, revolute, densely white to rusty-woolly beneath; flowers regular, 5-merous,
in terminal clusters; corolla white, of separate petals, 3–6 mm long; capsule
longer than wide, 5-celled, splitting upward from the base; rare in bogs and
peaty wetlands; NE and NW; flr. Jun; OBL; ❧.

Leiophyllum R.Hedw.

Leiophyllum buxifolium (Berg) Elliot   Sand-myrtle
Evergreen shrub to 1 m tall with glabrous twigs; leaves alternate, opposite, or
whorled, ovoid to elliptic, entire, glabrous, green below; inflorescence in terminal clusters; corolla white, of separate petals, 3–4 mm long; capsule ovoid, 3–4
mm long, 2–3(5)-celled; very rare in open rocky places; NE; believed to be extirpated; flr. May; FACU–; ❧.

Leucothoe D.Don

Leucothoe racemosa (L.) A.Gray   Fetter-bush, swamp dog-hobble
Deciduous shrub to 3 m tall; flower buds conspicuous over the winter; leaves
ovoid to elliptic, toothed, hairy on the veins beneath; inflorescence a 1-sided
raceme; calyx subtended by 2 bracts; corolla white, tubular, 7–9 mm long; capsule globe shaped, 4–5 mm wide, 5-celled; rare in wet woods and thickets; SE
and SC; flr. late May–early Jun; FACW; ❧. [syn. *Eubotrys racemosa* (L.) Nutt.]

Lyonia Nutt.

Deciduous shrubs with twigs lacking a true terminal bud; leaves alternate,
toothed or entire; flowers regular, 5-merous, clustered; corolla globe-shaped or
tubular; fruit a 5-celled capsule.

Leiophyllum buxifolium

Ledum groenlandicum

Lyonia mariana

Lyonia ligustrina

Lyonia mariana, fruits

A. leaves toothed; buds appressed, scales 2(3); corolla 2–5 mm long; capsule 2–4 mm long .... L. ligustrina
A. leaves entire; buds divergent, scales 4 or more; corolla 6–13 mm long; capsule 4–6 mm long .... L. mariana

Lyonia ligustrina (L.) DC.   Maleberry
Deciduous shrub to 3 m tall; twigs pubescent or not; buds appressed, scales mostly 2(3); leaves elliptic, variably and often obscurely toothed, usually pubescent below; inflorescence a panicle; corolla white, globe-shaped, 2–5 mm long; capsule globe-shaped, 2–4 mm long; frequent in dry to wet woods and thickets; throughout, except in the northernmost counties; flr. Jun; FACW.

Lyonia mariana (L.) D.Don   Staggerbush
Deciduous shrub to 2 m tall; twigs usually glabrous; buds divergent, scales 4 or more; leaves elliptic, entire, often glandular dotted and glabrous or pubescent on the veins below; flowers in whorled clusters; corolla white, tubular, 6–13 mm long; capsule ovoid, 4–6 mm long; rare in dry woods and serpentine barrens; SE; flr. Jun; FAC–; 🍂.

Menziesia Sm.

Menziesia pilosa (Michx.) Juss.   Minniebush
Deciduous shrub to 2 m tall; twigs pubescent with a true terminal bud; leaves elliptic to oval, ciliate with chaffy scales on midrib below, tip mucronate; flower parts in 4s; corolla greenish-yellow to dull yellow, lobes often reddish, urn-shaped, 6–10 mm long; capsule ovoid, 4–6 mm long, 4-celled; rare in dry to wet woods and stream banks; SW and E; flr. late May; FAC–; 🍂.

Oxydendrum arboreum

Oxydendrum DC.

Oxydendrum arboreum (L.) DC.   Sourwood

Tree to 20 m tall with glabrous twigs; leaves lanceolate to elliptic, irregularly toothed or nearly entire; inflorescence a terminal panicle; corolla white, cylindric, 5–8 mm long; capsule ovoid, 4–7 mm long, 5-celled; rare in dry woods and slopes; SW; flr. Aug; UPL; 🍂.

Rhododendron L.

Deciduous or evergreen shrubs; true terminal bud present, often enlarged; leaves alternate, ciliate (in deciduous species) or entire; flowers clustered, parts in 5s; corolla regular or irregular, funnel-shaped or bell-shaped with spreading lobes; capsule longer than wide, 5-celled. Previous reports of *R. canescens* (Michx.) Sweet in southeastern Pennsylvania (Rhoads and Klein, 1993) were based on misidentified specimens.

A. leaves evergreen, mostly 8–15 (or more) cm long, the margins hairless .... *R. maximum*
A. leaves deciduous, mostly <8 cm long, the margins usually ciliate ("azaleas")
    B. flowers appearing before or with the leaves
        C. corolla rose-purple (white), very irregular; stamens 10 .... *R. canadense*
        C. corolla pink, white, orange, or yellow, regular or nearly so; stamens 5
            D. flowers yellow or orange .... *R. calendulaceum*
            D. flowers pink or white
                E. plant to 7.5 dm tall, forming thickets; gland-tipped hairs of corolla lobes in rows or lines .... *R. atlanticum*
                E. plant 1–3 m tall; gland-tipped hairs of the corolla lobes scattered, or absent

F. pedicel, ovary, and usually the corolla tube with nonglandular hairs .... R. periclymenoides
F. pedicel, ovary, and the corolla tube with gland-tipped hairs .... R. prinophyllum
  B. flowers appearing after the leaves are fully expanded
    G. pedicels with short hairs and long gland-tipped hairs; midrib of lower leaf surfaces with long stiff hairs; calyx lobes mostly 0.5–2 mm long..... .... R. viscosum
    G. pedicels with long hairs only, or glabrous; midrib glabrous beneath or nearly so; calyx lobes mostly 2–5 mm long .... R. arborescens

Rhododendron arborescens (Pursh) Torr.   Smooth azalea
Deciduous shrub to 3 m tall with glabrous, often whitened twigs; leaves glabrous or nearly so beneath, often whitened; flowers fragrant; corolla white, the tube with gland-tipped hairs; pedicels with long hairs (sometimes gland-tipped), or glabrous; ovary and capsule with gland-tipped hairs; occasional in woods, thickets, and stream banks; mostly SW; flr. May; FAC.

Rhododendron atlanticum (Ashe) Rehder   Dwarf azalea
Deciduous shrub to 0.75 m tall, forming thickets; twigs glabrous or bristly; leaves bristly on the midrib beneath; flowers fragrant; corolla white, the lobes with gland-tipped hairs in rows or lines; pedicel, ovary and capsule with bristly, sometimes gland-tipped, hairs; rare in moist woods; York Co.; flr. May; FAC; ❦.

Rhododendron calendulaceum (Michx.) Torr.   Flame azalea
Deciduous shrub to 3 m tall; bud scales usually glabrous; twigs and lower leaf surfaces with rather dense short, soft hairs; flowers only slightly if at all fragrant; corolla yellow or orange; pedicel, ovary, and capsule with short hairs and long (often gland-tipped) hairs; very rare in thickets; Somerset Co.; flr. May; believed to be extirpated; ❦.

Rhododendron canadense (L.) Torr.   Rhodora
Deciduous shrub to 1 m tall with whitened twigs; leaves often whitened, revolute, pubescent below, the longer hairs rusty or brownish; corolla rose-purple (white), with two long lobes and three short lobes; capsule asymmetrical at the base and "hump backed," often whitened; rare, but often locally abundant in bogs, peaty wetlands, and barrens; NE; flr. late May, before the leaves; FACW.

Rhododendron canadense, flower

Rhododendron canadense, capsule (×1)

Rhododendron maximum L.   Rosebay
Evergreen shrub to 5 m tall with stout twigs bearing large terminal flower buds; leaves leathery, entire; corolla white (pink), the upper lobe(s) usually green-spotted; stamens 10; pedicel, ovary, and capsule with gland-tipped hairs; common on stream banks, cool slopes, and swamps; throughout; flr. late Jun–Jul; FAC.

Rhododendron periclymenoides (Michx.) Shinners   Pinxter-flower
Deciduous shrub to 3 m tall; bud scales usually glabrous; twigs glabrous, bristly, or rarely with short soft hairs; leaves glabrous, bristly on midrib, or rarely with dense short soft hairs beneath; flowers slightly if at all fragrant; corolla pink or white, the tube usually without gland-tipped hairs; pedicel, ovary, and capsule with long nonglandular hairs; common in dry woods, thickets, and stream

banks; throughout; flr. May; FAC. [syn: R. nudiflorum (L.) Torr.] Hybridizes with
R. prinophyllum and R. viscosum.

Rhododendron prinophyllum (Small) Millais   Mountain azalea
Deciduous shrub to 3 m tall; twigs, bud scales, and lower leaf surfaces with
rather dense short soft hairs; flowers fragrant; corolla pink (white); corolla tube,
pedicel, ovary, and capsule all with gland-tipped hairs; frequent in dry woods,
thickets, and rocky slopes; throughout; flr. late May–early Jun; FAC.

Rhododendron viscosum (L.) Torr.   Swamp azalea
Deciduous shrub to 3 m tall; leaves with long stiff hairs on the midrib beneath
and sometimes whitened; flowers fragrant; corolla white, the tube with gland-
tipped hairs; pedicel, ovary, and capsule with short hairs and long gland-tipped
hairs; frequent in wet woods and swamps; mostly E; flr. Jun; FACW+.

Rhododendron
viscosum

Rhododendron
viscosum, flower

Vaccinium L.

Deciduous or evergreen shrubs lacking a true terminal bud; leaves alternate,
toothed or entire; flowers regular, 4- or 5-merous, solitary or in racemes; ovary
inferior; fruit a many-seeded berry.

A. stems creeping or trailing, leaves evergreen, usually <1.5 cm long and 7 mm
   wide; flower parts in 4s; fruit red
   B. leaves oblong-elliptic; pedicel with bracts above the middle; corolla lobes
      6–10 mm long; fruit 10–20 mm .... *V. macrocarpon*
   B. leaves ovate to ovate-oblong; pedicel with bracts at or below the middle;
      corolla lobes mostly 4–6 mm long; fruit 5–12 mm .... *V. oxycoccos*
A. stems upright, the leaves deciduous and often >1.5 cm long and 7 mm wide;
   flower parts in 5s; fruit blue, black, purplish or greenish
   C. twigs and branchlets relatively smooth (but usually pubescent), not warty
      or pebbly; base of pedicel with a leaf-like bract; corolla open bell-shaped,
      stamens exposed .... *V. stamineum*
   C. twigs and branchlets with a warty or pebbly appearance (under magnifica-
      tion); base of pedicel naked or with a scale-like bract; corolla urn-shaped,
      cylindrical, or tubular, stamens included
      D. plants 1 m or more tall .... *V. corymbosum*
      D. plants 1 m or less, often forming thickets
         E. leaves entire, the lower surface densely pubescent; twigs densely pu-
            bescent .... *V. myrtilloides*
         E. leaves toothed or entire, the lower surface glabrous or slightly pu-
            bescent; twigs glabrous or slightly pubescent
            F. leaves toothed, mostly twice or more as long as wide, the lower
               surface green or less frequently pale; twigs reddish or brownish-
               gray .... *V. angustifolium*
            F. leaves entire or toothed, mostly less than twice as long as wide,
               the lower surface pale; twigs green .... *V. pallidum*

Vaccinium angustifolium Aiton   Low sweet blueberry
Deciduous shrub to 0.75 m tall; twigs warty-dotted, mostly glabrous; leaves elliptic, finely toothed, mostly twice or more as long as wide, green or pale below;
corolla white or pink, tubular, 4–8 mm long; berry blue, sweet, 5–12 mm; common in dry woods and barrens; throughout; flr. May, frt. late Jun–Jul; FACU–.

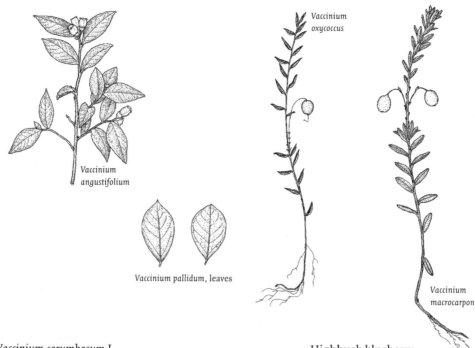

Vaccinium angustifolium

Vaccinium oxycoccus

Vaccinium pallidum, leaves

Vaccinium macrocarpon

Vaccinium corymbosum L.   Highbush blueberry
Deciduous shrub to 4 m tall; twigs warty-dotted, glabrous or pubescent; leaves elliptic, mostly entire, glabrous or pubescent; corolla white or pink, tubular, 5–12 mm long; berry blue, sweet, 5–12 mm; common in dry to wet woods, thickets, and stream banks; throughout; flr. May, frt. Jul–Aug; FACW–. [syn: *V. atrococcum* (A.Gray) A.Heller and *V. caesariense* Mack. in part]

Vaccinium macrocarpon Aiton   Cranberry
Trailing evergreen shrub; twigs glabrous; leaves oblong-elliptic, entire, pale or whitened below; flower parts in 4s; corolla pink, the lobes strongly recurved, 6–10 mm long; berry red, sour, 10–20 mm; occasional in bogs, peaty wetlands, and seepy places; throughout; flr. Jun, frt. Sep–Oct; OBL.

Vaccinium myrtilloides Michx.   Sour-top blueberry
Deciduous shrub to 0.75 m tall with densely pubescent twigs; leaves elliptic, entire, densely pubescent beneath; corolla white or pink, tubular, 4–6 mm long; berry blue, sour, 4–10 mm; infrequent in wet thickets and barrens; mostly N; flr. May, frt. Jul–Aug; FAC.

Vaccinium oxycoccos L.   Small cranberry
Trailing evergreen shrub; twigs glabrous; leaves ovate to ovate oblong, entire, revolute, whitened below; flower parts in 4s; corolla pink, the lobes strongly recurved, 4–6 mm long; berry red, sour, 5–12 mm; infrequent in bogs and peaty wetlands; mostly N; flr. Jun, frt. Sep–Oct; OBL.

Vaccinium pallidum Aiton   Lowbush blueberry
Deciduous shrub to 1 m tall; twigs green, warty-dotted, mostly glabrous; leaves elliptic to oval, mostly less than twice as long as wide, entire or toothed, paler below; corolla white or pink, tubular, 4–8 mm long; berry blue, sweet, 4–10 mm; common in dry woods and barrens; throughout; flr. May, frt. Jul–Aug.

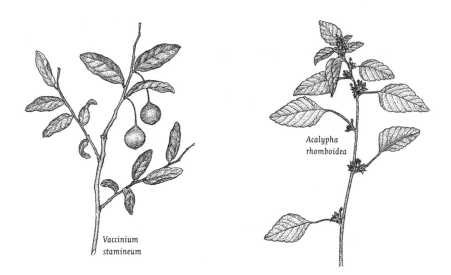

Vaccinium
stamineum

Acalypha
rhomboidea

Vaccinium stamineum L.   Deerberry

Deciduous shrub to 2 m tall; twigs usually pubescent, not warty-dotted; leaves elliptic, entire, usually pubescent below; base of the pedicel with a leaf-like bract; corolla white, open bell-shaped, 4–8 mm long, lobes spreading; stamens exposed; berry green or purplish, sour and rather thick-skinned; common in dry woods, openings, and barrens; throughout; flr. May–Jun, frt. Sep–Oct; FACU.

EUPHORBIACEAE Spurge Family

Annual or perennial herbs, some with milky sap; leaves alternate or opposite, simple; stipules present or absent; flowers unisexual, regular; petals mostly absent; sepals distinct or nearly so, or absent; ovary superior, mostly of 3 fused carpels; styles distinct or united below, often branched; fruit a 3-lobed capsule.

A. sap milky; the greatly reduced flowers (1 pistillate and several to many staminate) enclosed in a cup-shaped involucre (cyathium) that mimics a single flower; calyx absent
  B. involucres with 1 marginal gland .... *Poinsettia*
  B. involucres with 4 or 5 marginal glands
    C. leaves usually alternate below (opposite in E. *lathyris*), their bases symmetrical; cyathia in a terminal umbel .... *Euphorbia*
    C. leaves all opposite, their bases asymmetrical; cyathia solitary or clustered in the upper axils .... *Chamaesyce*
A. sap not milky; flowers not enclosed in an involucre; calyx present
  D. flowers in small axillary clusters; leaves entire .... *Phyllanthus*
  D. flowers in terminal or axillary spikes or racemes; leaves mostly toothed or lobed
    E. leaves palmately lobed .... *Ricinus*
    E. leaves not palmately lobed
      F. plants with stellate hairs or scales; stamens 3–8
        G. fruit 2- or 3-locular, dehiscent; lower leaf surface with stellate hairs .... *Croton*

G. fruit 1-locular, indehiscent; lower leaf surface with flattened
   scales .... *Crotonopsis*
F. plants with simple hairs or none; stamens 8–20 .... *Acalypha*

Acalypha L.

Annual herbs; leaves alternate, with 2 prominent lateral veins from the base, crenate or serrate; stipules present; flowers small, in axillary or terminal spikes or racemes subtended by a leafy bract that enlarges in fruit; staminate flowers with 4 sepals and 4–8 stamens; pistillate flowers, which are located at the base of the staminate spike, with 3–5 sepals; ovary 3-locular, styles deeply fringed.

A. bracts 5–9-lobed; petiole ½ to as long as the leaf blade
   B. fruit 3-seeded .... *A. rhomboidea*
   B. fruit 2-seeded .... *A. deamii*
A. bracts 10–15-lobed; petioles <¼ as long as the blade
   C. lobes of the bracts ovate to deltoid, glandular .... *A. gracilens*
   C. lobes of the bracts linear to oblong, usually not glandular .... *A. virginica*

Acalypha deamii (Weath.) Ahles   Three-seeded mercury
Very similar to *A. rhomboidea* but with only 2 seeds per fruit; known from only a single location in Allegheny Co.; believed to be extirpated; 🌿.

Acalypha gracilens A.Gray   Slender mercury
Stems erect, 1–4 dm, pubescent with incurved hairs; petioles about ¼ as long as the leaf blades; blades linear to oblong or narrowly elliptic; pistillate bracts with 9–15 ovate or triangular lobes; staminate spike conspicuously longer than the bract; fruits with 3 seeds; occasional in dry fields, open woods, and shaly slopes; SE; flr. Aug–early Sep.

Acalypha rhomboidea Raf.   Three-seeded mercury
Stems erect, 2–6 dm, glabrous or with lines of incurved hairs; leaves ovate to rhombic, toothed; petiole more than ½ as long as the blade; pistillate bracts 5–9-lobed, usually stipitate glandular; staminate spikes scarcely longer than the bracts; fruit with 3 seeds; common on wooded slopes, roadsides, fields and waste ground; throughout; flr. Jul–mid-Oct.

Acalypha rhomboidea, inflorescence (×2)

Acalypha virginica L.   Three-seeded mercury
Stem erect, 2–6 dm, pubescent with incurved hairs; leaf blades lance-ovate, shallowly crenate; petioles ⅓ to ½ as long as the blade; pistillate bracts deeply cleft into 10–15 linear segments; the staminate spikes equaling or slightly exceeding the bracts; frequent in dry or moist soil of fields, wooded slopes, stream banks, and waste ground; S; flr. Jul–Oct.

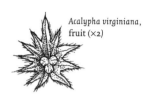

Acalypha virginiana, fruit (×2)

Chamaesyce S.F.Gray

Prostrate or erect, low-growing annuals with milky sap; leaves opposite, elliptic-ovate to lanceolate, the base of the blade distinctly asymmetrical; margins serrate or entire; stipules present, toothed; flowers in axillary cyathia, the glands with or without appendages. Often included in *Euphorbia.*

A. young stems and leaves glabrous
    B. leaves entire
        C. appendages conspicuous, ovate or ovate-oblong .... *C. serpens*
        C. appendages no longer than the glands, or lacking .... *C. polygonifolia*
    B. leaves serrulate, at least at the tip .... *C. serpyllifolia*
A. young stems pubescent at least in lines, young leaves pubescent at least on one side
    D. stems ascending to erect, simple below .... *C. nutans*
    D. stems prostrate or spreading, branched from the base
        E. ovary and capsule pubescent .... *C. maculata*
        E. ovary and capsule glabrous .... *C. vermiculata*

Chamaesyce maculata,
fruit (×10)

Chamaesyce maculata (L.) Small   Spotted spurge, milk-purslane
Mat-forming annual with stems to 4 dm tall; leaves dark green often with a reddish blotch, ovate or ovate-oblong, 5–15 mm long; involucre slit on one side; ovary and fruit hairy; styles bifid, ¼ to ⅓ their length; fruit 1.5 mm; frequent in dry, disturbed ground and pavement cracks; throughout; flr. Jun–early Oct. The very similar *C. prostrata* (Aiton) Small, with styles bifid to the base, has occurred on ballast.

Chamaesyce nutans (Lag.) Small   Eyebane
Annual with erect to ascending stems to 8 dm tall, puberulent when young, at least in a line, becoming glabrous; leaves oblong or oblong-ovate, asymmetrical at the base, 1–3.5 cm long, serrulate; fruit 2–2.5 mm, strongly 3-lobed, glabrous; frequent in dry disturbed ground, waste areas, and cultivated fields; throughout except the northernmost counties; flr. Jul–mid Oct.

Chamaesyce polygonifolia (L.) Small   Seaside spurge
Glabrous, mat-forming annual, stems branched, to 2.5 dm; leaves linear to nar-

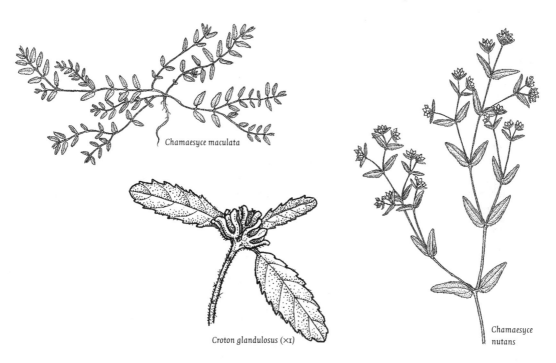

Chamaesyce maculata

Croton glandulosus (×1)

Chamaesyce nutans

rowly oblong, 8–16 mm long, entire; involucre 1–1.4 mm with broadly oval glands with or without appendages; ovary and capsule smooth; capsule 3–5.5 mm long; very rare on dunes and sand plains; NW; flr. Jul–Aug; ✿.

Chamaesyce serpens (Kunth) Small   Round-leaved spurge
Glabrous annual with prostrate, freely branched stems 1–4 dm with short internodes; leaves nearly round, 2–7 mm, entire; stipules forming a fringed, scale-like structure; appendages of the glands of the cyathium very small; capsule 1–1.5 mm, 3-angled; rare on ballast, waste ground, and margins of streamlets; SE; flr. Jul–Oct; native farther west and south; FACW.

Chamaesyce serpyllifolia (Pers.) Small   Thyme-leaved spurge
Prostrate to ascending annual, glabrous throughout; stems to 3 dm; leaves oblong or obovate, 3–14 mm long, serrulate above the middle, usually with a red blotch along the midrib; cyathia solitary, glands with small appendages; capsule ovoid, 1.3–1.9 mm long, glabrous; known from a single site in Tioga Co.; flr. Jul–Oct; native farther west.

Chamaesyce vermiculata (Raf.) House   Hairy spurge
Hairy annual with prostrate to ascending stems to 4 dm; leaves ovate to lanceolate, 0.5–2 cm, serrate; fruit 1.5–2 mm, glabrous; frequent in dry open soil and waste ground including walkways, railroad ballast, and roadsides; throughout; flr. Jun–Sep.

Croton L.

Monoecious annuals with stellate pubescence; leaves mostly alternate; stipules present; flowers in a dense terminal inflorescence, unisexual; calyx 5–7-lobed; petals as many as and as long as the sepals in staminate flowers, reduced or absent in pistillate flowers; stamens 8–20; ovary 3-locular; styles distinct or nearly so, branched.

A. leaves toothed, with 1 or 2 glands at the base of the blade .... *C. glandulosus*
A. leaves entire or wavy-margined, lacking glands
    B. calyx lobes 5; styles with 6 stigmatic tips; fruit pendulous ....
    .... *C. lindheimerianus*
    B. calyx lobes 6–9; styles with 12–24 stigmatic tips; fruit erect ..... *C. capitatus*

Croton capitatus Michx.   Hogwort, woolly croton
Stems to 1 m tall, sparingly branched above; inflorescence 1–3 cm; staminate flowers with 5 sepals, 5 petals, and 10–14 stamens; pistillate flowers with 6–9 sepals and a total of 12–24 style branches; railroads and waste ground; scattered; flr. Aug–Sep.

Croton glandulosus L.   Croton
Stems 2–6 dm tall, usually branched, densely stellate-hairy and somewhat glandular; leaves 3–7 cm, with rounded teeth, acuminate tips, and 2 whitish glands on the lower surface near the junction with the petiole; stipules tiny; staminate flowers with 4 or 5 calyx lobes, white petals and 7–9 stamens; pistillate flowers with a 5-lobed calyx, minute petals, and 3 bifid styles; rare on railroads, riverbanks, and waste ground; native as far north as VA, adventive in PA; flr. Jul–

Sep. *C. punctatus* Jacq., with lower leaf surfaces silvery flecked with brown, has occurred on ballast.

Croton lindheimerianus Scheele   Croton
Densely stellate-pubescent, grayish-white annual; leaves elliptic-ovate, rounded at the tip, 1–4 cm long; petioles 0.5–3 cm; pistillate flowers with 5 sepals; styles bifid resulting in a total of 6 stigmatic tips; rare in waste ground and pastures; SE; flr. Aug–Oct; native farther south.

Crotonopsis Michx.

Crotonopsis elliptica Willd.   Elliptical rushfoil
Annual, 1–4 dm tall; leaves alternate below, opposite above, linear-lanceolate, 1–3 cm long, stellate hairy above and with silvery fringed scales beneath; flowers in terminal or axillary clusters, unisexual; staminate flowers with a 5-lobed calyx, 5 petals, and 5 stamens; pistillate flowers with 5 petal-like glands opposite the calyx lobes; rare in sandy soil of open woods; SE; flr. Jul–Sep; believed to be extirpated; ✿.

Euphorbia L.

Herbs with milky sap; leaves alternate or opposite, entire or finely serrate; inflorescence often umbel-like; flowers greatly reduced and enclosed in a cup-like involucre forming a cyathium that simulates a single flower, each cyathium containing several staminate flowers, each consisting of a single stamen, surrounding a single, stalked pistillate flower; the margin of the cyathium usually bearing glands, sometimes with petal-like marginal appendages; ovary and capsule 3-lobed. The sap of some species may be irritating.

A. glands of the involucre with 5 conspicuous white petal-like appendages
    B. leaves just below the inflorescence with wide white margins; blades 1.5–3 cm wide .... E. *marginata*
    B. leaves just below the inflorescence not white-margined, narrower............
    .... E. *corollata*
A. glands of the involucre without conspicuous white petal-like appendages
    C. stems spreading to ascending; leaves opposite; cyathia solitary, on long pedicels from the nodes .... E. *ipecacuanhae*
    C. stems erect, ending in an umbel of 3 or more rays; leaves mostly alternate (opposite in E. *lathyris*)
        D. glands of the involucre oval, elliptic, or rounded; leaves entire or finely serrulate
        E. seeds smooth
            F. leaves 5–10 cm, entire; involucre 3 mm ....E. *purpurea*
            F. leaves 1–5 cm, finely serrulate; involucre 1.5 mm
                G. involucre glabrous; styles separate to the base .... E. *obtusata*
                G. involucre villous; styles united at the base .... E. *platyphyllos*
        E. seeds reticulate... E. *helioscopia*
        D. glands of the inflorescence crescent-shaped with the concave side outward; leaves entire
        H. stem leaves strictly opposite, aligned in 4 vertical ranks .... E. *lathyris*
        H. stem leaves all or mostly alternate

I. stem leaves linear, 1–3 mm wide .... *E. cyparissias*
I. stem leaves wider
   J. rays of the umbels 3–5; seeds pitted
      K. floral leaves wider than long, oblanceolate to obovate; capsule smooth, not winged .... *E. commutata*
      K. floral leaves longer than broad
         L. stem leaves sessile or nearly so; capsule not winged .... *E. falcata*
         L. stem leaves petioled, obtuse to retuse; capsule winged . ... *E. peplus*
   J. rays of the umbels >5; seeds smooth
      M. principal stem leaves <1 cm wide, 1-nerved .... *E. esula*
      M. principal stem leaves 1–2 cm wide, pinnately veined .... *E. lucida*

Euphorbia commutata Engelm.   Wood spurge
Glabrous perennial with branching stems 2–4 dm tall from a decumbent base; leaves alternate, obovate to oblanceolate, 2–4.5 cm long, entire, short petioled; primary umbel with 3–4 rays; involucre with 3 or 4 crescent-shaped glands with slender horns; capsule 3 mm long, globose-ovoid, smooth; rare in open woods and wooded slopes on limestone; SC and SW; flr. May–Jul; FACU.

Euphorbia commutata, cyathium (×3)

Euphorbia corollata L.   Flowering spurge
Erect, glabrous to villous perennial; stems to 1 m tall, branched above the middle; stem leaves alternate, linear to elliptic, 3–6 cm, those of the inflores-

Euphorbia corollata, cyathium (×2)

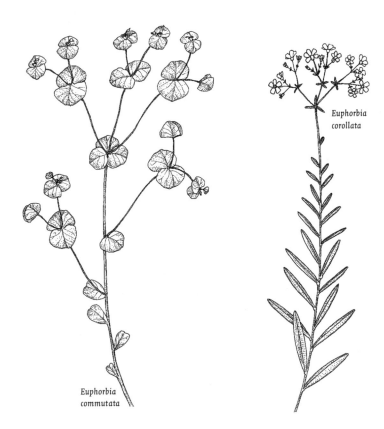

Euphorbia commutata

Euphorbia corollata

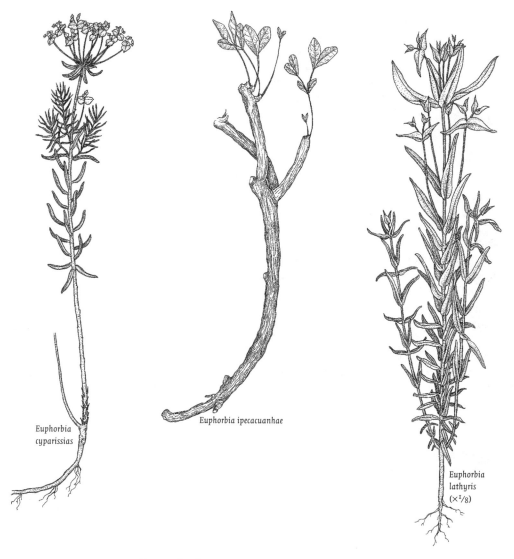

Euphorbia
cyparissias

Euphorbia ipecacuanhae

Euphorbia
lathyris
(×1/8)

cence smaller and often opposite; involucres bearing conspicuous white appendages on the 5 glands; occasional in dry, open woods, shale barrens, fields, and sandy waste ground; throughout, except for the northernmost counties; flr. Jul–Sep.

Euphorbia
cyparissias,
cyathium (×3)

Euphorbia cyparissias L.   Cypress spurge, cemetery-plant
Glabrous perennial with erect stems to 3 dm tall, colonial by creeping rhizomes; leaves numerous, alternate, linear, 1–3 cm long; rays of the umbel numerous, often 10 or more; involucre with 4 crescent-shaped glands lacking appendages; fruit 3 mm long and granular-roughened, but seldom produced; formerly cultivated and frequently naturalized on roadsides, railroads, old fields, and disturbed open woods; throughout; flr. late Apr–Jul; native to Eurasia.

Euphorbia esula L.   Leafy spurge, wolf's-milk
Glabrous perennial with erect, lightly branching stems 3–7 dm tall, from horizontal rhizomes; stem leaves alternate, linear to lance-linear, 3–8 cm, 1-nerved;

rays of the primary umbel 7–15; involucre with 4 or 5 crescent-shaped glands lacking appendages; fruit 3–3.5 mm, finely granular; an occasional weed of roadsides and waste ground; throughout; flr. May–Aug; native to Eurasia.

Euphorbia falcata L.   Spurge
Glabrous annual with branched stems 1–4 dm tall; stem leaves alternate, oblanceolate, entire, 0.5–3 cm, acute, falling early; rays of the primary umbel 3–5; involucre with crescent-shaped glands each with 2 horns; capsule smooth, 2 mm long; seeds deeply pitted and furrowed; rare on dry, shaly roadside slopes; SE and NW; flr. Jul–Sep; native to Europe.

Euphorbia helioscopia L.   Wartweed
Glabrous annual to 6 dm tall, unbranched; stem leaves alternate, sessile, spatulate, 1.5–5 cm, finely serrate, falling early; rays of the primary umbel 5; involucres with 4 brownish or yellowish-green, elliptic glands without appendages; fruit 3 mm long, smooth; rare and scattered in dry open soil, gardens, and waste ground; flr. May–Jul; native to Europe.

Euphorbia ipecacuanhae L.   Wild ipecac
Perennial with a long, thick taproot; stems to 3 dm tall, spreading to ascending and repeatedly branched; leaves opposite, red or green, ovate to obovate; involucres solitary and long-pediceled from the nodes; glands of the involucre with minute green margins; capsule 4–5 mm; rare in sandy or gravelly soil of the coastal plain; SE; flr. Apr–Jun; believed to be extirpated; ❦.

Euphorbia ipecacuanhae, cyathium (×3)

Euphorbia lathyris L.   Caper spurge, mole-plant
Glabrous and glaucous biennial; stem to 1.2 m tall, simple to the inflorescence; leaves opposite, linear or lance-linear, sessile, forming 4 vertical ranks; primary umbel with 2–4 rays; involucral glands with 2 horns; fruit 1 cm long; planted to repel moles and occasionally escaped to roadside banks, cultivated fields, and waste ground; S; flr. May–Aug; native to Europe.

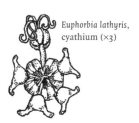

Euphorbia lathyris, cyathium (×3)

Euphorbia lucida Waldst. & Kit.   Leafy spurge
Rhizomatous perennial to 1 m tall; stem leaves alternate; lance-oblong or oblong, 5–10 cm long by 1–2 cm wide, sessile and somewhat clasping; rays of the umbel >5; involucral glands with 2 horns; fruit smooth, but deeply grooved, 3.5–4 mm long; rare in dry fields; NE; flr. Jun–Jul; native to Europe.

Euphorbia marginata Pursh   Snow-on-the-mountain
Erect, softly hairy annual; stems 3–8 dm tall, branched above; stem leaves alternate, sessile, broadly ovate to elliptic, 4–20 cm, those just below the inflorescence with broad white margins or even entirely white; involucres hairy, the 5 glands bearing white, petal-like appendages; cultivated and occasionally escaped to vacant lots, dumps, railroad cinders, and roadsides; flr. Jun–Sep; native farther west; UPL.

Euphorbia obtusata Pursh   Blunt-leaved spurge
Glabrous annual; stems erect, 2–7 dm tall, rarely branching below the umbel; leaves alternate, oblong-oblanceolate, 1.5–4 cm, finely serrulate, sessile with somewhat clasping bases; rays of the primary umbel usually 3; involucres glabrous, bearing 4 or 5 reddish or orange-brown, elliptical glands; styles separate

to the base; fruit 3–3.5 mm long, warty; rare in rich woods and stream banks; SC and NW; flr. Apr–May; believed to be extirpated; FACU–; .

Euphorbia peplus L.   Petty spurge
Much branched, glabrous annual, 1–3 dm tall; stem leaves alternate, petioled, ovate to obovate, 1–2 cm long, entire; rays of the umbel 3; involucre with 4 yellow, crescent-shaped glands with 2 slender horns; fruit 2 mm long with 2 ridges on each valve; occasional in yards, gardens, and pavements; S; flr. Jun–Sep; native to Eurasia.

Euphorbia platyphyllos L.   Broad-leaved spurge
Branching annual, 3–7 dm tall; stem leaves alternate, oblanceolate, 2–5 cm, finely serrulate, acute; primary umbel with 5 or more rays; involucres 1.5 mm, finely villous; styles united at the base; fruits 3 mm long, warty; rare on shores and waste ground; W; flr. Jul–Sep; native to Europe.

Euphorbia purpurea (Raf.) Fernald   Glade spurge
Robust perennial, stem erect, to 1 m tall from a short, stout rhizome; stem leaves alternate, elliptic to narrowly lance-oblong, entire, 5–10 cm long, those subtending the umbel shorter and broader; rays of the primary umbel 5–8; involucres 3 mm, glabrous; fruit 6–8 mm long, warty; rare in swamps or moist thickets on rich soils; SE and SC; flr. May–Jul; FAC; .

Phyllanthus L.

Phyllanthus caroliniensis Walter   Carolina leaf-flower
Erect, glabrous, branched annual; stems wiry, 1–4 dm tall; leaves alternate and 2-ranked on the stem, entire, oblong-obovate, 1–2 cm long; stipules present; monoecious; flowers in axillary clusters containing 1 staminate and several pistillate, sessile or very short-stalked; stamens 3; ovary 3-celled; styles 3, each

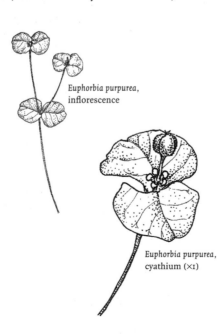

Euphorbia purpurea, inflorescence

Euphorbia purpurea, cyathium (×1)

Euphorbia purpurea (×¼)

Poinsettia dentata

once-branched; rare in moist, sandy soil, stream banks, and ravines; SE; flr. Aug–Sep; 🌣.

Poinsettia Graham

Poinsettia dentata (Michx.) Klotzsch & Garcke   Spurge

Erect, usually branched, hairy annual, 2–6 dm tall; leaves mostly opposite, alternate near the base, petiolate, coarsely toothed; inflorescence terminal, crowded, subtended by reduced foliar leaves; involucre 3 mm long with 3–5 lobes and a single gland; styles split half their length; fruit 5 mm thick, smooth; occasional on wooded slopes, railroad banks, roadsides, and waste ground; mostly S; flr. Jul–Sep. [syn: *Euphorbia dentata* Michx.]

Ricinus L.

Ricinus communis L.   Castor-bean

Stout, glabrous annual with large, palmately lobed leaves 1–4 dm wide; monoecious, flowers in racemes; stamens numerous, with branching filaments; styles 3, each divided and plumose; capsule 3-lobed, prickly; cultivated and occasionally naturalized in waste ground and floodplains; SE; flr. Jul–Sep; native to Africa.

FABACEAE Pea or Bean Family

Herbaceous or woody plants with alternate, pinnately compound or trifoliate (occasionally palmately compound or even simple) leaves and more or less prominent stipules; petiole and petiolules with basal swellings (pulvini); flowers perfect, borne in racemes, spikes, or heads, often showy; sepals 5 (occasionally 4 by fusion), generally forming a short tube basally; corolla 5-merous, typically papilionaceous (consisting of an upper standard or banner, 2 lateral wings, and the 2 lower petals partially fused to form a keel that encloses the stamens and pistil); stamens 5 or more often 10, frequently with 9 filaments forming a sheath around the pistil and the 10th partly or wholly separate (diadelphous); pistil 1, simple; fruit a legume, typically dehiscent, but sometimes indehiscent or jointed.

A. plants woody
  B. leaves palmately compound .... *Cytisus*
  B. leaves pinnately compound (if trifoliate, the terminal leaflet stalked)
    C. twining vines
      D. leaves trifoliate .... *Pueraria*
      D. leaflets >3 .... *Wisteria*
    C. trees or shrubs
      E. leaves pinnately trifoliate .... *Lespedeza*
      E. leaflets >3
        F. stems with stipular spines, bristles, or stiff glandular hairs ....
        .... *Robinia*
        F. spines, bristles, or stiff glandular hairs not present
          G. large shrubs; flowers purple; fruits <1 cm long .... *Amorpha*
          G. trees; flowers white; fruits >1 cm long .... *Sophora*

A. plants herbaceous

    H. leaves simple .... *Crotalaria*

    H. leaves compound

        I. leaves palmately compound .... *Lupinus*

        I. leaves pinnately compound or trifoliate

            J. terminal leaflet replaced by a tendril

                K. stems not winged .... *Vicia*

                K. stems winged or angled .... *Lathyrus*

            J. tendrils not present

                L. twining vines

                    M. high-climbing, semiwoody vine .... *Pueraria*

                    M. scrambling or trailing vines <2 m

                        N. calyx with only 1 well-developed lobe, leaflets 3–7 .... *Apios*

                        N. calyx with 4–5 well-developed lobes; leaves all trifoliate

                            O. calyx 5-lobed .... *Phaseolus*

                            O. calyx 4-lobed

                                P. inflorescence a raceme

                                    Q. a pair of small bracteoles present at the base of the calyx; calyx teeth unequal .... *Galactia*

                                    Q. bracteoles not present, each flower subtended by a single ovate bract; calyx teeth nearly equal .... *Amphicarpa*

                              P. inflorescence a dense head or umbel ..... *Strophostyles*

                L. erect, reclining, or prostrate herbs, nontwining

                  R. leaves trifoliate

                      S. terminal leaflet stalked (pinnately trifoliate)

                        T. fruit 1-seeded or separating into 2–5 1-seeded segments

                            U. fruit 1-seeded; leaflets without stipels

                              V. leaflets toothed

                                W. inflorescence a short raceme or subcapitate cluster; flowers purple or yellow ...... *Medicago*

                                W. inflorescence a slender, elongate raceme; flowers yellow or white .... *Melitotus*

                              V. leaflets entire

                                X. stipules lance-ovate, brown, persistent; calyx lobes blunt .... *Kummerowia*

                                X. stipules slender, often falling early; calyx lobes acuminate .... *Lespedeza*

                        U. fruit separating into 2–5 single-seeded segments; leaflets with stipels .... *Desmodium*

                      T. fruit with >1 seed, not segmented

                        Y. flowers in a dense head

                            Z. flowers white, yellow or pink-purple, corolla persistent .... *Trifolium*

                            Z. flowers blue-violet, corolla falling after flowering, ... *Medicago*

                      Y. flowers not in a dense head

                          AA. flowers 1–3 in axillary clusters .... *Clitoria*

                          AA. flowers in elongate axillary racemes

BB. flowers yellow or white; fruit ovate or round, 2–5 mm, indehiscent .... *Melilotus*

BB. flowers purple; fruit flat, 3–6 cm by 1 cm wide, dehiscent .... *Phaseolus*

S. terminal leaflet not stalked (palmately trifoliate)

CC. leaflets toothed .... *Trifolium*

CC. leaflets entire

DD. plants glabrous; stipules soon falling; flowers in racemes .... *Baptisia*

DD. plants hairy; stipules always present; flowers in heads or spikes .... *Stylosanthes*

R. leaves pinnately compound, leaflets >3

EE. inflorescence an umbel

FF. flowers yellow; leaflets 5 .... *Lotus*

FF. flowers pink or white; leaflets 11–25 .... *Coronilla*

EE. inflorescence a spike or raceme

GG. fruit 1-seeded, indehiscent .... *Aeschynomene*

GG. fruit dehiscent, normally containing >1 seed

HH. leaves and stem covered with long white hairs .... *Tephrosia*

HH. leaves and stem glabrous to thinly pubescent

II. racemes 1-sided; fruit cylindrical, 3–4 cm, not inflated... .... *Galega*

II. racemes not 1-sided; fruit inflated, 1–2 cm ..... *Astragalus*

Aeschynomene L.

Aeschynomene virginica (L.) Britton, Stearns & Poggenb.   Sensitive joint-vetch
Stems erect to 1 m tall or more, sparsely pustulate-hairy above; leaflets numerous; flowers yellow with red veins, 10–15 mm, in few-flowered axillary racemes; fruits 2–7 cm; very rare in freshwater tidal marshes; SE; believed to be extirpated; flr. Jun–Oct; OBL; ✥.

Amorpha L.

Amorpha fruticosa L.   False indigo
Deciduous shrub with pinnately compound leaves and small purple flowers in dense terminal racemes; the 2 lower calyx lobes longer and more narrow than the upper 3; corolla consisting of the 5–6-mm-long standard only, wings and keel absent; fruit glandular, 5–6 mm long; frequent in alluvial soils along streams and rivers and other low, moist ground; mostly SE; apparently not originally native to Pennsylvania, introduced from farther west and south and now widely naturalized; flr. late May–Jun; FACW.

Amorpha fruticosa, fruit (×2)

Amphicarpa Elliott

Amphicarpa bracteata (L.) Fernald   Hog-peanut
Slender, annual, twining vine to 1.5 m; leaflets 3, with stipels; flowers in axillary racemes or panicles, each pedicel subtended by a bract; calyx slightly irregular, 4-lobed; stamens 10, diadelphous; style glabrous; fruit flat, 3-seeded; in addition

Amphicarpa bracteata, flower (×2)

Amorpha fruticosa (×¼)

Amphicarpa
bracteata

Apios americana

apetalous flowers, which develop into subterranean 1-seeded fruits, are produced on slender peduncles from the lower stem; common in moist woods and floodplains throughout; flr. late May–early Sep; FAC.

Apios Medik.

Apios americana,
flower

Apios americana Medik.   Ground-nut, wild bean
Twining, rhizomatous, tuber-producing, perennial vine; leaflets 3–7; flowers in axillary racemes; corolla purple-brown, 10–13 mm; fruit linear, several-seeded, the valves coiling after dehiscence; common in moist woods, floodplains, and thickets; throughout; flr. late Jun–early Sep; FACW.

Astragalus L.

Perennial herbs with branching stems; leaves pinnate, with numerous leaflets; flowers in axillary racemes, white or yellowish; calyx cylindrical or campanulate, 5-lobed; stamens 10, diadelphous; fruits erect, inflated, 1–2 cm.

A. fruits 1–1.5 cm, unilocular .... *A. neglectus*
A. fruits 1.5–2 cm, 2-locular by intrusion of the sutures .... *A. canadensis*

Astragalus canadensis L.   Milk-vetch

Astragalus canadensis, fruit (×1)

Rhizomatous; stems erect, to 1.5 m tall; stipules connate; leaflets 15–35; fruits 1–1.5 cm, nearly terete and partitioned longitudinally to form 2 locules; rare in rocky roadside banks, limestone ledges, and shale barrens; SW; flr. late Jun–early Aug; FAC; ❧.

Astragalus neglectus (Torr. & A.Gray) Sheldon   Cooper's milk-vetch
Tap-rooted; stems erect, 3–9 dm tall, hollow; stipules free; leaflets 11–23; fruit sessile, 1.5–2 cm, inflated; very rare in gravelly thickets and roadsides; NE and C; believed to be extirpated; flr. Jun–Jul; FACU; ❧.

Baptisia Vent.

Glabrous, rhizomatous, perennial herbs with erect branched stems; leaves trifoliate, turning black when dry; flowers in conspicuous terminal racemes; stamens 10, distinct; fruits subcylindric to somewhat flattened, distinctly stalked above the calyx.

A. flowers blue; fruits 3–5 cm long .... B. *australis*
A. flowers yellow; fruits 8–15 mm long .... B. *tinctoria*

Baptisia australis (L.) R.Br.   Blue false indigo
Stems to 1.5 m tall; stipules persisting at least until flowering; flowers blue-violet, 2–2.7 cm, in racemes up to 4 dm long; fruits 3–5 cm on a stipe slightly longer than the calyx; occasional in open woods, stream banks, and sandy floodplains; W; flr. May–Jun; native, but also cultivated; ❧.

Baptisia tinctoria (L.) Vent.   Wild indigo
Stems to 1 m tall; leaves dark bluish-green; stipules falling early; racemes numerous; flowers yellow, 8–13 mm; fruits thick and lenticular, 8–15 mm long; frequent in dry, open woods and clearings in sandy, acidic soils; throughout; flr. Jun–Aug.

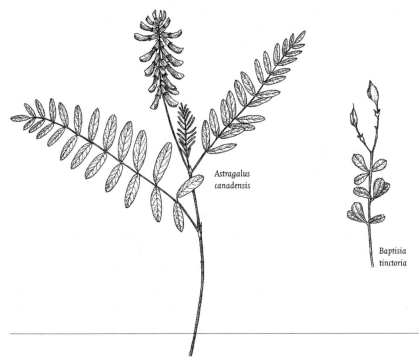

Astragalus canadensis

Baptisia tinctoria

Clitoria L.

Clitoria mariana L.   Butterfly pea
Glabrous, twining, perennial vine to 1 m; leaves trifoliate; flowers 1–3 in axillary clusters, inverted with the banner on the bottom and the keel above; corolla pale blue or pinkish, 4–6 cm; each flower subtended by a pair of small bracts; calyx irregular, the 2 upper lobes somewhat fused and shorter than the lower; stamens 10, monadelphous below; fruit stipitate, flattened, dehiscent; rare in dry, open areas on sandy soils; SE; flr. Jul–Sep; 🌿.

Coronilla L.

Coronilla varia, fruits

Coronilla varia L.   Crown-vetch
Sprawling to ascending perennial 3–10 dm tall; leaves sessile, with 11–25 leaflets; flowers pinkish (or white) in long-peduncled, axillary umbels; fruits linear, 4-angled; planted extensively along highways; throughout; flr. Jun–Nov; native to S. Europe.

Crotalaria L.

Crotalaria sagittalis L.   Rattlebox
Annual; stems erect, 1–4 dm tall, simple or branched above; leaves simple, lanceolate to linear; stipules decurrent along the stem; flowers in racemes, yellow; calyx irregular; stamens 10, monadelphous below the middle; fruits inflated, 1.5–3 cm; occasional in dry, sandy, or gravelly soil of woods, old fields, or roadsides; SE and SC; flr. Jul–Sep.

Coronilla varia

Clitoria mariana

Cytisus L.

Cytisus scoparius (L.) Link   Scotch-broom

Unarmed shrub to 2 m tall with stiff, green, angled branches; leaves mostly tri-foliate (upper may be unifoliate); leaflets obovate, 5–10 mm; flowers yellow, soli-tary or paired in upper axils to form a terminal raceme; fruits 2–3 cm, with long hairs along the suture; cultivated and occasionally escaped to railroad embank-ments and dry, sandy waste ground; flr. May–Jun; native to Europe.

Desmodium L.

Perennial herbs with trifoliate leaves with stipules that are sometimes persistent but more often fall early; leaflets entire, with or without stipels; stems and leaves variously villous, pilose or covered with short hooked hairs; flowers in terminal (and axillary) panicles (or racemes) with bracts subtending each branch or flower cluster and smaller, nonpersistent bracts subtending each flower; corolla papilionaceous, violet-purple or greenish, fading quickly; stamens 10, all united, or 9 fused and 1 free (diadelphous); fruits with 2–6 sections that separate into indehiscent, 1-seeded, triangular or rounded segments that adhere to rough sur-faces.

A. stems decumbent or trailing
    B. leaflets suborbicular or obovate; fruit with elliptic segments ....
    .... D. *rotundifolium*
    B. leaflets ovate or rhombic; fruit with triangular segments ...... D. *humifusum*
A. stems erect
    C. flowering stems leafless .... D. *nudiflorum*
    C. inflorescences borne on leaf-bearing stems
        D. stamens monadelphous; leaves clustered beneath the inflorescence ....
        .... D. *glutinosum*
        D. stamens diadelphous (9 united and 1 free); leaves spaced along the stem
            E. axis of the inflorescence with long straight hairs as well as shorter hooked hairs
                F. stipules narrow, falling early; leaves short-petioled to subsessile
                .... D. *canadense*
                F. stipules broad, some persistent; leaves (except the uppermost) with well-developed petioles .... D. *canescens*
            E. axis of the inflorescence with only short hooked hairs, or only a few longer straight hairs
                G. some or most of the stipules persistent .... D. *cuspidatum*
                G. stipules falling early
                    H. fruits with 3–5 segments, angled on the lower edge; or plants strongly villous
                        I. plants conspicuously velvety-villous and grayish in appear-ance, especially on the leaf undersides
                            J. fruits straight above, with 4–5 segments that are ob-tusely angled below; leaflets 1–1.5 times as long as wide
                            .... D. *viridiflorum*
                            J. upper margins of the fruits convex, the 2–4 segments rounded or obtusely angled below .... D. *nuttallii*

I. plants not conspicuously villous; stems pilose and with short hooked hairs, or glabrate; leaflets glabrous or appressed pubescent beneath

    K. fruit segments 2–4, with straight upper margins and rounded or obtusely angled lower edges; bracts villous .... D. nuttallii

    K. fruit segments 3–5, with straight upper margins and obtusely angled lower edges; bracts not villous

        L. corolla 8–10 mm; stem and leaves glabrous; leaflets pale beneath.....
        .... D. laevigatum

        L. corolla 5–7 mm; stems pubescent or nearly glabrous; leaflets green or only slightly pale beneath

            M. leaflets 3–8 times as long as wide; leaflets with short (<0.5 mm) hairs or none .... D. paniculatum

            M. leaflets 1–3 times as long as wide; leaflets with long (>0.5 mm) spreading hairs

                N. stem and petioles with long straight hairs or none ....
                .... D. perplexum

                N. stem and petioles sparsely or densely covered with short, hooked hairs .... D. glabellum

H. fruits with 2–3 segments, rounded on the lower edge; plants glabrate or pubescent but not strongly villous

    O. leaflets 5–10 times as long as wide .... D. sessilifolium

    O. leaflets 1–3.5 times as long as wide

        P. corolla 6–7 mm long; leaflets villous and grayish beneath .... D. nuttallii

        P. corolla 3.5–6 mm long; leaflets not grayish-villous

            Q. terminal leaflet longer and narrower than the lateral leaflets ....
            .... D. obtusum

            Q. terminal leaflet similar to the lateral leaflets

                R. stem and leaflets glabrous or nearly so; pedicels 8–15 mm ....
                .... D. marilandicum

                R. stem pilose; pedicels 3–8 mm .... D. ciliare

Desmodium canadense (L.) DC.   Showy tick-trefoil
Stems erect, branched, 5–10 dm tall, with straight and hooked hairs, at least above; stipules linear, falling early; leaflets ovate-lanceolate or lanceolate, with incurved hairs beneath; inflorescences densely flowered; corolla 8–11 mm, blue-violet; calyx 3.5–5 mm; fruits with 3–5 rounded segments; frequent in open woods; throughout; flr. Jul–Sep; FAC.

Desmodium canescens (L.) DC.   Hoary tick-trefoil
Stems erect, branched, 0.5–2 m, pilose and with some hooked hairs; stipules ovate, persistent; leaflets ovate, 1.5–2 times as long as wide, with short hooked hairs on the veins; inflorescence branched; corolla 9–13 mm, purple; fruit convex above and obtusely angled below, with 4–6 segments; occasional in dry, open woods and fields; mostly S; flr. Jun–Oct.

Desmodium
ciliare, fruits

Desmodium
ciliare, leaf

Desmodium ciliare (Muhl. ex Willd.) DC.   Tick-clover, tick-trefoil
Stems slender, erect, 4–15 dm tall, with long spreading hairs; stipules narrow, 2–4 mm, falling early; petioles hairy, leaflets broadly ovate to elliptic-ovate, blunt, 1.5–2.5 cm, hairy or glabrous; corolla 4–5 mm, lavender-purple; calyx 1.5–2 mm; fruits with 1–2(3) rounded segments; occasional in dry sandy woods and edges; SE and SC; flr. Aug–Sep.

Desmodium cuspidatum (Muhl. ex Willd.) Loudon   Tick-clover, tick-trefoil
Stems stout, erect to 2 m tall, glabrate or slightly pilose; stipules lanceolate, persistent; petioles 5–8 cm; leaflets glabrate or with a few spreading hairs beneath, lance-ovate to ovate, the terminal one 6–12 cm long and sharply acuminate; inflorescence branched; corolla 8–12 mm long, purple; calyx 3–4 mm; fruit with 4–6 segments, rounded above and obtusely angled beneath; occasional in rich, rocky woods and banks; throughout; flr. Jul–Sep.

Desmodium glabellum (Michx.) Kuntze   Tall tick-trefoil
Stems erect or ascending, 3–10 dm, branched above, densely covered with short hooked hairs; stipules 4–6 mm, falling early; leaflets pubescent beneath, broadly to narrowly ovate, 1.5–3 times as long as wide; stipules narrow, falling early; corolla 6–7 mm, lilac to purple; calyx 2–3 mm; fruit with 3–5 segments, convex above and obscurely angled below; rare in dry, wooded roadside banks and open woods; SE and W; flr. Jun–Aug; 🌿.

Desmodium glutinosum (Muhl. ex Willd.) A.W.Wood   Sticky tick-clover
Stems erect, unbranched, 3–10 dm tall; leaves clustered toward the top of the stem with the long panicle extending above; stipules narrow, falling early; leaflets broadly ovate and abruptly acuminate, 7–13 cm long and 4–12 cm wide, lacking stipels; corolla pink-purple, 5–7 mm; stamens monadelphous; calyx 2 mm; fruits with 2–3 segments, concave above and rounded beneath; common in rich woods; throughout; flr. Jun–Aug.

Desmodium humifusum (Muhl. ex Bigelow) Beck   Tick-trefoil
Stems prostrate, 1–2 m, with curved and straight hairs; stipules ovate to lance-acuminate, falling early or semipersistent; leaflets ovate or rhombic, hairy on both sides, sometimes with a few hooked hairs; inflorescences terminal and axillary; corolla 8–9.5 mm long, purple; calyx 2.5–3 mm; fruits with 3 or 4 segments, convex above, obtusely rounded beneath; rare in dry, sandy woods; scattered; flr. Aug–Sep; 🌿.

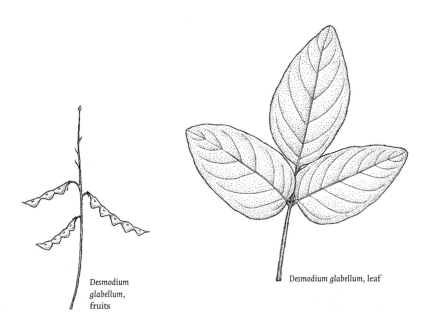

Desmodium
glabellum,
fruits

Desmodium glabellum, leaf

Desmodium laevigatum (Nutt.) DC. — Smooth tick-clover

Stems ascending to erect, 3–10 dm, simple or branched, glabrous or nearly so; stipules ovate-lanceolate, 5–8 mm; leaflets ovate or ovate-lanceolate, glabrous or puberulent, glaucous beneath; corolla 8–10 mm long, lavender to purple; calyx 3–4 mm; fruits with 3–5 segments, convex to straight above and obtusely angled to rounded beneath; rare in dry sandy woods and roadsides; mostly SE; flr. Jul–Sep; ✿.

Desmodium marilandicum (L.) DC. — Maryland tick-clover

Stems slender, ascending to erect, 3–10 dm, unbranched below the inflorescence; stipules slender, falling early; leaflets ovate or oval, 1.5–2.5 cm, glabrous or slightly hairy beneath; corolla 4–6 mm long, lavender to red-violet; calyx 2–3 mm; fruit with 2–3 rounded segments; occasional in dry, open, upland woods, fields, and edges; SE and SC; flr. Jul–Sep.

Desmodium nudiflorum (L.) DC. — Naked-flowered tick-trefoil

Flowering stem erect, leafless, to 7 dm; leaves crowded at the top of a separate 1–5 dm stem; stipules narrow, falling early; leaflets ovate, lacking stipels, the terminal one 4–10 cm long; corolla rose-purple, 6–8 mm, pink or white; fruit with 2 or 3 segments, concave above and rounded below; common in rich, deciduous woods and edges; throughout; flr. Jul–Sep.

Desmodium nuttallii (Schindl.) Schub. — Nuttall's tick-trefoil

Stems ascending or erect, 3–10 dm, branched from the base, villous, with or without hooked hairs; leaflets ovate to narrowly ovate, the terminal leaflet 1.5–2 times as long as wide, velvety-pubescent beneath; stipules narrow, falling early; corolla 6–7 mm long, purple or pink; fruits with 2–4 segments, convex (or occasionally straight) above and rounded or obtusely angled below; rare in dry woods and edges; SE; flr. Jul–Sep; ✿.

Desmodium obtusum (Muhl. ex Willd.) DC. — Tick-trefoil

Stems rigid, ascending to erect, frequently clustered, 5–15 dm, sparsely to densely covered with fine hooked hairs and with longer hairs at the nodes; stipules triangular-subulate, falling early; leaflets elliptic-ovate to lanceolate, 5–7 cm long and 2.5–4 times as long as wide, thick and veiny, hairy beneath and rough above; inflorescence glandular-pubescent; corolla 4–5 mm, pink-purple; fruits with 2 or 3 rounded segments; rare in dry, open woods on sandy soils; mostly SE; flr. Aug; ✿. [syn: D. rigidum (Elliot) DC.]

Desmodium paniculatum (L.) DC. — Tick-trefoil

Stems slender, erect or ascending and eventually sprawling, 3–10 dm, branched above, glabrate; stipules narrow, falling early; leaflets lanceolate or oblong, 3–8 times as long as wide, glabrous or appressed-hairy beneath; corolla 6–7 mm; calyx 2–3 mm; fruit with 3–6 segments, convex above and obtusely angled below; common in clearings and edges of moist or dry woods; throughout; flr. Jul–Sep; UPL.

Desmodium perplexum Schub. — Tick-trefoil

Desmodium perplexum, stem (×1½)

Stems erect, ascending, or spreading, clustered, 3–10 dm, with coarse straight hairs and sometimes also fine hooked hairs; stipules narrow, falling early; leaflets elliptic-ovate to narrowly ovate, 1.5–3 times as long as wide, sparsely pu-

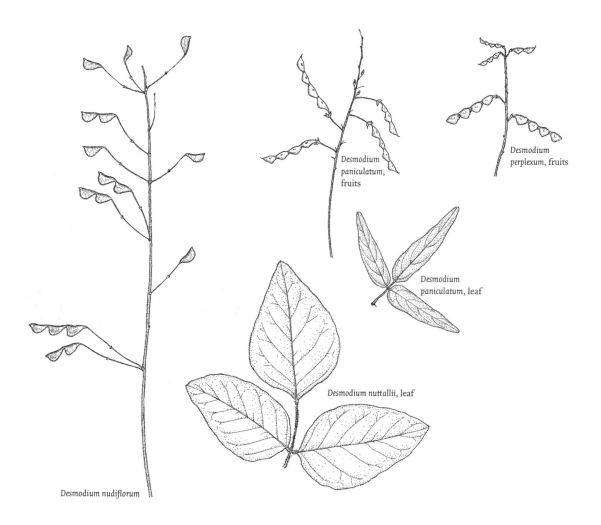

Desmodium paniculatum, fruits

Desmodium perplexum, fruits

Desmodium paniculatum, leaf

Desmodium nuttallii, leaf

Desmodium nudiflorum

bescent above, with few to many long, spreading hairs beneath; corolla 6–8 mm long, lavender or purple; fruits with 3–5 segments, convex above, obtusely angled below; common in dry or moist, open woods; throughout except at the highest elevations; flr. Jun–Sep.

Desmodium rotundifolium DC.   Round-leaved tick-trefoil
Stems prostrate, 1–2 m, branched, pilose; stipules ovate, persistent; leaflets rounded or broadly ovate, the terminal one 3–5.5 cm long; inflorescence generally unbranched; corollas 9–11 mm, pinkish; calyx 2.5–4 mm; fruits with 3–6 rounded segments; occasional in dry, open woods; throughout; flr. Jun–Sep.

Desmodium sessilifolium (Torr.) Torr. and A.Gray   Sessile-leaved tick-trefoil
Stems ascending to erect, 5–10 dm, mostly unbranched, finely hairy; leaves sessile or nearly so; stipules lanceolate, falling early; leaflets oblong to linear-oblong, densely hairy beneath; corolla lavender to reddish-purple, 5 mm long; fruit with 2(3) segments, nearly straight above and rounded below; very rare in dry, open woods; SE and NC; believed to be extirpated; flr. Aug–Sep; ❦.

Desmodium viridiflorum (L.) DC.   Velvety tick-trefoil
Stems ascending or erect, 3–10 dm, finely pubescent with some hooked hairs;

stipules triangular, falling early; leaflets broadly ovate or deltoid-ovate, 1–1.5 times as long as wide, densely velvety-pubescent beneath; inflorescence 2–5 dm; corolla 7–8 mm long, pinkish-green; calyx 2–3 mm; fruit with 4 or 5 segments, straight or convex above and obtusely angled below; very rare in abandoned fields and other dry, open places; SE; flr. Aug–Sep; 🌿.

Galactia P. Browne

Twining or trailing perennial herbs; leaves pinnately trifoliate; flowers in short racemes, small, purplish, each subtended by a pair of small bracts; calyx 4-lobed; stamens 10, diadelphous; fruit flat, few-seeded, twisted after dehiscence.

A. stems prostrate, twining only at the tip; flowers 12–18 mm .... G. regularis
A. stems twining and climbing; flowers 7–14 mm .... G. volubilis

Galactia regularis (L.) Britton, Stearns & Poggenb.   Eastern milk-pea
Stems trailing, twining only at the tips; leaflets elliptic to ovate-oblong, the terminal one 2–5 cm, glabrous to somewhat hairy; racemes few-flowered, 3–6 cm long; flowers 12–18 mm; fruits 3–5 cm, hairy; very rare in dry sandy soil; SE; believed to be extirpated; flr. Aug; 🌿.

Galactia volubilis,
flower (×2)

Galactia volubilis (L.) Britton   Downy milk-pea
Stems twining and climbing, to 1.5 m, hairy; leaflets ovate or ovate-oblong; raceme 1 dm, with the flowers well separated; flowers 7–10 mm; fruits 2–4 cm, densely hairy; very rare in dry thickets and edges; SE; believed to be extirpated; flr. Jul–Aug; 🌿.

Galega L.

Galega officinalis L.   Goat's-rue
Upright to sprawling perennial herb with stems to 1 m; leaves pinnate with 11–15 elliptic to oblong leaflets; racemes axillary, 1–2 dm, with numerous white or pale

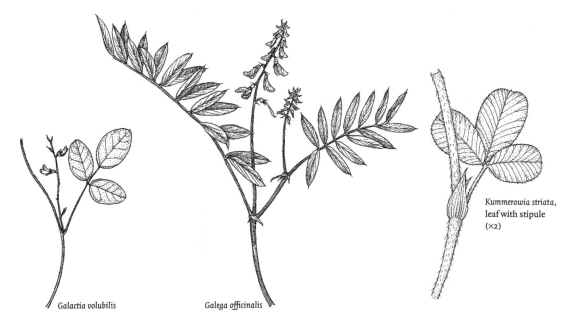

Galactia volubilis   Galega officinalis

Kummerowia striata,
leaf with stipule
(×2)

purple flowers; fruits linear, subterete, 3–5 cm; naturalized in moist open meadows and stream banks at a few scattered sites; flr. mid Jun–Sep; native to Eurasia; designated as a federal noxious weed.

Kummerowia Schindler

Very similar to *Lespedeza* with which it is sometimes combined, annuals with trifoliate leaves and broad, glabrous, membranous, persistent stipules; flowers of
2 types, those with a closed corolla (cleistogamous) are self-pollinating; fruit a
loment.

A. hairs of the stem pointing downward .... *K. striata*
A. hairs of the stem pointing upward .... *K. stipulacea*

Kummerowia stipulacea (Maxim.) Schindl.   Korean lespedeza
Stems to 6 dm with sparse spreading or upward pointing hairs; petioles 2–10
mm; stipules broad and conspicuous; flowers bluish-purple; fruit 3 mm long;
occasional in dry open soil of pastures, roadsides and waste ground, planted for
erosion control and widely naturalized; SE and SC; flr. late Apr–early Oct; native
to Asia. [syn: *Lespedeza stipulacea* Maxim.]

Kummerowia striata (Thunb.) Schindl.   Japanese clover
Stems wiry, branching, 1–3 dm; stipules ovate, 2–4 mm long, brown and scarious; petioles 1–2 mm; leaflets oblong or obovate, 6–14 mm long, ciliate along
the margins; flowers pinkish; fruit barely longer than the calyx; occasional in
fields, railroad banks, waste ground, and shale barrens; SE and SC; flr. Jul–Aug;
native to Asia. [syn: *Lespedeza striata* (Thunb.) Hook. and Arn.]

*Kummerowia
striata*, fruit
(×3)

Lathyrus L.

Sprawling vines with angled or winged stems; leaves with 2–16 leaflets and terminating in a tendril; inflorescences axillary on long peduncles; flowers 2–many
borne on one side near the end of the peduncle; corolla strongly papilionaceous,
pink, purple, or white; stamens diadelphous; style flattened, pubescent on the
inner side; fruit dehiscent.

A. leaflets 2
   B. stipules with 2 basal lobes, symmetrical .... *L. japonicus*
   B. stipules with 1 basal lobe, asymmetrical
      C. stem not winged; flowers red-purple .... *L. tuberosus*
      C. stem winged; flowers violet, pink, or white
         D. fruit hairy; calyx lobes similar .... *L. hirsutus*
         D. fruit glabrous; calyx lobes unequal .... *L. latifolius*
A. leaflets 4–12
   E. basal lobes of the stipules rounded; flowers yellowish-white ....
   .... *L. ochroleucus*
   E. basal lobes of the stipules sharp; flowers red-purple (or white)
      F. stems not winged; leaflets 8–12; racemes with 10–20 flowers ....
      .... *L. venosus*
      F. stems winged; leaflets 4–8; racemes with 2–6 flowers .... *L. palustris*

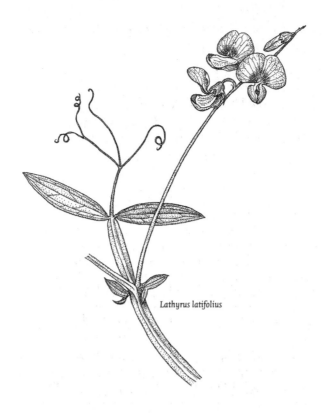

Lathyrus latifolius

Lathyrus japonicus Willd.   Beach pea
Decumbent to suberect, rhizomatous perennial to 1 m; leaflets 6–12; stipules broadly ovate with basal lobes; calyx irregular; corolla purple; very rare in sandy or gravelly shores, sandplains, and dunes; NW; flr. Jun–Sep; ❦. Ours is var. *glaber* (Ser.) Fernald. [syn: *L. maritimus* (L.) Bigelow]

Lathyrus latifolius L.   Perennial sweetpea
Trailing or climbing, rhizomatous perennial; stems to 2 m, broadly winged; leaflets 2, on a winged petiole; raceme with 4–10 flowers; corolla purple, pink or white, 1.5–3 cm; fruit 6–10 cm frequently planted and/or naturalized in roadsides, old fields, and waste ground; throughout; flr. Jun–Oct; native to Europe.

Lathyrus ochroleucus Hook.   Wild pea, vetchling
Glabrous, rhizomatous perennial to 8 dm; stipules rounded at the base; leaflets 6–10; racemes with 5–10 yellowish-white flowers; rare in dry, upland woods and slopes; mostly N; flr. May–Jul; ❦.

Lathyrus palustris L.   Marsh pea or vetchling
Slender, rhizomatous perennial; stems to 1 m, winged; stipules with pointed basal lobes; leaflets 4–8; raceme with 2–6 red-purple flowers; rare in shores, moist meadows, sandplains, swamps, and thickets; scattered; flr. Jun–Aug; FACW+; ❦.

Lathyrus tuberosus L.   Field pea, yellow vetchling
Rhizomatous, tuber-bearing perennial; stems to 8 dm, 2-angled but not winged; leaflets 2; stipules lanceolate; racemes crowded, with 2–10 red-purple flowers;

rarely naturalized in fields, meadows, roadsides, and alluvium; scattered; flr. Jun–Jul; native to Europe and W. Asia.

Lathyrus venosus Muhl. ex Willd.   Veiny pea or vetchling
Stout, rhizomatous perennial to 1 m; stipules narrow, with pointed basal lobes; leaflets 8–12; racemes dense; flowers purple; rare in sandy or rocky shores, wooded slopes, and railroad banks; scattered; flr. May–Jun; FACW; 🍃.

Lespedeza Michx.

Deciduous shrubs or perennial herbs with small trifoliate leaves; stipules linear, falling early or persisting; leaflets without stipels; flowers purple or yellowish, axillary, solitary or in spikes or racemes; each flower subtended at the base by 2 or 4 small bractlets; many species have both petaliferous and apetalous flowers; stamens 10, diadelphous; fruit an oval to elliptic loment surrounded by a persistent 5-lobed calyx.

A. woody to semiwoody shrubs; stipules persistent; flowers purple; apetalous flowers numerous, in axillary clusters
    B. racemes loose, usually drooping; calyx lobes longer than the tube............
    .... .L. thunbergii
    B. racemes compact, upright; calyx lobes about half as long as the tube ....
    ....L. bicolor
A. perennial herbs; stipules often falling early; flowers purple or yellowish; apetalous flowers few and mingled with the others, or none
    C. flowers purple, in axillary racemes or clusters
        D. plants trailing or procumbent
            E. stem and petioles glabrous or with appressed hairs............. L. *repens*
            E. stem and petioles with dense spreading hairs .... L. *procumbens*
        D. plants erect or ascending
            F. peduncles of the petaliferous flowers longer than their subtending leaves; keel 1–2 mm longer than the wings .... L. *violacea*
            F. peduncles of the petaliferous flowers shorter than their subtending leaves; keel equal to or shorter than the wings
                G. leaves thinly hairy on both sides, or glabrous above, the 2 upper calyx lobes fused for at least 1/3 their length
                    H. leaflets linear, 2–5 mm wide ....L. *virginica*
                    H. leaflets oval or rounded, 6–20 mm wide .... L. *intermedia*
                G. leaves densely velvety beneath; the 2 upper calyx lobes barely fused at the base ....L. *stuevei*
    C. flowers yellowish-white, solitary, in small axillary clusters, spikes, racemes, or dense terminal heads
        I. flowers solitary or in axillary clusters of 2–3; upper calyx lobes fused for about half their length; wings and keel equal .... L. *cuneata*
        I. flowers in dense spikes, racemes, or heads; upper calyx lobes not fused; wings longer than the keel
            J. peduncles shorter than the subtending leaves; calyx lobes 6–10 mm
            .... L. *capitata*
            J. peduncles longer than the subtending leaves; calyx lobes 3–7 mm
                K. leaflets oval to obovate, 1/2–3/4 as wide as long ....L. *hirta*
                K. leaflets linear to narrowly oblong, up to 1/3 as wide as long.........
                ....L. *angustifolia*

Lespedeza angustifolia (Pursh) Elliot   Narrow-leaved bush-clover
Stems 5–12 dm, branched only near the top; leaflets linear or narrowly oblong,
2.5–6 cm long and 2–6 mm wide; flowers yellowish, in dense axillary racemes;
fruit nearly circular, shorter than the stipules; very rare in moist, open, sandy soil
of an abandoned gravel pit; SE; flr. Aug–Sep; FAC; ✥.

Lespedeza bicolor Turcz.   Bicolor lespedeza
Shrub or semishrub to 3 m tall; stipules narrow and persistent; leaflets elliptic,
1–4 cm; flowers purple, 9–13 mm, in numerous, strongly ascending, axillary racemes that give the appearance of a leafy, terminal panicle; planted for erosion
control and occasionally naturalized in old fields and open woods; flr. Aug–early
Oct; native to Japan.

Lespedeza capitata Michx.   Round-headed bush-clover
Stems stiff, erect, 6–15 dm, usually simple or with a few branches in the upper
portion; leaflets narrowly oblong to elliptic, 2.5–5 cm long and 5–18 mm wide,
the lower surfaces grayish due to appressed silky hairs; flowers clustered near
the stem tip in dense terminal and axillary heads on peduncles shorter than the
subtending leaves; corolla yellowish-white; fruit ovate to oblong about half as
long as the calyx lobes; occasional in dry old fields and thin woods; throughout;
flr. Jul–Sep; FACU–. Hybridizes with L. *virginica*.

Lespedeza cuneata,
flower (×2)

Lespedeza cuneata (Dum.Cours.) G.Don   Bush-clover
Stems stiff, erect to 1 m with ascending branches; leaves numerous; leaflets narrowly wedge-shaped, grayish-pubescent beneath; flowers in axillary clusters,
whitish or purple-veined; fruit 2 mm long; fields and grassy roadsides, planted
for erosion control; flr. Jul–Oct; native to eastern Asia.

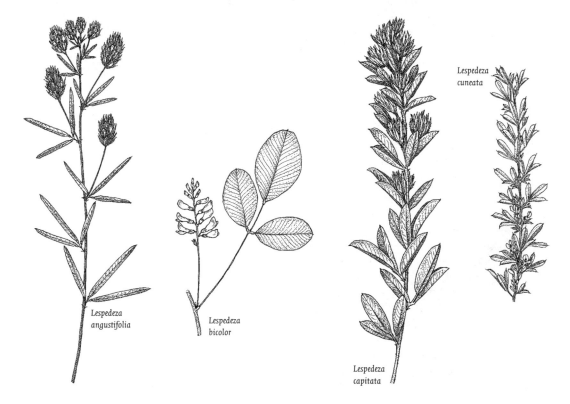

Lespedeza
cuneata

Lespedeza
angustifolia

Lespedeza
bicolor

Lespedeza
capitata

Lespedeza intermedia

Lespedeza thunbergii

Lespedeza thunbergii, flower (×2)

Lespedeza hirta, fruit (×2)

Lespedeza intermedia, fruit (×2)

Lespedeza hirta

Lespedeza hirta (L.) Hornem.   Bush-clover
Stems 6–12 dm with fine, spreading hairs, branched above; leaflets oval to nearly circular, obtuse to somewhat notched at the ends, longer than the petioles; flowers in dense axillary racemes on peduncles that are longer than the subtending leaves; corolla yellowish or sometimes purple-spotted; fruit oval, about as long as the calyx; frequent in dry, open soils; throughout; flr. Jun–Sep. Hybridizes with L. intermedia and L. virginica.

Lespedeza intermedia (S.Watson) Britton   Bush-clover
Stems erect, 3–9 dm, finely pubescent to glabrous, branched above; leaflets oval to oblong, obtuse or truncate at the tip; flowers in short-stalked axillary clusters; corolla violet-purple; fruit about 4 mm long, exceeding the calyx; common in dry, open, rocky woods and thickets; throughout; flr. Jul–Sep.

Lespedeza procumbens Michx.   Trailing bush-clover
Stems trailing or reclining, 3–10 dm, with densely spreading hairs; leaflets downy-pubescent, oval or elliptic, rounded at both ends and often notched at the tip; petaliferous flowers 2–8 at the ends of peduncles that are longer than the leaves; corolla violet-purple; fruit 3–7 mm long, oval to nearly circular; frequent in sandy or rocky soils of dry woods, fields, or roadsides; mostly S; flr. Jun–early Oct. Hybridizes with L. virginica.

Lespedeza repens (L.) W.Bartram   Creeping bush-clover
Stems slender, branched, prostrate or reclining, to 6 dm, glabrous or with appressed hairs; leaflets oval or obovate, 6–16 mm long, obtuse; flowers in long-peduncled clusters; corolla violet to purple, 4–6 mm long; fruit oval to circular, 3 mm long; frequent in wooded banks or edges in dry, sterile, acidic soils; mostly S; flr. Jun–Sep.

Lespedeza stuevei Nutt.   Tall bush-clover
Stems erect to spreading, rarely branched; 3–12 dm; leaves numerous, covered with fine spreading pubescence; leaflets oval to oblong, densely pubescent beneath; flowers in sessile axillary racemes; corolla violet to purple; fruit ovate to nearly circular, much longer than the calyx; very rare in dry, open woods and edges in sterile soil; SE; believed to be extirpated; flr. Aug–Sep; ✺.

Lespedeza thunbergii (DC.) Nakai   Thunberg's bush-clover
Shrub or semishrub 1–3 m tall; stipules slender, persistent; leaflets 1–4 cm, ellip-

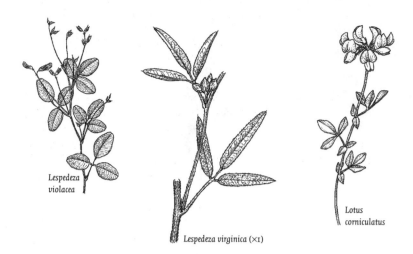

Lespedeza
violacea

Lespedeza virginica (×1)

Lotus
corniculatus

tic or lance-elliptic; glabrous above, strigose beneath; flowers showy, purple, 12–17 mm, in spreading or drooping axillary racemes; rarely escaped from cultivation to dry fields and open woods; SE; flr. Aug–Oct; native to Asia.

Lespedeza violacea (L.) Pers.   Bush-clover
Stems erect or ascending, 3–9 dm, branched and sparsely pubescent; leaves well separated on the stem; leaflets oval-oblong, rounded at both ends with appressed hairs beneath; flowers in few-flowered, axillary racemes or panicles on slender peduncles that are longer than the subtending leaves; corolla violet to purple, 6–10 mm long; fruit 4–6 mm long; occasional in dry, upland woods, thickets, and openings; mostly S; flr. Aug–Sep.

Lespedeza virginica (L.) Britton   Slender bush-clover
Stems slender, erect, with few if any branches, 3–11 dm tall; leaves numerous, on long slender petioles; leaflets linear to narrowly oblong; 1.2–3.5 cm long by 2–5 mm wide; flowers in dense sessile clusters in the upper leaf axils; corolla violet purple, 4–8 mm long; fruit ovate or broadly oval, 4 mm long; occasional in dry fields, stony banks, and serpentine barrens; mostly S; flr. Jul–Sep.

Lotus L.

Lotus corniculatus L.   Bird's-foot trefoil
Glabrous perennial with spreading to ascending stems to 6 dm; leaves 5-foliate, the lowest pair of leaflets basal; stipules absent; flowers yellow in long-peduncled umbels; frequently planted along roadsides or other open, disturbed areas, also cultivated as a forage crop; flr. May–Sep; native to Europe; FACU–.

Lupinus L.

Lupinus perennis L.   Blue lupine
Erect, pubescent perennial, 2–6 dm tall; leaves palmately compound with 7–11 leaflets; flowers blue (occasionally white) in erect racemes above the leaves; rare on stream banks, open fields, woods edges, and roadsides in sandy, acidic soils; E and C; flr. Apr–Jul; ❧.

Medicago L.

Herbaceous; leaves trifoliate; leaflets serrulate, the terminal one stalked; stipules fused with the petioles at the base; flowers in globose to short-cylindric heads, papilionaceous; calyx tube campanulate; fruits indehiscent.

A. flowers yellow; fruits 1-seeded, kidney-shaped .... M. lupulina
A. flowers blue-violet; fruits several-seeded, coiled .... M. sativa

Medicago lupulina L.   Black medic
Annual or biennial with prostrate or ascending stems to 8 dm; flowers yellow, 2–4 mm; fruits nearly black, 2–3 mm, kidney-shaped, 1-seeded; common in roadsides and waste places; throughout; flr. May–Oct; native to Eurasia.

Medicago lupulina,
fruit (×3)

Medicago sativa L.   Alfalfa
Perennial with ascending stems to 1 m; flowers blue-violet, 6–12 mm; fruits coiled; widely cultivated as a forage crop and occasionally escaped in roadsides, railroad embankments, and vacant lots; throughout; flr. May–Sep; native to Eurasia.

Medicago sativa, fruit (×3)

Medicago
sativa

Lupinus perennis

Melilotus Mill.

Erect biennial (or annual) herbs with trifoliate leaves and yellow or white flowers in numerous axillary racemes; calyx teeth subequal; stamens 10, diadelphous; style glabrous; fruit indehiscent; in addition to the species treated below, *Melilotus altissima* Thuill., with yellow flowers and pubescent, veiny, 2-seeded fruits, has been collected in Lehigh and Northampton Cos.

A. corolla white, 4–5 mm long .... *M. alba*
A. corolla yellow, 5–7 mm long .... *M. officinalis*

Melilotus alba Medik.   White sweet-clover
Stems 1–3 m tall; racemes 5–20 cm; flowers white; common in roadsides and old fields; throughout; flr. early May–Nov; native to Eurasia; FACU.

Melilotus officinalis (L.) Pall.   Yellow sweet-clover
Stems 0.5–1.5 m tall; racemes 5–15 cm; flowers yellow, 5–7 mm; common in fields, waste ground, and roadsides; throughout; flr. May–Nov; native to Eurasia; FACU.

Phaseolus L.

Phaseolus polystachios (L.) Britton, Stearns & Poggenb.   Wild bean
Herbaceous perennial twining vine to 3–4 m; leaves pinnately trifoliate; leaflets broadly ovate; flowers in a slender raceme, purple, 10–12 mm; keel petal coiled; fruit flat, 3–6 cm, coiled after dehiscence; infrequent in moist woods, roadside banks and waste ground; S; flr. Jul–Sep.

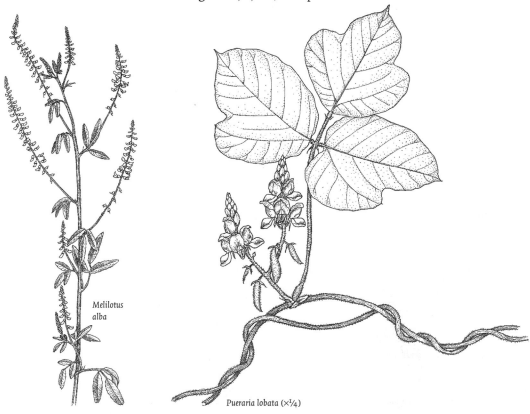

Melilotus alba

Pueraria lobata (×¼)

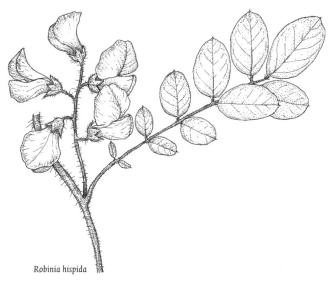

Robinia hispida

Pueraria DC.

Pueraria lobata (Willd.) Ohwi   Kudzu

Vigorous growing, half-woody twining vine that drapes itself extensively over trees and shrubs to a height of 10–20 m; leaves trifoliate; leaflets broadly ovate, 10–15 cm long, often with 2 or 3 lobes; inflorescence 1–2 dm, flowers purple; rarely forming mature seed here and usually dying to the ground in winter; planted for ornament or erosion control and occasionally escaped to waste ground and woods edges; flr. Jul–Oct; native to eastern Asia; designated as a noxious weed in Pennsylvania.

Robinia L.

Deciduous trees or shrubs with alternate, odd-pinnate leaves, setaceous stipules or stipular spines, and showy, white or pink flowers in axillary racemes; calyx bilabiate; corolla papilionaceous; stamens 10, diadelphous; fruit elongate, flat, many-seeded.

A. trees with stipular spines and otherwise puberulent to subglabrous twigs; flowers white; fruit glabrous .... *R. pseudoacacia*
A. shrubs or small trees with bristly or glandular twigs; flowers pink; fruit hispid
 B. twigs covered with sessile or short-stalked glands; leaflets 13–25 .... *R. viscosa*
 B. twigs covered with bristly, glandular hairs; leaflets 7–13 .... *R. hispida*

Robinia hispida L.   Bristly or mossy locust, rose-acacia

Rhizomatous shrub, 1–2 m tall, stems, peduncles, and calyces densely to sparsely covered with bristly, glandular hairs; leaflets 7–13, ovate-oblong, 3–6 cm; racemes 3–10 flowered; flowers rose-pink to pink-purple, 2.5–3 cm; fruits densely hispid; rare in dry, open woods, slopes and roadsides; flr. May–Jun; native to the southern mountains, VA to AL, but planted in Pennsylvania.

Robinia pseudoacacia L.   Black locust

Upright tree to 25 m tall; paired stipular spines to 1 cm; leaflets 7–23, oval or elliptic, 2–6 cm long, entire; flowers 1.5–2.5 cm, white, in drooping, many-flowered racemes, fragrant; fruit 5–10 cm; common in open woods, floodplains, thickets, and fencerows; throughout; flr. May–Jun; native, at least in the west; FACU–.

Robinia viscosa Vent.   Clammy locust
Large shrub or small tree to 5 m tall (or more); twigs and peduncles covered with large sessile or short-stalked glands; leaflets 13–25, lance-ovate to oval, flowers pink, 2.5 cm, in short, suberect racemes; fruit covered with glandular hairs; infrequent in dry, open ground, thin woods, or slopes; E; flr. Jun–Jul; native, but also planted.

Sophora L.

Sophora japonica L.   Japanese pagoda-tree
Deciduous tree to 20 m tall with green branches and alternate, odd-pinnate leaves; leaflets 7–17, entire; flowers yellowish-white in loose panicles; fruits 5–7.5 cm long, indehiscent, somewhat constricted between the seeds; cultivated and occasionally naturalized in disturbed woods; flr. Jul–early Aug; native to China.

Strophostyles Elliot

Trailing or twining vines with hairy stems; leaves trifoliate, the leaflets with stipels; flowers pink-purple or white, sessile in small heads on long axillary peduncles; each flower closely subtended by a pair of bracteoles; stamens diadelphous; style bearded along the upper side; fruits cylindrical or slightly flattened.

A. leaflets narrowly oblong; calyx tube and bracteoles densely hairy ....
....S. leiosperma
A. leaflets ovate; calyx tube and bracteoles glabrous or with only a few appressed hairs
    B. bracteoles lanceolate, at least as long as the calyx tube; seeds 6–10 mm ....
    .... S. helvola
    B. bracteoles ovate or oblong, blunt, to ½ as long as the calyx tube; seeds 3–6 mm .... S. umbellata

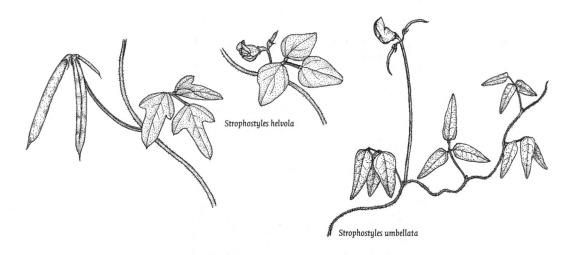

Strophostyles helvola

Strophostyles umbellata

Strophostyles helvola (L.) Elliot   Wild bean
Annual, stems to 1 m tall; leaflets ovate, frequently with 1 or 2 lateral lobes or with concave lateral margins; flowers several on 0.5–3-dm peduncles; corolla pink-purple; fruits 4–9 cm; seeds 5–10 mm; rare in sandy fields and railroad embankments; mostly SE; flr. Jul–Sep; FACU.

Strophostyles leiosperma (Torr. & A.Gray) Piper   Wild bean
Annual; leaflets narrowly oblong to lanceolate; flowers 5–8 mm; fruit 2–4 cm; seeds 2.5–3 mm; rare in open, sandy ground and railroad ballast; S; flr. Jul; native farther west.

Strophostyles umbellata (Muhl. ex Willd.) Britton   Wild bean
Similar to S. *helvola* but perennial; leaflets never lobed; bracteoles at least as long as the calyx tube; seeds 3–5 mm; rare in clearings and fields, sandy soils, and serpentine barrens; SE; flr. Jul–Sep; FACU; ❦.

Stylosanthes Swartz

Stylosanthes biflora (L.) Britton, Stearns & Poggenb.   Pencil-flower
Stiffly erect, pubescent perennial with wiry stems and trifoliate leaves; upper leaves crowded at the base of the flower clusters; stipules fused in a tube around the stem; flowers in small axillary or terminal clusters, yellow, 6–8 mm long; rare in sandy fields, riverbanks, and dry rocky or shaly slopes; S; flr. Jul–Sep; ❦.

Tephrosia Pers.

Tephrosia virginiana (L.) Pers.   Goat's-rue
Hairy perennial with clustered, mostly unbranched, stems 2–7 dm tall; leaflets 15–25; flowers in terminal racemes; standard yellowish-white, wings and keel pinkish-purple; occasional in dry woods and openings in sandy, acidic soils; mostly S; flr. Jun–Jul.

Stylosanthes biflora

Trifolium L.

Herbs with trifoliate leaves, stipules fused to the petioles; leaflets finely toothed; flowers in dense globose to cylindrical heads; individual flowers sessile or short-stalked; calyx bell-shaped or tubular, the lobes with slender tips; corolla persisting dry after flowering; 9 of the stamens united, 1 separate; fruit 1–6-seeded, often shorter than the persistent calyx, indehiscent or opening along only 1 suture.

A. flowers yellow
    B. leaflets all sessile .... *T. aureum*
    B. terminal leaflet stalked
        C. flowers 3.5–5 mm; petiolule of the terminal leaflet 1–3 mm ....
        .... *T. campestre*
        C. flowers 2.5–3.5 mm; petiolule of the terminal leaflet 1 mm at most ......
        .... *T. dubium*
A. flowers white, pink, or purple
    D. individual flowers sessile or nearly so
        E. flower heads ovoid to cylindrical, longer than wide

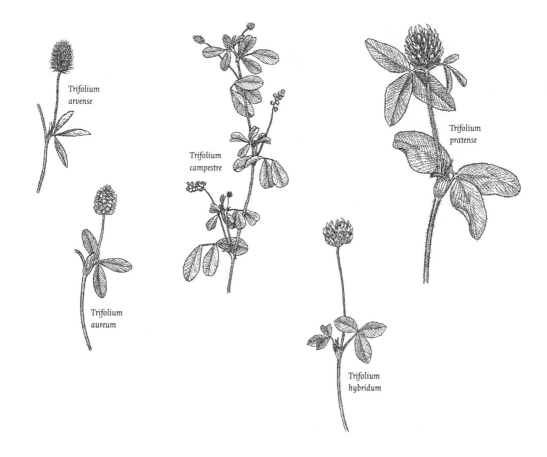

Trifolium arvense

Trifolium aureum

Trifolium campestre

Trifolium pratense

Trifolium hybridum

      F. corolla bright crimson, longer than the calyx lobes ..... T. incarnatum
      F. corolla whitish or pinkish, shorter than the calyx lobes ..... T. arvense
    E. flower heads globose, nearly as broad as long
        G. heads subtended by a pair of opposite leaves; flowers deep magenta
                .... T. pratense
        G. heads on naked peduncles, not subtended by a pair of leaves; flow-
                ers purplish .... T. resupinatum
   D. individual flowers with a pedicel at least 2 mm long
      H. calyx hairy or bristly
        I. leaflets narrowly oblong; stems prostrate .... T. virginicum
        I. leaflets obovate; stems erect or ascending .... T. reflexum
      H. calyx glabrous or nearly so
        J. leaflets notched at the tip; stem stoloniferous .... T. repens
        J. leaflets rounded at the tip; stems erect or ascending ...... T. hybridum

Trifolium arvense,
flower (×3)

Trifolium arvense L.   Rabbit's-foot clover
Grayish-pubescent, branched annual, 1–4 dm tall; leaflets oblanceolate, toothed
only at the tip; heads ovoid to cylindric, densely flowered; sepals with slender
tips longer than the corolla; occasional in dry fields and roadsides; mostly S; flr.
Jun–early Oct; native to Eurasia and N. Africa.

Trifolium aureum Pollich   Large yellow hop-clover
Erect, nearly glabrous, branched annual or biennial, 2–5 dm tall; leaflets all

sessile; flowers yellow, sessile in a dense ovoid or short cylindric head 1–2 cm long and 1–1.5 cm in diameter; common in dry fields, waste ground, and roadsides; throughout; flr. Jun–Nov; native to Eurasia.

Trifolium campestre Schreb.   Low hop-clover
Branched annual, 1–4 dm tall; the terminal leaflet with a 1–3-mm stalk; flowers yellow in a globose to short cylindric head 8–15 mm long and 8–12 mm in diameter, with 20–30 flowers; individual flowers 3.5–5 mm; frequent in roadsides, old fields, and waste places; mostly S; flr. May–Oct; native to Europe.

Trifolium dubium Sibth.   Little hop-clover
Very similar to *T. campestre* but with the stalk of the terminal leaflet <1 mm and the heads 5–10 mm long with 5–15 flowers; individual flowers 2.5–3.5 mm; infrequent in roadsides, old sand pits, and lawns; scattered; flr. May–Oct; native to Europe; UPL.

Trifolium hybridum L.   Alsike clover
Ascending perennial, 3–8 dm tall; heads globose, densely flowered, on peduncles longer than the leaves; flowers pink and white turning brown with age, distinctly stalked; common in fields, clearings, and roadsides; throughout; flr. May–Oct; native to Eurasia; FACU.

Trifolium hybridum, flower (×1)

Trifolium incarnatum L.   Crimson or Italian clover
Erect, soft-hairy annual; leaves long-petioled; heads cylindrical or ovoid, 2–6 cm long; flowers crimson, sessile, 8–15 mm long; roadsides and waste ground, spreading from cultivation; flr. May–Sep; native to Europe.

Trifolium pratense L.   Red clover
Ascending, pubescent perennial to 8 dm tall; leaves long-petioled; heads globose, 2–3 cm long, subtended by a pair of trifoliate leaves; flowers magenta-pink, 12 mm long, sessile; widely grown as a forage plant and extensively naturalized in pastures, fields, and roadsides; throughout; flr. May–Oct; native to Europe; FACU–.

Trifolium reflexum L.   Buffalo clover
Ascending annual or biennial, 1–5 dm tall, not stoloniferous; heads globose, 2.5–4.5 cm in diameter; corolla with a red or white standard and white wings and keel; very rare in woodland edges, fields, and roadsides; SE and SW; believed to be extirpated; flr. May–Jun; 🌿.

Trifolium repens L.   White clover
Creeping, glabrous perennial with stems prostrate and rooting at the nodes; leaflets often notched at the tip; flowers 7–11 mm, distinctly stalked; corolla white or pink-tinged; common in lawns, fields, and roadsides; throughout; flr. May–Oct; native to Europe; FACU–.

Trifolium resupinatum L.   Persian clover
Ascending or decumbent, glabrous annual to 3 dm tall; heads globose, 1–1.5 cm in diameter; flowers inverted with the standard on the lower side and the keel upward, purplish; rare in lawns, fields, roadsides, or stream banks; scattered; flr. May–Sep; native to Europe; UPL.

Trifolium virginicum Small   Kate's-mountain clover
Prostrate perennial from a taproot; stems pubescent to 1–2 dm tall; leaflets linear to oblanceolate; heads globose, 2–3 cm in diameter; corolla yellowish-white, 10–12 mm long; rare in shale barrens; SC; flr. May–Aug; 🌿.

Vicia L.

Herbaceous vines with pinnate leaves bearing a tendril in place of the terminal leaflet; stem not winged; flowers axillary in racemes or few-flowered clusters; calyx lobes 5, nearly equal; corolla violet or white; standard obovate to almost round; wings fused to the middle of the keel; stamens 10, 9 fused and 1 free; style pubescent at the summit; fruit flat or rounded, dehiscent, with 2–many seeds.

A. flowers few, sessile (or nearly so) in the upper axils .... *V. sativa*
A. flowers solitary or few on long peduncles or in long-stalked racemes
    B. flowers 1–8 on long peduncles; corolla white or whitish, 3–7 mm
        C. calyx lobes unequal; fruit 4-seeded, glabrous .... *V. tetrasperma*
        C. calyx lobes equal; fruit 2-seeded, hairy .... *V. hirsuta*
    B. flowers in racemes of 8–many; corolla >7 mm long, violet
        D. racemes mostly shorter than the subtending leaves, 3–10 flowered
            E. flowers blue-violet; stipules serrate .... *V. americana*
            E. flowers white with a blue-tinted keel; stipules entire ....*V. caroliniana*
        D. raceme as long as or longer than the subtending leaf, flowers 10–20 (or more); corolla violet
            F. calyx swollen on the upper side at the base, the pedicel thus appearing lateral .... *V. villosa*
            F. calyx not swollen on the upper side at the base, the pedicel appearing basal .... *V. cracca*

Vicia americana Muhl. ex Willd.   Purple vetch
Climbing perennial with stems to 1 m, glabrous or nearly so; stipules palmately toothed; leaflets 8–18; raceme 3–9-flowered, shorter than the subtending leaves; flowers blue-violet; occasional in moist gravelly shores, thickets, meadows, or roadside banks; mostly E; flr. May–Jul; FACU–.

Vicia caroliniana Walter   Wood vetch
Trailing perennial with slender stems to 1.5 m; leaflets 10–20, mucronate; stipules entire; raceme loose, shorter than the subtending leaf, bearing 8–10 white flowers; occasional in rich woods and thickets, often on limestone; scattered throughout; flr. Apr–Jun; FACU–.

Vicia cracca L.   Canada pea, cow vetch
Trailing perennial with angular, pubescent stems to 2 m; leaflets 10–24, mucronate; stipules entire; raceme 1-sided, many-flowered; flowers blue-violet (occasionally white), 9–13 mm long; naturalized in fields, roadsides, or floodplains; flr. May–Aug; native to Eurasia.

Vicia hirsuta (L.) S.F.Gray   Vetch
Slender annual with climbing or decumbent stems 3–6 dm; leaflets 12–16; pe-

duncles 1–3 cm, bearing 3–8 whitish flowers; calyx lobes nearly equal; fruit flattened, hairy, 2-seeded; rare in roadsides and waste places; SE; flr. Apr–Jun; native to Europe.

Vicia sativa L.   Common vetch
Slender, ascending or climbing annual to 1 m; leaflets 8–16; stipules serrate with a glandular spot beneath; flowers sessile or nearly so, paired in the upper axils, violet; fruits 3.5–7.5 cm long; rare in cultivated fields and roadsides; mostly SE; flr. May–Aug; native to Europe; FACU–; 2 varieties:

A. leaflets 4–10; corolla 1–1.8 cm long .... ssp. *sativa*
A. leaflets 2–5; corolla 1.8–2.5 cm long .... ssp. *nigra* (L.) Ehrh.

Vicia tetrasperma (L.) Schreb.   Slender vetch
Slender annual with climbing stems to 5 dm; leaflets 4–10; peduncles 1–3 cm long bearing 1–6, pale purple to whitish flowers; calyx lobes unequal; fruits flat, glabrous, 4-seeded; naturalized in moist meadows, roadsides, and moist areas on serpentine barrens; E; flr. May–Aug; native to Eurasia.

Vicia villosa Roth   Hairy or winter vetch
Hairy annual or biennial to 1 m tall; leaflets 10–20; racemes long-peduncled, 1-sided, bearing 10–40 violet flowers; calyx swollen on the upper side at the base; spreading from cultivation to fields and roadsides; throughout; flr. May–Sep; native to Europe; 2 varieties:

A. plants conspicuously villous; dorsal calyx lobe 2–4 mm long .... ssp. *villosa*
A. plants glabrate or inconspicuously pubescent; dorsal calyx lobe 1–2 mm long
.... ssp. *varia* (Host) Corb.

Wisteria Nutt.

Twining woody vines with alternate, deciduous, odd-pinnate leaves and showy, violet-blue, strongly papilionaceous flowers in drooping racemes that appear before the leaves have fully expanded; fruit linear, flattened, dehiscent.

A. racemes 15–40 cm long; ovary and fruit pubescent
   B. leaflets 15–19 .... *W. floribunda*
   B. leaflets 7–13 .... *W. sinensis*
A. racemes 4–15 cm long; ovary and fruit glabrous .... *W. frutescens*

Wisteria floribunda (Willd.) DC.   Japanese wisteria
Vigorous twining vine; leaflets 15–19, ovate-elliptic, 3–7 cm long; flowers in racemes to 40 cm; fruit velvety-pubescent, to 10–15 cm; a popular landscape ornamental that occasionally runs wild in roadside thickets or abandoned gardens; SE; flr. Apr–Jun; native to Japan.

Wisteria frutescens (L.) Poir.   American wisteria
Stout climbing vine to 35 m; leaflets 9–15, somewhat pubescent, ovate to ovate-lanceolate, 2.5–5 cm long; flowers in compact racemes to 10 cm; fruit glabrous, 5–10 cm; rare in alluvial woods and riverbanks; flr. Apr–Jul; native VA to FL, Pennsylvania collections probably represent introductions.

*Vicia
tetrasperma*

Wisteria sinensis, leafless stem

Wisteria sinensis (Sims) Sweet   Chinese wisteria
Climbing vine; leaflets 7–13, ovate-acuminate or ovate-lanceolate, short-stalked, 5–7 cm long; flowers in racemes 15–30 cm; fruit velvety-pubescent, 10–15 cm; a landscape ornamental occasionally naturalized in disturbed woods or abandoned nurseries or gardens; flr. May–Jul; native to China.

FAGACEAE Beech Family
John Kunsman

Deciduous trees and shrubs; leaves simple, alternate, the margins lobed, toothed, or entire; flowers unisexual, the staminate in catkins, the pistillate solitary or in spikes; fruit a nut, enclosed by a scaly cup or a spiny bur.

A. leaves lobed, toothed, or entire; twigs with a cluster of buds at the ends; nut (acorn) enclosed by a scaly cup; staminate catkins slender and elongate, drooping .... *Quercus*
A. leaves toothed; twigs with a single bud at the ends; nut(s) enclosed by a spiny bur; staminate catkins globe-shaped, or slender, elongate and ascending
    B. nut sharply triangular, the bur weakly spiny; staminate catkins globe-shaped, drooping; buds slender and elongate with 6 or more scales ....
    .... *Fagus*
    B. nut angled or rounded, the bur sharply spiny; staminate catkins elongate, ascending; buds ovoid with 2–3 scales .... *Castanea*

Castanea Mill.

Trees or shrubs; true terminal bud absent; bud scales 2–3; leaves elliptic-oblong, coarsely toothed with parallel lateral veins; staminate catkins fragrant, ascending; nut(s) enclosed within a sharply spiny bur; flr. Jun, after the leaves are well grown.

A. mature leaves green and glabrous below; branches and buds glabrous; nuts usually 2–4 per bur... *C. dentata*
A. mature leaves whitish- or grayish-pubescent below
    B. shrub or small tree to 5 m tall; nut usually 1 per bur .... *C. pumila*
    B. tree to 20 m tall; nuts 2 or 3 per bur .... *C. mollissima*

Castanea dentata (Marshall) Borkh.   American chestnut
Shrub or small tree to 5 m tall (formerly much larger before the introduction of the chestnut blight); mature twigs and buds glabrous or nearly so; leaves glabrous or nearly so, tip acuminate, teeth often incurved; nuts usually 2–4 per bur; common in dry woods and thickets; throughout.

Castanea mollissima Blume   Chinese chestnut
Tree to 20 m tall distinguished from *C. dentata* by its pubescent branches (at least when young) and pubescent leaf undersides; nuts 2 or 3 per bur; cultivated and sometimes persisting; native to Asia.

Castanea pumila Mill.   Chinquapin
Shrub or small tree to 5 m tall; mature twigs and buds usually pubescent; leaves

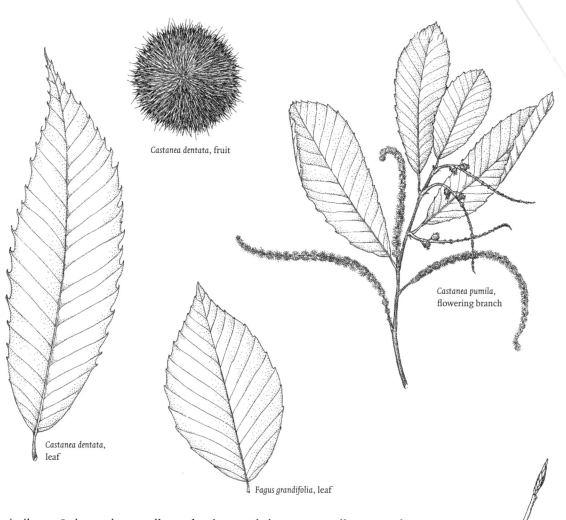

Castanea dentata, fruit

Castanea pumila,
flowering branch

Castanea dentata,
leaf

Fagus grandifolia, leaf

similar to *C. dentata* but smaller and pubescent below; nut usually one per bur; occasional in dry woods; SC.

Fagus L.

Fagus grandifolia Ehrh.   American beech
Tree to 25 m tall with smooth gray bark; true terminal bud present; bud scales 6 or more; twigs with nearly circular stipule scars; leaves toothed, with parallel lateral veins, thin in texture; the staminate catkins globe-shaped, drooping; nuts usually paired, enclosed within a weakly spiny 4-parted bur; common in moist woods; throughout; flr. before or with the leaves; FACU.

Quercus L.

Trees or shrubs with stellate pith; true terminal bud present, usually surrounded by a cluster of axillary buds; bud scales 6 or more; leaves mostly lobed or toothed, but sometimes entire; the staminate catkins very slender and drooping; nut (acorn) enclosed by a scaly cup; flr. before or with the leaves. Identification of oaks is often difficult due to variation within and hybridization between species.

Fagus grandifolia,
fruit

Fagus grandifolia,
winter twig

A. tips of the leaves or tips of the teeth or lobes ending in a bristle; acorns maturing in 2 growing seasons (1 season in *Q. acutissima*), immature acorns often present on the twigs in winter; "red oaks"

    B. leaves entire, usually 4.0 cm or less wide

        C. leaves mostly 2–4 cm wide, densely pubescent beneath ..... *Q. imbricaria*

        C. leaves mostly 1–2 cm wide, usually glabrous or nearly so beneath ....
        .... *Q. phellos*

    B. leaves toothed or lobed, often >4 cm wide

        D. lower surface of mature leaves glabrous or with tufts of hairs in the vein axils, or if pubescent throughout, then the hairs easily removed by light rubbing with a finger

        E. leaves lanceolate, sharply but coarsely toothed............. *Q. acutissima*

        E. leaves broader, lobed

            F. mature buds densely pubescent (usually whitish or grayish) throughout their length; scales around rim of the acorn cup loose at their tips .... *Q. velutina*

            F. mature buds glabrous, or densely pubescent only on the upper half or near the tip; scales around rim of the acorn cup appressed

            G. sinuses between lobes of the leaf relatively shallow, mostly $\frac{1}{2}$ or less the distance to the midrib .... *Q. rubra*

            G. sinuses between lobes relatively deep, >$\frac{1}{2}$ (often $\frac{3}{4}$ or more) the distance to the midrib

                H. upper scales of mature buds densely pubescent and contrasting with the glabrous or sparsely pubescent lower scales; acorn cup covering about $\frac{1}{2}$ of the nut; upland species .... *Q. coccinea*

                H. upper scales of mature buds mostly glabrous, similar to the lower scales; acorn cup covering $\frac{1}{4}$–$\frac{1}{3}$ of the nut; lowland species

                    I. bud scales red-brown, relatively lustrous; nut about 1 cm long; leaf lobes 5–7, tending to narrow to the tip with relatively few teeth and secondary lobes .... *Q. palustris*

                    I. bud scales brownish or grayish, relatively dull; nut 1.5–2.5 cm long; leaf lobes mostly 7, tending to enlarge to the tip with relatively numerous teeth and secondary lobes .... *Q. shumardii*

        D. lower surface of mature leaves densely and closely pubescent, the hairs crusty-granular or soft and felt-like, not easily removed by light rubbing with a finger

        J. shrub; leaf blades mostly <10 cm long, white or light gray below ....
        .... *Q. ilicifolia*

        J. tree; leaf blades usually >10 cm long, green, yellow-green, rusty, or grayish below

            K. leaves widest near the tip, often with 3 shallow lobes, the hairs of the lower surface crusty-granular; acorn cup covering $\frac{1}{2}$ of the nut... *Q. marilandica*

            K. leaves widest below to above the middle, with 3–7 mostly narrow, often curved lobes, the hairs of the lower surface relatively soft and felt-like; acorn cup covering $\frac{1}{3}$–$\frac{1}{2}$ of the nut .... *Q. falcata*

A. tips of leaves or tips of teeth or lobes rounded, acute, or with a short cusp; acorns maturing in one growing season; "white oaks"
 L. leaves lobed
  M. leaves acute, obtuse, or rounded at the base
   N. leaves glabrous beneath .... *Q. alba*
   N. leaves pubescent beneath (hairs may not be evident without magnification)
    O. leaves variously lobed, the upper half tending to be coarsely toothed; scales around the rim of the acorn cup elongate and curly, the cup covering ½ or more of the nut .... *Q. macrocarpa*
    O. leaves, or many of them, tending to be cross-shaped; scales around the rim of the acorn cup scaly, the cup covering ⅓–½ of the nut .... *Q. stellata*
  M. leaves auricled at the base .... *Q. robur*
 L. leaves coarsely toothed (occasional leaves of *Q. bicolor* may be lobed)
  P. stalk of acorn 2 cm or more; lower surface of leaves densely pubescent and soft to the touch, usually whitish or pale gray; in poorly drained habitats .... *Q. bicolor*
  P. stalk of acorn <2 cm; lower leaf surface glabrous or variously pubescent (hairs often not visible without magnification), green, pale green, or somewhat whitened; in well-drained habitats
   Q. shrub; leaf teeth mostly 4–7 on each margin .... *Q. prinoides*
   Q. tree; leaf teeth 8 or more on each margin
    R. leaves with rounded teeth, the hairs of the lower surface (under magnification) mostly 2–4 branched, branches ascending; mature bark dark gray, deeply ridged and furrowed; acorn cup funnel-shaped, the nut 2–3 cm long .... *Q. montana*
    R. leaves with acute or incurved teeth, the hairs of the lower surface (under magnification) mostly 6- or more branched, the branches appressed or spreading; mature bark light gray, scaly; acorn cup bowl-shaped, the nut 1–2 cm long .... *Q. muhlenbergii*

Quercus acutissima

Quercus acutissima Carruth.   Sawtooth oak

Tree to 20 m tall; mature twigs finely pubescent to glabrate; buds slender, sharp-pointed, hairy; leaves lanceolate, with coarse sharp teeth bearing long bristles; acorn cup heavily fringed; occasionally escaped from cultivation to fallow fields, also planted by the Pennsylvania Game Commission; native to Asia.

Quercus alba L.   White oak

Tree to 30 m tall with light gray bark; mature twigs glabrous; buds red-brown, blunt, mostly glabrous, to 5 mm long; leaves with 5–11 rounded lobes, glabrous and paler below; nut 1.5–3 cm long, ¼–⅓ covered by a warty, bowl-shaped cup; common in dry to moist woods; throughout; FACU. Hybridizes with *Q. bicolor, Q. macrocarpa,* and *Q. montana.*

Quercus bicolor Willd.   Swamp white oak

Tree to 30 m tall with gray ridged bark, conspicuously peeling on young branches; mature twigs glabrous; buds brown, blunt, mostly glabrous, to 5 mm long; leaves with 4–8 coarse teeth (or shallow lobes) on each margin, softly hairy

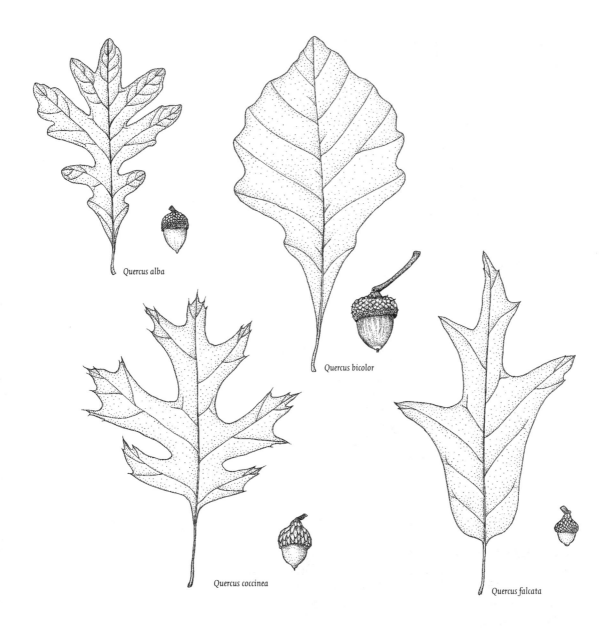

Quercus alba

Quercus bicolor

Quercus coccinea

Quercus falcata

and usually whitened below; nut 2–3 cm long, ⅓–½ covered by a bowl-shaped cup with a stalk 2 cm or more; frequent in swamps and low woods; mostly S; FACW+. Hybridizes with *Q. montana*.

Quercus coccinea Münchh.   Scarlet oak
Tree to 30 m tall with gray ridged bark; mature twigs glabrous; buds red-brown, bluntish, pubescent above the middle, to 6 mm; leaves with 7–9 lobes, sinuses deep, glabrous or with axillary tufts of hairs below; nut 1.5–2 cm long, ½ covered by a scaly bowl-shaped cup; common in dry woods; throughout, except the northernmost counties. Hybridizes with *Q. ilicifolia* and *Q. rubra*.

Quercus falcata Michx.   Southern red oak
Tree to 25 m tall with gray bark; mature twigs usually pubescent; buds red-

brown, pointed, usually pubescent, to 6 mm long; leaves with 3–7 variable but usually narrow and curved lobes, base broadly rounded, softly pubescent below; nut 1–1.5 cm long, ⅓ covered by a scaly saucer-shaped cup; rare in dry to moist woods; SE; FACU–; 🌿.

Quercus ilicifolia Wangenh.   Scrub or bear oak
Shrub to 4 m tall, forming thickets; mature twigs usually pubescent; mature buds red-brown, bluntish, pubescent or not, to 5 mm; collateral buds often present; leaves with 5–7 triangular lobes, softly white or gray-pubescent below; nut 1–1.5 cm, ½ covered by a scaly bowl-shaped cup; frequent in dry thickets and barrens; throughout, except the far N and W. Hybridizes with *Q. marilandica, Q. rubra,* and *Q. velutina.*

Quercus imbricaria Michx.   Shingle oak
Tree to 20 m tall with gray bark; mature twigs glabrous; mature buds brownish, pointed, mostly glabrous, to 4 mm; leaves entire, densely pubescent below; nut 1–2 cm long, ⅓–½ covered by a scaly bowl-shaped cup; dry to moist woods; frequent W, occasional E; FAC. Hybridizes with *Q. marilandica, Q. palustris, Q. rubra,* and *Q. velutina.*

Quercus macrocarpa Michx.   Bur or mossycup oak
Tree to 30 m tall with gray, ridged bark; branchlets sometimes corky-winged; twigs usually pubescent; mature buds brownish, bluntish, pubescent or not, to 6 mm; leaves variable, tending to be lobed on the bottom half and coarsely toothed on the upper half, pubescent below; nut 2–4 cm long, ½ or more covered by a bowl-shaped cup with elongate and curly scales; occasional in dry to moist woods, usually on limestone; mostly SC and SW; FAC–. Hybridizes with *Q. muhlenbergii.*

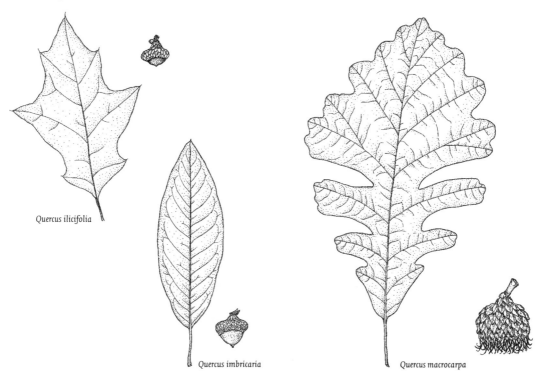

Quercus ilicifolia

Quercus imbricaria

Quercus macrocarpa

Quercus marilandica Münchh.   Blackjack oak
Tree to 10 m tall with rough bark; mature twigs usually pubescent; mature buds red-brown, pointed, densely pubescent, to 10 mm; leaves thick, variable, but mostly broadest above the middle with 3 lobes, crusty-granular below; nut 1–2 cm long, ½ covered by a loose-scaly bowl-shaped cup; rare on dry wooded slopes and serpentine barrens; SE. Hybridizes with *Q. phellos* and *Q. velutina*.

Quercus montana Willd.   Chestnut oak
Tree to 25 m tall with gray, deeply furrowed bark; mature twigs glabrous; mature buds brownish, mostly glabrous, pointed, to 8 mm; leaves with 8–16 rounded teeth on each margin, green and slightly pubescent below; nut 2–3 cm long, ⅓–½ covered by a warty, funnel-shaped cup; common in dry woods and rocky slopes; throughout; UPL.

Quercus muhlenbergii Engelm.   Chinquapin or yellow oak
Tree to 20 m tall with gray bark; mature twigs usually glabrous; mature buds brownish, pointed, pubescent or not, to 6 mm; leaves with 8–16 pointed and often incurved teeth on each margin, pale and pubescent below; nut 1–2 cm long, ⅓–½ covered by a scaly bowl-shaped cup; infrequent in dry to moist woods, usually on limestone; S; UPL.

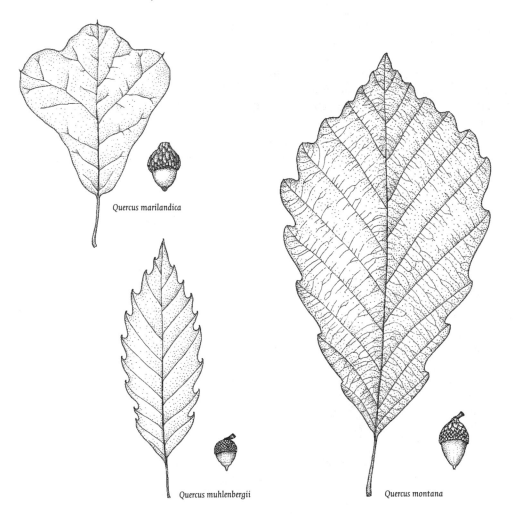

Quercus marilandica

Quercus muhlenbergii

Quercus montana

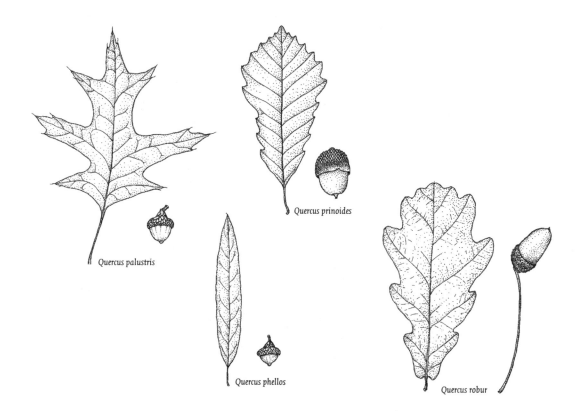

Quercus palustris

Quercus prinoides

Quercus phellos

Quercus robur

Quercus palustris Müenchh.   Pin oak
Tree to 20 m tall with gray bark and drooping lower branches; mature twigs glabrous; mature buds red-brown, pointed, glabrous, to 5 mm long; leaves with 5–7 tapering lobes, each with a few teeth and separated by deep sinuses, glabrous or with axillary tufts of hairs below; nut about 1 cm long, ¼ covered by a scaly saucer-shaped cup; frequent in swamps and low woods; throughout, except in the northernmost counties; FACW.

Quercus phellos L.   Willow oak
Tree to 30 m tall with gray bark; mature twigs glabrous; mature buds brownish, pointed, glabrous, to 4 mm long; leaves entire, mostly 1.0–2.0 cm wide, glabrous or nearly so below; nut 1–1.5 cm long, ¼–½ covered by a scaly saucer-shaped cup; rare in moist to wet woods; SE; FAC+; ❧. Hybridizes with *Q. rubra*.

Quercus prinoides Willd.   Dwarf chestnut oak
Shrub to 4 m tall, forming thickets; mature twigs pubescent or not; mature buds brownish, pubescent or not, bluntish, to 4 mm; leaves with 3–7(8) teeth on each margin, pale and pubescent below; nut 1–2 cm long, ⅓–½ covered by a scaly bowl-shaped cup; occasional in dry thickets and barrens; throughout, except in the northernmost counties.

Quercus robur L.   English oak
Tree to 25 m tall with gray bark; mature twigs glabrous; leaves with 6–10 rounded lobes, auricled at the base, paler and glabrous below; nut 1.5–2.5 cm long, ⅓ covered by a scaly cup; cultivated and very rarely escaped; native to Europe.

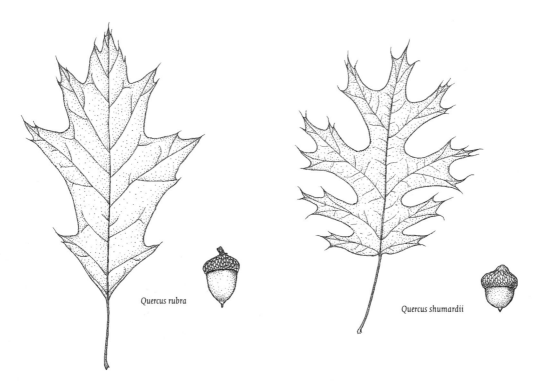

Quercus rubra

Quercus shumardii

Quercus rubra L.   Northern red oak
Tree to 30 m tall with gray, ridged bark; mature twigs glabrous; mature buds
red-brown, pointed, glabrous or pubescent at the tip, to 7 mm; leaves with 7–11
tapering lobes with few teeth, sinuses relatively shallow, glabrous or with axillary tufts of hairs and paler beneath; nut 2–3 cm long, ¼–⅓ covered by a scaly
saucer-shaped cup; common in dry to moist woods; throughout; FACU–.

Quercus shumardii Buckley   Shumard oak
Tree to 30 m tall with gray bark; mature twigs glabrous; mature buds grayish or
brownish, dull, glabrous, pointed, to 8 mm; leaves with 7–9 lobes that widen toward the tip and with numerous teeth, sinuses deep, axillary tufts of hairs beneath; nut 1.5–2.5 cm long, ¼–⅓ covered by a scaly saucer-shaped cup; rare in
moist to wet woods; SC; FAC+; 🌿.

Quercus stellata Wangenh.   Post oak
Tree to 20 m tall with gray bark; mature twigs pubescent; mature buds red-
brown, bluntish, pubescent, to 6 mm; leaves thick and leathery with 5–7 lobes,
often cross-shaped, pubescent below; nut 1–2 cm long, ⅓–½ covered by a scaly
bowl-shaped cup; occasional in dry woods, rocky slopes, and serpentine barrens; mostly SE; UPL.

Quercus velutina Lam.   Black oak
Tree to 30 m tall with gray, ridged bark; mature twigs pubescent or not; mature
buds with dense whitish or grayish, hairs often angled, pointed, to 12 mm;
leaves dark green, variable in lobing and depth of sinuses, the pubescence, if
present, tending to be easily removed; nut 1–2 cm long, ½ covered by a scaly
bowl-shaped cup with loose scales around the rim; common in dry woods;
throughout, except in the northernmost counties.

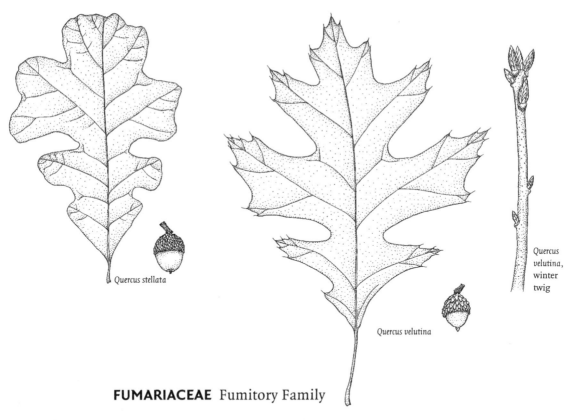

Quercus stellata

Quercus velutina

Quercus velutina, winter twig

FUMARIACEAE Fumitory Family

Annual, biennial, or perennial herbs, glabrous (rarely pubescent) and often glaucous, juice clear; leaves basal or basal and cauline, usually much dissected; flowers perfect, irregular, and bilaterally symmetrical, at least one outer petal spurred or with a sac-like extension projecting rearward, stamens 6 in two groups of 3; inflorescence a raceme or panicle; fruits mostly elongate capsules, 2-valved; all species contain isoquinoline alkaloids and should be considered toxic.

A. all leaves basal .... *Dicentra*
A. at least some leaves cauline
  B. plant a climbing vine .... *Adlumia*
  B. plant not vine-like
    C. flowers dark red to red-purple (without any yellow); fruit globose, about 2.5 mm across, 1-seeded .... *Fumaria*
    C. flowers yellow or bicolored pink to pink-purple and yellow; fruit a several-seeded capsule, 1–4 cm long .... *Corydalis*

Adlumia Raf.

Adlumia fungosa (Aiton) Greene ex Britton, Stearns & Poggenb.

Allegheny-vine, climbing fumitory

Biennial vine, forming a basal rosette the first year and developing a climbing stem up to 5 m long the second year; leaves pinnately compound and much divided, the leaflets much reduced upward along the elongate, twining rachis; inflorescences axillary, cyme-like racemes, up to 30-flowered; flowers white or pink to purple, corolla 1.5–2 cm long, flowers persisting in fruit; fruits about as

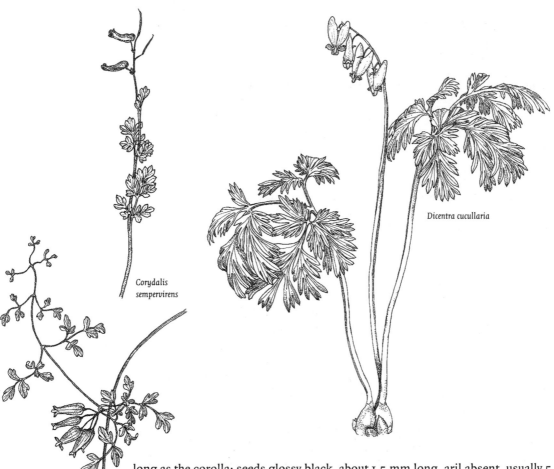

Corydalis
sempervirens

Dicentra cucullaria

Adlumia fungosa

long as the corolla; seeds glossy black, about 1.5 mm long, aril absent, usually 5 per capsule; moist, rocky slopes and woodlands; occasional throughout; flr. Jun–Sep.

Corydalis Vent.

Annual or biennial, stems hollow and semisucculent; both basal and cauline leaves generally present, glabrous and glaucous, cauline leaves somewhat to only slightly reduced upward; flowers irregular, the upper petal only with a sac-like projection.

A. stems erect, 3–8 dm tall; flowers pinkish to pink-purple with yellow tips ......
.... C. sempervirens

A. stems erect when young but soon trailing or prostrate to ascending; flowers yellow
   B. mature flowers 6–9 mm long, the upper petal with a 3–4-lobed or toothed crest .... C. flavula
   B. mature flowers 12–20 mm long, the upper petal not crested .... C. aurea

Corydalis aurea Willd.   Golden corydalis
Stems 1.5–6 dm tall, branching above; basal leaves pinnately compound with 5–7 major lobes, each lobe further dissected, cauline leaves similar and only slightly reduced upward; inflorescence a terminal or axillary raceme; capsules

drooping at maturity, 1.5–3 cm long; seeds shiny black, without a marginal ridge, aril about 1.5 mm long; very rare along roadsides; flr. May–Jul; ✺.

Corydalis flavula (Raf.) DC.   Yellow fumewort, yellow harlequin
Stems 1–4 dm tall, much branched above; leaves similar to *C. aurea*; inflorescence a terminal raceme; capsules spreading or drooping at maturity, 1.5–2 cm long; seeds similar to *C. aurea* but with a prominent marginal ridge, aril about 1 mm long; occasional in moist open woods, slopes, and edges; mostly S; flr. Apr–May; FACU.

Corydalis sempervirens (L.) Pers.   Rock harlequin
Stems 3–8 dm tall, sometimes branched above; basal leaves with 3–5 major lobes, each lobe further dissected, petiolate, lower cauline leaves similar to basal, upper cauline leaves substantially reduced and sessile; inflorescences racemes or panicles terminating the branches, capsules erect at maturity, 2.5–4.5 cm long; seeds compressed, shiny black, aril about 0.5 mm long; frequent in dry, rocky woods and on rock outcrops; throughout; flr. May–Sep.

Dicentra Bernh.

Perennials arising from rhizomes or clusters of tubers; leaves glabrous, twice compound, the leaflets often much dissected; inflorescence terminal on a scape 1–5 dm tall; two outer petals spurred or with sac-like extensions, white to purple; fruit a slender capsule.

A. inflorescence a panicle; flowers pink to purple .... D. eximia
A. inflorescence a raceme; flowers white to cream-colored
    B. petal spurs more or less acute, divergent; capsules about 1 cm long ....
    .... D. cucullaria
    B. petal spurs more or less rounded, not or scarcely divergent; capsules about
    1.5 cm long .... D. canadensis

Dicentra canadensis (Goldie) Walp.   Squirrel-corn
Flowering scape to about 2.5 dm tall; leaves generally 1–3 dm long, often somewhat glaucous; both leaves and flowering scape arising from a small cluster of yellow tubers each 6–12 mm across; outer petals saccate, corolla about 1–1.5 cm long, white to cream-colored; fruits 1.4–1.7 cm long by 4–6 mm wide; seeds very glossy black, 2.2–2.6 mm long, aril pale, wing-like; frequent in rich, moist woods; throughout; flr. Apr–May.

Dicentra
canadensis,
flower

Dicentra cucullaria (L.) Bernh.   Dutchman's-breeches
Similar in habit to *D. canadensis*; leaves similar but often not as glaucous; leaves and flowering scape arising from a cluster of small pink tubers each 4–6 mm across; outer petals spurred, corolla about 1.5–2 cm long, white to cream-colored, sometimes tinged with pink; fruits about 1 cm long by 4 mm wide; seeds glossy black, 1.7–1.9 mm long, aril composed of several filaments; frequent in rich woods; throughout; flr. Apr–May.

Dicentra eximia (Ker Gawl.) Torr.   Wild bleeding-heart
Flowering scape 2–5 dm tall; leaves 4–6 dm long, glaucous; leaves and scape arising from a scaly rhizome, without tubers as in *D. canadensis* or *D. cucullaria*;

outer petals sac-like, corolla 1.5–2 cm long, pink to purple; fruits 1.5–2 cm long by 5–8 mm wide; seeds dull brown, aril cream to golden, nearly as large as the seed; very rare in rich woods and on cliffs, but also often cultivated; scattered throughout; flr. Jun–Jul; ✤.

Fumaria L.

Fumaria officinalis L.   Common fumitory, earth-smoke
Stems lax, much-branched; racemes dense, many-flowered; sepals 2, small; corolla about 8 mm long; cultivated sites and waste ground; occasional throughout; flr. May–Sep; native to Eurasia.

Dicentra eximia,
inflorescence

GENTIANACEAE Gentian Family
James S. Pringle

Annual, biennial, or perennial herbs; leaves sessile, opposite, whorled, or rarely alternate, sometimes also basal, simple, entire; flowers in terminal and axillary inflorescences, perfect, regular, 4- or 5-merous; sepals united or rarely absent; petals united; stamens fused to the corolla tube, alternate with the corolla lobes and of the same number; pistil 1, of 2 carpels; fruit a capsule.

A. leaves reduced to scales, scarcely wider than stem diameter .... *Bartonia*
A. leaves larger
    B. corolla lobes alternating with and connected by wedge-shaped appendages
    .... *Gentiana*
    B. corolla lobes not separated by wedge-shaped appendages
        C. corolla lobes each with a conspicuous nectary pit; pit openings distinctly fringed .... *Swertia*
        C. nectaries none or indistinct, or, if in pits on the corolla, the pit openings inconspicuous, not fringed
            D. flowers subtended by 2 separate, leaf-like bracts but without a calyx of 4 or 5 basally united sepals .... *Obolaria*
            D. flowers with a calyx of 4 or 5 basally united sepals
                E. margins of the corolla lobes fringed .... *Gentianopsis*
                E. corolla lobes not fringed
                    F. corollas violet or blue-violet (whitish), narrowly funnelform, lobes ascending... *Gentianella*
                    F. corollas pink or rose-violet (white), rotate or salverform, lobes abruptly spreading.
                        G. corollas rotate, lobes >4 times as long as the tube.....*Sabatia*
                        G. corollas salverform, lobes shorter than or about as long as tube .... *Centaurium*

Bartonia Muhl. ex Willd.

Annuals; plants slender, 1–4 dm tall, yellow-green, lower parts often purplish; leaves opposite or alternate, minute, scale-like; inflorescence paniculate; flowers 4-merous; corolla funnelform, white to yellowish green, the lobes longer than the tube, appendages none.

A. leaves opposite or nearly so; corolla lobes erose, apex rounded to abruptly acute, mucronate .... *B. virginica*
A. leaves mostly alternate; corolla lobes entire, apex acute to acuminate, not mucronate .... *B. paniculata*

Bartonia paniculata (Michx.) Muhl.   Screwstem
Decumbent to erect; inflorescence with variable, often arched-ascending, branches; corolla often purple-tinged, 3.0–6.2 mm; anthers 0.3–0.6 mm; rare in bogs and peaty lake margins; SE; flr. Aug–Oct; OBL; ✽.

Bartonia virginica (L.) Britton, Stearns & Poggenb.   Bartonia
Erect; inflorescence with strongly ascending branches; corolla 2.3–4.4 mm; anthers 0.5–1.2 mm; occasional in bogs, shores, and wet, open woods; throughout, but more frequent southward; flr. Jul–Sep; FACW.

Bartonia
virginica

Centaurium Hill

Annuals or biennials; cauline leaves elliptic-oblong (lower) to lanceolate (upper); flowers in cymes; 4 or 5-merous; corolla purplish-pink (white) with a yellow throat, salverform, the lobes shorter than the tube, appendages none.

A. flowers sessile, directly subtended by a pair of bractlets; corolla lobes 5–6.5 mm .... *C. erythraea*
A. flowers on pedicels 2–5 mm above the subtending bractlets; corolla lobes 2.5–5 mm .... *C. pulchellum*

Centaurium erythraea Rafn   Common centaury
Erect biennial 2–5 dm tall; basal leaves usually present at flowering time, obovate to elliptic, to 7 cm by 2 cm, cauline 0.8–5 cm by 1–8 mm; corolla 10–17 mm long, lobes 5–7 mm; very rare in fields, roadsides, other open, disturbed sites; E; flr. Jul–Sep; native to Eurasia.

Centaurium pulchellum (Sw.) Druce   Lesser centaury
Branched annual to 2 dm tall; only cauline leaves present at flowering time, 1–3 cm by 1–5 mm or rarely wider; corolla 10–15 mm long, lobes 2.5–5 mm; occasional in moist to wet, open places; SE; flr. Jul–Sep; native to Europe.

Gentiana L.

Perennials with clustered stems to about 1 m; leaves opposite; flowers few to many in dense clusters at the stem tip and upper axils, 5-merous; corolla cylindric to narrowly campanulate, remaining closed or opening only slightly, the corolla lobes much shorter than the tube, true lobes (with midvein) alternating with wedge-shaped appendages.

A. corolla appendages variously bifid, notched, lacerate, or fringed, generally as long as or longer than wide
  B. corolla appendages conspicuously longer than lobes, notched and shallowly toothed at summit .... *G. andrewsii*
  B. corolla appendages shorter than or about as long as lobes; lobes well-developed, semicircular to ovate or obovate
    C. calyx lobes broadly ovate to orbicular or obovate; corollas remaining

tightly closed; corolla lobes about as long as the appendages....G. *clausa*
  C. calyx lobes lanceolate to oblanceolate; corollas opening at least slightly,
     or only loosely closed; corolla lobes longer than the appendages
    D. leaves linear to elliptic, widest near the middle; calyx lobes shorter
       than or about as long as the tube; corolla lobes somewhat incurved,
       usually <2 mm longer than the appendages .... G. *saponaria*
    D. leaves ovate, widest near the base; calyx lobes longer than the tube;
       corolla lobes spreading, usually 2–4 mm longer than the appendages
       ....G. *catesbaei*
A. corolla appendages obliquely low-triangular, wider than long, not cleft, lac-
   erate or fringed, but sometimes with a few low teeth
  E. leaves linear to narrowly lanceolate, generally <15 mm wide and >6 times
     as long as wide .... G. *linearis*
  E. leaves ovate to obovate, at least the upper leaves >15 mm wide and/or <6
     times as long as wide
    F. leaves ovate; calyx lobes ovate, decurrently keeled; corollas white to pale
       yellow .... G. *alba*
    F. leaves elliptic to obovate; calyx lobes linear to oblong-lanceolate or ob-
       lanceolate, not keeled; corollas usually suffused with violet.... G. *villosa*

Gentiana alba Muhl. ex Nutt.   Yellowish or pale gentian
Leaves ovate; calyx lobes ovate-triangular, decurrently keeled; corolla whitish to
pale yellow, slightly open; lobes broadly ovate-triangular, 4–6 mm; appendages
obliquely triangular, erose to shallowly lacerate; very rare on wooded hillsides;
widely scattered; believed to be extirpated; flr. Sep; FACU; ✿.

*Gentiana
andrewsii,
flower*

*Gentiana
andrewsii,
split corolla*

Gentiana andrewsii Griseb.   Prairie closed or bottle gentian
Leaves lanceolate to ovate; calyx lobes lanceolate to ovate; corolla blue (white),
completely closed; lobes reduced to a minute point <1 mm long; appendages oblong, shallowly bifid, truncate, erose, conspicuous at the summit of intact flowers; occasional in moist to wet fields, open woods, and swamps in calcareous
soils or on diabase; mostly S; flr. late Jul–early Nov; FACW.

Gentiana catesbaei Walter   Catesby's or Coastal Plain gentian
Leaves ovate; calyx lobes lanceolate, sometimes more or less foliaceous; corolla
blue, slightly to fully open, lobes deltoid-ovate, 5–10 mm; appendages deeply
cleft into 2 more or less triangular, lacerate segments; very rare in moist open
woods and clearings; SE; believed to be extirpated; flr. Sep–Oct; OBL; ✿.

Gentiana clausa Raf.   Meadow closed or bottle gentian
Leaves ovate; calyx lobes broadly ovate to orbicular or obovate; corolla blue
(white), completely closed; lobes broadly ovate-triangular to semicircular, 0.75–
2 mm, more or less hooded and largely concealing the appendages in an intact
flower; appendages oblong, more or less deeply bifid, summit erose; frequent in
moist, open woods, stream banks, and meadows in noncalcareous soils;
throughout; flr. Aug–Oct; FACW.

*Gentiana
clausa, flower*

*Gentiana
clausa, split
corolla*

Gentiana linearis Froel.   Narrow-leaved or bog gentian
Leaves linear to lanceolate; calyx lobes linear to oblong; corolla blue (violet or
white), loosely closed or slightly open; lobes semicircular, 2.5–5 mm; append-

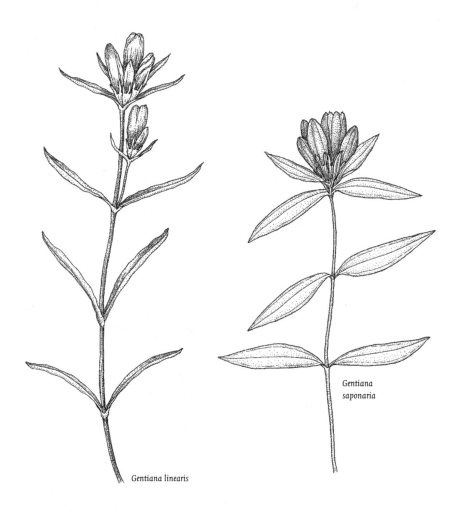

Gentiana linearis

Gentiana saponaria

ages obliquely triangular, entire or shallowly erose; occasional in bogs, moist barrens, and wet meadows; NE and SW in the mountains; flr. Jul–Sep; OBL; 🍂.

Gentiana saponaria L. Soapwort gentian
Leaves linear to elliptic; calyx lobes narrowly oblanceolate; corolla blue, slightly open; lobes ovate-triangular, 3–7 mm; appendages deeply cleft into 2 more or less triangular, lacerate segments; rare in moist open woods, roadsides, and swamps; SE and SW; flr. Sep–Oct; FACW; 🍂.

Gentiana saponaria, split corolla

Gentiana saponaria, flower

Gentiana villosa L. Striped gentian
Leaves obovate or upper leaves elliptic; calyx lobes linear to oblanceolate; corolla greenish-white, more or less suffused with violet, or grayish-violet throughout, slightly open; lobes ovate-triangular, 4–10 mm; appendages obliquely triangular, erose; rare in dry, open woods and serpentine barrens; SE and SC; flr. Sep–Oct; 🍂.

Gentianella Moench

Gentianella quinquefolia (L.) Small Stiff gentian
Annual to 8 dm tall; leaves opposite, ovate, 0.5–8 cm; flowers in dense terminal and axillary clusters, 4- merous; calyx 2–8 mm, lobes subulate to linear-oblong;

corolla narrowly funnelform, violet (white), 10–23 mm; lobes more or less erect or incurved, ovate-triangular; appendages none; occasional in moist open woods, springy slopes, and stream banks; NE and NC, and at higher elevations along the Allegheny Front; flr. Jul–Oct; FAC.

Gentianopsis Ma

Annuals or biennials, to about 6 dm tall; leaves opposite; flowers solitary on 5–20-cm peduncles arising from the stem apex and upper axils, 4-merous; corolla blue (white); lobes about as long as the tube, oblong-obovate, margins fringed; appendages none.

A. leaves lanceolate to ovate, >1 cm wide; margins of the corolla lobes deeply fringed on the sides and apex .... *G. crinita*
A. leaves linear to lanceolate, <1 cm wide; margins of the corolla lobes fringed on the sides, merely toothed at the apex .... *G. virgata*

Gentianopsis crinita, flower

Gentianopsis virgata, flower

Gentianopsis crinita (Froel.) Ma   Eastern fringed gentian
Leaves lanceolate to ovate; margins of the corolla lobes deeply fringed on the apex and sides; occasional in wet meadows, swamps, fens, stream banks, and other moist, open sites on calcareous soils; SE, C, and NW; flr. Aug–Oct; OBL.

Gentianopsis virgata (Raf.) Holub   Narrow-leaved fringed gentian
Very similar to *G. crinita* but the leaves <1 cm wide, and the margins of the corolla lobes fringed on the sides, but merely toothed at the apex; very rare on moist, calcareous shores near Lake Erie; NW; native, but probably extirpated; flr. Sep–Oct; FACW+; ❦. [*G. procera* (Holm) Ma of Rhoads and Klein, 1993; there is disagreement regarding the correct name for this species.]

Obolaria L.

Obolaria virginica L.   Pennywort
Perennial, pale green to somewhat purplish; stems to 1.7 dm tall, solitary or few together, mostly simple; leaves opposite, fan-shaped to spatulate-obovate or orbicular, the upper 4–16 mm by 3–11 mm, the lower scale-like; flowers terminal and axillary, solitary or in 3s; 4-merous; calyx none (a pair of bracts below each flower alternatively interpreted as 2 separate sepals); corolla narrowly campanulate, white to pale violet, 6–15 mm long, lobes slightly longer than the tube; occasional in moist deciduous woods, especially on diabase; S; flr. Mar–Jun.

Sabatia Adanson

Perennials, biennials, or annuals, variable in size; leaves opposite and often also basal; flowers solitary or in cymes, 5-merous; corolla rotate, pink (white), with a yellow eye, the lobes much longer than the tube, appendages none.

A. primary branching mostly opposite; calyx lobes narrowly oblong-lanceolate or foliaceous .... *S. angularis*
A. branching all or mostly alternate; calyx lobes linear .... *S. campanulata*

Sabatia angularis (L.) Pursh   Common marsh-pink; rose-pink
Biennial, primary branching mostly opposite, secondary mostly alternate; basal

leaves oblong-spatulate to ovate-orbicular, cauline ovate to lanceolate, 1–4 cm by 5–30 mm; calyx lobes narrowly oblong-lanceolate or rarely foliaceous; occasional in open woods, fields, marshes, and moist meadows; mostly S; flr. Jul–Oct; FAC+.

Sabatia campanulata (L.) Torr.   Slender marsh-pink
Perennial, branches all or mostly alternate; leaves cauline only, linear to narrowly lanceolate; calyx lobes narrowly linear, 7–15 mm; very rare in marshes and wet fields; SE; believed to be extirpated; flr. Jul–Oct; FACW; .

Swertia L.

Swertia caroliniensis (Walter) Kuntze   American columbo; green gentian
Stems simple, to 3 m tall; plants flowering once after several years in a rosette stage; rosette and basal leaves elliptic to narrowly obovate, 2–4.5 dm long, cauline leaves in whorls of 3–5, oblong-lanceolate, 1.6–3.2 dm; flowers in a terminal and axillary paniculate inflorescence, 4-merous; corolla rotate, 2–3 cm across, pale yellowish-green with purple dots, the lobes each bearing a large elliptical gland surrounded by a long fringe; rare in open deciduous woods on calcareous soils; NW; flr. May–Jun; .

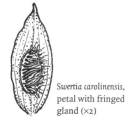

Swertia carolinensis, petal with fringed gland (×2)

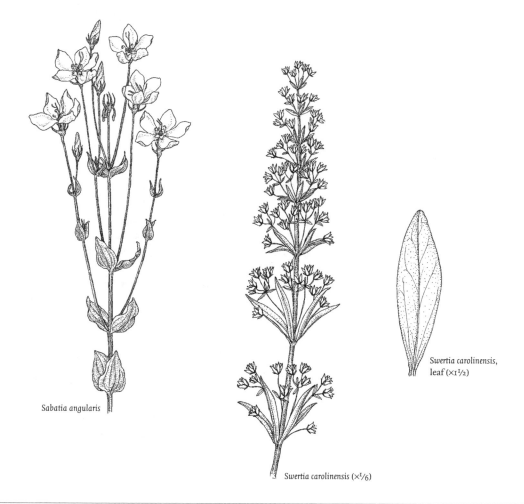

Sabatia angularis

Swertia carolinensis (×1/6)

Swertia carolinensis, leaf (×1½)

GERANIACEAE Geranium Family

Annual or perennial herbs with lobed, dissected, or compound leaves with stipules; flowers perfect, regular, 5-merous; petals distinct, alternating with nectar glands; stamens 5 or 10; ovary superior, of 5 fused carpels; style 1 with 5 stigmas; placentation axile; fruit an elongate capsule, the valves separating from the central column.

A. leaves palmately lobed or compound; fertile stamens 10 (5 in G. pusillum) .... Geranium
A. leaves pinnately compound; fertile stamens 5 .... Erodium

Erodium L.

Erodium cicutarium (L.) L'Hér. ex Soland.   Red-stem filaree

Winter annual or biennial; stems initially short with mostly basal leaves, later diffusely branched to 4 dm; leaves pinnately compound with several deeply lobed, sessile leaflets; inflorescence umbel-like, with a long peduncle and 2–8 flowers; sepals with short awns; petals pinkish; fruit 2–4 cm, the valves separating from the central axis from the top down; occasional in weedy fields, roadsides and waste ground; mostly SE; flr. late Apr–Nov; native to Europe. E. moschatum (L.) L'Hér. (white-stem filaree), which differs by its larger size (stems to 6 dm) and stalked leaflets, has been collected rarely.

Geranium L.

Herbs with opposite or basal, palmately lobed or divided leaves; flowers in pairs at the ends of axillary branches; petals pink, purple, or white; fertile stamens 10 (5 in G. pusillum), the filaments broadened below; fruit an elongate, beaked capsule, the valves separating by coiling upward from the base.

A. petals 12–20 mm; anthers 2 mm or more .... G. maculatum
A. petals <12 mm; anthers 1 mm or less
    B. leaves compound, the terminal leaflet distinctly stalked ...... G. robertianum
    B. leaves deeply lobed but not compound
        C. outer sepals blunt or acute but not awn-tipped
            D. stamens 5; capsule pubescent... G. molle
            D. stamens 10; capsule glabrous .... G. pusillum
        C. outer sepals with a 0.7–3-mm awn
            E. fruiting pedicels shorter than to less than twice as long as the calyx
                F. fruits with long upward-pointing hairs .... G. carolinianum
                F. fruits with spreading, often glandular hairs ....G. dissectum
            E. fruiting pedicels more than twice as long as the calyx
              G. pedicels spreading hairy, not glandular.................... G. sibiricum
              G. pedicels glandular-villous or strigose
                H. pedicels glandular-villous .... G. bicknellii
                H. pedicels with closely appressed, downward-pointing hairs ....
                .... G. columbinum

Geranium bicknellii Britton   Cranesbill

Erect annual or biennial, 2–4 dm tall; leaves cleft nearly to the base into 3–5 segments, which are again deeply incised; flowers in pairs on glandular-villous pedicels; sepals with a conspicuous subulate tip; petals 7–9 mm; fruit 2.5 cm; rare in dry, open woods, clearings, and rocky ledges; scattered; flr. Jun–Aug; ❦.

Geranium carolinianum L.   Wild geranium
Freely branched, hairy annual to 6 dm tall, often with a reddish tinge; leaves round or kidney-shaped, deeply divided into 5–9 toothed lobes, those of the stem alternate; pedicels not much longer than the calyx; sepals tipped with a short awn; petals white or pinkish, barely longer than the sepals; fruit 2–5 cm with long upward-pointing hairs; occasional in dry ground of fields, open woods, roadsides, and railroad rights-of-way; mostly S; flr. May–Aug.

Geranium columbinum L.   Long-stalked cranesbill
Annual or biennial; stems creeping to ascending, 1.5–8 dm; leaves kidney-shaped, 3–5-lobed nearly to the base, cauline leaves opposite; peduncles bearing 2 elongate pedicels with appressed, downward-pointing pubescence; sepals with a subulate tip; petals purple, slightly longer than the sepals; fruit 2.5 cm; occasional in dry, rocky, or shaly slopes or old fields; flr. May–Oct; native to Europe.

Geranium dissectum L.   Cut-leaved cranesbill
Spreading to erect, hairy annual to 6 dm tall; leaves round-cordate, shallowly 5-lobed below, smaller but more deeply lobed above; peduncles 2-flowered; pedicels about as long as the calyx; flowers 6 mm wide; sepals with a short awn; petals purple, about as long as the sepals; fruit 15–20 mm; rare and scattered in fields and waste ground; flr. May–Aug; native to Eurasia.

Geranium maculatum L.   Wood geranium
Perennial with 5–7-lobed basal leaves with long petioles; flowering stems erect, 5–7 dm, with a single pair of short-petioled leaves; flowers 2.5–4 cm wide; se-

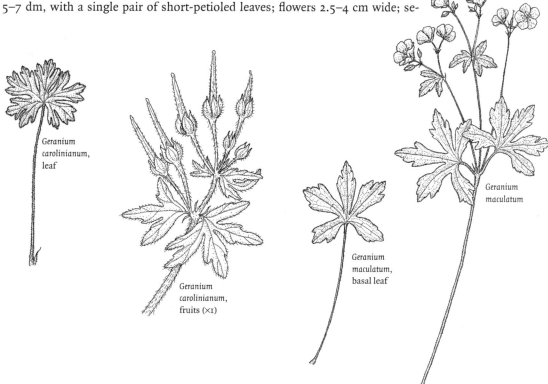

Geranium carolinianum, leaf

Geranium carolinianum, fruits (×1)

Geranium maculatum, basal leaf

Geranium maculatum

pals awn-tipped; petals pink; fruits erect, 3–4 cm; common in woods, roadsides, and fields; throughout; flr. May–early Jul; FACU.

Geranium molle L.   Dove's-foot cranesbill
Branched, pubescent annual; stems spreading or ascending, 2–5 dm; basal leaves round in outline, 5–9-lobed, upper leaves progressively reduced and alternate; flowers numerous, dark purple; calyx and pedicels densely hairy and glandular; sepals acute; petals deeply notched; fruit 9–13 mm; rare and scattered on roadsides, canal banks, and railroad tracks; flr. May–Aug; native to Europe.

Geranium pusillum L.   Slender cranesbill
Annual or biennial with spreading or ascending stems 1–5 dm; leaves long-petioled, kidney-shaped to orbicular in outline and deeply 5–7-lobed; flowers reddish-purple, 5–9 mm wide; sepals callus-tipped; stamens 5; fruit 6–9 mm long, finely pubescent; scattered in roadsides, fields, and waste ground; flr. Jun–Aug; native to Europe.

Geranium robertianum, flower (×1)

Geranium robertianum L.   Herb-robert
Branched, spreading annual or biennial, aromatic; stem to 6 dm, swollen at the nodes; leaves opposite, palmately compound, the terminal leaflet distinctly petioled, the leaflets pinnately lobed; flowers pink or purple; sepals awn-tipped; petals 9–13 mm; common in moist, wooded, rocky slopes and ravines; mostly E; flr. May–Oct.

Geranium sibiricum L.   Siberian cranesbill
Annual or short-lived perennial with weak, spreading stems to 1 m; leaves kid-

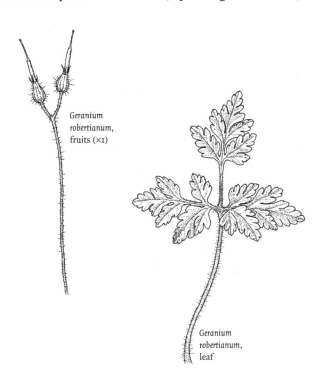
Geranium robertianum, fruits (×1)

Geranium robertianum, leaf

ney-shaped, with 3–5 coarsely toothed lobes; peduncles with 1 or 2 flowers on villous pedicels; sepals with a short awn; petals barely longer than the sepals; fruit 1.5–2 cm; cultivated and rarely escaped to roadsides and waste ground; flr. Jul–Sep; native to Eurasia.

GROSSULARIACEAE Gooseberry Family

Deciduous shrubs with alternate, simple, but often deeply lobed leaves; flowers 5-merous, regular, with a prolonged saucer-shaped, campanulate, or tubular hypanthium; stamens 5, alternating with 5 staminoids; stigma 2-lobed, fruit a capsule or berry with numerous seeds.

A. leaves palmately veined and lobed; ovary inferior .... *Ribes*
A. leaves pinnately veined, serrulate; ovary superior .... *Itea*

Itea L.

Itea virginica L.   Virginia-willow, tassel-white
Shrub to 3 m tall, twigs green and finely hairy when young; leaves elliptic to oblong-lanceolate, 4–10 cm long, glabrous; the small white flowers in terminal racemes; fruit an elongate, bilocular capsule with a persistent style; very rare in moist coastal plain woods; SE; flr. May–Jun; OBL; 🍂.

Itea virginica,
fruit

Itea virginica,
stem showing
pith (×3)

Ribes L.

Erect or sprawling shrubs, unarmed or with spines at the nodes and sometimes also bristly internodes; leaves, stems, and fruits sometimes glandular or glandular-bristly; leaves crowded on short lateral branches, palmately 3–7-lobed; flowers in small clusters or racemes; hypanthium tube prolonged beyond the inferior ovary; fruit a berry.

A. stems with spines at at least some of the nodes
  B. ovary and fruit glabrous
    C. stamens longer than the sepals
      D. floral bracts glabrous or sparsely glandular .... *R. rotundifolium*
      D. floral bracts ciliate with glandular and nonglandular hairs present...
      .... *R. missouriense*
    C. stamens just equaling the sepals .... *R. hirtellum*
  B. ovary and fruit glandular-bristly
    E. flowers solitary or in clusters of 2–3; pedicels not jointed at the summit
    .... *R. cynosbati*
    E. flowers in a definite raceme of 5 or more; pedicels jointed at the summit
      F. hypanthium broadly campanulate, pubescent .... *R. uva-crispa*
      F. hypanthium broadly saucer-shaped, glabrous .... *R. lacustre*
A. stems not spiny
  G. ovary and fruit bristly .... *R. glandulosum*
  G. ovary and fruit not bristly

      H. leaves with resinous glands beneath .... R. *americanum*
      H. leaves without glands
          I. flowers yellow-green; fruit red
             J. lateral leaf lobes directed forward.................................. R. *triste*
             J. lateral leaf lobes spreading .... R. *rubrum*
          I. flowers bright yellow; fruit black .... R. *odoratum*

Ribes americanum Mill.   Wild black currant
Erect unarmed shrub to 2 m tall; leaves with glandular dots on the lower surface; inflorescence a drooping, many-flowered raceme with lanceolate bracts; fruit smooth, black; occasional in moist woods, marshes and thickets; throughout; flr. late Apr–May; FACW.

Ribes cynosbati L.   Prickly gooseberry, dogberry
Shrub with spines at the nodes and frequently also bristles; leaves pubescent but not glandular; inflorescence few-flowered; fruits prickly and glandular when immature, pale red; frequent in thin, moist, rocky woods; Allegheny Plateau; flr. May–Jun.

Ribes glandulosum Grauer   Skunk currant
Reclining or sprawling shrub, stems unarmed; leaves glabrous or sparsely hairy on the veins beneath; fruit glandular-bristly, dark red; occasional in swamps, bogs, wet woods, and moist rocky slopes; N and at high elevations along the Allegheny Front; flr. May–Jun; FACW.

Ribes hirtellum Michx.   Northern wild gooseberry
Stems with a few spines and bristles or unarmed; leaves softly hairy to glabrous, not glandular; fruit smooth, greenish-purple; infrequent in calcareous marshes, swamps, rocky woods, and cliffs; scattered; flr. May–Jun; FAC.

Ribes lacustre (Pers.) Poir.   Bristly black or swamp currant
Stems with spines at the nodes and numerous bristles; leaves deeply lobed, glabrous; racemes spreading or drooping; fruit glandular-bristly, dark purple to black; rare in swamps and cool, wet woods; mostly NE and NW; flr. May–Jun; FACW; ✿.

Ribes missouriense Nutt. ex Torr. & A.Gray   Missouri gooseberry
Stems with stout spines at the nodes; leaves each with 2 deep sinuses, hairy beneath; pedicels and bracts with glandular and nonglandular hairs; fruit smooth, red to purple; very rare in rich woods; C and SE; flr. late Apr–May; ✿.

Ribes odoratum H.L.Wendl.   Buffalo currant
Stems erect, unarmed; leaves ciliate and finely puberulent beneath; flowers yellow, fragrant; fruit smooth, black; cultivated and occasionally escaped to meadows and waste ground; mostly S; flr. late Apr–May; native to the Great Plains.

Ribes rotundifolium Michx.   Wild gooseberry
Stems arching or trailing, with short (or no) spines; leaves glabrous or with tiny hairs; peduncles glabrous; bracts glabrous or sparsely glandular; fruits glabrous, purple or greenish; frequent in rocky upland woods; throughout; flr. late Apr–May.

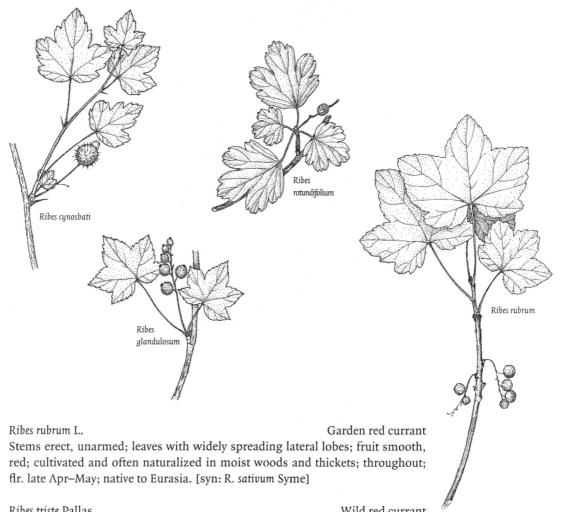

Ribes cynosbati

Ribes
rotundifolium

Ribes
glandulosum

Ribes rubrum

Ribes rubrum L.   Garden red currant
Stems erect, unarmed; leaves with widely spreading lateral lobes; fruit smooth, red; cultivated and often naturalized in moist woods and thickets; throughout; flr. late Apr–May; native to Eurasia. [syn: R. *sativum* Syme]

Ribes triste Pallas   Wild red currant
Stems straggling, unarmed; leaves glabrous or hairy beneath; racemes many-flowered, drooping, the axis and pedicels glandular; fruit smooth, red; rare in wet, rocky woods, swamps, and cliffs; N; flr. Jun–Jul; OBL; 🍃.

Ribes uva-crispa L.   European garden gooseberry
Stems erect, with stout spines, also bristly; leaves to 6 cm wide; fruit pubescent and often with gland-tipped bristles; cultivated and rarely escaped; flr. May–Jun; native to Eurasia. Ours is var. *sativum* DC.

HALORAGACEAE Water-milfoil Family

Rooted, aquatic, herbaceous plants with opposite, whorled, or alternate leaves mostly pinnately divided into filiform segments (leaves reduced in M. *tenellum* and on the emergent stem tips of some other species); flowers small, sessile in the axils of leaves or bracts, regular, 3–4-merous, perfect or unisexual; calyx tube fused to the inferior ovary; stamens 1–8; ovary 1–4-celled; fruit several nutlets or a drupe.

A. at least some of the leaves whorled; flowers 4-merous .... *Myriophyllum*
A. all leaves alternate
    B. all leaves reduced to minute bracts; flowers 4-merous ....
    .... *Myriophyllum tenellum*
    B. leaves well-developed, submersed leaves once pinnate to pinnatifid; emersed leaves simple; flowers 3-merous .... *Proserpinaca*

Myriophyllum L.

Monoecious, aquatic herbs with finely, pinnately divided leaves, at least some of which are whorled (except *M. tenellum*, which has very small, alternate bracts throughout); stem tips emergent in some species and bearing reduced, bract-like leaves; flowers in the axils of upper leaves or bracts, 4-merous, unisexual, the upper staminate, the lower pistillate; perianth 4-lobed, often absent in the pistillate flowers; stamens 4 or 8; fruit 4-lobed, separating into 4 nutlets when mature; some species form dense, leafy vegetative buds, or turions, which remain dormant over the winter.

A. stems erect from a slender rhizome; leaves reduced to small scattered bracts
    .... *M. tenellum*
A. stems flexuous; submersed leaves pinnately divided into filiform segments
    B. submersed leaves partly whorled and partly alternate or scattered
        C. flowers (and fruits) borne in the axils of reduced bract-like leaves; fruits smooth .... *M. humile*
        C. flowers (and fruits) borne in the axils of ordinary foliage leaves; fruits tuberculate .... *M. farwellii*
    B. submersed and emersed leaves all whorled
        D. leaves (or bracts) subtending the flowers deeply pinnately lobed or laciniate
            E. emergent leaves (or bracts) 2.5–3.5 cm long .... *M. verticillatum*
            E. emergent leaves (or bracts) <2 cm long .... *M. aquaticum*
        D. leaves (or bracts) subtending the flowers entire or serrulate
            F. floral bracts serrulate, much longer than the flowers ....
            .... *M. heterophyllum*
            F. floral bracts entire, slightly longer than, equal to, or shorter than the flowers
                G. leaf segments 4–14 per side; stem diameter uniform ....
                .... *M. sibiricum*
                G. leaf segments 5–25 per side; stem thicker below the inflorescence
                .... *M. spicatum*

Myriophyllum aquaticum (Vell.) Verdc.   Parrot's-feather
Leaves all whorled with 10–18 segments on each side, the lower segments much shorter; dioecious, only the pistillate form known here; flowers in emergent spikes in the axils of the scarcely reduced upper leaves; rare in ponds; E and SC; native to South America; OBL.

Myriophyllum farwellii Morong   Farwell's water-milfoil
Wholly submersed; leaves alternate or subopposite; flowers in the axils of foliage leaves; fruit 2–2.5 mm long, each segment with 2 low, irregular longitudinal ridges, turions produced; rare in lakes and ponds; NE; OBL; 🌿.

Myriophyllum heterophyllum Michx.   Broad-leaved water-milfoil
Leaves whorled, pinnately divided into capillary segments; spikes emersed; floral bracts whorled, lanceolate to oblong, sharply serrulate; fruit subglobose,
rough; rare in still water of ponds and lakes; widely scattered; OBL; 🌿.

Myriophyllum humile (Raf.) Morong   Water-milfoil
Submersed leaves whorled below but mostly alternate or subopposite; flowers in
the axils of submersed leaves or in short emersed spikes with pinnately divided
to entire bracts that are longer than the flowers; fruits 0.7–1.2 mm long,
rounded on the back and smooth; occasional in lakes and ponds; mostly E; OBL.

Myriophyllum sibiricum Komarov   Northern water-milfoil
Stem width uniform throughout; leaves mostly in whorls of 4, fan-shaped in
outline with 4–14 segments per side; floral bracts shorter than to rarely equaling
the fruits; turions produced; rare in rivers, lakes, ponds, and marshes, mostly in
water <1 m deep; W and S; 🌿. [syn: *M. exalbescens* Fernald]

Myriophyllum spicatum L.   Eurasian water-milfoil

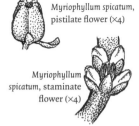

Myriophyllum spicatum,
pistilate flower (×4)

Stem width increasing below the emergent inflorescence; leaves mostly in
whorls of 4, broadly ovate to elliptic in outline with 5–25 segments per side;
flora bracts as long or slightly longer than the fruits; turions not produced; an
increasingly common invader of lakes and rivers, abundant in water 1–3 m deep;
throughout; native to Eurasia; OBL.

*Myriophyllum
spicatum,* staminate
flower (×4)

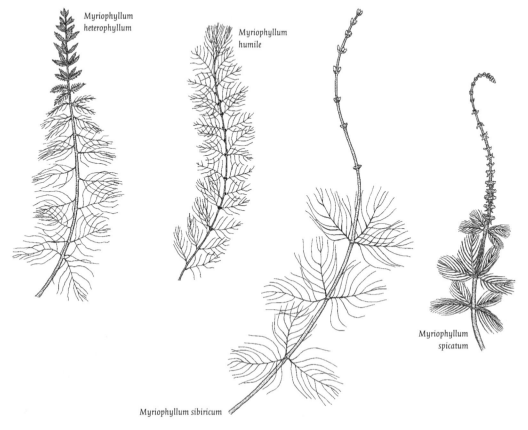

*Myriophyllum
heterophyllum*

*Myriophyllum
humile*

*Myriophyllum
spicatum*

Myriophyllum sibiricum

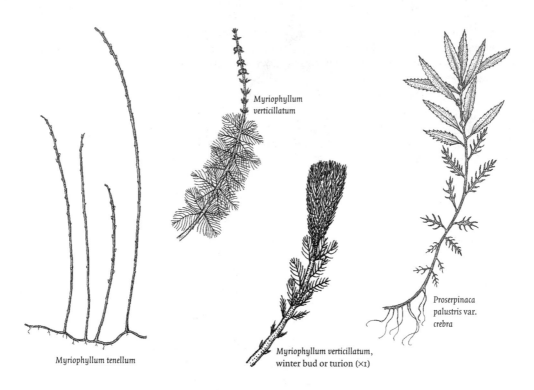

Myriophyllum verticillatum

Myriophyllum tenellum

Myriophyllum verticillatum, winter bud or turion (×1)

Proserpinaca palustris var. crebra

Myriophyllum tenellum Bigelow   Slender water-milfoil

Stems erect, simple, to 2 dm, from slender rhizomes, forming a loose turf; leaves alternate, reduced to a few tiny bracts that subtend the flowers; rare in shallow water of lakes and ponds; NE; OBL; 🍂.

Myriophyllum verticillatum L.   Whorled water-milfoil

Stem width uniform throughout; leaves in whorls of 4–5, with 9–17 segments per side; floral bracts 1–10 times the length of the fruits, pinnate or pectinate, never entire; turions produced; rare in shallow (<1 m deep) water of ponds or marshes; NW; OBL; 🍂.

Proserpinaca L.

Perennial herbs of shallow water or wetlands; stems prostrate and often submergent at the base with ascending, emergent tips; leaves all alternate, pinnately lobed or toothed; flowers axillary on emergent portions of the stems; calyx lobes 3; petals lacking; stamens 3; fruit 3-sided and 3-seeded, nondehiscent.

A. leaves on the emergent portion of the stem merely serrate; submersed leaves pinnatifid .... P. palustris
A. leaves all pinnatifid, including those subtending the flowers and fruits ....
.... P. pectinata

Proserpinaca palustris L.   Common mermaid-weed

Submerged leaves sessile, 2–6 cm long, deeply divided into linear segments; emergent leaves simply serrate; fruits with concave sides and sharp or winged angles; swamps, bogs, ponds, and marshes; OBL; 2 varieties:

A. fruits 4–6 mm wide, angles wing-margined .... var. *palustris*
very rare; SE and NW.
A. fruits 2–4 mm wide, angles not winged .... var. *crebra* Fernald & Griscom
occasional; E and NW.

Proserpinaca palustris var. palustris, fruit (×3)

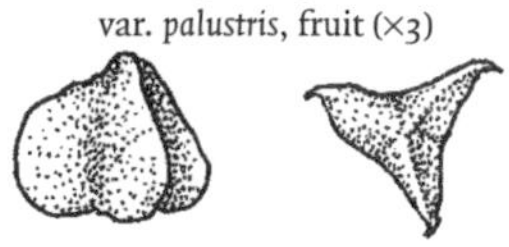

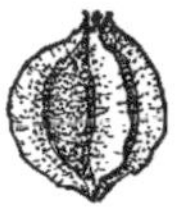

Proserpinaca palustris var. crebra, fruit (×3)

Proserpinaca pectinata Lam. — Comb-leaved mermaid-weed

All leaves deeply pinnately lobed including those on the emergent, flowering portions of the stems; fruits obtusely angled with flat sides; rare in swamps or bogs; SE; believed to be extirpated; OBL.

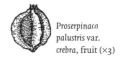

HAMAMELIDACEAE Witch-hazel Family

Deciduous trees or shrubs with alternate, simple leaves and deciduous stipules; flowers perfect or unisexual, 4–5-merous or stamens numerous, ovary partly inferior, bicarpellate; fruit a woody capsule.

A. leaves palmately lobed; flowers unisexual, in dense heads; petals lacking ...... *Liquidambar*
A. leaves coarsely toothed; flowers perfect, in small axillary clusters; petals present .... *Hamamelis*

Hamamelis L.

Hamamelis virginiana L. — Witch-hazel

Deciduous shrub to 5 m tall, blooming in the fall as the leaves are dropping; leaves 5–15 cm, obovate or oblong with coarse rounded teeth; flowers perfect, regular with 4 strap-shaped, yellow petals, fruit a woody capsule with 2 locules, each containing a single seed that is forcibly expelled when the capsule opens; common in moist, rocky woods; throughout; flr. Oct–Nov; FAC–.

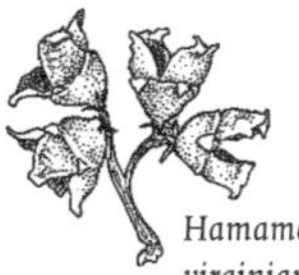

Hamamelis virginiana, fruit

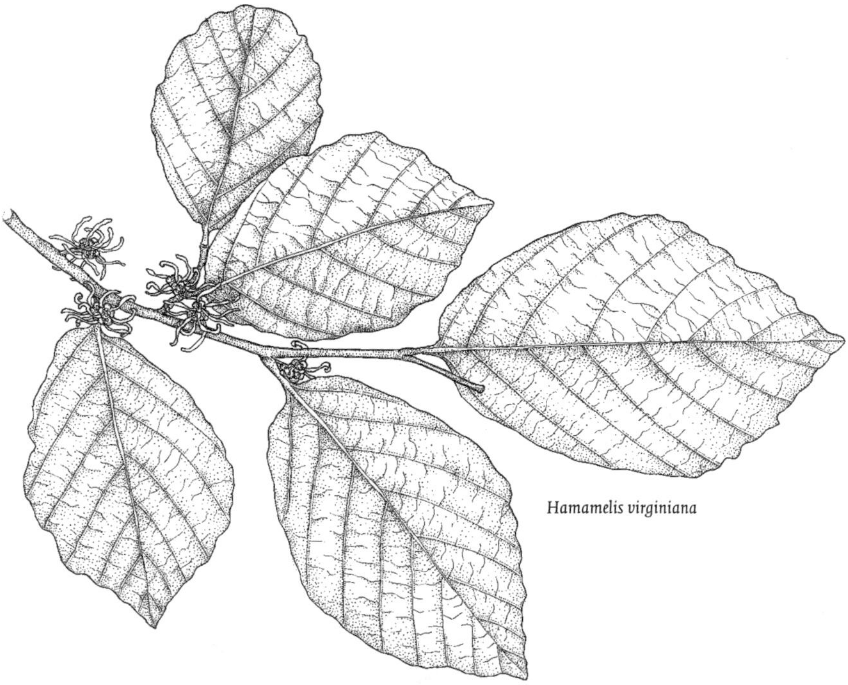

Hamamelis virginiana

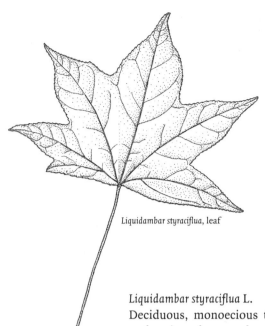
Liquidambar styraciflua, leaf

Liquidambar L.

Liquidambar styraciflua L.   Sweetgum
Deciduous, monoecious tree to 40 m tall; smaller branches sometimes with corky wings; leaves palmately veined and lobed, the 5–7 lobes triangular and divergent; flowers tiny, clustered in dense heads, staminate heads in an erect raceme, pistillate flowers in a single, long-peduncled head at the base of the inflorescence; fruit a spherical head of fused woody capsules; rare in low, wet coastal plain woods; SE; also cultivated; flr. late Apr; FAC.

HIPPOCASTANACEAE Horse-chestnut Family

Deciduous trees or shrubs with opposite, palmately compound leaves and terminal racemes of showy flowers; flowers irregular, 4–5-merous; stamens 6–8, an inner whorl of 5 and a partial outer whorl; ovary superior with a long style and 3-lobed stigma; fruit a dehiscent capsule often with a single large seed.

Aesculus L.

Characteristics of the family.

A. winter buds sticky; leaves mostly with 7 leaflets .... *A. hippocastanum*
A. winter buds not sticky; leaves mostly with 5 leaflets
    B. trees; flowers yellow-green, yellow, or reddish; stamens shorter than to somewhat longer than the petals
        C. bark light grayish-brown, smooth; fruit not prickly; leaflets widest above the middle; petals yellow or reddish, of varying lengths, as long as or longer than the stamens .... *A. flava*
        C. bark dark brownish-black, scaly; fruit prickly; leaflets widest at the middle; petals yellowish-green, similar in length, shorter than the stamens
        .... *A. glabra*
    B. stoloniferous shrub; flowers white; stamens more than twice as long as the petals .... *A. parviflora*

Aesculus flava Sol.   Yellow buckeye, sweet buckeye

Tree to 30 m tall; leaflets 5, drooping, widest above the middle, inflorescence 1–1.5 dm; petals 4, yellow or reddish; stamens 7 or 8, as long as or only slightly longer than the petals; fruit smooth, 5–8 cm; occasional in low alluvial woods; SW; flr. early May; native, but also escaped from cultivation.

Aesculus glabra Willd.   Ohio buckeye

Tree to 20 m tall; leaflets 5(7), widest at about the middle, inflorescence 1–1.5 dm; petals 4, greenish-yellow; stamens 7, up to twice the length of the corolla; fruit spiny, 3–4 cm; occasional in moist woods and bottomlands; SW; flr. late Apr–early May; native, but also escaped from cultivation; FACU+.

Aesculus hippocastanum L.   Horse-chestnut

Tree to 25 m tall with stout resinous winter buds; leaflets 7–9, wedge-shaped or obovate, irregularly serrate; inflorescence 2–3 dm; petals 5, white with red or yellow markings; fruit spiny, 5 cm; cultivated and occasionally escaped to railroad banks or waste ground; flr. late May; native to Eurasia.

Aesculus parviflora Walter   Bottlebrush buckeye

Spreading, stoloniferous shrub to 4 m tall; leaflets 5–7, elliptic to oblong-

Aesculus flava (×¼)

Aesculus hippocastanum (×¼)

Aesculus hippocastanum,
winter twig

Aesculus parviflora,
inflorescence (×¼)

obovate, crenate-serrulate; inflorescence 2–3 dm; petals white; stamens 7, much longer than the petals; fruit smooth, 2.5–4 cm; cultivated and occasionally naturalized in disturbed urban woods; flr. Jun; native SC to FL.

HYDRANGEACEAE Hydrangea Family

Deciduous shrubs with opposite, simple leaves lacking stipules; flowers 4–5-merous, regular, perfect; ovary partly to completely inferior; stamens 8–many; fruit a capsule.

A. leaves strongly roughened (scabrous) above .... *Deutzia*
A. leaves not strongly roughened above
    B. ovary and fruit 2- or rarely 3-celled; stamens 8–10; sterile flowers usually present .... *Hydrangea*
    B. ovary and fruit usually 4-celled; stamens 20–40; all flowers perfect ....
    .... *Philadelphus*

Deutzia Thunb.

Deutzia scabra Thunb.   Deutzia

Shrub to 2.5 m tall; branches with brown, exfoliating bark; leaf blades ovate to oblong-lanceolate, 3–8 cm, acute or acuminate, crenate-denticulate with stellate pubescence on both sides and very rough to the touch above; petioles about 2 mm; upright panicles 6–12 cm long; flowers white or pinkish, 1.5–2 cm, stamens nearly as long as the petals; capsule subglobose, 5 mm; cultivated and occasionally persisting on roadside banks or waste ground; mostly SE; flr. Jun; native to Asia.

Hydrangea L.

Deciduous shrubs with opposite, petiolate leaves and numerous small flowers in a flat-topped or pyramidal inflorescence that often includes sterile flowers with enlarged sepals; calyx and corolla 4–5-merous, regular; ovary inferior or half inferior with 2 or 3 styles; fruit a 2- or 3-celled capsule.

A. inflorescence flat-topped or slightly domed; ovary inferior; sepals of sterile flowers white .... *H. arborescens*
A. inflorescence pyramidal; ovary half inferior; sepals of sterile flowers at first white, becoming bronzy red in age .... *H. paniculata*

Hydrangea arborescens L.   Sevenbark, wild hydrangea

Straggling shrub to 2 m tall; leaves ovate-oblong, 6–20 cm, acuminate, serrate; inflorescence 5–10 cm wide; flowers all fertile or only a few sterile; petals 5, white; ovary inferior, 2-celled; frequent in rich woods, slopes, and stream banks; throughout; flr. late Jun–early Jul; FACU.

Hydrangea arborescens

Hydrangea paniculata Siebold   Peegee hydrangea
Shrub to 5 m tall; leaves ovate to elliptic, 5–12 cm; inflorescence pyramidal, 1.5–2.5 dm; fertile flowers with half inferior, 2- or 3-celled ovary; sterile flowers with 4 enlarged and persistent sepals, which are at first white, changing to bronzy red; cultivated and occasionally persisting at abandoned home sites or disturbed woods; SE and SW; flr. Aug; native to Asia; FAC.

Philadelphus L.

Erect, deciduous shrubs with opposite, simple leaves and showy white flowers; inflorescence a raceme or few-flowered cyme; sepals and petals 5; stamens numerous; ovary inferior or half inferior; fruit a 4-valved capsule.

A. flowers in a 5–9-flowered raceme
    B. bark brown, exfoliating; leaves glabrous beneath except on the veins; flowers fragrant .... *P. coronarius*
    B. bark light gray, nonexfoliating; leaves densely pubescent beneath; flowers not fragrant .... *P. pubescens*
A. flowers solitary or in 3-flowered cymes .... *P. inodorus*

Philadelphus coronarius L.   Mock-orange
Shrub to 3 m tall with brown, exfoliating bark; leaf blades ovate to ovate-oblong, 4–8 cm, acuminate, sparsely denticulate, glabrous except on the veins and vein axils beneath; flowers in 5–7-flowered racemes, creamy-white, fragrant, 2.5–3.5 cm across; frequently cultivated and occasionally spreading to banks, roadsides, or alluvial woods; mostly S; flr. late May–early Jun; native to Eurasia.

Philadelphus inodorus L.   Mock-orange
Shrub to 3 m tall with white cup-shaped flowers 4–5 cm across; leaf blades ovate, 2.5–10 cm, acute or short acuminate, entire or sparsely denticulate, 3–5 veined at the base, glabrous above and hairy in the vein axils beneath; petiole 2–3 mm; flowers solitary or in few-flowered cymes, not fragrant; cultivated and escaped at a few scattered sites; flr. May; native farther south. Ours is var. *grandiflorus* A.Gray. [syn: *P. grandiflorus* Willd.]

Philadelphus pubescens Loisel.   Mock-orange
Shrub to 3 m tall with light gray, nonexfoliating bark; leaves ovate to elliptic, 4–10 cm, acuminate, remotely dentate to almost entire, glabrous above, densely grayish-pubescent beneath; flowers 3–4 cm across, in 5–9-flowered, leafy racemes, not fragrant; cultivated and occasionally spreading; mostly S; flr. late May–early Jun; native farther south.

HYDROPHYLLACEAE Waterleaf Family

Herbaceous plants with regular, perfect, 5-merous flowers with a tubular corolla; stamens 5, inserted on the corolla tube alternate with the lobes; ovary superior, bicarpellate but 1-locular; style 1, stigmas 2; leaves palmately or pinnately lobed or compound.

A. flowers several to many in a terminal inflorescence
    B. inflorescence coiled, elongating in flower; principal leaves <8 cm wide ....
    .... *Phacelia*
    B. inflorescence not elongating greatly at flowering; principal leaves >8 cm
    wide .... *Hydrophyllum*
A. flowers solitary, opposite the leaves and in few-flowered terminal clusters ....
    .... *Ellisia*

Ellisia L.

Ellisia nyctelea L   Waterpod, Aunt Lucy
Annual to 4 dm tall with pinnately lobed leaves, the lobes coarsely toothed; flowers solitary, opposite the leaves and in small terminal clusters; sepals triangular, free nearly to the base; corolla campanulate, whitish; rare on damp, shady banks and rich alluvial woods; SE; flr. May; FACU; ✿.

Hydrophyllum L.

Perennial or biennial herbs with a 5-merous tubular corolla; stamens and styles exerted; filaments and ovary hairy; leaves pinnately or palmately lobed or divided.

A. leaves pinnately lobed or divided
    B. stem glabrous or with sparse appressed hairs; leaf segments 3–7 ....
    .... *H. virginianum*
    B. stem conspicuously hairy; basal leaves with at least 7 segments ....
    ....*H. macrophyllum*
A. leaves (at least the uppermost) palmately lobed
    C. upper stem and inflorescence mostly glabrous or with a few stout hairs;
    calyx with a few scattered hairs, and without a lobe in the sinuses ....
    .... *H. canadense*
    C. plants hairy, especially above; calyx bristly hairy with a small reflexed lobe
    in each sinus .... *H. appendiculatum*

Hydrophyllum appendiculatum Michx.   Waterleaf
Hairy biennial to 5 dm tall with palmately lobed upper leaves, basal or lower leaves somewhat pinnately lobed or with a few separate leaflets at the base; inflorescence fairly compact in flower, more open in fruit; flowers lavender or pinkish, drying blue; stamens barely exerted; calyx bristly-hairy with a small reflexed lobe in each sinus; rare on wooded slopes, thickets, and stream banks; SW; flr. May.

Hydrophyllum canadense L   Canadian waterleaf
Rhizomatous perennial; stems 3–5 dm tall, mostly glabrous or with a few stout hairs; leaves palmately 3–5-lobed, the largest basal leaves sometimes with a few small, widely separated leaflets at the base; corolla white to pink-purple; calyx lacking a lobe in the sinuses; occasional on rocky wooded slopes, ravines, and moist woods; throughout; flr. early Jun–early Jul; FACU.

Ellisia nyctelea

Hydrophyllum appendiculatum, flower (×1)

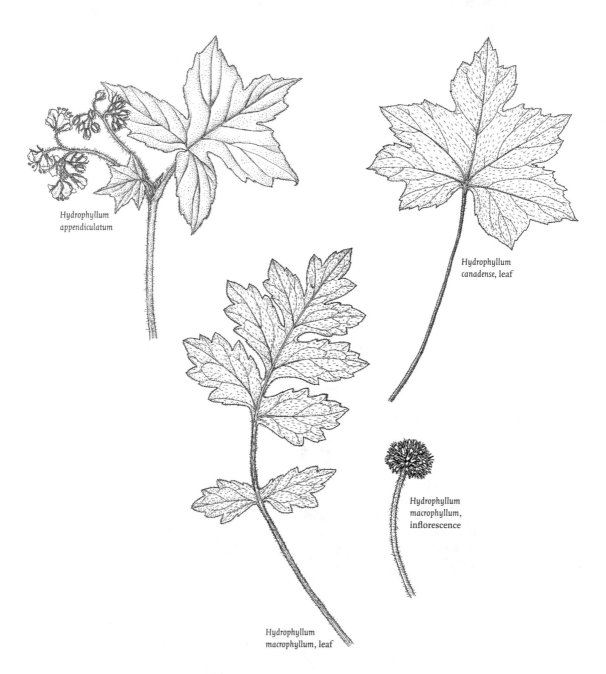

Hydrophyllum appendiculatum

Hydrophyllum canadense, leaf

Hydrophyllum macrophyllum, inflorescence

Hydrophyllum macrophyllum, leaf

Hydrophyllum virginianum, flower (×1)

Hydrophyllum macrophyllum Nutt. Large-leaved waterleaf
Hairy perennial, 3–7 dm tall; leaves pinnately divided nearly to the midvein into 7–9 coarsely toothed segments, lower leaves sometimes with a few remote basal segments; inflorescence very dense; corolla white or pinkish; stamens and style much longer than the corolla; rare in mesic, calcareous woods; SC and SW; flr. early Jun; ❦.

Hydrophyllum virginianum L. Virginia waterleaf
Rhizomatous perennial; stems 3–8 dm tall with sparse, appressed hairs above; leaves pinnately divided almost to the midrib, sometime with a few distinct leaflets at the base on the larger basal leaves; leaf blade segments coarsely toothed

and often marked with whitish blotches; flowers white; stamens and styles extending well beyond the corolla lobes; common in moist woods, thickets, and stream banks; throughout; flr. May; FAC.

Phacelia A.L.Juss.

Annual or biennial, mostly hairy herbs with alternate, pinnately lobed or compound leaves; inflorescence a helicoid cyme that elongates in flowering; flowers blue or white; fruit a capsule.

A. corolla lobes fringed; calyx lobes subequal .... *P. purshii*
A. corolla lobes entire; calyx lobes unequal .... *P. dubia*

Phacelia dubia (L.) Trel.   Scorpion-weed, small flowered phacelia
Annual; stem simple or diffusely branching; leaves pinnate or pinnatifid, mostly 3(5)-lobed; corolla white to lilac (drying blue) with entire lobes; calyx lobes unequal; infrequent on shaly wooded slopes and cliffs; SE and SC; flr. May–early Jun.

Phacelia purshii Buckley   Miami-mist
Annual or biennial; stem erect or ascending, simple or branched, 1.5–5 dm tall; leaves pinnately 5–12-lobed; corolla bluish, lobes fringed; calyx lobes nearly equal, linear to linear-lanceolate with divergent hairs; rare in rich, moist woods and creek banks; mostly SW; flr. early May–early Jun.

*Phacelia
purshii*

*Hydrophyllum
virginianum*

JUGLANDACEAE Walnut Family

Deciduous trees to 30–40 m tall with alternate, pinnately compound leaves with resinous glands on the surface; stipules lacking; staminate flowers in long, slender, drooping catkins; pistillate flowers solitary or in short spikes; fruit a 2-valved hard nut enclosed by a dehiscent or indehiscent husk; flr. before the leaves emerge.

A. terminal leaflet the largest; husk of fruit more or less dehiscent into 4 valves; pith homogenous .... *Carya*
A. leaflets uniform or the median lateral leaflets the largest; husk of fruit indehiscent; pith chambered .... *Juglans*

Carya Nutt.

Trees to 30 m tall with alternate, pinnately compound leaves with 5–11 serrate leaflets; bark tight, somewhat scaly, or exfoliating in long strips; pith continuous; staminate flowers in slender, elongate catkins borne in peduncled groups of 3; the hard-shelled, subglobose to oblong or obovoid nut enclosed in a partially to completely dehiscent husk of varying thickness.

A. buds bright yellow; bud scales 4–6, valvate; leaflets 7–9(11); husk winged along the sutures .... *C. cordiformis*
A. buds brown or gray; bud scales >6, imbricate; leaflets 5–7(9); husk not winged
    B. terminal buds <1 cm long; twigs glabrous; leaflets mostly glabrous beneath .... *C. glabra*
    B. terminal buds 1–2.7 cm; twigs pubescent; leaflets pubescent beneath, at least on the veins
        C. mature bark deeply furrowed but not shaggy; outer bud scales deciduous; husk of fruit <6 mm thick .... *C. tomentosa*
        C. bark exfoliating in long shaggy strips; outer bud scales persistent; husk of fruit to 12 mm thick
            D. leaflets 7–9, dense lower surface pubescence persisting at maturity; fruit oblong, 5–8 cm long, nut flattened; apices of outer bud scales prolonged into stiff, spreading points .... *C. laciniosa*
            D. leaflets 5, lower surface pubescence mostly limited to veins and tufts of hairs in the teeth at maturity; fruit subglobose to ovoid, 3.5–5 cm long, nut not flattened; apices of outer bud scales prolonged, but not stiff and spreading .... *C. ovata*

Carya cordiformis,
fruit

Carya cordiformis (Wang.) K.Koch   Bitternut hickory
Bark scaly; winter buds bright yellow; leaflets mostly 7–9, pubescent beneath, often somewhat falcate, the terminal one long-cuneate at the base and sessile; fruit 2.5–3.5 cm, husk thin, winged along the sutures above the middle and splitting halfway to the base; nut 1.5–3 cm, slightly angled, pointed at top, bitter; common in moist woods and stream banks; throughout; FACU+.

Carya glabra (Mill.) Sweet   Pignut hickory
Gray bark furrowed but tight; twigs glabrous; terminal bud <1 cm long, lanceolate, acuminate with mostly deciduous outer bud scales; leaflets usually 5; petioles, rachis, and lower surface of leaflets glabrous or pubescent only on the

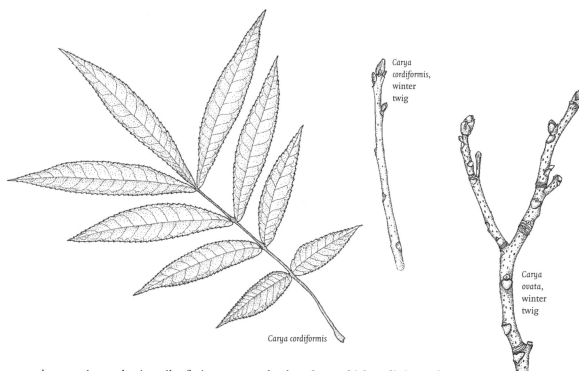

Carya
cordiformis,
winter
twig

Carya cordiformis

Carya
ovata,
winter
twig

larger veins and vein axils; fruit 2–3.5 cm; husk 2–6 mm thick, splitting only at the apex; nut round or compressed, thick-shelled, astringent; common in upland forests; throughout, except in the northernmost counties; FACU–. *C. glabra* is a highly variable species, which here includes what has been called *C. ovalis* (Wang.) Sarg.

Carya laciniosa (F.Michx.) Loudon   Shellbark hickory
Shaggy gray bark exfoliating in long strips; twigs hairy; terminal bud >1 cm long with persistent outer bud scales; leaflets 7–9, permanently pubescent beneath; fruit 3.5–7 cm; husk 6–12 mm thick, splitting to the base; nut 3–6 cm, strongly compressed, prominently angled, wedge-shaped at the base, very thick-shelled, edible; occasional in moist, rich bottomlands and floodplains; mostly S; FAC.

Carya
laciniosa,
nut

Carya ovata (Mill.) K.Koch   Shagbark hickory
Shaggy gray bark exfoliating in long strips; twigs hairy; terminal bud 1 cm or more long with persistent outer bud scales; leaflets usually 5, at maturity glabrate beneath except on the veins and margins where the teeth often bear subapical tufts of hairs; the obovate terminal leaflet much larger than the laterals; fruit 3.5–5 cm; husk 3–12 mm thick, splitting to the base; nut slightly compressed, rounded at the base, angled, sharp-pointed above, thin-shelled, edible; common in moist woods; throughout; FACU.

Carya
ovata,
fruit

Carya tomentosa (Lam. ex Poir.) Nutt.   Mockernut hickory
Bark dark gray and deeply furrowed, but not shaggy; twigs reddish-brown, hairy; outer bud scales of terminal buds deciduous exposing the grayish-brown, silky tomentose inner bud scales; leaflets 7–9; leaves persistently pubescent beneath and on the petioles and rachis with curly fascicled hairs; fruit 3.5–5 cm; the 2–6-mm-thick husk splitting to the base; nut 1.5–3 cm, slightly compressed, rounded at the base, edible; frequent in moist open woods and slopes; mostly S.

Carya
tomentosa,
fruit

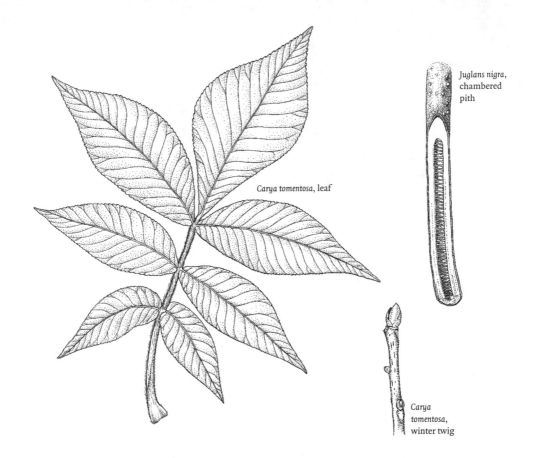

Carya tomentosa, leaf

Juglans nigra, chambered pith

Carya tomentosa, winter twig

Juglans L.

Deciduous trees to 40 m tall with alternate, odd-pinnate, compound leaves; pith diaphragmed; staminate catkins visible during the winter; the nut enclosed in a fibrous, indehiscent husk.

A. pith dark brown; upper margin of the leaf scars not notched, with a prominent hairy fringe all the way across; fruit oblong-ovoid, pointed .... *J. cinerea*

A. pith tan; upper margin of the leaf scars notched, sometimes with a pad of hairs at the center, but not hairy all the way across; fruit subglobose, not at all pointed ....*J. nigra*

Juglans cinerea, nut

Juglans cinerea L.   Butternut
Tree to 30 m tall; bark gray, fissured but with smooth ridges; pith dark brown, chambered; leaflets 11–17, oblong-lanceolate, acuminate, length <2.5 times the width; fruit 4–7 cm, oblong and somewhat pointed; nuts short cylindric, rough; occasional in rich woods; throughout, but declining due to *Sirococcus* canker disease; FACU+.

Juglans nigra L.   Black walnut
Tree to 40 m tall; bark dark brown, rough; pith tan, chambered; leaflets 11–23, lanceolate, acuminate, length >2.5 times the width; fruit 5–8 cm thick, subglobose; nut very rough and only slightly, if at all, flattened; common in open woods, meadows, and floodplains in moist, alluvial soils; throughout, except at the highest elevations; FACU.

LAMIACEAE Mint Family

Herbaceous plants with simple, often aromatic leaves and mostly square stems; leaves opposite; flowers perfect, ovary superior and 4-lobed or 4-parted, corolla tubular and often irregular; fruit a cluster of four nutlets, separating at maturity; inflorescences appearing terminal or axillary, consisting of whorls or pairs of flowers at the nodes, the individual flowers often subtended by bracteoles.

A. anther-bearing stamens 2
  B. calyx irregular, appearing 2-lipped; the upper lip entire or 3-lobed, the lower lip 2-lobed
    C. flowers in axillary clusters .... *Hedeoma*
    C. flowers in terminal inflorescences
      D. inflorescence a dense head-like cluster (dense subterminal clusters sometimes also present) .... *Blephilia*
      D. inflorescence variable but not head-like
        E. inflorescence a panicle or raceme; flowers yellow .... *Collinsonia*
        E. inflorescence a raceme or spike; flowers white, blue, or purplish .... *Salvia*
  B. calyx essentially regular, not appearing 2-lipped
    F. inflorescences mostly terminal (axillary clusters sometimes present)
      G. calyx <4 mm long; corolla <1 cm long .... *Cunila*
      G. calyx >4 mm long; corolla >1 cm long .... *Monarda*
    F. inflorescences mostly axillary (small terminal clusters sometimes present) .... *Lycopus*
A. anther-bearing stamens 4
  H. calyx with a distinct projection on the upper side .... *Scutellaria*
  H. calyx without such a projection
    I. corolla regular or nearly so, its lobes of nearly equal length (lower corolla lobe usually deflexed in *Trichostema*)
      J. calyx distinctly 2-lipped
        K. stamens extending 7–15 mm beyond the corolla .... *Trichostema*
        K. stamens not extending or extending <5 mm beyond the corolla... .... *Perilla*
      J. calyx regular or nearly so
        L. inflorescence spike-like, with all flowers on one side (secund) ... . *Elsholtzia*
        L. inflorescences variable but never secund
          M. inflorescences dense, head-like terminal clusters .... *Pycnanthemum*
          M. inflorescences terminal or axillary but not dense and head-like
            N. inflorescence a loose, leafy panicle .... *Trichostema*
            N. inflorescences of many-flowered terminal or axillary clusters .... *Mentha*
    I. corolla conspicuously lipped, not at all regular
      O. corolla distinctly 1-lipped; the upper lip greatly reduced or absent
        P. upper lip absent; lower lip 5-lobed .... *Teucrium*
        P. upper lip small and 2-lobed; lower lip 4-lobed .... *Ajuga*
      O. corolla distinctly 2-lipped
        Q. stamens not extending beyond the corolla tube, not visible without splitting the tube

R. calyx lobes 10 .... *Marrubium*

R. calyx lobes 5 .... *Sideritis*

Q. stamens extending beyond the corolla tube, visible without splitting the tube

  S. calyx regular or nearly so

    T. upper lip of corolla hooded

      U. calyx lobes prolonged into short, stiff spines

        V. lower lip of corolla with 2 yellow, conical projections at its base...

        ....*Galeopsis*

        V. lower lip of corolla without such projections .... *Leonurus*

      U. calyx lobes not prolonged into spines

        W. inflorescence essentially terminal; whorls of flowers subtended by bracts or greatly reduced foliage leaves

          X. flowers in opposite pairs at nodes of the inflorescence

            Y. leaves cordate at base .... *Meehania*

            Y. leaves not cordate at base .... *Physostegia*

          X. flowers several to many at nodes of the inflorescence

            Z. plants generally decumbent, <5 dm tall .... *Lamium*

            Z. plants generally erect, 5–10 dm tall .... *Stachys*

        W. inflorescence essentially axillary; whorls of flowers subtended by normal foliage leaves or somewhat reduced foliage leaves

          AA. flowers with distinct pedicels .... *Glechoma*

          AA. flowers sessile .... *Lamium*

    T. upper lip of corolla more or less flattened

      BB. stamens clearly extending beyond the corolla

        CC. inflorescences spike-like .... *Agastache*

        CC. inflorescences head-like .... *Pycnanthemum*

      BB. stamens not clearly extending beyond the corolla .... *Nepeta*

  S. calyx distinctly 2-lipped

    DD. calyx lobes prolonged into short, stiff spines .... *Leonurus*

    DD. calyx lobes not prolonged into spines

      EE. inflorescence essentially axillary; whorls of flowers subtended by normal foliage leaves or somewhat reduced foliage leaves

        FF. lobes of the upper calyx lip wider than long .... *Melissa*

        FF. lobes of the upper calyx lip longer than wide

          GG. individual clusters of flowers sessile

            HH. flowers purple .... *Acinos*

            HH. flowers yellow .... *Lamiastrum*

          GG. individual clusters of flowers pedunculate ....

          .... *Calamintha*

      EE. inflorescence essentially terminal; whorls of flowers subtended by bracts or greatly reduced foliage leaves

        II. stamens extending beyond the corolla lips .... *Thymus*

        II. stamens not extending beyond the corolla lips

          JJ. upper calyx lip 1-lobed, lower lip 4-lobed ....

          .... *Dracocephalum*

          JJ. upper calyx lip 3-lobed, lower lip 2-lobed

            KK. inflorescence with many bristle-like bracteoles ...

            .... *Clinopodium*

            KK. inflorescence with broadly rounded bracteoles...

            .... *Prunella*

Acinos Mill.

Acinos arvensis (Lam.) Dandy   Mother-of-thyme
Finely pubescent annual, 1–2 dm tall; leaves 6–12 mm long, petioles 1–2 mm long; generally 1–3 flowers in the upper axils; calyx 5–6 mm long with a bulge on the lower side; flowers pale purple with white marks on the lower lip; rare on calcareous hillsides and railroad banks; scattered, mostly E; flr. Jun–Sep; native to Europe. [syn: *Satureja acinos* (L.) Scheele]

Agastache J.Clayton

Erect perennial herbs to about 1.5 m tall; inflorescence spike-like; nutlets minutely pubescent at one end.

A. leaves densely pubescent and distinctly whitened beneath .... *A. foeniculum*
A. leaves glabrous or only slightly pubescent beneath
    B. calyx lobes 1–1.5 mm long .... *A. nepetoides*
    B. calyx lobes 2–2.5 mm long .... *A. scrophulariifolia*

Agastache foeniculum (Pursh) Kuntze   Anise or blue giant-hyssop
Plant to about 1 m tall, branching above; leaves generally ovate, up to 9 cm long, coarsely serrate, rounded or truncate at the base, densely pubescent with very small hairs beneath; inflorescence up to 15 cm long; flowers blue; cultivated and rarely escaped; mostly S; flr. Jul–Aug; native to western U.S.

Agastache nepetoides (L.) Kuntze   Yellow giant-hyssop
Plants 1–1.5 m tall, branching above; leaves ovate or ovate-lanceolate, to about 15 cm long, coarsely serrate, rounded or subcordate at the base, finely pubescent beneath, petioles up to 6 cm long; inflorescence to 2 dm long; flowers greenish-yellow; frequent in rich woods, moist thickets, fields, and roadsides; mostly S; flr. Aug–Sep; FACU.

Agastache nepetoides,
flower (×1½)

Agastache scrophulariifolia (Willd.) Kuntze   Purple giant-hyssop
Plants very similar in appearance and habit to *A. nepetoides*; inflorescence up to 15 cm long; flowers purple; occasional in rich woods, moist thickets, and roadsides; throughout; flr. Aug–Sep.

Ajuga L.

Perennial herbs of variable habit; flowers in whorls of 2–6 in the axils of bracts, forming a terminal spike; flowers blue.

A. plants with abundant runners, forming mats .... *A. reptans*
A. plants erect, runners absent .... *A. genevensis*

Ajuga genevensis L.   Bugleweed
Erect plant to about 3 dm tall; leaves ovate to oblong-spatulate, 2–5 cm long, the lower short-petioled and the upper sessile; calyx densely villous, 6–8 mm long; cultivated and rarely escaped to lawns and fallow fields; mostly E; flr. Apr–Jun; native to Europe or northern Asia.

Agastache nepetoides

Ajuga reptans

Ajuga reptans, flower (×1)

Ajuga reptans L.   Carpet bugleweed
Spreading plant with flowering stems erect to about 3 dm; leaves similar to *A. genevensis*; calyx sparsely villous, 4–6 mm long; escaped from cultivation and naturalized in fields, wooded banks, and roadsides; throughout; flr. May–Jun; native to Eurasia.

Blephilia Raf.

Erect perennials, 4–8 dm tall, sometimes sparsely branched toward the top; flowers in densely crowded whorls forming a terminal spike-like head, or sometimes with dense subterminal whorls; flowers pale purple with darker purple spots.

A. stems with very short, curved and appressed hairs .... *B. ciliata*
A. stems with straight, stiff hairs, 1–2 mm in length .... *B. hirsuta*

Blephilia ciliata (L.) Benth.   Wood-mint
Leaves lanceolate to ovate, 3–6 cm long, generally entire or nearly so to occasionally few-toothed; whorls of inflorescence 3–5 and densely crowded (the lower whorl occasionally separated); frequent on wooded slopes, calcareous hillsides, and swamps; mostly W and S; flr. May–Jul.

Blephilia hirsuta (Pursh) Benth. — Wood-mint

Leaves ovate-lanceolate to broadly ovate, 4–8 cm long, generally serrate; whorls of inflorescence 3–5 and generally separated by internodes (the upper two whorls often crowded); frequent in moist woods and swamps; mostly W; flr. May–Aug; FACU–.

Calamintha Mill.

Perennial herbs of variable habit; leaves generally entire or with a few low teeth; inflorescence an elongate panicle, each flower cluster subtended by progressively reduced foliage leaves.

A. leaves linear, 1–2 cm long .... *C. arkansana*
A. leaves ovate or ovate-oblong, generally <1 cm long .... *C. nepeta*

Calamintha arkansana (Nutt.) Shinners — Calamint

Stoloniferous plant with stiffly erect branches, simple or sometimes branched, glabrous except for a pubescent area at each node; calyx glabrous, 4–6 mm long; corolla pale purple, 8–15 mm long; rare on calcareous shores along Lake Erie; flr. May–Aug; FACU.

Calamintha nepeta (L.) Savi — Basil-thyme, calamint

Much branched herb to about 1 m tall, stems pubescent; leaves broadly ovate, 10–20 mm long, shallowly crenate-serrate to nearly entire with no more than 5 teeth on each side; calyx glabrous, 4–6 mm long; corolla pale purple to white, 9–12 mm long; Bucks and Philadelphia Cos.; flr. Jun–Sep; native to Europe. Ours is subsp. *glandulosa* (Req.) P.W.Ball [syn: *Satureja calamintha* (L.) Scheele subsp. *glandulosa* (Req.) Gams]

Clinopodium L.

Clinopodium vulgare L. — Wild basil

Perennial, 2–6 dm tall, often branched toward the top; leaves ovate or ovate-oblong to ovate-lanceolate, 2–4 cm long, generally entire or with a few low teeth, the lower leaves petioled and the upper sessile to petioled; inflorescence a many-flowered head-like cluster terminating the main stem and often the side branches (head-like clusters also present at the first node on vigorous growth); flowers pale purple, pink, or white; common in open woods, fields, and roadsides; throughout; Jun–Sep; a plant of circumboreal distribution, but those in Pennsylvania probably native to Europe. [syn: *Satureja vulgaris* (L.) Fritsch]

Clinopodium vulgare

Clinopodium vulgare, flower (×1)

Collinsonia L.

Collinsonia canadensis L. — Horse-balm, stoneroot

Erect perennial herb from a woody rhizome, to 12 dm tall; leaves ovate or ovate-oblong, the largest up to 2 dm long on petioles to 1 dm long, serrate, progressively smaller and on shorter petioles upward, the uppermost essentially sessile; inflorescence a terminal panicle; flowers yellow, the calyx sometimes weakly irregular (appearing nearly regular), the corolla 1.2–1.5 mm long; frequent in rich woods and on wooded floodplains; throughout; flr. Jul–Sep; FAC+.

Collinsonia canadensis, flower (×1)

Collinsonia
canadensis

Cunila
origanoides

Collinsonia canadensis, rhizome

Cunila L.

Cunila origanoides,
flower (×1½)

Cunila origanoides (L.) Britton   Common dittany, stone-mint
Much-branched perennial herb, 2–4 dm tall, glabrous; leaves ovate to deltoid-ovate, 2–4 cm long, few-toothed, nearly sessile; inflorescence branching tri-chotomously from the terminal node or arising from the upper axils; flowers purple to white, the corolla nearly regular, the stamens extending far beyond the corolla lobes; frequent in dry, open woods, shaly slopes, and serpentine out-crops; mostly S; flr. Aug–Oct.

Dracocephalum L.

Dracocephalum parviflorum Nutt.   Dragonhead
Erect, sometimes branched perennial, 2–8 dm tall, glabrous to finely pubescent; leaves lanceolate to ovate-lanceolate, 3–8 cm long, sharply serrate; inflorescence spike-like, 2–10 cm long; flowers blue, the corolla about as long as the calyx; very rare in yards and waste ground; mostly S; flr. May–Jul; native farther N and W, introduced or adventive in Pennsylvania; FACU–; 🍂.

Elsholtzia Willd.

Elsholtzia
ciliata, flower (×3)

Elsholtzia ciliata (Thunb.) Hyl.   Elsholtzia
Erect or ascending annual, 3–5 dm tall; leaves ovate to ovate-lanceolate, 3–7 cm

long, tapering to the base, crenate-serrate; inflorescence 2–5 cm long; flowers pale blue; on stream banks, roadsides, and open woods; SE; rare but probably spreading throughout; flr. Jul–Sep; native to Asia.

Galeopsis L.

Galeopsis bifida Boenn.   Hemp-nettle
Stems hispid (often densely), 3–8 dm tall; leaves lanceolate to ovate, 5–10 cm long, crenate-serrate, pubescent on both sides; flowers variegated, usually with white and pink; occasional in moist or swampy woods, lake margins, and roadsides; mostly N; flr. Jun–Sep; native to Eurasia. [syn: *G. tetrahit* L.]

Glechoma L.

Glechoma hederacea L.   Gill-over-the-ground, ground-ivy
Stems creeping, to 1 m long; leaves rotund to reniform, 1.5–3 cm wide, crenate; calyx 5.5–9 mm long; bractlets subulate, shorter than calyx; flowers purple; common in fields, disturbed woods, roadsides, gardens, and waste ground; throughout; flr. Apr–Jun; native to Eurasia; FACU.

Glechoma hederacea,
flower (×1)

Hedeoma Pers.

Small, strong-scented annuals; flowers blue, in few-flowered axillary clusters; corolla rather weakly 2-lipped, scarcely if at all exceeding the calyx.

A. leaves lanceolate to ovate or obovate .... *H. pulegioides*
A. leaves linear .... *H. hispidum*

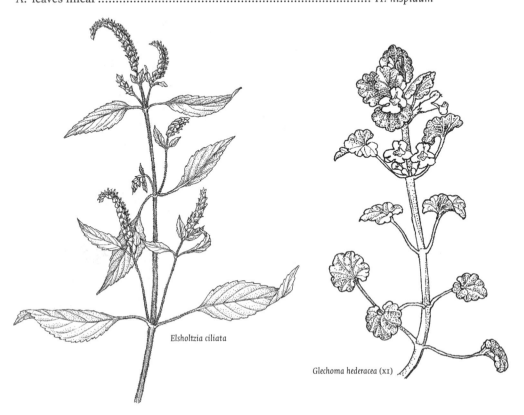

Elsholtzia ciliata

Glechoma hederacea (×1)

Hedeoma hispidum Pursh   Mock pennyroyal

Plant 5–20 cm tall; leaves linear to oblong-linear, 1–2 cm long, sessile, entire; calyx about 5 mm long, the upper lip lobed halfway or more to the base, the lower lower divided to the base; rare in fallow fields; Berks and Lancaster Cos.; flr. May–Aug; native farther north and west.

Hedeoma pulegioides,
calyx (×1½)

Hedeoma pulegioides (L.) Pers.   American pennyroyal, pudding-grass

Plant 1–4 dm tall; leaves lanceolate to ovate or obovate, 1–3 cm long, petioled, entire or serrulate; calyx similar to H. *hispidum* except the upper lip lobed ⅓ to ½ way to the base; common in dry fields, pastures, woods, and roadsides; throughout; flr. Jul–Sep.

Lamiastrum Heist. ex Fabr.

*Lamiastrum
galeobdolon,*
flower (×1)

*Lamiastrum
galeobdolon,*
calyx (×1)

Lamiastrum galeobdolon (L.) Ehrend. & Polatschek   Yellow archangel

Erect perennial; stems 1.5–6 dm tall, sparsely to densely hairy; leaves ovate to orbicular, 3–8 cm long, coarsely toothed, truncate to subcordate at base; calyx 7–10 mm long, the lobes about ¼ as long as the tube; corolla 15–25 mm long, bright yellow with brown markings; Chester Co., possibly elsewhere; native to Eurasia.

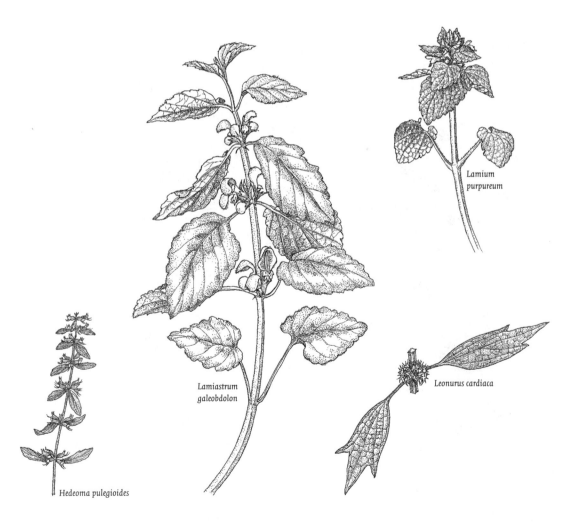

*Lamium
purpureum*

*Lamiastrum
galeobdolon*

Leonurus cardiaca

Hedeoma pulegioides

Lamium L.

Annuals or perennials, often spreading or decumbent; flowers white to red or purple, in axillary clusters of 6–12, or the clusters crowded into an interrupted terminal spike.

A. corolla <2 cm long
    B. upper leaves sessile or clasping ....L. amplexicaule
    B. upper leaves petiolate .... L. purpureum
A. corolla 2 cm or more long
    C. corolla generally white; lateral lobes of the corolla with 2–3 teeth ....
    .... L. album
    C. corolla pink to purple (although rarely white); lateral lobes of the corolla
       with 1 tooth ....L. maculatum

Lamium album L.   Snowflake, white dead-nettle
Perennial, erect or ascending from a decumbent base; stems pubescent, 2–5 dm tall; leaves ovate to deltoid, 3–10 cm long, crenate; calyx 10–13 mm long, the lobes longer than the tube; cultivated and rarely escaped to moist thickets; Berks and Bucks Cos.; flr. Apr–Sep; native to Eurasia.

Lamium amplexicaule L.   Henbit
Annual, weakly ascending or decumbent; stems 1–4 dm long; leaves subrotund, 1–3 cm wide, deeply crenate; calyx 5–7 mm long, densely villous, the lobes about as long as the tube; common in wooded slopes, fields, roadsides, fencerows, and shale barrens; throughout; flr. Mar–Nov; native to Eurasia.

Lamium maculatum L.   Spotted dead-nettle
Perennial, erect or ascending from a decumbent base; stems 2–6 dm tall; leaves ovate to deltoid, usually with a white stripe along the midrib; calyx 8–10 mm long, the lobes shorter than to about as long as the tube; common in wooded slopes, floodplains, cultivated fields, roadsides, and waste ground; throughout; flr. Apr–Oct; native to Eurasia.

Lamium purpureum L.   Purple dead-nettle
Annual, similar to *L. amplexicaule* in habit and size; calyx 5–7 mm long, the lobes about as long as the tube; common in wooded slopes, fields, and roadsides; throughout; flr. Apr–Oct; native to Eurasia.

Lamium purpureum,
flower (×1)

Leonurus L.

Erect, strong-scented annuals or perennials, 1–1.5 m tall; stems finely pubescent; leaves dentate to lobed; flowers pink to white, in dense, crowded clusters forming an interrupted terminal spike.

A. leaves palmately 3–7-lobed; corolla 8–12 mm long .... L. cardiaca
A. leaves entire to coarsely toothed but not lobed; corolla 5–7 mm long ....
....L. marrubiastrum

Leonurus cardiaca L.   Common motherwort
Perennial; larger leaves ovate to suborbicular, sharply toothed as well as lobed, upper leaves progressively smaller and narrower, uppermost often merely 3-

Leonurus cardiaca,
flower (×1½)

toothed; calyx 6–8 mm long, the lobes about as long as the tube; corolla pale pink; frequent in stream banks, roadsides, railroad cinders, fields, and woods; throughout; flr. Jun–Aug; native to Asia.

Leonurus marrubiastrum L.   Motherwort
Annual or biennial; leaves ovate-lanceolate to ovate, serrate to crenate; corolla scarcely surpassing the calyx; occasional in old fields, vacant lots, and roadsides; mostly SE; flr. Jun–Sep; native to Eurasia.

Lycopus L.

Perennials, usually stoloniferous, often decumbent, generally inhabiting wet areas; flowers very small, white, in dense axillary clusters; flr. Jun–early Sep.

A. calyx lobes <1 mm long, obtuse or acute
    B. stem pubescent; stolons without tubers ....*L. virginicus*
    B. stem generally glabrous or minutely pubescent; stolons with white tubers
    ....*L. uniflorus*
A. calyx lobes 1–2 mm long, sharp-pointed
    C. leaves with stiff, appressed hairs above ....*L. europaeus*
    C. leaves glabrous or only slightly pubescent above
        D. calyx lobes acute to acuminate but not subulate-tipped .... *L. rubellus*
        D. calyx lobes subulate-tipped ....*L. americanus*

Lycopus americanus Muhl. ex W.Bartram   Water-horehound
Stolons short, without tubers; leaves lanceolate to nearly linear, 3–8 cm long, upper nearly entire, lower serrate to pinnatifid; corolla barely surpassing the calyx; common on shaded hillsides, fields, moist thickets, wet ditches, and swamps; throughout; OBL.

Lycopus europaeus L.   European water-horehound
Similar to L. *americanus*; leaves generally ovate in outline, upper and lower serrate; calyx lobes about 2 mm long; rare in canal banks and moist waste ground; mostly SE; native to Europe; OBL.

Lycopus rubellus Moench.   Gypsy-wort, water-horehound
Stolons long, bearing tubers; leaves lanceolate to elliptic, 5–10 cm long, sharply acute to acuminate; calyx lobes 1.3–2 mm long; corolla surpassing the calyx by about 1 mm; rare in bogs, stream banks, pond margins, and wet ditches; scattered throughout; OBL; ❦.

Lycopus uniflorus Michx.   Bugleweed, water-horehound
Stolons long; leaves lanceolate to oblong, 2–6 cm long, acute to short-acuminate, teeth widely spaced; calyx lobes broadly triangular; common in swampy meadows, bogs, lake margins, and floodplains; throughout; OBL.

Lycopus virginicus L.   Bugleweed, water-horehound
Stolons long; leaves lanceolate to ovate or elliptic, 5–12 cm long, acuminate, coarsely serrate; calyx lobes ovate to triangular-ovate; frequent in moist woods, stream banks, swamps, and wet ditches; throughout; OBL.

Lycopus virginicus,
flower (×2)

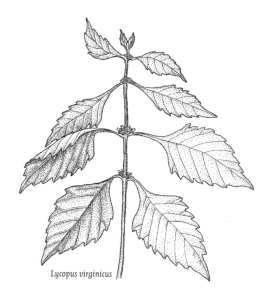

Lycopus virginicus

Marrubium L.

Marrubium vulgare L.   Common horehound
Strong-scented perennial; erect, 4–6 dm tall; stems white-hairy; leaves ovate, 3–
5 cm long, rugose, white-hairy; flowers in dense axillary clusters; flowers about
6 mm long; calyx lobes 2–3 mm long, hooked at the apices in fruit; cultivated
and occasionally escaped to fields and roadsides; mostly S; flr. Jun–Aug; native
to Eurasia.

Meehania Britton

Meehania cordata (Nutt.) Britton   Heart-leafed meehania
Decumbent to trailing perennial with erect flowering stems 1–2 dm tall; leaves
broadly ovate, 3–6 cm long, long-petioled, shallowly crenate; terminal spikes
few-flowered, 3–5 cm long; calyx tube about 1 cm long; upper calyx lobes 5–6
mm long, lower lobes about 1 mm shorter; flowers blue; occasional on banks
and wooded slopes; SW; flr. May–Jul; 🌿.

Melissa L.

Melissa officinalis L.   Lemon-balm
Citrus-scented perennial; stems 4–8 dm tall; leaves ovate to deltoid, primary
leaves 4–7 cm long, coarsely crenate; flowers white to pale blue, in few-flowered
axillary clusters; calyx 7–10 mm long, the lobes about 2/3 as long as the tube; cultivated and occasionally escaped to woods and roadsides; mostly S; flr. Jun–Aug;
native to Asia.

Mentha L.

Strong-scented rhizomatous or stoloniferous perennials with erect stems; leaves
serrate; flowers blue to lavender, in dense axillary clusters or in terminal spikes
or heads; flr. Jun–Sep. Most species of *Mentha* are capable of hybridizing with

each other resulting in plants with characters intermediate between those of the parents. Those hybrids that have been collected in Pennsylvania have been included in the key, however, it is entirely possible that other hybrids might be found.

A. inflorescence essentially terminal; whorls of flowers subtended by bracts or greatly reduced foliage leaves
- B. inflorescence head-like, consisting of 1–3 whorls densely packed at the end of the stem .... M. aquatica
- B. inflorescence spike-like, consisting of 3 or more whorls extending down the stem
  - C. leaves densely pubescent beneath
    - D. principal leaves more than twice as long as wide .... M. longifolia
    - D. principal leaves less than twice as long as wide
      - E. inflorescence 5–6 mm across .... M. x rotundifolia
      - E. inflorescence 7–8 mm across .... M. x villosa
  - C. leaves glabrous or with a few scattered hairs beneath
    - F. principal leaves sessile or with petioles not exceeding 3 mm long ...
      .... M. spicata
    - F. principal leaves with petioles 4–15 mm long .... M. x piperata
A. inflorescence essentially axillary; whorls of flowers subtended by normal foliage leaves or somewhat reduced foliage leaves
- G. leaves with 2–3 lateral veins on each side of the midvein .... M. pulegium
- G. leaves with 4 or more lateral veins on each side of the midvein
  - H. plants generally pubescent
    - I. calyx 1.5–2.5 mm long .... M. arvensis
    - I. calyx 2.5–4 mm long .... M. x verticillata
  - H. plants essentially glabrous or with sparse pubescence .... M. x gentilis

Mentha aquatica L.   Water mint
Stems 3–6 dm tall, sparsely to densely pubescent; leaves broadly ovate, 3–7 cm long, rounded to subcordate at base; calyx and pedicels pubescent; very rare in stream banks; SE and SW; native to Eurasia; FACW+.

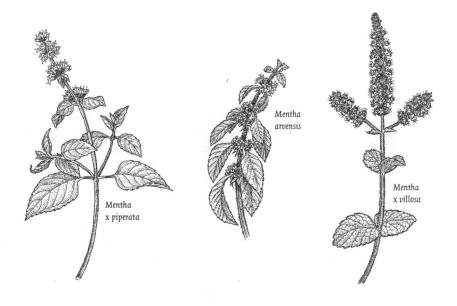

Mentha
x piperata

Mentha
arvensis

Mentha
x villosa

Mentha arvensis L.   Field mint

Stems 3–6(10) dm tall; leaves lanceolate to broadly ovate, 3–7 cm long, weakly serrate; calyx sparsely pubescent to long villous; frequent in moist banks, wet meadows, and swamps; throughout; FACW.

Mentha arvensis,
flower (×2)

Mentha x gentilis   Red mint

Cultivated and frequently naturalized in gravelly shores, stream banks, and roadsides; throughout; native to Eurasia; FACW. [*M. arvensis x spicata* of Rhoads and Klein, 1993]

Mentha longifolia (L.) L.   Horse mint

Stems finely pubescent, to 1 m tall; leaves sessile to subsessile, oblong-lanceolate, 2–7 cm long, serrate with appressed teeth; calyx pubescent; cultivated and rarely escaped to stream banks; mostly S; native to Eurasia; FACU.

Mentha x piperata   Peppermint

Frequently naturalized in wet fields, moist thickets, swamps, and stream banks; throughout; native to Eurasia; FACW+. [*M. aquatica x spicata* of Rhoads and Klein, 1993]

Mentha x piperata,
flower (×2)

Mentha pulegium L.   Pennyroyal

Stems 1–4 dm tall; leaves narrowly elliptic to ovate-oblong, 1–2 cm long, entire or few-toothed, densely pubescent; calyx finely villous; cultivated, also occasionally occurring on ballast or waste ground; mostly S; native to Europe.

Mentha x rotundifolia   Apple mint, pineapple mint

Similar to *M. longifolia*; leaves broadly elliptic to round-ovate, crenate-dentate; cultivated and rarely escaped to roadsides; mostly S; native to Eurasia; FACW. [*M. longifolia x suaveolens* of Rhoads and Klein, 1993]

Mentha spicata L.   Spearmint

Stems 1–5 dm tall, generally glabrous; leaves oblong lanceolate, 2–6 cm long, sharply serrate, rounded to obtuse at base, mostly glabrous; calyx tube glabrous, the lobes pilose; naturalized in stream banks, swamps, and wet pastures; throughout; native to Europe; FACW+.

Mentha x verticillata   Mint

Rare in ballast and waste ground; mostly SE; native to Eurasia. [syn: *M. sativa* L.; *M. aquatica x arvensis* of Rhoads and Klein, 1993]

Mentha x villosa   Apple mint, wooly mint

Similar to *M. longifolia*; leaves round-oval to round-oblong, coarsely dentate; occasional in fields, springy slopes, and roadsides; throughout; native to Eurasia. [syn: *M. alopecuroides* (Hull) Briq.; *M. spicata x suaveolens* of Rhoads and Klein, 1993]

Monarda L.

Erect perennials with showy flowers aggregated into head-like clusters in the axils of the upper leaves, or terminal on the branches; inflorescences subtended by leaf-like bracts; flowers subtended by linear to subulate bractlets.

A. stamens not extending beyond the lobes of the corolla .... M. punctata
A. stamens extending beyond the lobes of the corolla
   B. corolla bright red .... M. didyma
   B. corolla reddish-purple or pink to white
      C. bracts of the inflorescence strongly tinged with purple .... M. media
      C. bracts of the inflorescence green
         D. upper lip of the corolla bearded at the tip .... M. fistulosa
         D. upper lip of the corolla not bearded at the tip .... M. clinopodia

Monarda didyma,
inflorescence

Monarda clinopodia L.   Bee-balm
Stems to 1 m tall, glabrous to sparsely pubescent; leaves ovate to lanceolate, 6–12 cm long, serrate, acuminate, rounded at base; corolla white to yellowish-white, 1.5–3 cm long; flowering heads 1.3–3 cm across (excluding corollas); frequent in moist woods, fields, and floodplains; throughout; flr. Jun–Jul.

Monarda didyma L.   Bee-balm, oswego-tea
Stems 7–15 dm tall, glabrous to sparsely pubescent especially at the nodes; leaves ovate to nearly ovate-lanceolate, 7–15 cm long, serrate, acuminate, rounded at base; corolla crimson, 3–4.5 cm long; flowering heads 2–4 cm across (excluding corollas); frequent in creek banks, floodplains, and moist woods, also cultivated; throughout; flr. Jul–Sep; FAC+.

Monarda fistulosa,
flower (×1)

Monarda fistulosa L.   Horsemint, wild bergamot
Stems 5–12 dm tall, generally pubescent above; leaves deltoid-lanceolate to ovate, serrate, acuminate; rounded to truncate or acute at base; corolla lavender to pink, 2–3 cm long; flowering heads 1.5–3 cm across (excluding corollas); flowers lavender to pink; throughout; common in fields, thickets, and roadsides; throughout; flr. Jun–Sep; UPL.

Monarda media Willd.   Bee-balm, purple bergamot
Stems to 1 m tall, glabrous to sparsely pubescent; leaves ovate to broadly lanceolate; bracteal leaves purplish; calyx usually purple, finely pubescent, glandular; corolla rose-red, 2–2.5 cm long; flowering heads 1.5–3 cm across (excluding corollas); occasional in low woods, stream banks, and floodplains; scattered throughout; flr. Jul–Aug.

Monarda punctata L.   Spotted bee-balm
Stems 3–10 dm tall, sparsely pubescent; leaves lanceolate to narrowly oblong, 2–8 cm long, pubescent; bracteal leaves pale green, often tinged with purple; calyx usually villous; corolla pale yellow with purple spots; rare in dry, open, sandy fields; scattered throughout; flr. Jun–Sep; UPL; 🍂.

Nepeta L.

Nepeta cataria,
flower (×1½)

Nepeta cataria L.   Catnip
Perennial, finely and densely pubescent throughout; stems erect to 1 m tall; leaves deltoid to deltoid-ovate, 3–8 cm long, coarsely crenate-dentate, truncate to subcordate at base; inflorescence terminal, 2–6 cm long, consisting of numerous small blue to white flowers; cultivated and frequently escaped to woods, fields, and roadsides; throughout; flr. Jul–Oct; native to Eurasia; FACU.

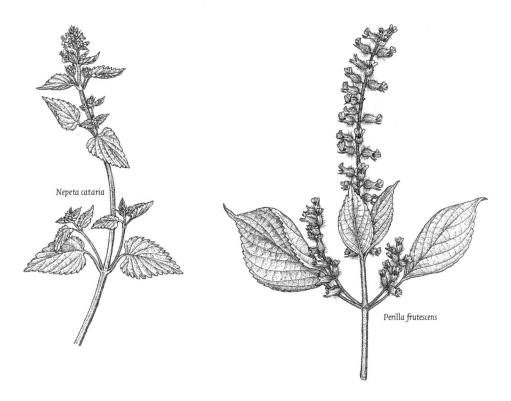

Nepeta cataria

Perilla frutescens

Perilla L.

Perilla frutescens (L.) Britton   Perilla
Stems mostly 10–40 cm; leaves ovate-oblong to broadly ovate, 8–15 cm long, long-petioled, generally purple below, cuneate at base; calyx 3 mm long at flowering, 9–12 mm long in fruit, strongly 2-lipped; flowers small, purplish to white, paired along the spike-like inflorescence; cultivated and occasionally naturalized in moist or shaded roadsides and disturbed woods; mostly SC and SE; flr. Aug–Sep; native to India; FACU+.

Perilla frutescens,
calyx (×1)

Physostegia Benth.

Physostegia virginiana (L.) Benth.   False dragonhead
Stems to 1 m tall; leaves lanceolate to oblanceolate or narrowly oblong, the larger ones 5–12 cm long, sharply serrate; upper leaves similar to lower but greatly reduced, often entire; flowers pinkish-purple, corolla 2.5–3 cm long; occasional on stream banks and moist shorelines, also cultivated and occasionally escaped; throughout; flr. Jul–Sep; FAC+.

Prunella L.

Low-growing perennials; stems creeping or decumbent with erect branches to 6 dm tall; inflorescences terminal, dense, and head-like; flowers violet to pink or white.

A. at least some leaves deeply dissected .... *P. laciniata*
A. all leaves entire to somewhat crenate .... *P. vulgaris*

Prunella laciniata L. Heal-all, self-heal
Stems densely pubescent; principal leaves pinnatifid; occasional in fields, open woods, and weedy roadsides; mostly SE; flr. Jun–Aug; native to Europe.

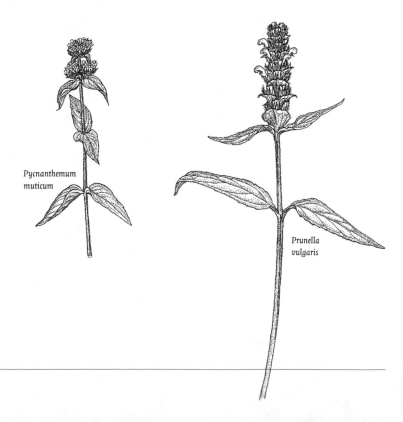

Prunella vulgaris L. Heal-all, self-heal
Stems to 6 dm long; leaves generally glabrous above; calyx 7–10 mm long, green or purple; corolla violet-blue to pink or white, 10–16 mm long; inflorescence spike-like, 2–5(8) cm long, 1.5–2 cm across; throughout; flr. May–Oct; FACU+; 2 varieties:

A. principal leaves lanceolate to oblong, cuneate at the base ....
.... var. *lanceolata* (Barton) Hultén
  frequent in fields, woods, floodplains, and roadsides.
A. principal leaves ovate to ovate-oblong, rounded at the base .... var. *vulgaris*
  rare in moist fields and fencerows; native to Europe.

Pycnanthemum Michx.

Erect, rhizomatous, perennials; flowers small, in crowded head-like clusters terminating the main stem and branches, or also in the axils of the upper leaves; corolla purple to white, the lower lip often spotted with purple; flr. Jul–Sep. A previous report of *P. pycnanthemoides* (Leavenw.) Fernald (Rhoads and Klein, 1993) was based on a misidentified specimen.

A. calyx lobes of equal or nearly equal length
  B. mature leaves 1.5–3 cm across .... *P. muticum*
  B. mature leaves rarely >1 cm across
    C. upper stems glabrous; leaves linear, mostly 2–3 mm wide ....
    .... *P. tenuifolium*

Prunella vulgaris,
flower (×1)

Pycnanthemum
muticum

Prunella
vulgaris

C. upper stems pubescent at least on the angles; leaves linear-lanceolate, most leaves 3 mm or more wide
  D. tips of calyx lobes prolonged into sharp tips .... *P. torrei*
  D. tips of calyx lobes not prolonged into sharp tips
    E. upper stems pubescent only on the angles .... *P. virginianum*
    E. upper stems pubescent both on the angles and on the sides ....
    .... *P. verticillatum*
A. calyx lobes clearly not all of equal length
  F. leaves glabrous below .... *P. clinopodioides*
  F. leaves whitened with fine pubescence below .... *P. incanum*

Pycnanthemum clinopodioides Torr. & A.Gray   Mountain-mint
Stems with both short curving and long spreading hairs; leaves lanceolate, the larger ones 1–2 cm wide, acuminate, serrate to nearly entire; rare on dry slopes; mostly SE; 🌿.

Pycnanthemum incanum (L.) Michx.   Mountain-mint
Stems to about 1 m tall; upper stems densely pilose with short spreading hairs and a few longer spreading hairs; leaves ovate to ovate-oblong, remotely toothed; flowering heads 15–35 mm across, rather loose; frequent in old fields, thickets, and barrens; throughout.

Pycnanthemum muticum (Michx.) Pers.   Mountain-mint
Stems 4–8 dm tall, minutely pubescent; leaves oblong to ovate-lanceolate, the larger ones 4–7 cm long, weakly serrate; flowering heads 8–15 mm across; occasional in moist woods, thickets, meadows, and swales; mostly SE; FACW.

Pycnanthemum muticum, flower (×2½)

Pycnanthemum tenuifolium Schrad.   Mountain-mint
Stems 5–8 dm tall, glabrous, much-branched above; leaves linear, 2–5 cm long, entire, glabrous; flowering heads very dense, 3–8 mm across; common in moist old fields, sandy stream banks, and floodplains; mostly S, scattered elsewhere; FACW.

Pycnanthemum torrei Benth.   Torrey's mountain-mint
Stems to 1 m tall, glabrous below, thinly pubescent above; leaves linear-lanceolate, 4–6 cm long, entire, glabrous above, pubescent below; flowering heads 6–15 mm across; rare in upland woods and thickets; mostly S; 🌿.

Pycnanthemum verticillatum (Michx.) Pers.   Mountain-mint
Stems to 1 m tall, densely pubescent above; leaves narrowly lanceolate, larger ones 3–5 cm long, few-toothed; flowering heads dense, 8–15 mm across; FAC; 2 varieties:

A. leaves glabrous below or pubescent only along the major veins ....
  .... var. *verticillatum*
  occasional in abandoned fields, swampy meadows, marshes, and woods; throughout.
A. leaves pubescent below .... var. *pilosum* (Nutt.) Cooperr.
  rare in rocky meadows; NE and SE.

Pycnanthemum virginianum (L.) Durand & Jacks. ex B.L.Rob. & Fernald

Mountain-mint

Stems to 1 m tall; leaves linear-lanceolate, glabrous above, pubescent on the veins below, the larger ones 3–6 cm long; flowering heads dense, 7–15 mm across; occasional in boggy fields, swamps, and moist woods; throughout, uncommon N; FAC.

Salvia L.

Annuals or perennials; flowers relatively showy, in continuous or interrupted terminal spike-like racemes.

A. upper lip of calyx 1-lobed and untoothed .... *S. reflexa*
A. upper lip of calyx 3-lobed or with 3 small teeth (visible at 10x)
    B. veins on underside of leaf prominently raised .... *S. pratensis*
    B. veins on underside of leaf not raised (sometimes even sunken)
        C. most leaves basal .... *S. lyrata*
        C. most leaves cauline
            D. upper lip of calyx about equal in length to the lower lip ....
            .... *S. verticillata*
            D. upper lip of calyx about half as long as the lower lip .... *S. nemorosa*

Salvia lyrata, flower (×1)

Salvia lyrata, split calyx (×1)

Salvia lyrata L.   Lyre-leaved sage
Perennial; stems 3–6 dm tall; principal leaves basally whorled, 1–2 dm long, deeply pinnately lobed into rounded segments; cauline leaves usually 1 pair (rarely 2–3 pairs), similar to basal leaves but much smaller; occasional in moist pastures, thickets, or woods; S; flr. May–Jun; UPL.

Salvia nemorosa L.   Woodland sage
Stems 3–7 dm tall, branched above; principal leaves cauline, ovate-lanceolate, 5–10 cm long, mostly glabrous above, pubescent below; very rare on railroad banks and cinders; SE; flr. Jun–Jul; native to Eurasia.

Salvia pratensis L.   Meadow sage
Stems 3–6 dm tall; principal leaves basal, long-petioled, ovate-oblong, 7–12 cm long, serrate-crenate; cauline leaves similar to basal but much smaller and fewer; rare in dry pastures and on waste ground; mostly SE; flr. Jun–Aug; native to Europe.

Salvia reflexa Hornem.   Lance-leaved sage
Annual; 3–6 dm tall, much-branched, sparsely pubescent; leaves lanceolate to linear, 3–5 cm long, entire to few-toothed, narrowed to base; rare in stream banks, old fields, roadsides, cinders, and quarry waste; mostly S; flr. Jun–Sep.

Salvia verticillata L.   Sage
Perennial; stems 4–8 dm tall; leaves broadly ovate-oblong to deltoid, 5–10 cm long, coarsely serrate; subcordate to truncate at base, sparsely pubescent on both sides; cultivated and occasionally escaped to roadsides; mostly SE; flr. Jul–Sep; native to Eurasia.

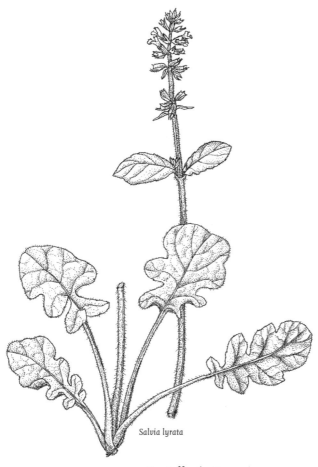

Salvia lyrata

Scutellaria L.

Perennials, mostly rhizomatous; calyx strongly 2-lipped, both lips entire; corolla 2-lipped, lower lip 3-lobed; flowers blue or violet to pink or white.

A. all flowers solitary in the axils of normal or slightly reduced foliage leaves
    B. corolla 15 mm or more long .... S. *galericulata*
    B. corolla 6–10 mm long
        C. leaves 1–1.5 cm long .... S. *leonardii*
        C. leaves 2–4.5 cm long .... S. *nervosa*
A. flowers in racemes subtended by bracts or substantially reduced foliage leaves
    D. corolla 5–8 mm long... S. *lateriflora*
    D. corolla >1 cm long
        E. leaf blades truncate or cordate at the base .... S. *saxatilis*
        E. leaf blades rounded or gradually narrowed at the base
            F. middle to upper leaves entire .... S. *integrifolia*
            F. all leaves variously toothed
                G. calyx with glandular hairs (and also somewhat pubescent) ....
                .... S. *elliptica*
                G. calyx variably pubescent but glandular hairs absent
                    H. racemes generally 3–5; the lowest pair of flowers of each raceme subtended by bracts .... S. *incana*
                    H. racemes generally 1 (rarely 3); the lowest pair of flowers subtended by normal foliage leaves .... S. *serrata*

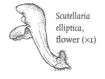
Scutellaria elliptica, flower (×1)

Scutellaria elliptica Muhl. ex Spreng. Hairy skullcap

Stems 3–6 dm tall, usually unbranched; middle and upper leaves ovate to deltoid, 4–7 cm long, crenate from widest point to apex, acute at base; inflorescences 3–10 cm long; flowers blue; flr. May–Aug; 2 varieties:

A. upper stem with short curved hairs .... var. *elliptica*
    occasional in open woods, wooded banks, and shale barrens; mostly SC and SE.
A. upper stem with glandular hairs 1–2 mm long ....
    .... var. *hirsuta* (Short & Peter) Fernald
    rare in open woods; mostly SW.

Scutellaria galericulata L. Common skullcap

Stems 3–6 dm tall, pubescent on the angles with recurved hairs; leaves ovate-oblong to lanceolate, largest ones 3–5 cm long, crenate-serrate, acute to acuminate at apex, rounded to subcordate at base; flowers blue; frequent in bogs, swamps, and marshy meadows; throughout, less common S; flr. Jun–Aug; OBL.

Scutellaria elliptica

Scutellaria lateriflora

Scutellaria leonardii

Scutellaria incana Biehler　　　　　　　　　　　　Downy skullcap

Stems to 1 m tall, usually minutely pubescent; leaves ovate to lanceolate, 5–10 cm long, crenate except near the base, obtuse to subcordate at base; flowers blue; frequent in rocky woods and roadsides; mostly C and W; flr. Jun–Aug.

Scutellaria integrifolia L.　　　　　　　　　　　　Hyssop skullcap

Stems 3–7 dm tall, finely pubescent; middle and upper leaves lanceolate to oblanceolate, 2–6 cm long, entire, obtuse at apex, tapering to sessile base; flowers blue to pink; occasional in swamps, bogs, moist woods, fields; mostly SC and SE; flr. May–Jul; FACW.

Scutellaria lateriflora L.　　　　　　　　　　　　Mad-dog skullcap

Stems 3–7 dm tall, glabrous to minutely pubescent on the angles; leaves ovate or deltoid to lanceolate, the larger ones 4–7 cm long, coarsely serrate; inflorescences 3–10 cm long; flowers blue, or rarely pink or white; common in wet woods, bogs, lake margins, stream banks, floodplains, and swampy pastures; throughout; flr. Jul–Sep; FACW+.

Scutellaria leonardii Epling　　　　　　　　　　　　Small skullcap

Stems several, 1–2 dm tall from a rhizomatous base, minutely pubescent on the angles; leaves ovate-lanceolate, 10–16 mm long, entire, glabrous to somewhat scabrous above, sessile; flowers blue; occasional in open woods and shores; mostly S and NW; flr. May–Jul.

Scutellaria nervosa Pursh　　　　　　　　　　　　Skullcap

Stems 3–7 dm tall, glabrous to minutely pubescent on the angles above, glandular pubescent below; leaves lanceolate to round-ovate, 2–4.5 cm long, crenate-serrate to nearly entire, acute to obtuse at apex, rounded to subcordate at base; flowers blue; occasional in moist wooded slopes, stream banks, and floodplains; mostly S; flr. May–Jun; FAC.

Scutellaria saxatilis Riddell　　　　　　　　　　　　Rock skullcap

Stems often decumbent, 2–5 dm long, glabrous or sparsely pubescent; leaves ovate to deltoid-ovate, larger ones 2–4 cm long, with a few rounded teeth, rounded to cordate at base; rare in low woods, rocky stream banks, and roadsides; mostly SW; flr. Jul–Aug; ❦.

Scutellaria serrata Andr.　　　　　　　　　　　　Showy skullcap

Stems 3–6 dm tall, generally glabrous; leaves few, ovate to elliptic, larger ones 5–11 cm long, crenate, broadly rounded to a cuneate base; flowers blue; very rare in rocky, humusy woods and floodplains; SE; believed extirpated; flr. May–Jun; ❦.

Sideritis L.

Sideritis romana L.　　　　　　　　　　　　Ironwort

Stems 1–3 dm tall, branched from the base, strongly hirsute; leaves spatulate to oblanceolate, 1–2.5 cm long, crenate-serrate; calyx 2-lipped, the upper lip 1-lobed, larger than the 4-lobed lower lip; rare in cultivated fields; flr. Jun–Aug; Lancaster Co., possibly elsewhere; native to Eurasia.

Stachys L.

Calyx nearly regular, 5–10-nerved; corolla strongly 2-lipped.

A. stem and undersides of leaves densely white-wooly .... *S. germanica*
A. stem and undersides of leaves glabrous to somewhat pubescent but not white-wooly
    B. stems pubescent on the sides and on the angles (visible at 10x)
        C. calyx lobes about half as long as the tube .... *S. nuttallii*
        C. calyx lobes ²∕₃ as long to as long as the tube .... *S. palustris*
    B. stems glabrous or with hairs only on the angles (visible at 10x)
        D. principal leaves sessile or nearly so .... *S. hyssopifolia*
        D. petioles of principal leaves 1–2 cm long .... *S. tenuifolia*

Stachys germanica L.   Hedge-nettle
Stems 3–10 dm tall; leaves lanceolate to oblong, 4–8 cm long, serrate; flowers red; rare along roadsides and woods edges; mostly S; flr. Jun–Aug; native to Europe.

Stachys hyssopifolia Michx.   Hedge-nettle
Stems 3–5 dm tall, often branched from the base, bearded at the nodes, glabrous to sparsely pubescent on the internodes; leaves linear to linear-oblong, 2–7 cm long, sessile to subsessile, entire to few-toothed; flowers mostly pink; SC and SE; flr. Jul–Sep; 🌿; 2 varieties:

A. stems hispid on the angles; leaves 5–15 mm wide .... var. *ambigua* A.Gray
    rare in fallow fields and on waste ground; FACW.
A. stems generally glabrous; leaves 2–9 mm wide .... var. *hyssopifolia*
    rare in fields and stream banks; FACW+.

Stachys nuttallii Shuttlew. ex Benth.   Nuttall's hedge-nettle
Stems 5–10 dm tall, with pustulate-based hairs on the sides and angles, and minutely glandular-pubescent; leaves ovate-lanceolate to oblong, 6–12 cm long, crenate-serrate, obtuse to rounded at base; very rare on wooded mountain slopes; SC and SW; flr. Jun–Jul; FAC; 🌿. [syn: *S. cordata* Riddell]

Stachys palustris L.   Hedge-nettle, woundwort
Stems 5–10 dm tall, seldom branched, pubescent on both the sides and the angles; leaves linear-oblong to ovate, 5–10 cm long, sharply serrate, acute at apex, obtuse to subcordate at base; flowers usually purplish; flr. Jul–Aug; OBL; 2 varieties:

A. calyx with glandular hairs mixed with short nonglandular hairs....var. *palustris*
    occasional in stream banks, moist fields, woods, and edges; throughout; native to Europe.
A. calyx with long (1.5–3 mm) hairs mixed with shorter hairs ....
    .... var. *pilosa* (Nutt.) Fernald
    very rare, Westmoreland Co.; native farther W.

Stachys tenuifolia, flower (×1)

Stachys tenuifolia Willd.   Creeping hedge-nettle
Stems to 1 m tall, usually unbranched, glabrous to densely hispid or scabrous on the angles only; leaves oblanceolate to oblong, 6–15 cm long, crenate-serrate, glabrous to densely hispid above, acuminate at apex, obtuse to subcordate at

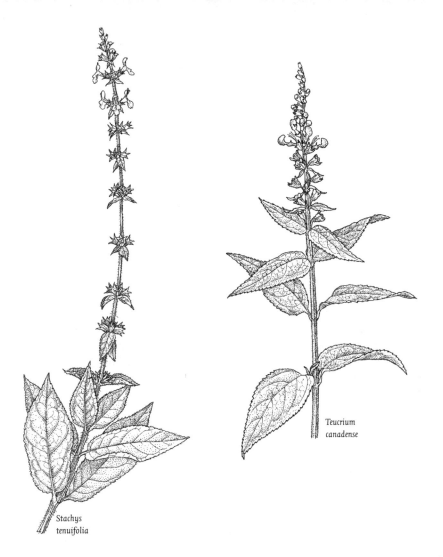

Stachys
tenuifolia

Teucrium
canadense

base; frequent in moist, wooded bottomland, stream banks, wet meadows, and fields; throughout; flr. Jun–Aug; FACW+. [includes S. *hispida* Pursh]

Teucrium L.

Teucrium canadense L.
Rhizomatous perennial; stems 5–10 dm tall, pubescent, rarely branched; leaves ovate-lanceolate to oblong, 5–12 cm long, dentate-serrate, acute to acuminate at apex, obtuse to rounded at base; flowers purplish to pink or cream-colored in spike-like racemes; flr. Jun–Aug.

Teucrium canadense,
flower (×1)

A. hairs in the inflorescence gland-tipped ....
.... var. *boreale* (E.P.Bicknell) Shinners; Northern germander, wood-sage
rare in ballast, wharves, swamps, and cultivated ground; scattered throughout; native farther N; FACW–. [syn: var. *occidentale* (A.Gray) E.M. McClint. &
Epling]
A. hairs in the inflorescence not gland-tipped ....
.... var. *virginicum* (L.) Eaton; Wild germander, wood-sage
common in floodplains, lake margins, moist fields, and fencerows; throughout; FACW.

Thymus L.

Thymus pulegioides L.   Creeping thyme
Low-growing perennial; stems much branched, 1–4 dm long; leaves linear to nearly rotund, 5–10 mm long; flowers purple; cultivated and occasionally escaped to roadside banks; throughout; flr. Jun–Sep; native to Europe.

Trichostema L.

Ovary shallowly 4-lobed; corolla lobes about equal, the lower one usually deflexed; flowers generally blue, rarely pink to white.

A. stamens extending <5 mm beyond the corolla .... *T. brachiatum*
A. stamens extending 7–15 mm beyond the corolla
    B. leaves linear with only a midvein evident .... *T. setaceum*
    B. leaves oblong to lanceolate with lateral veins evident .... *T. dichotomum*

Trichostema brachiatum L.   False pennyroyal
Stems 2–4 dm tall, much-branched, finely pubescent, glandular in the inflorescence; leaves entire, 1–3-nerved, acute at apex, narrowed to base; flowers pale blue; corolla about 5 mm long, scarcely longer than the calyx; occasional in open woods, rocky shores, shale barrens, and dry slopes; S; flr. Aug–Sep. [syn: *Isanthus brachiatus* (L.) Britton, Stearns & Poggenb.]

Trichostema dichotomum, flower (×1)

Trichostema dichotomum L.   Blue-curls
Stems up to 7 dm tall, much-branched, densely glandular short-hairy, especially in the inflorescence; leaves 2–5 cm long, narrowed to the base; corolla 4–6 mm long; frequent in open woods, fields, rock outcrops, barrens, and dry roadsides; throughout; flr. Aug–Sep.

Trichostema setaceum Houtt.   Narrow-leaved blue-curls
Stems 1–3 dm tall, finely pubescent; leaves linear, glabrous or nearly so; corolla 6–10 mm long; very rare on dry, sandy banks and shaly slopes; Berks and Montour Cos.; flr. Aug–Sep; ✿.

LARDIZABALACEAE Lardizabala Family

Woody climbers with alternate, palmately compound leaves; monoecious; sepals 3 or 6; petals lacking; stamens 3 or 6; ovaries 3–12, distinct; fruit an oblong berry that eventually splits open along one side; seeds numerous.

Akebia Decne.

Characteristics of the family.

Akebia quinata (Houtt.) Decne.   Five-leaf akebia
High-climbing, woody twining vine; leaves evergreen, alternate, palmately compound with 5 oval leaflets; flowers purplish-brown, in axillary racemes, the lower pistillate, 2–3 cm wide, the upper smaller and staminate; locally abundant in disturbed woods, naturalized from cultivated sources; mostly SE; flr. May; native to Asia.

LAURACEAE Laurel Family

Deciduous, aromatic trees or shrubs with alternate, simple, lobed or unlobed, entire leaves lacking stipules; dioecious; flowers small, unisexual, regular, perigynous; perianth of 6 greenish-yellow tepals fused at the base to form a hypanthium; stamens 9; ovary superior, unilocular; fruit a drupe.

A. trees to 20 m tall; leaves variable with 0–3 lobes; inflorescence terminal; fruit dark blue .... *Sassafras*
A. shrubs to 3 m tall; leaves never lobed; inflorescences lateral; fruit red ....
.... *Lindera*

Lindera Thunb.

Lindera benzoin (L.) Blume   Spicebush
Shrub to 3 m tall; leaves elliptic or obovate, pinnately veined; flowers small, greenish-yellow, in dense, nearly sessile clusters along the branches; fruits red, 6-10 mm, ellipsoid; common in moist woods; throughout, except in the northernmost counties; flr. Mar–Apr, before the leaves, frt. Sep–Oct; FACW–.

Sassafras Nees

Sassafras albidum (Nutt.) Nees   Sassafras
Deciduous tree to 20 m tall; leaves variable with 0–3 lobes; inflorescence a raceme or panicle; flowers greenish-yellow; fruit 1 cm, blue; common in old fields, hedgerows, and woods edges; throughout; flr. late Apr–May, with the leaves; frt. Sep–Oct; FACU–.

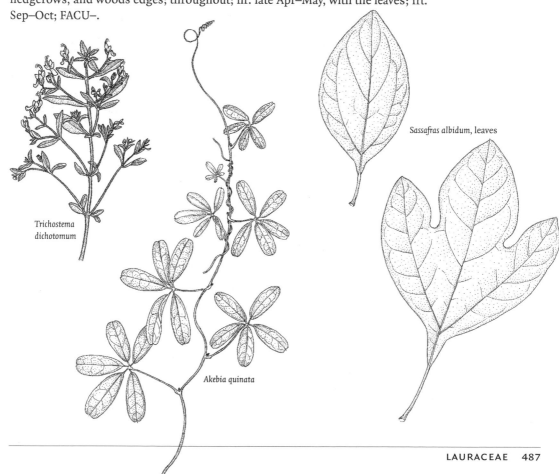

Trichostema dichotomum

Sassafras albidum, leaves

Akebia quinata

LENTIBULARIACEAE Bladderwort Family

Aquatic herbs with linear to finely dissected leaves usually bearing small traps that catch and digest tiny aquatic organisms; flowers on emergent, leafless peduncles; calyx slightly irregular; corolla strongly bilabiate and spurred, the upper lip 2-lobed, the lower 3-lobed; stamens 2, inserted on the corolla tube; ovary superior, with 2 fused carpels; fruit a capsule; in addition to chasmogamous flowers described above, some species bear cleistogamous flowers under water on pedicels arising from the nodes of the stems.

Utricularia L.

Characteristics of the family; plants entirely suspended or partially buried in sand, peat, or muck; flowers yellow or pink-purple, solitary or in short racemes, each flower subtended by a bract, sometimes additional empty bracts present below the inflorescence; dense, leafy winter buds or turions are formed in some species.

A. basal portions buried in the substrate; leaves few, erect, linear; traps few, mostly hidden in the substrate
    B. floral bracts tubular; flowers solitary; corolla purple .... U. resupinata
    B. floral bracts not tubular; flowers 1–5; corolla yellow
        C. flowering stems very slender with a zigzag inflorescence; pedicels longer than the bracts .... U. subulata
        C. stems stout; pedicels shorter than the bracts .... U. cornuta
A. plants anchored or suspended; leaves numerous, arising from the nodes of elongate, floating stolons, or just above the surface of the submerged substrate, more or less finely, dichotomously divided into flattened or capillary segments; traps well-developed
    D. leaves whorled; traps terminal on the leaf segments; flowers purple ....
    .... U. purpurea
    D. leaves alternate or crowded at the base; traps lateral on the leaf segments; flowers yellow
        E. peduncle with a whorl of spongy, inflated floats .... U. radiata
        E. peduncle not bearing a whorl of floats
            F. divisions of the leaves flat, with a visible midrib; floral bracts auriculate at the base
                G. traps on separate stems; leaf divisions spine-toothed; pedicels erect in fruit .... U. intermedia
                G. traps on the leaves; leaf divisions not spine-toothed; pedicels curved downward in fruit... U. minor
            F. divisions of the leaves filiform, without a distinct midrib; bracts not auriculate
                H. peduncles without extra bracts below the lowest flower ....
                .... U. geminiscapa
                H. peduncles with 1–5 empty bracts below the lowest flower
                    I. upper corolla lip larger than the lower; bracts tapering to a clasping base .... U. gibba
                    I. upper corolla lip smaller than the lower; bracts not clasping ...
                    .... U. macrorhiza

Utricularia cornuta Michx.   Horned bladderwort

Perennial, lower portions buried in the substrate; traps numerous on stolons, rhizoids, and leaves; peduncles 0.5–1.5 mm thick; flowers 1–6, on filiform 1–2 mm pedicels; empty bracts numerous, similar to the floral bracts; corolla yellow; rare in shallow water of marshes, ponds, and ditches; mostly NE; flr. Jul–Aug; OBL; ✤.

Utricularia geminiscapa Benj.   Bladderwort

Suspended perennial; leaves very numerous, divided into numerous capillary segments; traps numerous to few, lateral or in the angles between leaf segments; inflorescences of 2 types; those bearing chasmogamous flowers erect, 5–25 cm long, 0.5–0.7 mm thick with 2–8 yellow flowers; floral bracts narrowly ovate, no empty bracts present; cleistogamous inflorescences below the water surface, 0.5–2 cm long with pedicels arising directly from the stolons and the corolla reduced or absent; rare in bogs, vernal ponds, and river margins; scattered, E and C; flr. Jul–Aug; OBL; ✤.

Utricularia gibba L.   Humped bladderwort

Rooted or suspended, often mat-forming annual or perennial; leaves sparsely dichotomously branched into capillary, slightly flattened, entire or sparsely denticulate ultimate segments; traps few, lateral; inflorescence 1–20 cm long; peduncle filiform, 0.4–0.8 mm thick; floral bracts semicircular, minutely glandular, a few empty bracts below; flowers 2–6 on erect or spreading pedicels 0.2–3 cm long; corolla yellow; rare in shallow water or exposed peat, sand, or mud flats; widely scattered; flr. Jul–Aug; OBL; ✤. [syn: *U. fibrosa* Walter in part]

Utricularia intermedia Hayne   Flat-leaved bladderwort

Perennial anchored in the substrate; leaves numerous, mostly above the submerged substrate, circular in outline with dichotomously branched segments that are flattened and have a distinct midrib; traps few, lateral on the segments, mostly on reduced leaves buried in the substrate; floral bracts broadly ovate, auriculate at the base; flowers 2–3 on erect pedicels; corolla yellow; rare in lakes and wet edges of exposed floating bog mats; NE; flr. Jul–Aug; OBL; ✤.

Utricularia macrorhiza LeConte   Common bladderwort

Suspended perennial; leaves numerous, finely dichotomously divided, the ultimate segments with or without short marginal teeth; traps numerous, of 2 sizes, lateral on the leaf segments; inflorescence 10–40 cm, peduncle 1–3 mm thick; bracts broadly ovate; flowers 3–14, pedicels filiform, 0.8–1.5 cm long; corolla yellow with reddish-brown streaks on the lower lip; frequent in lakes, ponds, swamps, marshes, and ditches; throughout; flr. late Jun–Aug; OBL. [Formerly included in the European *U. vulgaris* L.]

Utricularia intermedia, leaf (×3)

Utricularia intermedia

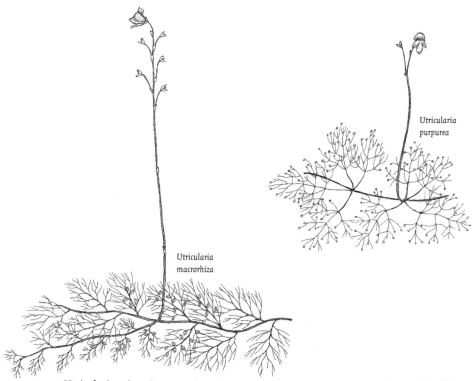

Utricularia purpurea

Utricularia macrorhiza

Utricularia minor L.   Lesser bladderwort

Perennial anchored in the substrate by non-green, reduced leaves and stolons; leaves above the substrate semicircular in outline and dichotomously divided into 7–22 flattened segments with midribs; traps more numerous on the buried portions, few or absent on those above the substrate; inflorescence 2.5–25 cm long, peduncle filiform, straight; bracts broadly ovate, auriculate at the base; flowers 2–6 on 4–8 mm pedicels; corolla yellow; rare in shallow water of lakes, ponds, and swamps; NE and NW; flr. Jun–Aug; OBL; ❦.

Utricularia purpurea Walter   Purple bladderwort

Suspended perennial with stolons to 6 dm or more; leaves numerous, whorled, repeatedly divided into capillary segments; traps numerous, terminal on the ultimate segments; inflorescence 3–20 cm with 1–3 flowers; floral bracts extending below the point of attachment, empty bracts not present; corolla pink-purple with a yellow blotch at the base of the lower lip; rare, but sometimes locally abundant in lakes and ponds; NE; flr. Jul–early Sep; OBL; ❦.

Utricularia radiata Small   Floating bladderwort

Suspended annual; leaves numerous, dichotomously divided into capillary segments; traps numerous, lateral; the peduncle bearing a whorl of spongy, inflated floats that hold it erect; flowers 1–7; corolla yellow with brown streaks at the base of the lower lip; very rare in shallow ponds and ditches; SE; believed to be extirpated; flr. Jun–Aug; OBL; ❦.

Utricularia resupinata B.D.Greene ex Bigelow   Northeastern bladderwort

Anchored, mat-forming perennial; leaves numerous from the peduncle base and nodes of the stolon; traps of 2 sizes (0.5 and 1 mm long), moderately numerous on stolons and lower part of the leaves; peduncle 0.5–1 mm thick, flowers pink-purple, solitary; floral bracts tubular, several empty bracts below the flower; very rare in swamps; NW; believed to be extirpated; flr. Jul–Aug; OBL; ✿.

Utricularia subulata L. Slender bladderwort
Stolons and leaves mostly embedded in the substrate; traps few; peduncle filiform, 2–20 cm; floral bracts ovate to elliptic, attached at or below the middle; flowers 1–12 on 4–15 mm pedicels; corolla yellow; very rare in wet soil or shallow water; SE; believed to be extirpated; flr. Jun–Aug; OBL.

LIMNANTHACEAE Meadow-foam Family

Small, weak-stemmed annuals with alternate, pinnate leaves; flowers 3-merous, petals distinct, smaller than the sepals; stamens 6, 3 with nectar glands at the base; ovary deeply 2-lobed with basally attached style.

Floerkea Willd.

Floerkea proserpinacoides Willd. False mermaid
Small, bright yellow-green, glabrous herbs with inconspicuous solitary flowers on long axillary pedicels; locally abundant in moist woods and floodplains; throughout; flr. Apr–May; FAC.

Floerkea proserpinacoides

Floerkea proserpinacoides, flower (×2)

LINACEAE Flax Family

Annual or perennial, upright herbs with numerous, narrow, sessile leaves that are alternate above and often opposite below; flowers in a branched terminal inflorescence, regular, perfect, 5-merous; sepals distinct or nearly so, overlapping; petals distinct; stamens same number as the petals and alternate with them, filaments expanded below; ovary superior, of 3–5 united carpels; styles distinct or partially united; fruit a globose, dehiscent capsule.

Linum L.

A. flowers blue
    B. stigmas capitate; margins of the inner sepals entire .... *L. perenne*
    B. stigmas linear or clavate; margins of the inner sepals ciliate
        C. stems usually solitary; capsule 6–9 mm .... *L. usitatissimum*
        C. stems several; capsule 4–6 mm .... *L. bienne*
A. flowers yellow
    D. leaves with a pair of minute stipular glands at the base .... *L. sulcatum*
    D. leaves without stipular glands

E. capsule ovoid, rounded to pointed at the top .... *L. intercursum*
E. capsule rounded and flattened at the top (depressed globose)
    F. margins of the inner sepals with stalked glands .... *L. medium*
    F. margins of the inner sepals glandless or with tiny sessile glands
        G. branches terete; inflorescence a broad, flat-topped corymb ....
        .... *L. virginianum*
        G. branches striate-angled; inflorescence elongate, panicle-like .....
        .... *L. striatum*

Linum bienne Mill.   Blue flax

Erect biennial or perennial; stems 0.6–6 dm tall, branched, clustered; leaves linear, acuminate, 1–3- veined; sepals with a conspicuous midvein, the inner ciliate, the outer entire; petals blue, 2–3 times as long as the sepals; capsule subglobose, 4–6 mm; occasionally escaped from cultivation to waste ground and ballast; native to Europe.

Linum intercursum, capsule (×2½)

Linum intercursum E.P.Bicknell   Sandplain wild flax

Perennial; stems 1–7 dm tall, branched above; leaves narrowly oblong to elliptic; flowers yellow, 10–15 mm wide; outer sepals entire, inner ones shorter and glandular-ciliate; capsule ovoid, pointed; rare in moist, clayey, open thickets and serpentine barrens; SE; flr. Jul–Aug; ❧.

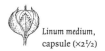
Linum medium, capsule (×2½)

Linum medium (Planch.) Britton   Yellow flax

Perennial; stems to 1 m tall; leaves linear to lance-elliptic, the upper ones with subulate tips; flowers on stiff upright branches; flowers yellow, 10–16 mm wide; inner sepals glandular-ciliolate; capsule flattened-globose; flr. Jul–Aug; FACU; 2 varieties:

A. leaves thick, blunt, 3–5 mm wide .... var. *medium*
    very rare in moist sand flats; NW.
A. leaves thin, 1.5–3.5 mm wide .... var. *texanum* (Planch.) Fernald
    occasional in moist to dry sandy fields or woodland openings; SE.

Linum perenne L.   Perennial flax

Perennial; stems 3–7 dm tall, clustered; leaves linear, 1-nerved or obscurely 3-nerved at the base; flowers blue, 2.5 cm wide; cultivated and rarely naturalized along roadsides; flr. Jun; native to Europe.

Linum striatum, flower (×2½)

Linum striatum, capsule (×2½)

Linum striatum Walter   Ridged yellow flax

Perennial; stems 3–9 dm tall, ridged or angled; leaves elliptic to oblanceolate or obovate, the lower ones opposite, the upper often alternate; inflorescence an elongate panicle; flowers yellow, 10–14 mm wide; sepals entire or the inner ones with a few sessile glands; capsule depressed globose; occasional in moist meadows and wet, open woods; throughout, except the northernmost counties; flr. Jul–Aug; FACW.

Linum sulcatum Riddell   Grooved yellow flax

Annual; stems 1.5–7.5 dm tall, winged and grooved; leaves mostly or all alternate, 3-nerved, with a pair of dark stipular glands at the base; flowers yellow, 1–2

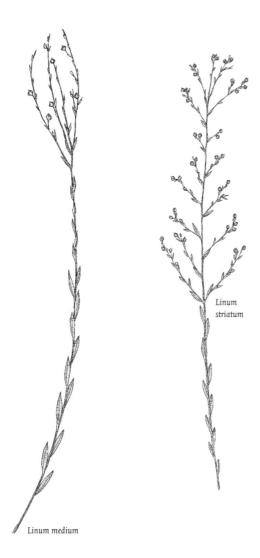

Linum striatum

Linum medium

cm wide; sepals all glandular serrate; capsule ovoid, pointed; rare in sandy bar-rens; scattered; ❧.

Linum usitatissimum L.   Common flax
Annual; stems to 1 m tall; leaves lance-linear, 3-veined; flowers blue, 2–3 cm wide; capsule subglobose, 6–10 mm; occasional on railroad banks and road-sides; mostly S; native to Europe.

Linum virginianum L.   Slender yellow flax
Perennial; stems 1.5–8 dm tall; leaves elliptic to oblanceolate to obovate, oppo-site below, alternate above, and with a single decurrent wing extending a short distance from the base of each leaf; flowers yellow, 8–14 mm wide; the inner se-pals with a few tiny sessile glands along the margins; capsule depressed glo-bose; frequent in dry, open woods, old fields, or shaly slopes; mostly S; flr. Jul–Aug; FACU.

LOGANIACEAE Logania Family

Deciduous shrubs with opposite, simple, leaves; flowers perfect, regular; corolla narrowly tubular with 4 spreading lobes; stamens attached to the corolla tube, same number as and alternate with corolla lobes; ovary superior; fruit a 2-valved, dehiscent capsule.

Buddleja L.

Buddleja davidii Franch.   Butterfly-bush, summer-lilac
Shrub to 3 m tall with long, slender terminal panicles of purple flowers; branches 4-angled; leaves narrowly ovate, 1–2 dm, acuminate, densely white tomentose beneath; inflorescences 1.5–2.5 dm long; cultivated and occasionally naturalized on roadsides or abandoned railroad beds; scattered; flr. Jul–Sep; native to China.

LYTHRACEAE Loosestrife Family

Terrestrial or semiaquatic herbs with opposite, alternate, or whorled simple leaves lacking stipules; flowers perfect, regular or irregular, strongly perigynous, 4–8-merous; the petals and sepals attached to the rim of the hypanthium, the sepals alternating with appendages; stamens as many or twice as many as the petals, attached within the hypanthium, the filaments often of 2 lengths; ovary superior, with several fused carpels; style and stigma 1; fruit a dry capsule.

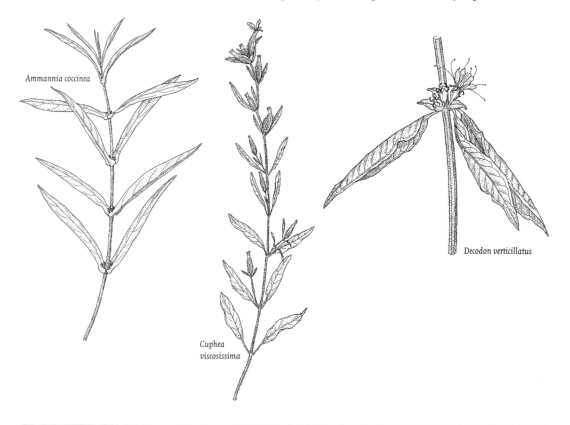

Ammannia coccinea

Cuphea viscosissima

Decodon verticillatus

A. hypanthium tubular, much longer than wide; sepals and petals each 6
    B. flowers irregular; petals unequal; hypanthium bulging at the base on one side .... *Cuphea*
    B. flowers regular; petals equal; hypanthium tube symmetrical at the base..... .... *Lythrum*
A. hypanthium cup-like, about as wide as long; sepals and petals each 4– 6
    C. petals 10–15 mm long, purple .... *Decodon*
    C. petals about 1 mm long, white or pink
        D. flowers clustered in the leaf axils; appendages between the sepal lobes small and narrow .... *Ammannia*
        D. flowers solitary in the leaf axils; appendages nearly the same size and shape as the sepals .... *Rotala*

Ammannia L.

Ammannia coccinea Rottb.   Toothcup
Upright to decumbent annual of wet soil; leaves opposite, sessile, linear or lance-linear; flowers in small clusters in the leaf axils, 4-merous; hypanthium globose to campanulate; sepals alternating with very tiny appendages; petals deep pink, 1.7–2.5 mm; capsule splitting irregularly; rare in wet, sandy or silty shores; SE; flr. Jul–Aug; OBL; ✺.

Ammannia coccinea, flower (×4)

Cuphea P.Browne

Cuphea viscosissima Jacq.   Blue waxweed
Erect, sparingly branched, glandular-hairy annual, 1.5–6 dm tall; leaves opposite, lanceolate to ovate, long-petioled; flowers solitary or paired in the upper axils, irregular, 6-merous; hypanthium nearly 1 cm long, swollen at the base on one side; petals red-purple, the upper 2 larger than the lower; stamens 11 or 12; ovary 2-celled, fruit opening along 1 side; occasional on dry open banks, fencerows, and fields; S; flr. Jul–Sep; FAC–.

Decodon J.F.Gmel.

Decodon verticillatus (L.) Elliot   Water-willow
Perennial with a woody base, stems 1–3 m, usually arched; leaves opposite or in whorls of 3 or 4, lanceolate; flowers in the upper axils, 4–6-merous, of 3 forms varying in the length of style and stamens; hypanthium cup-shaped; sepals triangular, alternating with linear appendages; petals purple, 1–1.5 cm long; stamens 8 or 10; occasional in bogs, swamps, and lake or pond margins in shallow water; E and NW; flr. Jul–Aug; OBL.

Lythrum L.

Upright herbs with opposite or alternate leaves; flowers 6-merous, of 2 or 3 forms varying in the length of styles and stamens; hypanthium cylindrical; petals purple to white; stamens 6 or 12; ovary 2-celled; fruit a capsule.

A. flowers in a dense terminal spike-like inflorescence .... *L. salicaria*
A. flowers solitary or paired in the leaf axils
    B. hypanthium sharply 12-winged; petals purple .... *L. alatum*
    B. hypanthium not winged; petals white to pale purple .... *L. hyssopifolia*

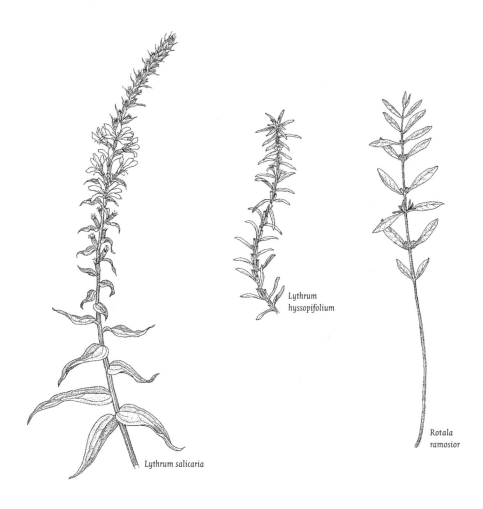

Lythrum hyssopifolium

Rotala ramosior

Lythrum salicaria

Lythrum alatum Pursh   Winged loosestrife
Erect, branched perennial; stem 4-angled, 4–8 dm tall; leaves sessile, linear-oblong to lance-ovate, all but the lowest alternate; flowers solitary in the upper axils; hypanthium with 12 narrow wings; appendages twice as long as the sepals; petals purple; rare in swamps, wet meadows, marshy shores, and ditches; SE and W; flr. Jun–early Sep; FACW+; ❧.

Lythrum hyssopifolia L.   Hyssop loosestrife
Pale green annual with wiry, angled stems 1–6 dm tall; leaves linear, mostly alternate; flowers solitary or in pairs in the axils, sessile, 4–6-merous; corolla pale purple or white; rare, but occasionally locally abundant in moist roadside ditches and shores; SE; flr. Jun–Sep; native to Europe; OBL.

Lythrum salicaria L.   Purple loosestrife
Erect perennial to 2 m, with a persistent woody base; leaves opposite or whorled, sessile, lanceolate to linear, the larger ones slightly cordate at the base; flowers in terminal, leafy-bracted spikes 1–4 dm long; appendages linear, twice as long as the sepals; petals purple; stamens 12; common in swamps, wet meadows, and shores; throughout; flr. Jul–Sep; native to Europe; FACW+; designated as a noxious weed in Pennsylvania.

Rotala L.

Rotala ramosior (L.) Koehne   Toothcup

Erect or prostrate, glabrous annual of wet soils; leaves opposite, linear-spatulate; flowers solitary in the leaf axils, 4-merous; hypanthium campanulate; sepals alternating with appendages of nearly the same size; petals white or pink,
falling early; stamens 4; ovary 4-celled; rare on wet, sandy shores and other
swampy open ground; S; flr. Jul–Sep; OBL; 🌿.

Rotala ramosior,
fruit (×2½)

MAGNOLIACEAE Magnolia Family

Deciduous or semi-evergreen, aromatic trees with alternate, simple, entire or
lobed, stipulate leaves; flowers large, perfect, regular, borne singly; perianth of 3
or more whorls of poorly differentiated tepals; stamens and pistils numerous,
spirally arranged on an elongate receptacle; ovaries superior; fruit an aggregate
of follicles or samaras.

A. leaves entire; tepals white or greenish-yellow; fruit a partly fused aggregate of
   follicles, each of which splits open to expose a red or orange seed ... *Magnolia*
A. leaves lobed; tepals green with orange blotches; fruit a cone-shaped cluster of
   distinct samaras .... *Liriodendron*

Liriodendron L.

Liriodendron tulipifera L.   Tuliptree, yellow-poplar

Deciduous tree to 45 m tall with a tall, straight trunk; terminal buds flattened;
bud scales valvate; leaves 4-lobed, truncate or broadly notched at the apex; tepals
erect, green with orange blotches; fruit a cone-like cluster of overlapping samaras that are dispersed individually; common in mesic forests; throughout, except
in the northernmost counties; flr. late May, after the leaves; FACU.

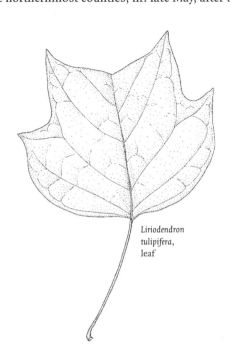

*Liriodendron
tulipifera,
leaf*

*Liriodendron
tulipifera, fruit*

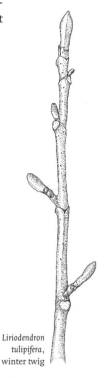

*Liriodendron
tulipifera,
winter twig*

Magnolia L.

Deciduous or semi-evergreen trees with large leaves and flowers; perianth un-differentiated, white or greenish-yellow; stamens and pistils numerous, spirally arranged on an elongate receptacle; fruit a knobby, cone-shaped cluster of partly fused, single-seeded follicles; the seeds, which have a red or orange aril, remain attached by a thread after dehiscence.

Summer key

A. leaves semi-evergreen, thick and leathery, strongly white-glaucous beneath.... .... M. virginiana
A. leaves promptly deciduous, thin, green beneath
    B. leaves in whorl-like clusters at the ends of the branches, to 6 dm long; pet-als 8–14 cm, white .... M. tripetala
    B. leaves scattered along the branch, not clustered
        C. large tree to 30 m; flowers appearing with the leaves, greenish-yellow.... .... M. acuminata
        C. smaller trees to 10 m; flowering before the leaves emerge
            D. flowers white .... M. kobus
            D. flowers purple .... M. soulangeana

Winter key

A. pith diaphragmed; leaves often persistent .... M. virginiana
A. pith not diaphragmed; leaves deciduous
    B. buds glabrous .... M. tripetala
    B. buds pubescent
        C. young twigs glabrous
            D. youngest twigs dark brown above, greenish on side away from sun . ... M. kobus
            D. youngest twigs uniformly red-brown .... M. acuminata
        C. young twigs pubescent .... M. soulangeana

Magnolia acuminata (L.) L.   Cucumber-tree
Deciduous tree to 25 m tall; leaves scattered, 10–25 cm, broadly oblong or ellip-tic; flowers with 6 erect, greenish-yellow tepals; fruit knobby, ellipsoid, 5–8 cm; frequent in rich upland forests; W; flr. May, with the leaves.

Magnolia kobus DC.   Kobus magnolia
Deciduous tree to 10 m tall; with pubescent winter buds; flowers white, 8–15 cm long; petals 6, thin and spreading; sepals 3; occasionally naturalized in dis-turbed woods; SE; flr. late Mar–early Apr, before the leaves; native to Japan.

Magnolia soulangeana Soul.-Bod.   Saucer magnolia
Deciduous tree to 10 m tall with densely hairy winter buds; leaves broadly ovate to obovate; flowers purple; occasionally naturalized in disturbed woods; SE; flr. late Mar–Apr, before the leaves; native to Asia.

Magnolia tripetala (L.) L.   Umbrella-tree
Deciduous tree to 10 m tall with large leaves in whorl-like clusters at the ends of the branches; tepals white, 8–14 cm, spreading; fruit 7–12 cm, ellipsoid-cylin-

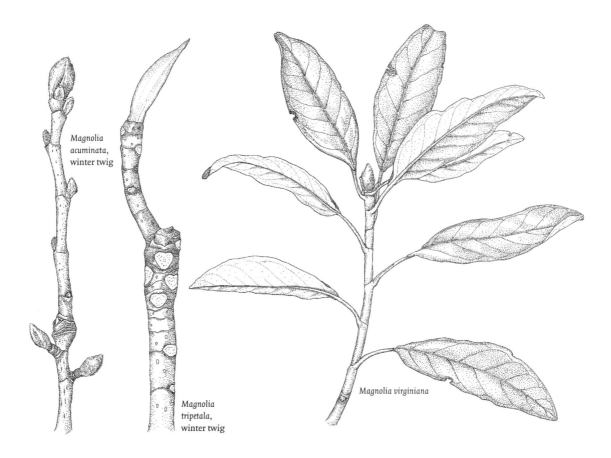

Magnolia acuminata, winter twig

Magnolia tripetala, winter twig

Magnolia virginiana

dric; rare on rich wooded slopes and floodplains; S; flr. May, with the leaves; native, but also apparently introduced and spreading rapidly in some locations; FACU; 🌿.

Magnolia virginiana L.   Sweetbay magnolia
Semi-evergreen tree to 20 m tall, often multistemmed; leaves leathery, white-glaucous and finely hairy beneath; flowers intensely fragrant; tepals creamy white, concave, 3–5 cm; fruit ellipsoid, knobby, 3–5 cm; rare in moist woods and swamps, in sandy-peaty soils; SE; flr. late May–Jun; FACW+; 🌿.

MALVACEAE Mallow Family

Herbaceous plants or shrubs with alternate, simple, usually palmately veined leaves with stipules; flowers regular, 5-merous, perfect, calyx often surrounded by an additional whorl of bracts; petals distinct or fused at the base; stamens numerous, their filaments forming a tube surrounding the style; fruit a 5-locular, dehiscent capsule or a ring of 10–20 loosely united, indehiscent nutlets.

A. bracts present below the calyx
   B. bracts 3 .... *Malva*
   B. bracts 6–12
      C. styles 5 .... *Hibiscus*
      C. styles numerous .... *Althaea*

A. bracts not present below the calyx
  D. leaves strongly cordate-acuminate, velvety on both surfaces .... *Abutilon*
  D. leaves ovate or palmately 3–5-lobed, pubescent but not velvety .... *Sida*

Abutilon Mill.

Abutilon theophrastii, fruit

Abutilon theophrastii Medik.   Butter-print, velvet-leaf
Velvety-pubescent annual, 6–12 dm tall; leaves cordate-acuminate; bracts not present below the calyx; flowers yellow, 1.2–2.5 cm wide; carpels 12–15, arranged in a ring with beak-like projections pointing outward; a frequent weed of cultivated fields, roadsides, and waste ground; mostly S; flr. Jul–early Oct; native to Asia; UPL.

Althaea L.

Althaea officinalis L.   Marsh-mallow
Perennial to 12 dm tall; leaves ovate to slightly cordate, velvety-pubescent; 6–9 narrowly lanceolate bracts present below the calyx; flowers pink, 3 cm wide; cultivated and sometimes escaped to wet ground; mostly S; flr. Jul–Sep; native to Europe; FACW+.

Hibiscus L.

Shrubs or herbs with large flowers; approximately 12 bracts present below the calyx; stamen tube with anthers along most of its length; style with 5 branches each with a capitate stigma; ovules several in each locule.

A. woody shrubs .... *H. syriacus*
A. stems herbaceous, although the base of the plant may be somewhat woody
  B. at least the upper leaves distinctly lobed
    C. leaves deeply 3-parted, the lobes again coarsely toothed or lobed ....
    .... *H. trionum*
    C. leaves with 1 or 2 basal lobes on each side .... *H. laevis*
  B. leaves not lobed or only slightly and irregularly lobed .... *H. moscheutos*

Hibiscus laevis All.   Halberd-leaved rose-mallow
Coarse perennial to 1–2 m tall; at least the upper leaves with widely spreading basal lobes; flowers pink with a purple center, 12–16 cm wide; rare in alluvial shores and marshes in shallow water; SE; flr. late Jul–early Sep; OBL. [syn: H. militaris Cav.]

Hibiscus moscheutos L.   Rose- or swamp-mallow
Coarse perennial to 1–2 m tall; leaves lanceolate to ovate, occasionally shallowly lobed below; flowers 10–20 cm wide, purplish, pink, or white with a darker center; occasional in swamps, marshes, and ditches in shallow water; SE and W; flr. late Jul–early Oct; OBL. [syn: H. palustris L.]

Hibiscus syriacus L.   Rose-of-Sharon
Branching upright shrub to 3 m tall; leaves deciduous, ovate, 3-lobed, toothed; flowers 6–8 cm wide, white, pink, red, or violet with a darker center; capsule 5-locular, pointed, stellate tomentose; seeds with hairy margins; cultivated and oc-

casionally spreading to empty lots, waste ground, or roadsides; mostly S; flr. late
Jul–early Sep; native to Asia.

Hibiscus trionum L.   Flower-of-the-hour
Hairy, branched annual to 5 dm tall; leaves deeply 3-lobed, the segments again
coarsely toothed or lobed; flowers 2.5–5 cm wide, pale yellow with a purplish
base; fruit a many-seeded capsule; a frequent weed of cultivated fields, stream
banks, and dry, rocky ground; mostly S; flr. late Jun–early Oct; native to Europe.

Malva L.

Erect or prostrate herbs with alternate, palmately veined leaves; flowers with 3
bracts below the calyx, petals obcordate; carpels 10–20 arrayed in a ring and
separating into 1-seeded nutlets at maturity; styles as many as the carpels.

A. upper leaves deeply lobed .... *M. moschata*
A. leaves round to kidney-shaped, only slightly if at all lobed
    B. petals reddish-purple; bracts oblong-ovate .... *M. sylvestris*
    B. petals white tinged with pink or purple; bracts narrowly lanceolate to
       linear
        C. plants erect, up to 2 m tall; flowers sessile or nearly so .... *M. verticillata*
        C. plants low, spreading; flowers stalked
            D. petals twice as long as the sepals; carpels smooth and rounded on
               the back .... *M. neglecta*
            D. petals only slightly if at all longer than the sepals; carpels flat and
               veiny on the back .... *M. rotundifolia*

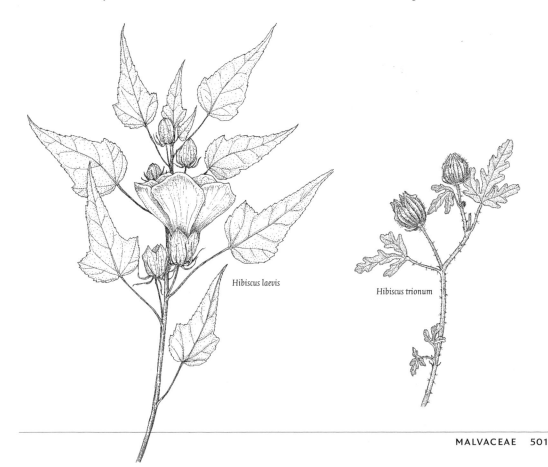

Hibiscus laevis

Hibiscus trionum

Malva moschata L.   Musk or rose mallow
Branched perennial, 3–6 dm tall; leaves deeply palmately lobed and toothed; flowers pink or white, 5–6 cm wide; carpels 15–20, rounded; naturalized in dry, sandy fields, roadsides, and fencerows; throughout; flr. May–early Sep; native to Europe; UPL.

Malva neglecta,
fruit (×1)

Malva neglecta,
flower (×1)

Malva neglecta Wallr.   Chesses, common mallow
Prostrate, branched biennial; leaves round or cordate, long-petioled, margins wavy to slightly lobed; flowers pale lavender or white, 1.4–2.4 cm wide; carpels 12–15, rounded; a common weed of gardens, roadsides, and waste ground; throughout; flr. early May–Aug; native to Eurasia and N. Africa.

Malva rotundifolia L.   Chesses
Similar to *M. neglecta* but smaller; petals barely if at all longer than the sepals; carpels 8–11; rare in ballast and waste ground; flr. May–Sep; scattered; native to Europe.

Malva sylvestris L.   Chesses, high mallow
Erect biennial, 4–10 dm tall; leaves rounded or kidney-shaped, 3–7-lobed; flowers in the upper axils; petals red-purple; cultivated and occasionally escaped to ballast, roadsides, and waste ground; scattered; flr. Jun–Sep; native to Eurasia.

Malva
neglecta, leaf

Sida
hermaphrodita

Malva verticillata L. Whorled mallow
Erect annual to 2 m tall; leaves 8–15 cm wide, kidney-shaped with 5–7 rounded lobes; flowers nearly sessile, in crowded axillary clusters; petals white or bluish; cultivated and rarely escaped to roadsides and meadows; scattered; flr. Jul–Sep; native to Asia.

Sida L.

Herbs with serrate or lobed leaves and small, axillary flowers; bracts not present below the calyx; calyx persistent, enclosing the fruit; stamen tube with anthers at the top; carpels 5–10, 1-seeded; styles with head-like stigmas; fruit separating into 1-seeded, beaked nutlets.

A. leaves palmately 3–7-lobed, petioles without spines; flowers white .... S. *hermaphrodita*
A. leaves ovate-lanceolate with a spine at the base of the petiole; flowers yellow . ... S. *spinosa*

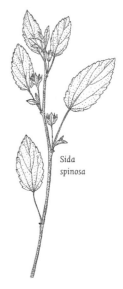

Sida hermaphrodita (L.) Rusby Virginia mallow
Perennial, 1–3 m tall, somewhat stellate-hairy when young, becoming glabrate; leaves deeply 3–7-lobed; flowers white, axillary, 1.8–2.5 cm wide; rare on stream banks; SC; flr. Jul–Oct; FAC; ✹.

Sida spinosa L. Prickly sida, false mallow
Branching annual 3–6 dm tall with a spine at the base of each leaf; leaves oblong or elliptic, petiolate; flowers clustered in the axils, pale yellow; a frequent weed of cultivated fields, roadsides, stream banks, waste ground, and ballast; S; flr. Jul–Sep; native to the tropics; UPL.

Sida spinosa

MELASTOMACEAE Melastome Family

Perennial herbs; leaves opposite, simple, 3-veined, often with ciliate margins; stipules not present; flowers perfect, regular except for the stamens, 4-merous, perigynous; sepals valvate; petals distinct; stamens 8, with a conspicuous spur-like appendage attached to each anther; fruit a capsule.

Rhexia L.

Characteristics of the family; stems 2–10 dm tall; petals light to dark pink; hypanthium tubular in flower, persisting in fruit, the basal portion swelling around the capsule but not fused to it, except at the base.

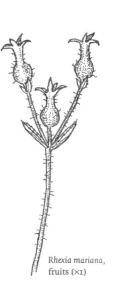

A. neck of the hypanthium as long or longer than the body; petals pale pink to white .... R. *mariana*
A. neck of the hypanthium mostly shorter than the body; petals dark pink .... R. *virginica*

Rhexia mariana L. Maryland meadow-beauty
Stems freely branched, very unequally and obscurely 4-sided; petals pale pink;

Rhexia mariana, fruits (×1)

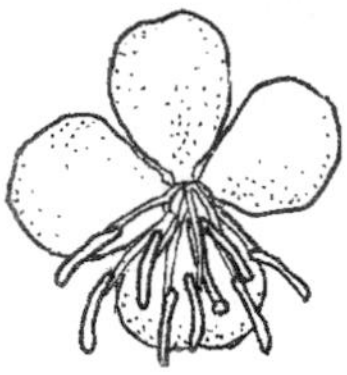

Rhexia virginica, flower (×1)

neck of the hypanthium as long or longer than the body in fruit; rare in moist, open sandy soils; SE; flr. Jul–Aug; OBL; .

Rhexia virginica L. — Meadow-beauty
Stems subequally 4-sided and somewhat winged, bristly at the nodes; petals dark pink; the neck of the hypanthium shorter than the body in fruit; occasional in moist open areas on sandy soils; SE, SW, and C; flr. Jul–early Sep; OBL.

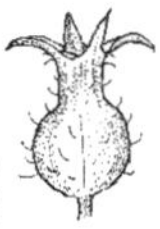

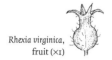

Rhexia virginica, fruit (×1)

MENISPERMACEAE Moonseed Family

Twining woody vines with alternate, simple, palmately veined leaves; dioecious; flowers small, 3-merous; flower parts all distinct; sepals and petals each 6–9, petals shorter than the sepals; stamens 12–24; pistils 2 or 3, superior; each ripening into a drupe.

Menispermum L.

Menispermum canadense L. — Moonseed
Twining vine, 2–5 m tall; leaves suborbicular, entire to slightly lobed, 10–15 cm wide, peltate near the margin; flowers whitish, in small axillary panicles; fruit dark blue, 6–10 mm; seed flattened and crescent-shaped with a roughened surface; frequent on moist riverbanks, floodplains, and edges; throughout; flr. Jun, frt. Sep–Oct; FACU.

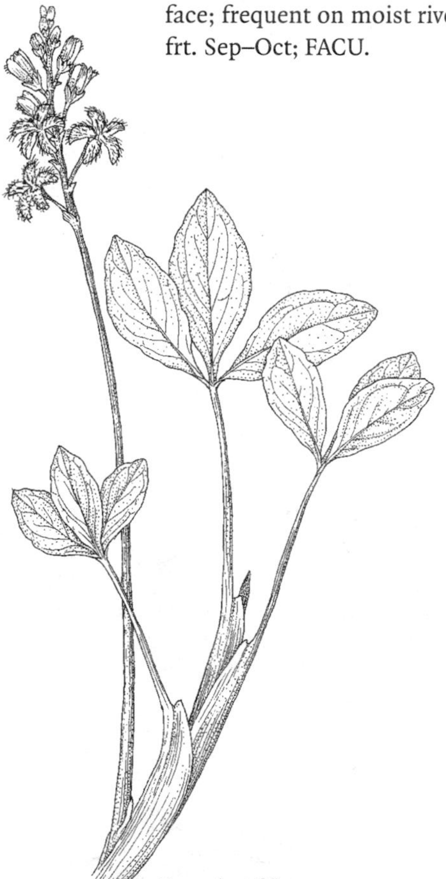

Menyanthes trifoliata

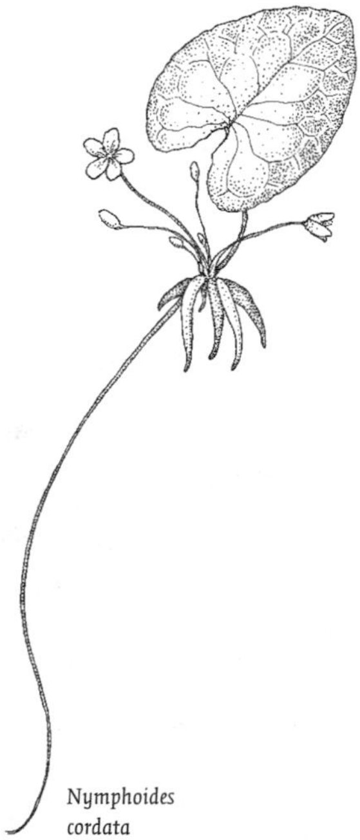

Nymphoides cordata

MENYANTHACEAE Buckbean Family

Aquatic or semi-aquatic herbs with alternate, floating or emergent leaves; flowers perfect, regular, 5-merous; petals fused; stamens attached to the corolla tube alternate with the lobes; ovary bicarpelate but unilocular, superior to half inferior; fruit a capsule or berry.

A. leaves trifoliate, emergent; flowers in a raceme .... *Menyanthes*
A. leaves simple, floating; flowers in an umbel .... *Nymphoides*

Menyanthes L.

Menyanthes trifoliata L.   Bogbean, buckbean
Glabrous, somewhat fleshy perennial with a coarse rhizome; leaves trifoliate with a long petiole and sheathing base; corolla whitish, densely hairy inside; ovary ⅓ inferior; fruit eventually dehiscent by 2 valves; occasional in bogs, sphagnous swamps, or shallow water of pond and lake margins; N and at higher elevations along the Allegheny Front; flr. May–Aug; OBL.

Nymphoides Hill

Aquatic herbs with floating leaf blades arising directly from the rhizome on long petioles or subtending the inflorescence near the tip of a long flowering stem; flowers in an umbel, perfect; corolla campanulate, deeply 5-lobed; ovary tapering to a nearly sessile, 2-lobed stigma; fruit a nondehiscent capsule.

A. flowering stem with a single leaf at the base of the inflorescence; flowers white; fruit not beaked .... *N. cordata*
A. flowering stem with 2 leaves at the base of the inflorescence; flowers yellow; fruit strongly beaked .... *N. peltata*

Nymphoides cordata (Elliot) Fernald   Floating-heart
Leaves deeply cordate, 3–7 cm; flowers white or creamy, usually interspersed in the inflorescence with stubby, thickened roots; corolla 5–8 mm, deeply 5-lobed with a yellow gland at the base of each lobe; fruit not beaked; very rare in lakes and ponds; NE; flr. Jul–Aug; OBL.

Nymphoides peltata (J.G.Gmel.) Kuntze   Waterfringe, yellow floating-heart
Leaves suborbicular, 5–15 cm; flowers bright yellow, corolla lobes fringed below; fruit strongly beaked; rare in ponds and lakes; SE; flr. Jul–Sep; native to Europe; OBL.

MIMOSACEAE Mimosa Family

Trees or herbs with alternate, bipinnately compound leaves with stipules; flowers regular, aggregated in dense heads; sepals 5, fused below to form a short tube; petals 5, funnelform; stamens commonly colored and conspicuously exerted; ovary 1, simple, superior; fruit a legume.

Albizia Durazz.

Albizzia julibrissin, leaflet

Albizia julibrissin Durazz.   Mimosa, silktree
Unarmed tree to 10 m tall with spreading branches, leaves bipinnately compound, 2–5 dm; flowers in dense heads, pink; calyx and corolla each tubular, 5-lobed; stamens numerous, 15 mm long, extending well beyond the corolla, some filaments fused below; fruit broadly linear, flat, with several seeds; cultivated and occasionally escaped to roadsides and woods edges; flr. Jun–early Aug; native to Asia.

MOLLUGINACEAE Carpetweed Family

Annual herbs; flowers perfect; sepals 5, distinct; petals absent; stamens 3–4 (5); ovary superior, 3–5 celled; fruit a many-seeded capsule.

Mollugo L.

Mollugo verticillata L.   Carpetweed
Prostrate or slightly ascending plant forming mats up to 4 dm across; leaves oblanceolate without a strongly differentiated petiole, in whorls of 3–8 at the nodes; flowers pale green to white, 2–5 on pedicels at the nodes; capsules ovoid, about 3 mm long; a common weed of waste ground, roadsides, and pavement cracks; throughout; native to tropical America; flr. Jun–Sep; FAC.

MONOTROPACEAE Indian-pipe Family

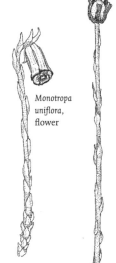

Monotropa uniflora, flower

Monotropa uniflora, fruit

Low-growing, non-green, mycotrophic plants with scale-like leaves; flowers perfect, regular; sepals 5 or absent; petals 4–6, distinct; stamens 8 or 10, anthers opening by longitudinal slits; ovary superior, 4- or 5-locular; fruit an ovoid capsule, opening from the top down.

Monotropa L.

Characteristics of the family; plants white, pinkish, yellowish, or red, 1–3 dm tall; flowers solitary or in a raceme, nodding at first, becoming erect in fruit.

A. plants white; flowers solitary .... *M. uniflora*
A. plants yellowish or red; flowers few to many in a raceme .... *M. hypopithys*

Monotropa hypopithys L.   Pinesap
Plant red, pink, or yellowish, 1–3 dm tall; flowers in a raceme; frequent in dry to moist woods; throughout; flr. Jul.

Monotropa uniflora L.   Indian-pipe
Plant waxy white, 1–2 dm tall, stems usually clustered; flowers solitary; common in dry to moist woods; throughout; flr. late Jun–early Aug; FACU–.

Mollugo
verticillata

Broussonetia
papyrifera, fruit

Broussonetia papyrifera, leaves

MORACEAE Mulberry Family

Deciduous trees with milky sap; leaves alternate, simple or variously lobed, entire or serrate; flowers unisexual; plants dioecious or occasionally monoecious; pistillate flowers in dense clusters coalescing in fruit; fruit an aggregation of drupelets (actually achenes surrounded by fleshy layers derived from the calyx).

A. stems thorny in the juvenile phase; leaves entire .... *Maclura*
A. stems lacking thorns; leaves serrate and frequently lobed
    B. bark furrowed; twigs and branches orange; pistillate flowers in cylindric catkins; fruit short-cylindric .... *Morus*
    B. bark smooth; twigs grayish; pistillate flowers in dense globose heads; fruit globose .... *Broussonetia*

Broussonetia L'Hér.

Broussonetia papyrifera (L.) Vent.   Paper-mulberry
Spreading deciduous tree to 15 m tall with smooth bark and pubescent twigs; dioecious; leaves serrate, variously lobed or simple, pubescent beneath; fruit globose, 2–3 cm, red; occasional in waste ground, woods margins, and urban land; mostly SE; flr. Jun, frt. Aug–Sep; native to Asia.

Maclura Nutt.

Maclura pomifera,
winter twig

Maclura pomifera (Raf.) C.K.Schneid. Osage-orange
Deciduous tree to 20 m tall with stout, 1.2-cm thorns in the juvenile phase, dioecious; leaves petiolate, lance-ovate, entire, glossy above; fruit spherical, to 12 cm in diameter, yellow-green with a convoluted surface; frequent in fencerows, roadsides and abandoned pastures; mostly S; flr. Jun, frt. Aug–Oct; native to the southwestern U.S.; UPL.

Morus L.

Deciduous trees with orange twigs; leaves variously lobed or occasionally simple, serrate; monoecious or dioecious; flowers in cylindrical catkins; fruit a short cylindrical aggregation of fleshy drupelets.

A. winter buds 5–8 mm; leaves pubescent beneath; fruit red then purplish-black .... *M. rubra*
A. winter buds 3–4 mm; leaves glabrous beneath except on the larger veins; fruit white, pink, or rarely purple .... *M. alba*

Morus alba, fruit, x 1

Morus alba L. White mulberry
Deciduous tree to 15 m tall with furrowed bark exposing orange inner layers and reddish-brown or orange twigs, dioecious or occasionally monoecious; leaves coarsely serrate, variously lobed, glabrous except on the main veins beneath; fruit 1–2 cm, white to pink or occasionally purple; common in fencerows, disturbed woods, and waste ground; mostly S; flr. May, frt. Jun; native to Asia; UPL.

Morus rubra L. Red mulberry
Deciduous tree to 20 m tall with scaly red or yellow-tinged bark; twigs green, becoming orange-brown; leaves coarsely serrate, variously lobed, pubescent beneath; dioecious; fruit 2–3 cm long, dark purple to black; frequent in rich, moist alluvial soils; mostly S; flr. May, frt. Jun–early Jul; FACU.

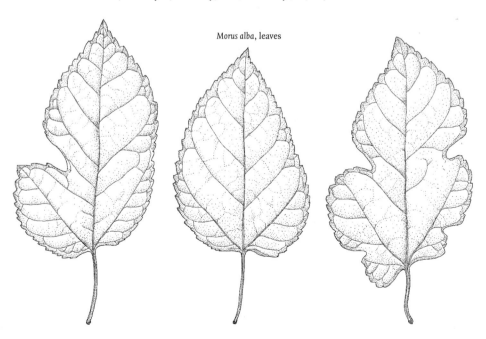

Morus alba, leaves

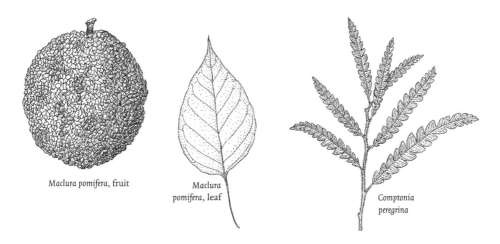

Maclura pomifera, fruit

Maclura pomifera, leaf

Comptonia peregrina

MYRICACEAE Bayberry Family

Deciduous, rhizomatous, aromatic shrubs with alternate, simple, pinnately veined leaves, with or without stipules; plants dioecious; flowers in unisexual catkins, lacking perianth; ovary bicarpellate; fruit a nutlet, achene, or small drupe.

A. leaves pinnately lobed, stipules present; fruit a glabrous nutlet surrounded by a bur .... *Comptonia*
A. leaves entire or with a few widely spaced teeth, stipules absent; fruit an achene or waxy drupelet, not enclosed by a bur .... *Myrica*

Comptonia L'Hér

Comptonia peregrina (L.) J.M.Coult.   Sweet-fern
Deciduous, rhizomatous shrub to 1.5 m tall, glandular; leaves 6–12 cm long, pinnately lobed; bracts and bracteoles of the pistillate inflorescence enlarging to form a bur that encloses the hard, glabrous nutlets; frequent in dry, sterile soil of open woods and barrens; throughout, except in the westernmost counties; flr. late Apr–early May.

Myrica L.

Deciduous, stoloniferous, glandular shrubs; leaves alternate, simple, entire or slightly toothed; plants dioecious; flowers in catkins; fruit an achene or small drupe subtended by 2–6 small bracts. Although previously reported (Rhoads and Klein, 1993), there is no clear evidence of *Myrica heterophylla* Raf. in PA.

A. fruit a waxy, subglobose drupelet, the 4–6 subtending bracts remaining small and inconspicious; plants of dry to moist, usually sandy soils ....
.... *M. pensylvanica*
A. fruit a flattened achene enclosed by 2 enlarged and thickened bracts; a plant of peaty, northern wetlands .... *M. gale*

Myrica gale L.   Sweet-gale
Deciduous shrub to 1.5 m tall; leaves oblanceolate, 3–6 cm long, minutely serrate; fruiting catkins 10–12 mm, the 2 subtending bracts of each flower becom-

ing enlarged and thickened and clasping the flattened achenes; rare in bogs and shallow water of lake and stream edges; NE; flr. May, before the leaves; OBL; 🌿.

Myrica pensylvanica Loisel.   Bayberry
Deciduous shrub to 2 m tall; leaves oblanceolate to obovate, 4–8 cm long, entire or with a few teeth toward the tip; flowers on previous year's wood; fruits globose, 3.5–5 mm, densely covered with grayish-white wax and short hairs; occasional in old fields, sand dunes, and open woods in dry to moist, sterile, sandy soils; SE and NW; flr. May; FAC. [Wilbur (1994) places M. pensylvanica in the segregate genus Morella as M. caroliniensis (Mill.) Small.]

NELUMBONACEAE Lotus-lily Family

Perennial aquatic herbs from large, fleshy rhizomes; leaves nearly circular, peltate, entire; flowers perfect, regular, the sepals and petals similar, stamens many; fruits nut-like, embedded in the surface of an enlarged flat-topped receptacle, each fruit occupying a separate chamber in the receptacle.

Nelumbo Adans.

Nelumbo lutea (Willd.) Pers.   American lotus
Leaf blades 3–7 dm across, floating or elevated above the surface of the water on long petioles; flowers solitary, on long peduncles often rising well above the surface of the water, pale yellow, 15–25 cm across; mature receptacle about 1 dm across; individual fruits about 1 cm in diameter; rare in ponds and other quiet water; S; flr. Jul–Sep; OBL; 🌿.

Nelumbo lutea,
fruit (×¼)

NYCTAGINACEAE Four-o'clock Family

Flowers perfect, subtended by an involucre of 5 united bracts resembling a calyx; corolla absent, but calyx corolla-like, tubular; ovary superior; stamens 3–5; style elongate; leaves opposite.

Mirabilis L.

Characteristics of the family; flr. and frt. summer; native to tropical America.

A. flower 1 per involucre; involucre enlarged in fruit, brown and papery ....
.... M. jalapa
A. flowers 3–5 per involucre; involucre not enlarged in fruit .... M. nyctaginea

Mirabilis jalapa L.   Four-o'clock, marvel-of-Peru
Fruit strongly 5-ribbed, <3 mm wide; involucre bell-shaped; calyx 30–50 mm long, narrowly funnel-shaped; fruit ovoid, glabrous to minutely short-hairy; cultivated and occasionally spreading to waste ground; mostly S.

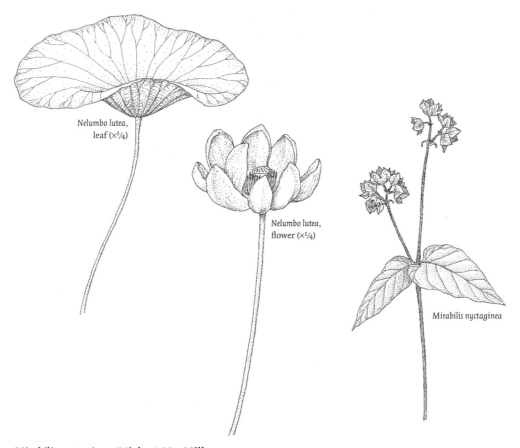

Nelumbo lutea,
leaf (×¹/₄)

Nelumbo lutea,
flower (×¹/₄)

Mirabilis nyctaginea

Mirabilis nyctaginea (Michx.) MacMill.
Heart-leaved umbrella-wort, wild four-o'clock
Fruit smooth to moderately ribbed or angled, often >3 mm wide; involucres cup-shaped; calyx about 10 mm long, broadly funnel-shaped; fruit club-shaped, hairy; dry soil and waste ground; throughout; FACU–.

Mirabilis nyctaginea,
inflorescence (×1)

NYMPHAEACEAE Water-lily Family

Aquatic perennials; leaves floating, emergent or submerged, arising from stout rhizomes, cordate to hastate at the base of the blade, very long-petioled; flowers solitary at the ends of long peduncles arising from the rhizomes, appearing at or slightly above the water surface, perfect, hypogynous, sepals 4–6, petals numerous, stamens numerous with some or all filaments flattened; fruit a many-seeded leathery berry.

A. flowers yellow to occasionally red or purple, up to 5 cm across .... *Nuphar*
A. flowers white to pinkish, 7–12 cm across .... *Nymphaea*

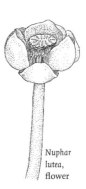

Nuphar
lutea,
flower

Nuphar Sm.

Nuphar lutea (L.) Sibth. & Sm. *sensu lato*   Spatterdock, yellow pond-lily, cow-lily
Leaves generally ovate in outline, blade 5–40 cm long; flowers yellow; sepals 5–

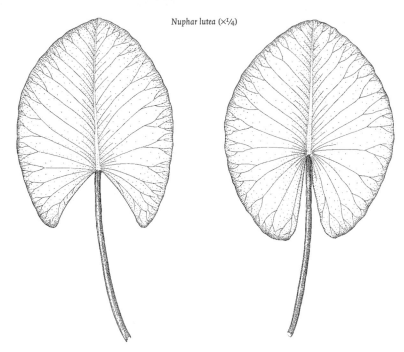

Nuphar lutea (×¼)

6, petal-like, much larger than the numerous strap-like petals; stamens resembling the petals, all filaments flattened; fruits ovoid with numerous locules, each locule with numerous small non-arilate seeds, maturing above the water surface; frequent on lake margins, in ponds, slow-moving streams, swamps, and tidal marshes; throughout; flr. May–Sep; OBL. This species has been split variously by some authors. [syn: *N. advena* (Aiton) W.T.Aiton in part, *N. variegata* Durand in part, *N. microphyllum* (Pers.) Fernald in part, *N. variegatum* Engelm. in part]

Nymphaea L.

Nymphaea odorata Aiton   Fragrant water-lily
Leaves generally rounded to elliptic in outline, blade 10–30 cm across, the sinus at the base of the blade very narrow; flowers white or rarely pink, fragrant, opening in early morning and closing by afternoon; sepals 4, green; petals 15–30, about as long as the sepals; stamens many, the inner ones with linear filaments and long anthers, the outer gradually becoming petal-like; fruits depressed globose with numerous locules, maturing below the water surface; seeds with a sac-like aril; occasional in quiet waters of lakes and ponds; throughout; flr. May–Sep; OBL.

OLEACEAE Olive Family

Deciduous or semi-evergreen trees or shrubs with opposite leaves without stipules; flowers regular, 4-merous, perfect or unisexual and very reduced, lacking calyx and corolla in some species; petals fused if present; stamens 2 (rarely 4), inserted on the corolla tube when a corolla is present; ovary superior, 2-locular; fruit a samara, drupe or capsule.

Summer key

A. trees; leaves pinnately compound; flowers unisexual, lacking a corolla; fruit a samara .... *Fraxinus*
A. shrubs; leaves simple; flowers perfect or unisexual, with a more or less showy corolla; fruit a capsule or drupe
   B. pith hollow or chambered between the nodes; leaves usually somewhat serrate; flowers yellow .... *Forsythia*
   B. pith solid; leaves entire; flowers white or lilac
      C. leaves >7 cm long; flowers from axillary buds; corolla lobes longer than the tube .... *Chionanthus*
      C. leaf blades <7 cm long; inflorescence terminal; corolla lobes shorter than the tube
         D. leaves ovate, truncate, or cordate at the base with an acuminate tip; flowers purple or occasionally white; anthers hidden within the corolla tube; fruit a capsule .... *Syringa*
         D. leaves elliptic-oblong, tapered at the base and tip; flowers white; anthers extending beyond the corolla tube; fruit a drupe ...... *Ligustrum*

Winter key

A. vein scars crowded in a curved line
   B. trees .... *Fraxinus*
   B. large shrubs .... *Chionanthus*
A. vein scar 1
   C. pith hollow or chambered between the nodes .... *Forsythia*
   C. pith solid throughout
      D. fruit a drupe .... *Ligustrum*
      D. fruit a capsule .... *Syringa*

Chionanthus L.

Chionanthus virginicus L.   Fringetree
Deciduous shrub or small tree to 10 m tall; leaves simple, petiolate; blades lance-elliptic to oblong, 8–15 cm, entire; dioecious; flowers in conspicuous 1–2-dm panicles; corolla with 4 narrow, 15–30 mm, drooping lobes; stamens 2; fruit a dark blue drupe; rare in moist woods; S; flr. May; FAC+; ✿.

Forsythia Vahl

Spreading deciduous shrubs with arching branches and numerous yellow, bell-shaped flowers appearing before the leaves; leaves simple, entire or toothed; petals united, corolla deeply 4-lobed; stamens 2, inserted on the corolla tube; fruit a 2-locular, many seeded capsule. Many forms are in cultivation based on the species below and their hybrid F. x *intermedia* Zabel.

A. branches suberect; pith chambered .... *F. viridissima*
A. branches arching; pith hollow except at the nodes .... *F. suspensa*

Forsythia suspensa (Thunb.) Vahl   Forsythia
Branches arching, to 3 m tall; pith hollow except at the nodes; leaves toothed or

lobed; cultivated and occasionally escaped or persisting; flr. Apr–early May; native to China.

Forsythia viridissima Lindl.   Forsythia

Branches suberect to 3 m tall; pith chambered; leaves entire or irregularly serrate above the middle; cultivated and occasionally escaped; flr. Apr–early May; native to China.

Fraxinus L.

Deciduous trees with opposite, pinnately compound leaves, stout twigs, and finely, reticulately ridged bark; flowers unisexual; corolla absent; calyx present or absent; plants dioecious and wind pollinated; fruit a samara; flr. late Apr–early May, before the leaves.

Summer key

A. leaflets sessile; calyx absent; samaras flat throughout, winged to the base ....F. nigra

A. leaflets definitely stalked; calyx present and persistent in fruit; samaras with a distinct subterete body and a flat wing
    B. petiolules of the middle and lower leaflets usually with a wing extending from the blade nearly to the rachis; wing of the samara extending halfway or more down the body .... F. pennsylvanica
    B. petiolules of the middle and lower leaflets wingless or nearly so; wing of the samara extending 1/3–1/2 the length of the subterete body
        C. leaflets pale or whitish and papillose beneath; twigs, petioles, and rachis usually glabrous .... F. americana
        C. leaflets green or tawny beneath, not papillose; twigs, petioles, and rachis pubescent .... F. profunda

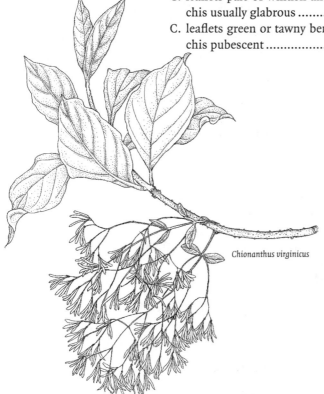

Chionanthus virginicus

Winter key

A. leaf scars mostly with a concave upper margin, indented by the bud ....
.... *F. americana*
A. leaf scars with a truncate or only slightly concave upper margin, not strongly
indented by the bud
    B. twigs pubescent
        C. widespread in moist woods and floodplains .... *F. pennsylvanica*
        C. only in vernal ponds and wet woods along Lake Erie .... *F. profunda*
    B. twigs glabrous
        D. terminal bud wider than long .... *F. pennsylvanica*
        D. terminal bud as long as or longer than wide .... *F. nigra*

Fraxinus americana L.   White ash

Fraxinus americana, fruit

Tree to 40 m tall with mostly glabrous twigs, petioles, and rachis; old leaf scars
concave; leaflets 5–9, stalked, oblong to ovate, entire to somewhat serrate, whitish and papillose beneath; petiolules not winged; samara linear to oblanceolate,
3–5 cm, the wing longer than and extending ⅓ of the length of the subterete
body; common in woods, fencerows, and old fields; throughout; FACU. Variety
biltmoreana (Beadle) J. Wright, with more or less pubescent twigs and petioles,
occurs in the southern part of the state.

Fraxinus nigra Marshall   Black ash
Tree to 25 m tall with glabrous twigs; leaflets 7–11, sessile, lanceolate to oblong,
long-acuminate, serrate; samara lanceolate to oblanceolate, 2.5–4 cm, flat,
winged to the base, lacking a calyx; occasional in swamps, wet woods, and
bottomlands; throughout; FACW.

Fraxinus pennsylvanica Marshall   Red ash

Fraxinus pennsylvanica, fruit

Tree to 25 m tall with twigs and leaves pubescent to glabrous, not papillose; leaf
scars only slightly if at all concave on the upper margin; leaflets 5–9, ovate to oblong or elliptic, acute or acuminate, extending down along the short petiolule at
the base; margin serrate to subentire; samara linear to spatulate, 4–7.5 cm, the
wing extending to the middle of the subterete body; common in alluvial woods,
stream banks, and moist fields; throughout; FACW.

Fraxinus profunda (Bush) Bush   Pumpkin ash
Tree to 40 m tall with densely pubescent twigs, petioles, and rachis; leaflets 5–9,
green beneath, lanceolate to oblong or elliptic, abruptly acuminate, entire; petiolules 8–15 mm, not winged; samaras linear-oblong to spatulate, 4–7.5 cm, the
wing extending to the middle of the terete body or beyond; rare in shallow woodland ponds and wet, wooded flats; NW; OBL; ❧.

Ligustrum L.

Deciduous or semi-evergreen shrubs with opposite, simple, entire leaves and
small, white perfect flowers in terminal panicles; corolla tubular with 4 spreading lobes; stamens 2; fruit a small black drupe.

A. twigs glabrous
    B. leaves oblong-ovate to lanceolate; filaments not extending beyond the co-
      rolla tube .... *L. vulgare*

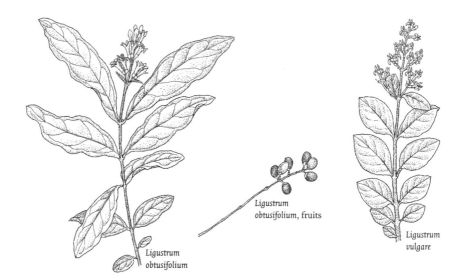

Ligustrum obtusifolium

Ligustrum obtusifolium, fruits

Ligustrum vulgare

B. leaves elliptic-ovate; filaments extending beyond the corolla tube............
.... L. *ovalifolium*
A. twigs pubescent or puberulent
  C. twigs densely pubescent
    D. leaves deciduous, elliptic to oblong-obovate; calyx pubescent; anthers reaching the middle of the corolla lobes .... L. *obtusifolium*
    D. leaves semi-evergreen, oval to oblong; calyx glabrous or slightly pubescent near the base; anthers not reaching the middle of the corolla lobes .... L. *amurense*
  C. youngest twig segments puberulent; calyx glabrous .... L. *vulgare*

Ligustrum amurense Carrière   Amur privet
Semi-evergreen shrub to 5 m tall; branches pubescent when young; leaves oval or oblong 1.5–5 cm, glabrous except midrib beneath; panicle to 5 cm; calyx glabrous or slightly pubescent near the base; corolla tube 3 times as long as the lobes, anthers not reaching the middle of the corolla lobes; cultivated and rarely spreading to thickets; SE; flr. Jun; native to China.

Ligustrum obtusifolium, flower (×2)

Ligustrum obtusifolium Siebold & Zucc.   Obtuse-leaved privet
Deciduous shrub to 3 m tall with pubescent twigs; leaves elliptic to oblong-ovate, 2.5–5 cm, acute or obtuse, glabrous or pubescent only on the midrib beneath; panicle to 3.5 cm; calyx pubescent; corolla tube 2–3 times longer than the lobes, anthers reaching the middle of the corolla lobes; cultivated and frequently naturalized in disturbed woods, thickets, hedgerows, and old fields; mostly S; flr. Jun, frt. Sep–winter; native to Japan.

Ligustrum ovalifolium Hassk.   California privet
Semi-evergreen shrub to 5 m tall with glabrous twigs; leaves elliptic-ovate or oblong, 2.5–6 cm, acute; panicle to 10 cm; calyx glabrous, corolla tube 2–3 times as long as the lobes; stamens longer than the corolla lobes; cultivated and occasionally escaped to roadsides and thickets; SE and SW; flr. Jul, frt. Nov; native to Japan.

Ligustrum vulgare L. Common privet

Semi-evergreen shrub to 5 m tall; twigs puberulent; leaves oblong-ovate, 2.5–6 cm, obtuse or acute; panicle to 6 cm; corolla tube same length as the lobes; stamens as long as the corolla; frequently cultivated and occasionally persisting on roadsides or abandoned home sites; mostly S; flr. Jun–early Jul, frt. Oct–Nov; native to Europe; FACU.

Ligustrum vulgare, flower (×2)

Syringa L.

Syringa vulgaris L. Common lilac

Deciduous shrub with simple, entire heart-shaped leaves and showy lilac or white flowers in terminal panicles; corolla lobes shorter than the tube; stamens 2, included within the corolla tube; fruit a 2-locular capsule; frequently cultivated and often persisting on roadsides and abandoned home sites; throughout; flr. May; native to Europe.

ONAGRACEAE Evening-primrose Family

Herbaceous plants with simple, opposite or alternate leaves; flowers mostly 4-merous, but some 2–3- or 5–6-parted, perfect, regular or irregular; ovary inferior; hypanthium tubular and frequently prolonged well beyond the end of the ovary; petals 2–9, distinct; stamens as many or twice as many as the petals and inserted at the top of the hypanthium tube; ovary 2–4-celled; fruit a dehiscent capsule or indehiscent and nut-like.

A. flowers 2-merous; leaves opposite .... *Circaea*
A. flowers 4(or more)-merous; leaves alternate (or occasionally opposite)
    B. flowers irregular; fruit indehiscent .... *Gaura*
    B. flowers regular; fruit dehiscent longitudinally or by a terminal pore
        C. sepals persistent on the fruit; hypanthium not prolonged ...... *Ludwigia*
        C. sepals falling after flowering; hypanthium mostly prolonged beyond the ovary
            D. hypanthium only slightly longer than the ovary; seeds with a tuft of hairs at the end .... *Epilobium*
            D. hypanthium elongate; seeds without a tuft of hairs .... *Oenothera*

Circaea L.

Rhizomatous, perennial herbs with opposite, petioled leaves; inflorescence a terminal raceme, sometimes with secondary lateral branches; flowers small, petals, sepals, and stamens each 2; fruit indehiscent, bur-like, covered with hooked bristles.

A. flowers well-spaced along the raceme; fruits 3.5–5 mm; pedicels spreading in flower; fruits equally bilocular and 2-seeded .... *C. lutetiana*
A. flowers clustered near the top of the peduncle; pedicels spreading-ascending in flower; fruits 2–3 mm, unilocular and 1-seeded .... *C. alpina*

Circaea alpina L.   Enchanter's-nightshade

Stems erect, 1–3 dm tall; leaves ovate, 2–6 cm, coarsely dentate, the petiole winged beneath; inflorescence with <15 white or pinkish flowers, crowded near the top on ascending pedicels; petals 1–2.5 mm, notched; fruit 2–3 mm, unilocular and 1-seeded; occasional in moist hemlock forests, rocky wooded slopes, and stream banks; N and at higher elevations along the Allegheny Front; flr. May–Oct; FACW. Ours is ssp. *alpina.* Hybridizes with *C. lutetiana* to form *Circaea* x *intermedia* Ehrh., a sterile hybrid that is intermediate between the 2 parental species and persists in vegetative colonies.

Circaea lutetiana L.   Enchanter's-nightshade

Stems erect, 3–7 dm tall; leaves oblong-ovate, 6–12 cm, shallowly denticulate; inflorescence with many, well-spaced, white flowers on spreading pedicels; petals 2.5–4 mm, lobed; fruit 3.5–5 mm, bilocular, 2-seeded; common in woods and floodplains; throughout; flr. Jun–Aug; FACU. Ours is ssp. *canadensis* (L.) Asch. & Magnus. [syn: *C. quadrisulcata* of authors, not (Maxim.) Franch. & Sav.]

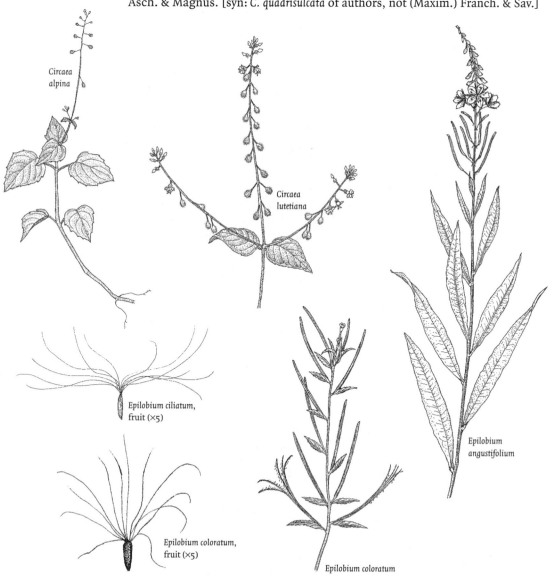

Circaea alpina

Circaea lutetiana

Epilobium ciliatum, fruit (×5)

Epilobium coloratum, fruit (×5)

Epilobium coloratum

Epilobium angustifolium

Epilobium L.

Annual or perennial herbs with alternate or opposite leaves; flowers in a terminal raceme, 4-merous, the hypanthium prolonged beyond the ovary for a short distance only; sepals not persistent in fruit; petals white, pink, or purple; stamens 8; style short; fruit an dehiscent linear capsule; seeds with a tuft of hairs at the end. Recent evidence indicates that the fireweeds are distinct and should be treated as a separate genus, *Chamerion*.

A. stigma 4-parted
    B. petals entire; leaves nearly entire, glabrous .... *E. angustifolium*
    B. petals notched; leaves toothed, hairy
        C. petals 10–15 mm, shallowly notched .... *E. hirsutum*
        C. petals 4–9 mm, deeply notched .... *E. parviflorum*
A. stigma not lobed or cleft
    D. leaves entire, <1 cm wide; stems without lines of hairs extending from the leaf bases
        E. hairs of the stem straight, spreading .... *E. strictum*
        E. hairs of the stem incurved
            F. leaves glabrous above .... *E. palustre*
            F. leaves finely hairy on the upper surface .... *E. leptophyllum*
    D. leaves toothed, usually >1 cm wide; stem with lines of hairs extending from the leaf bases
        G. seeds beakless with a tuft of brown hairs .... *E. coloratum*
        G. seeds with a short beak, hairs white .... *E. ciliatum*

Epilobium angustifolium L.   Fireweed
Erect perennial with a simple stem 1–3 m tall, glabrous except in the inflorescence; leaves alternate, sessile, lanceolate to lance-linear, crowded; flowers many in a terminal raceme, purplish-pink; hypanthium not prolonged beyond the ovary; stigma deeply 4-lobed; occasional in woods edges and recent clearings, in open sandy ground; scattered throughout; flr. Jun–Sep; FAC.

Epilobium ciliatum Raf.   Willow-herb
Perennial with basal rosettes and an erect, branching stem 1–1.5 m; stem pubescent, at least in lines; leaves opposite or the uppermost alternate, with a few scattered teeth along the margin; flowers numerous, white or pink, with notched petals; seeds longitudinally ribbed, with a tuft of white hairs; occasional in moist, springy soil and wet rocks; throughout, except the southernmost counties; flr. Jun–Sep; FAC–.

Epilobium coloratum Biehler   Purple-leaved willow-herb
Erect, branched perennial to 1 m tall, stem with lines of incurved hairs; leaves opposite below the inflorescence, narrowly lanceolate, irregularly serrate, often reddish or purplish; flowers numerous; petals 4–5 mm, pink or white; capsule 3–5 cm; seeds with a tuft of brown hairs; common in moist fields, marshes, shores, and floodplains; throughout; flr. Jul–early Sep; FACW+.

Epilobium hirsutum L.   Hairy willow-herb
Rhizomatous perennial with erect, branched, hairy stems 5–12 dm tall; leaves mostly opposite, lanceolate and somewhat clasping, sharply serrulate, hairy on both sides; flowers solitary in the axils of alternate upper leaves, reddish-purple,

2.5 cm; petals shallowly notched; stigma 4-lobed; capsule 5–8 cm; infrequent in wet fields, marshes, and ditches; scattered; flr. Jul–Sep; FACW; native to Europe.

Epilobium leptophyllum, stem (×5)

Epilobium leptophyllum Raf.   Willow-herb
Erect perennial 2–10 dm tall from slender stolons, pubescent with incurved hairs; leaves linear to linear-lanceolate, 1.5–7 cm long by 1–7 mm wide, entire or revolute; flowers in the upper axils, petals pink to white, 4–6 mm, entire; capsule to 5 cm; hairs of the seeds pale brown; occasional in marshes and boggy pastures; mostly E and C; flr. Apr–early Sep; OBL.

Epilobium palustre L.   Marsh willow-herb
Perennial from slender stolons; stems erect, 1–5 dm tall, simple or branched; puberulent in the inflorescence, but the hairs more sparse below and often in lines; leaves opposite, or alternate above, lanceolate or lance-linear, 2–7 cm long by 2–15 mm wide, entire or revolute, glabrous or with a few scattered hairs along the midrib; flowers in the upper axils; petals white to pink or lilac, 4–6 mm, notched; capsule to 5 cm; rare in bogs and wooded swamps; NE; flr. Jul–Sep; OBL; ❦.

Epilobium parviflorum Schreb.   Willow-herb
Similar to E. hirsutum, but smaller; stems to 8 dm tall; leaves 2–8 cm, sessile but not clasping, the upper alternate; petals 4–9 mm, deeply notched; rare in moist shores; NW; flr. Aug–Sep; native to Europe.

Epilobium strictum, stem (×5)

Epilobium strictum Muhl.   Downy willow-herb
Perennial from slender rhizomes; stems erect, 3–6 dm tall, simple or slightly branched; leaves lanceolate to lance-linear, entire, revolute, 2–4 cm long by 3–8 mm wide, all but the lowest alternate; young stems, upper surfaces of leaves, pedicels, and fruits covered with straight, spreading hairs; petals pink, 5–8 mm, notched; hairs of the seeds pale brown; rare in calcareous marshes, meadows, and thickets; SE, SC, and NW; flr. Jul–Sep; OBL; ❦.

Gaura L.

Gaura biennis L.   Gaura
Erect, branched biennial to 2 m tall, hairy and glandular, especially in the inflorescence; leaves lanceolate, to 12 cm; inflorescence a many-flowered terminal spike; flowers irregular, the 4 white to pink petals all on one side; hypanthium tube not much longer than the ovary; stamens 8; fruit indehiscent; frequent in moist meadows, floodplains, stream banks, and roadside thickets; throughout, except the northernmost counties; flr. Jul–early Oct; FACU.

Ludwigia L.

Wetland herbs with alternate or opposite leaves and flowers in the upper leaf axils; hypanthium tube not prolonged beyond the ovary; sepals persistent in fruit; petals yellow or lacking; ovary cylindrical or with winged angles; fruit a dehiscent capsule opening longitudinally or by terminal pores.

Epilobium
strictum

Gaura biennis,
flower (×2)

Ludwigia alternifolia

A. stamens 4; fruit up to 1 cm long
    B. leaves opposite; flowering stems prostrate or creeping .... L. *palustris*
    B. leaves alternate; flowering stems erect
        C. flowers with a definite pedicel; petals conspicuous .... L. *alternifolia*
        C. flowers sessile or nearly so; petals absent or very small
            D. bracteoles 2–5 mm, attached at or above the base of the ovary ....
            .... L. *polycarpa*
            D. bracteoles 1 mm, attached at or below the base of the ovary ....
            .... L. *sphaerocarpa*
A. stamens 8 or 10; fruit 2–4 cm long
    E. flowers 4-merous; stem winged .... L. *decurrens*
    E. flowers 5(6)-merous; stem not winged
        F. flowering stems creeping or floating; petals 1–1.5(2) cm long ....
        ....L. *peploides*
        F. flowering stems more or less erect; petals 1.5–2.5 cm long ....
        .... L. *hexapetala*

Ludwigia alternifolia L.   Seedbox, false loosestrife
Erect branching perennial, 4–12 dm tall; leaves alternate, lanceolate; flowers
solitary in the leaf axils, on 1–5-mm pedicels, 4-merous; petals yellow, about as
long as the sepals; fruit 4-sided and somewhat wing-margined; common in

swampy fields and wet woods; throughout, except in the northernmost counties; flr. Jun–Aug; FACW+.

Ludwigia decurrens Walter   Upright primrose-willow
Erect, branched, glabrous annual to 2 m tall; stems square and usually winged due to the decurrent leaf bases; leaves alternate, lanceolate to linear; flowers 4-merous, stamens 8, petals 8–12 mm; fruit 4-angled or winged; rare in sandy shores; SE; flr. Jul–Sep; OBL; 🥬.

Ludwigia hexapetala (Hook. & Arn.) Zardini, H.Gu & P.H.Raven   Water-primrose
Rhizomatous perennial with some slender, often floating, leafy stems; leaves alternate, lanceolate to oblanceolate, with hairy petioles up to 1 cm; flowers borne on erect or ascending stems 3–6 dm tall; petals 1.5–2.5 cm, yellow; fruit cylindrical, 2–4 cm tall; rare in shallow water of streams and canals; SE; flr. Jul–Sep; OBL. [syn: L. uruguayensis (Cambess.) H.Hara]

Ludwigia palustris (L.) Elliot   Marsh- or water-purslane
Perennial with stems prostrate, creeping or partially floating; leaves opposite; flowers sessile in the leaf axils, 4-merous, lacking petals; fruit 4-sided, with rounded angles; common in swamps, wet meadows, muddy shores, stream banks, and ditches; throughout; flr. Jul–Sep; OBL.

Ludwigia peploides (Kunth) P.H.Raven   Primrose-willow
Perennial with stems prostrate or floating and bending upward at the tips, rooting at the nodes; leaves alternate, lanceolate, oblanceolate, or obovate; flowers 5-merous, petals yellow, 1–1.5(2) cm; stamens 10; fruit cylindrical, 2–4 cm; rare, but sometimes locally abundant in shallow water or silty, muddy shores; SE; flr. Jul–Sep; OBL. Ours is ssp. *glabrescens* (Kuntze) P.H.Raven.

Ludwigia polycarpa Short & Peter   Seedbox, false loosestrife
Stoloniferous perennial, 2–10 dm tall, branched; stem 4-angled, glabrous; leaves lanceolate or lance- linear; flowers sessile in the axils of the leaves, 4-merous; petals very small or absent; stamens 4; fruit 4-sided with rounded angles, the bracteoles attached at or above the base; rare in wet meadows and swales; SC; flr. Jul–Sep; OBL; 🥬.

Ludwigia
peploides,
fruit

Ludwigia peploides,
flower

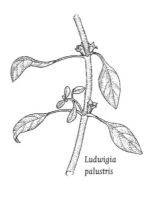

Ludwigia
palustris

Ludwigia sphaerocarpa Elliot   Spherical-fruited seedbox

Stoloniferous, branched perennial to 1 m tall; leaves lanceolate to linear; flowers in the axils of reduced leaves, subsessile, 4-merous, lacking petals; fruit subglobose with bracteoles attached at or below the base; very rare in coastal plain swamps; believed to be extirpated; flr. Jul–Sep; OBL; ❧.

Oenothera L.

Herbs with alternate narrow leaves; flowers in a terminal spike or raceme or solitary in the axils, mostly lasting only 1 day, 4-merous; hypanthium tube much longer than the ovary; sepals fused but with free tips, splitting and becoming reflexed at anthesis; petals 4, yellow (white or pink in *O. speciosa*); stamens 8; ovary inferior, 4-locular with elongate style and 4-lobed stigma; capsule winged or rounded, cylindrical, tapered, or club-shaped.

A. ovary rounded or only slightly 4-angled
  B. petals white or pink .... *O. speciosa*
  B. petals yellow
    C. some of the leaves coarsely toothed to pinnatifid; fruit linear ....
    .... *O. laciniata*
    C. leaves entire or shallowly dentate; fruit tapering from base to tip
      D. petals 3–6.5 cm long; style longer than the stamens
        E. cauline leaves 4–10 mm wide; tip of the inflorescence curved .....
        .... *O. argillicola*
        E. cauline leaves >1.5 cm wide; tip of the inflorescence erect
          F. stems, hypanthium, and ovary pubescent .... *O. glazioviana*
          F. stems, hypanthium, and ovary appearing glabrous to the naked eye .... *O. grandiflora*
      D. petals 0.7–2.5 cm long; style shorter than to as long as the stamens
        G. apex of the inflorescence erect
          H. inflorescence glabrous to sparsely glandular-puberulent ....
          .... *O. nutans*
          H. inflorescence conspicuously pubescent
            I. ovary lacking bulbous-based hairs .... *O. villosa*
            I. ovary with bulbous-based hairs .... *O. biennis*
        G. apex of the inflorescence curved
          J. leaves grayish or dull green; plant with long silky hairs ....
          .... *O. oakesiana*
          J. leaves bright green; plant with erect pubescence or appearing glabrous to the naked eye .... *O. parviflora*
A. ovary and fruit strongly 4-angled or winged
  K. leaves pinnatifid, all basal .... *O. triloba*
  K. leaves entire or toothed, cauline leaves present
    L. petals 3–10 mm; style 1–10 mm; apex of the inflorescence nodding ......
    .... *O. perennis*
    L. petals 15–30 mm; style 10–20 mm; apex of the inflorescence erect
      M. stems densely covered with long, spreading hairs; fruit somewhat club-shaped to elliptic .... *O. pilosella*
      M. stems with sparse straight or incurved hairs or glabrous; fruit distinctly club-shaped .... *O. fruticosa*

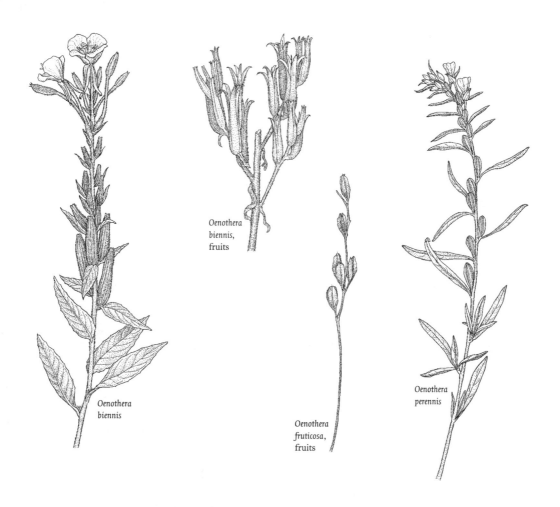

Oenothera
biennis,
fruits

Oenothera
biennis

Oenothera
fruticosa,
fruits

Oenothera
perennis

Oenothera argillicola Mack. Shale-barren evening-primrose
Erect or ascending biennial with branched stems to 4 dm tall, glabrous to somewhat hairy in the inflorescence; leaves dark green, glossy, sometimes puberulent; inflorescence curved at the tip; sepals yellowish-green flushed with red, the
free tips 3–9 mm; petals 2–4 cm; capsule 2–4 cm; rare in shale barrens; SC; flr.
Jul–Sep; ❧.

Oenothera biennis L. Evening-primrose
Erect, glabrous to finely canescent biennial 0.5–2 m tall; leaves 5–22 cm , lanceolate to oblong, entire or slightly toothed; inflorescence an erect, simple or
branched terminal spike with leafy bracts; sepals 1.2–2.2 cm, yellowish-green,
the free tips 1.5–3 mm; petals 1–2.5 cm; fruit 1.5–4 cm, stout and tapering to the
tip; common in cultivated fields, waste ground and roadsides; throughout; flr.
Jun–early Oct; FACU–.

Oenothera fruticosa L. Sundrops
Erect, variously hairy to nearly glabrous perennial 2–9 dm tall; leaves ovate to
nearly linear, entire or very slightly toothed; inflorescence mostly erect; petals
1.5–3 cm; capsules strongly 4-angled or winged, club-shaped, 4–15 mm; frequent in fields, meadows, and roadsides; throughout, except at the highest elevations, mostly S; flr. late May–Aug; FAC; 2 subspecies:

A. inflorescence and fruits with gland-tipped hairs .... ssp. *fruticosa*
[syn: *O. tetragona* Roth var. *longistipata* (Pennell) Munz, *O. fruticosa* L. var. *linearis*
(Michx.) S.Watson]

A. inflorescence and fruits not glandular .... ssp. *glauca* (Michx.) Straley
[syn: *O. tetragona* Roth var. *tetragona, O. tetragona* var. *latifolia* (Rydb.) Fernald]

Oenothera glazioviana Micheli　　　　　　　　　　　　Evening-primrose
Similar to *O. biennis* but with broader leaves and larger flowers; sepals reddish,
their free tips 5–8 mm; petals 3–5 cm; cultivated and rarely naturalized in waste
ground; E; flr. Aug–Oct; native to Europe.

Oenothera grandiflora L'Hér.　　　　　　　　　　　　Evening-primrose
Erect biennial 10–30 dm tall; stems green above, red below, with scattered hairs
including a few pustulate ones below the inflorescence; leaves bright green, elliptic or lanceolate; inflorescence usually with secondary branches below the
main spike, tip of the inflorescence erect; sepals 2.2–4.6 cm, yellowish-green,
often flushed with red, the free tips 2–9 mm; petals 3–4.5 cm; capsule 1.5–3.5
cm; cultivated and rarely escaped; N; flr. Jul–Oct; native to the southeastern U.S.

Oenothera laciniata Hill　　　　　　　　　Cut-leaved evening-primrose
Puberulent or sparsely hairy annual with decumbent or erect stems 1–6 dm tall;
leaves coarsely toothed to pinnatifid, sessile or short-petioled below; flowers
few, sessile in the upper axils; petals 5–18 mm, yellow becoming reddish with
age; fruit linear, 2–3 cm, usually curved; rare in dry sandy or gravelly soil; SE and
W; flr. Jun–Nov; FACU–.

*Oenothera
laciniata,
leaf*

Oenothera nutans G.F.Atk. & Bartlett　　　　　　　　Evening-primrose
Erect biennial; stem 3–20 dm tall, with scattered hairs, sometimes glandular-puberulent in the inflorescence; leaves dark green, elliptic to lanceolate, glabrous
or with scattered hairs; inflorescence erect, unbranched; sepals yellowish-green
with reddish ends, 1–2.3 cm, the free tips 1.5–6 mm; petals 1.4–2.5 cm; capsules
1.2–3.6 cm; frequent in old fields, roadsides, and fallow land; throughout; flr.
Jun–Oct.

Oenothera oakesiana (A.Gray) Robins ex S.Watson & Coult.　　　Evening-primrose
Grayish-green, silky-hairy biennial; stem 1–6 dm tall, erect or procumbent;
leaves narrowly oblanceolate or elliptic, remotely dentate; inflorescence simple,
the tip curved; sepals greenish or yellow with red flecks or stripes; petals pale
yellow, 0.7–2 cm; capsule 1.5–4 cm; rare in railroad ballast and alluvium; E; flr.
Jul–Oct; FACU–; 🍂. [syn: *O. parviflora* L. var. *oakesiana* (Robbins) Fernald]

Oenothera parviflora L　　　　　　　　　　　　　　Evening-primrose
Similar to *O. biennis* but with intermingled long and short hairs, some of them
gland-tipped; leaves bright green; tip of the inflorescence nodding; free tips of
the sepals 1–3 mm; petals 1–2 cm; fruit 1.5–4 cm, only slightly tapering; occasional in old fields, roadsides, and railroad embankments; mostly E, but scattered elsewhere; flr. Jul–Oct; FACU–.

Oenothera perennis L　　　　　　　　　　　　　　　　　Sundrops
Perennial with erect, simple, clustered stems 1–6 dm tall; leaves 3–6 cm, oblanceolate to elliptic, tapered to a short petiole; flowers in the axils of reduced upper

*Oenothera perennis,
fruit (×1)*

leaves; tip of the inflorescence nodding in bud; petals 5–10 mm, notched; fruits 4–7 mm, tapering to the base, strongly 4-angled; common on shaly slopes, moist or dry fields, pastures, and roadsides; throughout; flr. late May–early Aug; FAC–.

Oenothera pilosella Raf.   Sundrops
Hairy, rhizomatous perennial; stems to 8 dm tall, simple or branched above; leaves 3–10 cm, acute or acuminate, sessile; inflorescence compact, erect; petals 1.5–2.5 cm; fruit club-shaped or ellipsoid; rare in open woods, meadows, and roadsides, mostly escaped from cultivation; flr. May–Jul; native farther west; FAC.

Oenothera speciosa Nutt.   White evening-primrose
Ascending to erect perennial 1–6 dm tall; leaves linear to lanceolate or oblanceolate, the lower ones often coarsely toothed or pinnatifid; flowers sessile in the upper axils; petals white or pink; fruit narrowly club-shaped; cultivated and rarely escaped; mostly E; flr. Jun–Jul; native farther west.

Oenothera triloba Nutt.   Evening-primrose
Tap-rooted winter annual or biennial; leaves basal, pinnatifid with a large terminal lobe; flowers axillary, petals 1–2.5 cm, pale yellow becoming whitish with age; fruit winged, 1–2 cm; very rare in dry, often calcareous soil; flr. May–Jun; native farther west.

Oenothera villosa Thunb.   Evening-primrose
Erect biennial; stem simple or branched, 5–20 dm tall; leaves dull green to grayish green, lanceolate or elliptic, dentate to subentire; inflorescence erect; sepals 0.9–1.8 cm, yellowish-green with red stripes, the free tips 0.5–3 mm; petals 0.7–2 cm; capsule 2–4.3 cm, tapering toward the apex; rare in fallow fields and waste ground in moist soil; scattered; flr. Jul; native to the Great Plains and farther west; FAC.

OROBANCHACEAE Broom-rape Family

Parasitic herbs lacking chlorophyll; flowers perfect, irregular; calyx 2–5-lobed; corolla tubular with 4–5 lobes; ovary superior, unilocular with 2–6 parietal placentae and a terminal style; capsule 2-valved, many seeded.

A. flowers in a dense, fleshy spike 2–3 cm in diameter .... *Conopholis*
A. flowering stems more slender
    B. stems usually branched; flowers in the axils of alternate scale-like leaves....
    .... *Epifagus*
    B. stems unbranched; flowers solitary or in a terminal spike .... *Orobanche*

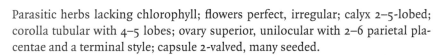

Conopholis Wallr.

Conopholis americana

Conopholis americana (L.) Wallr.   Squawroot, cancer-root
Stout herb with unbranched stems to 1.5 dm tall with fleshy, overlapping scale-like leaves below and a dense, fleshy spike of flowers above; pale brown or yellowish throughout and appearing somewhat like a cluster of elongate pine cones

upon emergence in the spring; corolla very irregular, the lower lip downward curved; parasitic on *Quercus* spp.; occasional in forests, mostly S and W; flr. May– early Jun.

Epifagus Nutt.

Epifagus virginiana (L.) W.Bartram   Beechdrops

Yellow or brownish-purple branched herb to 3 dm tall, with scale-like, alternate leaves and axillary flowers; parasitic on the roots of *Fagus grandifolia* and occurring with the host; frequent in forests; throughout; flr. Jul–Oct.

Epifagus virginiana, flower (×1)

Orobanche L.

Small glandular-hairy herbs lacking leaves; calyx campanulate 4–5-lobed; corolla with a curved tube much longer than the lobes; parasitic on the roots of many other plants.

A. flowers apparently solitary .... *O. uniflora*
A. flowers several to many in a dense spike-like raceme .... *O. minor*

Orobanche minor Sm. ex Sowerby   Small broom-rape

Plant 1.5–5 dm tall; flowers in a more or less elongate spike, sessile; calyx split to the base; corolla white or yellowish marked with purple; parasitic on *Trifolium* species; very rare in yards, vacant lots, and roadsides; native to Eurasia; designated as a federal noxious weed.

Orobanche uniflora L.   Broom-rape, cancer-root

Aboveground flowering stems (actually pedicels) erect, 6–20 cm tall, each bearing a single white to violet flower 2 cm long; calyx lobes 5, slightly longer than the tube; frequent in forests; throughout; flr. May; FACU.

Epifagus virginiana

OXALIDACEAE Oxalis Family

Low perennial herbs with alternate, trifoliate leaves with or without stipules; inflorescence umbel-like, or flowers solitary; flowers regular, perfect, 5-merous; petals distinct; stamens 10, their filaments joined at the base; ovary superior, of 5 fused carpels; styles distinct; placentation axile; fruit an erect, more or less elongate, dehiscent capsule.

Oxalis L.

Characteristics of the family.

A. leaves and flowering stalks arising from a bulbous or rhizomatous base; flowers yellow, white or purple
  B. flowers yellow .... *O. corniculata*
  B. flowers white or purple
    C. leaves and peduncles arising from a rhizome; flowers solitary ....
    .... *O. acetosella*

C. leaves and peduncles arising from a bulb; flowers several per peduncle
.... *O. violacea*
A. leaves and flowering stalks arising from the axils of an erect stem; flowers yellow
    D. petals 12–20 mm .... *O. grandis*
    D. petals 5–11 mm
        E. stipules absent .... *O. stricta*
        E. stipules present
            F. stems creeping, rooting at the nodes; stipules broad, leaves often purplish .... *O. corniculata*
            F. stems erect, not creeping; stipules narrow; leaves green .... *O. dillenii*

Oxalis acetosella L.   Northern wood-sorrel
Rhizomatous, leaves all basal, long-petioled; flowers solitary on peduncles about as tall as the leaves; petals white with pink veins; frequent in rich, moist woods, bogs, and swamps; N and at high elevations along the Allegheny Front; flr. May–Aug; FAC–. [syn: *O. montana* Raf.]

Oxalis corniculata L.   Creeping yellow wood-sorrel
Stems trailing and rooting freely; leaves often purplish; flowers yellow; seeds brown; an occasional weed of roadsides, pavement, gardens, and greenhouses; flr. May–Sep; native farther south; FACU.

Oxalis dillenii Jacq.   Southern yellow wood-sorrel
Stems tufted, erect to decumbent but not creeping or rhizomatous; petals yellow, 4–10 mm; fruit 1.5–2.5 cm; seeds brown with whitish ridges; common in rich woods, roadsides, and waste ground; throughout except in the northernmost counties; flr. May–Sep. Ours is ssp. *filipes* Small.

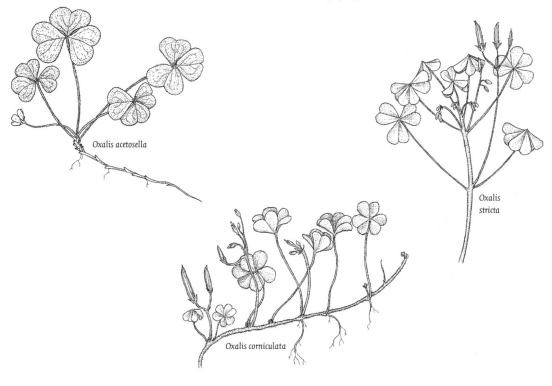

Oxalis acetosella

Oxalis stricta

Oxalis corniculata

Oxalis grandis Small   Great yellow wood-sorrel
Rhizomatous; stems 3–10 dm tall, erect; leaflets with purple margins; petals yellow, 12–20 mm; fruit 8–10 mm; rare in rich woods, wooded banks, and shale
barrens; SW; flr. late May–Jul.

Oxalis stricta L.   Common yellow wood-sorrel
Rhizomatous; stems prostrate to erect to 5 dm; leaflets 1–2 cm wide; stipules absent; flowers yellow; petals 4–9 mm; fruit 8–15 mm; common in lawns, gardens,
and fields; throughout; flr. May–Sep; UPL. [syn: *O. europaea* Jord.]

Oxalis violacea L.   Violet wood-sorrel
Leaves and flowering stems arising from a scaly, brown bulb; peduncles much
taller than the leaves, bearing a cluster of rose-violet flowers; occasional in dry,
open woods and shaded banks; S; flr. Apr–Jun.

PAPAVERACEAE Poppy Family

Annual, biennial, or perennial herbs, juice generally milky or colored; leaves
basal or alternate, often deeply divided; flowers mostly solitary or in few-flowered
inflorescences, regular, mostly perfect, hypogynous, usually showy; petals 4 or
more; sepals present but often falling early; stamens many; fruit a capsule,
sometimes conspicuously flattened on top into a stigmatic disk; most species
contain isoquinoline alkaloids traditionally used in medicine and all should be
considered toxic.

A. leaf basal and single .... *Sanguinaria*
A. cauline leaves present
  B. stems and leaves spiny .... *Argemone*
  B. stems and leaves not spiny, although sometimes coarsely hairy
    C. juice clear .... *Eschscholzia*
    C. juice milky or colored
      D. inflorescence a terminal panicle .... *Macleaya*
      D. flowers solitary or in few-flowered umbels
        E. flowers yellow; capsule dehiscing longitudinally along the sides
          F. capsule elliptic, pubescent or bristly .... *Stylophorum*
          F. capsule linear, glabrous .... *Chelidonium*
        E. flowers white to red, orange or purple; capsule dehiscing by slits
        or pores just below a prominent, flattened stigmatic disk ....
        .... *Papaver*

Argemone L.

Argemone mexicana L.   Mexican poppy, prickly poppy
Coarse, often branched annual or biennial, all parts bristly, up to 1.5 m tall, juice
yellow; both basal and cauline leaves present, sessile or clasping, lobed and
toothed with each tooth ending in a bristle; flowers 3–6 cm across, yellow or
cream-colored; capsules spiny, 25–45 by 10–20 mm, elliptic; occasional on ballast and waste ground; scattered E; flr. May–Sep; native to tropical America.

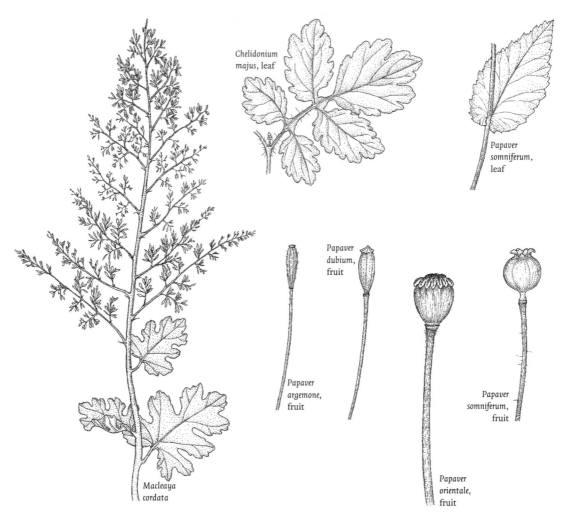

Chelidonium
majus, leaf

Papaver
somniferum,
leaf

Papaver
dubium,
fruit

Papaver
argemone,
fruit

Papaver
somniferum,
fruit

Papaver
orientale,
fruit

Macleaya
cordata

Chelidonium L.

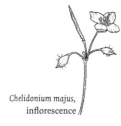

Chelidonium majus,
inflorescence

Chelidonium majus L.   Greater celandine, swallowwort
Biennial, 3–8 cm tall, juice yellow-orange; both basal and cauline leaves present, deeply pinnately lobed and variously toothed; flowers about 2 cm across, 4-petaled, most often bright yellow, sometimes cream-colored to orange-yellow; capsules glabrous, 20–40 by 2–3 mm, straight or slightly curved; common on moist soils, roadsides and garden sites; throughout; flr. Mar–May; native to Eurasia; UPL.

Eschscholzia Cham.

Eschscholzia californica Cham.   California poppy
Annual or perennial, 2–7 dm tall, juice clear, stems somewhat ribbed; leaves much divided, the ultimate segments linear, up to 6 cm long, often glabrous and glaucous but sometimes densely hairy; flowers 4-petaled, 4–12 cm across, color variable, usually yellow, pink, or orange, sometimes white; capsules linear, 30–80 by 1–3 mm, glabrous, straight or slightly curved; occasionally escaping from cultivation; mostly SE, scattered elsewhere; flr. summer; native to the west coast of North America.

Macleaya R.Br.

Macleaya cordata (Willd.) R.Br.   Plume-poppy, tree celandine
Perennial up to nearly 3 m tall, stems glabrous and glaucous; leaves up to 30 cm long on the lower part of plant, reduced upward, 5–11-lobed, each lobe with smaller rounded lobes, glabrous above and densely short-hairy below; flowers small, petals absent; inflorescence a much branched terminal panicle up to about 50 cm long; capsules 8–18 mm long, flattened, ovoid; occasional in railroad banks and cultivated and waste ground; scattered throughout; flr. Jun–Jul; native to eastern Asia.

Macleaya cordata, fruit (×1)

Papaver L.

Annual, biennial, or perennial herbs with milky juice; both basal and cauline leaves present in biennial or perennial plants; flowers long-peduncled and solitary; capsules topped by a prominent stigmatic disk and dehiscing by small slits or pores just below the disk.

A. at least some cauline leaves with clasping bases .... *P. somniferum*
A. none of the cauline leaves with clasping bases
    B. capsule covered with bristles .... *P. argemone*
    B. capsule glabrous or nearly so
        C. capsule about 2 times as long as wide .... *P. dubium*
        C. capsule about as long as wide
            D. pedicels with coarse, appressed hairs; plants generally 1 m or more tall; petals 4–6 cm long .... *P. orientale*
            D. pedicels with weak, spreading hairs; plants generally less <1 m tall; petals 1.5–3.5 cm long .... *P. rhoeas*

Papaver argemone L.   Long rough-fruited poppy
Plant to about 5 dm tall, sparsely branched, stems slender; leaves once or twice pinnately divided; flowers dark red to red-orange; capsules narrowly ellipsoid, about 15 mm long, bristly; rare on waste ground and ballast; flr. Apr–May; native to southern Europe.

Papaver dubium L.   Long-pod poppy, smooth-fruited poppy
Plant 3–6 dm tall, sparsely branched, stems slender; leaves pinnately divided; flowers red to white; capsules narrowly obovoid, glabrous; occasionally escaping from cultivation; mostly S; flr. May–Jul; native to Europe.

Papaver orientale L.   Oriental poppy
Plant to 1.3 m tall, rarely branched; leaves pinnately lobed and coarsely toothed, each tooth ending in a stiff hair; flowers generally some shade of orange; capsules 1.5–2.5 cm long, ovoid to obovoid, glabrous and somewhat glaucous; occasionally persisting where cultivated; mostly S; flr. May–Jun; native to Asia.

Papaver rhoeas L.   Corn poppy, field poppy
Plant 1–10 dm tall, sparsely branched; leaves once to twice pinnately divided; flowers red, white, or purple; capsules ovoid to subglobose, glabrous; occasionally escaping from cultivation; mostly S; flr. May–Sep; native to Eurasia and northern Africa.

Papaver somniferum L.   Opium poppy
Plants to 1 m or more tall, fairly stout, rarely branched; leaves coarsely toothed or
shallowly lobed; flowers red, purple, or white; capsules subglobose, 2–6 cm
long, glabrous; occasionally escaping from cultivation; mostly S, scattered elsewhere; flr. May–Jul; native to Eurasia.

Sanguinaria L.

Sanguinaria canadensis L.   Bloodroot, red puccoon
Flowering stem 5–15 cm tall, glabrous; leaf solitary, basal, shallowly 3–9-lobed,
broadly reniform, up to 20 cm long and 30 cm wide at maturity; both flowering
stem and leaf arising from a branching rhizome; juice orange-red; flower solitary, white to rarely pinkish with golden-yellow anthers; capsule narrowly elliptic, 3–5 cm long, glabrous; common in rich woods and on roadside banks;
throughout; flr. Apr–early May; UPL.

Stylophorum Nutt.

Stylophorum diphyllum (Michx.) Nutt.   Celandine-poppy
Plant 3–5 dm tall, somewhat branched, juice yellow-orange; most leaves basal,
5–7-lobed, 10–25 cm long, long-petioled, cauline leaves similar to basal but
smaller, generally 1 pair only; flowers yellow, in few-flowered terminal umbels;
capsules 2–3 cm long, ellipsoid; flr. Apr–May; generally native to areas farther
west, all collections made in Pennsylvania are probably escapes from cultivation.

PASSIFLORACEAE Passion-flower Family

Herbaceous perennial vines climbing by tendrils; leaves alternate, palmately
lobed; stipules present; flowers perfect, regular, on long peduncles in the leaf
axils, perigynous; sepals and petals each 5, attached to the rim of a hypanthium,
which also bears a fringed corona; stamens 5, the filaments united to form a
sheath around the style; ovary 1, superior, stalked; stigmas 3; fruit a berry, 1 cm
thick.

Passiflora L.

Passiflora lutea L.   Passion-flower
Vines to 3 m tall; leaves shallowly 3-lobed; flowers 1.5–2.5 cm wide, greenish-
yellow; fruit purple; rare in moist stream bank thickets; mostly SW; flr. Jul; 🌿.

PEDALIACEAE Sesame Family

Herbs with stalked, branched trichomes and other hairs; leaves simple, lacking
stipules, opposite or the upper alternate; flowers irregular, bilabiate, somewhat
spurred or saccate at the base; stamens 4, inserted on the corolla tube; ovary superior, bicarpellate, unilocular with 2 parietal placentae; fruit a capsule.

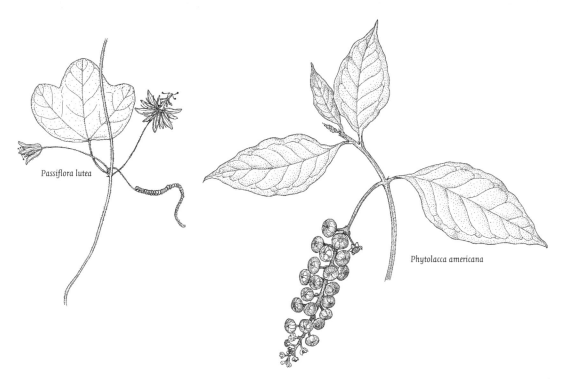

Passiflora lutea

Phytolacca americana

A. sprawling, vine-like plants; fruit indehiscent, bearing several long appendages ....*Proboscidea*
A. erect plants; fruit opening by terminal pores, without appendages ... *Sesamum*

Proboscidea Schmidel

Proboscidea louisianica (Mill.) Thell.   Unicorn-plant
Sprawling-ascending, glandular pubescent annual to about 2 m tall; leaves long-
petioled, orbicular to reniform-cordate, the upper often alternate; racemes 8–20-
flowered; corolla 3.5–5.5 cm, dull white or yellowish spotted with purple; fruit
1–2 dm long, an indehiscent capsule with curved appendages; cultivated and occasionally escaped to fields and waste ground; mostly S; flr. Jun–Sep; native to
the southeastern U.S.; FACU.

Sesamum L.

Sesamum indicum L.   Sesame
Hairy, erect herb to 1.5 m tall; leaves lanceolate to ovate, entire or coarsely
toothed; flowers solitary, axillary on ascending pedicels; corolla pink or yellowish-white, 2–3 cm; fruit a columnar, 4-angled capsule opening at the summit;
cultivated and occasionally naturalized in urban waste ground; SE; native to
India.

PHYTOLACCACEAE Pokeweed Family

Coarse perennial herbs with alternate, entire leaves; flowers in racemes, perfect,
regular, 5-merous; sepals 5; stamens 10; ovary superior, carpels united except
for the styles; fruit a berry.

Phytolacca L.

Phytolacca americana L.   Pokeweed

Branching perennial to 3 m tall; flowers greenish-white, racemes drooping in fruit; berries juicy, dark purple; common in forest openings, roadsides, gardens, and open ground; throughout; flr./frt. Jun–Oct; FACU+.

PLANTAGINACEAE Plantain Family

Annual or perennial herbs with basal leaves (except *P. psyllium*); leaves linear to broadly ovate, conspicuously parallel-veined and often ribbed; inflorescence a dense spike of small, 4-merous, perfect flowers on a leafless stem, each flower subtended by a bract; corolla of fused petals, chaffy, and persistent; stamens 2 or 4, inserted on the corolla tube; ovary superior, bilocular; fruit a capsule.

Plantago L.

Characteristics of the family.

A. leaves cauline, opposite; inflorescences on axillary peduncles .... *P. psyllium*
A. leaves basal; inflorescences on leafless scapes arising from the base of the plant
  B. leaves linear or narrowly lanceolate
    C. bracts conspicuous, much longer than the flowers they subtend ....
       .... *P. aristata*
    C. bracts not conspicuous .... *P. pusilla*
  B. leaves lanceolate to broadly ovate
    D. inflorescence extending for ½ or more of the length of the flowering scapes
      E. leaves and scape glabrous or inconspicuously short-hairy; stamens 4
        F. petioles green throughout; calyx lobes obtuse; capsule ovoid .....
           .... *P. major*
        F. petioles purple at the base; calyx lobes acute; capsule ellipsoid....
           .... *P. rugelii*
      E. leaves and scape densely and conspicuously hairy; stamens 2 ....
         .... *P. virginica*
    D. inflorescence extending <½ the length of the flowering scape
      G. leaves >6 times as long as wide; sepals 3 .... *P. lanceolata*
      G. leaves <5 times as long as wide; sepals 4 .... *P. media*

Plantago
aristata,
inflorescence

Plantago aristata Michx.   Bristly plantain, buckhorn

Hairy annual or short-lived perennial, leaves linear to narrowly oblanceolate; flowering stems to 3.5 dm, the lowest bracts 2 cm long, the upper progressively shorter; occasional in dry sandy fields, shale barrens, railroad cinders, and other dry, open ground; mostly S; flr. Jun–Aug; native to western North America.

Plantago lanceolata L.   English plantain, ribgrass

Perennial with lanceolate or oblanceolate, strongly ribbed leaves 6–30 mm wide; inflorescence narrowly conical at first, elongating and becoming cylindrical in

Plantago
aristata

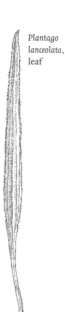

Plantago
lanceolata,
leaf

Plantago
lanceolata,
inflorescence
(×¼)

fruit; stamens conspicuously exert; common in lawns, roadsides, old fields, clearings, and waste ground; throughout; flr. May–early Oct; native to Europe; UPL.

Plantago major L.   Broad-leaved plantain, whiteman's-foot
Very similar to *P. rugelii* and differing mainly in the more pubescent leaves and scape, green petioles, and higher line of dehiscence on the capsule; common in lawns, gardens, and waste ground; throughout; flr. Jun–Sep; native to Europe; FACU.

Plantago major,
capsule (×2½)

Plantago media L.   Hoary plantain, lamb's-tongue
Perennial; leaves elliptic, obovate or oblanceolate, tapering at each end; flowering stem 2–4 dm, strigose above; spike conical, becoming cylindric; bracts and sepals similar in length; occasional in lawns, waste ground, and ballast; flr. Jul; native to Eurasia.

Plantago psyllium L.   Flaxseed plantain, fleawort
Annual with leafy stems to 6 dm tall; leaves numerous, linear, opposite; peduncles axillary; spikes dense, nearly 1 cm thick; very rare on sandy shores, railroad tracks, and other dry waste ground; scattered; flr. Jun–Oct; native to the Mediterranean region.

Plantago pusilla Nutt.   Dwarf plantain
Annual, leaves linear, to 8 cm; flowering stems to 10 cm tall, spike about ½ the length of the scape, flowers well-spaced; rare on dry, sandy, or rocky open ground; SE; flr. May–Jun; UPL.

Plantago rugelii Decne.   Broad-leaved or Rugel's plantain
Perennial; leaves thin, broadly oval or elliptic, glabrous or slightly pubescent; petioles glabrous and usually purple at the base; flowering stems to 3 dm,

Plantago rugelii,
capsule (×2½)

mostly glabrous, flowers extending >½ the length of the scape, dense above to more loosely arranged toward the base; line of dehiscence on the capsules well below the middle; common in gardens, lawns, meadows, wet pastures, roadside banks and waste ground; throughout; flr. Jul–Aug; FACU.

Plantago virginica L.   Dwarf plantain, pale-seeded plantain
Densely gray-hairy annual or biennial; leaves obovate to oblanceolate, 5–15 cm long; flowering stems bearing a dense, elongate spike that extends >½ its length; floral bracts narrow, shorter than the calyx; occasional in fields, low meadows, open banks, and waste ground along railroads; mostly S; flr. May–Jul; UPL.

PLATANACEAE Planetree Family

Deciduous trees with alternate, simple, palmately veined and lobed leaves; base of the petiole surrounding and covering the winter bud; stipules present; monoecious; flowers tiny, clustered in dense unisexual heads; fruit a spherical head of achenes each with a tuft of hairs, individual achenes dispersed by the wind during late winter and early spring.

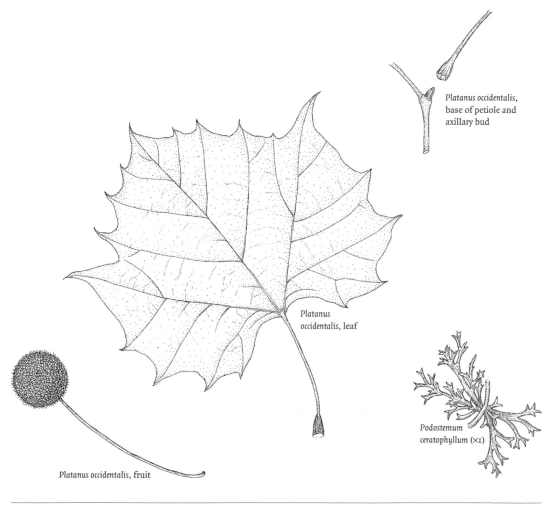

Platanus occidentalis, base of petiole and axillary bud

Platanus occidentalis, leaf

Podostemum ceratophyllum (×1)

Platanus occidentalis, fruit

Platanus L.

A. bark white; fruiting heads borne singly .... *P. occidentalis*
A. bark yellowish-gray; fruiting heads 2–3 per peduncle .... P. x *acerifolia*

Platanus x *acerifolia* Willd.   London planetree
Very similar to the native sycamore but with blotchy, yellowish-gray, exfoliating bark, and multiple fruiting heads per peduncle; frequently planted in urban areas and occasionally escaped. [syn: *P. occidentalis* x *orientalis*]

Platanus occidentalis L.   Sycamore, buttonwood, planetree
Deciduous tree to 50 m tall with blotchy, brownish-gray to chalky white, exfoliating bark; leaves palmately veined, with 3–5 broad lobes, margins serrate; stipules foliaceous; flowers in long-peduncled heads; common on riverbanks, low woods, floodplains, and alluvial soils; throughout; flr. Apr, before the leaves; FACW–.

Platanus occidentalis,
achene (×2)

PLUMBAGINACEAE Leadwort Family

Perennial herbs with basal leaves; flowers perfect, regular; calyx and corolla each tubular and 5-lobed; stamens 5; ovary superior, 1-celled; styles 5; fruit dry and indehiscent, surrounded by the persistent calyx.

Armeria Willd.

Armeria maritima (Miller) Willd.   Thrift
Leaves linear and evergreen; flowers pink, in a dense, capitate head with an involucre of membranous subtending bracts; styles fused at the base; naturalized and locally abundant on a mountaintop in the vicinity of Lehigh Gap in zinc and lead contaminated soil, also cultivated; flr. Jul–Aug; native farther north and in Eurasia.

*Armeria
maritima*

PODOSTEMACEAE Riverweed Family

Reduced, nearly thalloid plants; stems 2–10 cm tall, branched, attached to the substrate by fleshy disks; leaves similar in appearance to the stems, alternate, 1–10 cm long, narrow and entire or forked forming numerous narrow segments, stipules with a free tip above the sheathing base; flowers axillary, perfect; perianth absent; ovary superior; stamens 2; fruit a ribbed capsule, 2.5–3 mm.

*Podostemum
ceratophyllum,*
fruit (×4)

Podostemum Michx.

Podostemum ceratophyllum Michx.   Riverweed
Characteristics as above; an occasional native growing attached to rocks in shallow, fast-moving water of rivers and streams; throughout, often locally abundant; OBL

POLEMONIACEAE Phlox Family

Herbaceous annuals or perennials with opposite or alternate leaves lacking stipules; flowers in showy panicles or cymes, perfect, regular, 5-merous; calyx and corolla tubular; corolla lobes convolute in bud; stamens 5, inserted on the corolla tube, sometimes at different heights; ovary superior, of 3 fused carpels; style 1; stigmas 3; fruit a dehiscent capsule.

A. leaves simple
  B. leaves, at least the lowest, opposite; corolla limb >1 cm wide .... *Phlox*
  B. leaves alternate, corolla limb <1 cm wide .... *Collomia*
A. leaves compound .... *Polemonium*

Collomia Nutt.

Collomia linearis Nutt.   Collomia
Annual to 3 dm tall with an erect, usually simple stem, alternate leaves, and flowers in leafy-bracted terminal clusters; corolla salverform, white to pink; rarely naturalized along railroad tracks and waste ground; scattered; native farther north and west.

Phlox L.

Herbaceous perennials, upright or creeping, with mostly opposite, entire leaves; flowers with a narrow corolla tube and abruptly spreading lobes; stamens inserted on the tube at uneven heights; capsule 3-valved with 1–4 seeds per locule.

A. plants semi-woody at the base, mat-forming; leaves subulate to linear, evergreen .... *P. subulata*
A. plants herbaceous, with more or less upright flowering stems; leaves lanceolate to ovate.
  B. style shorter than the corolla tube or calyx
    C. leaves linear to lanceolate; flowering stems erect or ascending, arising in a cluster from a root crown; sterile shoots not produced; corolla tube pubescent on the outside .... *P. pilosa*
    C. leaves ovate to lance-ovate; horizontal stems rooting at the nodes and producing sterile shoots as well as upright flowering stems; corolla tube glabrous .... *P. divaricata*
  B. style as long or longer than the corolla tube or calyx
    D. leaves with distinct lateral and sub-marginal connecting veins ....
       .... *P. paniculata*
    D. leaves without distinct lateral veins
      E. stoloniferous with basal, horizontal leafy shoots present at flowering; sterile shoots with spatulate, evergreen leaves .... *P. stolonifera*
      E. nonstoloniferous, basal shoots upright, with narrow, non-evergreen leaves
        F. flowering stems bearing 6 or more pairs of leaves below the cylindric inflorescence .... *P. maculata*
        F. flowering stems bearing 4–5 pairs of leaves below the flat-topped, corymbiform inflorescence .... *P. ovata*

Phlox divaricata L. Wild blue phlox, wild sweet-william

Perennial with semi-evergreen leaves and decumbent sterile shoots forming roots at the nodes; flowers pale blue to occasionally violet or white; corolla with notched lobes and a glabrous tube; calyx glandular pubescent; frequent in rich deciduous woods; mostly W and S; flr. late Apr–early Jun; FACU. The midwestern ssp. *laphamii* (A.W.Wood) Wherry, an occasional garden escape, is distinguished from the native ssp. *divaricata* by its unnotched corolla lobes.

Phlox maculata L. Meadow phlox

Stems 3–8 dm tall, glabrous and usually finely streaked or spotted with purple; flowers deep pink to rose-purple, in a cylindrical inflorescence; calyx and corolla tube glabrous; frequent in wet meadows, abandoned fields, and thickets; throughout; flr. May–early Aug; FACW.

Phlox ovata L. Mountain phlox

Stems 3–5 dm tall, decumbent at base, glabrous, with only 4 or 5 pairs of leaves below the inflorescence; flowers few, deep pink, calyx and corolla tube glabrous; rare in openings and edges in dry, sandy woods; SC and SE; flr. late May–Jul; 🍂.

Phlox pilosa,
flower (×1)

Phlox paniculata L. Summer phlox

The tallest of our native species, to 2 m tall; corolla deep pink to white, tube hairy; calyx puberulent or glabrous; frequent in thickets, hillsides and stream banks; throughout, often in calcareous soils; flr. Jul–Oct; FACU; native, but also cultivated and frequently escaped.

Phlox pilosa L. Prairie or downy phlox

Flowering stems 3–6 dm tall, clustered from a definite root crown; stem, upper leaves, calyx, and corolla tube pubescent to villous, calyx also frequently stipitate-glandular; flowers pink; rare in moist meadows; SE; flr. late May–Jun; FACU; 🍂.

Phlox stolonifera Sims. Creeping phlox

Stoloniferous, bearing decumbent leafy basal shoots with broadly spatulate, obovate, evergreen leaves; flowering stems upright, 1–4 dm tall; corolla violet to rose-purple; calyx glandular-pubescent; frequent in rich woods; SW; flr. Apr–May.

Phlox subulata L. Moss-pink

Mat-forming with small, stiff, tufted, evergreen leaves; corolla deep pink to violet or white, often with a contrasting eye; occasional on dry slopes, rocky ledges, and serpentine barrens; throughout; flr. Apr–Jun. Subspecies *brittonii* (Small) Wherry, which is smaller and has gland-tipped hairs in the inflorescence, is known from a few shale barrens.

Polemonium L.

Perennial herbs with alternate, pinnately compound leaves and 5-merous, sympetalous blue flowers; stamens 5, nearly equal, inserted on the corolla tube.

Phlox pilosa

Phlox subulata

Phlox stolonifera

Polemonium reptans

A. stamens shorter than to as long as the corolla
    B. stems 1.5–5.5 dm tall, diffuse and low branching; pedicel as long as the calyx .... *P. reptans*
    B. stems simple, to 1 m tall; pedicel shorter than the calyx .... *P. caeruleum*
A. stamens longer than the corolla .... *P. van-bruntiae*

Polemonium caeruleum L.   Jacob's-ladder
Stems simple, erect to 1 m tall, leafy; inflorescence and upper stem glandular pubescent; flowers blue; stamens about the same length as the corolla; style slightly longer; cultivated and occasionally escaped; flr. Jun–Jul; native to Europe.

Polemonium reptans,
flower (×1)

Polemonium reptans L.   Spreading Jacob's-ladder
Stem solitary or clustered, erect to reclining, 1.5–5.5 dm tall, branched; flowers light blue, only the style extending beyond the corolla; frequent in low moist woods and wooded floodplains; throughout, except in the northeastern counties; flr. Apr–Aug; FACU.

Polemonium van-bruntiae Britton   Jacob's-ladder
Stems erect, mostly solitary, to 1 m tall; basal leaves long-petioled, overlapping; the upper well separated and nearly sessile; flowers blue, with stamens and style extending well beyond the corolla; pedicel shorter than the calyx; rare in sphagnous glades, swamps, or marshes; NE and SW; flr. Jun–Jul; FACW; 🍃.

POLYGALACEAE Milkwort Family

Annual, biennial, or perennial herbs with milky sap; leaves alternate or whorled, simple, entire; inflorescence a terminal or axillary raceme, or head-like, or the flowers solitary or few from the stem apex; flowers perfect, strongly irregular; sepals 5, the 2 inner ones larger; petals 3, all fused to the filaments to form a tube, the lower petal keel-shaped with a fringe-like crest; stamens 6 or 8, the filaments forming a sheath split along the upper side; ovary superior, 2-locular, with a terminal unequally 2-lobed style; fruit a capsule.

Polygala L.

A. flowers 1–4 from the stem tip, 1.5–2.3 cm; leaves few .... P. paucifolia
A. flowers numerous, in a raceme or dense head-like inflorescence; leaves many
    B. flowers yellow or orange; sepals decurrent on the pedicels .... P. lutea
    B. flowers rose-purple to white or greenish
        C. at least some of the leaves whorled
            D. raceme 8–20 mm thick, cylindrical .... P. cruciata
            D. raceme 2–4.5 mm thick, tapered to the tip.................. P. verticillata
        C. leaves all alternate
            E. stems clustered, arising from a thick, knotty base
                F. flowers white; cleistogamous flowers not produced .... P. senega
                F. flowers rose-purple to white; cleistogamous flowers produced in
                    subterranean racemes .... P. polygama
            E. stems solitary; flowers purple, pink or green
                G. stem glaucous; leaves linear-subulate and falling early ....
                .... P. incarnata
                G. stem not glaucous; leaves broader, flat, persistent
                    H. raceme 5–6 mm thick, tapering to the tip .... P. nuttallii
                    H. raceme 6–14 mm thick, rounded at the tip
                        I. raceme remaining dense and crowded .... P. sanguinea
                        I. raceme loosening as the inflorescence matures ....
                        .... P. curtissii

Polygala cruciata L.   Cross-leaved milkwort
Erect annual, 1–3 dm tall, with simple or lightly branched stems; leaves in whorls of 3 or 4, linear to narrowly elliptic; raceme cylindric, 1–6 cm by 1–1.5 cm thick; flowers rose-purple to greenish; rare in boggy pastures and mountain bogs; SE and SW; flr. late Jul–Sep; FACW+; 🌿.

Polygala curtissii A.Gray   Curtis's milkwort
Annual, stems erect, 1–4 dm tall; leaves alternate, linear to linear-oblong or narrowly oblanceolate; raceme dense, cylindric, 1–2 cm by 8–13 mm thick; flowers rose-purple; very rare in dry, open serpentine barrens; SE; flr. Jul–early Oct; 🌿.

Polygala incarnata L.   Pink milkwort
Annual with a simple or sparingly branched stem, 2–6 dm tall; leaves alternate, erect, linear; racemes dense, 1–4 cm long and 10–15 mm thick; flowers pale rose-purple; very rare on serpentine barrens; SE; flr. Jul–Aug; UPL; 🌿.

Polygala
paucifolia

Polygala
sanguinea

Polygala
verticillata var.
verticillata

Polygala lutea L. Yellow milkwort
Biennial or perennial frequently with clustered stems, 1–4 dm tall; leaves alternate, obovate or oblong-obovate; flowers orange to yellow, in a dense head-like raceme; very rare in moist sandy soil; SE; believed to be extirpated; flr. late Jun–Aug; FACW+; ✿.

Polygala nuttallii Torr. & A.Gray Nuttall's milkwort
Very similar to *P. curtissii* but the raceme only 5–6 mm thick; rare in open woods, peaty thickets, and sphagnum bogs; E; flr. Jul–Oct; FAC; ✿.

Polygala paucifolia Willd. Fringed milkwort, bird-on-the-wing
Rhizomatous, colonial perennial; stems 8–15 cm tall; leaves few, alternate, the upper 3–6 larger; flowers 1–4, rose-purple, 1.3–1.9 cm; common in rich rocky woods and wooded slopes; throughout, except in the westernmost counties; flr. May–Jun; FACU.

Polygala polygama Walter Racemed or bitter milkwort
Biennial with clustered stems 1–2.5 dm tall from a decumbent base; lower leaves spatulate or obovate, upper linear-oblanceolate; raceme loose and open, 2–10 cm; flowers rose-purple to white; rare in abandoned fields and wooded bogs; mostly SC; flr. Jun–Jul; UPL; ✿.

Polygala sanguinea L. Field or rose milkwort
Erect annual, 1–4 dm tall; stems simple or branched above; leaves alternate, linear to linear-oblong; raceme very dense and head-like, 1–4 cm by 6–14 mm thick; flowers rose-purple; frequent in open areas on moist, sterile, acidic soils; throughout; flr. late Jun–Sep; FACU.

Polygala senega L. Seneca snakeroot
Perennial with clustered, unbranched stems, 1–5 dm tall; leaves alternate, linear; racemes dense, 1.5–4 cm by 5–8 mm thick; flowers white; rare on rocky, wooded limestone slopes; scattered; flr. Jun–Jul; FACU; 2 varieties:

A. leaves 0.3–1.5 cm wide .... var. *senega*
A. leaves 1.5–3.5 cm wide .... var. *latifolia* Torr. & A.Gray

Polygala verticillata L. Whorled milkwort
Erect, branched annual, 1–4 dm tall; leaves linear, whorled below and alternate above; racemes cylindric-conical, continuous, to 4 cm long but appearing shorter because the lower flowers mature and fall before the uppermost have opened; flowers greenish-white or pinkish; flr. late Jun–early Oct; UPL; 3 varieties:

A. leaves mostly alternate ....var. *ambigua* (Nutt.) Wood
    occasional in moist meadows, marshes, and ravines; S.
A. leaves whorled at all the primary nodes
    B. pedicels 0.5–1 mm; sepals purplish....................................var. *verticillata*
        common in dry, open woods, old fields, and roadsides; throughout.
    B. pedicels 0.1–0.3 mm; sepals white or greenish............var. *isocycla* Fernald
        occasional in waste ground and along railroad tracks; S.

POLYGONACEAE Smartweed Family
Richard S. Mitchell

Perennial or annual herbs, vines, or shrubs with simple, alternate leaves; nodes often swollen and with tubular, funnelform, or oblique sheathing stipules (ocreae) and inflorescence bracts (ocreolae); flowers perfect or plants monoecious, dioecious, or polygamous; flowers borne in small fascicles at the nodes, or commonly in panicles that resemble racemes or spikes (rarely singly); flowers with a single superior ovary; petals absent; stamens 4–12, often 8; perianth (calyx) fused at base into a cup or tube, its lobes 4–6 in a single series, or in 2 whorls of 3, green or often petaloid and white to brightly colored; fruit an achene or berry.

A. achenes winged; plant robust, with a basal rosette of cordate-ovate leaves that are 15–40 cm broad, borne on fleshy, reddish petioles .... *Rheum*
A. achenes not winged (although the calyx surrounding them may be); plants herbaceous, vining, or large and bamboo-like, but not as above
  B. flowers consistently one per bract, reflexed downward on short pedicels; inflorescences wiry spikes on very slender, erect plants, with needle-like, jointed leaves that drop early .... *Polygonella*
  B. flowers 2 or more per bract, often in fascicles (sometimes drooping); plants various, but leaves mostly persisting through the season
    C. calyx of two distinct whorls of 3 lobes each; inner perianth lobes often greatly enlarging in fruit to become papery or leathery valves with wing-like (or spiny) margins; calyx valves sometimes with swollen basal grains .... *Rumex*
    C. calyx a single whorl with 5 lobes that usually retain their texture and color in fruit; lacking grains, but sometimes laterally winged
      D. achene not strongly exserted from the calyx; usually <5 mm long ....
      ....*Polygonum*
      D. achene often strongly exserted from the calyx; 5–8 mm long
        E. leaves broadest at the base, cordate to truncate; flowers in loose panicles .... *Fagopyrum*
        E. leaves narrowed at the base; flowers in spike-like racemes or dense head-like clusters .... *Polygonum*

Fagopyrum Mill.

Annual herbs with broadly triangular, petiolate leaves and oblique ocreae; flowers perfect, white or pink; tepals 5, fused toward the base; stamens 8; achene 3-angled, longer than the calyx when mature; flr. Jun–Sep.

A. flowers creamy to white, 3–4 mm long, borne in corymb-like panicles at or near the stem tips; achenes mostly 5–7 mm long .... *F. esculentum*
A. flowers greenish, 2–3 mm long, borne in raceme-like panicles, mostly in the leaf axils; achenes mostly 5.5 mm long or less .... *F. tataricum*

Fagopyrum esculentum Moench   Buckwheat
Erect to spreading annual herb from a slender taproot; leaves triangular-hastate on slender petioles; ocreae oblique, often split along one side; inflorescences

corymb-like, branched panicles; ocreolae not ciliate; flowers creamy to white; achene with 3 rounded angles, dark brown, 5–7 mm long, strongly exserted from the fruiting calyx; cultivated and escaping to roadsides, fields, and waste places; native to Asia. [*F. sagittatum* Hill of Rhoads and Klein, 1993]

Fagopyrum tataricum (L.) Gaertn.   India-wheat
Erect to spreading annual herb from a slender taproot; leaves broadly hastate; ocreae tubular to lacerate; inflorescences raceme-like panicles of fascicles borne mostly in the upper leaf axils; ocreolae not ciliate; flowers greenish; achene 3-angled, exserted from the fruiting calyx, dull brown, and about 5 mm long; sometimes cultivated and rarely escaped; native to Eurasia.

Polygonella Michx.

Polygonella articulata

Polygonella articulata (L.) Meisn.   Jointweed
Erect annual with wiry stems from a small taproot; leaves linear and dropping early; inflorescences terminal and subterminal from upper nodes, extremely slender, with overlapping floral bracts and a jointed appearance; flowers white or pink-tinged with 5 subequal lobes, borne singly in the axils of the bracts and pendulous on short, strongly reflexed pedicels; fruit a 3-angled achene; very rare in sandy, open places; SE; flr. late Aug–Oct; ✿.

Polygonum L.

Annual or perennial herbs with alternate, simple, entire, leaves and well-developed ocreae; inflorescences terminal or axillary; flowers borne in small clusters, spike-like racemes, or rarely, dense heads; flowers perfect or unisexual, in fascicles subtended by sheathing ocreolae; tepals mostly 5, fused at the base; stamens 3–8; fruit a lens-shaped or 3-angled achene enclosed by the persistent perianth, or a fleshy berry (in *P. perfoliatum*); most species flower continuously through the summer and early fall.

A. stems and petioles with sharp spines or recurved prickles
    B. fruit a fleshy, blue berry; ocreae conspicuous, spreading funnel-like collars .... *P. perfoliatum*
    B. fruit a hard-walled achene; ocreae inconspicuous, sheathing, narrowly funnelform or oblique
        C. leaves hastate; flowers in small racemes; achene lens-shaped .... *P. arifolium*
        C. leaves sagittate; flowers in dense heads; achene 3-angled .... *P. sagittatum*
A. stems and petioles lacking sharp spines or prickles (soft bristles are present at the nodes in *P. cilinode*)
    D. vines or bamboo-like shrubs; fruiting calyx winged or keeled
        E. plants erect, bamboo-like and somewhat woody
            F. leaves truncate at the base, with cuspidate tips, mostly <10 cm long .... *P. cuspidatum*
            F. leaves cordate at the base, gradually tapered toward acuminate tips, often 12–20 cm long ....*P. sachalinense*
        E. plants vine-like; climbing or trailing
        G. calyx lobes strongly winged in fruit

        H. inflorescences pale throughout, the creamy ocreolae caudate-tipped .... *P. aubertii*

        H. inflorescences with green axes and greenish to scarious-brown ocreolae, their margins truncate to oblique, not caudate .... *P. scandens*

    G. calyx lobes keeled, but not strongly winged

        I. nodes with soft, reflexed bristles just below the stipules; styles 3, not fused .... *P. cilinode*

        I. nodes lacking bristles below the stipules; styles fused to near their tips .... *P. convolvulus*

D. annual or perennial herbs; fruiting calyx not winged or keeled

  J. leaf blade with 2 obvious, longitudinal fold marks paralleling the midrib ....*P. tenue*

  J. leaf blade not creased

    K. inflorescences dense terminal heads, each flower subtended by a long bract .... *P. nepalense*

    K. inflorescences spike-like racemes or axillary clusters

      L. flowers borne in loose fascicles in the leaf axils, scattered at the nodes of the main stems or along short shoots throughout the upper and middle portions of the plant

        M. leaves of mature plants of 2 sizes, those of the axillary short shoots much reduced

          N. larger leaves narrowly lanceolate, mostly 8–15 times longer than broad; achenes often strongly exserted from the fruiting calyx ....*P. ramosissimum*

          N. larger leaves oval to broadly lanceolate, mostly 2–7 times longer than broad; achenes usually totally enclosed in the fruiting calyx

            O. leaves of the main stems elliptic-ovate with rounded tips, mostly 1–2 cm broad; foliage and flowers yellow-green, the calyx lobes conspicuously boat-shaped at the tips ....*P. erectum*

            O. leaves of the main stems ovate-lanceolate, often with acute tips, mostly 0.5–1 cm broad, gray-green; flowers white (pink- or green-tinged); calyx lobes not boat-shaped ....*P. aviculare*

        M. leaves of mature plants relatively uniform in size, often gradually reduced upward on the stem

          P. leaves narrowly lanceolate to linear, pale green, sometimes reddish or yellowish; plants wiry, erect-ascending ....*P. bellardii*

          P. leaves oval to narrowly oblong, dark green or bluish to gray-green; young shoots sometimes ascending as spring or winter sprouts, but mature plants mat-forming, mounding, or sprawling

            Q. calyx lobes divided only about a third of the flower length, narrowly boat-shaped, constricted to form a neck above the achene .... *P. achoreum*

            Q. calyx lobes divided half the flower length or more, not constricted into a neck above the achene

R. calyx tube relatively symmetrical, cup-like, lacking veiny pouches, the lobes with rounded tips, not conspicuously boat-shaped; a common, mound-forming weed .... *P. arenastrum*

R. calyx somewhat asymmetrical, with heavily-veined, lateral pouches, the lobes boat-shaped at the tips; plants creeping, usually on gravelly shores, rare .... *P. buxiforme*

L. flowers borne in spikes at the branch tips (inflorescence usually leafless)

S. styles conspicuous, thick, unfused, hooked at their tips, forming a springy dispersal device; flowers reflexed, borne in remote clusters of 2–3 on a long, arching spike .... *P. virginianum*

S. styles slender, fleshy, and often fused at base; flowers in contiguous or interrupted spikes

T. ocreae of the middle and upper leaves with a ring of bristles along the upper margin

U. perennials

V. flowers dotted with numerous golden glands (that are darker and more obvious on drying)

W. ocreolae almost totally lacking cilia at margins; robust plants with dense, contiguous spikes; leaves mostly 2–4 cm broad..... ....*P. robustius*

W. ocreolae with ciliate margins; slender plants, often with interrupted spikes; leaves rarely >2.5 cm broad .... *P. punctatum*

V. flowers lacking golden, punctate glands (very rarely with a few pale, discoid structures)

X. achenes 3-angled

Y. flowers creamy (to greenish); spikes often clustered at the stem tips .... *P. setaceum*

Y. flowers pink (rarely white); spikes usually single or paired .... *P. hydropiperoides*

X. achenes lens-shaped .... *P. amphibium*

U. annuals

Z. flowers with golden, glandular dots (that darken and are more obvious on drying)

AA. achene surfaces dull brown, powdery; ocreolae mostly overlapping .... *P. hydropiper*

AA. achene surfaces shiny, dark brown to black; ocreolae bracts often remote .... *P. punctatum*

Z. flowers without glandular dots

BB. leaves ovate to broadly lanceolate, the larger ones 5–10 cm broad, cordate .... *P. orientale*

BB. leaves ovate-lanceolate to rhombic, rarely over 4 cm broad, not cordate

CC. upper stems covered with stalked glands (well below peduncles); achenes consistently lens-shaped; spikes somewhat curved or lax and drooping .... *P. careyi*

CC. upper stems lacking stalked glands; achenes 3-angled (or both types); spikes not lax

DD. bristles of the ocreolae 2–4(5) mm long often extending beyond the tips of the flowers .... *P. caespitosum*

DD. bristles of the ocreolae short or lacking .... *P. persicaria*

T. ocreae entire or shattering, lacking bristles along the upper margin
   EE. perennial plants with rhizomes or stolons that root at the nodes; floating
     aquatics or large emergents with solitary (or paired) terminal spikes ....
     .... *P. amphibium*
      FF. spikes solitary or rarely paired; perianth pink or reddish ....
      .... *P. amphibium*
      FF. spikes several; perianth white .... *P. densiflorum*
   EE. annual plants with taproots, usually bearing many spikes at the tips of
     lateral branches
      GG. calyx with conspicuous, raised, anchor-shaped veins, the lobes
       greenish to pink-tinged with a silvery sheen; spike often arched or
       drooping; ocreolae and ocreae obliquely angled at the top ....
       .... *P. lapathifolium*
      GG. calyx without raised, anchor-shaped veins, the lobes pale pink to
       hot rose; spikes stout, not lax; ocreolae and ocreae truncate ....
       .... *P. pensylvanicum*

Polygonum achoreum S.F.Blake   Homeless knotweed
Prostrate, mat-forming annual with oval to elliptic, bluish-green leaves; ocreae
4–10 mm, becoming brown and ragged; flowers in axillary fascicles; perianth
bluish or gray-green, fused for the basal two-thirds and constricted above the
achene; rare in weedy locations; SE; FACU.

Polygonum amphibium L.   Water smartweed
Robust perennial with oval to lanceolate leaves; stems erect-ascending or lax and
trailing on water; rhizomes and stolons stout, rooting at the nodes; plants functionally dioecious, with pink to rose-red flowers borne in a single terminal spike;
achenes lens-shaped; lakes, shores and ditches; throughout; 2 varieties (and intermediates):

A. leaves broadly lanceolate-acuminate, not floating; spikes narrowly cylindric,
   4–15 cm long, with glandular peduncles .... var. *emersum* Michx.
   emergent aquatic or terrestrial plant; frequent throughout; OBL.
A. floating leaves present, ovate to narrowly elliptic; spikes stout, 1–4 cm long
   with glabrous peduncles .... var. *stipulaceum* Coleman;
   rare stoloniferous aquatic; not usually flowering; mostly N; OBL; 🍂.

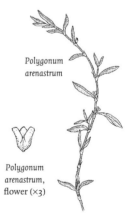

*Polygonum
arenastrum*

*Polygonum
arenastrum,*
flower (×3)

Polygonum arenastrum Jord. ex Boreau   Doorweed, knotgrass
Caespitose, much-branched annual rosettes from a slender taproot; leaves subequal, elliptic to oblong, bluish- to gray-green; flowers tiny, white (green- or red-
tinged), abundant in small fascicles at the nodes; a common weed of sunny, disturbed ground, especially in urban settings; SE; native to Eurasia.

Polygonum arifolium L.   Halberd-leaf tearthumb
Vine with hastate leaves on well-developed petioles; stems, petioles and peduncles armed with reflexed prickles; inflorescences small, terminal, and axillary, stalked racemes of white to greenish-pink flowers; common in swamps,
wet meadows, and marshes; throughout; OBL.

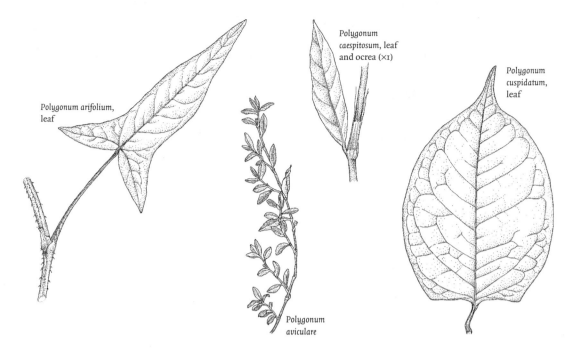

Polygonum arifolium, leaf

Polygonum caespitosum, leaf and ocrea (×1)

Polygonum cuspidatum, leaf

Polygonum aviculare

Polygonum aubertii L.Henry   Silver-lace vine

Vine with oblong to truncate-sagittate leaves; petioles and stems slender, lacking prickles; inflorescences profusely branched panicles, bearing many hundreds of flowers, pale throughout, including the central axis and small, caudate
bracts; fruiting calyx cream-colored, club-shaped, with well-developed wings;
cultivated and infrequently naturalized on roadsides; native to Asia.

Polygonum aviculare, flower (×3)

Polygonum aviculare L.   Knotweed

Slender annual, moderately branched, ascending from a wiry taproot; leaves
gray-green, ovate to lanceolate, of two sizes on mature plant, those of the slender, axillary short shoots much reduced; flowers tiny, white (green- or red-
tinged), abundant in small fascicles at the nodes, especially on the short shoots
in late summer; a common weed of fields and roadsides, especially in rural areas; native to Europe; FACU.

Polygonum bellardii All.   Needle-leaf knotweed

Strongly ascending, wiry annual from a slender taproot; leaves subequal, narrowly lance-elliptic to linear, yellow-green or reddish-tinged; flowers white or
red-tinged, in small fascicles at the nodes; achenes flattened, 3-angled; occasional in disturbed, sandy, or gravelly areas; native to Eurasia. [*P. neglectum* Bes-
ser of Rhoads and Klein, 1993]

Polygonum buxiforme Small   Knotweed

Annual with tough, wiry, depressed stems branching laterally from the apex of a
tough, twisted taproot; leaves subequal, gray-green, oblong, somewhat fleshy;
flowers white (green- or red-tinged), veiny, and pouched, borne sparingly in fascicles at the nodes; rare on gravelly shores, ballast, and disturbed places; SE.

Polygonum caespitosum Blume   Low smartweed

Slender annual with weakly ascending stems; leaves rhombic, subsessile on very short petioles; stipules and bracts tubular with extremely long bristles on the margins; spikes terminal, slender, of pink to rose-purple, nonglandular flowers; achenes 3-angled; common in woods, fields, and waste ground; throughout; native to Asia; FACU. Ours is predominately var. *longisetum* (DeBruyn) Stewart, a single Pennsylvania specimen is apparently var. *caespitosum*, with red flowers and shorter bristles.

Polygonum careyi Olney   Pinkweed, smartweed
Ascending, robust, annual from a taproot; leaves lanceolate; stipules tubular with bristled margins; inflorescence a lax or slightly drooping spike of pink to dark rose flowers with ciliate bracts and densely glandular peduncles; upper stems also densely glandular; achenes lens-shaped; a rare, scattered native of sandy, open woodlands and disturbed places, particularly after fire; FACW; 🌿.

Polygonum cilinode Michx.   Fringed bindweed
Vine with ovate-cordate leaves, the basal lobes somewhat truncate; stems often reddish with oblique, sheathing stipules and a fringe of reflexed bristles at the base of each node; inflorescences slender panicles of greenish-cream to white flowers borne in leaf axils near the stem apex; calyx lobes obscurely keeled but not winged; achene 3-angled; frequent in rocky woods, thickets and stream courses; throughout.

Polygonum convolvulus L.   Black bindweed, nimble-will
Vine with ovate-cordate to sagittate leaves; stems often reddish, with oblique ocreae; nodes lacking reflexed bristles; slender, few-flowered panicles borne in leaf axils near plant tips; flowers creamy, the fruiting calyx obscurely keeled but not winged; achenes 3-angled; a common weed in sunny lots and fallow fields; throughout, especially thriving around human disturbance; native to Europe; FACU.

Polygonum cuspidatum Siebold & Zucc.   Japanese knotweed, Mexican bamboo
Perennial, arching, bamboo-like shrubs to 3 m tall, with strongly jointed zigzag stems from thick, spreading rhizomes; leaves ovate, up to 15 cm, with truncate bases and cuspidate tips; inflorescences copious, slender-branched panicles in the axils of upper and middle leaves; calyx lobes winged, greenish-white; achenes 3-angled; an aggressive weed of river and stream banks, railroad rights-of-way, and waste ground; throughout; native to Japan; FACU–.

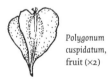

Polygonum cuspidatum, fruit (×2)

Polygonum densiflorum Meissner   Smartweed
Rhizomatous perennial to 1.5 m tall; stems rooting at the lower nodes; leaves lanceolate, 10–25 cm long; ocrea smooth, entire; spikes to 10 cm, slender; flowers overlapping; perianth white; rare in swamps and shallow water; SE; OBL.

Polygonum erectum L.   Erect knotweed
Bushy annual from a taproot; leaves of two sizes, those of the lateral short shoots much reduced; larger stem leaves elliptic-ovate up to 6 by 3 cm; foliage and flowers bright-green to yellowish; flower fascicles borne mostly in the axils of small leaves along short shoots; lobes half the calyx length or more, not constricted into a neck above; achene 3-angled; occasional on river banks and moist slopes; mostly S, once frequent, but seriously declining; FACU.

Polygonum hydropiper, flowers (×2)

Polygonum hydropiper L.   Smartweed, water-pepper

Annual with reddish branches, erect-ascending from a taproot; leaves lanceolate; ocreae bristle-fringed and pubescent; spikes slender, often interrupted and sometimes with a few reduced leaves; fruiting calyx creamy, greenish to reddish or bronze-tinged, copiously punctate with golden-brown glands; achenes both 3-angled and lens-shaped; common in wet meadows, swamps, stream margins and other moist open places; native to Europe; OBL.

Polygonum hydropiperoides Michx.   Mild water-pepper, water smartweed

Perennial with ascending to erect stems from stolons and rhizomes; leaves broadly to narrowly lanceolate; ocreae bristle-fringed; inflorescences terminal spikes with fringed ocreolae; calyx lacking golden-brown punctate glands, but rarely with a few pale, flat glands; achenes dark, lustrous, sharply 3-angled; 2 varieties:

A. undersurfaces of leaves lacking glands; fruiting calyx pale pink, almost always enclosing the achene .... var. *hydropiperoides* common on lake shores and streams; throughout; OBL.

A. undersurfaces of leaves with numerous pale, flat glands (sometimes also present on flowers); fruiting calyx greenish-white (or red-tinged), the lobes often incurved to expose the achene at maturity.. .... var. *opelousanum* (Ridd.) Stone very rare in coastal wetlands; OBL.

Polygonum lapathifolium L.   Willow-weed, dock-leaf smartweed

Polygonum lapathifolium, flower (×3)

Stout, much-branched annual from a taproot; leaves broadly to narrowly lanceolate, often with a central purple spot; ocreae and ocreolae oblique, lacking bristles at margins; spikes often lax and drooping; flowers greenish or pink-tinged with a silvery sheen, lacking glands; calyx lobes of most flowers with prominent, raised, anchor-shaped veins; achene lens-shaped with one indented side; a common weed of streets, roadsides, waste places, open woods, and clearings; throughout; native to Europe; FACW+.

Polygonum nepalense Meisn.   Knotweed

Lax annual to 4 dm tall, leaves ovate or deltoid, abruptly narrowed to the winged petiole, which is often expanded and clasping at the base; ocreae bristly at the base; peduncles stipitate-glandular; inflorescence head-like; flowers greenish-white or pinkish with bracts equaling or exceeding the flowers; achenes lens-shaped; rare and scattered on river banks and urban waste ground; native to Asia.

Polygonum orientale L.   Prince's feather, kiss-me-over-the-garden-gate

Tall, erect annual from a thick taproot; leaves broadly ovate, large, up to 15 by 25 cm with acuminate tips; ocreae with marginal bristles, sometimes flared at the margin; spikes terminal, stiff to somewhat drooping; ocreolae fringed; flowers pink to deep rose, the lobes not glandular; achenes lens-shaped, indented on one or both sides; escaped from cultivation into waste places and vacant lots; throughout; native to India; FACU–.

Polygonum pensylvanicum, flower (×3)

Polygonum pensylvanicum L.   Pinkweed, smartweed

Coarse, erect-ascending annual from a stout taproot; leaves broadly to narrowly

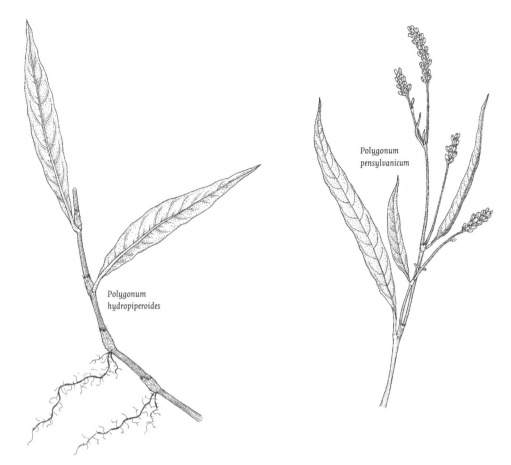

Polygonum pensylvanicum

Polygonum hydropiperoides

lanceolate; ocreae lacking marginal bristles; spikes thick to narrowly cylindric, not drooping, terminal and in upper leaf axils on glandular (rarely glabrous) peduncles; ocreolae oblique, their margins glabrous or ciliate; fruiting calyx pink, lacking glands, enclosing the fruit; achenes lens-shaped, somewhat flattened and indented on one or both sides; common in seasonally moist ditches, meadows, fields and waste places; throughout; FACW.

Polygonum perfoliatum L.   Mile-a-minute weed
Slender, annual vine with reflexed prickles and barbs; leaves roughly triangular with barbed midribs; ocreae conspicuous, shallow, saucer-like collars at the nodes; inflorescences axillary or terminal panicles with discoid ocreolae similar to the ocreae; flowers creamy-white, the lobes of the calyx wingless; fruit a small, juicy blue berry; climbing or sprawling in thickets, open woodlands, meadows, fields, and roadsides; S; native to Asia; designated as a noxious weed in Pennsylvania; FAC.

Polygonum persicaria L.   Lady's-thumb, heart's-ease
Erect-ascending, much-branched annual from a taproot; leaves rhombic to lanceolate, often with a dark purple central spot; ocreae truncate, with a few bristles at the margin; spikes numerous at branch tips, short- to long-cylindric, usually not drooping; ocreolae with weakly ciliate margins; calyx lobes pink (or greenish-tinged), lacking prominent, raised veins; achenes both 3-angled and lens-shaped; a ubiquitous weed of open places; native to Europe; FACW.

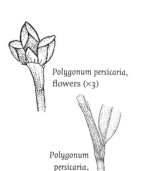

Polygonum persicaria, flowers (×3)

Polygonum persicaria, ocrea (×1)

Polygonum punctatum Elliot   Dotted or water smartweed
Annual or perennial; leaves lance-ovate to rhombic; ocreae with bristles at the margin; inflorescences cylindric or much-interrupted, slender spikes borne at branch tips or from leaf axils; ocreolae with small bristles at the margin; flowers white (greenish) with golden, punctate glands (drying brown); achenes both lens-shaped and 3-angled; a common native of muddy shores and shallow water; throughout; OBL 2 varieties:

A. annual from a small taproot or weakly rooted at a few nodes; spikes very slender and interrupted, terminal and axillary, sometimes with reduced leaves interspersed .... var. *confertiflorum* (Meisn.) Fassett
A. perennial from a rhizome; spikes cylindric, leafless, mostly borne at branch tips .... var. *punctatum*

Polygonum ramosissimum Michx.   Knotweed
Slender, erect-ascending annual from a taproot; leaves narrowly lanceolate, of two sizes, those of stem tips and short-shoots much-reduced; flowers in small fascicles, borne mostly in the axils of the small upper leaves and lateral short-shoots; fruiting calyx yellow-green to brownish, deeply divided, often exposing the achene, bearing mostly 3-angled fruit, but sometimes with paler, very large, lens-shaped achenes; rare in coastal ballast, shores, and waste places; native to western North America; FAC; 🌱.

Polygonum robustius (Small) Fernald   Large dotted or water smartweed
Robust perennial from stout stolons and rhizomes; leaves broadly lanceolate; ocreae with stiff, marginal bristles; spikes cylindric, uninterrupted, borne at or just below shoot tips; ocreolae almost totally lacking marginal cilia; flowers white (greenish) with numerous golden, punctate glands that dry brown; achenes 3-angled (very rarely lens-shaped); rare in swamps, lake shores, and streams; SE and NW; OBL.

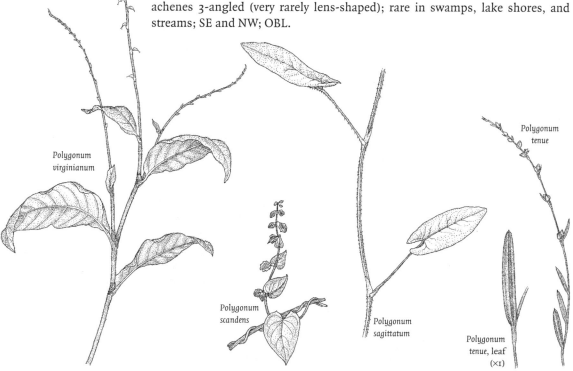

Polygonum virginianum

Polygonum scandens

Polygonum sagittatum

Polygonum tenue

Polygonum tenue, leaf (×1)

Polygonum sachalinense F.W.Schmidt ex Maxim.   Giant knotweed
Giant perennial, bamboo-like shrub, to 4 m tall; stems jointed, zigzag branching; leaves cordate, to 20 by 30 cm, with acute to acuminate tips; panicles borne
in profusion at the middle and upper nodes; flowers greenish-creamy, the calyx
lobes strongly winged; achenes 3-angled; escaped from cultivation to disturbed
sites such as railroad embankments, ditches, old lots, and roadsides; native to
Asia; UPL.

Polygonum sagittatum L.   Tearthumb, scratch-grass
Sprawling, vine-like annual with recurved prickles on the stems and leaves;
leaves narrowly sagittate; inflorescences long-peduncled, terminal or axillary;
flowers borne in condensed heads, calyx green and pink to red or whitish, not
winged, enclosing the 3-angled achene; common in bogs, marshes, and wet
meadows; throughout; OBL.

Polygonum scandens L.   Climbing false buckwheat
Climbing or sprawling vine; leaves ovate to hastate with cordate to truncate
bases; inflorescences racemes, borne terminally and on short shoots in leaf axils; flowers greenish to creamy-white; calyx lobes strongly winged; achenes
sharply 3-angled; woods, thickets, fencerows, and waste places; throughout; 3
varieties:

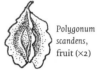

*Polygonum
scandens,
fruit (×2)*

A. calyx 10–15 mm long with flat wings .... var. *scandens*
    common throughout; FAC.
A. calyx 7–10 mm long
    B. calyx weakly or only partially winged ....
        .... var. *cristatum* (Engelm. & A.Gray) Gleason
        SE.
    B. calyx with flat, well-developed wings .... var. *dumetorum* (L.) Gleason
        rare and scattered; native to Europe.

Polygonum setaceum Baldwin ex Elliot   Swamp smartweed
Robust, erect-ascending perennial from stolons and rhizomes; leaves broadly
lanceolate; ocreae strigose, with stiff, marginal bristles; spikes usually several at
the stem apex, with strongly ciliate ocreolae; flowers creamy (not white or pink),
lacking glands, the fruiting calyx enclosing the 3-angled achene; rare in swamp
forest margins, shores, and shallow water; NW and SE; OBL; 🍂. Ours is var. *inter-
jectum* Fernald.

Polygonum tenue Michx.   Slender knotweed
Wiry, erect, annual from a taproot; leaves lance-linear, with a longitudinal fold
on each side of the midrib; flowers in slender, much-interrupted, leafy spikes at
the branch tips; fruiting calyx deeply cut, greenish-creamy to tan, enclosing the
3-angled achene; occasional on dry, rocky hillsides and serpentine and shale
barrens; E and SC.

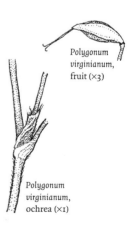

*Polygonum
virginianum,
fruit (×3)*

*Polygonum
virginianum,
ochrea (×1)*

Polygonum virginianum L.   Jumpseed
Robust, erect-ascending perennial from knotty rhizomes; leaves ovate to elliptic-
lanceolate, to 7 by 15 cm; ocreae with marginal bristles; inflorescences terminal
and subterminal, with fascicles of 2–3 flowers scattered remotely along long,
slender spikes; flowers reflexed on short pedicels; calyx greenish to white or

pink tinged; achene lens-shaped, exserted from the calyx, bearing two springy, persistent, hooked styles that serve as a dispersal device; common in moist woodlands, floodplains, and clearings; throughout; FAC.

Rheum L.

Rheum rhababarum L.   Rhubarb
Robust, perennial herb, to 1 m tall or more; central flowering stalk from a stout rootstock with a basal rosette of crinkled deltoid to cordate leaves on stout, fleshy petioles; stalks often striped or blushed with red; inflorescence a terminal, open panicle of greenish-pink flowers; achenes exserted, 8–10 mm long, brown with pale wings; commonly cultivated and occasionally escaping to open, disturbed ground; flr. May; native to Europe. [R. rhaponticum L. of Rhoads and Klein, 1993]

Rumex L.

Perennial (or annual) herbs with alternate and/or basal leaves, ocreae, and erect flowering stems; flowers perfect or unisexual; inflorescence a terminal panicle of whorled fascicles of small flowers; perianth of 2 whorls of 3 tepals each; stamens 6; ovary superior, 3-lobed with 3 styles; fruit an achene closely enclosed by the 3 inner tepals or valves, 1 or more of which usually bears a swollen grain at the base of the midrib; flr. early to midsummer, frt. midsummer to fall.

A. leaves hastate or sagittate, to spatulate or marginally lobed; dioecious
    B. fruiting calyx cap-like, closely conforming to the achene, lacking wing-like valves .... R. acetosella
    B. fruiting calyx with veiny, wing-like valves
        C. leaves primarily clustered in a basal rosette; calyx valves lacking basal grains .... R. hastatulus
        C. leaves mostly borne on the stem; some mature calyx valves with small basal grains .... R. acetosa
A. leaves not hastate, sagittate or marginally lobed; plants with bisexual flowers or unisexual and the plants monoecious
    D. valve margins of the fruiting calyx with deeply cut teeth and/or spines
        E. only 1 calyx valve per flower bearing a grain; valve margins mostly toothed, with a few small spines
            F. largest basal leaves fiddle-shaped, often conspicuously larger than the rest; pedicels jointed near the middle; flower clusters usually distant on the spike.. R. pulcher
            F. largest basal leaves ovoid to cordate, not fiddle-shaped; pedicels jointed obscurely near the base; flower clusters frequently contiguous .... R. obtusifolius
        E. all 3 calyx valves with basal grains; valve margins with conspicuous, slender spines... R. maritimus
    D. valve margins of the fruiting calyx with smooth or wavy margins, or with very small teeth, but not spiny
        G. main stem with well-developed lateral branches, some of which may flower and fruit; plants often with a pale, glaucous or olive sheen
            H. pedicels several times the length of the fruiting calyx, straight; dangling from their immediate bases at maturity to form skirt-like fringes of fruit below the nodes .... R. verticillatus

H. pedicels shorter, from one half to twice the length of the fruiting calyx, often curved; not dangling by their bases in fruit
    I. calyx valves deltoid, truncate at the base, all 3 usually bearing a basal grain .... *R. salicifolius*
    I. calyx valves ovate, tapered toward the base; 1 or 2 bearing a grain .... *R. altissimus*
G. main stem lacking well-developed lateral branches; plants usually dark green (or infused with red)
    J. leaf margins and blades strongly crisped and convoluted; fruiting pedicels jointed near the middle .... *R. crispus*
    J. leaf margins and blades not strongly crisped; fruiting pedicels jointed near the base
        K. all 3 calyx valves bearing basal grains; an emergent aquatic .... *R. orbiculatus*
        K. only 1 or 2 calyx valves bearing basal grains; a plant of dry sites.... .... *R. patientia*

Rumex acetosa L.   Garden sorrel
Dioecious perennial from slender rhizomes; leaves mostly cauline, sagittate, borne on long petioles; stem erect with a terminal panicle of many small flowers; fruiting calyx with 3 reddish, wing-like valves, only one of which usually develops a starchy grain; cultivated and occasionally escaped to fallow fields and waste ground; SE; native to Europe; FACU.

Rumex acetosella L.   Sheep sorrel, sourgrass
Small, dioecious perennial from slender rhizomes; leaves borne primarily in basal rosettes, hastate to linear-spatulate; stem lax to erect, with a slender, terminal panicle of many small, reddish-green flowers, the achene closely invested in a wingless calyx; a ubiquitous weed of pastures, lawns, roadsides, and barrens; throughout; native to Eurasia; UPL.

Rumex altissimus A.W.Wood   Pale, tall or peach-leaf dock
Erect, branching perennial to 1.5 m tall, with ovate to oblanceolate, pale green leaves on short petioles; inflorescence a moderately dense, terminal panicle; flowers bisexual; fruiting calyx with brownish, ovate to cordate valves, only one of which usually bears a starchy grain; frequent in rich, alluvial soils of river banks and wetland margins; throughout; FACW–.

Rumex crispus L.   Curly dock
Erect perennial to 1 m tall with a basal rosette, the strong central axis also with petioled, narrowly oblong, contorted leaves with strongly crisped margins; inflorescence a moderately dense, terminal panicle; flowers bisexual; fruiting calyx with 3 light brown to ebony, broadly ovate valves, all of which usually bear a starchy basal grain; a common weed of roadsides, waste places, and fields; throughout; native to Europe; FACU.

Rumex hastatulus Baldwin ex Elliot   Heart or red sorrel
Slender, dioecious perennial with erect stems and a taproot; leaves both cauline and in a basal rosette, often narrowly hastate; inflorescence a slender, terminal panicle of small, pale flowers; fruiting calyx with 3 ovate valves that do not develop starchy basal grains; rare in meadows; SE; FACU–; ✿.

Rumex acetosella

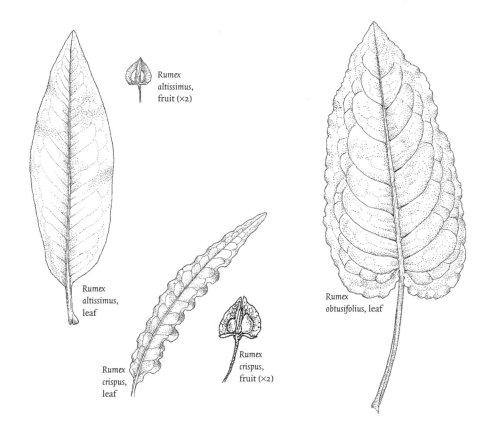

Rumex
altissimus,
fruit (×2)

Rumex
altissimus,
leaf

Rumex
crispus,
leaf

Rumex
crispus,
fruit (×2)

Rumex
obtusifolius, leaf

Rumex
obtusifolius,
fruit (×2)

Rumex maritimus L.   Golden dock

Small, bushy annual or biennial with linear to lance-elliptic, short-petioled leaves; flowers bisexual, small and yellow-green, borne in dense, globose clusters of fascicles at the nodes; fruiting calyx with 3 golden-brown valves, each bearing a conspicuous basal grain and 4–6 spiny teeth at the margin; rare in sandy and gravelly waste ground; SE; FACW. Ours is var. *fueginus* (Phil.) Dusen.

Rumex obtusifolius L.   Bitter dock

Leathery perennial from tough rootstock; main stem erect-ascending to 1.2 m tall, subtended by a rosette, but also bearing large oblong to lanceolate leaves upward; leaves dark green, infused with red along the veins, the upper ones crisped; flowers borne in fascicles in relatively compact panicles near the plant apex; fruiting calyx with golden to dark red-brown valves, at least one of which bears a plump basal grain; valve margins with well-developed teeth and small spines; common in fields, woods and roadsides; native to Europe; FACU–.

Rumex orbiculatus A.Gray   Great water-dock

Erect, monoecious plant to 2.5 m tall from a stout, perennial rootstock, bearing large, oblong-lanceolate leaves that are progressively smaller and narrower upward; inflorescence an elongate panicle of small fascicles; fruiting calyx with veiny reddish-brown valves that are orbicular and somewhat truncated at the base, each with a narrow basal grain; occasional on wet shores, or emergent from shallow water in rivers and along swampy banks; mostly NW, scattered elsewhere; OBL.

Rumex patientia L.   Monk's rhubarb
Robust plant to 3 m tall, the stem arising from a stout perennial rootstock; leaves large (to 3.5 × 1.5 dm), oblong, the margins generally not crisped; inflorescence a stout, terminal panicle; flowers bisexual; fruiting calyx with tan to dark brown oval valves, only one usually bearing a small, starchy grain near its base; cultivated and occasionally escaped to fields or waste ground; native to Europe.

Rumex pulcher L.   Fiddle-dock
Leathery plant from a tough perennial rootstock; the main stem erect, 2–8 dm tall, with relatively small leaves compared to the basal rosette, the larger leaves of which are fiddle-shaped, red-infused, and somewhat crisped; inflorescence of spike-like panicles of remote fascicles; valves of the fruiting calyx ovate with truncate bases and toothed margins, brown and veiny, each with a basal grain; very rare in waste ground; SE; native to Europe; FACW–.

Rumex salicifolius Weinm.   Willow-leaf dock
Erect-ascending monoecious perennial to 1 m tall; the main stem branching, sometimes with several elongate panicles at the branch tips; leaves lanceolate, gray-green, mostly cauline; fruiting calyx with deltoid valves, rusty brown with smooth margins (or tiny teeth), each valve usually bearing an elongate basal grain; occasional in waste dumps, shores, and ballast; scattered; native farther west; FACU. Ours is var. *mexicanus* (Meisn.) Hitchc. [R. *triangulivalvis* (Danser) Rech.f. of Rhoads and Klein, 1993]

Rumex verticillatus L.   Swamp dock
Robust, tap-rooted perennial to 1.5 m tall; leaves flat, lanceolate, or lance linear, narrowed to the base; inflorescence of whorls of flowers (or fruits) on long, slender dangling pedicels; valves broadly triangular, as wide as long, obtuse, truncate at the base, all 3 bearing basal grains; rare in swamps, floodplains, and swales in wet soil or shallow water; mostly NW; OBL.

PORTULACACEAE Purslane Family

Herbaceous, often succulent annuals or perennials; flowers perfect, regular; sepals 2; petals 5; stamens 3–many, if as many as the petals, then opposite the petals; fruit a capsule, dehiscent by longitudinal valves or opening transversely, usually many-seeded.

A. stamens 3 or 5; leaves opposite
    B. cauline leaves 2, often arising from the stem at ground level and appearing basal .... *Claytonia*
    B. cauline leaves more than 2, appearing along stem .... *Montia*
A. stamens 6 or more; leaves alternate or crowded near base of plant
    C. leaves crowded near the base of the plant, more or less rounded in cross section .... *Talinum*
    C. leaves more or less evenly distributed along the stems, not rounded in cross section .... *Portulaca*

Claytonia L.

Perennial herbs arising from rounded tubers; one pair of opposite cauline leaves arising below the inflorescence; petals 5, white or white tinged with pink; stamens 5 and opposite the petals; inflorescence a loose raceme; fruit a 3-valved capsule with 3–6 seeds, opening by inrolling; flr. late Mar–early May.

A. leaves at least 6 times as long as wide, scarcely or not at all tapering to a petiole ....*C. virginica*

A. leaves 3–6 times as long as wide, clearly tapering to a petiole .... *C. caroliniana*

Claytonia caroliniana Michx.　　　　　　　　　　　Carolina spring-beauty
Leaves mostly 3–6 cm long and 10–15 mm wide, narrowly ovate or oblanceolate, blade narrowing to a distinct petiole; raceme with 2–11 flowers, the bract subtending the lowest pedicel scarious; occasional on moist, rocky, wooded slopes; mostly N and W; FACU.

Claytonia virginica L.　　　　　　　　　　　　　　Spring-beauty
Leaves mostly more than 7 cm long and 2–10 mm wide, linear to linear-lanceolate, blade not narrowing significantly at base; raceme with 5–19 flowers, the bract subtending the lowest pedicel leaf-like; common in moist woods and meadows, frequently on alluvial soils; throughout; FAC.

Montia L.

Montia chamissoi (Ledeb. ex Spreng.) Greene　　　Chamisso's miner's-lettuce
Creeping or ascending perennial, sending out long slender runners; leaves spatulate to obovate, 2–5 cm long; racemes axillary or terminal, 2–7-flowered or flowers sometimes solitary, petals pink, sepals obovate, about 3 mm long; very rare on moist, rocky ledges and river banks; extreme NE; flr. Jun–Jul; 🌿.

Montia chamissoi, flower (×1½)

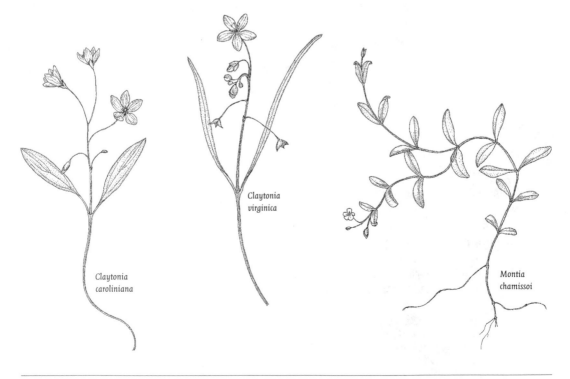

Claytonia caroliniana

Claytonia virginica

Montia chamissoi

Portulaca L.

Succulent, much-branched annual herbs, often forming extensive mats; leaves alternate, the upper ones crowded near the inflorescence; petals 5; stamens 8–many; fruit a capsule, opening transversely, many-seeded.

A. leaves spatulate to obovate, flattened; plants entirely glabrous .... *P. oleracea*

A. leaves linear, more or less rounded in cross section; stems densely hairy at the nodes .... *P. grandiflora*

Portulaca grandiflora Hook.   Moss-rose

Flowers 2–4 cm wide, white, yellow, pink, and various shades of red; otherwise with the characteristics of the genus; cultivated and occasionally lingering in abandoned gardens or rubbish dumps; mostly SE and SW, scattered elsewhere; flr. summer; native to Argentina.

Portulaca oleracea L.   Purslane

Flowers 5–10 mm wide, sessile, yellow; otherwise with the characteristics of the genus; a common weed of gardens and fields; throughout; flr. summer; FAC.

Talinum Adans.

Talinum teretifolium Pursh   Round-leaved fameflower

Succulent perennial to about 2 dm tall when in flower; leaves clustered at the base of plant; inflorescence a long-peduncled cyme with pink flowers that remain open for only a few hours in full sunshine; fruit a capsule, 5–6 mm long; seeds minutely roughened; serpentine barrens; SE; flr. late Jun–Jul; 🌿.

Talinum
teretifolium

PRIMULACEAE Primrose Family

Herbs with simple opposite, whorled, or basal leaves that lack stipules; flowers regular, perfect, 5-merous; petals fused to form a tube at least at the base; stamens 5, opposite the corolla lobes; ovary superior (half inferior in *Samolus*), unilocular with free central placentation; fruit a capsule.

A. floating aquatics with deeply dissected leaves; flowering stems inflated .... Hottonia

A. terrestrial plants or emergent aquatics; stems of the inflorescence not inflated; leaves not dissected
  B. leaves all or mostly in a basal rosette
    C. corolla showy, rose-pink (or rarely white), lobes strongly reflexed .... *Dodecatheon*
    C. corolla tiny, white, lobes not reflexed .... Samolus
  B. leaves all or mostly cauline, alternate, opposite, or whorled
    D. leaves opposite or whorled
      E. leaves in a single terminal whorl...................................... *Trientalis*
      E. leaves opposite or in several whorls
        F. flowers yellow .... Lysimachia
        F. flowers scarlet to dull red .... Anagallis

D. leaves alternate
    G. flowers tiny, white; ovary half inferior .... *Samolus*
    G. flowers yellow; ovary superior .... *Lysimachia*

Anagallis L.

Anagallis arvensis L.   Scarlet pimpernel
Spreading or erect, branched annual, 1–3 dm tall; leaves opposite, entire, elliptic or ovate, 1–2 cm; flowers solitary in the axils, on slender pedicels; corolla scarlet or dull red; a common weed of lawns, gardens, and waste ground; mostly S; flr. Jun–early Oct; native to Eurasia; UPL.

Dodecatheon L.

Dodecatheon meadia L.   Shooting-star
Glabrous perennial with a rosette of basal, oblong to oblanceolate leaves with entire margins; flowering stem leafless and erect, 2–6 dm tall; flowers in a terminal umbel; corolla divided nearly to the base and strongly reflexed, rose purple to lavender or occasionally white; stamens more or less fused, filaments short, anthers erect; style about as long as the stamens; capsule opening by 5 terminal valves; rare in open wooded slopes, bluffs, and meadows on limestone; SE and SC; flr. late Apr–May; FACU; ❦. *Dodecatheon* in Pennsylvania includes populations sometimes distinguished as *D. radicatum* Greene [syn: *D. amethystinum* (Fassett) Fassett] on the basis of thinner capsule walls, smaller overall stature, and the lack of a red tinge at the base of the leaves; however, these distinctions are not consistent.

Hottonia L.

Hottonia inflata Elliot   American featherfoil
Floating, aquatic herb with finely dissected submergent leaves and flowers in whorls on the inflated, jointed stems of the terminal inflorescence; corolla tubular, shorter than the 5-lobed calyx; very rare in ponds and ditches; E; believed to be extirpated; flr. Jun–Jul; OBL; ❦.

Lysimachia L.

Herbs with simple, opposite or whorled leaves (alternate in L. *clethroides*) lacking stipules, punctate-dotted in some species; flowers axillary or in terminal racemes or panicles, mostly 5-merous (6-merous in L. *thyrsiflora*); corolla yellow (white in L. *clethroides*), often marked with dots or lines; corolla tube very short, lobes spreading; stamens attached near the throat of the corolla tube, filaments united in some species; capsule splitting longitudinally.

A. leaves alternate; corolla white .... *L. clethroides*
A. leaves opposite or whorled; corolla yellow
    B. plants prostrate, rooting at the nodes .... *L. nummularia*
    B. plants erect
        C. leaves in whorls of 3–5
            D. flowers axillary .... *L. quadrifolia*

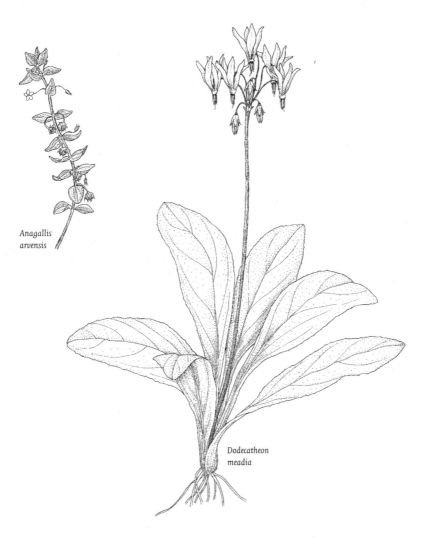

Anagallis arvensis

Dodecatheon meadia

D. flowers in a terminal inflorescence, although the lowest ones may be axillary

   E. calyx lobes red-margined; corolla lobes entire .... *L. vulgaris*

   E. calyx lobes not red-margined; corolla lobes glandular-ciliate ......

   .... *L. punctata*

C. leaves opposite

  F. leaves not punctate-dotted

    G. leaves ovate to lance-ovate, >1.5 cm wide

      H. petiole ciliate its entire length .... *L ciliata*

      H. petiole ciliate below the middle, glabrous above ..... *L. hybrida*

    G. leaves narrowly lanceolate to linear, <1.5 cm wide

      I. leaves pinnately veined; margins ciliate or scabrous, not revolute

        J. rhizome short and stout .... *L. hybrida*

        J. rhizome long and slender, stolon-like .... *L. lanceolata*

      I. leaves with obscure lateral veins; margins entire and revolute...

      .... *L. quadriflora*

F. leaves punctate-dotted
- K. flowers in short, dense racemes in the axils of the middle leaves...
....*L. thyrsiflora*
- K. inflorescence terminal
  - L. flowers all subtended by linear bracts .... *L. terrestris*
  - L. lower flowers from the axils of foliage leaves .... *L. x producta*

Lysimachia ciliata L.   Fringed loosestrife

Rhizomatous perennial with upright stems 4–13 dm tall; leaves ovate to lanceolate, the petioles ciliate for their entire length; flowers solitary in the upper axils, yellow, 1.2–2.5 cm wide; filaments distinct; common in low, moist ground of fields, stream banks and floodplains; throughout; flr. Jun–Aug; FACW.

Lysimachia clethroides Duby   Loosestrife

Stems erect to 1 m tall; leaves opposite, ovate-lanceolate with revolute margins; flowers in a slender terminal spike, corolla white; cultivated and occasionally spreading from gardens; flr. Jul–Aug; native to Asia.

Lysimachia hybrida Michx.   Lance-leaved loosestrife

Stems erect to 1.5 m tall with short branches from the upper nodes; leaves opposite, linear-lanceolate, petiole ciliate near the base, glabrous above; inflorescence terminal and from the upper leaf axils; flowers yellow, 1.5–2.5 cm broad; corolla lobes ciliate; filaments distinct; rare in swamps, wet meadows, fens, and pond margins; SE; flr. Jun–Aug; OBL; ✿.

Lysimachia lanceolata Walter   Loosestrife

Stems erect, from slender rhizomes, 2–9 dm tall, with branches about as long as the leaves; leaves opposite, linear to narrowly lanceolate; flowers in the upper axils, yellow, 1.2–2 cm wide; corolla lobes ciliate and cuspidate; filaments distinct; occasional on stream banks, sandy or rocky shores, or moist fields; SW and SC; flr. Jul–Aug; FAC.

Lysimachia ciliata

Lysimachia nummularia L.   Moneywort, creeping-charlie
Creeping, mat-forming perennial; leaves opposite, short-petioled, oval to nearly round; flowers 2–3 cm wide, solitary in the leaf axils; corolla yellow dotted with dark red; a common weed of lawns, meadows, wet woods, and floodplains; throughout; flr. late May–Oct; native to Europe; FACW–.

Lysimachia x producta (A.Gray) Fernald   Loosestrife
A fertile hybrid, similar to L. *terrestris* but with floral bracts longer and wider; open thickets and moist bottomland; scattered. [syn: L. *quadrifolia x terrestris*]

Lysimachia punctata L   Spotted loosestrife
Stems erect, 6–9 dm tall, rarely branched; leaves in whorls of 3 or 4, ovate to ovate-lanceolate, punctate; flowers yellow, 2–3 cm wide, in axillary whorls at the top of the stem; corolla lobes with glandular-ciliate margins; filaments united at the base; cultivated and occasionally naturalized in wet meadows and roadsides; E; flr. May–Jul; native to Eurasia; OBL.

Lysimachia quadriflora Sims   Four-flowered loosestrife
Stems 2–9 dm tall, erect from a short rhizome, branching from the upper nodes; leaves linear, tapering to the sessile base, margins revolute; flowers in terminal clusters on the stem and upper branches; corolla yellow; filaments distinct; very rare in calcareous meadows and fens; C; flr. Jul–Aug; FACW+; ❦.

Lysimachia quadrifolia L.   Whorled loosestrife
Erect, unbranched perennial from long rhizomes; leaves in whorls of (3)4(6), punctate, narrowly or broadly lanceolate; flowers axillary on slender, 2–5-cm pedicels; corolla 1–1.5 cm wide, yellow with dark lines; filaments united at the base; common in dry, open woods; throughout; flr. May–early Aug; FACU–.

Lysimachia terrestris (L.) Britton, Stearns & Poggenb.   Swamp-candles
Rhizomatous perennial with erect, glabrous, often branched stems 4–8 dm tall; leaves opposite, lanceolate, punctate; elongate bulblets forming in the axils late in the season; flowers yellow, in a dense terminal raceme; common in swamps and bogs; throughout; flr. Jun–Aug; OBL.

Lysimachia thyrsiflora L.   Tufted loosestrife
Stems erect, 3–7 dm tall, from long rhizomes; leaves opposite, narrowly lanceolate to linear; inflorescence several short, dense racemes from the axils of the middle leaves; flowers mostly 6-merous, yellow; corolla lobes linear, marked with black; filaments distinct; rare in bogs and swamps; NE and NW; flr. May–Jun; OBL.

Lysimachia vulgaris L.   Garden loosestrife
Stems erect to 1.2 m tall, from long, stolon-like rhizomes; leaves whorled or opposite, lanceolate or lance-ovate, punctate; inflorescence a terminal panicle or raceme; calyx lobes red-margined; corolla yellow, lobes entire; filaments united about half their length; cultivated and occasionally naturalized in swales or moist shores; mostly E; flr. Jun–Aug; native to Eurasia; FAC+.

*Samolus
parviflorus*

*Trientalis
borealis*

Chimaphila maculata

Samolus L.

Samolus parviflorus Raf.   Water pimpernel
Perennial herb, 1–3 dm tall; leaves basal and alternate, spatulate to obovate, entire; inflorescence a diffuse raceme of tiny white flowers; calyx campanulate,
lobes shorter than the tube; corolla about 2 mm across, deeply 5-lobed; ovary
partly inferior; capsule globose, opening by 5 valves as far as the adherent calyx
tube; rare on muddy stream banks and shores, especially on diabase; S and NW;
flr. Jun–Oct; OBL; ❦. [syn: *S. floribundus* Kunth]

Trientalis L.

Trientalis borealis Raf.   Starflower
Rhizomatous, perennial herb to 2 dm tall with a terminal whorl of leaves subtending 1–3 small white flowers on slender, 2–5-cm erect pedicels; leaves lance-
acuminate, 4–10 cm long, somewhat variable in size on a single plant; flowers 5-
merous; common in moist woods and bogs; throughout; flr. May–Jun; FAC.

PYROLACEAE Shinleaf Family

Low-growing, glabrous, perennial herbs or subshrubs; leaves evergreen, simple, basal, alternate, opposite, or whorled; stipules not present; inflorescence erect; flowers 5-merous; sepals distinct or barely fused at the base; petals distinct; stamens 10; anthers opening by pores; ovary 5-locular; stigma 5-lobed; fruit a capsule.

A. subshrubs with cauline leaves; inflorescence an umbel or corymb .... Chimaphila
A. herbaceous plants with leaves mostly basal; inflorescence a raceme
    B. raceme one-sided; styles straight .... Orthilia
    B. raceme not one-sided; styles curved .... Pyrola

Chimaphila Pursh

Rhizomatous, evergreen subshrubs somewhat woody at the base; leaves thick and leathery, sharply toothed; flowers 5-merous, in sparse umbels or corymbs; pollen grains in clusters of 4; flr. late Jun–Jul.

A. leaves lanceolate, striped with white .... C. maculata
A. leaves oblanceolate, not marked with white .... C. umbellata

Chimaphila maculata (L.) Pursh   Pipsissewa, spotted wintergreen
Stems 1–3 dm tall; leaves few, clustered, lanceolate, dark green with whitish stripes along the veins; flowers solitary or 2–5 in an umbel, creamy-white; common in wooded areas; throughout, except in the northernmost counties.

Chimaphila umbellata (L.) W.Bartram   Pipsissewa, prince's-pine
Stems 1–3 dm tall; leaves scattered to nearly whorled, uniformly green; flowers 4–8 in a corymb, white or pinkish; occasional in upland woods or barrens; scattered; throughout.

Orthilia Raf.

Orthilia secunda (L.) House   One-sided shinleaf, wintergreen
Leaves basal or separated by distinct internodes, elliptical or broadly ovate; inflorescence a 1-sided raceme, 1–2 dm; petals 5, white or greenish with 2 basal tubercles; style elongate, straight; pollen grains single; occasional in dry to moist woods; mostly E; flr. late Jun–Jul; FAC. [syn: *Pyrola secunda* L.]

Pyrola L.

Rhizomatous herbs with a few broad, petiolate, basal leaves; flowers in an erect raceme, 5-merous; stamens 10, anthers inverted in the flower, each opening by 2 pores; ovary 5-celled, style curved; pollen grains in clusters of 4; capsule opening from the base upward; flr. Jun–early Jul.

A. calyx lobes much longer than broad .... P. americana
A. calyx lobes as broad or broader than long
    B. leaves 1–3 cm; sepals broadly ovate, obtuse, or rounded .... P. chlorantha
    B. leaves 3–7 cm; sepals triangular, shortly acuminate .... P. elliptica

Pyrola americana Sweet   Wild lily-of-the-valley

Leaves firm and glossy, blades elliptic to nearly circular; flowering stalks 1.5–3 dm tall; petals white or creamy, 6–10 mm; frequent in woods; throughout; FAC. [syn: *P. rotundifolia* L. var. *americana* (Sweet) Fern.]

Pyrola chlorantha Swartz   Wintergreen

Leaves thick and leathery, blades obovate to nearly circular, broadly rounded at both ends; scapes 1–3 dm with 2–13 flowers; petals white with greenish veins; infrequent in dry woods; E and C; UPL.

Pyrola elliptica Nutt.   Shinleaf, wild lily-of-the-valley

Leaves thin and dull, blades oblong to elliptic, decurrent along the petiole; flowering stalks 1.5–3 dm, with 3–21 flowers; petals white or creamy with greenish veins; occasional in dry or moist woods on rich soils; throughout; UPL.

RANUNCULACEAE Buttercup Family
Carl S. Keener

Perennial or annual herbs, rarely woody climbers or low shrubs; leaves simple to variously compound or dissected, alternate, or rarely opposite or whorled, usually basal and cauline; flowers solitary or in racemes, cymes, or panicles, radially

Pyrola elliptica

or bilaterally symmetric, perfect or sometimes unisexual; sepals 3–5(20), distinct, often petaloid and colored, occasionally spurred; petals (sometimes called honey leaves or staminodia) usually 5–10, nectariferous and colored, sometimes reduced or absent; stamens many, occasionally petal-like; carpels 1–many, superior, usually distinct; fruits achenes or follicles, rarely a berry, utricle, or capsule, often aggregated into globose to cylindrical heads. Identification of specimens usually requires both flowers and fruit.

A. flowers bilaterally symmetric, the upper sepal distinctly unlike the others; fruits follicles

    B. upper sepal hooded or helmet-shaped, not spurred; petals 2(5), entirely concealed by the hooded sepals .... *Aconitum*

    B. upper sepal spurred, not hooded; petals 2 or 4 with at least the limbs exerted

        C. carpel 1; petals 2, united .... *Consolida*

        C. carpels 2 or more; petals 4, distinct .... *Delphinium*

A. flowers radially symmetric; sepals mostly equal; fruits various

    D. leaves opposite; styles of mature fruits usually elongate and feathery .... *Clematis*

    D. leaves whorled, alternate, or basal; styles of mature fruits not elongate and feathery

        E. fruits achenes or utricles; carpels with 1 ovule

            F. sepals or sepal-like bracts green; petals or petal-like sepals present, usually showy

                G. petals (or petaloid sepals) yellow .... *Ranunculus*

                G. petals (or petaloid sepals) white, pink, purplish, or blue

                    H. plants terrestrial .... *Hepatica*

                    H. plants aquatic.. *Ranunculus*

            F. sepals white or variously colored or absent; petals absent

                I. leaves compound, the leaflets distinctly stalked; flowers often unisexual .... *Thalictrum*

                I. leaves simple, often deeply parted or, if compound, the leaflets scarcely stalked; flowers usually perfect

                    J. leaves alternate; fruit a dehiscent utricle .... *Trautvetteria*

                    J. leaves basal, opposite, or whorled; fruit an achene

                        K. leaves all basal .... *Hepatica*

                        K. a whorl of leaves present below the inflorescence...*Anemone*

        E. fruits follicles, or rarely berries or capsules; carpels with >1 ovule

            L. small shrubs; inflorescence a drooping panicle .... *Xanthorhiza*

            L. herbaceous; flowers solitary or in racemes

                M. largest leaves simple, variously toothed to deeply palmately lobed

                    N. sepals 3; ovules 2; fruit a berry; rhizomes yellow internally .... *Hydrastis*

                    N. sepals >3; ovules >2; fruit a follicle; rhizomes not yellow

                        O. petals absent; sepals usually bright yellow; leaves simple, merely shallowly toothed .... *Caltha*

                        O. petals present; sepals various; leaves palmately lobed

                          P. cauline leaves 2–3, whorled; rhizomes tuberous .... *Eranthis*

                          P. cauline leaves several, alternate; rhizomes more or less fibrous, nontuberous .... *Trollius*

M. largest leaves compound or finely dissected
    Q. leaves finely dissected into narrow segments; carpels more or less united...
    .... *Nigella*
    Q. leaves variously compound with generally broad, toothed or lobed leaflets; carpels not united
        R. petals distinctly spurred; perianth variously colored .... *Aquilegia*
        R. petals absent or, if present, not spurred; perianth usually white or cream
            S. plants <1.5 dm tall; sepals conspicuous, petal-like, persistent; flowers solitary or few .... *Coptis*
            S. plants >3 dm tall; sepals small, not petal-like, soon falling; flowers many, in racemes
                T. fruit a follicle; flowering racemes >15 cm long; carpels 1–8 ....
                .... *Cimicifuga*
                T. fruit a berry; flowering racemes <5 cm long; carpel 1 .... *Actaea*

Aconitum L.

Erect to reclining, slender, perennial herbs; leaves simple, alternate, palmately lobed or cleft, upwardly reduced, petiolate; inflorescences paniculate or racemose, terminal, few-flowered, bracteate; flowers bilaterally symmetric, bisexual, showy; sepals 5, lower 2 narrowly oblong, lateral 2 broadly rounded or reniform, uppermost (galea) helmet-shaped; petals 2(5), upper 2 clawed, with a nectary blade concealed within the helmet, lower minute or absent; carpels usually 3–5; follicles oblong; seeds several. All parts of the plant (but especially the roots) contain poisonous alkaloids.

A. perianth blue; rachis or pedicels or both pilose, the hairs straight and spreading; rootstocks tuberous .... *A. uncinatum*
A. perianth white to cream; rachis and pedicels with short, stiff incurved hairs; rootstocks not tuberous .... *A. reclinatum*

Aconitum reclinatum A.Gray   White monkshood
Similar to *A. uncinatum*; leaves subglabrous to strigose beneath; galea elongate-conic, 1.5–2.3 cm high, about ½ as wide, beaked in front; rare in shaded ravines in rich, moist mountain woods; SW; flr. Jun–Sep; 🌿.

Aconitum uncinatum L.   Blue monkshood
Stems 0.3–2.5 m, glabrous; leaves reniform or cordate to ovate in outline, 3–5(7) lobed, 4–10 cm wide, firm, glabrous; segments rhombic-ovate, entire to variously lobed, toothed or incised; sepals 5, deep to pale blue, galea erect, rounded-conic, 16–25 mm long, nearly as wide, prominently beaked in front; follicles ellipsoid, 8–21 mm long, pubescent to glabrate; rare in rich, moist deciduous woods; SW; flr. Aug–Oct; 🌿.

Actaea L.

Erect, more or less glabrous, perennial herbs 3–8 dm tall, from hard rootstocks; leaves 1–3 times ternate to pinnate, the divisions ultimately 1–5-foliolate, petiolate; inflorescences racemose, terminal, <5 cm long at anthesis, on long peduncles; flowers radially symmetric, perfect, numerous, small, whitish; sepals 3–5, whitish, soon falling, broadly obovate; petals 4–10, white, clawed, spatulate to obovate; carpel 1, stigma broad, 2-lobed; berries subglobose; seeds 3–16. Plants poisonous due to glycosides or essential oils.

A. mature pedicels >1 mm thick; stigmas at flowering 1–2 mm broad, sessile; petals with truncate or cleft tips; berries white (rarely red) .... *A. pachypoda*
A. mature pedicels <1 mm thick; stigmas at flowering 0.5–1 mm broad, subsessile; petals with acute or obtuse tips; berries red (or often white) ....
.... *A. rubra*

Actaea pachypoda Elliot   Doll's-eyes, white baneberry
Leaflets broadly ovate or obovate to ovate-oblong, to 11 cm long, usually glabrous beneath, deeply incised to sharply toothed, occasionally lobed; racemes 1–2 cm at flowering, elongating to 1 dm in fruit; peduncles to 2.6 dm; petals 3–7, soon deciduous, clawed, 3–5 mm long, usually truncate or cleft at the tip; stigmas 1.1–2.1 mm broad, during anthesis, as broad as the ovary; berries 6–10 mm long; pedicels often reddish, 1–2 mm thick, 1–2 cm long; frequent in rich open woods and thickets; throughout; flr. May–Jun; UPL.

Actaea rubra (Aiton) Willd.   Red baneberry
Similar to *A. pachypoda*; leaflets commonly with hooked hairs on the veins beneath; petals with acute to obtuse tips; stigmas 0.6–1 mm broad at flowering, usually much narrower than the ovary; pedicels <1 mm thick, 1.3–3 cm long; occasional in rich woods and thickets; N; flr. May–Jun; UPL.

Anemone L.

Erect perennial herbs; basal leaves usually deeply palmately divided or compound, long-petioled; involucral bracts leaf-like, usually similar to the basal leaves, opposite or whorled; flowers solitary or in cymes on elongated peduncles, radially symmetric, perfect; sepals few–many, petal-like; petals absent; carpels numerous; fruiting heads globose to long-cylindric; achenes orbicular to fusiform, nearly glabrous to densely woolly, beaked.

A. flowering stems usually branched, 2-or-more-flowered; plants usually >4 dm tall at flowering
  B. involucral leaves sessile; achenes nearly glabrous to finely pubescent, forming loose globose heads .... *A. canadensis*
  B. involucral leaves petioled; achenes densely pubescent, forming woolly, compact, ovoid to cylindric heads
    C. involucral leaves usually 3; styles greenish; achene heads usually >1 cm thick .... *A. virginiana*
    C. involucral leaves >3; styles crimson; achene heads 1 cm or less thick.....
    .... *A. cylindrica*
A. flowering stems simple, usually 1-flowered; plants usually <1 dm tall at flowering .... *A. quinquefolia*

Anemone canadensis L.   Canadian anemone
Stems 2–7 dm tall, villous to glabrate; basal leaves usually deeply 5–7-parted, orbicular in outline, 0.5–1.7 dm wide; segments 1–3-cleft, sharply toothed, basally

cuneate; petioles to 3.5 dm long; primary involucral bracts usually 3, whorled, similar to basal leaves, sessile or subsessile; secondary to tertiary involucral bracts opposite, similar, sessile; sepals 5, white, 1–2.3 cm long; fruiting heads globose, 1–2 cm broad; occasional in calcareous or sandy low ground, meadows, and thickets; scattered; flr. May–early Jul; FACW.

Anemone cylindrica A.Gray　　　　　　　　　Thimbleweed, long-headed anemone
Similar to but more densely pubescent than *A. virginiana*; basal leaves variously parted, cleft, or incised; involucral leaves 3–10, similar to the basal ones; flowers 2–6, long pedunculate; sepals 5, greenish-white, 0.8–1.2 cm long; fruiting heads 2–3 cm long, up to 1 cm thick; very rare on dry open slopes and calcareous fields; Centre and Erie Cos.; flr. Jun–Aug; 🌿.

Anemone quinquefolia L.　　　　　　　　　　Wood anemone, windflower
Stems 0.9–2(3) dm tall; involucral leaves whorled, compound, glabrate to sparingly pilose, petiolate; leaflets 3(5), sessile or short-petiolate; terminal leaflets prominently cleft, incised, or lobed; lateral leaflets usually 2-parted or -divided; basal leaves similar, long-petiolate; flowers solitary; sepals 5(8), whitish, ovate to obovate, 0.8–1.5(2) cm long; fruiting heads subglobose; common in moist open woods and thickets; throughout, although less N; flr. Apr–May; FACU.

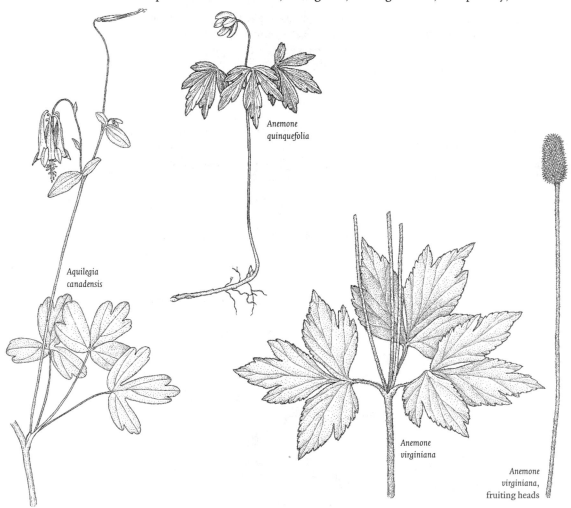

Anemone quinquefolia

Aquilegia canadensis

Anemone virginiana

Anemone virginiana, fruiting heads

Anemone virginiana L.   Thimbleweed, tall anemone

Stems loosely pubescent; basal leaves compound or deeply 3-parted, to 15 cm wide; segments or leaflets rhombic-obovate in outline, variously parted, cleft, lobed or incised; petioles to 2.8 dm; involucral leaves similar, lower whorled, upper opposite; flowers (1)3–5(9), long pedunculate; sepals 5, whitish, ovate to obovate, 0.8–2 cm long, pubescent outside; fruiting heads oblong 1.5–3.0 cm long, 1–1.5 cm thick; common in dry rocky open woods, slopes, and thickets; throughout; flr. late May–Aug; FACU.

Anemone virginiana, achene (×1½)

Aquilegia L.

Erect, slender, branching, perennial herbs from a persistent semi-woody base; basal leaves ternately 1–3 times compound; leaflets variously cleft or lobed, stalked; cauline leaves alternate, similar to the basal ones, upwardly reduced; inflorescence a few-flowered panicle; flowers radially symmetric, perfect, pendulous on long pedicels; sepals 5, colored, short-clawed; petals 5, each flat-limbed and spurred; carpels 5, stigmas linear, styles elongate; follicles erect, short-beaked; seeds many.

A. perianth red, yellow, or both; stamens conspicuously exerted; spurs straight to curved, not strongly hooked distally .... *A. canadensis*

A. perianth blue, purple, pink, or white; stamens scarcely exserted; spurs strongly hooked distally .... *A. vulgaris*

Aquilegia canadensis L.   Wild columbine

Stems 1.5–8 dm tall, branched above; leaflets cuneate-obovate, 1–4.5 cm long, glabrous above, glaucous and glabrous to pilose beneath; flowers on 5–10 cm pedicels; sepals red, ovate to lance-ovate, 1–1.8(2.1) cm long, short-clawed; petal blades yellow, 6–8 mm long, spurs red, 2–2.5 cm long, swollen at the tips; follicles ellipsoid, 1.5–2.2(3) cm long, usually pubescent; common in rich rocky woods and slopes, cliffs, ledges, pastures, and roadside banks; throughout; flr. Apr–Jun; FAC.

Aquilegia canadensis, fruit

Aquilegia vulgaris L.   Columbine

Foliage similar to *A canadensis*; sepals lance-ovate, 1.8–2.5 cm long; petal blades oblong-obovate, 10–13 mm long, spurs 1.5–2.2 cm long; stamens about as long as the sepals; cultivated and occasionally established along roadsides, waste places, and fields; scattered throughout; flr. Apr–Jun; native to Europe.

Caltha L.

Caltha palustris L.   Marsh-marigold

Erect to decumbent, glabrous, succulent, perennial herb; stems hollow, to 4 dm or more, branched above; roots fleshy from a short rootstock; basal leaves rounded to ovate-reniform or cordate, 4–15 cm wide, entire or crenate to dentate; petioles to 20 cm; cauline leaves alternate, similar to the basal leaves, reduced upward; inflorescences axillary or terminal; flowers perfect, 1–3 per bract; sepals 5(9), yellow to whitish, oval-orbicular to narrowly obovate, 0.8–2 cm long, conspicuously nerved; carpels (3)5–many, styles short, stigmatose; follicles divergent, 8–14 mm long, short-beaked; seeds many; common in wet open woods and meadows, marshes, swamps, and bogs; throughout; flr. Apr–early Jul; OBL.

Cimicifuga Wernisch.

Tall, coarse, long-lived, perennial herbs from knotty rhizomes; leaves basal and cauline, alternate, ternately 1–3 times compound, the divisions ultimately 1–5-foliolate, petiolate; leaflets 3–many, oblong-ovate to broadly cordate, often lobed, dentate or serrate to incised, the usually symmetric terminal leaflets larger than the asymmetric lateral ones, petiolulate; inflorescences racemose, simple or paniculately branched, elongate, bracteate; flowers perfect, numerous; sepals usually 4–5, white, petal-like; petals or staminodia present or rarely absent, 1–8, clawed, usually notched at the tip; follicles 1–8, ovoid to oblong or obovoid; seeds 4–15. This genus is included in *Actaea* by some authors.

A. carpels 1(2) per flower, sessile; follicles firm-walled .... *C. racemosa*
A. carpels 3–8 per flower, stalked; follicles thin, papery .... *C. americana*

Cimicifuga
racemosa,
inflorescence

Cimicifuga americana Michx.   American bugbane
Similar to *C. racemosa*; sepals 5; petals 2, ovate, about 3 mm long, 2-horned; follicles obovoid, flattened, 0.8–1.7 cm long, glabrous, obliquely beaked; seeds 5–8, in 1 row; rich moist wooded slopes and coves in the mountains; SW and SC; flr. Aug; ✤.

Cimicifuga racemosa (L.) Nutt.   Black cohosh, black snakeroot
Stems 0.7–2.5 m tall, pubescent to glabrate, solitary; terminal leaflets ovate-obovate in outline, 6–16 cm wide, palmately 3-lobed with 3 prominent veins arising basally, glabrous to sparsely ciliate beneath, dentate to deeply serrate, incised or lobed (rarely dissected), bases cuneate to subcordate; panicles usually with 4–9 slender racemes 1–6 dm long; sepals 4; petals (1)4(8), oblong, 3 mm long, usually bifid; follicles ovoid, about 0.7 cm long, pubescent; seeds 8–10, in two rows; common in rich moist woods, wooded slopes, ravines, and thickets; throughout; flr. Jun–early Aug.

Clematis L.

Perennial, herbaceous to semiwoody vines; leaves simple, pinnate, bipinnate or 1–2 times ternate; flowers solitary or in cymose-paniculate inflorescences, perfect or unisexual; sepals usually 4, valvate, variously colored, thin to thick and leathery; petals absent; petal-like staminodia sometimes present; carpels numerous; achenes light brown, compressed; styles persistent in fruit, elongate and feathery.

A. flowers numerous in cymose-paniculate inflorescences; sepals whitish; filaments glabrous
  B. flowers unisexual; leaves mostly trifoliate; leaflets coarsely toothed or lobed; anthers <1.5 mm long .... *C. virginiana*
  B. flowers perfect; leaves mostly pinnately 5-foliate; leaflets entire, seldom cleft; anthers 1.5 mm or more long .... *C. terniflora*
A. flowers usually solitary; sepals variously colored; filaments pubescent
  C. sepals thin, spreading, not converging; leaves trifoliate .... *C. occidentalis*
  C. sepals thick, usually erect and more or less converging; leaves simple to pinnately compound .... *C. viorna*

Clematis occidentalis (Hornem.) DC.   Purple clematis
Leaves trifoliate; leaflets lance-ovate to subcordate, uncleft to 3-lobed, acuminate, margins entire to deeply crenate-serrate; flowers solitary, campanulate, peduncles elongate; sepals 4, reddish-violet, lance-ovate to oblong-elliptic, 2.5–6
cm long, thin, sparsely pilose outside, margins densely villous; carpels 40–120;
styles white, 3–7 cm long; uncommon in open woods, banks, rocky calcareous
slopes and cliffs; E and C; flr. late Apr–Jun.

Clematis terniflora DC.   Sweet autumn clematis
Leaves pinnately 5-foliate or rarely trifoliate; leaflets elliptic to ovate or subcordate, to 5(8) cm long, coriaceous, glabrous, margins usually entire; inflorescences cymose-paniculate, axillary, and terminal; flowers bisexual, numerous,
fragrant; sepals 4, whitish, spreading, elliptic to obovate, 6–16 mm long, margins tomentose; styles whitish, to 3 cm long; thickets, fencerows, and waste
places; S; flr. Aug–Sep; native to Japan; FACU–.

Clematis viorna L.   Leather-flower
Leaves pinnately to bipinnately compound; rachis finely to sparsely pubescent,
tendril-like toward the tip; leaflets simple to 3-lobed or 3-foliolate, elliptic to
lanceolate to ovate or rarely cordate, to 8 cm or more long, green and soft pilose
to villous or glabrate beneath; flowers nodding, ovoid-campanulate; peduncles
axillary, 1–7-flowered, usually thinly pubescent, 2-bracteate near the middle; sepals dull purple, ovate to oblong-lanceolate, 1–2.5 cm long, puberulent outside,
margins white-tomentose distally, tips spreading to recurving; styles tawny, 3–6
cm long, spreading to recurved; very rare on rich wooded banks and thickets; S;
flr. May–Jul; 🌿.

Clematis virginiana L.   Virgin's-bower
Leaves trifoliate or rarely 5-foliolate; leaflets ovate to elliptic-ovate, 2–10 cm
long, sparingly pilose to glabrate beneath, acuminate, entire to coarsely serrate
or lobed, bases rounded to subcordate; inflorescences cymose-paniculate, axil-

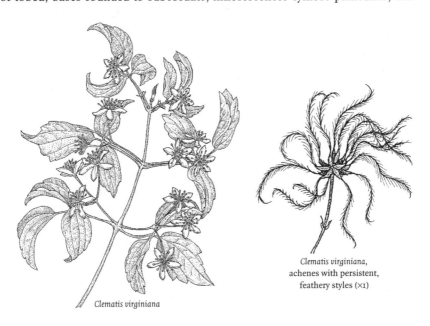

Clematis virginiana,
achenes with persistent,
feathery styles (×1)

Clematis virginiana

lary; flowers unisexual, numerous; sepals spreading, dull white, oval to oblong-spatulate, 4–12 mm long, marginally tomentose; sterile stamens usually present in pistillate flowers; styles white, 2.5–4 cm long; common in low woods, stream banks, thickets, and waste places; throughout; flr. Jul–late Aug; FAC.

Consolida (DC.) A.Gray

Consolida ajacis (L.) Schur   Garden larkspur
Erect, glabrous to glabrate, annual herb with simple or few-branched stems to 8(10) dm; leaves palmately dissected into numerous linear to filiform segments up to 3 mm wide; flowers in loose, elongate racemes, naked or with leafy bracts; flowers bisexual, showy; sepals 5, blue or purple to pink or white, 4 of the sepals suborbicular or reniform to ovate, 0.9–1.8 cm long, 1 sepal with a slender curved spur 1.2–2 cm long; petals apparently 2, united, bearing a nectariferous spur protruding into the sepal spur; follicles ovoid, 1.5–2 cm long, pubescent; seeds black, angled, scaly; occasionally established along roadsides, old fields, and waste places; S; flr. late Jun–early Sep; native to Europe. [C. ambigua of Rhoads and Klein, 1993.] C. regalis A.Gray, a rare escape occurring on waste ground, differs by its entire (vs. dissected) lower bracts. All species contain poisonous alkaloids.

Coptis Salisb.

Coptis trifolia (L.) Salisb.   Goldthread
Low, acaulescent, subglabrous perennial herb; rhizomes filiform, bright yellow; leaves basal, evergreen, trifoliate; leaflets 1.5–3.5 cm long, usually shallowly lobed and serrate to lobulate with mucronulate teeth, sessile or short-petiolate; flowers solitary on erect peduncles, perfect, small; sepals 5–7, whitish, deciduous, narrowly elliptic-lanceolate to oblanceolate, 5–8 mm long; petals 5–7, whitish, fleshy, 3–4 mm long, nectariferous in the hollow tip; carpels 3–9; follicles divergent, ovoid, 6–10 mm long; seeds several; common in rich, damp mossy woods, bogs, and swamps, often associated with Tsuga canadensis; N, and at higher elevations along the Allegheny Front; flr. Apr–May; FACW.

Delphinium L.

Erect to lax, perennial herbs with simple stems; leaves basal and cauline, alternate, the largest palmately divided into 3–5 variously lobed or parted segments; flowers in terminal racemes or panicles, perfect, bluish-purple to rarely white or pink; sepals 5, 4 ovate to obovate, upper (posterior) 1 spurred; petals 4, upper 2 irregular (cleaver-shaped), nectariferous, the spurs extending into the spurred sepal and the limbs exerted, lower 2 clawed; carpels usually 3; follicles erect to widely spreading, arcuate-oblong, beaked; seeds several, light brown to black, indistinctly wing-angled. Plants (especially the seeds and young plants) contain poisonous alkaloids.

A. racemes short, open, 5–15-flowered, follicles soon widely divergent, 1.3–1.8 cm long; stems soft, herbaceous, lax, mostly <4 dm tall .... D. tricorne
A. racemes elongate, 12–50-flowered; follicles erect, usually closely parallel, 0.8–1 cm long; stems firm, erect, usually >4 dm tall .... D. exaltatum

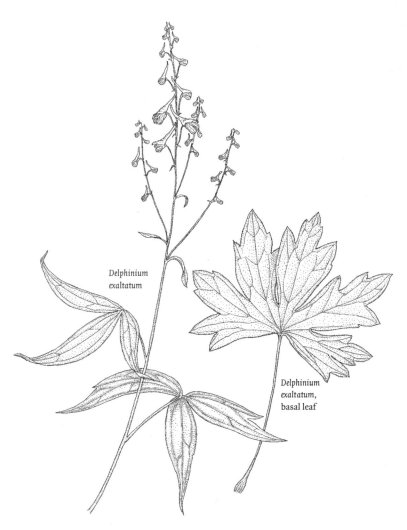

Delphinium
exaltatum

Delphinium
exaltatum,
basal leaf

Delphinium exaltatum Aiton   Tall larkspur

Stems erect, slender, glaucous, glabrous below, puberulent above; roots elongate, woody; leaves usually all cauline at flowering; inflorescences 12–50-flowered; sepals light blue or purplish to white, ovate; lower 2 petals oblong, upper 2 oblique; follicles short-oblong, 0.8–1 cm long, puberulent; rare in rich shaded woods and on rocky limestone bluffs; SW and SC; flr. Jul–Aug; 🍂.

Delphinium tricorne Michx.   Dwarf larkspur

Stems erect to lax, puberulent to glabrate; roots tuberous; leaves basal and cauline; inflorescences 5–15-flowered; sepals blue or violet to whitish, oval-oblong; lower 2 petals widely ovate, upper 2 clavate; follicles arcuate-oblong, 1.3–1.8 cm long, glabrous; locally abundant in rich moist woods, thickets, river bluffs, and calcareous slopes; SW and SC; flr. late Apr–Jun.

Eranthis Salisb.

Eranthis hyemalis (L.) Salisb.   Winter aconite

Low, glabrous, perennial herb 5–15 cm tall; basal leaves deeply palmately divided, petiolate; cauline leaves 3, whorled, sessile, closely subtending the single terminal flower; flowers perfect, cup-shaped, showy; sepals 5–8, yellow, 1.5–1.8 cm long; petals 6, yellow, tubular, 2-lipped, about 4 mm long; follicles 0.8–1.4 cm long, distinctly beaked; seeds few, subglobose; cultivated and occasionally escaped or persistent in rich moist bottomlands; SE; flr. Feb–late Apr; native to Europe.

Hepatica Mill.

Hepatica nobilis Mill.   Liverleaf

Tufted perennial herb 5–19 cm tall with short rhizomes; leaves persistent, often purplish beneath, widely cordate-reniform in outline, to 11 cm wide, 3-lobed; petioles to 17 cm long, pilose; flowers solitary on a villous to pilose peduncle; bracts 3, calyx-like, closely subtending the sepals, 6–9 mm long; flowers perfect; sepals petaloid, 5–12, blue, white, or pink, 6–16 mm long; carpels numerous; achenes lance-ovoid, 3–4 mm long, villous; *Hepatica* is occasionally lumped with *Anemone*, but is here treated as a separate genus; 2 varieties:

A. leaves cleft to about the middle of the blade; apex of leaf lobes and bracts rounded to obtuse; sepals typically bluish (sometimes white or pink) .... var. *obtusa* (Pursh) Steyerm. common in rich woods and dry upland slopes; throughout, except N; flr. late Mar–early May. [syn: *H. americana* (DC.) Ker Gawl., *Anemone americana* (DC.) H.Hara.]

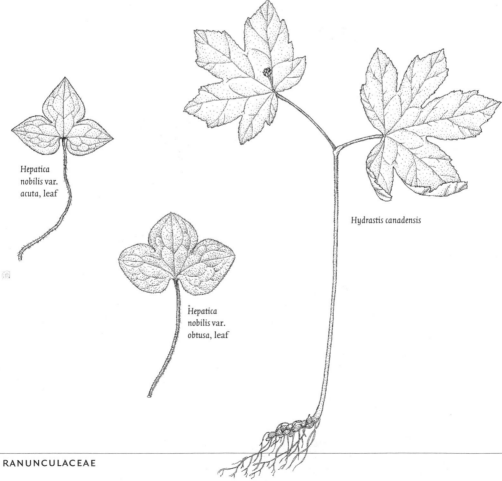

Hepatica nobilis var. acuta, leaf

Hepatica nobilis var. obtusa, leaf

Hydrastis canadensis

A. leaves parted to below the middle of the blade; apex of leaf lobes and bracts acute; sepals typically whitish .... var. *acuta* (Pursh) Steyerm. common in upland woods and rocky slopes, often in calcareous soil; throughout, except SE; flr. early Apr–May. [syn: *H. acutiloba* DC., *Anemone acutiloba* (DC.) G.Lawson.]

Hydrastis L.

Hydrastis canadensis L.   Goldenseal
Erect, pubescent, perennial herb, 1.5–3 dm tall, pilose; leaves usually 3, 1 basal and 2 cauline, broadly cordate-reniform in outline, 0.8–2.9 dm wide, 3–7-parted, segments variously lobed and singly to doubly serrate; inflorescence solitary; peduncle 0.3–2.5 cm, arising from the base of the sessile upper leaf; flowers perfect, small; sepals greenish-white, ovate, 3.5–7 mm long, soon falling; petals absent; carpels 5–15, stigmas 2-lipped, styles short; berries dark red, 5–6 mm long, in globose heads; rare, but sometimes locally abundant, in rich moist woods; mostly SE and SW, declining due to overcollecting; flr. Apr–early Jun; ❦.

Nigella L.

Nigella damascena L.   Fennel-flower, love-in-a-mist
Erect, glabrous annual, stems 1–7 dm tall, often upwardly branched; involucral bracts similar to the upper cauline leaves, closely subtending the flower and curling upward; flowers solitary, terminal or axillary, perfect, showy; sepals 5, bluish, petaloid, persistent, 0.8–2.5 cm long; petals 5–10, clawed, lobed at the tips, shorter than the sepals; stamens longer than the petals; carpels 5–10, united to the apex; capsules with long erect styles, dehiscing apically; cultivated and occasionally escaping to waste places or sometimes persisting about old home sites; flr. Jun–Jul; native to Europe.

Ranunculus L.

Terrestrial to aquatic, perennial, biennial, or annual herbs with erect to procumbent stems 0.2–12 dm or more tall, branching or simple; leaves simple and entire to compound and finely dissected, usually variously lobed or divided; inflorescences cymose-racemose, cymose paniculate, or flowers solitary; flowers bisexual, often showy; sepals 3–5; petals usually 5, yellow or rarely white, obovate and short-clawed, with a basal nectariferous scale or pit on the adaxial surface; carpels usually numerous; achenes aggregated into globose to cylindric heads, the achenes turgid to discoid, smooth to papillate, glabrous to hairy or spiny, usually beaked. Both flowers and fruit are essential for proper identification. Some species can cause dermatitis, and most species are poisonous if ingested.

A. sepals 3(4); petals 7–12; achenes pubescent, beakless; leaves cordate ....
.... *R. ficaria*
A. sepals (3)5(6); petals typically 5; achenes smooth to spiny or pubescent, usually with well-developed beaks; leaves seldom cordate
    B. petals dull, white; achenes roughly transverse-ridged; aquatic herbs
        C. leaves floating, shallowly lobed; receptacles glabrous .... *R. hederaceus*

  C. leaves usually submersed, finely dissected; receptacles hispid ....
   .... *R. aquatilis* var. *diffusus*
 B. petals usually glossy, yellow; achenes usually not transverse-ridged; terrestrial or aquatic herbs
  D. leaves all simple, not lobed or deeply divided; margins entire to denticulate or serrulate
   E. petals 1–3, 1–2 mm long; achenes <1.5 mm long .... *R. pusillus*
   E. petals 5, 2–7 mm long; achenes >1.4 mm long
    F. petals mostly as long as the sepals; cauline leaves >1 cm wide; achene beaks >1 mm long .... *R. ambigens*
    F. petals twice as long as the sepals; cauline leaves <1 cm wide; achene beaks <0.5 mm long .... *R. flammula*
  D. leaves (especially the lower cauline ones) variously lobed to divided or compound; margins various
   G. basal leaves rarely deeply lobed, distinctly unlike the deeply parted cauline leaves; achenes turgid, ovoid, 1–2.5 mm long, without pronounced marginal rims
    H. achene beaks 0.1–0.3 mm long; petals >½ the length of the sepals; sepals glabrous to sparsely long-villous
     I. plants usually glabrous; basal leaves 1–6(10) cm wide, reniform to cordate; receptacles usually villous .... *R. abortivus*
     I. plants villous, at least basally; basal leaves 1–2.5 cm wide, truncate to cuneate (rarely cordate) at base; receptacles usually glabrous .... *R. micranthus*
    H. achene beaks 0.6–1 mm long; petals <½ the length of the sepals; sepals hirsute .... *R. allegheniensis*
   G. basal leaves mostly deeply parted or compound, usually similar to the smaller cauline leaves; achenes various, 2–5 mm long, with or without pronounced marginal rims
    J. achenes markedly spiny, papillose or tuberculate .... *R. sardous*
    J. achenes smooth, rarely pubescent or papillose
     K. achenes usually turgid, the marginal rims scarcely evident, corky-thickened below; marsh or aquatic plants
      L. petals 2–5 mm long; achenes essentially beakless (0.1 mm long); plants terrestrial to palustrine, without submersed leaves .... *R. sceleratus*
      L. petals 7–15 mm long; achene beaks well-developed (0.6–1.5 mm long); plants aquatic with finely dissected leaves....
       .... *R. flabellaris*
     K. achenes generally flattened, discoid, marginal rims usually pronounced, not corky-thickened below; normally terrestrial plants
      M. petals 2–4 mm long, about equaling the sepals
       N. achene beaks markedly recurved; largest leaves merely deeply dissected; fruiting heads globose .... *R. recurvatus*
       N. achene beaks straight to slightly curved; largest leaves ternately compound; fruiting heads cylindrical ....
        .... *R. pensylvanicus*
      M. petals 5–18 mm long, distinctly longer than the sepals
       O. achene beaks recurved or hooked, usually <1.5 mm long, stigmatose laterally

P. sepals spreading
  Q. stems repent, rooting at the lower nodes; basal leaves ternately compound; receptacles hispid .... *R. repens*
  Q. stems erect, never rooting; basal leaves deeply 5-parted; receptacles glabrous .... *R. acris*
P. sepals tightly reflexed (bent backward along the peduncles)
  R. plants cormose perennials; petals 8–14 mm long; achenes smooth, unevenly thick, the apex much thicker than the base, the faces asymmetrically obovate to orbicular, variously bulged .... *R. bulbosus*
  R. plants soft-based annuals; petals 5–8(10) mm long; achenes smooth to variously protuberant, more or less uniformly thick throughout or slightly bulged in the center, the faces generally orbicular, usually flat.... .... *R. sardous*
O. achene beaks straight to flexuous, (1)1.5–3 mm long, stigmatose apically
  S. later basal leaves generally cordate-ovate in outline, usually wider than long, 3-lobed to trifoliate; stem pubescence hispid, spreading or appressed to glabrous; petals rounded to obovate, widest above the middle
    T. stems erect or repent and stolon-like, 1–3 mm thick, the longest aerial shoots usually <5 dm long at time of fruiting; sepals spreading to reflexed; petals typically <1 cm long; plants generally of dry habitats .... *R. hispidus*
    T. stems repent, often strongly stoloniferous and rooting at the nodes, 2–8 mm thick, the longest aerial shoots usually >5 dm long at time of fruiting; sepals usually spreading; petals typically >1 cm long; plants of wet habitats .... *R. caricetorum*
  S. later basal leaves generally ovate-oblong in outline, often longer than wide, usually pinnately divided; stem pubescence silky-appressed to spreading; petals linear oblong, usually widest below the middle .... *R. fascicularis*

Ranunculus abortivus L.   Small-flowered crowfoot
Annual or biennial, 1–5 dm tall; basal leaves simple, reniform, 1–6(10) cm wide, crenate to crenulate, or 3-lobed to 3-foliolate; cauline leaves and bracts alternate, deeply divided or parted (rarely merely shallowly toothed or lobed); flowers solitary; sepals 3–4 mm long; petals 5, pale yellow, 2.5–3.5 mm long; achenes discoid-obovoid, 1.4–1.6 mm broad, glabrous; beaks 0.1–0.2 mm long, straight to slightly curved; fruiting receptacles fusiform-cylindroid, to 4 mm long, usually villous; common in rich low woods, low fields, and moist waste places; throughout; flr. Mar–Jun; FACW–. Highly variable in leaf morphology and pubescence.

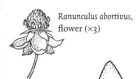

Ranunculus abortivus, flower (×3)

Ranunculus abortivus, petal with nectar gland (×10)

Ranunculus acris L.   Tall or common meadow buttercup
Perennial, 5–10 dm tall; basal leaves simple, 5–10 cm wide, palmately 5–7-parted; cauline leaves alternate, similar to the basal, upwardly reduced; inflorescences paniculate to racemose; sepals 4–7 mm long; petals 5, yellow to whitish, 8–14 mm long; styles stigmatose laterally; achenes discoid-obovoid, about 2 mm broad, smooth, glabrous, margins sharply keeled, beaks flat, deltate, 0.3–0.6 mm long, typically recurved; fruiting receptacles pear-shaped, 2.5 mm long, glabrous; common in pastures, meadows, clearings, and waste places; throughout, but less so S; flr. May–Sep; native to Europe; FAC+.

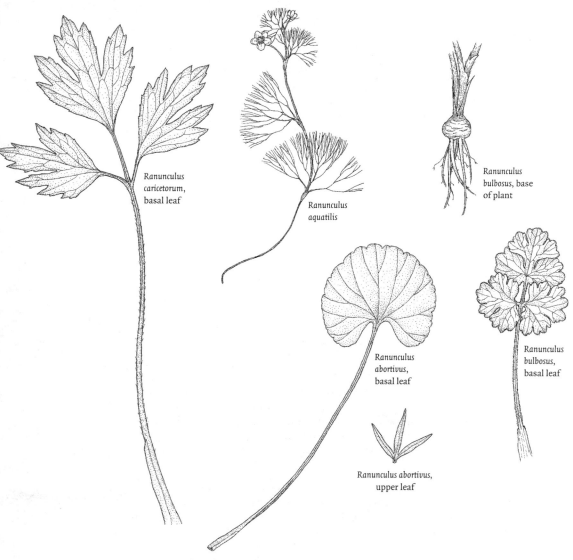

Ranunculus
caricetorum,
basal leaf

Ranunculus
aquatilis

Ranunculus
bulbosus, base
of plant

Ranunculus
abortivus,
basal leaf

Ranunculus
bulbosus,
basal leaf

Ranunculus abortivus,
upper leaf

Ranunculus alleghensis Britton   Allegheny or mountain crowfoot
Similar to but generally more robust than *R. abortivus*; sepals yellowish, 2.5–3.5
mm long; petals 5, yellow, 1–1.5 mm long; achenes discoid-suborbicular, 1.4–1.6
mm broad, smooth, glabrous; beaks 0.6–1 mm long, usually strongly recurved;
frequent in rich woods and on calcareous slopes; SW and SC; flr. May–Jun; FAC.

Ranunculus ambigens S.Watson   Water-plantain spearwort
Decumbent to suberect, rhizomatous perennial; stems 0.3–1 m long, hollow,
with slightly swollen nodes; basal leaves absent; cauline leaves simple, alternate,
lanceolate to ovate-lanceolate or rarely subcordate-ovate, the largest blades 6–14
cm long, denticulate; petioles winged, up to 6 cm long; flowers solitary; sepals
spreading, about 4 mm long; petals 5(6), yellow, clawed, 5–7 mm long; achenes
cuneate-obovate, about 2 mm long and 1.4 mm broad, smooth to finely reticulate, glabrous, margins narrowly keeled, beaks subulate, to 1.3 mm long, deflexed; fruiting receptacles cylindroid-obovoid, 3–5 mm long, glabrous; uncommon in low wet ground, swamps, and muddy ditches; scattered; flr. May–Aug;
OBL.

Ranunculus aquatilis L.   White water-crowfoot

Usually submerged, aquatic annual or perennial, stems 2–6(20) dm long; submerged leaves finely dissected into capillary segments, 1–4 cm long; flowers solitary; sepals 2.5–3.5 mm long; petals 5, whitish with yellowish bases, 3.5–5.5 mm long; achenes obovoid, about 1–1.5 mm long, transverse-ridged, glabrous to hispid, beaks nearly absent, to 1.2 mm; fruiting receptacles subglobose, 1–2 mm long, hirsutulous; uncommon in lakes, ponds, rivers, and streams; scattered; flr. May–Jul; OBL; 🍃. A very variable species, ours is var. *diffusus* With. [includes *R. longirostris* and *R. trichophyllus* of Rhoads and Klein, 1993].

Ranunculus bulbosus L.   Bulbous buttercup

Ranunculus bulbosus, achene (×5)

Perennial, stems 3–7 dm tall from a cormose base to 1.5 cm thick; basal leaves pinnately compound, 1–5 cm wide; leaflets 3, the divisions variously lobed and toothed; cauline leaves alternate, similar to the basal, greatly reduced and less divided upward; inflorescences paniculate, racemose, or rarely 1-flowered; sepals usually tightly reflexed, about 7 mm long; petals 5, bright yellow, 10–14 mm long; styles stigmatose laterally; achenes obliquely obovoid, 2–3 mm broad, smooth, glabrous, margins sharply keeled, beaks stout, about 1.5 mm long, recurved; fruiting receptacles ovoid, 4 mm long, pubescent; common in lawns, fields, roadsides, and pastures; mostly SE, less frequent elsewhere; flr. late Apr–early Sep; native to Europe; UPL.

Ranunculus caricetorum Greene   Northern swamp or marsh buttercup

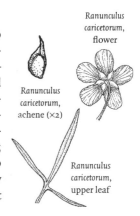
Ranunculus caricetorum, flower

Ranunculus caricetorum, achene (×2)

Ranunculus caricetorum, upper leaf

Palustrine perennial; stems 3–10 dm long, 2–8 mm thick, becoming stoloniferous and rooting at the lower nodes; basal leaves all ternately compound, 2–14(20) cm wide, leaflets 2–3-lobed to deeply cleft, the segments variously lobed to sharply incised or toothed; basal stipules 2–4 cm long; cauline leaves alternate, the lower similar to basal, upper reduced, entire to variously cleft; inflorescence irregularly cymose-paniculate; sepals 6–11 mm long; petals 5–8(10), yellow, 8–14 mm long; styles stigmatose apically; achenes 15–30, obovoid, 2–2.5 mm broad, smooth, glabrous, margins prominently keeled, the central keel up to 0.3 mm wide, beaks flat, stout, 1.5–3 mm long, hispidulous; common in low woods, swamps, marshes, meadows, and alluvial thickets; throughout, except NE; flr. Apr–Jun.

Ranunculus fascicularis Muhl. ex J.M.Bigelow   Early or tufted buttercup

Tufted perennial, 1–2.5 dm tall; basal leaves usually pinnately 3–5-divided, 2–4 cm wide, usually longer than wide; cauline leaves alternate, reduced, entire to variously cleft; inflorescences irregularly cymose or 1-flowered; sepals 6–9 mm long; petals 5–7(10), yellow, 8–12 mm long, usually widest below the middle; styles stigmatose apically; achenes 10–35, obovoid, 1.8–3 mm broad, smooth, glabrous, margins scarcely keeled, beaks slender, 1.3–3.3 mm long, straight; fruiting receptacles conical, 1.7–6 mm long, hispidulous; rare in thin dry woods and exposed calcareous slopes and ledges; scattered; flr. Apr–May; FACU; 🍃. This species is sometimes confused with *R. hispidus* var. *hispidus*.

Ranunculus ficaria L.   Lesser celandine, pilewort

Ranunculus ficaria, flower

Erect to reclining perennial from a cluster of tuberous roots; stems 1–3 dm tall, branching, producing small aerial tubers at the nodes; basal leaves simple, broadly cordate, 1–5 cm long and 1–4.5 cm wide, entire or sinuate to broadly crenately toothed; inflorescences; sepals 5–10 mm long; petals 7–12, yellow, 8–

15 mm long; achenes obovoid, ca. 2.5 mm long, but rarely maturing; fruiting receptacles pear-shaped to globose, 1–2 mm long, glabrous; an invasive weed of low, open woods, floodplains, meadows, and waste places; SE, scattered elsewhere; flr. Apr–May; native to Eurasia; FAC. A form with double flowers and silvery blotches on the leaves occurs in some areas.

Ranunculus flabellaris Raf.   Yellow water-crowfoot
Submersed aquatic perennial; stems 3–7 dm long, rooting at the nodes; submersed leaves alternate, finely decompound, to 12 cm broad, the segments filiform to ribbon-like; emergent leaves 3-parted, broadly reniform, the segments cleft to lobed; flowers solitary or in cymose-racemose inflorescences; sepals 5–8 mm long, deciduous; petals 5–8, yellow, 7–19 mm long; achenes obovoid, about 1.3 mm broad, smooth to rugose, glabrous, margins corky-thickened basally, beaks curved, 1–2 mm long, broad and flat; fruiting receptacles ovoid-cylindroid, 5–7 mm long, pubescent; uncommon in shallow water and on muddy shores; NW, NE, and SC; flr. May; OBL; 🌿.

Ranunculus flammula L.   Creeping spearwort
Erect to prostrate perennial; stems up to 5 dm long, usually rooting at the nodes; leaves lanceolate to oblanceolate or filiform, to 10 mm wide, entire or slightly toothed; flowers solitary; sepals 1–3 mm long; petals 5–6, yellow, 3–5 mm long; achenes 1.2–1.6 mm long, beaks lanceolate to linear, more or less straight, 0.1–0.6 mm long; fruiting receptacles glabrous; scattered; flr. Jun–Aug; FACW; 🌿; 2 varieties:

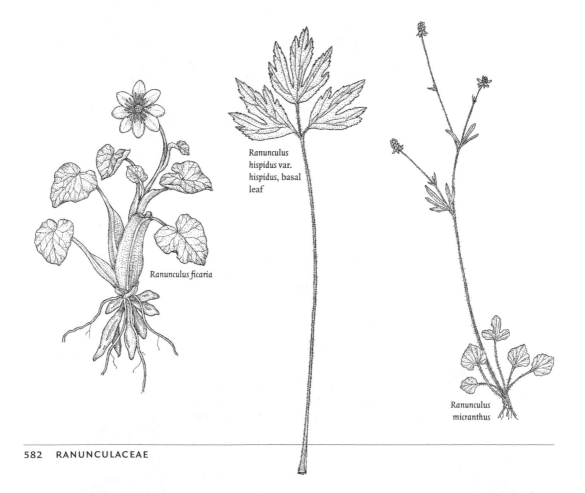

Ranunculus ficaria

Ranunculus hispidus var. hispidus, basal leaf

Ranunculus micranthus

A. leaves 2–8 mm wide, sepals 2–3 mm long ....
.... var. *ovalis* (J.M.Bigelow) L.D.Benson
rare in muddy ground, gravelly or sandy shores, and shallow water.
A. leaves 0.4–1 mm wide; sepals 1–2 mm long .... var. *reptans* (L.) E.Mey.
rare in springy thickets and shallow water. [R. *flammula* var. *filiformis* (Michx.)
DC. of Rhoads and Klein, 1993]

Ranunculus hederaceus L. Long-stalked crowfoot
Aquatic annual or perennial; stems floating, 0.5–3 dm long, rooting at the lower
nodes; leaves simple, floating, broadly kidney-shaped, 5–13 mm wide, shallowly
3–5-lobed; inflorescences 1-flowered; sepals 2–2.5 mm long; petals 5, whitish,
2–3.5 mm long; achenes obovoid, 1.2–1.6 mm long, transverse-ridged, glabrous, margins prominent, beaks about 0.1 mm long; fruiting receptacles 1–1.5
mm long, glabrous; very rare in shallow water; Lancaster Co.; believed to be extirpated; flr. Jun; OBL; ✿.

Ranunculus hispidus Michx. Hairy buttercup

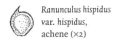

Ranunculus hispidus var. hispidus, achene (×2)

Perennial, stem 1–4 dm tall, freely branching from the base; basal leaves simple,
3-cleft or ternately compound, 2–11 cm wide; basal stipules 2–3 cm long, narrow; cauline leaves alternate, reduced above, entire to variously cleft; inflorescences irregularly cymose or 1-flowered; sepals 3.5–6 mm long; petals 5–8(10),
yellow, 5–10(13) mm long; styles stigmatose apically; achenes 5–35, obovoid,
2.5–3.5 mm broad, smooth, glabrous, margins narrowly to sharply wing-keeled,
the keels to 1 mm wide, beaks slender, 1–2 mm long, straight to curved; fruiting
receptacles cylindrical, 4–8 mm long, sparsely hispidulous; flr. Apr–May; 2 varieties:

A. stems erect, often hispid; basal leaves usually both simple and trifoliate; sepals spreading or reflexed from the base; achenes 12–35; achene margins narrowly winged, the wings less than 0.3 mm wide .... var. *hispidus*
common in dry rocky to rich moist oak and oak-hickory woods; S, and scattered to the N; FAC.
A. stems repent, often stoloniferous and typically more or less glabrous; basal
leaves usually trifoliate; sepals usually reflexed 1 mm above the base; achenes
5–15; achene margins conspicuously winged, the wings 0.4–1 mm wide ....
.... var. *nitidus* (Chapm.) T.Duncan
infrequent in low woods, thickets, and marshes, usually in floodplains; scattered; FACW+. [syn: R. *septentrionalis* Poir., R. *carolinianus* DC.]

Ranunculus micranthus Nutt. Small-flowered crowfoot

Ranunculus micranthus, achene (×5)

Perennial, 1.5–4 dm tall; basal leaves simple, 1–2.5(3.5) cm wide, crenulate to
crenately lobed, often basally truncate, or deeply 3-lobed to 3-foliolate; cauline
leaves and bracts alternate, deeply cleft; inflorescences 1-flowered; sepals 2.5–
3.5 mm long; petals 5, yellow, 2–3.5 mm long; achenes discoid-ovoid, 1.2–1.4
mm broad, smooth, glabrous, beaks 0.2–0.3 mm long, slender, curved; fruiting
receptacles cylindroid, to 6 mm long, usually glabrous; occasional in rich
woods, rocky hillsides, and calcareous banks; S; flr. May; FACU.

Ranunculus pensylvanicus L.f. Bristly crowfoot
Copiously hispid or hirsute annual or perennial, 3–10 dm tall; basal leaves pinnately compound, 9–12 cm wide; leaflets deeply toothed or incised; cauline

leaves alternate, similar to the basal ones; inflorescences cymose-paniculate; sepals 4–5 mm long; petals 5, yellow, 2–3(5) mm long; achenes ovoid, 2.5 mm long, smooth, glabrous, margins keeled, beaks flat-subulate, about 1 mm long, straight to slightly curved; fruiting receptacles cylindroid, 8–12 mm long, pubescent; uncommon in marshes, wet woods, and meadows; scattered; flr. May–Jul; OBL.

Ranunculus pusillus Poir.   Low spearwort
Palustrine annual, 1–5 dm tall; basal leaves simple, lance-ovate to suborbicular, the blades 0.6–5 cm long, entire to denticulate; cauline leaves alternate, similar to the basal ones, progressively narrower and shorter upward; inflorescences irregularly cymose-paniculate; sepals usually 1–2 mm long; petals 1–3(5), yellow, 1–1.2 mm long; achenes oblong-obovoid, about 1 mm long and 0.7 mm broad, smooth or slightly to markedly papillate, glabrous, margins inconspicuous, beaks short, 0.1–0.2 mm long; fruiting receptacles pear-shaped, 1.5–3 mm long, glabrous; rare in low wet ground, swamps, ditches, and shallow pools; SE and SC; flr. May–Jun; OBL; ❦.

Ranunculus recurvatus Poir.   Hooked crowfoot
Perennial, stems 1.5–5 dm tall from a bulbous base; basal leaves simple, cordate-reniform in outline, 3–9 cm wide, 3–5 lobed or parted, strigose above and below; cauline leaves alternate, similar to the basal ones; flowers solitary or in cymose-racemose inflorescences; sepals lance-ovate, 4–5 mm long, soon deciduous; petals 5, pale yellow, 2.5–3 mm long; achenes discoid-obovoid, 1.5 mm broad, minutely reticulate-pitted, glabrous, with sharply keeled, sparsely hirsute margins, beaks slender, 1.2–1.4 mm long, hooked; fruiting receptacles pear-shaped, 3 mm long, hairy toward the tips; common in rich, low woods; throughout; flr. May–Jul; FAC+.

Ranunculus
recurvatus,
flower (×2)

Ranunculus repens L.   Creeping buttercup
Prostrate to suberect perennial; stems 1–8 dm long, commonly stoloniferous and rooting at the nodes; basal leaves pinnately compound, 2–8 cm wide; leaflets 2–3 cleft to rarely divided; cauline leaves alternate, similar to the basal ones; inflorescences irregularly cymose-paniculate or rarely flowers solitary; sepals 5–7 mm long; petals 5, bright yellow, 7–13 mm long; styles stigmatose laterally; achenes discoid-obovoid, 2 mm broad, minutely reticulate-pitted, glabrous, margins prominently keeled, beaks stout, 1 mm long, curved; fruiting receptacles subglobose-ovoid, 3 mm long, pubescent; common in low meadows, stream banks, and waste places; throughout, S; flr. May–late Aug; native to Europe; FAC. A very variable species, a double-flowered form occasionally escapes from cultivation.

Ranunculus
sardous,
achene (×5)

Ranunculus sardous Crantz   Hairy or Sardinian buttercup
Erect to suberect annual, 1–6 dm tall; basal leaves simple, 3-parted to pinnately compound, 2–4(10) cm wide; segments variously lobed and incised; cauline leaves alternate, generally similar to the basal leaves; flowers solitary; sepals 3–5 mm long, deciduous; petals 5, yellow, 5–10 mm long; achenes discoid, 2–3 mm long and broad, sparsely protuberant to smooth, glabrous, margins conspicuously rimmed, beaks deltate, 0.2–0.5(1) mm long, hooked to straight; fruiting receptacles pear-shaped, 2 mm long, pubescent; rare in low wet fields and waste

places; scattered; flr. May–early Oct; native to Europe; UPL. Frequently confused with *R. bulbosus*.

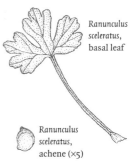

Ranunculus
sceleratus,
basal leaf

Ranunculus sceleratus L.   Celery-leaved or cursed crowfoot
Fleshy palustrine annual, stems 1–5(10) dm tall, inflated; basal leaves simple, deeply 3-parted, reniform, 1.5–10 cm wide, the segments lobed to deeply cleft; cauline leaves alternate, similar to the basal leaves, the segments linear to oblanceolate or obovate, apically toothed or entire; flowers solitary; sepals 2–3 mm long; petals 5, pale yellow, 2–3(5) mm long; achenes 40–300, obovoid, about 0.6 mm broad, smooth to rugulose, glabrous, margins spongy-thickened basally, beaks essentially absent; fruiting receptacles cylindroid, 3–9 mm long, sparingly pubescent; occasional in marshes, wet ditches, lake margins, and stream banks; SE; flr. Apr–May; OBL.

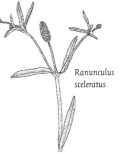

Ranunculus
sceleratus,
achene (×5)

Ranunculus
sceleratus

Thalictrum L.

Erect to procumbent perennial herbs; leaves ternately 1–4 times compound, alternate; basal and lower cauline leaves with basally dilated petioles; leaflets green and glabrous above, green to glaucous and glabrous or minutely pubescent beneath, cordate-reniform to obovate; flowers in panicles or umbels, perfect or sometimes unisexual, small, usually numerous, whitish to greenish-yellow or purplish; sepals 4–10, 1–15 mm long, petal-like, and usually soon deciduous; carpels usually numerous, separate and crowded on small receptacles, stigmas capitate or variously elongate, styles short or absent; achenes ribbed, flat to turgid.

A. inflorescence an umbel; plants usually <2(3) dm tall .... *T. thalictroides*
A. inflorescence a panicle; plants usually >2 dm tall
    B. leaflets 3(or more)-lobed apically, the lobes often crenate; filaments colored, filiform; plants dioecious, rarely polygamous
       C. achenes sessile; upper cauline leaves often long-petioled ...... *T. dioicum*
       C. achenes stipitate, the stipes wing-angled; upper cauline leaves sessile or subsessile .... *T. coriaceum*
    B. leaflets entire or 3-lobed apically, the lobes entire (rarely crenate); filaments usually white, more or less club-shaped; plants polygamous, polygamo-dioecious, or sometimes dioecious
       D. achenes and lower surfaces of leaflets with stipitate or sessile glands....
       .... *T. revolutum*
       D. achenes and lower surfaces of leaflets glabrous (rarely papillose) or pubescent, but not glandular
          E. leaflets more or less finely pubescent beneath
             F. stigmas 0.5–2.5 mm long, about ½ the length of the mature achene body, often coiled distally; filaments rigid, ascending, prominently club-shaped; anthers usually <1.5 mm long ....
             .... *T. pubescens*
             F. stigmas 1.5–4 mm long, nearly as long as the mature achene body, more or less filiform; filaments flexuous, filiform, scarcely dilated toward the tips; anthers 1.2–3.5 mm long.....*T. dasycarpum*
          E. leaflets glabrous (rarely papillose) beneath

G. anthers 0.5–1.6 mm long; stigmas 0.6–3 mm long; largest leaflets usually cordate basally, the lobes rounded to obtuse; flowers often perfect .... *T. pubescens*

G. anthers 1.3–3.6 mm long; stigmas 1–5 mm long; largest leaflets rounded to cuneate basally, rarely cordate, the lobes usually sharply acute or sometimes obtuse to rounded; flowers rarely perfect

H. leaflets coriaceous, minutely spiny to whitish-papillose beneath (at least on the veins), the veins usually prominent; stipes often >1 mm long; plants of dry open woods, brushy banks, thickets and barrens .... *T. revolutum*

H. leaflets membranous to firm, not minutely spiny or whitish-papillose beneath, the veins relatively inconspicuous; stipes usually <1 mm long; plants of damp thickets, swamps and wet meadows .... *T. dasycarpum*

Thalictrum coriaceum (Britton) Small　　　　　　Thick-leaved meadow-rue
Dioecious herb; stems 0.6–2 m tall from a typically bright yellow nonrhizomatous or rarely rhizomatous caudex; leaves ternately 1–4 times compound; upper subsessile, petioles of lower leaves well-developed, clasping the stem; leaflets 1–7.5 cm wide, membranous, glabrous beneath, apex 3–9-lobed or toothed; inflorescences pyramidal, loosely branched; sepals white to purplish, 1.5–3.5 mm long; achenes obliquely ovoid, bodies 3–6.5 mm long; stipes more or less wing-angled, 0.7–2.5 mm long; rare in rich rocky woods, thickets, and moist alluvium; SC; flr. late May–Jun; FACU; ✿. [syn: *T. steeleanum* B.Boivin].

Thalictrum dasycarpum Fisch. & Avé-Lall.　　　　　　Purple meadow-rue
Similar to *T. pubescens* in habit; plant usually dioecious, often with purplish stems; inflorescences paniculate, more or less pyramidal; sepals lanceolate, 3–5 mm long; stigmas filiform, 1.5–4.7 mm long, about equaling the length of the achene body, often dehiscent when fruits mature; achenes ovoid to fusiform, ribbed, bodies 4–6 mm long, narrowed to a sessile or subsessile base; very rare in wet alluvial thickets, swamps, and wet meadows; Warren and Forest Cos.; flr. late May–Jun; FACW; ✿.

Thalictrum dioicum L.　　　　　　Early meadow-rue
Dioecious herb, 3–8 dm tall; leaves ternately 1–4 times compound; petioles of basal and upper cauline leaves 3–12 cm long; leaflets 1–4.5 cm wide, 3–12 lobed at the tip; sepals greenish to purple, ovate or obovate to oval, 1.8–4 mm long; stamens pendulous, anthers 2–4 mm long, filaments colored; achenes ellipsoid, bodies 3.5–5 mm long, short-stipitate; common in rich rocky woods, ravines, and alluvial terraces; throughout; flr. Apr–May; FAC.

Thalictrum
pubescens,
leaflet (×1)

Thalictrum pubescens,
flower (×2)

Thalictrum pubescens Pursh　　　　　　Tall meadow-rue
Herb, 0.5–3 m tall; leaves ternately and pinnately decompound, basal and lower cauline leaves petiolate, upper cauline sessile; leaflets 0.5–5.5 cm wide, tip entire to 2–3(5)-lobed or toothed; inflorescences racemose to paniculate, more or less rounded; flowers unisexual or the pistillate ones with a few stamens; sepals white to purplish, elliptic rounded, 2–3.5 mm long; anthers 0. 5–1.5(2.1) mm long, blunt or only slightly apiculate, filaments white to purplish, ascending, rigid, distinctly club-shaped, 1.5–7 mm long; stigmas often coiled distally, 0.6–

2.5 mm long, about ½ the length of the achene body, usually persistent; achenes ellipsoid, ribbed, bodies 3–5 mm long; common in rich woods, low thickets, swamps, wet meadows, and stream banks; throughout; flr. Jun–Aug; FACW+. A very variable species. [syn: *T. polygamum* Muhl]

Thalictrum revolutum DC.   Skunk or purple meadow-rue
Usually dioecious herb, 0.5–2 m tall; leaves ternately 1–4 times compound, upper sessile, lower petiolate; leaflets grayish- or brownish-green, 0.5–4 cm wide, lower surfaces with glands, minutely spiny or whitish papillose, tip entire or 2–3(5)-toothed, margins strongly revolute; inflorescence racemose to paniculate, elongate; sepals whitish, ovate to oblanceolate, (2)3–4 mm long; anthers (0.7) 1.2–2.7 mm long, blunt to apiculate, filaments white, more or less flexuous, upwardly slightly club-shaped, 2.5–7.8 mm long; carpels 8–16; stigmas linear filiform, (1)1.5–3.3 mm long; achenes lanceolate to ellipsoid, ribbed bodies 1.5–5 mm long; frequent in dry open woods, brushy banks, thickets, and barrens; mostly S; flr. May–Jul; UPL.

Thalictrum revolutum, leaflet (×1)

Thalictrum revolutum, achene (×2½)

Thalictrum thalictroides (L.) A.J.Eames & B.Boivin   Rue anemone, windflower
Herb, 1–3 dm tall; basal leaves ternately twice-compound; involucral leaves usually ternately compound, subsessile, otherwise similar to the basal leaves; leaflets 0.8–3 cm wide, 3-lobed at the tip; flowers perfect, (1)2–several; sepals usually 5, white to pinkish, 5–15 mm long; anthers 0.4–0.7 mm long, filaments narrowly club-shaped distally, 3–4 mm long; achenes fusiform, bodies 3–4.5 mm long, strongly ribbed; stipes 0.1–0.4 mm long; common in wooded banks and thickets; throughout, except at the highest elevations; flr. Apr–Jun; FACU–. [syn: *Anemonella thalictroides* (L) Spach.]

Trautvetteria Fisch. & C.A.Mey.

Trautvetteria caroliniensis (Walter) Vail   Carolina tassel-rue
Erect, leafy, glabrous to glabrate, perennial herb; stem 0.5–1.5 m tall from a short rootstock; basal leaves palmately 5–11-lobed, 1–3(4) dm wide, segments variously lobed, margins lacerate to serrate, base cordate; cauline leaves alter-

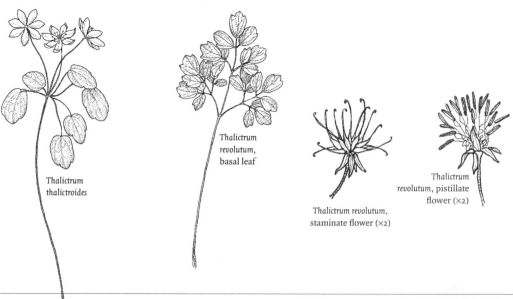

Thalictrum thalictroides

Thalictrum revolutum, basal leaf

Thalictrum revolutum, staminate flower (×2)

Thalictrum revolutum, pistillate flower (×2)

nate, similar to the basal ones, upwardly reduced; inflorescences corymbose, many-flowered, long-pedunculate; pedicels densely covered with hooked hairs; flowers perfect, small; sepals 3–5, soon falling, white, 3–5 mm long; carpels numerous; fruits (dehiscent utricles) inflated, ellipsoid to obovoid, angled, 3–4 mm long, membranous, glabrous, facially nerved, styles persistent, recurved; seed 1; very rare in wooded seepage slopes, stream banks, and bogs; SW; flr. Jun–Jul; FACW–; ✹.

Trollius L.

Trollius laxus Salisb.   Spreading globe-flower
Erect, glabrous, perennial herb; stem 1–5 dm tall, from short rhizomes; basal leaves palmately 3–7 parted with 3-lobed coarsely toothed segments; cauline leaves 1–3, similar to the basal ones, upwardly reduced; flowers perfect, showy, solitary at ends of branches; sepals 5–7, petal-like, 1–2 cm long; petals 10–15, yellowish, oblong, 3–5 mm long, 1-lipped, shorter than the stamens; follicles elongate with persistent beaks 8–12 mm long; seeds several; rich moist calcareous meadows, swamps, and open woods; rare; mostly in extreme E and NW; flr. Apr–May; OBL; ✹.

Xanthorhiza Marshall

Xanthorhiza simplicissima Marshall   Shrub yellowroot
Colonial, weak, pubescent shrub with deep yellow wood; stems simple to few-

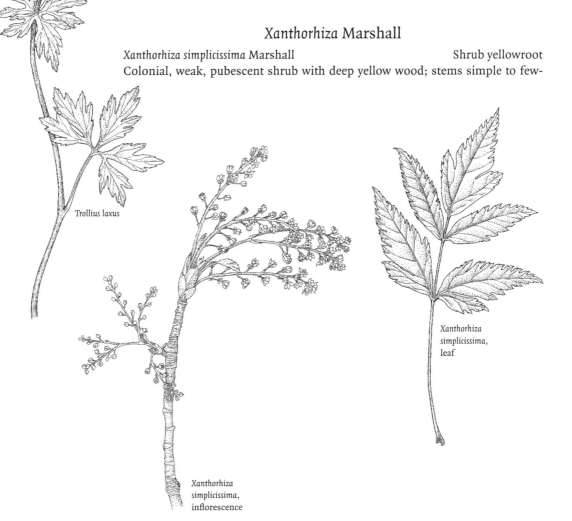

Trollius laxus

Xanthorhiza simplicissima, inflorescence

Xanthorhiza simplicissima, leaf

branched, 2–7 dm tall; roots yellow; leaves pinnately compound, alternate; leaflets 3–5, to 9 cm long, variously cleft, incised, or serrate; inflorescence paniculate, drooping; flowers perfect or sometimes unisexual; sepals 5, dark purple, about 3 mm long; petals 5, clawed, nectary transversely oblong, bilobed; stamens 5(10); carpels (2)5–10, styles short; follicles yellowish, compressed, 3–4 mm long; sometimes naturalizing in woods; SE; flr. Apr–May; native to the southeastern U.S.; FACW.

RESEDACEAE Mignonette Family

Herbs with alternate leaves; stipules reduced to a pair of glands; flowers in a dense raceme, irregular; sepals and petals each 4–6; stamens 10–25, attached to a 1-sided, hypogenous disk; ovary of 3–4 fused carpels; fruit a 3–6-lobed capsule splitting open at the top.

Reseda L.

Flowers greenish-white to greenish-yellow; sepals nearly equal; petals unequal, the upper one larger; flr. Jun–early Aug.

A. leaves mostly entire
    B. pedicels twice as long as the calyx; petals with numerous filiform appendages .... *R. odorata*
    B. pedicels short; petals merely toothed or cleft .... *R. luteola*
A. leaves pinnately divided
    C. leaves regularly pinnatifid; flowers greenish-white .... *R. alba*
    C. leaves irregularly pinnatifid; flowers greenish-yellow .... *R. lutea*

Reseda alba L.   White mignonette
Erect, glabrous, glaucous annual or perennial to 8 dm tall; leaves deeply pinnately divided into linear segments; flowers greenish-white, fragrant; petals 5 or 6, with 3 appendages; ovary and fruit 4-lobed; rare on ballast and waste ground; SE; native to S. Europe.

Reseda lutea L.   Yellow mignonette
Biennial or perennial with a decumbent to erect stem to 8 dm tall; leaves oblanceolate, deeply divided into linear segments; sepals 6; petals 6, greenish-yellow, irregularly cleft and with appendages; occasional in waste ground, often on limestone; SE and SC; native to Europe.

Reseda luteola L.   Dyer's rocket
Biennial, stem erect to 8 dm tall; leaves linear to lanceolate or oblanceolate, 4–10 cm long with a short petiole; raceme narrow, pedicels short; flowers 2–4 mm wide; sepals 4; petals 4 or 5, the upper lobed, the lower toothed; capsule 3-lobed; rare on waste ground and roadsides, formerly cultivated for dye; SE; native to Europe.

Reseda odorata L.   Mignonette
Annual or biennial, decumbent-ascending, to 6 dm tall; leaves oblanceolate to

Reseda lutea

obovate, entire or with a few basal lobes; flowers fragrant; petals greenish- or yellowish-white with numerous filiform appendages; cultivated and rarely escaped to waste ground; SE; native to N. Africa.

RHAMNACEAE Buckthorn Family

Shrubs and small trees with simple, alternate or opposite leaves; stipules present (though sometimes minute); flowers small, stamens and petals (when present) attached to a disk lining the floral cup; flowers perfect or occasionally dioecious, 4–5-merous; stamens as many as and alternate with sepals; fruit a capsule or a berry-like drupe.

A. leaves pinnately veined; flowers solitary or in few-flowered clusters; fruit fleshy, berry-like .... *Rhamnus*
A. leaves strongly 3-veined from top of petiole; flowers numerous in panicles of clustered umbels, fruit dry, 3-lobed .... *Ceanothus*

Ceanothus L.

Ceanothus americanus L.   New Jersey tea
Shrub to 1 m tall; leaves alternate, margins toothed, 3–8 cm, narrowly to broadly ovate; flowers white, perfect, 5-merous; inflorescences on long peduncles to 2 dm on lower branches, often subtended by 1–3 reduced leaves; fruit 3-lobed, capsule-like, the lobes separating when ripe, 5–6 mm diameter; wooded bluffs, roadside banks, and shaly slopes; throughout; flr. summer.

Rhamnus L.

Shrubs or small trees; leaves pinnately veined, alternate or opposite, entire or toothed; flowers small, greenish, solitary or in few-flowered umbels in axils of lower leaves of current season's growth, 4–5 merous, perfect or dioecious; fruit a fleshy berry-like drupe containing 2–4 seeds.

Summer key

A. leaves opposite or subopposite; twigs often spine-tipped
    B. leaves 3–7 cm, papery, with 3 lateral veins on each side of the midrib; fruit 4-seeded .... R. *cathartica*
    B. Leaves 6.5–13 cm, leathery, with 4 or more lateral veins on each side of the midrib; fruit 2-seeded ....R. *davurica*
A. leaves alternate; twigs not spine-tipped
    C. leaves entire; 2–8 flowers per inflorescence ....R. *frangula*
    C. leaves toothed; 1–3 flowers per inflorescence
        D. leaf margins serrulate (12–15 teeth per cm); sepals, petals, and stamens 4; fruit 2-seeded ....R. *lanceolata*
        D. leaf margins crenate-serrate (7–8 teeth per cm); sepals and stamens 5, petals absent; fruit 3-seeded ....R. *alnifolia*

Ceanothus
americanus

Rhamnus cathartica

Winter key

A. leaf scars opposite or subopposite .... R. *cathartica* or R. *davurica*
    (difficult to distinguish in winter condition)
A. leaf scars alternate
    B. terminal buds 2–5 mm long, covered by glabrous scales
        C. terminal bud about 2 mm long, brown ....R. *lanceolata*
        C. terminal bud 4–5 mm long, red-brown ....R. *alnifolia*
    B. terminal buds 5–10 mm long, naked, pubescent ....R. *frangula*

Rhamnus alnifolia L'Hér.   Alder-leaved buckthorn
Shrub to 1 m tall; leaves alternate, margins unevenly toothed, 4.5–11.5 cm, elliptic or ovate, apex acuminate to obtuse; dioecious; flowers 5-merous, 1–3 per inflorescence; fruit black, 3-seeded, 6 mm diameter; fens, calcareous marshes,
and wet thickets; NE and NW; flr. May–June; OBL.

Rhamnus cathartica L.   Buckthorn
Shrub or small tree to 6 m tall; leaves opposite, margins crenulate-serrate, 3–7
cm, ovate to elliptic; some branches ending in spines; usually dioecious; flowers
4-merous, 10–15 per inflorescence; fruit glassy-black, 4-seeded, 5 mm diameter;
open woods, pastures, fencerows, and roadside banks; flr. May–Jun; native to
Europe.

Rhamnus davurica Pall.   Buckthorn

Large shrub or rarely a small tree to 7 m tall; leaves opposite, margins crenulate-serrate, 6.5–13 cm, oblong to elliptic, somewhat leathery; twigs thick, spiny; dioecious; flowers 4-merous, 10–15 per inflorescence; fruit blue-black, 2-seeded, 6–8 mm diameter; cultivated and occasionally spreading to old fields and roadside thickets; flr. late Apr–May; native to N. China.

Rhamnus frangula L.   Alder buckthorn

Shrub or small tree to 2–5 m tall; leaves alternate, entire, 5–8 cm, oblong or usually obovate-oblong, 7–9 lateral veins on each side of the midrib; flowers perfect, 5-merous, 2–8 per inflorescence; fruit red changing to purple-black, 2–3-seeded, 6–10 mm diameter; cultivated and escaped to fields, thickets, roadside banks, aggressively colonizing wet habitats; flr. May–June; native to Europe.

Rhamnus lanceolata Pursh   Lanceolate buckthorn

Shrub to 1–2 m tall; leaves alternate, margins serrulate (12–15 teeth per cm), 3–8 cm, elliptic, apex acuminate, 6–9 lateral veins on each side of the midrib; dioecious; flowers 4-merous, 2–3 per inflorescence; fruit black, 2-seeded, 6–7 mm diameter; boggy fields, stream banks, and calcareous woods; mostly SC; flr. late Apr–May; 🌱.

ROSACEAE Rose Family

A diverse family including trees, shrubs, and herbs; stems armed or unarmed; leaves alternate (except *Rhodotypos*), simple or compound, usually serrate; stipules present (except *Spiraea*); flowers regular and usually perfect, perigynous or epigynous, mostly 5-merous (occasionally 4-merous), hypanthium shallowly saucer-shaped to narrowed at the top and enclosing the ovaries; petals distinct; carpels 1–many, distinct or united; stamens often numerous; fruit an achene, follicle, hip, drupe, aggregation of drupelets, or pome.

A. plants woody
  B. leaves simple
    C. leaves opposite; flowers 4-merous .... *Rhodotypos*
    C. leaves alternate; flowers 5-merous
      D. branches armed
        E. thorns located at the nodes
          F. small trees or large shrubs; flowers solitary and pedicellate or in simple or compound cymes, opening after the leaves .... *Crataegus*
          F. low shrubs; flowers nearly sessile, solitary or in clusters of 2–5, opening before the leaves .... *Chaenomeles*
        E. some branches ending in thorns; flowers in small, umbel-like clusters
          G. ovary inferior; fruit a pome
            H. styles fused at the base; mouth of the hypanthium open; fruit without stone cells .... *Malus*
            H. styles separate to the base; mouth of the hypanthium closed around the styles; fruit containing stone cells .... *Pyrus*

    G. ovary superior; fruit a drupe .... *Prunus*
D. branches unarmed
    I. shrub; young branches and petioles densely glandular-pubescent........
        ....*Rubus odoratus*
    I. trees or shrubs; young branches not densely glandular-pubescent
        J. petioles usually bearing a pair of glands near the base of the leaf
            blade; bark with prominent, horizontal lenticels; fruit a drupe........
            .... *Prunus*
        J. petioles without glands; bark without horizontal lenticels; fruit a
            pome or a cluster of follicles or achenes
            K. leaves with small dark glands or scales along the upper surface
                of the midrib .... *Aronia*
            K. leaves without scales on the midrib
                L. leaves palmately lobed; bark papery, exfoliating .... *Physocarpus*
                L. leaves serrate to entire, some or all of the leaves not lobed;
                    bark not exfoliating
                    M. branches and twigs green, smooth .... *Kerria*
                    M. branches and twigs brown or gray
                        N. ovary superior, fruit a cluster of follicles; stipules (or
                            stipule scars) not present .... *Spiraea*
                        N. ovary inferior, fruit a pome; stipules (or stipule scars)
                            present
                            O. flowers (and fruits) mostly in racemes; fruits <1 cm
                                thick .... *Amelanchier*
                            O. flowers (and fruits) not in racemes; fruits small or
                                large
                                P. leaf margins crenate or serrate; fruits glabrous
                                    Q. fruit pear-shaped or the surface covered with
                                        small punctate dots, flesh containing stone
                                        cells .... *Pyrus*
                                    Q. fruit not pear-shaped or punctate, lacking
                                        stone cells
                                        R. flowers and fruits solitary, fascicled, or in
                                            umbels .... *Malus*
                                        R. flowers and fruits in corymbs or panicles....
                                            .... *Photinia*
                                P. leaf margins entire
                                    S. fruits 1 cm or less in diameter, glabrous.........
                                        .... *Cotoneaster*
                                    S. fruits 7–10 cm in diameter, pubescent ....
                                        .... *Cydonia*
B. at least some of the leaves compound
    T. branches armed
        U. fruit a hip .... *Rosa*
        U. fruit a cluster of coherent drupelets on a convex receptacle .... *Rubus*
    T. branches unarmed
        V. shrubs; leaflets 5–7, flowers yellow .... *Potentilla fruticosa*
        V. shrubs or trees; leaflets 9–21; flowers white
            W. inflorescence flat-topped; fruit a small orange pome .... *Sorbus*
            W. inflorescence elongate; fruit a cluster of dehiscent follicles....*Sorbaria*

A. plants herbaceous
  X. leaves simple, crenate .... *Dalibarda*
  X. leaves compound or at least strongly pinnatifid
    Y. leaves trifoliate or palmately compound
      Z. low-growing or trailing plants spreading by rhizomes or stolons; the leaves all petiolate
        AA. leaflets mostly 5, occasionally 3 or 7 .... *Potentilla*
        AA. leaflets 3
          BB. calyx with bractlets between the sepals thus appearing 10-lobed
            CC. bractlets entire, similar to the sepals; receptacle enlarging and turning red and becoming sweet and edible in fruit .... *Fragaria*
            CC. bractlets with 3–5 teeth at the tips, larger than the sepals; receptacle enlarging somewhat and turning red, but not becoming sweet and edible ... *Duchesnea*
          BB. calyx 5-lobed, without bractlets between the sepals .... *Waldsteinia*
      Z. plants with upright stems, not stoloniferous or trailing
        DD. the lowest leaves with petioles <2 cm .... *Porteranthus*
        DD. basal or lower cauline leaves with petioles >2 cm
          EE. style shorter than the achene, neither hooked nor persistent .... *Potentilla*
          EE. basal portion of the style persistent as a hooked appendage on the achene .... *Geum*
    Y. leaves pinnate or 2 or 3 times compound or dissected
      FF. leaves 2 or 3 times compound .... *Aruncus*
      FF. leaves pinnate
        GG. hypanthium with a double row of hooked prickles at the top, enclosing the achenes in fruit .... *Agrimonia*
        GG. hypanthium without hooked prickles
          HH. plants more or less hairy, individual flowers >8 mm wide, in a few-flowered, open corymb
            II. style exceeding the achene, the basal portion persistent as a hooked appendage on the achene .... *Geum*
            II. style shorter than the achene, neither hooked nor persistent .... *Potentilla*
          HH. plants glabrous, individual flowers 2–8 mm wide, numerous, in a dense spike or panicle
            JJ. plants glaucous; terminal leaflet similar to the lateral ones; petals lacking .... *Sanguisorba*
            JJ. plants not glaucous; terminal leaflet broadly 3-lobed; petals present .... *Filipendula*

Agrimonia L.

Herbaceous, more or less hairy perennials with upright, mostly unbranched stems; leaves pinnately compound, the larger leaflets interspersed with very reduced ones along the rachis; stipules large; flowers in spike-like racemes; petals yellow; hypanthium top-shaped and crowned with a double row of hooked prickles; fruit consisting of 2 achenes enclosed within the persistent hypanthium that, in addition to the prickles, has, at maturity, a beak-like top formed by the sepal lobes.

A. leaflets (excluding the tiny intermediate ones) mostly 11 or more, each 3–4 times as long as wide .... *A. parviflora*
A. leaflets 3–9(11), each not more than twice as long as wide
    B. axis of the inflorescence glandular; lower leaf surface hairy only on the veins
        C. stems 2–3 mm thick, glabrous or with very sparse hairs; fruit with few bristles .... *A. rostellata*
        C. stems 4 mm or more thick, with long spreading hairs; fruit with many bristles .... *A. gryposepala*
    B. axis of the inflorescence not glandular; lower leaf surfaces densely pubescent
        D. leaflets glandular dotted beneath .... *A. striata*
        D. leaflets not glandular beneath
            E. larger leaves with 5–9 leaflets; stem and floral axis with ascending or incurved hairs ....*A. pubescens*
            E. larger leaves with 3–5 well-developed leaflets; stem and floral axis with spreading hairs .... *A. microcarpa*

Agrimonia gryposepala Wallr.   Agrimony, harvest-lice
Stems stout, 0.3–1.5 m tall, hairy; leaflets 5–9 (excluding the tiny ones), nearly glabrous beneath with resin droplets; flowers 8–12 mm across; fruits 6–8 mm long, the outer bristles reflexed; common in woods, fields, and floodplains; throughout; flr. Jul–Sep; FACU.

Agrimonia microcarpa Wallr.   Small-fruited agrimony
Stems slender, simple, pubescent, to 6 dm tall; leaflets mostly 3, pilose beneath; fruits 4 mm; occasional in woods; mostly SE; flr. Jul–Aug.

Agrimonia parviflora Aiton   Southern agrimony
Stems stout to 1.2 m tall, hairy; leaves with 11–15 elliptic-lanceolate, sharply serrate leaflets interspersed with smaller ones; flowers 6–10 mm wide; fruits 4–6 mm long, the outer bristles spreading, the inner erect; frequent in bogs, moist woods, and thickets; mostly S; flr. Jul–early Sep; FACW.

Agrimonia parviflora, fruits (×2)

Agrimonia pubescens Wallr.   Downy agrimony
Stems to 1 m tall, hairy; leaflets 5–9, oblong or elliptic, velvety-pubescent beneath; flowers 6–8 mm wide; axis of the inflorescence hairy but not glandular; fruit 4–5 mm long; frequent in rich, rocky woods, woods edges, and slopes; throughout except in the northernmost counties; flr. Jul–Sep.

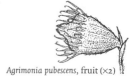

Agrimonia pubescens, fruit (×2)

Agrimonia rostellata Wallr.   Woodland agrimony
Stems slender, nearly glabrous, 2–10 dm tall; leaflets elliptic to obovate with resin droplets beneath and few hairs; axis of the inflorescence glandular; flowers 4–5 mm wide; fruit 7–9 mm long; occasional in woods, fields, and thickets; throughout; flr. Jul–Sep; FACU.

Agrimonia rostellata, fruit (×2)

Agrimonia
parviflora (×¼)

Agrimonia
rostellata

Agrimonia pubescens

Agrimonia striata Michx.   Roadside agrimony

Stems to 1 m or more tall; stout and hairy; leaflets 7–9, the lower surfaces bearing resin droplets and hairy on the veins; flowers 6–10 mm wide; fruits 6–8 mm long, strongly ridged; frequent on stream banks, open floodplains, or woods edges; mostly N; flr. Jul–Aug; FACU–.

Amelanchier Medik.
Ann Robinson

Early spring flowering shrubs or small trees; leaves simple and serrate; flowers mostly in racemes; sepals 5, erect to reflexed, persistent; petals 5, white (or very rarely pinkish); stamens mostly 20, shorter than the petals; fruit a 10-seeded pome. The taxonomy of *Amelanchier* is complicated by wide variability observed in the field resulting from hybridization, polyploidy, aneuploidy, and apomixis.

A. flowers 1–4, commonly in erect or semi-erect fascicles; leaves not folded in
   bud; fruit pear-shaped; angles of the leaf apex and base commonly acute ......
   .... *A. bartramiana*

A. flowers more numerous, in semi-erect or flexed racemes; leaves folded length-
    wise in bud; fruit round, ovoid, or pear-shaped; angle of the leaf apex and base
    variable
    B. mature leaves averaging >1.5 times longer than wide; outer scales of over-
        wintering buds tannish-brown; fruit round to ovoid
        C. angle of the apex of mature leaves usually <90%; sepals on fruit flexed
            at their base
            D. leaves at flowering time folded or barely beginning to unfold; leaf
                undersurface tomentose to densely appressed-pilose, although some
                hairs are shed early, hairs always present on leaf undersurface at base
                of midrib at leaf abscission; young leaves green .... A. arborea
            D. leaves at flowering time beginning to unfold; amount of hair on the
                leaf undersurface variable, but undersurface completely glabrous by
                the time of leaf abscission, and frequently before; young leaves fre-
                quently copper-colored, red pigment also evident in fall coloration...
                ....A. laevis
        C. angle of the apex of mature leaves usually >90%; sepals on fruit semi-
            erect to erect .... A. canadensis
    B. mature leaves averaging <1.5 times longer than wide; outer scales of over-
        wintering buds maroon; fruit pear-shaped
        E. summit of ovary glabrous at flowering; sepals on fruit erect, spreading,
            or less commonly revolute from the middle .... A. obovalis
        E. summit of the ovary densely appressed pilose at flowering; sepals on
            fruit spreading or revolute from the middle
            F. mature leaves coarsely serrate-dentate, with 3 or more secondary
                veins on each side entering directly into the teeth of the mature leaf;
                floral hypanthium broadly saucer-shaped .... A. sanguinea
            F. mature leaves less coarsely serrate, secondary veins more frequently
                forking prior to the leaf margin with tertiary veins entering directly
                into the teeth; floral hypanthium cup-shaped
                G. leaf apex commonly obtuse; leaf base rounded, less commonly
                    subcordate; leaf blades oval; leaves just unfolding at flowering
                    time; petal length 5–9 mm .... A. stolonifera
                G. leaf apex commonly broadly subacute; leaf base subcordate; leaf
                    blades oval-oblong; leaves ½ expanded at flowering time; petal
                    length 7–10 mm .... A. humilis

Amelanchier arborea (F.Michx.) Fernald   Shadbush, serviceberry, Juneberry
Erect shrub or small tree to 15 m tall; leaves folded at flowering, densely pubescent beneath when young, becoming glabrous later except for some hairs at the
base of the midrib; petals 10–15 mm long; ovary glabrous to somewhat pubescent at the summit; fruit dark red, dryish in texture; edges, upland woods, and
rocky bluffs; frequent throughout; flr. mid Apr–early May, usually 10–14 days
earlier than *A. humilis*, *A. obovalis*, *A. sanguinea* or *A. stolonifera*, frt. Jun; FAC–.

Amelanchier bartramiana (Tausch) M.Roem.
                    Mountain Juneberry, oblong-fruited serviceberry
Shrub to 2 m tall, often growing in clumps; leaves glabrous, not folded in bud,
the midrib broad and conspicuous on the upper side; inflorescence cymose with
at least the lowest of the 1–4 flowers in a leaf axil; ovary tomentose at the summit
forming a cap of hairs at the base of the styles; fruit pear-shaped, dark purple;

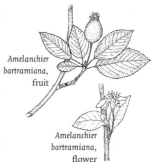

*Amelanchier
bartramiana,
fruit*

*Amelanchier
bartramiana,
flower*

rare in sphagnum bogs and peaty thickets; NE and NC; flr. Apr–early May, frt. Jun; FAC; 🌿.

Amelanchier canadensis (L.) Medik.   Shadbush, serviceberry, Juneberry
Shrub or small tree to 8 m tall, often growing in fastigiate clumps; leaves folded at flowering and densely tomentose beneath, becoming glabrescent, elliptic, elliptic-ovate to obovate, to 3–8 cm long, veins branching and anastomosing toward the margins; racemes short and erect; pedicels and hypanthium tomentose; petals 7–10 mm long; ovary glabrous at the summit; fruit dark purple to black, juicy and sweetish; rare in moist woods and swamps; SE; flr. mid Apr–mid May, frt. Jun–early Jul; FAC.

Amelanchier humilis Wiegand   Low Juneberry or serviceberry
Colonial shrub 0.3–8 m tall; leaves ½ expanded at flowering, densely tomentose beneath becoming glabrescent, oval-oblong or less commonly oval, rarely obovate-cuneate, serrate, veins forking and entering the teeth; raceme erect, densely tomentose; petals 7–10 mm long; fruit black with a waxy bloom, sweet and juicy; rare in dry, open ground and on bluffs; scattered; flr. Apr–mid May, frt. Jun–early Jul; FACU; 🌿.

Amelanchier laevis Wiegand   Smooth shadbush or serviceberry
Erect shrub or tree to 13 m tall; leaves elliptic to ovate-oblong, 4–6 cm long, finely serrate, rounded or subcordate at the base, beginning to unfold and with a distinct reddish hue at flowering, glabrous, lateral veins curving and anastomosing toward the margins; pedicels glabrous; petals 10–22 mm long; ovary glabrous at the summit; fruit dark purple-red, sweet and juicy; frequent in rocky woods, thickets and roadside banks; throughout; flr. mid Apr–mid May, frt. Jun–early Jul.

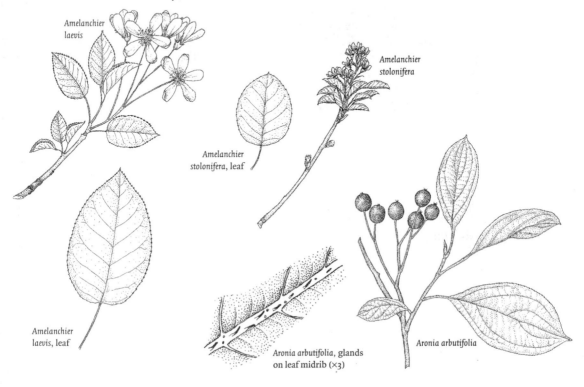

Amelanchier
laevis

Amelanchier
stolonifera

Amelanchier
stolonifera, leaf

Amelanchier
laevis, leaf

Aronia arbutifolia, glands
on leaf midrib (×3)

Aronia arbutifolia

Amelanchier obovalis (Michx.) Ashe — Coastal Juneberry or shadbush

Colonial shrub 0.3–1.5 m tall; leaves barely evident at flowering, densely tomentose beneath when young, becoming glabrescent, broadly elliptic to subrotund, 3–5 cm long, sharply serrulate nearly to the base, veins branching and anastomosing toward the margins; flowers in dense spike-like racemes to 3 cm long; pedicels and hypanthia tomentose; petals 6–7 mm long; ovary glabrous at the summit; fruit pear-shaped; very rare in open woods and thickets; SE; flr. mid Apr–mid May, frt. Jun–early Jul; FACU; .

Amelanchier sanguinea (Pursh) DC. — Roundleaf Juneberry or shadbush

Erect shrub or small tree 3(6) m tall, usually clumped; leaves half grown at flowering, tomentose beneath, becoming glabrous, oblong to subrotund, coarsely toothed, mostly above the middle, lateral veins extending into the teeth; flowers in long drooping racemes; petals 11–15 mm long; ovary tomentose at the summit; rare and scattered in open woods, rocky slopes and barrens; flr. mid Apr–late May; frt. Jun–early Jul; .

Amelanchier stolonifera Wiegand — Low Juneberry or shadbush

Colonial shrub 0.3–1.5 m tall; leaves just unfolding at flowering, densely pubescent beneath, becoming glabrous, commonly oval, rarely oblong-oval, or orbicular, 2–5 cm long, finely toothed, veins branched and anastomosing near the margins; petals 5–9 mm long; ovary tomentose at the summit; occasional in woods, old fields, fencerows, and serpentine barrens; throughout; flr. Apr–early May, frt. Jun; FACU. [syn: *A. spicata* (Lam.) K.Koch]

Amelanchier stolonifera, fruit

Aronia Medik.

Erect, colonial shrubs to 3 m tall; leaves simple, alternate, deciduous; leaf blades broadly oblanceolate to elliptic, finely serrate, glabrous above except for small dark scales along the midrib; flowers small, white, in a cluster; petals 5, broadly ovate, spreading; stamens 20; ovary inferior; fruit a small pome; flr. late Apr–May. In addition to the 2 species below, a probable hybrid, with variable, intermediate characters, has sometimes been recognized as *A. prunifolia* (Marsh.) Rehder.

A. branchlets, pedicels, and lower leaf surfaces pubescent; fruits red .... *A. arbutifolia*

A. branchlets and lower leaf surfaces glabrous or nearly so; fruit black .... *A. melanocarpa*

Aronia arbutifolia (L.) Elliot — Red chokeberry

Upright shrub varying in height from 0.5 to 3 m depending on habitat; leaves dull green, turning red in autumn; branchlets and lower leaf surfaces tomentose; fruits red, persistent; frequent in swamps, bogs, and moist woods; throughout; FACW.

Aronia melanocarpa (Michx.) Elliot — Black chokeberry

Upright shrub with lustrous green leaves turning brown in autumn; branchlets and lower leaf surfaces nearly glabrous; fruits black, not persistent; occasional in swamps, bogs, wet or dry woods, and barrens; throughout; FAC.

Aruncus Adans.

Aruncus dioicus (Walter) Fernald   Goat's-beard
Herbaceous, dioecious perennial, 1–2 m tall, with numerous small white flowers in a 1–3-dm terminal panicle; leaves 2–3 times compound; leaflets oblong to broadly ovate, doubly serrate; fruit a cluster of follicles; occasional in rich woods and wooded roadsides; SW; flr. Jun–Jul; FACU.

Chaenomeles Lindl.

Chaenomeles speciosa (Sweet) Nakai   Flowering quince
Thorny deciduous shrub with alternate, simple leaves, conspicuous stipules, and showy red-orange, pink, or white flowers; cultivated and occasionally persisting at old home sites; flr. Mar–Apr, before the leaves; native to China. [syn: *C. lagenaria* Koidz.]

Cotoneaster Medik.

Cotoneaster apiculata Rehder & Wilson   Cranberry cotoneaster
Deciduous shrub to 2 m tall; leaves suborbicular or orbicular-ovate, 1.5 cm, glabrous and shining above, sparsely hairy beneath; flowers in small clusters, pink; fruit red, subsessile, 7–10 mm; cultivated and rarely naturalized; scattered; flr. Jun; native to China.

Crataegus L.

Thorny trees or shrubs with alternate, simple leaves that are coarsely toothed and usually more or less lobed; flowers perfect, regular, solitary (in *C. uniflora*) or more often in simple or compound cymes; petals 5, white; stamens 10 or 20; fruit a greenish or red pome with 1–5 nutlets and a persistent calyx (deciduous in *C. phaenopyrum*) that is sometimes prominently elevated. Crataegus is a difficult genus, complicated by hybridization, polyploidy and apomixis. Numerous "species" have been named and later relegated to synonymy. In the absence of a recent definitive study of the Pennsylvania material, we have chosen to follow Gleason and Cronquist (1991) and recognize broad species complexes. For more complete synonymy see Rhoads and Klein (1993).

A. leaves with some of the lateral veins extending to the sinuses as well as the points of the lobes
    B. leaves sharply acute or acuminate, cordate at the base; calyx not persistent in fruit .... *C. phaenopyrum*
    B. leaf apex rounded to acute, base not cordate; calyx persistent in fruit........ .... *C. monogyna*
A. leaves with the lateral veins running only to the points of the lobes or teeth
    C. shrubs 1–2 m tall; flowers mostly solitary (or 2–3 per cluster) ....*C. uniflora*
    C. trees or large shrubs; flowers in compound cymes
        D. leaves of the flowering branches narrowly tapered or wedge-shaped at the base
            E. leaves of the flowering branches not lobed, mostly obovate to oblong-elliptic
                F. branches of the inflorescence glabrous

G. nutlets 1–2, shallowly or not at all pitted .... *C. crus-galli*

G. nutlets 2–3, deeply pitted .... *C. succulenta*

  F. branches of the inflorescence pubescent

    H. nutlets 3–5, smooth or only slightly pitted .... *C. punctata*

    H. nutlets 2–3, deeply pitted .... *C. calpodendron*

    E. leaves of the flowering branches lobed, rarely obovate

      I. petiole and leaf base glandular; fruit hard and dry ...... *C. intricata*

      I. petiole not glandular; fruit becoming soft or succulent

        J. leaves obtuse or acute at the tip; nutlets not pitted ....
.... *C. rotundifolia*

        J. leaves abruptly acuminate; nutlets with a shallow pit............
.... *C. brainerdii*

  D. leaves of the flowering branches broadly rounded, truncate, or cordate at the base

    K. lower surfaces of the leaves, inflorescence branches, and fruits pubescent .... *C. mollis*

    K. mature leaves and fruits mostly glabrous; inflorescence glabrous or villous

      L. flowers 2.2–2.6 cm wide; leaves of vegetative branches usually cordate .... *C. dilatata*

      L. flowers <2 cm wide; leaves of vegetative branches rounded to truncate or subcordate

      M. sepals deeply glandular-serrate .... *C. coccinea*

      M. sepals entire or slightly glandular-serrate

        N. young leaves glabrous; fruiting calyx prominent and elevated .... *C. pruinosa*

        N. young leaves with short hairs above; fruiting calyx small and sessile .... *C. flabellata*

Crataegus brainerdii Sarg.   Brainerd's hawthorn

Shrub or tree to 8 m tall; twigs glabrous, thorny; leaves elliptic to ovate with shallow lateral lobes; flowers 1.3–1.8 cm wide; cyme glabrous; fruit red; nutlets 2–5, acute, shallowly pitted on the inner surface; rare in moist bottomlands; scattered; flr. May–Jun, frt. Aug–Sep; ✿.

Crataegus calpodendron (Ehrh.) Medik.   Pear or blackthorn hawthorn

Very similar to *C. succulenta* but with twigs villous to tomentose when young and larger leaves that are short villous above and usually also beneath; rare in woods, thickets, and low meadows; mostly SE; flr. May–Jun, frt. Aug–Sep.

Crataegus coccinea L.   Red-fruited hawthorn

Large shrub or tree to 10 m tall; twigs glabrous or villous when young; leaves ovate or suborbicular with 3–5 pairs of lateral lobes, glabrous, or short-hairy above at least when young; petiole often glandular; sepals deeply glandular-serrate; fruit bright red, 0.7–1.3 cm thick, becoming succulent; occasional in open woods, fields, roadsides, and stream banks; flr. Apr–May, frt. Aug–Sep.

Crataegus crus-galli L.   Cockspur hawthorn

Large shrub or tree to 10 m tall with widely spreading branches; leaves mostly obovate, unlobed or only slightly so, dark green and glossy; glabrous; flowers

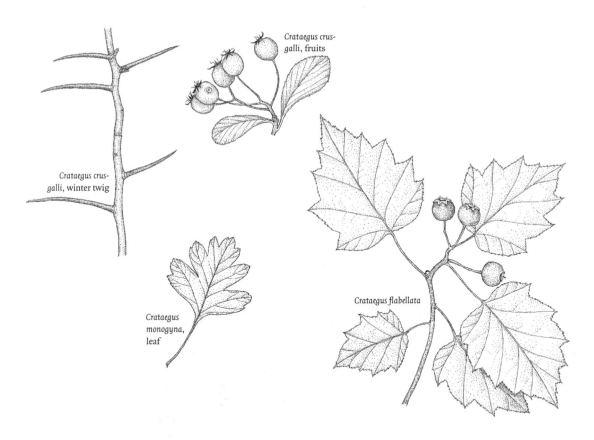

Crataegus crus-galli, fruits

Crataegus crus-galli, winter twig

Crataegus monogyna, leaf

Crataegus flabellata

0.8–1.8 cm wide; pedicels glabrous; fruit red to green, remaining hard and dry; nutlets 1–2; occasional in woods, meadows, roadsides, and thickets; flr. May–Jun, frt. Aug–Sep; FACU.

Crataegus dilatata Sarg.   Broadleaf hawthorn
Shrub or small tree to 7 m tall armed with stout, 3–5-cm-long, dark purple thorns; leaves broadly ovate or deltoid, cordate at the base, sharply and deeply serrate with 4–5 pairs of triangular lateral lobes, short-hairy at least on the upper surface when young or even at maturity; rare in pastures, thickets, and hillsides; C; flr. May–Jun, frt. Aug–early Sep; 🍂.

Crataegus flabellata (Spach) G.Kirchn.   Fanleaf hawthorn
Shrub or small tree to 8 m tall with glabrous twigs; leaves broadly ovate to deltoid, rounded to obtuse or truncate at the base, short-hairy when young and villous along the veins beneath; flowers 1.3–3 cm wide in glabrous or villous compound cymes; fruit red, becoming soft or succulent; nutlets 3–5; frequent in open woods, fencerows, abandoned fields, and roadsides; throughout; flr. Apr–May, frt Aug–Sep.

Crataegus intricata Lange *sensu lato*   Biltmore hawthorn
Irregularly branched shrub or small tree to 8 m tall, with slender, 2.5–5-cm long thorns; leaves with several pairs of shallow lateral lobes and glandular petioles; flowers 1.3–2 cm wide in a 4–10-flowered cyme containing numerous bracts with conspicuous stipitate margins; fruits greenish, orange, or red with thin flesh

and 3–5 nutlets; infrequent in woods, pastures, thickets, and barrens; S; flr. May–Jun, frt. Aug–Sep.

Crataegus mollis (Torr. & A.Gray) Scheele   Downy hawthorn
Tree to 12 m tall with wide-spreading branches with few thorns or sometimes thornless; young twigs villous; leaves densely short-hairy above and more or less tomentose beneath when young, becoming less hairy at maturity, those of the vegetative shoots deeply lobed; flowers 1.6–2.3 cm wide in a tomentose compound cyme; fruit red, often with pale dots, hairy at least near the ends; very rare in open areas on limestone; SE and NW; FACU; flr. late Apr–May, frt. Aug–Sep; 🌿.

Crataegus monogyna Jacq.   English hawthorn
Large shrub or tree to 12 m tall; leaves wedge-shaped to truncate at the base, deeply 3–5-lobed; fruit red, 5–8 mm thick, usually containing only a single nutlet; occasionally escaped from cultivation to roadsides and waste ground; E and W; flr. Jun–Jul, frt. Aug–winter; native to Eurasia.

Crataegus monogyna, fruit

Crataegus phaenopyrum (L.f.) Medik.   Washington hawthorn
Tree to 12 m tall; leaves ovate or cordate, acuminate, with 1–4 pairs of lobes; flowers 1–1.3 cm wide; fruit scarlet, 4–5 mm thick; cultivated and occasionally escaped to roadsides, hedgerows, and open ground; S; native, but also frequently planted and occasionally escaped; flr. May–Jun, frt. Aug–winter; FAC.

Crataegus pruinosa (H.L.Wendl.) K.Koch *sensu lato*   Frosted hawthorn
Shrub or small tree to 8 m tall with very thorny branches; leaves rounded or truncate to subcordate at the base, broadly ovate with shallow lateral lobes, glabrous; flowers 1.2–2 cm wide in nearly simple, glabrous cymes; sepals entire or nearly so; fruit red or greenish with prominent elevated calyx; occasional in open woods and thickets; throughout; flr. May–Jun, frt. Oct.

Crataegus punctata Jacq.   Dotted or white hawthorn
Shrub or small tree to 10 m tall with thorns 4–6 cm long; leaves thick and firm, dull above with impressed veins; leaves of the flowering branches mostly obovate and unlobed or only obscurely lobed, those of the vegetative branches larger and more or less lobed; flowers 1–2 cm wide, on pubescent pedicels; sepals entire or glandular-serrate; fruit dull or bright red or yellow with 3–5 nutlets; occasional in woods, thickets, and alluvial banks; throughout; flr. May, frt. Oct.

Crataegus rotundifolia Moench   Fireberry hawthorn
Shrub or small tree to 7 m tall with glabrous or villous thorny twigs; leaves glabrous or short-villous when young, with impressed veins above, elliptic to broadly ovate or obovate, shallowly lobed, glabrous or slightly glandular; flowers 1.2–2 cm wide, in glabrous to slightly villous compound cymes; fruit red, becoming soft or succulent, with 3–5 nutlets; occasional in rocky pastures, open woods, and roadsides; throughout; flr. May, frt. Aug–Oct.

Crataegus succulenta Schrad. ex Link   Long-spined or fleshy hawthorn
Tree to 8 m tall, twigs glabrous or slightly hairy when young, armed with stout, 3–4.5-cm-long thorns; leaves elliptic to ovate or oblong-ovate, dark green and glossy, glabrous at maturity; flowers 1–1.7 cm wide; fruit bright red 0.7–1.2 cm

wide; nutlets 2–3, deeply pitted; infrequent in woods, thickets, banks, fence-rows, and meadows; S; flr. May–Jun, frt. Sep–Oct.

Crataegus uniflora Münchh. One-fruited hawthorn
Shrub 1–2 m tall with slender branches; leaves obovate to broadly elliptic or sub-orbicular; flowers 1 (or rarely 2) together on short-pubescent pedicels; fruit greenish-yellow to dull red; rare in open woods and dry slopes; S; flr. May–Jun, frt. Aug–Sep.

Cydonia Mill.

Cydonia oblonga Mill. Quince
Unarmed, deciduous tree to 6 m tall with alternate, simple, entire leaves and stipules; flowers solitary, white or pinkish; ovary inferior; fruit a yellow, tomentose, somewhat pear-shaped pome 7–10 cm across; cultivated and occasionally escaped to woods, hedgerows, thickets, and old fields; flr. May–Jun; native to Eurasia.

Dalibarda L.

Dalibarda repens L. Dewdrop
Low-growing, perennial herb with creeping stems; leaves cordate, 3–5 cm broad with hairy petioles; flowers solitary on slender peduncles; petals 5, white; stamens numerous; sepals 5–6; fruit a cluster of dry, achene-like drupes; occasional in bogs, peaty barrens, and cool, mossy woods; mostly N and at higher elevations along the Allegheny Front; flr. May–Jun; FAC.

Duchesnea Sm.

Duchesnia indica, flower

Duchesnea indica (Andr.) Focke Indian strawberry
Low-growing, spreading, perennial herb; leaves trifoliate; flowers solitary from the nodes of the decumbent spreading stems or stolons, with 5 yellow petals and 5 sepals alternating with 5 larger, 3-toothed bracts; receptacle enlarging and turning red in fruit (but not becoming sweet and edible), nearly covered with numerous achenes; extensively naturalized in woods, lawns and waste ground; SE; flr. Mar–Sep; native to Asia; FACU–.

Filipendula Adans.

Erect perennials to 1–2 m tall with large flat-topped panicles of small pink or white flowers; leaves pinnately compound with tiny leaflets interspersed with the larger ones, the terminal leaflet deeply cleft into several serrate segments; stipules present; fruits 1-seeded, indehiscent.

A. leaves glabrous beneath; flowers pink; fruits straight .... *F. rubra*
A. leaves tomentose beneath; flowers white; fruits twisted .... *F. ulmaria*

Filipendula rubra (Hill) B.L.Rob. Queen-of-the-prairie
Moist meadows, thickets, and roadsides; scattered; native, but also escaped from cultivation; flr. Jun–Aug; FACW; 🌢.

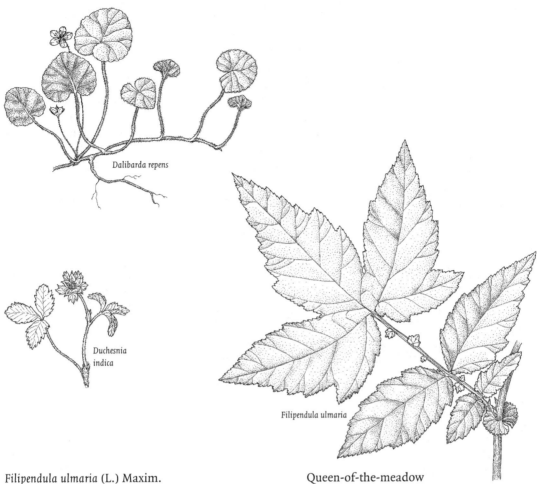

Dalibarda repens

Duchesnia indica

Filipendula ulmaria

Filipendula ulmaria (L.) Maxim.   Queen-of-the-meadow
Cultivated and occasionally escaped to swamps or woods edges; flr. late Jun–Jul; native to Eurasia.

Fragaria L.

Low-growing perennials spreading by stolons; leaflets 3, coarsely serrate; flowers in stalked clusters; petals 5, white; stamens numerous; carpels numerous on a convex receptacle that enlarges and becomes red and edible in fruit with the individual achenes distributed on the surface. The cultivated strawberry (*F. ananassa* Duchesne), with larger fruit and petals 10–15 mm long, occasionally persists in fields and waste ground.

A. central tooth of the terminal leaflet smaller than the 2 adjacent teeth; achenes in depressions in the surface of the fruit .... *F. virginiana*
A. central tooth of the terminal leaflet not smaller than the adjacent teeth; achenes on the surface of the fruit, not in pits .... *F. vesca*

Fragaria vesca L.   Sow-teat strawberry
Foliage yellow-green, terminal leaflet usually sessile; peduncles at anthesis shorter than the leaves but eventually exceeding them; pedicels unequal; occa-

sional in deciduous woods and wooded slopes; throughout; flr. Apr–May, frt. Jun; 2 subspecies:

A. long hairs of the petioles and peduncles spreading or retrorse........ ssp. *vesca* native to Europe.

A. hairs on the petioles and peduncles ascending or appressed .... ssp. *americana* (Porter) Staudt.

Fragaria virginiana Mill.   Wild strawberry
Foliage dark bluish green, terminal leaflet usually distinctly petiolate; peduncles shorter than the leaves even at maturity; pedicels of nearly equal length; common in old fields, meadows, and other dry, open ground; throughout; flr. Apr–Jun, frt. Jun; FACU. Several varieties have been described, but their distinctiveness is unclear.

Geum L.

Herbaceous perennials with basal rosettes of pinnately compound leaves; flowers in terminal corymbs, petals white or yellow; 5 outer bractlets alternating with the sepals in some species; fruit a cluster of achenes each with a persistent hooked style that functions in seed dispersal. In addition to the taxa treated below, G. *urbanum* L., which differs from G. *alleppicum* in its shorter petals and fewer achenes, was collected once in Pittsburgh.

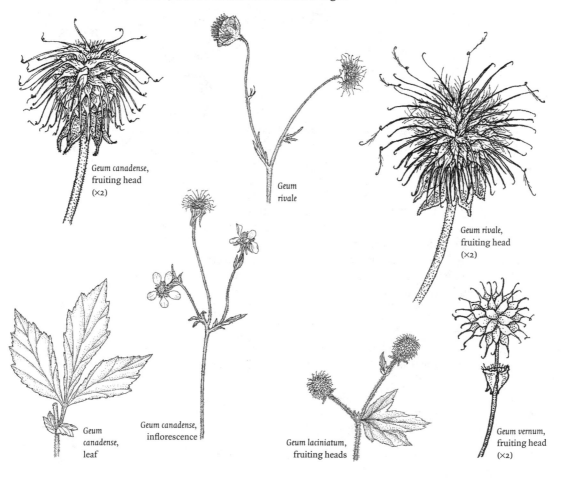

Geum canadense,
fruiting head
(×2)

Geum
rivale

Geum rivale,
fruiting head
(×2)

Geum
canadense,
leaf

Geum canadense,
inflorescence

Geum laciniatum,
fruiting heads

Geum vernum,
fruiting head
(×2)

A. fruiting head elevated above the calyx on a stipe; bractlets lacking.....*G. vernum*
A. fruiting head sessile; bractlets present
    B. sepals purple; terminal section of the style plumose ....*G. rivale*
    B. sepals green; terminal section of the style not plumose
        C. petals bright yellow ....*G. aleppicum*
        C. petals white or creamy-white to pale yellow
            D. pedicels hirsute; receptacle glabrous ....*G. laciniatum*
            D. pedicels minutely puberulent, sometimes with a few longer hairs; receptacle densely hairy
                E. petals pure white, nearly as long as the sepals .... *G. canadense*
                E. petals creamy-white, distinctly shorter than the sepals ....
                ....*G. virginianum*

Geum aleppicum Jacq.   Yellow avens
Erect, hirsute stems to 1 m tall; leaves variable, the principal leaflets often interspersed with minute ones; petals yellow, about as long as the sepals; as many as 200 achenes per fruiting head; receptacle glabrous or short-hispid; occasional in woods, moist fields, swamps, and roadsides; throughout except SW; flr. Apr–Jul; FAC.

Geum canadense Jacq.   White avens
Stems slender, 4–10 dm tall, glabrous to sparsely pubescent below to densely puberulent above; basal leaves long-petioled, with 3 leaflets, cauline leaves progressively reduced, the uppermost simple and sessile; petals white, as long or longer than the sepals; fruiting head 10–15 mm, the receptacle densely hairy; common in woods and roadsides; throughout; flr. May–Aug; FACU+.

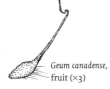

Geum canadense, fruit (×3)

Geum laciniatum Murray   Herb-bennet, rough avens
Stems 4–10 dm tall, covered with deflexed hairs; lower leaves long-petioled, pinnately lobed and incised, the upper trifoliate or 3-lobed; pedicels long-hairy; petals creamy white, distinctly shorter than the sepals; fruiting head 12–18 mm broad, receptacle glabrous; frequent in woods, swamps, bogs, and wet ditches; throughout; flr. Jun–Jul; FAC+.

Geum laciniatum, fruit (×3)

Geum rivale L.   Water avens, purple avens
Stems 3–6 dm tall, sparsely hairy; basal leaves 3- or 5-foliate the terminal leaflets 3-lobed, additional lateral leaflets remote and irregular, cauline leaves much smaller; flowers usually nodding; sepals purple, petals yellowish with purple veins; the distal portion of the style plumose; occasional in bogs, calcareous marshes, and peaty meadows; mostly N; flr. May–Jun; OBL.

Geum rivale, fruit (×3)

Geum vernum (Raf.) Torr. & A.Gray   Spring avens
Erect or ascending perennial to 6 dm tall; most leaves pinnate or trifoliate; sepals reflexed, lacking intervening bractlets; petals yellow, 1–2 mm long; fruiting head elevated above the sepals on a 1–2-mm stipe; occasional in rich woods and ravines; S; flr. late Apr–early May; FACU.

Geum vernum, fruit (×3)

Geum virginianum L.   Cream-colored avens
Very similar to *G. canadense*, but with larger, more coarsely toothed leaves and the stem hirsute below; petals greenish-white and ½–⅓ as long as the sepals; occa-

sional in rich woods, open swamps, ravines, and bluffs; mostly S; flr. Jun–Aug; FAC–.

Kerria DC.

Kerria japonica DC.   Japanese kerria
Deciduous shrub to 2.5 m tall with green twigs and branches; leaves alternate, simple, acuminate, doubly-serrate with narrow, deciduous stipules; flowers yellow, 2.5–5 cm across, solitary at the ends of leafy lateral shoots; pistil of 5–8 distinct ovaries; fruit a small, dry achene; cultivated and occasionally escaped; flr. May–Jun; native to Japan. A double-flowered form is sometimes seen.

Malus Mill.

Deciduous trees to 10–15 m tall; leaves alternate, simple or somewhat lobed, serrate; flowers in umbel-like clusters on short, lateral branch spurs that sometimes end in thorns; hypanthium well-developed; sepals and petals 5; stamens numerous; ovary inferior, 3–5 locular; fruit a pome. Hybridization and the presence of numerous horticultural cultivars can make identification difficult. [Sometimes included in the genus *Pyrus* L.]

Key to flowering material

A. leaves conduplicate in bud, at least some leaves lobed or notched (although these may not be evident at flowering time); lateral branchlets often spine-tipped; anthers reddish
    B. pedicels and hypanthium glabrous or slightly villous; sepals persistent .... *M. coronaria*
    B. pedicel and hypanthium pubescent; sepals deciduous .... *M. floribunda*
A. leaves convolute in bud, unlobed; branchlets not spine-tipped, anthers yellow
    C. leaves pubescent or villous beneath; fruit 2 cm or greater in diameter; sepals persistent
        D. hypanthium and pedicels tomentose .... *M. pumila*
        D. hypanthium and pedicels glabrous to villous .... *M. prunifolia*
    C. leaves becoming glabrate beneath; fruit 1 cm or less in diameter; sepals deciduous .... *M. baccata*

Key to fruiting material

A. fruit 2 cm or more in diameter; sepals persistent
    B. buds, petioles, undersides of leaves, and pedicels tomentose; fruit >5 cm in diameter .... *M. pumila*
    B. buds, leaves, and pedicels glabrous to villous; fruit 2–3 cm in diameter
        C. winter buds glabrous; fruit yellowish-green, 2–3 cm in diameter .... *M. coronaria*
        C. winter buds silky-hairy; fruit yellow or red, 2 cm in diameter .... *M. prunifolia*
A. fruit at most 1 cm in diameter; sepals deciduous
    D. pedicels glabrous .... *M. baccata*
    D. pedicels pubescent .... *M. floribunda*

Malus baccata (L.) Borkh.   Siberian crabapple

Twigs, leaves, and hypanthium glabrous or subglabrous; flowers white; fruit yellow or red, 1 cm or less in diameter, calyx lobes not persistent; cultivated and occasionally escaped to waste ground or cinders; flr. Apr–early May, frt. Sep–Oct; native to Eurasia.

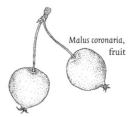

Malus baccata, fruit (×1)

Malus coronaria (L.) Mill.   Sweet crabapple

Branches thorny; mid- to late season leaves often with a few triangular lobes or coarse teeth, glabrous beneath or sparsely villous when young; flowers pink fading to white; anthers pink; hypanthium glabrous or slightly villous; fruit subglobose, greenish, 2–3 cm in diameter, with persistent sepals; frequent in woods, old fields, and thickets; throughout; flr. May, frt. Aug–Oct. [Includes *M. lancifolia* and *M. glaucescens* of Rhoads and Klein, 1993]

Malus coronaria, fruit

Malus floribunda Siebold ex Van Houtte   Showy crabapple

Leaves pubescent beneath when young, becoming glabrate, sharply serrate; pedicels and hypanthium pubescent; flowers rose-red fading white; fruits red, 1 cm in diameter, calyx deciduous; cultivated and rarely escaped; scattered; native to Asia.

Malus prunifolia (Willd.) Borkh.   Chinese crabapple

Young branches pubescent; young leaves pubescent beneath but becoming glabrescent; flowers white; calyx glabrous to villous; fruits yellow or red, 2.5 cm in diameter, with persistent calyx; cultivated and rarely escaped, parks, waste ground, and railroad ballast; scattered; flr. May, frt. Aug–Oct; native to Asia.

Malus pumila Mill.   Apple

Young growth densely pubescent; leaves becoming glabrous above but remaining pubescent beneath; flowers white or light pink; hypanthium and pedicels tomentose; fruit globose, 6–12 cm in diameter, with persistent calyx lobes; cultivated for the edible fruit and frequently persisting at abandoned farms or orchards, hedgerows, or roadsides; flr. late Apr–early May, just before or with the leaves, frt. late Aug–Oct; native to Eurasia. [syn: *Pyrus malus* L.]

Photinia Lindl.

Photinia villosa DC.   Photinia

Shrub or small tree to 3 m tall; leaves deciduous, alternate, simple, finely serrulate, long-hairy beneath, glabrous above; flowers in a short panicle or corymb, white, 12 mm wide; pedicels hairy; fruit a red pome with 1–4 seeds, about 1 cm in diameter; cultivated and rarely escaped in urban woods and floodplains; SE; flr. late May–Jun; native to Asia. *P. parviflora* (Pritz) Schneid., another Asian species with glabrous pedicels and an umbel-like inflorescence, has also been found.

Physocarpus Raf.

Physocarpus opulifolius (L.) Raf.   Ninebark

Deciduous shrub to 3 m tall with erect or spreading stems and papery, exfoliating bark; leaves alternate, simple, palmately 3–5-lobed, petiolate; stipules present but deciduous; flowers white, in small umbel-like corymbs at the ends of short lateral branches, perigynous, with 3–5 distinct carpels; fruit a cluster of

Physocarpus opulifolius, fruit (×1)

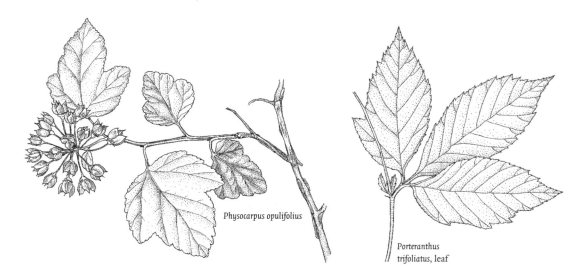

Physocarpus opulifolius

Porteranthus
trifoliatus, leaf

follicles; frequent on moist cliffs, wet woods, and sandy or rocky stream banks; throughout, except the northernmost counties; flr. May–Jul; FACW–.

Porteranthus Britton

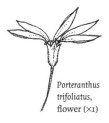

Porteranthus
trifoliatus,
flower (×1)

Porteranthus trifoliatus (L.) Britton   Bowman's-root, mountain Indian-physic
Perennial herb with upright stems to 1 m tall; leaves sessile or the lower ones with a short petiole, palmately trifoliate; leaflets sharply and irregularly serrate, tapering at both ends; stipules narrowly linear to lanceolate and soon falling; flowers white, 5-merous, perigynous with a narrowly campanulate hypanthium; fruit a cluster of hairy follicles; frequent in upland woods; throughout; flr. late May–Jul.

Potentilla L.

Herbaceous, frequently more or less hairy, perennials or small shrubs with compound leaves and conspicuous stipules; flowers with 5 (occasionally 4) sepals alternating with the same number of similar bractlets thus appearing to have a 8–10-parted calyx; petals (4)5, yellow, or in a few species white, creamy, or red-purple; fruit a cluster of achenes on a convex, dry receptacle surrounded by the persistent calyx.

A. low-growing shrubs .... *P. fruticosa*
A. erect or creeping herbaceous perennials (although some species are somewhat woody at the base)
    B. plants aquatic or semi-aquatic; stem glabrous, rooting at the lower nodes; petals red-purple .... *P. palustris*
    B. plants of terrestrial habitats; stem variously hairy; petals yellow or white
        C. leaflets densely silvery tomentose beneath
            D. leaves pinnately compound .... *P. anserina*
            D. leaves palmately compound .... *P. argentea*
        C. leaflets variously hairy but not silvery tomentose beneath
            E. leaves trifoliate or palmately compound
                F. leaflets 3-toothed at the apex, otherwise entire; flowers white ....
            .... *P. tridentata*

F. leaflets with more than 3 teeth; flowers yellow
  G. flowers solitary on naked peduncles arising from the nodes of trailing stems or stolons
    H. flowers 10–15 mm wide; anthers 0.6–1.0 mm long
      I. terminal leaflet ovate with teeth extending at least half way to the base; the first flower arising from the second node on the stem .... *P. simplex*
      I. terminal leaflet obovate, teeth extending less than half way to the base; the first flower arising from the first node on the stem .... *P. canadensis*
    H. flowers 18–25 mm wide; anthers >1 mm long ...... *P. reptans*
  G. flowers few to many in a cymose inflorescence; stems generally upright
    J. flowers mostly 4-merous .... *P. anglica*
    J. flowers 5-merous ....
      K. leaflets mostly 7, 5 in reduced upper leaves .... *P. recta*
      K. leaflets 3–5
        L. principal leaves 3-foliate .... *P. norvegica*
        L. principal leaves 5-foliate .... *P. intermedia*
E. leaves pinnately compound, mostly with 7 or more leaflets
  M. petals yellow .... *P. paradoxa*
  M. petals creamy to almost white .... *P. arguta*

Potentilla anglica Laichard.   English cinquefoil
Perennial with trailing to suberect stems (or stolons); leaves palmately 3 or 5 foliate; flowers solitary on axillary pedicels, yellow, 12–18 mm wide, mostly 4-merous; known from only a single site in Wayne Co.; flr. May–Jun; native to Europe.

Potentilla anserina L.   Silverweed
Low-growing with long stolons that root at the nodes; leaves pinnately compound, leaflets densely silvery-tomentose beneath; flowers yellow, 1.5–2.5 cm wide, solitary on axillary peduncles; very rare in moist, sandy or gravelly shores or ballast; mainly NW; flr. May–Jun; ❦.

Potentilla argentea L.   Hoary or silvery cinquefoil
Freely branched with upright or ascending leafy stems 1–5 dm tall; leaves palmately 5-foliate, leaflets silvery or white-tomentose beneath; flowers in a terminal inflorescence, 7–10 mm wide; petals yellow; calyx white-tomentose; occasional in dry, open ground; flr. May–Aug; native to Eurasia; UPL.

Potentilla arguta Pursh   Tall cinquefoil
Erect, simple, glandular-pubescent stems 3–10 dm tall; leaves pinnately compound, leaflets 7–11 or fewer above; flowers in a terminal cyme, creamy to white, 12–18 mm wide; occasional on dry rocky ledges, fields, and woods; scattered throughout; flr. May–Jul; UPL.

Potentilla canadensis L.   Cinquefoil
Low-growing, creeping, with stolons rooting at the nodes; leaves palmately 5-foliate, teeth of the central leaflet confined mainly to the distal half; flowers solitary on axillary peduncles, yellow, 10–15 mm wide, the lowest flower developing

in the axil of the first well-developed leaf; common in dry, open woods and fields; throughout; flr. Apr–Jul.

Potentilla fruticosa L. Shrubby cinquefoil

Bushy shrub to 1 m tall with pinnate leaves consisting of 5 or 7 narrow, entire leaflets; flowers yellow, 2–3 cm wide, solitary or a few together; ovaries and achenes villous; very rare in calcareous swamps; NE; native, but also frequently cultivated; flr. Jun–Aug; FACW; ✿.

Potentilla intermedia L. Downy cinquefoil

Stems erect, hairy, 3–7 dm tall; leaves palmately 5-foliate; flowers in a leafy, terminal cyme, yellow, 8–10 mm wide; infrequent on roadsides and waste ground; flr. May–Aug; native to Eurasia.

Potentilla norvegica L. Strawberry-weed

Stout, hirsute, leafy annual or short-lived perennial to 6 dm tall; leaves with 3 elliptic to obovate leaflets; flowers yellow, about 1 cm wide, in a branched, terminal inflorescence; common on roadsides and waste ground; throughout; flr. Jun–Aug; FACU. Ours is ssp. *monspeliensis* (L.) Aschers. & Graebn.

Potentilla palustris (L.) Scop. Marsh cinquefoil

Stems coarse, decumbent or ascending, 2–6 dm tall; leaves long-petioled, pinnately compound with 5 or 7 leaflets; inflorescence leafy, flowers red-purple, 2 cm wide, sepals longer than the petals; achenes smooth on a spongy receptacle; rare in swamps, bogs, and peaty lake margins where it often grows as an emergent aquatic; NE and NW; flr. Jun–Jul; OBL.

Potentilla paradoxa Nutt. Bushy cinquefoil

Diffusely branched annual or short-lived perennial; stems spreading to ascending, 2–5 dm tall; leaves pinnate with 7–11 leaflets; inflorescence terminal, much-branched, flowers yellow, 5–7 mm wide; rare on moist, sandy shores; NW; flr. Jul–Sep; OBL; ✿.

Potentilla recta L. Sulfur cinquefoil

Erect, leafy, hairy stems 2–7 dm tall; leaves palmately 5- or 7-foliate; flowers in a terminal cyme, pale yellow, 1.5–2.5 cm wide; common in dry fields and waste ground; throughout; flr. May–Oct; native to Europe.

Potentilla reptans L. Creeping cinquefoil

Similar to *P. canadensis* and *P. simplex* but the flowers 1.8–2.5 cm wide; occasional on waste ground and ballast; SE; flr. May–Oct; native to Eurasia.

Potentilla simplex Michx. Old-field cinquefoil

Very similar to *P. canadensis*, except that the central leaflet is toothed nearly to the base and the lowest flower occurs in the axil of the second well-developed leaf; common in dry woods, fields, meadows, and roadsides; throughout; flr. May–Jul; FACU–.

Potentilla tridentata Aiton Three-toothed cinquefoil

Stems erect from a somewhat woody base, 5–25 cm tall; leaves trifoliate, leaflets entire except for 3 teeth at the tip; flowers in a few-flowered, terminal cyme,

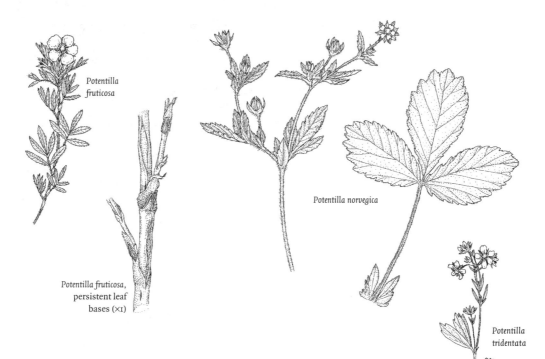

Potentilla fruticosa

Potentilla norvegica

Potentilla fruticosa,
persistent leaf
bases (×1)

*Potentilla
tridentata*

white, 6–10 mm wide; rare on dry, exposed, rocky balds and mountain tops; flr. Jun–Jul; ❧.

Prunus L.

Deciduous trees or shrubs with scaly bark and thorn-tipped branchlets or smoother bark with conspicuous horizontal lenticels; leaves alternate, simple, serrate; petiole or base of the leaf blade often bearing glands; flowers perigynous, 5-merous; hypanthium cup-shaped, deciduous; petals 5, white (or pink); ovary 1, superior; fruit a 1-seeded, fleshy drupe. *P. subhirtella* Miq., an Asian cherry with pendulous branches and doubly serrate leaves, has also been found.

A. flowers and fruits sessile or subsessile; fruit tomentose or velvety
    B. leaves lanceolate, folded at the midrib and somewhat falcate .... *P. persica*
    B. leaves broadly ovate to obovate
        C. fruits 2.5–4 cm in diameter, orange .... *P. armeniaca*
        C. fruits about 1 cm in diameter, red .... *P. tomentosa*
A. flowers and fruits pedicellate; fruit glabrous
    D. inflorescence a raceme
        E. racemes with <12 flowers .... *P. mahaleb*
        E. racemes with 20 or more flowers
            F. leaves with a conspicuous hairy zone along the midrib toward the base beneath; leaf teeth rounded at the tips; calyx lobes persistent in fruit .... *P. serotina*
            F. leaves lacking hairs along the midrib beneath; leaf teeth with fine hair-like tips; calyx lobes deciduous in fruit
                G. petals as long as or only ⅓ longer than the stamens ....
                .... *P. virginiana*
                G. petals twice as long as the stamens .... *P. padus*
    D. flowers in umbel-like clusters

H. bark smooth with prominent horizontal lenticels; branchlets never thorny; stone globose

    I. prostrate or upright shrubs; leaves oblanceolate or obovate, entire toward the base .... *P. pumila*

    I. trees or large shrubs; leaves oblong-lanceolate to oblong-ovate or obovate, toothed to the base

        J. leaves finely serrulate, glabrous; fruits 6 mm in diameter .... *P. pensylvanica*

        J. leaves serrate or doubly serrate; fruits 1.5–2 cm in diameter

            K. leaves glabrous beneath .... *P. cerasus*

            K. leaves pubescent beneath at least on the veins .... *P. avium*

H. bark rough and scaly, lenticels not prominent; branchlets sometimes ending in thorns; stone flattened, with lateral ridges (plums)

    L. leaf teeth glandless; sepals lacking marginal glands

        M. leaves ovate, oval, or obovate, acute to obtuse; lower leaf surfaces, pedicels, and hypanthium pubescent .... *P. maritima*

        M. leaves lanceolate to ovate, long acuminate; lower leaf surfaces slightly pubescent to glabrate; pedicels and hypanthium glabrous

            N. leaves abruptly long-acuminate; petals 8–15 mm long .... *P. americana*

            N. leaves gradually acuminate; petals 4–8 mm long .... *P. allegheniensis*

    L. leaf teeth gland-tipped (or with a scar where the gland had been); calyx without marginal glands .... *P. angustifolia*

Prunus alleghaniensis Porter   Allegheny plum

Straggling shrub or tree to 5 m tall; branchlets at first pubescent, becoming glabrate and reddish-brown, thorny; leaves slightly pubescent or glabrate beneath; flowers white, 1–1.5 cm wide; fruit dark purple, 1 cm in diameter; rare on rocky bluffs, shale barrens, roadsides, and floodplains; mostly SC; flr. Apr–early Jun, frt. Aug; UPL.

Prunus americana Marshall   Wild plum

Coarse shrub or tree to 8 m tall often with spine-tipped branchlets; leaves abruptly long-acuminate, sharply and often doubly serrate; petioles without glands; flowers white, 1.8–3 cm wide; fruit red to yellow, 2–3 cm in diameter;

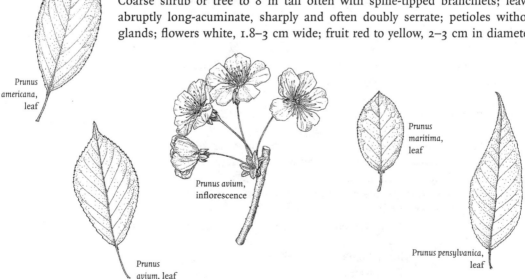

Prunus americana, leaf

Prunus avium, leaf

Prunus avium, inflorescence

Prunus maritima, leaf

Prunus pensylvanica, leaf

frequent on wooded slopes, hedgerows and roadside thickets throughout, except in the northernmost counties; flr. late Apr–May, frt. late Jul–Sep; FACU–.

Prunus angustifolia Marshall   Chickasaw plum
Shrub or small tree to 4 m with glabrous, reddish-brown, spiny-tipped branchlets; leaves glabrous with gland-tipped teeth, often folded along the midvein; petioles with glands at the summit; flowers opening before the leaves, white, 6–9 mm wide; calyx lobes without glands; fruit red or yellow, 1–2 cm; very rare in roadside thickets; flr. Apr, frt. Jun; native, or perhaps escaped from cultivation.

Prunus americana, flower bud (×1)

Prunus armeniaca L.   Apricot
Small tree with reddish-brown bark and glabrous twigs; leaves round-ovate, abruptly short-pointed, serrate; petiole with conspicuous glands; flowers sessile, pinkish to white, 2–2.5 cm wide, before the leaves; fruit pubescent, orange to yellow, sweet; cultivated and occasionally persisting at abandoned home sites; flr. Apr–May, frt. Jun–Jul; native to Asia.

Prunus avium (L.) L.   Sweet cherry
Tree to 20 m tall with spreading, ashy-gray branches; leaves to 1.5 dm long, abruptly acuminate, pubescent on the nerves beneath, coarsely and doubly serrate; flowers white, 2.5–3.5 cm broad; fruit dark reddish-purple, 2–2.5 cm in diameter, sweet; common in forests, wooded edges and fencerows; mostly E of the Allegheny Front; flr. May with the leaves, frt. late Jun–Jul; native to Eurasia.

Prunus cerasus L.   Pie or sour cherry
Tree to 10 m tall with ascending branches; leaves 5–8 cm long, glabrous; flowers white, 2.5 cm wide; fruit red, sour; cultivated and occasionally naturalized in woods edges, thickets, or waste ground; flr. Apr–May, frt. Jun; native to Eurasia.

Prunus mahaleb L.   Mahaleb cherry
Shrub or small tree to 10 m tall with open branching and aromatic bark; leaves round-ovate, broadly rounded to cordate at base; inflorescence a 4–10-flowered, corymbiform raceme; flowers white, 1.5 cm wide, fragrant; fruit black, 6–10 mm in diameter; occasionally naturalized in roadsides, woods, or waste ground; flr. Apr–May, frt. Jul; native to Europe.

Prunus maritima Marshall   Beach plum
Straggling shrub or small tree to 2.5 m tall; branchlets at first pubescent, becoming glabrate; leaves ovate, elliptic, or obovate, pubescent beneath, with crenate teeth and basal glands; flowers white, 1.2–2 cm wide, opening before the leaves; fruit blue-purple, 1.3–2.5 cm in diameter; very rare on dry roadside banks and hedgerows; SE; flr. late Apr, before the leaves, frt. Aug; ❦.

Prunus maritima, flower (×1)

Prunus padus L.   European bird cherry
Tree to 10 m tall, similar to *P. virginiana* but with flowers to 1.3 cm wide on drooping or spreading racemes; fruit black; cultivated and occasionally escaped; flr. Apr–May, frt. Jul–early Aug; native to Europe.

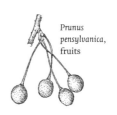
Prunus pensylvanica, fruits

Prunus pensylvanica L.f.   Pin or fire cherry
Upright tree or coarse shrub to 12 m tall with thin, reddish–brown bark; leaves oblong-lanceolate, finely serrate; flowers white, 1.2–1.6 cm wide, fruit light red,

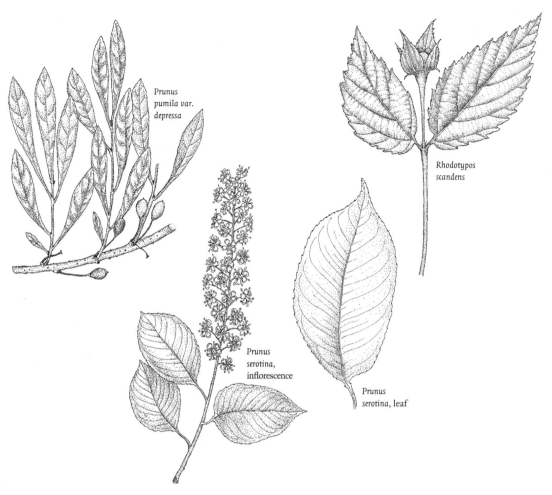

Prunus pumila var. depressa

Rhodotypos scandens

Prunus serotina, inflorescence

Prunus serotina, leaf

5–7 mm in diameter with thin acidic flesh; frequent in dry woods and openings; mostly N and W; flr. May, with the leaves, frt. late Jul–Aug; FACU–.

Prunus persica (L.) Batsch   Peach
Tree to 8 m tall with glabrous twigs; leaves oblong-lanceolate, folded along the midrib and often somewhat curved; flowers solitary, sessile, pink; fruit fleshy, pubescent, yellow or red, sweet; cultivated and occasionally escaped; flr. Apr–early May, before the leaves, frt. Jul–Aug; native to Asia.

Prunus pumila L.   Sand cherry
Prostrate or upright shrub to 1.5 m tall; leaves glabrous, oblanceolate to spatulate, 2.5–5 cm long; closely serrate; flowers white, 1 cm wide; fruit purple-black, 0.8 cm in diameter; ✵; 3 varieties:

A. stems prostrate .... var. *depressa* (Pursh) Gleason
    rare, but sometimes locally abundant on alluvial islands and sandy or gravelly shores; N; flr. Apr–May, frt. Aug.
A. stems upright
    B. leaves oblong or oblong-obovate; youngest twigs minutely pubescent ......
        .... var. *susquehanae* (Hortulan ex Willd.) Jaeger
      rare on dry, exposed rock outcrops and mountain tops; E; flr. Apr–May, frt. Aug. [syn: *Prunus susquehanae* Willd.]

B. leaves oblanceolate; youngest twigs glabrous .... var. *pumila*
  very rare, known only from Presque Isle; flr. May–Jun.

Prunus serotina Ehrh.   Wild black cherry
Tree to 30 m tall but often smaller in the east; branchlets reddish-brown, bark
aromatic; leaves lance-oblong, acuminate, crenate-serrate with blunt callous-
tipped teeth; flowers white, in a slender, elongate raceme; fruit black; common
in forests and fencerows; throughout, but especially well developed on the Allegheny Plateau where it is a major timber species; flr. May–early Jun, after the
leaves, frt. Aug; FACU.

Prunus tomentosa Thunb.   Nanking cherry
Small tree or large shrub with very pubescent young growth; leaves broadly oval
to obovate, unequally serrate, dull and veiny above and tomentose beneath; flowers 1 or 2 together or clustered on spurs, nearly sessile; fruits light red, slightly
hairy; rarely naturalized in sandy soil; NW; flr. Apr, before the leaves; native to
Asia.

Prunus tomentosa, leaf

Prunus tomentosa, fruit

Prunus virginiana L.   Choke cherry
Shrub or small tree, bark not aromatic; leaves ovate or obovate, sharply serrulate; petiole with glands at the apex; flowers in a raceme; fruit red; frequent in
rocky woods, edges, roadside banks, and borders of swamps; throughout; flr.
May–early Jun, with the leaves, frt. Aug–Sep; FACU.

Prunus virginiana,
flower (×1½)

Pyrus L.

Deciduous trees to 15 m; leaves simple, alternate; petals 5, white; sepals 5; ovary
inferior; carpels 2–5; fruit a globose or pear-shaped fleshy pome; the flesh containing stone cells.

A. styles 5; fruit diameter >4 cm; calyx lobes persistent in fruit .... *P. communis*
A. styles 2–3; fruit diameter 1 cm or less; calyx lobes not persistent ....
  .... *P. calleryana*

Pyrus calleryana Decne.   Callery pear
Leaves ovate, glossy, and somewhat leathery, glabrous; margins crenate; flowers
about 2 cm wide, petals white; frequently cultivated, especially as a street tree,
and occasionally escaped to roadsides, old fields, and disturbed woods; flr. late
Mar–Apr, before the leaves; native to China.

Pyrus communis L.   Pear
Leaf blades elliptic, glabrous at maturity; margins serrate to nearly entire, flowers 2–3 cm across, in small clusters; petals white; fruit a large fleshy pome containing stone cells; cultivated for the edible fruit and occasionally persisting at
abandoned home sites or orchards; flr. Apr, frt. Aug–Sep; native to Eurasia.

Rhodotypos Siebold & Zucc.

Rhodotypos scandens (Thunb.) Makino   Jetbead
Deciduous shrub to 2 m tall with opposite, simple, doubly serrate leaves; flowers
2–3 cm across with 4 white petals and 4 large, toothed sepals; ovaries distinct,

Rhodotypos
scandens,
fruits

superior; fruit a cluster of black, 1-seeded dry drupelets; cultivated and occasionally escaped to roadsides and disturbed woods; flr. Apr–May; native to Japan.

Rosa L.

Deciduous, prickly and/or bristly shrubs with alternate odd-pinnate leaves and
stipules adnate to the base of the petiole; flowers relatively large and showy, solitary or in corymbs or panicles, perigynous; petals 5–many; stamens very numerous; fruit a hip.

A. margins of stipules fringed or laciniate
  B. branches prostrate or trailing, sometimes sprawling over other shrubs;
     leaves glabrous .... R. wichuraiana
  B. branches upright-arching to climbing or scrambling; petiole, rachis, and
     lower surface of leaflets pubescent .... R. multiflora
A. margins of stipules entire or with sessile or stipitate glands, but not fringed
  C. styles united in a column and protruding from the orifice of the hypan-
     thium .... R. setigera
  C. styles distinct, only slightly or not at all protruding
    D. current year twigs tomentose as well as densely bristly/prickly............
       .... R. rugosa
    D. current year twigs glabrous or nearly so although stipitate glands,
       bristles and/or prickles may be present
      E. flowers solitary at the ends of the branches, the pedicel not sub-
         tended by a bract .... R. gallica
      E. flowers solitary or clustered, when solitary with a bract at the base
         of the pedicel
        F. hypanthium, and usually the pedicel, stipitate-glandular
          G. hypanthium and pedicels both stipitate-glandular at least at
             flowering; rachis and undersides of the leaflets pubescent
            H. leaflets finely serrate, teeth 0.5 mm; infrastipular prickles
               stout, usually strongly decurved .... R. palustris
            H. leaflets coarsely serrate, teeth to 1 mm; infrastipular prick-
               les straight to slightly decurved
              I. infrastipular prickles slender, straight; internodal prick-
                 les usually numerous .... R. carolina
              I. infrastipular prickles stout, flattened toward the base
                 and often somewhat decurved; internodal prickles rarely
                 present .... R. virginiana
          G. pedicels, rachises, and leaflets completely glabrous ....
             .... R. cinnamomea
        F. hypanthium, and usually the pedicel, glabrous
          J. stems unarmed or with a few slender bristles on the lower in-
             ternodes .... R. blanda
          J. stems bearing infrastipular and/or intranodal prickles
            K. stems with stout, hooked prickles only; leaflets single-
               toothed .... R. canina
            K. stout, usually curved prickles and more slender bristles
               both usually present; leaflets double-toothed
              L. styles short-hairy; sepals persistent .... R. eglanteria
              L. styles glabrous; sepals soon falling .... R. micrantha

Rosa blanda Aiton   Meadow rose
Upright shrub, 1–2 m tall; stems unarmed or with a few prickles near the base; leaflets 5 or 7, coarsely serrate; flowers solitary or few, 5–6 cm wide, single, petals pink; hypanthium glabrous; sepals persistent in fruit; occasional in fence-rows and hedges; scattered; flr. Jun.

Rosa canina L.   Dog rose
Shrub with arching stems 1–3 m long; prickles numerous, stout, flattened, and hooked; bristles absent; leaflets glabrous and without glands beneath; pedicels and hypanthium glabrous; flowers pink, 5–6 cm wide; sepals not persistent; cultivated and occasionally escaped to roadsides and old fields; flr. May–Jun; native to Europe.

Rosa carolina L.   Pasture rose
Upright, lightly branched shrubs to 1 m tall with large, solitary, single pink flowers with 5 petals, 1.5–2 cm long; stems armed with slender, straight infrastipular prickles that are terete to the base and numerous smaller internodal prickles; sepals not persistent in fruit; frequent in fields, rocky banks, shale barrens and other dry, open ground; throughout; flr. May–Jul; UPL.

Rosa cinnamomea L.   Cinnamon rose
Colonial shrub with stems 1–2 m tall armed with stout, curved infrastipular prickles and a few internodal prickles; leaves glaucous and pubescent beneath; flowers pink, often double, 2–3 cm wide; cultivated and occasionally escaped to dry, open fields; mostly N; flr. Jun; native to Eurasia. [syn: *R. majalis* Herrm.]

Rosa eglanteria L.   Sweetbrier
Stems erect, 1–3 m tall with numerous stout, flattened, hooked prickles and a few straight bristles; leaflets glandular-pubescent beneath; hypanthium glabrous; styles villous or woolly; calyx lobes persistent; occasional in dry fields, shaly hillsides, and roadsides; throughout; flr. Jun–Jul; native to Europe.

Rosa gallica L.   French rose
Erect, colonial shrub to 1.5 m tall; stem with stout, hooked prickles and bristles; flowers solitary at the ends of the branches, semi-double, deep pink to scarlet; occasionally persisting at old home sites or roadsides; flr. Jun; native to Europe. Many cultivated forms.

Rosa micrantha Borrer ex Sm.   Sweetbrier
Similar to *R. eglanteria* but the styles glabrous, prickles numerous, bristles absent; rarely naturalized on roadsides and old fields; scattered; flr. Jun; native to Europe; FACU.

Rosa multiflora Thunb. ex Murray   Multiflora rose
Vigorous shrub with arching-ascending branches to 3 m long; stipules conspicuously fringed; leaflets 5–11; flowers in many-flowered terminal panicles; petals white or slightly pinkish; hips 5 mm in diameter; frequently naturalized in disturbed woods, pastures, old fields, roadsides, and thickets; throughout; flr. late May–Jun; native to Asia; designated as a noxious weed in Pennsylvania; FACU.

Rosa carolina

Rosa multiflora

Rosa palustris

Rosa wichuriana

Rosa palustris Marshall   Swamp rose
Colonial shrub to 2 m tall with stout, decurved infrastipular prickles but lacking internodal prickles and bristles; leaflets finely toothed, flowers to 5–7 cm wide with 5 pink petals; frequent in swamps and marshes; throughout; flr. May–Jun; OBL.

Rosa rugosa Thunb.   Rugosa rose
Stems 1–2 m tall, pubescent and densely bristly, infrastipular prickles also present; leaflets 7 or 9, rugose above; pedicels glandular, bristly, and pubescent; petals 5, pink or occasionally white, 3–5 cm; hips 2–3 cm thick, dark red; cultivated and occasionally escaped; scattered; flr. May–Sep; native to eastern Asia; FACU–.

Rosa setigera, flower

Rosa setigera Michx.   Prairie rose
A large shrub with arching, 2–4-m-long branches bearing stout, curved infrastipular and internodal prickles but no bristles; leaflets 3 or 5, coarsely toothed; stipules, petioles, pedicels and calyx all glandular; flowers in terminal clusters, single; petals pink, 2–3 cm long; occasional in sandy old fields, open thickets, roadsides, and fencerows; flr. Jun–Jul; native farther west, perhaps adventive here; FACU.

Rosa virginiana Mill. Wild or pasture rose

Upright shrub to 2 m tall; infrastipular prickles straight to somewhat curved, broad and flattened toward the base; internodal prickles few, mostly at the base; flowers 5–7 cm across, single, pink; sepals reflexed in anthesis and eventually deciduous in fruit; this species is often difficult to distinguish from R. *carolina* due to the occurrence of numerous intermediates; frequent in pastures, fields, open woods, thickets, and roadsides, usually in moist soil; flr. May–early Jul; FAC; ◍.

Rosa wichuraiana Crép. Memorial rose

Stems prostrate or sprawling, with small leaflets; flowers double, 2.5 cm wide, light or dark pink; naturalized in fields, roadsides, floodplains, and waste ground; scattered; flr. Jun–Jul; native to Asia. Most of the material included here is cv. Dorothy Perkins, a R. *wichuraiana* hybrid, and an early cultivated climbing rose much planted in the past.

Rubus L.

Erect, arching or sprawling shrubs or trailing vines with prickly stems; leaves alternate, simple or more often, compound; leaflets with serrate margins; flowers with a small, flat hypanthium; sepals 5, spreading or reflexed; petals 5; stamens many; pistils numerous, distinct, inserted on a convex receptacle; fruit a cluster of adherent drupelets. The raspberries and blackberries produce biennial stems that are unbranched and vegetative with mainly compound leaves the first year (primocanes) and flowering branches often bearing some simple leaves the second (floricanes). Fruits of the blackberries and dewberries are dark purple-black and remain firmly attached to the receptacle when ripe, those of the raspberries separate from the receptacle at maturity and are red or black as noted.

The taxonomy of *Rubus*, especially the blackberries, is complicated by hybridization, polyploidy and apomixis. More study is needed to resolve the status of the many species that have been described over the years. We have chosen to follow the treatment of Gleason and Cronquist (1991), and recognize broad species complexes, in the absence of a definitive taxonomic treatment. See Rhoads and Klein (1993) for a more complete listing of synonymy of the taxa that have been collected in Pennsylvania.

A. leaves all simple .... R. *odoratus*
A. principal leaves compound
    B. stems trailing, with short erect flowering branches
        C. stems unarmed except for narrow-based, stiff bristles or glandular hairs
            D. horizontal stems nearly herbaceous, bearing few if any bristles ......
            .... R. *pubescens*
            D. horizontal stems definitely woody, bearing numerous small-based
            bristles .... R. *hispidus*
        C. stems bearing stout-based, often hooked prickles, lacking bristles or
        glandular hairs
            E. inflorescence a raceme, flowers subtended by stipules only ....
            .... R. *recurvicaulis*
            E. flowers solitary or few, all but the terminal 1 or 2 subtended by
            simple or trifoliate leaves

F. terminal leaflet of the trifoliate leaves ovate, with subcordate or broadly rounded base and acute to long-acuminate tip .... R. flagellaris

F. terminal leaflet of the trifoliate leaves oblong or obovate with cuneate base and abruptly and inconspicuously acuminate tip .... R. enslenii

B. stems erect, arching, or sprawling, only occasionally decumbent

G. leaflets deeply laciniate, green beneath .... R. laciniatus

G. leaflets not laciniate

H. leaflets densely gray or white-tomentose beneath

I. stems strongly arching and frequently rooting at the tips; fruit separating from the receptacle at maturity

J. stems densely covered with long purple bristles and glandular hairs; fruit orange-red, enclosed by the sepals until ripe .... R. phoenicolasius

J. stems purple, conspicuously glaucous, bearing broad-based prickles only; fruit black, never enclosed by the sepals .... R. occidentalis

I. stems erect to sprawling, not rooting at the tips; fruit remaining attached to the receptacle

K. stems erect .... R. cuneifolius

K. stems sprawling

L. prickles straight; stems glabrous; petals pink or red .... R. bifrons

L. prickles curved, stems canescent toward the tip; petals white to pinkish .... R. discolor

H. leaflets glabrous to pubescent beneath, but not densely gray or white-tomentose

M. fruits red, separating from the receptacles

N. leaves of the floricanes 3-foliate or simple; stems with slender-based prickles and glandular bristles .... R. idaeus

N. leaves of the floricanes pinnately 5–foliate; stems with broad-based prickles but not glandular or bristly .... R. illecebrosus

M. fruits black, remaining attached to the receptacles

O. stems not strongly angled, bearing stiff bristles and/or slender-based prickles .... R. setosus

O. stems strongly angled, bearing stout, broad-based, hooked prickles

P. pedicels, axis of the inflorescence, and petioles with numerous glandular hairs .... R. allegheniensis

P. pedicels and other parts lacking glandular hairs or with only a few

Q. leaves glabrous beneath; prickles few and small or lacking on old canes .... R. canadensis

Q. leaves softly pubescent beneath; canes strongly armed ....R. pensilvanicus

Rubus allegheniensis Porter   Common blackberry

Stems erect, 0.5–2 m tall or more, with nearly straight prickles with an elongate base, and also frequently glandular; prickles smaller and hooked in the inflores-

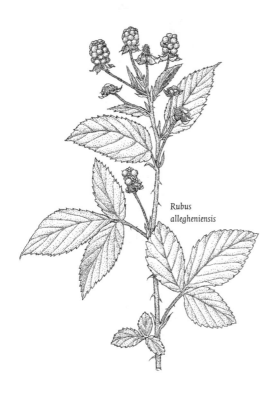

*Rubus
allegheniensis*

cence; principal leaves 5-foliate with smaller, hooked prickles present on the petioles and midveins; inflorescence an elongate, many-flowered raceme; pedicels tomentose and glandular; lower flowers subtended by leaves, the upper by stipules only; common in old fields, open woods, and clearings; throughout; flr. May–Jun, frt. Jul–Aug; FACU-.

Rubus bifrons Vest ex Tratt.   Blackberry
Very similar to R. *discolor* but stems glabrous; prickles of stem straight and spreading with subulate tips; leaflets 3 or 5; petals pink to red; very rare in dry, open ground; SE; native to Europe.

Rubus canadensis L.   Smooth blackberry
Similar to R. *pensilvanicus* but stems with only a few weak prickles and leaflets glabrous beneath; occasional in cool moist woods, rocky slopes, and thickets, especially at higher elevations; flr. May–Jun, frt. Jul–Aug.

Rubus cuneifolius Pursh   Sand blackberry
Stems erect to 1 m tall, with numerous stout-based, straight or hooked prickles; leaves of the primocanes 3- or 5-foliate with a prickly petiole; leaflets densely gray-tomentose beneath; terminal leaflet obovate, sharply serrate above but entire from the widest point to the wedge-shaped base; inflorescence of 1–3 flowers mostly subtended by a leaf with an expanded blade; rare in dry open thickets and roadsides, in sandy soil; SE; flr. May–Jun, frt. Jul; UPL; 🍃.

Rubus discolor Weihe & Nees   Himalaya-berry
Stems coarse, scrambling, to 2 m long, canescent toward the tips; prickles strongly flattened, curved; primocane leaves mostly 5-foliate; leaflets grayish-

tomentose beneath, bright green and glabrous to pilose above; petals white or pinkish; fruit a blackberry; cultivated and rarely spreading to roadside thickets and waste ground; SE; flr. Jun–Jul, frt. Aug; native to Europe; UPL.

Rubus enslenii Tratt.   Southern dewberry
Stems prostrate, rooting at the tips, with slender to stout, recurved, broad-based prickles; floricane leaves trifoliate or simple, dull above; flowers white; occasional on sandy banks and oak-hickory woods; mostly S; flr. May–Jun, frt. Jul–Aug.

Rubus flagellaris Willd.   Prickly or northern dewberry
Stems prostrate, rooting at the tips, with broad-based, hooked prickles but lacking bristles or glands; flowering branches upright, bearing 1–5 flowers, each (except perhaps the uppermost) subtended by a trifoliate or simple leaf; common on rocky or shaly slopes, cliffs, or fields; throughout; flr. May–Jul, frt. Aug; FACU.

Rubus hispidus L.   Swamp dewberry
Stems trailing, rooting at the tips, with slender-based, slightly reflexed prickles and shorter glandular hairs; leaves 3–5-foliate, shiny above; flowering branches erect; inflorescence a raceme; flowers white; frequent in bogs, swamps, moist woods, thickets, and barrens; throughout; flr. May–early Aug, frt. Aug–Sep; FACW.

Rubus idaeus L.   Red raspberry
Stems to 2 m tall, erect or spreading, with slender-based prickles and stiff glandular bristles; leaflets 3 or 5, gray-pubescent beneath; flowers in small clusters

Rubus hispidus

Rubus idaeus

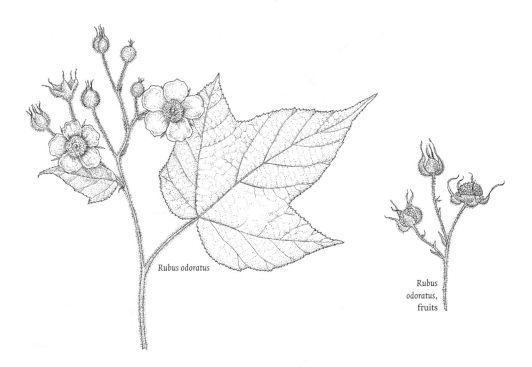

Rubus odoratus

Rubus odoratus, fruits

or solitary in the axils of upper, often simple leaves; pedicels with glandular hairs and bristles; petals white or greenish, shorter than the sepals; fruit red; frequent in rocky woods, clearings, and thickets, especially at higher elevations; N and W; flr. Jun–Aug, frt. Aug–Sep; FAC–. Ours is var. *strigosus* (Michx.) Maxim.

Rubus illecebrosus Focke   Strawberry raspberry
Stems with a few wide-based prickles; leaves with 5–9 sessile, pinnate leaflets; flowers 3–4 cm wide; petals white; fruit red; naturalized in thin woods, roadsides, and fields at a few scattered sites; S; flr. Jun–Jul, frt. Aug; native to Japan.

Rubus laciniatus Willd.   Cut-leaved blackberry
Stems coarse, sprawling to 2 m or more in length, with numerous stout-based prickles; leaves evergreen, mostly 5-foliate; leaflets deeply cleft to twice compound; flowers numerous, in a panicle; petals pinkish or white; fruit elongate, 1–1.5 cm thick; spreading from cultivation to roadsides, fields, sandy woods, and waste ground; mostly SE and SW; flr. Jun–Jul, frt. Jul–Aug; native to Europe; UPL.

Rubus occidentalis L.   Black-cap, black raspberry
Stems conspicuously glaucous, purple, erect or arching and rooting at the tips, with stout-based straight or hooked prickles but not glandular; leaves compound with 3 or 5 leaflets, or simple on the upper portion of the flowering stems; leaflets irregularly serrate, gray-tomentose beneath; petals white, shorter than the sepals; fruit a dark purple-black raspberry; common in sandy or rocky woods, wooded slopes, old fields, or thickets; throughout; flr. May–Jun, frt. late Jun–Jul.

Rubus odoratus L.   Purple-flowering raspberry, thimbleberry
Unarmed shrub to 1–2 m tall, densely covered with glandular hairs and sticky to

the touch; leaves simple with 3–5 triangular lobes; flowers 5–6 cm wide with rose-purple petals; fruit red, a shallowly cupped raspberry; frequent on moist, shaded cliffs, ledges, and rocky wooded slopes; throughout; flr. May–Aug, frt. Jul–Sep.

Rubus pensilvanicus Poir.   Blackberry
Stems stout, 1–3 m tall, with broad-based, straight or slightly reflexed prickles; leaves 3- or 5-foliate, with shorter hooked prickles on the petioles, petiolules, and even the midveins; leaflets pubescent beneath; raceme short, few-flowered; petals white; common in thickets, rocky banks, woods, fields, and waste ground; throughout; flr. May–Jun, frt. Jul–Aug.

Rubus phoenicolasius Maxim.   Wineberry
Stems arching, to 2 m long, densely covered with purple, glandular hairs and bristles; leaflets 3, densely white tomentose beneath; flowers in a panicle; petals white; fruit an orange-red raspberry, enclosed by the long-acuminate, glandular-hairy sepals until ripe, common on roadsides, banks, woods, and thickets; mostly S; flr. Jun, frt. Jul–Aug; native to Asia.

Rubus pubescens Raf.   Dwarf raspberry
Stems creeping at or near the soil surface, upright branches arising from near the base or along the horizontal stems; unarmed or with a few bristles only; leaves primarily 3-foliate; leaflets acute or acuminate and sharply toothed, especially above the middle, bases of the leaflets wedge-shaped and entire; inflorescences 1–3-flowered; pedicels with a few stipitate glands; petals white or pale pink; fruit a dark red raspberry; occasional in bogs, swampy woods, or moist slopes; mostly N; flr. May–Jun, frt. Jul; FACW.

Rubus recurvicaulis Blanch.   Dewberry
Much like R. *flagellaris* in habit, but the inflorescence a raceme with the several flowers subtended by stipules only, although the lowest may be axillary to an expanded, petiolate leaf; occasional in dry, rocky, or sandy soil; throughout; flr. Jun, frt. Jul.

Rubus setosus Bigelow   Bristly blackberry
Stems erect, to 1 m tall, with bristles or slender, soft prickles but lacking stout, broad-based prickles; leaves of the primocanes 3- or 5-foliate, with bristly petioles; inflorescence few- to many-flowered with glandular pedicels; petals white; rare in damp thickets and swamps; scattered; flr. May–Jun, frt. Jul–Aug; FACW; 🌿.

Sanguisorba L.

Glabrous perennial herbs from thick rhizomes; leaves alternate, pinnately compound; leaflets 7–17, ovate or elliptic, serrate; flowers small, white, in dense spikes or heads; sepals 4, petaloid; petals none; stamens 4–many, prominently exerted; pistils 1 or 2; fruit 1 or 2 achenes enclosed by the hypanthium.

A. inflorescence cylindrical, principal leaflets >2 cm long .... S. *canadensis*
A. inflorescence ovoid or globose; principal leaflets <2 cm long .... S. *minor*

Sanguisorba canadensis L.   American burnet

Stems erect, to 1.5 m tall; inflorescence 3–12 cm long; flowers all perfect; occasional in moist meadows, swamps, bogs, and floodplains; throughout; flr. May–Sep; FACW+.

Sanguisorba minor Scop.   Salad burnet

Stems erect to 7 dm tall; inflorescence 8–20 mm long; the lower flowers perfect or staminate, the upper pistillate; cultivated and occasionally escaped to open rocky ground, roadsides, or ballast; scattered; flr. May–Jul; introduced from Eurasia.

Sorbaria (Ser.) A.Braun

Sorbaria sorbifolia (L.) A.Braun   False spiraea

Deciduous shrub to 2 m tall forming dense colonies; leaves odd-pinnate; leaflets 13–21, lanceolate, doubly serrate; flowers small, white, in elongate terminal panicles; fruits dehiscent follicles; cultivated and occasionally escaped to roadsides, moist stream banks, or thickets; scattered throughout; flr. Jun–Jul; native to Asia.

Sorbus L.

Deciduous trees or large shrubs with alternate, odd-pinnate leaves with serrate leaflets; flowers small, white, in 6–15-cm-wide, flat-topped clusters; fruits small, orange-red pomes in large flat-topped clusters.

A. twigs, leaves, inflorescence branches, and hypanthium glabrous; winter buds glutinous
    B. leaflets long-acuminate, 3–5 times as long as wide; fruit 4–7 mm thick ....
       .... *S. americana*
    B. leaflets short-acuminate or acute, 2–3 times as long as wide; fruit 7–10 mm
       thick .... *S. decora*
A. twigs, lower leaf surfaces, inflorescence branches, hypanthiums and winter buds densely white villous; buds not glutinous .... *S. aucuparia*

Sorbus americana Marshall   American mountain-ash

Shrub or tree to 10 m tall with glabrous twigs and glutinous winter buds; leaflets gradually acuminate, 5–9 cm long; fruit 4–7 mm thick; occasional on rocky slopes, bogs and swamps; N and W; flr. Apr–Jun; frt. Jul–Sep; FACU.

Sorbus aucuparia L.   European mountain-ash

Tree to 10 m tall with densely white-villous twigs, leaf under-surfaces, and inflorescences; leaflets oblong, acute, or obtuse, 3–5 cm long; fruit 10 mm thick; cultivated and occasionally escaped; N and W; flr. May, frt. Sep; native to Europe.

Sorbus decora (Sarg.) Schneid.   Showy mountain-ash

Shrub or tree to 10 m tall; leaflets oblong or oblong elliptic, short-acuminate or acute, 3.5–8 cm long; fruit 7–10 mm thick; very rare on rocky slopes; N; flr. May, frt. Sep–Oct; FAC; 🌿.

Sanguisorba
canadensis,
inflorescence

Sanguisorba
canadensis,
leaf

Spiraea L.

Deciduous shrubs with alternate, simple leaves lacking stipules; flowers perigynous, 5-merous, white or pink, in terminal or axillary clusters; pistils distinct; fruit a cluster of follicles.

A. flowers in elongate terminal panicles
    B. leaves mostly glabrous; flowers white to pinkish; follicles glabrous
        C. leaves 2–3 times as long as wide; branches of the inflorescence and the hypanthium glabrous .... S. alba
        C. leaves 3–4 times as long as wide; inflorescence branches and hypanthium puberulent .... S. latifolia
    B. leaves villous to densely tomentose beneath; flowers pink; follicles pubescent
        D. leaves densely tomentose beneath .... S. tomentosa
        D. leaves villous beneath .... S. x billardii
A. flowers axillary or in flattened corymbiform inflorescences that are wider than long
    E. flowers in compound branched corymbs
        F. leaves acute, obtuse, or rounded; flowers white or rarely pink
            G. leaves twice as long as wide; leaf base obtuse to subcordate .... S. betulifolia
            G. leaves wider than long; leaf base wedge-shaped .... S. virginiana
        F. leaves long-acuminate; flowers pink .... S. japonica
    E. flowers in sessile, axillary umbels .... S. prunifolia

Spiraea alba DuRoi   Meadow-sweet
Shrub to 2 m tall with dull brown or yellow-brown twigs; leaves 3–7 cm long, obovate, 3–4 times as long as wide, finely toothed, mostly glabrous; flowers in a dense terminal panicle, white or pinkish, 4–7 mm wide with spreading sepals; branches of the inflorescence and the hypanthium puberulent; fruit glabrous; frequent in bogs, swamps, and moist meadows; throughout; flr. Jun–Sep; FACW+.

Spiraea betulifolia Pallas   Dwarf spiraea
Simple or little-branched shrub to 1 m tall; leaves broadly ovate to obovate, 4–7 cm long, irregularly toothed and rounded at the tip, obtuse to subcordate at the base; inflorescence terminal, broadly rounded, 3–10 cm wide; flowers white (or pink), 4–5 mm wide; fruit glabrous, 2–3 mm; very rare on rocky, wooded slopes; SC; flr. Jun; ✤. [Ours is var. corymbosa (Raf.) Maxim.]

Spiraea x billardii Hérincq   Spiraea
Erect shrub to 2 m tall, branches brown-pubescent; leaves oblong to lance-oblong, 5–7.5 cm long, acute, sharply and doubly serrate except toward the base, villous to gray-tomentose beneath; inflorescence a dense panicle to 20 cm long; flowers bright pink; cultivated and occasionally escaped to roadsides or stream banks; flr. Jun–Jul. [Syn: S. douglasii x salicifolia]

Spiraea japonica L.f.   Japanese spiraea
Upright shrub to 1.6 m tall; leaves ovate to ovate-oblong, 2.5–7 cm long, acuminate, doubly and sharply serrate; inflorescence terminal, flat-topped; flowers pink; cultivated and occasionally escaped; mostly S; flr. Jun–Sep; native to Japan; FACU–.

Spiraea latifolia (Aiton) Borkh.   Meadow-sweet
Very similar to S. *alba* but twigs purple-brown or red-brown; axis and branches
of the inflorescence and the hypanthium glabrous; and the leaves mostly 2–3
times as long as wide; frequent in bogs, moist woods, barrens, and swamps;
mostly E; flr. Jun–Sep; FAC+.

Spiraea prunifolia Siebold & Zucc.   Bridal-wreath spiraea
Erect shrub 2–3 m tall; leaves ovate or ovate-oblong, finely serrate; flowers in axillary umbels, small, white, often double; cultivated and occasionally persisting
at abandoned home sites or along roadsides; mostly S; flr. Apr–May; native to
Asia.

Spiraea tomentosa L.   Hardhack, steeple-bush
Stems unbranched, erect, to 1 m tall; leaves ovate or oblong, 3–5 cm long,
densely white or brownish tomentose beneath; inflorescence a terminal panicle,
5–15 cm long; flowers pink, 3–4 mm wide; fruit pubescent; frequent in wet
meadows, moist old fields, bogs, and swamps; throughout; flr. Jul–Sep; FACW–.

Spiraea virginiana Britton   Appalachian spiraea
Simple or lightly branched shrub to 1 m tall; leaves broadly oblanceolate, 3–6 cm
long, acute, mucronate, wedge-shaped at the base, entire or with a few low
teeth, glaucous beneath; inflorescence terminal, 3–8 cm wide; flowers white;
rare on rocky slopes and stream banks; known only from a single site in Fayette
Co.; flr. Jun–Jul; FACU; 🌿.

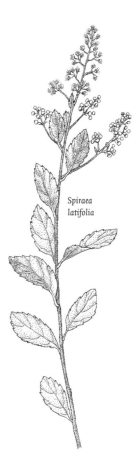

*Spiraea
latifolia*

Waldsteinia Willd.

Waldsteinia fragarioides (Michx.) Tratt.   Barren strawberry
Rhizomatous, perennial herb; leaves trifoliate, basal, with an elongate petiole;
leaflets serrate to somewhat lobed, the lateral ones asymmetrical; flowers several, on basal peduncles about as long as the leaves; petals yellow; stamens numerous, carpels 2–6, distinct, enclosed by the hypanthium except for the protruding styles; fruit a cluster of achenes; occasional in moist, rich woods, and
pastures; throughout; flr. Apr–Jul.

*Waldsteinia
fragarioides,
inflorescence*

*Waldsteinia
fragarioides, leaf*

RUBIACEAE Madder Family

Herbaceous or woody plants with opposite or whorled, simple leaves with or without stipules; flowers regular, (3)4–8-merous; petals fused; stamens inserted on the corolla tube, as many as the lobes and alternate with them; ovary inferior, 2–5-carpellate; styles 1 or 2, slender with lobed or capitate stigmas; fruit a nutlet, capsule, or berry; ovules 1–many per locule.

A. shrubs with whorled or opposite leaves and spherical heads of flowers .... *Cephalanthus*
A. herbs with inflorescence various, but mostly not head-like
  B. principal leaves opposite
    C. bases of the leaves connected by a stipular sheath bearing slender bristles .... *Diodia*
    C. bases of the leaves not connected by a bristly stipular sheath, although small, inconspicuous stipules may be present
      D. flowering stems erect, fruit a capsule with many seeds per locule .... *Houstonia*
      D. flowers on horizontal, creeping, leafy stems; fruit a berry crowned with 2 sets of calyx lobes .... *Mitchella*
  B. principal leaves whorled
    E. calyx lobes not present; flowers mostly in open, branched cymes .... *Galium*
    E. calyx lobes present; flowers in terminal heads surrounded by an involucre .... *Sherardia*

Cephalanthus occidentalis

Cephalanthus L.

Cephalanthus occidentalis L.   Buttonbush
Densely branched, deciduous shrub to 3 m tall with shaggy, exfoliating bark; leaves in whorls of 3 or opposite, simple; blades ovate-oblong to lance-oblong, 8–15 cm; flowers and fruits in 3-cm spherical heads that also include sterile bractlets between the flowers; flowers 4-merous, ovary inferior, bilocular, topped by an hypanthium; calyx lobes short; corolla white, 5–8 mm, funnel-form, with short lobes; fruit splitting into several indehiscent nutlets; frequent in swamps, bogs, lake margins, and other low, wet ground; throughout; flr. Jun–early Aug; OBL.

Diodia L.

Diodia teres Walter   Rough buttonweed
Branching annual herb to 2–8 dm tall; leaves stiff, narrow, connected across the node by a stipular sheath that bears long bristles; flowers in the upper axils, small, whitish; sepals 4; corolla with a slender tube and 4 lobes; ovary with 2 single-seeded locules and a capitate style; fruit splitting into 2 hairy nutlets with persistent sepals; occasional in railroad tracks, roadsides, old fields, and sandy open ground; S; flr. Jul–Oct.

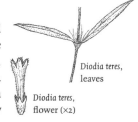

Diodia teres, leaves

Diodia teres, flower (×2)

Galium L.

Erect to reclining or matted herbs with 4-angled stems; leaves in numerous whorls of 4–8, sometimes reduced in size or number in the inflorescence; flowers in small axillary clusters, or more often in branched terminal and axillary cymes; corolla rotate, 3- or 4-lobed; stamens shorter than the corolla; ovary 2-lobed; fruit a fleshy berry or more often dry, smooth or bristly, each carpel indehiscent.

A. principal stem leaves in whorls of 4
   B. fruits smooth
      C. leaves 1-nerved
         D. flowers yellow-green, on very short axillary peduncles .... G. pedemontanum
         D. corolla white
            E. corolla 4-lobed; nodes short-bearded .... G. obtusum
            E. corolla 3-lobed; nodes not bearded
               F. flowers solitary on peduncles .... G. trifidum
               F. flowers 2–3 per peduncle .... G. tinctorium
      C. leaves 3-nerved
         G. flowers white; leaves with stiff, prickle-like hairs beneath .... G. boreale
         G. flowers purple; leaves without stiff, prickle-like hairs .... G. latifolium
   B. fruits hairy or bristly
      H. at least some of the flowers sessile or subsessile along the branches of the inflorescence
         I. leaves oval or elliptic, broadest near the middle; flowers greenish..... .... G. circaezans
         I. leaves lanceolate, broadest well below the middle; flowers purple (or greenish) .... G. lanceolatum

H. flowers all pediceled

    J. principal leaves oval to round or oblong; hairs of the fruits hooked
       .... *G. pilosum*

    J. principal leaves lance-linear; hairs of the fruit not hooked...*G. boreale*

A. principal stem leaves in whorls of 5 or more

  K. fruits smooth, granular, or warty but lacking hairs or bristles

    L. flowers yellow .... *G. verum*

    L. flowers white or greenish

      M. leaves acute or with a cuspidate tip

        N. stems more or less erect from a decumbent base; leaves 6–8 per node; flowers many in a highly branched, conspicuous terminal panicle .... *G. mollugo*

        N. stems weak and matted, scrambling to ascending; inflorescence 2–3 times branched, few-flowered

          O. leaves linear, antrorsely scabrous on the margins or entire ....
            .... *G. concinnum*

          O. leaves elliptic or oblanceolate, retrorsely scabrous on the margins and midveins beneath .... *G. asprellum*

      M. leaves blunt, obtuse, or rounded

        P. corolla 4-lobed

          Q. cymes repeatedly branched, forming a many-flowered terminal panicle; nodes not bearded .... *G. palustre*

          Q. cymes once or twice branched, bearing 2–4 flowers; nodes short-bearded

            R. leaves soon reflexed, 1–2.5 mm wide .... *G. labradoricum*

            R. leaves ascending or spreading, >2.5 mm wide ... *G. obtusum*

        P. corolla 3-lobed

          S. leaves mostly in 4s but occasionally 5s; flowers solitary at the ends of peduncles 1 cm or more long .... *G. trifidum*

          S. leaves in whorls of 4–6; flowers 2–3 per peduncle; peduncles mostly <1 cm .... *G. tinctorium*

  K. fruits with hooked bristles

    T. stems and/or leaves retrorsely scabrous on the angles, midribs and margins, at least below; corolla <4 mm wide

      U. corolla 1–2 mm wide; fruits 2–4 mm; annual .... *G. aparine*

      U. corolla 2–4 mm wide; fruits 1.5–2 mm; perennial .... *G. triflorum*

    T. stems with retrorse hairs at the nodes but otherwise glabrous; corolla 4–7 mm wide .... *G. odoratum*

Galium aparine L.   Bedstraw, cleavers
Annual; stems to 1 m tall, weak and scrambling, retrorsely hooked or scabrous on the angles; leaves in whorls of 6 or 8, narrow, 1-nerved, cuspidate, retrorsely ciliate on the margins and the midrib beneath and with hooked hairs above; inflorescence 3–5-flowered on axillary peduncles; corolla greenish-white, 4-lobed, 1–2 mm wide; fruit with hooked bristles or rarely smooth; common in woods, stream banks, and roadsides; throughout; flr. Apr–Jul; FACU.

Galium asprellum Michx.   Rough bedstraw
Perennial; stems scrambling to 2 m tall, retrorse-scabrous on the angles; leaves in 6s (or those on the branches in 4s or 5s), narrowly elliptic or oblanceolate, sharply acute or cuspidate, retrorsely scabrous on the margins and midrib be-

low; inflorescence a small, loose terminal panicle rarely >2 cm; corolla white, 4-lobed, 3 mm wide; fruit smooth, 3 mm; frequent in swamps, bogs, stream banks, and wet thickets; throughout; flr. Jun–Oct; OBL.

Galium boreale L.   Northern bedstraw
Erect, stiffish perennial, 2–8 dm tall; stems short-bearded at the nodes; leaves in 4s, glabrous or scaberulous, lance-linear, 1.5–4.5 cm, 3-nerved with a minute rounded tip; flowers numerous, in terminal panicles; corolla white, 3.5–7 mm wide; fruit 2 mm, glabrous or with short straight hairs; occasional on rocky slopes, low fields, fens, and roadside banks; throughout; flr. Jun–Sep; FACU.

Galium circaezans Michx.   Wild licorice

Galium circaezans, fruit (×3)

Erect or ascending perennial to 6 dm tall; stems more or less pubescent; leaves in 4s, oval-elliptic or ovate-oblong, 2–5 by 1–2.5 cm, obtuse, 3–5 nerved, glabrous to thinly or densely hairy; inflorescence from the upper axils, simple or with few branches; flowers sessile or nearly so; corolla greenish; fruit 3 mm, with hooked bristles; frequent in rich woods, calcareous slopes, and bluffs; throughout; flr. Jun–Aug; UPL. Plants with more densely hairy leaves have been called var. *hypomalacum* Fernald.

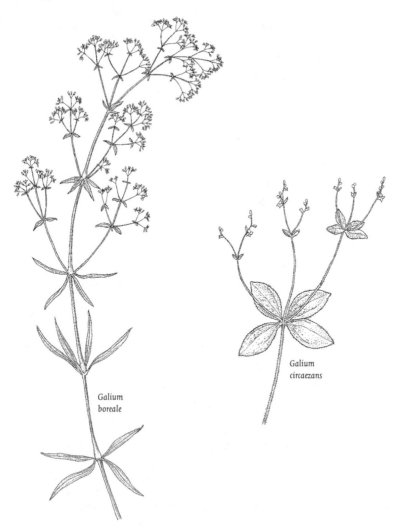

Galium boreale

Galium circaezans

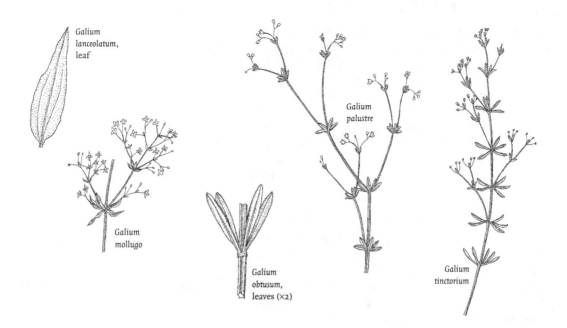

Galium lanceolatum, leaf

Galium palustre

Galium mollugo

Galium obtusum, leaves (×2)

Galium tinctorium

Galium concinnum Torr. & A.Gray   Shining bedstraw

Perennial; stems 2–5 dm tall, slender, spreading, sparsely retrorse-scabrous on the angles; leaves in 6s (or 4s on the branches), linear or linear-elliptic, 1–2 cm, sharply acute or cuspidate and antrorsely scabrous on the margins; inflorescence terminal and from the upper axils; corolla 4-lobed, white, 2.5–3 mm wide; fruit smooth; occasional on wooded slopes; primarily S and W; flr. Jun–Aug; UPL.

Galium labradoricum (Wieg.) Wieg.   Bog bedstraw

Perennial; stems slender, to 4 dm tall, bearded at the nodes but otherwise glabrous; leaves in 4s, deflexed, linear-oblanceolate, 8–15 by 1–2.5 mm; peduncles mostly terminal, 3-flowered; corolla white 2–3 mm wide, 4-lobed; fruit smooth, 2–3 mm; very rare in sphagnum bogs and moist banks; scattered; flr. Jun–Aug; OBL; ✹.

Galium lanceolatum Torr.   Wild licorice

Very similar to G. circaezans but the leaves lanceolate, broadest well below the middle, 3–8 by 1–2.5 cm with a long-tapering acute or acuminate tip; corollas purple or greenish; frequent in woods; throughout; flr. May–Jul.

Galium latifolium Michx.   Purple bedstraw

Perennial; stems to 6 dm tall, smooth or hairy at the nodes; leaves in 4s, lanceolate or lance-ovate, 3–6 cm long, 3-nerved; inflorescences from the upper nodes, branched; corolla purple; fruit smooth or granular, 3–4 mm; rare in woods, rocky slopes, and roadsides; SC; flr. Jun–Jul; ✹.

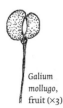

Galium mollugo, fruit (×3)

Galium mollugo L.   White bedstraw, wild madder

Perennial to 1.2 m tall, sprawling to ascending from a decumbent base, glabrous or short hairy below; leaves in 6s or 8s, scabrous-ciliate on the margins but otherwise glabrous, 1-nerved, cuspidate; flowers numerous, in terminal panicles; corolla white, 2–5 mm wide; fruit glabrous; common in fields, roadsides, and waste ground; throughout; flr. May–Nov; native to Eurasia.

Galium obtusum Bigelow   Cleavers, marsh bedstraw
Matted perennial; stems 2–8 dm tall, smooth on the angles and short-bearded at
the nodes; leaves in 4s, linear to lanceolate or oblanceolate, 10–30 by 1–6 mm,
obtuse to emarginate at the tip; midrib and margin usually scabrous or hispid;
inflorescences terminal with 3–5 flowers on short pedicels; corolla white, 4-
lobed; fruit 4–5 mm, tuberculate; occasional in wet woods, bogs, and swamps; S
and W; flr. May–Jul; FACW+.

Galium odoratum (L.) Scop.   Sweet woodruff
Perennial; stems to 5 dm tall, erect, retrorsely hispid at the nodes, otherwise glabrous; leaves in whorls of 6–10, oblanceolate or narrowly elliptic, cuspidate,
antrorsely ciliate on the margins and midrib beneath, otherwise glabrous; inflorescences terminal; corolla white, 4–7 mm wide; fruit 3 mm, with hooked
bristles; cultivated and rarely escaped to rich, dry woods; S; flr. May–Jun; native
to Europe.

Galium palustre L.   Ditch or marsh bedstraw
Perennial; stems 4–6 dm tall, minutely retrorse-scabrous on the angles, nodes
not bearded; leaves in whorls of 5 or more, linear to narrowly oblanceolate,
blunt, antrorsely scabrous on the margins; inflorescence repeatedly branched,
many flowered; corolla white, 4-lobed, 4 mm wide; fruit smooth, 2 mm; occasional on stream banks, marshes, or wet ditches; mainly N; flr. Jun; OBL.

Galium pedemontanum All.   Bedstraw
Annual to 3.5 dm tall with spreading hairs and recurved prickles on the stems;
leaves in 4s, 3–11 by 2–4 mm, 1-nerved with revolute margins; peduncles much
shorter than the subtending leaves; flowers yellow or yellow-green, 0.5–1 mm
wide; fruit 1 mm, glabrous; rare in disturbed ground; S; flr. Apr–May; native to
Europe.

Galium pilosum Aiton   Bedstraw, cleavers
Erect or ascending perennial to 1 m tall; leaves in 4s, elliptic to oval, 1–2.5 cm,
firm, 3-nerved; inflorescences from the upper axils, each peduncle forked 2 or 3
times, the flowers terminating the branches; corollas greenish-white to purplish; fruit with hooked bristles; occasional in old fields, shale barrens, and dry,
open sandy ground; especially S; flr. Jun–late Aug.

Galium tinctorium L.   Bedstraw, cleavers
Very similar to *G. trifidum* but with leaves in whorls of 4–6 and peduncles
straight, <1 cm long, and bearing 2 or 3 flowers; common on moist, wooded
slopes, wooded floodplains, stream banks, and swales; throughout; flr. Jun–
Aug; OBL.

Galium trifidum L.   Small bedstraw
Perennial; stems weak and reclining, much branched and retrorse hispid on the
angles to nearly glabrous; leaves mostly in 4s but occasionally in whorls of 5 or
6, linear, blunt, 1-nerved; peduncles >1 cm, flexuous, bearing 1 or 2 flowers each;
corolla whitish, 1–1.5 mm wide, 3-lobed; fruit glabrous; rare in moist woods,
thickets, and swales; scattered; flr. Jun; FACW+; ❦.

Galium obtusum,
flower (×3)

*Galium
tinctorium,*
flower (×3)

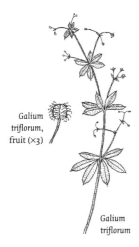

Galium
triflorum,
fruit (×3)

Galium
triflorum

Galium triflorum Michx.   Sweet-scented bedstraw

Perennial; stems to 8 dm tall, prostrate or scrambling, usually retrorsely scabrous on the angles below, but sometimes smooth; leaves in 6s (or 4s on smaller branches), narrowly elliptic to oblanceolate, cuspidate, 1-nerved, retrorsely scabrous on the margins and midrib beneath, otherwise glabrous, vanilla scented; inflorescences axillary and terminal; peduncles simple and 3–5(or more)-flowered; corolla greenish-white, 2–3 mm wide; fruit with hooked bristles; common in rocky woods, shaded hillsides, and roadside banks; throughout; flr. Jun–early Sep; FACU.

Galium verum L.   Yellow bedstraw

Perennial; stems erect, 4–10 dm tall; leaves in whorls of 8–12, linear, 1-nerved, acute, with inrolled margins, shiny and hairy above, densely pubescent beneath; flowers bright yellow, in a dense terminal panicle; occasionally naturalized in fields, roadsides, waste ground, and ballast; flr. May–Sep; native to Eurasia; 2 varieties:

A. inflorescence interrupted; flowers lemon yellow, not fragrant ....
.... var. *wirtgenii* (F.W.Schultz) Oborny
      occasional; throughout. [*G. wirtgenii* F.W.Schultz of Rhoads and Klein, 1993]
A. inflorescence not interrupted; flowers golden yellow, fragrant .... var. *verum*
      rare; E.

Houstonia L.

Low perennial herbs with opposite, or basal, entire leaves and somewhat sheathing stipules; flowers blue to purple, 4-merous, distylous (except *H. serpyllifolia*); corolla with a slender tube and spreading lobes; stamens 4, inserted on the corolla tube; ovary 2-celled, partly inferior; style slender, stigma 2-lobed; fruit a dehiscent capsule. Treatment follows Terrell (1996).

A. flowers solitary on long pedicels
    B. plants erect or spreading .... *H. caerulea*
    B. plants prostrate or creeping .... *H. serpyllifolia*

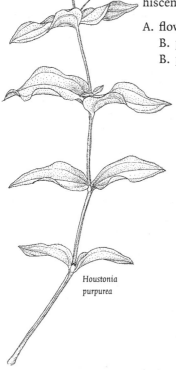

Houstonia
purpurea

Houstonia
caerulea

Houstonia
longifolia

A. flowers in few-flowered cymes
  C. lower or median cauline leaves broadly ovate to lanceolate, 3–35 mm wide
    .... H. purpurea
  C. lower or median cauline leaves filiform, linear, oblanceolate, or obovate,
    0.3–8.5 mm wide
    D. stems with 3–6 internodes, the lower internodes longer than the upper;
      lower or median cauline leaves oblanceolate or obovate ....H. canadensis
    D. stems with 3–13 internodes, the lower internodes not longer than the
      upper; cauline leaves linear to elliptic .... H. longifolia

Houstonia caerulea L.   Bluets, Quaker-ladies
Plant with a persistent basal rosette and slender flowering stems to 2 dm tall
bearing widely spaced pairs of reduced leaves; flowers solitary on long pedicels,
nodding or erect; corolla light blue with a yellow center; common in meadows,
open woods, and edges; throughout; flr. Apr–Jun or later; FACU. [*Hedyotis caerulea* (L.) Hook. of Rhoads and Klein, 1993]

Houstonia canadensis Willd. ex Roem. & Schult.   Fringed bluets
Clumped or rhizomatous plant with a persistent rosette and erect flowering
stems to 2.5 dm tall; basal leaves obovate to spatulate, cauline ones oblanceolate, obovate, or elliptic, 1-nerved; flowers in few-flowered cymes from the upper nodes; corolla white to light purple; rare on dry, rocky slopes; SW; flr. May–Jun. [*Hedyotis canadensis* (Willd.) Fosberg of Rhoads and Klein, 1993]

Houstonia longifolia Gaertn.   Long-leaved bluets
Clumped or rhizomatous, forming a winter rosette; stems erect or ascending,
much branched, to 5 dm tall; median cauline leaves linear to narrowly elliptic, 1-
nerved; flowers in terminal cymes; corolla white to purple, glabrous to puberulent inside; frequent in dry, sandy fields, shale barrens, and dry, thinly wooded
slopes; SW and SC; flr. May–Jul. [*Hedyotis longifolia* (Gaertn.) Hook. of Rhoads
and Klein, 1993, *Hedyotis purpurea* (L.) Torr. var. *tenuifolia* (Nutt.) Fosberg in part]

Houstonia purpurea L.   Purple or southern bluets
Tufted with slender rhizomes; stems erect or ascending, to 4 dm tall, leafy,
branching from the upper nodes; median cauline leaves broadly ovate to ovate-
lanceolate, 3–7 nerved, pubescent to glabrous; flowers in few- to many-flowered
cymes, heterostylous; corolla purple, puberulent to densely pubescent inside;
very rare in open woods and roadsides; SW and SC; flr. Apr–Aug; ✿. Ours is var.
purpurea. [*Hedyotis purpurea* (L.) Torr. var. *purpurea* of Rhoads and Klein, 1993]

Houstonia serpyllifolia Michx.   Creeping or thyme-leaved bluets
Perennial with prostrate, creeping stems rooting at the nodes and forming mats;
leaf blades ovate to suborbicular, those of the flowering stems narrower; flowers
solitary on slender, erect pedicels; corolla light blue with yellow blotches at the
base of each lobe; very rare on moist streambanks or pastures; SW; flr. Apr–Jul;

Mitchella repens, in fruit

FAC; ✿. [*Hedyotis michauxii* Fosberg of Rhoads and Klein, 1993]

Mitchella L.

Mitchella repens L.   Partridge-berry
Creeping, mat-forming, evergreen perennial; leaves opposite, nearly round, 1–2

cm in diameter; flowers white, paired; fruits red, each pair fused to form a double berry crowned by 2 sets of calyx lobes; common in moist woods; throughout; flr. Jun; FACU.

Sherardia L.

Sherardia arvensis L.   Field-madder

Branched annual to 4 dm tall with whorled leaves; stem square, roughly hairy; inflorescence dense, head-like surrounded by an involucre of 8 lanceolate, partly fused leaves; corolla pink or blue, with a slender tube and 4 lobes; fruit dry, 2-ribbed, with persistent sepals; infrequent in cultivated fields, lawns, and pavement; mostly S; flr. May–Jul; native to Eurasia.

RUTACEAE Rue Family

Deciduous, aromatic trees or shrubs (some prickly or thorny) or herbaceous perennials; leaves alternate or opposite, compound, with translucent glandular spots, lacking stipules; flowers perfect or unisexual, 3–5-merous, regular; ovary superior, with 2–8 fused or distinct carpels; stamens as many as the petals and alternate with them or twice as many; fruit various, a fleshy follicle, samara, drupe, capsule, or hesperidium.

A. herbs; leaves simple to 2–3 times compound .... *Ruta*
A. trees or shrubs; leaves trifoliate or pinnately compound
   B. trees or shrubs, branches without thorns
      C. shrub; leaves trifoliate; fruit a samara .... *Ptelea*
      C. trees; leaves odd-pinnate
         D. inner bark bright yellow; axillary buds enclosed by the base of the petiole; fruit a berry .... *Phellodendron*
         D. inner bark not yellow; axillary buds not enclosed by the base of the petiole; fruit a dehiscent capsule .... *Euodia*
   B. shrubs; branches bearing thorns or prickles
      E. branches brown, prickly; leaves pinnately compound; fruit 5 mm .... *Zanthoxylum*
      E. branches green with a single large thorn at each node; leaves trifoliate; fruit 3.5–5 cm .... *Poncirus*

Euodia J.R.Forst. & G.Forst.

Euodia
hupehensis,
winter twig

Euodia hupehensis Dode   Bee-bee tree

Deciduous tree 15–20 m tall with opposite, pinnate leaves and exposed axillary buds; plants dioecious; flowers unisexual, whitish, in large terminal clusters; fruit a 5-merous dehiscent capsule opening to reveal dark red-purple fleshy seeds; occasional in disturbed woods and roadsides; SE; flr. early Aug, frt. Sep–Oct; native to Asia. In addition, *E. daniellii* (Benn.) Hemsl., which is a smaller tree with subsessile leaflets may also be present.

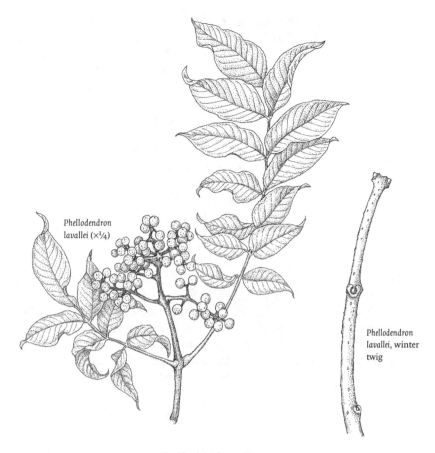

Phellodendron
lavallei (×¼)

Phellodendron
lavallei, winter
twig

Phellodendron Rupr.

Deciduous trees with opposite, odd-pinnate leaves and axillary buds that are enclosed by the base of the petiole; bark corky and ridged or thin and fissured, inner bark bright yellow; dioecious; flowers small, yellowish-green, in terminal panicles; petals and sepals each 5–8; staminate flowers with 5 or 6 stamens, pistillate with a superior, 5-celled ovary and 5-lobed stigma; fruit a fleshy black berry, 8–12 mm wide, with 5 seeds; flr. early Jun, frt. Sep–Jan.

A. leaf undersides pubescent, at least on the veins, pale green or grayish-green
    B. bark thick and corky; leaflets broadly cuneate at base .... P. lavallei
    B. bark thin; leaflets truncate or subcordate at base .... P. japonicum
A. leaf undersides glabrous and somewhat glaucous .... P. sachalinense

Phellodendron japonicum Maxim.   Japanese corktree
Tree to 10 m tall with thin, slightly fissured bark; leaflets 9–13, dull green above, villous beneath; inflorescence 5–8 cm across; fruits 10–12 mm, black; cultivated and rarely naturalized in disturbed woods; SE; native to Japan.

Phellodendron lavallei Dode   Corktree
Tree to 15 m tall with thick, corky bark; leaflets 5–13, dull yellowish green above, pubescent beneath at least on the veins; inflorescence 6–8 cm wide; fruit 8 mm; cultivated and occasionally naturalized in disturbed woods and roadsides; SE; native to Japan.

Phellodendron sachalinense (F.Schmidt) Sarg. Corktree

Tree to 15 m tall with slightly fissured bark that separates into thin plates; leaflets 7–11, dull green above, glabrous or nearly so beneath; inflorescence 6–8 cm high; fruit 1 cm; rarely naturalized on roadsides; SE; native to Asia.

Poncirus Raf.

Poncirus trifoliata L. Hardy or trifoliate orange

Deciduous shrub to 3 m tall with stiff, angled, green branches bearing a single sharp, flattened thorn at each node; leaves trifoliate with a somewhat winged petiole; flowers in the axils of the thorns, white, fragrant, perfect; fruit orange-like, 3–5 cm; cultivated and occasionally escaped to stream banks and roadside thickets; E; flr. May, before the leaves, frt. Sep–Nov; native to Asia.

Ptelea L.

Ptelea trifoliata L. Hoptree, wafer-ash

Ptelea trifoliata,
bud scar (×1)

Shrub or small tree with alternate, trifoliate leaves and a nearly circular, dry, winged fruit (samara); leaflets sessile, ovate or elliptic; inflorescence 4–8 cm wide; flowers small, greenish-white, perfect or unisexual; sepals, petals and stamens each 3–5; ovary superior, of 2 fused carpels; styles 2; fruit with the wing surrounding the body; rare in old fields, stream banks, and alluvial thickets; S and NW; flr. late May–early Jun, frt. Jul–Sep; FAC; 🍂.

Zanthoxylum
americanum (×¼)

Ptelea
trifoliata

Ruta L.

Ruta graveolens L.   Common rue

Perennial herb to 1 m tall with a semiwoody base; leaves glaucous, simple to 2 or 3 times pinnate, alternate; flowers yellow, perfect, 4–5-merous; fruit a 4- or 5-lobed, many-seeded capsule; cultivated and occasionally escaped to roadsides, fields, and pastures; scattered; flr. Jun; native to Europe.

Zanthoxylum L.

Zanthoxylum americanum Mill.   Prickly-ash

Upright shrub or small tree to 8 m tall with prickly stems and alternate, pinnately compound leaves; leaflets 5–11, sessile, oblong to elliptic; plants dioecious; flowers in axillary clusters, yellow, unisexual; ovaries 3–5, distinct; fruit a fleshy follicle with the 1–2 seeds sometimes dangling on a thread; occasional on stream banks, river bluffs, and wet woods, usually on limestone or diabase; S; flr. late Apr–early May, before the leaves, frt. Jul–Nov; FACU.

SALICACEAE Willow Family

Deciduous trees and shrubs with alternate, simple leaves with stipules; dioecious; flowers small, unisexual, borne in catkins; sepals transformed into a disk or nectaries; petals absent; stamens 1 or more; ovary of 2–4 fused carpels, superior; fruit a thin-walled, 1-locular capsule that splits open to release seeds bearing conspicuous tufts of hairs.

A. leaves generally 2 or more times as long as wide, with strictly pinnate venation; winter buds with 1 bud scale; flowers with nectaries; stamens 1–9 .... Salix
A. leaves about as wide as long, with a strong pair of basal secondary veins; winter buds with 3 or more bud scales; flowers with a disk; stamens (5)8–40 .... Populus

Populus L.
James E. Eckenwalder

Trees predominantly with ovate, deltoid, or orbicular leaves that sometimes have 1 or more glands at the base of the blade; terminal buds present; catkins pendulous, flowering before the leaves, on shoots of the previous year; floral bracteoles falling early.

Summer key

A. petiole markedly flattened near its apex
    B. leaf margin minutely ciliate, especially on the teeth
        C. petiole densely pubescent .... *P. x jackii*
        C. petiole glabrous
            D. large leaves notably broader than long, without glands at the base of the blade .... *P. nigra*
            D. large leaves no broader than long, with 0–6 glands

  E. leaves truncate to subcordate, with 2–6 glands at the base of the
   blade .... *P. deltoides*

  E. leaves broadly cuneate, with 0–1 glands .... *P. x canadensis*

 B. leaf margin without cilia

  F. teeth 3–15 on each side, coarse .... *P. grandidentata*

  F. teeth >15 on each side, fine

   G. leaves hairy beneath, with 2 cup-shaped glands at the base of the
    blade .... *P. grandidentata*

   G. leaves glabrous beneath, without glands at the base of the blade .....
    .... *P. tremuloides*

A. petiole round or only slightly flattened top to bottom

 H. petiole glabrous to short hairy .... *P. balsamifera*

 H. petiole tomentose

  I. leaves densely white tomentose beneath, leaves of long shoots distinctly
   palmately 5-lobed .... *P. alba*

  I. leaves grayish tomentose beneath, leaves of long shoots coarsely and irregularly toothed .... *P. x canescens*

Key to flowering specimens

A. floral bracteoles shallowly dentate to laciniate, ciliate

 B. bracteoles dentate

  C. bracteoles shallowly dentate .... *P. alba*

  C. bracteoles dentate for half their length .... *P. x canescens*

 B. bracteoles laciniate

  D. rachis of catkins pubescent .... *P. grandidentata*

  D. rachis of catkins glabrous .... *P. tremuloides*

A. floral bracteoles laciniate, but not ciliate

 E. rachis of catkins pubescent

  F. carpels 2, or plant staminate .... *P. balsamifera*

  F. carpels 3–4; no staminate plants .... *P. x jackii*

 E. rachis of catkin glabrous

  G. both sexes common; stamens 30–40; carpels 4 .... *P. deltoides*

  G. usually staminate; stamens 12–30; carpels 2–3

   H. stamens 20–30; carpels 3 .... *P. x canadensis*

   H. stamens 12–20; carpels 2 .... *P. nigra*

Winter key

A. buds resinous

 B. resin yellow or tan on brown buds .... *P. deltoides*

 B. resin orange or red on reddish-brown buds

  C. resin with balsamic fragrance when warm

   D. first-year twigs hairy, especially at the nodes .... *P. x jackii*

   D. first-year twigs glabrous to sparsely hairy .... *P. balsamifera*

  C. resin, if fragrant, not balsamic

   E. lateral buds appressed to the shoots .... *P. nigra*

   E. lateral buds divergent .... *P. x canadensis*

A. buds not resinous

 F. buds glabrous .... *P. tremuloides*

 F. buds hairy

  G. buds with short, straight hairs .... *P. grandidentata*
  G. buds tomentose
    H. first-year twigs tomentose ....*P. alba*
    H. first-year twigs sparsely hairy to glabrate .... *P. x canescens*

Populus alba L.   White poplar
Colonial tree to 25 m tall; young twigs and buds white tomentose; leaves dark green above, densely white tomentose beneath at first, those of long shoots palmately 5-lobed; columnar clones staminate (Bolleana poplar); stamens 6–10; pistillate clones with a spreading crown; ovaries and capsules narrowly ovoid, 2-valved, usually not maturing; capsules 3–5 mm long when mature; cultivated, persisting, and spreading at old garden sites; mostly S; native to Eurasia.

Populus balsamifera L.   Balsam poplar
Colonial tree to 30 m tall; young twigs glabrous or sparsely minutely hairy; buds resinous, with balsamic fragrance; leaves dark green above, glaucous white and marked with red resin stains beneath, generally ovate, with minute teeth along most of the margin; stamens about 20–30; ovaries and capsules ovoid, 2-valved; capsules 5–8 mm long; rare in cool, seasonally wet soils and bog margins; throughout; 🌿.

Populus x *canadensis* Moench   Carolina poplar
Single-trunked tree to 30 m tall; young twigs glabrous; buds resinous; leaves dark green above, light green and marked with yellowish-orange resin stains beneath, broadly ovate, often noticeably longer than wide, with regular, graded teeth to 2 mm deep; stamens 15–30; pistillate trees rare; ovaries and capsules ovoid, 2–3 valved; capsules 6–9 mm long; cultivated and rarely persisting; mostly S; a hybrid of Eurasian origin. [syn: *P. deltoides* x *nigra*]

Populus x *canescens* (Aiton) Sm.   Gray poplar
Colonial tree to 25 m tall; young twigs and buds thinly tomentose to glabrate; leaves dark green above, thinly grayish-tomentose to glabrate beneath, those of long shoots coarsely and irregularly toothed; staminate plants rare, stamens 8–15; ovaries and capsules narrowly ovoid, 2-valved, usually not maturing; capsules 3–5 mm long; cultivated and rarely persisting or spreading; SE; native to Eurasia. [syn: *P. alba* x *tremula*]

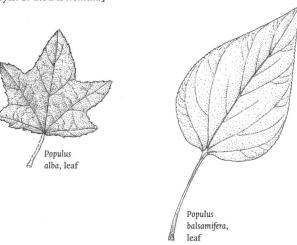

*Populus
alba*, leaf

*Populus
balsamifera,*
leaf

Populus deltoides Marsh.   Eastern cottonwood

Noncolonial tree to 40 m tall but often dividing into 2 or more trunks near the base; young twigs glabrous; buds resinous, moderately fragrant; leaves about as broad as long, pale to bright green above and beneath with graded crenate teeth to 5 mm deep along the middle portion; stamens 30–40; ovaries and capsules ovoid, 3–4-valved; capsules 8–11 mm long; occasional in floodplains and swamps; scattered; FACU–.

Populus grandidentata Michx.   Bigtooth aspen

Colonial tree to 25 m tall; young twigs glabrous; buds with dense short hairs; leaves dark green above, greenish-white beneath, with 5–15 coarse, pointed teeth up to 5 mm deep on each side; stamens 6–12; ovaries and capsules narrowly ovoid, 2-valved; capsules 2–5 mm long; common in early successional forests; throughout; FACU–.

Populus x jackii Sarg.   Balm-of-Gilead

Sparingly colonial tree to 30 m tall; young twigs densely short-hairy; buds with abundant balsamic resin; leaves dark green above, greenish white and marked with red resin stains beneath, heart-shaped, with minute teeth along most of the margin; staminate trees not found; ovaries and capsules ovoid, with 2 or 3 valves; capsules usually not maturing; cultivated and persisting sparsely throughout; native to the north, but origin of cultivated clone unknown. [syn: *P. balsamifera* x *deltoides*]

Populus nigra L.   Black poplar

Sparingly colonial tree to 30 m tall, most individuals columnar (cv. Italica, Lombardy poplar); young twigs glabrous; buds resinous, moderately fragrant; leaves

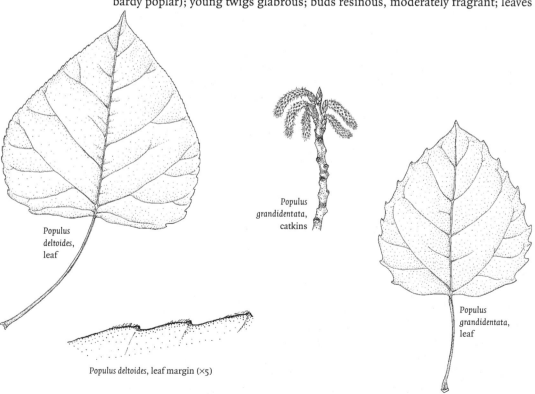

Populus
deltoides,
leaf

Populus
grandidentata,
catkins

Populus
grandidentata,
leaf

Populus deltoides, leaf margin (×5)

dark green above, paler but not glaucous beneath, with slightly graded crenate teeth to 1.5 mm deep along most of the margin; leaves of long shoots distinctly broader than long, nearly triangular; usually staminate, stamens 20–30; ovaries and capsules, when present, broadly ovoid, 2-valved; capsules 5–9 mm long; cultivated, persisting and spreading modestly; E; native to Eurasia.

Populus tremuloides Michx.   Quaking aspen

Colonial tree to 20 m tall; young twigs and buds glabrous, the buds shiny; leaves dark green above, slightly glaucous beneath, often nearly round, with 15–40 low rounded teeth on each side; stamens 6–12; ovaries and capsules narrowly ovoid, 2-valved; capsules 2–4.5 mm long; common in dry to wet, open woods; throughout, particularly following disturbance.

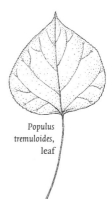

Populus
tremuloides,
leaf

Salix L.
George W. Argus

Shrubs or trees; buds with 1 external bud scale the margins of which are usually fused but occasionally free and overlapping; stems mostly erect and flexible at the base; leaves alternate (except S. purpurea), often with leafy stipules; mature leaf blades linear, lanceolate, ovate, or broadly obovate; leaf margins entire or toothed; flowers in catkins that are sessile or borne on short, leafy branchlets, flowering as the leaves emerge (exceptions noted below); flowers lacking a perianth but with a nectar gland present at the base; each flower subtended by a bract that is usually dark-colored and persistent after flowering; staminate flowers with 2–9 stamens; pistillate flowers with a single bicarpellate ovary borne on a stipe or sessile; style present or lacking; stigmas 2; fruit a capsule that splits open to release the seeds, each of which is surrounded by a tuft of hairs.

Summer key

A. all, or at least some, of the leaves opposite or subopposite .... S. purpurea
A. all leaves alternate
    B. bud scale margins free and overlapping; buds sharply pointed
        C. leaf blades not glaucous beneath, shiny above .... S. nigra
        C. leaf blades glaucous beneath, dull or highly glossy above
            D. leaves dull above, broadly to narrowly lanceolate, 3–6 times as long as wide; stipules lacking or rudimentary, leaf-like only on vigorous or late-season shoots .... S. amygdaloides
            D. leaves highly glossy above; leaves lanceolate to very narrowly so, 5–10 times as long as wide; stipules leaf-like .... S. caroliniana
    B. bud scale cap-like, its margins fused; buds blunt
        E. leaves not glaucous beneath
            F. stipules leaf-like .... S. lucida
            F. stipules lacking or rudimentary
                G. branches yellow-, gray-, or red-brown, epidermis flaky; petioles deeply grooved, the margins not touching; secondary veins of leaf blades flat on both surfaces .... S. serissima
                G. branches brownish, epidermis not flaky; petioles deeply grooved but with the margins covering the groove; secondary veins of leaf blades protruding on both surfaces .... S. pentandra
        E. leaves glaucous beneath or the lower surface obscured by dense hairs

H. trees
    I. leaves with stomata on the lower surface only, upper surface not minutely gray-dotted
        J. leaf tips acute to obtuse; petioles puberulent; branchlets red-brown .... *S. bebbiana*
        J. leaf tips acuminate; petioles glabrous or tomentose; branchlets gray- or yellow-brown .... *S. caprea*
    I. leaves with stomata on both surfaces, upper surface minutely gray-dotted
        K. branches strongly pendulous .... *S. babylonica* hybrids
        K. stems erect or drooping
            L. leaves dull above, remaining silky in age; branches flexible to somewhat brittle at the base, epidermis flaky............................. *S. alba*
            L. leaves shiny or highly glossy above (sometimes dull in S. x rubens); soon glabrate; branches highly brittle at the base, epidermis not flaky
                M. leaves serrulate to serrate, base acute to wedge-shaped; petioles pilose or villous ....*S. x rubens*
                M. leaves coarsely serrate, base obtuse, rounded or broadly wedge-shaped; petioles glabrous or puberulent .... *S. fragilis*
H. shrubs
    N. juvenile leaves glabrous or sparsely hairy
        O. leaf margin entire; stipules absent or rudimentary; petioles glabrous or puberulent; bog plants forming colonies by layering .... *S. pedicellaris*
        O. leaf margin toothed; stipules leaf-like; petioles sparsely to densely hairy; not bog plants or colonial
            P. leaf margin bluntly serrulate to crenate, sometimes almost entire; branch epidermis not flaky, branchlets red-brown .... *S. myricoides*
            P. leaf margin sharply serrate; branch epidermis flaky, branchlets yellow-brown .... *S. eriocephala*
    N. juvenile leaves more densely hairy
        Q. leaves linear, 6.5–31 times as long as wide; margins remotely spinulose-serrulate; blades with stomata on both surfaces (minutely gray-dotted above) .... *S. exigua*
        Q. leaves narrowly oblong to elliptic, oblanceolate and broadly obovate, sometimes ligulate, 1.5–11 times as long as wide; margins entire, glandular-dotted, or variously toothed but not as above; blades with stomata on the lower surfaces only (not gray-dotted above)
            R. leaf margins strongly revolute
                S. branchlets glabrescent, villous, tomentose, or velvety; branches red-brown, epidermis not flaky; leaves pilose or velvety becoming glabrate with wedge-shaped bases .... *S. humilis*
                S. branchlets woolly, branches gray-brown, epidermis flaky; leaves densely woolly to tomentose with acute bases .... *S. candida*
            R. leaf margins slightly revolute or flat
                T. leaf margins serrate or serrulate
                    U. stipules leaf-like; juvenile leaves pilose or sparsely villous; leaf bases obtuse or cordate .... *S. eriocephala*
                    U. stipules absent or rudimentary; juvenile leaves densely silky; leaf bases acute to wedge-shaped

V. leaves densely short-silky beneath; juvenile leaves densely short-silky; branches highly brittle at the base; leaf bases wedge-shaped ... *S. sericea*

V. leaves glabrous, long-silky or glabrate beneath; juvenile leaves moderately densely long-silky; branches flexible at the base; leaf bases acute .... *S. petiolaris*

T. leaf margins entire, remotely or irregularly serrate, crenate or undulate

    W. branchlets woolly; leaves tomentose above .... *S. candida*

    W. branchlets glabrescent or variously hairy but not woolly; leaves glabrous, glabrescent, pubescent, pilose, or short-silky above

        X. hairs of lower leaf surfaces appressed

            Y. leaf margins entire, blades ligulate or very narrowly elliptic .... *S. petiolaris*

            Y. leaf margins crenate to undulate, blades narrowly elliptic, elliptic, oblanceolate, obovate, or broadly obovate

                Z. juvenile leaves pilose; mature leaf margins flat .... *S. discolor*

                Z. juvenile leaves tomentose or long-silky; mature leaves slightly revolute .... *S. atrocinerea*

        X. hairs of lower leaf surfaces divergent, or erect

            AA. branchlets tomentose, thinly glaucous .... *S. aurita*

            AA. branchlets pubescent, pilose, villous or velvety, not glaucous

                BB. petioles puberulent; branchlets red-brown; mature leaf margins flat .... *S. bebbiana*

                BB. petioles glabrescent, tomentose or velvety; branchlets yellow- or gray-brown; mature leaf margins slightly revolute

                    CC. stipules lacking or falling early, except on vigorous shoots; wood beneath the bark smooth .... *S. caprea*

                    CC. stipules conspicuous, persistent; wood beneath the bark with prominent ridges

                        DD. leaves tomentose beneath with white or gray hairs .... *S. cinerea*

                        DD. leaves tomentose or silky beneath with a mixture of white and rust-colored hairs .... *S. atrocinerea*

Key to plants with pistillate flowers

A. ovaries tomentose or silky

    B. catkins emerging with the leaves or throughout the season

        C. floral bracts deciduous after flowering, tip ragged; ovary beak abruptly tapering to the style .... *S. exigua*

        C. floral bracts persistent, tip entire; ovary beak gradually tapering or slightly bulged below the style

            D. ovaries tomentose; catkins densely or moderately densely flowered; branchlets woolly .... *S. candida*

            D. ovaries silky; catkins loosely flowered; branchlets glabrous, villous, or velvety

                E. floral bracts dark brown; ovary pear-shaped; beak short, gradually tapering to the style .... *S. myricoides*

                E. floral bracts tawny; ovary inverse club-shaped; beak long and slender, slightly bulged below the style .... *S. bebbiana*

B. catkins emerging before the leaves
   F. juvenile leaves glabrous or sparsely hairy
      G. catkins opposite or subopposite; ovaries inverse top-shaped; nectary longer than the stipe; juvenile leaves glabrous or pubescent............ .... S. *purpurea*
      G. catkins alternate; ovaries inverse club-shaped, pear-shaped, or ovate; nectary shorter than the stipe; juvenile leaves pilose, tomentose, or silky .... S. *discolor*
   F. juvenile leaves densely hairy
      H. ovaries ovate; juvenile leaves densely short-silky; floral bracts widest at the base; branches highly brittle at the base .... S. *sericea*
      H. ovaries pear- or inverse club-shaped; juvenile leaves tomentose, sometimes sparsely so; floral bracts widest at the middle or top; branches flexible at the base
         I. ovaries sparsely or moderately densely hairy; plants colonial by layering .... S. *humilis*
         I. ovaries very densely hairy; plants not layering
            J. stipules falling early; wood beneath the bark smooth...S. *caprea*
            J. stipules persistent; wood beneath the bark with ridges
               K. leaves rugose; branchlets tomentose; hairs of the ovary wavy or crinkled .... S. *aurita*
               K. leaves flat or not conspicuously rugose; branchlets pubescent, pilose, villous, or velvety; hairs of the ovary straight
                  L. catkins densely flowered; ovaries pear-shaped; leaves with white or gray hairs .... S. *cinerea*
                  L. catkins loosely to moderately densely flowered; ovaries inverse club-shaped; leaves with a mixture of white and rust-colored hairs .... S. *atrocinerea*
A. ovaries glabrous
   M. floral bracts persistent after flowering
      N. trees; stems strongly pendulous .... S. *babylonica* hybrids
      N. shrubs; stems erect
         O. floral bracts tawny to light rose; styles 0.1–0.2 mm long .... S. *pedicellaris*
         O. floral bracts brown, black, or bicolor; styles 0.3–1.6 mm long
            P. catkins loosely flowered; ovary beak gradually tapering to the style; stigmas slender-cylindric .... S. *myricoides*
            P. catkins moderately to very densely flowered; ovary beak slightly bulged below the style; stigmas with two plump lobes .... S. *eriocephala*
   M. floral bracts deciduous after flowering
      Q. bud scale margins free and overlapping; buds sharply pointed
         R. leaves not glaucous beneath; branches highly brittle at the base .... S. *nigra*
         R. leaves glaucous beneath; branches flexible, or somewhat brittle at the base
            S. leaves ligulate to lanceolate or narrowly so, 5–10 times as long as wide, tip acuminate; stipules prominent and persistent .... S. *caroliniana*

    S. leaves very narrowly elliptic to lanceolate or oblanceolate, 3–6 times as long as wide, tip acuminate to caudate; stipules rudimentary, leaf-like only on later leaves and vigorous shoots .... *S. amygdaloides*

Q. bud scale margins fused; buds blunt and cap-like

  T. stipules leaf-like

    U. leaves not glaucous beneath; stipules strongly resinous, tip obtuse or rounded; margins of leaves on flowering branchlets serrulate or at least gland-dotted .... *S. lucida*

    U. leaves glaucous beneath; stipules not strongly resinous, tip caudate to acute; margins of leaves on flowering branchlets entire

      V. petioles glabrous or puberulent; leaves glossy and glabrescent above .... *S. fragilis*

      V. petioles pilose, villous, or silky; leaves dull and glabrescent or silky above

        W. stipule tip acuminate; ovary beak gradually tapering to the style; petioles pilose or villous; leaves glabrescent .... *S. x rubens*

        W. stipule tip acute; ovary beak slightly bulged below the style; petioles long-silky; leaves remaining silky .... *S. alba*

  T. stipules lacking or rudimentary

    X. leaf margins remotely spinulose-serrulate, blades linear; plants forming colonies by root shoots; petioles not glandular at the distal end .... *S. exigua*

    X. leaf margins serrate or serrulate, leaf blades narrowly oblong, narrowly elliptic, elliptic, lanceolate, or narrowly ovate; plants not colonial; petioles glandular-dotted or -lobed at the distal end

      Y. leaf margins coarsely serrate; branchlets pubescent, pilose, or glabrescent; leaves sparsely silky or glabrescent beneath, glabrous above .... *S. fragilis*

      Y. leaf margins serrulate; branchlets glabrous; leaves glabrous on both surfaces

        Z. catkins 16–35 by 11–22 mm, <3 times as long as wide; floral bracts moderately densely hairy all over; ovary beak slightly bulged below the style .... *S. serissima*

        Z. catkins 25–65 by 7–13 mm, 3 or more times as long as wide; floral bracts sparsely hairy mainly at the base; ovary beak gradually tapering to the style .... *S. pentandra*

Salix alba L.   White willow

Tree to 25 m tall with flexible to somewhat brittle branches; branchlets yellowish to gray- or red-brown, hairy; juvenile leaves very densely silky; petioles glandular-dotted or -lobed; leaf blades narrowly oblong to lanceolate, silky to glabrescent, upper surface dull, base acute to wedge-shaped, margins serrulate, apex acute to acuminate; catkins on short leafy branchlets; floral bracts tawny, the pistillate ones deciduous; ovaries glabrous; cultivated and occasionally escaping to roadsides and old fields; mostly S; native to Eurasia; FACW.

Salix amygdaloides Andersson   Peach-leaved willow

Tree to 20 m tall with flexible to somewhat brittle branches; branchlets yellow-, red-, or gray-brown, glabrous; bud scale margins free and overlapping; juvenile

Salix amygdaloides

leaves glabrous to pilose with white or rust-colored hairs; stipules rudimentary, leaf-like on later leaves; petioles glandular-dotted; leaf blades elliptic, lanceolate, or oblanceolate, glabrous, upper surface dull, base acute to rounded, margins serrulate, tip acuminate to caudate; catkins on short leafy branchlets; floral bracts tawny, the pistillate ones deciduous; stamens 3–7; ovaries glabrous; rare in swamps, bogs, and wet shores; mostly NW; FACW.

Salix atrocinerea L.   Rusty willow
Shrub to 12 m tall with gray- or yellow-brown, hairy branchlets; juvenile leaves tomentose or silky with white or rust-colored hairs; leaf blades narrowly elliptic to obovate or oblanceolate, lower surface tomentose or villous to glabrescent with white or rust-colored hairs, upper surface dull or shiny, pilose to glabrescent, base acute or rounded, margins crenate to undulate, tip acute to rounded; catkins sessile, or on short leafy branchlets; ovaries short-silky; rare on alluvial soils; E; flr. before the leaves emerge; native to Europe.

Salix aurita L.   Eared willow
Shrub to 3 m tall with red- or yellow-brown, sparsely hairy branchlets; juvenile leaves very densely tomentose with white, or sometimes rust-colored, hairs; leaf blades obovate to elliptic, pilose to glabrescent, upper surface dull or shiny, base wedge-shaped, obtuse or acute, margins remotely or irregularly serrate and crenate, tip abruptly acute to rounded; catkins sessile or on short leafy branchlets; ovaries silky; rarely naturalized in disturbed urban woods; SE; flr. before the leaves emerge; native to Europe.

Salix babylonica L.   Weeping willow
Tree to 20 m tall with strongly pendulous, yellowish or yellow-brown branches; petioles glandular-dotted or not; leaf blades linear to narrowly lanceolate, lower surface glabrous or silky to glabrescent, upper surface shiny or dull, silky to glabrescent, base wedge-shaped, margins serrulate, tip gradually acuminate; catkins on short leafy branchlets; floral bracts tawny; ovaries glabrous; cultivated and occasionally naturalized; mostly S; native to China; FACW–; represented by 2 hybrids:

A. leaf margins serrulate; juvenile leaves pubescent or silky ....
.... *S.* x *pendulina* Wender.
   [syn: *S. babylonica* x *fragilis*]
A. leaf margins remotely or irregularly serrate; juvenile leaves glabrous or sparsely silky .... *S.* x *sepulcralis* Simonk.
   [syn: *S. alba* x *babylonica*]

Salix bebbiana Sarg.   Long-beaked or gray willow
Shrub or tree to 10 m tall with flexible branches that are red-brown and hairy; juvenile leaves tomentose to silky; stipules leaf-like; leaf blades elliptic to oblanceolate or obovate, lower surface tomentose or silky to glabrescent, upper surface dull, pubescent to glabrescent, base acute to obtuse, margins entire or crenate, tip abruptly acute to obtuse; staminate catkins sessile or on short leafy branchlets; pistillate catkins on short, leafy branchlets; floral bracts tawny; ovaries silky; occasional in upland deciduous woods, moist or dry thickets, and edges; throughout, especially N; flr. just before the leaves emerge; FACW.

Salix candida Flüggé ex Willd.   Hoary or sage-leaved willow

Shrub to 1 m tall, sometimes layering; branchlets yellow- or gray-brown, densely hairy; juvenile leaves very densely tomentose; leaf blades very narrowly elliptic to oblanceolate; usually very densely tomentose or woolly beneath, upper surface dull or shiny, base acute, margins entire to undulate revolute, tip acute; catkins on short leafy branchlets; floral bracts tawny or brown; ovaries tomentose; rare in fens and wet meadows on calcareous soils; NE and NW; OBL; ❦.

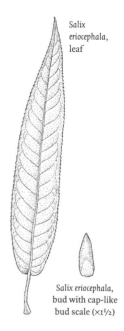

Salix candida, capsule (×1½)

Salix candida, leaves

Salix caprea L.   Goat willow

Shrub or tree to 15 m tall with yellow- or gray-brown, hairy branchlets; juvenile leaves densely tomentose with white or rust-colored hairs; stipules leaf-like; leaf blades broadly oblong to obovate; lower surface sparsely tomentose or pubescent, upper surface dull or shiny, pubescent to glabrescent, base rounded to cordate or acute to wedge-shaped, margins crenate, undulate or entire, tip abruptly acuminate; catkins sessile; ovaries silky; cultivated and occasionally spreading to thickets and roadsides; mostly SE; flr. before the leaves emerge; native to Europe.

Salix caroliniana Michx.   Carolina willow

Tree to 10 m tall with somewhat brittle branches; branchlets yellow- or red-brown, glabrous or hairy; bud scale margins free and overlapping; juvenile leaves glabrous or moderately densely tomentose with white or rust-colored hairs; petioles glandular-dotted or lobed; leaf blades ligulate to lanceolate or narrowly so, glabrous or glabrescent, upper surface highly glossy, base acute, rounded to cordate, margins serrulate, tip acuminate; catkins on short leafy branchlets; floral bracts tawny, the pistillate ones deciduous; stamens 4–7; ovaries glabrous; rare in stream banks, shores, and low woods; mostly SW and SC; OBL; ❦.

Salix cinerea L.   Gray or pussy willow

Shrub to 7 m tall with yellow-brown, hairy branchlets; juvenile leaves tomentose; leaf blades elliptic to oblanceolate or obovate, lower surface tomentose, upper surface dull or shiny, pubescent to glabrescent, base rounded; margins crenate, undulate or entire, tip acute; catkins sessile; ovaries silky; cultivated and rarely escaped; SE; flr. before the leaves emerge; native to Eurasia.

Salix discolor Muhl.   Pussy willow

Shrub to 4 m tall with yellowish to dark brown, hairy branchlets; juvenile leaves pilose with white or rust-colored hairs; stipules foliaceous or rudimentary; leaf blades narrowly elliptic to oblanceolate or obovate, pilose to glabrescent, upper surface dull or shiny, base obtuse, acute or wedge-shaped, margins crenate, undulate or entire; tip subacuminate; catkins sessile; ovaries silky; common in swamps and moist or wet woods; throughout; flr. before the leaves emerge; FACW.

Salix eriocephala, leaf

Salix eriocephala Michx.   Diamond willow

Shrub to 6 m tall with yellow-brown, hairy to glabrescent branchlets; juvenile leaves glabrous or pilose; leaf blades narrowly elliptic to obovate, lower surface glabrous or glabrescent, upper surface shiny or dull, glabrous or pilose, base cordate or obtuse, margin serrate, tip acute to acuminate; staminate catkins sessile, flr. before the leaves emerge; pistillate catkins on short leafy branchlets,

Salix eriocephala, bud with cap-like bud scale (×1½)

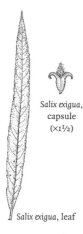

Salix exigua, capsule (×1½)

Salix exigua, leaf

flr. with the leaves; ovaries glabrous; common on shores and bottomlands; throughout; FACW. [syn: *S. cordata* Muhl., *S. rigida* Muhl.]

Salix exigua Nutt.   Sandbar willow
Shrub or tree to 6 m tall, forming colonies by root shoots; branchlets yellow- or red-brown, hairy or glabrescent; juvenile leaves long-silky; stipules rudimentary or lacking; leaf blade linear, lower surface villous or silky to glabrescent, upper surface shiny, villous or pilose to glabrescent, base wedge-shaped, margin remotely spinulose-serrulate, tip acuminate; catkins on short leafy branchlets; floral bracts tawny, the pistillate ones deciduous after flowering; ovaries silky or glabrous; frequent on sandy or gravelly bars or shores; flr. throughout the season; OBL. Ours is ssp. *interior* (Rowlee) Cronquist. [syn: *S. interior* Rowlee]

Salix fragilis L.   Crack or brittle willow
Tree to 20 m tall with yellow-brown, hairy to glabrescent branchlets; stipules leaf-like or rudimentary; petioles glandular-dotted or lobed; leaf blade lanceolate to very narrowly elliptic, lower surface very sparsely silky to glabrescent, upper surface shiny or highly glossy, base obtuse to rounded, margin coarsely serrate, tip acuminate to caudate; catkins on short, leafy branchlets; floral bracts tawny, the pistillate ones deciduous after flowering; ovaries glabrous; infrequently cultivated and rarely escaping to roadsides and woods edges; native to Europe; FAC+.

Salix humilis Marshall   Upland willow
Shrub to 3 m tall with erect or decumbent, layering stems; branchlets yellow- to red-brown, densely hairy to glabrescent; juvenile leaves tomentose to glabrescent with white or rust-colored hairs; stipules leaf-like, rudimentary, or lacking; leaf blades narrowly oblong, elliptic to broadly obovate, lower surface usually densely tomentose with white or rust-colored hairs, upper surface shiny, pilose or velvety to glabrescent, base wedge-shaped, margins entire, crenate, or undulate, tip acute to obtuse; catkins sessile; ovaries silky; flr. before the leaves emerge; FACU; 2 varieties:

A. stipules leaf-like .... var. *humilis*
    common in dry thickets and barrens; throughout.
A. stipules lacking .... var. *tristis* (Aiton) Griggs
    occasional in moist barrens and thickets; throughout. [syn: *S. humilis* var. *microphylla* (Andersson) Fernald]

Salix lucida

Salix lucida Muhl.   Shining willow
Shrub or tree to 6 m tall with yellow-, gray-, or red-brown, glabrous or densely hairy branchlets; juvenile leaves glabrous, villous, or silky with white or rust-colored hairs; stipules leaf-like, strongly resinous; petioles glandular-dotted or lobed; leaf blades narrowly elliptic to lanceolate; lower surface glabrous or pilose with white or rust-colored hairs, not glaucous, upper surface glossy, glabrous, or villous, base acute to rounded, margins serrulate; tip caudate; catkins on short leafy branchlets; floral bracts tawny, the pistillate ones deciduous; stamens 3–6; ovaries glabrous; occasional in swamps, low ground, and wet shores; throughout; FACW. Ours is ssp. *lucida*.

Salix myricoides Muhl. Broad-leaved willow

Shrub to 3.5 m tall with flexible to highly brittle branches; branchlets red-brown, glabrous to hairy; juvenile leaves glabrous or pubescent with white or rust-colored hairs; leaf blades narrowly oblong to oblanceolate, lower surface strongly glaucous, glabrous to glabrescent, midrib villous, upper surface shiny, base rounded to cordate, margins serrulate to crenate or entire, tip acute to acuminate; catkins on short leafy branchlets; ovaries glabrous; rare on stream banks and swamps; scattered; FAC; .

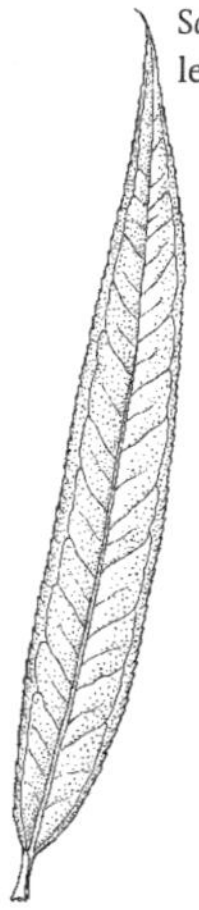

Salix nigra, leaf

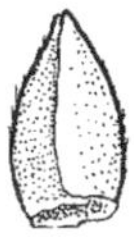

Salix nigra, bud with overlapping bud scale margins (×5)

Salix nigra Marshall Black willow

Tree to 20 m tall or more with highly brittle branches; branchlets gray- or red-brown, glabrous or hairy; bud scale margins free and overlapping; juvenile leaves glabrous or pilose with white or rust-colored hairs; petioles glandular-dotted; leaf blades very narrowly elliptic to lanceolate, glabrous or pilose, lower surface not glaucous, upper surface shiny, base acute, margins serrulate, tip acuminate; catkins on short leafy branchlets; floral bracts tawny, the pistillate ones deciduous; stamens 4–6; ovaries glabrous; common in swamps, wet meadows, and rich alluvial soils; throughout; FACW+.

Salix pedicellaris Pursh Bog willow

Shrub to 1.5 m tall, erect or decumbent, layering; branchlets yellow- or red-brown, glabrous or sparsely hairy; juvenile leaves glabrous to puberulent with white or rust-colored hairs; stipules rudimentary; leaf blades narrowly oblong to elliptic or oblanceolate, glabrous or puberulent, upper surface dull, base acute to rounded, margin entire, tip rounded to acute; catkins on short leafy branchlets; floral bracts tawny; ovaries glabrous; very rare in fens and bogs; NW; OBL; .

Salix pentandra L. Bay-leaved willow

Shrub or tree to 5 m tall with red-brown to brownish, glabrous branchlets; juvenile leaves glabrous; stipules rudimentary; petioles glandular-dotted or lobed; leaf blades narrowly elliptic to lanceolate, glabrous, lower surface not glaucous, upper surface highly glossy; base rounded or wedge-shaped, margins serrulate, tip acuminate; catkins on short leafy branchlets; floral bracts tawny, the pistillate ones deciduous; stamens 4–9; ovaries glabrous; cultivated and rarely escaping; mostly E; native to Eurasia.

Salix petiolaris Sm. Slender willow

Shrub to 6 m tall with yellowish or yellow-green, hairy branchlets; juvenile leaves moderately densely silky with white or rust-colored hairs; stipules lacking or rudimentary; leaf blades ligulate or very narrowly elliptic, glabrous to silky with white or rust-colored hairs, upper surface dull or shiny, base acute, margins entire or serrulate, tip acute; staminate catkins on short leafy branchlets or sessile; pistillate catkins on short leafy branches; ovaries silky; rare in meadows and swales; scattered; flr. just before the leaves emerge; OBL; .

Salix purpurea L. Basket willow

Shrub or tree to 5 m tall with flexible to somewhat brittle branches; branchlets yellow- to olive-brown or violet tinged, glabrous; leaves opposite or subopposite; juvenile leaves glabrous or pubescent with white or rust-colored hairs; stipules lacking; leaf blades narrowly oblong to oblanceolate, glabrous, upper surface dull to sublusterous; base rounded to wedge-shaped; margins serrulate or entire, tip acute to acuminate; staminate catkins sessile, flr. before the leaves emerge; pistillate catkins sessile or on short leafy branchlets, flr. before or with the leaves; floral bracts black or bicolor; stamens apparently 1; ovaries silky; cultivated and occasionally naturalized in low ground; mostly S; native to Europe.

Salix x *rubens* Schrank Willow
Tree to 15 m tall with erect or drooping, highly brittle branches; branchlets red-brown to golden-yellow, hairy; juvenile leaves silky or glabrous; leaf blades narrowly elliptic, glabrous or sparsely silky to glabrescent, upper surface shiny or dull, base acute to wedge-shaped, margins serrate to serrulate, tip acuminate; catkins on short leafy branchlets; floral bracts tawny, the pistillate ones deciduous; ovaries glabrous; cultivated and often naturalized along roadsides and streams; native to Europe. [syn: *S. alba* x *fragilis*]

Salix *sericea* Marshall Silky willow
Shrub to 4 m tall with highly brittle branches; branchlets yellow-green to mottled yellow-brown, velvety; juvenile leaves very densely silky with white or rust-colored hairs; stipules rudimentary or lacking; petioles glandular-dotted or not; leaf blades ligulate to narrowly elliptic, lower surface densely silky, upper surface dull, pubescent to glabrescent, base wedge-shaped, margins serrulate; tip short acuminate; catkins sessile; ovaries silky; common in swamps, bogs, stream banks, and low woods; throughout; flr. before the leaves emerge; OBL.

Salix *serissima* (Bailey) Fernald Autumn willow
Shrub to 5 m tall with yellow- to red-brown, glabrous branchlets; juvenile leaves glabrous; stipules lacking or rudimentary; petioles glandular-dotted; leaf blades narrowly oblong to lanceolate or narrowly ovate, glabrous, lower surface not glaucous or thinly so, upper surface highly glossy, base acute to rounded, margins serrulate, tip acuminate; catkins on short leafy branchlets; floral bracts tawny; stamens 3–7; ovaries glabrous; rare in fens and wet meadows on calcareous soils; scattered; flr. after the leaves emerge; OBL; 🌿.

SANTALACEAE Sandalwood Family

Deciduous shrubs or herbaceous plants which, although green and photosynthetic, are root parasites on other species; leaves alternate (in our species), simple, entire, without stipules; flowers perfect or unisexual, regular, 4–5-merous; ovary inferior, topped by a nectar disk; fruit a nut or drupe.

A. herb; flowers in a terminal branching inflorescence .... *Comandra*
A. shrub; flowers in terminal racemes .... *Pyrularia*

Comandra Nutt.

Comandra *umbellata* (L.) Nutt. Bastard toadflax
Rhizomatous, glabrous, perennial herb, 1–4 dm tall, parasitic on roots of oaks (*Quercus* sp.); leaves 2–5 cm long, numerous, alternate, pale green, lanceolate to elliptic, sessile or nearly so; flowers perfect with a well-developed hypanthium, white; fruit 4–6 mm thick; frequent in dry, open oak woods; throughout; flr. May–Jul; FACU–.

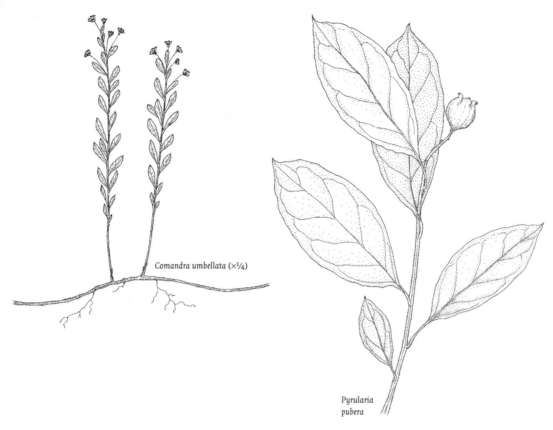

Comandra umbellata (×¼)

Pyrularia
pubera

Pyrularia Michx.

Pyrularia pubera Michx.   Buffalo-nut, oil-nut
Deciduous shrub to 5 m tall, parasitic on the roots of trees and shrubs; leaves
alternate, simple, entire, petiolate, 5–15 cm long; flowers perfect or staminate,
small, greenish; fruit subglobose, 1.5–3 cm, with a large stone and thin outer
layer; rare in woods and thickets; SW; flr. Jun, frt. Sep–Nov; UPL; 🍂.

SAPINDACEAE Soapberry Family

Deciduous trees or herbaceous vines with alternate, pinnately compound leaves
lacking stipules and small flowers in a terminal panicle; flowers perfect or unisexual, slightly irregular, 4–5-merous; ovary superior; stamens 8; fruit a 3-lobed
inflated capsule.

A. deciduous tree .... *Koelreuteria*
A. herbaceous vine .... *Cardiospermum*

Cardiospermum L.

Cardiospermum halicacabum L.   Balloon-vine
Herbaceous annual vine to 3 m tall with alternate, twice compound leaves; flowers white, 3 cm, slightly irregular, in few-flowered, tendril-bearing corymbs;
fruit an inflated, 3-lobed, 2.5-cm capsule; cultivated and rarely escaped to bal-

last, railroad sidings, or urban land; flr. Jul–Sep; native to the tropics and sub-tropics; FACU.

Koelreuteria Laxm.

Koelreuteria paniculata Laxm.   Golden rain-tree
Deciduous tree with stout, coarse branches, alternate, odd-pinnate leaves and terminal panicles of small yellow flowers followed by conspicuous inflated, papery fruits; cultivated, seedlings occasionally becoming established in waste ground and along roadsides; SE and SW; flr. Jul–Aug, frt. Sep–Oct; native to Asia.

Koelreuteria paniculata, fruit

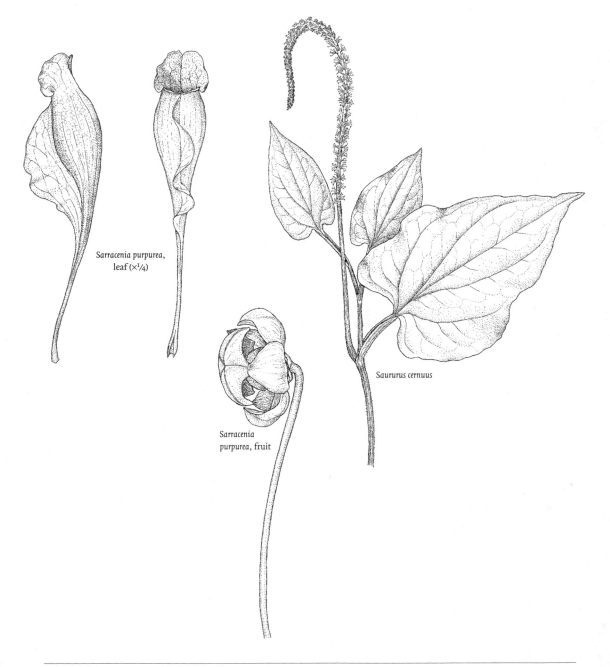

Sarracenia purpurea, leaf (×¼)

Sarracenia purpurea, fruit

Saururus cernuus

SARRACENIACEAE Pitcher-plant Family

Carnivorous, perennial herbs with basal leaves in the form of tubular "pitchers" that collect rain water and serve as traps for insects; flowers perfect, solitary on naked stalks; sepals and petals each 5; petals distinct; stamens many; ovary superior; fruit a capsule.

Sarracenia L.

Sarracenia purpurea L.   Pitcher-plant

Leaves 1–2 dm, curved-ascending at the base, winged on the lower side and with a pronounced lip; flowers nodding, petals maroon or red; style expanded in a 5-lobed, umbrella-like structure that persists long after the petals have dropped; frequent in sphagnum bogs; NE and NW and at high elevations along the Allegheny Front; flr. Jun; OBL.

SAURURACEAE Lizard's-tail Family

Aquatic, perennial herbs with alternate, simple, entire and stipulate leaves; flowers perfect, hypogynous, perianth absent, stamens 6–8, carpels 3–4 with one style per carpel; inflorescence a spike; fruits indehiscent, 1-seeded.

Saururus L.

Saururus cernuus L.   Lizard's-tail

Plants 5–12 dm tall, sometimes branched, often in large colonies; leaves cordate to cordate-ovate, 6–15 cm long, petioles shorter than the blade and sheathing the stem at the nodes; spikes terminal, slender, 6–15 cm long, nodding near the tip even before anthesis; flowers white, very small, each subtended by a tiny bractlet, fragrant; fruits 2–3 mm in diameter, strongly wrinkled when dry; occasional in swamps and shallow water along the edges of streams and rivers; NW and S; flr. Jun–Aug; OBL.

SAXIFRAGACEAE Saxifrage Family

Herbaceous perennials with simple, alternate, opposite, or basal leaves with small, inconspicuous stipules or none; flowers regular or irregular, perfect, perigynous; sepals mostly 4–5, appearing as lobes of the hypanthium; petals the same number as the sepals or absent; stamens 4–10; carpels 2–5(7), fused below but usually with individual styles or stylal beaks above; fruit dry, dehiscent.

A. ovary 5–7-celled; stamens 10 .... *Penthorum*
A. ovary of 2–4 carpels; stamens 4–10
    B. stems creeping, branched; petals lacking; flowers 4-merous ....
    .... *Chrysosplenium*
    B. stems erect, unbranched; petals present; flowers 5-merous
        C. flowers solitary on scape-like stems .... *Parnassia*
        C. flowers in a raceme or panicle
            D. stamens 10

> E. petals deeply fringed .... Mitella
> E. petals entire or nearly so
>> F. leaves pinnately veined, entire or toothed but not lobed; styles 2 .... Saxifraga
>> F. leaves palmately 3–5-lobed and toothed; style 1 .... Tiarella
> D. stamens 5 .... Heuchera

Chrysosplenium L.

Chrysosplenium americanum Schwein ex Hooker   Golden saxifrage
Succulent, semiaquatic herb with decumbent stems; leaves ovate to circular, opposite below, alternate above; flowers axillary, sessile; sepals 4, yellowish or purplish; petals none; stamens 4–10, inserted on a conspicuous disk; anthers red-orange; ovary bicarpellate; styles 2; frequent in wet woods, springs, seeps, and cold swamps; throughout; flr. late Apr–Jun; OBL.

Heuchera L.

Flowering stems, 3–7 dm tall; leaves all or mostly basal, round-cordate and toothed and lobed; inflorescence a panicle of small greenish flowers on a long slender peduncle; sepals, petals, and stamens each 5; hypanthium saucer-shaped or tubular and partially fused to the base of the ovary; carpels 2; flr. late Apr–Jun. Heuchera sanguinea Engelm., the cultivated coralbells, with red flowers, has rarely spread to waste ground.

A. calyx lobes equal; stamens distinctly longer than the petals .... H. americana
A. calyx lobes unequal; stamens as long as or only slightly longer than the petals .... H. pubescens

Heuchera
americana,
flower (×1½)

Heuchera americana L.   Alum-root
Hypanthium fused to the ovary for more than half its length; calyx lobes equal; petals greenish, white or pink; stamens extending well beyond the corolla; frequent in rich woods, rocky slopes, and shaly cliffs; mostly S; native.

Heuchera pubescens Pursh   Alum-root
Free portion of the hypanthium longer than the basal fused portion; calyx lobes unequal; petals white, pink, or purple; stamens and styles barely longer than the corolla; occasional in rocky woods, banks, and shale barrens; SC.

Mitella L.

Erect herbs with leaves mostly basal, palmately veined; the flowering stem with 1 or 2 leaves midway to the base; inflorescence a raceme of small white or yellow-green flowers, each subtended by a bract; flowers regular, perigynous, 5-merous; petals deeply fringed; stamens 10; carpels 2, fused to form a 2-lobed, superior ovary with 2 short styles.

A. flowering stem bearing a pair of opposite leaves; petals white ...... M. diphylla
A. flowering stem with a single leaf, or none; petals yellow-green .... M. nuda

Mitella diphylla,
flower (×2)

Mitella diphylla L.   Bishop's-cap, miterwort
Stems 1–4 dm tall, bearing a single pair of sessile leaves below the inflorescence;

most leaves basal, ovate-rotund and shallowly 3–5-lobed, with a cordate base; flowers white, nodding, 4–5 mm wide, well separated along the axis of the raceme; the valves of the fruit spreading and funnel-like after dehiscence revealing the shiny black seeds; frequent in rich, moist woods; throughout; FACU.

Mitella nuda L.   Bishop's-cap, miterwort
Stems 0.5–2 dm tall, bearing 1 or no leaves below the inflorescence; most leaves basal, round or kidney-shaped, cordate, crenate, and only obscurely lobed; flowers greenish-yellow, 6–7 mm wide; seeds black and shining; rare in swamps and moist, mossy woods; NE and NW; flr. May–Jun; FACW–; 🌿.

Parnassia L.

Parnassia glauca Raf.   Grass-of-Parnassus
Glabrous herbs to 2.5–5 dm tall; leaves mostly basal, long-petioled, kidney-shaped, entire; flowers solitary on erect stems that bear a single clasping leaf near the base; sepals 5, slightly fused at the base; petals 5, white with green veins; ovary 1; stigmas 4; rare in boggy meadows or seeps on calcareous soils; E and W; flr. Aug–Sep; OBL; 🌿.

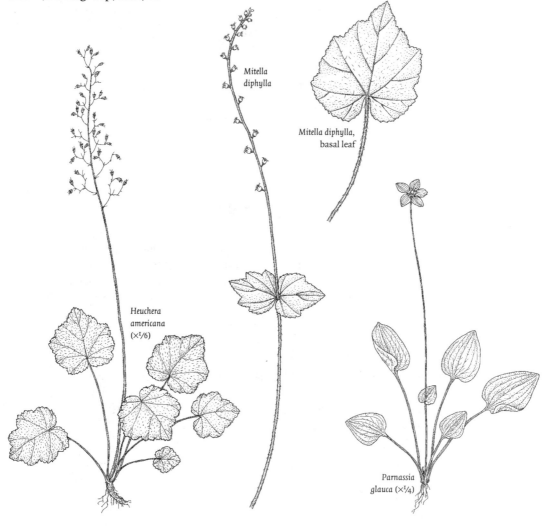

Mitella diphylla

Mitella diphylla, basal leaf

Heuchera americana (×1/6)

Parnassia glauca (×1/4)

Penthorum sedoides

Saxifraga
virginiensis

Tiarella
cordifolia

Penthorum sedoides,
flower (×2)

Penthorum sedoides,
fruit (×2)

Penthorum L.

Penthorum sedoides L.   Ditch stonecrop

Stems erect, 1.5–8 dm tall; leaves alternate, sessile, narrowly elliptic with serrate margins; inflorescence a branched cyme; flowers greenish-yellow; petals small or none; carpels 5–7, united to the middle and forming a capsule with spreading, tapered lobes; common in low, wet ground and ditches; throughout; flr. Jul–early Sep; OBL.

Saxifraga L.

Erect herbs with leaves in a basal rosette; inflorescence a panicle borne on a leafless stem, the branches subtended by small bracts; flowers regular, 5-merous; hypanthium fused to the base of the ovary; carpels 2, fused at the base to form a 2-celled ovary that becomes a 2-beaked fruit.

A. leaves coarsely dentate, ovary superior .... *S. micranthidifolia*
A. leaves shallowly crenate-dentate to entire; ovary partly inferior
    B. petals white; leaves to 8 cm long; plants of dry, rocky banks or woodlands
    .... *S. virginiensis*

B. petals greenish; leaves 1–2 dm long; plants of wet places ..... *S. pensylvanica*

Saxifraga micranthidifolia (Haw.) Steud.   Mountain or lettuce saxifrage
Flowering stems, 3–9 dm tall; leaves oblong to oblanceolate, 1–3 dm long, coarsely dentate and tapering to a winged petiole; panicle large, much-branched; flowers white with a yellow spot at the base of the petals; calyx lobes reflexed; ovary superior, carpels barely fused at the base; rare in shaded stream-beds and seepage areas; S; flr. May–Jun; OBL; 🐾.

Saxifraga pensylvanica L   Swamp saxifrage
Flowering stems, 2–8 dm tall; leaves leathery, 1–2 dm long, oval to spatulate-oblong with broad, clasping petioles; inflorescence an open panicle; flowers greenish; calyx lobes reflexed; carpels free except at the base; frequent in wet woods, bogs, and swamps; throughout; flr. May–Jun; OBL.

Saxifraga virginiensis Michx.   Early saxifrage
Flowering stems 1–3 dm tall; leaves oval or obovate, 1–8 cm long, narrowed to a margined petiole; leaf margins crenate or dentate; panicle narrow; flowers white; calyx lobes erect; carpels distinct except at the base where the ovary is fused to the hypanthium; common in rock crevices and dry slopes; throughout, except in the northernmost counties; flr. late Apr–May; FAC-.

Saxifraga virginiensis, flower (×3)

Tiarella L.

Tiarella cordifolia L.   Foamflower
Rhizomatous, stoloniferous herb with stems 1–3.5 dm tall; leaves basal, palmately veined and shallowly 3–5-lobed; flowers in a dense raceme, 5-merous, regular, barely perigynous; hypanthium small; petals and sepals white; stamens 10; carpels 2, unequal, fused only at the base, but sometimes appearing as 1; common in moist, rocky, wooded slopes; throughout except in the SE; flr. May–Jun; FAC–.

Tiarella cordifolia, flower (×1½)

SCROPHULARIACEAE Snapdragon Family

Herbaceous plants with opposite or alternate leaves lacking stipules; flowers with distinct calyx and corolla; sepals fused, at least at the base, usually regular but sometimes irregular; petals fused, at least at the base; corolla nearly regular to more often slightly or strongly irregular, often bilabiate, sometimes spurred; fertile stamens 2, 4, or 5, borne on the corolla tube, sterile stamens or staminodes often present also; ovary superior, 2-locular; fruit a capsule.

A. corolla nearly regular, lobes spreading radially
   B. corolla 4-lobed, fertile stamens 2
      C. leaves whorled .... *Veronicastrum*
      C. leaves opposite or alternate .... *Veronica*
   B. corolla 5-lobed; fertile stamens 4 or 5; capsule not flattened
      D. corolla white or yellow, tube much shorter than the lobes; fertile sta-mens 5 .... *Verbascum*
      D. corolla blue, tube longer than the lobes; fertile stamens 4 .... *Buchnera*

A. corolla distinctly irregular
    E. corolla with a spur at the base
        F. stems erect
            G. flowers in terminal racemes .... *Linaria*
            G. flowers solitary in the leaf axils
                H. plants glandular-hairy; leaves linear-spatulate ...... *Chaenorrhinum*
                H. plants hairy but not glandular; leaves broadly ovate to triangular or hastate .... *Kickxia*
        F. stems prostrate or trailing
            I. leaves palmately veined and lobed .... *Cymbalaria*
            I. leaves pinnately veined, entire or with a few coarse teeth at the base .... *Kickxia*
    E. corolla lacking a basal spur
        J. foliage leaves alternate or basal
            K. plants aquatic; leaves basal, linear with only a slightly expanded blade; flowers solitary on basal peduncles
                L. leaves in pairs from a slender stolon, fertile stamens 2 .... *Glossostigma*
                L. leaves in clusters from the nodes of the stolon; fertile stamens 4 .... *Limosella*
            K. plants terrestrial; leaves alternate, with well-developed blades; flowers in terminal spikes or racemes
                M. foliage leaves evenly pinnately lobed or toothed .... *Pedicularis*
                M. foliage leaves entire or with 3–5 irregular, linear lobes
                    N. flowers subtended and nearly hidden by conspicuous scarlet-tipped bracts .... *Castilleja*
                    N. flowers not subtended by red-tipped bracts .... *Digitalis*
        J. foliage leaves opposite or whorled
            O. foliage leaves in whorls of 4–7 .... *Veronicastrum*
            O. foliage leaves opposite
                P. prostrate or creeping plants of tidal mudflats; calyx 4-lobed .... *Micranthemum*
                P. plants erect; calyx 4- or 5-lobed
                  Q. fertile stamens 2
                    R. plants mostly glabrous; flowers without extra bracts below the sepals .... *Lindernia*
                    R. entire plant glandular pubescent; flowers with 2 additional bracts just below the sepals .... *Gratiola*
                  Q. fertile stamens 4
                    S. upper lip of the corolla arched and concave, or straight and folded, so as to enclose the anthers
                    T. leaves palmately veined, no more than twice as long as wide .... *Euphrasia*
                    T. leaves pinnately veined, many times longer than wide
                      U. leaves entire or with a few irregular teeth near the base .... *Melampyrum*
                      U. leaves regularly and evenly pinnatifid .... *Pedicularis*
                  S. upper lip of the corolla not enfolding the anthers, although the stamens may be completely hidden within the corolla tube

V. sepals distinct nearly to the base
  W. bracteal leaves only gradually reduced .... *Leucospora*
  W. bracteal leaves abruptly smaller than the foliage leaves below them
    X. each flower subtended by 2–3 large sepal-like bractlets .... *Chelone*
    X. flowers not subtended by 2–3 large bractlets, although each pedicel may have a small subtending bractlet
      Y. corolla tube without a prominent basal bulge
        Z. corolla white, lavender or bluish, 1.5–3.5 cm long; sterile stamen as long as or longer than the fertile ones, hairy near the tip .... *Penstemon*
        Z. corolla greenish or brownish, 7–11 mm long; staminode a flap-like projection on the upper corolla lobe .... *Scrophularia*
      Y. corolla tube with a prominent basal bulge .... *Antirrhinum*
V. sepals distinctly connate, forming a lobed tube
  AA. corolla strongly 2-lipped (bilabiate)
    BB. corolla bicolored, lower lobes blue and upper lobes white .... *Collinsia*
    BB. corolla not bicolored, although there may be a patch of yellow at the throat
      CC. erect; flowers in the axils of opposite foliage leaves .. *Mimulus*
      CC. creeping; flowers in the axils of alternate bracteal leaves .... *Mazus*
  AA. corolla not strongly 2-lipped, nearly campanulate
    DD. corolla yellow .... *Aureolaria*
    DD. corolla pink or lavender .... *Agalinis*

Agalinis Raf.

Simple or branched herbs; leaves opposite (becoming alternate above), sessile, mostly linear; flowers in the axils of the upper leaves or in terminal racemes; calyx regular, 5-lobed; corolla pink or purple, slightly irregular, campanulate, with 5 spreading lobes; stamens 4; parasitic on the roots of various woody and herbaceous plants.

A. stems hairy; upper leaves with narrow spreading lobes at the base .... *A. auriculata*
A. stems glabrous; leaves entire, never lobed
  B. pedicels shorter than the calyx
    C. corolla 2–4 cm; calyx lobes up to half as long as the tube .... *A. purpurea*
    C. corolla 1.5–2 cm; calyx lobes <½ to as long as the tube ..... *A. paupercula*
  B. pedicels longer than the calyx
    D. plant dark green, often tinged with purple; calyx tube with obscure longitudinal veins or veinless .... *A. tenuifolia*
    D. plant pale green; calyx tube distinctly net veined .... *A. decemloba*

Agalinis auriculata

Agalinis auriculata (Michx.) S.F.Blake   Eared false foxglove
Upright, mostly unbranched, hairy perennial with sessile, lanceolate leaves, some of which have narrow, spreading basal lobes; corollas purple, 2–2.5 cm; very rare in moist meadows, fields, roadsides and waste ground; SE; flr. late Aug–Sep; ❧. [syn: *Tomanthera auriculata* (Michx.) Raf.]

Agalinis decemloba (Greene) Pennell   Blue Ridge false foxglove
Stems glabrous, pale green, 2–7 dm tall; leaves linear, rough pubescent above; flowers pink with faint yellow lines; rare on serpentine barrens and dry roadside banks; SE; apparently extirpated; flr. Sep; FACU; ❦. [syn: A. obtusifolia Raf. in part]

Agalinis paupercula (A.Gray) Britton   Small-flowered false foxglove
Very similar to A. purpurea but lobes of the calyx nearly as long as the tube and the corolla only 1.5–2 cm long; rare in moist, open, sandy ground and pond shores; mostly S; flr. Aug–Sep; FACW+; ❦.

Agalinis purpurea, calyx and capsule (×1)

Agalinis purpurea (L.) Pennell   False foxglove
Glabrous annual, 3–7 dm tall with spreading branches; leaves linear, 2–4 mm wide; flowers in elongate racemes; pedicels 1–4 mm, usually shorter than the calyx; corolla purple; occasional in moist sandy fields, rocky shores, and serpentine barrens; SE; flr. late Jul–Sep; FACW–.

Agalinis tenuifolia, calyx and capsule (×1)

Agalinis tenuifolia (Vahl) Raf.   Slender false foxglove
Glabrous, branching annual, 2–6 dm tall with linear leaves 1–3 mm wide; pedicels 1–2 cm at anthesis; corolla pink, 1–2 cm, the upper lip arching over the stamens; frequent in moist to dry, sandy fields, meadows, and open woods; mostly SE and SW; flr. Aug–early Oct; FAC.

Antirrhinum L.

Antirrhinum majus L.   Snapdragon
Erect glabrous herb with narrow, alternate leaves and flowers in a terminal raceme; corolla bilabiate with a bulge or pouch at the base extending between the 2 lower sepals, the mouth of the corolla tube closed by a prominent palate; stamens 4; calyx regular, deeply 5-lobed; capsule with 2 unequal locules; widely cultivated and occasionally escaping to waste ground; flr. Jun–Oct; native to Europe.

Aureolaria Raf.

Erect, branching herbs with opposite, entire to deeply lobed leaves; flowers large, yellow, solitary in the axils of the upper, only slightly reduced leaves; corolla irregular with a somewhat oblique tube, the lobes shorter than the tube; stamens 4; parasitic on the roots of oaks and other trees.

A. glandular hairs present on pedicels, calyx, outside of corolla, and capsules..... ....*A. pedicularia*
A. glandular hairs not present
  B. plants finely pubescent; capsules rusty-pubescent .... *A. virginica*
  B. plants and capsules glabrous
    C. stem green; leaves rarely lobed .... *A. laevigata*
    C. stem glaucous (whitened); leaves deeply lobed .... *A. flava*

Aureolaria flava (L.) Farw.   Yellow false foxglove
Perennial to 1–2 m tall; stem glabrous and glaucous; leaves deeply pinnately

Agalinis
purpurea

Agalinis
tenuifolia

Aureolaria pedicularia

lobed, reduced upward; axis of the inflorescence, pedicels and calyx glabrous; dry, open woods and rocky thickets; flr. late Jun–Oct; 2 varieties:

A. calyx lobes 2–5 mm .... var. *flava*
widespread.
A. calyx lobes 5–15 mm .... var. *macrantha* Pennell
very rare; NW.

Aureolaria laevigata (Raf.) Raf.   False foxglove
Perennial, 0.5–1.5 m tall; stems glabrous but not glaucous; leaves lance-ovate, entire or with a few lobes; upper leaves sometimes serrate; pedicels and calyx glabrous; occasional in rocky, open woods; mostly SC and SW; flr. Jul–Oct.

Aureolaria pedicularia (L.) Raf.   Cut-leaf false foxglove
Annual to 1 m tall; leaves pinnatifid with 5–8 pairs of deeply cleft leaflets; pedicels and calyx glandular-hairy; corolla yellow with brown markings; occasional in dry, open woods and edges; mostly SE and SC; flr. Aug–Sep.

Aureolaria virginica (L.) Pennell   Downy false foxglove
Pubescent perennial, 0.5–1.5 m tall; lower leaves lobed below the middle, the upper leaves progressively reduced; occasional in dry, open deciduous woods; throughout; flr. late Jun–Sep.

Buchnera L.

Buchnera americana L.   Bluehearts
Erect, unbranched perennial herb, 3–8 dm tall; sessile, opposite leaves; flowers
nearly regular, corolla purple, 2 cm long with a long slender tube and 5 spreading lobes; stamens 4, shorter than the tube; parasitic on the roots of other
plants; rare in moist, sandy ground of serpentine barrens; SE; believed to be extirpated; flr. Jul–Sep; FACU; ✿.

Castilleja Mutis

Castilleja coccinea (L.) Spreng.   Indian paintbrush
Erect annual, 2–6 dm tall; leaves alternate, entire, or with 3–5 lobes; flowers in a
dense terminal spike, each flower subtended by a showy crimson-tipped bract;
corolla yellowish-green, the upper lip arched and enclosing the 4 stamens; parasitic on the roots of other plants; rare in moist meadows on limestone and diabase; S; flr. Apr–Jun; FAC; ✿.

Chaenorrhinum Reichb.

Chaenorrhinum minus (L.) Lange   Dwarf snapdragon, lesser toadflax
Erect, freely branched, glandular-hairy annual to 3 dm tall; leaves alternate, linear to narrowly spatulate, entire; flowers solitary in the axils; corolla bilabiate,
spurred at the base, blue-purple with a yellow palate; stamens 4; occasional
in cinders, roadsides, railroad tracks, and waste ground; flr. May–Sep; native to
Europe.

Chelone L.

Chelone glabra L.   Turtlehead
Erect glabrous perennial, 5–8 dm tall, with opposite leaves; flowers white or
pinkish, in dense terminal spikes, each flower subtended by 2–3 large bractlets
below the sepals; corolla tubular and bilabiate, the throat nearly closed by a
raised palate; stamens 4, inserted at the base of the corolla tube; frequent in wet
wood, stream banks, and swamps; throughout; flr. late Jul–Sep; OBL.

Chelone glabra

Collinsia Nutt.

Collinsia verna Nutt.   Blue-eyed-Mary
Upright annual to 4 dm tall with opposite leaves and flowers in whorls of 2–8 from the upper nodes; flowers strongly bilabiate and bicolored, the upper lip white, lower blue; rare in floodplain forests and alluvial thickets; SW; flr. late Apr–May; FAC–; 🌿.

Cymbalaria Hill

Cymbalaria muralis Gaertn., B.Mey. & Schreb.   Kenilworth-ivy
Trailing annual or short-lived glabrous perennial, stems rooting at the nodes; leaves palmately veined and lobed, alternate; flowers solitary on long pedicels in the axils; corolla violet with a yellow patch on the throat and a 2–3-mm spur; frequently naturalized in crevices in rock walls and other masonry; flr. Mar–Dec; native to the Mediterranean region.

Digitalis L.

Erect, leafy, unbranched biennials or perennials with alternate leaves and an elongate terminal raceme; corollas tubular, irregular with an open throat and 5 spreading lobes; stamens 4. Poisonous.

A. middle lobe of the lower lip of the corolla much longer than the others ....
.... *D. lanata*
A. middle lobe of the lower lip of the corolla only slightly longer than the others
   B. plant glabrous except in the inflorescence; corolla pale yellow or whitish,
     1–2.5 cm long .... *D. lutea*
   B. leaves and stem pubescent; corolla purple to white, 4–5 cm long ....
     .... *D. purpurea*

Digitalis lanata Ehrh.   Woolly foxglove
Biennial or perennial; stems 5–10 dm tall; leaves narrowly lanceolate to oblong-lanceolate, sessile, entire; corolla pale yellow to whitish with brown or violet lines, middle lobe of the lower lip 8–13 mm long; cultivated and rarely escaped to roadsides or open ground; flr. Jul; native to Europe.

Digitalis lutea L.   Yellow foxglove
Glabrous perennial, 5–8 dm tall; leaves sessile, oblong-oblanceolate; corolla yellow, the 5 lobes nearly equal in length; cultivated and occasionally escaped to roadsides or open ground; flr. Jul; native to Europe.

Digitalis purpurea L   Foxglove
Pubescent biennial to 1.5 m tall; leaves pubescent at least beneath, tapering to a somewhat winged petiole; corolla purple or white with darker purple spots within; cultivated and occasionally naturalized on wooded roadsides and meadows; flr. May–Jul; native to Europe.

Euphrasia L.

Euphrasia stricta J.P.Wolff ex J.F.Lehm.   Eyebright
Annual herb, 1–4 dm tall, with ovate to rotund, palmately veined leaves, foliage

leaves opposite, bracteal leaves subtending the flowers mostly alternate; calyx 4-lobed; corolla pale lavender with darker lines and a yellow patch at the throat, 2-lipped, the upper lip arched over the 4 stamens and shallowly 2-lobed or notched, the lower lip 3-lobed and spreading; rare on gravelly roadsides and in sterile fields; flr. Jun–Oct; native to Europe.

Glossostigma L.

Glossostigma diandrum,
flower (×10)

Glossostigma diandrum (L.) Kuntze   Mudmat
Submergent or emergent mat-forming, rooted aquatic, 1–2 cm tall; leaves basal, in pairs at the nodes of slender stolons, narrow, strap-shaped without a well-defined blade; flowers solitary on pedicels about as long as the leaves, corolla 1 mm wide, pinkish, irregular; stamens 2 or, in deeper water, flowers apparently cleistogamous on very short pedicels; naturalized in shallow water or on exposed mud flats at the margin of a lake; flr. Jul–Sep; native to New Zealand and Australia.

Gratiola L.

Low glandular-pubescent plants of moist or wet soils, 1–3 dm tall; leaves opposite, sessile; the solitary, long-pediceled, axillary flowers each subtended by 2 strap-shaped bracts just below the 5-lobed calyx; corolla yellow or whitish, the 2 stamens inserted on the corolla tube.

A. flowers bright yellow; leaves lanceolate to ovate .... G. aurea
A. corolla lobes white, the tube yellowish; leaves lanceolate to oblanceolate ......
.... G. neglecta

Gratiola aurea Muhl. ex Pursh   Goldenpert, hedge-hyssop
Perennial, glandular pubescent at least in the upper portions; leaves linear to narrowly ovate; corolla bright yellow, 8–13 mm; very rare on moist stream banks and sandy pond shores; E; flr. Jul–Aug; OBL; ✿.

Gratiola neglecta Torr.   Hedge or mud-hyssop
Annual, glandular-pubescent throughout; leaves 2–5 cm, lanceolate to oblanceolate, coarsely toothed; corolla 8–10 mm, tube yellowish, lobes white; frequent on stream banks, muddy shores, bogs, wet fields, and ditches; throughout; flr. May–Oct; OBL.

Kickxia L.

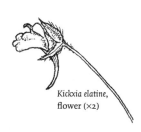

Kickxia elatine,
flower (×2)

Kickxia elatine (L.) Dumort.   Cancerwort
Hairy annual, upright or with prostrate branches; leaves broadly ovate or triangular-hastate, short-petioled, opposite below, alternate above; flowers solitary on long pedicels in the upper axils; corollas 6–8 mm, spurred, yellowish-green, with a purple blotch on the upper lip; occasional in waste ground and gravel; SE; flr. Jun–Oct; native to the Mediterranean region; FAC. *Kickxia spuria* (L.) Dumort., with round-ovate leaves that are cordate at the base, was collected early on ballast.

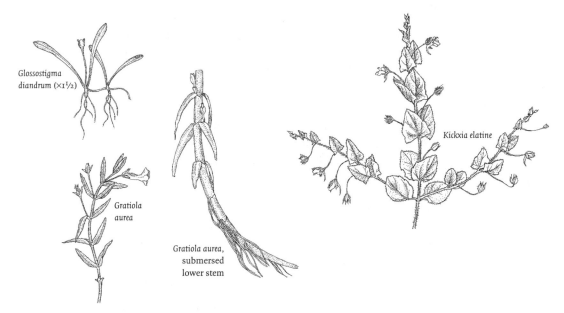

Glossostigma diandrum (×1½)

Gratiola aurea

Gratiola aurea, submersed lower stem

Kickxia elatine

Leucospora Nutt.

Leucospora multifida (Michx.) Nutt.   Narrowleaf paleseed
Erect, branched, pubescent annual herb with opposite or whorled, pinnatifid leaves and solitary or paired axillary flowers; corolla pale lavender, tubular; stamens 4, inserted near the middle of the corolla tube; rare in waste ground and ballast; SE; flr. Jun–Sep; native farther west, apparently adventive here; OBL. [syn: *Conobea multiflora* (Michx.) Benth.]

Limosella L.

Limosella australis R.Br.   Awl-shaped mudwort
Small aquatic or emergent herb with prostrate stems or stolons producing clusters of 5–10 linear leaves at the nodes; flowers solitary on peduncles arising from the leaf clusters; corolla white, 3.5–4 mm; stamens 4; tidal mudflats; SE; believed to be extirpated; flr. Aug–Sep; OBL; ❦.

Linaria Mill.

Erect glabrous herbs with numerous, alternate leaves; flowers in a terminal raceme; calyx deeply 5-lobed; corolla strongly bilabiate and spurred; stamens 4. In addition to the taxa treated below, *Linaria supina* (L.) Chaz., which is glandular-pubescent in the inflorescence and has yellow flowers tinged with purple, is represented by several early ballast collections.

A. flowers blue (or occasionally white) .... *L. canadensis*
A. flowers yellow with an orange palate
    B. leaves linear, narrowed to a short petiole, gray-green and glaucous ....
    .... *L. vulgaris*
    B. leaves ovate to lance-ovate, sessile, and clasping at the base.... *L. genistifolia*

Linaria canadensis (L.) Dum.Cours.   Old-field toadflax
Slender, upright annual to 5 dm tall with linear cauline leaves and short, prostrate, leafy, basal shoots; flowers light blue; occasional in sandy fields, serpentine barrens, and railroad banks; mostly SE; flr. May–Jul.

Linaria genistifolia (L.) Mill.   Toadflax
Stout, upright perennial, 4–12 dm tall; leaves ovate-clasping; flowers in an elongate terminal raceme; corolla bright yellow with an orange-bearded palate, spurred; cultivated and occasionally escaped to railroad banks and roadsides; flr. Jun; native to the Mediterranean region. Ours is ssp. *dalmatica* (L.) Maire & Petitm.

Linaria vulgaris Hill   Butter-and-eggs
Glaucous perennial, 3–8 dm tall, with numerous narrow, gray-green leaves; corolla yellow with an orange palate and a spur; fields, roadsides, shale barrens, railroad tracks, and waste ground; common throughout; flr. Jun–Nov; native to Eurasia.

Lindernia All.

Lindernia dubia (L.) Pennell   False pimpernel
Branching, annual herb to 2 dm tall, mostly glabrous; leaves opposite, entire or few-toothed; flowers solitary in the upper axils, on long pedicels; corolla irregular, white to pale violet, 5–10 mm, tubular, with 5 spreading lobes; fertile stamens 2; 3 varieties:

A. leaves subtending the flowers tapering at the apex; pedicels 0.5–2 cm
    B. lower pedicels shorter than the subtending leaves .... var. *dubia*
        common on muddy and swampy shores; throughout; flr. May–Sep; OBL.
    B. lower and upper pedicels longer than the subtending leaves ....
       .... var. *anagallidea* (Michx.) Cooperr.
       occasional in moist meadows, stream banks, and ditches; SE; flr. May–Sep; OBL.
A. leaves subtending the flowers rounded at the tip; pedicels 1–5 mm ....
    ....var. *inundata* (Pennell) Pennell
    wet shores and tidal mud flats; SE; flr. Jul–Oct; OBL.

Mazus Lour.

Low herbs with basal or opposite leaves below becoming alternate above; flowers in few-flowered terminal racemes; corolla violet or white, 2-lipped; stamens 4.

A. semierect, pubescent annual; corolla 7–10 mm... M. *pumilus* (Burm.f.) Steenis
A. creeping, glabrous perennial; corolla 15 mm .... M. *miquelii* Makino

Mazus miquelii Makino   Mazus
Glabrous, creeping perennial; cultivated and occasionally escaped to lawns, gardens, and paving cracks; SE; flr. Apr–Jul; native to Asia.

Mazus pumilus (Burm. f.) Steenis   Japanese mazus
Sprawling short-hairy annual to 15 cm tall; cultivated and occasionally escaped to lawns or moist alluvial shores; SE; flr. May–Sep; native to Asia; FACU.

Melampyrum L.

Melampyrum lineare Desr.   Cow-wheat

Annual herbs with simple or lightly branched stems to 4 dm tall; leaves opposite, linear or lance-ovate, entire, upper bracteal leaves subtending the flowers often toothed near the base; corolla white with a yellow palate, 6–12 mm, tubular, and bilabiate, the upper lip straight and folded enclosing the 4 stamens; parasitic on the roots of other plants; frequent in acidic woods, bog margins, and barrens; throughout; flr. Jun–Sep; FACU. Plants with more deeply toothed upper bracts have been recognized as var. *pectinatum* (Pennell) Fernald, as opposed to the more common var. *americanum* (Michx.) P.Beauv.

Melampyrum
lineare

Micranthemum Michx.

Micranthemum micranthemoides (Nutt.) Wettst.   Nuttall's mud-flower

Small, glabrous annual with creeping stems and ascending branches 5–20 cm tall; leaves opposite, elliptic, 2–5 mm, entire; flowers solitary and axillary, corolla 2 mm; stamens 2, inserted near the corolla throat; rare on tidal mudflats; SE; believed to be extinct; flr. Sep–Oct; OBL; ✤.

Mimulus L.

Perennial herbs of moist soils; leaves opposite; flowers yellow or blue, bilabiate, solitary in the upper axils; stamens 4, inserted midway on the corolla tube, stigma 2-lobed.

A. flowers blue
     B. leaves sessile; pedicel longer than the calyx .... M. ringens
     B. leaves petiolate; pedicel shorter than the calyx .... M. alatus
A. flowers yellow
     C. calyx lobes nearly equal; leaf blades pinnately veined .... M. moschatus
     C. calyx lobes conspicuously unequal; leaf blades palmately veined ....
     .... M. guttatus

Mimulus alatus Aiton   Winged monkey-flower

Very similar to *M. ringens*, but leaves tapering to a 1–2-cm winged petiole; pedicels only 2–10 mm at anthesis; frequent in swamps, wet meadows, and shores; mostly S; flr. Jul–Sep; OBL.

Mimulus guttatus Fisch. ex DC.   Monkey-flower

Erect or ascending herb to 5 dm tall, glandular-pubescent in the inflorescence, glabrous elsewhere; corolla yellow with orange spots in the throat, 4 cm long; cultivated and very rarely escaped to wet roadsides; C; flr. Jun–Jul; native to western North America; OBL.

Mimulus moschatus Douglas ex Lindl.   Muskflower

Glandular-pubescent, creeping to ascendent herb; leaves opposite, entire or remotely denticulate; flowers on slender, 1–2-cm pedicels in the axils of upper pairs of leaves; corolla yellow, 1.7–2.2 cm long, open at the throat; wet shores, seeps, and springy swales; occasional; mostly E; flr. Jul–Aug; OBL.

Mimulus ringens L.   Allegheny monkey-flower

Erect, glabrous perennial to 1.3 dm tall; sessile, opposite leaves and 4-angled stems; calyx angled, the tubular portion longer than the lobes; flowers solitary in the upper axils on 2–4.5-cm pedicels; corolla blue, 2–3 cm, strongly 2-lipped; common in wet, open ground of swamps, meadows, and shores; throughout; flr. Jun–Aug; OBL.

Pedicularis L.

Hemiparasitic perennials with pinnately lobed or toothed leaves; flowers in leafy-bracted, terminal spikes or racemes; corolla strongly irregular, the upper lip arched over and enclosing the 4 stamens.

A. spring blooming; leaves chiefly basal .... *P. canadensis*
A. fall blooming; leaves mostly cauline .... *P. lanceolata*

Pedicularis canadensis L.   Wood-betony, forest lousewort

Perennial herb, 1.5–4 dm tall; leaves chiefly basal, deeply pinnately lobed; flowers in a dense, leafy terminal spike; corolla yellow to purple, upper lip distinctly longer than the lower; parasitic on the roots of a wide variety of woody and herbaceous plants; frequent in woods, old fields, and edges; throughout; flr. Apr–May; FACU.

Pedicularis lanceolata Michx.   Swamp lousewort, wood-betony

Erect perennial herb, 3–8 dm tall; leaves mostly opposite but becoming alternate above, shallowly pinnately lobed or toothed; flowers in a terminal raceme and from the upper axils, corolla yellow, upper lip barely longer than the lower; parasitic on the roots of other plants; rare in swamps, boggy meadows, and swales; mostly S; flr. Aug–Sep; FACW; 🍃.

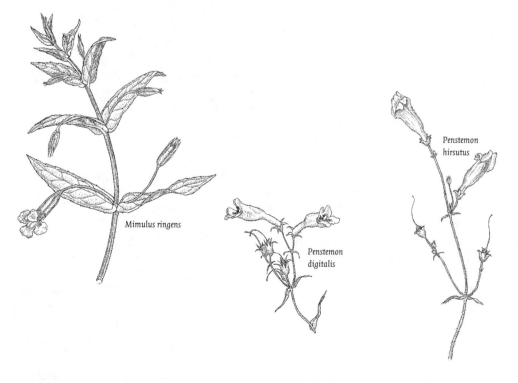

Mimulus ringens

Penstemon digitalis

Penstemon hirsutus

Penstemon Schmidel

Erect perennial herbs with sessile, opposite leaves and a terminal raceme or panicle; corolla 2-lipped, the tube much longer than the lobes; calyx deeply 5-lobed; fertile stamens 4, the sterile filament as long as or longer than the fertile ones and bearded near the tip, hence the common name "beard-tongue."

A. corolla tube closed by a palate created by the upward arching lower lip; stem hirsute .... *P. hirsutus*
A. corolla tube open; stem variously glabrous or finely pubescent
    B. corolla with a basal tube and an abruptly wider, dilated throat
        C. outside of the corolla white or just slightly tinged with purple at the base; anthers hairy ....*P. digitalis*
        C. outside of the corolla pale lavender or purple; anthers glabrous
            D. tips of the calyx lobes acute but flat; corolla 15–20 mm long .... *P. laevigatus*
            D. tips of the calyx lobes linear-subulate; corolla 20–35 mm long ....*P. calycosus*
    B. corolla tube only gradually dilated upward
        E. corolla tube white, the lower lip with purple lines .... *P. pallidus*
        E. corolla tube pale violet or blue-purple .... *P. canescens*

Penstemon calycosus Small   Beardtongue
Stems 0.5–1.3 m tall, sparsely pubescent in the inflorescence, glabrous below; sepals linear subulate at the tip, 5–12 mm long; corolla purple, 2–3.5 cm; anthers glabrous; occasional in swampy woods, old fields, and roadsides; SE; flr. Jun–Jul; native farther west and south; UPL.

Penstemon canescens (Britton) Britton   Beardtongue
Stems 4–8 dm tall, often clustered, finely whitish-pubescent or sometimes also with spreading hairs to 1 mm; corolla pale violet-purple externally, white inside, 2–3 cm long, throat flattened and strongly ridged within; rare on dry, rocky, wooded slopes; SC; 🌱.

Penstemon digitalis Nutt. ex Sims   Tall white beardtongue
Stems to 1.5 m tall, glabrous or sometimes finely pubescent; sepals ovate with acuminate tips; corolla white or occasionally faintly tinged with purple at the base or with purple lines within, 1.5–3 cm long; anthers bearing stiff hairs; common in meadows, old fields, and roadsides; throughout; flr. Jun–early Aug; FAC.

Penstemon digitalis, corolla

Penstemon hirsutus (L.) Willd.   Northeastern beardtongue
Stems usually clustered, 4–8 dm tall, hirsute below, often glandular-pubescent in the inflorescence; corolla pale violet-purple with whitish lobes, 2.3–3.2 cm long, tube closed by the upward arching palate; frequent in dry, open, rocky slopes, dry fields, and roadside banks; throughout, except at the highest elevations; flr. late May–early Aug.

Penstemon hirsutus, flower

Penstemon laevigatus (L.) Aiton   Eastern beardtongue
Similar to P. *digitalis;* sepals 3–6 mm long, lance-ovate with elongate, but flat, tips; corolla pale violet-purple on the outside, 1.5–2.2 cm long; anthers glabrous; rare on wooded hillsides, moist meadows, and roadside banks; mostly W; flr. Jun–Jul; FACU; 🌱.

Penstemon pallidus Small   Eastern white beardtongue
Stems 3–7 dm tall, pubescent throughout and glandular in the inflorescence; leaves pubescent on both sides; corolla white with fine purple lines, 1.6–2.2 cm long; infrequent in meadows and old fields; SE; flr. May–Jun; native farther west and south; FACU.

Scrophularia L.

Upright, leafy, perennial herbs, glabrous or sparsely glandular above; leaves opposite, ovate or ovate-lanceolate, petiolate; flowers in a terminal panicle; calyx only slightly fused at the base; corolla purple-brown or greenish, the 2-lobed upper lip extending beyond the 3-lobed lower lip; stamens 4, the single staminode reduced to a flap-like projection attached to the upper lip of the corolla.

A. staminode dark purple or brown, longer than wide; larger leaf blades often cordate .... *S. marilandica*
A. staminode yellowish-green, wider than long; leaf blades ovate-lanceolate, truncate to rounded at the base, never cordate .... *S. lanceolata*

Scrophularia lanceolata, flower

Scrophularia lanceolata, upper lip of corolla with staminode

Scrophularia lanceolata Pursh   Lanceleaf figwort
Stems to 2 m tall, lower leaves usually long-petioled, blades ovate-cordate; corolla 7–11 mm, lower lobe yellowish-green; staminode yellowish-green wider than long; occasional in low woods, thickets, stream banks, and moist roadsides; throughout; flr. late May–Aug; FACU+.

Scrophularia marilandica, upper lip of corolla with staminode

Scrophularia marilandica, flower

Scrophularia marilandica L.   Eastern figwort
Stems to 3 m tall; corolla 5–8 mm; staminode purple or brown, longer than wide; frequent in alluvial woods, river banks, moist shores and roadsides; throughout; flr. Jun–Aug; FACU–.

Verbascum L.

Upright biennial herbs with leafy stems to 2 m tall arising from a basal rosette; variously densely tomentose or stipitate-glandular; flowers in terminal spikes or racemes, corolla nearly regular, 5-lobed, with a very short tube; stamens 5, inserted on the corolla tube, some or all of the filaments bearing long hairs.

A. leaves green and glabrous on both sides or with simple glandular hairs; calyx and pedicels glandular pubescent
   B. flowers solitary at the nodes of the inflorescence .... *V. blattaria*
   B. flowers in clusters of 2–5 at the lowest nodes of the inflorescence ....
     .... *V. virgatum*
A. plant densely covered with branching hairs, except perhaps on the upper leaf surface, and appearing gray or yellowish-gray; calyx and pedicels mostly not glandular
   C. inflorescence very dense, rarely branched
     D. leaves decurrent forming wings on the stem extending to the next node below ....*V. thapsus*
     D. leaves not decurrent, or only slightly so .... *V. phlomoides*
   C. inflorescence freely branched
     E. basal leaves not lobed; inflorescence not glandular .... *V. lychnitis*
     E. basal leaves lobed; inflorescence glandular .... *V. sinuatum*

Verbascum blattaria L.   Moth mullein

Biennial to 1.2 m tall, glandular pubescent in the inflorescence, leaves and lower stem glabrous; corolla yellow or white, filament hairs purple or purple and white; common in fields, roadsides, railroad embankments, and waste ground; throughout; flr. May–Jul; native to Eurasia; UPL.

Verbascum lychnitis L.   White mullein

Sparsely gray-tomentose biennial to 1.5 m tall; leaves cuneate at the base, sessile but not decurrent, green above, whitish-tomentose beneath; inflorescence freely branched; flowers white or creamy; all 5 filaments bearing white hairs; rare on roadsides, shaly banks, railroad cuts, and vacant lots; mostly SE; flr. Jun–Aug; native to Eurasia.

Verbascum phlomoides L.   Mullein

Gray or yellowish-tomentose biennial to 1.2 m tall; leaves elliptical to lance-ovate, sessile but not decurrent, densely tomentose above and below; inflorescence simple; corolla white or yellow, only the 3 upper filaments bearing white hairs; rare and scattered on roadsides and disturbed open ground; flr. Jul–Aug; native to Europe.

Verbascum sinuatum L.   Mullein

Densely yellowish-gray-tomentose biennial to 1 m tall; glandular in the freely branched inflorescence; filament hairs violet; rare in ballast; SE; flr. Jul–Aug; native to Europe.

Verbascum thapsus L.   Common mullein, flannel-plant

Densely yellowish-gray-tomentose biennial to 2 m tall; leaves decurrent forming broad wings on the stem; inflorescence dense, usually unbranched; corolla yellow, white hairs present on the upper filaments; common in fields, roadsides, shale barrens, railroad embankments, and dry waste ground; throughout; flr. Jun–Aug; native to Europe.

Verbascum virgatum Stokes   Mullein

Very similar to *V. blattaria* but glandular pubescent throughout and the flowers in clusters of 2–5 at the lower nodes of the inflorescence; very rare and scattered in urban waste ground; flr. Jun–Jul; native to Europe.

Verbascum thapsus

Verbascum blattaria, leaf

Verbascum blattaria

Veronica
anagalis-aquatica

Veronica arvensis

Veronica L.

Herbaceous perennial herbs of terrestrial or semiaquatic environments; leaves simple, opposite (except those subtending flowers, which tend to be alternate); flowers only slightly irregular, blue, violet, or white, often with darker veins; calyx and corolla usually 4-merous, occasionally 5-merous; stamens 2, on the corolla; fruit a 2-locular, more or less flattened, heart-shaped capsule surrounded by the persistent sepals.

A. flowers in terminal racemes or solitary in the axils of the alternately arranged upper bracteal leaves
    B. plants completely glabrous .... *V. peregrina*
    B. stems, leaf blades, sepals, and fruits variously pubescent and/or glandular pubescent
        C. flowers sessile or with pedicels at most 3–4 mm
            D. flowers crowded in a spike-like inflorescence; leaves sharply serrate
            .... *V. longifolia*
            D. inflorescence more raceme-like; leaves very shallowly toothed to nearly entire
                E. leaves oblong to linear-oblong, with a single midvein ....
                .... *V. peregrina*
                E. leaves elliptic to broadly ovate, palmately veined
                    F. flowers white with purple veins; fruit shallowly cordate with an elongate style .... *V. serpyllifolia*
                    F. flowers blue; fruit deeply cordate with style length about equaling the depth of the notch .... *V. arvensis*
        C. flowers (and fruits) on slender pedicels >1 cm, equaling or exceeding the subtending leaves
            G. leaves shallowly 3–5-lobed; fruit barely notched .... *V. hederifolia*
            G. leaves toothed, but not lobed; fruit distinctly cordate
                H. mature pedicels 15–40 mm; corolla 8–11 mm wide

          I. leaf blades mostly <1 cm wide; plants creeping, rooting at the nodes .... *V. filiformis*

          I. leaf blades often >1 cm wide; plants loosely ascending .... *V. persica*

       H. mature pedicels 6–15 mm; corolla 4–8 mm wide

          J. fruit sparsely glandular hairy .... *V. agrestis*

          J. fruit densely covered with both short nonglandular and longer glandular hairs .... *V. polita*

A. flowers in axillary racemes; the leaves all opposite

   K. plants variously pubescent, of terrestrial upland habitats

      L. leaves pinnately lobed to dissected; calyx 5-lobed .... *V. austriaca*

      L. leaves merely toothed; calyx 4-lobed

         M. leaves sessile; corolla 9–11 mm wide, 4-lobed .... *V. chamaedrys*

         M. leaves with a short petiole; corolla 3–4 mm, 5-lobed ...... *V. officinalis*

   K. plants glabrous, of shallow water or moist habitats

      N. leaves all with a short petiole

         O. leaves broadest near the base; style 2.5–3.5 mm .... *V. americana*

         O. leaves broadest at or above the middle; style 1.8–2.2 mm .... *V. beccabunga*

      N. leaves, at least the middle and upper ones, sessile

         P. fruit plump, scarcely notched; leaves 1.5–3 times as long as wide .... *V. anagallis-aquatica*

         P. fruit flattened, distinctly notched; leaves 4–10 times as long as wide .... *V. scutellata*

Veronica agrestis L.   Field speedwell
Prostrate annual with ovate or rotund, 1–2-cm leaves with a short petiole; flowers solitary in the axils of the upper, alternate leaves; corolla blue, 4–8 mm wide; pedicels 6–10 mm, elongating in fruit; fruit with a deep, narrow notch; rare on roadsides, urban waste ground, and ballast; flr. Apr–Jul; native to Eurasia.

Veronica americana (Raf.) Schwein. ex Benth.   American brooklime or speedwell
Rhizomatous perennial with erect stems to 1 m tall; leaves opposite, all with short petioles; blades lanceolate to lance-ovate, 2–4 times as long as wide; flowers in axillary racemes; corolla light bluish-violet, 5–10 mm wide; fruit scarcely notched; style 2.5–3.5 mm; frequent on moist banks, wet ditches, and stream edges; throughout; flr. May–Oct; OBL.

Veronica anagallis-aquatica L.   Brook-pimpernel, water speedwell
Rhizomatous biennial or short-lived perennial with erect stems to 1 m tall; leaves opposite, the upper sessile and somewhat clasping and the lower often sub-petiolate; flowers in axillary racemes; corolla light blue, 5–8 mm wide; fruit scarcely notched; occasional in wet fields, ditches, and stream edges in shallow water; mostly S; flr. May–Sep; OBL.

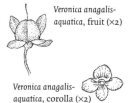
Veronica anagalis-aquatica, fruit (×2)

Veronica anagalis-aquatica, corolla (×2)

Veronica arvensis L.   Corn speedwell
Low growing, pubescent annual with short curved and longer glandular hairs; leaves palmately 3–5 veined; flowers blue, solitary on pedicels 1 mm or less, in the axils of reduced upper bracteal leaves; frequent in disturbed soils of woods, roadside banks, meadows, and lawns; flr. Apr.–Sep; native to Eurasia.

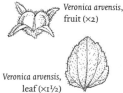
Veronica arvensis, fruit (×2)

Veronica arvensis, leaf (×1½)

Veronica austriaca L.   Speedwell

Erect glabrous or hairy perennial to 6 dm tall; leaves pinnately lobed, sessile; flowers in axillary racemes; corolla deep blue, 12 mm wide; calyx mostly 5-merous; cultivated and rarely escaped to fields and roadsides; E; flr. Jun; native to Europe.

Veronica beccabunga L.   European brooklime

Very similar to *V. americana* but with leaves more elliptic or obovate; corolla violet-blue, 5–7 mm wide; fruit not notched; rare in ditches and wet shores; SE; flr. Sep–Nov; native to Eurasia; OBL.

Veronica chamaedrys L.   Bird's-eye, germander speedwell

Low-growing, rhizomatous perennial with hairy, opposite, pinnately veined leaves and 2 rows of hairs on the stem; flowers in axillary racemes; corolla blue, 9–11 mm wide; fruit rarely produced; rare and scattered on wooded slopes, roadsides, lawns, and cultivated ground; flr. Apr–Jun; native to Eurasia; UPL.

Veronica filiformis Sm.   Creeping speedwell

Creeping, rhizomatous perennial, rooting at the nodes; leaves <1 cm, cordate-orbicular to reniform, slightly toothed; flowers blue, on long, slender pedicels in the axils of upper alternate leaves; corolla 8 mm wide; fruit broadly cordate with style greatly exceeding the notch; occasional in lawns and open ground; SE; flr. Apr–Jun; native to Eurasia.

Veronica hederifolia, fruit (×1.5)

Veronica hederifolia L.   Ivy-leaved speedwell

Hairy annual with weak, trailing stems; leaves broader than long with 3–5 lobes; flowers white with faint blue veins, 3 mm wide, on pedicels 1–1.5 mm long in the axils of the mostly alternate leaves; only a few pairs of opposite leaves near the base of the stem; fruits plump, barely indented at the style; occasional in open woods, floodplains, roadside banks, and ballast; flr. Apr–Jun; native to Eurasia.

Veronica longifolia L.   Speedwell

Upright perennial to 1 m tall with sharply serrate opposite leaves; flowers in a dense, spike-like terminal raceme, pale violet-blue, 8–10 mm wide; fruit barely flattened; cultivated and occasionally escaped to roadsides and railroad banks; flr. May–Sep; native to Eurasia.

Veronica officinalis L   Common speedwell, gypsyweed

Creeping hairy perennial with opposite, elliptic or elliptic-ovate, petiolate leaves; flowers in axillary racemes; corolla light blue-violet, 4–8 mm wide; fruit subcordate; style 2.5–4.5 mm; woods, roadsides, old fields, shale barrens, and lawns; common; throughout; flr. May–Sep; native to Europe; FACU–.

Veronica peregrina, fruit (×1.5)

Veronica peregrina L.   Neckweed, purslane speedwell

Glabrous, upright annual to 2 dm tall; leaves oblong or linear-oblong, sparsely denticulate; flowers white, 3–4 mm wide, solitary in the axils of the upper alternate leaves; fruit flattened, cordate with style not exceeding the length of the notch; common in open woods, alluvial banks, gardens, and waste ground; throughout; flr. Apr–Aug; FACU–. Most of our plants are ssp. *peregrina*, the more western ssp. *xalapensis* (Kunth) Pennell, which is distinguished by the presence of glandular pubescence, is found occasionally.

Veronica hederifolia

Veronica peregrina

Veronica officinalis

Veronica persica

Veronica persica Poir.   Bird's-eye speedwell

Low growing, hairy annual with broadly elliptic or ovate leaves 1–2 cm; flowers solitary in the axils of alternate upper leaves; pedicels 1.5–4 cm long; corolla blue, 8–11 mm wide; fruit strongly cordate, sparsely glandular pubescent; style 1.8–3 mm; occasional in lawns, roadsides, cultivated fields, and waste ground; mostly S; flr. Apr–Oct.; native to Eurasia.

Veronica persica, fruit (×1.5)

Veronica polita Fr.   Speedwell

Very similar to *V. persica* with densely pubescent fruit with short nonglandular and longer glandular hairs; rare in lawns, fields, roadsides, disturbed ground, and ballast; mostly SE; flr. Apr–Aug; native to Eurasia.

Veronica polita, fruit (×1.5)

Veronica scutellata L.   Marsh or narrow-leaved speedwell

Sprawling, rhizomatous perennial to 4 dm tall, glabrous or occasionally hairy; leaves all opposite, linear to lanceolate; flowers in axillary racemes; corolla bluish, 6–10 mm wide; fruit flattened, broadly notched; style 2–4 mm; frequent in swamps, wet woods, ditches, and swales; throughout; flr. May–Aug; OBL.

Veronica serpyllifolia,
fruit (×2)

Veronica serpyllifolia L   Thyme-leaved speedwell

Low-growing, spreading perennial rooting at the nodes; stems and pedicels densely short-hairy; leaf blades mostly glabrous except on the margins; flowers white with purple veins, 5 mm wide; fruit glandular-hairy, flattened, somewhat cordate with long style; common in lawns, fields, meadows, and open woods; throughout; flr. May–Aug; native to Europe; FAC+.

Veronicastrum Fabr.

Veronicastrum
virginicum,
flower (×2)

Veronicastrum virginicum (L.) Farw.   Culver's-root

Upright perennial herb to 2 m tall; leaves in whorls of 4–7, variously glabrous or villous beneath; inflorescence a dense terminal spike of small flowers; corolla white or pink, stamens 2, inserted on the corolla tube and extending beyond it; occasional in moist meadows, thickets, and swamps; mostly SE and SW; flr. late Jun–Sep; FACU.

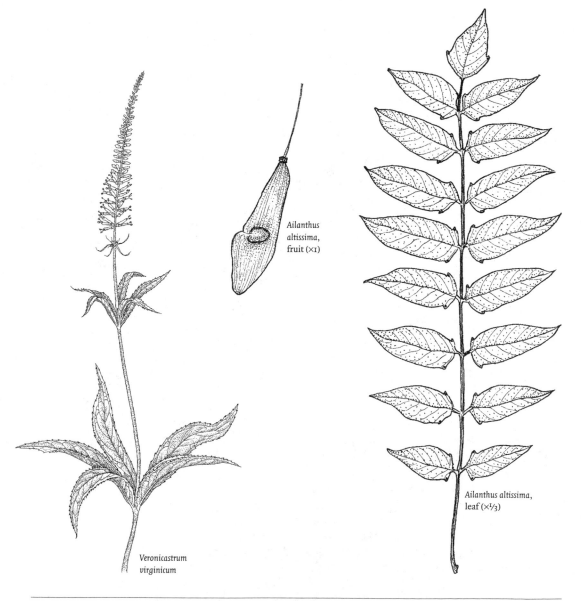

Ailanthus
altissima,
fruit (×1)

Ailanthus altissima,
leaf (×⅓)

Veronicastrum
virginicum

SIMAROUBACEAE Quassia Family

Deciduous, dioecious trees with alternate, pinnately compound leaves and smooth gray bark; flowers unisexual, small, regular, 5-merous; ovary superior; stamens 10, alternate with the petals; fruit a samara.

Ailanthus Desf.

Ailanthus altissima (Mill.) Swingle   Tree-of-heaven

Tree to 25 m tall with coarse twigs and odd-pinnate leaves to 1 m long; leaflets 11–41, with one or more glandular teeth near the base; dioecious; inflorescences terminal, 1–2 dm; flowers 5 mm wide, greenish, malodorous; samaras winged at both ends, produced in large, conspicuous, reddish clusters; widely naturalized in disturbed woods, roadsides, fencerows, vacant lots, and railroad rights-of-way; flr. Jun–early Jul, frt. Aug–winter; native to Asia.

SOLANACEAE Nightshade Family

Herbs or shrubs with alternate leaves; inflorescence cymose, or flowers axillary and solitary; flowers perfect, 4- or 5-merous; corolla regular, tubular to rotate, the lobes strongly reflexed in some genera; sepals forming a basal tube, sometimes enlarging to loosely enclose the fruit; stamens 4 or 5, fused to the corolla tube; anthers opening by longitudinal slits or apical pores; ovary 1, superior, 2–5-celled; fruit a capsule or berry.

A. woody shrubs or trailing vines with a semiwoody base; fruit a red berry
    B. spiny shrubs .... *Lycium*
    B. trailing vines lacking spines .... *Solanum*
A. herbaceous annuals or perennials; fruit a capsule or a black, green, or yellow berry
    C. flowers mostly >1 per node, in peduncled inflorescences or on axillary peduncles .... *Solanum*
    C. flowers axillary, 1 per node
       D. calyx enlarging after flowering and loosely enclosing the fruit
          E. corolla white, greenish or yellow; calyx lobed only at the tip ....*Physalis*
          E. corolla blue; calyx lobes extending nearly to the base ....*Nicandra*
       D. calyx not enlarging and covering the fruit
          F. calyx tube 4–10 cm long; fruit a spiny capsule .... *Datura*
          F. calyx tube shorter, or sepals separate; capsule not spiny .... *Petunia*

Datura L.

Erect annual herbs with large, entire to coarsely toothed leaves; flowers 8–20 cm long, solitary, with a conspicuous funnel-shaped corolla with 5 flaring lobes and an elongate tube; calyx cylindric or prismatic, 5-lobed, not persistent in fruit; anthers opening by longitudinal slits; fruit a spiny capsule. All species are poisonous.

A. leaves coarsely toothed throughout .... *D. stramonium*
A. leaves entire except for an occasional single tooth .... *D. meteloides*

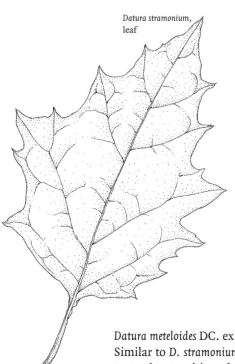

Datura stramonium, leaf

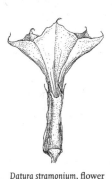

Datura stramonium, flower

Lycium barbarum

Datura meteloides DC. ex Dunal   Downy thorn-apple, Indian-apple
Similar to *D. stramonium* in habit; leaves entire or with a single tooth; flowers to 20 cm long; cultivated and occasionally escaped to ballast, railroad tracks or fallow ground; S; flr. Aug–Oct; native to China.

Datura stramonium, fruit

Datura stramonium L.   Jimsonweed
Erect, branched, glabrous annual to 1.5 m tall; leaf blades to 1.5 dm long, coarsely toothed throughout; flowers 8–10 cm long; the corolla flaring to 3–5 cm wide, white or pinkish; capsule erect, 3–5 cm, usually covered with short prickles; common in cultivated fields, roadsides, and waste ground; S; flr. Aug–Oct; geographic origin uncertain, widely distributed in warm-temperate regions; designated as a noxious weed in Pennsylvania.

Lycium L.

Lycium barbarum, flower (×1)

Lycium barbarum L.   Matrimony-vine
Woody shrub with alternate leaves and spines at the nodes; flowers axillary, solitary or in small clusters; calyx irregularly 4–5-lobed, splitting as the fruit develops; corolla purplish, 4–5-lobed, tubular or funnel-shaped; anthers opening by longitudinal slits; fruit a red berry; occasional in waste places and roadsides; S; flr. Apr–Oct; native to Eurasia. [syn. *L. halimifolium* Mill.]. *Lycium chinense* Mill., which has a shorter corolla tube and leaves wider at the base, has been collected occasionally.

Nicandra Adanson

Nicandra physalodes (L.) Gaertn.   Apple-of-peru, shoofly-plant
Glabrous annual to 1.5 m tall; leaves 1–2 dm, coarsely and unevenly toothed; flowers solitary in the axils; corolla blue; the dryish berry enclosed in a leafy, enlarged calyx; cultivated and occasionally naturalized in old fields or along railroads; mostly S; flr. Jul–Oct; native to Peru.

Petunia A.L.Juss.

Petunia x hybrida Vilm.   Petunia

Clammy-pubescent, annual herb with alternate, simple leaves; flowers solitary in the axils, 3–6 cm long with broadly flaring, pink-violet corolla; calyx deeply 5-lobed; widely cultivated in a wide variety of forms and colors; occasionally seeding into waste ground and paving cracks; scattered throughout; flr. Jul–Oct; parental species of South American origin.

Physalis L.

Annual or perennial herbs; leaves alternate or opposite, entire to toothed or lobed; flowers mostly solitary at the nodes, drooping; corolla yellowish with 5 dark blotches within (or whitish in *P. alkekengi*), broadly funnel- or bell-shaped, only shallowly or not at all lobed; calyx expanding to form a loose, membranous covering around the green, purplish, or orange, pulpy berry.

A. corolla whitish; calyx bright orange or red at maturity .... *P. alkekengi*
A. corolla yellow; calyx green at maturity
    B. stems glabrous or with a few upward pointing hairs near the top
        C. pedicels <1 cm, even in fruit .... *P. philadelphica*
        C. pedicels 1–3 cm .... *P. subglabrata*
    B. stems hairy
        D. hairs of the stem recurved, none glandular .... *P. virginiana*
        D. hairs of the stem spreading, glandular hairs present
            E. flowers 6–10 mm long; mature fruiting calyx <2.5 cm long; pedicels <1 cm .... *P. pubescens*
            E. flowers 12–18 mm long; mature fruiting calyx 2.5–3.7 cm long; pedicels >1.5 cm .... *P. heterophylla*

Physalis alkekengi L.   Chinese-lantern

Erect, unbranched, rhizomatous perennial, 4–6 dm tall; leaves ovate; flowers white; calyx enlarged and bright orange-red in fruit; cultivated and occasionally escaped to woods, roadsides, stream banks, or vacant lots; scattered throughout; flr. Jun–Jul; native to Eurasia.

Physalis heterophylla Nees   Clammy ground-cherry

Erect, glandular-hairy, rhizomatous perennial 2–9 dm tall; leaves ovate, irregularly toothed to entire, pedicels 1 cm in flower, 3 cm in fruit; corolla 1.2–2 cm; fruiting calyx ovoid, 3–4 cm; frequent in fields, sandy or cindery open ground, and cultivated areas; throughout; flr. Jun–Sep.

Physalis philadelphica Lam.   Tomatilla

Nearly glabrous, branching annual, 2–6 dm tall; leaves ovate or rhombic, 2–6 cm; pedicels 3–5 mm in flower and fruit; corolla 7–15 mm; berry purplish, edible; cultivated and occasionally occurring on ballast, wharves, and waste ground; SE; flr. Jun–Oct; native to Mexico.

Physalis pubescens L.   Hairy ground-cherry

Widely branching, glandular-hairy annual to 6 dm tall; leaves ovate, 3–10 cm; pedicels 5–10 mm in flower, to 2 cm in fruit; corolla 6–10 mm; mature fruiting calyx 2–4 cm; rare in woods, fields, and gardens; flr. Jun–Oct; FACU–; 2 varieties:

A. leaves with 5–8 teeth per side .... var. *pubescens*
   rare; W.
A. leaves with 3–4 teeth per side .... var. *integrifolia* (Dunal) Waterf.
   occasional; S.

Physalis subglabrata Mack. & Bush   Ground-cherry
Erect, branched, rhizomatous perennial, 4–8 dm tall, mostly glabrous except for the youngest parts; leaves 4–10 cm, abruptly narrowed to a long petiole; pedicels 1–3 cm in flower and fruit; corolla 11–17 mm; fruiting calyx ovoid to short cylindric, 3–4 cm; frequent in fields, waste ground, hedgerows and limestone uplands; throughout, except the northernmost counties; flr. Jul–Sep.

Physalis virginiana Mill.   Virginia ground-cherry
Rhizomatous perennial, 3–6 dm tall with ascending branches; pubescence of short recurved hairs; leaves ovate to narrowly lanceolate, the blade decurrent on the long petiole; pedicels 1–2 cm in flower and fruit; fruiting calyx 5-angled; berry orange; rare in ballast and waste ground; widely scattered.

Solanum L.

Erect, annual or perennial herbs or semiwoody trailing vines; leaves alternate, entire to pinnatifid or pinnate; flowers in a cymose inflorescence arising at or between the nodes; corolla rotate or broadly bell-shaped, deeply 5-lobed; anthers fused to form a ring around the style, opening by terminal pores; fruit a berry. All species are poisonous.

A. plants spiny or prickly
   B. calyx with few if any prickles, not enclosing the fruit .... *S. carolinense*
   B. calyx prickly, fully enclosing the fruit when mature .... *S. rostratum*
A. plants not spiny or prickly
   C. trailing or sprawling vines with a semiwoody base; fruit red at maturity....
      .... *S. dulcamara*
   C. erect herbs; fruit black
      D. inflorescence raceme-like; fruit dull with adherent sepals ...... *S. nigrum*
      D. inflorescence umbel-like; fruit shiny with reflexed sepals ....
         .... *S. americanum*

Solanum americanum Mill.   Black nightshade
Very similar to *S. nigrum*, distinguished by the more umbel-like inflorescence, smaller anthers, erect fruiting pedicels, and shiny fruit with reflexed sepals; occasional in fields, woods, and disturbed ground; flr. Jun–Oct; FACU–. [*S. nigrum* L. in part of Rhoads and Klein, 1993]

Solanum carolinense L.   Horse-nettle
Erect, spiny, branching, rhizomatous perennial to 1 m tall; leaves ovate with 2–5 large teeth or lobes, spiny along the major veins and stellate-hairy above and below; corolla pale violet to white, 2 cm wide; fruit yellow; common in fields, roadsides, and sandy stream banks; throughout; flr. Jun–Sep; UPL.

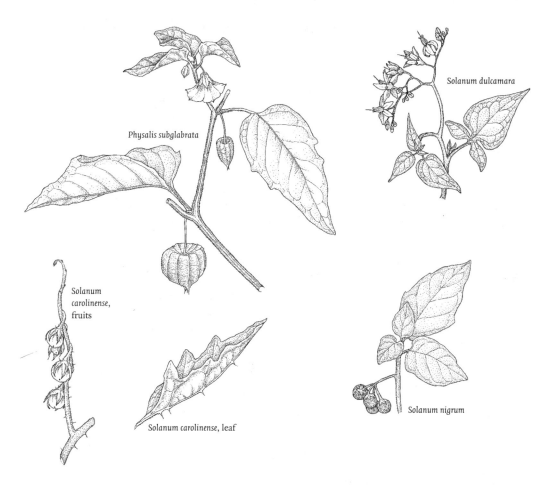

Physalis subglabrata

Solanum dulcamara

Solanum carolinense, fruits

Solanum carolinense, leaf

Solanum nigrum

Solanum dulcamara L.   Trailing bittersweet or nightshade
Nearly glabrous, rhizomatous perennial with a semiwoody base and trailing, vine-like branches to 3 m tall, malodorous when bruised; leaves highly variable, simple to lobed; inflorescence 10–25-flowered; corolla purple to occasionally white; anthers yellow; fruit 8–11 mm, bright red; common in moist, open woods, bogs, wet thickets, and waste ground; flr. May–Aug; native to Eurasia; FACU. The more pubescent var. *villosissimum* Desv. is rare in the SE.

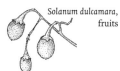

Solanum dulcamara, fruits

Solanum nigrum L.   Black nightshade
Annual with erect, mostly glabrous, branching stems 1.5–6 dm tall; leaves ovate to deltoid, irregularly toothed or nearly entire; inflorescence raceme-like; corolla white or faintly bluish, 5–10 mm wide; fruit globose, 8 mm, dull black, with adherent sepals on recurved pedicels; common in fields, woods, roadsides, and moist disturbed ground; mostly S; flr. Jun–Oct; native to Europe.

Solanum rostratum Dunal   Buffalo-bur
Coarse, branching, spiny and stellate-hairy annual, 3–10 dm tall; leaves deeply pinnately lobed; inflorescence a short pedunculate, 3–15-flowered raceme; flowers light yellow, 2–3 cm wide; berry completely enclosed by the prickly calyx; rare in cultivated ground and stream banks; scattered; flr. Jul–Sep; native to the Great Plains.

STAPHYLEACEAE Bladdernut Family

Deciduous shrubs with opposite, compound leaves and deciduous stipules; flowers regular, perfect, 5-merous; sepals and petals distinct; stamens alternate with the petals and attached on the nectar disk; ovary and fruit 3-lobed; fruit a thin-walled, inflated capsule.

Staphylea L.

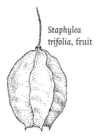

Staphylea
trifolia, fruit

Staphylea trifolia L.   Bladdernut

Erect shrub to 5 m tall with trifoliate leaves and small greenish-white flowers in drooping terminal panicles; leaflets elliptic, 5–10 cm, serrate; terminal leaflet with a long petiolule, lateral leaflets sessile or nearly so; inflorescence 4–10 cm; flowers with long pedicels, campanulate; fruit papery, inflated, 5 cm; frequent in moist rocky woods and stream banks; throughout, except in the northernmost counties; flr. May, frt. Aug–Oct; FAC.

Staphylea trifolia

Halesia carolina

STYRACACEAE Storax Family

Deciduous trees with simple, alternate leaves lacking stipules; flowers regular, perfect, 4- or 5-merous; corolla campanulate, white; stamens 8–10, fruit dry, indehiscent, 1-seeded.

A. leaves evenly serrulate; corolla lobes 4; ovary inferior; fruit winged ..... *Halesia*
A. leaf margins irregularly serrate to crenate; corolla lobes 5; ovary superior; fruit
  not winged .... *Styrax*

Halesia J.Ellis ex L.

Halesia carolina L.   Carolina silverbell
Tree to 15 m tall; leaves 10–15 cm, oblong or elliptic, acuminate, serrulate; flowers in small clusters on slender pedicels; corolla white, bell-shaped, 2–2.5 cm broad; fruit 2.5–3.5 cm, with 4 wings; cultivated and occasionally naturalized in disturbed woods or edges; flr. May; native from WV south. [syn: *H. tetraptera* Ellis]

Styrax L.

Styrax japonicus Siebold & Zucc.   Japanese styrax
Tree to 8 m tall; leaf blade elliptic, glabrous except for sparse stellate pubescence on the veins, margin irregularly serrate to crenate; flowers white, in small clusters along the branches, drooping; fruit ovoid, densely gray stellate-tomentose, about 1 cm; cultivated and occasionally naturalized in disturbed, urban woods; flr. May; native to Asia.

SYMPLOCACEAE Sweetleaf Family

Deciduous shrubs or small trees with simple, alternate leaves lacking stipules; flowers regular, 5-merous, half-inferior; calyx tube fused to the ovary only as far as the sinuses; petals fused only at the base; stamens 12–many; ovary 2–5-locular; style 1; fruit a drupe.

Symplocos Jacq.

Symplocos paniculatus Thunb.   Sapphire-berry
Deciduous shrub to 10 m tall; leaves obovate; flowers in axillary panicles, white; stamens numerous, distinct; fruit a bright blue fleshy drupe; cultivated and occasionally naturalized in disturbed woods and floodplains; SE; flr. May; native to Asia.

THYMELAEACEAE Mezereum Family

Woody or herbaceous plants with alternate (or opposite), simple, entire leaves lacking stipules; flowers perfect (in our species), regular, 4–5-merous, perigynous with a tubular hypanthium; ovary superior, 2-locular; stamens 8; fruit a drupe.

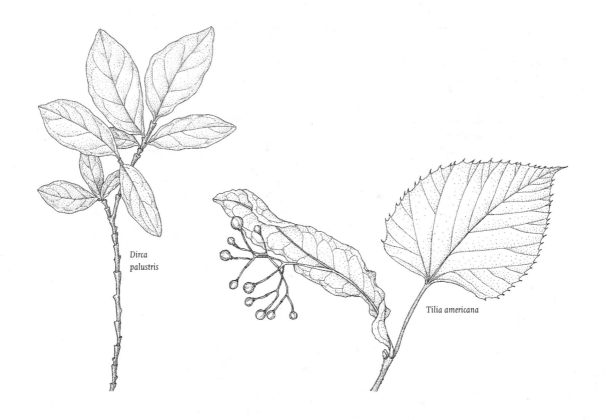

Dirca
palustris

Tilia americana

Dirca L.

Dirca palustris L. Leatherwood

Deciduous shrub to 1.5 m tall with alternate, simple, entire leaves and distinctive swollen nodes that have a jointed appearance; flowers small, yellow, perfect, regular, perigynous; stamens 8, inserted at the summit of the hypanthium tube and extending well beyond; fruit a drupe, 12–15 mm long, pale yellowish-green; occasional in rich deciduous woods and thickets; scattered throughout; flr. Apr; FAC.

TILIACEAE Linden Family

Deciduous trees with alternate, simple, palmately veined leaves with stipules; peduncle of the inflorescence subtended by and fused part way to a narrow, elongate, somewhat leaf-like bract; flowers white, fragrant, perfect, 4-merous; fruit densely hairy, indehiscent, nut-like, containing 1–2 seeds.

Tilia L.

Tilia americana L. Basswood

Tree to 40 m tall; leaves broadly ovate, 7–15 cm, serrate, asymmetrical at the base; fruit subglobose, 6–8 mm; flr. late Jun–early Jul; FACU; 2 varieties:

A. leaves green beneath, glabrous to sparsely stellate-hairy, especially in the vein axils .... var. *americana* frequent; throughout.

A. leaves permanently pubescent beneath with fine white or brown stellate hairs .... var. *heterophylla* (Vent.) Loudon occasional; mostly SW and SC.

TRAPACEAE Water-chestnut Family

Rooted aquatics with a submersed stem bearing opposite or subopposite, highly dissected leaves; the terminal portion of the stem bearing a cluster of simple floating or slightly emergent leaves on elongate petioles with a swollen portion in the middle that functions as a float; flowers perfect, regular, 4-merous; sepals forming a short hypanthium and becoming hardened and spine-like in fruit; petals 4; stamens 4; fruit indehiscent, 1-seeded.

Trapa L.

Trapa natans L.   Water-chestnut

Stem to 1 m tall or more, often breaking and thus the terminal portion floating freely; the floating or emergent leaves 2–5 cm wide, coarsely serrate; flowers axillary; petals white; fruit 3–4 cm wide, with 4 stout spines; locally abundant in a few scattered ponds and lakes; native to Eurasia; OBL.

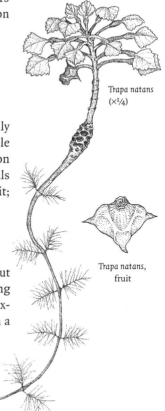

Trapa natans (×¼)

Trapa natans, fruit

ULMACEAE Elm Family

Trees or rarely shrubs with simple, alternate, double- or occasionally single-toothed, 2-ranked leaves; buds with imbricate scales; flowers small, perfect or monoecious, 4–5-merous, petals absent; fruit a single samara or a drupe.

A. leaves distinctly asymmetrical at base
    B. lateral veins anastomosing near margin .... *Celtis*
    B. lateral veins extending into teeth .... *Ulmus*
A. leaves not distinctly asymmetrical at base
    C. leaves sharply serrate, the teeth with acuminate tips .... *Zelkova*
    C. leaves serrate-dentate, the teeth with acute tips .... *Ulmus*

Celtis L.

Deciduous trees with alternate, simple leaves; monoecious; pistillate flowers solitary or in pairs from the upper axils; staminate flowers lower; calyx 5-parted, persistent on the fruit; stamens 5–6; ovary superior, 1-locular, with 2 recurved styles; fruit a drupe with a large stone surrounded by a thin layer of dryish pulp; flr. late Apr–early May, frt. Aug–Oct.

A. leaves serrate to below the middle; fruiting pedicels longer than the petioles .... *C. occidentalis*

A. leaves entire or with only a few scattered teeth above the middle; fruiting pedicels about as long as the petioles .... *C. tenuifolia*

Celtis occidentalis L.   Hackberry, sugarberry

Medium to large tree to 35 m tall with light gray, irregularly ridged or warty bark; leaves lance-ovate to broadly ovate or deltoid, serrate, asymmetrical at the base with one side strongly cordate, acuminate to long-acuminate at the tip, pubescent beneath and veiny and scabrous above; pedicel usually distinctly longer than the subtending petiole; fruit ellipsoid to subglobose, 7–13 mm, dark red to black, sweet; dry to moist woods, rocky slopes, rich banks, and bottomlands; throughout; FACU.

Celtis tenuifolia Nutt.   Dwarf hackberry

Shrub or small irregularly branched tree to 5 m tall; leaves ovate to deltoid, firm, dark green above, pubescent beneath, the base oblique to slightly cordate, nearly symmetrical; pedicel no longer than the subtending petiole; fruit subglobose, 5-9 mm, orange, reddish, or brown, not sweet; shale banks, dry wooded hillsides, and limestone cliffs; mostly SE and SC.

Ulmus L.

Leaves double-toothed (single-toothed in *U. pumila* and *U. parvifolia*); flowers perfect, in short racemes or pendant fascicles; fruit a single samara.

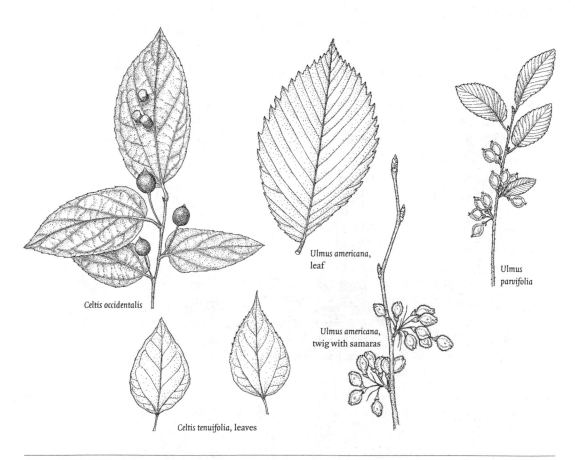

Celtis occidentalis

Ulmus americana, leaf

Ulmus parvifolia

Celtis tenuifolia, leaves

Ulmus americana, twig with samaras

Summer key

A. leaves and twigs very roughly scabrous; buds rusty-tomentose .... *U. rubra*
A. leaves and twigs glabrous or only slightly scabrous; buds glabrous or somewhat pale-pubescent
   B. leaves mostly single-toothed
      C. leaf blades 5–8 cm long, acuminate at the apex, mostly symmetrical at the base .... *U. pumila*
      C. leaf blades 2–5 cm long, acute to occasionally acuminate at the apex, mostly asymmetric at the base .... *U. parvifolia*
   B. leaves mostly double-toothed
      D. leaves 5–8 cm long, often scabrous above; branches often with corky wings .... *U. procera*
      D. leaves 8–15 cm long, glabrous to only slightly scabrous above; branches not corky-winged .... *U. americana*

Flowering key

A. flowering in early spring before the leaves
   B. flowers on slender, drooping pedicels .... *U. americana*
   B. flowers on short pedicels, not drooping
      C. calyx pubescent
         D. stigmas pink to red; anthers reddish .... *U. rubra*
         D. stigmas white; anthers dark brown .... *U. procera*
      C. calyx glabrous .... *U. pumila*
A. flowering in late summer–fall .... *U. parvifolia*

Winter key

A. bud scales 2-ranked
   B. buds densely rusty-brown-hairy; inner bark mucilaginous .... *U. rubra*
   B. buds not rusty-brown hairy; inner bark not mucilaginous
      C. bud scales with darker margins .... *U. americana*
      C. bud scales uniformly colored
         D. bud scales dark brown-black, slightly pubescent, margins ciliate ....
         .... *U. procera*
         D. bud scales light brown, glabrous to slightly pubescent, margins not ciliate .... *U. pumila*
A. bud scales not 2-ranked .... *U. parvifolia*

Ulmus americana L.   American or white elm
Large tree to 40 m tall; leaves ovate-oblong to somewhat obovate; flowers fascicled, somewhat pendant; fruit elliptic, to 1 cm long, deeply notched at the apex, surface glabrous, the margin of the wing densely ciliate; stream banks and floodplains; throughout, but large trees are very rare due to the effects of Dutch elm disease; flr. late Mar–early Apr, before the leaves, frt. Apr–May; FACW–.

Ulmus parvifolia Jacq.   Chinese elm
Small tree to 15–25 m tall; bark smooth and often mottled; leaves elliptic to ovate or obovate; flowers in axillary clusters; fruit elliptic-ovate, about 1 cm long, notched at apex, surface glabrous, the margins of the wings glabrous; widely cultivated and escaping; flr. Aug–Sep, frt. Sep–Oct; native to Asia.

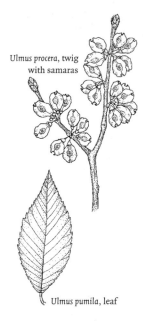

Ulmus procera, twig
with samaras

Ulmus pumila, leaf

Ulmus procera Salisb. English elm

Large tree to 40–50 m tall; leaves ovate or broad-elliptic; flowers short-stalked, in axillary clusters; fruit suborbicular, to 1.2 cm across, with a small closed notch at apex, the seed very close to the notch; widely cultivated and escaping; flr. early spring before the leaves appear; native to Europe.

Ulmus pumila L. Siberian elm

Tree to 25 m tall or sometimes a large shrub; leaves elliptic to elliptic-lanceolate; flowers very short-stalked; fruit suborbicular, to 1.5 cm long, with a closed notch at the apex, seed close to the middle of the fruit; cultivated and sometimes escaped; flr. early spring; native to Asia.

Ulmus rubra Muhl. Red elm, slippery elm

Tree to 20 m tall; leaves obovate to oblong and very rough above; flowers short-stalked in very dense clusters; fruit suborbicular to broadly elliptic, to 2 cm long, slightly notched at the apex, tawny-pubescent at the center; moist woods, stream banks, and floodplains; throughout; flr. late Mar–Apr, frt. Apr–May; FAC–.

Zelkova Spach.

Zelkova serrata (Thunb.) Makino Japanese zelkova

Tree to 30 m tall; leaves ovate to lanceolate-ovate, coarsely single-toothed, 3–8 cm long; flowers solitary or in few-flowered inflorescences; fruit a drupe, broader than long, about 4 mm across; often cultivated and escaped; flr. late Apr–May; native to Asia.

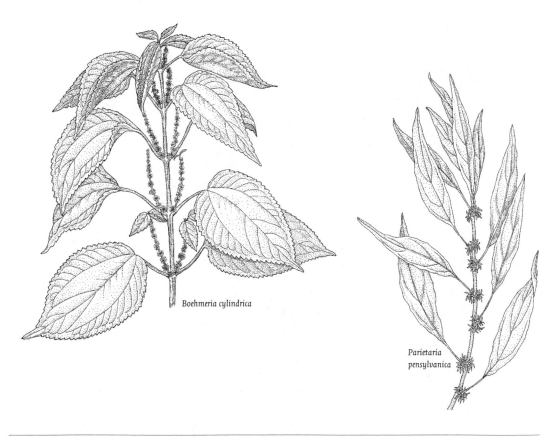

Boehmeria cylindrica

Parietaria
pensylvanica

URTICACEAE Nettle Family

Erect, mostly unbranched herbs with alternate or opposite, simple leaves and often with stinging hairs; stipules usually present (absent in *Parietaria*); flowers small, mostly unisexual, in simple or branched axillary clusters; sepals 3–5, distinct or united at the base; petals none; stamens 4 or 5; ovary superior, 1-celled, style 1; fruit an achene.

A. stinging hairs present
    B. leaves alternate .... *Laportea*
    B. leaves opposite .... *Urtica*
A. stinging hairs not present
    C. leaves alternate ....*Parietaria*
    C. leaves mostly opposite
        D. flowers in unbranched axillary spikes; achenes completely enclosed by the calyx .... *Boehmeria*
        D. flowers in axillary panicles or clusters; achenes longer than the calyx....
        .... *Pilea*

Boehmeria Jacq.

Boehmeria cylindrica (L.) Sw.   False nettle
Erect perennial, 4–10 dm tall, lacking stinging hairs, dioecious; leaves opposite, or rarely alternate, long- petioled; blades ovate or lance-ovate, 3-nerved, coarsely serrate, acute or acuminate; flowers tiny, forming continuous or interrupted, leaf-tipped spikes in the upper axils; common in moist shady ground, wet woods, and stream margins; throughout; FACW+. Plants growing in the sun have smaller, narrower leaves and have been called var. *drummondiana* (Wedd.) Wedd.

Laportea Gaudich.

Laportea canadensis (L.) Wedd.   Wood nettle
Rhizomatous, monoecious, perennial herb, 5–10 dm tall; leaves alternate, long-petioled, coarsely serrate, with numerous stinging hairs; blades ovate, acute, pinnately veined; a single 2-lobed stipule present; flowers in loose, spreading, branched terminal and axillary inflorescences, unisexual; staminate flowers in the lower axils and 5-merous, pistillate in the terminal inflorescence and 4-merous; common in low, moist woods and stream banks; throughout; FAC.

Parietaria L.

Parietaria pensylvanica Muhl. ex Willd.   Pellitory
Erect hairy annual, 1–4 dm tall; leaves alternate, lanceolate, 3-nerved, entire; stipules not present; flowers unisexual or perfect, in axillary clusters, subtended and exceeded by hairy bracts; calyx of pistillate flowers tubular at the base, 4-lobed; ovary ovoid, stigma sessile; achene smooth and shiny; common in dry woods, roadside banks, and waste ground; mostly S.

Pilea Lindl.

Glabrous, succulent, translucent annuals lacking stinging hairs, monoecious or dioecious; leaves opposite, 3-nerved, coarsely toothed, long-petioled; stipules inconspicuous; staminate flowers with a 4-lobed calyx; calyx of the pistillate flowers 3-lobed, persistent in fruit.

A. mature achenes green or tan marked with purple .... *P. pumila*
A. mature achenes dark olive to purple or black .... *P. fontana*

Pilea fontana (Lunell) Rydb.   Lesser clearweed
Very similar to *P. pumila*; achenes broader at the base, dark olive or purple to nearly black and covered with broad, shallow warts; rare in spring heads, wet shores, and swamps; scattered; FACW+.

Pilea pumila, seed (×5)

Pilea pumila (L.) A.Gray   Clearweed, coolwort, richweed
Stems 1–5 dm tall; leaves long-petioled, shining; achene pale green with raised purple spots; common in cool, moist, shady areas; throughout; FACW.

Urtica L.

Urtica dioica L.   Stinging or great nettle
Erect, unbranched, rhizomatous perennial to 2 m tall, with numerous stinging hairs, dioecious or monoecious; leaves opposite, serrate, tip acute or acuminate; stipules linear; flowers in panicles from the upper axils, 4-merous; FACU; 2 subspecies:

A. dioecious; leaves cordate at the base; stinging hairs present on both sides of the leaves .... ssp. *dioica*
widely naturalized in floodplains and other low, moist areas; mostly S; native to Europe.

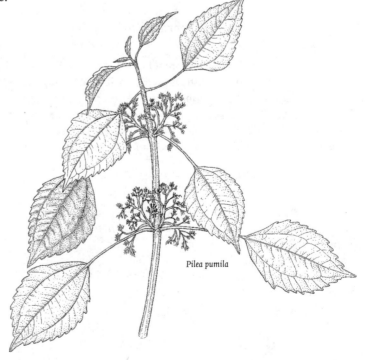

Pilea pumila

A. monoecious; leaves not cordate at the base; stinging hairs on the lower leaf surface only .... ssp. *gracilis* (Aiton) Selander
common on floodplains and thickets; throughout.

In addition *U. urens* L., an annual from Europe with the inflorescences shorter than the petioles, has been collected occasionally in the SE.

VALERIANACEAE Valerian Family

Upright herbs with opposite leaves and no stipules; flowers small, in terminal, head-like cymes, perfect or unisexual, regular or slightly irregular, mostly 5-merous; sepals tiny and bristle-like or absent; corolla tubular; stamens 3, attached to the corolla tube; ovary inferior, 3-celled but only 1 cell fertile, the others smaller; fruit dry, indehiscent.

A. cauline leaves pinnately divided; calyx lobes 9–20 .... *Valeriana*
A. cauline leaves simple; calyx lobes tiny or absent .... *Valerianella*

Valeriana L.

Perennials with strongly scented roots; basal leaves undivided, cauline leaves pinnately lobed or divided; corolla tube swollen near the base; the 3 stamens longer than the corolla; calyx becoming bristle-like in fruit.

A. basal leaves mostly simple; corolla tube 12–16 mm .... *V. pauciflora*
A. leaves all pinnate; corolla tube 0.5–4 mm................................... *V. officinalis*

Valeriana officinalis L.   Garden heliotrope
Stems 5–15 dm tall from a short rhizome; leaves pinnately divided, the segments toothed; inflorescence open; flowers white or pinkish, fragrant; cultivated and occasionally escaped to woods, fields, and roadsides; scattered throughout; flr. May–early Aug; native to Eurasia.

Valeriana pauciflora Michx.   Valerian
Stems 3–8 dm tall from a slender rhizome, glabrous; basal leaves broadly cordate and long-petioled; cauline leaves short-petioled, pinnate, the terminal lobe larger than the lateral ones; inflorescence dense at first, becoming open-pyramidal; flowers small, pale pink; rare in moist wooded stream banks and floodplain forests; SW and SE; flr. May–Jun; FACW.

Valerianella Miller

Annual or biennial herbs with forking stems; leaves simple, mostly sessile, the lowest sometimes fused; flowers very small, white or pale blue; calyx minute or lacking; corolla regular or nearly so; stamens 3 (rarely 2), longer than the corolla; fruit dry, 3-celled, but 1-seeded; flr. late Apr–Jun.

A. corolla bluish; fruit with a conspicuous corky mass on the back of the seed-bearing cell.. *V. locusta*
A. corolla white or tinged with pink; fruit without a corky mass

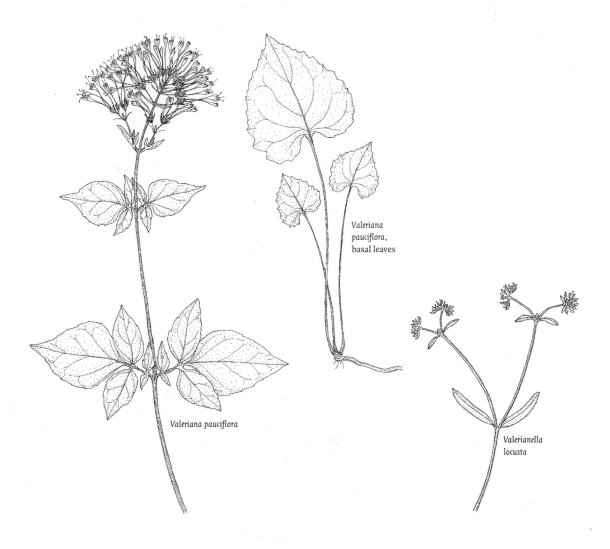

Valeriana pauciflora, basal leaves

Valeriana pauciflora

Valerianella locusta

B. fruits suborbicular, the inner face saucer-shaped .... V. umbilicata
B. fruits longer than broad, the inner face not saucer-shaped ....
.... V. chenopodiifolia

Valerianella chenopodiifolia (Pursh) DC.   Goose-foot corn-salad
Plants 3–6 dm tall, glabrous; basal leaves spatulate with wavy margins; upper
leaves oblong to lanceolate with acute tips; inflorescence dense; flowers white,
2–4 mm long; fruit triangular in cross-section, 4 mm long and 2.5 mm thick, the
fertile cell wider than the 2 empty ones but not corky-thickened; occasional in
mesic woods, floodplains, old fields, and roadsides; mostly SW.

Valerianella locusta (L.) Betcke   Corn-salad, Lamb's-lettuce
Stems 1–5 dm tall, clustered, slightly pubescent; basal leaves entire, those on the
stem slightly toothed near the base; bracts of the flower clusters ciliate on the
margins; fruit 2–4 mm long, laterally flattened with a thick corky mass on the
back of the seed-bearing cell; occasional in moist open ground and roadside
ditches; S; native to Europe.

Valerianella
locusta,
flower (×5)

Valerianella umbilicata (Sullivant) Wood   Corn-salad
Stems 0.3–1 m tall; bracts of the inflorescence ciliate only at the tips; flowers
white; fruits globose, 2–2.5 mm long, grooved on the top and without corky
thickening; occasional in moist meadows, swampy woods, and roadsides; SE
and SW; FAC.

VERBENACEAE Vervain Family

Herbs or shrubs with opposite leaves and square stems, nonaromatic (except
Vitex); flowers perfect; corolla tubular, slightly irregular, 4- or 5-merous; stamens 4, inserted on the corolla tube; ovary superior, shallowly 4-lobed or
unlobed and single-seeded; stigmas mostly 2-lobed; fruit nutlets, an achene, or
a drupe. In addition to the species described below, *Clerodendrum trichotomum*
Thunb., an ornamental shrub from Japan with a white corolla, red calyx, and
blue fruit, was collected once from a naturalized site along a river bank.

A. nonaromatic herbs with simple or pinnatifid leaves
    B. fruit 2 or 4 nutlets that separate at maturity
        C. fruit 4 nutlets; corolla 5-lobed, slightly irregular .... *Verbena*
        C. fruit 2 nutlets; corolla 4-lobed and 2-lipped .... *Phyla*
    B. fruit an achene .... *Phryma*
A. aromatic shrub with palmately compound leaves .... *Vitex*

Phryma L.

Phryma leptostachya L.   Lopseed
Upright, perennial herb with ovate leaves; flowers in open racemes terminating
the stem and arising from the upper axils; flowers purple; fruit an achene tipped
by the persistent hooked style; frequent in rich woods, rocky limestone slopes
and swamps; mostly S; flr. Jul–Aug; FACU–.

*Phryma
leptostachya,
fruit (×3)*

Phyla Lour.

Phyla lanceolata (Michx.) Greene   Fog-fruit
Trailing perennial with ascending flowering stems; leaves opposite, lance-elliptic or ovate, toothed to below the middle; flowers pink, in long-peduncled,
dense, head-like spikes arising from the leaf axils; rare in moist, low ground of
river and stream banks; SE and NE; flr. July–Sep.; OBL. [syn: *Lippia lanceolata*
Michx.]

Phryma leptostachya

Verbena L.

Perennial herbs with opposite leaves and flowers in spikes terminating the stem and upper branches; calyx tubular, unequally 5-lobed; corolla tubular, slightly 2-lipped; stamens 4, in 2 unequal pairs, shorter than the corolla tube; style 2-lobed at the tip; ovary shallowly 4-lobed, eventually separating into 4 nutlets.

A. leaves singly or doubly serrate
    B. spikes slender and elongate, flowers, and even more so the fruits, widely separated; corolla white .... *V. urticifolia*
    B. flowers and fruits close and overlapping; corolla blue or purple
        C. leaves and stem glabrous or short-hairy
            D. leaves lance-ovate, acuminate, coarsely serrate throughout and sometimes 2-lobed at the base .... *V. hastata*
            D. leaves lanceolate to lance-obovate, obtuse, serrate to below the middle, basal portion entire .... *V. simplex*
        C. leaves densely pubescent, especially beneath .... *V. stricta*
A. leaves pinnatifid or deeply incised, at least at the base
    E. bracts of the individual flowers greatly exceeding the calyx .... *V. bracteata*
    E. bracts of the flowers shorter than to equaling the calyx
        F. spike 2–4 cm thick, short, flowers densely overlapping; limb of the corolla >10 mm wide .... *V. canadensis*
        F. spike <10 mm thick, elongate, flowers becoming well separated in fruit; limb of the corolla <5 mm wide .... *V. officinalis*

Verbena bracteata Lag. & Rodr.   Prostrate vervain
Hairy annual or perennial with prostrate to arching stems 1–5 dm long; leaves pinnately incised or 3-lobed; flowers bluish or purple, in a dense terminal spike; bracts of the individual flowers conspicuous; rare on ballast, railroad embankments, and sidewalks; mostly SE; flr. Aug; native farther west, but mainly adventive here; UPL.

Verbena canadensis (L.) Britton   Rose verbena
Decumbent perennial, stems trailing and rooting at the lower nodes; flowers blue, purple, or white, in dense showy spikes; rare in ballast, quarry waste, and sidewalk cracks; flr. Jun; native farther west, but adventive here. [*Glandularia canadensis* (L.) Nutt. of Rhoads and Klein, 1993]

Verbena hastata, flower

Verbena hastata L.   Blue vervain, Simpler's-joy
Erect short-hairy perennial to 1.5 m tall; leaves coarsely serrate, sometimes with 2 basal lobes; flowers blue, in erect clustered spikes that are densely flowered except sometimes toward the base; common in moist meadows, floodplains, wet ditches, and roadsides; throughout; flowering mid-Jul–Aug; FACW+. Hybridizes with *V. urticifolia* to form *V. x engelmannii* Moldenke.

Verbena officinalis L.   European vervain
Slender annual to 6 dm tall; stem glabrous; leaves once or twice pinnatifid; spikes elongate, slender; flowers widely spaced, not overlapping; calyx glandular-hairy; corolla blue or violet; rare in waste ground; flr. Jul–Sep; native to Europe; FACU–.

Verbena simplex Lehm.   Narrow-leaved vervain
Erect perennial to 7 dm tall; leaves lance-obovate to narrowly oblong, toothed to
entire toward the base; flowers deep lavender or purple, densely clustered except
in the lowest portions of the slender, solitary spikes; infrequent in shale barrens,
fields, railroad banks, and roadside ditches; SE and SC; flr. mid-Jun–Aug.

Verbena stricta Vent   Hoary vervain
Erect perennial to 1.2 m tall; stems nearly round; leaves sharply serrate, densely
whitish-pubescent beneath; flowers deep blue or purple, in dense, thick spikes;
occasional in old fields and abandoned lots; flr. late Jul–Sep; native farther west,
adventive here.

Verbena urticifolia L.   White vervain
Erect annual or perennial; leaves broadly lanceolate to oblong-ovate, coarsely
doubly-serrate; flowers white, widely spaced in slender clustered terminal
spikes; moist fields, meadows and waste ground; flr. Jul–Sep; FACU; 2 varieties:

A. leaves hirtellous or glabrous; nutlets 2 mm, corrugated on the back ....
.... var. *urticifolia*
common; throughout.
A. leaves densely velutinous; nutlets 1.5 mm, smooth ....
.... var. *leiocarpa* Perry & Fernald
occasional; S.

Verbena hastata

Verbena urticifolia

Vitex L.

Vitex agnus-castus L.   Chaste-tree
Deciduous shrub or small tree to 5 m tall; leaves palmately compound with 5–9
narrow leaflets; twigs and undersides of leaflets densely pubescent; flowers in an
elongate terminal inflorescence; corolla purple; cultivated and rarely escaped to
fencerows; flr. Jul–Aug; native to Eurasia.

VIOLACEAE Violet Family
Harvey E. Ballard, Jr.

Perennial or annual herbs; leaves various; stipules broadly lanceolate, acute, entire to serrate; flowers white, yellow, blue to purple, or multicolored, bilaterally
symmetrical and commonly with a spur enveloping nectaries; petals and sepals
5, petals distinct; pistil compound and 3-carpellate, ovary superior, style 1; stamens 5; fruit a 3-valved capsule; cleistogamy common.

A. plants commonly 5 dm or more tall in flower; flowers 2–4 mm long, greenish-white; capsule >2 cm .... *Hybanthus*
A. plants <4 dm tall in flower; flowers typically 5–15 mm; corolla white, cream,
   yellow, blue, purple, or multi-colored; capsule <2 cm .... Viola

Hybanthus Jacq.

Hybanthus concolor (T.F.Forst.) Spreng.   Green-violet
Perennial herb 4–8 dm tall; leaves oblanceolate, entire to coarsely and remotely

Hybanthus concolor

serrate; corolla very small, greenish-white, the bottom petal slightly longer than lateral and upper ones; capsules green, drying to tan; seeds black; frequent in rich mesic or alluvial woods; S; flr. May–Jun; FACU–.

Viola L.

Herbs to 4 dm tall, perennials except as noted; some species with leaf-bearing stems, others stemless (either stoloniferous or strictly rosette-forming) with all leaves basal; leaves in most species elliptic to reniform and rounded to cordate at the base; margins regularly serrate or crenate to lobed or dissected; corolla variously colored but not greenish-white, the bottom petal distinctly shorter than the others; sepals often bearing auricles posteriorly; capsules green, tan, or heavily purple-flecked; seeds variously colored, <2 mm except in *V. odorata*.

A. plants with ascending or erect stems
  B. corolla cream or blue to purple, without yellow center
    C. corolla solid creamy-white; spur to 4 mm, <¼ of the length of the entire spurred petal (blade plus spur); sepal margins ciliate .... *V. striata*
    C. corolla light blue to pale purple; spur commonly >5 mm, >¼ of the length of the whole spurred petal; sepal margins not ciliate
      D. corolla solid pale blue; lateral petals bearded within .... *V. labradorica*
      D. corolla lavender with dark purple "eyespot" around the throat; lateral petals not bearded within .... *V. rostrata*
  B. corolla yellow throughout, or white, cream or purple with a yellow center
    E. annuals or biennials with weak, slender taproots <1 mm thick; stipules deeply lobed, herbaceous, and leaf-like
      F. leaf blades and terminal segment of stipules with 0, 1(2) crenations .... *V. bicolor*
      F. leaf blades and terminal segment of stipules with 3 or more crenations
        G. lateral petals distinctly shorter than the sepals (rarely surpassing them by up to 2 mm); corolla cream-white with a yellow center.... .... *V. arvensis*
        G. all petals distinctly longer than the sepals; corolla yellow-orange to purplish with a yellow center, the uppermost 2 petals often blue-black at the tip or throughout .... *V. tricolor*
    E. perennials with rhizomes >3 mm thick; stipules entire or lobed, often partly or completely membranous and not leaf-like
      H. corolla white with a yellow center (purplish on the underside); stipules long-tapering, narrowly and sharply acute at the tip .... *V. canadensis*
      H. corolla completely yellow; stipules ovate, acute, or obtuse at the tip
        I. blade of most or all stem leaves as broad as long or broader, obtuse or abruptly acute at the tip .... *V. pubescens*
        I. blades of all stem leaves much longer than broad, long-tapering to a narrowly acute tip
          J. blades of stem leaves arrow-shaped, truncate to cordate at the base .... *V. hastata*
          J. blades of stem leaves lance-ovate, rounded to tapering at the base .... *V. tripartita*

A. plants stemless; some species also producing stolons that commonly root at the nodes

K. corolla yellow; stipules entire .... *V. rotundifolia*

K. corolla white to blue; stipules glandular-ciliate or shallowly lacerate

    L. bottom of the style tipped by a long, slender, recurved conic hook; leaves densely short-pubescent throughout; capsules densely short-pubescent.. *V. odorata*

    L. bottom of the style merely expanded into a short scoop-shaped tip; leaves sparsely hairy on the upper surfaces, otherwise glabrous; capsules glabrous

    M. spur much longer than thick, 3 mm or more long

        N. lateral petals bearded within; leaf blades reniform, shallowly cordate at the base, the inner edges of the basal lobes never overlapping; stipules free .... *V. appalachiensis*

        N. lateral petals glabrous within; leaf blades narrowly ovate, deeply cordate at the base, the inner edges of the basal lobes often overlapping; stipules adnate for about half of their length... *V. selkirkii*

    M. spur shorter than thick, 2 mm or less long

        O. corolla entirely white (often some petals with brown to purple lines)

            P. leaf blades >1.5 times as long as broad

                Q. leaf blades narrowly cuneate at the base; margins minutely denticulate, each tooth tipped with a small but distinct black gland... *V. lanceolata*

                Q. leaf blades subtruncate to subcordate at the base; margins crenate, teeth lacking a conspicuous black gland.............. .... *V. primulifolia*

            P. leaf blades <1.5 times as long as broad (often broader than long)

                R. leaf blades strictly glabrous on both sides, the underside not distinctly paler than the upper; base shallowly and broadly cordate, lobes divergent................... *V. macloskeyi*

                R. leaf blades typically at least sparsely pubescent on both sides, the underside distinctly paler than the upper; the base moderately to deeply cordate, inner edges of the lobes frequently overlapping

                    S. plant stoloniferous; rhizomes horizontal .... *V. blanda*

                    S. plant not stoloniferous; rhizomes vertical (although sometimes twisted) .... *V. renifolia*

        O. corolla blue to violet (rarely white with a purple center)

            T. leaf blades lobed or dissected

                U. leaf blades dissected nearly to the base into slender, linear segments

                    V. corolla frontally flattened, tips of the stamens exposed; lateral petals glabrous within; stipules long-adnate to the petioles ....*V. pedata*

                    V. corolla with petals oriented forward in life, tips of the stamens not exposed; lateral petals bearded within; stipules not strongly adnate .... *V. brittoniana*

U. leaf blades coarsely incised or lobed, the sinuses extending about halfway from the margin to the base, the segments triangular to ovate or reniform

    W. leaf blades much longer than broad; coarse teeth or linear-oblong lobes restricted to the lowest ¼–⅓ of the blade; sepals long-tapering, sharply acute at tip; auricles well-developed, 2–6 mm in fruit .... *V. sagittata*

    W. leaf blades scarcely if at all longer than broad; lobes triangular to broadly ovate, reaching to or beyond the middle of the blade margin; sepals oblong, obtuse, or rounded at the tip; auricles short, <1 mm, even in fruit

        X. early spring (outermost, smallest) and late summer (innermost, largest) leaf blades mostly unlobed; central division of midseason blades unlobed, the lateral divisions lobed .... *V. palmata*

        X. all leaf blades lobed; central lobe of midseason blades deeply lobed like the lateral divisions .... *V. subsinuata*

T. leaf blades merely serrate along the margins

  Y. most or all leaf blades distinctly longer than broad

    Z. foliage densely to moderately pubescent; leaf blades oblong to elliptic-ovate, broadly tapering to subcordate at the base; most petioles shorter than the blades in flower, less than twice as long as the blades in fruit .... *V. sagittata*

    Z. foliage glabrous or essentially so; leaf blades ovate, moderately to deeply cordate at the base; petioles at least twice as long as the blades in later flower and fruit

        AA. lateral petals bearded within with long thread-like hairs; spurred petal densely bearded within; auricles <1 mm in fruit; seeds light to medium brown .... *V. affinis*

        AA. lateral petals bearded within with short knob-shaped hairs; spurred petal glabrous within; auricles 3–10 mm in fruit; seeds olive-black .... *V. cucullata*

  Y. most or all leaf blades as broad as long or broader

    BB. sepals long-tapering, sharply acute at the tip; lateral petals bearded within with short, knob-shaped hairs; auricles 3–10 mm in fruit .... *V. cucullata*

    BB. sepals oblong to lance-ovate, obtuse to rounded at the tip; lateral petals bearded within with long, thread-like hairs; auricles <2 mm in fruit

        CC. spurred petal moderately to densely bearded within; leaves glabrous except for scattered hairs on the upper leaf surfaces near the margins, commonly purple-flushed but with distinctly paler veins .... *V. hirsutula*

        CC. spurred petal glabrous or with a few hairs within; leaves commonly moderately to densely long-pubescent throughout, uniformly green .... *V. sororia*

Viola affinis Leconte   LeConte's violet
Stemless; rhizome ascending, 4–6 mm thick; leaves simple, blades narrowly ovate-triangular, acuminate at the tip and cordate at the base, glabrous except for scattered small hairs on upper surface; margins serrate; capsules purple-spotted on prostrate or declined peduncles; seeds light brown; frequent in rich, moist (especially alluvial) woods; throughout, except NC; flr. Apr–Jun; FACW. [syn: *V. sororia* Willd. in part]

Viola appalachiensis L.K.Henry   Henry's violet

Stemless; producing surficial stolons; rhizome prostrate to ascending, 2–4 mm thick; leaves simple, blades broadly ovate to reniform, rounded at the tip and shallowly cordate at the base, margin crenate, glabrous except for scattered small hairs on the upper surface; stipules free, sparingly lacerate; corolla purple; spurred petal glabrous, spur 2–3 times as long as thick; capsules green, on ascending peduncles; seeds light brown; rare in bogs and stream banks in rich, moist woods; W; flr. Apr–Jun; FACU; ✿.

Viola arvensis Murray   European field pansy

Annual; stems to 3.5 dm tall; taproot erect, 1–2 mm thick; leaves simple, blades obspatulate to narrowly obovate, acute to rounded at the tip, and tapering to rounded at the base; stipules palmately lobed near the base; frequent in fields and along railroad tracks; mostly SE; flr. Apr–Sep; native to Europe.

Viola bicolor Pursh   Field pansy

Annual; stems to 2.5 dm tall; taproot erect, 1–2 mm thick; leaves simple, blades obspatulate to orbicular, rounded at the tip and tapering to rounded at the base; stipules pinnately lobed near the base; corolla pale blue with a yellow center; petals twice as long as the sepals; fields, dry open woods, and floodplain terraces; mostly S; flr. Apr–May. [V. rafinesquii Greene of Rhoads and Klein, 1993]

Viola blanda Willd.   Sweet white violet

Stemless; producing naked or leafy subsurface stolons; rhizome prostrate to ascending, 1–3 mm thick; leaves simple, blades broadly ovate, obtuse to rounded at the tip and cordate at the base, nearly glabrous to densely pubescent; margins low-serrate; capsules purple-spotted, on prostrate or declined peduncles; seeds medium brown; frequent in moist woods and swamps; throughout; flr. Apr–May; FACW. [syn: V. incognita Brainerd]

Viola brittoniana Pollard   Coast violet

Viola
brittoniana,
leaf

Stemless; rhizome ascending, 4–6 mm thick; leaves deeply twice-ternately dissected, blade outline ovate, rounded at the tip and cordate at the base, glabrous; segment margins subentire to remotely serrate; corolla purple; spurred petals bearded within with long thread-like hairs; capsules green, on erect peduncles; seeds brown; very rare in moist sandy woods and flats; extreme SE; flr. Apr–May; FAC; ✿. [V. pedatifida G.Don ssp. brittoniana (Pollard) McKinney of Rhoads and Klein, 1993]

Viola canadensis L.   Canada violet

Viola canadensis,
corolla

Stems to 3 dm tall; rhizome ascending, 3–5 mm thick; leaves simple, blades ovate, acuminate at the tip and cordate at the base, glabrous to moderately short-pubescent; margins serrate; stipules free, membranous, entire; lateral petals bearded with short knob-shaped hairs; spurred petal glabrous, spur short and broadly rounded; capsules green, on ascending peduncles; seeds ivory; frequent in moist woods and swamps; throughout, except SE; flr. late Apr–Jul.

Viola cucullata Aiton   Blue marsh violet

Viola cucullata,
dissected corolla

Stemless; rhizome ascending, 4–6 mm thick; leaves simple, blades narrowly ovate-triangular to ovate, acute at the tip and cordate at the base, glabrous; margins serrate; flowers on peduncles exceeding the leaves, corolla pale purple with darker eyespot; capsules green, on erect peduncles; seeds olive-black; frequent in bogs, meadows, and swamps; throughout; flr. Apr–Jul; FACW+.

Viola hastata Michx.   Halberd-leaved yellow violet
Stems to 3 dm tall; rhizome ascending, 3–5 mm thick; leaves simple, glabrous; margins serrate; stipules free, membranous, entire; lateral petals bearded with short knob-shaped hairs; spurred petal glabrous, spur short and broadly rounded; capsules green, on ascending peduncles; seeds tan; frequent in rich deciduous woods; W; flr. late Apr–May; UPL.

Viola hirsutula Brainerd   Southern wood violet
Stemless; rhizome ascending, 4–6 mm thick; leaves simple, blades broadly ovate to orbicular, obtuse at the tip and cordate at the base; margin serrate; stipules free, membranous, glandular-ciliate; corolla purple; lateral and spurred petals bearded within with long thread-like hairs; capsules purple-spotted, on prostrate peduncles; seeds brown; frequent in dry, rich deciduous woods, often near streams; mostly S; flr. Apr–May.

Viola labradorica Schrank   American dog violet
Stems to 2 dm tall; rhizome ascending, 2–4 mm thick; leaves simple; blades orbicular to ovate, apiculate at the tip and cordate at the base, glabrous except for scattered small hairs on upper surface; margins crenate; stipules free, sparingly lacerate; spurred petal glabrous; spur 2–3 times as long as thick; capsules green, on ascending peduncles; seeds light brown; frequent in moist woods and swamps; throughout; flr. May–Jul; FACW. [*V. conspersa* Reichenb. of Rhoads and Klein, 1993]

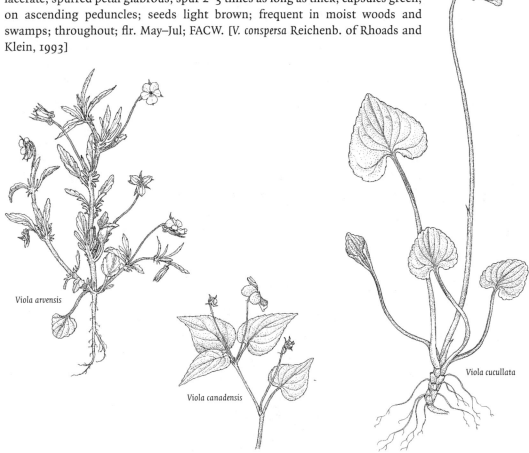

Viola arvensis

Viola canadensis

Viola cucullata

Viola lanceolata L.　　　　　　　　　　　　　　　　Lance-leaved violet
Stemless; producing leafy surficial stolons with cleistogamous capsules; rhizome prostrate to ascending, 1–2 mm thick; leaves simple; blades lanceolate; capsules green, on erect peduncles; seeds olive-black; frequent in moist sandy shores, flats, and bogs; throughout; flr. late Mar–Jun; OBL. Ours is var. *lanceolata*.

Viola macloskeyi F.E.Lloyd　　　　　　　　　　　　Sweet white violet
Stemless; producing leafy surficial stolons with cleistogamous capsules; rhizome prostrate to ascending, 1–2 mm thick; leaves simple; blades narrowly ovate to orbicular; margin subentire to shallowly crenate; capsules green, on erect peduncles; seeds olive-black; frequent in bogs, swamps, and wet woods; throughout; flr. Apr–Jul; OBL. Ours is ssp. *pallens* (Banks ex Ging.) M.S.Baker. [syn: *V. pallens* (Banks ex Ging.) Brainerd]

Viola odorata L.　　　　　　　　　　　　　　　　English or sweet violet
Stemless; producing subsurface stolons throughout much of the year; rhizome prostrate to ascending, 2–4 mm thick; leaves simple; blades ovate to orbicular, rounded at the tip and shallowly cordate at the base, rugose above; margins crenate; stipules free, herbaceous, lacerate; corolla white to pale purple; lateral petals glabrous or bearded with thread-like hairs; spurred petal glabrous; spur well-developed, 2–3 times as long as thick; capsules black-spotted, on prostrate or declined peduncles; seeds ivory; occasional in lawns, fencerows, or city woodlots; mostly S; flr. Apr–May; native to Europe.

Viola odorata, flower with
hooked stigma (×2)

Viola palmata L.　　　　　　　　　　　　　　　　Early blue violet
Stemless; rhizome ascending, 4–6 mm thick; leaves lobed; blade outline ovate, acute to obtuse at the tip and cordate at the base, glabrous; segment margins

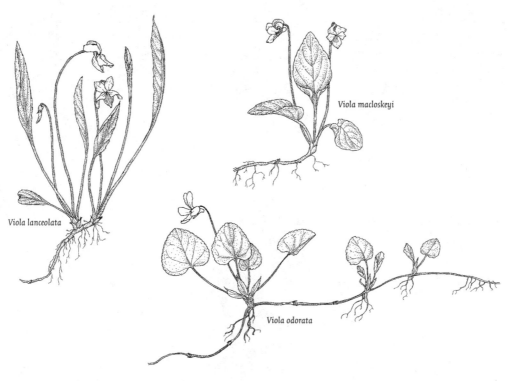

Viola lanceolata

Viola macloskeyi

Viola odorata

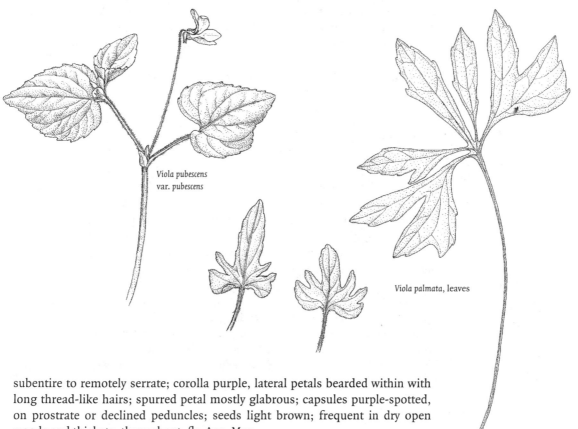

Viola pubescens
var. pubescens

Viola palmata, leaves

subentire to remotely serrate; corolla purple, lateral petals bearded within with long thread-like hairs; spurred petal mostly glabrous; capsules purple-spotted, on prostrate or declined peduncles; seeds light brown; frequent in dry open woods and thickets; throughout; flr. Apr–May.

Viola pedata L.   Birdfoot violet
Stemless; rhizome barrel-like, erect, 4–8 mm thick; leaves deeply ternately dissected, lateral divisions dissected again; blade ovate in outline, rounded at the tip and cordate at the base, glabrous to ciliate; segment margins subentire to remotely serrate; corolla uniformly pale blue or the upper petals purple-black; spurred petals glabrous within; capsules green, on erect peduncles; seeds light brown; occasional on sandy or rocky barrens or dry forested slopes; SE and SC; flr. May–Jun.

Viola primulifolia L.   Primrose violet
Stemless; producing leafy stolons with cleistogamous capsules; rhizome prostrate to ascending, 1–2 mm thick; leaves simple; blades lance-ovate; capsules green, on erect peduncles; seeds olive-black; occasional in meadows and swamps; mostly E, but scattered elsewhere; flr. Apr–Jun; FAC+.

Viola pubescens Aiton   Downy yellow violet
Stems to 3.5 dm tall; leaves simple, rhizome ascending, 3–5 mm thick; leaf blades ovate to reniform, acute to broadly obtuse at the tip and truncate or broadly tapering to cordate at the base, nearly glabrous to densely villous; margins serrate; stipules free, largely or completely herbaceous, entire; lateral petals bearded with short knob-shaped hairs; spurred petal glabrous, spur short and broadly rounded; capsules green, on ascending peduncles; seeds tan; frequent throughout; flr. May–Jun; 2 varieties:

also smooth next page to 122 Rhoad

Viola pubescens var.
pubescens, dissected
flower

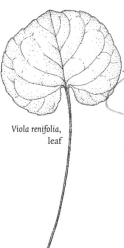

Viola renifolia,
leaf

A. foliage densely pubescent; stems 1–3, erect throughout; basal leaves 1 or none; blades of stem leaves reniform, broadly obtuse at the tip, broadly tapering to truncate at the base .... var. *pubescens* frequent in dry woods; throughout; FACU.

A. foliage glabrous or nearly so; stems usually several, spreading; basal leaves 2–3; blades of stem leaves ovate, acute at the tip, cordate at the base .... var. *scabriuscula* Nutt. ex Torr. & A.Gray frequent in moist woods and swamps; throughout. [syn: *V. eriocarpa* Schwein. of Rhoads and Klein, 1993]

Viola renifolia A.Gray   Kidney-leaved violet
Stemless; rhizome erect (although sometimes contorted), 1–2 mm thick; leaves simple, blades reniform or orbicular to broadly ovate, obtuse to rounded at the tip and cordate at the base, glabrous or pubescent with long silky hairs on one or both surfaces; margin low-serrate; capsules purple-spotted on prostrate or declined peduncles; seeds medium brown; rare in dry cool woods, rock crevices, or hummocks in swamps, especially on limestone soils; NE; flr. May–Jul; FACW; ☙.

Viola rostrata,
dissected flower

Viola rostrata Pursh   Long-spurred violet
Stems to 2.5 dm tall; rhizome ascending, 2–4 mm thick; leaves simple; blades ovate, acute at the tip and cordate at the base, glabrous; margins serrate; stipules free, moderately lacerate; all petals glabrous; spur at least as long as the petal blade; capsules green on ascending peduncles; seeds light brown; frequent in rich woods; throughout; flr. Apr–early Jun; FACU.

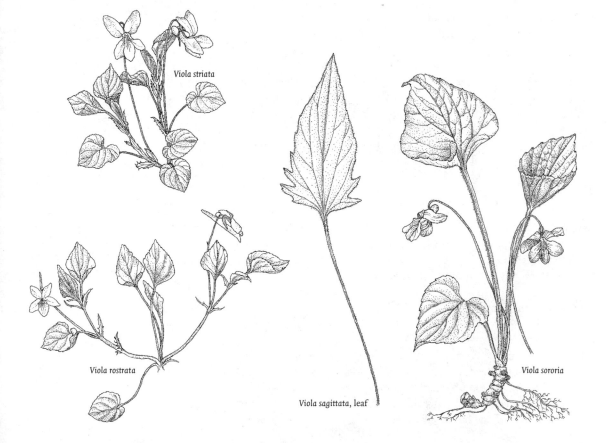

Viola striata

Viola rostrata

Viola sagittata, leaf

Viola sororia

Viola rotundifolia Michx.   Round-leaved violet

Stemless; rhizome ascending, 4–6 mm thick; leaves simple, mostly flat against the ground; blades narrowly ovate to orbicular, broadly rounded at the tip and shallowly cordate at the base, the lobes often overlapping, sparsely hairy on upper surfaces; margins crenate; stipules free; lateral petals bearded within with short knob-shaped hairs; spurred petal glabrous; spur short and broadly rounded; capsules black-spotted, on prostrate or declined peduncles; seeds ivory; frequent in cool, moist woods and banks; throughout, except SW and SC; flr. Apr–May; FAC+.

Viola sagittata Aiton

Stemless; rhizome ascending, 4–6 mm thick; leaves simple; blades narrowly ovate-triangular to ovate, acute at the tip and cordate at the base, glabrous; margins serrate; corolla pale purple with darker eyespot; lateral petals bearded with short knob-shaped hairs; spurred petal glabrous within; capsules green, on erect peduncles; seeds olive-black; frequent in dry woods, fields, and edges; throughout; flr. Apr–Jun; FACW; 2 varieties

A. foliage glabrous to moderately short-pubescent; leaves erect; blades oblong-lanceolate to long-triangular, commonly shallowly to deeply lobed at the base with linear segments; petioles much longer than blades ....
....var. *sagittata*—arrow-leaved violet
A. foliage densely short-pubescent; leaves prostrate or weakly ascending; blades elliptic to ovate, at most coarsely serrate at the base; petioles distinctly shorter than the blades at flowering time, up to as long in fruit ....
.... var. *ovata* (Nutt.) Torr. & A.Gray—ovate-leaved violet

Viola selkirkii Pursh ex Goldie   Great-spurred violet

Stemless; rhizome ascending, 2–3 mm thick; leaves simple, blades ovate, obtuse to acutish at the tip and deeply cordate at the base, sparsely hairy on upper surfaces; margin crenate; corolla purple; spur thick, well-developed; capsules black-spotted, on prostrate or declined peduncles; seeds medium brown; rare in dry, cool woods and rock crevices, especially on limestone soils; NE; flr. May–Jul; ❧.

Viola
selkirkii,
leaf

Viola selkirkii,
flower (×1)

Viola sororia Willd.   Common blue violet

Stemless; rhizome ascending, 4–6 mm thick; leaves simple; blades broadly ovate, acutish to obtuse at the tip and cordate at the base, glabrous to densely villous, especially on lower surfaces; margins serrate; lateral petals bearded within with long thread-like hairs; capsules purple-spotted, on prostrate peduncles; seeds brown; common in dry to moist woods, swamps, and thickets; throughout; flr. Mar–Jun; FAC. [syn: *V. papilionacea* auct. non Pursh]

Viola sororia,
dissected flower

Viola striata Aiton   Striped violet

Stems to 3 dm tall; rhizome ascending, 2–4 mm thick; leaves simple; blades ovate, acuminate at the tip and cordate at the base, glabrous except for scattered small hairs on the upper surface; margin crenulate; stipules free, moderately to heavily lacerate; lateral petals bearded with thread-like hairs, occasionally the others as well; spur 2–3 times as long as thick; capsules green, on ascending peduncles; seeds light brown; frequent in alluvial woods and alkaline swamps; throughout; flr. Apr–Jun; FACW.

Viola striata,
dissected flower

Viola subsinuata Greene   Violet

Stemless; rhizome ascending, 4–6 mm thick; leaves lobed; blade ovate in out-line, acute to obtuse at the tip and cordate at the base, glabrous; segment margins subentire to remotely serrate; lateral petals bearded within with long thread-like hairs; spurred petal mostly glabrous; capsules purple-spotted, on prostrate or declined peduncles; seeds light brown; frequent in rich open woods; mostly S; flr. Apr–May. [syn: *V. palmata* auct. non L.]

Viola tricolor L.   Johnny-jump-up

Annual; stems to 3.5 dm tall; taproot erect, 1–3 mm thick; leaves simple, blades obspatulate to orbicular, acute to rounded at the tip and tapering or rounded at the base; stipules palmately lobed near the base; occasional in fields, abandoned gardens, or dumps; scattered; flr. spring–summer; native to Europe. [includes the garden pansy, *V.* x *wittrockiana* Gams.]

Viola tripartita Elliot   Three-parted violet

Stems to 4 dm tall; rhizome ascending, 3–5 mm thick; leaves simple, blades ovate-lanceolate, acuminate at the tip and narrowly tapering to rounded at the base, glabrous to moderately villous; margins remotely serrate; stipules free, membranous, entire; lateral petals bearded with short knob-shaped hairs; spurred petal glabrous; spur short and broadly rounded; capsules green, on ascending peduncles; seeds tan; very rare in rich deciduous woods; SW; flr. late Apr–May; 🌿. Ours is var. *glaberrima* (DC.) R.M.Harper, which lacks lobed leaves.

VISCACEAE Mistletoe Family

Greatly reduced, brownish-green, evergreen subshrubs, parasitic on the branches of trees; leaves opposite, simple, and entire, or reduced to scales; dioecious; flowers regular, perianth greatly reduced; fruit a drupe.

A. stems mostly <2 cm; leaves tiny, brown, scale-like .... *Arceuthobium*
A. stems to 4 dm; leaves green, 2–5 cm .... *Phoradendron*

Arceuthobium M.Bieb

Arceuthobium pusillum Peck   Dwarf mistletoe

Inconspicuous parasite on *Picea mariana*; stems 0.5–1.5 cm long; leaves minute brown scales; flowers axillary, appearing as short lateral branches; rare in sphagnum bogs where the host is found; flr. Jun–Jul; NE; 🌿.

Phoradendron Nutt.

Phoradendron leucarpum (Raf.) Reveal & M.C.Johnst.   Christmas mistletoe

Parasite on *Nyssa sylvatica* and other deciduous trees; stems 2–4 dm long, branching freely; leaves leathery, oblong, 2–6 cm; flowers 3-merous, in 1–5-cm spikes that are solitary or clustered; fruit white, 4–5 mm; very rare; SE; believed to be extirpated; frt. Oct–Dec; 🌿.

VITACEAE Grape Family

Woody vines climbing by tendrils sometimes equipped with adhesive disks; leaves alternate with tendrils or inflorescences opposite them; leaves simple or compound, palmate, stipulate; flowers small, yellowish or greenish, regular, perfect or unisexual, 5-merous; ovary superior, 2-locular, free from to partly embedded in the nectar disk; style simple with a 2-lobed, capitate stigma; stamens same number as the petals and opposite them; petals distinct or remaining attached at the tips; fruit a berry with 1–4 bony seeds.

A. leaves compound .... *Parthenocissus*
A. leaves simple
    B. inflorescence a panicle; pith brown, interrupted at the nodes .... *Vitis*
    B. inflorescence cymose; pith white, continuous through the nodes
        C. scrambling vines; tendrils lacking adhesive disks .... *Ampelopsis*
        C. vines climbing by means of adhesive disks .... *Parthenocissus*

Ampelopsis Michx.

Ampelopsis brevipedunculata (Maxim.) Trautv.   Porcelain-berry
Climbing or scrambling vine with tendrils or flowers opposite the palmately-lobed leaves; tendrils without adhesive disks; petals falling separately; fruit a bright blue berry; cultivated and occasionally naturalized in disturbed woods, fields, floodplains, and roadside thickets; SE and SW, a serious weed in some areas; flr. May–Aug, frt. Sep–Oct; native to Asia.

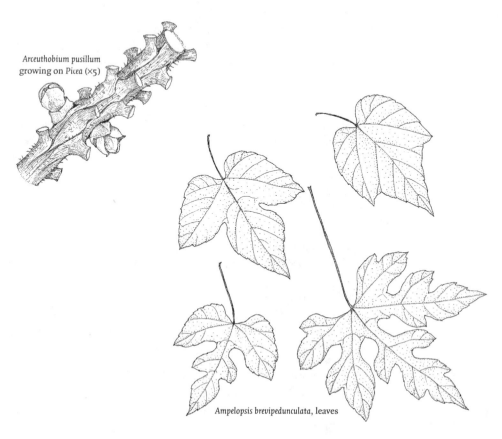

Arceuthobium pusillum growing on Picea (×5)

Ampelopsis brevipedunculata, leaves

Parthenocissus Planch.

Woody vines climbing by tendrils bearing adhesive disks; leaves simple or compound, palmate; flowers 5-merous, mostly perfect; petals falling separately; fruit a bluish-black, 1–4-seeded berry; flr. May–Jun, frt. Sep–Oct.

A. leaves compound with 5 leaflets throughout
    B. tendrils with 5–12 branches and adhesive disks at the tips; inflorescence with a central axis .... P. quinquefolia
    B. tendrils with 3–5 branches mostly lacking adhesive disks; inflorescence dichotomously branched, lacking a central axis .... P. inserta
A. leaves mostly simple, or 3-foliate on flowering branches .... P. tricuspidata

Parthenocissus inserta (Kern.) Fritsch   Grape woodbine
Very similar to *P. quinquefolia* but tendrils with only 3–5 branches and few, if any, adhesive disks; leaflets more strongly toothed; inflorescence branching dichotomously, without a central axis; occasional in woods, fields, and alluvial thickets; mostly N; FACU. [syn. *P. vitacea* (Knerr) A.Hitchc.]

Parthenocissus quinquefolia (L.) Planch.   Virginia-creeper, woodbine
Climbing vine adhering by adhesive disks at the tips of the much branched tendrils; leaf long-petioled, 5-foliate, palmately compound; leaflets 6–12 cm long, obovate, coarsely toothed above the middle, turning bright red in the fall; inflorescence terminal or from the upper axils, with a definite central axis; fruits purplish-black, 6 mm; common in woods, fields, and edges; throughout; FACU.

Parthenocissus tricuspidata (Sieb. & Zucc.) Planch.   Boston ivy
Vine climbing by means of tendrils tipped with numerous adhesive disks; leaves cordate, mostly simple and 3-lobed but trifoliate on flowering branches; cultivated and occasionally escaped; mostly SE; native to China and Japan.

Vitis L.

Woody vines with shredding or peeling bark, climbing by means of twining tendrils; leaves simple, with a tendril or inflorescence opposite the leaf at most nodes; flowers functionally unisexual, (plants polygamodioecious) with a hypogynous disk composed of 5 nectar glands alternate with the 5 stamens; petals distinct at the base but joined at the tip and falling as a unit; fruit a juicy berry with 1–4 seeds; flr. May, frt. Aug–Nov.

A. mature leaves glaucous beneath; nodes often glaucous .... V. aestivalis
A. mature leaves not glaucous beneath; nodes not glaucous
    B. tendrils and/or panicles sometimes produced from 3 or more successive nodes
    C. leaves tomentose beneath
        D. leaves moderately tomentose beneath; fruits few, in a simple globular cluster .... V. labrusca
        D. leaves densely tomentose beneath; fruits many, in a branched cluster .... V. x labruscana
    C. leaves at maturity merely pubescent or slightly cobwebby on the veins beneath .... V. novae-angliae

B. tendrils or panicles present at only 2 consecutive nodes and missing at every third node, or totally lacking
- E. tendrils none or only opposite the uppermost leaves; leaf blade kidney-shaped .... *V. rupestris*
- E. tendrils present except at every third node, leaf blade cordate or cordate-ovate
  - F. leaves pubescent beneath with short straight hairs when young, glabrous at maturity
    - G. leaves hardly lobed; leaf margins with obtuse teeth, not ciliolate; berries black .... *V. vulpina*
    - G. lobes of the leaves well-developed, pointing forward; leaf margins with acute teeth, ciliolate; berries glaucous .... *V. riparia*
  - F. leaves pubescent beneath with long matted hairs, at least when young .... *V. cinerea*

Vitis aestivalis Michx.   Summer grape, pigeon grape
Stems with tendrils or inflorescences absent at every 3rd node; young stems terete, glabrous or sparsely pilose at anthesis, glaucous at the nodes; leaves glaucous and persistently floccose-tomentose at least in the vein axils beneath; blade broadly cordate-ovate to suborbicular, with narrow to broad basal sinus, shallowly to deeply 3- or 5-lobed; petiole glabrous or very sparsely pilose; fruit dark purple or black, 5–10 mm; common in upland woods and wooded slopes; throughout; FACU.

Vitis cinerea (Engelm. in A.Gray) Engelm. ex Millardet   Possum grape
Stems angled, cobwebby-tomentose becoming glabrescent; leaves unlobed or shallowly 3-lobed; petioles and lower leaf surface permanently pilose; fruit black, 4–7 mm, in dense clusters, becoming sweet after frost; rare in lowland woods; SW; FACW; ✚. Ours is var. *baileyana* (Munson) Comeaux. [syn. *V. baileyana* Munson]

Vitis labrusca L.   Fox grape
Stems regularly with tendrils or inflorescences at 3 or more successive nodes; underside of leaves permanently covered with a dense rusty tomentum that conceals the entire surface; leaf blades firm with prominent veins, round-cordate, with coarsely dentate margin; fruit 15 mm in diameter, becoming sweetish; frequent in rocky woods, moist thickets, and stream banks; throughout, except in the northernmost counties; FACU.

Vitis x labruscana Bailey   Grape
Similar to *V. labrusca* but with generally less tomentose leaves, larger fruit clusters, and bigger berries; cultivated and occasionally spreading to roadsides and waste ground. A collective name for cultivars of hybrid origin with *V. labrusca* as one parent, such as Concord.

Vitis novae-angliae Fernald   New England grape
Stems frequently bearing tendrils or inflorescences from 3 or more successive nodes; leaves and stems tomentose when young but nearly glabrous at maturity; blades round-ovate, acuminate, unlobed, or with angled shoulders; leaf teeth short and wide; fruit black, glaucous, 12–17 mm, acid; rare in moist mountain

woods, ravines, and roadside thickets; scattered; 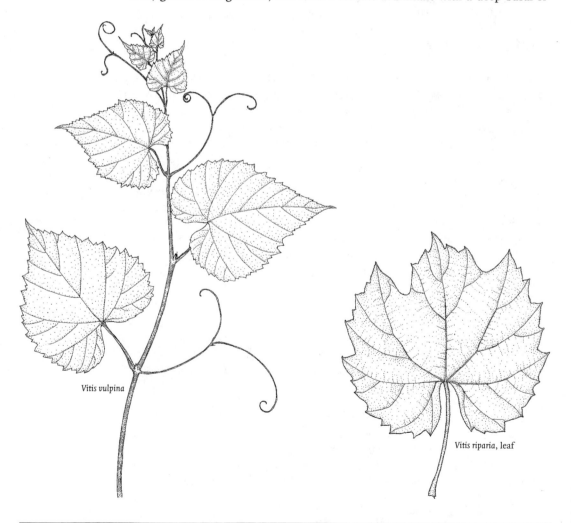. [*V. labrusca* x *riparia* of Rhoads and Klein, 1993]

Vitis riparia Michx.   Frost grape
Stems with tendrils or inflorescences lacking at every 3rd node; young stems glabrous or pubescent becoming glabrate; leaf blades cordate-ovate with a prolonged apex and 2 forward-pointing lateral lobes, coarsely serrate with ciliolate margins, pubescent beneath when young; fruit black, 6–12 mm, strongly glaucous, acidic; frequent on riverbanks and alluvial thickets; throughout; FACW.

Vitis rupestris Scheele   Sand grape
Stems usually lacking tendrils at all but the uppermost nodes, therefore merely prostrate or reclining, not climbing; leaf blades reniform to depressed ovate, often folded, coarsely toothed with a broad basal sinus, glabrous or sparsely hairy along the major veins; fruit 6–10 mm, black, sweet; very rare on riverbanks; S; UPL; .

Vitis vulpina L.   Frost grape
Stems lacking tendrils or inflorescences at every 3rd node; young branches terete, glabrous or glabrate; leaf blades cordate-acuminate with a deep basal si-

Vitis vulpina

Vitis riparia, leaf

nus, unlobed or with angled shoulders, firm, glabrous or with pubescence only in the vein axils beneath; inflorescence 10–15 cm long and loosely flowered; fruits shiny black, 5–10 mm, becoming sweet after frost; occasional in woody thickets, rocky slopes, roadsides, and sand dunes; mostly S; FAC. *V. vinifera* L., the cultivated wine grape, which occurs as a rare escape in our region, differs from *V. vulpina* by the deeply 3- or 5-lobed leaf blades that are somewhat pubescent beneath.

ZYGOPHYLLACEAE Creosote-bush Family

Herbs with opposite, pinnate leaves and stipules; flowers regular, perfect, 4- or 5-merous; stamens 2–3 times as many as the petals; nectar disk or separate nectaries present; ovary superior, of 5 fused carpels with axile placentation; fruit a capsule.

Tribulus L.

Tribulus terrestris L.   Puncture-weed, caltrop
Much branched, mat-forming annual to 5 dm tall; leaves opposite, odd-pinnate; flowers solitary in the axils, 8–10 mm wide; fruit 1 cm thick with 5 spiny segments; rare on ballast and railroad tracks; SE and SW; flr. Jul–Oct; native to the Mediterranean region.

ANGIOSPERMS, MONOCOTS

ACORACEAE Sweet Flag Family

Rhizomatous perennials with erect, sword-shaped leaves; flowers tiny, 3-merous, perfect, crowded on an elongate spadix that diverges laterally from the leaf-like stem; tepals 6; stamens 6; ovary superior; spathe extending above the spadix and appearing very much like a leaf.

Acorus L.

Two species with characteristics as above and differing mainly by the prominence of the midrib of the leaf.

A. midrib of the leaf more prominent than the other veins .... *A. calamus*
A. midrib of the leaf equal in prominence to the other veins .... *A. americanus*

Acorus americanus (Raf.) Raf.   Sweet flag
Rare in shallow water of ponds; NW; flr. May–Aug; OBL; 🌱.

Acorus calamus L.   Sweet flag
Common in wet meadows, freshwater tidal marshes, stream edges, and ditches, where it forms dense, and sometimes extensive, colonies; flr. May–Aug; throughout; native to Europe; OBL.

Acorus calamus,
leaf section

AGAVACEAE Agave Family

Coarse, evergreen herbs; leaves all basal, linear, stiff, and spine-tipped; inflorescence an erect panicle; flowers with 6 creamy white, petaloid tepals, 6 stamens, and a 3-lobed superior ovary; fruit a dehiscent capsule.

Yucca L.

Yucca flaccida Haw.   Adam's needle
Leaves numerous, to 8 dm long, fibrous along the margins; inflorescence 1–3 m tall; capsules cylindric, 2–4 cm long; occasional in dry sandy soil near old home sites; flr. Jul–Aug; native farther south. [syn: *Y. filamentosa* L.]

Acorus calamus (×^{1}/$_{10}$)

ALISMATACEAE Water-plantain Family

Aquatic or emergent, perennial (or annual in the case of S. *calycina*) herbs, often rhizomatous; leaves all basal, with a petiolar sheath and a broad blade, or narrow and strap-shaped without an expanded blade or a defined petiole; flowers in a scapose inflorescence; regular, perfect or unisexual; petals and sepals each 3, distinct; pistils numerous, distinct; ovaries superior; stamens 6–many; fruit a cluster of achenes.

A. inflorescence a panicle with numerous perfect flowers; stamens 6 ....*Alisma*
A. flowers mostly unisexual, in a raceme of 1–12 whorls of 3 flowers each; stamens 7–many .... *Sagittaria*

Alisma L.

Upright emergent aquatics; leaves all basal, long-petioled, with ovate or elliptic blades; flowering stem 1–10 dm tall with whorled branches and pedicels bearing numerous small, white, perfect flowers; stamens 6, carpels 10–28 in a single whorl; achenes flattened but not winged.

A. flowers 7–13 mm wide; fruiting heads 4–7 mm ....*A. triviale*
A. flowers 4–5 mm wide; fruiting heads 3–4 mm .... *A. subcordatum*

Alisma subcordatum L. Water-plantain
Flowers 4–5 mm wide; fruiting heads 3–4 mm wide; common in ditches, lake and pond margins, and muddy shores; throughout; flr. late Jul–late Aug; OBL. [syn: *A. plantago-aquatica* L. var. *parviflorum* (Pursh) Torr.]

Alisma triviale Pursh Water-plantain
Very similar to *A. subcordatum* but flowers and fruiting heads larger; rare in shallow water of ditches, lake margins, and stream edges; mostly W; OBL; ✿. [syn: *A. plantago-aquatica* L. var. *americana* Schult. & Schult.]

Sagittaria L.

Emergent or wholly submersed aquatics; leaves highly variable, upright with emergent, expanded, often arrow-shaped blades, narrow and stiffly upright, or ribbon-like with the tips floating (in deeper water); plants monoecious, flowers mostly unisexual, in a raceme of 1–12 whorls with staminate flowers above, pistillate below; petals white; sepals persistent; carpels aggregated in a globose head; achenes flattened, winged on the margins and sometimes also on the face.

A. sepals appressed in fruit; pedicels 3–5 mm in diameter. ....*S. calycina*
A. sepals reflexed in fruit; pedicels more slender
  B. leaves rarely or never with sagittate blades
    C. stamens 7–15; plants low growing, stoloniferous, mat-forming
      D. leaves 5–10 cm long, stiffly erect or spreading................ *S. subulata*
      D. leaves 30–90 cm long, ribbon-like, the terminal portions floating ...
      .... *S. filiformis*
    C. stamens 12 to many; plants not mat-forming
      E. pistillate flowers sessile .... *S. rigida*
      E. pistillate flowers with definite pedicels .... *S. graminea*

Alisma subcordatum (×¼)

B. sagittate leaf blades almost always present
    F. beak of the achene spreading to nearly horizontal; sides of the achene without ridges .... *S. latifolia*
    F. beak of the achene ascending; sides of the achenes ridged ... *S. australis*

Sagittaria australis (J.G.Sm.) Small   Appalachian arrowhead
Perennial; leaf blades triangular-ovoid with broad basal lobes, petiole with 5 wings; flowers in 5–12 whorls; achenes 2.3–3.2 mm; the beak 0.7–1.7 mm, often incurved above the ascending base; occasional in alluvial meadows, wet woods, and backwater pools; SE and SW; flr. late Jul–early Sep.; OBL. [syn: *S. engelmanniana* J.G.Sm. var. *longirostra* (Micheli) Bogin]

Sagittaria australis,
achene (×2½)

Sagittaria calycina Engelm.   Long-lobed arrowhead
Annual; leaves strap-shaped with a blunt tip or with an expanded, arrow-shaped blade and elongated and very spongy petiole; flowers in 3–12 whorls, the upper staminate, the lower ones perfect; pedicels recurved and sepals appressed in fruit; rare on tidal mudflats; SE; flr. Jul–early Sep; OBL; ❦. [syn: *S. montevidensis* Cham. & Schltdl.]

Sagittaria filiformis J.G.Sm.   Arrowhead
A deep-water, nontidal form of *S. subulata* with 30–90-cm, ribbon-like leaves with floating ends and equally long or longer flexuous flowering stems; very rare in lakes and ponds; SE but probably extirpated; OBL; ❦. [*S. stagnorum* Small of Rhoads and Klein, 1993]

Sagittaria graminea Michx.   Grass-leaved arrowhead
Erect and emersed or more lax, submersed perennial; leaves often bladeless and acute, or with an expanded, elliptic-ovate blade, rarely with basal lobes; flowers in 2–12 whorls, pediceled; achenes 1.2–3 mm; occasional in shallow water, mudflats, and tidal shores; E; flr. Jun–early Aug; OBL. Ours is var. *graminea*.

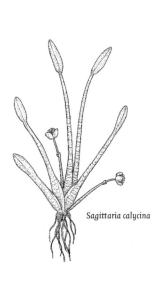

Sagittaria calycina,
sagittate leaf

Sagittaria calycina

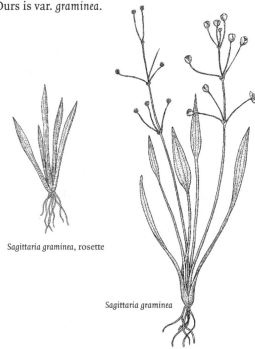

Sagittaria graminea, rosette

Sagittaria graminea

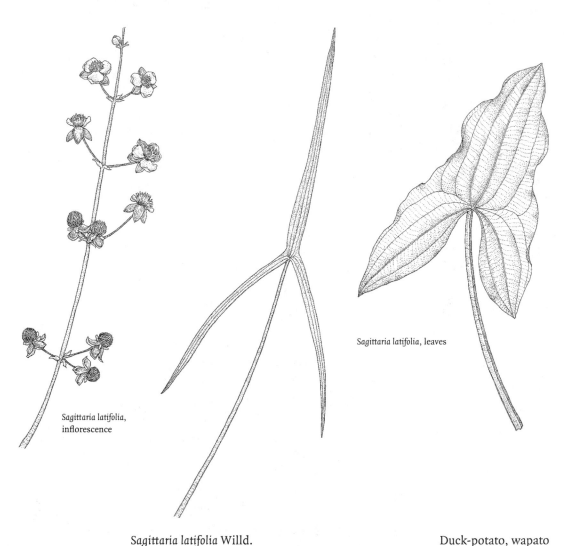

Sagittaria latifolia, inflorescence

Sagittaria latifolia, leaves

Sagittaria latifolia, achene (×2¹/₂)

Sagittaria latifolia Willd. Duck-potato, wapato
Erect perennial from large, edible tubers; leaves sagittate with blade and basal lobes highly variable in width; flowers in 2–15 whorls, occasionally with branches present at the lowest whorls; achenes 2.5–4 mm with the beak at right angles to the body; swamps, wet shores, and shallow water of ponds and streams; flr. Jul–early Sep; OBL; 2 varieties:

A. plants glabrous .... var. *latifolia*; common; throughout.
A. plants pubescent .... var. *pubescens* (Muhl. ex Nutt.) Sm.; occasional; S.

Sagittaria rigida Pursh Arrowhead
Perennial, emersed or submersed; leaves mostly with an expanded, elliptic to lanceolate blade, basal lobes absent or poorly developed; flowering stem 1–8 dm, with 2–8 whorls of flowers, the stem usually bent near the lowest whorl, pistillate flowers and fruiting heads nearly sessile; achenes obovate to oblong, 2–3 mm; frequent in pond margins, mudflats, and stream edges; throughout; flr. Jul–late Aug; OBL.

Sagittaria subulata (L.) L.Buch Subulate arrowhead
Low-growing, stoloniferous, mat-forming perennial; leaves mostly without ex-

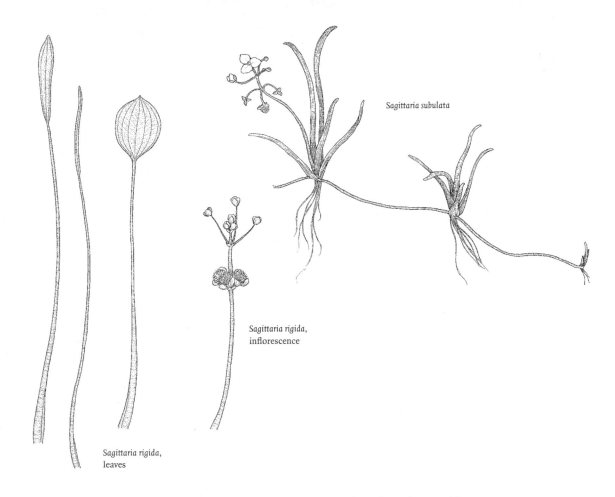

Sagittaria subulata

Sagittaria rigida,
inflorescence

Sagittaria rigida,
leaves

panded blades or just slightly wider at the tips, 5–10 cm long by 1–6 mm wide, obtuse; flowering stem erect, few-flowered; pedicels recurved; achenes 1.5–2.5 mm; rare, but sometimes locally abundant on tidal shores and mudflats; SE; flr. Jun–early Sep; OBL; ✿.

ARACEAE Arum Family

Perennial herbs with expanded, mostly net-veined leaves with a sheathing petiole arising from a corm or tuber; flowers tiny, perfect or unisexual, in a densely crowded, fleshy spike (spadix) that is usually subtended by and often surrounded by an erect sheathing spathe; sepals 0, 4, or 6; stamens 4–10; ovary superior, but often embedded in the spadix; fruit a berry.

A. spathe conspicuous
    B. leaves compound, or deeply lobed
        C. spadix not fused to the spathe; leaves not producing bulblets....*Arisaema*
        C. spadix fused to the spathe at the base; leaves producing bulblets at the
            base of the leaflets ....*Pinellia*
    B. leaves simple
        D. spathe white, flat and open ....*Calla*
        D. spathe green or purplish, surrounding the spadix

E. spadix globose; leaves cordate or rounded, smelling of skunk when bruised .... *Symplocarpus*
E. spadix elongate; leaves sagittate, not malodorous .... *Peltandra*
A. spathe lacking an expanded blade at flowering .... *Orontium*

Arisaema C.Mart.

Mostly dioecious perennials arising from a corm, with 1–2 trifoliate or palmately compound leaves with sheathing bases; flowering stem bearing a single inflorescence with flowers only at the base of the spadix; flowers unisexual; plants monoecious or dioecious; spathe with a tubular lower portion and a spreading or arching tip; fruit a cluster of red berries.

A. leaflets 3; spathe arching over the blunt spadix .... *A. triphyllum*
A. leaflets 5–7; spadix elongate, extending well beyond the spathe ....
.... *A. dracontium*

Arisaema dracontium (L.) Schott   Green-dragon
Plant to 1 m tall with a single palmately compound leaf; the terminal portion of the spadix extending well beyond the erect, tapered spathe; occasional in low woods, floodplains, and swamps; throughout; flr. May; FACW.

Arisaema triphyllum (L.) Schott   Jack-in-the-pulpit
Plant to 1 m tall usually with 2 long-petioled, trifoliate leaves; terminal portion of the spathe arched over the shorter, blunt-tipped spadix; the smaller plants usually staminate, larger pistillate; flr. late Apr–Jun; FACW–; 3 subspecies:

A. lateral leaflets strongly asymmetrical .... ssp. *triphyllum*
    common in moist woods, swamps, and bogs; throughout.
A. lateral leaflets not strongly asymmetrical
    B. back of the spathe strongly ridged .... ssp. *stewardsonii* (Britton) Huttl.
        scattered; throughout.
    B. back of the spathe not ridged .... .ssp. *pusillum* (Peck) Huttl.
        scattered; mostly S and W.

Calla L.

Calla palustris L.   Wild calla
Emergent aquatic with petioled, cordate leaves arising from a creeping rhizome; spathe open, white; spadix globose, shorter than the spathe; occasional in bogs and swamps; N; flr. May–Aug; OBL.

Orontium L.

Orontium aquaticum L.   Goldenclub
Emergent aquatic with entire, oblong to elliptic leaves 1–3 dm long; flowers bright yellow, on an elongate, exposed spadix; spathe covering the spadix when young but gone by flowering time or reduced to an inconspicuous basal bract; infrequent in swamps, lakes, ponds, streams, ditches, and wet shores; E and SC; flr. late Apr–May; OBL; ❧.

Peltandra Raf.

Peltandra virginica (L.) Schott & Endl.   Arrow-arum, tuckahoe
Emergent aquatic from a basal root crown; leaves strongly veined, oblong to tri-angular with divergent basal lobes; peduncles shorter than the leaves and bend-ing downward after flowering, spathe green, tightly rolled, enclosing the spadix; occasional in swamps, stream or lake edges, and freshwater tidal marshes; E and W; flr. May; OBL.

Pinellia Ten.

Pinellia ternata (Thunb.) Ten. ex Breitenb.   Pinellia
Perennial arising from a corm and spreading vegetatively by rhizomes and bulb-lets that are produced at the base of the leaflets; the single leaf trifoliate; spadix fused to the spathe at the base and with an elongate tip that extends well beyond the spathe; a rare weed of gardens; SE; flr. May–Jun; native to Japan.

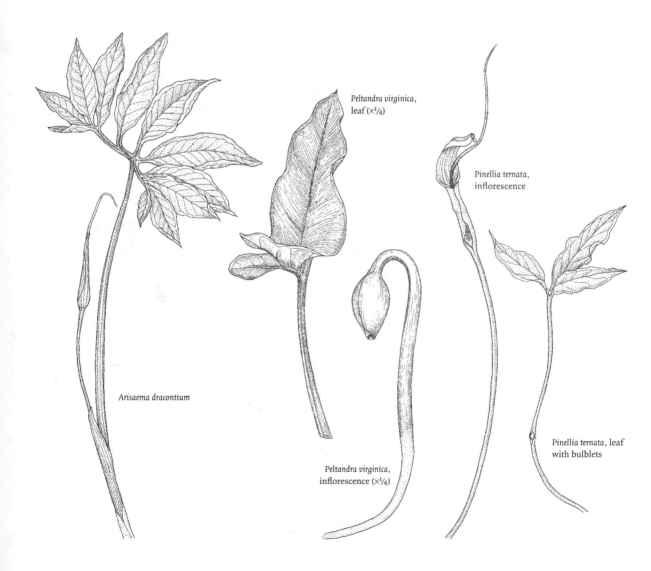

Peltandra virginica,
leaf (×¼)

Pinellia ternata,
inflorescence

Arisaema dracontium

Peltandra virginica,
inflorescence (×¼)

Pinellia ternata, leaf
with bulblets

Symplocarpus Salisb.

Symplocarpus foetidus (L.) Salisb. ex Nutt.   Skunk cabbage
Skunky-smelling (when bruised) wetland plant with large, bright green, net-veined leaves; flowering in the very early spring before the leaves, the mottled-purple, hood-shaped spathe often poking through the ice; spadix globose, enclosed by the spathe; common in moist woods, swamps and bogs; throughout; flr. Feb–Apr; OBL.

BUTOMACEAE Flowering-rush Family

Rhizomatous, emergent aquatic herbs with an erect stem terminated by a cymose umbel subtended by 3 bracts; leaves parallel-veined with an expanded, sheathing base; flowers regular, perfect; sepals 3; petals 3, pink; stamens 9; carpels 6, fused at the base into a ring; fruit a cluster of follicles.

Butomus L.

Butomus umbellatus L.   Flowering-rush
Leaves basal, erect, linear, to 1 m, emersed, floating, or even submersed in deeper water; flowering stem to 1.5 m or more; flowers numerous, 2–2.5 cm wide, sepals and petals persistent; rare along pond margins and in marshes; NW; flr. late Jun–early Aug; native to Eurasia; OBL.

Commelina communis var.
communis

Symplocarpus foetidus,
inflorescence

COMMELINACEAE Spiderwort Family

Herbs with alternate leaves that taper to sheathing bases, the upper leaves becoming sheathing bracts subtending the flowers; flowers perfect, regular, or irregular; petals 3, lavender, blue, or white; fertile stamens 3 or 6; ovary superior, 2–3-celled; fruit a capsule.

A. corolla irregular; fertile stamens 3, filaments not hairy .... *Commelina*
A. corolla regular; fertile stamens 6, filaments hairy .... *Tradescantia*

Commelina L.

Branching herbs, ascending or trailing; leaves narrowed to a short sheathing petiole; flowers subtended by a spathe-like bract, irregular; petals 3, the lower one usually smaller than the other 2; stamens 3, filaments not hairy; flowers remaining open for a single day or less.

A. margins of the floral bracts not fused
   B. lower petal white; leaf sheaths 1–2 cm long .... *C. communis*
   B. all 3 petals blue; leaf sheaths 0.5–1 cm long.............................. *C. diffusa*
A. margins of the floral bracts fused in the lower third
   C. lower petal blue, nearly as large as the others .... *C. virginica*
   C. lower petal white, definitely smaller than the others .... *C. erecta*

Commelina communis L.   Asiatic dayflower
Weak-rooted annual, stem spreading and rooting at the nodes; leaves lanceolate or ovate, sheaths glabrous at the top; flowers with the 2 larger petals blue, the third smaller and white; capsule 2-celled; moist, shaded ground of gardens, woods, roadsides, and stream banks; flr. Jul–Sep; native to Asia; FAC–; 2 varieties:

A. sheaths glabrous on the margins and summits .... var. *communis*
   common; mostly S.
A. sheaths ciliolate .... var. *ludens* (Miquel) C.B.Clarke
   scattered; mostly E.

Commelina diffusa Burm.f.   Creeping dayflower
Annual, diffusely branched and rooting from the lower nodes; all 3 petals blue; capsule 3-celled; very rare in waste ground; flr. Jul–Sep; native to Eurasia; FACW.

Commelina erecta L.   Erect or slender dayflower
Erect perennial to 1.2 m tall from thickened roots; sheaths of the principal leaves white-ciliate, prolonged at the summit into flaring auricles; upper petals blue, the lower one smaller and white; rare in dry sandy soil and rocky banks; SE; flr. late Jun–Sep; believed to be extirpated; 🌿.

Commelina virginica L.   Virginia dayflower
Rhizomatous perennial from thickened roots; stems erect to 1.2 m tall; leaf sheaths with a fringe of red bristles at the top but not prolonged into distinct auricles; floral bracts clustered toward the summit, sessile or short peduncled; petals all blue, nearly equal in size; very rare in moist shores; SE; flr. Jul–Sep; believed to be extirpated; FACW; 🌿.

Tradescantia virginiana (×¼)

Tradescantia L.

Perennial herbs with upright leafy stems; leaves elongate, keeled, sheathing at the base; flowers regular, showy, in umbels subtended by leaf-like bracts; sepals 3; petals 3, blue or purple; stamens 6, the filaments hairy; ovary and capsule 3-celled.

A. sepals and pedicels pubescent; plants not glaucous .... *T. virginiana*
A. sepals glabrous or pubescent at the top only; plants glabrous and glaucous....
.... *T. ohiensis*

Tradescantia ohiensis Raf.   Spiderwort
Plant glabrous and glaucous, 3–9 dm tall; leaves glabrous except the sheaths that are often hairy; pedicels glabrous; sepals with hairs only at the tip; alluvial woods and waste ground; SE and W; flr. May–Jul; native, but also cultivated and occasionally naturalized; FAC; 🌿.

Tradescantia virginiana L.   Spiderwort, widow's-tears
Plant 3–9 dm tall, leaves thin and glabrous; floral bracts usually shorter and broader than the leaves; umbels terminal, solitary; pedicels hairy, sepals green and hairy throughout; occasional on wooded slopes, shale outcrops, moist fields, roadsides, and railroad ballast; S; flr. late Apr–Jun; native, but also cultivated and naturalized; FACU.

CYPERACEAE Sedge Family

Alfred E. Schuyler (except *Carex*)

Herbaceous plants with 3-angled (or rounded) stems; leaves with a narrow, elongate blade and a tubular basal sheath, the blade or sheath sometimes absent; flowers perfect or unisexual, arranged in 1 or more spikelets or spikes, each flower in the axil of a subtending scale; perianth reduced to 1 or more (often 6) bristles, or lacking; stamens 1–3; ovary superior; style 2- or 3-branched; fruit a 3-angled or flattened achene.

A. flowers unisexual; pistillate flowers and fruits enclosed in a sac (perigynium) with a small opening in the tip through which the style protrudes
  B. leaves lacking sheaths or midribs; base of the inflorescence lacking bracts or leaves .... *Cymophyllus*
  B. leaves with sheaths, blades with midribs; base of the inflorescence with bracts or leaves .... *Carex*
A. flowers perfect or unisexual; pistillate flowers and fruits not enclosed in a sac (perigynium)
  C. flowers unisexual; staminate and pistillate flowers in different spikelets; achenes globose, usually whitish and bony in texture at maturity, having a disk (hypogynium) at the base .... *Scleria*
  C. flowers bisexual (sometimes unisexual); in similar spikelets; achenes various, usually not whitish and bony in texture at maturity, lacking a disk at the base
    D. flowers in spikelets with 2-ranked scales
      E. stems round and hollow; inflorescences axillary; flowers and achenes with bristles at the base; achene summit tapering to a persistent style .... *Dulichium*
      E. stems 3-angled and solid; inflorescences terminal; bristles lacking at the base of flowers and achenes; achene summit tapering to the base of the nonpersistent style .... *Cyperus*
    D. flowers in spikelets with spirally arranged scales
      F. numerous elongate bristles present at the base of the flowers and becoming a conspicuous cottony tuft in fruit .... *Eriophorum*
      F. 0–10 straight or twisted rigid bristles present at the base of flowers and fruits
        G. achene summit with a persistent tubercle
          H. leaves consisting of bladeless sheaths; inflorescence a single spikelet without subtending bracts .... *Eleocharis*
          H. leaves with elongate blades; inflorescence having few to many spikelets subtended by bracts
            I. style 3-branched; achene 3-angled; tubercle minute, <½ as wide as the achene .... *Bulbostylis*
            I. style 2-branched; achene 2-angled; tubercle prominent, >½ as wide as the achene .... *Rhynchospora*
        G. achene summit tapering to the style without forming a persistent tubercle
          J. leaf blades <1 mm wide; achenes <1 mm long (*Bulbostylis* may key here if the minute tubercle was not recognized) ....
            .... *Lipocarpha*
          J. leaf blades >2 mm wide or achenes >1 mm long

K. spikelets with 1 perfect flower and 1 achene ....*Cladium*
K. spikelets with many perfect flowers and many achenes
  L. inflorescence a single spikelet, subtended by a scale-like bract; plants of dry woods .... *Trichophorum*
  L. inflorescence having 1–many spikelets, subtended by leaf-like bracts or with one of the bracts appearing as a continuation of the culm; plants of wet or periodically wet places
    M. inflorescence subtended by leaf-like bracts
      N. plants lacking tubers; scales glabrous; achenes <2 mm long
        O. leaves >3 mm wide; stems >2 mm thick; styles about same width from base to tip ...... *Scirpus*
        O. leaves <3 mm wide; stems <2 mm thick; styles broadened at the base................... *Fimbristylis*
      N. plants with tubers; stems sharply angled to the base; scales pubescent; achenes >2 mm long......
      .... *Schoenoplectus*
    M. lowermost inflorescence bract appearing as a continuation of the culm......................... *Schoenoplectus*

Bulbostylis Kunth

Bulbostylis capillaris (×1)

Bulbostylis capillaris, achene (×10)

Bulbostylis capillaris (L.) C.B.Clarke   Sandrush
Tufted, lacking rhizomes; leaves basal with capillary blades; spikelets fewer than 10, mostly sessile at the tips of the stems, subtended by 1 or more erect bracts; open dry rocky, gravelly, or sandy substrates; throughout; frt. Aug–Oct; FACU. [syn: *Stenophyllus capillaris* (L.) Britton]

Carex L.

Paul E. Rothrock

Herbaceous perennials; root stock fibrous and shoots tufted (cespitose), or rhizomatous with solitary or tufted shoots; shoots mostly 3-sided; leaves 3-ranked, the blades linear, usually narrow; leaf sheaths forming a closed cylinder, often with hyaline zone or band on ventral face (i.e., side away from the leaf blade); spikes 1 to many per shoot, round or angular (not flattened) in cross section, usually subtended by a leafy or scale-like bract; plants usually bisexual, flowers strictly unisexual; sepals and petals lacking; male flower consisting of 3 stamens subtended by a scale; female flower consisting of a pistil enclosed in sac-like perigynium; perigynia each subtended by a single scale; stigmas protruding through the terminal orifice of the perigynium; styles deciduous or persistent, slender throughout their length, never forming a tubercle; fruit a brown or black achene.

Recognition of species from this large genus relies heavily on characters of the mature perigynia; flowering or vegetative material should clearly be avoided. Working the keys is greatly aided by 10× magnification, fine forceps, and an accurate millimeter scale. A small number of diagnostic characters are best discerned with even higher magnification (20–30×). In the field be sure to note whether the plants grow in isolated tufts or tend to form colonies through rhizomes.

Key to subgenera

A. inflorescences with male spikes only
  B. shoots in tufts (cespitose); habitat a fen or wet sedgy meadow ....*C. sterilis*, p. 805
  B. shoots solitary, forming a colony by coarse rhizomes; habitat a salted highway right-of-way... *C. praegracilis*, p. 783
A. inflorescences with female spikes, bisexual spikes, or both male and female spikes
  C. spikes 1 per fertile shoot (all the flowers attached to the main stem); stigmas 3 and achenes 3-angled (triangular in cross section) .... subgenus *Carex*, p. 729
  C. spikes 2 or more on each fertile shoot; stigmas and achenes various
    D. stigmas 2 and achenes 2-angled (lenticular in cross section); spikes typically all sessile, similar in appearance, and bisexual (rarely unisexual as in *C. praegracilis* and *C. sterilis*); lower sheaths brown, green, blackened, but never red-tinged .... subgenus *Vignea*, p. 780
    D. stigmas usually 3 and their achenes 3-angled (stigmas 2 and achenes 2-angled in sect. *Racemosae*, species with sheathing lower bracts and pedunculate lower spikes, and sect. *Phacocystis*, species with terminal male spikes and cylindrical lateral spikes); spikes (at least the lower ones) pedunculate, often clearly differentiated into terminal male spikes and lateral spikes that are female or nearly so; lower sheaths sometimes red-tinged .... subgenus *Carex*, p. 729

Subgenus *Carex*

Inflorescences of 1–many spikes; lower bracts sheathing or sheathless; spikes unisexual, the terminal spikes often male, or less often bisexual but typically very different in form from the lateral spikes; lateral spikes usually pedunculate, cylindrical, oblong, ovoid, to globose; achenes 3-sided and stigmas 3 (except in sections *Phacocystis* and *Racemosae*).

Key to sections of *Carex* subgenus *Carex*

A. spikes 1 per fertile shoot (all the flowers attached to the main stem)
  B. spikes often unisexual; perigynia usually minutely pubescent .... sect. *Acrocystis*
  B. spikes bisexual; perigynia glabrous
    C. spikes female above, male at base; perigynia tightly packed together (and thus hiding the lower 3/4 of the perigynium body), the body obovoid or obconic .... sect. *Squarrosae*
    C. spikes male above, female at base; perigynia loosely arranged (most or all of the perigynium body evident), the body variously shaped
      D. perigynia spreading or reflexed at maturity, narrowly lanceolate, 6 mm or more long; pistillate scales soon deciduous, acute .... sect. *Leucoglochin*
      D. perigynia ascending, obovoid, ellipsoid, or globose, <6 mm long; pistillate scales persistent

E. perigynia with beaks >½ the length of the perigynium body; lowest pistillate scales longer (often 2 times or more) than the perigynia, frequently bract-like or even foliaceous .... sect. *Phyllostachyae*

E. perigynia essentially beakless; lowest pistillate scales mostly shorter than the perigynia, always scale-like, thin, and nonfoliaceous

    F. achenes occupying the lower half of the ellipsoid perigynium body; perigynia 2.5–5 mm long; plants of wet habitats .... sect. *Polytrichoideae*

    F. achenes filling the obovoid perigynium body; perigynia 4.5–6 mm long; plants of limestone bluffs .... sect. *Firmiculmes*

A. spikes 2 or more on each fertile shoot

G. stigmas 2, achenes 2-angled (lenticular or planoconvex in cross section)

    H. longest lateral spikes 2.5–10 cm tall, erect or more often drooping; lowest bracts more or less sheath-less; fertile shoots 2–15 dm tall .... sect. *Phacocystis*

    H. longest lateral spikes 0.4–2.5 cm tall, more or less erect; lowest bracts long-sheathing (>4 mm long); fertile shoots 5–55 cm tall .... sect. *Racemosae*

G. stigmas 3, achenes 3-angled (triangular in cross section)

I. leaves and/or sheaths pubescent or scabrous (be sure to check the lower leaf surface and circumference of the leaf sheath near its summit)

    J. lowest bracts more or less sheathless (sheaths 0–3 mm long); perigynia beakless or beaked

        K. perigynia with beaks >0.5 mm long; perigynium bodies filled to the summit by the achene; terminal spikes male....sect. *Halleranae*

        K. perigynia essentially beakless; perigynium bodies not tightly filled with the achene, at least the summit empty; terminal spikes male or more often female above, male at base

            L. lateral spikes 1–2.5 mm wide, the lower ones borne on slender, drooping peduncles; perigynia glabrous .... sect. *Hymenochlaenae* (*C. aestivalis*)

            L. lateral spikes 2–6.5 mm wide, the lower ones erect or spreading; perigynia glabrous or pubescent .... sect. *Porocystis*

    J. lowest bracts long-sheathing (>3 mm long); perigynia beaked

    M. beaks sharply bidentate, the teeth 0.8–3 mm long; plants rhizomatous; styles persistent, continuous in texture and color with the achene .... sect. *Carex*

    M. beaks entire or shallowly bidentate; plants tufted; styles deciduous, jointed near the base

        N. spikes erect, the terminal one male; perigynia 3–9 per spike, nerves numerous (about 25 or more), closely spaced, and often impressed (giving a corrugated appearance) .... sect. *Griseae* (*C. hitchcockiana*)

        N. spikes drooping, the terminal one with at least a few perigynia above; perigynia usually >9 per spike, their nerves 2–12 and raised .... sect. *Hymenochlaenae*

I. leaves *and* sheaths glabrous

O. lower bracts with obvious sheaths (>3 mm long) but rudimentary blades (0–10 mm long)

P. leaf blades 8–32 mm wide; female spikes 8–30 mm long ....
.... sect. *Careyanae* (*C. plantaginea*)
P. leaf blades 0.3–4 mm wide; female spikes <8 mm long
  Q. basal sheaths, bract sheaths, and male spikes red-tinged; leaf
  blades flat, 1.5–4 mm wide .... sect. *Clandestinae*
  Q. basal sheaths, bract sheaths, and male spikes brown or pale;
  leaf blades filiform, 0.3–1 mm wide .... sect. *Albae*
O. lower bracts sheathless or sheathing and with well-developed blades
(rarely bracts are apparently lacking)
  R. perigynia pubescent and distinctly beaked
    S. larger female spikes <1.5 cm long; perigynium bodies filled by
    the achene .... sect. *Acrocystis*
    S. larger female spikes 1.5 cm long or longer; perigynium bodies
    not tightly filled by the achene, at least the summit empty
      T. spikes 1.5–4.5 mm wide, the lower ones nodding on slen-
      der peduncles; perigynia strongly ascending ....
      .... sect. *Hymenochlaenae* (*C. debilis*)
      T. spikes 5–41 mm wide, the lower ones more or less erect;
      perigynia spreading
        U. perigynia >5 mm long
          V. female spikes globular, 2.6–4.1 cm wide; plants
          tufted .... sect. *Lupulinae* (*C. grayi*)
          V. female spikes more or less cylindrical, 8–15 mm
          wide; plants rhizomatous ....
          .... sect. *Carex* (*C. trichocarpa*)
        U. perigynia <5 mm long
          W. leaf blades 5–18 mm wide, upper surface with 2
          prominent midlateral veins; perigynia sparsely pu-
          bescent, not obscuring the 2 ribs and 2–6 smaller
          nerves .... sect. *Anomalae*
          W. leaf blades 0.7–5 mm wide, lateral veins all of simi-
          lar diameter; perigynia densely pubescent, obscuring
          the many nerves .... sect. *Paludosae*
  R. perigynia glabrous, beaked or indistinctly beaked
    X. lowest pistillate scales longer (often 2 times or more) than the
    perigynia, frequently bract-like or even foliaceous; bracts ap-
    parently lacking, the peduncles arising from sheaths of basal
    leaves .... sect. *Phyllostachyae*
    X. lowest pistillate scales mostly shorter than the perigynia, al-
    ways scale-like, thin, and nonfoliaceous; bracts present, pe-
    duncles arising from bracts or sometimes from sheaths of
    basal leaves
      Y. beaks 0.7 mm long or longer and sharply, stiffly bidentate;
      perigynia 4–20 mm long; styles usually persistent, continu-
      ous in texture and color with the achene (styles deciduous,
      jointed near the base in *Hymenochlaenae* and *Spirostachyae*)
        Z. terminal spikes female above (with 35 or more perig-
        ynia), male at base .... sect. *Squarrosae*
        Z. terminal spikes male or with only a few perigynia
          AA. lateral spikes drooping on long, slender peduncles

BB. perigynia with 2–3 strong ribs and 0–8 weak nerves; pistillate scales obtuse to acuminate or smooth awned .... sect. *Hymenochlaenae*

BB. perigynia coarsely 6–20-nerved; pistillate scales with long, rough awns .... sect. *Vesicariae*

AA. lateral spikes erect or somewhat spreading on short, erect peduncles

CC. perigynia subulate or narrowly lanceolate (4–14 times as long as wide), the beaks gradually tapered

DD. beak teeth erect or spreading; leaf blades 6–14(17) mm wide .....
.... sect. *Folliculatae*

DD. beak teeth sharply reflexed; leaf blades 1.5–5 mm wide ....
.... sect. *Collinsiae*

CC. perigynia obovoid, ellipsoid, or ovoid (<4 times as long as wide), the beaks abruptly narrowed from perigynium body

EE. largest perigynia 10–20 mm long; terminal male spikes 1 ....
.... sect. *Lupulinae*

EE. largest perigynia 4–10 mm long; terminal male spikes 1–6

FF. lateral spikes subglobose to ellipsoid, 1.2–2 times as long as wide

GG. leaves wiry and strongly involute, 0.8–2.5 mm wide; female spikes 3–15 flowered; perigynia ascending, broadly ovoid, 2.5–3.2 mm wide ....
.... sect. *Vesicariae (C. oligosperma)*

GG. leaves more or less flat, 1.5–5 mm wide; female spikes 15–many flowered; perigynia horizontally spreading or reflexed, ellipsoid to obovoid, 1–2 mm wide ....
.... sect. *Spirostachyae*

FF. lateral spikes cylindrical, mostly 2 or more times as long as wide

HH. perigynium bodies obovoid or obconic; pistillate scales longer than the perigynia and rough-awned; plants tufted .... sect. *Squarrosae (C. frankii)*

HH. perigynium bodies ovoid-lanceolate, ovoid, or ellipsoid (widest at or below the middle); pistillate scales mostly shorter than the perigynia, acute to rough-awned; plants tufted or rhizomatous

II. pistillate scales obtuse to acuminate; perigynium walls thin and papery .... sect. *Vesicariae*

II. pistillate scales rough-awned; perigynium walls thin and papery to thick and firm

JJ. male spikes 1 per inflorescence; perigynia inflated, the walls thin and papery, nerves coarse, 7–25; plants tufted (except for the rare *C. schweinitzii*) .... sect. *Vesicariae*

JJ. male spikes 2–4 per inflorescence; perigynia with thick, firm walls, nerves fine (sometimes obscure), 14–25; plants rhizomatous ....
.... sect. *Paludosae*

Y. beaks lacking or short (i.e., <0.7 mm long) or, if >0.7 mm long, then more or less entire at apex; perigynia 2–7 mm long (longer in several species with lateral spikes on long drooping peduncles); styles usually deciduous, jointed near the base

    KK. perigynia distinctly beaked (at least ¼ as long as perigynium body) *and* horizontally spreading or reflexed; lateral spikes erect, densely subglobose to ellipsoid (1.2–2 times as long as wide); plants of wet calcareous habitat .... sect. *Spirostachyae*

    KK. perigynia beakless or ascending; lateral spikes erect, spreading or drooping, usually cylindrical; plants of various wet to dry habitats

        LL. lower bracts sheathless or sheaths only 1–3 mm long

            MM. perigynia beakless or with obscure beaks <0.5 mm long

                NN. terminal spikes usually male; lateral spikes drooping on slender peduncles; roots covered by a brown felt.......... .... sect. *Limosae*

                NN. terminal spikes female above, male at base; lateral spikes erect; roots without brown coating

                    OO. pistillate scales awned or narrowly acuminate, longer than the perigynia; perigynia ellipsoid, finely papillose (20×); plants long-stoloniferous.... .... sect. *Atratae*

                    OO. pistillate scales short-mucronate or acute, mostly shorter than the perigynia; perigynia obovoid, smooth; plants tufted .... sect. *Shortianae*

            MM. perigynia with distinct, usually slender beaks (in *C. prasina* the beak is broadly triangular, gradually tapered, more or less flattened-bent), 0.5 mm long or longer

                PP. leaf blades wiry and strongly involute, 0.8–2.5 mm wide; perigynia 2.5–3.2 mm wide; styles persistent, continuous in texture and color with the achene .... sect. *Vesicariae* (*C. oligosperma*)

                PP. leaves thin (thick in *C. tonsa* with leaf blades up to 2.5–5 mm wide) and flat, V-, or W-shaped in cross section; perigynia 1–2.6 mm wide; styles deciduous, jointed near the base

                    QQ. fertile shoots of varying lengths with at least some spikes more or less sessile and hidden among leaf bases; spikes ovoid, <2 times as long as wide; leaf blades hard and firm .... sect. *Acrocystis* (*C. tonsa*)

                    QQ. fertile shoots all elongate above the leaf bases, the lower spikes drooping on long, slender peduncles; spikes more or less cylindrical (2–7 times as long as wide); leaf blades flexible and thin .... sect. *Hymenochlaenae*

        LL. lower bracts long-sheathing, >3 mm long

            RR. lateral spikes (especially the lower ones) drooping on long, slender peduncles

SS. perigynia more or less beakless, many-nerved and conspicuously 3-angled at maturity; female spikes 0.6–2(3) cm long, the lowest often borne in the axils of basal leaves; terminal spikes always male .... sect. *Careyanae*

SS. perigynia beaked and usually few-nerved or beakless and round in cross section; female spikes 1–8 cm long, the lowest always borne in the axils of cauline bracts; terminal spikes female above, male below or all male .... sect. *Hymenochlaenae*

RR. lateral spikes erect or spreading on usually short peduncles

TT. nerves of perigynia numerous (about 25 or more), closely spaced, and often impressed (giving a corrugated appearance); awns of pistillate scales rough or ciliate .... sect. *Griseae*

TT. nerves of perigynia several–25, slightly raised; awns of pistillate scales absent or smooth

UU. lowest female spikes borne in the axils of basal leaves; perigynia sharply 3-angled at maturity (the faces flat to slightly convex) ... . sect. *Careyanae*

UU. lowest female spikes borne in the axils of cauline bracts; perigynia obtusely 3- angled or round

VV. perigynia convex to rounded at base, broadly ellipsoid to subglobose in overall shape and nearly round in cross section, deep olive-brown at maturity; apex of perigynia broadly conical, terminating in a minute beak .... sect. *Granulares*

VV. perigynia concave, acute or tapering at base, lanceolate, fusiform, or obovoid in overall shape and obtusely 3-sided, greenish, pale brown, or straw-colored at maturity; apex of perigynia often with a slender or sometimes strongly bent beak

WW. plants tufted, fertile shoots easily compressed and relatively weak, sometimes nearly wing-edged; leaf blades 1–52 mm wide; perigynia smooth .... sect. *Laxiflorae*

WW. plants rhizomatous (caution: they may be deep-seated), fertile shoots firm and slender; leaf blades 1.5–6(7) mm wide; perigynia with many to few fine papillae (30×), especially near or on the beak .... sect. *Paniceae*

Section *Acrocystis*

Plants of mostly low stature, tufted to rhizomatous; leaves and sheaths more or less glabrous; bracts sheathless; spikes 1–several per inflorescence, the terminal one normally male; perigynia pubescent (except sometimes glabrous in *C. tonsa*), 2-ribbed and stipitate; beaks bidentate; the achenes 3-sided, filling the perigynium body; styles deciduous; stigmas 3.

A. fertile shoots all elongate (5–40 cm long) and of more or less uniform length, few or no inflorescences on very short peduncles or hidden at the base of the plant

B. perigynium bodies (excluding stipitate base and beak) about as wide as long, subglobose to ovoid
- C. plants without elongate rhizomes; widest leaves (especially overwintering leaves) ca. 3–5 mm wide, their ligules conspicuously longer than wide; bracts beneath middle female spikes with lobes at base of awn ... . *C. communis*
- C. plants with elongate, red-tinged rhizomes; widest leaves 1.3–3 mm wide, their ligules wider than long; bracts beneath middle female spikes tapering to the awn
  - D. beaks 0.2–0.8 mm long, 0.1–0.5 times as long as the body; peduncles usually smooth below the inflorescence .... *C. pensylvanica*
  - D. beaks 0.9–2 mm long, 0.5–1 times as long as the body; peduncles usually strongly scabrous (20×) below the inflorescence .... *C. lucorum*
B. perigynium bodies (excluding stipitate base and beak) much longer than wide, ellipsoid
- E. widest leaves (especially overwintering leaves) ca. 3–5 mm wide, their ligules conspicuously longer than wide; bracts beneath middle female spikes with lobes at base of awn .... *C. communis*
- E. widest leaves <3 mm wide, their ligules wider than long; bracts beneath middle female spikes tapering to the awn
  - F. shoots loosely tufted or with slender rhizomes; lowest female spikes on short peduncle ca. 1 mm long .... *C. novae-angliae*
  - F. shoots densely tufted; lowest female spikes sessile
    - G. fertile shoots firm, more or less erect and surpassing the leaves; spikes (especially the lowest one) separated; staminate scales obtuse to broadly acute, the midrib ending below the apex .... *C. albicans*
    - G. fertile shoots lax, arching, and often shorter than the leaves; spikes usually crowded; staminate scales sharply acute, the midrib prominent .... *C. emmonsii*
A. fertile shoots of varying lengths, many inflorescences on very short peduncles (0–5 cm long), some or all often hidden at the base of the plant
- H. pistillate scales purplish, rarely pale; fertile shoots bearing multiple spikes on a single peduncle .... *C. nigromarginata*
- H. pistillate scales green to reddish-brown; fertile shoots bearing 2–4 peduncles, the longest bearing a male spike and female spikes, the shorter ones bearing female spikes
  - I. beaks 0.4–1 mm long, 0.2–0.45 times as long as the perigynium body . ... *C. umbellata*
  - I. beaks 1–2 mm long, 0.45–1 times as long as the perigynium body .... *C. tonsa*

Carex albicans Spreng.   Sedge

Carex albicans,
perigynium (×3)

Carex albicans,
pistillate scale (×3)

Tufted; fertile shoots more or less erect, firm, 5–50 cm tall, usually longer than the leaves; leaf blades 0.5–2.5 mm wide, ligules wider than long; inflorescences 0.8–3 cm long; spikes sessile, the lowest usually separated; scales of male spikes each with midrib weak or absent below the apex; pistillate scales white hyaline or reddish-brown-tinged, lanceolate, and equal to or longer than the perigynium body; perigynia 2.3–3.2 by 0.7–1.1 mm, the bodies ellipsoid; beaks 0.5–0.9 mm long; dry, wooded slopes; common S, occasional N.

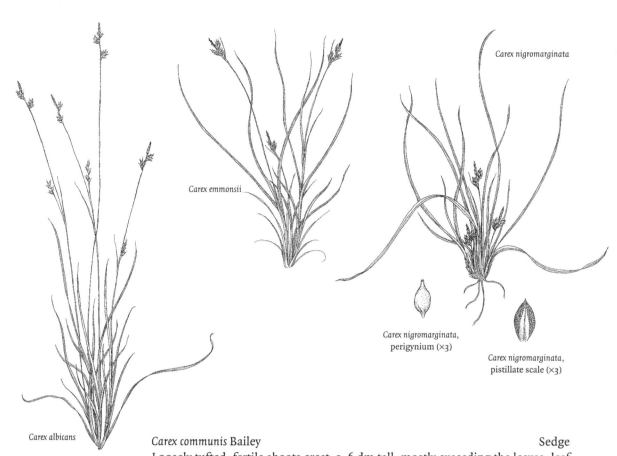

Carex nigromarginata

Carex emmonsii

Carex nigromarginata,
perigynium (×3)

Carex nigromarginata,
pistillate scale (×3)

Carex albicans

Carex communis,
perigynium (×3)

Carex communis,
pistillate scale (×3)

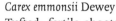

Carex emmonsii,
perigynium (×3)

Carex emmonsii,
pistillate scale (×3)

Carex lucorum,
perigynium (×3)

Carex lucorum,
pistillate scale (×3)

Carex communis Bailey — Sedge

Loosely tufted; fertile shoots erect, 1–6 dm tall, mostly exceeding the leaves; leaf blades (1.3)2–5 mm wide, ligules longer than wide; inflorescences 2–5 cm long, bracts from middle of inflorescences scariously lobed at base of awns; spikes aggregated or the lower more or less separated, the lowest often peduncled (1–7 mm long); pistillate scales reddish- or purple-brown-tinged, ovate and as long as to longer than the perigynium body; perigynia 2.5–3.4 by 1.1–1.4 mm, the bodies ovoid to subglobose; beaks 0.4–0.9 mm long; dry to mesic woods and clearings; common; throughout.

Carex emmonsii Dewey — Sedge

Tufted; fertile shoots lax and slender, 3–25 cm tall, usually surpassed by the leaves; leaf blades 0.5–2.2 mm wide, ligules wider than long; inflorescences 0.8–2 cm long; spikes aggregated, sessile; scales of male spikes each with prominent midrib and sharply pointed apex; pistillate scales white-hyaline or reddish-brown-tinged, lanceolate, longer than the perigynium body; perigynia 1.8–3 by 0.7–1 mm, the bodies ellipsoid; beaks 0.4–0.7 mm long; dry, acidic woods; frequent; mostly E.

Carex lucorum Link — Sedge

Rhizomatous; fertile shoots erect, 1–5.5 dm tall, usually longer than the leaves and strongly scabrous below the inflorescence; leaf blades 0.7–3(3.4) mm wide, ligules wider than long; inflorescences 1–5 cm long, bracts from middle of inflorescence tapering to an awn; spikes aggregated or the lower more or less separated, sessile; pistillate scales reddish-brown- or purple-tinged, ovate or lance-ovate, and longer than the perigynium body; perigynia 2.8–3.8 by (1)1.2–1.7 mm,

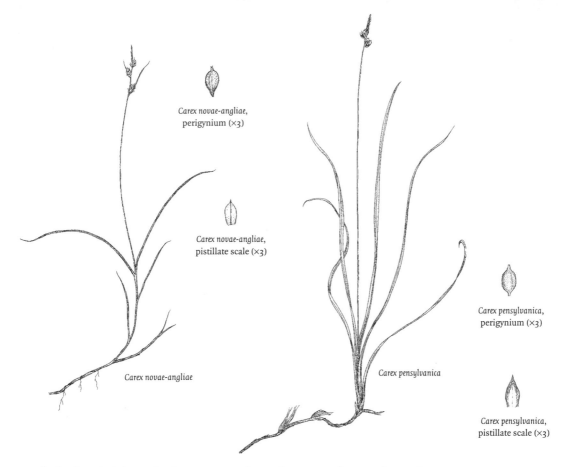

Carex novae-angliae,
perigynium (×3)

Carex novae-angliae,
pistillate scale (×3)

Carex novae-angliae

Carex pensylvanica

Carex pensylvanica,
perigynium (×3)

Carex pensylvanica,
pistillate scale (×3)

the body subglobose; beaks 0.9–2 mm long; dry, open often sandy woods; rare to infrequent N and W, infrequent elsewhere.

Carex nigromarginata Schwein. Sedge
Tufted with short, ascending rhizomes; fertile shoots erect or lax, 2–25 cm tall, ranging (often on the same plant) from much shorter to about as long as the leaves; leaf blades 1.5–4 mm wide, ligules wider than to as wide as long; inflorescences 0.5–1.5 cm long; spikes aggregated, sessile; pistillate scales purplish, rarely pale, ovate, and longer than the perigynium body; perigynia 2.8–3.6 by 0.8–1.3 mm, the body ellipsoid; beaks 0.8–1.2 mm long; dry wooded slopes; occasional; SE and SC; UPL.

Carex novae-angliae Schwein. Sedge
Loosely tufted or with slender rhizomes; fertile shoots slender and weak, 5–40 cm tall, usually longer than the leaves; leaf blades 0.7–1.5 mm wide, ligules mostly wider than long; inflorescences 1.5–4 cm long; lower spikes widely separated and often on short (about 1 mm) peduncles; pistillate scales reddish-brown, lance-ovate, and longer than the perigynium body; perigynia pale, 2–2.6 by 0.7–1 mm, the body ellipsoid; beaks 0.4–0.7 mm long; wet woods and thickets; locally frequent; N; FACU.

Carex pensylvanica Lam. Sedge
Rhizomatous; fertile shoots erect, 1–4.5 dm tall, usually longer than the leaves and smooth below the inflorescence; leaf blades 0.5–3(3.4) mm wide, ligules wider than long; inflorescences 1–4.5 cm long, bracts from middle of inflores-

cence tapering to an awn; spikes aggregated or the lower more or less separated, sessile; pistillate scales reddish-brown- or purple-tinged, ovate and longer than the perigynium body; perigynia 2.2–3.2 by 1.1–1.5(1.7) mm, the body subglobose; beaks 0.2–0.8(0.9) mm long; dry, open woods and wooded slopes; common; throughout.

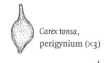

Carex tonsa,
perigynium (×3)

Carex tonsa,
pistillate scale (×3)

Carex tonsa (Fernald) E.P.Bicknell   Sedge

Tufted, sometimes forming dense mats; fertile shoots erect 2–15 cm tall. much exceeded by the leaves; leaf blades mostly 1.5–5 mm wide, ligules much wider than to as wide as long; inflorescences ca. 0.5–3 cm long; spikes aggregated but borne on 2–4 separate peduncles emerging from the leaf sheath; pistillate scales green to reddish-brown, ovate and about as long as the perigynia; perigynia 3–4 by 1–1.2 mm, the body subglobose to ellipsoid; beaks 1–2 mm long; dry woods, hillsides, and railroad banks; occasional; mostly E; 2 varieties:

A. leaves thick, stiff, the widest blades 2.5–5 mm wide; perigynia usually glabrous .... var. *tonsa*
A. leaves thin, relatively soft, the widest blades 1.5–3 mm wide; perigynia pubescent .... var. *rugosperma* (Mack.) Crins

Carex umbellata Willd.   Sedge

Tufted, sometimes forming dense mats; fertile shoots erect, 2–15 cm tall, much exceeded by the leaves; leaf blades 1.5–3 mm wide, ligules much wider than to as

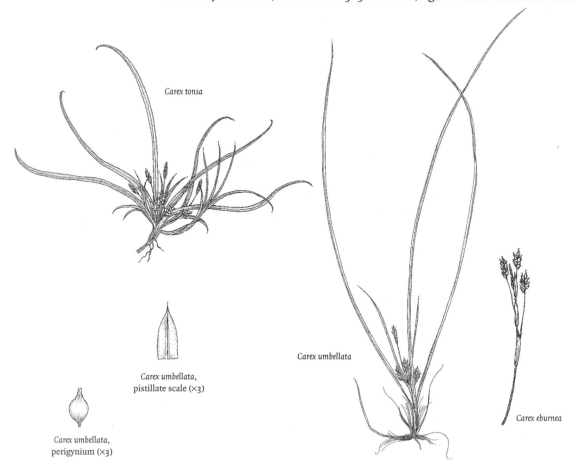

Carex tonsa

Carex umbellata,
pistillate scale (×3)

Carex umbellata

Carex umbellata,
perigynium (×3)

Carex eburnea

wide as long; inflorescences ca. 0.5–1.5 cm long; spikes aggregated but borne on 2–4 separate peduncles emerging from the leaf sheath; pistillate scales green to reddish-brown, ovate, and longer than the perigynium body; perigynia 2.5–3.5 by 0.9–1.2 mm, the body ellipsoid; beaks 0.4–1 mm long; dry, open woods and fields; frequent; mostly SE.

Section *Albae*

Carex eburnea Boott   Ebony sedge

Rhizomes long and wiry; fertile shoots 1–3.5 cm tall, filiform; leaves glabrous, filiform, 0.3–1 mm wide; lower bracts bladeless, sheathing, 3–10 mm long in vigorous shoots; terminal spikes sessile, male, overtopped by the long-peduncled female spikes; pistillate scales obtuse or broadly acute; perigynia glabrous, 1.6–2.1 by 0.7–1 mm, sharply 3-angled; beak short, ca. 0.3 mm long; achenes 3-sided, filling the body of the perigynia; styles deciduous; stigmas 3; limestone ledges; rare; Erie, Centre, Huntingdon, Lehigh, and Northampton Cos.; FACU; 🌿.

Carex eburnea,
perigynium (×3)

Carex eburnea, pistillate
scale (×3)

Section *Anomalae*

Carex scabrata Schwein.   Sedge

Rhizomes long, tough; fertile shoots 2.5–8 dm tall; leaves 5–18 mm wide, ribbon-like with 2 prominent midlateral veins on upper surface, blades and sheaths glabrous; bracts sheathless, the longest ones about 2 times as long as the inflorescence; terminal spikes staminate, short-peduncled; pistillate scales acute to short-awned, margins ciliate; perigynia pubescent, 2-ribbed with 2–6 intermediate nerves, 2.5–4 by 1.2–2 mm; beaks 0.8–2 mm long, distinctly bidentate; achenes 3-sided; styles deciduous; stigmas 3; wet woods, streambanks, and swamps; common; throughout; OBL.

Section *Atratae*

Carex buxbaumii Wahlenb.   Brown sedge

Long stoloniferous; fertile shoots 2.5–10 dm tall, the base filamentous and red-tinged; leaves glabrous, the blades 1.5–4 mm wide; bracts sheathless; terminal spikes female above, male at base; pistillate scales dark-purple-tinged (except for the pale midrib), mostly awned or narrowly acuminate, as long as or longer than the perigynia; perigynia elliptical, finely 10- or more-nerved, finely papillose (20×), 2.5–3.6 by 1–1.8 mm; beaks lacking to about 0.2 mm long and obscurely bidentate; achenes 3-sided; styles deciduous; stigmas 3; calcareous swamps, swales, and meadows; infrequent; S and Lycoming Co.; OBL; 🌿.

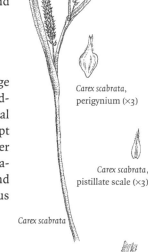

Carex scabrata,
perigynium (×3)

Carex scabrata,
pistillate scale (×3)

Carex scabrata

Carex scabrata,
fragment of stem

Section *Carex*

Rhizomatous; shoot bases mostly reddish and fibrillose; vegetative shoots with leaves mostly clustered at apex; hyaline band of leaf sheaths not red-dotted (though sometimes tinged with red above); lower bracts sheathing; terminal 1–6 spikes male, elevated above lateral spikes; lateral spikes female, mostly cylindrical, erect; pistillate scales acute to rough-awned; perigynia ovoid to ovoid-

Carex buxbaumii,
perigynium (×3)

Carex buxbaumii,
pistillate scale (×3)

Carex buxbaumii

lanceolate, glabrous or pubescent, many-ribbed and thick-walled, much larger than the achene, round or obtusely 3-sided in cross section; beaks long (about 1.7–4.6 mm); beak teeth long (about 1–3 mm); styles usually persistent and continuous with the 3-sided achenes; stigmas 3. Key based on draft by A. A. Reznicek.

A. ventral hyaline band of leaf sheaths pubescent at summit
  B. perigynia glabrous, (7)8–10 mm long; beaks smooth, longest teeth 1.5–3 mm long .... C. atherodes
  B. perigynia usually pubescent, 4.8–7.5 mm long; beaks scabrous, teeth 0.8–1.7 mm long .... C. hirta
A. ventral hyaline band of leaf sheaths glabrous or scabrous at summit
  C. vegetative shoots hard, center solid with tissue; longest ligules 2–12(17) mm long; perigynia pubescent, their longest beak teeth (0.8)1.1–2.3(2.7) mm long .... C. trichocarpa
  C. vegetative shoots soft, with a large hollow center; longest ligules (6)11–45 mm long; perigynia glabrous, their longest beak teeth 1.5–3 mm long .....
  .... C. atherodes

Carex atherodes,
perigynium (×3)

Carex atherodes,
pistillate scale (×3)

Carex atherodes Spreng.   Awned sedge
Fertile shoots 3–15 dm tall; ventral hyaline band of leaf sheaths glabrous to pubescent and pale at summit; leaf blades glabrous on upper surface, pilose on lower surface, 3.5–10 mm wide; ligules to (6)11–45 mm long; perigynia glabrous, (7)8–10 by 1.5–2.7 mm; beaks smooth, 2.5–4.6 mm long, their teeth 1.5–3 mm long; open seepy slopes; rare; Erie and Lawrence Cos.; OBL; ❧.

Carex hirta L.   Sedge
Fertile shoots (1)2–10 dm tall; ventral hyaline band of leaf sheaths pubescent and pale at summit; leaf blades pilose, 2–6 mm wide; ligules 2–12(17) mm long; perigynia pubescent, 4.8–7.5 by 1.8–2.5 mm; beaks scabrous, 2.5–4.6 mm long, their teeth 0.8–1.7 mm long; introduced into waste places, dry fields, meadows, and sandy alluvium; Philadelphia region; native to Europe.

Carex trichocarpa Schkuhr   Sedge
Fertile shoots 5–12 dm tall; ventral hyaline band of leaf sheaths glabrous and red-tinged at summit; leaf blades glabrous, 3–7.5 mm wide; ligules 2–12(17) mm long; perigynia pubescent, 6–10.2 by 1.9–3.2 mm; beaks 2.5–4.6 mm long, their teeth (0.8)1.1–2.3(2.7) mm long; marshes and wet meadows; frequent; throughout except SW; OBL.

Section *Careyanae*

Tufted; lower bracts with well-developed sheaths; terminal spikes male, lateral spikes female, lowest female spikes borne in the axil of basal leaf; pistillate scales acute to awned, shorter than the perigynia; perigynia glabrous, sharply 3-sided (the faces flat to slightly convex), many-nerved; beaks lacking or short and tapered, erect or slightly bent; achenes 3-sided, filling the body of the perigynia; styles deciduous; stigmas 3.

A. base of plant purple; male spikes purple-tinged
  B. perigynia 5–7 mm long; bracts with well-developed blades ...... C. careyana

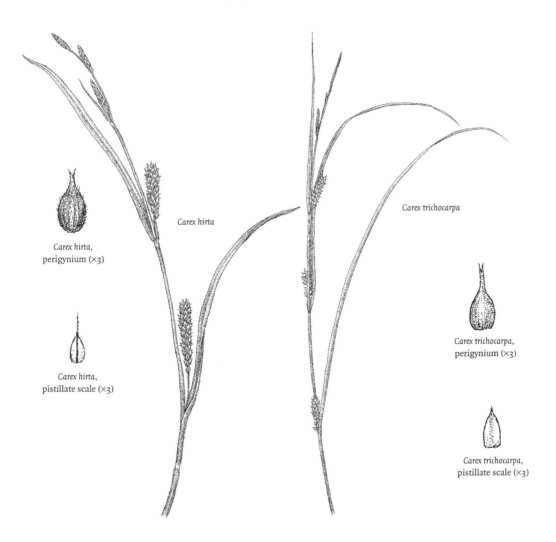

Carex hirta, perigynium (×3)

Carex hirta, pistillate scale (×3)

Carex hirta

Carex trichocarpa

Carex trichocarpa, perigynium (×3)

Carex trichocarpa, pistillate scale (×3)

B. perigynia 4–5 mm long; bracts with essentially bladeless sheaths ....
.... *C. plantaginea*
A. base of plant greenish or greenish-white; male spikes pale or whitish
  C. blades of lower bracts much shorter than the inflorescence; leaves of sterile shoots (7)10–28 mm wide .... *C. platyphylla*
  C. blades of lower bracts mostly as long as or longer than the inflorescence; leaves of sterile shoots 2–10(12) mm wide
    D. staminate spikes sessile, inconspicuous or nearly so, hidden by the uppermost pistillate bracts and 1–2 pistillate spikes; lower bracts long, surpassing the inflorescence by 5–10 cm .... *C. abscondita*
    D. staminate spikes on evident peduncles 2–95 mm long, conspicuous, not hidden by pistillate bracts or adjacent pistillate spike; lower bracts shorter than to 5(7) cm longer than the inflorescence
      E. widest leaf blades 2.5–6 mm wide; lateral spikes female, without staminate flowers at their base .... *C. digitalis*
      E. widest leaf blades 6–10(12) mm wide, glaucous; lateral spikes with 1–2 staminate flowers or empty scales at the base .... *C. laxiculmis*

Carex abscondita Mack.   Sedge

Fertile shoots 6–30 cm tall, greenish or greenish-white at base; leaf blades 3–4.5 mm wide; blades of lower bracts long, usually 5–10 cm longer than the inflorescence and often hiding the spikes; male spikes sessile, pale; female spikes 6–12 mm long, the upper 2 approximate; perigynia 2.5–4.2 by 1.4–2 mm; moist to wet woods; coastal plain and scattered in eastern ⅔ of the state; FAC.

Carex careyana Dewey   Carey's sedge

Fertile shoots 3–6.5 cm tall, purple at base; leaf blades of sterile shoots 4–18 mm wide, those of fertile shoots narrower; blades of lower bracts much shorter than the inflorescence; 1–4 times as long as their sheaths; male spikes purple-tinged, their peduncles 8–80 mm long; female spikes 7–18 mm long; perigynia 5–7 by 2.5–3 mm; rich, calcareous woods; rare; Allegheny, Bedford, Fayette, and McKean Cos.; ❧.

Carex digitalis Willd.   Sedge

Fertile shoots 7–55 cm tall, greenish or greenish white at base; leaf blades deep-green, 1–6 mm wide; blades of lower bracts about as long as to longer than the inflorescence; male spikes pale, their peduncles 4–87 mm long; lateral spikes completely female, 6–18 mm long, the lowermost borne on slender drooping peduncles; perigynia 2–3.2 by 1.2–1.8 mm; dry woods; common; throughout; UPL.

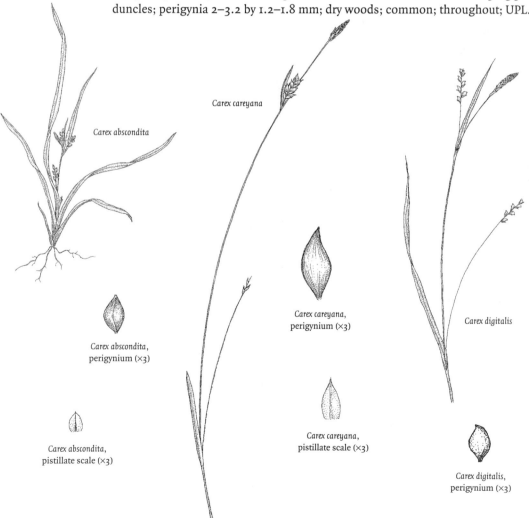

Carex careyana

Carex abscondita

Carex careyana,
perigynium (×3)

Carex digitalis

Carex abscondita,
perigynium (×3)

Carex careyana,
pistillate scale (×3)

Carex abscondita,
pistillate scale (×3)

Carex digitalis,
perigynium (×3)

Carex laxiculmis Schwein. Sedge
Fertile shoots 9–40 cm tall, greenish or greenish-white at base; leaf blades 6–10(12) mm wide; blades of lower bracts shorter than to as long as the inflorescence; male spikes pale, their peduncles 2–95 mm long; lateral spikes female with 1–2 staminate flowers (or empty scales) at the base, 6–20 mm long, peduncles slender, weak; perigynia 2.5–4 by 1.5–2 mm; 2 varieties:

Carex laxiculmis var. laxiculmis, perigynium (×3)

A. leaves glaucous green .... var. *laxiculmis*
    rich, dry woods; common; throughout.
A. leaves deep-green .... var. *copulata* (Bailey) Fernald
    rich calcareous woods; occasional; NE, SC, SW.

Carex laxiculmis var. laxiculmis, pistillate scale (×3)

Carex plantaginea Lam. Plantain sedge
Fertile shoots 2.5–5.5 dm tall, purple at base; leaf blades 8–32 mm wide; bracts purple-tinged, the sheaths essentially bladeless; male spikes purple-tinged, their peduncles 0–50 mm long; female spikes 8–30 mm long; perigynia 4–5 by 1.6–2 mm; woods and wooded slopes; frequent N, SC, and SW, rare SE.

Carex plantaginea, perigynium (×3)

Carex plantaginea, pistillate scale (×3)

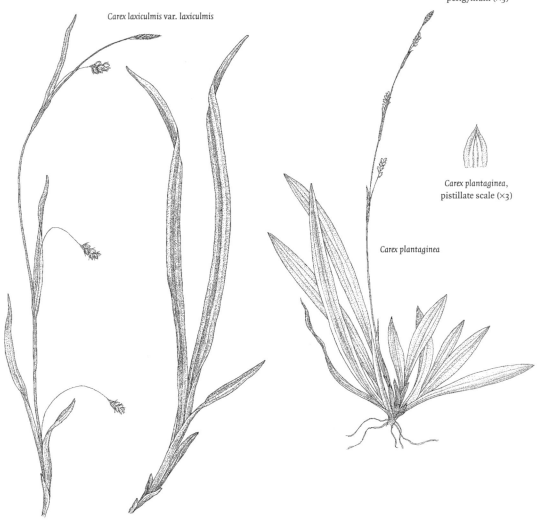

Carex laxiculmis var. laxiculmis

Carex plantaginea

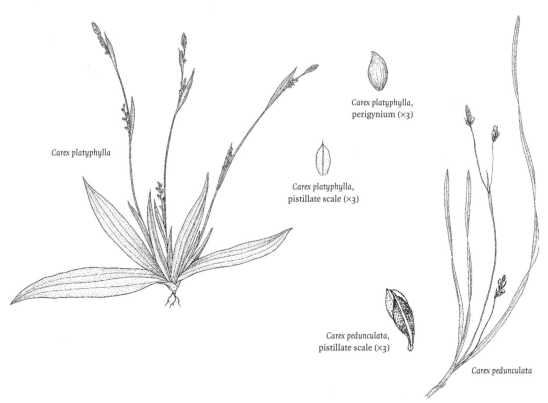

Carex platyphylla,
perigynium (×3)

Carex platyphylla

Carex platyphylla,
pistillate scale (×3)

Carex pedunculata,
pistillate scale (×3)

Carex pedunculata

Carex platyphylla Carey   Broad-leaf sedge
Fertile shoots 1.5–4 dm tall, greenish or greenish-white at base; leaf blades of
sterile shoots 7–30 mm wide, those of fertile shoots up to 5 mm wide; blades of
lower bracts much shorter than the inflorescence; male spikes pale, their peduncles 1–8 mm long; female spikes 6–16 mm long; perigynia 2.8–4 by 1.6–2
mm; rich woods and wooded slopes; frequent; throughout.

Section *Clandestinae*

Bracts sheathing, red-tinged, their blades 10 mm long or less; terminal spikes
male or with a few perigynia at the base, red-tinged; perigynia ovoid to obovoid,
minutely pubescent to glabrous; beaks lacking or indistinct, 0–0.5 mm long;
achene 3-sided, filling the perigynium body; styles deciduous; stigmas 3.

A. terminal spikes with a few perigynia at base; pistillate scales truncate and
    awned; lowest spikes on capillary peduncles arising from leaf sheaths ....
    .... *C. pedunculata*
A. terminal spikes male; pistillate scales acute or obtuse; lowest spikes arising
    from bracts .... *C. richardsonii*

Carex pedunculata Willd.   Sedge
Loosely tufted; fertile shoots 5–30 cm tall, often much exceeded by leaves, red-
tinged at the base; leaf blades 2–3 mm wide; terminal spikes usually with a few
perigynia basally; lowest lateral spikes borne on long, capillary peduncles arising from leaf sheaths; pistillate scales truncate and awned, red-tinged, about
as long as the perigynia; perigynia sparsely pubescent or glabrous, 3.4–4.5 by
1.4–1.7 mm; rich rocky, wooded slopes or swampy woods; rare SW, occasional
elsewhere.

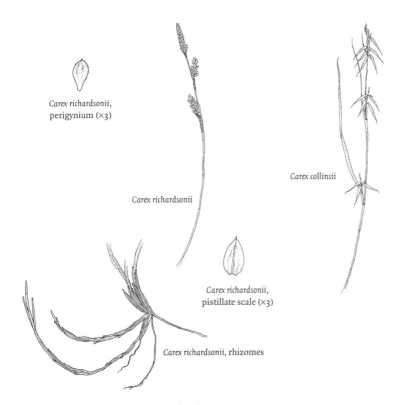

Carex richardsonii,
perigynium (×3)

Carex richardsonii

Carex collinsii

Carex collinsii,
perigynium (×3)

Carex richardsonii,
pistillate scale (×3)

Carex richardsonii, rhizomes

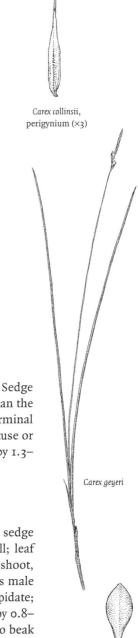

Carex geyeri

Carex geyeri,
perigynium (×3)

Carex geyeri, pistillate
scale (×3)

Carex richardsonii R.Br.   Sedge
Rhizomatous; fertile shoots 1.2–3.5 dm tall, about as long as or longer than the leaves, brown or red-tinged at the base; leaf blades 1.5–4 mm wide; terminal spikes male; lowest lateral spikes arising from bracts; pistillate scales obtuse or acute, red-tinged, longer than the perigynia; perigynia pubescent, 2–3.2 by 1.3–1.5 mm; serpentine barrens; Chester Co.; UPL; ❦.

Section *Collinsiae*

Carex collinsii Nutt.   Collin's sedge
Tufted; fertile shoots slender, relatively weak and arching, 1.3–8 dm tall; leaf blades 1.5–5 mm wide, flat; lower bracts long-sheathing; spikes 2–5 per shoot, terminal ones male, 5–10 mm long, sessile or pedunculate; lateral spikes male above, 1–8 perigynia at base; pistillate scales persistent, awned or cuspidate; perigynia glabrous, deep-green, spreading or reflexed at maturity, 9–13 by 0.8–1.3 mm; beaks very gradually tapered, distance from summit of achene to beak apex 6.5–9.2 mm, the teeth abruptly reflexed; achenes 3-sided, the styles persistent; stigmas 3; sphagnum bogs and swampy woods; rare; Carbon, Monroe, Pike, and Schuylkill Cos.; OBL; ❦.

Section *Firmiculmes*

Carex geyeri Boott   Geyer's sedge
Short-rhizomatous; fertile shoots 1–4 dm tall; leaf blades 1–3.5 mm wide, lowest ones bladeless or blades greatly reduced; bracts lacking; spikes 1 per fertile shoot, male above, 1–3 perigynia at base; pistillate scales not foliaceous, usually

Carex folliculata

Carex folliculata,
perigynium (×1½)

longer and wider than the perigynia; perigynia obovoid, about 4.5–6 by 1.8–2.6 mm; beaks lacking; achenes 3-sided, filling the body of the perigynia; styles deciduous; stigmas 3; dry, wooded limestone bluff; Centre Co.; 🍂.

Section *Folliculatae*

Carex folliculata L.   Sedge

Tufted; fertile shoots relatively stout, 3–12 dm tall; leaf blades glabrous, 6–14(17) mm wide, flat; lower bracts long-sheathing; terminal spikes male; lateral spikes female; pistillate scales awned or cuspidate; perigynia glabrous, yellow-green, lanceolate, spreading, 10–15 by 2–3.4 mm; beaks very gradually tapered, distance from summit of achene to beak apex 5–8.5 mm; achenes 3-sided; styles persistent; stigmas 3; bogs, swamps, and wet woods; frequent except in extreme SW; OBL.

Section *Granulares*

Carex granularis Willd.   Sedge

Tufted; fertile shoots 2–9 dm tall, brownish at base; leaf blades 2.5–9 mm wide; lower bracts long-sheathing; terminal spikes male, sessile or short-peduncled; lateral spikes erect, cylindrical, (4)5–6 mm thick, the perigynia densely arranged; perigynia glabrous with many raised nerves, more or less round in cross section, 2.5–3.6 by 1.5–2.5 mm, deep-olive brown at maturity; beaks minute or indistinct; achenes 3-sided, not filling the body of the perigynia; styles deciduous; stigmas 3; FACW+; 2 varieties:

A. perigynia 2.5–4 mm long, 1.5–2.5 mm wide, wide-spreading; female spikes (4)5–6 mm thick .... var. *granularis*
calcareous wet meadows, swales and woods; frequent to common S, scattered N.

A. perigynia 2–2.5(2.8) mm long, 1–1.5 mm wide, ascending; female spikes 3–5 mm thick.. var. *haleana* (Olney) Porter
wet meadows, swales, and wet cliffs of limestone regions; scattered; mostly S.

Section *Griseae*

Tufted; lower bracts with well-developed sheaths; leaves and bracts glabrous (except scabrous in *C. hitchcockiana*); terminals spikes male, lateral spikes female; pistillate scales mostly rough-awned; perigynia glabrous, inflated (closely enveloping the achene in *C. hitchcockiana*), round or obtusely triangular in cross section, the many veins becoming impressed giving the surface a corrugated appearance; achenes 3-sided; styles deciduous; stigmas 3. Treatment based on Naczi (1999).

A. shoot bases red
   B. perigynia spirally arranged in the spikes; sheaths of lowest bracts loose and enlarged upward; ligules 2–13 mm long
      C. perigynia more or less round in cross section, the largest 2–2.6 mm wide; achene bodies 2.2–3 mm long (excluding stipitate base and

apiculum formed by style), 5–10 times as long as the stipitate base ......
.... *C. grisea*

  C. perigynia obtusely triangular in cross section, the largest 1.5–2(2.2) mm wide; achene bodies 1.8–2.4 mm long (excluding stipitate base and apiculum formed by style), 3–6 times as long as the stipitate base ....
.... *C. amphibola*

B. perigynia arranged in a single plane giving the fresh spike a flattened appearance; sheaths of lowest bracts tight; ligules seldom reaching 3.5 mm long

  D. perigynia 2–2.6 times as long as wide, the bodies usually abruptly contracted to a distinct beak 0.5–1 mm long; longest female spikes 4–8(10)-flowered ....*C. oligocarpa*

  D. perigynia 2.5–3.3 times as long as wide, the bodies gradually tapered to an indistinct beak 0.1–0.4 mm long; longest female spikes (5)7–14-flowered .... *C. planispicata*

A. shoot bases brown

  E. sheaths of leaves and bracts scabrous; perigynia 4.5–5.6(6.2) mm long with prominent (0.5)0.8–1.3-mm-long beaks .... *C. hitchcockiana*

  E. sheaths of leaves and bracts glabrous; perigynia <4.5 mm long or, if longer, then beaks obscure or lacking, 0–0.2 mm long

    F. leaves and bracts glaucous green; awns of pistillate scales 0–1 mm long
.... *C. glaucodea*

    F. leaves and bracts deep or light green; awns of pistillate scales usually 1.2–3.8 mm long

      G. perigynia 18–40 per spike, 2.8–3.6(4.3) mm long, 17–25-nerved; peduncles of lateral spikes rough .... *C. conoidea*

      G. perigynia 5–16 per spike, mostly 4–5 mm long, 40–65-nerved; peduncles of lateral spikes smooth

        H. perigynia more or less round in cross section, the largest 2–2.6 mm wide; achene bodies 2.2–3 mm long (excluding stipitate base and apiculum formed by style), 5–10 times as long as the stipitate base .... *C. grisea*

        H. perigynia obtusely-triangular in cross section, the largest 1.5–2(2.2) mm wide; achene bodies 1.8–2.4 mm long (excluding stipitate base and apiculum formed by style), 3–6 times as long as the stipitate base .... *C. amphibola*

Carex granularis var. granularis

Carex granularis var. granularis, perigynium (×3)

Carex granularis var. granularis, pistillate scale (×3)

Carex amphibola Steud.   Sedge
Fertile shoots 1.4–8 dm tall, their base red or brown; leaf blades glabrous, 3.3–6.8 mm wide; bract sheaths loose and enlarged upward; perigynia 5–16 per spike, spirally arranged, 4.2–5 by 1.5–2(2.2) mm; beaks 0–0.2 mm long; dry to moist woods, meadows, and swales; common SE and SC, occasional elsewhere; FAC. [*C. amphibola* Steud. var. *rigida* (Bailey) Fernald of Rhoads & Klein, 1993]

Carex amphibola, perigynium (×3)

Carex conoidea Willd.   Sedge
Fertile shoots 1.2–8 dm tall, their bases brown; leaf blades glabrous, 1.3–4 mm wide; bract sheaths tight and minutely serrulate on edges; perigynia 8–25 per spike, spirally arranged, 2.8–3.6(4.3) by 1.2–1.7 mm; beaks 0–0.2 mm long; meadows, and swales; mostly SE; FACU.

Carex conoidea, perigynium (×3)

Carex conoidea, pistillate scale (×3)

Carex glaucodea,
perigynium (×3)

Carex glaucodea,
pistillate scale (×3)

Carex glaucodea Tuck.   Sedge

Fertile shoots 1–5 dm tall, their bases brown; leaf blades glaucous-green, glabrous, 4.3–10.8 mm wide; bract sheaths loose and enlarged upward; perigynia 10–45 per spike, spirally arranged, 3.2–4 by 1.5–2.5 mm; beaks 0–0.3 mm long; dry to moist calcareous woods or fields; frequent; mostly S.

Carex grisea Wahlenb.   Sedge

Fertile shoots 1.6–9 dm tall, their bases red or brown; leaf blades glabrous, 3.6–9 mm wide; bract sheaths loose and enlarged upward; perigynia 5–16 per spike, spirally arranged, (4)4.3–5(5.6) by 2–2.6 mm; beaks 0–0.2 mm long; dry to moist woods, meadows, and swales; occasional; throughout; FAC.

Carex hitchcockiana,
perigynium (×3)

Carex hitchcockiana,
pistillate
scale (×3)

Carex hitchcockiana Dewey   Sedge

Fertile shoots 1.2–7 dm tall, their bases brown; leaf blades scabrous, 2.4–5.4(6.5) mm wide; bract sheaths tight; perigynia 3–9 per spike, spirally arranged, 4.5–5.6(6.2) by 1.9–2.3 mm; beaks (0.5)0.8–1.3 mm long; rocky, limestone woods and slopes; infrequent; mostly S.

Carex oligocarpa Willd.   Sedge

Fertile shoots 1.4–6 dm tall, their bases red; leaf blades glabrous, 2.2–6.5 mm wide; bract sheaths tight; perigynia 4–8(10) per spike, distichous (i.e., in a single plane), 4.2–5 by 1.6–2 mm; beaks abruptly contracted, (0.2)0.4–1 mm long; dry limestone woods, slopes, and thickets; locally frequent; mostly SE and SC.

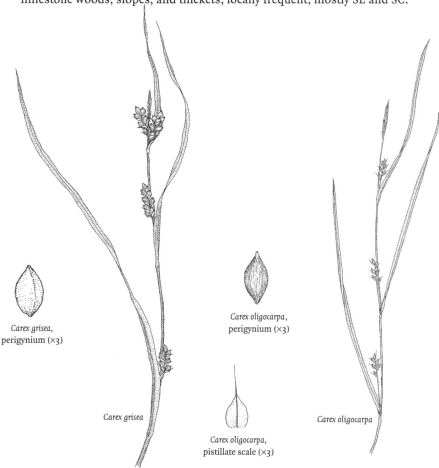

Carex grisea,
perigynium (×3)

Carex grisea

Carex oligocarpa,
perigynium (×3)

Carex oligocarpa,
pistillate scale (×3)

Carex oligocarpa

Carex planispicata Naczi   Sedge

Fertile shoots 1.4–6 dm tall, their bases red; leaf blades glabrous, 2.2–6.5 mm wide; bract sheaths tight; perigynia (5)7–14 per spike, distichous, 4.2–5 by 1.6–1.8 mm; beaks indistinct, 0–0.2 mm long; dry to moist woods; locally frequent; mostly SE and SC; FAC. [*C. amphibola* Steud. var. *amphibola* of Rhoads and Klein, 1993]

Carex planispicata,
perigynium (×3)

Carex planispicata,
pistillate scale
(×3)

Section *Halleranae*

Carex hirtifolia Mack.   Sedge

Loosely tufted or short-rhizomatous; fertile shoots 3–8 dm tall; leaves and sheaths pubescent, the blades 2.5–9 mm wide; bracts essentially sheathless; terminal spikes male, sessile or short-peduncled; lateral spikes female; pistillate scales cuspidate to awned; perigynia pubescent, 3.4–4.3 by 1.2–1.6 mm, sharply 3-angled; beaks slender, 0.8–1.3 mm long; achenes 3-sided, filling the body of the perigynia; style deciduous; stigmas 3; dry woods; common SE, scattered elsewhere.

Section *Hymenochlaenae*

Mostly tufted; lower bracts long-sheathing to sheathless; terminal spikes female above and male at base or all male; lateral spikes cylindrical or oblong, the lower ones on slender drooping peduncles; perigynia glabrous (except in *C. debilis* var. *pubera*); achenes 3-sided, not filling the summit of the perigynia; styles deciduous; stigmas 3.

Carex hirtifolia

A. sheaths and leaf blades glabrous
  B. perigynia spreading and their beaks abruptly contracted, slender, 1.8–3.8(5) mm long; base of plants pale or brown, densely fibrous, short-rhizomatous .... *C. sprengelii*
  B. beaks spreading to ascending and beakless or with gradually tapering beaks 0.5–1.5 mm long; base of plants reddish or brown, at most only slightly fibrous, tufted
    C. lower bract sheaths 0–1.2 cm long; shoots brown at base; perigynia with 2 sharp ribs (otherwise nerveless), their beaks often symmetrical and nearly as long as the body .... *C. prasina*
    C. lower bract sheaths (1)1.5–8 cm long; shoots reddish at base; perigynia 3–8-nerved, beakless or with shorter, symmetrical beaks
      D. beaks lacking; perigynia 2.4–3.6 mm long, round in cross section....
      .... *C. gracillima*
      D. beaks present; perigynia 3.2–9.3 mm long, 3-angled in cross section
        E. leaf blades of vegetative shoots 4–10 mm wide, distinctly wider than those of fertile shoots; perigynia stipitate, 3.2–4.8 mm long, their achenes more or less sessile; pistillate scales cuspidate or awned .... *C. arctata*

Carex hirtifolia,
perigynium (×3)

        E. leaf blades of vegetative shoots 1.8–4.7 mm wide, similar in width to those of fertile shoots; perigynia sessile, 4.7–9.3 mm long, their achenes on a definite slender stipe 0.5–1.3 mm long; pistillate scales obtuse to acuminate, rarely awned .... *C. debilis*
A. sheaths and/or leaf blades pilose

Carex hirtifolia,
pistillate scale (×3)

F. perigynia 2.5–3 mm long, beakless; leaf blades 1.5–3 mm wide ....
.... *C. aestivalis*

F. perigynia 3.5–6 mm long, with short, shallowly bidentate beaks; leaf
blades 3–8 mm wide

   G. pistillate scales long-attenuate to long-awned, about as long as or
longer than the perigynia; lateral spikes female throughout and borne
on peduncles shorter than the spike .... *C. davisii*

   G. pistillate scales acute to obtuse (sometimes with a short awn), shorter
than the perigynia; lateral spikes with several male flowers (or their
empty scales) at the base and borne on peduncles much longer than the
spike .... *C. formosa*

Carex aestivalis Schwein.   Sedge

Fertile shoots 3–6 dm tall, reddish at base; leaf sheaths and blades pilose, the
blades 1.5–3 mm wide; sheaths of lower bracts 0–5 cm long; terminal spikes female above and male at base; lateral spikes female throughout, 20–40 by 1–2.5
mm; pistillate scales obtuse to awned; perigynia ascending, obscurely 3-sided in
cross section, 2.5–3 by 0.8–1.3 mm, nerves 7–10; beaks lacking; achenes sessile
or on very short (about 0.3 mm) stipe; dry woods; infrequent; N and SC. Putative
hybrids between *C. aestivalis* and *C. gracillima* have been reported for 6 widely
scattered counties.

Carex aestivalis

Carex arctata Boott   Sedge

Fertile shoots 2–10 dm tall, reddish at base; leaf sheaths and blades glabrous,
the blades of fertile shoots 2.5–4 mm wide, those of vegetative shoots 4–10 mm
wide; sheaths of lower bracts 2–8 cm long; terminal spikes male; lateral spikes
female throughout, 25–80 by 2–5 mm; pistillate scales cuspidate or awned;
perigynia ascending to spreading, short (about 0.5 mm) stipitate, sharply 3-
angled in cross section, 3.2–4.8 by 1–1.6 mm, nerves 3–8; beaks gradually contracted, 0.5–1.4 mm long; achenes sessile; rich, dry to moist woods and clearings; locally frequent; N and Cambria Co.; OBL.

Carex aestivalis,
perigynium (×3)

Carex davisii Schwein. & Torr.   Sedge

Fertile shoots 3–9 dm tall, reddish at base; leaf sheaths and blades pilose, the
blades 3–8 mm wide; sheaths of lower bracts 2–8 cm long; terminal spikes female above and male at base; lateral spikes female throughout, 15–35 by 3.5–6.5
mm; pistillate scales long-attenuate to long-awned; perigynia more or less ascending, sessile, obscurely 3-sided in cross section, 4.5–6 by 2–2.5 mm, nerves
6–12; beaks usually abruptly contracted, 0.4–1 mm long; achenes on very short
stipes about 0.2–0.4 mm long; rich woods and stream banks; frequent SE, rare
elsewhere; FAC–.

Carex aestivalis,
pistillate scale (×3)

Carex debilis Michx.   Sedge

Fertile shoots 2.5–12 dm tall, reddish at base; leaf sheaths and blades glabrous,
the blades 1.8–4.7 mm wide; sheaths of lower bracts 1–5 cm long; terminal
spikes male; lateral spikes female throughout, 25–60 by 1.5–4.5 mm; pistillate
scales obtuse to acuminate, sometimes awned; perigynia glabrous, strongly ascending, sessile, obscurely 3-angled in cross section, 6.8–9.3 mm by 1–1.8 mm,
2–3 strong and 4–8 weaker nerves; beaks about 0.8–2.3 mm long; achenes on
short stipes about 0.5–1.3 mm long; FAC; 3 varieties:

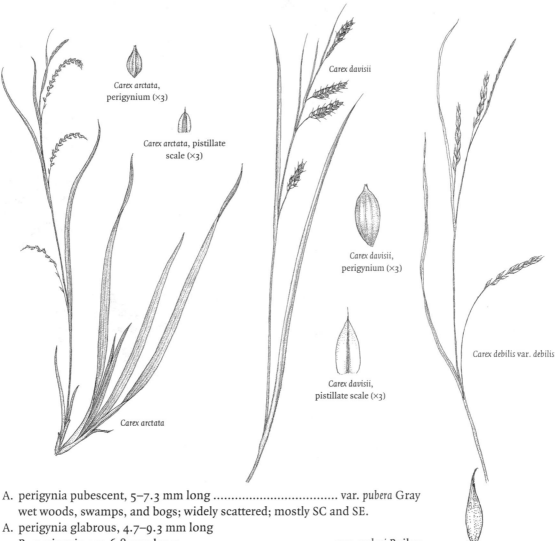

Carex arctata,
perigynium (×3)

Carex arctata, pistillate
scale (×3)

Carex arctata

Carex davisii

Carex davisii,
perigynium (×3)

Carex davisii,
pistillate scale (×3)

Carex debilis var. *debilis*

Carex debilis var. *debilis,*
perigynium (×3)

Carex debilis var. *rudgei,*
perigynium (×3)

Carex formosa,
perigynium (×3)

Carex formosa,
pistillate scale (×3)

A. perigynia pubescent, 5–7.3 mm long .... var. *pubera* Gray
　　wet woods, swamps, and bogs; widely scattered; mostly SC and SE.
A. perigynia glabrous, 4.7–9.3 mm long
　　B. perigynia 4.7–6.8 mm long .... var. *rudgei* Bailey
　　　　moist, open or rocky woods; common; throughout.
　　B. perigynia 6.8–9.3 mm long .... var. *debilis*
　　　　swamps, thickets, and low woods; frequent; mostly SE.

Carex formosa Dewey   Handsome sedge
Fertile shoots 3–8 dm tall, reddish at base; leaf sheaths and blades pilose, the
blades 3–8 mm wide; sheaths of lower bracts 1.5–6 cm; terminal spikes largely
male with a few perigynia above; lateral spikes with several male flowers (or
empty scales) at the base, 15–30 by 3.5–6 mm; pistillate scales acute to obtuse;
perigynia ascending, sessile, obscurely 3-angled in cross section, 3.5–5 by 1.6–
2.2 mm, 2–3 strong nerves; beaks gradually tapered, 0.3–0.6 mm long; achenes
sessile; dry calcareous woods; rare; Centre Co.; FAC; ✿.

Carex gracillima Schwein.   Sedge
Fertile shoots 2–10 dm tall; reddish at base; leaf sheaths and blades glabrous,
the blades 2.8–7(9) mm wide; sheaths of lower bracts 1.5–8 cm long; terminal
spikes female above and male at base or sometimes all male; lateral spikes female throughout, 10–50 by 2–4 mm; pistillate scales obtuse to cuspidate;

Carex gracillima,
perigynium (×3)

Carex gracillima,
pistillate scale (×3)

perigynia ascending, sessile, round in cross section, 2.4–3.6 by 1–1.7 mm, nerves 4–8; beaks lacking; achenes sessile; dry to moist woods and meadows; common; throughout; FACU.

Carex prasina Wahlenb.   Sedge

Fertile shoots 2.5–8 dm tall, mostly brownish at base; leaf sheaths and blades glabrous, the blades 2.5–4 mm wide; sheaths of lower bracts 0–1.2 cm long; terminal spikes all male or female above and male at base; lateral spikes female throughout, 20–60 by 3.5–6.5 mm; pistillate scales cuspidate or awned; perigynia spreading, sessile, sharply 3-angled in cross section, 3–4 by 1–1.9 mm, 2–3-nerved at edges otherwise essentially nerveless; beaks gradually tapered and twisted, nearly as long as the perigynium body, about 0.5–1.5 mm long; achenes sessile; wet woods, thickets, and stream banks; common; throughout; OBL.

Carex sprengelii Spreng.   Sedge

Rhizomatous; fertile shoots 3–10 cm tall, dull-brown or pale and densely fibrous at base; leaf sheaths and blades glabrous, the blades 2–4 mm wide; sheaths of lower bracts 0–0.5 cm; terminal spikes male; lateral spikes female throughout or male above, 10–35 by 8–11 mm; pistillate scales awned to acute; perigynia

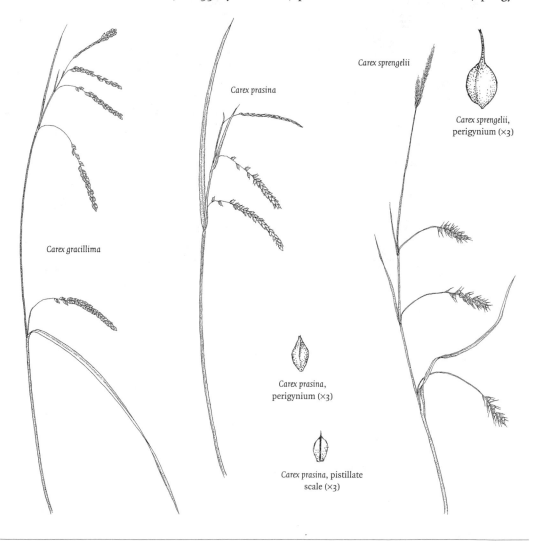

Carex gracillima

Carex prasina

Carex sprengelii

Carex sprengelii,
perigynium (×3)

Carex prasina,
perigynium (×3)

Carex prasina, pistillate
scale (×3)

spreading, sessile, round in cross section, 5–7(8) by 1.3–2.5 mm, 2–3 ribs (otherwise essentially nerveless); beaks abruptly contracted and about as long as the perigynium body, 1.8–3.8(5) mm; achenes sessile; rich alluvial thickets, stream banks, moist limestone outcrops, and grassy banks; scattered; E; FACU; 🍂.

Section *Laxiflorae*

Tufted; lower bracts with well-developed sheaths; terminal spikes male, lateral spikes female, lowest female spikes borne in the axils of cauline bracts; perigynia glabrous, stipitate, obtusely 3-sided (the faces convex below); achenes 3-sided, filling the body of the perigynia; styles deciduous; stigmas 3.

In order to evaluate color at the base of the plant, it is critical to collect or observe the lowest scaly sheaths at the soil line.

A. perigynia with only 2 distinct nerves (and sometimes several additional faint ones on each face), often sharply angled above .... *C. leptonervia*

A. perigynia with 6–7 conspicuous nerves on each face and (except for *C. styloflexa*) obtusely angled above

  B. beaks short and usually strongly bent (almost at right angles to axis of perigynium); edges of lower bract sheaths and culms serrulate

    C. bracts 5–20 mm wide; widest leaves of sterile shoots 12–52 mm wide, forming a more or less vase-shaped tuft; pistillate scales mostly obtuse .... *C. albursina*

    C. bracts 0.5–6 mm wide; widest leaves 3–15 mm wide, sterile shoots forming an elongated base; pistillate scales mucronate or awned

      D. plants pale brown or greenish-white at base; peduncles of male spikes 0–17 mm long; staminate scales pale at maturity .... *C. blanda*

      D. plants purple or wine-red at base; peduncles of male spikes 5–106 mm long; staminate scales orange-tinged at maturity ....
.... *C. gracilescens*

  B. beaks tapering and slightly angled or straight; edges of lower bract sheaths and culms usually smooth

    E. female spikes 5–15 mm long, the lower on capillary, spreading or drooping peduncles; perigynia fusiform, becoming sharply angled above .... *C. styloflexa*

    E. female spikes 10–50 mm long, the lower on erect, stiff peduncles; perigynia obovoid (or fusiform in *C. striatula*)

      F. plants sparingly purple or wine-red at base; perigynia <2 times as long as wide, abruptly contracted into a very short beak ....
.... *C. ormostachya*

      F. plants brownish or pale at base; perigynia usually >2 times as long as wide and tapered to a prominent straight or slightly angled beak

        G. largest perigynia <4 mm long, obovoid; male spikes on peduncles <22 mm long .... *C. laxiflora*

        G. largest perigynia 4–5 mm long, fusiform; male spikes on peduncles 22–120 mm long .... *C. striatula*

Carex albursina,
pistillate scale (×3)

Carex albursina E.Sheld.   Sedge
Fertile shoots 1–6.5 cm tall, their bases brown, green, or sometimes purple; leaf blades wide, those of sterile shoots 12–52 mm wide; bracts 5–20 mm wide, the edges of their sheaths serrulate; male spikes sessile; female spikes 6–36 mm

Carex albursina,
perigynium (×3)

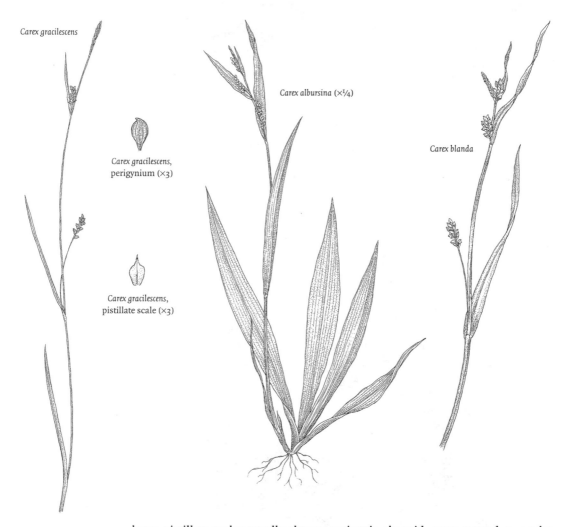

Carex gracilescens

Carex alibursina (×¼)

Carex blanda

Carex gracilescens,
perigynium (×3)

Carex gracilescens,
pistillate scale (×3)

Carex blanda,
perigynium (×3)

Carex blanda, pistillate
scale (×3)

long; pistillate scales usually obtuse; perigynia obovoid, many-nerved, 3–4.2 by 1.8–2.1 mm; beaks short, abruptly bent; dry to moist, rich woods; occasional; throughout, mostly SW.

Carex blanda Dewey   Sedge
Fertile shoots 1.5–5.5 dm tall, pale brown or greenish-white at base; leaf blades 1–10(15) mm wide; bracts 2–6 mm wide, the edges of their sheaths serrulate; peduncle of male spike 0–17 mm long; staminate scales pale; female spikes 15–18 mm long; pistillate scales mucronate to awned; perigynia obovoid, many-nerved, 2.4–2.6 by 1–1.2 mm; beaks short, abruptly bent; dry to moist woods, thickets, and meadows; common SE and SC, occasional elsewhere; FAC.

Carex gracilescens Steud.   Sedge
Fertile shoots 1.3–8 dm tall, purple or wine-red at base; leaf blades 1–5(7) mm wide; bracts 0.5–3 mm wide, the edges of their sheaths serrulate; peduncles of male spike 5–106 mm long; staminate scales orange-tinged; female spikes 5–27 mm long; pistillate scales mucronate to awned; perigynia obovoid, many-nerved, 2.8–3 by 1.5–1.8 mm; beaks short, abruptly bent; rich woods; common; throughout.

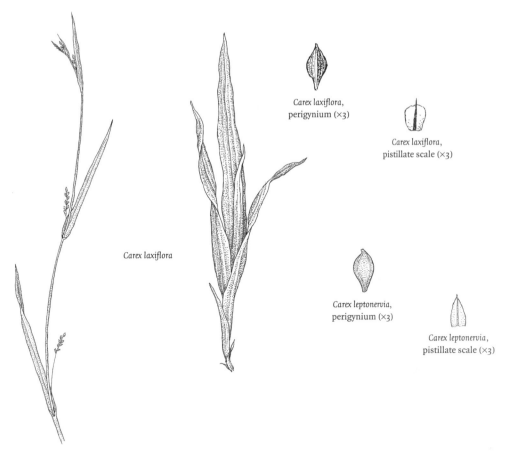

Carex laxiflora

Carex laxiflora,
perigynium (×3)

Carex laxiflora,
pistillate scale (×3)

Carex leptonervia,
perigynium (×3)

Carex leptonervia,
pistillate scale (×3)

Carex laxiflora Lam. — Sedge

Fertile shoots 1.5–5.5 dm tall, pale or greenish-white at base; leaf blades of sterile shoots 7–26 mm wide, those of fertile shoots up to 8 mm wide; bracts 0.7–7 mm wide, the edges of their sheaths smooth or rarely serrulate; peduncle of male spike 10–22 mm long; female spikes 10–33 mm long; pistillate scales mucronate or awned; perigynia obovoid, many-nerved, 2.6–4 by 1.2–1.6 mm; beaks gradually tapered, straight or slightly bent; rich woods; common; throughout; FACU.

Carex leptonervia (Fernald) Fernald — Sedge

Fertile shoots 1.2–4.5 dm tall, brownish or pale at base; leaf blades 3–10 mm wide; bracts 0.5–6 mm wide, the edges of their sheaths serrulate; peduncle of male spike about 2 mm long; female spikes 9–22 mm long; pistillate scales awned or obtuse; perigynia more or less obovoid, 2 distinct veins (and sometimes additional faint ones), 2.2–3.2 by 1–1.5 mm; beaks straight or slightly bent, often with sharply angled edges; moist woods and thickets; frequent; mostly N and SC; FACW.

Carex ormostachya Wiegand — Spike sedge

Fertile shoots 2.5–4.5 dm tall, sparingly purple or wine-red at base; leaf blades (2)3.5–12 mm wide; bracts 1–5 mm wide, the edges of their sheaths smooth; peduncle of male spike 2–26 mm long; female spikes 8–28 mm long; pistillate scales mostly awned; perigynia obovoid, many-nerved, 2.2–3 by 1.2–1.6 mm;

Carex ormostachya,
perigynium (×3)

Carex ormostachya,
pistillate scale (×3)

beaks very short, abrupt, straight or slightly bent; rocky woods and shale barrens; scattered; throughout; 🌱.

Carex striatula,
perigynium (×3)

Carex striatula,
pistillate scale (×3)

Carex striatula Michx.   Sedge

Fertile shoots 3–6.5 dm tall, pale or greenish-white at base; leaf blades 3–19 mm wide; bracts 0.5–7 mm wide, the edges of their sheaths smooth; peduncle of male spike 4–120 mm long; female spikes 20–62 mm long; perigynia fusiform, many-nerved, 3.4–5 by 1.2–2 mm; beaks straight or slightly bent; rich, open woods; frequent; mostly SE.

Carex styloflexa Buckley   Sedge

Fertile shoots 2.5–8.5 dm tall, pale or greenish-white at base; leaf blades 2.5–8.7 mm wide; bracts 1–3 mm wide, the edges of their sheaths smooth; peduncle of male spike 0–90 mm long; female spikes 5–15 mm long; perigynia fusiform, many-nerved, 3.5–5.5 by 1.2–1.6 mm; beaks long, straight or slightly bent, 2 strong nerves forming green, sharply angled edges; wet woods; frequent; mostly SE; FACW–.

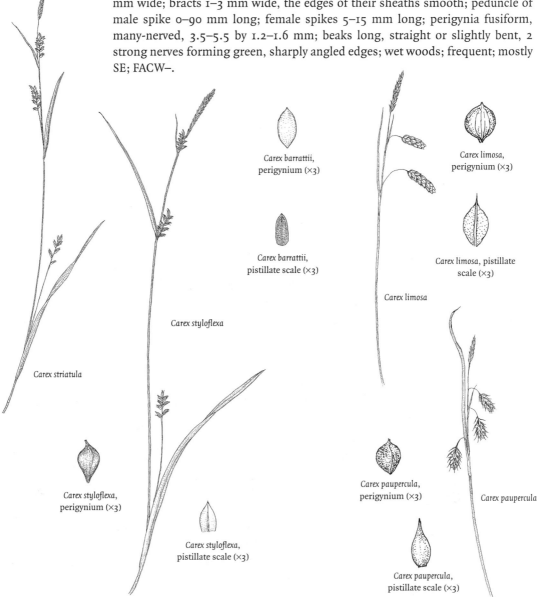

Carex striatula

Carex styloflexa

Carex styloflexa,
perigynium (×3)

Carex styloflexa,
pistillate scale (×3)

Carex barrattii,
perigynium (×3)

Carex barrattii,
pistillate scale (×3)

Carex limosa,
perigynium (×3)

Carex limosa, pistillate
scale (×3)

Carex limosa

Carex paupercula,
perigynium (×3)

Carex paupercula,
pistillate scale (×3)

Carex paupercula

Section *Limosae*

Roots covered by brownish felt; sheaths and leaf blades glabrous; lower bracts more or less sheathless; terminal spikes male; lateral spikes female or with a few basal male flowers, drooping on slender peduncles; perigynia finely papillose (20×); beaks lacking; achenes 3-sided, loosely filling the perigynia; styles deciduous; stigmas 3.

A. plants densely to loosely tufted; pistillate scales long-acuminate, much longer than the perigynia; male spikes usually 0.5–1.5 cm long .... *C. paupercula*
A. plants strongly rhizomatous; pistillate scales obtuse to acute, shorter than to about as long as the perigynia; male spikes usually 1.5–5.5 cm long
    B. female spikes oblong-cylindric, 0.5–2.7 cm long, rarely with male flowers at apex; pistillate scales as wide as or wider than the perigynia .... *C. limosa*
    B. female spikes linear-cylindric, usually 2–5.5 cm long, often with male flowers at apex; pistillate scales much narrower than the perigynia ....
    .... *C. barrattii*

Carex barrattii Schwein. & Torr.   Barratt's sedge
Strongly rhizomatous; fertile shoots 3–9 dm tall; leaf blades 2.5–6 mm wide, flat; male spikes 2.5–5.5 cm long; female spikes linear-cylindric, (1)2–5.5 cm long, often with a few male flowers at apex; pistillate scales persistent, obtuse to acute, somewhat shorter and much narrower than the perigynia; perigynia 2.5–3.5 by 1.2–1.8 mm; wet woods of Coastal Plain; Delaware Co.; probably extirpated; OBL; ✹.

Carex limosa L.   Mud sedge
Strongly rhizomatous; fertile shoots 2–6 dm tall; leaf blades 0.5–3 mm wide, involute; male spikes (1)1.5–5 cm long; female spikes oblong-cylindric, 0.5–2.7 cm long, rarely with male flowers at apex; pistillate scales persistent, obtuse to acute, about as long and as wide as the perigynia; perigynia 2.5–4 by 1.4–2.2 mm; sphagnum bog mats and hummocks; locally frequent NE, rare elsewhere; OBL; ✹.

Carex paupercula Michx.   Bog sedge
Densely or loosely tufted; fertile shoots 1–8 dm tall; leaf blades 2–4 mm wide, flat; male spikes 0.5–1.5(2) cm long; female spikes 0.5–2 cm long, occasionally with male flowers at base; pistillate scales deciduous before perigynia, long-acuminate, much longer and narrower than the perigynia; perigynia 2.4–3.7 by 1.5–2.6 mm; sphagnum bogs and boggy woods; rare; NE and NC; OBL; ✹.

Section *Leucoglochin*

Carex pauciflora Lightf.   Few-flowered sedge
Rhizomes long, slender; fertile shoots slender but stiff, 1–6 dm tall; leaf blades 1–1.5 mm wide, involute; bracts none; spikes one per shoot, male above and 2–6 perigynia at base; pistillate scales soon deciduous, acute; perigynia glabrous, light-green, reflexed at maturity, 6–7.2 by 0.8–1.3 mm; beaks very gradually tapered, distance from summit of achene to beak apex 2.8–4 mm; achenes 3-sided; styles persistent; stigmas 3; open sphagnum bogs; rare; N; OBL; ✹.

Carex pauciflora

Carex pauciflora, perigynium (×3)

Section *Lupulinae*

Plants coarse, 2–13 dm tall; leaves and sheaths glabrous; terminal spikes male, the female spikes globose to cylindric, sessile or stalked; perigynia glabrous (or pubescent in some populations of *C. grayi*), (9)10–20 mm long, membranous, much inflated and coarsely 13–25-nerved; styles persistent and continuous with the 3-sided achenes; stigmas 3. Treatment based on Reznicek and Ball (1974).

A. sheaths of uppermost leaf absent or <1.5 cm long; female spikes globose to ovoid, the beaks 1.5–4.2 mm long; achenes elliptic or obovate
    B. perigynia 1–12 per spike and ascending-spreading, lustrous, rounded at the base .... *C. intumescens*
    B. perigynia 8–30 per spike and radiating out in all direction, cuneate at the base .... *C. grayi*
A. sheaths of uppermost leaf usually 1.7 cm or longer; female spikes oblong-cylindric (definitely longer than wide), the beaks 4.5–10 mm long; achenes angularly rhombic
    C. achenes with smoothly curved angles (not knobbed) and flat to slightly concave faces .... *C. lupulina*
    C. achenes with prominently knobbed angles and strongly concave faces .....
    .... *C. lupuliformis*

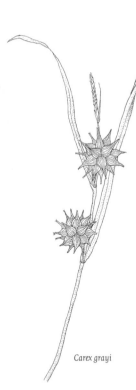

Carex grayi

Carex grayi Carey   Sedge
Fertile shoots 2.5–9 dm tall, solitary or tufted; leaf blades 4–11 mm wide, upper leaf sheaths absent or up to 1.5 cm long; peduncles of male spike 0.5–6 cm long; female spikes 1–2 per fertile shoot, globular, 2.5–4.2 by 2.6–4.1 cm; perigynia 8–35 per spike and radiating out in all directions, ovoid with a cuneate base, 12.5–20 by 4–8 mm, dull (sometimes finely pubescent); beaks poorly defined, 1.5–3 mm long; achenes ellipsoid to obovoid, faces convex, angles not thickened; swamps and wet woods with basic clay soils; rare NE, occasional to frequent elsewhere; FACW+.

Carex intumescens Rudge   Sedge
Fertile shoots 1.5–8 dm tall, solitary or tufted; leaf blades 3.5–8 mm wide, upper leaf sheaths absent or up to 1.5 cm long; peduncles of male spike 0.5–4 cm long; female spikes 1–4 per fertile shoot, ovoid to obovoid, 1–2.7 by 1–2.8 cm; perigynia 1–12 per spike and spreading to ascending, lanceolate with a rounded base, (9)10–16.5 by 2.5–6.5 mm, lustrous; beaks poorly defined 2–4.2 mm long; achenes ellipsoid to obovoid, faces convex, angles not thickened; swamps, wet meadows, and wet woods; common; throughout; FACW+.

Carex lupuliformis Sartwell   False hop sedge
Fertile shoots 5–12 dm tall, rhizomatous; leaf blades 6–13 mm wide, upper leaf sheaths 3–21 cm long; peduncles of male spike 1–12 cm; female spikes 2–6 per fertile shoot, usually cylindric, 2–8 by 1.5–3 cm; perigynia 8–75 per spike and ascending-spreading, 12–18.5 by 3.8–6 mm, shiny; beaks conic, 6–9 mm long; achenes rhomboid, faces strongly concave, angles knobbed; calcareous marshes and wet woods, sometimes in shallow water; rare; mostly SC and SE; FACW; ❦.

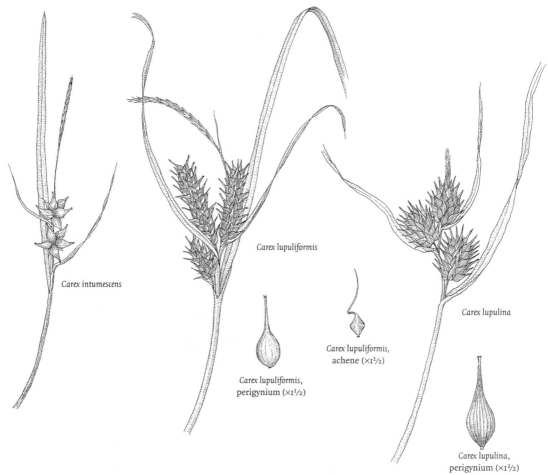

Carex intumescens

Carex lupuliformis

Carex lupulina

Carex lupuliformis,
perigynium (×1½)

Carex lupuliformis,
achene (×1½)

Carex lupulina,
perigynium (×1½)

Carex lupulina,
achene (×1½)

Carex lupulina Willd.   Sedge

Fertile shoots 2–13 dm tall, rhizomatous; leaf blades 4–15 mm wide, upper leaf sheaths 1.5–25 cm long; peduncles of male spike 0.5–6 cm long; female spikes 2–5 per fertile shoot, ovoid to cylindric, 1.5–6.5 by 1.3–3 cm; perigynia 8–80 per spike and ascending, 11–19 by 3–6 mm, shiny; beaks conic, 4.5–7 mm long; achenes rhomboid, faces flat to concave, angles merely curved; swamps, wet ditches, and wet woods; common; throughout; OBL.

Section *Paludosae*

Rhizomatous; shoot bases mostly reddish (or brownish in *C. hyalinolepis*) and fibrillose; vegetative shoots with leaves clustered at base; ventral hyaline band of leaf sheaths often red-dotted; lower bracts sheathless or sheathing; terminal 1–6 spikes male, elevated above lateral spikes; lateral spikes female, cylindrical to ovoid, erect; pistillate scales long-awned or sometimes acute; perigynia ellipsoid to ovoid, glabrous or pubescent, many-nerved and thick-walled, much larger than the achene, round or obtusely 3-sided in cross section; beaks 0.5–1.7(2) mm long; beak teeth 0.3–0.8(1) mm long; styles persistent and continuous with the 3-sided achenes; stigmas 3. Key based on draft by A. A. Reznicek.

A. perigynia glabrous
  B. longest ligules 2–12 mm long, <2 times as long as wide; lower leaves with well developed blades .... *C. hyalinolepis*
  B. longest ligules 12–40 mm long, much longer than wide; lower leaves bladeless or blades much reduced .... *C. lacustris*
A. perigynia pubescent
  C. beaks hyaline, friable, and not regularly bidentate at the apex; peduncles of male spikes 2–20 mm long .... *C. vestita*
  C. beaks firm, bidentate with teeth 0.2–0.7 mm long; peduncles of male spikes mostly 20–90 mm long
    D. leaves flat, 2.2–5(8.5) mm wide; edges of culms sharply angled and usually rough .... *C. pellita*
    D. leaves involute or triangular-channeled at base, tapering to a long filiform tip, 0.7–2 mm wide; edges of culms obtusely angled and smooth .... *C. lasiocarpa*

Carex hyalinolepis,
perigynium (×3)

Carex hyalinolepis Steud.   Shoreline sedge
Fertile shoots 4–10 dm tall; edges of flowering culms sharply 3-angled, angles smooth; leaf blades 3.5–15 mm wide, those of lower leaves well-developed; ligules 2–12 mm long, <2 times as long as wide; male peduncles 12–40 mm long; perigynia glabrous, 5.5–8 by 1.7–2.6 mm; beaks 1.2–2.5 mm long, their teeth 0.3–0.9 mm long; damp banks; Philadelphia Co.; probably extirpated; OBL; 🍂.

Carex lacustris,
perigynium (×3)

Carex lacustris Willd.   Sedge
Fertile shoots 6–13 dm tall; edges of flowering culms obtusely angled and smooth; leaf blades 4.5–16 mm wide, those of lower leaves much reduced or lacking; ligules 12–40 mm long, much longer than wide; male peduncles 20–100 mm long; perigynia glabrous, 4.8–7 by 1.5–2.3 mm; beaks about 1–1.8 mm long, their teeth 0.4–0.9 mm long; swamps, wet meadows, swales, and shores; throughout except SW; OBL.

Carex lasiocarpa var.
americana, perigynium (×3)

Carex lasiocarpa Ehrh.   Many-fruited sedge
Fertile shoots 3–12 dm tall; edges of flowering culms obtusely angled and smooth; leaf blades 0.7–2 mm wide, involute or triangular-channeled at base, tapering to a long filiform tip, blades of lower leaves lacking or much reduced; ligules 0.5–2.5 mm long; male peduncles (15)20–50 mm long; perigynia pubescent, 2.8–4.5 by 1.5–2.1 mm; beaks 0.5–1.2 mm long, their teeth 0.3–0.6 mm long; sphagnum bogs; locally frequent NE, rare elsewhere; OBL; 🍂. Ours are var. *americana* Fernald.

Carex pellita, perigynium
(×3)

Carex pellita Willd.   Sedge
Fertile shoots 3–10 dm tall; edges of flowering culms sharply 3-angled and serrulate above; leaf blades flat, 2.2–5(8.5) mm wide, those of lower leaves much reduced; ligules 1.5–3.5 mm long; male peduncles 20–90 mm long; perigynia densely pubescent, 2.5–4.2 by 1.3–2.4 mm; beaks 0.8–1.3 mm long, their teeth 0.3–0.7 mm long; swamps, wet meadows, and shores; common; mostly SC and SE; OBL.

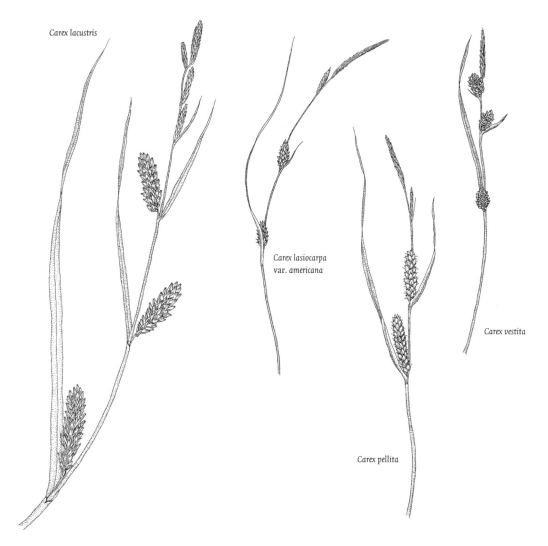

Carex lacustris

Carex lasiocarpa
var. *americana*

Carex vestita

Carex pellita

Carex vestita,
perigynium (×3)

Carex vestita Willd.   Sedge
Fertile shoots 3–10 dm tall; edges of flowering culms sharply 3-angled and serrulate above; leaf blades 2.5–5 mm wide, those of lower leaves much reduced;
ligules 2.8–5 mm long; male peduncles 2–20 mm long; perigynia pubescent, 3–
4.2 by 1.1–1.8 mm; beaks 0.7–1.7 mm long, their apices hyaline, whitish, only
obscurely bidentate; dry or moist sandy woods, clearings, and barrens; locally
frequent; E.

Section *Paniceae*

Rhizomatous; leaves glabrous; lower bracts sheathing; terminal spike male (terminal 1–3 in *C. polymorpha*); lateral spikes usually entirely female, erect, and cylindrical, 0.5–4 cm long; pistillate scales rounded, obtuse, or awned, becoming
brown- or purple-tinged; perigynia minutely to strongly papillose, with raised
nerves, and often 2-ribbed; beaks short to long; achenes 3-sided; styles deciduous; stigmas 3.

A. perigynia contracted into an elongated, cylindrical beak 0.5–1.8(2.2) mm long; perigynia heavily papillose .... *C. polymorpha*

A. perigynia tapered or contracted to a minute beak <0.5 mm long; perigynia sparsely or minutely papillose

    B. lower sheaths bladeless and strongly tinged with reddish-purple; plants of woodland habitats, sometimes forming extensive closed colonies of vegetative shoots from superficial rhizomes; perigynia tapering to a stipitate base 0.5–1.2 mm long .... *C. woodii*

    B. lower sheaths with blades, brownish or faintly and irregularly tinged with reddish-purple; plants of wetland habitats with widely scattered vegetative shoots and deep rhizomes; perigynia rounded or acute at base

        C. achenes 1.7–2.2(2.5) mm wide; ligules 0.4–1.2 times as long as wide; perigynia more or less spreading, tightly arranged and yellow-green ... . *C. meadii*

        C. achenes 1.2–1.6(1.8) mm wide; ligules mostly 1.2–1.8 times as long as wide; perigynia appressed or ascending, loosely arranged, dark green . ... *C. tetanica*

Carex meadii, perigynium (×3)

Carex meadii Dewey   Mead's sedge

Fertile shoots 1.5–6 dm tall; lower sheaths bearing a blade, brownish; leaf blades 2–5(7) mm wide, their ligules 0.4–3.6 mm long (0.4–1.2 times wider than long); inflorescences 4–25 cm long; spikes unisexual, the perigynia more or less spreading and tightly arranged; perigynia yellow-green, 2.3–5 by 1.7–2.5 mm, minutely papillose; beaks minute and bent; achenes 1.7–2.2(2.5) mm wide; wet meadows on diabase or serpentinite; rare; SE; FAC; 🌿.

Carex polymorpha, perigynium (×3)

Carex polymorpha Muhl.   Variable sedge

Fertile shoots 3–6 dm tall; lower sheaths bladeless, brown- to red-purple-tinged; leaf blades 2.5–6 mm wide, their ligules 2–10 mm long (longer than wide); inflorescences 6–19 cm long; terminal 1–3 spikes male, lower spikes all female or male distally, the perigynia densely arranged; perigynia green to light brown, 4–5.5 by 1.5–2.5 mm, with low, broad papillae; beak 1–2.2 mm long, purple-tinged, its orifice hyaline and oblique; achenes 1.4–2.3 mm wide; thin woods and barrens in sandy-peaty soils; rare; E; FACU; 🌿.

Carex tetanica, perigynium (×3)

Carex tetanica, pistillate scale (×3)

Carex tetanica Schkuhr   Wood's sedge

Fertile shoots 1.5–6.5 dm tall; lower sheaths bearing a blade, brownish- or purple-tinged; leaf blades 1.5–5 mm wide, their ligules 1–6 mm long (1.2–1.8 times as long as wide); inflorescences 4–32 cm long; spikes unisexual, the perigynia appressed or ascending and often loosely arranged; perigynia dark green, 2.5–4 by 1.2–2.2 mm, minutely papillose, tapering to a minute beak; achenes 1.2–1.6(1.8) mm wide; calcareous wet meadows and swales; scattered; mostly S; FACW; 🌿.

Carex woodii Dewey   Sedge

Fertile shoots 3–7 dm tall; lower sheaths bladeless, purple-tinged; leaf blades 2.5–4 mm wide, their ligules 0.5–2 mm long (wider than long); inflorescences 5–13 cm long; spikes unisexual, the perigynia ascending and loosely arranged; perigynia green to brown, 2.5–4 by 1–2 mm, stipitate, minutely papillose only near beak; beaks short, tapered, and strongly bent; achenes 1.2–1.4 mm wide; rocky limestone woods and rich, wooded slopes; occasional; W; UPL.

Carex woodii,
perigynium (×3)

Section *Phacocystis*

Plants (2)3.5–15 dm tall, coarse; leaves and sheaths glabrous or scabrous; bracts sheathless; terminal spike(s) male; lateral spikes female, 2.5–10 cm long, cylindrical, erect or drooping; beaks usually short and entire; achenes lenticular; styles deciduous; stigmas 2. Treatment largely based on Stanley (1985, 1989).

A. spikes erect; plants often rhizomatous; pistillate scales awnless
  B. lowest bracts equaling or surpassing inflorescence
    C. fertile shoots usually surrounded by dried-up leaves of previous year; leaf sheaths lacking a pinnate network of persistent veins, the hyaline band glabrous ....*C. aquatilis*
    C. fertile shoots with conspicuous bladeless sheaths at base; leaf sheaths regularly displaying a conspicuous pinnate network of persistent veins, the hyaline band scabrous .... *C. stricta*

Carex woodii

Carex meadii

Carex meadii, rhizomes

Carex polymorpha

Carex polymorpha,
rhizomes

B. lowest bracts mostly shorter than inflorescence
- D. pistillate scales generally longer than the perigynia, acute or acuminate; pistillate spikes more or less truncate at the base, lowest ones 1–5 cm long; perigynia broadly elliptic or obovate, somewhat inflated and essentially nerveless .... *C. haydenii*
- D. pistillate scales shorter than the perigynia, obtuse or acute; pistillate spikes tapering at the base (due to more loosely arranged perigynia), lowest ones 1.5–11 cm long; perigynia elliptic, flattened, 0–5-nerved
  - E. leaf sheaths regularly displaying a conspicuous pinnate network of persistent veins, the hyaline band scabrous; mouth of sheaths concave; ligules longer than wide .... *C. stricta*
  - E. leaf sheaths lacking a pinnate network of persistent veins, the hyaline band glabrous; mouth of sheaths convex; ligules about as wide as long .... *C. emoryi*

A. spikes (at least the lowest) drooping; plants usually lacking rhizomes; pistillate scales awned or awnless
- G. pistillate scales not awned, black; lower bracts about as long as or shorter than the inflorescence; achenes not indented .... *C. torta*
- G. pistillate scales awned, brown; lower bracts generally much longer than the inflorescence; achenes generally indented
  - H. lower sheaths glabrous; perigynia obovate, inflated at maturity .... *C. crinita*
  - H. lower sheaths scabrous; perigynia elliptic, not inflated at maturity
    - I. perigynia smooth, 0–1 nerve on each face; achenes indented, oblong-ovoid; pistillate scales tapering at summit to awn....*C. gynandra*
    - I. perigynia distinctly fine-papillose (at 10×), often 2–4 nerves on each face; achenes indented, orbicular; pistillate scales with an awn arising from a truncate or retuse apex .... *C. mitchelliana*

Carex aquatilis

Carex aquatilis Wahlenb.   Water sedge

Rhizomatous; fertile shoots 4–12 dm tall, usually surrounded at base by dried-up leaves from previous year; leaf sheaths glabrous; leaf blades 2.5-8 mm wide; ligules about as wide as long; lowest bracts as long as to much longer than the inflorescence (5–40 cm long); female spikes erect, 1–11.5 cm; pistillate scales brown or purple-brown, acute and about as long as the perigynia; perigynia el-

Carex aquatilis,
perigynium (×3)

liptic or obovate, 2–3.6 by 1.3–2.3 mm, yellow-brown (sometimes with red-brown dots), nerves lacking; achenes not indented; marshy swales; rare; NW, NC, and SE; OBL; ⚘.

Carex aquatilis,
pistillate scale (×3)

Carex crinita Lam.   Short-hair sedge

Tufted; fertile shoots 7–15 dm tall; leaf sheaths glabrous; leaf blades 5–13 mm wide; ligules longer than wide; lowest bracts much longer than the inflorescence (17–55 cm long); female spikes drooping, 2.5–10 cm; pistillate scales with a long awn (1–7.5 mm long) arising from a truncate or retuse summit; perigynia inflated and obovate, 2–3.2 by 1.5–2 mm, 0–1 nerve on each face; achenes indented on one side; OBL.

A. pistillate scales 2–3 times as long as the perigynia; perigynia <3.2 mm long; achenes indented on one side .... var. *crinita*
    moist to wet woods, thickets, marshes, ditches, and stream banks; common; throughout.

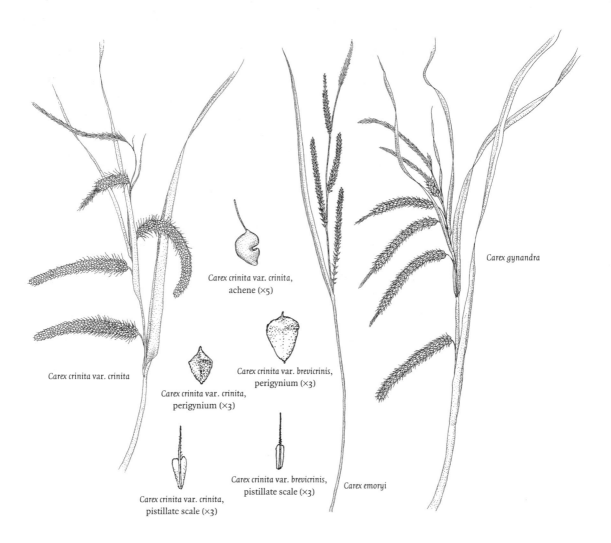

Carex crinita var. crinita,
achene (×5)

Carex crinita var. crinita

Carex crinita var. crinita,
perigynium (×3)

Carex crinita var. brevicrinis,
perigynium (×3)

Carex crinita var. crinita,
pistillate scale (×3)

Carex crinita var. brevicrinis,
pistillate scale (×3)

Carex emoryi

Carex gynandra

Carex emoryi,
perigynium (×3)

Carex emoryi,
pistillate scale (×3)

Carex gynandra,
perigynium (×3)

Carex gynandra,
pistillate scale (×3)

A. pistillate scales mostly 1–2 times as long as the perigynia; perigynia up to 4.5 mm long; achenes lacking indentations .... var. *brevicrinis* Fernald moist to wet woods; SE; ❦.

Carex emoryi Dewey   Sedge
Rhizomatous; fertile shoots 3.5–12 dm tall; leaf sheaths glabrous; leaf blades 3–8 mm wide; ligules wider than long; lowest bracts about as long as inflorescence (5–35 cm long); female spikes erect, 2.5–10 cm; pistillate scales red-brown, acute or obtuse, and shorter than the perigynia; perigynia elliptic, 1.7–3.2 by 1–2.1 mm, yellow-brown, 0–3 nerves on upper face and 3–5 nerves on lower face; achenes not indented; stream banks, swales, and marshes; infrequent; throughout; OBL.

Carex gynandra Schwein.   Sedge
Tufted; fertile shoots 6.5–16.5 dm tall; sheaths scabrous; leaf blades 4–13 mm wide; ligules longer than wide; lowest bracts much longer than the inflorescence (17–60 cm long); female spikes drooping, 2.5–10 cm; pistillate scales tapering to a long awn (1.5–5.8 mm long); perigynia generally smooth, flattened, and elliptic, 2.4–4.2 by 1.4–2.1 mm, 0–1 nerve on each face; achenes indented on one side; marshy areas and wet woods; common; throughout; OBL.

Carex haydenii,
perigynium (×3)

Carex haydenii,
pistillate scale (×3)

Carex haydenii Dewey

Cloud sedge

Tufted or occasionally producing rhizomes; fertile shoots 3.5–11.5 dm tall; leaf sheaths glabrous, lowest ones sometimes with a network of persistent veins; leaf blades 3–4 mm wide; ligules longer than wide; lowest bracts about as long as the inflorescence (3–16 cm long); female spikes erect, 1–5 cm; pistillate scales brown to dark red-brown, acute to acuminate and mostly longer than the perigynia; perigynia broadly elliptic or obovate, inflated, 1.5–2.3 by 1.2–2 mm, olive green (often with red-brown dots), nerveless or with 2–3 faint nerves; achenes not indented; swamps and wet meadows; scattered; mostly E; OBL; ✹.

Carex mitchelliana,
perigynium (×3)

Carex mitchelliana,
pistillate scale (×3)

Carex mitchelliana Curtis

Mitchell's sedge

Tufted; fertile shoots 4–13 dm tall; leaf sheaths scabrous; leaf blades 2.5–8 mm wide; ligules longer than wide; lowest bracts about as long as or longer than the inflorescence (7.5–33 cm long); female spikes drooping, 2–8 cm; pistillate scales (especially lower ones) with an awn (0.5–4 mm long) arising from a truncate or retuse summit; perigynia fine-papillose, ovate to broadly ovate, 2.5–4.1 by 1.2–2.2 mm, usually 2–4 nerves on each face; achenes generally not indented; swamps, wet meadows, and stream banks; rare; Crawford Co.; OBL; ✹.

Carex stricta,
perigynium (×3)

Carex stricta,
pistillate scale (×3)

Carex stricta Lam.

Tussock sedge

Tufted, often forming tussocks; fertile shoots 4.5–14.5 dm tall; leaf sheaths scabrous (especially on hyaline band) and displaying a pinnate network of persistent veins; leaf blades 3–6 mm wide; ligules longer than wide; lowest bracts 0.5–1.5 times as long as the inflorescence (3–35 cm long); female spikes erect, 1.5–11 cm; pistillate scales pale red-brown, acute or obtuse and shorter than the perigynia; perigynia ovate or elliptic, 1.7–3.4 by 0.8–1.8 mm, yellow-brown (often with red-brown dots), 0–5 nerves on each face; achenes not indented; swamps, stream banks, and wet meadows; common SE, occasional elsewhere except NC; OBL.

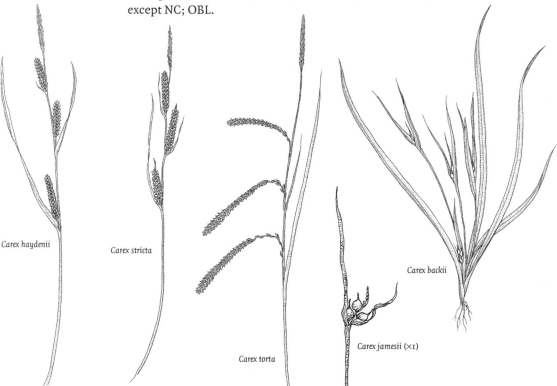

Carex haydenii

Carex stricta

Carex torta

Carex jamesii (×1)

Carex backii

Carex torta Boott ex Tuck.   Sedge

Tufted; fertile shoots 2–9 dm tall; leaf sheaths glabrous; leaf blades 3–5 mm wide; ligules longer than wide; lowest bracts about as long as the inflorescence (2–15 cm long); upper female spikes erect, lower ones drooping, 3–9 cm; pistillate scales black, obtuse, and shorter than the perigynia; perigynia usually ovate, 2.3–4.9 by 1.1–1.8 mm, green, lacking nerves, its upper portion and the beak often twisted; achenes not indented; floodplains and rocky stream banks; common; throughout; FACW.

Carex torta,
perigynium (×3)

Carex torta,
pistillate scale (×3)

Section *Phyllostachyae*

Plants low, tufted; leaves glabrous and overtopping fertile culms; bracts apparently lacking; spikes male above and female at base; staminate scales often small and encircling the rachis; lowest pistillate scales mostly longer than the perigynia or even leafy, upper scales smaller; perigynia glabrous, 2-ribbed; beaks 1.5–3 mm long; achenes 3-sided, filling the body of the perigynia; styles deciduous; stigmas 3. Key based on Catling et al. (1993).

A. lowest pistillate scales 2.4–4.5 mm wide, much wider than perigynia (essentially concealing them), entirely green and leafy; male flowers 2–4 per spike; widest leaves 2.8–5.5 mm wide .... *C. backii*

A. lowest pistillate scales 1.2–2.5(3) mm wide, not more than 1.5 times as wide as perigynia (spreading and not concealing them), their margins hyaline at the base; male flowers 5–25 per spike; widest leaves 2–3.5 mm wide

    B. perigynia (3)4–9 per spike, their bodies obovoid-oblong and gradually tapering to a stout, triangular beak; staminate scales obtuse to acute; achenes 1.5–2 times as long as wide .... *C. willdenowii*

    B. perigynia 2–3(4) per spike, their bodies subglobose and abruptly contracted to a slender beak; staminate scales more or less truncate; achenes 1–1.5 times as long as wide .... *C. jamesii*

Carex backii Boott   Back's sedge, Rocky Mountain sedge

Fertile shoots 5–30 cm tall; widest leaf blades 2.8–5.5 mm wide; male flowers (or their empty scales) 2–4; staminate scales obtuse to cuspidate; lowest pistillate scales 2.4–4.5 mm wide, foliaceous, appressed against and essentially concealing the perigynia; perigynia 2–5 per spike, 4.5–6 by 1.8–2.5 mm, the body more or less ellipsoid; beaks 1.5–2.5 mm long, stout; achenes 1.3–1.5 times as long as wide; rocks at base of sandstone ledge; rare; Lackawanna Co.; ✿.

Carex backii,
perigynium (×3)

Carex jamesii Schwein.   Sedge

Fertile shoots 5–30 cm tall; widest leaf blades 1.8–3 mm wide; male flowers 5–10 (or more); staminate scales truncate; lowest pistillate scales 1.2–2.5(3) mm wide, margins hyaline at the base, spreading, 2 or more times as long as the perigynia but not concealing them; perigynia 2–3(4) per spike, 5–5.8 by 1.8–2.4 mm, the body subglobose and stipitate; beaks 1.8–3 mm long, abruptly contracted, slender; achenes 1–1.5 times as long as wide; limestone woods; infrequent; S.

Carex jamesii,
perigynium (×3)

Carex willdenovii Schkuhr ex Willd.   Sedge

Fertile shoots 5–25 cm; widest leaf blades 2–3 mm wide; male flowers 5 to many; staminate scales obtuse to acute; lowest pistillate scales 1.2–2.5(3) mm

Carex willdenovii,
perigynium (×3)

wide, margins hyaline at the base, spreading, mostly longer than the perigynia but not concealing them; perigynia (3)4–9 per spike, 4.3–5.5 by 1.3–1.8 mm, the body obovoid-oblong; beak 1.8–2.4 mm long, stout-triangular; achenes 1.5–2 times as long as wide; dry, open rocky woods; frequent SC and SE, rare NW and SW; UPL.

Section Polytrichoideae

Carex leptalea,
perigynium (×3)

Carex leptalea Wahlenb.   Sedge
Tufted; leaves and sheaths glabrous, the leaf blades soft, lax, 0.5–1.3 mm wide; inflorescences consisting of a solitary spike 0.5–1.5 cm long, bractless; spikes male above, female at base; staminate scales tightly clasping the rachis, their margins united at the base; perigynia glabrous, appressed, narrowly ellipsoid, 2.5–5 by 0.9–1.4 mm; beaks lacking; achenes 3-sided, occupying the lower half of perigynia; styles deciduous; stigmas 3; sphagnum bogs, swampy woods, and swales; rare SW, frequent to occasional elsewhere; OBL.

Section Porocystis

Tufted; sheaths and usually leaves pubescent, the blades 1.5–4.5 mm wide; bracts more or less sheathless; terminal spikes female above, male at base (except all male in *C. pallescens*); perigynia pubescent or glabrous; beaks lacking; achenes 3-sided, not filling the bodies of the perigynia; styles deciduous; stigmas 3.

A. terminal spikes male throughout; perigynia glabrous and faintly nerved ....
....*C. pallescens*
A. terminal spikes female above and male at base; perigynia glabrous with sharp nerves (often lighter in color than perigynium wall) or pubescent
  B. perigynia glabrous; spikes 3.5–6.5 mm thick
    C. leaf blades glabrous (except perhaps at the base); ventral band of sheaths more or less glabrous, especially at the summit.....*C. caroliniana*
    C. leaf blades pubescent; ventral band of sheaths pubescent, often densely so at summit
      D. perigynia flattened on upper face; pistillate scales shorter than the perigynia; anthers 1.2–2 mm long .... *C. hirsutella*
      D. perigynia inflated, almost round in cross section; pistillate scales usually longer than the perigynia; anthers 2–3 mm long ..... *C. bushii*
  B. perigynia pubescent; spikes 2–4.5 mm thick
    E. lateral spikes (especially lowest one) 5–16(20) mm long, perigynia closely spaced throughout; fertile shoots often shorter than the leaves; anthers 0.7–1.3 mm long .... *C. swanii*
    E. lateral spikes (especially lowest one) 16–40 mm long, lower perigynia loosely spaced; fertile shoots usually exceeding the leaves; anthers 1.4–2.3 mm long .... *C. virescens*

Carex bushii,
perigynium (×3)

Carex bushii Mack.   Sedge
Fertile shoots 3–9 dm tall, shorter than to exceeding the leaves; leaf blades and sheaths pilose; lateral spikes 5–14 by 4.5–6.5 mm (widths exclude pistillate scales), the terminal spikes much longer (up to 25 mm long); pistillate scales

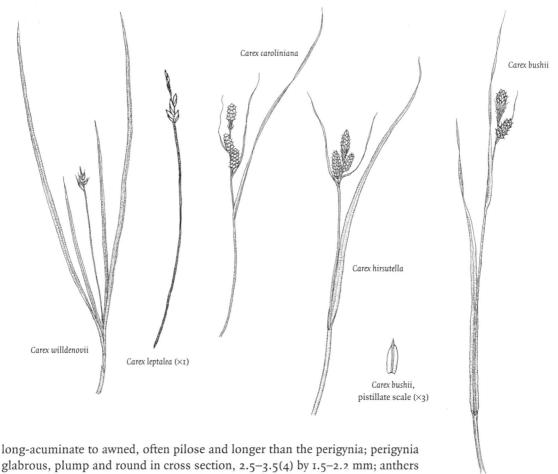

Carex caroliniana

Carex bushii

Carex hirsutella

Carex willdenovii

Carex leptalea (×1)

Carex bushii,
pistillate scale (×3)

long-acuminate to awned, often pilose and longer than the perigynia; perigynia glabrous, plump and round in cross section, 2.5–3.5(4) by 1.5–2.2 mm; anthers 2–3 mm long; dry to moist open woods, thickets, and fields; frequent S, rare N; FACW.

Carex caroliniana Schwein.   Sedge

Fertile shoots 2.5–8 dm tall, exceeding the leaves; leaf blades glabrous except at the base, leaf sheaths soft-hairy dorsally, the ventral band more or less glabrous; lateral spikes 8–20 by 3.5–5 mm; pistillate scales obtuse to short-cuspidate, shorter than the perigynia; perigynia glabrous, plump and more or less round in cross section, 2–2.5(3) by 1.3–2.2 mm; anthers 1.2–2 mm long; wet meadows, thickets and woods; locally frequent; S; FACU.

 Carex caroliniana,
perigynium (×3)

Carex caroliniana,
pistillate scale (×3)

Carex hirsutella Mack.   Sedge

Fertile shoots 2–9 dm tall, shorter than to exceeding the leaves; leaf blades and sheaths soft pubescent; lateral spikes 6–16 by 3.5–5 mm; pistillate scales short-cuspidate to obtuse, shorter than the perigynia; perigynia glabrous, upper side flattened, 2–3 by 1.3–2 mm; anthers 1.2–2 mm long; dry fields, open woods, and serpentine barrens; common; throughout.

 Carex hirsutella,
perigynium (×3)

Carex hirsutella,
pistillate scale (×3)

Carex pallescens L.   Sedge

Fertile shoots 2–8 dm tall, shorter than to exceeding the leaves; leaf blades and sheaths pubescent; lateral spikes 5–22 by 4–5.8 mm; pistillate scales acute to

 Carex pallescens,
perigynium (×3)

Carex pallescens,
pistillate scale (×3)

cuspidate, shorter than to slightly longer than the perigynia; perigynia glabrous, ellipsoid and more or less round in cross section, 2–2.8 by 1–1.5 mm; anthers 1.3–2.2 mm long; moist woods and meadows; locally frequent; mostly N.

Carex swanii,
perigynium (×3)

Carex swanii,
pistillate scale (×3)

Carex swanii (Fernald) Mack.   Sedge

Fertile shoots 2–6 dm tall, mostly shorter than the leaves; leaf blades and sheaths pubescent; lateral spikes 5–16(20) by 3–4.5 mm, the perigynia tightly arranged throughout; pistillate scales acute to awned, shorter than the perigynia; perigynia pubescent, obovoid or broadly ellipsoid, 2–2.3 by 1.2–1.5 mm; anthers 0.7–1.3 mm long; dry woods, thickets, and fields; common; throughout; FACU.

Carex virescens,
perigynium (×3)

Carex virescens,
pistillate scale (×3)

Carex virescens Willd.   Sedge

Fertile shoots 3.5–10 dm tall, exceeding the leaves; leaf blades and sheaths pubescent; lateral spikes 16–40 by 2–4 mm, lower spikes with a tapering base due to more loosely arranged perigynia; pistillate scales acute to awned, shorter than the perigynia; perigynia pubescent, ellipsoid, 1.8–2.5 by 0.9–1.4 mm; anthers 1.4–2.3 mm long; dry woods and thickets; common SE, frequent to occasional elsewhere.

Section *Racemosae*

Rhizomatous; sheaths and leaves glabrous; lower bracts long sheathing; lateral spikes more or less erect, 4–25 by 3–5 mm, lower ones pedunculate; perigynia plump, finely papillose (visible at 20×), elliptic in cross section; beaks lacking; achenes lenticular, more or less filling the bodies of the perigynia; styles deciduous; stigmas 2.

Carex swanii

Carex aurea

Carex pallescens

Carex virescens

Carex shortiana

Carex garberi

A. mature perigynia golden-orange when fresh (when dry becoming brownish or dusty white, especially if immature); pistillate scales mostly ¾ the length of the perigynia or less, widely spreading at maturity, reddish-yellow or pale; terminal spikes all male or rarely with a very few perigynia at apex .... *C. aurea*

A. mature perigynia dusty white when fresh or dry; pistillate scales mostly 0.75–1 times as long as the perigynia, ascending at maturity, brownish-red-tinged; terminal spikes with few to many perigynia above, male at base ...... *C. garberi*

Carex aurea Nutt.   Golden-fruited sedge
Fertile shoots 5–55 cm tall; terminal spikes male or rarely with a very few perigynia at apex; pistillate scales reddish-yellow or pale, mostly acute or cuspidate, spreading and distinctly shorter than the perigynia; perigynia golden-orange when fresh (brownish or dusty white when dry), 2–3 by 1.3–2.1 mm; moist, calcareous slumps and seeps; rare; Erie Co.; FACW; ✿.

Carex aurea, perigynium (×3)

Carex aurea, pistillate scale (×3)

Carex garberi Fernald   Elk sedge
Fertile shoots 5–40 cm tall; terminal spikes with few to many perigynia above, male at base; pistillate scales brownish-red-tinged, mostly blunt to acute, ascending and about as long as the perigynia; perigynia dusty white, 2–3 by 1.1–1.6 mm; sandy swales and calcareous gravel; rare; Erie Co.; FACW; ✿.

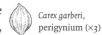
Carex garberi, perigynium (×3)

Carex garberi, pistillate scale (×3)

Section *Shortianae*

Carex shortiana Dewey   Sedge
Tufted; fertile shoots 2–9 dm tall; leaf blades and sheaths glabrous, the blades 4–10 mm wide; bracts more or less sheathless; spikes 3–6 per inflorescence, densely cylindrical, 3.8–5.2 mm wide, female above and male at base or all female (especially lower spikes); pistillate scales short-mucronate or acute; perigynia squarrose, obovoid, 2–3 by 1.5–2.4 mm, glabrous but finely transverse corrugated, nerveless; beaks essentially lacking, 0.2 mm long or less; achenes 3-sided; styles deciduous; stigmas 3; calcareous wet meadows and swamps and rich woods; infrequent; SC and SW; FAC; ✿.

Carex shortiana, perigynium (×5)

Carex shortiana, pistillate scale (×5)

Section *Spirostachyae*

Tufted; lower bracts short- to long-sheathing; terminal spikes male (occasionally female above); female spikes subglobose to ellipsoid, at least the upper one(s) sessile; pistillate scales acute or short-cuspidate; perigynia glabrous, horizontally spreading or reflexed, strongly many-nerved; achenes 3-sided; styles deciduous; stigmas 3.

A. perigynia 2–3 mm long, horizontally spreading; beaks about 0.25–0.5 times as long as the perigynium body .... *C. viridula*

A. perigynia 3–6.2 mm long, at least the lower ones reflexed; beaks about 0.5–0.8 times as long as the perigynium body
    B. perigynia 4.5–6.2 mm long, the beaks minutely and sparsely serrulate toward the tip; pistillate scales becoming brown- or red-tinged .... *C. flava*
    B. perigynia 3–4.5 mm long, the beaks smooth; pistillate scales pale or slightly brown-tinged .... *C. cryptolepis*

Carex cryptolepis,
perigynium (×3)

Carex cryptolepis,
pistillate scale (×3)

Carex cryptolepis Mack.   Northeastern sedge

Fertile shoots 1.5–6 dm tall; leaf blades 1.5–3.5 mm wide; spikes 7–10 mm thick; pistillate scales pale or slightly brown-tinged; perigynia reflexed, 3–4.5 by 1–2 mm; beaks smooth, 1–2 mm long, about 0.5–0.8 times as long as the perigynium body; wet calcareous meadows and lake shores; rare; NE and NW; OBL; ✿.

Carex flava,
perigynium (×3)

Carex flava L.   Yellow sedge

Fertile shoots 1–8 dm tall; leaf blades (2.5)2.8–5 mm wide; spikes 8–12 mm thick; pistillate scales becoming brown- or red-tinged; perigynia reflexed, (4.5)5–6.2 by 1.2–2 mm; beaks minutely and sparsely serrulate (especially toward the tip), (1.1)1.5–2.6 mm long, about 0.5–0.8 times as long as the perigynium body; calcareous wet meadows and fens; rare; NE and NW; OBL; ✿.

Carex viridula,
perigynium (×3)

Carex viridula,
pistillate scale (×3)

Carex viridula Michx.   Green sedge

Fertile shoots 5–40 cm tall; leaf blades 1–3 mm wide; pistillate scales pale or reddish-brown-tinged; spikes 4–7 mm thick; perigynia horizontally spreading but not reflexed, 2–2.8 by 0.7–1.1 mm; beaks smooth, 0.3–0.8 mm long, about 0.25–0.5 times as long as the perigynium body; wet calcareous sand flats; rare; Erie Co.; OBL; ✿.

Section *Squarrosae*

Tufted; terminal spikes either all male or female above and male at base; spikes 1–7 per inflorescence; perigynia glabrous and papery, very tightly packed, their bodies obovoid or obconic, 2–many-nerved; beaks long, slender, abruptly contracted from the perigynium body; achenes 3-sided, loosely enveloped by the perigynium; style persistent and about the same texture as the achene; stigmas 3.

A. pistillate scales with rough awns, longer than the perigynia; terminal spikes usually male; edges of inflorescence axis and peduncles smooth and rounded
.... *C. frankii*

A. pistillate scales obtuse to short awned, shorter than the perigynia, often hidden among the perigynia; terminal spikes female above, male at base; edges of inflorescence axis and peduncles rough and sharply angled

   B. spikes rarely more than 1, subglobose to ellipsoid (1.2–1.8 times as long as wide), the beaks in the midregion widely spreading; widest leaves 3–4.7(5.7) mm wide; styles strongly sinuous .... *C. squarrosa*

   B. spikes 1–3 (rarely more), the terminal one ellipsoid to cylindrical (1.8–4 times as long as wide), the beaks in the midregion ascending; widest leaves (3.7)4.7–7.7 mm wide; styles more or less straight .... *C. typhina*

Carex viridula

Carex frankii Kunth   Sedge

Fertile shoots 1–8 dm tall; leaf blades 3–10 mm wide; bracts long-sheathing; spikes 2–7, the terminal spikes male throughout or sometimes female above; female spikes cylindrical, mostly 2–3 times as long as wide, 1–4.5 by 0.8–1.2 cm; pistillate scales with long, rough awns exceeding the perigynia; perigynia 4–5.7 by 2–2.8 mm, 12–20-nerved; beaks 1.2–2 mm long; achenes 1.7–2.2 mm long; styles straight; swamps, wet woods, streambanks, and ditches; frequent; mostly S; OBL.

Carex frankii,
pistillate scale (×3)

Carex frankii, achene (×3)

Carex squarrosa L. Sedge

Fertile shoots 3–9 dm tall; leaf blades 2–4.7(5.7) mm wide; bracts reduced and sheathless; spikes usually 1 (occasionally 2–4), the terminal spikes female above, male at base; female portion of spikes subglobose to ellipsoid, 1.2–1.8 times as long as wide, 1.7–3 by 1.4–2 cm; pistillate scales acute or short-awned, shorter than the perigynia; perigynia 6.3–8 by 2.8–3.8 mm, 2-ribbed and faintly nerved; beaks 2.4–3.5 mm long; achenes slenderly ellipsoid (2.5–3 by 1–1.3 mm), 2.2 or more times as long as wide; styles strongly sinuous; swamps, wet woods, meadows, and ditches; common; mostly S; FACW.

Carex squarrosa, achene (×3)

Carex typhina Michx. Cat-tail sedge

Fertile shoots 3–9 dm tall; leaf blades (2.5)4.7–7.7 mm wide; bracts essentially sheathless; spikes 1–3 (rarely up to 6), the terminal spikes female above, male at base; female portion of spikes ellipsoid to cylindrical, 1.8–4 times as long as wide, 1.5–4.5 by 1–1.7 cm; pistillate scales more or less obtuse, much shorter than the perigynia; perigynia 3.8–5.2 by 1.8–2.8 mm, 8–12-nerved; beaks 1.8–2.8 mm long; achenes broadly ellipsoid (1.9–2.6 by 1.2–1.4 mm), 1.5–2.2 times as long as wide; styles more or less straight; calcareous bottomlands, swamps, and wet woods; rare; throughout; FACW+; ✿.

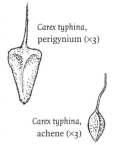

Carex typhina, perigynium (×3)

Carex typhina, achene (×3)

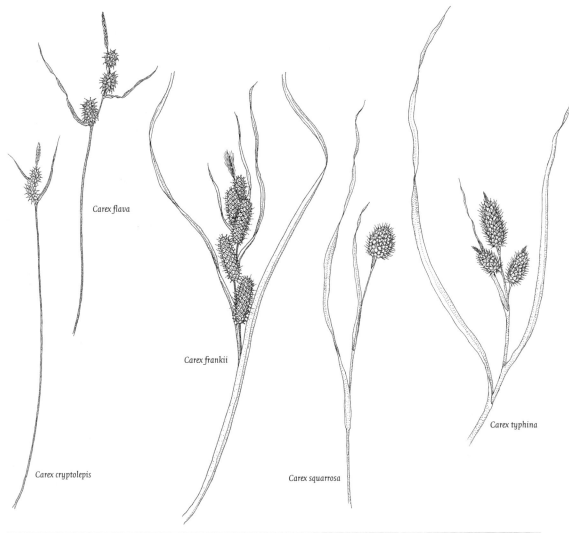

Carex flava

Carex frankii

Carex cryptolepis

Carex squarrosa

Carex typhina

Section *Vesicariae*

Plants relatively tall and coarse; bracts mostly sheathless or short-sheathing; male spikes 1, or less often 2–4, terminal; lower spikes female, erect-ascending to nodding, linear-cylindric (rarely subglobose as in *C. oligosperma*); pistillate scales acute, acuminate, or rough-awned; perigynia glabrous, 3.8–10 mm long, often inflated, coarsely 6–25-nerved; beaks (0.7)1–4.8 mm long, bidentate, the teeth 0.2–1.2 mm long; achenes 3-sided; styles persistent; stigmas 3. Treatment based on draft by A.A. Reznicek.

A. pistillate scales obtuse to acuminate, the scale margins smooth
    B. leaves wiry and strongly involute, 0.5–2.5 mm wide; female spikes sub-globose, 3–15 flowered; peduncles round or obtusely 3-sided, smooth .... *C. oligosperma*
    B. leaves flat, U-, V-, or W-shaped in cross section, 1.5–12(15) mm wide; female spikes oblong-cylindric to cylindric, 15–many-flowered; peduncles round to sharply 3-sided, often rough on angles
        C. widest perigynia (4)4.5–7 mm wide, their beaks 2.4–4.8 mm long; achenes asymmetrical, deeply indented on one side .... *C. tuckermanii*
        C. widest perigynia (2)2.5–3.5(4.5) mm wide, their beaks 1–4.5(4.8) mm long; achenes symmetrical, not indented
            D. lowest bracts (ignore spikes from long sheaths near base of plant) mostly 3–9 times as long as the inflorescence; male spikes usually 1 and only slightly if at all elevated above summit of crowded female spikes; lower perigynia reflexed .... *C. retrorsa*
            D. lowest bracts shorter than to 2.5 times as long as the inflorescence; male spikes 2–4 and elevated above summit of the separated female spikes; lower perigynia ascending or spreading
                E. beaks 2.4–4.2(4.8) mm long, usually finely scabrous (15×) at least near the tip and on the teeth; widest leaves 1.8–4.3(5) mm wide... .... *C. bullata*
                E. beaks 1–2.7 mm long, smooth; widest leaves 1.5–15 mm wide
                    F. plants forming colonies from long-creeping rhizomes; widest leaves 4.5–15 mm wide; ligules about as wide as long; basal sheaths usually spongy-thickened and little or not at all red-tinged .... *C. utriculata*
                    F. plants tufted; widest leaves 1.8–6.5 mm wide; ligules usually much longer than wide; basal sheaths not spongy-thickened, often tinged with reddish-purple .... *C. vesicaria*
A. pistillate scales with a prominent scabrous awn, the scale margins often ciliate
    G. plants forming colonies from elongate, creeping rhizomes; staminate scales acute to acuminate, the margins essentially smooth except at the very tip; perigynia 7–11-nerved .... *C. schweinitzii*
    G. plants densely or loosely tufted, rhizomes (if any) <10 cm long; staminate scales often with a distinct scabrous awn, the margins sometimes ciliate; perigynia 6–25-nerved
        H. perigynia 6–12-nerved, the nerves separated nearly to the beak apex, the body broadly ellipsoid to globose, (1.8)2–4.2 mm wide; achenes rough-papillate (15×)
            I. spikes (1.2)1.5–2.2 cm thick, usually <2.5 times as long as wide; perigynia mostly 6.5–10.8 mm long, their beaks 2.5–5.9 mm long; widest leaves 4–11.5 mm wide .... *C. lurida*

I. spikes 0.8–1.5 cm thick, usually 2.5–3.5 times as long as wide; perigynia mostly 4.8–6.5 mm long, their beaks 2.2–4 mm long; widest leaves 2.4–4(5) mm wide .... *C. baileyi*

H. perigynia 12–25-nerved, the nerves (except for the 2 prominent laterals) confluent at or below the middle of the beak, the body ellipsoid to lance-ovoid, 1.1–2.2 mm wide; achenes smooth

J. perigynia ascending or spreading at maturity (lowest few sometimes reflexed), more or less inflated and rounded in cross section, herbaceous, many nerves separated by more than 3 times their width; longest beak teeth 0.3–0.9 mm .... *C. hystericina*

J. perigynia more or less reflexed at maturity, compressed-triangular and uninflated in cross section, leathery, strongly and closely nerved; longest beak teeth 0.7–2.1(2.8) mm

K. female spikes 1.1–1.8 cm thick; beak teeth strongly outcurved, the longest 1.3–2.1(2.8) mm .... *C. comosa*

K. female spikes 0.9–1.2 cm thick; beak teeth more or less erect, the longest 0.7–1.2(1.4) mm .... *C. pseudocyperus*

Carex baileyi Britton   Sedge

Tufted; fertile shoots 2–6.5 dm tall, obtusely angled and smooth below the inflorescence; leaf blades 1.8–4(5) mm wide; ligules about as long as wide; lowest bracts longer than the inflorescence; male spikes solitary, peduncles short; female spikes 1–4 by 0.8–1.5 cm, the lowest ones nodding; staminate and pistillate scales with a long, rough awn; perigynia 4.8–6.5(7.6) by 1.8–2.7 mm, the body 5–9-nerved and strongly inflated; beaks 2.2–4 mm long, their teeth 0.1–0.6 mm long; achenes finely papillate, not notched; swamps and wet woods; frequent; mostly N and SC; OBL.

Carex bullata Willd.   Bull sedge

Rhizomatous; fertile shoots 3–8.5 dm tall, sharply triangular below inflorescence; leaf blades flat or channeled, 1.5–4.3 mm wide; ligules wider than long; lowest bracts shorter than or exceeding inflorescence; males spikes 1–3, peduncles long; female spikes 1–5 by 1–2 cm, erect; pistillate scales obtuse to acuminate; perigynia spreading, 5.9–10.2 by 2.4–3.3 mm, the body 9–12-nerved; beaks rough near the tip, 2.4–4.2(4.8) mm long, their teeth 0.5–1 mm long; achenes smooth; swamps; rare; E; OBL; ❦.

Carex comosa Boott   Sedge

Tufted; fertile shoots 5–12 dm tall, sharply triangular below the inflorescence; leaf blades more or less flat, 5–16 mm wide; ligules much longer than wide; lowest bracts up to several times longer than the inflorescence; male spikes solitary, peduncles short; female spikes 2–5.5 by 1.1–1.8 cm, nodding; pistillate scales long, rough-awned; perigynia reflexed, (4.8)5.5–8.7 by 1.1–1.8 mm, the body 14–22-nerved and flattened-triangular in cross section, scarcely inflated, thick-walled; beaks (including teeth) 2–3.8 mm long, smooth, their teeth strongly spreading, 1.3–2.1(2.8) mm long; achenes smooth; swamps, marshes and swales; frequent NE and NW, occasional SC and SE; OBL.

Carex hystericina Willd.   Sedge

Tufted; fertile shoots 2–10 dm tall, rough triangular below the inflorescence; leaf blades flat, 2.5–8.5 mm wide; ligules often longer than wide; lowest bracts

Carex baileyi

Carex baileyi,
perigynium (×3)

Carex bullata,
perigynium (×3)

Carex comosa,
perigynium (×3)

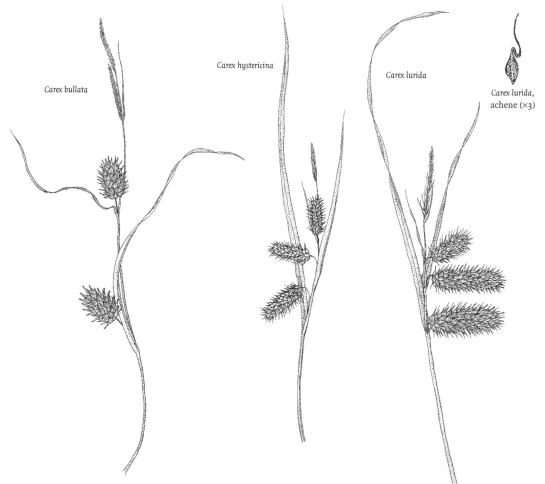

Carex bullata

Carex hystericina

Carex lurida

Carex lurida,
achene (×3)

Carex hystericina,
perigynium (×3)

Carex lurida,
perigynium (×3)

somewhat longer than the inflorescence; male spikes solitary, sometimes elevated above the female spikes; female spikes 1–4(6) by 1–1.5 cm, the lower ones nodding; pistillate scales with a long, rough awn; perigynia ascending or spreading, 4.5–7.3 by 1.4–2.1 mm, the body 13–20-nerved, more or less round in cross section, inflated and thin-walled; beaks 1.7–2.8 mm long, their teeth erect, 0.3–0.9 mm long; achenes smooth; marshes, swamps, and swales; frequent; S and NW; OBL.

Carex lurida Wahlenb.   Sedge
Tufted; fertile shoots 1.5–11 dm tall, usually obtuse-angled and smooth below the inflorescence; leaf blades 2.4–11.5 mm wide; ligules longer than wide; lowest bracts much longer than the inflorescence; male spikes solitary, peduncles short; female spikes 1–6.5 by 1.4–2.2 cm the lower ones often nodding; staminate and pistillate scales with a long, rough awn; perigynia (6)6.5–10.8 by (1.8)2–3.5(4.2) mm, the body 7–12-nerved and strongly inflated; beaks 2.5–5.9 mm long their teeth erect or spreading, 0.2–0.8 mm long; achenes finely papillate, not notched; swamps, wet meadows, and bogs; common; throughout; OBL.

Carex oligosperma Willd.   Few-seeded sedge
Rhizomatous; fertile shoots 3–9 dm tall, triangular and more or less rough below the inflorescence; leaf blades wiry and strongly involute, 0.5–2.5 mm wide;

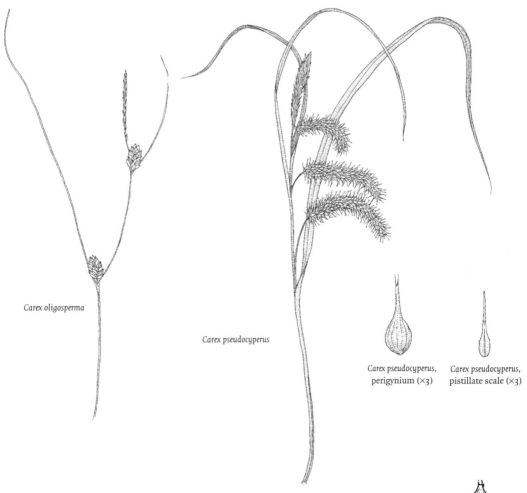

Carex oligosperma

Carex pseudocyperus

Carex pseudocyperus,
perigynium (×3)

Carex pseudocyperus,
pistillate scale (×3)

ligules much wider than long; lowest bracts shorter than to longer than the inflorescence; male spikes solitary, peduncles long; female spikes subglobose, 3–15-flowered, 0.7–2 by 0.6–0.9 cm; pistillate scales acute to acuminate; perigynia ascending, 4–6.7 by 2.5–3.4 mm, several-nerved, thick-walled, inflated; beaks 0.3–0.9 mm long, their teeth short, 0.1–0.3 mm long; achenes smooth; sphagnum bogs; rare; NC and NE; OBL; ❦.

Carex oligosperma,
perigynium (×3)

Carex pseudocyperus L.   Cyperus-like sedge
Tufted; fertile shoots 3.5–10 dm tall, rough triangular below the inflorescence; leaf blades flat with revolute margins, 4–13 mm wide; ligules usually much longer than wide; lowest bracts much longer than the inflorescence; males spikes solitary, peduncles short; female spikes 2–6 by 0.8–1.2 cm, nodding; pistillate scales long, rough-awned; perigynia reflexed, 3.4–6.1 by 1–1.7 mm, the body 12–20-nerved, flattened-triangular in cross section, scarcely inflated, thick-walled; beaks smooth, 1.2–2.2 mm long, their teeth more or less erect, 0.7–1.2(1.4) mm long; achenes smooth; calcareous swamps and swales; rare; NW; OBL; ❦.

Carex retrorsa Schwein.   Backward sedge
Loosely tufted; fertile shoots 1–10.5 dm tall, obtusely angled and smooth below the inflorescence; leaf blades 3–10 mm wide, more or less flat, their sheaths

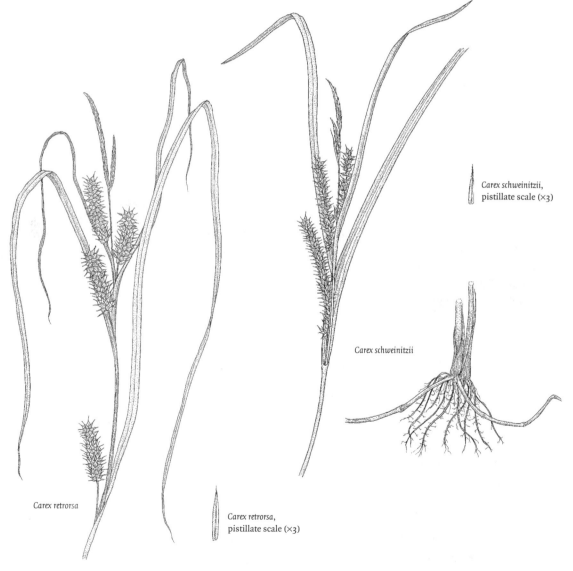

Carex retrorsa,
perigynium (×3)

Carex schweinitzii,
perigynium (×3)

spongy; ligules longer than to about as long as wide; lower bracts 2.5–9 times the length of the inflorescence; male spikes usually 1, peduncles short; pistillate scales acuminate; female spikes 1.5–5(8) by 1.2–2 cm; perigynia spreading or reflexed, 6–10 by (1.6)2.1–3.4 mm, the body 6–13-nerved and much inflated; beaks smooth or serrulate, 2.1–4.5 mm long, their teeth 0.3–1.1 mm long; achenes smooth; marshes, swales, wet thickets; scattered; mostly N and SC; FACW+; ✿.

Carex schweinitzii Schwein.   Schweinitz' sedge

Rhizomes long, slender; fertile shoots 2–6.5 dm tall, sharply triangular below inflorescence; leaf blades 4–11 mm wide; ligules wider than long; lowest bracts much longer than the inflorescence; male spikes usually solitary, peduncles short; female spikes 2.5–7.5(8) by 0.8–1.3(1.5) cm, spreading or erect; staminate scales awnless or cuspidate; pistillate scales with long, rough awns; perigynia 4.2–7 by 1.3–1.8 mm, the body 7–11-nerved and inflated; beaks 1.4–2.4 mm long, their teeth 0.2–0.5 mm long; achenes smooth; calcareous marshes and stream banks; rare; SC and Monroe Co.; OBL; ✿.

Carex schweinitzii,
pistillate scale (×3)

Carex schweinitzii

Carex retrorsa

Carex retrorsa,
pistillate scale (×3)

Carex tuckermanii Dewey   Sedge

Short rhizomatous; fertile shoots 4–12 dm tall, rough triangular below the inflorescence; leaf blades flat, 2–5 mm wide; ligules about as long as to longer than wide; male spikes usually 2, elevated above female spikes; female spikes 2–6 by 1.2–1.8 cm; pistillate scales acute or acuminate; perigynia erect or spreading, 7.5–12.5 by 4.5–6.7 mm, the body 7-12-nerved and inflated; beaks smooth, 2.4–4.8 mm long, their teeth 0.7–1.9 mm long; achenes deeply notched on one side; swampy woods and bogs; infrequent; mostly W; OBL.

Carex utriculata Boott   Sedge

Colonial with long, deep-seated rhizomes; fertile shoots 2.5–10 dm tall, obtusely angled and most often smooth below the inflorescence; leaf blades more or less flat, (3)4.5–15 mm wide, the lower sheaths spongy (septate-nodulose); ligules shorter than wide; lowest bracts 1–2.5 times as long as the inflorescence; male spikes 2–4, somewhat elevated above the female spikes; female spikes 2–15 by 0.9–2 cm; pistillate scales acute to short awned; perigynia widely spreading, (3.2)4–8.6 by 1.7–3 mm, the body 9–15-nerved and inflated; beaks 1–2.7 mm long, their teeth 0.2–1.3 mm long; achenes smooth; swamps. bogs, and lake shores; locally frequent NE, occasional to rare elsewhere; OBL.

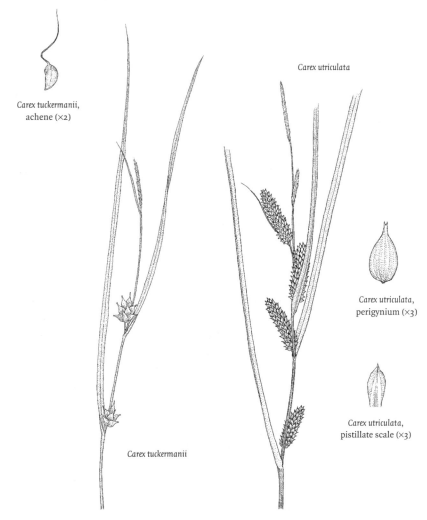

Carex utriculata

Carex tuckermanii,
achene (×2)

Carex utriculata,
perigynium (×3)

Carex utriculata,
pistillate scale (×3)

Carex tuckermanii

Carex vesicaria,
pistillate scale (×3)

Carex vesicaria

Carex vesicaria,
perigynium (×3)

Carex vesicaria L. Sedge

Tufted with stout, short-creeping rhizomes; fertile shoots 1.5–10.5 dm tall, acutely angled and rough below the inflorescence; leaf blades flat, 1.8–6.5 mm wide, their sheaths not spongy, often purple-tinged; ligules longer than wide; lowest bracts 1–2.5 times as long as the inflorescence; male spikes 2–4, elevated above the female spikes; female spikes 2–7.5 by 0.7–1.5 cm, erect; pistillate scales acute to long-acuminate; perigynia ascending, (3.6)4–7.5(8.2) by 1.7–3.5(4.5) mm, the body 7–12-nerved and inflated; beaks 1.1–2.6 mm long, their teeth 0.3–0.9 mm long; achenes smooth; swamps, wet meadows, ditches, and wet woods; locally frequent SC, SE, and NW, scattered or rare elsewhere; OBL.

Subgenus *Vignea*

Inflorescences of 2–many spikes; lower bracts sheathless; spikes usually bisexual and not strongly dimorphic (except for *C. praegracilis* and *C. sterilis* with occasional male inflorescences or male spikes); lateral spikes sessile, usually ovoid to globose, occasionally oblong to almost cylindrical; achenes 2-sided and stigmas 2.

Key to Sections of *Carex* subgenus *Vignea*

A. shoots in tufts (i.e., cespitose), plants lacking coarse-elongate rhizomes (rarely having compactly creeping rootstocks or slender stolons)
  B. male flowers (or their remnant filaments and/or empty scales) at the apex of some or all spikes; inflorescences compound (i.e., the lowest nodes with 2 or more spikes or with spike-bearing branches) in some common species
    C. perigynia <6 per spike, more or less round in cross section, and essentially beakless; plants of very delicate stature (<4 dm tall) and interconnected by stolons .... sect. *Dispermae*
    C. perigynia 5 or more per spike, planoconvex, clearly beaked; plants generally of coarse stature (3–12 dm tall)
      D. inflorescences simple (unbranched, one spike per node), spikes seldom >10 .... sect. *Phaestoglochin*
      D. inflorescences compound, spikes usually 10 to many
        E. inflorescence peduncles stout (i.e., about 1.5 mm or more thick at 3 cm below the inflorescence), soft and loosely cellular (flattening under pressure), often wing-angled; perigynia more or less truncate at the base due to spongy thickenings .... sect. *Vulpinae*
        E. inflorescence peduncles more slender, firm, and more or less wiry (due to prominent veins), sometimes angled on the edges but not winged; perigynia rounded to cuneate at the base
          F. widest leaves 3.5–10 mm wide; perigynia 3–4.5(5.6) mm long .... some individuals of sect. *Phaestoglochin*
          F. widest leaves 4 mm wide or less; perigynia 2–3.2 mm long
            G. leaf sheaths more or less smooth on ventral surface; pistillate scales obtuse to short-pointed; bracts short, inconspicuous, or absent .... sect. *Heleoglochin*
            G. leaf sheaths cross-puckered on ventral surface; pistillate scales awned; bracts (at least the lower ones) very slender and longer than their adjacent spike .... sect. *Multiflorae*

B. males flowers found at the base of the spikes (especially the terminal one); inflorescences rarely compound

    H. perigynia with winged margins, somewhat scale-like, not spongy-thickened at the base .... sect. *Ovales*

    H. perigynia round- or thin-edged, often plump, spongy, corky, or wrinkled at the base at maturity

        I. achenes more or less filling the bodies of the perigynia; perigynia elliptic with rounded edges, 1.8–3.7 mm long .... sect. *Glareosae*

        I. achenes occupying only the upper ½–⅔ of the perigynium body; perigynia triangular-ovate, ovate, or lanceolate, their edges thin, 2–5.5 mm long

            J. perigynia appressed-ascending, 4–5.5 mm long .... sect. *Deweyanae*

            J. perigynia spreading or reflexed, 2–4 mm long ..... sect. *Stellulatae*

A. shoots solitary, forming a colony by coarse, elongating rhizomes or from old prostrate shoots

    K. fertile shoots arising from nodes of last year's prostrate shoots

        L. perigynia strongly flattened, distinctly wing-edged; inflorescences more or less oblong, 1–6 cm long; plants associated with non-sphagnum wetlands .... some individuals of C. *longii* and C. *tribuloides* in sect. *Ovales*

        L. perigynia plump, planoconvex, not wing- or sharp-edged; inflorescences ovoid, 0.5–1.8 cm long; plants of sphagnum bogs .... sect. *Chordorrhizae*

    K. fertile shoots arising from subterranean rhizomes

        M. perigynia 4.7–6 mm long, thin-winged above; beaks sharply bidentate, their teeth often >0.5 mm long; leaf sheaths with hyaline band on ventral surface .... sect. *Ammoglochin*

        M. perigynia 2–4.5 mm long, thin but not winged; beaks obscurely bidentate; leaf sheaths with hyaline band or green and veined on ventral surface

            N. upper sheaths green and veined on ventral surface, the summit prolonged above the base of the leaf blade; inflorescences 3–6 cm long; plants of freshwater swamps .... sect. *Intermediae*

            N. upper sheaths with hyaline band on ventral surface, the summit truncate; inflorescences 1.5–3.5 cm long; plants introduced in saline roadside habitats .... sect. *Divisae*

Section *Ammoglochin*

Carex siccata Dewey   Sedge

Rhizomes long-creeping, shoots arising singly or in small clumps; fertile shoots 2–8 dm tall; leaf blades 1–3 mm wide; leaf sheaths with broad, smooth hyaline band on ventral surface; inflorescences simple, (1.5)2–3.5 cm long, their peduncles firm, not wing-angled; spikes male at the base, all female, all male, or even male at the apex; pistillate scales acute to short-pointed; perigynia planoconvex, more or less winged above, 4.4–6 by 1.6–2 mm; beaks abruptly contracted, sharply bidentate, 1.7–2.6 mm long; sandy open woods and fields; infrequent; mostly SE; FAC+; 🍂.

Carex siccata, perigynium (×3)

Carex siccata, pistillate scale (×3)

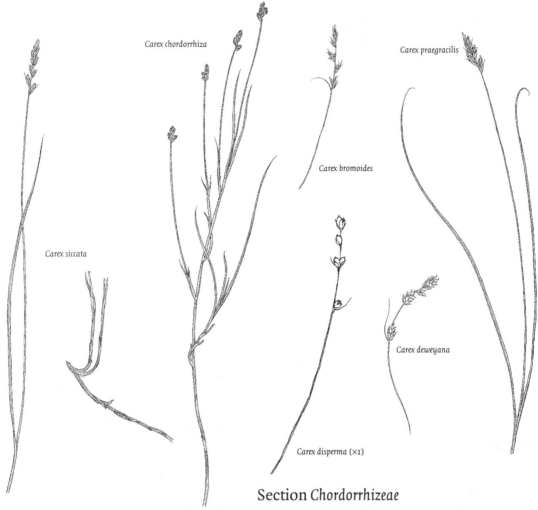

Carex chordorrhiza

Carex praegracilis

Carex bromoides

Carex siccata

Carex deweyana

Carex disperma (×1)

Carex chordorrhiza, pistillate scale (×3)

Section *Chordorrhizeae*

Carex chordorrhiza, perigynium (×3)

Carex chordorrhiza L.f.   Creeping sedge
Fertile shoots 1–4 dm tall, becoming prostrate and sending up new shoots from the nodes; leaf blades 1–2.5 mm wide; inflorescences simple, 0.5–1.8 cm long, peduncles firm, not wing-angled; spikes male at apex, few-flowered and aggregated; pistillate scales acute or acuminate, mostly longer than the perigynium body; perigynia thickly planoconvex, 2–3.8 by 1.2–1.8 mm, lustrous brown, nerved on both faces; beaks abrupt and short, 0.3–0.8 mm long; sphagnum bog; rare; Tioga Co.; OBL; ✹.

Section *Deweyanae*

Tufted; leaf sheaths not cross-puckered on ventral surface; inflorescences simple, their peduncles firm, not wing-angled; terminal spikes male at the base (or sometimes with a separate, predominately male spike at its base); perigynia planoconvex, appressed-ascending, the body narrowly elliptical to lanceolate, strongly spongy at the base; achenes occupying only the upper $1/2$–$2/3$ of the perigynium body.

A. leaves 1–2.5 mm wide; perigynia 0.8–1.3 mm wide and strongly nerved (especially on outer face); plants of wetlands .... *C. bromoides*
A. leaves 2–5 mm wide; perigynia 1.3–1.7 mm wide, more or less nerveless on both faces; plants of mesic woods .... *C. deweyana*

Carex bromoides Willd. Sedge
Tufted; fertile shoots 2.5–8 dm tall; leaves 1–2.5 mm wide; inflorescences 2–5.5 cm long; spikes lance-cylindric; pistillate scales often orange- or brown-tinged, obtuse to acuminate (or short-awned); perigynia 4–5.5 by 0.8–1.3 mm, nerved on both faces; beaks tapering, serrulate, 1.2–2.2 mm long; swamps and wet woods; frequent; throughout; FACW.

Carex bromoides,
perigynium (×3)

Carex bromoides,
pistillate scale (×3)

Carex deweyana Schwein. Sedge
Fertile shoots 2–10 dm tall; leaves 2–5 mm wide; inflorescences 2–5 cm long; spikes ovoid or oblong; pistillate scales pale, acute to awned; perigynia (3.5)4–5(5.5) by 1.3–1.7 mm, inner face nerveless, sometimes obscurely nerved at base of outer face; beaks tapering, serrulate, 1.2–2 mm long; rich often rocky woods; frequent; N; FACU.

Carex deweyana,
perigynium (×3)

Section *Dispermae*

Carex disperma Dewey Soft-leaved sedge
Loosely tufted, slightly stoloniferous; fertile shoots 1–4 dm tall; leaf blades 0.7–1.5 mm wide; inflorescences simple, 1–2.8 cm long, their peduncles firm, not wing-angled; spikes male at apex, the lower one separated, perigynia 1–6 per spike; bracts bristle-like usually <1 cm long; pistillate scales acuminate or short-pointed; perigynia ellipsoid, more or less round in cross section, spongy at the base, 2.2–3 by 1.2–1.7 mm; beaks abruptly contracted, minute (<0.25 mm long); bogs and wet acidic woods; infrequent; N; FACW+; ✿.

Carex disperma,
perigynium (×5)

Carex disperma,
pistillate scale (×5)

Section *Divisae*

Carex praegracilis Boott Freeway sedge
Rootstock long-creeping, stout, shoots arising singly or in small clumps; fertile shoots 2–7.5 dm tall; leaf blades 1.5–3 mm wide; leaf sheaths with smooth hyaline band on ventral surface; inflorescences simple, 1.5–3.5 cm long, their peduncles firm, not wing-angled; spikes male at apex or sometimes entirely male; pistillate scales acute to short-pointed; perigynia planoconvex, spongy at base, ovate, 3–3.8 by 1.2–1.7 mm, the inner face nerveless; beaks tapered, 0.9–1.4 mm long; rare; Erie Co.; UPL; native to western U.S.

Carex praegracilis,
perigynium (×3)

Carex praegracilis,
pistillate scale (×3)

Section *Glareosae*

Tufted (loosely stoloniferous in *C. trisperma*); inflorescences simple, their peduncles firm, not wing-angled; spikes male at base or sometimes all female; perigynia planoconvex, elliptic, their margins rounded or only slightly thin-edged, the body spongy at the base and more or less filled by the achene.

A. spikes 1–3, each containing only 1–5 perigynia; lowest bracts bristle-like, many times longer than the spike; perigynia 2.7–3.7 mm long ..... *C. trisperma*
A. spikes 3–many, each containing 5–35 perigynia; lowest bracts short or not apparent, at most about 2 times as long as the spike; perigynia 1.8–2.8 mm long
  B. leaves deep-green, the widest ones 1–2.5 mm wide; spikes 3–7 mm long, subglobose; perigynia 5–10 per spike and loosely spreading ....
  .... *C. brunnescens*

B. leaves gray-green, the widest ones 2–4 mm wide; spikes (4)6–12 mm long, ovoid to oblong; perigynia 10–35 per spike and appressed-ascending .... *C. canescens*

Carex brunnescens, perigynium (×3)

Carex brunnescens, pistillate scale (×3)

Carex brunnescens (Pers.) Poir.   Sedge
Fertile shoots 7–70 cm tall; leaf blades 1–2.5 mm wide, deep-green; inflorescences 1.5–5 cm long; lowest bracts up to 2 times as long as its spike; spikes 5–10 per inflorescence, subglobose, 3–7 mm long, each containing 5–10 perigynia; pistillate scales rounded to acute; perigynia loosely spreading, 2–2.6 by 0.9–1.4 mm; beaks serrulate, 0.3–0.6 mm long; borders of bogs and acidic woods; frequent NE and NW, local SW; FACW.

Carex canescens var. *canescens*, perigynium (×3)

Carex canescens L.   Sedge
Fertile shoots 1–8 dm tall; leaf blades 1.5–4 mm wide, gray-green; inflorescences 2–15 cm long; lowest bracts up to 2 times as long as its spike; spikes 4–8 per inflorescence, ovoid to oblong, (4)6–12 mm long, each containing 10–35 perigynia; pistillate scales acute to short-awned; perigynia appressed-ascending, 1.8–2.8 by 1–1.5 mm; beaks smooth (or serrulate at base), ca. 0.1–0.4 mm long; OBL; 2 varieties:

A. inflorescences 2–7 cm long, their spikes approximate (the lowest ones only 0.5–2.5 cm apart) .... var. *canescens*

Carex canescens var. *canescens*, pistillate scale (×3)

boggy areas; rare; throughout except SE.
A. inflorescences 6–15 cm long, their spikes distant (the lowest ones 2–4 cm apart) .... var. *disjuncta* Fernald sphagnum bogs and wet meadows, woods, and shores; E and NW, rare elsewhere.

Carex trisperma, perigynium (×3)

Carex trisperma Dewey   Sedge
Fertile shoots 2–5 dm tall, slender; leaf blades 0.5–2 mm wide, deep-green; inflorescences 1–5.5 cm long; lowest bracts mostly 3–10 cm long, very slender and bristle-like; spikes 1–3 per inflorescence, widely spaced, each containing 1–5 perigynia; pistillate scales acute; perigynia ascending, 2.7–3.7 by 1–1.7 mm; beak smooth, 0.4–0.7 mm long; bogs and wet woods; frequent, mostly N; OBL.

Section *Heleoglochin*

Tufted; leaf sheaths not cross-puckered on ventral surface; inflorescences compound, their peduncles firm, not wing-angled; bracts short or absent; spikes male at apex or all female; perigynia planoconvex, the base spongy, margins thin.

A. sheaths copper-tinged at mouths; inflorescences interrupted (generally 2 or more internodes exposed); perigynia dull yellow to brown, lanceolate to elliptic (2–3 times as long as wide), appressed .... *C. prairea*
A. sheaths pale or sometimes with red dots; inflorescences more or less dense (0–1 internodes exposed); perigynia lustrous olive to dark brown, broadly elliptic (1.6–2.2 times longer than wide), spreading .... *C. diandra*

Carex diandra, perigynium (×5)

Carex diandra Schrank   Lesser panicled sedge
Fertile shoots 3–7 dm tall; leaf sheaths pale or sometimes with red dots at mouth; leaf blades 1–2.5(3) mm wide, not overtopping the fertile shoots; inflo-

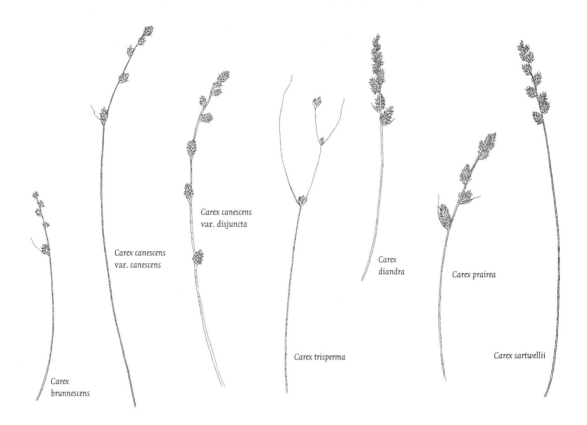

Carex canescens
var. disjuncta

Carex canescens
var. canescens

Carex
diandra

Carex prairea

Carex trisperma

Carex sartwellii

Carex
brunnescens

rescences 2.5–5(8) cm long, the spikes aggregated; pistillate scales obtuse to acute, about as long as the perigynia; perigynia spreading, 2.2–2.6 by 1–1.3 mm, broadly elliptic, lustrous olive or dark brown; beaks serrulate, tapering or abruptly contracted, 0.5–1 mm long; bog hummocks and pond margins; rare; extreme NE and NW; OBL; ❦.

Carex diandra,
pistillate scale (×5)

Carex prairea Dewey   Prairea sedge
Fertile shoots 5–10 dm tall; leaf sheaths copper-tinged at mouth; leaf blades 2–3 mm wide, not overtopping the fertile shoots; inflorescences 4–8(10) cm long, the upper spikes aggregated, the lower ones separated; pistillate scales acute to acuminate, about as long as the perigynia; perigynia appressed, 2–3 by 0.9–1.3 mm, elliptic or lanceolate, dull yellow to brown; beaks tapering, 0.8–1.1 mm long; wet calcareous marshes and fens; scattered; throughout; FACW; ❦.

Carex prairea,
perigynium (×5)

Carex prairea,
pistillate scale (×5)

Section Intermediae

Carex sartwellii Dewey   Sartwell's sedge
Rootstocks long-creeping, shoots arising singly or in small clumps; fertile shoots 4–10 dm tall; leaf blades 2–4.5 mm wide; leaf sheaths green and veined on ventral surface; inflorescences simple, (2)3–6 cm long, peduncles firm, not wing-angled; spikes male at apex, sometimes all male or female; pistillate scales obtuse to short-pointed; perigynia planoconvex, 2.8–4 by 1.4–1.7 mm; beaks abruptly contracted, 0.7–1 mm long; swamps; rare; NW; OBL; ❦.

Carex sartwellii,
perigynium (×3)

Carex sartwellii,
pistillate scale (×3)

Carex annectens

Carex annectens,
perigynium (×5)

Section *Multiflorae*

Tufted; leaf sheaths (at least lower ones) cross-puckered on ventral surface; inflorescences frequently compound, peduncles firm, not wing-angled; spikes typically 10–many, male at apex or all female; perigynia planoconvex, somewhat spongy at base, sharp-edged, yellow or brown, beaked. Key provided by A.A. Reznicek.

A. perigynia abruptly contracted into beak 0.25–0.55 times as long as the body; larger perigynia 1.5–2.4 mm wide; fertile shoots exceeding leaves at maturity ....*C. annectens*
A. perigynia tapered into beak 0.5–1 times as long as the body; larger perigynia 1–1.9 mm wide; fertile shoots often shorter than to more or less equaling the leaves even at maturity ....*C. vulpinoidea*

Carex annectens,
pistillate scale (×5)

Carex annectens (E.P.Bicknell) E.P.Bicknell   Sedge
Fertile shoots 4–9 dm tall; leaves 2–4 mm wide, shorter than the fertile shoot; inflorescences 2–7(10) cm long; pistillate scales awned; perigynia ovate, 2.5–3.2 by 1.5–2.4 mm; beaks abruptly contracted, 0.6–1 mm long; dry or seasonally moist fields, open woods, and ditches; common; throughout; FACW. [syn: *C. vulpinoidea* Michx. var. *ambigua* F. Boott.]

Carex vulpinoidea,
perigynium (×5)

Carex vulpinoidea Michx.   Sedge
Fertile shoots 3–9 dm tall; leaves 2–4 mm wide, overtopping the fertile shoots; inflorescences 3–10(15) cm long; pistillate scales awned; perigynia ovate, 2–2.7 by 1–1.9 mm; beaks tapered, 0.5–1 mm long; swampy places and ditches; common; throughout; OBL.

Carex vulpinoidea,
pistillate scale (×5)

Section *Ovales*

Mostly tufted (occasionally in *C. longii* and *C. tribuloides* fertile shoots may arise from last year's prostrate shoots); leaf sheaths smooth or finely papillose dorsally, not cross-puckered on ventral surface; inflorescences simple, peduncles firm, not wing-angled; terminal spikes male at base, the lateral spikes male at base or occasionally all pistillate; perigynia mostly yellowish- or reddish-brown, flat to planoconvex, the margins narrowly to broadly thin winged and serrulate; beaks conspicuous.

The species in section *Ovales* are the most difficult to identify in the entire genus. In learning these species, special care must be taken to collect plants with mature fruit. Careful measurements at 10× magnification (or higher) are recommended. In analyzing the perigynia, avoid those with atypical size and shape at the very base or in the upper ¼ of a spike. Differences in the habitat and the form of sterile shoots can also be very helpful in learning to recognize the different species.

A. pistillate scales reaching only to the base or middle of the beaks (not covering the beaks of mature perigynia), the apex awned in a few species; plants of wet to dry habitats
   B. perigynia 2 mm wide or less
      C. pistillate scales obtuse or merely acute, sometimes inconspicuous
         D. leaf sheaths loose and expanded near the summit, uniformly pale or greenish, sharp-edged or bearing narrow wings continuous with the midrib and/or edges of the leaf blade; widest blades 3–7 mm wide;

perigynia often thin and scale-like, the wings often conspicuously narrowed (even indented) near the middle of the body to form a cuneate base; leaves of vegetative shoots spreading and spaced along distal ½ of the shoot

E. perigynia stiffly spreading or recurved, the spikes more or less globose; pistillate scales usually hidden between perigynia (1.6–2.3 mm long); summit of leaf sheaths and/or ligules often rust-tinged .... C. cristatella

E. perigynia loosely spreading, ascending, or appressed, the spikes subglobose to ovoid-oblong; pistillate scales evident (2–3 mm long); summit of leaf sheaths and ligules not colored

F. inflorescences stiffly oblong, spikes overlapping; perigynia more than 30, beaks appressed-ascending; leaf sheaths firm at summit .... C. tribuloides

F. inflorescences flexuous, the lower spikes usually separated; perigynia 15–40, loosely spreading or ascending; leaf sheaths friable at summit .... C. projecta

D. leaf sheaths usually tight, the edges more or less rounded, unwinged; widest blades 1–4(4.2) mm wide (or blades wider in C. normalis in which the sheaths generally have green veins contrasting with white intervenal areas); perigynia usually plumply planoconvex, the wings typically tapered evenly to the round or sometimes cuneate base; vegetative shoots with leaves ascending and clustered at apex

G. perigynia narrowly lanceolate (2.5–4 times as long as wide), the distance from beak tip to top of achene mostly 2.2–5 mm (about ⅗ the length of the perigynia)

H. inflorescences dense to open or arched, the lowest internodes 2–17 mm long; perigynia 1.2–2 mm wide; achenes 0.7–1.1 mm wide .... C. scoparia

H. inflorescences dense, the lowest internodes 2–3(5) mm long; perigynia 0.9–1.3 mm wide; achenes 0.6–0.8 mm wide ....
.... C. crawfordii

G. perigynia obovate, circular, ovate, or lanceolate (<2.5 times as long as wide), the distance from beak tip to top of achene 0.8–2.2 mm (about ½ or less the length of the perigynia)

I. perigynia lanceolate, ovate, or round, widest at or below the middle of the body

J. inflorescences dense, more or less head-like, lowest internodes <6 mm long

K. perigynia nerveless on inner face, 1.2–2 mm wide, rust-colored; plants of wet habitat .... C. bebbii

K. perigynia nerved on inner face or if nerveless then 1.8–3 mm wide, not rust-colored; plants of woods and fields

L. leaf blades 2.5–6.5 mm wide, the mouth of sheaths prolonged up to 2 mm above base of leaf blades; perigynia lanceolate (about 2–2.5 times as long as wide) and usually greenish- or greenish-brown-tinged .... C. normalis

L. leaf blades 1.5–4 mm wide, the mouth of sheaths concave or at most shortly prolonged above base of leaf blades; perigynia broadly elliptic (mostly 1.5–2 times as long as wide), quickly turning pale- or yellowish-brown .... *C. molesta*

 J. inflorescences usually open at least proximally, lowest internodes >6 mm long

  M. perigynia more or less circular, abruptly narrowed to beak; leaves 1–3.5 mm wide; lower pistillate scales (and adjacent staminate scales) acute .... *C. festucacea*

  M. perigynia ovate or lanceolate, tapering to beak; leaves 1.3–6.5 mm wide; lower pistillate scales (and adjacent staminate scales) obtuse

   N. inflorescences stiff, the lower spikes usually overlapping; fertile shoots with 3–7 leaves, the blades 2.5–6.5 mm wide; sheaths loose and often with green veins contrasting with pale or white intervenal areas; perigynia spreading, usually greenish- or greenish-brown-tinged .... *C. normalis*

   N. inflorescences open, often flexuous and nodding, the lower spikes usually distant; fertile shoots with 3–4 leaves, the blades 1.3–2.5(3) mm wide, the sheaths tight and evenly colored; perigynia appressed or ascending, becoming pale- or reddish-brown............
.... *C. tenera*

 I. perigynia obovate, widest above the middle of the body

  O. beaks appressed-ascending, triangular, and gradually tapered from the body; pistillate scales obtuse; styles straight
.... *C. longii*

  O. beaks spreading, slender, and abruptly contracted from the body; pistillate scales acute; styles with a strong S-loop at the base .... *C. albolutescens*

C. pistillate scales acuminate with a subulate or awn-like tip

 P. perigynia generally ascending, narrowly lanceolate (2.5–4 times as long as wide).. go to couplet H

 P. perigynia spreading, broadly elliptic (<2.5 times as long as wide) ...
.... *C. straminea*

B. perigynia >2 mm wide

 Q. pistillate scales obtuse or merely acute

  R. perigynia lanceolate, ovate, elliptic, or circular, widest at or below the middle of the body; leaf sheaths various, some with a prominent hyaline band ventrally; achenes 0.9–1.7 mm wide

   S. perigynia thin (the thickness of the achene typically more or less equally visible on both faces of the perigynium), 4.5–6.5 mm long and with translucent coppery wings, 5 or more sharp nerves on the inner face .... *C. bicknellii*

   S. perigynia planoconvex, 2.5–4.8 mm long, the wings white hyaline, the inner face 0–5-nerved

    T. perigynia lanceolate (about 2–2.5 times as long as wide), greenish- or greenish-brown-tinged, the beaks gradually tapered; leaves 2.5–6.5 mm wide .... *C. normalis*

    T. perigynia broadly ovate to circular (about 1–2 times as long as wide), quickly turning pale- or yellow-brown, the beaks often abruptly contract from body; leaves 1–3.5 mm wide

U. spikes on larger inflorescences 2–4 (rarely more), rounded at the base, the terminal one lacking a conspicuous male base; inflorescences 1.3–3 cm long (the lowest internodes generally 1.5–6 mm long); perigynium bodies elliptic to ovate (rarely more or less circular; 1–1.6 times as long as wide) .... C. molesta

U. spikes on larger inflorescences (4)5–7 or more, tapered at the base, the terminal one with a conspicuous male base; inflorescences typically 2.5–4.5(6) cm long (the lowest internodes generally 5–13 mm long); perigynium bodies broadly ovate to circular (mostly 0.9–1.2 times as long as wide)

V. larger perigynia 3.2–4.7(5) mm long, 2.5–3.3(3.5) mm wide, the inner face usually nerveless; achenes 1.3–1.7 mm wide .... C. brevior

V. larger perigynia 2.5–4(4.2) mm long, 1.5–2.4(2.6) mm wide, the inner face mostly 2–4-nerved; achenes 0.9–1.2 mm wide .... C. festucacea

R. perigynia obovate, widest above the middle of the body; leaf sheaths green-nerved ventrally nearly to the summit; achenes 0.7–1.2 mm wide

W. beaks spreading, slender, and abruptly contracted from the body; pistillate scales acute; styles with a strong S-loop at the base .... C. albolutescens

W. beaks appressed-ascending, triangular, and gradually tapered from the body; pistillate scales obtuse or acute; styles straight or occasionally bent

X. perigynia nerveless on inner face; widest leaves 3–6 mm wide, their sheaths truncate at summit and extending 0–3 mm above base of leaf blade .... C. cumulata

X. perigynia nerved on inner face; widest leaves 2–4 mm wide, their sheaths concave at summit and not prolonged above base of leaf blade .... C. longii

Q. pistillate scales acuminate with a subulate tip or awn up to 0.8 mm long

Y. widest leaves 2.5–6 mm wide; inflorescences stiff, spikes approximate or densely aggregated; perigynia strongly obovate, yellow-brown, and very faintly nerved, 2.5–4 mm wide .... C. alata

Y. widest leaves 1.5–3 mm wide; inflorescences arched, spikes separated; perigynia broadly elliptical or sometimes weakly obovate, rust-colored, and sharply 4–7-nerved on each face, 1.9–2.8 mm wide .... C. straminea

A. pistillate scales nearly as long as or longer than mature perigynia, usually concealing the beaks of the perigynia, the apex acute or obtuse but never awned; plants of dry habitats

Z. inflorescences arched or flexuous, up to 8 cm long, the lowest spikes usually remote

AA. perigynia erect-ascending, finely granular-papillose, the inner face strongly and evenly nerved; spikes (4)7–15, the distal ones crowded; achenes 1–1.2 mm wide .... C. argyrantha

AA. perigynia ascending to spreading, smooth, the inner face often nerveless or with few curved nerves; spikes 3–7(11), the distal ones more or less separated; achenes 1.2–1.7 mm wide .... C. foenea

Z. inflorescences stiff, dense to open, up to 5 cm long, the lowest spikes aggregated or approximate

BB. pistillate scales distinctly narrower than the body of the perigynia; perigynia thin (ca. 0.5 mm thick); achenes <1.2 mm wide .... *C. ovalis*

BB. pistillate scales about as wide as the body of perigynia; perigynia plump (about 0.8 mm thick); achenes 1.7–2 mm wide .... *C. adusta*

Carex adusta,
perigynium (×3)

Carex adusta,
pistillate scale (×3)

Carex adusta Boott   Crowded sedge

Fertile shoots 2.5–8 dm tall; leaf sheaths with hyaline summit prolonged up to 1.5 mm above base of leaf blade; leaf blades 2–3.8 mm wide; inflorescences dense or open, 2–4.5 cm, the spikes acute to rounded at base and tip; pistillate scales about as long and wide as perigynia, acute; perigynia plump, 4–5 by 2–2.5 mm, the body ovate to broadly ovate, inner face essentially nerveless; beaks contracted, 1–1.5 mm long; achenes 1.7–2 mm wide; dry acidic soil in open woods; Northampton Co.; probably extirpated; 🌿.

Carex alata,
perigynium (×3)

Carex alata,
pistillate scale (×3)

Carex alata Torr.   Broad-winged sedge

Fertile shoots 3–14 dm tall; leaf sheaths green-ribbed ventrally nearly to summit; leaf blades 2–6 mm wide; inflorescences dense to open, 2–6.5 cm long, the spikes with rounded or short-acute bases and acute to obtuse tips; pistillate scales reaching base of beak, sharply acuminate or with awn up to 0.8 mm long; perigynia yellow-brown, 4–5.5 by 2.5–4 mm, the body obovate and very faintly nerved on inner face; beaks abruptly contracted, 0.8–1.8 mm long; achenes 0.9–1.1 mm wide; swamps, peaty shores, wet thickets, and woods, usually on calcareous soils; rare; NW and lower Delaware River; 🌿.

Carex albolutescens,
perigynium (×3)

Carex albolutescens,
pistillate scale (×3)

Carex albolutescens Schwein.   Sedge

Fertile shoots 2.5–12 dm tall; leaf sheaths green-ribbed ventrally nearly to summit; leaf blades 2–3.5 mm wide; inflorescences open, 1.5–4 cm long, the spikes with acute bases and rounded tips; pistillate scales reaching bases of beaks, acute; perigynia green or light brown, 2.6–4.5 by 1.5–2.7 mm, the body obovate and with 4–many nerves on each face; beaks abruptly contracted, 0.9–1.5 mm long, spreading; achenes 0.7–1 mm wide, their styles strongly sinuous near the base; acidic swamps, low woods, and thickets; occasional; mostly S; FACW.

Carex argyrantha,
perigynium (×3)

Carex argyrantha,
pistillate scale (×3)

Carex argyrantha Tuck.   Sedge

Fertile shoots 3–10 dm tall; leaf sheaths very finely papillose, well-defined hyaline band on ventral surface; leaf blades 2–4.5 mm wide; inflorescences crowded distally, arched or nodding, 2.5–6.5 cm long, the spikes with attenuate bases and rounded tips; pistillate scales about as long as the perigynia, usually covering the beaks, acute; perigynia matte silvery-green, 3–4.5 by 1.8–2.3 mm, the body elliptic and with 5–8 sharp, parallel nerves on each face; beaks tapered, 0.7–1 mm long; achenes 1–1.2 mm wide; dry and rocky (typically sandstone) woods and clearings; frequent E, occasional W.

Carex bebbii,
perigynium (×3)

Carex bebbii,
pistillate scale (×3)

Carex bebbii (Bailey) Fernald   Bebb's sedge

Fertile shoots 2–9 dm tall; sheaths hyaline ventrally; leaf blades 1.5–4.2 mm wide; inflorescences mostly dense, 1–3 cm long, the spikes rounded at base; pistillate scales reaching bases of beaks, acute or acuminate; perigynia rust colored, 2.5–3.8 by 1.2–2 mm wide, the body ovate or elliptic and with 0–3 faint nerves on the inner face; beaks tapered, 0.7–1.2 mm long; achenes 0.6–0.8 mm wide; calcareous or neutral wet meadows, moist sand flats, and shores; rare; NW, Centre, Huntingdon, and Monroe Cos.; OBL; 🌿.

Carex bicknellii Britton   Bicknell's sedge

Fertile shoots 3–12 dm tall; sheaths very finely papillose, the hyaline mouth often extending 1–2 mm above the base of leaf blade; leaf blades 2.2–4 mm wide; inflorescences open, 2–5.5 cm long, the spikes acute or tapered at base and rounded at tip; pistillate scales reaching base of beaks, more or less acute; perigynia 4.5–6.5 by 2.7–4 mm, the hyaline wings becoming copper-tinged, the body ovate or circular, 5 or more conspicuous nerves on each face; beaks gradually to abruptly narrowed, 1.2–2 mm long, sharply bidentate; achenes 1.2–1.5 mm wide; dry woods, thickets, fields, and serpentine barrens; rare; SE; 🌿.

Carex bicknellii,
perigynium (×3)

Carex bicknellii,
pistillate scale (×3)

Carex brevior (Dewey) Mack. ex Lunell   Sedge

Fertile shoots 2.5–11.5 dm tall; sheaths with well-defined ventral hyaline band sometimes prolonged up to 2 mm above the base of leaf blade; leaf blades 1.5–3.5 mm wide; inflorescences open, (1.5)2.5–4.5(5) cm long, the spikes (3)4–7, mostly tapered at base and acute-rounded at tip; pistillate scales reaching to middle of beaks, often brown-tinged, acute or obtuse; perigynia 2.6–4.7(5) by 2–3.3(3.5) mm, the body more or less circular, usually nerveless on the inner face; beaks abruptly contracted, 0.8–1.5 mm long, often sharply bidentate; achenes 1.3–1.7 mm wide; dry, open thickets, banks, fields and roadsides; infrequent; mostly SE; UPL; 🌿.

Carex brevior,
perigynium (×3)

Carex brevior,
pistillate scale (×3)

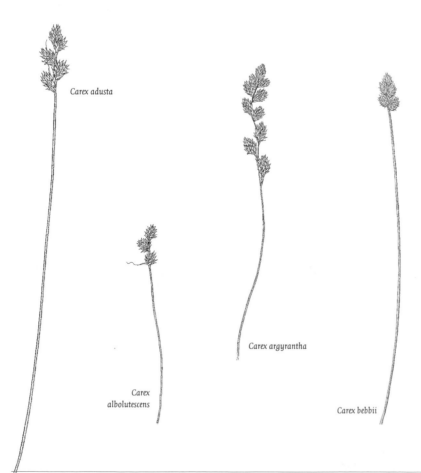

Carex adusta

Carex argyrantha

Carex albolutescens

Carex bebbii

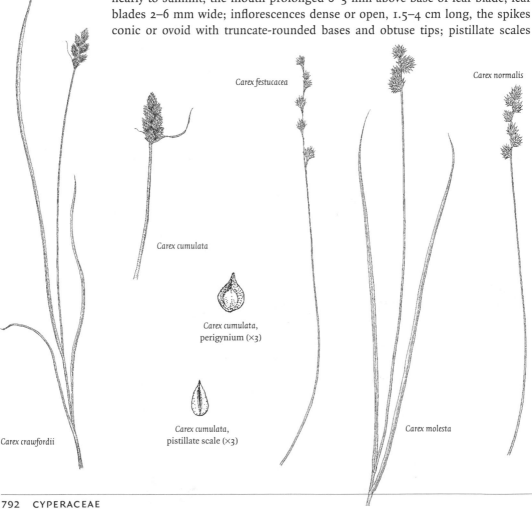

Carex crawfordii,
perigynium (×3)

Carex crawfordii,
pistillate scale (×3)

Carex cristatella,
perigynium (×3)

Carex cristatella,
pistillate scale (×3)

Carex crawfordii Fernald   Crawford's sedge

Fertile shoots 2.5–6 dm tall; leaf blades 2–4 mm wide; inflorescences dense, 1.8–3 cm long, the spikes acute-truncate at base and tip; pistillate scales reaching to middle of beaks, acuminate to subulate; perigynia 3.4–4.7 by 0.9–1.3 mm, the body narrowly lanceolate, gradually tapering to base and to beak, the inner face with 0–3 faint nerves; beaks about 1–1.6 mm long; achenes 0.6–0.8 mm wide; wet or dry, open, sandy soil; rare; Pike Co.; FAC; ✿.

Carex cristatella Britton   Sedge

Fertile shoots 3–10 dm tall; leaves of vegetative shoots spreading and evenly spaced along distal ½; sheaths loose and somewhat winged above, the mouth hyaline and often rust-tinged; leaf blades 3–7.5 mm wide; inflorescences dense or proximally open, 2–4.5 cm long, the spikes more or less globose; pistillate scales inconspicuous among perigynia; perigynia spreading to recurved, 2.7–4 by 1–1.8 mm, the body ovate–elliptic but abruptly narrowed near base, 2–6 conspicuous nerves on each face; beaks tapering, 0.8–1.2 mm long; achenes 0.6–0.8 mm wide; wet meadows, marshes, thickets, and stream banks; frequent SE, occasional elsewhere; FACW.

Carex cumulata (Bailey) Mack.   Sedge

Fertile shoots 2–8 dm tall; sheaths very finely papillose, green-ribbed ventrally nearly to summit, the mouth prolonged 0–3 mm above base of leaf blade; leaf blades 2–6 mm wide; inflorescences dense or open, 1.5–4 cm long, the spikes conic or ovoid with truncate-rounded bases and obtuse tips; pistillate scales

Carex festucacea

Carex normalis

Carex cumulata

Carex cumulata,
perigynium (×3)

Carex cumulata,
pistillate scale (×3)

Carex molesta

Carex crawfordii

reaching base of beaks, obtuse or acute; perigynia appressed, 3–4.2 by 2–3.2 mm wide, the body obovate, nerveless on the inner face; beaks broadly triangular and gradually tapering, 0.8–1.3 mm long; achenes 1–1.2 mm wide; mostly dry, sandy, or rocky soil of barrens, abandoned railroad beds, and thickets; scattered; NE; FACU.

Carex festucacea Schkuhr ex Willd.   Sedge

Fertile shoots 4.5–10 dm tall; sheaths green or with pale intervenal areas, the mouth prolonged ventrally 0–2 mm above the base of leaf blade; leaf blades 1–3.5 mm wide; inflorescences open and erect, 2.5–6 cm long, the spikes (3)5–7(10), acute–attenuate at base and rounded at tip; pistillate scales lanceolate, reaching base of beaks, acute; perigynia spreading, yellowish-brown, 2.5–4(4.2) by 1.5–2.4(2.6) mm, the body circular–elliptic, nerves on inner face (0)2–4; beaks abruptly contracted, 0.7–1.3 mm long; achenes 0.9–1.2 mm wide, the styles straight or bent; moist, open woods or thickets; frequent; mostly SC and SE; FAC.

Carex festucacea,
perigynium (×3)

Carex festucacea,
pistillate scale (×3)

Carex foenea Willd.   Fernald's hay sedge

Fertile shoots 2–12 dm tall; sheaths very finely papillose and with a well-defined ventral hyaline band; leaf blades 2–4 mm wide; inflorescences open, arched, or nodding, 1.5–8 cm long, the spikes attenuate at base and usually round at tip; pistillate scales about as long as the perigynia and usually covering the beaks, sometimes brown-tinged, acute; perigynia spreading or loosely ascending, 3.3–5 by 1.7–2.7, the body ovate, the inner face nerveless or sometimes with up to 5 curved nerves; beaks tapering, 1–1.5 mm long; achenes 1.2–1.7 mm wide; dry, gravelly or sandy banks; rare (perhaps extirpated); Centre and Wayne Cos.; ❦.

Carex foenea,
perigynium (×3)

Carex foenea,
pistillate scale (×3)

Carex longii Mack.   Long's sedge

Fertile shoots 3–12 dm tall; sheaths often very finely papillose, green-ribbed ventrally nearly to summit; leaf blades 2–4 mm wide; inflorescences usually open, 1–4.5 cm long, the spikes rounded or obtuse at base and obtuse to broadly acute at tip; pistillate scales reaching to middle of beaks, obtuse; perigynia appressed, 3–4.5 by 1.6–2.8 mm, the body obovate and with 5 or more conspicuous nerves on each face; beaks broadly triangular or gradually tapering, 0.8–1.3 mm long; achenes 0.7–1 mm wide, their styles straight; wet or seasonally wet sandy soils in swamps, thickets, and meadows; scattered; E; OBL; ❦.

Carex longii,
perigynium (×3)

Carex longii,
pistillate scale (×3)

Carex molesta Mack. ex Bright   Sedge

Fertile shoots 3.5–11 dm tall; sheaths green or rarely with pale intervenal areas and a narrow hyaline band ventrally near summit; leaf blades 1.5–3.5 mm wide; inflorescences stiff, 1.3–3 cm long, the spikes 2–4(5), somewhat aggregated and with rounded bases and tips; pistillate scales reaching base of beaks, acute; perigynia 3.2–4.8 by 1.8–3 mm, the body elliptic to ovate (rarely more or less circular), (0)3–6-nerved on the inner face; beaks tapering, 0.9–1.5 mm long; achenes 1–1.3 mm wide; dry to wet often heavy soils in fields, roadsides, and open woods; frequent SE, infrequent SC and SW.

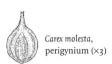

Carex molesta,
perigynium (×3)

Carex molesta,
pistillate scale (×3)

Carex normalis Mack.   Sedge

Fertile shoots 3.5–14 dm tall; sheaths with green veins and pale or white intervenal areas, the prominent ventral Y-shaped hyaline area prolonged up to 2 mm above the base of leaf blades; leaf blades 3–7 per shoot, 2.5–6.5 mm wide;

Carex normalis,
perigynium (×3)

Carex normalis,
pistillate scale (×3)

inflorescences usually stiff and open, 1.5–5(7) cm long, the spikes with rounded (sometimes tapering) bases and rounded tips; pistillate scales reaching base of beaks, obtuse or acute; perigynia green-tinged, 2.7–4.1 by 1.3–2.3 mm, the body lanceolate or elliptic, 4 or more nerves on each face; beaks gradually tapering, 1–1.6 mm long; achenes 0.9–1.2 mm wide; open, often moist, woods and meadows; common; throughout; FACU.

Carex ovalis Gooden.   Sedge

Carex ovalis,
perigynium (×3)

Carex ovalis,
pistillate scale (×3)

Fertile shoots 3.5–8.5 dm tall; sheaths with ventral hyaline band; leaf blades 1.5–4 mm wide; inflorescences 1.5–4 cm long, the spikes aggregated with tapering bases and rounded tips; pistillate scales reddish-brown, about as long as and concealing the perigynia, acute or acuminate; perigynia 3.4–5.2 by 1.3–2.2 mm, the body narrowly elliptic, 0–5 nerves on the inner face; beaks tapering or abruptly narrowed, distal 0.3 mm usually cylindric and entire; achenes 0.9–1.2 mm wide; Philadelphia Co.; FAC; native to Europe. [*C. leporina* L. of Rhoads & Klein, 1993]

Carex projecta Mack.   Sedge

Carex projecta,
perigynium (×3)

Carex projecta,
pistillate scale (×3)

Fertile shoots 5–9 dm tall; leaves of vegetative shoots spreading and evenly spaced along distal ½; sheaths loose and somewhat winged above, the ventral hyaline area narrow, untinged at the summit; leaf blades 3–7 mm wide; inflorescences open at least proximally, lax or flexuous, 2.5–6 cm long, the subglobose spikes with about 15–40 perigynia and acute bases; pistillate scales reaching base of beaks, obtuse to acute; perigynia loosely spreading or ascending, 2.9–4.5 by 1.2–1.8 mm, the body lanceolate or elliptic and often contracted below middle, 3 or more conspicuous nerves on each face; beaks gradually tapering, 0.9–1.5 mm long; achenes 0.6–1 mm wide; wet woods; occasional; throughout; FACW.

Carex scoparia Schkuhr ex Willd.   Broom sedge

Carex scoparia,
perigynium (×3)

Carex scoparia,
pistillate scale (×3)

Fertile shoots 2–10 dm tall; sheaths tight and lacking sharp edges; leaf blades 1.4–3.5 mm wide; inflorescences variable, dense to open and even arched, 1.5–6 cm long, the spikes acute at base and more or less acute or sometimes rounded at tip; pistillate scales reaching base of beaks, acute to long acuminate; perigynia 4–6.8 by 1.2–2 mm, the body lanceolate, approximately 5 nerves on each face; beaks gradually tapered, 1.2–2.4 mm long; achenes 0.7–1.1 mm wide; mostly moist open ground; common; throughout; FACW.

Carex straminea Willd.   Sedge

Carex straminea,
perigynium (×3)

Carex straminea,
pistillate scale (×3)

Fertile shoots 3.5–10 dm tall; sheaths green-ribbed ventrally nearly to summit; leaf blades 1.5–3 mm wide; inflorescences open and usually arched, 3.5–8 cm long, the spikes with attenuated bases and rounded tips; pistillate scales reaching base of beaks, sharply acuminate or with awns up to 0.8 mm long; perigynia rust-colored, widely spreading, 4–5.6 by 1.8–2.8 mm wide, the body broadly elliptic–obovate and with 5 or more nerves on each face; beaks abruptly narrowed, 1.2–2 mm long; achenes 0.8–1 mm wide; marshes, swales, and wet woods; locally frequent E, scattered W; OBL.

Carex tenera Dewey   Sedge

Fertile shoots 2–9 dm tall; sheaths tight, the ventral hyaline band extending up to 0.5–1.5 mm above base of leaf blade; leaf blades 3–5 per shoot, 1.3–2.5 mm

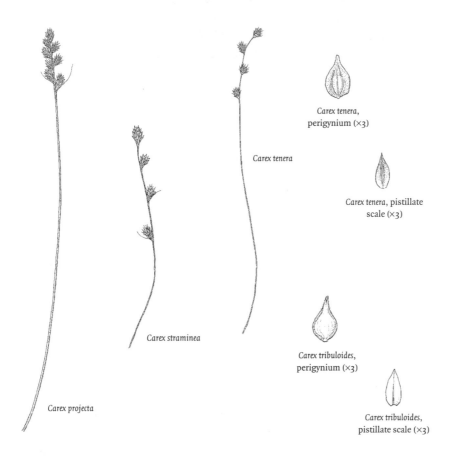

Carex tenera, perigynium (×3)

Carex tenera, pistillate scale (×3)

Carex tenera

Carex straminea

Carex tribuloides, perigynium (×3)

Carex projecta

Carex tribuloides, pistillate scale (×3)

wide; inflorescences open, typically flexuous or nodding above lowest spike, 2–5 cm long, the spikes acute to attenuate at base and obtuse–rounded at tip; pistillate scales ovate, reaching middle of beaks, obtuse or acute; perigynia appressed-ascending, pale- or reddish-brown, 2.8–4.2 by 1.4–2 mm, the body broadly elliptic or ovate, nerves on the inner face (0)3–5; beaks gradually to abruptly narrowed, 0.8–1.6 mm long; achenes 0.8–1.1 mm wide; dry–moist open woods and meadows; occasional; throughout except SW; FAC.

Carex tribuloides Wahl.   Sedge
Fertile shoots 5–11 dm tall; leaves of vegetative shoots spreading and evenly spaced along distal ½; sheaths loose and somewhat winged above, green-ribbed ventrally except for a Y-shaped, whitish hyaline area at summit; leaf blades 2.5–7 mm wide; inflorescences stiff, somewhat open proximally and dense above, 2–6 cm long, the spikes acute-rounded at base and rounded at tip; pistillate scales reaching base of beaks, acute or obtuse; perigynia ascending or appressed, 3–5.2 by 1–1.7 mm, the body ovate-elliptic and somewhat contracted below the middle, usually 3–5 sharp nerves on the inner face; beaks gradually tapered, 1–1.8 mm long; achenes 0.6–0.9 mm wide; wet woods and marshes; common; throughout; FACW+.

Section Phaestoglochin

Mostly tufted, sometimes with short creeping rootstocks, fertile shoots erect or angled with the ground but not prostrate and rooting; leaf sheaths bearing a smooth (sometimes cross-puckered in *C. sparganioides*) hyaline band on ventral surface; inflorescences usually simple, their peduncles stiff, the edges serrulate or occasionally smooth and sharp-edged but not distinctly winged; bracts in our species typically shorter than the inflorescence; spikes 3–10, male at apex; pistillate scales obtuse, acute, or awned; perigynium bodies sharp-edged, lanceolate to circular (or even more or less deltoid), planoconvex (except almost biconvex in *C. retroflexa*), beaked.

A. widest leaves 1–4.5 mm wide; sheaths tight, usually uniformly pale brown or green, cross-veins obscure or lacking
    B. inner face of mature perigynia spongy at base, the beaks widely spreading or reflexed at maturity; spikes often with only a few perigynia and widely separated
        C. margins of perigynia (especially the beaks) serrulate; pistillate scales acute or obtuse, much shorter than the perigynium body, persistent; spikes (especially the lowest) separated
            D. widest leaves 0.9–1.7 mm wide; base of fertile shoots 0.7–1.4 mm wide
                E. stigmas mostly coiled around the beak tip; base of perigynia cuneate, the distance from the base of perigynium to base of achene (often evident on outer perigynium face) 0.1–0.5 mm ....*C. appalachica*
                E. stigmas reflexed, otherwise more or less straight; base of perigynia truncate, the distance from the base of perigynium to base of achene (often evident on outer perigynium face) 0.5–0.9 mm... .... *C. radiata*
            D. widest leaves 1.7–2.8 mm wide; base of fertile shoots more than 1.4 mm wide
                F. stigmas straight to slightly twisted, slender (0.03–0.06 mm thick) .... *C. radiata*
                F. stigmas mostly coiled, thick (0.07–0.10 mm thick) .... *C. rosea*
        C. margins of perigynia smooth; pistillate scales acuminate or short-pointed, almost as long as the perigynium body, soon deciduous; spikes somewhat aggregated
            G. perigynia 1.3–1.8 mm wide, the lower 1/2–2/3 becoming spongy; widest leaf blades 1.5–3 mm wide .... *C. retroflexa*
            G. perigynia 1–1.3 mm wide, the lower 1/4 becoming spongy; widest leaf blades 1–1.5 mm wide .... *C. texensis*
    B. perigynia of uniform thickness and texture to the base, ascending or spreading; spikes bearing numerous perigynia and often aggregated
        H. inflorescences 0.7–2 cm long, the spikes in dense ovoid-oblong to globular heads
            I. lower perigynia ovate-deltoid; pistillate scales obtuse, acute or sometimes short-awned (<1 mm long); beaks very short, their margins usually smooth or sparsely serrulate .... *C. leavenworthii*
            I. lower perigynia ovate; pistillate scales mostly awned (awns reaching 3.2 mm long); beaks elongate, their margins clearly serrulate
                J. leaves smooth on their upper surface; ligules mostly longer than wide; pistillate scales (excluding awns) only 0.5–0.75 times as long as the perigynium body .... *C. cephalophora*
                J. leaves finely papillate (evident at 10× or higher magnification) on

upper surface; ligules wider than long; pistillate scales (excluding awns) about as long as the perigynium body or longer ....
....*C. mesochorea*

    H. inflorescences >1.5 cm long, oblong to elongate, usually interrupted revealing individual spikes or internodes

        K. ligules prolonged, much longer than wide; perigynia 4–5.5 mm long; inflorescences 1.5–4.5 cm long .... *C. spicata*

        K. ligules short, rarely longer than wide; perigynia 2.5–4 mm long (or up to 4.5 mm long in the very rare *C. divulsa*, which has elongate inflorescences)

            L. inflorescences 4.5–8 cm long; pistillate scales acuminate; perigynia 3.5–4.5 mm long ....*C. divulsa*

            L. inflorescences 1.5–4 cm long; pistillate scales short-pointed or awned; perigynia 2.5–4 mm long

                M. pistillate scales pale or greenish and (excluding awn) about as long as the perigynium body .... *C. muhlenbergii*

                M. pistillate scales reddish-brown-tinged and (excluding awn) about ¾ as long as the perigynium body .... *C. muricata*

A. widest leaves 3.5–10 mm wide; sheaths loose or baggy (rarely tight in *C. aggregata*), mottled green with white spots or with pale intervenal areas and contrasting darker veins, cross-veins evident

    N. mouth of leaf sheaths concave, thickened, not friable; widest leaf blades 3.5–6 mm wide; pistillate scales acuminate and about as long as the perigynium body .... *C. aggregata*

    N. mouth of leaf sheaths straight or concave but not thickened, friable; widest leaf blades 5–8.5(10) mm wide; pistillate scales mostly obtuse or acute (rarely short-awned) and about 0.5–0.75 times as long as the perigynium body

        O. inflorescences 3–10 cm long, the spikes (except the uppermost) distinctly separated .... *C. sparganioides*

        O. inflorescences 1.5–3.5 cm long, the spikes densely aggregated ....
....*C. cephaloidea*

Carex aggregata Mack.   Sedge

Fertile shoots 4–10 dm tall, from short creeping rootstocks; widest leaf blades 3.5–6 mm wide, sometimes papillate (visible at 10×) on upper-proximal surface; sheaths loose or tight, green with white mottling or veins green and intervenal areas white, the mouth concave, callused, not friable; ligules usually wider than long; inflorescences 2.3–5 cm long, aggregate except for lowest spike; pistillate scales acuminate, about as long as the perigynium body; perigynia spreading, 3.5–4.5 by 2–3 mm, the inner face more or less flat and nerveless; beaks serrulate, 1.2–2 mm long; moist woods, meadows, and ditches; common SE, scattered SC and SW; FACU.

Carex aggregata, perigynium (×3)

Carex aggregata, pistillate scale (×3)

Carex appalachica Webber & Ball   Sedge

Fertile shoots 2–6 dm tall, 0.7–1.3 mm thick; widest leaf blades 0.9–1.7 mm wide; sheaths tight and uniformly greenish; ligules wider than to about as wide as long; inflorescences 2–8 cm long, with usually <30 perigynia, the spikes separated; pistillate scales usually obtuse, shorter than the perigynium body; perigynia 2–3.4 by 0.8–1.3 mm, the inner face spongy at the base; beaks serrulate, their stigmas slender (0.05–0.08 mm thick) and tightly coiled; rocky upland woods; frequent; mostly N.

Carex appalachica, perigynium (×3)

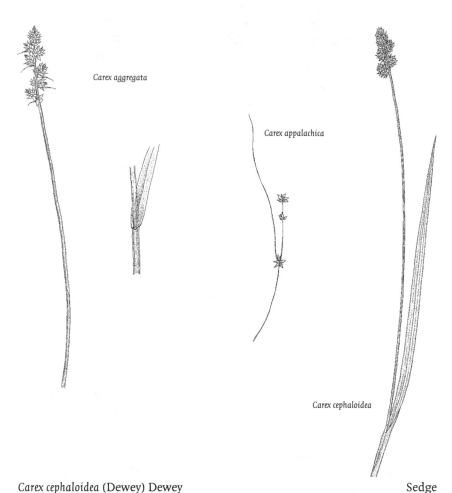

Carex aggregata

Carex appalachica

Carex cephaloidea

Carex cephaloidea,
perigynium (×3)

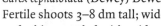

Carex cephaloidea,
pistillate scale (×3)

Carex cephaloidea (Dewey) Dewey   Sedge

Fertile shoots 3–8 dm tall; widest leaf blades 5–8 mm wide; sheaths baggy, green and white mottled with evident cross-veins, friable; ligules longer than wide; inflorescences 1.5–3.5(4) cm long, densely aggregated; pistillate scales obtuse to acute, about 0.5–0.75 times as long as the perigynium body; perigynia 3.2–4(4.5) by 1.5–2.4 mm, the inner face more or less flat or with raised rim 0.2 mm wide or less; beaks serrulate, 0.8–1.7 mm long; rich, dry to moist woods, thickets, and stream banks; occasional; throughout; FAC+.

Carex cephalophora,
perigynium (×3)

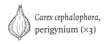

Carex cephalophora,
pistillate scale (×3)

Carex cephalophora Willd.   Sedge

Fertile shoots 2.5–6.5 dm tall; widest leaf blades 2–4.5 mm wide, the upper surface smooth; sheaths tight and usually uniformly greenish; ligules mostly longer than wide; inflorescences 0.7–2 cm long, the spikes forming a dense ovoid to globular head; pistillate scales mostly awned (excluding awns) about 0.5–0.75 times as long as the perigynium body; perigynia (2)2.4–3.4(3.8) by 1.2–2 mm, widest just below the middle of the body, the inner face more or less flat and nerveless; beaks with a double row of fine serrations, 0.7–1.4 mm long; dry woods, thickets, and fields; common; throughout; FACU.

Carex divulsa Stokes   Sedge

Fertile shoots 3–10 dm tall, from short creeping rhizomes; widest leaf blades 2–3(4) mm wide; sheaths tight and usually uniformly greenish; ligules about as long as wide; inflorescences (3)4.5–8 cm long, the spikes separated; pistillate

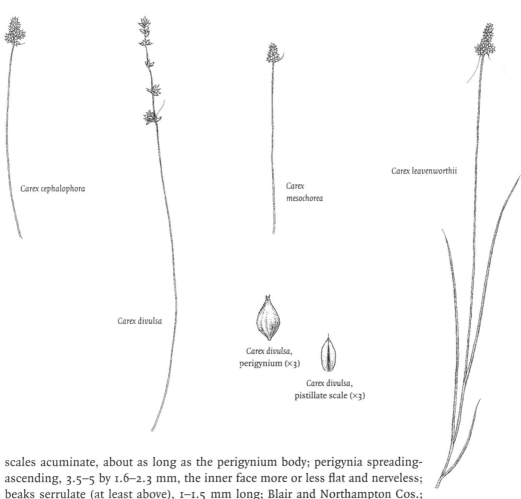

Carex cephalophora

Carex
mesochorea

Carex leavenworthii

Carex divulsa

Carex divulsa,
perigynium (×3)

Carex divulsa,
pistillate scale (×3)

scales acuminate, about as long as the perigynium body; perigynia spreading-ascending, 3.5–5 by 1.6–2.3 mm, the inner face more or less flat and nerveless; beaks serrulate (at least above), 1–1.5 mm long; Blair and Northampton Cos.; native to Europe.

Carex leavenworthii Dewey   Sedge
Fertile shoots 1–6 dm tall; leaf blades 1–3(3.5) mm wide; sheaths tight, often green and white mottled; ligules wider than long; inflorescences 0.7–2 cm long, the spikes forming a dense, ovoid-oblong head; pistillate scales obtuse, acute, or short-awned, about 0.75–1 times as long as the perigynium body; perigynia 2–3.2 by 1.2–2 mm, widest near the more or less truncate base (especially the lowest perigynia in a spike), the inner face more or less flat and nerveless; beaks often smooth or sparsely serrulate, 0.4–0.8 mm long; dry woods, fields, pastures, and clearings; frequent; mostly SE.

Carex leavenworthii,
perigynium (×3)

Carex leavenworthii,
pistillate scale (×3)

Carex mesochorea Mack.   Midland sedge
Fertile shoots 2.5–10 dm tall; widest leaf blades 2–4.5 mm wide and finely papillate on upper surface (visible at 10×); sheaths tight and uniformly greenish; ligules wider than long; inflorescences 1–1.9 cm long, the spikes forming a dense ovoid head; pistillate scales awned and (excluding awns) about as long as or exceeding the perigynium body; perigynia 2.7–3.5 by 1.6–2.4 mm, widest just below the middle of the body, the inner face more or less flat and nerveless; beaks serrulate, 0.5–1 mm long; dry, open woods, fields, and roadsides; scattered; S; FACU.

Carex mesochorea,
perigynium (×3)

Carex mesochorea,
pistillate scale (×3)

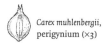

Carex muhlenbergii,
perigynium (×3)

Carex muhlenbergii,
pistillate scale (×3)

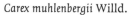

Carex muhlenbergii Willd.   Sedge

Fertile shoots 2–10 dm tall, sometimes rhizomatous; widest leaf blades 2–4 mm wide and papillate on upper surface (visible at 10×); sheaths tight and uniformly greenish; ligules about as long as wide; inflorescences 1.5–4 cm long, oblong, somewhat aggregated to interrupted; pistillate scales short-pointed or awned, about as long as the perigynium body; perigynia 2.5–4 by 1.6–3 mm, the inner face more or less flat and either nerveless or sharply nerved; beaks serrulate, 0.7–1.2 mm long; dry woods, thickets, roadsides, and open sandy habitat; frequent; mostly E.

Carex muricata,
perigynium (×3)

Carex muricata,
pistillate scale (×3)

Carex muricata L.   Sedge

Fertile shoots 3–7 dm tall; widest leaf blades 1.8–3.5 mm wide; sheaths tight and uniformly greenish; ligules about as wide as long; inflorescences 1.8–3.5 cm long, aggregated above, lower spike separated; pistillate scales short-pointed, reddish-brown-tinged, about 0.75–1 times as long as the perigynium body; perigynia 3–4 by 1.7–2.1 mm, the inner face more or less flat and nerveless; beaks serrulate, 0.6–1.2 mm long; Northampton Co.; native to Europe.

Carex radiata,
perigynium (×3)

Carex radiata (Wahlenb.) Small   Sedge

Fertile shoots 2–8 dm tall; widest leaf blades 0.8–1.8 mm wide; sheaths tight and uniformly greenish; ligules wider than long; inflorescences 2.5–7 cm long and with usually <30 perigynia, the spikes separated; pistillate scales obtuse to acute, shorter than the perigynium body; perigynia 2.6–3.8 by 1–1.5 mm, the inner face spongy at the base; beaks serrulate and 0.6–1 mm long, their stigmas slender, reflexed but otherwise more or less straight; dry to moist woods; common; throughout.

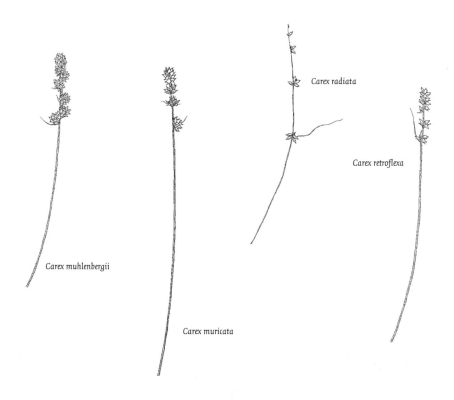

Carex radiata

Carex retroflexa

Carex muhlenbergii

Carex muricata

Carex retroflexa Schkuhr   Sedge

Fertile shoots 2–4 dm tall; widest leaf blades 1.5–3 mm wide; sheaths tight and uniformly greenish; ligules wider than long; inflorescences 1.2–3 cm long, more or less aggregated or lowest spike separated; pistillate scales acuminate or short-awned, about as long as the perigynium body and deciduous; perigynia 2.2–3 by 1.3–1.8 mm, the inner face spongy at the base; beaks with smooth margins, 0.6–1 mm long; dry, open, rocky woods; frequent; mostly S.

Carex rosea Willd.   Sedge

Fertile shoots 2–9 cm tall; widest leaf blades 1.8–2.8 mm wide; sheaths tight and uniformly greenish; ligules wider than to about as wide as long; inflorescences 2.5–6.5 cm long with about 25–50 perigynia, the spikes separated; pistillate scales obtuse to acute, shorter than the perigynium body; perigynia 2.6–4 by 1.1–1.8 mm, the inner face spongy at the base; beaks serrulate and 0.7–1.3 mm long, their stigmas relatively stout (0.07–0.1 mm thick) and tightly coiled; dry to moist woods; common; throughout.

Carex sparganioides Willd.   Sedge

Fertile shoots 4–10 dm tall; widest leaf blades 5–8.5(10) mm wide; sheaths loose, green and white mottled with cross-veins, friable; ligules about as wide as long; inflorescences 3–10 cm long, the lower spikes widely separated; pistillate scales acuminate to obtuse, occasionally short-awned, about 3/4 as long as the perigynium body; perigynia 3–4.2 by 1.8–2.2 mm, the inner face with a raised rim (>0.2 mm wide), otherwise more or less flat and nerveless; beaks serrulate, 0.8–1.6 mm long; rich, dry woods and meadows; common SE, occasional elsewhere; FACU.

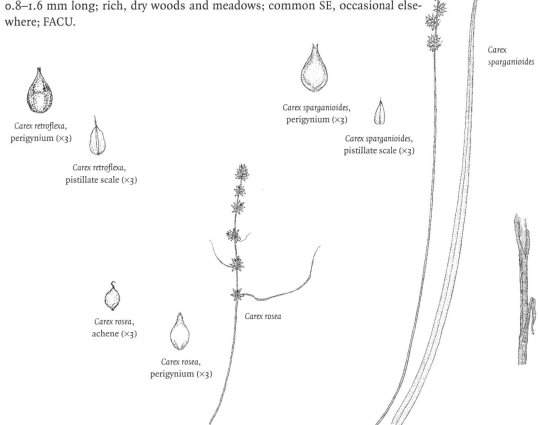

Carex retroflexa, perigynium (×3)

Carex retroflexa, pistillate scale (×3)

Carex rosea, achene (×3)

Carex rosea, perigynium (×3)

Carex rosea

Carex sparganioides, perigynium (×3)

Carex sparganioides, pistillate scale (×3)

Carex sparganioides

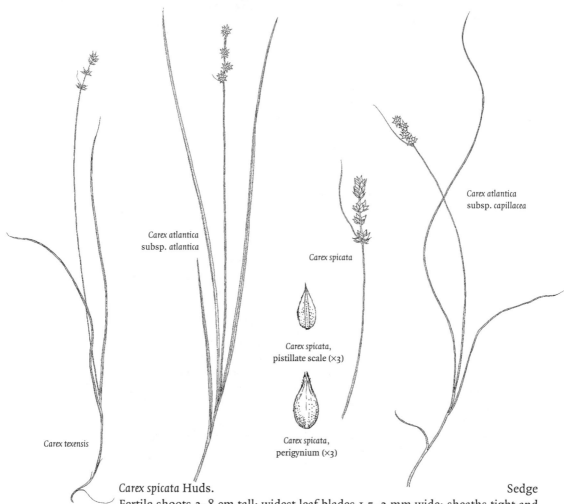

Carex texensis

Carex atlantica
subsp. atlantica

Carex spicata

Carex atlantica
subsp. capillacea

Carex spicata,
pistillate scale (×3)

Carex spicata,
perigynium (×3)

Carex texensis,
perigynium (×3)

Carex texensis,
pistillate scale (×3)

Carex spicata Huds.   Sedge

Fertile shoots 2–8 cm tall; widest leaf blades 1.5–3 mm wide; sheaths tight and uniformly greenish; ligules much longer than wide; inflorescences 1.5–4.5 cm long, interrupted; pistillate scales mostly acuminate, slightly longer than the perigynium body and often reddish-brown-tinged; perigynia 4–5.5 by 2.2–2.8 mm, the inner face flat and nerveless; beaks serrulate, 1.2–1.8 mm long; dry fields and roadsides; mostly SE; native to Europe.

Carex texensis (Torr.) Bailey   Sedge

Fertile shoots 1–4.5 dm tall; widest leaf blades 1–1.5 mm wide; sheaths tight and uniformly greenish; ligules wider than long; inflorescences 0.6–3 cm long, more or less aggregated or lower 1–3 spikes separated; pistillate scales acuminate or short-awned, about as long as the perigynium body, deciduous; perigynia 2.2–3.2 by 1–1.3 mm, the inner face spongy at the base; beaks with smooth margins, 0.7–1.4 mm long; rare; Bradford Co.; native to central U.S.

Section *Stellulatae*

Tufted; leaf sheaths not cross-puckered on ventral surface; inflorescences simple, peduncles firm, not wing-angled; spikes usually male at the base but sometimes all female or all male; perigynia planoconvex and thick-edged, the base spongy; beaks usually serrulate, bidentate, spreading or reflexed; achenes occupying the upper ½ or ⅔ of perigynium body.

Two characteristics important for keying these species are perigynia length/

width ratio and leaf width. In Stellulatae the lowest perigynia within a spike should be chosen for measurements since they show the greatest difference between species and the least variation within a species. In determining leaf width, always choose the largest one, especially any old leaves from the vegetative shoots of the previous season. The present treatment is based on Reznicek and Ball (1980).

A. beaks with smooth margins, 0.2–0.6 mm long; body of perigynia widest near the middle, elliptic or elliptic-ovate .... *C. seorsa*
A. beaks serrulate, 0.4–2 mm long; body of perigynia widest below the middle, lanceolate, ovate, or deltoid
   B. widest leaves 0.8–2.7 mm wide
      C. terminal spikes usually all male or all female (occasionally with several male flowers at base); anthers 1–2.2(2.35) mm long; plants of calcareous swamps and fens .... *C. sterilis*
      C. terminal spikes with tapered staminate base 1.5–16 mm long; anthers 0.6–1.6 mm long; plants mostly of acidic bogs and other noncalcareous wet habitats
         D. lower perigynia of spikes 1.8–3.2 times as long as wide, 2.7–4 mm long; beaks 0.85–2 mm long, 0.4–0.85 times as long as the body ....
         .... *C. echinata*
         D. lower perigynia of spikes 1–2 times as long as wide, mostly 2–3 mm long; beaks <0.95 mm long, about 0.2–0.5 times as long as the body
            E. inner face of perigynia mostly nerveless over achene; perigynia often convex-tapered from widest point to beak .... *C. interior*
            E. inner face of perigynia 1–12-nerved over achene; perigynia cuneate or even concave-tapered from widest point to beak ....
            .... *C. atlantica*
   B. widest leaves 2.8–5 mm wide
      F. lower perigynia of spike 1.1–1.7 times as long as wide (2.1–3 mm wide) and 1–12-nerved over achene on inner face; spikes aggregated to separated .... *C. atlantica*
      F. lower perigynia of spike 1.5–2.5 times as long as wide (1.4–2 mm wide), the inner face mostly nerveless; spikes aggregated .... *C. wiegandii*

Carex atlantica subsp. atlantica, perigynium (×5)

Carex atlantica Bailey   Bog sedge
Fertile shoots 1–11 dm tall; leaves usually equaling to slightly shorter than fruiting shoots; inflorescences 1.5–5.5 cm long; spikes aggregated to separated, the terminal one long-staminate at base; pistillate scales ¾ to as long as the perigynium body, usually acute and pale; perigynia broadly ovate, triangular-ovate (occasionally broadly elliptic), inner face with 1–12 nerves over the achene; beaks serrulate, 0.45–1.25 mm long, 0.23–0.5 times as long as the body; anthers 0.6–1.5 mm long; 2 varieties:

Carex atlantica subsp. atlantica, pistillate scale (×5)

A. widest leaf blades 1.5–4(4.5) mm wide; inflorescences 1.5–5.5 cm long; perigynia 2.3–3.8 by 2.1–3 mm, 1.1–1.7 times as long as wide .... ssp. *atlantica* bogs and swamps; common SE, frequent to occasional elsewhere; FACW+.
A. widest leaf blades 0.6–1.6 mm wide; inflorescences 0.8–2.0 cm long; perigynia 2–3 by 1.5–2 mm, 1.3–1.7 times as long as wide ....
.... ssp. *capillacea* (Bailey) Reznicek swamps, bogs, and shores; occasional to infrequent; NE, SE, SC, and NW; OBL.

Carex atlantica subsp. capillacea, perigynium (×5)

Carex atlantica subsp. capillacea, pistillate scale (×5)

Carex echinata,
perigynium (×5)

Carex echinata,
pistillate scale (×5)

Carex echinata Murray   Prickly sedge

Fertile shoots 1–9 dm tall; leaves shorter than to equaling fruiting shoots, the widest blades 1–2.7 mm wide; inflorescences 1–7.5 cm long; spikes aggregated to widely separated, the terminal one long-staminate at base; pistillate scales about as long as the perigynium body, acute, becoming brown-tinged; perigynia lanceolate to ovate, 2.7–4 by 0.8–2 mm, 1.8–3.2 times as long as wide, inner face with 0–12 nerves over achene; beaks long and serrulate, 0.85–2 mm, 0.4–0.85 times as long as the body; anthers 0.8–1.8 mm long; acidic wetlands on Sphagnum, peat, or wet sandy substrates; frequent E, occasional to rare W; OBL.

Carex interior,
perigynium (×5)

Carex interior,
pistillate scale (×5)

Carex interior Bailey   Sedge

Fertile shoots 1–9.5 dm tall; leaves often shorter than fruiting shoots, the widest blades 1–2.4(2.7) mm wide; inflorescences 0.8–4 cm long; spikes aggregated to separated, the terminal one long-staminate at base; pistillate scales distinctly shorter than perigynium body, obtuse or acute, becoming brown-tinged; perigynia ovate to broadly ovate, 1.95–3(3.3) by 1.1–1.8 mm, 1.4–2(2.2) times as long as wide, inner face nerveless (or occasionally up to 6-nerved) over achene; beaks setulose-serrulate, 0.4–0.95 mm long, 0.18–0.44 times as long as the body; anthers 0.6–1.4 mm long; non-acidic sedge meadows, swales, and swamps; frequent NW, SC, and SE, rare elsewhere; OBL.

Carex seorsa,
perigynium (×5)

Carex seorsa,
pistillate scale (×5)

Carex seorsa Howe   Sedge

Fertile shoots 1.5–7.5 dm tall; leaves shorter than fruiting shoots, the widest blades 2–4 mm wide; inflorescences 2–7 cm long; spikes separated, the terminal one long-staminate at base; pistillate scales shorter than perigynium body, acute, pale; perigynia elliptic to elliptic-ovate, 1.8–3 by 1–1.6 mm, 1.2–2.1 times as long as wide, inner face with up to 6 nerves over achene; beaks with smooth margins, 0.2–0.6 mm long, 0.1–0.3 times as long as the body; anthers 1–1.9 mm long; wet woods and swamps; occasional; NW and SE; FACW.

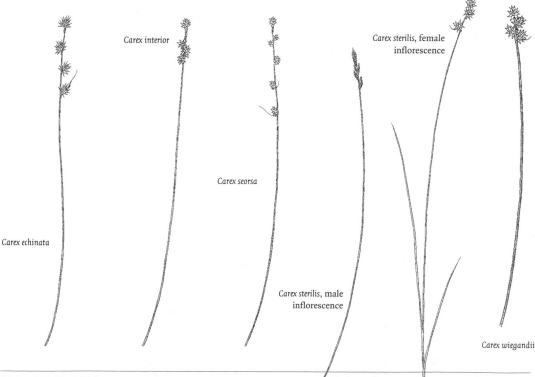

Carex echinata

Carex interior

Carex seorsa

Carex sterilis, male
inflorescence

Carex sterilis, female
inflorescence

Carex wiegandii

Carex sterilis Willd. Atlantic sedge

Fertile shoots 1–7.5 cm tall; leaves shorter than fruiting shoots, the widest blades 1.6–2.6 mm wide; inflorescences 1–4.5 cm long; terminal spikes staminate, pistillate or mixed; pistillate scales longer than the perigynium body, acute, becoming brown-tinged; perigynia ovate to deltoid, 2.1–3.8 by 1.2–2.2 mm, 1.4–2.3(2.7) times as long as wide, inner face with 0–10 nerves over achene; beaks densely serrulate, 0.65–1.6 mm long, 0.45–0.8 times as long as the body; anthers 1–2.2(2.35) mm long; calcareous swamps and fens; scattered; mostly S; OBL; ✤.

Carex sterilis,
perigynium (×5)

Carex sterilis,
pistillate scale (×5)

Carex wiegandii Mack. Wiegand's sedge

Fertile shoots 1–10.5 dm tall; leaves shorter than fruiting shoots, the widest blades 2.5–5 mm wide; inflorescences 0.9–3 cm long; spikes aggregated, the terminal one long-staminate at base; pistillate scales shorter than the perigynium body, obtuse to acute, becoming brown-tinged; perigynia broadly ovate, 2.6–3.7 by 1.4–2 mm, 1.5–2.5 times as long as wide, inner face mostly nerveless or with up to 10 faint nerves over achene; beaks serrulate, 0.55–1.1 mm long, 0.25–0.55 times as long as the body; anthers 0.7–1.3 mm long; sphagnum bogs, openings, or thickets; rare; NC; OBL; ✤.

Carex wiegandii,
perigynium (×5)

Carex wiegandii,
pistillate scale (×5)

Section *Vulpinae*

Tufted; fertile shoots erect; leaf sheaths bearing a smooth or cross-puckered hyaline band on ventral surface; inflorescences composed of numerous, densely aggregated spikes, the peduncles stout, soft, and loosely cellular, wing-angled; spikes either male at the apex or all female; pistillate scales acuminate or awned; perigynia planoconvex, spongy at base; beaks typically long.

A. perigynia ovate, (2.5)2.8–5 mm long, the inner face more or less nerveless; beaks contracted and about 0.5–1 times as long as the perigynium body; hyaline band of leaf sheaths often with red dots near the summit ......*C. conjuncta*
A. perigynia lanceolate, 3.8–6 mm long, the inner face nerved; beaks gradually tapering and about as long as the perigynium body or longer; leaf sheaths lacking red dots on the ventral band
   B. leaf sheaths smooth ventrally, the mouth thickened and either truncate or concave .... *C. laevivaginata*
   B. leaf sheaths cross puckered ventrally, the mouth thin, friable, and prolonged at the summit .... *C. stipata*

Carex conjuncta

Carex conjuncta Boott Sedge

Fertile shoots 5–12 dm tall; leaf blades mostly 5–10 mm wide; hyaline band of leaf sheaths cross-puckered, friable, often with red dots, prolonged at the summit; inflorescences 2–7.5 cm long; perigynia greenish, ovate, 3.4–5 by 1.6–2 mm, the inner face nerveless or several-nerved at the base, the outer face with several strong nerves; beaks contracted, 1–1.7 mm long; moist woods and meadows; frequent SE, scattered NE, SC, SW, and Erie Co.; FACW.

Carex conjuncta,
perigynium (×3)

Carex conjuncta,
pistillate scale (×3)

Carex laevivaginata (Kük.) Mack. Sedge

Fertile shoots 3.5–8 dm tall; leaf blades 2.5–7 mm wide; hyaline band of leaf sheaths smooth, thick and either truncate or concave at summit; inflorescences

Carex laevivaginata,
perigynium (×3)

Carex laevivaginata,
pistillate scale (×3)

2–6.5 cm long; perigynia yellowish-green or brown, lanceolate, 4.3–6 by 1.5–2 mm, both faces strongly nerved; beaks gradually tapered, 1.5–3.3 mm long; swamps and stream banks; frequent; mostly S and NW; OBL.

Carex stipata var. maxima,
perigynium (×3)

Carex stipata Willd.   Sedge

Fertile shoots 4–10 dm tall; leaf blades 3.5–15(17) mm wide; hyaline band of leaf sheaths often cross-puckered, thin, and prolonged at the summit; inflorescences 3–10 cm long; perigynia yellowish-green or brown, lanceolate, 3.8–6 by 1.3–2 mm, strongly nerved on both faces; beaks gradually tapered; 2 varieties:

Carex stipata var. maxima,
pistillate scale (×3)

A. perigynia 4–5 mm long; widest leaves 4–8 mm wide; beaks about as long as the perigynium body .... var. *stipata* common; throughout.

A. perigynia 5–6 mm long; widest leaves 8–15(17) mm wide; beaks about 2 times as long as the perigynium body .... var. *maxima* Chapm. scattered; throughout.

Cladium R.Br.

Cladium mariscoides,
achene (×5)

Cladium mariscoides (Muhl.) Torr.   Twig-rush

Rhizomatous, stems solitary or a few together, often forming large colonies; culms to 1 m tall, leaf blades 1–3 mm wide, channeled at the base, flat in the middle and rounded at the tip; inflorescence cymose with small terminal clusters of spikelets; spikelet with a single perfect flower, a staminate or abortive flower below, and an empty scale at the base; rare in marshes, floating bog mats, and shallow lake margins; NE and scattered elsewhere; OBL; 🌿.

Cymophyllus Mack. ex Britton

Cymophyllus fraserianus (Ker Gawl.) Kartesz & Ghandi   Fraser's sedge

Cespitose perennial with evergreen basal leaves that lack sheaths; flowering stems arising from the center of the whorl of leaves, each surrounded at the base by a tubular leaf that later splits and elongates into a sheathless blade lacking a midvein; flowers in a single spike, staminate above and pistillate below; pistillate flowers enclosed in white perigynia; very rare on rich, wooded slopes; SC; flr./frt. Apr–Jul; 🌿.

Cyperus L.

Annual or perennial herbs with solid, angled stems; leaves with closed basal sheaths and a well-developed elongate blade; spikelets few or many in dense heads or elongate spikes that are clustered in a simple or compound terminal umbel; florets perfect, each subtended by a scale; stamens 1–3; style branches 2 or 3; achenes lens-shaped or 3-angled; flr./frt. mid summer–early fall. Many species show a wide range in the height of individual plants ranging from a few centimeters to several decimeters depending on growing conditions.

A. styles 3-parted, achenes 3-angled
    B. spikelets in obvious spikes with a visible elongate axis
        C. spikelets flattened, scales 1–4 mm long, achenes usually <2 mm long

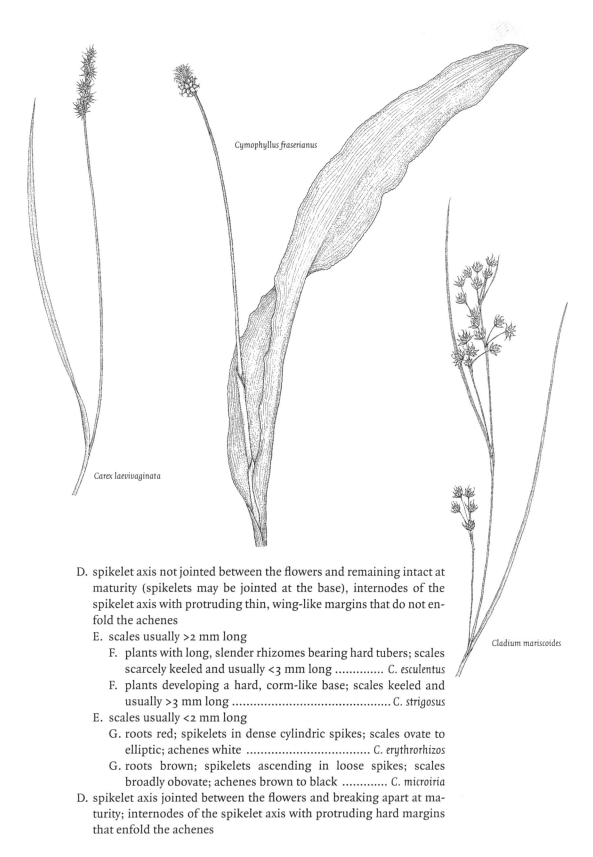

Cymophyllus fraserianus

Carex laevivaginata

Cladium mariscoides

D. spikelet axis not jointed between the flowers and remaining intact at maturity (spikelets may be jointed at the base), internodes of the spikelet axis with protruding thin, wing-like margins that do not enfold the achenes

  E. scales usually >2 mm long

    F. plants with long, slender rhizomes bearing hard tubers; scales scarcely keeled and usually <3 mm long .... *C. esculentus*

    F. plants developing a hard, corm-like base; scales keeled and usually >3 mm long .... *C. strigosus*

  E. scales usually <2 mm long

    G. roots red; spikelets in dense cylindric spikes; scales ovate to elliptic; achenes white .... *C. erythrorhizos*

    G. roots brown; spikelets ascending in loose spikes; scales broadly obovate; achenes brown to black .... *C. microiria*

D. spikelet axis jointed between the flowers and breaking apart at maturity; internodes of the spikelet axis with protruding hard margins that enfold the achenes

H. scales obviously overlapping those above on the same side of the spikelet; achenes ellipsoid or slenderly obovoid .... *C. odoratus*

H. scales not or scarcely reaching those above on the same side of the spikelet; achenes linear-oblong and somewhat curved (banana-shaped) .... *C. engelmannii*

C. spikelets nearly round in cross section, scales 4–6 mm long, achenes usually >2 mm long

    I. culms smooth; only the lowest spikelets reflexed, most radiating horizontally and the uppermost ascending

        J. spikelets loosely arranged along the axis; scales scarcely overlapping those on the same side of the spikelet; achenes up to 3 mm long .... *C. refractus*

        J. spikelets densely arranged along the axis; scales obviously overlapping those on the same side of the spikelet; achenes 2–2.6 mm long .... *C. lancastriensis*

    I. culms scabrous or rough-hairy near the tips; uppermost spikelets radiating horizontally, all others reflexed

        K. inflorescence branches and upper surfaces of bracts rough or hairy .... *C. plukenetii*

        K. inflorescence branches and upper surfaces of bracts smooth .... *C. retrofractus*

B. spikelets in dense heads or short head-like spikes with the spike axis obscured by the dense arrangement of spikelets

    L. scales with obviously outwardly curved tips

        M. scales 7–9-nerved with a distinct awn at the tip .... *C. squarrosus*

        M. scales 3-nerved, somewhat acuminate but not awn-tipped .... *C. acuminatus*

    L. tips of the scales not outwardly curved (sometimes with scarcely outwardly curved awns)

        N. scales <1 mm long .... *C. difformis*

        N. scales >1 mm long

            O. heads with numerous densely aggregated spikelets that are <2 mm wide; scales strongly overlapping with appressed tips

                P. heads globose or globose-ellipsoid; spikelets becoming dark brown; achenes to 2.2 mm long........................... *C. echinatus*

                P. heads cylindric; spikelets pale green or brown; achenes at most 1.5 mm long .... *C. retrorsus*

            O. heads with few to many loosely arranged spikelets (sometimes appearing densely aggregated in variants of C. lupulinus) that are usually >2 mm wide; scales with free tips giving the spikelet a dentate margin

                Q. plants with a short, hard rhizomes with hard, knotty tubers; spikelets deciduous

                    R. culms scabrous near the tip; scales with awns 0.5–1 mm long .... *C. schweinitzii*

                    R. culms smooth; scales with awns <0.5 mm long

                        S. scales oblong-elliptic; achenes about ½ as wide as long .... *C. lupulinus*

                        S. scales about as wide as long; achenes ⅔ or more as wide as long .... *C. houghtonii*

Q. plants with long slender rhizomes or lacking apparent rhizomes; spikelet axis persisting
    T. plants with long slender rhizomes; scales with up to 7 nerves concentrated in the central portion .... *C. dentatus*
    T. plants lacking rhizomes; scales with >7 nerves extending nearly to the margins .... *C. compressus*
A. styles 2-parted, achenes 2-angled
  U. spikelets readily discernible in spikes or spike-like heads, >5 mm long at maturity and many-flowered
    V. plants with rhizomes; leaves >5 mm wide; spikelets in obvious spikes with an elongate axis >2 cm long .... *C. serotinus*
    V. plants usually tufted and lacking rhizomes; leaves <5 mm wide; spikelets in heads or short spikes with an axis <2 cm long
      W. scales broadly ovate; achenes black with very fine transverse lines ... . *C. flavescens*
      W. scales ovate to narrowly oblong; achenes brown and lacking transverse lines
        X. scales blunt with appressed tips, usually tinged with purple or brown
          Y. dark pigments of scales concentrated at the tips and extending downward along the margins; styles split nearly to the base and persisting in fruit .... *C. diandrus*
          Y. dark pigments of scales more concentrated below the tips and extending from the midribs to the margins; styles split to about the middle .... *C. bipartitus*
        X. scales with a prolonged midrib giving the spikelet a serrated margin, usually yellow
          Z. scales 2–3.5 mm long; achenes obovoid, 1.2–1.4 mm long .... *C. filicinus*
          Z. scales 1.5–2.3 mm long; achenes oblong, 0.8–1 mm long .... *C. polystachyos*
  U. spikelets strongly overlapping in dense, head-like spikes, <5 mm long at maturity, 1-flowered
    AA. plants with long rhizomes; spikelets usually >3 mm long; achenes up to 1.8 mm long .... *C. brevifolioides*
    AA. plants usually tufted and lacking rhizomes; spikelets usually <3 mm long; achenes up to 1.4 mm long .... *C. tenuifolius*

Cyperus acuminatus Torr. & Hook.   Short-pointed flatsedge
Annual; culms tufted, 0.5–4 dm tall; spikelets in dense globose clusters, strongly flattened; scales 1.3–2 mm, ovate with recurved acuminate tip; achenes 0.6–1 mm; rare in damp, sandy, disturbed ground; SE; native as far north as VA, adventive here; OBL.

Cyperus bipartitus Torr.   Umbrella sedge
Annual; similar to *C. diandrus*, but the rays of the umbel longer (to 1.2 dm); styles 2-branched to about the middle and scales strongly tinged with purple below the tips and along the margins almost to the base; frequent in wet, sandy ground and gravelly or muddy shores; throughout; FACW+.

Cyperus acuminatus (×1)

Cyperus acuminatus, spikelet (×2)

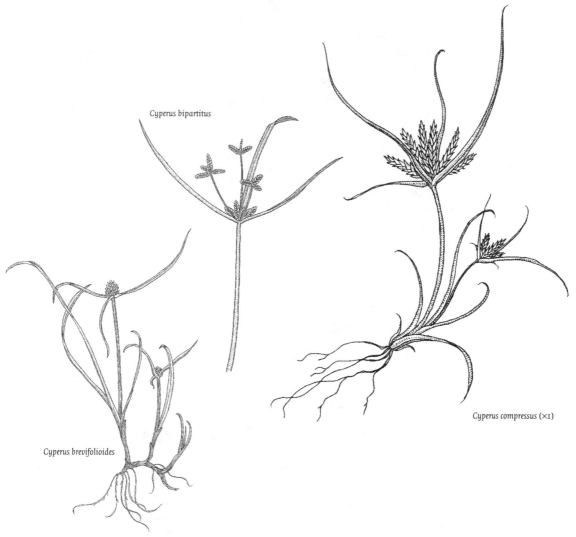

Cyperus bipartitus

Cyperus compressus (×1)

Cyperus brevifolioides

Cyperus brevifolioides Thieret & Delahoussaye Umbrella sedge
Rhizomatous perennial often forming a turf; culms 1–3 dm tall; leaves 2–4 mm wide; inflorescence a solitary, round-ovoid spike that is sessile at the top of the stem; spikelets very numerous, 3.5–4.5 mm long, 1-flowered; achenes flattened, obovate, 1–1.4 mm; locally abundant in sandy dredge spoil, moist turf, tidal river banks, and urban waste ground; SE; native to Asia.

Cyperus compressus L. Umbrella sedge
Annual; culms tufted, 1–3 dm; leaves 1–2 mm wide; spikes sessile and sometimes also at the ends of 1–4 rays; spikelets 3–10, radiating from a short axis; scales with pale, nerveless margins; achenes 1–1.4 mm; rare in sandy waste ground and ballast; SE; FAC+.

Cyperus dentatus,
spikelet (×2)

Cyperus dentatus Torr. Umbrella sedge
Perennial with slender rhizomes ending in tubers; culms 1–6 dm, stiffly erect; leaves rigid and keeled; spikelets in subglobose clusters, or sometimes replaced by leafy tufts or bulblets; 2.5 mm wide, scales ovate with the nerves clustered near the midrib and a broad nerveless margin; achenes obovoid, 1 mm or less; occasional on moist, sandy stream banks and dredge spoil; SE and C; FACW+.

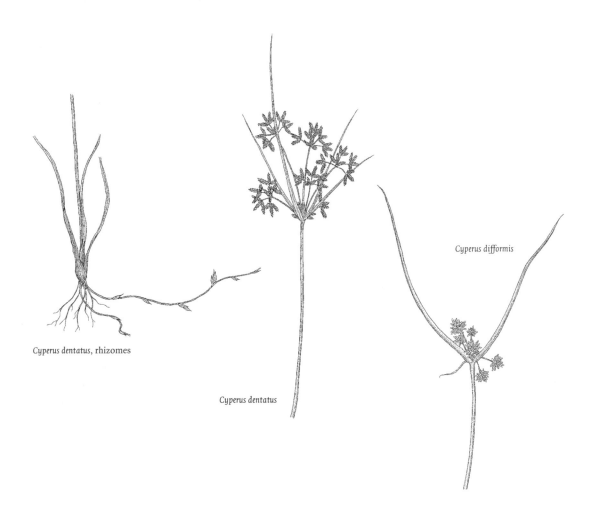

Cyperus dentatus, rhizomes

Cyperus dentatus

Cyperus difformis

Cyperus diandrus Torr.   Umbrella sedge
Annual; culms slender, tufted, or solitary, 0.2–4.5 dm; inflorescence a single radiating cluster of spikelets, or with several rays to 6 cm long, bearing smaller
clusters; spikelets reddish-brown to greenish, very flat, 6–32-flowered; scales 2–
2.7 mm long, ovate with purple-brown tips and margins; stamens 2 or 3; styles
split nearly to the base and persisting in fruit; very rare on moist stream banks,
bogs, and marshes; scattered; E and NW; FACW; ❦.

Cyperus difformis L.   Flatsedge
Annual; culms soft, smooth, tufted, 2–7 dm tall; heads globose, dense, 8–12 mm
thick; spikelets 10–20-flowered; scales ovate-rotund, broader than long with
reddish or brownish sides; achenes light yellowish, about as long as the scales;
very rare on sandy pond margins and rock crevices in shallow water; SE; native
to the Old World tropics; OBL.

Cyperus difformis,
spikelet (×5)

Cyperus echinatus (L.) A.Wood   Umbrella sedge
Perennial; culms 3–10 dm tall; leaves flat, 3–8 mm wide; inflorescence of dense
globose spikes, several sessile and others terminating the 2–10 elongate rays;
spikelets very numerous and closely packed, radiating out in all directions,
scales strongly overlapping with appressed tips; flowers 1–3; achenes 1.8–2.2
mm long; occasional in dry to moist soil of woods and fields; S; FACU.

Cyperus engelmannii,
spikelet (×2)

Cyperus engelmannii Steud.   Engelmann's flatsedge
Annual, similar to *C. odoratus,* with which it often grows, but differing by the non-overlapping scales and linear-oblong, somewhat curved achenes; rare in moist shores, tidal mudflats, and mixed emergent marshes; SE and NW; FACW; 🌿.

Cyperus erythrorhizos Muhl.   Redroot flatsedge
Annual with red roots; culms 1.5–9 dm tall, tufted; leaf sheaths purple at the base; inflorescence with numerous spikes, some sessile and others clustered at the tips of the rays; spikelets 3–12 mm long with 6–many flowers; scales 1.2–1.5 mm long, with prominent green keels; achenes pearly white; occasional on moist shores, swamps, and wet ground; S and NW; FACW+.

Cyperus erythrorhizos,
spikelet (×2)

Cyperus esculentus L.   Yellow nutsedge
Perennial with slender rhizomes ending in small, hard tubers; culms 1–7 dm tall, leaves mostly basal, 3–8 mm wide; inflorescence of several rays that are often branched at the top as well as several sessile spikes; spikelets with 8–30 flowers, spikelet axis persistent, winged but not enclosing the achenes; frequent, and often a serious weed in moist ground of fields, meadows, lawns, and gardens; throughout; FACW.

*Cyperus
esculentus,*
spikelet (×2)

Cyperus filicinus Vahl   Umbrella sedge
Annual; culms slender, soft, tufted, or solitary, 0.1–3.5 dm tall; leaves 1–3 mm wide; inflorescence sessile at the tip of the stem or with 1–6 rays to 10 cm long; spikelets radiating; scales yellowish, narrowly oblong, 2–3.5 mm long with pro-

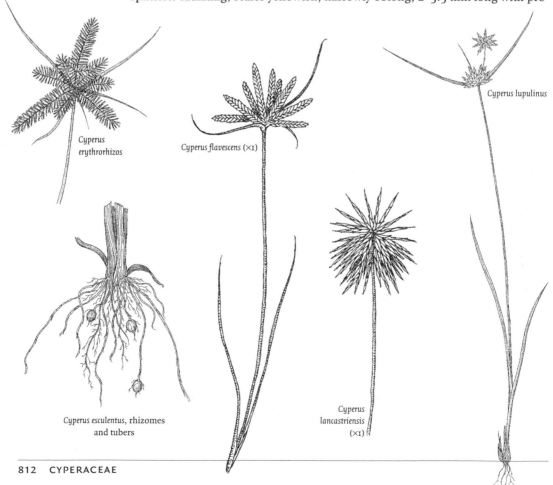

*Cyperus
erythrorhizos*

Cyperus flavescens (×1)

Cyperus lupulinus

Cyperus esculentus, rhizomes
and tubers

*Cyperus
lancastriensis*
(×1)

longed midribs giving the spikelet a serrated margin; stamens 2; achenes flattened, obovoid, 1.2–1.4 mm long; very rare on sandy stream banks and ballast; SE.

Cyperus flavescens L.   Umbrella sedge
Annual; culms tufted or solitary, 0.5–4 dm tall; inflorescence a single sessile cluster of spikelets or with several short rays each bearing a short spike; spikelets very flat, many-flowered; scales broadly ovate with thin green keels, about 1.7–2.5 mm long; stamens 3; style deeply 2-cleft; achenes black, suborbicular; occasional in fields, ditches, and moist, open ground; mostly S; OBL.

Cyperus flavescens, spikelet (×2)

Cyperus houghtonii Torr.   Houghton's flatsedge
Perennial; similar to *C. schweinitzii* but the culms smooth; scales rotund with minute pointed tips <0.5 mm long; rare on railroad ballast and dry, sandy soil; scattered; 🍃.

Cyperus lancastriensis Porter   Umbrella sedge
Perennial with a short rhizome with corm-like enlargements; culms smooth; leaves 4–10 mm wide; rays of the umbel well-developed; spikes 1–3 cm long, fairly dense, unbranched; spikelets 3–8-flowered, the lowest reflexed, those in the middle radiating horizontally and the uppermost ascending; achenes 2–2.6 mm long, held between the clear wings of the spikelet axis; rare on dry slopes, open woods, and fields; SE; FACU; 🍃.

Cyperus lancastriensis, spikelet (×2)

Cyperus lupulinus (Sprengel) Marcks   Umbrella sedge
Perennial from a short, tuberous-knotty rhizome; culms 1–5 dm tall, smooth; leaves 1–3.5 mm wide, flat or folded; inflorescence a single sessile cluster of spikelets or occasionally also with 1–2 rays bearing smaller clusters; spikelets crowded, 2.5–4 mm wide; scales oblong-elliptic, with minute awn-tips <0.5 mm long; frequent in dry woods, fields, waste ground, and roadsides; mostly SE and SC, scattered elsewhere; UPL.

Cyperus lupulinus, spikelet (×2)

Cyperus microiria Steud.   Umbrella sedge
Annual; culms to 6 dm tall, tufted; spikes 2–4 cm long, loose and usually branched, some nearly sessile, others terminating more or less elongate rays; scales truncate with green, pointed tips to 0.4 mm long; rare in damp, sandy, disturbed ground; SE; native to Asia.

Cyperus odoratus L.   Umbrella sedge, rusty flatsedge
Annual; culms erect, 0.1–8 dm tall; leaves flat; spikelets in an elongate spike; spikelet axis breaking into short segments with broad clasping wings that enclose the achenes; scales distinctly overlapping the next above on the same side of the spikelet; achenes 1.5–2 mm long, ellipsoid or slenderly obovoid; occasional in moist meadows, wet sandy or gravelly flats, and riverbanks; SE and NW; FACW.

Cyperus odoratus, achene (×5)

Cyperus odoratus, spikelet (×2)

Cyperus plukenetii Fernald   Plukenet's flatsedge
Very similar to *C. retrofractus*; culms smooth; spikes dense and only slightly longer than thick; spikelets 6–8 mm with 1–2 flowers; wings of the spikelet axis clasping the achenes for ½ their length; very rare in dry sandy soil; Philadelphia Co.; believed to be extirpated.

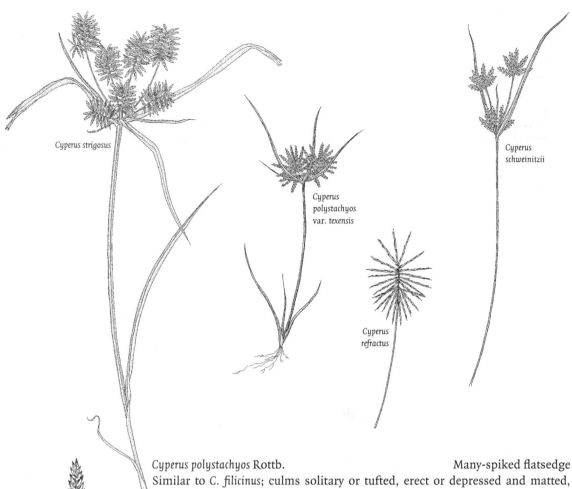

Cyperus strigosus

Cyperus polystachyos var. texensis

Cyperus refractus

Cyperus schweinitzii

Cyperus polystachyos var. texensis, spikelet (×2)

Cyperus refractus, spikelet (×2)

Cyperus polystachyos Rottb. Many-spiked flatsedge
Similar to *C. filicinus*; culms solitary or tufted, erect or depressed and matted, 0.1–2 dm tall; spikelets in a single sessile cluster or with several rays up to 4 cm long bearing smaller clusters; spikelets yellowish; scales 1.5–2.3 mm long, narrowly elliptic-ovate; achenes 0.8–1 mm long; very rare, tidal river banks; SE; believed to be extirpated; FACW; ✿. Ours is var. *texensis* (Torr.) Fernald.

Cyperus refractus Engelm. Reflexed flatsedge
Perennial with short, knotty rhizomes; culms 3–8 dm tall, smooth; leaves flat; spikes very loose and open, terminating elongate rays, not branched; spikelets slender, 1–3 cm long, 2–6-flowered, those below the middle reflexed at maturity; achenes linear, 2.5–3 mm long; rare on sandy, alluvial banks and dry woods; SE; FACU+; ✿.

Cyperus retrofractus (L.) Torr. Rough flatsedge
Perennial with a hard, corm-like base; culms rough, tufted, 3–9 dm tall; leaves 3.5–8 mm wide; spikes not branched; spikelets 1–4-flowered, the uppermost spikelets radiating horizontally, lower ones strongly reflexed; wings of the spikelet axis clasping the achenes for ¾ of their length; very rare in dry, sandy soil; known only from early collections from Philadelphia and Lancaster Cos.; believed to be extirpated; ✿.

Cyperus retrorsus Chapm. Retrorse flatsedge, umbrella sedge
Perennial; culms slender, 3–10 dm tall; leaves 3–5 mm wide; spikes short cylindric, in a simple umbel; spikelets very numerous, crowded, radiating horizontally, scales strongly overlapping with appressed tips, 1–3-flowered; achenes 1–1.5 mm; rare in dry, sandy soil and ballast; SE; FAC–; 🌿.

Cyperus schweinitzii Torr. Schweinitz' flatsedge
Perennial with short, knotty rhizomes; culms 1–10 dm tall, rough above and on the angles; leaves 2–8 mm wide; inflorescence usually of 1 sessile spike and several on more or less elongate rays; spikelets crowded, ascending; flattened, 3–4.5 mm wide; scales broadly ovate-elliptic to rotund with conspicuous awn-tips 0.5–1 mm long; rare in dry or moist sand flats or dunes; SE and NW; FACU; 🌿.

Cyperus schweinitzii, spikelet (×2)

Cyperus serotinus Rottb. Umbrella sedge
Perennial with soft rhizomes; culms to 1 m tall, leafy at the base; leaves 6–10 mm wide; spikelets in obvious spikes with axes >2 cm long, lustrous reddish-brown, 10–30-flowered; styles 2; achenes flattened; rare (but locally abundant) on tidal river banks; SE; OBL; native to Eurasia.

Cyperus squarrosus L. Umbrella sedge
Annual; culms tufted, 0.5–2 dm tall; leaves 0.5–2 mm wide; spikes in dense, head-like clusters; bracts of the spikelets with conspicuous recurved tips ending in short awns; occasional in moist soil of river banks, stream margins, wet ditches, and waste ground; mostly E and SC; FACW+.

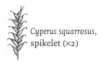

Cyperus squarrosus, spikelet (×2)

Cyperus strigosus L. False nutsedge
Perennial with a hard, corm-like base; culms 0.1–1 m tall, usually tufted; leaves flat, 0.1–1.2 cm wide; inflorescence simple to compound, the somewhat open spikes usually branched at the base; spikelets, 4–14-flowered; scales golden with green midribs, keeled; spikelet axis winged, achenes linear, 1.5–2 mm, much shorter than the scales; common in moist fields, woods, swamps, and stream banks; throughout; FACW.

Cyperus strigosus, spikelet (×2)

Cyperus strigosus, achene (×5)

Cyperus strigosus, scale (×5)

Cyperus tenuifolius (Steud.) Dandy Thin-leaved flatsedge
Annual; culms slender, tufted, 0.5–3 dm tall; leaves 1–3 mm wide; spikelets numerous, in 1–3 sessile, ovoid spikes subtended by 2 or 3 bracts, 1-flowered; rare on moist, sandy, or muddy shores; FACW; native farther south, apparently adventive in Pennsylvania.

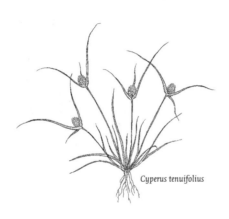

Cyperus tenuifolius

Dulichium Pers.

Dulichium arundinaceum (L.) Britt.   Three-way sedge

Culms erect to 10 dm tall, round, hollow, and somewhat jointed; leaves numerous, in 3 vertical rows; spikelets in flattened spikes that are solitary in the upper leaf axils; flowers perfect; stamens 3; style 2-branched; achene surrounded by bristles; common in bogs, swamps, marshes, lake margins, and ditches; throughout; flr. Jul–Sep, frt. late Aug–early Oct; OBL.

Dulichium arundinaceum, achene (×3)

Eleocharis R.Br.

Annual or perennial, aquatic or terrestrial plants with erect, solitary, or clustered culms; leaves reduced to basal, bladeless sheaths; inflorescence a single terminal spikelet with spirally-overlapping scales; flowers perfect; stamens 1–3; styles 2- or 3-branched; achene flat or 3-sided, topped by an expanded tubercle and surrounded at the base by bristles; flr./frt. mid- to late summer.

A. spikelet about the same diameter as the stem or only slightly thicker; scales persistent
  B. stems 4-sided .... E. quadrangulata
  B. stems triangular .... E. robbinsii
A. spikelet distinctly thicker than the stem; scales deciduous
  C. tubercle confluent with the achene except for being a slightly different texture
    D. scales 2.5–5 mm long; plants 1 dm or more tall
      E. spikelets with (5)10–20 florets; stems 4–10 dm tall .... E. rostellata
      E. spikelet with 3–9 florets; stems 1–3 dm tall .... E. pauciflora
    D. scales 1.5–2 mm long; plants <1 dm tall .... E. parvula
  C. tubercle usually separated from the achene by a narrow constriction, also differentiated by shape and texture
    F. plants with obvious, short to elongate, hard or cord-like rhizomes
      G. spikelets densely flowered, ellipsoid to cylindric; scales about 2 mm long; achenes with prominent keeled angles; bristles none ....
      .... E. tricostata
      G. spikelets closely to loosely flowered, ellipsoid, ovoid, or narrowly ovoid; scales >2 mm long; achenes with blunt angles; bristles few and delicate or none
        H. stems 4–8-angled or strongly compressed; achenes 3-angled with a rough surface
          I. culms strongly compressed but not angled .... E. compressa
          I. culms 4–8-angled (roll between fingers to feel the edges)
            J. stems 4–5-angled; achenes olive-colored .... E. tenuis
            J. stems 6–8-angled; achenes yellow to dull orange ....
            .... E. elliptica
        H. culms nearly round or somewhat flattened; achenes lens-shaped with a smooth surface
          K. 2 empty scales surrounding the base of the spikelet; achenes to 1.7 mm long .... E. palustris
          K. a single empty scale surrounding the base of the spikelet; achenes to 1.4 mm long .... E. erythropoda
    F. plants with inconspicuous short, hard rhizomes hidden by the culmbases and roots, thread-like rhizomes, or no apparent rhizomes

Dulichium arundinaceum (×¼)

L. styles usually 2-branched, less often 3-branched; achenes 2-angled
    M. stems densely tufted; rhizomes hard and inconspicuous, or lacking; tubercles broadly triangular to depressed, 2/3 or more as wide as the achenes
        N. spikelets ovoid to ovoid-cylindric; tubercles broadly triangular, >1/4 as long as the achenes; bristles usually extending beyond the tubercles .... *E. obtusa*
        N. spikelets slender-cylindric; tubercles depressed, <1/4 as long as the achenes; bristles usually not reaching the tubercles .... *E. engelmannii*
    M. stems tufted or scattered; rhizomes delicate and thread-like, or lacking; tubercles < 2/3 as wide as the achenes
        O. rhizomes delicate and thread-like; sheaths loose; spikelets mostly acute; achenes olive or dark brown, tubercles conical .... *E. olivacea*
        O. rhizomes lacking; spikelets obtuse; achenes shiny dark purple to black, tubercles depressed .... *E. caribaea*
L. styles usually 3-branched; achenes 3-angled or nearly round with obscure angles
    P. plants with rhizomes
        Q. plants with delicate thread-like rhizomes; spikelets flattened; achenes with prominent longitudinal and fine horizontal ribs .... *E. acicularis*
        Q. plants with short, hard rhizomes; spikelets cylindric; achenes with prominent keeled angles, but not ribbed .... *E. tricostata*
    P. plants cespitose, lacking rhizomes
        R. scales tough with a leathery texture; achene body and tubercle about the same size .... *E. tuberculosa*
        R. scales soft or chaffy; achene body much larger than the tubercle
            S. culms round or nearly so; achenes brown or yellowish-olive; plants of calcareous substrates .... *E. intermedia*
            S. culms 4-angled; achenes pale greenish or whitish; plants of acidic substrates .... *E. microcarpa*

Eleocharis acicularis (L.) Roem. & Schult.   Needle spike-rush
Perennial; culms in tufts or close together along thread-like rhizomes, 3–12 cm tall, often occurring as extensive stands of short (or sometimes long and hair-like) sterile stems when submerged in water; achenes with prominent longitudinal ridges; frequent in shallow water and wet shores of lakes, ponds, and streams; throughout; OBL.

Eleocharis acicularis,
achene (×10)

Eleocharis caribaea (Rottb.) S.F.Blake   Spike-rush
Tufted annual; culms 3–20 cm tall; rhizomes lacking; sheaths tight; achenes with depressed or saucer-like tubercles; very rare in damp sandy depressions; Presque Isle; 🍂.

Eleocharis acicularis

Eleocharis compressa var. compressa, spikelet (×3)

Eleocharis compressa var. compressa, achene (×10)

Eleocharis elliptica, achene (×10)

Eleocharis compressa Sull.   Flat-stemmed spike-rush

Perennial with hard rhizomes; culms slender, compressed and often twisted, 1.5–8 dm tall; spikelets 3–10 mm long, the scales with scarious, acuminate tips; achenes golden to brown, 1–1.5 mm long; very rare in wet, sandy ground, and stream banks; NW and SE; OBL; 🍃. Ours is var. *compressa*.

Eleocharis elliptica Kunth   Slender spike-rush

Similar to E. *tenuis*; stems 6–8-angled; spikelets larger with blackish scales; achenes yellow to dull orange; rare in moist, calcareous, sandy flats, fens, and swales; scattered; FACW+; 🍃. [syn: E. *tenuis* (Willd.) Schultes var. *borealis* (Svenson) Gleason]

Eleocharis engelmannii Steud.   Spike-rush

Closely resembling E. *obtusa*, but differing by having longer spikelets with more acute tips as well as the other characters in the key; occasional in shallow water and moist soil; mostly SE and SC; FACW+.

Eleocharis erythropoda Steud.   Spike-rush

Similar to E. *palustris* but with the scales broadly rounded or obtuse; spikelet encircled at the base by a single empty scale; occasional on tidal shores, stream banks, wet meadows, and swales, often on calcareous or diabase substrates; throughout; OBL.

Eleocharis intermedia (Muhl.) Schult.   Matted spike-rush

Densely tufted annual, lacking rhizomes; culms very slender, usually of unequal lengths ranging from 5 to 25 cm; achenes smooth with sharply triangular tu-

Eleocharis compressa var. compressa

Eleocharis elliptica, spikelet (×3)

Eleocharis intermedia, achene (×10)

Eleocharis intermedia

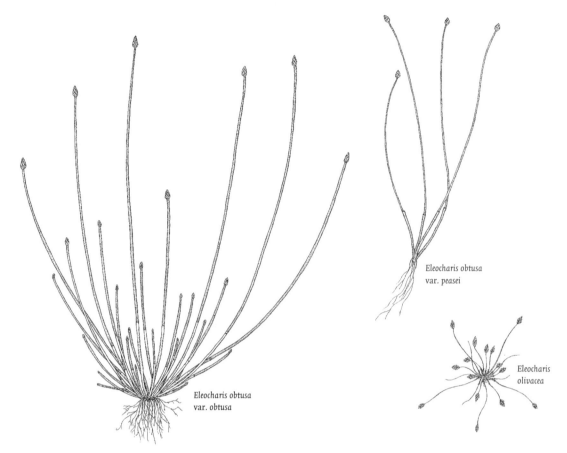

Eleocharis obtusa
var. peasei

Eleocharis
olivacea

Eleocharis obtusa
var. obtusa

bercles; rare in marshes, wet meadows, and stream banks, on calcareous substrates; mostly SE and NW; FACW+; 🍃.

Eleocharis microcarpa Torr.   Spike-rush
Tufted annual to 3 dm tall; culms 4-angled; sheath oblique; spikelet 2–7 mm long with several flowers; scales rounded at the tip and pale; styles 3-branched; achenes 3-angled, 0.6–0.8 mm long with a tiny tubercle; rare on acidic substrates; collected once in Somerset Co.; OBL.

Eleocharis obtusa (Willd.) Schult.   Spike-rush
Densely tufted perennial; rhizomes short and hard (difficult to detect); spikelets often thimble-like and compact with numerous closely appressed scales; OBL.

Eleocharis obtusa
var. *obtusa*,
achene (×10)

A. bristles present .... var. *obtusa*
     common in shallow water, along the margins of lakes, ponds, and wet depressions and in marshes, wet meadows, and ditches; throughout.
A. bristles absent .... var. *peasii* Svenson
     rare on tidal shores and mudflats; SE; 🍃.

Eleocharis olivacea Torr.   Capitate spike-rush
Perennial; culms in tufts or scattered along thread-like rhizomes, 3–15 cm tall; sheaths loose; spikelet 2–7 mm long; scales with dark brown edges; achenes lens-shaped, 1 mm long, with a conical tubercle; occasional in bogs and sandy-peaty depressions; E and NW; OBL; 🍃. [syn: *E. flavescens* (Poiret) Urban]

Eleocharis olivacea,
achene (×10)

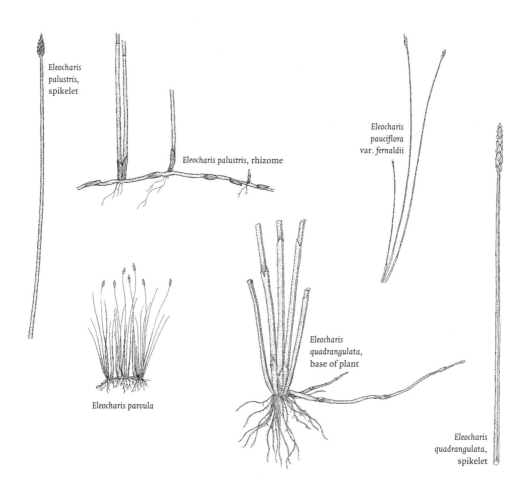

Eleocharis palustris, spikelet

Eleocharis palustris, rhizome

Eleocharis pauciflora var. fernaldii

Eleocharis quadrangulata, base of plant

Eleocharis parvula

Eleocharis quadrangulata, spikelet

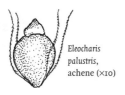

Eleocharis palustris, achene (×10)

Eleocharis palustris (L.) Roem. & Schult.   Creeping spike-rush
Culms arising singly from a perennial rhizome, 1–10 cm tall; spikelets 5–40 mm long, surrounded at the base by 2–3 empty scales; styles 2-branched; achenes lenticular, 2–3 mm long, roughened; bristles surrounding the achene retrorsely barbed; forming large colonies at the margins of lakes and streams, and in bogs, swamps, and marshy swales; throughout; OBL.

Eleocharis parvula (Roem. & Schult.) Link ex Buffon & Fingerh.   Dwarf spike-rush
Perennial from slender rhizomes; culms 2–6 cm tall, very slender; spikelet 2.5–4 mm long; styles 3-branched; achenes 3-angled, about 1 mm long including the short, thickened tubercle; very rare on tidal shores and mudflats; SE; OBL; ✤.

Eleocharis pauciflora var. fernaldii, achene (×10)

Eleocharis pauciflora (Lightf.) Link   Spike-rush
Rhizomatous perennial with slender clustered culms 1–3 dm tall and rarely more than 1 mm thick; spikelet 4–8 mm long, 3–9-flowered; styles 3-branched; achenes 1.9–2.6 mm long including the tubercle, rounded or planoconvex with a roughened surface; very rare in wet, calcareous sand at Presque Isle; ✤. Ours is var. *fernaldii* Svenson.

Eleocharis quadrangulata, achene (×5)

Eleocharis quadrangulata (Michx.) Roem. & Schult.   Four-angled spike-rush
Perennial; culms robust, to 1 m tall, sharply 4-sided; spikelet barely thicker than the stem; rare in lake margins, swamps, and ponds; W and SE; OBL; ✤.

Eleocharis robbinsii Oakes   Robbins's spike-rush
Perennial with erect, emergent flowering culms and, flexuous, capillary underwater or floating sterile stems; spikelet no thicker than the stem, 4–8-flowered;
styles 3-branched; achenes biconvex; very rare in shallow water of glacial lakes
and ponds; NE; OBL; ✿.

Eleocharis rostellata (Torr.) Torr.   Beaked spike-rush
Perennial with clustered stems 4–10 dm tall; stems flattened toward the top,
some of them bending to the ground and rooting at the tip; spikelets slender, 8–
13 mm long; styles 3-branched; achenes rounded-trigonous to planoconvex,
1.9–2.8 long including the prominent tubercle; very rare in calcareous swamps;
W; OBL; ✿.

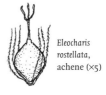

Eleocharis rostellata, achene (×5)

Eleocharis tenuis (Willd.) Schult.   Slender spike-rush
Culms tufted or close together on elongate cord-like rhizomes; sheaths truncate,
reddish below; spikelet ellipsoid to ovoid 3–10 mm long; scales obtuse or acute,
with dark margins; achenes unequally 3-angled, roughened; bristles very small
or lacking; FACW+.

A. culms delicately capillary with 4–5 angles, not winged; rarely >3 dm tall
  B. tubercle pyramidal, ⅕ as high as the achene; surface of the achene reticu-
     late without warty thickenings .... var. *tenuis*
     frequent in moist fields, swamps, bogs, and wet ditches; throughout.
  B. tubercle depressed, about ⅛ as high as the achene; surface of the achene
     reticulate with warty thickenings .... var. *verrucosa* (Svenson) Svenson
     rare in moist, open ground; scattered; ✿.

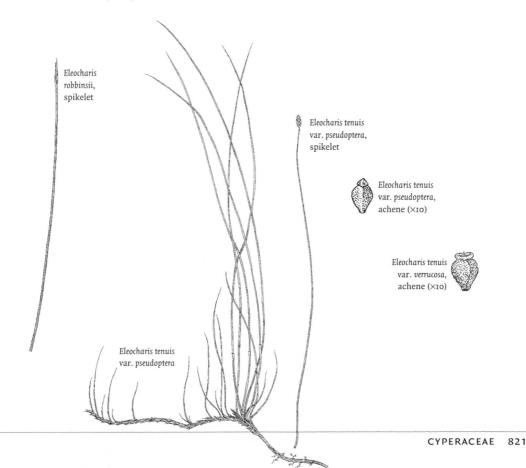

Eleocharis robbinsii, spikelet

Eleocharis tenuis var. *pseudoptera,* spikelet

Eleocharis tenuis var. *pseudoptera,* achene (×10)

Eleocharis tenuis var. *verrucosa,* achene (×10)

Eleocharis tenuis var. *pseudoptera*

A. culms firm with 1 or more of the angles winged, 3–9 dm tall ....
.... var. *pseudoptera* (Weath.) Svenson
rare in moist meadows, hayfields, and damp areas, especially on limestone or
serpentine; mostly SE.

Eleocharis tricostata Torr.   Three-ribbed spike-rush
Perennial with a short rhizome; culms 2–6 dm tall, rounded or slightly compressed; sheaths truncate with mucronate tips; spikelet 6–16 mm with numerous closely overlapping scales, styles 3-branched; achenes 0.8–1 mm long, yellowish, lacking bristles; very rare in wet, sandy-peaty meadows; Delaware Co.
but believed to be extirpated; OBL; ❦.

*Eleocharis
tuberculosa,
achene (×10)*

Eleocharis tuberculosa (Michx.) Roem. & Schult.   Long-tubercled spike-rush
Tufted annual with flattened culms 2–8 dm tall; sheaths oblique; spikelets ovoid,
5–15 mm with straw-colored or dark scales; achenes 3-angled, 1.2–1.5 mm long;
tubercle about as long and as thick as the achene; bristles present; very rare in a
sphagnous bog; SE but believed to be extirpated; OBL; ❦.

Eriophorum L.

Perennial herbs of wet habitats with narrow grass-like leaves with basal sheaths,
upper sheaths sometimes lacking blades; spikelets solitary and terminal or in a
cymose or head-like cluster subtended by bracts; flowers perfect, each in the axil
of a scale; perianth consisting of numerous elongate bristles that form a conspicuous cottony tuft 2–4 cm long in the mature spikelets.

A. sheaths of the upper stem lacking blades; inflorescence a single spikelet without subtending leaf-like bracts; lower scales of the spikelet sterile ....
.... *E. vaginatum*
A. sheaths of the upper stem bearing blades; inflorescence subtended by one or
more leaf-like bracts and with 2–many spikelets; lower scales of the spikelets
fertile
B. primary inflorescence bract divergent from the stem tip; spikelets in a
dense cluster; frt. Aug–Sep .... *E. virginicum*
B. primary inflorescence bract erect or nearly so; spikelets solitary at the tips
of inflorescence branches (sometimes a few in a cluster); frt. May–Jul
C. inflorescence usually with 1 leaf-like bract at the base (other bracts lacking an obvious blade); blades of stem leaves usually V-shaped and about
the same width from base to tip
D. blade of upper stem leaf usually shorter than the sheath; scales usually blackish; frt. May .... *E. gracile*
D. blade of upper stem leaf usually longer than the sheath; scales usually reddish brown; frt. Jun .... *E. tenellum*
C. inflorescence usually with >1 leaf-like bract at the base; blades of stem
leaves nearly flat except for the abruptly narrowed elongate tip ....
.... *E. viridicarinatum*

Eriophorum gracile Koch ex Roth   Slender cotton-grass
Spreading with soft rhizomes, sometimes forming floating mats; stems often
spreading and reclining; leaf blades up to 1.5 mm wide; scales blackish; very rare
in bogs and peaty depressions; NE and NW, former sites in the SE apparently are
gone; frt. May; OBL; ❦.

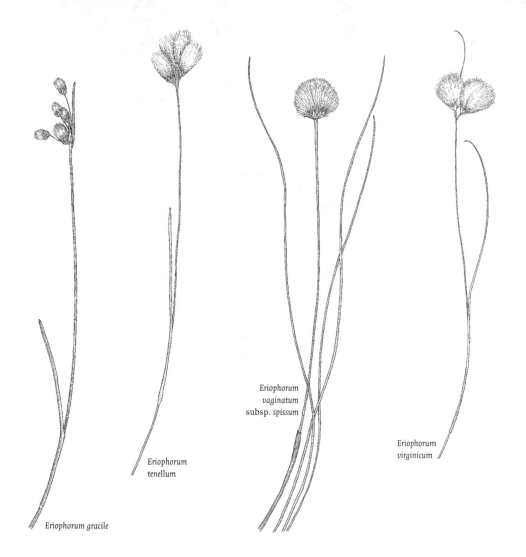

Eriophorum
vaginatum
subsp. spissum

Eriophorum
tenellum

Eriophorum
virginicum

Eriophorum gracile

Eriophorum tenellum Nutt.   Rough cotton-grass
Closely resembling but generally coarser than E. *gracile* and blooming later; scales reddish-brown; very rare in bogs, peaty depressions, and peaty swamps; NE and NW, former sites in the SE apparently are gone; frt. Jun; OBL; 🌿.

Eriophorum vaginatum L.   Cotton-grass
Forming dense tussocks; stems each with a single terminal spikelet; occasional in bogs and peaty swamps; N; frt. May; OBL. Ours are ssp. *spissum* (Fernald) Hulten.

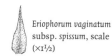

Eriophorum vaginatum
subsp. *spissum*, scale
(×1½)

Eriophorum virginicum L.   Tawny cotton-grass
Tufted with slender rhizomes; leaf blades to 4 mm wide; spikelets several, in a dense cluster; scales brown; bristles tawny; widespread in bogs, peaty meadows, and peaty swamps; throughout; frt. Aug–Sep; OBL.

Eriophorum
virginicum,
scale (×1½)

Eriophorum viridicarinatum (Engelm.) Fernald   Thin-leaved cotton-grass
Tufted with slender rhizomes; leaf blades 2–6 mm wide; scales blackish; rare in bogs, wet meadows, and swamps; NE and NW; frt. May–Jun; OBL; 🌿.

Fimbristylis Vahl

Annual (or perennial) herbs with slender stems and a few grass-like leaves at the base; spikelets several to many in a dense cluster or a simple or compound umbel-like inflorescence subtended by sheathless bracts; scales spirally arranged; flowers perfect, perianth none; stamens 1-3; style 2- or 3-branched, the lower, unbranched part flattened and sometimes fringed; achenes lens-shaped or 3-angled.

Fimbristylis
annua, achene
(×10)

A. styles 3-branched, not fringed; achenes 3-angled, <0.8 mm long..................
.... F. autumnalis
A. styles 2-branched, fringed; achenes 2-angled, >0.8 mm long
  B. stem to 3 dm tall; not rhizomatous; stamens 1 or 2 .... F. annua
  B. stems to 1 m tall; rhizomes present; stamens 3 .... F. puberula

Fimbristylis annua (All.) Roem. & Schult.   Annual fimbry
Annual; differing from *F. autumnalis* by having <10 ovate spikelets; rare in moist depressions on serpentine barrens; SE; flr./frt. Jul–Oct; FACW–; ❧. [syn: *F. bald-winiana* (Schult.) Torr.]

Fimbristylis
autumnalis,
achene (×10)

Fimbristylis autumnalis (L.) Roem. & Schult.   Slender fimbry
Tufted annual to 2 dm tall; leaves basal with blades to nearly 3 mm wide; spikelets narrowly ovate, numerous, in umbel-like inflorescences; occasional in stream banks, pond margins, or other moist sandy or peaty substrates; S; flr./frt. Aug–Oct; FACW+.

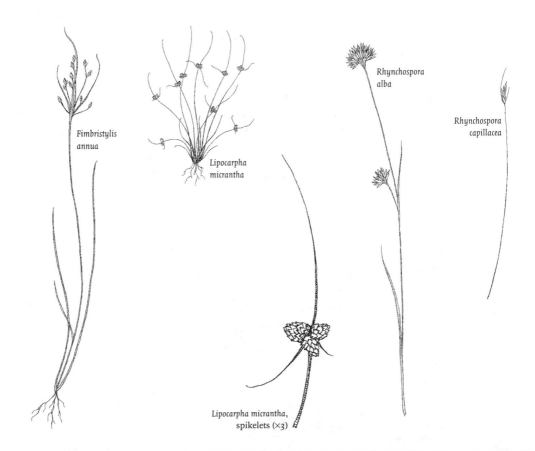

Fimbristylis
annua

Lipocarpha
micrantha

Rhynchospora
alba

Rhynchospora
capillacea

Lipocarpha micrantha,
spikelets (×3)

Fimbristylis puberula (Michx.) Vahl   Hairy fimbry

Perennial from short, knotty rhizomes; stems in small tufts, to 1 m tall; edges of
the leaves rolled inward; spikelets 5–10 mm long, in a compound inflorescence;
styles 2-branched; achenes lens-shaped; very rare in the margin of a swamp;
Lancaster Co. but believed to be extirpated; flr./frt. May–Jul; OBL; ✿.

Lipocarpha R.Br.

Lipocarpha micrantha (Vahl) G.C.Tucker   Common hemicarpa

Tufted small annual lacking rhizomes; leaves basal with capillary blades; spikelets 1–3, sessile; flowers perfect, each subtended by a scale and an additional pair
of colorless scales, 1 above and persistent along the rachilla, and 1 below and
falling with the achene; moist sand; known only from Erie and Northumberland
Cos.; frt. Jul–Oct; FACW+; ✿. [syn: Hemicarpha micrantha (Vahl) Britton]

Rhynchospora Vahl

Perennial herbs of wet habitats; stems erect, leafy; spikelets in 1 or more dense
glomerules that are arranged in terminal and axillary inflorescences; scales spirally overlapping, the lowest ones usually empty; flowers perfect or some staminate; stamens 1–3; styles 2-branched; achenes lens-shaped with persistent tubercles; bristles present or absent; frt. mid-summer–early fall.

A. bristles downwardly toothed or barbed, or smooth
  B. spikelets white, becoming whitish-brown; bristles 9–12, hairy at the base
  .... R. alba
  B. spikelets brown; bristles 6, not hairy at the base
    C. leaves flat; glomerules spreading, fan-shaped; achenes obovoid with
    whitish margins .... R. capitellata
    C. leaves with inrolled margins; glomerules ellipsoid; achenes oblong-
    ellipsoid to narrowly obovate .... R. capillacea
A. bristles upwardly toothed or barbed
  D. achenes appearing smooth (at 10×)
    E. achene obovate, prolonged at the base and broadest well above the
    middle .... R. fusca
    E. achene elliptic or subrotund, not prolonged at the base, broadest at the
    middle .... R. gracilenta
  D. achenes appearing roughened (at 10×) .... R. globularis

Rhynchospora alba (L.) Vahl   White beak-rush

Stems erect to 7 dm tall, clustered, longer than the leaves; spikelets white or
whitish-brown, in 1–3 glomerules, inflorescences from the lower axils long-
peduncled; achenes flattened, pear-shaped, narrowed at the base; occasional in
bogs and swamps; throughout; OBL.

Rhynchospora
alba, achene
(×5)

Rhynchospora capillacea Torr.   Capillary beak-rush

Cespitose and tussock-forming with very slender stems; leaves narrow, the uppermost exceeding the stems in length; spikelets in 1 or 2 glomerules, the terminal glomerule with 2–10 erect spikelets, the lateral glomerules smaller and
subsessile; achenes narrowly elliptic; bristles 6; rare in fens and calcareous
swamps; OBL; ✿.

Rhynchospora
capillacea,
achene (×5)

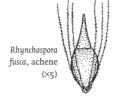

Rhynchospora capitellata, achene (×5)

Rhynchospora capitellata (Michx.) Vahl Beak-rush

Erect, cespitose with stems 3–8 dm tall; spikelets brown, in 2 or more glomerules the terminal one to 1.5 cm thick, lateral glomerules smaller; inflorescences increasingly long-peduncled toward the base; frequent in boggy meadows, open woods, abandoned gravel pits, and vernal ponds; mostly SE, but scattered elsewhere; OBL.

Rhynchospora fusca, achene (×5)

Rhynchospora fusca (L.) Ait. f. Brown beak-rush

Rhizomatous; stems 1.5–4 dm tall; leaves very slender with inrolled edges; inflorescences with 1–3 glomerules, the lower glomerule long-peduncled; spikelets dark brown; achenes obovate, light brown, smooth; very rare in bogs and glacial lake margins; NE but believed to be extirpated; OBL; 🌿.

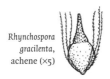

Rhynchospora globularis, achene (×5)

Rhynchospora globularis (Chapm.) Small Beak-rush

Cespitose, stems erect or ascending, 3–9 dm tall; leaves flat, 2–4 mm wide; inflorescences with 1–4 glomerules, the lateral ones peduncled; spikelets brown, broadly ovoid; achenes ovoid to subrotund with roughened surfaces and pyramidal tubercles; rare on sandy shores and in swamps and sphagnum bogs; S; FACW; 🌿.

Rhynchospora gracilenta, achene (×5)

Rhynchospora gracilenta Gray Beak-rush

Stems very slender, cespitose, to 1 m tall; leaves to 2 mm wide; inflorescences with 2–4 glomerules, the terminal glomerule 6–12 mm wide; spikelets dark reddish-brown; very rare in sphagnum bogs; known from a single site in Lancaster Co., but believed to be extirpated; OBL; 🌿.

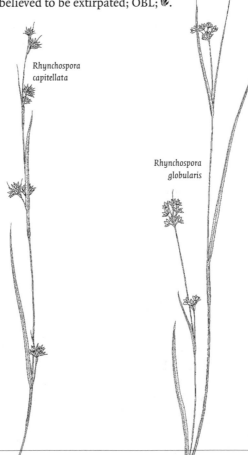

Rhynchospora capitellata

Rhynchospora globularis

Schoenoplectus (Rchb.) Palla

Rhizomatous perennials; leaves basal or cauline; spikelets 1–15, sessile or with short pedicels and subtended by an erect bract that appears as a continuation of the stem, the inflorescence thus appearing lateral, or the inflorescence terminal and subtended by several flat, leaf-like bracts; scales numerous, spirally overlapping; flowers perfect, each in the axil of a scale; stamens 3; styles 2- or 3-branched; achenes lacking a tubercle; bristles present or occasionally lacking. Previously included in *Scirpus*.

A. plants lacking tuberous rhizomes; leaves at or near the base of the stems, blades usually reduced and inconspicuous, inflorescence bract appearing as a continuation of the stem; spikelets with glabrous scales
  B. plants lacking rhizomes, culms tufted
    C. inflorescence bracts mostly <¼ as long as the stems, scales broadly elliptic and <1.4 times as long as wide, achenes obviously curved along the vertical axis of the ventral surface .... S. *purshianus*
    C. inflorescence bracts mostly >¼ as long as the stems, scales elliptic to obovate and mostly >1.4 times as long as wide, achenes nearly flat along the vertical axis of the ventral surface .... S. *smithii*
  B. plants rhizomatous with scattered culms
    D. leaves with elongate, flaccid, capillary underwater blades, inflorescence with 1 spikelet .... S. *subterminalis*
    D. leaf blades reduced to short projections from the sheaths or firm and elongate, rarely developing underwater; inflorescence with >1 spikelet (rarely 1)
    E. stems round
      F. inflorescences with comparatively stiff branches, spikelets mostly in small clusters at the tips of the inflorescence branches; scales dull gray-brown with brown spots, awns at the tips of the scales contorted and >0.5 mm long (frequently broken on herbarium specimens) .... S. *acutus*
      F. inflorescence with arching branches; spikelets solitary or in clusters of 2 or 3; scales pale gray to orange-brown; awns straight or nearly so, <0.8 mm long
        H. culms soft; spikelets mostly in clusters of 2 or 3; scales orange-brown; awns about 0.25 mm long; styles 2-parted and achenes 2-angled .... S. *tabernaemontani*
        H. culms hard; spikelets mostly solitary; scales pale gray; awns >0.25 mm long; styles 3- parted; achenes 3-angled....
        .... S. *heterochaetus*
    E. stems 3-angled
      G. rhizomes flaccid; leaf blades triangular-channeled; scales scarcely notched at the tip, awns lacking or inconspicuous ....
      .... S. *torreyi*
      G. rhizomes firm; leaf blades V-shaped; scale tips with conspicuous notches and awns .... S. *pungens*
A. plants with tuberous rhizomes; leaves scattered along the stems, with well-developed V-shaped blades; inflorescence bracts similar to leaves; spikelets with pubescent scales .... S. *fluviatilis*

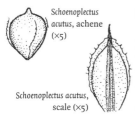

Schoenoplectus
acutus, achene
(×5)

Schoenoplectus acutus,
scale (×5)

Schoenoplectus acutus (Muhl. Ex Bigel.) Love & Love

Great or hard-stemmed bulrush
Rhizomatous, colonial; stems 1–3 m tall, round, and firm; leaves few, with short blades, mostly near the base of the stem; inflorescence appearing lateral, stiff-branched; spikelets numerous, in small clusters at the tips of stiff inflorescence branches; rare in shallow water of lake and pond margins; NW; frt. Jun–Aug; OBL; ❧. [syn: *Scirpus acutus* Muhl. ex Bigelow]

Schoenoplectus fluviatilis (Torr.) Strong   River bulrush

Stems erect, 0.7–2 m tall from a tuberous-thickened rhizome, sharply angled and leafy nearly to the top; leaf sheath convex at the top; inflorescence terminal, subtended by several leafy bracts; spikelets 1.5–4 cm long, solitary or clustered, some sessile and others at the tips of slender inflorescence rays; rare on moist, sandy shores and marshes, tidal or nontidal; scattered; frt. Jun–Aug; OBL; ❧. [syn: *Scirpus fluviatilis* (Torr.) A.Gray]

Schoenoplectus heterochaetus (Chase) Sojak   Slender bulrush

Similar to *S. acutus*, but more slender; stems hard; inflorescence more lax, all the spikelets pedicellate; lake and stream margins; Monroe Co. but believed to be extirpated; frt. Aug; OBL; ❧. [syn: *Scirpus heterochaetus* Chase]

Schoenoplectus pungens (Vahl) Palla   Chairmaker's rush, three-square

Rhizomatous, forming large colonies; stems to 1.5 m tall, sharply triangular; leaves few with short blades; inflorescence of 1–6 sessile spikelets appearing lateral due to the erect subtending bract; occasional on tidal shores and in

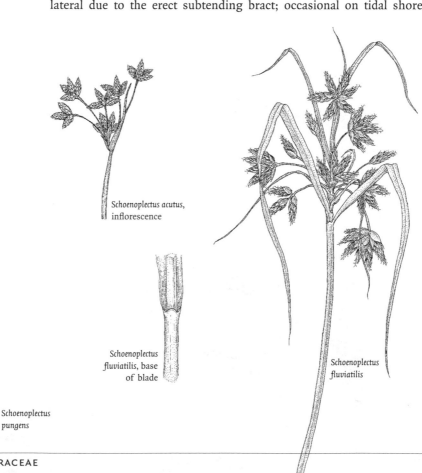

Schoenoplectus acutus,
inflorescence

Schoenoplectus
fluviatilis, base
of blade

Schoenoplectus
fluviatilis

Schoenoplectus
pungens

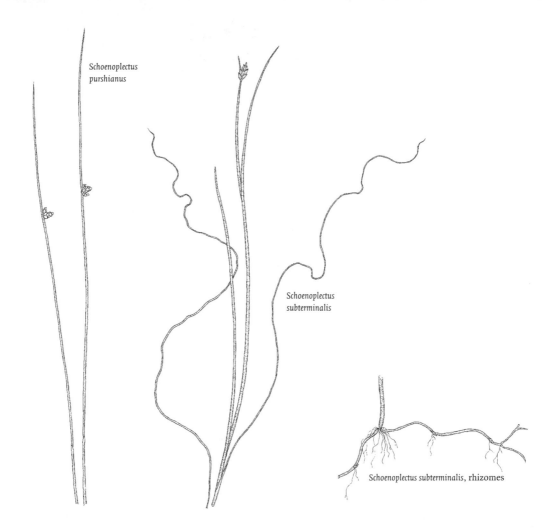

Schoenoplectus
purshianus

Schoenoplectus
subterminalis

Schoenoplectus subterminalis, rhizomes

shallow water and marshes; throughout; frt. May–Aug; FACW+. [syn: *Scirpus americanus sensu auct.* non Pers., *Scirpus pungens* Vahl]

Schoenoplectus purshianus (Fernald) Strong   Bulrush

Schoenoplectus
purshianus,
achene (×5)

Densely tufted and lacking rhizomes; spikelets sessile and appearing lateral; achenes (of Pennsylvania plants) usually with well-developed bristles; rare in moist or wet depressions, ditches, and shores; mostly E and S, scattered elsewhere; frt. Jul–Sep; OBL. [syn: *Scirpus purshianus* Fernald]

Schoenoplectus smithii (A.Gray) Sojak   Smith's bulrush

Similar to S. *purshianus* except that achenes are flatter on the ventral surface and have more truncate tips; Pennsylvania plants of S. *smithii* lack bristles while Pennsylvania plants of S. *purshianus* have them; rare in freshwater intertidal marshes and moist lake shores; SE and NW; frt. Jul–Sep; OBL; ❧. [syn: *Scirpus smithii* A.Gray]

Schoenoplectus subterminalis (Torr.) Sojak   Water bulrush

Schoenoplectus
subterminalis,
achene (×5)

Aquatic with submersed, capillary leaves; flowering stems extending above the water surface, each bearing a single spikelet subtended by an erect bract; rare in quiet water of lakes, ponds, and slow-moving boggy streams; NE, NW, and SC; frt. Jul–Sep; OBL; ❧. [syn: *Scirpus subterminalis* Torr.]

Schoenoplectus
torreyi, achene
(×5)

Schoenoplectus tabernaemontani,
inflorescence

Schoenoplectus tabernaemontani (Gmel.) Palla   Great or soft-stemmed bulrush
Rhizomatous, colony-forming; stems erect, round, 0.5–2.5 m tall; sheaths usually lacking blades; inflorescence a cluster of short, drooping branches from the stem tip, subtended by a short erect, involucral bract; spikelets solitary or clustered at the tips of the inflorescence branches; frequent in shallow water of swamps, lake and pond margins, wet ditches, and mudflats; throughout; frt. Jun–Aug; OBL. [syn: *Scirpus validus* Vahl]

Schoenoplectus torreyi (Olney) Palla   Torrey's bulrush
Rhizomatous, colonial perennial; stems erect, 5–10 dm tall, sharply 3-angled; leaves several, often longer than the stem; spikelets 1–4, sessile and clustered, inflorescence appearing lateral due to the erect, 6–15 cm long subtending bract; rare in shallow water of lake and pond margins; N; frt. Jul–Sep; OBL; 🍂. [syn: *Scirpus torreyi* Olney]

Schoenoplectus
torreyi (×⅛)

Scirpus L.

Herbs of wet habitats; stems round or 3-angled; leaves narrow, with basal sheaths, blades sometimes reduced or lacking; flowers in spikelets, each in the axil of a scale; spikelets numerous, in a compound terminal inflorescence; florets perfect, each in the axil of a scale; stamens 3; style with 2 or 3 branches; achene lacking a tubercle; bristles usually present.

A. bristles smooth, without teeth along the margins, strongly contorted, and exceeding the achenes when extended (if bristles are lacking, or are short and nearly straight, see S. *georgianus*).
  B. stem leaves lacking ligules; scales usually with prominent green midribs, mature bristles mostly contained within scales; achenes >1 mm long ....
    .... S. *pendulus*
  B. culm leaves with low narrow ligules; scales usually with inconspicuous midribs, mature bristles exceeding the scales and giving the inflorescence a woolly appearance; achenes <1 mm long (the species under this heading are difficult to distinguish and heavy reliance is placed on time of achene maturation)
    C. spikelets usually solitary at tips of pedicels (rarely in clusters with pedicels scarcely developed); scales usually blackish; frt. late Jun–early Jul
      .... S. *atrocinctus*
    C. spikelets solitary at tips of pedicels or in clusters with pedicels scarcely developed; scales reddish-brown, brown, or sometimes blackish; frt. mid Jul–Sep

D. spikelets usually solitary at the tips of pedicels; scales usually pale brown; frt. mid–late Jul .... *S. pedicellatus*

D. spikelets solitary at the tips of pedicels or in clusters with pedicels scarcely developed; scales reddish-brown to dark brown (sometimes blackish); frt. early Aug–Sep; widespread variable species ....
.... *S. cyperinus*

A. bristles with teeth along the margins, nearly straight to strongly contorted, shorter than to greatly exceeding the achenes when extended

  E. leaf sheaths red-tinged near the base; styles mostly 2-parted; achenes 2-angled .... *S. microcarpus*

  E. leaf sheaths green near the base (except S. *expansus*, which has lower leaf sheaths red-tinged near the base); styles mostly 3-parted; achenes 3-angled

    F. culms with 10–20 leaves; spikelets broadly ovate; scales reddish-brown, about as broad as long excluding tips; bristles contorted, more than twice as long as the achene when extended .... *S. polyphyllus*

    F. culms with 2–10 leaves; spikelets broadly ovate to narrowly ovate; scales brown or black, mostly longer than broad excluding tips, bristles straight or nearly so, shorter than to slightly longer than the achenes

      G. inflorescence branches arching at maturity, antrorsely scabrous; bristle teeth thick-walled and sharp-pointed, densely arranged almost to base of bristle

        H. plants spreading with long reddish rhizomes; lower leaf sheaths red-tinged near the base; scales usually broadest below the middle; achenes with poorly developed receptacles from which the bristles readily detach .... *S. expansus*

        H. plants cespitose with short brownish rhizomes; leaf sheaths green throughout; scales broadest about the middle; achenes with well-developed receptacles and firmly attached bristles ......
.... *S. ancistrochaetus*

      G. inflorescence branches stiff and branching at divergent angles, smooth or slightly scabrous near tips; bristle teeth thin-walled and with rounded tips, restricted to upper ⅔ of the bristle

        I. bristles 0–3, shorter than achenes, teeth (if present) concentrated near bristle tips .... *S. georgianus*

        I. bristles usually 5 or 6, shorter than to slightly longer than achenes, teeth extending downward from all or at least some of the bristles

          J. lower leaf blades and sheaths usually with prominent transverse septae between veins; spikelets ovate or narrowly ovate; scales brownish, longer bristles frequently exceeding achenes
.... *S. atrovirens*

          J. lower leaf blades and sheaths with inconspicuous transverse septae between veins; spikelets broadly ovate or ovate; scales blackish, longer bristles mostly shorter than or equaling achenes .... *S. hattorianus*

Scirpus ancistrochaetus Schuyler   Northeastern bulrush

Cespitose with short tough rhizomes; culms often forming bulblets at the nodes late in the season; branches of inflorescence arching at maturity; rare in intermittently wet or inundated depressions; C and NE; frt. Jul; OBL; 🌿. This species is most often confused with S. *atrovirens* from which it differs by having longer

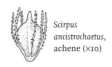

Scirpus
ancistrochaetus,
achene (×10)

achenes (1.1–1.7 mm vs. 1–1.3 mm). [syn: S. *atrovirens* Willd. var. *atrovirens* in part]

Scirpus atrocinctus Fernald   Blackish wool-grass
Similar to S. *pedicellatus*, but having blackish scales and fruiting earlier; mostly NE, rare elsewhere; frt. late Jun–early Jul; FACW+. [syn: S. *cyperinus* (L.) Kunth in part]

Scirpus atrovirens Willd.   Black bulrush
Similar in overall appearance to S. *georgianus* and S. *hattorianus*, but generally in

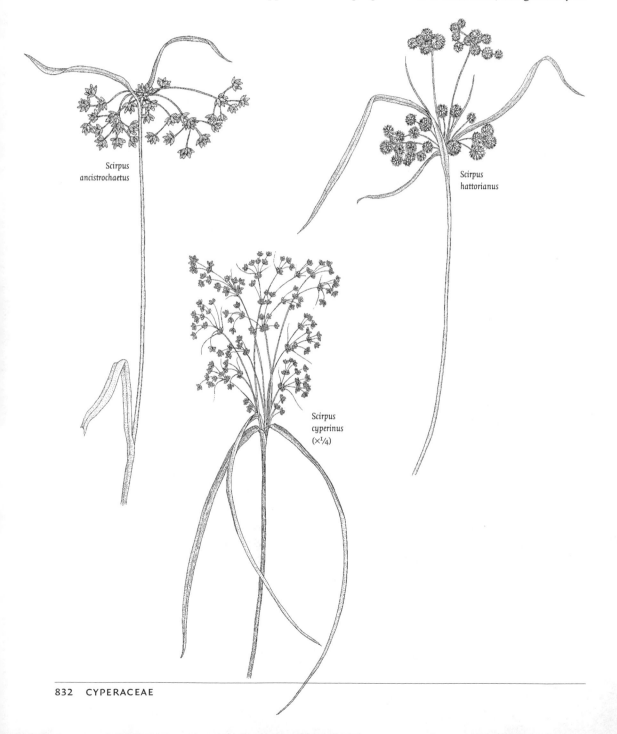

Scirpus
ancistrochaetus

Scirpus
hattorianus

Scirpus
cyperinus
(×¼)

wetter habitats; frequent in marshes, wet meadows, swales, and ditches; throughout, but rare NC; frt. late Jun–Jul; OBL.

Scirpus cyperinus (L.) Kunth   Wool-grass
Forming dense tussocks with short, tough, branching rhizomes; spikelets in clusters or solitary at tips of inflorescence branches; bristles highly contorted and exceeding the achenes at maturity; extremely variable; common in marshes, moist meadows, swales, shores, and ditches, often in early successional habitats; throughout; frt. Aug–Sep; FACW+.

Scirpus cyperinus,
achene (×10)

Scirpus cyperinus,
spikelet (×2)

Scirpus expansus Fernald   Wood bulrush
Spreading with long reddish rhizomes; lower sheaths red-tinged; styles 3; achenes 3-angled; marshes, wet meadows, and swales; mostly E, scattered elsewhere; frt. Jul–early Aug; OBL. [syn: *S. sylvaticus* L.]

Scirpus georgianus R.M.Harper   Bulrush
Resembling *S. hattorianus* and *S. atrovirens*; cespitose with short, tough rhizomes; inflorescence branches stiff and divergent with clusters of spikelets at the tips; frequent in marshes, moist meadows, swales, and ditches; mostly S; frt. Jun–Jul; OBL. [syn: *S. atrovirens* Willd. var. *atrovirens* in part, *S. atrovirens* Willd. var. *georgianus* (R.M.Harper) Fernald]

Scirpus georgianus,
achene (×10)

Scirpus hattorianus Makino   Bulrush
Similar in gross appearance to *S. georgianus* and in similar habitats, but generally more frequent N while *S. georgianus* is more frequent S; marshes, moist meadows, swales, and ditches; throughout; frt. Jun–early Jul; OBL. [syn: *S. atrovirens* Willd. var. *atrovirens* in part]

Scirpus hattorianus,
achene (×10)

Scirpus microcarpus C.Presl   Bulrush
Having reddish rhizomes and red-tinged sheaths similar to *S. expansus*, but differing from it and all other Pennsylvania species by having mostly 2-branched styles and 2-angled achenes; rare in marshes, moist meadows, swales, and ditches; mostly N, and scattered elsewhere; frt. Jun–Jul; OBL.

Scirpus pedicellatus Fernald   Wool-grass, stalked bulrush
Similar to *S. cyperinus*, but with spikelets solitary at the tips of inflorescence branches, pale brown scales, and an earlier fruiting time; rare in lowland alluvial wetlands and stream valleys; NW and NC; frt. Jul; OBL; ✤.

Scirpus pendulus Muhl.   Bulrush
Cespitose with short tough rhizomes; inflorescence drooping with solitary spikelets at the branch tips; achenes usually >1 mm long; occasional in marshes, moist meadows, swales, and ditches, often on somewhat calcareous substrates; mostly S and NW; frt. Jun–Jul; OBL.

Scirpus polyphyllus,
achene (×10)

Scirpus polyphyllus Vahl   Bulrush
Distinguished from all Pennsylvania species by the numerous stem leaves (usually >10) and the long contorted bristles with retrorse teeth above the middle; frequent in swamps and along wooded streams and other shaded, wet habitats; throughout; frt. Jul–Aug; OBL.

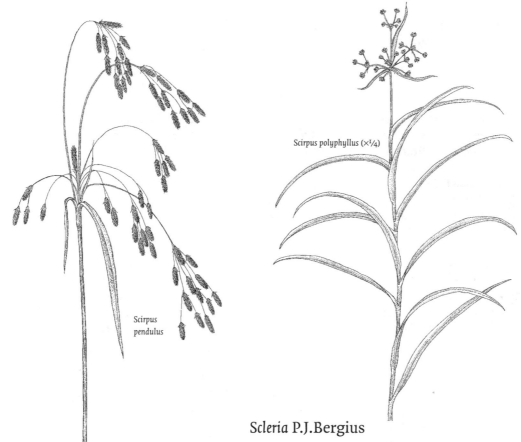

Scirpus polyphyllus (×¼)

Scirpus
pendulus

Scleria P.J.Bergius

Herbs with 3-angled stems; monoecious; inflorescences small, compact cymes; the staminate spikelets few-flowered, the pistillate spikelets each with several empty scales below the single fertile one; stamens 1–3; achene bony, white, with a disk-like base (hypogynium); frt. summer.

A. achenes smooth and shining
 B. leaf blades 4–8 mm wide; achenes 2.5–3.5 mm long .... *S. triglomerata*
 B. leaf blades 1–3 mm wide; achenes 1.5–2 mm long .... *S. minor*
A. achenes rough-textured (appearing like a miniature golf ball)
 C. hypogynium with 6 tubercles .... *S. pauciflora*
 C. hypogynium without tubercles
  D. hypogynium calyx-like, with 3 lobes appressed to the achene ....
  .... *S. muhlenbergii*
  D. hypogynium not lobed .... *S. verticillata*

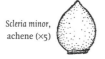

Scleria minor,
achene (×5)

Scleria minor (Britton) Stone   Small nut-rush
Rhizomatous perennial; stems slender, to 8 dm tall; leaf blades 1–3 mm wide; cymes 1–3; achene white, smooth with minute pointed tip; very rare in sphagnum bogs and swamps; SE; frt. Jun–Jul; FACW; ❦.

Scleria muhlenbergii, achene (×5)

Scleria muhlenbergii Steud.   Reticulated nut-rush
Annual or perennial with short slender rhizomes; stems to 8 dm tall; leaf blades 2–4 dm wide; cymes 2–4; achenes white or grayish, the surfaces ridged and pitted; rare in moist, sandy meadows and boggy pastures; SE; frt. Aug–Oct; OBL; ❦.
[syn: *S. reticularis* Michx.]

Scleria pauciflora Willd.   Few-flowered nut-rush

Perennial from short rhizomes; stems 2–5 dm tall; leaf blades 1–3 mm wide; cyme usually solitary; achenes white, looking like a miniature golf ball; hypogynium with 6 tubercles; rare, but sometimes locally abundant in dry, open woods and serpentine barrens; SE; frt. Jun–Sep; FACU+; ✿.

Scleria pauciflora, achene (×5)

Scleria triglomerata Michx.   Whip-grass, nut-rush

Rhizomatous perennial, stems to 1 m tall; leaf blades 4–8 mm wide; cymes 1–3; achenes white or grayish, blunt or with minute pointed tips; rare in sphagnum bogs, swampy meadows, and moist serpentine barrens; mostly SE, scattered elsewhere; frt. Jun–Sep; FAC; ✿.

Scleria triglomerata, achene (×5)

Scleria verticillata Willd.   Whorled nut-rush

Annual; stems slender, 2–6 dm tall; leaf blades 1 mm wide; cymes 2–8; achene white, the surfaces rough; very rare in moist, calcareous meadows, bogs, and fens; SE and NW; frt. Jul–Sep; OBL; ✿.

Trichophorum Pers.

Trichophorum planifolium (Sprengel) Palla   Club-rush

Densely cespitose with short rhizomes, often forming mounds; leaves up to 2 mm wide; stems each with a single terminal spikelet subtended by a scale-like bract with a projecting awn; rich, dry, rocky woods; widespread, but mostly SE; frt. May–Jun.

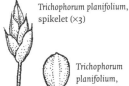

Trichophorum planifolium, spikelet (×3)

Trichophorum planifolium, achene (×5)

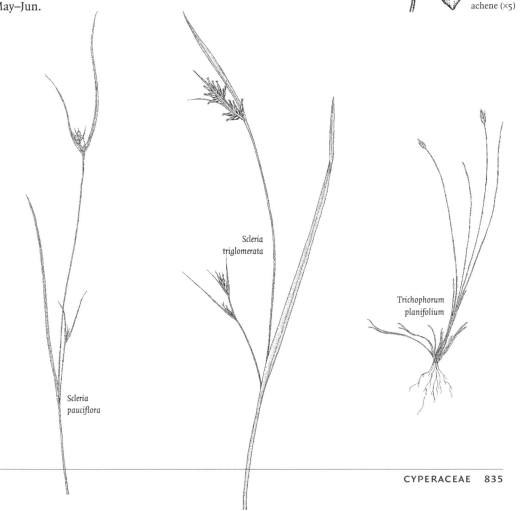

Scleria pauciflora

Scleria triglomerata

Trichophorum planifolium

DIOSCOREACEAE Yam Family

Twining, climbing herbaceous vines from a thickened rhizome; leaves alternate (sometimes opposite or whorled), with a broad, cordate or hastate blade with 3–13 curved convergent veins; plants dioecious; flowers 3-merous; tepals 6, fused at the base; stamens 6; ovary inferior, 3-locular; fruit a capsule; seeds flat, broadly winged.

Dioscorea L.

Flowers greenish-yellow, in axillary inflorescences; capsules winged, with 3 papery valves.

A. leaves broadly cordate, the sides concave above the spreading basal lobes; aerial tubers produced .... *D. batatas*
A. leaves cordate-ovate, with convex sides; aerial tubers not produced
    B. leaves nearly all alternate, occasionally the lowest whorled; rhizomes 5–10 mm thick .... *D. villosa*
    B. many of the leaves, especially toward the base, in whorls of 4–7; rhizomes 10–15 mm thick .... *D. quaternata*

Dioscorea batatas Decne.   Cinnamon-vine, Chinese yam
Climbing vine, 1–5 m tall from a large underground tuber, the stems often bearing small aerial tubers in the leaf axils; leaves alternate, opposite, or in whorls, broadly cordate with spreading basal lobes and concave sides; occasionally naturalized in disturbed woods and waste ground; mostly SE; flr. Jul; native to China.

Dioscorea quaternata (Walter) J.F.Gmel.   Wild yam
Very similar to *D. villosa*, but less strongly twining, with thicker rhizomes and more frequently whorled leaves; frequent in dry, rocky, wooded slopes and thickets; SC and SW; flr. Jun, frt. late Jul–fall; FACU.

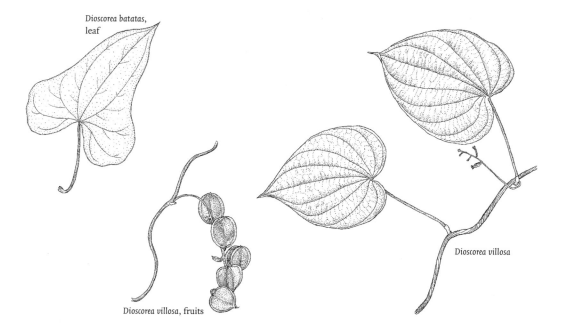

Dioscorea batatas,
leaf

Dioscorea villosa

Dioscorea villosa, fruits

Dioscorea villosa L.   Wild yam, colic-root
Stems glabrous, to 5 m tall, twining; leaves mostly alternate, cordate-ovate with 7–11 veins, abruptly acuminate; staminate inflorescence widely branched, with 1–4 flowers per node; pistillate inflorescence a spike 5–10 cm long with flowers solitary at each node; frequent in woods, thickets, and rocky slopes; throughout; flr. late Jun–early Jul, frt. Aug–Sep; FAC+.

ERIOCAULACEAE Pipewort Family

Aquatic (usually submergent) perennial herbs with a basal rosette of short, narrow leaves; monoecious; flowering stems leafless; flowers very small, crowded in a dense, white or grayish terminal head in which each flower is subtended by a bract; calyx of 2 keeled sepals; petals 2, each with a gland at the tip; staminate flowers with 4–6 stamens fused to the corolla tube; ovary superior; flr. Jul–Sep.

Eriocaulon L.

A. low plants rarely exceeding 2 dm; heads 3–6 mm across
     B. perianth and receptacular bracts densely white-hairy .... *E. aquaticum*
     B. perianth and bracts glabrous or only slightly hairy .... *E. parkeri*
A. coarser plants 2–10 dm; heads 7–15 mm across .... *E. decangulare*

Eriocaulon aquaticum

Eriocaulon aquaticum (Hill) Druce   Seven-angle pipewort, white-buttons
Leaves 3–10 cm long; flowering stems 3–20 cm (taller in deep water), 4–7-angled; receptacle glabrous; flowers densely white-hairy; rare in shallow water and peaty shores of northern lakes; NE; OBL. [syn: *E. septangulare* With.]

Eriocaulon decangulare L.   Ten-angle pipewort
Leaves 1–5 dm long with obtuse tips; flowering stems 2–8 dm, 10–12-angled; receptacle hairy between the flowers; very rare in glacial lakes; NE; believed to be extirpated; OBL; ❧.

Eriocaulon parkeri B.L.Rob.   Parker's pipewort
Leaves 1–6 cm long; flowering stems 2–10 cm; receptacle glabrous; flowers glabrous or fringed with minute hairs; very rare on freshwater tidal mudflats; SE; believed to be extirpated; OBL; ❧.

HYDROCHARITACEAE Frog's-bit Family

Rooted or free-floating, submergent aquatic perennials; flowers perfect or unisexual; sepals 3, green; petals 3 or lacking; stamens 3–many; ovary inferior, of 3–6 loosely joined carpels; styles lobed or bifid; flowers on long peduncles or the staminate flowers breaking free and floating.

A. leaves basal, elongate, and ribbon-like .... *Vallisneria*
A. leaves cauline, not more than 4 cm long
    B. principal leaves in whorls of 3 .... *Elodea*
    B. principal leaves in whorls of 4–8
        C. leaf margins very finely serrulate; leaves crowded, overlapping; tubers
            not produced .... *Egeria*
        C. leaf margins distinctly spine-toothed; internodes as long as the leaves;
            plants tuber-bearing... *Hydrilla*

Egeria Planch.

Egeria densa Planch.   Brazilian waterweed
Similar to *Elodea* but with leaves 4–8 per node and larger; dioecious, both staminate and pistillate flowers raised to the water surface on slender peduncles;
spathes 2–4-flowered; petals longer than the sepals; occasionally naturalized in
ponds and lakes; SE; flr. Jun–Oct; native to South America; OBL.

Elodea Michx.

Leafy, rooted or free-floating stems; leaves sessile, minutely serrulate, mostly 3
per node, or opposite below; flowers unisexual or perfect, raised to the water
surface on slender peduncles, or the staminate flowers breaking free and floating; petals 3, white, not much longer than the sepals; stamens 3 or 9, anthers
opening explosively; pollen floating; ovary of 3 carpels; sexual dimorphism occurs in the dioecious species making it difficult to distinguish between staminate plants of E. *nuttallii* and pistillate E. *canadensis* when flowers are not present;
in addition, late season shoots are often denser with shorter, frequently recurved
leaves; flr. late Jun–early Aug.

A. flowers unisexual; stamens 9
    B. leaves (1.5)1.8–3 mm wide by (6)8–12 mm long; staminate flowers remain-
        ing attached to the plant, spathe 7–10 mm long .... E. *canadensis*
    B. leaves 0.9–1.5(2) mm wide by 5–9(12) mm long; staminate flowers sepa-
        rating from the plant and floating at anthesis, spathe 2–4 mm long ....
        .... E. *nuttallii*
A. flowers perfect; stamens 3 .... E. *schweinitzii*

Elodea canadensis Rich. ex Michx.   Common waterweed, ditch-moss
Flowers unisexual, elevated to the surface on slender peduncles; petals white, 2–
4.5 mm; shallow, mostly calcareous water of rivers, creeks, lakes, and ponds;
throughout; pistillate plants are common, but staminate forms were known
from only a few sites and may be extirpated; OBL.

Elodea nuttallii (Planch.) St.John   Waterweed
Very similar to E. *canadensis* but the leaves narrower, the leaves of pistillate plants
longer and flaccid, those of staminate plants shorter and recurved; staminate
flowers sessile, breaking loose and floating to the water surface; petals 1–2 mm
or lacking; frequent in shallow water of rivers, streams, ponds, and freshwater
tidal mudflats; throughout; OBL.

Elodea schweinitzii (Planch.) Casp.   Schweinitz's waterweed
Differing from the above mainly by the perfect flowers and stamen number re-

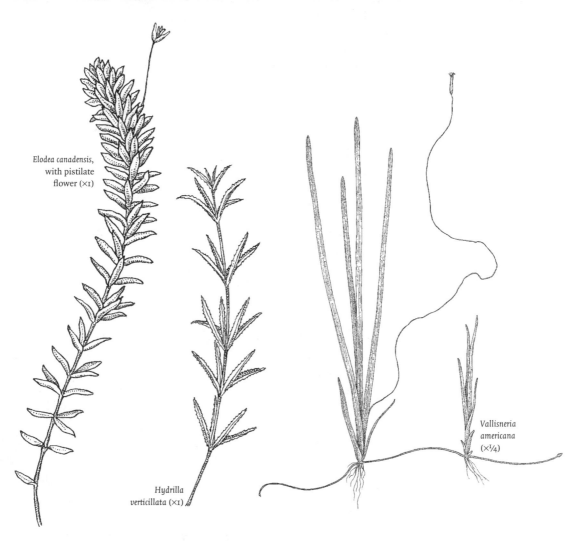

Elodea canadensis,
with pistilate
flower (×1)

Hydrilla
verticillata (×1)

Vallisneria
americana
(×¼)

duced to 3; very rare, known only from Northampton Co.; believed to be extirpated; OBL; ❦.

Hydrilla Rich.

Hydrilla verticillata (L.f.) Royle   Hydrilla
Similar to *Elodea* but with leaves in whorls of 4–8 and leaf margins with easily visible spine-tipped teeth; internodes generally as long as the leaves; plants tuber-
bearing; flowers unisexual; stamens 3; rare, but becoming locally abundant
along the margins of rivers and lakes; SE; native to Eurasia; OBL; designated as a
federal noxious weed.

Vallisneria L.

Vallisneria americana Michx.   Water-celery, tape-grass
Stoloniferous, rooted, submergent aquatic; leaves 3–10 mm wide and to 2 m
long; dioecious, staminate flowers in a short-stalked spathe, breaking loose and
floating to the surface; pistillate flowers solitary on an elongate peduncle that
coils after fertilization, retracting the elongate, indehiscent fruit; sepals, petals,
and stigmas each 3; frequent in rivers, streams, and lakes; throughout; flr. Jul–
Oct; OBL.

IRIDACEAE Iris Family

Perennial herbs; leaves simple, narrow, parallel-veined, sheathing at the base and with the blade folded and fused so that only the lower surface is exposed; inflorescence terminal, with 1 or 2 subtending bracts forming a spathe; flowers regular or nearly so, perfect, 3-merous; sepals 3, petal-like; petals 3; stamens 3, opposite the sepals; filaments distinct or fused; ovary inferior, style 3-lobed; fruit a 3-locular capsule.

A. style branches expanded and petal-like, covering the stamens .... *Iris*
A. style branches neither petal-like nor covering the stamens
    B. flowers orange-red with dark markings, 3–5 cm wide; filaments distinct
    .... *Belamcanda*
    B. flowers blue or white, up to 2 cm wide; filaments fused .... *Sisyrinchium*

Belamcanda Adans.

Belamcanda chinensis (L.) DC.   Blackberry-lily
Stems 3–6 dm tall; leaves sword-shaped; inflorescence widely branched; flowers orange with darker spots; perianth parts distinct; stamens attached to the base of the tepals, filaments distinct; capsule valves falling at maturity to expose clusters of fleshy black seeds; cultivated and occasionally escaped to roadside banks and open woods; SE; flr. Jun–Jul; native to Asia.

Iris L.

Erect perennial herbs with sword-shaped leaves and fleshy rhizomes; flowering stalks bearing 1–several showy flowers; sepals spreading or reflexed; petals erect or arching; stamens inserted at the base of the sepals, hidden by the 3 arching, petaloid stigma lobes; fruit an elongate 3-locular capsule.

A. plants <2 dm tall at flowering; petals and sepals nearly equal in length
    B. leaves >1 cm wide; sepals strongly crested (with hairy median lines) ....
    .... *I. cristata*
    B. leaves no more than 1 cm wide; sepals not crested .... *I. verna*
A. plants >2 dm tall at flowering; sepals longer and wider than the petals
    C. sepals bearded .... *I. germanica*
    C. sepals not bearded
        D. flowers yellow; capsules 6-angled .... *I. pseudoacorus*
        D. flowers blue; capsules 3-angled
            E. leaves no more than 1 cm wide; capsules sharply angled ....
            .... *I. prismatica*
            E. leaves >1 cm wide; capsule bluntly angled
                F. sepal blade hairy at the base; style branches auriculate at the base;
                   capsule ovoid .... *I. virginica*
                F. sepal blade not hairy at the base; style branches not auriculate;
                   capsule cylindrical .... *I. versicolor*

Iris cristata Sol.   Dwarf crested iris
Stems to 1 dm tall from a branched, creeping rhizome; leaves 6–20 mm wide and 6–25 cm long; flowers pale blue-violet or white; sepals spreading or recurved with a wide yellow band of hairs (the crest) down the center from the middle to the base; rare on wooded slopes and stream banks; SW and SC; flr. May; ❧.

Iris germanica L.   Bearded iris

Stems to 8 dm tall from stout rhizomes; leaves erect, glaucous; sepals recurved, dark purple and bearded on the median line; petals lighter purple, erect-arching; commonly cultivated and occasionally persisting on roadsides, or near abandoned homesteads; flr. May–Jun; native to Europe. The yellow-flowered *I. flavescens* DC. and other color forms may occasionally persist also.

Iris prismatica Pursh   Slender blue flag

Stems to 9 dm tall from slender, creeping rhizomes; leaves 3–7 mm wide and 5–7 dm long; flowers blue-violet; capsule sharply angled, 5–8 cm long; rare in moist meadows and sandy or gravelly shores; SE; flr. late May–Jun; OBL; 🌿.

Iris pseudoacorus L.   Yellow iris

Clump forming, stems to 1 m tall from short, thick rhizomes; leaves erect, 8–20 mm wide; flowers bright yellow; capsules 5–8 cm long, 6-angled; frequent in marshes, shallow water, or moist shores; flr. late May–Jul; native to Europe; OBL.

Iris verna L.   Dwarf iris

Stems to 1.5 dm tall from slender rhizomes; leaves 3–8 mm wide and to 3 dm long; flowers blue, perianth segments barely equal, sepals yellow and pubescent at the base; rare in dry to moist, acidic, sandy soils; SC; flr. May; 🌿. Ours is var. *smalliana* Fernald ex M.E.Edwards

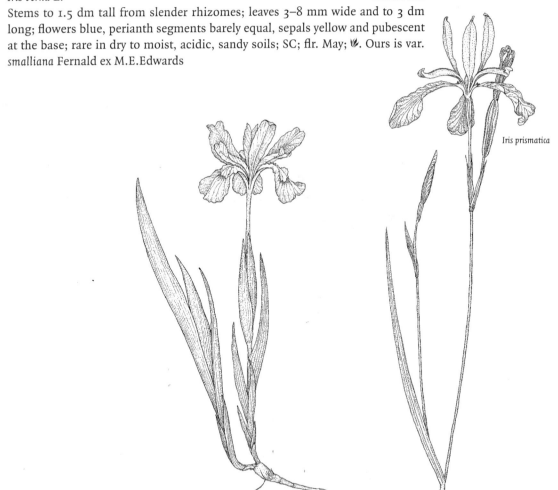

Iris cristata

Iris prismatica

Iris versicolor L.   Northern blue flag

Clump forming, stems 6–9 dm tall from thick, short rhizomes; leaves erect, 1–3 cm wide, glaucous; flowers blue-violet; sepals with a greenish-yellow blotch at the base of the blade; capsules cylindrical, bluntly 3-angled; frequent in wet meadows, bogs, and marshes; throughout; flr. May–Jul; OBL.

Iris virginica L.   Southern blue flag

Very similar to *I. versicolor*, differing by the softer, arching leaves; bright yellow, hairy blotch at the base of the blade of the sepals and ovoid capsules; rare in shallow water of pond margins; NW; flr. May–Jul; OBL; 🍂.

Sisyrinchium L.

Low, tufted perennial herbs with fibrous roots, basal, grass-like leaves, and flat, winged stems; flowers in umbel-like clusters from a sheathing spathe formed of 2 bracts; perianth blue (or rarely white), rotate, the lobes distinct or only slightly fused at the base; filaments fused; capsules globose, 3-lobed.

A. spathes sessile
    B. spathe bracts nearly equal in length, margins of the outer bract free to the base.. S. albidum
    B. spathe bracts strongly unequal, margins of the outer bract fused for 2–5 mm at the base
        C. stem barely winged, 1–1.5 mm wide .... S. mucronatum
        C. stem flattened and distinctly winged, 1.5–3 mm or more wide ....
        .... S. montanum
A. spathes pedunculate
    D. stem broadly winged, 3–4 mm wide .... S. angustifolium
    D. stem narrowly winged, 0.5–2 mm wide
        E. old leaf bases persistent, dry, brown, and fibrous; plants blackening when dry .... S. fuscatum
        E. old leaf bases not persistent; plant remaining green when dry ....
        .... S. atlanticum

Sisyrinchium angustifolium

Sisyrinchium albidum Raf.   Blue-eyed-grass

Stems erect to 4 dm tall, distinctly winged, 1.5–4 mm wide; leaves 1.5–4 mm wide; spathes paired at the top of a simple stem, sessile; flowers white with a yellow eye; very rare in dry, sandy, open soil; native mostly farther west, possibly adventive here; believed to be extirpated; FAC; 🍂.

Sisyrinchium angustifolium Mill.   Blue-eyed-grass

Stems erect or spreading, to 5 dm tall, broadly winged, 3–4 mm wide; leaves 1.5–6 mm wide; peduncles winged, 2–15 cm from the axil of the single stem leaf; spathe bracts nearly equal in length, the edges of the outer bract fused for 2 mm or more at the base; flowers pale blue; common in damp soil in meadows, floodplains, fields, and open woods; throughout; flr. May–Jul; FACW–.

Sisyrinchium atlanticum E.P.Bicknell   Eastern blue-eyed-grass

Stems erect or spreading, narrowly winged, 0.5–2 mm wide; leaves up to 2 mm wide; spathes on long peduncles from the axil of the single stem-leaf; spathe bracts nearly equal in length, the edges of the outer bract fused at the base for

2.5–4 mm; flowers pale blue; rare in moist to dry, sandy, open ground of fields and thin woods; SE; flr. May–Jun; FACW; ❦.

Sisyrinchium fuscatum E.P.Bicknell   Sand blue-eyed-grass
Stems erect, to 5 dm tall, narrowly winged; leaves 1–3.5 mm wide; spathes peduncled from the axil of the stem leaf; spathe bracts nearly equal in length, the edges of the outer bract fused for 2.5–4 mm at the base; flowers violet; very rare in dry, open, sandy ground of fields and thin woods; believed to be extirpated; flr. May–Jun; FACU; ❦. [syn: *S. arenicola* E.P.Bicknell]

Sisyrinchium montanum Greene   Blue-eyed-grass
Stems erect, to 5 dm tall, winged, 1.5–3 mm wide; leaves 2–3 mm wide; spathes solitary on a simple stem, sessile; spathe bracts strongly unequal in length, the edges of the outer bract fused for 2–5 mm at the base; flowers violet; occasional on stream banks, woods, and old fields; NE and scattered elsewhere; flr. May–Jul; FAC. Ours is var. *crebrum* Fernald.

Sisyrinchium mucronatum Michx.   Blue-eyed-grass
Stems to 4.5 dm tall, barely winged, 1–1.5 mm wide; leaves 1–2 mm wide; spathes mostly solitary, sessile; spathe bracts strongly unequal in length; flowers violet (rarely white); frequent in dry fields, roadsides, and open woods; throughout, except NW; flr. May–Jun; FAC+.

JUNCACEAE Rush Family

Perennial (except *J. bufonius*) herbs with narrow leaves with terete or flat blades; stems unbranched below the inflorescence; inflorescence terminal or appearing lateral due to an erect involucral bract; flowers perfect, regular, with 3 scarious sepals and 3 similar petals; stamens 3 or 6; ovary superior, 3-celled; fruit a dehiscent capsule.

A. plants glabrous; ovules and seeds several to many per capsule .... *Juncus*
A. leaves long-hairy along the margins; ovules and seeds 3 per capsule .... *Luzula*

Sisyrinchium montanum

Juncus L.

Herbaceous plants mostly of moist or wet habitats; leaves basal or cauline, linear, flat, tubular, or reduced to a bladeless sheath in some species; auricles at the juncture of the sheath and blade conspicuous in some species; inflorescence terminal or appearing lateral; flowers perfect, 3-merous, regular, subtended by a pair of bracts (prophyllate) in some species; sepals and petals both present but similarly lanceolate and dry in texture; stamens 3 or 6; ovary superior, completely or partially 3-locular; fruit a dehiscent capsule; seeds many; flr./frt. summer.

A. leaves lacking or reduced to bladeless sheaths; inflorescence appearing lateral
   (because the erect subtending bract appears to be a continuation of the stem)
   B. inflorescence about midway between the base and the (apparent) tip of the
      stem
      C. capsule as long as or longer than the perianth .... *J. filiformis*
      C. capsule shorter than the perianth .... *J. gymnocarpus*
   B. inflorescence much closer to the tip

    D. stamens 3; capsule broadly rounded .... *J. effusus*
    D. stamens 6; capsule more or less pointed at the tip
       E. tepals about half the length of the capsule (1.5–2.2 mm) ....
        .... *J. gymnocarpus*
       E. tepals as long as the capsule or nearly so (2.6–4.7 mm)
         F. stems arising in rows from long rhizomes .... *J. arcticus*
         F. stems clustered from a short rhizome .... *J. inflexus*
A. leaves with blades present; inflorescence clearly terminal
   G. leaves round in cross section with rigid internal cross partitions at intervals
   H. seeds with lighter colored, tail-like tips at both ends
     I. seeds 1.3–1.8 mm long, the tails accounting for more than half the length .... *J. canadensis*
     I. seeds 0.7–1.2 mm long, the tails accounting for less than half the total length
       J. tepals lance-oblong, obtuse to subacute .... *J. brachycephalus*
       J. tepals lance-subulate
         K. inflorescence 2–6 times as long as broad .... *J. brevicaudatus*
         K. inflorescence not more than twice as long as wide ....
          .... *J. subcaudatus*
   H. seeds blunt, lacking tails
     L. flowers solitary or in 2s .... *J. pelocarpus*
     L. flowers in dense clusters of 3–many
       M. flowers in hemispherical to obpyramidal heads; tepals lanceolate to ovate
       N. stamens 3
         O. heads 1 cm thick; capsule shorter than the perianth ....
          .... *J. brachycarpus*
         O. heads <1 cm thick; capsule about equaling or longer than the perianth
           P. capsule slightly shorter to about equaling the perianth
            ....*J. acuminatus*
           P. capsule longer than the perianth .... *J. debilis*
       N. stamens 6
         Q. cauline leaf 1(2), 2–4 mm thick, overtopping the inflorescence .... *J. militaris*
         Q. cauline leaves 2 or more, 0.5–1.5 mm thick, not overtopping the inflorescence
           R. inflorescence less than twice as long as wide; tepals acute .... *J. articulatus*
           R. inflorescence more than twice as long as wide; tepals obtuse or rounded .... *J. alpinoarticulatus*
       M. flowers in spherical or globose heads; tepals lance-linear or linear-subulate
       S. stamens 6; bract longer than the inflorescence
         T. sepals and petals similar in length; capsule longer than the perianth .... *J. nodosus*
         T. sepals longer than the petals; capsule about the same length as the sepals .... *J. torreyi*
       S. stamens 3; bract shorter than the inflorescence

  U. capsule longer than the perianth; tepals all about the same length .... *J. scirpoides*

  U. capsule shorter than the perianth; sepals distinctly longer than the petals ....*J. brachycarpus*

G. leaves flat (sometimes becoming involute when dry), rounded and channeled on the upper side, or terete, but without cross partitions

 V. inflorescence ⅓ the total height of the plant

  W. capsule shorter than the perianth ....*J. bufonius*

  W. capsule longer than the perianth................................. *J. pelocarpus*

 V. inflorescence <⅓ the total height of the plant

  X. flowers each subtended by 2 small bracteoles in addition to the bractlet at the base of the pedicel

   Y. sepals obtuse, shorter than to barely equaling the capsule; 1 or 2 stem leaves present ....*J. gerardii*

   Y. sepals acute, as long or longer than the capsule; leaves nearly all basal

    Z. leaves flat (often rolling inward when dry)

     AA. auricles of the leaf sheaths elongate, tongue-shaped .....
.... *J. tenuis*

     AA. auricles of the leaf sheaths rounded

      BB. flowers 2.5–3.5 mm long, secund ....*J. secundus*

      BB. flowers 4–6 mm long, not secund .... *J. dudleyi*

    Z. leaves terete or channeled above

     CC. capsule shorter than to about as long as the perianth ....
.... *J. dichotomus*

     CC. capsule longer than the perianth ....*J. greenei*

  X. flowers not subtended by bracteoles other than the bractlet at the base of the pedicel

   DD. plants 2–5 dm with 5–20 heads; leaf blades 1–3 mm wide ....
....*J. marginatus*

   DD. plants 6–12 dm with 20–100 heads; leaf blades 4–6 mm wide
.... *J. biflorus*

Juncus acuminatus

Juncus acuminatus,
capsule (×2½)

Juncus acuminatus Michx.   Sharp-fruited rush
Stems erect, clustered, 2–8 dm tall; leaves terete with cross partitions; inflorescence more than twice as long as wide with 5–20 hemispherical to nearly spherical heads each with 10–50 flowers; stamens 3; capsule 3-angled, about as long as the perianth; bud-like galls often replacing some of the flowers; common in wet meadows, swamps, marshes, and stream banks; throughout; OBL.

Juncus alpinoarticulatus Chaix in Vill.   Alpine rush
Similar to *J. articulatus* but smaller; stems 0.5–3 dm; tepals 1.5–2.5 mm, obtuse or rounded; capsule obtuse or rounded; rare in moist, sandy, calcareous shores and seeps; NW; OBL; ✤. Ours is ssp. *nodulosus* (Wahlenb.) Hämet-Ahti. [syn: *J. alpinus* Vill.]

Juncus arcticus Willd.   Baltic or wire rush
Unbranched stems, 4–8 dm tall, arising in rows from long rhizomes; leaves reduced to bladeless sheaths; inflorescence appearing lateral, the subtending bract ⅕–⅓ the entire height of the plant; flowers prophyllate, the inner bracteole

broadly round-ovate; tepals lanceolate, acuminate, with a dark stripe on each side of the midrib; anthers 6; capsule slightly longer than the perianth; rare in calcareous swamps and shores; scattered; OBL; ❧. Ours is var. *littoralis* (Engelm.) Boivin. [syn: *J. balticus* Willd. var. *littoralis* Engelm.]

Juncus articulatus L.   Jointed rush
Stems erect, 1–6 dm tall from a coarse rhizome; leaves cauline, terete with cross partitions; inflorescence not more than twice as long as wide with few to many heads of 3–10 flowers; tepals 2.5–3 mm long, lance-subulate; stamens 6; capsule sharply 3-angled, longer than the perianth and tapering to the tip; rare in bogs, swamps, swales, and mud flats; scattered; OBL.

Juncus biflorus Elliot   Grass rush
Stems 6–12 dm tall, solitary (but close together) from a rhizome, bulbous thickened at the base; leaves basal and cauline, blades 4–6 mm wide with 5 main veins; flowers 20–100 per head; sepals lanceolate; stamens 3; capsule broadly rounded, chestnut brown when mature, about as long as the perianth; rare in moist, open woods, boggy fields, gravel pits, and ditches; SE and SC; FACW; ❧.

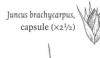

Juncus brachycarpus, capsule (×2½)

Juncus brachycarpus Engelm.   Short-fruited rush
Stems erect from a stout rhizome, 3–8 dm tall; leaves terete and cross-partitioned; inflorescence open or compact, of 3–10 spherical, 1 cm thick, many-flowered heads; stamens 3; capsules 3-angled, shorter than the perianth; very rare in moist open ground; SC; FACW; ❧.

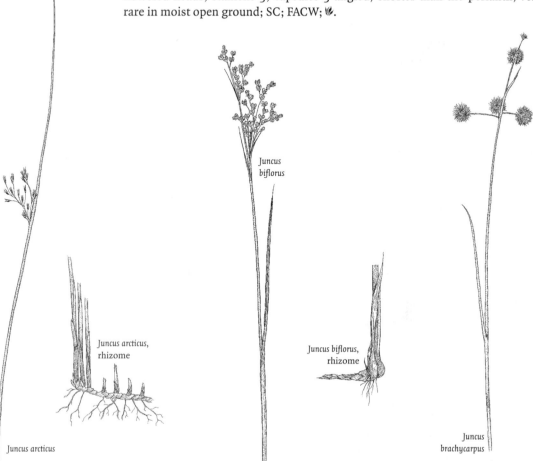

Juncus
biflorus

Juncus arcticus,
rhizome

Juncus biflorus,
rhizome

Juncus arcticus

Juncus
brachycarpus

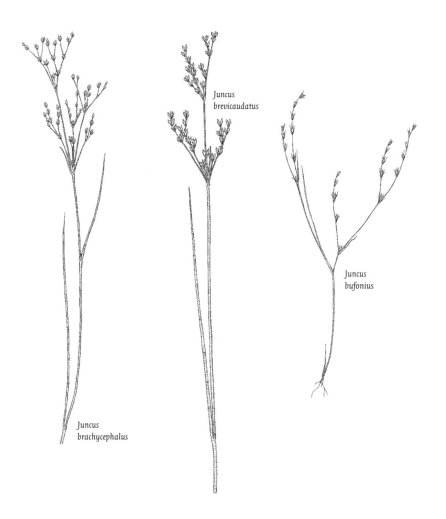

Juncus
brevicaudatus

Juncus
bufonius

Juncus
brachycephalus

Juncus brachycephalus (Engelm.) L.Buch   Small-headed rush

Stems erect, densely clustered, 3–7 dm tall; leaves terete with cross partitions; inflorescence of 2–5-flowered heads on spreading-ascending branches; stamens 3(6); capsule 3-angled, longer than the perianth; seeds with short white tails at both ends; rare and scattered on muddy or sandy, calcareous shores, clayey seeps, and springy or boggy fields; mostly W; OBL; .

Juncus brachycephalus,
capsule (×2½)

Juncus
brachycephalus,
seed (×10)

Juncus brevicaudatus (Engelm.) Fernald   Narrow-panicled rush

Stems erect, densely clustered, 1–5 dm tall; leaves erect, terete with cross partitions; inflorescence >3 times as long as wide, heads few to many, 2–7-flowered; sepals shorter than the petals; capsule 3-angled, much longer than the perianth; seeds with tails occupying 2⁄5 the total length; occasional in moist shores, swales, and ditches; mostly NE and NC; OBL.

Juncus brevicaudatus,
capsule (×2½)

Juncus
brevicaudatus,
seed (×10)

Juncus bufonius L.   Toad rush

Annual with slender, branched stems to 3 dm tall; leaves basal and cauline, flat or channeled above; flowers scattered singly or in few-flowered heads along the upper half of the stem, prophyllate; stamens 6; capsule shorter than the perianth; frequent on muddy river banks, moist roadside ditches, and other low, open ground that is seasonally wet; scattered throughout; FACW.

Juncus bufonius,
capsule (×2½)

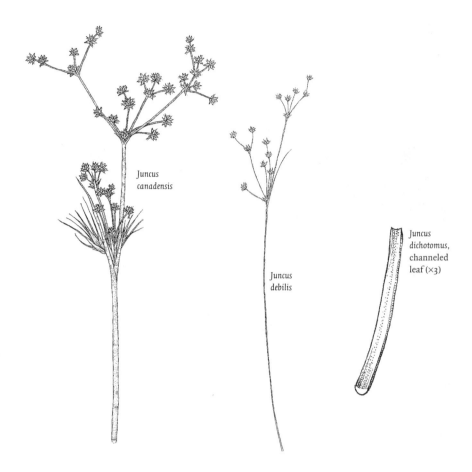

Juncus canadensis

Juncus debilis

Juncus dichotomus, channeled leaf (×3)

Juncus canadensis, capsule (×2½)

Juncus canadensis, seed (×10)

Juncus canadensis J.Gay ex Laharpe   Canada rush

Stems stiffly erect, clustered, 3–12 dm tall, leaves basal and cauline, terete with cross partitions at intervals; inflorescence longer than the subtending bract; flowers in globose heads of 5–40 flowers; stamens 3; capsule acutely 3-angled, slightly longer than the perianth; seeds with slender white tails more than half the total length; bud-like galls sometimes replacing some of the flowers; occasional in swamps, marshes, bogs, stream banks, pond margins, and swales; throughout; OBL.

Juncus debilis, seed (×20)

Juncus debilis, capsule (×2½)

Juncus debilis A.Gray   Weak rush

Stems slender, erect, clustered, 1–4 dm tall; leaves terete with cross partitions; inflorescence ⅓–½ the height of the plant, loosely branched; heads with 2–5 flowers; tepals narrowly scarious-margined; stamens 3; capsule 3-angled, distinctly longer than the perianth; infrequent on stream banks, mudflats, shores, and ditches; SE and SC; OBL; 🌿.

Juncus dichotomus, capsule (×2½)

Juncus dichotomus Elliot   Forked rush

Stems clustered, 3–9 dm tall; leaves basal, terete and channeled above, ⅓–½ the height of the stem; inflorescence longer than the subtending bract, the branches somewhat secund; stamens 6; capsule rounded at the top with a pointed tip, shorter than the perianth; rare in moist, sandy old fields, open woods, and gravel pits; mostly SE and SC; FACW–; 🌿.

Juncus dudleyi Wiegand   Rush

Stems clustered, 3–8 dm tall; leaves basal, to half the length of the stems; blades flat or inrolled, auricles rounded; inflorescence terminal, 1–7 cm long, shorter than the subtending bract, sparsely flowered; stamens 6; capsule oblong-ovoid to obovoid, shorter than the perianth; occasional in wet fields, stream banks, swales, and ditches; mostly S.

Juncus effusus L.   Soft rush

Stems densely clustered, to 1 m tall or more; leaves reduced to basal sheaths; inflorescence apparently lateral, many-flowered, the subtending bract erect, 10–25 cm long; flowers prophyllate; stamens 3; capsule obtuse to truncate, slightly shorter to slightly longer than the perianth; common in swamps, moist fields, floodplains, shores, and ditches; throughout; 2 varieties:

Juncus effusus var. *solutus* capsule (×2½)

A. stem with 10–25 coarse longitudinal ridges ....
.... var. *pylaei* (Laharpe) Fernald & Wiegand; FACW+
A. stem smooth to finely striate .... var. *solutus* Fernald & Wiegand; OBL

Juncus filiformis L.   Thread rush

Slender, leafless stems arising in rows from a long rhizome, the apparently lateral inflorescence subtended by an erect bract ½ to as long as the true stem; flowers prophyllate; stamens 6; rare in bogs and sandy shores; NE and NW; FACW; 🍂.

Juncus filiformis, capsule (×2½)

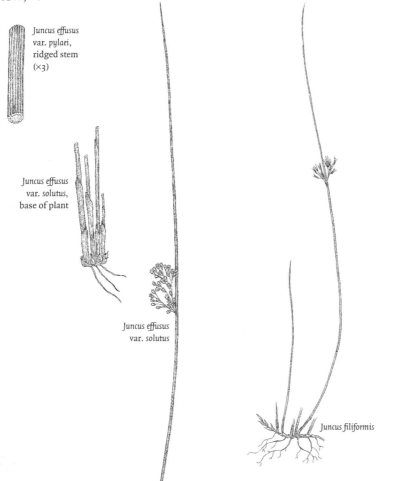

Juncus effusus var. *pylaei*, ridged stem (×3)

Juncus effusus var. *solutus*, base of plant

Juncus effusus var. *solutus*

Juncus filiformis

Juncus gerardii Loisel. Blackfoot rush, black-grass
Stems from long, slender rhizomes, 2–6 dm tall with 1 or 2 cauline leaves; inflorescence terminal, 2–8 cm, usually exceeding the subtending bract; flowers prophyllate; stamens 6; fruit rounded above with a short point, equaling or barely exceeding the tepals; occasional on ballast, waste ground, or moist roadsides where deicing salts are used; native in coastal areas but adventive in Pennsylvania; mostly E; FACW+.

Juncus greenei Oakes & Tuck. Greene's rush
Stems clustered from a short rhizome, 2–8 dm; leaves basal, filiform, subterete; inflorescence 2–5 cm, flowers prophyllate; capsule ovoid-cylindric, truncate, longer than the tepals; very rare on sandstone cliffs; NE; believed to be extirpated; FAC; 🌿.

Juncus gymnocarpus Coville Coville's or Pennsylvania rush
Stems arising in rows from long rhizomes; leaves reduced to reddish-purple, bristle-tipped basal sheaths; inflorescence apparently lateral, few-flowered; flowers prophyllate; stamens 6; capsule longer than the perianth; rare in sphagnous swamps, seeps, and springheads; mostly in or near Schuylkill Co.; OBL.

Juncus greenei,
channeled
leaf (×3)

Juncus greenei,
capsule (×2½)

Juncus gymnocarpus,
capsule (×2½)

Juncus
marginatus

Juncus
militaris
(×¼)

Juncus
pelocarpus

Juncus
secundus

Juncus inflexus L.   Meadow rush

Stems 3–8 dm tall, clustered from a short rhizome; leaves reduced to bladeless, bristle-tipped sheaths; inflorescence appearing lateral, the subtending bract $\frac{1}{4}$–$\frac{1}{3}$ as long as the true stem; flowers prophyllate; the capsule slightly shorter to slightly longer than the perianth; rare in moist bottomland and waste ground; scattered; native to Eurasia and N. Africa; FACW.

Juncus marginatus Rostk.   Grass-leaved rush

Stems clustered, 2–5 dm tall, bulbous thickened at the base; leaves basal and cauline, blades soft and flat 1–4 mm wide; inflorescence of globose heads of 2–12 flowers; petals with a green midrib and hyaline margin; stamens 3; capsule obovoid, rounded at the tip and about as long as the perianth; frequent in moist fields, ditches, swamps, and roadsides; throughout; FACW.

Juncus marginatus, capsule (×2½)

Juncus militaris Bigelow   Bayonet rush

Stems stout, stiffly erect, 5–10 dm tall, from a rhizome; the single terete, cross-partitioned leaf located halfway between the base and the inflorescence; the rhizome also producing many long, capillary leaves when submersed; inflorescence with ascending branches bearing heads of 5–13 flowers; stamens 6; capsule 3-angled, tapering to a conspicuous beak; rare in shallow water of lakes and ponds; NE; OBL; ✹.

Juncus nodosus L.   Knotted rush

Stem slender, erect, 1.5–4 dm tall, arising singly from a slender rhizome that is tuberous-thickened at intervals; leaves terete with cross partitions; inflorescence dense or open, bearing 2–10 spherical heads of 5–25 flowers; stamens 6; capsule sharply 3-angled, as long or longer than the perianth; occasional in moist fields, bogs, marshes, shores, and swales, often on calcareous soils; scattered; OBL.

Juncus nodosus, capsule (×2½)

Juncus nodosus, seed (×20)

Juncus pelocarpus B.Mey.   Brown-fruited rush

Stems slender, erect, 1–5 dm tall, from a rhizome; leaves terete and obscurely cross-partitioned; inflorescence much-branched; heads with 1–3 flowers, some of the flowers often replaced by bulbils; stamens 6; capsule narrowly ovoid, longer than the perianth; rare in bogs and marshes with seasonally fluctuating water levels; NE; OBL.

Juncus pelocarpus, capsule (×2½)

Juncus scirpoides Lam.   Sedge rush, scirpus-like rush

Stems 3–8 dm tall, erect from stout rhizomes; leaves terete with cross partitions; inflorescence of spherical, many-flowered heads on short (or sometimes longer) branches; stamens 3; fruit as long or longer than the perianth, 3-angled and tapering to a prominent beak; rare in moist, sandy, or peaty soil; SE; FACW; ✹.

Juncus secundus P.Beauv. ex Poir.   Rush

Stems 3–6 dm tall, clustered; leaves basal, about $\frac{1}{3}$ as long as the stem, narrowly linear, flat or inrolled, auricles rounded; inflorescence terminal with ascending, and often incurved, branches each with 3–8 secund, prophyllate flowers; stamens 6; fruit ovoid or short cylindric; occasional on upland slopes, rocky ledges, serpentine barrens, and roadside banks, often in clayey soil; mostly SE and SC; FACU.

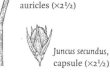

Juncus secundus, auricles (×2½)

Juncus secundus, capsule (×2½)

Juncus subcaudatus,
seed (×10)

Juncus subcaudatus,
capsule (×2½)

Juncus subcaudatus (Engelm.) Coville & S.F.Blake   Rush
Stems clustered, 3–8 dm tall, erect or spreading; leaves terete with cross partitions; inflorescence with 3–25 hemispherical heads; capsule 3-angled; seeds
with short white tails at each end; frequent in swamps, wet fields, ditches,
swales, and bogs; throughout; OBL.

Juncus tenuis Willd.   Path rush
Stems clustered, 1–8 dm tall; leaves basal, ⅓–½ the height of the stem, 1–1.5
mm wide, flat becoming inrolled; auricles elongate; inflorescence terminal, often exceeded by the subtending bract; flowers prophyllate; stamens 6; capsule
obtuse to truncate, shorter than the tepals; common in moist to dry, sometimes
heavily compacted soil of woods, fields, waste ground, and paths; throughout;
FAC–. Ours is var. *tenuis*.

Juncus tenuis,
capsule (×2½)

Juncus torreyi,
capsule (×2½)

Juncus torreyi,
seed (×20)

Juncus torreyi Coville   Torrey's rush
Stems stout, erect, 4–10 dm tall, arising singly from a tuberous-thickened rhizome; leaves terete with cross partitions; inflorescence with spherical heads of
25–100 flowers on short branches; sepals longer than the petals; stamens 6; capsule sharply 3-angled, about as long as the perianth; rare on muddy or sandy
shores, strip mine areas, swales, or ditches; scattered; FACW; 🍂.

Luzula DC.

Tufted plants; leaves with a closed sheath and a flat blade with loose hairs along
the margins; inflorescence terminal, paniculate or umbellate, each flower subtended by a pair of small bracts; ovary and capsule unilocular with 3 seeds; flr./
frt. Apr–Jul.

A. flowers mostly solitary at the ends of the inflorescence branches ....
.... L. acuminata
A. flowers in glomerules
  B. branches of the inflorescence strongly ascending
    C. plants producing basal tubers .... L. bulbosa
    C. plants not producing basal tubers .... L. multiflora
  B. some of the shorter branches of the inflorescence widely spreading or
    reflexed.. L. echinata

Luzula bulbosa

Luzula acuminata Raf.   Hairy woodrush
Plant loosely tufted, 1–4 dm tall; leaves 5–12 mm wide; inflorescence of simple,
filiform rays each bearing a single flower; frequent in swampy woods and floodplains; throughout; FAC.

Luzula bulbosa (A.W.Wood) Rydb.   Woodrush
Plant to 2–4 dm tall, in small tufts with glossy white tubers at the base; leaves
flat except toward the tip, 2–6 mm wide; umbel with ascending rays 0.5–7 cm
long; rare in fields, woods borders, serpentine barrens, and roadsides; S; FACU;
🍂.

Luzula echinata (Small) F.J.Herm.   Common woodrush
Tufted plant, 1–4.5 dm tall; leaves 4–7 mm wide; inflorescence loose, rays of the

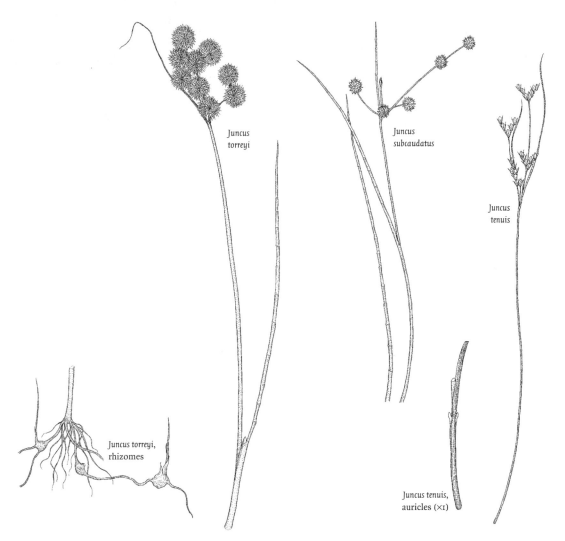

Juncus
torreyi

Juncus
subcaudatus

Juncus
tenuis

Juncus torreyi,
rhizomes

Juncus tenuis,
auricles (×1)

umbel 1–7 cm long, all but the central one spreading or reflexed; frequent in moist, rocky woods, and wet meadows; S; FACU.

Luzula multiflora (Ehrh.) Lej.   Field woodrush
Densely tufted plant to 9 dm tall; leaves 1–7 mm wide; umbel of 1–10 erect or ascending rays 0.5–6 cm long; common in woods, swamps, and floodplains; throughout; FACU.

JUNCAGINACEAE Arrow-grass Family

Rhizomatous, perennial herbs; leaves basal with an evident ligule at the juncture of the linear blade and the open sheath; flowers perfect, in a bractless, terminal raceme; perianth of 6 tepals; stamens 6, the filaments fused to the base of the tepals; ovary superior, of 3 carpels fused to a central axis, but separating, except at the tip, in fruit; fruit a cluster of 3 follicles.

Triglochin L.

Triglochin palustre L.   Marsh or slender arrow-grass

Emergent aquatic, 2–4 dm tall; leaves 1–2 mm wide; axis of the inflorescence with 3 broad wings; fruits narrowly club-shaped; very rare on moist, sandy shores; NW; believed to be extirpated; flr. May–Jul; OBL; 🌿.

LEMNACEAE Duckweed Family

Tiny, free-floating, aquatic plants consisting of a flattened thallus (frond) with or without one or more simple filamentous roots from the underside; reproducing mainly by budding followed by fragmentation; flowers borne in pouches on the margin or the upper surface, very rarely seen.

A. plants without roots
    B. fronds thin, flat, elongate or strap-shaped .... Wolffiella
    B. fronds thick, ellipsoid or globose, tapered or pointed on top, rounded below .... Wolffia
A. plants with 1 or more roots
    C. each frond with a single root; fronds 1–5-nerved .... Lemna
    C. each frond with 2 or more roots; fronds 5–12-nerved .... Spirodela

Lemna L.

Fronds flat, single or attached in clusters, ovate to oblong with 1–5 obscure nerves radiating from a point near one end; each frond with a single root.

A. fronds remaining attached to the parent plant by a stalk .... L. trisulca
A. fronds without a stalk, forming small colonies but eventually separating completely from the parent plant
    B. fronds nonsymmetrical, somewhat curved, 1-nerved .... L. valdiviana
    B. fronds rotund to elliptic, symmetrical, 3-nerved
        C. root sheath winged at the base; root <3 cm, tip sharply pointed ....
        .... L. perpusilla
        C. root sheath not winged; root >3 cm, tip rounded
            D. fronds with several equal papillae along the midline on the upper side; forming small, obovate to circular, rootless, dark green to brown turions that sink to the bottom .... L. turionifera
            D. fronds with 1 larger papilla or none; fronds not forming rootless turions
                E. papilla very prominent; fronds often red underneath ... L. obscura
                E. papilla not very prominent; fronds never red underneath ....
                .... L. minor

Lemna minor, top view (×2½)

Lemna minor, side view (×2½)

Lemna minor L.   Duckweed

Fronds rotund to elliptic, 2.5–6 mm, flat to slightly convex, symmetrical, 3-veined, solitary or in clusters of 2–5, each with a single root from the undersurface; common in still water of lakes, ponds, streams, swamps, and ditches; throughout, except at the highest elevations; OBL.

Lemna obscura (Austin) Daubs Little water duckweed
Similar to L. *minor*, but smaller; rare in shallow water; SE; believed to be extirpated; OBL; 🌿.

Lemna perpusilla Torr. Duckweed
Fronds asymmetrical and somewhat curved, with a prominent papilla at the apex of the veins and smaller papillae along the midrib; rare in ponds, bogs, and marshes; scattered; OBL; 🌿.

Lemna trisulca L. Star duckweed
Fronds oval or oblong, tapering to a 4–16-mm-long stalk that remains attached to the parent plant; rare, forming tangled colonies on lakes, ponds, bogs, swamps, marshes, and streams; mostly S; OBL.

Lemna turionifera Landolt Winter duckweed
Very similar to L. *minor* but forming rootless turions that overwinter; very rare in lakes, ponds, swamps, and marshes; W; OBL; 🌿.

Lemna valdiviana Phil. Pale duckweed
Fronds flat, narrowly oblong, 2.5–5 mm, with 1 faint nerve or nerveless; very rare, floating or on submerged debris in shallow water; believed to be extirpated; OBL; 🌿.

Spirodela Schleid.

Fronds disk-shaped, 2.5–10 mm, with 3–15 faint nerves radiating from a point near one end; roots 2–12 from the underside of the nodal point.

A. roots 5–12; fronds 4–10 mm .... S. *polyrhiza*
A. roots 2–5; fronds 2.5–5 mm .... S. *punctata*

Spirodela polyrhiza
(×2½)

Spirodela polyrhiza (L.) Schleid. Greater duckweed, water-flaxseed
Fronds round to obovate, 4–10 mm, green above and purple beneath with a cluster of 5–12 elongate roots; occasional in ponds, lakes, swamps, and margins of sluggish streams; throughout; OBL.

Spirodela punctata (G.Mey.) C.H.Thomps. Eastern water-flaxseed
Similar to S. *polyrhiza* but smaller, fronds 2.5–5 mm; roots fewer; very rare in ponds; N; OBL; 🌿.

Wolffia Horkel.

Fronds minute, up to 1.5 mm; globose, floating with the upper surface level with the water surface; no veins or roots present, but the upper surface papillate in some species.

A. fronds flattened above
    B. upper surface with a prominent papilla near the middle .... W. *brasiliensis*
    B. upper surface with a raised papilla at one end .... W. *borealis*
A. fronds rounded above, without an obvious papilla .... W. *columbiana*

Wolffia borealis (Engelm.) Landolt   Dotted watermeal
Fronds broadly ovoid, slightly asymmetrical, 0.5–1.5 mm tapering to a raised papilla at one end; very rare in ponds and swamps; widely scattered; OBL; 🌿.

Wolffia brasiliensis,
top view (×10)

Wolffia brasiliensis Wedd.   Pointed watermeal
Fronds symmetrical, with a prominent central papilla, 0.1–1.2 mm; a frequent native of lakes, ponds, and margins of slow moving streams; throughout; OBL.

Wolffia brasiliensis,
side view (×10)
showing waterline

Wolffia columbiana H.Karst.   Watermeal
Fronds broadly ellipsoid to globose, asymmetrical, 0.8–1.4 mm, rounded above and floating low in the water; occasional in lakes, ponds, marshes, ditches, and bogs; throughout; OBL.

Wolffiella Hegelm.

Wolffiella gladiata (Hegelm.) Hegelm.   Bog-mat
Fronds thin and flat, linear or strap-shaped, 4–14 mm; forming a tangled, partially submergent mass on open water channels in swamps; rare; NW; OBL; 🌿.

LILIACEAE Lily Family
John Kunsman

Rhizomatous or bulbous perennial herbs; leaves simple, basal (the plants often scapose or nearly so) or cauline; flowers regular; perianth segments 4 or 6; ovary superior or inferior; fruit a capsule or berry.

A. ovary inferior
   B. flowers in a raceme .... *Aletris*
   B. flowers occurring singly, in pairs, or in an umbel
      C. pedicels, outer surface of perianth, and usually the leaves pubescent.....
      .... *Hypoxis*
      C. pedicels, perianth, and leaves glabrous
         D. base of perianth segments united into a corona .... *Narcissus*
         D. base of perianth segments not united into a corona .... *Galanthus*
A. ovary superior
   E. perianth united for ⅕ or more of its length
      F. perianth <3 cm long
         G. leaves cauline; flowers axillary .... *Polygonatum*
         G. leaves basal or nearly so; flowers in a terminal raceme
            H. perianth blue (rarely white); leaves linear .... *Muscari*
            H. perianth white, greenish, or yellowish; leaves lanceolate, oblong, or elliptic
               I. perianth smooth on the outside; pedicels >5 mm long; leaves usually 2(3) .... *Convallaria*
               I. perianth roughened on the outside; pedicels mostly 1–3(5) mm long; leaves usually 4 or more .... *Aletris*
      F. perianth 3 cm or more long
         J. perianth orange or yellow; leaves linear .... *Hemerocallis*
         J. perianth blue, lilac, or white; leaves lanceolate to ovate, petioled ....
         .... *Hosta*

E. perianth of separate segments
    K. plant with an onion or garlic odor when bruised .... *Allium*
    K. plant with a neutral odor when bruised
        L. apparent leaves (actually stems) needle-like, <1 mm wide; true leaves scale-like .... *Asparagus*
        L. leaves >1 mm wide
            M. at least some of the stem leaves whorled
                N. leaves in a single whorl of 3; perianth of 3 green sepals and 3 white or colored petals .... *Trillium*
                N. leaves in 2 or more whorls, or in a single whorl of 4 or more; perianth of 6 colored segments
                    O. perianth 3 cm or more long; fruit a capsule .... *Lilium*
                    O. perianth 1.5 cm or less long; fruit a berry .... *Medeola*
            M. stem leaves alternate or opposite, or leaves all basal
                P. flowers solitary or in pairs (plant may have several single or paired flowers), often in the axils of the stem leaves
                    Q. leaves 2, mottled .... *Erythronium*
                    Q. leaves (2)3 or more, unmarked
                        R. perianth 3 cm or more long
                          S. leaves 2–3(4), thick and fleshy; flower erect ... *Tulipa*
                          S. leaves 4 or more, thin; flower ascending to drooping
                              T. perianth yellow, to 4.5 cm long; filaments much shorter than the anthers .... *Uvularia*
                            T. perianth orange-red or white, 5 cm or more long; filaments longer than the anthers .... *Lilium*
                        R. perianth <3 cm long
                          U. perianth yellow; fruit a capsule .... *Uvularia*
                          U. perianth white, yellow-green, pink, or rose; fruit a berry
                            V. perianth yellow-green; flowers terminal .... *Disporum*
                            V. perianth white, pink, or rose; flowers axillary ..... *Streptopus*
                P. flowers in a raceme, panicle, or umbel
                W. perianth 4 cm or more long .... *Lilium*
                W. perianth <4 cm long
                  X. flowers in an umbel
                    Y. fruit a berry; umbel bractless or subtended by leaves
                      Z. leaves basal or nearly so; umbel peduncled .... *Clintonia*
                      Z. leaves cauline; umbel sessile or nearly so .... *Disporum*
                  Y. fruit a capsule; umbel subtended by papery bract(s) .... *Allium*
                X. flowers in a raceme or panicle
                  AA. perianth segments 4 .... *Maianthemum*
                  AA. perianth segments 6
                      BB. perianth segments with 1 or 2 glandular spots

CC. perianth segments green or whitish, clawed at the base; inflorescence axis pubescent .... *Melanthium*

CC. perianth segments white, gradually narrowed at the base; inflorescence axis glabrous .... *Zigadenus*

BB. perianth segments without glandular spots

DD. flowers in a panicle

EE. perianth segments 1–3 mm long; fruit a berry .... *Smilacina*

EE. perianth segments 4–15 mm long; fruit a capsule

FF. perianth green, 10–15 mm long; leaves elliptic ..... *Veratrum*

FF. perianth white, 4–10 mm long; leaves linear .... *Stenanthium*

DD. flowers in a raceme

GG. styles 3

HH. pedicels 8–12 mm long, with bracts; basal leaves linear .... *Amianthium*

HH. pedicels 2–3 mm long, without bracts; basal leaves narrowly elliptic to oblanceolate .... *Chamaelirium*

GG. style 1

II. leaves cauline, lanceolate or wider; fruit a berry ..... *Smilacina*

II. leaves basal, linear; fruit a capsule

JJ. perianth segments white with a green stripe on the back .... *Ornithogalum*

JJ. perianth segments blue or violet, or if white, without a green stripe .... *Camassia*

Aletris L.

Aletris farinosa L.   Colic-root

Rhizomatous herb to 1 m tall; stem glabrous; leaves basal or nearly so, lanceolate; inflorescence a raceme; pedicels 1–3(5) mm long, shorter than the bracts; perianth white, mostly united, 5–8(10) mm long, roughened on the outside; ovary partly inferior; fruit a beaked capsule; rare in moist clearings; mostly SE; flr. May–Jul; FAC; ❀.

Allium L.

Bulbous herbs with onion or garlic odor when bruised; leaves basal or cauline, flat, keeled, or terete; inflorescence an umbel with a subtending bract(s); perianth segments separate; fruit a capsule.

A. leaves lanceolate to elliptic, mostly 2 cm or more wide, withering by flowering time .... *A. tricoccum*

A. leaves linear, mostly 2–15 mm wide, present at flowering time

B. umbel nodding .... *A. cernuum*

B. umbel erect

C. leaves terete and hollow

D. leaves 2–5 mm wide, basal and cauline; perianth 3–5 mm long .... *A. vineale*

D. leaves 1–3 mm wide, all basal; perianth 7–14 mm long .... *A. schoenoprasum*

C. leaves flat or keeled

E. bracts usually 1.5 cm or less in length; bulb coat fibrous reticulate .... *A. canadense*

E. bracts 2 cm or more long; bulb coat membranous .... *A. oleraceum*

Allium canadense L. Wild onion

Herb to 6 dm tall; bulb coat fibrous-reticulate; leaves basal or nearly so, flat, 2–5 mm wide; umbel erect, the 2 bracts 1.5 cm or less long; perianth white to pink, 5–9 mm long; flowers often replaced by bulblets; frequent in moist woods and stream banks; mostly S; flr. May–Jun; FACU.

Allium cernuum Roth Nodding onion

Herb to 6 dm tall; bulb coat membranous; leaves basal, flat, 2–6(8) mm wide; umbel nodding; perianth white, pink, or rose, 4–6 mm long; frequent on dry slopes, woods borders, and rocky banks; mostly SW and SC; flr. Jun–Aug.

Allium oleraceum L. Field garlic

Herb to 6 dm tall; bulb coat membranous; leaves cauline, flat, 2–4 mm wide; umbel erect, the 2 bracts elongate, 2 cm or more long; perianth green to purplish; cultivated and rarely escaped; SE; flr. Jul–Aug; native to Europe.

Allium schoenoprasum L. Chives

Herb to 5 dm tall; bulb coat membranous; leaves basal, terete, 1–3 mm wide; umbel erect; perianth pink to purple, 7–14 mm long; cultivated and rarely escaped; flr. May–Jun; native to Eurasia; FACU.

Allium tricoccum Aiton Ramp, wild leek

Herb to 5 dm tall; bulb coat slightly fibrous; leaves basal, flat, lanceolate to elliptic, petioled, (1)2–8 cm wide, withering by flowering time; umbel erect; perianth white, 4–7 mm long; frequent in rich woods and moist slopes; throughout; flr. Apr–Jul; FACU+.

Allium vineale L. Field garlic

Herb to 1 m tall; bulb coat membranous; leaves cauline, terete, 2–5 mm wide; umbel erect; perianth white, pink, or purple, 3–5 mm long; flowers often replaced by bulblets; common in disturbed ground and open woods; mostly S; flr. Jun–Jul; native to Europe; FACU–.

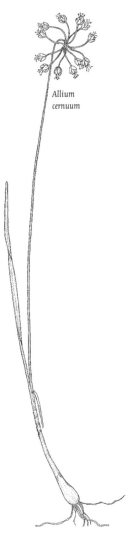

Allium
cernuum

Amianthium A.Gray

Amianthium muscaetoxicum (Walter) A.Gray Fly-poison

Bulbous herb to 1 m tall; stem glabrous; leaves mostly basal, linear, to 2.5 cm wide; inflorescence a raceme; pedicels mostly 8–12 mm, much longer than the bracts; perianth segments white (green in age), separate, 3–6 mm; fruit a 3-beaked capsule; frequent in woods and barrens; E, especially in the mountains; flr. May–Jul; FAC. Poisonous.

Amianthium
muscaetoxicum,
fruit (×1)

Asparagus L.

Asparagus officinalis L. Garden asparagus

Rhizomatous herb to 2 m tall; stem glabrous, branched, the branchlets needle-like and resembling leaves, 1 mm or less wide; true leaves reduced to scales; flowers axillary, single or paired; perianth segments yellowish to greenish-white, separate or nearly so, 3–6 mm long; fruit a red berry; frequent in woods borders, roadsides, and disturbed ground; throughout; flr. May–Jul; native to Europe; FACU.

Asparagus officinalis

Camassia Lindl.

Camassia scilloides (Raf.) Cory   Wild hyacinth
Bulbous herb to 6 dm tall; scape glabrous; leaves basal, linear, to 2 cm wide; inflorescence a raceme; lower pedicels 1–2 cm long, shorter than the bracts; perianth segments pale blue (white), separate, 7–18 mm long, 3–5-veined; fruit a globe-shaped capsule; rare in moist woods; W; flr. Apr–May; FAC; 🌿.

Chamaelirium Willd.

Chamaelirium luteum (L.) A.Gray   Devil's-bit
Rhizomatous herb to 1 m tall; stem glabrous; leaves basal and cauline, reduced in size upward, narrowly elliptic to oblanceolate; inflorescence a raceme; pedicels 2–3 mm long, without bracts; dioecious; perianth segments white or yellow, separate, 2–3 mm long; fruit an elliptic capsule; infrequent in dry woods, clearings, and barrens; throughout; flr. May–Jul; FAC.

Clintonia Raf.

Rhizomatous herbs; leaves 2–5, basal or nearly so, elliptic; inflorescence an umbel without bracts; perianth segments separate; fruit a berry; flr. May–Jun.

A. perianth yellow, 12–18 mm long; mature fruit blue .... *C. borealis*
A. perianth white, 6–10 mm long; mature fruit black .... *C. umbellulata*

Clintonia borealis (Aiton) Raf.   Blue bead-lily
Herb to 4 dm tall; flowers 2–5(8) per umbel; perianth yellow, 12–18 mm long; styles 10–15 mm long; mature fruit blue; frequent in moist woods; N and at higher elevations along the Allegheny Front; FAC.

Clintonia umbellulata (Michx.) Morong   Speckled wood-lily
Herb to 4 dm tall; flowers 6–25 per umbel; perianth white with purple spots, 6–10 mm long; styles 5–8 mm long; mature fruit black; frequent in moist woods; W.

Convallaria L.

Convallaria majalis L.   Lily-of-the-valley
Rhizomatous herb to 2 dm tall; scape glabrous; leaves appearing basal, elliptic, 2(3); inflorescence a raceme, the flowers nodding, fragrant; perianth white, united nearly to the tip, 6–10 mm long; fruit a red berry; occasionally naturalized in woods and disturbed ground; throughout; flr. Apr–early Jun; native to Europe.

Disporum Salisb.

Disporum lanuginosum (Michx.) G.Nicholson   Yellow mandarin
Rhizomatous herb to 1 m tall; stem branched, often pubescent; leaves cauline, alternate, ovate to lanceolate; flowers terminal, occurring singly, in pairs, or in an umbel; perianth segments yellowish-green, separate, recurved, 13–22 mm long; fruit a red berry; frequent in moist woods; W; flr. May–early Jun.

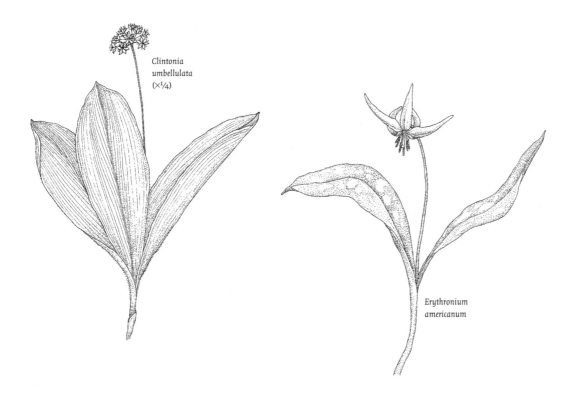

Clintonia
umbellulata
(×¹⁄₄)

Erythronium
americanum

Erythronium L.

Bulbous herbs; leaves appearing basal, paired, spotted/mottled, glabrous, lance-elliptic; flower single, nodding; perianth segments separate, recurved; fruit a capsule; flr. Apr–May.

A. perianth yellow; stigmas erect .... E. americanum
A. perianth white; stigmas spreading or recurved .... E. albidum

Erythronium albidum Nutt.   White trout-lily
Herb to 20 cm tall; leaves 2, spotted/mottled; perianth white, often with faint bluish or pinkish tinge, 1.5–4 cm long; stigmas spreading or recurved; infrequent in moist woods and rich slopes, especially on limestone; W and SC; FACU; .

Erythronium americanum Ker Gawl.   Yellow trout-lily
Herb to 20 cm tall; leaves 2, spotted/mottled; perianth yellow, 1.5–4 cm long; stigmas erect; common in moist woods and rich slopes; throughout.

Galanthus L.

Galanthus nivalis L.   Snowdrop
Herb to 3 dm tall; leaves linear to narrowly oblong; perianth white, the outer segments 12–25 mm long, the inner segments 6–11 mm long and with a green patch at the apex; cultivated and occasionally escaped in disturbed woodlands; SE; native to Europe.

Hemerocallis L.

Herbs with tuberous roots; leaves basal, linear; scape forked into two raceme-like branches; individual flowers blooming for a single day; perianth partly united, the recurved lobes longer than the tube; fruit a capsule.

A. perianth orange ....*H. fulva*
A. perianth yellow .... *H. lilioasphodelus*

Hemerocallis fulva (L.) L.   Orange day-lily
Herb to 1 m tall; leaves to 3 cm wide; flowers not fragrant; perianth orange, 6 cm or more long; fruit not developing; common in woods borders, clearings, roadsides, and disturbed ground; throughout; flr. Jun–early Aug; native to Asia; UPL. A double-flowered form occurs in some areas.

Hemerocallis lilioasphodelus L. emend. Hyl.   Yellow day-lily
Herb to 1 m tall; leaves to 1.5 cm wide; flowers fragrant; perianth yellow, 6 cm or more long; cultivated and occasionally escaped; flr. May–Jul; native to Europe.

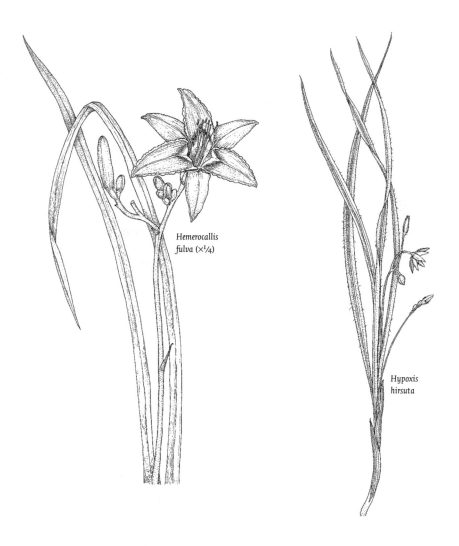

Hemerocallis fulva (×¼)

Hypoxis hirsuta

Hosta Tratt.

Rhizomatous herbs; leaves basal or nearly so, lanceolate to ovate, with distinct petioles; inflorescence a raceme; perianth partly united; fruit a capsule.

A. perianth 3–6 cm long
    B. perianth abruptly widening toward the tip; leaves ovate to cordate-ovate .... H. ventricosa
    B. perianth gradually widening toward the tip; leaves lanceolate to ovate-lanceolate .... H. lancifolia
A. perianth 10 cm or more long .... H. plantaginea

Hosta lancifolia (Houtt.) Engl.   Narrow-leaved plantain-lily
Herb to 7 dm tall; leaves lanceolate to ovate-lanceolate, to 5 cm wide; flowers not fragrant, horizontal to nodding; perianth lilac or lavender, united <½ its length, gradually widening toward the tip, 3–6 cm long; cultivated and rarely escaped; flr. Jul–Sep; native to Asia.

Hosta plantaginea (Lam.) Asch.   Fragrant plantain-lily
Herb to 7 dm tall; leaves ovate to cordate-ovate, to 15 cm wide; flowers fragrant, ascending; perianth white, united >½ its length, 10 cm or more long; cultivated and rarely escaped; mostly SE; native to Asia.

Hosta ventricosa (Salisb.) Stearn   Blue plantain-lily
Herb to 7 dm tall; leaves ovate to cordate-ovate, to 12 cm wide; flowers not fragrant, horizontal to nodding; perianth blue, united <½ its length, abruptly widening toward the tip, 3–6 cm long; cultivated and occasionally escaped; mostly SE; flr. Jun–Jul; native to Asia.

Hypoxis L.

Hypoxis hirsuta (L.) Coville   Yellow star-grass
Pubescent herb to 4 dm tall; leaves basal or nearly so, linear, to 8 mm wide; inflorescence umbel-like; perianth segments yellow, separate, 5–10 mm long, pubescent on the outer surface; ovary inferior; fruit a capsule; frequent in dry woods, clearings, and barrens; throughout, except in the northernmost counties; flr. May–Sep; FAC.

Lilium L.

Bulbous herbs; leaves cauline, alternate, or whorled; flowers single or in an inflorescence; perianth segments separate; fruit a capsule.

A. flowers erect; perianth segments clawed at the base .... L. philadelphicum
A. flowers horizontal or nodding; perianth segments gradually tapering to the base
    B. leaves in one or more whorls (alternate leaves may also be present)
        C. perianth segments strongly recurved; leaves usually smooth below .... L. superbum
        C. perianth segments spreading to recurved; leaves usually minutely roughened on the margins and on the veins below .... L. canadense
    B. leaves all alternate .... L. lancifolium

Lilium canadense

Lilium
philadelphicum

Lilium canadense L.   Canada lily
Herb to 2 m tall; stem glabrous; leaves all or in part whorled, lanceolate, usually
minutely roughened on the margins and on the veins below; flowers nodding;
perianth yellow, orange, or red, spotted, the segments spreading to recurved,
mostly 4.5 cm or more long; infrequent in moist woods and clearings; throughout; flr. late Jun–early Jul; FAC+. Plants with wider leaves and consistently red
perianth have been called ssp. *editorum* (Fernald) Wherry.

Lilium lancifolium Thunb.   Tiger lily
Herb to 1 m tall; stem pubescent, at least above; leaves alternate, lanceolate, often with bulbs in the axils; flowers nodding; perianth orange, spotted, the segments strongly recurved and pubescent on the lower midrib, mostly 4.5 cm or
more long; cultivated and occasionally escaped; native to Asia.

Lilium philadelphicum L.   Wood lily
Herb to 1 m tall; stem glabrous; leaves all or in part whorled, lanceolate; flowers
erect; perianth orange or red (yellow), spotted, the segments distinctly clawed
toward the base, mostly 4.5 cm or more long; occasional in open woods, borders, and clearings; throughout, except the far W; flr. late Jun–early Jul; FACU+.

Lilium superbum L.   Turk's-cap lily
Herb to 2.5 m tall; stem glabrous; leaves all or in part whorled, lanceolate, smooth or nearly so below; flowers nodding; perianth orange or orange-red, spotted, the segments strongly recurved, mostly 4.5 cm or more long; frequent in moist clearings and bottomlands; throughout; flr. late Jul–early Aug; FACW+.

Maianthemum Weber

Maianthemum canadense Desf.   Canada mayflower
Rhizomatous herb to 15 cm tall; leaves 2–3, alternate, cauline, ovate, base cordate-clasping; inflorescence a raceme; perianth of 2 sepals and 2 petals, white, separate, 2–3 mm long; fruit a red berry; common in dry to moist woods; throughout; flr. May–early Jun; FAC–.

Medeola L.

Medeola virginiana L.   Indian cucumber-root
White-tuberous herb to 1 m tall; leaves cauline in 2 whorls; inflorescence a sessile umbel, or the flowers single or paired; perianth segments yellowish, separate, recurved, 6–9 mm long; fruit a dark purple or black berry; frequent in moist woods; throughout; flr. late May–early Jun.

Medeola
virginiana (×¼)

Lilium superbum

Melanthium L.

Rhizomatous herbs; stem often pubescent above; leaves basal and cauline; inflorescence a panicle; perianth segments greenish, with a pair of glandular spots, clawed; fruit a capsule with winged seeds.

A. blade portion of perianth segments deltoid-ovate, about equaling the clawed portion .... M. latifolium
A. blade portion of perianth segments oblong-ovate, mostly 2 times or more as long as the clawed portion .... M. virginicum

Melanthium latifolium Desr.   Bunchflower
Herb to 2 m tall; leaves linear to oblanceolate, the larger ones 3–6 cm wide; pedicels 8 cm or more long, longer than the bracts; perianth 4–6 mm long, the blade portion of the segments deltoid-ovate, margin often crisped, about equaling the claw; infrequent in dry to moist woods and rocky slopes; mostly E and C; flr. Jul–Aug; FACU. [syn: M. hybridum Walter]

Melanthium
virginicum (×1)

Melanthium virginicum L.   Bunchflower
Herb to 2 m tall; leaves linear, the larger ones 1–3 cm wide; pedicels 8 cm or more long, longer than the bracts; perianth 4–8 mm long, the blade portion of the segments oblong-ovate, margin flat, about twice (or more) as long as the claw; occasional in moist woods, seepages, and damp clearings; mostly S; flr. Jun–Aug; FACW+; ❦.

Muscari Mill.

Bulbous herbs; scapes glabrous; leaves basal; inflorescence a raceme with nodding flowers; perianth united nearly to the tip; fruit a capsule.

A. perianth globe-shaped; leaves linear-oblanceolate .... M. botryoides
A. perianth ovoid to cylindric; leaves linear .... M. neglectum

Muscari botryoides (L.) Mill.   Grape-hyacinth
Herb to 4 dm tall; leaves linear-oblanceolate, 3–12 mm wide; perianth globe-shaped, blue or violet, 3–5 mm long; capsule 4–6 mm long; frequent in moist disturbed ground; mostly S; flr. Mar–May; native to Europe.

Muscari neglectum Ten.   Grape-hyacinth
Herb to 3 dm tall; leaves linear, 2–8 mm wide; perianth ovoid to cylindric, blue or violet, 4–7 mm long; capsule 8–10 mm long; cultivated and rarely escaped; SE; flr. Apr–May; native to Europe.

Narcissus L.

Bulbous herbs; scape glabrous; leaves basal, linear; flower solitary; perianth united at the base with a saucer-like extension or corona; ovary inferior; fruit a capsule; flr. Apr–May.

A. corona as long as or longer than the perianth segments ...... N. pseudonarcissus
A. corona <⅕ as long as the perianth segments .... N. poeticus

Narcissus poeticus L.   Poet's narcissus
Herb to 4 dm tall; leaves to 1.2 cm wide; perianth segments usually white, to 2.5 cm long, much longer than the yellow or white, often red-margined, corona; cultivated and rarely escaped; native to Europe.

Narcissus pseudonarcissus L.   Daffodil
Herb to 4 dm tall; leaves to 1.2 cm wide; perianth yellow, the segments to 2.5 cm long, equaling or shorter than the often fluted or frilled corona; cultivated and occasionally escaped; native to Europe.

Ornithogalum L.

Bulbous herbs; scape glabrous; leaves basal, linear; inflorescence a raceme; perianth white with a green stripe on the back of each lobe, separate or nearly so; filaments winged; fruit a capsule.

A. perianth 1–2 cm long; leaves 2–6 mm wide; lower pedicels >1 cm, longer than the bracts .... *O. umbellatum*
A. perianth mostly 2–2.5 cm long; leaves 8–15 mm wide; lower pedicels 1 cm or less, shorter than the bracts .... *O. nutans*

Ornithogalum nutans L.   Star-of-Bethlehem
Herb to 5 dm tall; leaves 8–15 mm wide; lower pedicels 1 cm or less long, shorter than the bracts; perianth white with green stripes on the back, 2–2.5(3) cm long;

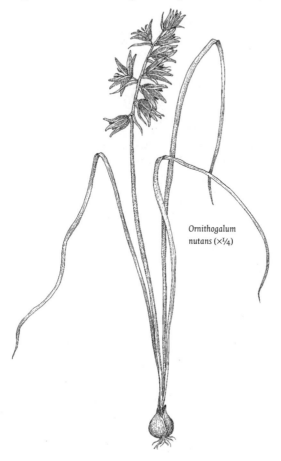

Ornithogalum nutans (×¼)

cultivated and occasionally escaped to roadsides and edges; mostly S; flr. Apr–May; native to Europe.

Ornithogalum umbellatum L. Star-of-Bethlehem
Herb to 3 dm tall; leaves 2–6 mm wide; lower pedicels >1 cm long, longer than the bracts; perianth white with green stripes on the back, 1–2 cm long; frequent in lawns and moist disturbed ground; mostly S; flr. Apr–Jun; native to Europe; FACU.

Polygonatum Mill.

Rhizomatous herbs; stem unbranched; leaves cauline, alternate, 2-ranked; flowers axillary, occurring singly, in pairs, or in an umbel; perianth united nearly to the tip; fruit a blue to black berry.

A. leaves glabrous below.. *P. biflorum*
A. leaves pubescent on the veins beneath .... *P. pubescens*

Polygonatum biflorum (Walter) Elliot Solomon's-seal
Herb to 2 m tall; leaves glabrous below; perianth white or greenish, 10–22 mm long; common in dry to moist woods; throughout; flr. Apr–Jun; FACU. Robust plants with larger leaves, flowers, and fruits have been called var. *commutatum* (Schult. f.) Morong or *P. canaliculatum* (Muhl.) Pursh.

Polygonatum pubescens (Willd.) Pursh Solomon's-seal
Herb to 1 m tall; leaves pubescent on the veins below; perianth white or greenish, 7–15 mm long; common in dry to moist woods; throughout; flr. Apr–May.

Smilacina Desf.

Rhizomatous herbs; stem unbranched; leaves cauline, alternate; inflorescence a raceme or panicle; perianth segments white, separate; fruit a red berry.

A. inflorescence a panicle; perianth segments 1–3 mm long .... *S. racemosa*
A. inflorescence a raceme; perianth segments 4–6 mm long
    B. leaves 2 or 3(4), glabrous beneath .... *S. trifolia*
    B. leaves 4 or more, usually pubescent beneath .... *S. stellata*

Smilacina racemosa (L.) Desf. False Solomon's-seal
Herb to 1 m tall; leaves 4 or more, usually pubescent beneath; inflorescence a panicle; perianth segments white, 1–3 mm long; common in dry to moist woods; throughout; flr. May.

Smilacina stellata (L.) Desf. Starflower
Herb to 6 dm tall; leaves 4 or more, usually pubescent beneath; inflorescence a more or less sessile raceme; perianth segments white, 4–6 mm long; occasional in moist to wet woods and shorelines; throughout; flr. late Apr–early May.

Smilacina trifolia (L.) Desf. False Solomon's-seal
Herb to 2 dm tall; leaves 2 or 3(4), glabrous beneath; inflorescence a stalked raceme; perianth segments white, 4–6 mm long; infrequent in bogs and peaty wetlands; N and SW; flr. May.

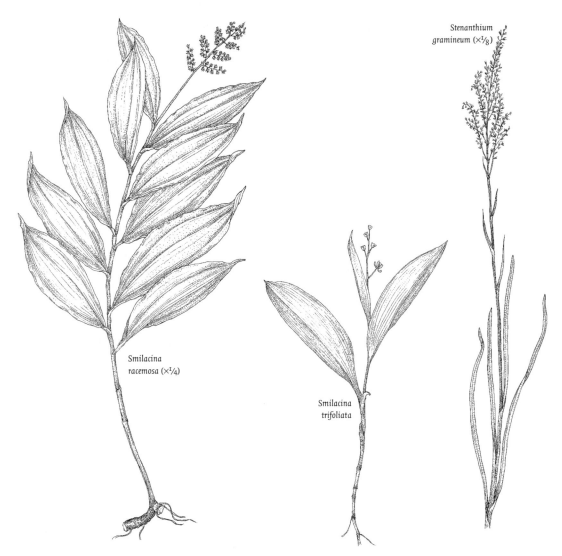

Stenanthium
gramineum (×⅛)

Smilacina
racemosa (×¼)

Smilacina
trifoliata

Stenanthium A. Gray

Stenanthium gramineum (Ker Gawl.) Morong   Featherbells

Bulbous herb to 1.5 m tall; stem glabrous; leaves basal and cauline, linear; inflorescence a panicle; pedicels 1–3(5) mm long, shorter than the bracts; perianth segments white (green), separate, 4–8(10) mm long, tapering to a slender pointed apex; fruit a cylindric capsule; infrequent in moist clearings; mostly W; flr. Jul–early Sep; FACW; ⚘.

Stenanthium
gramineum,
flower (×1)

Streptopus Michx.

Rhizomatous herbs; stem usually branched; leaves cauline, alternate; flowers axillary, mostly occurring singly, often nodding; perianth segments separate; fruit a red berry.

A. perianth segments pink or rose, spreading at the tip; leaf margins ciliate ......
....*S. roseus*

A. perianth segments greenish or whitish, spreading from near the middle; leaf margins glabrous .... *S. amplexifolius*

Streptopus amplexifolius (L.) DC. Twisted-stalk

Herb to 1 m tall; nodes usually glabrous; leaves ovate to lanceolate, clasping, the margins glabrous or nearly so; perianth segments greenish or whitish, spreading from near the middle; rare on seepy outcrops (often near waterfalls) and cool slopes; NE; flr. Jun–Jul; FAC+; ❦.

Streptopus roseus Michx. Rose mandarin

Herb to 6 dm tall; nodes often pubescent; leaves ovate to lanceolate, sessile to slightly clasping, the margins ciliate; perianth segments pink or rose, spreading at the tip; frequent in moist woods and stream banks; N and at higher elevations along the Allegheny Front; flr. May–Jul; FAC–. Plants with ciliate pedicels have been called var. *perspectus* Fassett.

Trillium L.

Rhizomatous herbs; stems unbranched; leaves in a single whorl of 3 just below the solitary flower; perianth of 3 green sepals and 3 colored petals; fruit a berry.

A. flower sessile; leaves usually mottled
  B. leaves sessile
    C. anther connective prolonged 2–4 mm, nearly as long as the filament; stigmas 1.5–2 times as long as the ovary .... *T. sessile*
    C. anther connective 1 mm, much shorter than the filament; stigmas up to as long as the ovary .... *T. cuneatum*
  B. leaves with petioles .... *T. recurvatum*
A. flower peduncled; leaves not mottled
  D. leaves sessile, or if narrowed to a petiole, then the ovary 6-angled or 6-winged; petals white, maroon, or pink (rarely yellow or green)
    E. stigmas of uniform width, straight or slightly curved; petals spreading from a constricted tube-like base, white (pink in age) .... *T. grandiflorum*
    E. stigmas widest at the base and tapering to the tip, recurved; petals mostly spreading from the point of attachment, white, maroon, or pink (rarely yellow or green)
      F. petals mostly 1–2 cm long; peduncle 1–2 cm, sharply declined with the flower held below the leaves .... *T. cernuum*
      F. petals mostly 2.5–4.5 cm long; peduncle mostly 4–6 cm long, divergent to declined with the flower usually held above the leaves
        G. petals maroon (occasionally white, yellow, or green); ovary maroon; filaments mostly ¼–½ as long as the anthers ..... *T. erectum*
        G. petals white (occasionally pink) ovary white or pink (occasionally darker); filaments mostly ¼ or less as long as the anthers ....
        .... *T. flexipes*
  D. leaves narrowed to a slender petiole; ovary 3-angled; petals white
    H. petals white, marked with purple at the base; tip of leaf acuminate; plant often 15 cm or more tall .... *T. undulatum*
    H. petals white (occasionally pink at the base); tip of leaf obtuse or blunt; plant usually <15 cm tall .... *T. nivale*

Trillium cernuum L. Nodding trillium

Herb to 4 dm tall; leaves ovate to rhombic, sessile or occasionally petioled, tip acuminate; peduncle (0.5)1–2(2.5) cm long, sharply declined with the flower held below the leaves; petals usually white, 1–2(2.5) cm long; anthers 3–7 mm long; ovary white; infrequent in moist woods; mostly SE; flr. Apr–May; FACW; ✿. Plants with wider and blunter petals and longer anthers have been called var. *macranthum* Eames & Wiegand.

Trillium cuneatum Raf.   Huger's trillium
Similar to T. *sessile* but differing by the stigmas shorter than to equal to the ovaries and the anther connective much shorter than the filament; collected at 2 sites in the SE, probably introduced; flr. late Apr–early May; native farther south.

Trillium erectum L.   Purple trillium, wakerobin
Herb to 4 dm tall; leaves ovate to rhombic, sessile or nearly so, tip acuminate; peduncle (2)4–6 cm long, divergent, erect, or declined, the flower usually held above the leaves; flowers often with an offensive odor; petals usually maroon (white, yellow, or green), (2)2.5–4.5(5) cm long; anthers 5–12 mm long; ovary maroon; common in moist woods; throughout except SE; flr. late Apr–early May; FACU–. White-flowered forms have been called var. *album* (Michx.) Pursh.

Trillium flexipes Raf.   Declined trillium
Herb to 4 dm tall; leaves ovate to rhombic, sessile or nearly so, tip acuminate; peduncle (2)4–6 cm long, divergent to declined, the flower held above or below the leaves; flowers usually without an offensive odor; petals usually white (pink), (2)2.5–4.5(5) cm long; anthers 6–15 mm long; ovary usually white or pink; occasional in moist woods; S; flr. late Apr–early May; FAC; ✿. Populations in the lower Susquehanna Valley may be a hybrid with T. *erectum*.

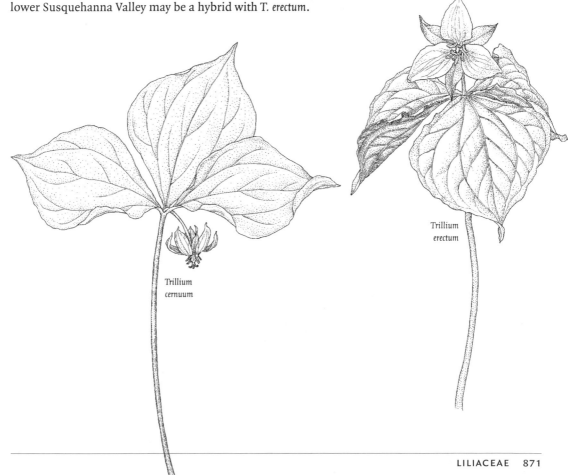

Trillium
cernuum

Trillium
erectum

Trillium grandiflorum (Michx.) Salisb.   Large-flowered trillium

Herb to 4 dm tall; leaves ovate to rhombic, sessile or nearly so, tip acuminate; peduncle (2)4–6 cm long, divergent to erect; petals white (pink in age), 3–6 cm long, obovate, spreading outward from a tube-like base; stigmas of uniform width; ovary pale; frequent in moist woods; mostly W; flr. late Apr–early May.

Trillium nivale Riddell   Snow trillium

Herb to 1.5 dm tall; leaves narrowly ovate to narrowly elliptic, base petioled, tip obtuse to blunt; peduncle 0.5–2 cm long, erect to divergent; petals white (occasionally pink at the base), 1.5–2.5(3) cm long; ovary 3-angled; very rare in moist woods; SW; flr. late Mar–Apr; ✿.

Trillium recurvatum Beck   Prairie trillium

Similar to T. sessile but differing in having petioled leaves, clawed petals and recurved sepals; introduced in Lancaster Co.; flr. May; native farther west; UPL.

Trillium sessile L.   Toadshade

Herb to 3 dm tall; leaves ovate, sessile, tip obtuse to acuminate, usually mottled; flowers sessile; petals usually maroon, 1.5–3.5 cm long; infrequent in moist woods; W and SC; flr. late Apr–early May; UPL.

Trillium undulatum Willd.   Painted trillium

Herb to 4 dm tall; leaves ovate, base petioled, tip acuminate; peduncle 1–2.5 cm long, erect to divergent; petals white, with a rose or purple triangle at the base, 1.5–2.5(3) cm long; ovary 3-angled; frequent in moist woods; N and at higher elevations along the Allegheny Front; flr. May–early Jun; FACU.

Tulipa L.

Tulipa sylvestris L.   Dutch-lily

Bulbous herb to 5 dm tall; stem glabrous; leaves 2–4, cauline, alternate, lanceolate, thick, and fleshy; flower solitary, erect; perianth segments yellow, separate, 3 cm or more long; fruit a capsule; cultivated and occasionally escaped; SE; flr. late Apr–early May; native to Europe.

Uvularia L.

Rhizomatous herbs; stem usually branched; leaves cauline, alternate; flowers axillary or terminal, occurring singly, usually nodding; perianth segments yellow, separate; fruit a capsule.

A. leaves sessile
    B. stem glabrous; ovary and capsule narrowed to a stalk-like base above the pedicel .... U. sessilifolia
    B. stem pubescent in a line below the leaf base; ovary and capsule sessile on the pedicel .... U. pudica
A. leaves perfoliate
    C. leaves glabrous below; perianth segments roughened on the inner surface .... U. perfoliata
    C. leaves usually pubescent below; perianth segments smooth on the inner surface .... U. grandiflora

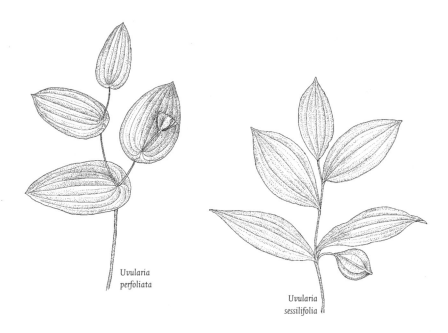

Uvularia
perfoliata

Uvularia
sessilifolia

Uvularia grandiflora Sm.   Bellwort
Herb to 7 dm tall; stem glabrous; leaves perfoliate, usually pubescent below; perianth yellow, 2.5–4.5 cm long, smooth on the inner surface; capsule broadest at the tip; infrequent in moist woods; W and NC; flr. Apr–May.

Uvularia perfoliata L.   Bellwort
Herb to 4 dm tall; stem glabrous; leaves perfoliate, glabrous below, usually glaucous; perianth yellow, 2–3 cm long, roughened on the inner surface; capsule broadest at the tip; common in moist woods; throughout; flr. May–Jul; FACU.

Uvularia pudica Michx.   Mountain bellwort
Herb to 4 dm tall; stem pubescent in a line below the leaf base; leaves sessile; perianth yellow, 1.5–3 cm long; style divided for ½ or more the length; capsule elliptic, without a stalk-like base above the pedicel; rare in moist woods; SW; flr. May; FACU; ✿. [U. puberula Michx. of Rhoads and Klein, 1993]

Uvularia sessilifolia L.   Bellwort
Herb to 4 dm tall; stem glabrous; leaves sessile; perianth yellow, 1.5–2.5 cm long; style usually divided for ⅓ or less its length; capsule elliptic, with a stalk-like base above the pedicel; common in dry to moist woods; throughout; flr. Apr–May; FACU–.

Veratrum L.

Veratrum viride Aiton   False hellebore
Rhizomatous herb to 1.5 m tall; stem pubescent; leaves cauline, alternate, elliptic, prominently veined, conspicuously sheathing the stem; inflorescence a panicle; pedicels 2–5 mm long, exceeded by the bracts; perianth segments green, separate, 10–15 mm long; fruit a capsule with winged seeds; frequent in wet woods, stream banks, and seepages; throughout; flr. May; FACW+.

Zigadenus Michx.

Zigadenus glaucus (Nutt.) Nutt. Camas

Bulbous herb to 6 dm tall; stem glabrous; leaves basal and cauline, linear, to 1.5 cm wide; inflorescence a raceme or panicle; lower pedicels 1 cm or more long, exceeding the bracts; perianth segments white, separate, glandular-spotted, 7–15 mm long; fruit a capsule; very rare on limestone ledges in Huntingdon Co.; flr. Aug; ✿.

NAJADACEAE Waternymph Family

Submersed, rooted, annual, aquatic herbs; leaves opposite or whorled, 1-nerved, with a serrate or serrulate blade and an expanded sheathing base; flowers solitary or few in the axils, unisexual, lacking a perianth; staminate flowers subtended by a ring or cup of scales; stamen 1; pistillate flowers lacking bracts, with a single, unilocular carpel bearing 2–4 elongate stigmas; fruit a cylindrical achene tapered at each end; flr./frt. mid- to late summer.

Najas L.

Characteristics of the family; freely branched, 1–8 dm tall, rooted, but not rhizomatous; the sheathing leaf base variously tapered or lobed at the top just below the toothed blade; fruits borne singly in the leaf axils; all species monoecious except N. *flexilis*, which is dioecious; flr./frt. Jul–Oct.

A. leaf teeth coarse, visible to the naked eye; stem and underside of leaf midvein prickly .... N. *marina*
A. leaf teeth finer, visible with a lens; leaves and stem not prickly
   B. tops of the leaf sheaths tapered
      C. style 1 mm or more long; seeds smooth and shining or only obscurely pitted .... N. *flexilis*
      C. style 0.5 mm or less in length; seeds dull and deeply pitted ....
      .... N. *guadalupensis*
   B. tops of the leaf sheaths lobe-like
      D. leaves recurved, with 7–15 teeth per side, lobes of the sheath fan-shaped and finely toothed .... N. *minor*
      D. leaves not recurved, thread-like, with 13–17 teeth per side, lobes of the sheath wedge-shaped and coarsely jagged .... N. *gracillima*

Najas flexilis (×1)

Najas gracillima (×1)

Najas flexilis (Willd.) Rostk. & Schmidt   Northern waternymph
Stem 0.5–5 dm tall, profusely branched above; leaf blade tapering to a long, slender point; lobes of the base tapering; seeds yellow to deep brown, glossy; occasional in lakes, ponds, creeks, and canals; throughout; OBL.

Najas gracillima (A.Braun) Magnus   Slender waternymph, bushy naiad
Stem 0.5–5 dm tall, sparingly branched; leaves lax, minutely serrulate, the base abruptly expanded and minutely toothed across the summit; seeds light brown, pitted; rare in shallow water of lakes, ponds, and reservoirs; scattered; OBL; ✤.

Najas guadalupensis (Spreng.) Magnus   Southern waternymph
Stem 1–8 dm tall, profusely branched; leaves minutely serrulate, tapering to the expanded base; seeds purple-tinged, pitted; rare in lakes, ponds, reservoirs, and freshwater tidal flats; scattered; OBL.

Najas marina L.   Holly-leaved naiad
Stem 0.5–4.5 dm tall, often prickly as are the undersides of the leaves; seeds reddish-brown, pitted; very rare in shallow, calcareous water; C; OBL; ✤.

Najas minor All.   Waternymph, naiad
Stems 1–2 dm tall, leaves becoming recurved in age, sheath abruptly expanded at the top, with coarse, jagged teeth; seeds purplish with quadrangular pits; rare in shallow water of lakes, ponds, and reservoirs; mostly SW; native to Europe; OBL.

ORCHIDACEAE Orchid Family

Herbaceous perennials with alternate, whorled or basal, parallel-veined leaves (or none), some species lacking chlorophyll and entirely mycotrophic; flowers irregular; sepals 3, distinct; petals 3, the 2 laterals similar and the lower one (the lip) distinctly different in shape; the single stamen (2 in *Cypripedium*) adnate to the style, forming a column with the anther(s), pollen dispensed in sac-like pollinia; ovary inferior; fruit a capsule with numerous tiny seeds.

A. leaves not present at flowering time
  B. flowers solitary; lip bearded .... *Arethusa*
  B. flowers in a spike or raceme; lip not bearded
    C. flowers with a conspicuous spur .... *Tipularia*
    C. flowers not spurred
      D. flowers white or yellowish; raceme spirally twisted .... *Spiranthes*
      D. flowers not white (except perhaps the lip); inflorescence not twisted
        E. lateral sepals free at the base; a single winter leaf produced ....
        .... *Aplectrum*
        E. lateral sepals fused at the base; leaves never present ... *Corallorhiza*
A. leaves present at flowering time
  F. lip conspicuously inflated and pouch-like .... *Cypripedium*
  F. lip not conspicuously pouch-like
    G. flowers 1–3, not in a raceme or spike
      H. leaves in a whorl of 5 or 6 just below the flowers .... *Isotria*
      H. leaves not whorled

I. flowers solitary (or occasionally 2) at the top of the stem; lip bearded
    J. leaves well developed before flowering; margin of the lip fringed .... *Pogonia*
    J. leaf slender, from a sheath, not visible until flowering time; margin of the lip wavy .... *Arethusa*
I. flowers several, axillary, nodding; lip not bearded.................. *Triphora*
G. flowers 2–many, in spikes or racemes
    K. flowers inverted so that the lip is above the sepals and petals ..... *Calopogon*
    K. flowers not inverted, lip not above the sepals and petals
        L. only a solitary leaf present
            M. flowers with a spur .... *Platanthera*
            M. flowers lacking a spur .... *Malaxis*
        L. leaves 2–several
        N. leaves basal
            O. flowers lacking a spur
                P. leaves several, in a rosette
                    Q. leaves ovate to lance-ovate, green with prominent white veins .... *Goodyera*
                    Q. leaves narrower, all green .... *Spiranthes*
                P. leaves 2
                    R. lip widest toward the tip ....*Liparis*
                    R. lip widest near the base .... *Malaxis*
             O. flowers with a spur
                S. flowers with a white lip and pink or pale purple lateral petals .... *Galearis*
                S. flowers green or greenish-white .... *Platanthera*
        N. leaves cauline
             T. a single pair of opposite leaves present midway between the flowers and the base .... *Listera*
             T. leaves alternate on the flowering stem
                U. raceme spirally twisted or merely secund................ *Spiranthes*
                U. raceme neither twisted nor secund
                    V. flowers with a spur
                      W. spur as long or longer than the lip, slender and elongate .... *Platanthera*
                      W. spur shorter than the lip, rounded .... *Coeloglossum*
                  V. flowers not spurred
                    X. lip saccate at the base and also crested .... *Epipactis*
                    X. lip neither saccate or crested .... *Malaxis*

Aplectrum Torr.

Aplectrum hyemale (Muhl. ex Willd.) Nutt.   Puttyroot
Stem 3–6 dm tall from a corm; leaf present in the fall and winter only, elliptic, longitudinally pleated and grayish-green with white veins; flowers in a short raceme on a leafless stalk, purplish; sepals and lateral petals similar, the petals extending forward over the column; lip broadly obovate, 3-lobed and with 3 ridges near the center; rare in moist, rich wooded slopes and bottomlands; mostly S; flr. May–Aug; FAC; 🌿.

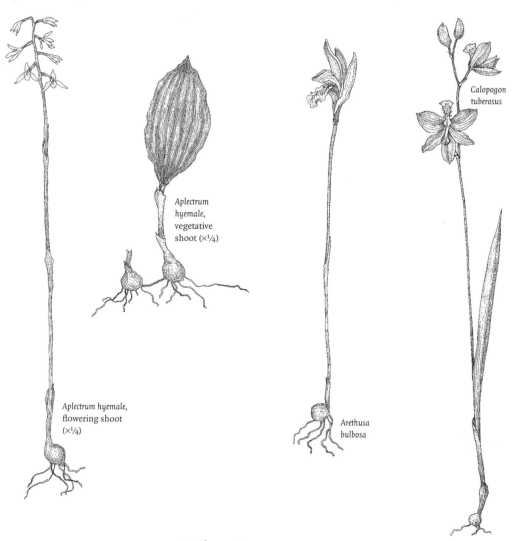

Aplectrum hyemale, vegetative shoot (×¼)

Aplectrum hyemale, flowering shoot (×¼)

Arethusa bulbosa

Calopogon tuberosus

Arethusa L.

Arethusa bulbosa L.   Dragon's-mouth

Stem 1–3 dm tall from a corm; the single narrow leaf arising after anthesis from the uppermost of several bracts on the stem; flowers solitary or very rarely 2; sepals and lateral petals magenta or pink; lip whitish with pink spots and a yellow crest; very rare in sphagnum bogs and seeps; scattered; flr. late May–Jun; OBL; 🍂.

Calopogon R.Br.

Calopogon tuberosus (L.) Britton, Stearns & Poggenb.   Grass-pink

Stem 3–7 dm tall from a corm; leaves 1 or 2, from near the base, narrow; flowers 3–15, in a loose raceme, rose-purple (or occasionally white); sepals and lateral petals spreading, sepals wider than the petals; lip uppermost, widest at the summit, with pink and yellow tipped hairs; occasional in bogs, fens and wet meadows; scattered; flr. Jun–Jul; FACW+.

Coeloglossum Hartm.

Coeloglossum viride (L.) Hartm.   Frog orchid, long-bracted green orchis
Stem 2–5 dm tall from fleshy roots; leaves several, alternate, reduced upward and becoming floral bracts, the lowest of which are longer than the flowers they subtend; inflorescence a spike; flowers green, sometimes tinged with purple; lip with upturned margins at the base and 3-toothed at the tip; the spur 2–3 mm long and rounded; rare in rich woods; scattered; flr. May–Aug; FACU; 🌿. Ours is var. *virescens* (Muhl. ex Willd.) Luer. [syn. *Habenaria viridis* (L.) R.Br.]

Corallorhiza Gagnebin.

Brownish or purplish, nonphotosynthetic plants with a cluster of knobby rhizomes; leaves reduced to sheathing scales; flowers in a terminal raceme; sepals and lateral petals narrow, usually extending forward over the column; the lateral sepals fused to the base of the column; lip oblong or rounded; column broad and boat-shaped; the single stamen producing 4 pollinia; fruits pendulous.

A. lip with lateral lobes or teeth
    B. plants >15 cm tall; lip 5–9 mm long; sepals and petals 3-nerved ....
    .... *C. maculata*
    B. plants <15 cm tall; lip 3–5 mm long; sepals and petals 1-nerved ... *C. trifida*
A. lip lacking lateral lobes or teeth, but sometimes notched at the tip
    C. sepals and petals >5 mm long; spring-flowering .... *C. wisteriana*
    C. sepals and petals <5 mm long; flowering in late summer or fall ....
    .... *C. odontorhiza*

Corallorhiza maculata, flower (×1)

Corallorhiza maculata (Raf.) Raf.   Spotted or large coralroot
Stem 2–6 dm tall, pinkish-purple with 10–40 flowers; sepals and lateral petals orange-brown marked with purple; spur present as a prominent swelling just above the ovary; lip 3-nerved, white with purple spots, bearing 2 small lateral lobes and with 2 short ridges on the surface; frequent in dry, deciduous or coniferous forests; throughout; flr. Jun–Sep; FACU.

Corallorhiza odontorhiza, flower (×1½)

Corallorhiza odontorhiza (Willd.) Nutt.   Autumn or small-flowered coralroot
Stem 1–2 dm tall, purplish, brownish, or greenish above, thickened and whitish at the base; flowers 5–15; sepals and lateral petals extending forward over the column, purplish; lip rounded, 3–4 mm long, white, often with a purple margin and 2 purple spots; occasional in rich woods on moist to dry soils; mostly SE; flr. Jun–Sep.

Corallorhiza trifida Châtel.   Early or northern coralroot
Stem 1–3 dm tall, yellowish or greenish with 5–15 flowers; sepals and lateral petals greenish-yellow to light purple, 4–5 mm long, 1-nerved, the lateral sepals extending forward close to the lip; spur lacking; lip 3.5–5 mm long, white, sometimes with a few purple spots, and with 2 short lateral lobes and 2 short ridges on the surface; rare in moist woods and bogs; scattered; flr. May–Jul; FACW.

Corallorhiza wisteriana, flower (×1)

Corallorhiza wisteriana Conrad   Wister's coralroot
Stem 1–4 dm tall, purple or reddish with 10–15 flowers; sepals and lateral petals extending forward over the column, greenish yellow marked with purple; lip broadly oval, notched at the tip, white dotted with purple; rare on rocky, wooded slopes on limestone and diabase; SE; flr. Apr–Jun; FAC; 🌿.

Cypripedium L.

Leaves basal or cauline, flowers solitary or few per plant; lip forming a conspicuous inflated pouch, sepals and lateral petals spreading; column bearing a fertile stamen on each side.

A. pouch pink
    B. sepals and lateral petals thin, brownish; leaves 2, basal .... *C. acaule*
    B. sepals and lateral petals white and petaloid; leaves >2, cauline ... *C. reginae*
A. pouch yellow or white (sometimes veined with purple)
    C. pouch yellow .... *C. calceolus*
    C. pouch white .... *C. candidum*

Cypripedium acaule Aiton   Pink lady's-slipper or moccasin flower
Leaves 2, basal; flowers solitary on a 1.5–4-dm scape; pouch pink; sepals and lateral petals lanceolate, greenish or brownish; frequent in moist to dry, acidic forests and bogs; throughout; flr. May–early Jul; FACU–.

Cypripedium calceolus L.   Yellow lady's-slipper or moccasin flower
Stem leafy, 2–8 dm tall with 1 or 2 terminal flowers; pouch yellow, often veined with purple; sepals and lateral petals greenish or purplish, often twisted; flr. Apr–Jun; 2 varieties:

A. lip 2–3 cm; petals 3–5 cm .... var. *parviflorum* (Salisb.) Fernald
    rare and scattered in moist woods and bogs, often on limestone; FACW–; ✿.
A. lip 3–5 cm; petals 5–8 cm .... var. *pubescens* (Willd.) Correll
    occasional in moist, rich, rocky woods and slopes; throughout; FAC+.

Cypripedium candidum Muhl. ex Willd.   Small white lady's-slipper
Stem 1.5–4 dm tall with 3–5 alternate leaves and a single terminal flower; pouch white; sepals and lateral petals greenish, the lateral petals often twisted; very

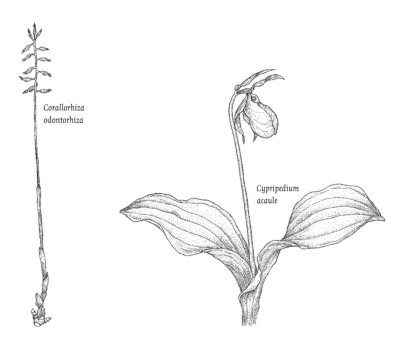

Corallorhiza odontorhiza

Cypripedium acaule

rare; known from only a single location in Lancaster Co.; believed to be extirpated; flr. May–Jun; OBL; 🌿.

Cypripedium reginae Walt. Large white or showy lady's-slipper
Stem leafy, 3–8 dm tall, bearing 1–3 flowers at the top; leaves veiny, ovate, 1–2 dm long; pouch streaked with pink or purple; sepals and lateral petals oval or oblong, white; rare in bogs, fens, and swampy woods; NW and C; flr. May–Jun; FACW; 🌿.

Epipactis Zinn

Epipactis helleborine (L.) Crantz Bastard hellebore
Stem to 8 dm tall from a rhizome; leaves cauline, sessile and clasping, reduced in size upward and becoming floral bracts, the lowest of which are longer than the flowers they subtend; flowers green veined with purple, in a many-flowered raceme; sepals and lateral petals lance-ovate; lip saccate at the base; naturalized along roadsides and forest edges; throughout; native to Europe; flr. Jun–Aug; UPL.

Epipactis helleborine,
flower (×1)

Epipactis helleborine

Goodyera pubescens

Galearis Raf.

Galearis spectabilis (L.) Raf.   Showy orchis
Stem 1–2 dm tall from fleshy roots; leaves 2, basal, ovate to broadly elliptic; inflorescence a loose raceme; flowers spurred; sepals and lateral petals pink or pale purple, converging to form a hood arching over the column; the lip white with a wavy margin; frequent in rich, deciduous forests; throughout; flr. Apr–Jul.

Goodyera R.Br.

Stems erect from basal rosettes of conspicuously white-veined leaves arising from creeping rhizomes, glandular-pubescent throughout; flowers in a terminal raceme, white; upper sepal and lateral petals fused at the margins to form a concave hood that arches forward over the column and lip; lip pouch-like at the base, narrowed to a straight or curved beak at the tip; column short; stamen 1, with 2 pollinia.

A. inflorescence densely cylindrical
   B. terminal part of the lip deflexed ....G. pubescens
   B. terminal part of the lip straight .... G. tesselata
A. inflorescence looser, spirally secund
   C. plants >20 cm tall; flower parts >5 mm long............................ G. tesselata
   C. plants <15 cm tall; flower parts <5 mm long .... G. repens

Goodyera pubescens (Willd.) R.Br.   Downy rattlesnake-plantain
Stems 2–4 dm tall with 4–14 bracts below the inflorescence, pubescent throughout, arising from rosettes of ovate to lance-ovate, dark green leaves marked with a network of white veins and a white stripes on each side of the midrib; inflorescence densely cylindrical; lip 3.5–5 mm long; frequent in dry to moist deciduous or coniferous forests; throughout; flr. Jun–Aug; FACU–.

Goodyera pubescens,
lip (×4)

Goodyera repens (L.) R.Br.   Lesser rattlesnake-plantain
Stems 1–2 dm tall with 2–4 bracts below the inflorescence; leaves ovate or oblong, the midrib green but the lateral and cross veins bordered with white; inflorescence loose, somewhat spiraled or 1-sided; lip 3–4 mm long, with a slender, deflexed beak; rare on moist forested slopes, mossy woods and swamps, usually under conifers; NE and C; flr. Jun–Sep; FACU+; ✿.

Goodyera tesselata Lodd.   Checkered rattlesnake-plantain
Stems 1.5–3.5 dm tall with 3–5 bracts below the inflorescence; leaves lacking broad white stripes along the midrib; lip 3–5.5 mm long, shallowly saccate; very rare in moist, coniferous or deciduous forests; NE; flr. Jul–early Sep; FACU–; ✿.

Goodyera tesselata, lip (×4)

Isotria Raf.

Stem erect, glabrous, arising from a cluster of fleshy roots, bearing a whorl of 5 leaves at the top; flowers 1 or 2, terminal; sepals linear, elongate; lateral petals shorter and wider; lip 3-lobed, with a prominent medial ridge; column extending forward over the lip; anther 1 with 2 pollinia.

A. sepals 3.5–6.5 cm; peduncle as long as the ovary or longer .... I. verticillata
A. sepals 1.2–2.5 cm; peduncle shorter than the ovary.................. I. medeoloides

Isotria medeoloides (Pursh) Raf.   Small whorled-pogonia
Stems glaucous, 1.5–2.5 dm tall; flowers light green, yellow-green, or whitish without darker markings; sepals 1.2–2.5 cm, spreading; very rare in dry, open oak forests; scattered; flr. late May–Jun; FACU; 🌿.

Isotria verticillata (Muhl. ex Willd.) Raf.   Whorled-pogonia
Stems 2–4 dm tall, slightly glaucous; sepals linear, 3.5–6.5 cm, greenish-brown to purple-brown, more than twice as long as the lateral petals, spreading; lateral petals greenish-yellow, fused, and extending forward over the column and lip; lateral lobes of the lip purple or purple-veined, terminal lobe yellowish-green; occasional in dry to moist, moderately acidic forests; throughout, except at the highest elevations; flr. late May–Jun; FACU.

Liparis Rich.

Stems erect from a corm or pseudobulb; leaves 2, basal; inflorescence a loose raceme subtended by a few scale-leaves on the otherwise leafless stem; sepals spreading, narrow; lateral petals linear, bending forward under the lip, rolled and twisted; lip broad; column 3–4 mm long, strongly curved, with lateral wings above; pollinia 4.

A. lip about 1 cm long, pale purple .... L. liliifolia
A. lip about 5 mm long, yellowish-green .... L. loeselii

Liparis liliifolia (L.) Rich. ex Lindl.   Lily-leaved twayblade
Stem 1–2.5 dm tall with 5–30 purplish flowers; leaves oval to elliptic, 5–15 cm long; sepals greenish-white, 1 cm, widely spreading; lateral petals very narrow, twisted, greenish to pale purple; projecting forward or deflexed and even crossing under the lip; lip pale purple, about 1 cm long and nearly as wide, almost flat; column 4 mm long; occasional in rich rocky forests, frequently on diabase; mostly S; flr. May–Jul; FACU–.

Liparis loeselii (L.) Rich.   Yellow twayblade
Stem 1–2.5 dm tall with 2–12 yellowish-green flowers; leaves lanceolate to elliptic, 5–15 cm long; sepals 5–6 mm long; lateral petals 5 mm long, reflexed; lip about 5 mm long, broadly ovate; column 3 mm long; rare in bogs, fens, wet meadows, and shaded ravines, especially on calcareous soils; scattered; flr. May–Jul; FACW.

Listera R.Br.

Stems erect with a terminal raceme of small flowers and a single pair of sessile, opposite leaves midway to the base; sepals and lateral petals similar, spreading or reflexed; lip longer than the lateral petals, projecting forward, deeply 2-lobed; column erect, shorter than the petals; the single anther bearing 2 pollinia.

A. lip deeply divided into 2 pointed lobes
     B. lip 3–5 mm long; pedicels and axis glabrous .... L. cordata
     B. lip 6–10 mm long; pedicels and axis glandular .... L. australis
A. lobes of the lip oblong or ovate, blunt... L. smallii

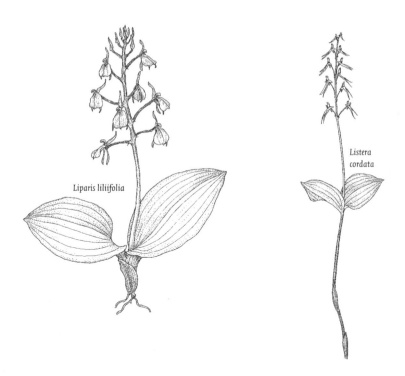

Liparis liliifolia

Listera
cordata

Listera australis Lindl.   Southern twayblade
Stems 1.5–3 dm tall, leaves ovate, 1–3.5 cm, rounded at the base; axis of the inflorescence glandular; flowers to 25, greenish-yellow to dull red; sepals and lateral petals ovate-oblong, reflexed, 2 mm long; lip 6–10 mm long with reflexed
lateral lobes at the base and the tip deeply divided into 2 slender, pointed lobes;
very rare in bogs; SE and NW; flr. May–Jun; FACW; ❧.

Listera cordata (L.) R.Br.   Heartleaf twayblade
Stem 1–2.5 dm tall; leaves 1–3 cm, broadly ovate or ovate-cordate; axis of the inflorescence glabrous; flowers to 25, green to purplish or reddish; lateral petals
2–2.5 mm long; lip 3–5 mm long, split to the middle or below into 2 linear
lobes; very rare in cool sphagnum bogs or mossy woods; mostly NE and SW; flr.
Jun–Jul; FACW+; ❧.

Listera cordata,
flower (×1)

Listera smallii Wiegand   Small's or kidney-leaved twayblade
Stem 1–3 dm tall; leaves broadly ovate to kidney-shaped, 1.5–4 cm; axis of the
inflorescence glandular; flowers 3–15; sepals and lateral petals lance-oblong, 3–
4 mm long, spreading or reflexed; lip whitish, broadly obovate, with 2 oblong
lobes, a minute tooth in the sinus, and a pair of spreading lateral teeth near the
base; column 1.5–3 mm long; very rare in damp, shady forests or bogs; SC; flr.
Jun–Jul; FACW; ❧.

Listera smallii,
flower (×1)

Malaxis Sol.

Small plants from a corm with an erect stem bearing a single leaf; flowers in a
terminal raceme which is crowded at first giving a corymbose appearance, but
elongates as the flowers mature; sepals larger than the lateral petals, lip broad,
lobed at the base; column 1 mm long or less; stamen 1, with 4 pollinia.

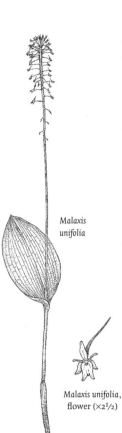

Malaxis
unifolia

Malaxis unifolia,
flower (×2½)

A. lip narrowly pointed, not lobed .... M. monophyllos
A. lip strongly lobed at the tip, with a short central tooth in the sinus
    B. flowers crowded at the apex of the inflorescence .... M. unifolia
    B. raceme elongate, slender .... M. bayardii

Malaxis bayardii Fernald   Adder's-mouth
Very similar to *M. unifolia* but with a slender, more elongate inflorescence; rare in dry, open upland forests and shale barrens; NE and SC; flr. Jul–Sep; 🌱.

Malaxis monophyllos (L.) Swartz   White adder's-mouth
Stem 1–2.5 dm tall; leaf elliptic with a sheathing base; inflorescence slender, tapering; sepals ovate to lance-ovate, 2–2.5 mm long, the central one erect, the lower deflexed; lateral petals oblanceolate, 1 mm long, spreading; lip greenish-white, deflexed, 2–3 mm long, broadly cordate and lobed at the base, pointed at the tip; very rare in wet woods, swamps and bogs; mostly N; flr. Jun–early Aug; FACW; 🌱. Ours is var. *brachypoda* (A.Gray) F.J.A.Morris.

Malaxis unifolia Michx.   Green adder's-mouth
Stem 1–3 dm tall; leaf oval or elliptic, sessile; sepals elliptic, obtuse, spreading; lateral petals linear, recurved; lip greenish, oblong and deeply lobed at the summit with a small tooth in the sinus; occasional in dry to moist forests; throughout; flr. Jun–Sep; FAC.

Platanthera Rich.

Plants glabrous, erect from a cluster of fleshy roots; leaves alternate and grading into bracts above, or basal; inflorescence a terminal raceme; sepals and lateral petals similar in shape and color, the petals fused to the upper sepal to form a hood over the column, lateral sepals spreading or recurved; lip narrow, ovate, or obovate, or 3-lobed, entire or variously fringed or toothed, and with a basal spur extending backward; column short; anther 1. Sometimes placed in the genus *Habenaria* Willd.

A. leaves basal
    B. stem lacking bracts below the inflorescence; lip tapering to a point, uncurved .... P. hookeri
    B. stem with 2 or 3 small bracts below the inflorescence; lip blunt, directed downward .... P. orbiculata
A. leaves cauline
    C. leaf solitary, sometimes with 1 or 2 small bract-like leaves above ....
    .... P. clavellata
    C. leaves several, gradually reduced upward
        D. lip entire, not fringed or lobed
            E. lip tapering to a narrow tip
                F. flowers greenish, lip lanceolate .... P. hyperborea
                F. flowers white, lip abruptly widened at the base .... P. dilatata
            E. lip broadly rounded at the tip with small lateral lobes at the base ....
            .... P. flava
        D. lip fringed and/or lobed
            G. lip deeply 3-lobed (the lobes in turn may be deeply fringed)
                H. flowers white or greenish

I. lateral petals strap-shaped .... *P. lacera*
I. lateral petals fan-shaped, broader above than at the base ....
.... *P. leucophaea*
  H. flowers magenta
    J. margins of the lobes slightly ragged. .... *P. peramoena*
    J. margins of the lobes distinctly fringed
      K. sepals and lateral petals 5–7 mm long; lateral lobes of the
        lip flat or curved backward .... *P. psycodes*
      K. sepals and lateral petals 6–10 mm long; lateral lobes of the
        lip curved forward .... *P. grandiflora*
 G. lip fringed, but not lobed
  L. flowers white .... *P. blephariglottis*
  L. flowers yellow to orange
    M. spur 5–9 mm long .... *P. cristata*
    M. spur 18–25 mm long .... *P. ciliaris*

Platanthera blephariglottis (Willd.) Lindl.   White fringed-orchid
Stem 4–10 dm tall with several alternate leaves; flowers white; sepals broadly ovate; lateral petals shorter and narrower, lip narrowly oblanceolate, 8–11 mm long, deeply fringed except at the narrowed base; spur slender and elongate; rare in sphagnum bogs and swamps; NE; flr. Jun–Aug; OBL; 🌿. Hybrids with *P. ciliaris* are known from several sites.

Platanthera ciliaris (L.) Lindl.   Yellow fringed-orchid
Stem 4–10 dm tall with several alternate leaves; flowers orange-yellow; sepals broadly oval and spreading; lateral petals shorter and narrower, toothed at the tip; lip linear oblong, 10–16 mm long, deeply fringed; spur 18–25 mm long; rare in bogs, moist meadows, and woods; E and S; flr. Jul–Aug; FACW; 🌿.

Platanthera clavellata (Michx.) Luer   Clubspur, or small green woodland orchid
Stem 1–4 dm tall with a single well-developed leaf at or below the middle; flowers white or greenish; the lip shallowly 3-toothed; spur 7–12 mm long, swollen at the tip and appearing lateral due to the twisted base of the flower; frequent in bogs, wet woods, and thickets; throughout; flr. Jun–Sep; FACW+.

Platanthera cristata (Michx.) Lindl.   Crested fringed-orchid
Stem 2–6 dm tall with several alternate leaves grading into bracts above; flowers yellow to orange; sepals rounded and spreading, lateral petals narrow and fringed; lip oblong, 7–10 mm long, deeply fringed; spur 5–9 mm long; known from a single site in Montgomery Co.; believed to be extirpated; flr. Jul–Aug; FACW+; 🌿.

Platanthera blephariglottis

Platanthera cristata, flower (×2)

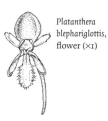

Platanthera blephariglottis, flower (×1)

Platanthera ciliaris, flower (×1)

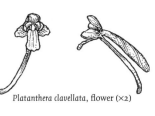

Platanthera clavellata, flower (×2)

Platanthera dilatata (Pursh) Lindl. ex Beck   Tall white bog-orchid
Stem to 1 m tall with up to 12 alternate leaves grading into bracts above; inflorescence 1–3 dm; flowers white; lateral petals arched forward with the upper sepal forming a hood over the column; lip 6–8 mm long, tapering from the wider base to a blunt tip; very rare in bogs, marshes, or wet meadows; NW; flr. Jun; FACW; 🌱.

Platanthera flava (L.) Lindl.   Tubercled rein-orchid
Stem 3–7 dm tall with several well-developed, lance-linear leaves; inflorescence 2–6 cm; flowers yellow-green; lip 3–4 mm long, deflexed with an irregular margin, a small lateral lobe on each side at the base, and a rounded projection (tubercle) on the upper surface just below the middle; occasional in swamps, bogs, or wet, open woods; throughout; flr. May–Aug; FACW. Ours is var. *herbiola* (R.Br.) Luer.

Platanthera grandiflora (Bigelow) Lindl.   Large purple fringed-orchid
Very similar to *P. psycodes*, but the flowers bigger; sepals 6–10 mm long; lip 18–25 mm long, its lateral lobes curved forward; rare in rich woods, thickets and meadows; throughout; flr. Jun–Aug; FACW.

Platanthera hookeri (Torr. ex A.Gray) Rydb.   Hooker's orchid
Stem to 2–4 dm tall, lacking bracts above the single pair of broadly rounded basal leaves; flowers sessile, yellowish-green; lip narrowly triangular, upcurved; spur 13–25 mm long, directed downward; rare in rich, well-drained deciduous woods; mostly N; flr. May–Jul; FAC; 🌱.

Platanthera hyperborea (L.) Lindl.   Tall green bog-orchid
Stem stout, 3–10 dm tall with alternate leaves grading upward into bracts; inflorescence 6–20 cm, compact; flowers green; lip lance-ovate, 4–7 mm long, blunt and gradually widened toward the base; spur about as long as the lip; very rare in wet meadows, bogs, and woods; NE and NW; flr. Jun–Aug; FACW; 🌱. Ours is var. *huronensis* (Nutt.) Farw.

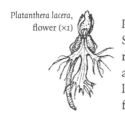
Platanthera lacera, flower (×1)

Platanthera lacera (Michx.) G.Don   Ragged fringed-orchid
Stem 3–8 dm tall with several alternate leaves grading into bracts above; inflorescence 5–15 cm; flowers greenish; sepals ovate; lateral petals narrower but about as long; lip deeply 3-lobed and each lobe deeply fringed; spur 14–21 mm long; frequent in open woods, moist meadows, bogs, and ditches; throughout; flr. Jun–Sep; FACW.

Platanthera leucophaea (Nutt.) Lindl.   Eastern prairie fringed-orchid
Stem 4–10 dm tall with alternate leaves; inflorescence 8–20 cm; flowers creamy white; sepals broadly oval; lateral petals slightly longer, spatulate; lip deeply 3-lobed and fringed; very rare in damp, calcareous meadows, and shores; NW; believed to be extirpated; flr. Jun–Jul; FACW+; 🌱.

Platanthera orbiculata, flower (×1)

Platanthera orbiculata (Pursh) Lindl.   Large round-leaved orchid
Stems 0.6–5 dm tall with a pair of broadly rounded basal leaves that lie flat on the ground and several small bracts below the inflorescence; flowers white or greenish; lip oblong-linear, pointing downward; spur 15–25 mm long; moist shady forests; flr. Jun–Aug; FAC; 2 varieties:

A. leaves 10 by 8 cm; stem to 40 cm with about 10 flowers .... var. *orbiculata*
   occasional; throughout, except in the SE.
A. leaves 20 by 15 cm; stem to 50 cm with about 20 flowers ....
   .... var. *macrophylla* (Goldie) Luer
   vary rare and scattered.

Platanthera peramoena (A.Gray) A.Gray   Purple fringeless orchid
Stem 3–10 dm tall, with alternate leaves, the upper reduced; inflorescence 6–18
cm; flowers magenta; sepals and lateral petals similar in length; lip deeply 3-
lobed with a ragged, but not fringed, margin; spur 22–31 mm long; rare in moist
meadows, low wet woods and ditches; S; flr. Jul–Aug; FACW; ✿.

Platanthera peramoena,
flower (×1)

Platanthera psycodes (L.) Lindl.   Purple fringed-orchid
Stem 3–15 dm tall, stout, with several alternate leaves; inflorescence dense and
cylindric; flowers magenta; sepals 5–7 mm long, broadly oval; lateral petals
oblong-spatulate and toothed; lip deeply 3-lobed and fringed, its lateral lobes
flat or curved backward; occasional in bogs, swamps, damp meadows, and open
woods; throughout, except SW; flr. Jun–Aug; FACW.

Platanthera psycodes,
flower (×1)

Pogonia Juss.

Pogonia ophioglossoides (L.) Ker Gawl.   Rose pogonia
Stem 1–4 dm tall from a short rhizome, with a lanceolate leaf near the middle
and a smaller one subtending the solitary (rarely 2), terminal, pink, flower; lip
veined with red, fringed, and bearded with pink and yellow hairs; occasional in
sphagnum bogs and boggy meadows; E, SW, and NW; flr. Jun–Jul; OBL.

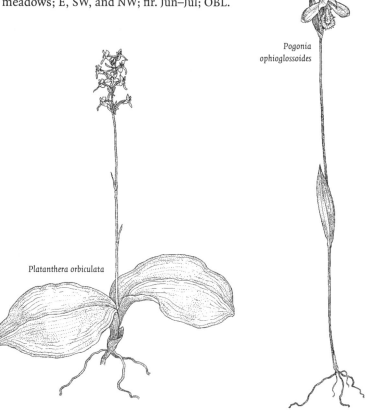

*Pogonia
ophioglossoides*

Platanthera orbiculata

Spiranthes Rich.

Stems erect; leaves narrow, mostly toward the base of the stem, alternate, with bladeless sheaths above; inflorescence a densely or loosely spiraled (or merely secund), spike-like raceme; flowers white or whitish; upper sepal (or all sepals) and lateral petals fused to form a hooded structure that projects forward over the lip and column; lip oblong or ovate, bearing 2 small swellings (tuberosities) at the base; anther on the back of the short column; pollinia 4.

A. lip constricted just above the middle and strongly expanded above; sepals and lateral petal fused to form a tubular hood .... S. romanzoffiana
A. lip not as above, if constricted near the middle then not expanded above
    B. sepals fused at the base; lip bright yellow .... S. lucida
    B. sepals distinct; lip whitish, pale yellow, or green
        C. axis of the inflorescence glabrous.. S. tuberosa
        C. axis of the inflorescence short-hairy or glandular
            D. axis of the inflorescence with dense, short, nonglandular hairs ....
            .... S. vernalis
            D. axis of the inflorescence sparsely to densely covered with short glandular-hairs
                E. upper sepal >7.5 mm long
                    F. leaves present at anthesis
                        G. base of the lip dilated, tuberosities small; flowers white ....
                        .... S. cernua
                        G. base of the lip not dilated, tuberosities large; flowers yellowish .... S. ochroleuca
                    F. leaves absent at anthesis .... S. magnicamporum
                E. upper sepal <7.5 mm long
                    H. inflorescence dense, the flowers in 3 or 4 vertical rows ....
                    ....S. ovalis
                    H. inflorescence mostly looser, the flowers clearly in a single spiral or merely secund
                        I. basal leaves erect, 5–10 times as long as wide; lip yellowish .... S. casei
                        I. basal leaves spreading, 1.5–3.5 times as long as wide; lip green ....S. lacera

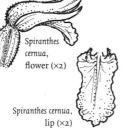

Spiranthes cernua, flower (×2)

Spiranthes cernua, lip (×2)

Spiranthes casei Catling & Cruise   Case's ladies'-tresses
Stems 2–4 dm tall; basal leaves lanceolate or lance-ovate, 5–15 cm; inflorescence 2–16 cm, loosely to densely flowered; flowers creamy or greenish-yellow, 5–7.5 mm long; lip fleshy, obovate or elliptic-ovate, the basal tuberosities about 1 mm long, longer than wide; very rare in dry open sandy soil; NC; flr. Aug–Sep; ✹.

Spiranthes cernua (L.) Rich.   Nodding ladies'-tresses
Stems 1–4 dm tall with narrow basal leaves to 30 cm; inflorescence 2–18 cm, dense, with glandular-pubescent axis; flowers white, in 3 or 4 vertical ranks; lip moderately recurved, yellowish-green at the center, basal tuberosities 1 mm long or more; common in moist, acidic soils of meadows, open woods and roadsides; throughout; flr. late Jul–Oct; FACW.

Spiranthes lacera (Raf.) Raf.   Northern slender ladies'-tresses
Stems slender, 1–4 dm tall; basal leaves ovate-elliptic or lanceolate, spreading to

Spiranthes
ochroleuca

form a rosette, but sometimes withering early; axis of the inflorescence sparsely and minutely glandular-hairy; flowers in a single loose or compact spiral, white with a green patch on the lip; occasional in open woods and grassy meadows; flr. Aug–Oct; FACU–; 2 varieties:

A. inflorescence loosely spiraled, lower flowers distinctly spaced .... var. *lacera* E and C.
A. inflorescence densely spiraled, flowers closely spaced ....
.... var. *gracilis* (Bigelow) Luer
mostly S.

Spiranthes lucida (H.H.Eaton) Ames   Shining ladies'-tresses
Stem 1–2.5 dm tall; basal leaves lance-oblong, to 12 cm; inflorescence 2–7 cm, dense, with minutely glandular-hairy axis; flowers in 3 vertical ranks, white with an orange-yellow patch on the lip; rare in moist shores and wet meadows, mostly on calcareous soils; S and W; FACW; flr. May–Jul; 🍂.

Spiranthes magnicamporum Sheviak   Great Plains ladies'-tresses
Very similar to S. *cernua* but with leaves absent at flowering time and lip not constricted near the center; very rare in wet meadows and fens on calcareous soils; SE; believed to be extirpated; flr. Sep–Oct; FACU–; 🍂.

Spiranthes ochroleuca (Rydb.) Rydb.   Yellow nodding ladies'-tresses
Very similar to S. *cernua* but flowers yellowish and the lip not dilated at the base;

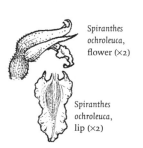

Spiranthes
ochroleuca,
flower (×2)

Spiranthes
ochroleuca,
lip (×2)

common in woods and roadsides; throughout; flr. late Jul–Oct. [syn: *S. cernua* (L.) Rich, in part]

Spiranthes ovalis, flower (×2)

Spiranthes ovalis Lindl.   October ladies'-tresses
Stems 1.5–3 dm tall with 2 or 3 basal leaves; inflorescence 2–6 cm, with glandular-puberulent axis; the flowers in 3 to 4 vertical ranks, white or the lip creamy; lip rounded at the end and narrowed at the middle, with short basal tuberosities; very rare in damp, humus-rich forests; SC; flr. Oct; FAC; ❧. Ours is var. *erostellata* Catling.

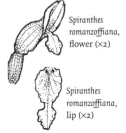

Spiranthes romanzoffiana, flower (×2)

Spiranthes romanzoffiana, lip (×2)

Spiranthes romanzoffiana Cham.   Hooded ladies'-tresses
Stems 1–4 dm tall with narrow basal leaves to 20 cm long; inflorescence 3–10 cm, dense, the flowers in 3 to 4 vertical ranks, white or creamy; sepals basally connate and fused to the lateral petals to form an open tubular hood; lip sharply deflexed, with an expanded tip, basal tuberosities very small; very rare in bogs or rich, open woods; NW; flr. Jul–Aug; OBL; ❧.

Spiranthes tuberosa Raf.   Slender ladies'-tresses
Stems slender 1.5–3 dm tall; basal leaves short-lived, ovate with a short petiole; inflorescence 2–8 cm with a glabrous axis, the flowers arranged in a single long spiral; flowers pure white, the lip broadly ovate, rounded or truncate at the tip; the basal tuberosities <1 mm long; very rare in grassy meadows and open woods; SE; flr. late Aug–Oct; FACU–; ❧.

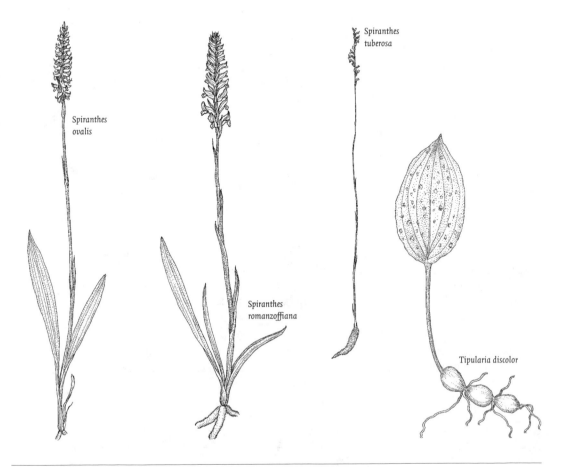

Spiranthes ovalis

Spiranthes romanzoffiana

Spiranthes tuberosa

Tipularia discolor

Spiranthes vernalis Engelm. & A.Gray   Spring ladies'-tresses
Stems 2–6 dm tall; basal leaves narrow, 5–25 cm long; inflorescence 3–15 cm, its
axis densely covered with pointed, nonglandular hairs; flowers in a single long
spiral, ivory or white; lip 5–8 mm long, white with a yellowish center, its basal
tuberosities 0.5–1 mm long, longer than thick; rare or moist, open sandy soils
and serpentine barrens; flr. May–Aug; FAC; 🌿.

Tipularia Nutt.

Tipularia discolor (Pursh) Nutt.   Cranefly orchid
Stem 2–5 dm tall, from a corm; the single leaf present only in the fall and winter,
ovate, green above and purple beneath; raceme without bracts; flowers greenish-
purple; the lip 3-lobed and spurred at the base; rare in deciduous forests and
stream banks; SE; flr. Jul–Aug; FACU; 🌿.

Tipularia discolor,
flower

Triphora Nutt.

Triphora trianthophora (Swartz) Rydb.   Nodding pogonia, three-birds orchid
Stem 1–3 dm tall, from a cluster of fleshy roots; leaves several, alternate; flowers
axillary, white or pale pink, short-lived; lip with 3 prominent green ridges; rare
in humus-rich, moist forests; SE; flr, Jul–Aug; UPL; 🌿.

POACEAE Grass Family

Flowers perfect or occasionally unisexual, without petals or sepals, each flower
subtended by a set of bracts, the lemma and palea, the flower with its bracts is
referred to as a floret; florets arranged 1–many in spikelets; each spikelet subtended by a pair of bracts, the glumes; spikelets variously arranged in panicles,
racemes or spikes; stamens 1–6 (mostly 3); pistil 1 with 1 ovary and 1 ovule, generally 2 (sometimes 1 or 3) styles with plumose stigmas; fruit a caryopsis; plants
usually herbaceous with 2-ranked leaves along an often hollow culm; leaves consisting of the usually flat blade, the sheath that surrounds the culm, and the
ligule, a membranaceous or hairy appendage at the base of the blade (sometimes
absent); plants annual or perennial, often rhizomatous or densely tufted, sometimes sod-forming.

Identification of grasses relies on the presence of flowers and, in some cases,
mature fruit. The dates given in the descriptions represent periods during which
intact inflorescences can be expected to be found; actual flowering times are
generally much shorter and toward the beginning of the range of dates given.

Key to groups of genera

A. culms woody, perennial; rarely flowering .... *Pseudosasa*
A. culms herbaceous, annual; generally regularly flowering
    B. tall, stout grasses, commonly to 2 m or more tall .... Key 1
    B. grasses of small to moderate size, usually <2 m tall
        C. spikelets concealed in a bur-like, spiny structure

   D. burs 8–15 mm long .... *Cenchrus*

   D. burs 3–4 mm long .... *Tragus*

  C. spikelets not concealed in a bur-like structure

    E. spikelets all unisexual, the staminate and pistillate ones grouped separately in the same or different inflorescences (monoecious), or on separate plants (dioecious) .... Key 2

    E. at least some spikelets with perfect flowers

      F. spikelets with 1–many florets, mostly laterally compressed or terete; reduced or sterile florets, if present, usually found above the fertile florets; mature spikelets usually (but not always) disarticulating above the glumes, the glumes remaining on the inflorescence

        G. inflorescence a panicle (sometimes contracted or spike-like, but the spikelets never sessile and 2-ranked)

          H. glumes lacking or reduced to a ring below the lemma and palea .... *Leersia*

          H. glumes present, although sometimes very small

            I. spikelets with 1 perfect floret only, no staminate, sterile, or reduced florets present .... Key 3

            I. spikelets with 2–many florets including perfect, staminate, reduced, or sterile florets

              J. glumes not equaling the lowermost floret .... Key 4

              J. second glume equaling or exceeding the lowermost floret (not equaling in *Sphenopholis* with obovate second glume), usually equaling the spikelet

                K. spikelets with 2–many florets, sterile florets, if present, above the fertile ones and similar to them (lower floret staminate in *Arrhenatherum*) .... Key 5

                K. spikelets containing 1 awnless perfect terminal floret, with 1 or 2 dissimilar sterile or staminate florets below it, these often hairy or reduced to minute scales .... Key 6

        G. inflorescence of 1 or more spikes, the spikelets sessile

          L. spike solitary, terminal, bearing 1–3 spikelets at each node, these 2-ranked on opposite sides of the usually zigzag, continuous or articulated rachis .... Key 7

          L. spikes several, digitate or racemose, bearing spikelets in 2 rows along their lower side .... Key 8

      F. spikelets with 1 perfect terminal floret and a sterile (rarely staminate) floret below, dorsally compressed; mature spikelets generally disarticulating below the glumes

        M. spikelets all alike, in panicles or spike-like racemes; glumes and sterile lemma membranaceous; fertile lemma and palea firm or coriaceous .... Key 9

        M. spikelets paired at each joint of a raceme, the members of each pair unequal, usually 1 fertile and sessile, the other staminate and pedicellate (sometimes fertile, sterile, or reduced to only the pedicel); glumes coriaceous; sterile lemma, fertile lemma, and palea thin and membranaceous .... Key 10

Key 1

A. at least some spikelets with perfect flowers
  B. spikelets with 2–many florets .... *Phragmites*
  B. spikelets with 1 floret only, or 1 perfect terminal floret and a sterile (rarely staminate) floret below
    C. spikelets with 1 perfect terminal floret and a sterile (rarely staminate) floret below .... *Panicum*
    C. spikelets with 1 floret only
      D. inflorescence with silky, grayish or silvery hairs concealing the spikelets .... *Miscanthus*
      D. inflorescence glabrous or pubescent, but not silky
        E. inflorescence consisting of 1 or more racemes of many pairs of spikelets
          F. spikelets of each pair equal and fertile .... *Erianthus*
          F. spikelets of each pair unequal, one greatly reduced and usually infertile .... *Andropogon*
        E. inflorescence a panicle of racemes, each of which is reduced to 1 or a few pairs of spikelets
          G. each sessile, perfect-flowered, awned spikelet accompanied by an awnless staminate one .... *Sorghum*
          G. each sessile spikelet accompanied by a hairy sterile pedicel, lacking a spikelet .... *Sorghastrum*
A. staminate and pistillate flowers separate, but in the same inflorescence
  H. inflorescence an open panicle .... *Zizania*
  H. inflorescence of 1–several unbranched spikes .... *Tripsacum*

Key 2

A. plants dioecious .... *Distichlis*
A. plants monoecious
  B. pistillate portion of the inflorescence 1–2-flowered, enclosed in a bony, bead-like structure .... *Coix*
  B. pistillate portion of the inflorescence several- to many-flowered
    C. pistillate portion of the inflorescence cylindrical, bony, the spikelets sunken into cavities in the rachis, at maturity this structure breaking into 1-seeded segments .... *Tripsacum*
    C. pistillate portion of the inflorescence not bony, the spikelets not sunken into cavities in the rachis .... *Zizania*

Key 3

A. inflorescence dense and spike-like, ovoid to cylindric, symmetrical, unlobed
  B. lemma 9 mm or more long .... *Ammophila*
  B. lemma much <9 mm long
    C. glumes awned .... *Phleum*
    C. glumes not awned
      D. lemmas longer than to about equaling the glumes
        E. lemmas longer than the glumes .... *Crypsis*
        E. lemmas about equaling the glumes .... *Alopecurus*
      D. lemmas much shorter than the glumes .... *Polypogon*

A. inflorescence loose and open, or if dense and spike-like, then asymmetrical, lobed or irregular
    F. lemma membranous or herbaceous at maturity, essentially like the glumes in texture, often with conspicuous nerves
        G. lemma surrounded at the base with a tuft of hairs, persisting on the inflorescence when the lemma is detached
            H. lemma with a short, delicate awn .... *Calamagrostis*
            H. lemma awnless
                I. leaf blades 3–8 mm wide .... *Calamovilfa*
                I. leaf blades 10–15 mm wide .... *Phalaris*
        G. lemma glabrous or pubescent, but not surrounded at the base by a tuft of hairs
            J. lemma awned
                K. awn attached below the apex of the lemma
                    L. spikelets disarticulating above the glumes .... *Agrostis*
                    L. spikelets disarticulating below the glumes .... *Cinna*
                K. awn attached at the apex of the lemma
                    M. body of the lemma <5 mm long; rachilla not prolonged behind the palea .... *Muhlenbergia*
                    M. body of the lemma 6–10 mm long; rachilla prolonged behind the palea as a naked bristle about 6 mm long .... *Brachyelytrum*
            J. lemma awnless
                N. lemma 3-nerved
                    O. first glume shorter than or rarely equaling the lemma, acuminate with a sharp tip; palea about equaling the lemma .... *Muhlenbergia*
                    O. first glume longer than the lemma, acute to acuminate but without a sharp tip; palea much shorter than the lemma or absent .... *Agrostis*
                N. lemma 1-nerved .... *Sporobolus*
    F. lemma hardened, distinctly unlike the glumes in texture when mature, rarely with conspicuous nerves
        P. lemma awnless .... *Milium*
        P. lemma awned
             Q. awn unbranched
                R. awn <3 times as long as the lemma .... *Oryzopsis*
                R. awn 4–10 times as long as the lemma
                  S. glumes 1–1.3 cm long, thin and translucent, usually purplish at the base .... *Piptochaetium*
                  S. glumes 2.5–4 cm long, firm and greenish .... *Stipa*
             Q. awn 3-branched .... *Aristida*

Key 4

A. lemmas 1–3-nerved, the nerves mostly prominent
    B. spikelets paired, one sessile and fertile, the other short-pedicellate and sterile; fertile lemmas 1-nerved .... *Cynosurus*
    B. spikelets all alike and fertile; lemmas 3-nerved
        C. lemma 2-lobed, the midnerve excurrent as a short point, the nerves prominently pubescent

D. plants perennial, with large exserted open panicles; palea not long-ciliate on the upper ½ .... Tridens

D. plants annual; terminal panicles small; palea long-ciliate on the upper ½ .... Triplasis

  C. lemma not lobed, the nerves glabrous

    E. mature spikelets stiff, with large, protruding, bottle-shaped grains .... Diarrhena

    E. mature spikelets not stiff, the grains neither protruding nor bottle-shaped

      F. lemmas not awned .... Eragrostis

      F. at least the upper lemma with a prominent awn .... Sphenopholis

A. lemmas 5–many-nerved or the nerves obscure

  G. sheaths united for at least ½ their length

    H. spikelets strongly flattened, in dense, one-sided clusters near the tips of the few, rigid panicles .... Dactylis

    H. spikelets not strongly flattened, in open or somewhat contracted panicles, not crowded in dense, one-sided clusters

    I. callus of the lemmas pilose or cobwebby .... Schizachne

    I. callus of the lemmas not pilose or cobwebby

      J. upper lemmas of the spikelet closely imbricated, forming a club-shaped sterile rudiment .... Melica

      J. upper lemmas of the spikelet gradually reduced, not forming a club-shaped rudiment

        K. lemmas awned .... Bromus

        K. lemmas not awned

          L. lemmas inflated, very broad .... Bromus

          L. lemmas not inflated, narrow .... Glyceria

  G. sheaths free except at the base

    M. lowermost floret of the spikelet sterile .... Chasmanthium

    M. lowermost floret of the spikelet fertile

    N. lemmas as broad as long, spreading perpendicular to the rachilla ... . Briza

    N. lemmas longer than broad, not spreading perpendicular to the rachilla

      O. lemmas rounded at apex, not awned

        P. lemmas prominently 5-nerved .... Torreyochloa

        P. lemmas obscurely nerved .... Puccinellia

      O. lemmas acute at apex or awned

        Q. callus of the lemmas pilose or cobwebby .... Poa

        Q. callus of the lemmas not pilose or cobwebby

          R. lemmas awned

            S. stamen 1; spikelets cleistogamous; plants annual .... Vulpia

            S. stamens 3; spikelets chasmogamous; plants perennial .... Festuca

          R. lemmas not awned

            T. lemmas keeled, pubescent on the nerves and keel, membranaceous .... Poa

            T. lemmas rounded on the back, glabrous, papery .... Festuca

Key 5

A. spikelets with 2 lemmas, one enclosing a perfect flower, the other enclosing a staminate flower
    B. lowest lemma awned from below the middle, the second from near the apex .... *Arrhenatherum*
    B. lowest lemma awnless, the second awned from just below the apex .... *Holcus*
A. spikelets with 2 or more lemmas, each enclosing a perfect flower
    C. awn arising at or below the middle of the body of the lemma .... *Deschampsia*
    C. awn absent, or terminal, or arising between the teeth of the lemma, or arising from the back of the lemma shortly below the apex
        D. glumes to 7 mm long
            E. panicle loose, nodding .... *Sphenopholis*
            E. panicle contracted and spike-like
                F. lemmas awnless .... *Koeleria*
                F. lemmas awned
                    G. spikelets disarticulating above the glumes .... *Trisetum*
                    G. spikelets disarticulating below the glumes .... *Sphenopholis*
        D. glumes 8.5 mm or longer
            H. glumes 7–11-nerved .... *Avena*
            H. glumes 1–3-nerved .... *Danthonia*

Key 6

A. the 2 lower florets as long or longer than the terminal fertile floret
    B. lower florets awnless, staminate .... *Hierochloe*
    B. lower florets awned, sterile .... *Anthoxanthum*
A. lower florets (1 or 2) <½ as long as the fertile floret .... *Phalaris*

Key 7

A. spikelets solitary at all nodes of the spike
    B. spikelets 2–several-flowered, not sunken into the rachis
        C. spikelets placed flat-wise to the rachis; both glumes developed
            D. glumes bristle-like .... *Secale*
            D. glumes lanceolate to oblong or ovate
                E. glumes lanceolate, acute, more than twice as long as broad; grain permanently enclosed by the lemma and palea
                    F. plants rhizomatous .... *Elytrigia*
                    F. plants not rhizomatous .... *Elymus*
                E. glumes broadly ovate, less than twice as long as broad; grain readily dropping from the floret at maturity .... *Triticum*
        C. spikelets placed edge-wise to the rachis, the first glume lacking except in the terminal spikelet .... *Lolium*
    B. spikelets 1–several-flowered, sunken into the thickened rachis ..... *Aegilops*

A. spikelets 2–several at a node
    G. spikelets 3 at a node
        H. rachis of the spikes disarticulating at maturity, the spikelets falling attached to the joints; lemma of central spikelet of each group usually <8 mm long .... *Critesion*
        H. rachis of the spikes not disarticulating; lemma of the central spikelet of each group >1 cm long .... *Hordeum*
    G. spikelets 2 at a node .... *Elymus*

Key 8

A. spikes closely aggregated at the summit of the culm
    B. spikelets with only 1 fertile lemma (1 or more sterile lemmas may be present)
        C. spikelets with only a fertile lemma (no sterile lemmas) .... *Cynodon*
        C. spikelets with 1 fertile and 1 or more sterile lemmas .... *Chloris*
    B. spikelets with 2 or more fertile lemmas .... *Eleusine*
A. spikes in a raceme, panicle or spike
    D. spikelets arranged on 2 sides of a 3-sided rachis
        E. spikelets densely crowded .... *Spartina*
        E. spikelets loosely arranged .... *Gymnopogon*
    D. spikelets inserted in 2 rows on 1 side of a narrow, flat rachis
        F. spikelets with 1 perfect floret .... *Bouteloua*
        F. spikelets with 2–3 perfect florets .... *Leptochloa*

Key 9

A. spikelets surrounded by an involucre of bristles
    B. spikelet or fruit falling leaving the bristles persistent on the panicle .... *Setaria*
    B. spikelet and the ring of bristles falling as a unit .... *Pennisetum*
A. spikelets not surrounded by an involucre of bristles
    C. spikelets arranged in 1–several slender, spike-like racemes
        D. rachilla joint thickened, forming a ring-like callus below the second glume, the spikelet appearing set in a shallow cup atop the pedicel ..... *Eriochloa*
        D. rachilla joint not thickened or not evident
            E. spikelets elliptic or narrowly ovate, acute; margins of the fertile lemma thin, not inrolled; fruit leathery and flexible .... *Digitaria*
            E. spikelets broadly elliptic or subcircular, usually obtuse; margins of the fertile lemma inrolled; fruit rigid .... *Paspalum*
    C. spikelets arranged in panicles
        F. spikelets not awned; fertile lemma obtuse to acute
            G. margins of the fertile lemma inrolled; first glume developed .... *Panicum*
            G. margins of the fertile lemma not inrolled; first glume minute or absent .... *Leptoloma*
        F. spikelets awned or strongly awn-tipped; fertile lemma apiculate .... *Echinochloa*

Key 10

A. pedicellate spikelet fertile, similar to the sessile one .... *Microstegium*
A. pedicellate spikelets reduced, staminate, neuter, or sometimes lacking
    B. pedicellate spikelet present in all pairs; rachis joints flattened; erect perennial grasses
        C. racemes solitary on each peduncle .... *Schizachyrium*
        C. racemes 2–several on each peduncle .... *Andropogon*
    B. pedicellate spikelet and its pedicel present only in the basal pairs of the raceme, or lacking; rachis filiform; low decumbent annuals, rooting at the nodes .... *Arthraxon*

Aegilops L.

*Aegilops cylindrica,
spikelet ×2*

Aegilops cylindrica Host   Jointed goatgrass
Annual; culms erect, to 6 dm tall; blades 2–3 mm wide; glumes keeled at one side and awned; lemmas mucronate at the tip or the uppermost awned, the awns very scabrous; spikelets 2–5-flowered, solitary at each joint of the axis, placed flatwise at each joint and fitting into the rachis, spikelets usually not reaching the one above on the same side thus exposing the rachis; waste ground, rubbish dumps, and fill; mostly SE; Jun–Jul; native to Eurasia.

Agrostis L.

Culms usually cespitose; spikelets 1-flowered, disarticulating above the glumes; glumes usually nearly equal, acute to acuminate, 1-nerved; lemma shorter than to nearly equaling the glumes, obscurely nerved; palea delicate, or reduced to absent in some species.

A. palea at least ½ as long as the lemma
    B. panicle branches naked to the base; leaf blades mostly 1–3 mm wide .... *A. capillaris*
    B. at least some of the panicle branches bearing spikelets to the base; leaf blades mostly 2–6(8) mm wide
        C. plants rhizomatous; the culms erect; panicle branches spreading at maturity .... *A. gigantea*
        C. plants not rhizomatous (although sometimes stoloniferous); the culms decumbent, rooting at the nodes; panicle narrow, contracted after flowering .... *A. stolonifera*
A. palea <¼ as long as the lemma or lacking
    D. panicle very diffuse, its branches branching only above the middle or toward the tip
        E. spikelets short-pediceled to subsessile, clustered near the tips of the panicle branches; flr. May–early June .... *A. hyemalis*
        E. spikelets on pedicels averaging 2 mm or more long, scattered along the panicle branches; flr July–September .... *A. scabra*
    D. panicle open, its branches branching from near or below the middle
        F. lemma awnless or with a very short straight awn from near the tip
            G. spikelets 2.7–3.5 mm long, clustered near the tips of the panicle branches .... *A. altissima*
            G. spikelets 1.8–2.8 mm long, scattered along the panicle branches .... *A. perennans*

F. lemmas bearing awns 3–10 mm long
    H. awns 3–4.5 mm long, bent or twisted .... *A. canina*
    H. lemma with a long flexuous awn (to about 10 mm) from just below
       the tip ....*A. elliottiana*

Agrostis altissima (Walter) Tuck.   Tall bentgrass
Perennial; similar to *A. perennans* but with shorter pedicels resulting in the spike-lets being aggregated in terminal spike-like clusters; sphagnum bogs, swales, or swamps; Montgomery Co.; believed to be extirpated; FACU; ✹. [syn: *A. perennans* (Walter) Tuck. var. *elata* (Pursh) A.Hitchc.]

Agrostis canina L.   Brown bent, velvet bent
Perennial; culms 2–6 dm tall; blades flat, 4–6 cm long by 1–3 mm wide; panicle 5–10 cm long, with scabrous branches; spikelets 1.9–2.5 mm long; lemma about 1.6 mm long with a bent awn attached at or near the middle; low, sandy or peaty soil; Bucks, Northampton, and Pike Cos.; May–Jul; native to Europe; FACU.

Agrostis caņina,
spikelet (×5)

Agrostis capillaris L.   Rhode Island bent
Perennial; culms 2–6 dm tall, loosely tufted, often with creeping stolons; blades 2–5 mm wide; ligule truncate, <1 mm long; panicle not contracted after flower-

Agrostis capillaris, spikelet (×5)

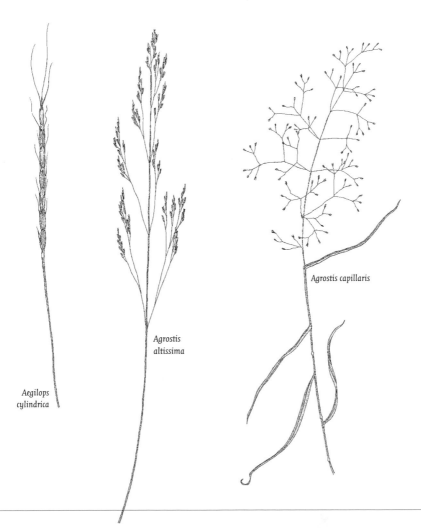

Aegilops
cylindrica

Agrostis
altissima

Agrostis capillaris

ing; spikelets 2–3 mm long; glumes scabrous on the midnerve; lemma 1.5–2.5 mm long; palea about ½ as long as the lemma; cultivated and occasionally established in dry, open ground along roadsides; throughout; Jun–Aug; native to Europe. [syn: *A. tenuis* Sibth.]

Agrostis elliottiana Schult.   Bentgrass
Annual; culms 1–6 dm tall; blades flat, 1–2 mm wide; panicle half or more as long as the entire plant; glumes 1.6–2 mm long, nearly glabrous on the midnerve; lemma 1.2–1.6 mm long, pilose at the base, truncate at the apex, with 2 minute setae; dry, open soil; Bucks and Montgomery Cos.; May–Jul; native south of Pennsylvania.

Agrostis gigantea Roth   Redtop
Perennial; culms to 1.5 m tall, often decumbent at the base, with strong creeping rhizomes; blades flat, 5–10 mm wide; panicle to 20 cm long, spreading in flower and usually contracting in fruit; cultivated and frequently established in moist soil of fields, roadsides, and waste ground; throughout; Jun–Aug; native to Europe; FACW–. [syn: *A. alba* L.]

Agrostis hyemalis (Walter) Britton, Stearns & Poggenb.   Hairgrass, ticklegrass
Perennial; culms 3–7 dm tall; leaves crowded toward the base of the plant,

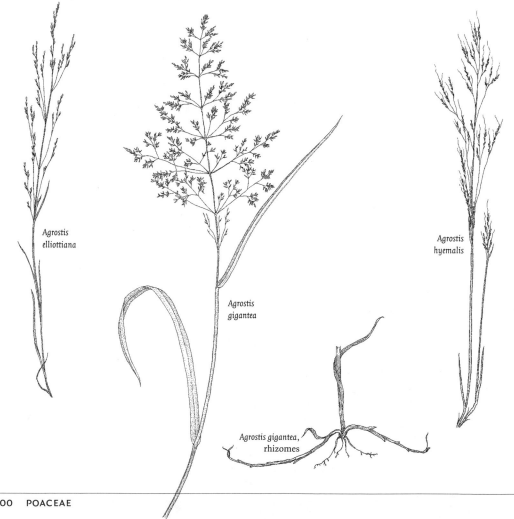

Agrostis hyemalis,
spikelet (×5)

*Agrostis
elliottiana*

*Agrostis
gigantea*

Agrostis gigantea,
rhizomes

*Agrostis
hyemalis*

Agrostis scabra

Agrostis perennans

Agrostis stolonifera

Agrostis stolonifera, rhizomes

blades involute, 1 mm or less wide (1–3 mm wide when unrolled); panicle 1–3 dm long, very diffuse, about ½ as long as the entire plant; spikelets 1.2–3 mm long; glumes scabrous on the midnerve; lemma 1–1.2 mm long; dry to moist, open, sterile soil; throughout but uncommon N; late May–early Jun; FAC.

Agrostis perennans (Walter) Tuck. Autumn bent, upland bent
Perennial; culms 5–10 dm tall; blades flat, 3–6 mm wide; panicle 10–25 cm long, branches scabrous; spikelets 1.8–2.8 mm long; glumes scabrous on the midnerve; lemma 1.3–2 mm long, usually awnless; dry, open ground or in light shade; throughout; Jul–Oct; FACU.

Agrostis perennans, spikelet (×5)

Agrostis scabra Willd. Fly-away grass, hairgrass, ticklegrass
Perennial; culms 3–8 dm tall; blades flat, 1–3 mm wide, scabrous; ligule 2–5 mm long; panicle 15–25 cm long, scabrous; spikelets 2–2.7 mm long, clustered near the ends of the panicle branches; glumes unequal, scabrous on the midnerve; lemma 1.5–1.7 mm long; moist, sandy-peaty ground and barrens; throughout but rare on the Coastal Plain and Piedmont; late Jun–Sep; FAC.

Agrostis scabra, spikelet (×5)

Agrostis stolonifera L. Carpet bentgrass, creeping bentgrass
Perennial; culms 2–5 dm tall, loosely tufted from a creeping rhizome; blades

Alopecurus
carolinianus

Ammophila
breviligulata

Andropogon
gerardii

flat, 1–3 mm wide; ligule 2–6 mm long; panicle 5–15 cm long, some branches bearing spikelets to the base; spikelets 2–2.5 mm long; glumes scabrous on the midnerve; lemma about 2/3 as long as the glumes; cultivated and established in wet meadows or on shores; throughout; flr. late Jun–early Sep; native to Europe; FACW. Ours is var. *palustris* (Huds.) Farw. [syn: *A. palustris* Huds.]

Alopecurus L.

Spikelets 1-flowered, disarticulating below the glumes; glumes equal, 3-nerved, strongly compressed, and keeled; lemma thin, 5-nerved, about as long as the glumes; palea usually absent.

A. spikelets 2–3 mm long
    B. awn not surpassing the glume or surpassing the glume by <1 mm............
    .... *A. aequalis*
    B. awn surpassing the glume by 1.5–3.5 mm .... *A. carolinianus*
A. spikelets 4–6 mm long .... *A. pratensis*

Alopecurus aequalis Sobol.   Short-awned foxtail
Annual or short-lived perennial; culms 2–5 dm tall, solitary or in small tufts; blades 1–4 mm wide; inflorescence 2–8 cm long by 3–5 mm wide; spikelets about 2 mm long; uncommon in swamps, ditches, and moist meadows; scattered throughout; late May–Jun; OBL; 🌿.

Alopecurus aequalis,
spikelet (×2½)

Alopecurus carolinianus Walter   Carolina foxtail, tufted foxtail
Annual; culms 1–6 dm tall, densely caespitose; inflorescence 2–5 cm long by 4–5 mm wide; spikelets 2–2.5 mm long; moist fields; restricted to the Coastal Plain and Piedmont; May–Sep; FACW.

Alopecurus pratensis L.   Meadow foxtail
Perennial; culms 4–8 dm tall; blades 2–6 mm wide; inflorescence 2–8 cm long by 5–10 mm wide; spikelets about 5 mm long; meadows and waste ground; mostly SE; May–Jul; native to Eurasia; FACW.

Alopecurus
pratensis,
spikelet (×2½)

Ammophila Host

Ammophila breviligulata Fernald   American beachgrass
Perennial; culms tufted, 7–10 dm tall, with strong, creeping rhizomes; blades involute, 4–8 mm wide when unrolled, scabrous above, glabrous below; inflorescence spike-like, 1–4 dm long; spikelets 1-flowered, 11–14 mm long; glumes nearly equal, first glume 1-nerved, second glume 3-nerved; lemma scabrous, with callus-hairs 1–3 mm long; sand dunes and beaches; NW and SE; Jul; FACU–; 🌿.

Ammophila
breviligulata,
spikelet (×1)

Andropogon L.

Perennials; spikelets paired at the joints of the rachis, one of each pair sessile and perfect, the other pediceled and staminate (sometimes sterile or greatly reduced); glumes of the fertile spikelet about equal, apparently without a midnerve; fertile lemma shorter than the glumes, usually awned.

A. pediceled spikelet staminate, similar to the sessile spikelet but awnless ....
.... *A. gerardii*
A. pediceled spikelet sterile, much reduced or reduced to only the pedicel
  B. common peduncle of sessile and pediceled spikelets 1 cm or less long; upper nodes glabrous or sparsely villous
    C. inflorescences terminal and axillary along the culm, not aggregated into a cluster .... *A. virginicus*
    C. inflorescences aggregated into a dense terminal cluster .... *A. glomeratus*
  B. common peduncle of sessile and pediceled spikelets >1 cm long; upper nodes densely villous at the summit .... *A. gyrans*

Andropogon gerardii Vitman   Big bluestem, turkeyfoot
Culms 1–3 m tall, in large clumps or occasionally sod-forming; blades 5–10 mm wide, lower ones villous, margins very scabrous; racemes 5–10 cm long, 2–6 on a long-exserted peduncle; sessile spikelet 7–10 mm long, appressed; awn 1–2 cm long, tightly twisted below; pediceled spikelet not much reduced; stream banks, roadsides, and moist meadows; throughout; Aug–early Oct; FAC–. [syn: *A. furcatus* Muhl.]

Andropogon
gerardii,
spikelet pair
(×2)

Andropogon glomeratus (Walter) Britton, Stearns & Poggenb.   Broom-sedge
Culms 5–15 dm tall, tufted, often villous on the upper nodes; blades 3–6 mm wide, often pilose; inflorescence compact, densely branched; sessile spikelets about 4–4.5 mm long, shorter than the sterile pedicel; awn straight, 1–2 cm long; rare in swamps and moist meadows; mostly SE; Aug–Oct; FACW+; ✺. [syn: *A. virginicus* L. var. *abbreviatus* (Hack.) Britton, Stearns & Poggenb.]

Andropogon gyrans Ashe   Elliott's beardgrass
Culms 4–9 dm tall, tufted, often branching above; upper blades usually crowded, 5–15 mm wide, usually brownish or copper-colored; inflorescence 4–5 cm long, feathery; sessile spikelets 4–5 mm long; awn loosely twisted, 1–1.5 cm long; dry or moist fields or open woods; SE; Sep–Oct; ✺. [syn: *A. elliotii* Chapm.]

Andropogon virginicus L.   Broom-sedge
Culms 5–15 dm tall, tufted, glabrous, or the upper nodes sometimes sparsely villous; blades 3–6 mm wide, often pilose on the sheath; sessile spikelets 3–5 mm long, shorter than the sterile pedicel; sterile spikelet usually absent; old fields, hillsides, and waste ground, in dry, sterile soil; mostly S; Aug–Oct; FACU.

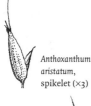

Andropogon virginicus, spikelet (×2½)

Anthoxanthum L.

Spikelets with 1 perfect terminal floret and 2 sterile lemmas, disarticulating above the glumes; glumes unequal, the first ovate and 1-nerved, the second lanceolate and 3-nerved, both longer than the lemmas; sterile lemmas about equal, the first awned near its apex, the second near its base; fertile lemma smaller than and enclosed by a sterile lemma.

A. culms unbranched; awns of the lower sterile lemmas straight ..... *A. odoratum*
A. culms branching from the lower nodes; awns of both sterile lemmas bent or twisted .... *A. aristatum*

Anthoxanthum aristatum Boiss.   Sweet vernalgrass
Annual; similar to *A. odoratum* but culms abruptly bent near the base and decumbent, 2–4 dm long; cultivated and occasionally escaped; scattered; May–Jun; native to Europe.

Anthoxanthum aristatum, spikelet (×3)

Anthoxanthum odoratum L.   Sweet vernalgrass
Perennial; culms erect, 3–7 dm tall, cespitose; blades 2–5 mm wide; inflorescence spike-like, 3–6 cm long; spikelets 8–10 mm long; first glume about 4 mm long, second glume 7–9 mm long; sterile lemmas pilose with golden hairs; common in open fields, meadows, and roadsides; throughout; Apr–early Jul; native to Eurasia; FACU.

Anthoxanthum odoratum, spikelet (×3)

Aristida L.

Spikelets 1-flowered, disarticulating above the glumes; glumes linear to lanceolate, usually 1-nerved; lemma bearing a long 3-branched awn.

A. lateral awns half as long as the central awn or shorter
    B. central awn spirally coiled at the base .... *A. dichotoma*
    B. central awn reflexed but not spirally coiled .... *A. longespica*
A. lateral awns about as long as the central awn

C. second glume 2–3 cm long; awns usually 4–6 cm long .... *A. oligantha*

C. second glume 0.5–1.5 cm long; awns usually 1.5–2.5 cm long ....
.... *A. purpurascens*

Aristida dichotoma Michx.   Poverty grass
Perennial; culms 2–5 dm tall; blades mostly involute, about 1 mm wide; terminal panicle few-branched, 3–8 cm long; lateral panicles much shorter, mostly enclosed in the subtending sheath; dry, open, or sterile sandy soil; Aug–early Oct; UPL; 2 varieties:

A. glumes markedly unequal, the second nearly reaching the middle of the lateral awns .... var. *curtissii* A.Gray
   rare; SC; ✹. [syn: *A. basiramea* Engelm. var. *curtissii* (Gray) Shinners]

A. glumes nearly equal, both surpassing the point of insertion of the awns of the lemma .... var. *dichotoma*
   throughout, but uncommon N.

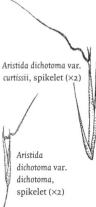

Aristida dichotoma var.
curtissii, spikelet (×2)

Aristida
dichotoma var.
dichotoma,
spikelet (×2)

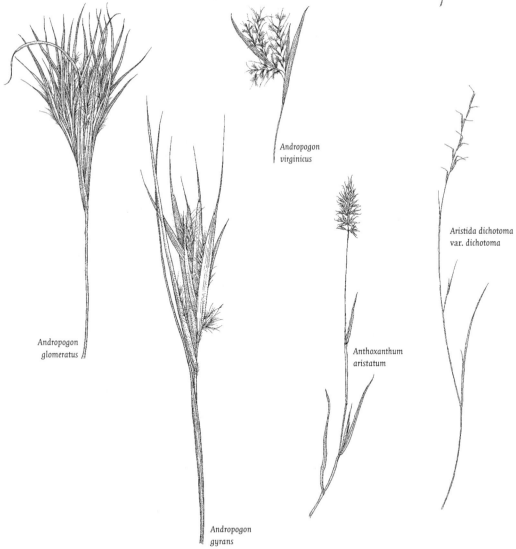

Andropogon
virginicus

Andropogon
glomeratus

Andropogon
gyrans

Anthoxanthum
aristatum

Aristida dichotoma
var. dichotoma

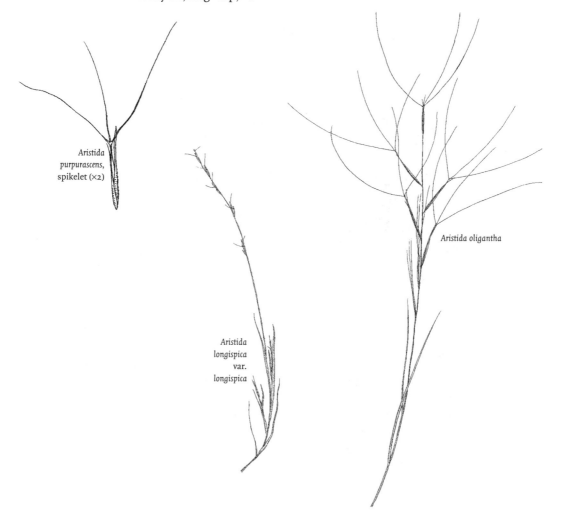

Aristida longispica
var. *geniculata*,
spikelet (×2)

Aristida longispica
var. *longispica*,
spikelet (×2)

Aristida longispica Poir. Slender threeawn
Annual; culms 2–4 dm tall, often branched from the lower nodes; blades often involute, 1–2 mm wide; terminal panicle very slender, 10–15 cm long; lateral panicles much reduced; serpentine barrens and other dry, sandy, or sterile soils; mostly SE; Aug–Sep; UPL; 🖑; 2 varieties:

A. lateral awns ⅕–⅓ as long as the central awn .... var. *longispica*
A. lateral awns ⅓–½ as long as the central awn ...... var. *geniculata* (Raf.) Fernald

Aristida oligantha Michx. Prairie threeawn
Annual; culms 2–4 dm tall, branched from the base and at most nodes; blades about 1 mm wide, flat or involute; terminal panicle few-flowered, 1–2 dm long, very lax; first glume 12–29 mm long, second glume slightly longer; lateral panicles few-flowered, dense; dry, open ground; mostly S; late Aug–Oct.

Aristida purpurascens Poir. Arrow-feather, three-awned grass
Perennial; culms 4–7 dm tall, cespitose from a knotty base; blades 1–2 mm wide; terminal panicle spike-like, 1–3 dm long; first glume 8–14 mm long, second glume 6.5–11.5 mm long; rare on serpentine barrens and other dry sandy soils; mostly SE; Aug–Sep; 🖑.

Aristida
purpurascens,
spikelet (×2)

Aristida oligantha

Aristida
longispica
var.
longispica

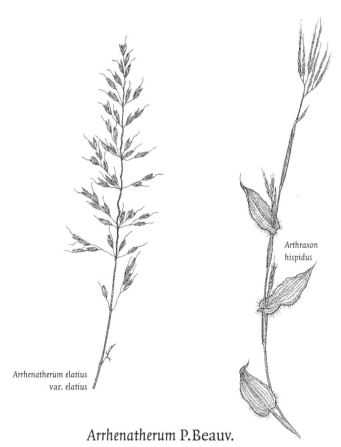

Arrhenatherum elatius
var. elatius

Arthraxon
hispidus

Arrhenatherum elatius var.
elatius, spikelet (×2)

Arrhenatherum P.Beauv.

Arrhenatherum elatius (L.) P.Beauv.   Tall oatgrass
Perennial; culms up to 2 m tall, cespitose, smooth to minutely pubescent on the
nodes; blades 4–8 mm wide, scabrous, sheaths smooth; panicle 1–3 dm long,
the short branches in fascicles; spikelets 2-flowered, the lower staminate, the
upper perfect, disarticulating above the glumes; first glume 4.5–8 mm long, second glume 6.6–10 mm long; roadsides, fields, waste ground, and wooded
slopes; common throughout; late May–Jun; native to Europe; FACU; 2 varieties:

A. awn of lower lemma 10–20 mm long, awn of second lemma 0–6 mm long ....
.... var. *elatius*
A. awns of both lemmas subequal .... var. *biaristatum* (Peterm.) Peterm.

Arthraxon P.Beauv.

Arthraxon hispidus (Thunb.) Makino   Grass
Annual; culms decumbent to creeping; blades 2–5 cm by 5–15 mm, cordate,
sheaths hirsute; racemes 2–4 cm long, several arranged digitately on a common
peduncle; fertile spikelets sessile, 3–5 mm long; sterile lemma usually awned;
moist meadows and waste ground; scattered; S; Aug–Oct; native to SE Asia; FAC.

Avena L.

Annuals; spikelets 2–3-flowered, disarticulating above the glumes and usually
between the lemmas; glumes nearly equal; lemma 2-toothed at the apex; often
awned; Jun–Sep.

A. lemmas glabrous; awns straight or lacking .... *A. sativa*
A. lemmas pubescent with white or brown hairs; awns bent or twisted ... *A. fatua*

Avena fatua L. Wild oats

Avena fatua,
spikelet

Culms 5–8 dm tall, smooth; blades 5–15 mm wide, somewhat scabrous, sheaths smooth; panicle very lax, the branches spreading horizontally; spikelets 3-flowered; glumes about 2.5 cm long; lemma about 2 cm long; awn 3–4 cm long; waste ground; scattered throughout; native to Europe.

Avena sativa,
spikelet

Avena sativa L. Oats
Culms 3–10 dm tall, branching from the base; spikelets 2-flowered; otherwise similar to *A. fatua*; cultivated and frequently escaped to fallow fields, roadsides, and waste ground; throughout; native to Europe.

Bouteloua Lag.

Bouteloua curtipendula (Michx.) Torr. Side-oats grama, tall grama
Perennial; culms 3–10 dm tall from slender rhizomes; blades 2–5 mm wide, scabrous on the margins; ligule a zone of short hairs; spikes 10–50 along 1 side of an axis 1–3 dm long, spikes 8–15 mm long, each bearing 3–6 spikelets; spikelets with 1 perfect floret and 1 or more rudimentary florets above, disarticulating above the glumes; glumes unequal, 1-nerved; fertile lemma usually slightly exceeding the glumes; serpentine barrens, dry calcareous clearings, and other dry, rocky, or sandy sites; mostly SC and SE; Aug–Sep; ❦.

Brachyelytrum
erectum,
spikelet (×1)

Brachyelytrum P.Beauv.

Brachyelytrum erectum (Schreb.) P.Beauv. Brachyelytrum
Perennial; culms 5–10 dm tall; blades 8–18 cm long by 8–16 mm wide, scabrous to pubescent, sheaths pubescent; panicle narrow, with few spikelets, 5–15 cm long; spikelets 1-flowered, disarticulating above the glumes; first glume <1 mm long or absent, second glume 1–4 mm long; awns 12–25 mm long; common in moist woods and thickets; throughout; Jun–Jul.

Briza L.

Briza media,
spikelet (×1½)

Briza media L. Quaking-grass
Perennial; culms 3–6 dm tall, loosely cespitose; blades 2–4 mm wide, scabrous on the margins; panicle 5–10 cm long, with erect branches; spikelets round-ovate, several-flowered, 4–5 mm long and about as wide, disarticulating above the glumes and between the lemmas; glumes 2.5–3.5 mm long; fallow fields and wet meadows; SE; May–Aug; native to Europe; FAC.

Bromus L.

Spikelets several- to many-flowered, disarticulating above the glumes and between the lemmas, oval to narrowly oblong, generally laterally flattened; glumes shorter than the lemmas, subequal; lemmas 5–9-nerved, 2-toothed (these often very short) at the apex, awned or awnless; sheath closed.

A. lemmas broad, rounded or tapered to the apex but without acuminate lateral teeth; callus not sharp and prolonged

  B. first glume mostly 1-nerved; glumes lanceolate to narrowly ovate

    C. plants rhizomatous; panicle branches and spikelets strongly ascending .... B. inermis

    C. plants not rhizomatous; panicle branches and spikelets nodding or drooping

      D. lemmas sparsely to densely pubescent on the back

        E. culms with 10–20 nodes; sheaths mostly longer than the internodes, bearing prominent auricles at the summit ..... B. altissimus

        E. culms with 3–7 nodes; sheaths shorter than the internodes and lacking auricles

          F. panicle mostly 9–12 cm long, contracted, nodding; anthers 1.5–2.6 mm long .... B. kalmii

          F. panicle mostly 15–20 cm long, open, lax; anthers 2.8–4.6 mm long .... B. pubescens

      D. lemmas pubescent only along the margins .... B. ciliatus

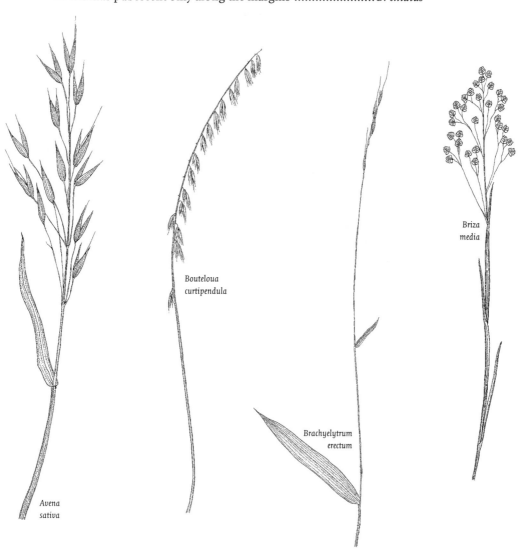

Avena
sativa

Bouteloua
curtipendula

Brachyelytrum
erectum

Briza
media

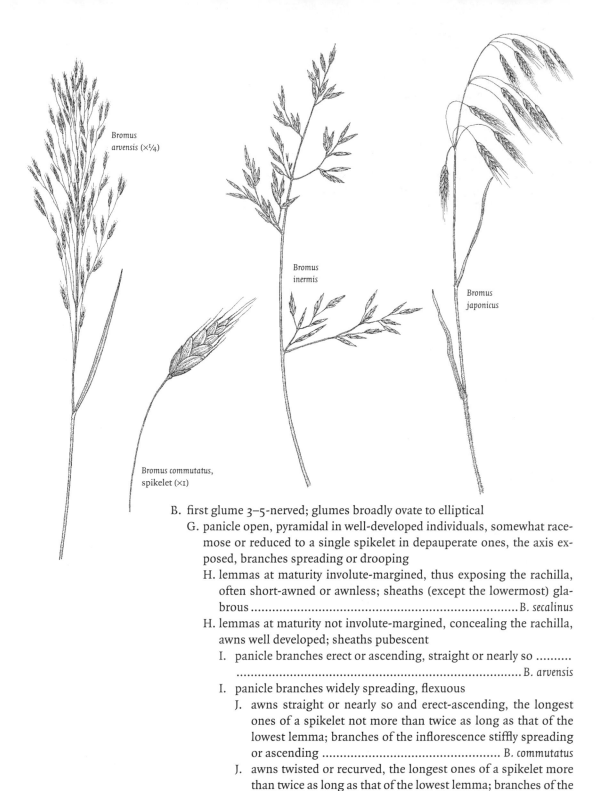

Bromus arvensis (×¼)

Bromus inermis

Bromus japonicus

Bromus commutatus, spikelet (×1)

B. first glume 3–5-nerved; glumes broadly ovate to elliptical
    G. panicle open, pyramidal in well-developed individuals, somewhat race-mose or reduced to a single spikelet in depauperate ones, the axis exposed, branches spreading or drooping
        H. lemmas at maturity involute-margined, thus exposing the rachilla, often short-awned or awnless; sheaths (except the lowermost) glabrous .... B. secalinus
        H. lemmas at maturity not involute-margined, concealing the rachilla, awns well developed; sheaths pubescent
            I. panicle branches erect or ascending, straight or nearly so .... B. arvensis
            I. panicle branches widely spreading, flexuous
                J. awns straight or nearly so and erect-ascending, the longest ones of a spikelet not more than twice as long as that of the lowest lemma; branches of the inflorescence stiffly spreading or ascending .... B. commutatus
                J. awns twisted or recurved, the longest ones of a spikelet more than twice as long as that of the lowest lemma; branches of the inflorescence lax and flexuous .... B. japonicus
    G. panicle ellipsoid, dense, usually not over 10 cm long, the branches shorter than the spikelets that overlap and conceal the axis
        K. lemmas glabrous .... B. racemosus
        K. lemmas pubescent .... B. hordeaceus

A. lemmas narrowly lanceolate, with a hard sharp callus and a bifid apex with
   acuminate teeth
   L. second glume 10 mm or less long; lemmas villous, 9–12 mm long, bearing
      awns 1–2 cm long .... B. tectorum
   L. second glume 12–15 mm long; lemmas scaberulous, bearing awns 2–3 cm
      long .... B. sterilis

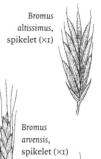

Bromus altissimus, spikelet (×1)

Bromus altissimus Pursh   Bromegrass
Perennial; similar to *B. ciliatus* in habit; rich alluvial thickets or woods; throughout; Aug–Sep; FACW. [syn: *B. latiglumis* (Shear) A.Hitchc.]

Bromus arvensis L.   Field chess
Annual; similar to *B. japonicus* in habit; disturbed soil; scattered; May–Jul; native
to southern Europe.

Bromus arvensis, spikelet (×1)

Bromus ciliatus L.   Fringed brome
Perennial; culms 6–12 dm tall, glabrous, often pubescent at the nodes; blades 5–
10 mm wide, glabrous to sparsely villous on one or both sides; panicle 1–2 dm
long, loose and open, branches up to 15 cm long; woods, clearings and meadows; N and in the mountains to Somerset Co.; May–Aug; FACW.

Bromus ciliatus, spikelet (×1)

Bromus commutatus Schrad.   Hairy chess
Annual; similar to *B. secalinus* in habit; fields, roadsides, and waste ground; common throughout; late May–early Jul; native to Europe.

Bromus hordeaceus L.   Soft chess
Annual; culms 4–12 dm tall; blades 3–6 mm wide, pubescent; spikelets 6–9-
flowered, usually pubescent; awns 6–9 mm long; fields and waste ground; scattered except SW; May–Jul; native to Europe. [syn: *B. mollis* L.]

Bromus hordeaceus, spikelet (×1)

Bromus inermis Leyss.   Smooth brome
Annual; culms 5–10 dm tall; blades 8–15 mm wide, glabrous; sheaths glabrous;
panicle 1–2 dm long, with 4–10 branches from each node; spikelets 7–11-
flowered, 15–30 mm long by about 3 mm wide; cultivated and escaped to fields,
roadsides, and waste ground; throughout; Jun–Jul; native to Europe.

Bromus inermis, spikelet (×1)

Bromus japonicus Thunb. ex Murray   Japanese chess
Annual; culms 3–9 dm tall, glabrous; blades 2–4 mm wide, densely villous;
panicle 1–2 dm long, branches spreading or drooping; spikelets 7–10-flowered,
glabrous or nearly so; roadsides and waste ground; throughout; late May–early
Jul; native to Eurasia; FACU–.

Bromus kalmii A.Gray   Bromegrass
Perennial; culms 5–10 dm long, solitary to loosely cespitose, often pubescent at
the nodes; blades 5–10 mm wide by 1–2 dm long; spikelets 6–11-flowered, 15–25
mm long, villous; awn 2–3 mm long; infrequent on rocky wooded slopes and dry
or moist woods; scattered; Jun–Jul; FACU; .

Bromus kalmii, spikelet (×1)

Bromus
pubescens,
spikelet (×1)

Bromus pubescens Muhl. ex Willd. Canada brome

Perennial; culms 6–15 dm tall; blades 8–15 mm wide, narrowed to base, glabrous to sparsely villous; panicle 1–2 dm long; spikelets 2–3 dm long; awn 2–8 mm long; dry to moist woods and thickets; throughout; Jun–Jul. [syn: *B. purgans* L.]

Bromus
racemosus,
spikelet (×1)

Bromus racemosus L. Soft chess

Annual; culms 4–9 dm tall, mostly glabrous; blades 3–6 mm wide, pubescent on both sides; sheaths finely villous; panicle 1–2 dm long, nodding at maturity; spikelets 5–10-flowered, mostly glabrous; awns 3–10 mm long; waste ground and disturbed sites; scattered throughout; May–Jul; native to Europe.

Bromus
secalinus,
spikelet (×1)

Bromus secalinus L. Cheat, chess

Annual; culms 3–6 dm tall; panicle 7–12 cm long, nodding; spikelets 1–2 cm long by 6–8 mm wide; grain fields and waste ground; throughout; late May–Jul; native to Europe.

Bromus sterilis L. Barren brome

Annual; culms 4–10 dm tall; blades 2–4 mm wide, glabrous; sheaths mostly glabrous; spikelets 5–7-flowered; roadsides and waste ground; throughout, but more common S; May–early Jul; native to Europe.

Bromus tectorum L. Downy chess

Annual; culms 3–7 dm tall, cespitose; blades 2–4 mm wide, softly pubescent; sheaths pubescent; spikelets to 3 cm long including glumes; palea conspicuously ciliate; dry, fallow fields and waste ground; throughout; May–Jun; native to Europe.

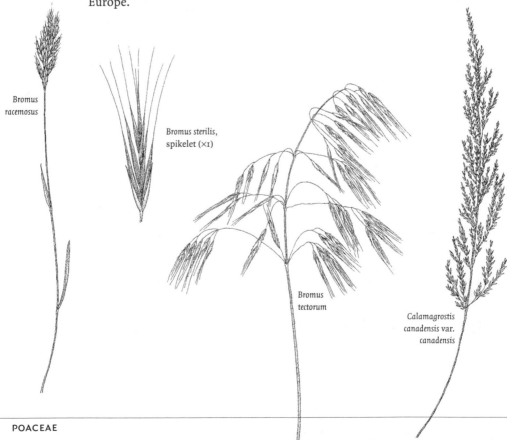

Bromus
racemosus

Bromus sterilis,
spikelet (×1)

Bromus
tectorum

Calamagrostis
canadensis var.
canadensis

Calamagrostis Adans.

Perennials; spikelets 1-flowered, disarticulating above the glumes; rachilla prolonged behind the palea as a short, hairy bristle; glumes about equal; lemma equaling or slightly shorter than the glumes.

A. awns bent or twisted below, laterally exserted from the glumes .... *C. porteri*
A. awns not bent or twisted, not exserted from the glumes
    B. callus hairs well exceeding the lemma; rachilla not prolonged behind the palea .... *C. epigejos*
    B. callus hairs up to about as long as the lemma; rachilla prolonged behind the palea as a hairy bristle
        C. awn inserted near the tip of the lemma; rachilla naked below with a prominent apical tuft of hairs which reach about to the apex of the floret; grain pubescent .... *C. cinnoides*
        C. awn inserted on the lower half of the lemma; rachilla hairy from base to tip; grain glabrous .... *C. canadensis*

Calamagrostis canadensis (Michx.) Beauv.   Canada bluejoint
Culms 5–15 dm tall, cespitose, from a creeping rhizome; blades 4–8 mm wide, sheaths glabrous; panicle 8–20 cm long, dense or open, but not contracted; spikelets 2–6 mm long; glumes about equal or the second sometimes slightly longer; lemma ¾ to as long as the glumes; wet meadows, bogs, and swamps; throughout; Jun–Aug; FACW+; 2 varieties:

A. spikelets 2.8 mm or more long; glumes acute to acuminate, distinctly longer than the floret; panicles mostly open .... var. *canadensis*
A. spikelets 2.2–2.8 mm long; glumes acute, about as long as the floret; panicles contracted .... var. *macouniana* (Vasey) Stebbins

Calamagrostis canadensis var. canadensis, spikelet (×3½)

Calamagrostis cinnoides (Muhl.) W.Bartram   Reedgrass
Culms 6–12 dm tall, glabrous; blades 5–10 mm wide, scabrous; panicle 1–2 dm long, dense and somewhat contracted, 2–5 cm wide; glumes usually curved outward, first glume 4.5–8 mm long, second usually somewhat shorter; lemma 3.7–6 mm long, awn inserted below the apex and only slightly surpassing the lemma; swamps or wet woods, in peaty or sandy soils; mostly E and S; Jul–early Sep; OBL.

Calamagrostis cinnoides, spikelet (×3½)

Calamagrostis epigejos (L.) Roth   Feathertop
Culms 10–15 dm from spreading rhizomes, forming dense patches; panicle dense and contracted, 25–35 cm long; glumes about equal, 5–6 mm long; lemma about ½ as long as the glumes, awned from below the middle; disturbed woods, roadsides, and waste ground; scattered; Jul–early Oct; native to Eurasia; FAC.

Calamagrostis epigejos, spikelet (×3½)

Calamagrostis porteri A.Gray   Porter's reedgrass
Culms to about 1 m; leaf sheaths pilose at the summit; panicle 1–2 dm long, some of its branches bearing spikelets to the base; first glume 3.5–5.5 mm long, second a little shorter; lemma 2.8–4.2 mm long; roadsides and open woods; mostly at elevations of 1000–2000 ft. in Bedford, Centre, Huntington, Snyder, and nearby counties; Jul–early Sep.

Calamagrostis porteri, spikelet (×3½)

Calamovilfa (A.Gray) Hack. ex Scribn. & Southw.

Calamovilfa longifolia, spikelet (×2½)

Calamovilfa longifolia (Hook.) Scribn.   Sandreed

Perennial; culms to 2 m tall, from long scaly rhizomes; blades 3–8 mm wide at the base, involute above; panicle 1–4 dm long, narrow with ascending branches; first glume 3.5–6 mm long, second a little longer; lemma nearly equaling the second glume, callus hairs about ½ as long as the lemma; planted and naturalized on beaches in Erie Co.; native to western and midwestern U.S.

Cenchrus L.

Annuals; burs enclosing 1–3 spikelets; first glume 1-nerved; second glume 3–5-nerved.

A. spines 45–75 per bur; burs (including spines) usually under 10 mm broad ....
.... *C. longispinus*

A. spines 15–40 per bur; burs (including spines) usually 10–17 mm broad ....
.... *C. tribuloides*

Cenchrus longispinus, spikelet (×1½)

Cenchrus longispinus (Hack.) Fernald   Sandbur

Culms 2–6 dm long, usually decumbent; burs pubescent; spikelets visible through the cleft in the bur; dry, sandy soil; mostly SE; Jul–Sep. [syn: *C. pauciflorus* Benth.]

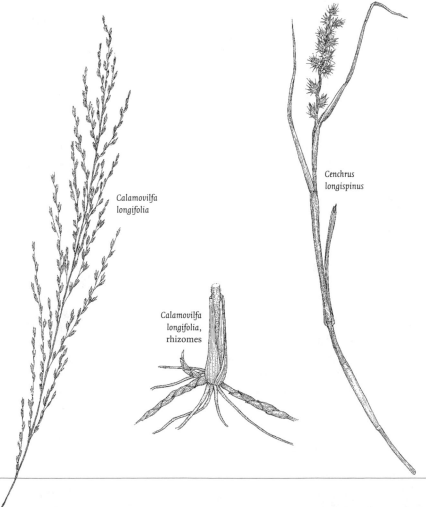

Calamovilfa longifolia

Cenchrus longispinus

Calamovilfa longifolia, rhizomes

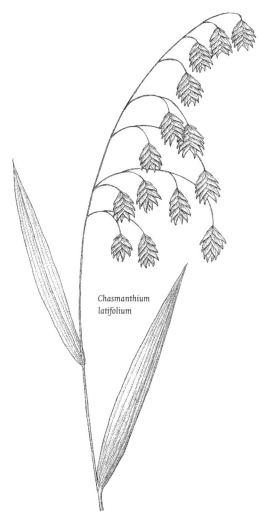

Chasmanthium
latifolium

Cenchrus tribuloides L.   Dune sandbur

Similar to *C. longispinus* in habit; burs densely villous; spikelets not visible; ballast ground, abandoned limestone quarries, and other dry, open, sandy sites; Jul–Sep; native east of Pennsylvania; UPL.

Cenchrus tribuloides,
spikelet (×1½)

Chasmanthium Link.

Perennials; spikelets 3–many-flowered, distinctly flattened, disarticulating above the glumes and between the lemmas; glumes generally shorter than the lemmas; lower lemma sterile.

A. panicle open, the branches drooping; spikelets 20–30 mm long, 8–12-flowered; largest leaf blades 15–25 mm broad .... *C. latifolium*

A. panicle strict, the branches ascending; spikelets 5–8 mm long, 3–4-flowered; largest leaf blades 3–7 mm broad .... *C. laxum*

Chasmanthium latifolium (Michx.) H.O.Yates   Sea-oats

Culms 10–15 dm from stout rhizomes; panicle 1–2 dm long; stream banks and alluvial woods; rare along lower Susquehanna River and at a few scattered sites S and W; Jul–Sep; FACU; ❦. [syn: *Uniola latifolia* Michx.]

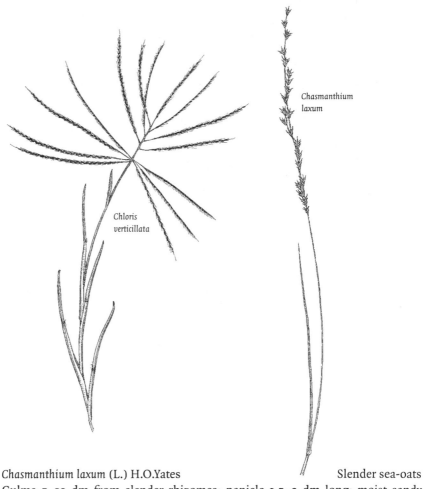

Chasmanthium laxum

Chloris
verticillata

Chasmanthium laxum,
spikelet (×2½)

Chasmanthium laxum (L.) H.O.Yates   Slender sea-oats
Culms 5–10 dm from slender rhizomes; panicle 1.5–3 dm long; moist sandy
soils of the Coastal Plain; extreme SE; Aug–Sep; FAC; 🍂. [syn: *Uniola laxa* Britton,
Stearns & Poggenb.]

Chloris Sw.

Chloris
verticillata,
spikelet
(×2)

Chloris verticillata Nutt.   Windmill-grass
Perennial; culms 1–4 dm tall, usually erect; leaves mostly near the base, blades
2–4 mm wide, usually <15 cm long; sheaths compressed; spikes numerous, 6–15
cm long; first glume 2–3 mm long, second glume 3–4 mm long; fertile lemma
2.4–2.8 mm long, its awn 5–9 mm long; fallow fields or lawns on sandy or
droughty soils; SE and SC; native to midwestern U.S.

Cinna L.

Perennials; spikelets 1-flowered, disarticulating below the glumes; glumes about
equal, linear to lanceolate, 1–3-nerved; lemma about equaling the second glume,
bearing a short straight awn from just below the apex; rachilla prolonged behind
the palea as a minute bristle.

A. spikelets 4.5–5.9 mm long; panicle usually rather dense .... *C. arundinacea*
A. spikelets 3.0–4.2 mm long; panicle usually lax .... *C. latifolia*

Cinna arundinacea L.   Wood reedgrass

Culms 10–15 dm tall, usually with 10–15 nodes; blades 6–12 mm wide, scabrous; panicle 1.5–3 dm long, often drooping; first glume about 4 mm long, 1-nerved, second about 5 mm long, 3-nerved; lemma usually a little longer than the first glume, awn usually <1 mm long; swamps and wet woods; throughout; Aug–Oct; FACW.

Cinna arundinacea, spikelet (×4)

Cinna latifolia (Trevis.) Griseb.   Drooping woodreed, wood reedgrass

Similar to *C. arundinacea* in habit, but usually with 3–7 nodes; blades up to 15 mm wide; panicle open; first glume about 3 mm long, second 1-nerved and slightly longer; awn up to 1.5 mm long; moist shores, damp woods, and bogs; throughout, but more frequent N; Jul–Aug; FACW.

Cinna latifolia, spikelet (×4)

Coix L.

Coix lacryma-jobi L.   Job's tears

Annual or short-lived perennial; culms to about 1 m; blades to 4 cm wide; cultivated and occasionally escaped to waste ground; Philadelphia and Lehigh Co.; Jul–Sep; native to tropical Asia; FACW.

Cinna arundinacea

Cinna latifolia

Critesion Raf.

Spikelets 1-flowered (rarely 2-flowered), disarticulating in groups of 3 attached to the joints of the rachis; rachilla prolonged behind the palea as a slender bristle.

A. awns very slender, 5–8 cm long .... *C. jubatum*
A. awns stout, <3 cm long
    B. glumes of the central spikelet glabrous to scabrous-margined, not ciliate
    .... *C. pusillum*
    B. glumes of the central spikelet ciliate-margined .... *C. murinum*

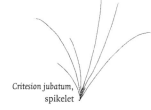

Critesion jubatum, spikelet

Critesion jubatum (L.) Nevski   Foxtail-barley
Perennial; culms 3–7 dm tall, cespitose, often decumbent at the base; principal blades 2–5 mm wide, often scabrous; upper sheath usually enclosing the base of the spike; spike 5–10 cm long, usually nodding; glumes about equal, 4–7 mm long; dry old fields, roadsides, and waste ground; throughout; late Jun–Aug; FAC. [syn: *Hordeum jubatum* L.]

Critesion murinum (L.) A.Löve   Wild-barley
Annual; culms 2–6 dm tall, usually branched from the base; glumes about equal, 12–30 mm long; ballast and waste ground; Lehigh Co.; native to Europe. [syn: *Hordeum murinum* L.]

Critesion pusillum, spikelet (×1)

Critesion pusillum (Nutt.) A.Löve   Little-barley
Annual; culms 2–4 dm tall, often decumbent; principal blades 2–6 mm wide; spike 2–6 cm long, somewhat flattened, enclosed in the upper sheath below; glumes about equal, 12–15 mm long; roadside banks; Bedford and Bucks Cos.; believed to be extirpated; May; FAC; 🍂. [syn: *Hordeum pusillum* Nutt.]

Crypsis Aiton.

Crypsis schoenoides, spikelet (×5)

Crypsis schoenoides (L.) Lam.   Grass
Annual; culms 1–4 dm long, often prostrate; inflorescences spike-like, ovoid-cylindric, 1–4 cm long by 6–10 mm wide, partially concealed in the sheaths of the subtending leaves; spikelets about 3 mm long; waste ground and ballast; mostly SE. [syn: *Heleochloa schoenoides* (L.) Host]

Cynodon Rich.

Cynodon dactylon, spikelet (×5)

Cynodon dactylon (L.) Pers.   Bermudagrass, wiregrass
Perennial; culms 1–3 dm tall, erect or ascending, spreading by stolons or rhizomes and forming mats; principal blades 2–4 mm wide; spikes usually 4–6, 3–5 cm long; spikelets disarticulating above the glumes; glumes nearly equal, the first curved, the second nearly straight; lemmas 2–2.5 mm long; palea nearly as long as the lemma; rachilla prolonged behind the palea as a slender bristle; cultivated and escaped; throughout, but more common S; Jul–Oct; native to Old World tropics; FACU.

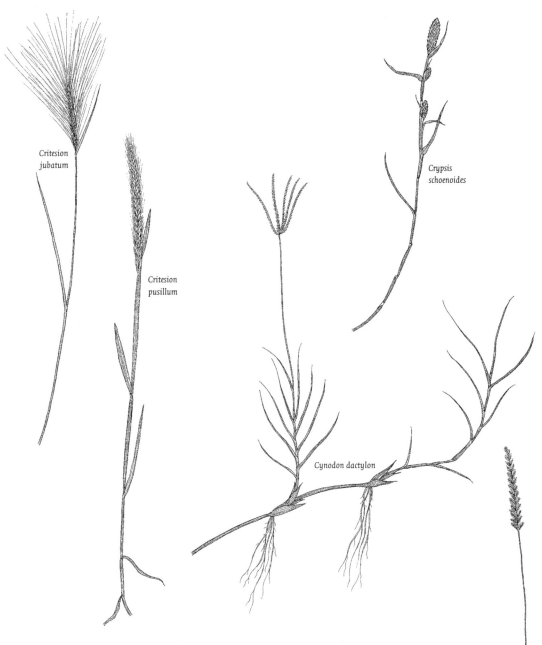

Critesion
jubatum

Critesion
pusillum

Crypsis
schoenoides

Cynodon dactylon

Cynosurus L.

Inflorescence spike-like or head-like, 1-sided; sterile spikelets consisting of 2 glumes and several lemmas; fertile spikelets 2–3-flowered, disarticulating above the glumes.

A. panicle narrow; awns 1 mm or less long .... *C. cristatus*
A. panicle ovoid; awns 1–4 cm long .... *C. echinatus*

Cynosurus cristatus L.   Crested dog's-tail
Perennial; culms 3–8 dm tall, densely cespitose; blades few, 1–3 mm wide; inflorescence 3–10 cm long, slender; lemmas 3–3.5 mm long; lawns, roadsides, and waste ground; mostly SE; Jun–Aug; native to Europe; UPL.

Cynosurus
cristatus

Cynosurus
cristatus,
spikelet (×3)

Cynosurus echinatus,
sterile spikelet (×1)

Cynosurus echinatus L.   Spiny dog's-tail
Annual; culms 2–4 dm tall; inflorescence head-like; 1–4 cm long; roadside banks, field margins, and disturbed soil; mostly SE; Jun–Jul; native to Europe.

Dactylis L.

*Dactylis
glomerata,*
spikelet (×1½)

Dactylis glomerata L.   Orchardgrass
Perennial; culms 6–12 dm tall, usually in large tussocks; blades 2–8 mm wide; sheaths scaberulous; ligules 5–7 mm long; panicle 5–20 cm long, few-branched, the branches stiff, as much as 10 cm long; spikelets 3–6-flowered; lemmas 5–8 mm long, usually ciliate on the keel; fields, meadows, and roadsides; common throughout; May–Jul; native to Europe; FACU.

Danthonia DC.

Perennials; spikelets 3–5-flowered, disarticulating above the glumes and between the lemmas; glumes equaling or exceeding the upper lemma; lemmas minutely 3-toothed at the apex, villous toward the base; palea shorter than the lemma, ciliate on the keel.

A. teeth of the lemmas 0.8–1.8 mm long, triangular-acuminate; inflorescence usually contracted and spike-like, branches ascending..................... *D. spicata*
A. teeth of the lemmas 2–4.5 mm long, bristle-tipped; inflorescence usually open, branches diverging .... *D. compressa*

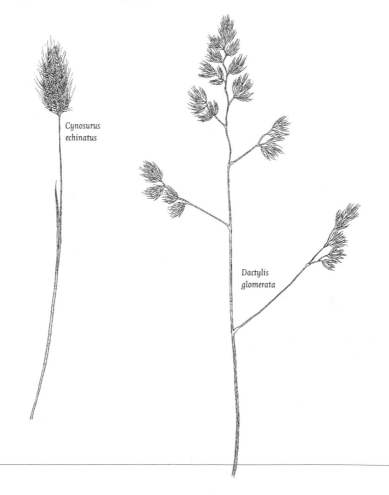

*Cynosurus
echinatus*

*Dactylis
glomerata*

Danthonia compressa Austin   Northern oatgrass
Culms 4–8 dm tall; blades usually flat, 2–4 mm wide, somewhat scabrous; sheaths glabrous; panicle 5–10 cm long, lower branches bearing 2–3 spikelets; dry, rocky woods and clearings; occasional throughout; Jun–Jul; FACU–.

Danthonia spicata (L.) P.Beauv. ex Roem. & Schult.   Poverty grass
Culms 2–6 dm tall; blades usually involute, 1–2 mm wide, glabrous or sparsely pilose; panicle 2–5 cm long, branches usually bearing 1 spikelet; dry, sandy or gravelly soil; common throughout; late May–early Sep.

Deschampsia P.Beauv.

Perennials; spikelets 2-flowered, disarticulating above the glumes and between the lemmas; rachilla prolonged behind the base of the upper lemma as a hairy bristle.

A. lemmas glabrous; awn straight or nearly so .... *D. cespitosa*
A. lemmas scabrous; awn strongly bent or twisted below .... *D. flexuosa*

Deschampsia cespitosa (L.) P.Beauv.   Tufted hairgrass
Culms 5–12 dm tall, loosely cespitose; blades flat or folded, 1–5 mm wide; ligule 3–12 mm long; lower panicle branches in whorls of 2–5; awns straight or nearly so, shorter than to barely surpassing the lemma; serpentine barrens, sandy shores, and thickets; mostly E; late May–Jun; FACW; 🌿.

Deschampsia cespitosa, spikelet (×2½)

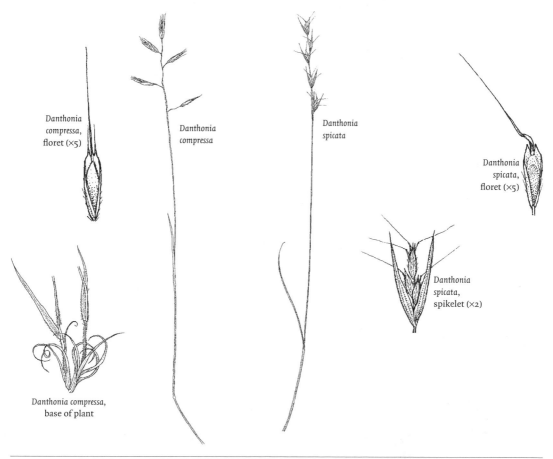

Danthonia compressa, floret (×5)

Danthonia compressa

Danthonia spicata

Danthonia spicata, floret (×5)

Danthonia spicata, spikelet (×2)

Danthonia compressa, base of plant

Deschampsia flexuosa, spikelet (×2½)

Deschampsia flexuosa (L.) Trin. Common hairgrass

Culms 3–10 dm tall; blades involute, 1–2 mm wide; ligule 1–2.5 mm long; panicle loose and open, up to 15 cm long, often nodding; awns twisted below the middle, surpassing the lemma by 1–3 mm; dry woods or rocky slopes; mostly E and S; late May–Jul; ✿.

Diarrhena P. Beauv.

Diarrhena obovata, spikelet (×1½)

Diarrhena obovata (Gleason) Brandenburg American beakgrain

Perennial; culms 5–12 dm tall; principal blades 2–4 dm long by 10–18 mm wide, mid-vein not centered; panicle 1–3 dm long, drooping; spikelets 3–5-flowered, strongly flattened, disarticulating above the glumes and between the lemmas; first glume 1-nerved, triangular, second 3–5 nerved, oblong, with a cuspidate tip; lemmas conspicuously 3-nerved; rich woods; SW; ✿. [syn: D. americana P. Beauv. var. obovata Gleason]

Digitaria Haller

Annuals; spikelets single or in groups of 2 or 3 on one side of an elongate rachis; first glume minute to absent; second glume 5–7-nerved, ⅓ to as long as the spikelet.

A. rachis of the spike wingless, triangular in cross section .... D. filiformis
A. rachis of the spike broadly winged, the wing at least as wide as the midrib
    B. second glume nearly as long as the spikelet, covering the fruit except at the tip; fruit brown-black at maturity .... D. ischaemum
    B. second glume about ½ as long as the spikelet, the fruit exposed nearly to the base; fruit pallid to gray-brown at maturity
        C. pedicels of the spikelets acutely triangular, scabrous on the angles
            D. spikelets appressed pubescent .... D. sanguinalis
            D. spikelets strongly ciliate .... D. ciliaris
        C. pedicels of the spikelets rounded, smooth .... D. serotina

Digitaria ciliaris, spikelet (×5)

Digitaria ciliaris (Retz.) Koeler Southern crabgrass

Similar to D. sanguinalis except as noted in the key; dry, sandy, disturbed soil; SE; Aug–Sep; native to Asia. [syn: D. sanguinalis (L.) Scop. var. ciliaris (Retz.) Parl.]

Digitaria filiformis (L.) Koeler Slender crabgrass

Culms 3–10 dm tall, branched from the base, erect or ascending; lower sheaths pilose, upper pilose to glabrous; racemes 2–6; spikelets in 2s or 3s; first glume absent, second more than ½ as long as the spikelet, usually pubescent; serpentine barrens and other dry, open sites; mostly SE; Aug–Oct.

Digitaria ischaemum, spikelet (×5)

Digitaria ischaemum (Schreb. ex Schweigg.) Schreb. ex Muhl. Smooth crabgrass

Culms 2–5 dm tall, branched and spreading from a decumbent base; leaves glabrous, often bluish or purplish; racemes 2–6, 4–8 cm long; spikelets 1.7–2.1 mm long, often purple; second glume and fertile lemma usually pubescent; disturbed soil, roadsides, and waste ground; common throughout; Aug–Oct; native to Eurasia; UPL.

*Digitaria
ciliaris*

*Digitaria
filiformis*

*Digitaria
sanguinalis*

*Digitaria
sanguinalis,*
spikelet (×5)

Digitaria sanguinalis (L.) Scop.   Northern crabgrass
Culms 3–6 dm long, much-branched, decumbent and rooting at the nodes;
blades mostly 4–10 cm long by 5–10 mm wide, pilose; racemes 3–6, 5–15 cm
long; spikelets 2.4–3 mm long; sterile lemma strongly 5-nerved; disturbed soil,
roadsides, lawns, and waste ground; common throughout; Jul–Oct; native to Europe; FACU–.

Digitaria serotina (Walter) Michx.   Dwarf crabgrass
Culms 2–5 dm long, decumbent and rooting at the nodes; blades 3–10 mm wide
by 3–8 cm long; racemes 2–6, often curved; spikelets 1.5–1.7 mm long; first
glume absent, second glume ⅓ to ½ as long as spikelet; sterile lemma 5-nerved;
waste ground and ballast; Philadelphia Co.; Aug–Oct; native east of Pennsylvania; FAC.

Distichlis Raf.

Distichlis spicata (L.) Greene   Seashore saltgrass

Perennial; culms 2–4 dm tall; blades rigid, involute, mostly 5–10 cm long; panicle ovoid, 2–5 cm long; pistillate spikelets usually 5-flowered; staminate spikelets 8–12-flowered; first glume 1.7–3 mm long, second 2.4–3.8 mm long; lemmas 3.3–4.5 mm long; paleas about equaling the lemmas; waste ground and ballast; Philadelphia Co.; believed to be extirpated; Aug–Oct; FACW+; ✿.

*Distichlis
spicata,
rhizomes*

Distichlis spicata

*Echinochloa
muricata*

*Echinochloa
crusgalli var.
crusgalli*

*Eleusine
indica*

Echinochloa P.Beauv.

Annuals; spikelets in thick racemes that are aggregated into a panicle; first glume 3-nerved, $\frac{1}{3}$–$\frac{1}{2}$ as long as the spikelet; second glume and sterile lemma as long as the spikelet, prominently nerved.

A. racemes few, distant, appressed or ascending, simple .... E. colona
A. racemes numerous, spreading or ascending and overlapping
    B. second glume with an awn usually 2–10 mm long; sterile lemma usually long-awned .... E. walteri
    B. second glume awnless; sterile lemma awned or awnless
        C. spikelets copiously hispid with stout spreading spines arising from yellowish papillae .... E. muricata
        C. spikelets hispid with slender, more or less appressed spines, their bases not prominently papillose, or merely pubescent, without spinulose hairs .... E. crusgalli

Echinochloa
colona

Echinochloa colona (L.) Link   Jungle-rice

Culms 1–6 dm long, often prostrate at the base; blades 2–6 mm wide; panicle slender, 5–12 cm long; racemes 1–2 cm long, to 1 cm apart; spikelets in 4 rows, 2.5–3 mm long, awnless; second glume and sterile lemma pubescent; dry, waste ground; Berks Co.; Jul–Sep; native to Old World tropics; FACW.

Echinochloa
colona,
spikelet (×5)

Echinochloa crusgalli (L.) P.Beauv.

Culms to 1.5 m tall, often branched from a decumbent base; blades 5–15 mm wide; racemes numerous, 2–4 cm long; first glume about $\frac{1}{2}$ as long as the spikelet; FACU; 2 varieties:

Echinochloa crusgalli
var. crusgalli,
spikelet (×2)

A. spikelets greenish or purple-tinged; sterile lemma often with a well developed awn; plants mostly 3–7 dm .... var. crusgalli—barnyard-grass fields, meadows, roadsides, and waste ground; common throughout; Aug–Oct; native to Eurasia.
A. spikelets grayish-brown to chocolate brown; sterile lemma usually awnless; plants mostly 7–15 dm .... var. frumentacea (Roxb.) W.Wight—billion-dollar grass, Japanese-millet cultivated and occasionally escaped; scattered throughout; Jul–Oct; native to Eurasia.

Echinochloa
crusgalli var.
frumentacea,
spikelet (×5)

Echinochloa muricata (P.Beauv.) Fernald   Barnyard-grass, cockspur

Similar in habit to E. crusgalli, differing as noted in the key; moist ground and alluvial shores; throughout; Aug–Sep; FACW+. [syn: E. pungens (Poir.) Rydb.]

Echinochloa
muricata,
spikelet (×5)

Echinochloa walteri (Pursh) A.Heller   Walter's barnyard-grass

Culms 1–2 m; sheaths papillose to hispid; panicle 1–3 dm long, often nodding, purplish; awn of sterile lemma to 3 cm long; tidal marshes and mudflats; restricted to the Coastal Plain; Aug–Sep; FACW+; ✹.

Eleusine Gaertn.

Eleusine indica (L.) Gaertn.   Goosegrass, wiregrass

Annual; culms 3–6 dm tall, spreading or ascending, branched from the base; sheaths compressed and keeled, papillose to pubescent at the summit; spikes

Eleusine indica,
spikelet (×2)

Elymus
canadensis var.
canadensis

Elymus hystrix

usually 3–8, 4–10 cm long; spikelets usually 3–6-flowered, disarticulating above the glumes; lemmas 2.5–4 mm long, strongly nerved near the keel; gardens, disturbed soil, and waste ground; becoming common S; Jun–Oct; native to Old World tropics; FACU–.

Elymus L.

Perennials; spikelets 1–7-flowered, usually in pairs (sometimes solitary, or 3 or more) at each joint of the rachis, disarticulating above the glumes and between the lemmas; glumes usually equal, 1–5-nerved, the midnerve nearer the adaxial margin; fruit adherent to the lemma and palea.

A. spikelets solitary at all (or nearly all) nodes of the spike .... *E. trachycaulus*
A. spikelets 2–many at all (or nearly all) nodes of the spike
 B. glumes bristle-like, reduced, usually much shorter than the lower floret or lacking; spikelets horizontally spreading at maturity .... *E. hystrix*
 B. glumes at least as long as the lower floret; spikelets ascending at maturity
  C. awns divergently curled when mature and dry; palea of lower floret 9–15 mm long .... *E. canadensis*
  C. awns straight when mature and dry; palea of lower floret 5–9 mm long
   D. glumes 1–2 mm wide above, hardened and bowed-out above the base .... *E. virginicus*
   D. glumes 0.4–0.8 mm wide above, not evidently bowed-out above the base
    E. leaf blades villous above; spikelets 1–2-flowered .... *E. villosus*
    E. leaf blades entirely glabrous; spikelets 2–4-flowered ... *E. riparius*

Elymus canadensis L.   Canada wild-rye
Culms 1 m tall or more, cespitose; blades 5–20 mm wide, glabrous to sparsely pilose; spike 10–15 cm long, often interrupted at the base; spikelets 2–7-flowered; glumes 15–30 mm long including the awns; lemmas 3–5 cm long including the awns; 2 varieties:

A. glumes usually 20–35 mm long; spikelets 2–5-flowered; inflorescences dense; culm blades 4–9, often involute, 5–15 mm wide .... var. *canadensis*
 alluvial shores and thickets; scattered throughout, especially along the larger rivers and their tributaries; late Jul–early Sep; FACU+.
A. glumes usually 15–20 mm long; spikelets 4–7-flowered; inflorescences lax; culm leaves 10–18, lax and flat, 13–20 mm wide ....
 ....var. *wiegandii* (Fernald) Bowden
 alluvial shores and thickets; scattered throughout, especially along the larger rivers and their tributaries; Aug–Sep; FACU+. [syn: *E. wiegandii* Fernald]

Elymus hystrix L.   Bottlebrush-grass
Culms 6–10 dm tall, usually solitary; blades 8–13 mm wide; spike 5–12 cm long, spike internodes 4–10 mm long; glumes varying from 0–16 mm long even on the same plant; lemmas 8–11 mm long; awn 1–4 cm long; moist, alluvial woods; throughout; Jun–early Aug. [syn: *Hystrix patula* Moench]

Elymus riparius Wiegand   Riverbank wild-rye
Culms 10–15 dm tall; blades mostly 8–15 mm wide, scabrous; spike 6–15 cm long by 2–3.5 cm wide, somewhat nodding; glumes 3-nerved, 17–30 mm long including the awns; lemmas 25–45 mm long including the awns; alluvial flats, meadows, and stream banks; throughout; Jul–Sep; FACW.

Elymus trachycaulus (Link) Gould ex Shinners   Slender wheatgrass
Culms 4–10 dm tall, cespitose; blades 4–10 mm wide; spikes 6–20 cm long; glumes 5–7-nerved; lemmas awnless or with an awn to 2 cm long; open woods, barrens and banks; infrequent in the Poconos and a few other scattered sites at higher elevations; Jun–Jul; FACU; 🍂. [syn: *Agropyron trachycaulum* (Link) Steud.]

Elymus riparius

Elymus riparius, spikelet (×1)

Elymus trachycaulus, spikelet (×1)

Elymus trachycaulus

Elymus villosus,
spikelet (×1)

*Elymus
virginicus,*
spikelet (×1)

*Elytrigia
pungens*

*Elytrigia
pungens,*
spikelet (×1)

Elymus villosus Muhl. ex Willd.   Wild-rye
Culms 5–10 dm tall, cespitose; principal blades 5–10 mm wide, softly villous on the upper side; spike 5–12 cm long, dense, declined; glumes 15–30 mm long including the awn, strongly 1–3-nerved above; lemmas 2–4 cm long including the awn; stream banks, moist woods, and marshes; rare N and W, more common SE; Jun–early Aug; FACU–.

Elymus virginicus L.   Virginia wild-rye
Culms 5–13 dm tall, stout, cespitose; principal blades 3–15 mm wide; spike 5–15 cm long, dense, usually straight and erect; glumes 4–5-nerved; moist woods, meadows, and stream banks; throughout; Jun–Sep; FACW–.

Elytrigia Desv.

Perennials, strongly rhizomatous; spikelets 4–11-flowered, disarticulating above the glumes and between the lemmas.

A. glumes flexible; leaf blades flat and lax, pubescent above, finely nerved ....
.... *E. repens*
A. glumes rigid; leaf blades mostly involute, stiff, glabrous, and glaucous, coarsely nerved ....*E. pungens*

Elytrigia pungens (Pers.) Tutin   Saltmarsh wheatgrass
Culms 5–10 dm tall; spikes 8–12 cm long; spikelets 7–11-flowered; lemmas awnless; ditches, rubbish heaps and ballast; adventive in Philadelphia Co.; native N of Pennsylvania; Jun–Aug; FACW. [syn: *Agropyron pungens* (Pers.) Roem. & Schult.]

Elytrigia repens (L.) Desv. ex Nevski   Quackgrass, witchgrass
Culms usually 5–10 dm tall; spikes 6–17 cm long; spikelets 4–8-flowered; lemmas awnless or with an awn up to 10 mm long; fields, roadsides, and waste ground; common throughout; late May–Jul; native to Eurasia. [syn: *Agropyron repens* (L.) P.Beauv.]

Eragrostis Wolf

Spikelets 2–many-flowered, mostly flattened, disarticulating above the glumes and between the lemmas, the lemma often remaining attached to the fruit; glumes usually unequal; lemmas 3-nerved, awnless, rounded or keeled on the back.

A. plants prostrate, forming mats, rooting at the lower nodes .... *E. hypnoides*
A. plants not prostrate (although sometimes decumbent) or mat-forming, not rooting at the nodes
   B. spikelets with 2–4 (rarely 5) florets
      C. panicle very diffuse, making up ⅔ or more of the height of the erect plant; pedicels of lateral spikelets 2–many times as long as spikelet .....
      .... *E. capillaris*
      C. panicle open but not diffuse, making up about ½ of the height of the erect plant; pedicels of the lateral spikelets usually less than twice the length of spikelet .... *E. frankii*
   B. spikelets with 6–many florets (rarely 5 in depauperate individuals)

D. plants with conspicuous tufts of white hairs in the throats of sheaths; perennial with hard, knotty bases .... E. *spectabilis*

D. plants without conspicuous tufts of hairs (although a few hairs are usually present) in the throats of sheaths; annuals with soft bases

    E. plants with conspicuous glandular dots on the keels of the lemmas and branches of panicles

        F. spikelets 2.5–3.5 mm wide .... E. *cilianensis*

        F. spikelets 1.3–2 mm wide .... E. *minor*

    E. plants lacking glandular dots

        G. spikelets diverging strongly from the panicle branches at maturity .... E. *pilosa*

        G. spikelets mostly closely appressed at maturity .... E. *pectinacea*

Eragrostis capillaris (L.) Nees   Lacegrass
Annual; culms 2–5 dm tall, densely cespitose; blades 2–4 mm wide; spikelets 2–3 mm long; lemmas 1.2–1.6 mm long; open woods, roadsides, and waste ground, in dry sandy soil; common SE, scattered elsewhere; late Jul–early Oct.

Eragrostis cilianensis (All.) F.T.Hubb.   Stinkgrass
Annual; culms 1–4 dm long, spreading from a decumbent base, densely cespitose; blades 3–12 cm long by 2–6 mm wide; spikelets 10–40-flowered; lemmas 2.1–2.6 mm long; fields, gardens, and waste ground; throughout, but more common SE; Jul–Oct; native to Europe; FACU.

Eragrostis frankii C.A.Mey. ex Steud.   Lovegrass
Annual; culms 1–3 dm tall, much-branched, densely cespitose; blades 1–3 mm wide; lemmas 1–1.5 mm long; moist stream banks; infrequent along the Delaware and Susquehanna Rivers, rare elsewhere; Sep; FACW.

Eragrostis hypnoides (Lam.) Britton, Stearns & Poggenb.   Creeping lovegrass
Annual; culms creeping, erect flowering portions 5–15 cm tall; blades 1–3 cm long by 1–3 mm wide; panicle 2–8 cm long; spikelets linear, 10–35-flowered; lemmas 1.5–2 mm long; wet shores and mudflats; occasional along the larger rivers and their tributaries; late Jun–Sep; OBL.

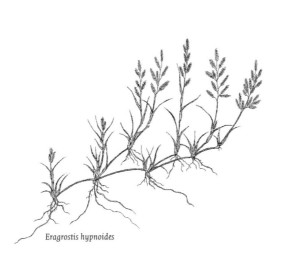

Eragrostis cilianensis

Eragrostis hypnoides

Eragrostis minor Host   Lovegrass

Annual; culms 1–4 dm long, often decumbent; blades 3–10 cm long by 2–5 mm wide; panicle 3–10 cm long, with spreading branches; spikelets 10–20-flowered, lemmas 1.7–1.9 mm long, glandular on the keel; dry roadsides and railroad embankments; scattered throughout; late Jun–Sep; native to Europe. [syn: *E. poaeoides* P.Beauv.]

Eragrostis pectinacea (Michx.) Nees   Carolina lovegrass

Annual; culms 1–5 dm tall, much-branched, densely cespitose, usually erect; blades 1–3 mm wide; panicle diffusely branched, often ½ as long as entire plant; spikelets 5–11-flowered; lemmas 1.5–1.8 mm long; gardens and waste ground; common throughout; Jul–early Oct; FAC.

Eragrostis pilosa (L.) P.Beauv.   India lovegrass

Annual; culms 1–6 dm tall, usually decumbent at the base, densely cespitose; blades 2–3 mm wide; panicle diffusely branched, 5–20 cm long; spikelets 5–9-flowered; lemmas 1.3–1.6 mm long; gardens, roadsides, and ballast; mostly SE, scattered elsewhere; Jul–Oct; native to the tropics; FACU.

Eragrostis spectabilis (Pursh) Steud.   Purple lovegrass, tumblegrass

Perennial; culms 3–6 dm tall, usually erect, cespitose; blades 3–6 mm wide, firm and stiff; panicle about ⅔ the length of the entire plant; spikelets 5–10-flowered, purple; lemmas 1.6–2.1 mm long, scabrous on the keel; dry, sandy fields and roadsides; common SE, rare elsewhere; Jun–Oct; UPL.

Eragrostis spectabilis,
spikelet (×2)

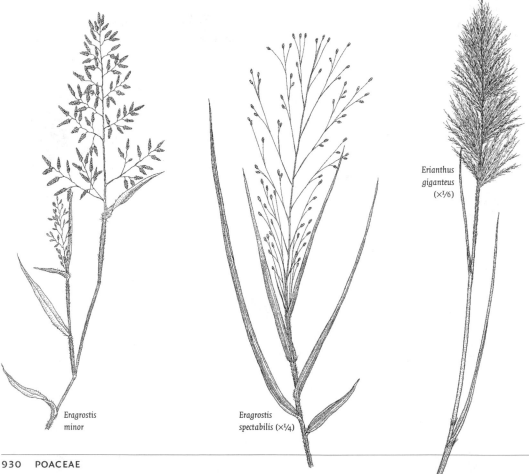

Erianthus giganteus (×⅙)

Eragrostis minor

Eragrostis spectabilis (×¼)

Erianthus Michx.

Erianthus giganteus (Walter) Muhl.   Giant beardgrass, sugarcane-plumegrass
Perennial; culms 1–3 m; smooth below, silky-hairy below the panicle, often hairy on the nodes; panicle narrow, 1.5–4 dm long, purplish; spikelets about 6 mm long; awns 15–25 mm long, not twisted; swamps and stream banks; SE; believed to be extirpated; FACW+; ✿. [syn: *E. saccharoides* (Walter) Muhl.]

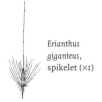

Erianthus giganteus, spikelet (×1)

Eriochloa Kunth

Eriochloa villosa (Thunb.) Kunth   Chinese cupgrass
Annual; culms 3–8 dm tall, tufted; blades 4–7 mm wide, pubescent; panicle 10–15 cm long, base enclosed in upper sheath, bearing several racemes each 1–2 cm long, rachis and pedicels pubescent; spikelets 3.5–4 mm long; fertile lemma with an awn about 0.8 mm long; open, disturbed ground; Berks Co. and probably elsewhere; native to eastern Asia.

Eriochloa villosa

Festuca L.

Perennials; spikelets 3–11-flowered, disarticulating above the glumes and between the lemmas; glumes narrow, unequal, usually shorter than the lemmas; paleas about equaling the lemmas.

A. leaf blades flat and lax, mostly 3 mm or more wide
    B. spikelets 10–20 mm long; leaf blades auriculate at the base
        C. auricles glabrous; internodes of the rachilla smooth or nearly so .... *F. pratensis*
        C. auricles ciliate; internodes of the rachilla scabrous .... *F. elatior*
    B. spikelets 8 mm or less long; leaf blades not auriculate at the base
        D. principle lowermost branches of the panicle bearing usually 2–7 spikelets along their outer half; leaf blades lax .... *F. obtusa*
        D. principle lowermost branches of the panicle bearing 8–20 spikelets clustered near their ends; leaf blades firm .... *F. paradoxa*
A. leaf blades mostly folded or involute, 2 mm or less wide
    E. leaf blades capillary; lemmas 2.4–3.4 mm long; awnless or with short awn-tips .... *F. tenuifolia*
    E. leaf blades narrow or involute, but not capillary; lemmas 4–7 mm long with awns usually 0.5 mm or more long
        F. basal sheaths soon disintegrating into fibers; plants often rhizomatous .... *F. rubra*
        F. basal sheaths persistent; plants never rhizomatous
            G. spikelets mostly 5–7 mm long .... *F. ovina*
            G. spikelets mostly 7–15 mm long .... *F. trachyphylla*

Festuca elatior L.   Fescue
Culms 8–10 dm tall, usually bent at the base, glabrous; blades 4–8 mm wide, glabrous; panicle 1–2 dm long, often nodding at tip; spikelets usually 7–8-flowered; first glume subulate, 2.5–4.5 mm long, second glume lanceolate, 3.5–7 mm long; lemmas 5.5–8 mm long; roadsides, fields, and open ground; throughout; late May–Aug; native to Europe; FACU–.

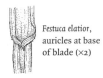

Festuca elatior, auricles at base of blade (×2)

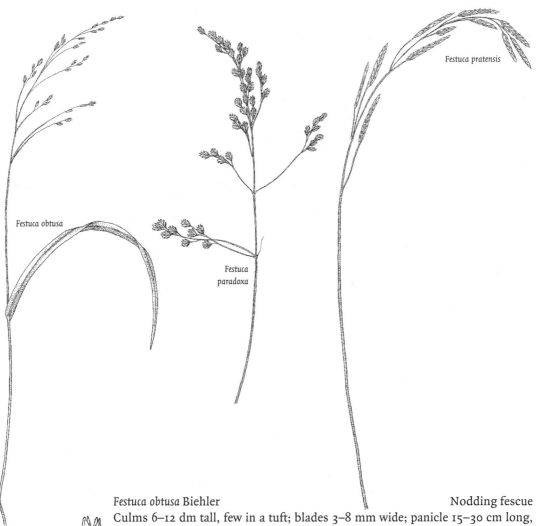

Festuca pratensis

Festuca obtusa

Festuca
paradoxa

Festuca obtusa,
spikelet (×2)

Festuca ovina,
spikelet (×2)

Festuca obtusa Biehler — Nodding fescue

Culms 6–12 dm tall, few in a tuft; blades 3–8 mm wide; panicle 15–30 cm long, branches bearing spikelets only above the middle; spikelets usually 3-flowered (rarely to 5-flowered); first glume subulate, 2.2–3.8 mm long, second glume ovate, 2.7–4 mm long; lemmas 3.3–4.7 mm long; moist woods and clearings; throughout; late May–Jun; FACU.

Festuca ovina L. — Sheep fescue

Culms 1–5 dm tall, densely cespitose, glabrous; blades about 0.5 mm wide; panicle very narrow, almost spike-like, 1–10 cm long; spikelets mostly 4–6-flowered; open woods, dry fields, and roadsides; throughout; May–Aug; native to Europe.

Festuca paradoxa Desv. — Cluster fescue

Similar to *F. obtusa* in general habit; spikelets 3–6-flowered, 5–8 mm long; first glume subulate, 2.4–4.2 mm long, second glume elliptic-oblong, 3.1–5.2 mm long; lemmas 3.6–5.2 mm long; moist, open ground and thickets; scattered S and W; May–Jul; FAC; ✿.

Festuca pratensis Huds. — Meadow fescue

Similar to *F. elatior* in habit; spikelets 7–9-flowered; lowest panicle branch less than ⅓ as long as the panicle; meadows, moist shores, roadsides, and disturbed ground; throughout; late May–Jul; native to Europe; FACU–.

Festuca rubra L.   Red fescue
Culms 3–10 dm tall, loosely tufted, often decumbent at the base; blades about 1 mm wide; lower sheaths reddish; panicle 5–20 cm long, usually narrow with ascending branches; spikelets 4–7-flowered; first glume subulate, 2.6–4.5 mm long, second glume 3.5–5.5 mm long; lemmas 4.8–6.1 mm long; dry woods, roadsides, waste ground, and ballast; scattered throughout; May–early Jul; native to Europe; FACU.

Festuca tenuifolia Sibth.   Hair fescue
Culms 3–10 cm tall, densely tufted; blades usually scabrous; panicle 10–20 cm long, narrow with ascending branches; spikelets 5–7-flowered; lemmas about 7 mm long; awns to 4 mm long; dry, open ground and oak woods; scattered throughout; late May–Jun; native to Europe. [syn: *F. capillata* Lam.]

Festuca
tenuifolia,
spikelet (×2)

Festuca trachyphylla (Hack.) Krajina   Hard fescue
Similar to *F. ovina*; blades somewhat scabrous, 0.5–1 mm wide; spikelets 4–8-flowered; dry, open soil; mostly SE; native to Europe. [*F. longifolia* of Rhoads and Klein, 1993]

Glyceria R.Br.

Perennials; spikelets 3–many-flowered, disarticulating above the glumes and between the lemmas; glumes unequal, shorter than the lemmas, often scarious at the apex and margins; lemmas awnless, rounded on the back, usually scarious at the apex, 5–9-nerved.

A. spikelets elongate-cylindric, 1–4 cm long
    B. lemmas acute, much exceeded by the paleas .... *G. acutiflora*
    B. lemmas obtuse, slightly exceeded by the paleas
        C. lemmas scaberulous on the nerves only; first glume 1–2 mm long; second glume 1.9–2.7 mm long .... *G. borealis*
        C. lemmas scaberulous over the surface; first glume 2–4 mm long; second glume 3.2–5.3 mm long .... *G. septentrionalis*

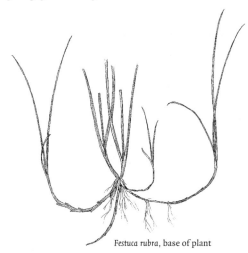
Festuca rubra, base of plant

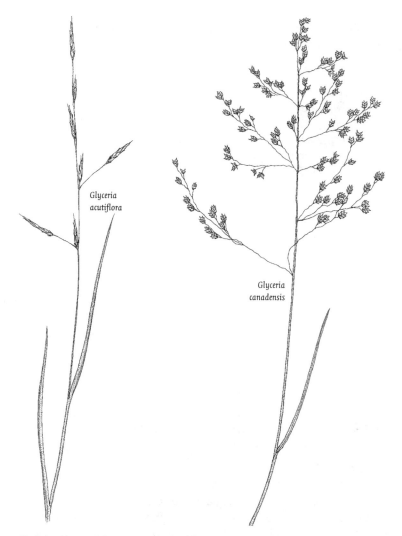

Glyceria
acutiflora

Glyceria
canadensis

A. spikelets ovate-oblong, somewhat flattened, 0.7 cm or less long
  D. panicle strict, the branches strongly ascending
    E. lemmas about 3.5 mm long; panicle dense, oblong-cylindric, 15 cm or less long .... G. obtusa
    E. lemmas 2–2.7 mm long; panicle slender, 15–36 cm long ..... G. melicaria
  D. panicle open, the branches spreading or drooping
    F. spikelets broadly ovate, 2–5 mm wide; lemmas not strongly nerved .... G. canadensis
    F. spikelets oblong-ovate, 1.2–2.1 mm wide; lemmas strongly nerved
      G. first glume 1 mm or less long .... G. striata
      G. first glume 1.4 mm or more long .... G. grandis

Glyceria acutiflora, spikelet (×1)

Glyceria acutiflora Torr.   Mannagrass
Culms to 1 m tall, often decumbent and rooting at the lower nodes; blades 3–6 mm wide; ligule 5–7 mm long; panicle 2–3 dm long, narrow, the branches erect; spikelets 5–12-flowered; first glume 2–3.3 mm long, second glume 4–5.5 mm long; lemmas 7–8.5 mm long; shallow water, muddy shores, and swamps; uncommon in mountains of central Pennsylvania, also NE and SE; late May–Aug; OBL.

Glyceria borealis (Nash) Batch. Northern mannagrass, small floating mannagrass
Culms to 1.2 m tall, decumbent at the base and rooting at the lower nodes; blades 2–5 mm wide; panicle 2–5 dm long, slender, the branches 8–12 cm long; spikelets 8–12-flowered; lemmas 3.1–3.9 mm long; shallow water of lakes and streams; mostly NE, scattered elsewhere; Jul; OBL; 🌿.

Glyceria borealis, spikelet (×2½)

Glyceria canadensis (Michx.) Trin.   Rattlesnake mannagrass
Culms to 1 m tall, solitary or a few in a tuft; blades 3–8 mm wide; panicle diffuse, with drooping branches bearing spikelets mostly toward the tips; spikelets ovate, 5–10-flowered, 4–8 mm long; lemmas 2.9–4 mm long; moist woods, marshes, swamps, and wet shores; throughout; late Jun–early Sep; OBL.

Glyceria canadensis, spikelet (×2½)

Glyceria canadensis x grandis
A hybrid found nearly throughout, distinguished from *G. canadensis* by having spikelets that are 3–5 mm long and 3–6-flowered; marshes, swamps, and wet woods; mostly in mountains NE; late Jun–early Sep; OBL. [syn: *G. canadensis* (Michx.) Trin. var. *laxa* (Scribn.) A.Hitchc.]

Glyceria grandis S.Watson   American mannagrass
Culms to 1.5 m tall, usually clustered; blades 8–12 mm wide; panicle 2–4 dm long, much-branched, usually nodding; spikelets 5–9-flowered; lemmas 2.1–2.7 mm long, distinctly 7-nerved, usually purple; shallow water or wet meadows; mostly N; Jun–Jul; OBL.

Glyceria grandis, spikelet (×2½)

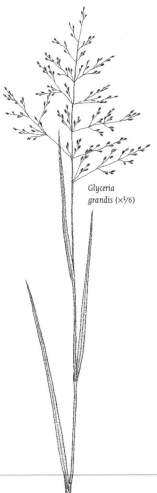

Glyceria grandis (×⅙)

Glyceria melicaria,
spikelet (×2½)

Glyceria melicaria (Michx.) F.T.Hubb. Slender mannagrass

Culms 5–10 dm tall; blades 2–5 mm wide; panicle very narrow, with 1–3 branches from each node; spikelets 2–4-flowered; wet woods, stream banks, and swamps; throughout except on the Coastal Plain; Jun–early Sep; OBL.

Glyceria obtusa,
spikelet (×2½)

Glyceria obtusa (Muhl.) Trin. Coastal mannagrass, blunt mannagrass

Culms 6–10 dm tall, often decumbent at the base; blades 3–8 mm wide; panicle bearing 3–8 branches from each node; spikelets 4–6-flowered; swamps, bogs and moist, sandy, peaty ground; rare on the Coastal Plain and scattered sites elsewhere; Jul–Aug; OBL; 🐾.

Glyceria
septentrionalis,
spikelet (×1)

Glyceria septentrionalis A.Hitchc. Floating mannagrass

Culms 10–15 dm tall, often rooting at the lower nodes; principal blades 6–10 mm wide; panicle 2–4 dm long, narrow; spikelets 8–16-flowered, 1–2 cm long; lemmas 3.7–5.3 mm long; wet meadows and shallow water of stream margins; mostly S; late May–Aug; OBL.

Glyceria striata,
spikelet (×4)

Glyceria striata (Lam.) A.Hitchc. Fowl mannagrass

Culms 5–12 dm tall; principal blades 2–5 mm wide; panicle 1–2 dm long, with spikelets mostly beyond the middle; spikelets 3–6-flowered, sometimes purplish; lemmas 1.4–2.1 mm long; wet woods, swamps, and bogs; common throughout; late May–Aug; OBL.

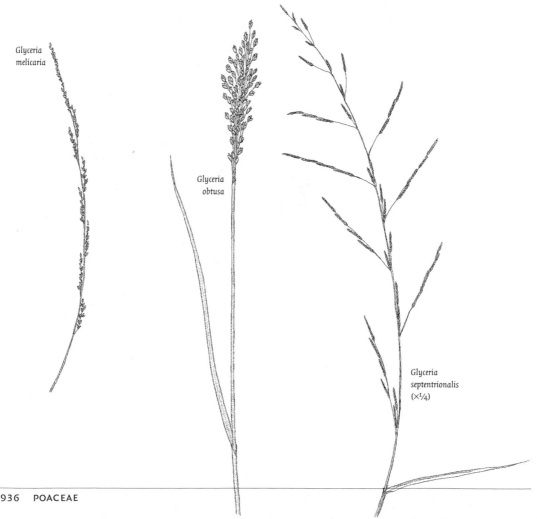

Glyceria
melicaria

Glyceria
obtusa

Glyceria
septentrionalis
(×¼)

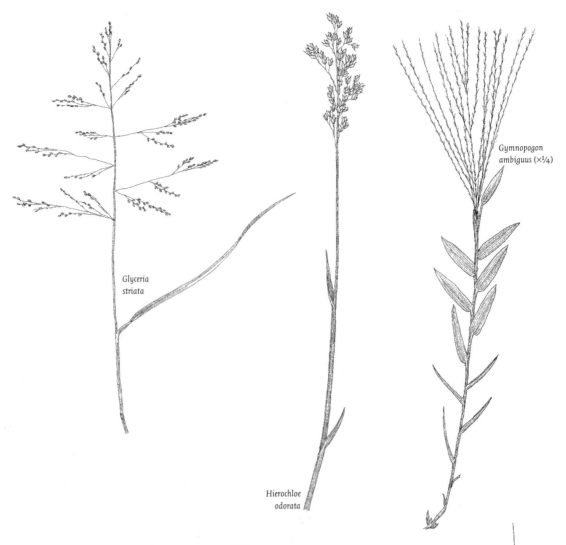

Glyceria
striata

Hierochloe
odorata

Gymnopogon
ambiguus (×¼)

Gymnopogon P.Beauv.

Gymnopogon ambiguus (Michx.) Britton, Stearns & Poggenb.

Broad-leaved beardgrass

Perennial; culms 3–6 dm tall, mostly solitary; leaves numerous, blades 6–12 mm wide by 5–10 cm long, rounded to subcordate at the base; panicle ⅓ to ½ as long as the entire plant; spikelets 1- flowered; glumes 4–6 mm long; lemma 3.5–4.3 mm long; awn 4.5–9 mm long; serpentine barrens; Lancaster Co.; believed to be extirpated; Jul–early Oct; ✤.

*Gymnopogon
ambiguus,*
spikelet
(×2½)

Hierochloe R.Br.

Hierochloe odorata (L.) P.Beauv.

Vanilla sweetgrass

Perennial; culms 3–6 cm; sheaths few, bladeless or with blades to 3 cm long; panicle 5–10 cm long, branches widely spreading to drooping; glumes 4–6 mm long; staminate lemmas usually pubescent; fertile lemma pubescent at apex; moist meadow or river shore; rare; W; May–Jun; FACW; ✤.

*Hierochloe
odorata,*
spikelet
(×2½)

Holcus
lanatus

Hordeum
vulgare

Koeleria
macrantha

Holcus L.

Holcus lanatus,
spikelet (×2½)

***Holcus lanatus* L.**   Velvetgrass
Perennial; culms 4–10 dm tall, softly hairy below, usually glabrous on the upper internode; blades 4–10 mm wide, pale green, softly pubescent; sheaths also softly pubescent; panicle dense, 5–15 cm long; spikelets disarticulating below the glumes; glumes 4–5 mm long; awn of second lemma hooked; meadows, old fields, river shores, and roadsides; common throughout; Jun–Aug; native to Europe; FACU.

Hordeum L.

Hordeum
vulgare,
spikelet

***Hordeum vulgare* L.**   Barley
Annual; culms 3–12 dm tall; principal blades 3–15 mm wide; spike 3–10 cm long excluding the awns, erect to slightly nodding; lemmas about 1 cm long; awn 6–15 cm long; cultivated and occasionally persisting on the edges of fields; throughout; late May–early Aug; native to Eurasia.

Koeleria Pers.

Koeleria macrantha,
spikelet (×2½)

***Koeleria macrantha* (Ledeb.) Schultes**   Junegrass
Perennial; culms 3–6 dm tall, cespitose, pubescent below the panicle; blades 1–3 mm wide, flat or involute when dry; panicle spike-like, 5–12 cm long by 1–2 cm

wide; spikelets usually 2-flowered, disarticulating above the glumes and between the lemmas; dry soil; Bradford Co.; believed to be extirpated; 🐾. [syn: K. *cristata* (L.) Pers.]

Leersia Sw.

Perennials; spikelets 1-flowered; lemma 3–5-nerved; palea about as long as the lemma, 3-nerved.

A. spikelets 4.2–5 mm long; panicles with 2 or more branches at some nodes ... . *L. oryzoides*
A. spikelets 2.9–3.9 mm long; panicles with 1 branch per node .... *L. virginica*

Leersia oryzoides (L.) Sw.   Rice cutgrass

Leersia
oryzoides,
spikelet
(×2½)

Culms 7–15 dm tall, often decumbent at the base; blades 6–12 mm wide by 15–20 cm long, rough-spiny on the margins; sheaths scabrous; ligules about 1 mm long; panicle 1–2 dm long, branches bearing spikelets in the upper ½ to ⅔; spikelets 3.8–6 mm long, in spike-like racemes about 1 cm long; marshes, bogs, and wet meadows; throughout; late Jul–Sep; OBL.

Leersia virginica Willd.   Cutgrass, whitegrass

Leersia
virginica,
spikelet
(×2½)

Culms 5–15 dm long, often decumbent and rooting at the base; blades 3–8 mm wide by 5–10 cm long, scabrous on the margins; ligule very short; panicle 5–20

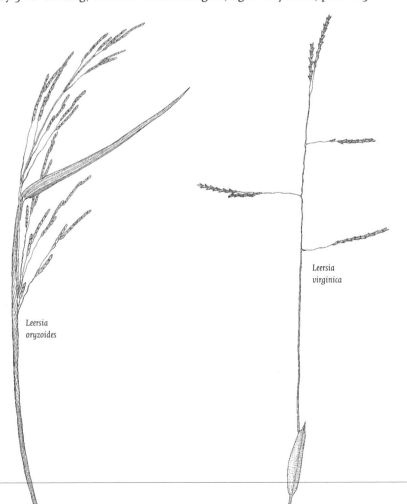

Leersia
oryzoides

Leersia
virginica

cm long, branches solitary, bearing spikelets from about the middle; spikelets in spike-like racemes 1–2 cm long; swamps or moist woods; throughout; Jul–Sep; FACW.

Leptochloa P.Beauv.

Annuals; spikelets 2–several-flowered, disarticulating above the glumes; glumes unequal, 1-nerved; lemmas 3-nerved; palea nearly as long as the lemma.

A. spikelets 1–2.5 mm long, 2–4-flowered; sheaths sparsely pilose ... *L. filiformis*
A. spikelets 4–10 mm long, 6–12-flowered; sheaths glabrous to scabrous ....
.... *L. fascicularis*

Leptochloa fascicularis (Lam.) A.Gray   Sprangletop
Culms 1–10 dm tall, erect to sometimes reclined, densely branched from the base; blades 1–3 mm wide, scabrous; sheaths usually smooth; lemmas 2-toothed at apex, awned from between the teeth, pubescent near the base; FACW; 2 varieties:

Leptochloa fascicularis var. acuminata, spikelet (×1½)

A. plant usually 5–10 dm .... var. *acuminata* (Nash) Gleason
    roadsides, railroad sidings, and waste ground, spreading along roadways; S
    and NW; probably introduced from southern North America.
A. plant usually <5 dm .... var. *maritima* (E.P.Bicknell) Gleason
    SE; a coastal plant, possibly adventive; ❦.

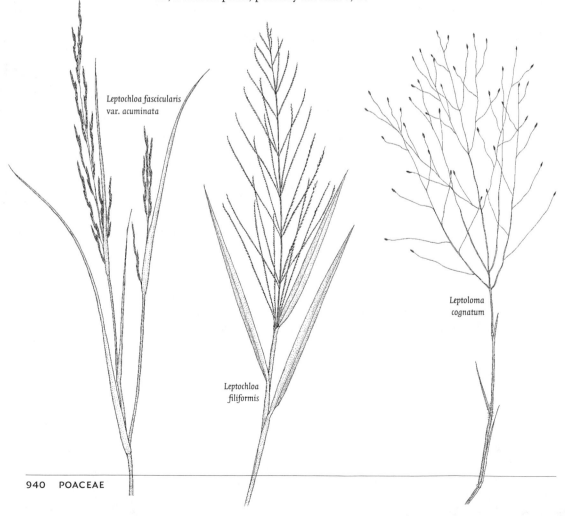

Leptochloa fascicularis var. acuminata

Leptochloa filiformis

Leptoloma cognatum

Leptochloa filiformis (Lam.) P.Beauv. Red sprangletop
Culms 3–10 dm tall, simple or branched from the base; larger blades 6–10 mm wide, scabrous on the margins; panicle 1–5 dm long, spikes numerous; lemmas 1.3–1.5 mm long, pubescent on the nerves near the base; wharves, railroad sidings, and rubbish dumps; mostly SE, but probably elsewhere; native to tropical America.

Leptochloa
filiformis,
spikelet (×5)

Leptoloma Chase

Leptoloma cognatum (Schult.) Chase Fall witchgrass
Perennial; culms 4–7 dm tall, tufted from a knotty base; blades 2–6 mm wide by 5–10 cm long; panicle very diffuse, usually ⅓ to ½ the entire length of the plant; spikelets 2.5–3 mm long, solitary at the ends of the peduncles; sandy, open ground; on or near the Coastal Plain and in Erie Co.; Jul–early Oct. [syn: *Digitaria cognatum* (Schultes) Chase]

Leptoloma
cognatum,
spikelet (×5)

Lolium L.

Spikelets several-flowered, disarticulating above the glumes and between the lemmas; Jun–Sep.

A. glume equaling or extending beyond the uppermost floret of the spikelet .... L. temulentum
A. glume not equaling the uppermost floret
 B. lemmas usually awnless; spikelets usually with 10 or fewer florets .... L. perenne
 B. lemmas usually with awns; spikelets usually with 10 or more florets .... L. multiflorum

Lolium
perenne

Lolium multiflorum Lam. Ryegrass
Annual or short-lived perennial; culms 3–8 dm tall, usually scabrous below the spike; principal blades 2–4 mm wide, scabrous above; joints of the spike rough on the side opposite the spikelets; spikelets 1–2 cm long; lowest lemma 5.5–8 mm long; cultivated and occasionally escaped; mostly SE; native to Europe. [syn: L. perenne var. italicum Parl.]

Lolium perenne L. Perennial ryegrass
Short-lived perennial; similar to L. multiflorum but smooth throughout; cultivated and frequently escaped; throughout; native to Europe; FACU–.

Lolium temulentum L. Darnel, poisongrass
Annual; culms 4–8 dm tall; principal blades 3–8 mm wide, scabrous above; spikelets 5–8-flowered; glume 12–22 mm long, 5–7-nerved; waste ground and ballast; mostly SE; native to the Mediterranean region.

Lolium
temulentum

Melica L.

Melica nitens Nutt. Tall melicgrass, three-flowered melicgrass
Perennial; culms 5–12 dm tall; blades 6–12 mm wide; panicle 1–2 dm long, few-

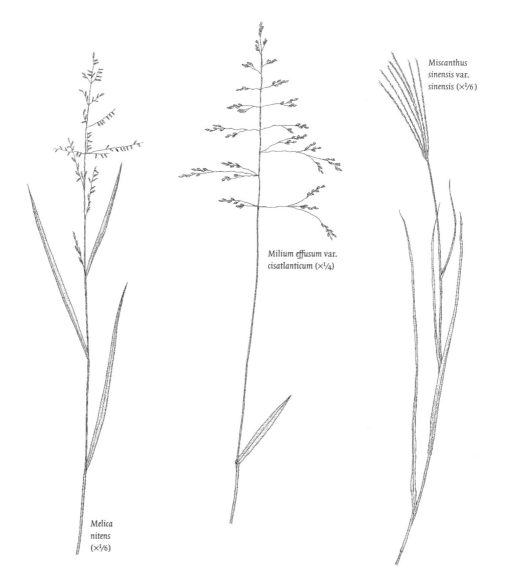

Miscanthus
sinensis var.
sinensis (×¹⁄₆)

Milium effusum var.
cisatlanticum (×¹⁄₄)

Melica
nitens
(×¹⁄₆)

Melica nitens,
spikelet (×1¹⁄₂)

branched; spikelets 2–3-flowered, 9–12 mm long; steep, rocky slopes and river banks; mostly along the lower Susquehanna and Casselman Rivers; late May–Jul; 🍂.

Microstegium Nees

Microstegium vimineum (Trin.) A.Camus   Stiltgrass

Annual or short-lived perennial; culms 6–10 dm long, decumbent and spreading; blades lanceolate, 5–10 mm wide by 3–8 cm long; panicle with 1–6 racemes; racemes 2–5 cm long; glumes about 5 mm long, awnless; lemmas shorter than the glumes, fertile lemma with an awn 4–8 mm long; moist ground of open woods, thickets, paths, clearings, fields and gardens; mostly SE, but spreading rapidly and probably elsewhere; Sep–early Nov; native to tropical Asia; FAC. [syn: *Eulaia viminea* (Trin.) Kuntze var. *variabilis* Kuntze]

Milium L.

Milium effusum L.   Milletgrass

Milium effusum var. cisatlanticum, spikelet (×2½)

Perennial; culms 1–4 mm long, often prostrate; principal blades lax, 10–20 cm long by 8–15 mm wide; panicle 10–20 cm long, its spreading branches bearing spikelets beyond the middle; glumes equal, about 3 mm long; lemma about as long as the glumes, white and shining at maturity; palea similar in texture to lemma; cool, rich woods; mostly N, but extending S in the mountains; late May–Jul. Ours is var. cisatlanticum Fernald.

Miscanthus Andersson

Miscanthus sinensis Andersson

Miscanthus sinensis var. sinensis, spikelet (×1½)

Perennials; culms 2–4 m tall, bunched; leaves mostly basal, to 1 m long, scabrous along the margins; panicle fan-shaped, of several simple racemes; cultivated and frequently escaped; mostly SE, but probably elsewhere; Aug–Oct; native to China; FACU; 2 varieties:

A. leaves not variegated .... var. sinensis—Japanese plumegrass, eulalia
A. leaves variegated ....var. zebrinus Beal—zebragrass

Muhlenbergia Schreb.

Perennials; spikelets 1-flowered, disarticulating above the glumes; glumes nearly equal, mostly 1-nerved, sometimes very reduced; lemma obscurely 3–5-nerved, rounded on the back; fruit permanently enclosed by the lemma; perennials, mostly rhizomatous.

A. plants with prominent creeping, scaly rhizomes
    B. glumes reduced, the first about ¼ as long as the floret, the second slightly
       longer ....M. frondosa x schreberi
    B. glumes at least ½ as long as the floret
       C. glumes broad below, abruptly tapering to an acuminate apex
          D. spikelets 1.7–2.5 mm long, awnless .... M. sobolifera
          D. spikelets (excluding awns) 3–4.5 mm long, awned....... M. tenuiflora
       C. glumes tapering from near the base to an acuminate or awned apex
          E. glumes about as long as the floret or shorter, the second 3.5 mm or
             less
             F. culms freely branching, with numerous axillary, often partly in-
                cluded panicles; internodes glabrous and shining ..... M. frondosa
             F. culms sparingly branching, the branches mostly elongated and
                bearing exserted terminal panicles
                G. panicle slender, contracted
                   H. panicle branches bearing spikelets mostly to their bases;
                      spikelets sessile or subsessile .... M. mexicana
                   H. longer panicle branches naked at their bases; spikelets on
                      pedicels at least as long as the glumes .... M. sylvatica
                G. panicle diffuse .... M. asperifolia
          E. glumes much exceeding the floret, 4.5 mm or more long ....
                .... M. glomerata

A. plants tufted or sprawling and rooting at the lower nodes, but not rhizomatous
 I. panicle diffuse, branches spreading
  J. lemmas awnless; spikelets about 1.5 mm long .... *M. uniflora*
  J. lemmas bearing awns 5–15 mm long; spikelets (excluding awns) 4–5.5 mm long .... *M. capillaris*
 I. panicle slender, branches appressed
  K. glumes minute, 0.1–0.3 mm long, or lacking .... *M. schreberi*
  K. glumes 1.7–2.8 mm long .... *M. cuspidata*

Muhlenbergia asperifolia (Nees & Meyen) Parodi   Scratchgrass
Culms 1–5 dm long, cespitose; spreading or decumbent at the base, scabrous below the nodes, branching mostly above the middle; principal blades 3–7 cm long by 2–4 mm wide; panicle slender, 2–5 cm long, its base often enclosed in the subtending sheath; glumes scabrous, 2–2.9 mm long; lemma 2.4–2.8 mm long, awnless; waste ground; Bedford and Blair Cos.; native to western and midwestern U.S.; FACW.

Muhlenbergia capillaris (Lam.) Trin.   Hairgrass, short muhly
Culms 5–10 dm tall, cespitose; principal blades 2–3 dm long by 2–4 mm wide; sheaths glabrous to somewhat scabrous; ligule 2–5 mm long; panicle usually ⅓–½ as long as the entire plant; spikelets purple; first glume 1.2–2.5 mm long, second a little longer; lemma 3–4.5 mm long excluding awns; river shores; Lancaster Co.; believed to be extirpated; Sep–Oct; FACU–; ✹.

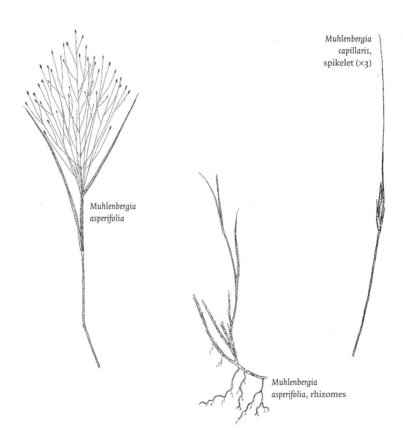

Muhlenbergia capillaris, spikelet (×3)

Muhlenbergia asperifolia

Muhlenbergia asperifolia, rhizomes

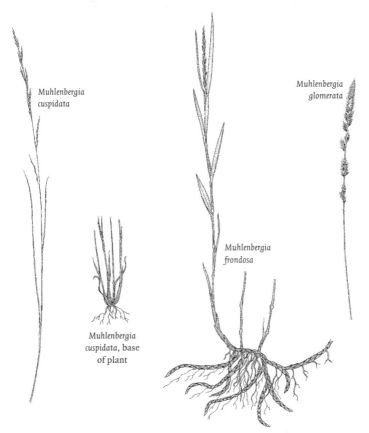

Muhlenbergia
cuspidata

Muhlenbergia
glomerata

Muhlenbergia
frondosa

Muhlenbergia
cuspidata, base
of plant

Muhlenbergia cuspidata (Torr.) Rydb.   Sharp-pointed muhly
Culms 2–7 dm tall, cespitose, strictly erect; blades flat or involute, 1–2 mm wide;
ligule usually <0.5 mm long; panicle spike-like, its branches 5–15 mm long;
glumes about equal; lemma 2.8–4.1 mm long, awnless; alluvial shores and rock
crevices; Northampton Co. along the Delaware River; native to western and
midwestern U.S., possibly adventive in Pennsylvania; 🍂.

Muhlenbergia
cuspidata,
spikelet (×5)

Muhlenbergia frondosa (Poir.) Fernald   Wirestem muhly
Culms often reclined, 5–10 dm long, much branched above the middle, glabrous
below the nodes; principal blades 5–10 cm long by 2–5 mm wide; panicles 3–8
cm long, often not surpassing the upper blades and partly enclosed by the subtending leaf sheath; glumes 1.6–3.1 mm long, nearly equal; lemmas 2.1–3.5 mm
long, usually awnless; moist, open woods and stream banks; common throughout; late Aug–early Oct; FAC.

Muhlenbergia
frondosa,
spikelet (×5)

Muhlenbergia frondosa x schreberi   Muhly
A hybrid with greatly reduced glumes similar to *M. schreberi* but overall plant size
similar to *M. frondosa*; Northampton and Delaware Cos., but may occur throughout where the parents are in close proximity. [syn: *M. curtisetosa* (Scribn.) Bush]

Muhlenbergia
frondosa x
schreberi,
spikelet (×5)

Muhlenbergia glomerata (Willd.) Trin.   Spike muhly
Culms 3–9 dm tall, often unbranched; principal blades 5–15 cm long by 2–5 mm
wide; panicle 3–10 cm long, spike-like, often interrupted at the base; glumes
about equal, awned; lemma about 3 mm long; rocky hillsides, sandy thickets,
and moist ground; mostly E, but scattered throughout; Jul–Oct; FACW.

Muhlenbergia
glomerata,
spikelet (×5)

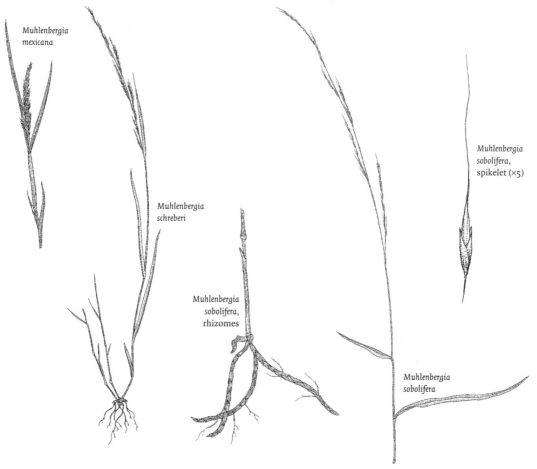

Muhlenbergia mexicana

Muhlenbergia schreberi

Muhlenbergia sobolifera, rhizomes

Muhlenbergia sobolifera, spikelet (×5)

Muhlenbergia sobolifera

Muhlenbergia schreberi, spikelet (×5)

Muhlenbergia mexicana (L.) Trin.   Muhly, satingrass
Culms 5–10 dm tall, often decumbent at the base, slightly pubescent below the nodes; principal blades 8–15 cm long by 3–6 mm wide; panicles slender, 6–12 cm long; glumes about equal, sharp-pointed or with a short awn; lemma 2.2–3.5 mm long excluding its sharp point or short awn; woods, rocky shores, swamps, and serpentine barrens; mostly SE, scattered elsewhere; Aug–early Oct; FACW.

Muhlenbergia schreberi J.F.Gmel.   Dropseed, nimble-will
Culms to about 1 m long, decumbent and much-branched below, erect above; principal blades 3–5 cm long by 2–4 mm wide; panicles slender, 3–10 cm long; first glume absent, second glume very small, mostly 0.3 mm or less long; lemma 1.9–2.5 mm long excluding its 2–4 mm long awn; woods, thickets, and waste ground; throughout; Aug–Oct; FAC.

Muhlenbergia sobolifera (Muhl.) Trin.   Creeping muhly
Culms to 1 m tall, erect, not densely cespitose; principal blades 10–15 cm long by 4–8 mm wide; panicles very slender, about 3 mm wide; glumes 1.2–2 mm long; lemma 1.7–2.3 mm long, usually awnless; dry, rocky slopes; mostly E, scattered elsewhere; late Jul–early Oct.

Muhlenbergia sylvatica (Torr.) Torr. ex A.Gray   Muhly, woodland dropseed
Culms 4–10 dm long, erect or spreading, freely branching, not densely ces-

pitose; principal blades 8–15 cm long by 3–5 mm wide; sheaths glabrous; panicles very slender, erect or nodding; glumes about equal, 1.8–3 mm long; lemma 2.2–3.8 mm long excluding its 4–11-mm-long awn; moist woods and shaded banks; occasional throughout; Aug–Sep; FAC+.

Muhlenbergia tenuiflora (Willd.) Britton, Stearns & Poggenb.

Muhly, woodland dropseed

Culms 5–12 dm tall, not densely cespitose, densely pubescent below the nodes; principal blades soft and lax, 8–15 cm long by 5–10 mm wide; panicles very slender, 8–15 cm long; glumes 1.7–2.9 mm long; lemma 2.6–3.5 mm long excluding its 4–10-mm-long awn; rocky, wooded slopes and along streams; occasional; mostly S; late Jul–Sep.

Muhlenbergia tenuiflora,
spikelet (×5)

Muhlenbergia uniflora (Muhl.) Fernald

Fall dropseed muhly

Culms 2–4 dm tall, densely cespitose; blades usually 1 mm or less wide; ligule 0.5–1.5 mm long; panicle slender, ¼–½ as long as the entire plant; glumes about ½ as long as the spikelet; lemma 1.2–2 mm long, awnless; marshes, bogs, and moist, sandy roadsides; rare at higher elevations, mostly NE; Aug–Sep; OBL; ⬥.

Muhlenbergia uniflora,
spikelet (×5)

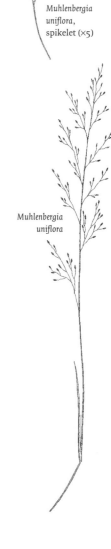

Muhlenbergia uniflora

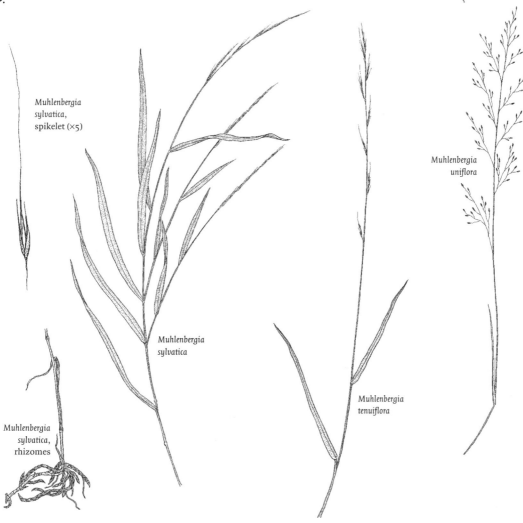

Muhlenbergia sylvatica,
spikelet (×5)

Muhlenbergia sylvatica

Muhlenbergia sylvatica,
rhizomes

Muhlenbergia tenuiflora

Oryzopsis Michx.

Perennials; spikelets 1-flowered, disarticulating above the glumes; glumes equal or nearly so, usually 7-nerved; lemma about equaling the glumes, its margins folded over the palea; awn strictly terminal on the lemma; palea similar to the lemma in texture at maturity.

A. spikelets 6–9 mm long; blades 5–15 mm wide
    B. lower leaf blades much shorter and narrower than the upper ones; mature fruit black .... *O. racemosa*
    B. upper leaf blades much reduced, most leaves basal; mature fruit straw-colored .... *O. asperifolia*
A. spikelets <5 mm long; blades 1–2 mm wide .... *O. pungens*

Oryzopsis asperifolia Michx.   Spreading ricegrass
Culms 3–7 dm long, loosely cespitose, often prostrate; basal leaves up to 4 dm long by 5–10 mm wide, scabrous above; upper blades <3 cm long; inflorescence 2–6 cm long, paired branches each bearing a single spikelet; glumes 7–8.5 mm long; lemma about as long as the glumes; awn 6–14 mm long; dry, sandy soil or rocky woods; mostly N, but extending to Bedford and Somerset Cos. in the mountains; May–early Jun.

Oryzopsis asperifolia

Oryzopsis pungens (Torr. ex Spreng.) A.Hitchc.   Slender mountain ricegrass
Culms 2–5 dm tall, densely cespitose; inflorescence 3–8 cm long, slender; glumes 3.5–4 mm long; lemma about as long as the glumes, pubescent; awn 1–2 mm long, straight or slightly bent; dry, sandy thickets and barrens; in the mountains of the NE; May; 🌿.

Oryzopsis pungens

Oryzopsis racemosa (Sm.) Ricker ex A.Hitchc.   Ricegrass
Culms 4–10 dm tall, loosely cespitose from a knotty base; upper blades 10–20 cm long by 8–15 mm wide; inflorescence 1–2 dm long, few-branched, bearing few spikelets; glumes 7–9 mm long; lemma somewhat shorter than the glumes; awn 12–22 mm long; open, rocky woods; mostly NE and at high elevations to Somerset Co., scattered SE; late Jun–Aug.

Panicum L.

Glumes usually very unequal, the first often minute, the second about as long as the spikelet and often prominently nerved; sterile lemma similar to the second glume, enclosing a palea and sometimes a staminate flower; fertile lemma usually nerveless, its margins inrolled over the palea.

Approximate flowering/fruiting dates are given in the descriptions for the species covered in Key 1. Unless otherwise stated, all other species produce vernal terminal panicles May–early Jul, and autumnal basal or axillary panicles later in summer or early fall.

A. plants blooming only once, usually in summer or fall; not forming winter rosettes, basal and culm leaves similar in appearance .... Key 1
A. plants blooming twice, bearing terminal panicles in spring and summer, with reduced axillary or basal panicles later; forming winter rosettes of short broad leaves unlike those of the culms (except species in Key 2)
    B. leaf blades of vernal culms elongate, 15 or more times as long as wide .....
      .... Key 2

B. leaf blades of vernal culms not elongate, most not more than 15 and usually less than 13 times as long as wide

  C. ligules prominent, densely ciliate, usually 2–5 mm long; pubescence of culms and sheaths, if present, of long, papillose-based hairs, not mixed with short, fine puberulence .... Key 3

  C. ligules usually inconspicuous, <1 mm long, or if longer, then the sheaths or culms bearing long hairs interspersed with short, fine puberulence

    D. spikelets with strong, conspicuous nerves .... Key 4

    D. nerves of spikelets not strong and conspicuous

      E. spikelets nearly spheroidal at maturity .... Key 5

      E. spikelets elliptic to obovate at maturity

        F. leaf blades 1–3 cm wide or wider, cordate at the base; spikelets 2.2–5 mm long .... Key 6

        F. leaf blades rarely >1 cm wide, usually not cordate at the base; spikelets not over 2.8 mm long

          G. culms glabrous .... Key 7

          G. culms puberulent or pubescent .... Key 8

Oryzopsis racemosa (×1/6)

Key 1

A. panicle with few, nearly simple primary branches; spikelets short-pedicellate, appressed and secund along the branches

  B. plants rhizomatous; spikelets falcate, set at an angle to the pedicels .... *P. anceps*

  B. plants lacking rhizomes; spikelets scarcely falcate, set straight on the pedicels

    C. ligule strongly ciliate, usually 2–3 mm long; leaf blades very elongate, narrow .... *P. longifolium*

    C. ligule membranaceous, erose, <1 mm long

      D. fruit with a pronounced stipe; spikelets slender, acuminate .... *P. stipitatum*

      D. fruit sessile; spikelets acute .... *P. rigidulum*

A. panicle much-branched, diffuse; spikelets mostly long-pedicellate, not secund on the branches

  E. spikelets smooth

    F. sheaths papillose-hispid

      G. spikelets <3.5 mm long; panicle branches slender

        H. spikelets slender-acuminate

          I. terminal panicle narrowly ellipsoid, usually not more than 1/3 the height of the plant; plants slender, usually 2.5–4 dm tall; pulvini of the lower panicle branches glabrous .... *P. flexile*

          I. terminal panicle broadly ovoid, as much as 1/2 the height of the plant; plants coarse, usually 4–10 dm tall; pulvini of the lower panicle branches hispid .... *P. capillare*

        H. spikelets acute or rarely short-cuspidate, never slender-acuminate

          J. plants large, coarse, usually 3.5–5.5 dm tall; larger leaf blades 5–11 mm wide, usually 10–20 cm long; panicles ellipsoid, not diffuse; spikelets appressed, subracemose along the panicle branches; pulvini of main panicle branches glabrous .... *P. gattingeri*

J. plants small, delicate, usually 1.5–4 dm tall; larger leaf blades 2–5 mm wide, 4–9 cm long; panicles ovoid, rather diffuse; spikelets tending to be clustered in 2s or 3s at the ends of the panicle branches

    K. pulvini of main panicle branches usually hispid .... *P. philadelphicum*

    K. pulvini of main panicle branches glabrous .... *P. tuckermanii*

G. spikelets 4.5 mm or more long; panicle branches stout, drooping .... *P. miliaceum*

F. sheaths not papillose-hispid

    L. first glume ¼–⅓ as long as the spikelet, truncate or broadly triangular; plants annual, lacking rhizomes .... *P. dichotomiflorum*

    L. first glume at least ½ the length of the spikelet, acuminate; plants rhizomatous perennials

        M. panicle diffuse or spreading..................................... *P. virgatum*

        M. panicle narrow, with appressed branches .... *P. amarum*

E. spikelets tuberculate .... *P. verrucosum*

Key 2

A. vernal blades very narrow, 20 or more times as long as wide; plants not forming distinct basal rosettes in autumn

    B. spikelets 3.2–4 mm long, the tips of the second glume and sterile lemma prolonged beyond the end of the fruit into an abrupt beak .... *P. depauperatum*

    B. spikelets 2.2–2.8 mm long, not beaked at the tip, the second glume and sterile lemma not prolonged beyond the end of the fruit..... *P. linearifolium*

A. vernal blades usually 15–20 times as long as wide; plants forming distinct basal rosettes in autumn

    C. autumnal phase branching sparingly from the middle and upper culm nodes .... *P. bicknellii*

    C. autumnal phase branching only from the base .... *P. laxiflorum*

Key 3

A. sheaths glabrous or the lower ones somewhat pubescent; blades glabrous, often ciliate at the base

    B. vernal panicle pyramidal, 1–2 times as long as wide .... *P. acuminatum*

    B. vernal panicle ellipsoid, 2–4 times as long as wide .... *P. spretum*

A. sheaths and usually blades strongly pubescent

    C. spikelets 1.4–1.9 mm long; culms, sheaths, and blades more or less pubescent, or the upper portions of the plant nearly glabrous; leaf blades mostly lax and spreading .... *P. acuminatum*

    C. spikelets 2–2.6 mm long; culms, sheaths, and blades heavily and conspicuously villous; leaf blades mostly ascending, rather firm ...... *P. villosissimum*

Key 4

A. herbage velvety throughout; culms bearing a viscid glabrous ring below the nodes .... *P. scoparium*

A. herbage more or less hispid, but not velvety to the touch; culms without a viscid ring below the nodes

    B. vernal panicles many-flowered, open, the branches divaricate .... *P. oligosanthes*

    B. vernal panicles few-flowered, narrow, the branches strongly ascending, or somewhat spreading at anthesis only

        C. leaf blades papillose-hispid on both surfaces, the margins prominently short-ciliate nearly to the tip .... *P. leibergii*

        C. leaf blades glabrous, scabrous-margined and with a few long cilia at the base .... *P. xanthophysum*

Key 5

A. vernal panicle narrowly ellipsoid or cylindrical, 2–4 times as long as wide; blade of the uppermost culm leaf 6–18 cm long, 9–28 mm wide; plants (2)4–9 dm .... *P. polyanthes*

A. vernal panicle broadly ovoid, less than twice as long as wide; blade of the uppermost culm leaf 2–6(9) cm long, 4–9(13) mm wide; plants 2–4(7.5) dm .... *P. sphaerocarpon*

Key 6

A. spikelets 2.2–3 mm long

    B. sheaths, especially those of the leaves of branches more or less papillose-hispid

        C. spikelets 2.7–3 mm long; leaf blades broadly lanceolate .... *P. clandestinum*

        C. spikelets 2.2–2.8 mm long; leaf blades linear-lanceolate ... *P. recognitum*

    B. sheaths not papillose-hispid

        D. leaf blades 12–25 mm wide; culms glabrous or nearly so, green .... *P. commutatum*

        D. leaf blades 5–12 mm wide; culms crisp-puberulent, usually purplish .... *P. boscii x commutatum*

A. spikelets 3.4–5 mm long

    E. nodes heavily bearded .... *P. boscii*

    E. nodes glabrous or nearly so

        F. leaf blades glabrous or nearly so, ciliate near the base; sheaths glabrous or nearly so .... *P. latifolium*

        F. leaf blades more or less prominently papillose-pubescent on both surfaces, short-ciliate nearly to the tip; sheaths prominently papillose-pilose .... *P. leibergii*

Key 7

A. plants erect, sometimes slender but the culms never weak and decumbent

    B. leaf blades velvety .... *P. annulum*

    B. leaf blades and sheaths essentially glabrous

        C. spikelets 1.5–1.7 mm long; nodes strongly bearded .... *P. microcarpon*

        C. spikelets 2 mm or more long; nodes glabrous or the lower slightly bearded

D. leaf blades erect or strongly ascending; spikelets pubescent ....
.... *P. boreale*

D. leaf blades spreading; spikelets glabrous or rarely puberulent

    E. spikelets 1.8–2.2 mm long, rounded at the tip; sheaths without whitish, flattened papillae .... *P. dichotomum*

    E. spikelets 2.1–2.7 mm long, pointed at the tip; sheaths bearing whitish, flattened papillae .... *P. yadkinense*

A. plants very slender; culms weak and decumbent; plants of swamps or bogs
.... *P. lucidum*

Key 8

A. sheaths glabrous; culms crisp-puberulent.................... *P. boscii x commutatum*

A. sheaths pubescent; at least some of the sheaths and internodes with long hairs interspersed with short, fine puberulence

    B. spikelets 1.3–1.9 mm long; plants mostly rather small and delicate

        C. spikelets 1.5–1.9 mm long; larger midculm leaf blades mostly 4–8 cm long by 4–8 mm wide .... *P. columbianum*

        C. spikelets 1.3–1.5 mm long; larger midculm leaf blades mostly 2–4 cm long by 2–4 mm wide .... *P. meridionale*

    B. spikelets 2–2.8 mm long; plants mostly rather robust ..... *P. commonsianum*

Panicum acuminatum, spikelet (×5)

Panicum acuminatum Sw.   Panic grass
Perennial; highly variable species; culms glabrous to papillose-pilose, or villous; blades glabrous or pubescent on both sides; autumnal culms spreading or prostrate, branching mostly from the middle nodes; dry woods, slopes and clearings; common throughout; FAC. [syn: *P. huachucae* Ashe in part; *P. implicatum* Scribn. in part; *P. lanuginosum* Elliot; *P. lindheimeri* Nash in part; *P. longiligulatum* Nash in part]

Panicum amarum var. *amarum*, spikelet (×5)

Panicum amarum Elliot
Perennial; culms 3–10 dm tall, usually solitary from extensive rhizomes; blades 5–12 mm wide by 10–30 cm long, rather thick, involute toward the tip; panicle ¼–⅓ the height of the entire plant; 2 varieties:

A. spikelets 6–8 mm long, the first glume ⅔ as long or more ....
.... var. *amarum*—Panic grass
ballast and shore of Lake Erie.

A. spikelets 4.7–5.8 mm long, the first glume ½ to ⅔ as long ....
.... var. *amarulum* (A.Hitchc. & Chase) P.G.Palmer—Beachgrass on sand and gravel fill along the Delaware River; Bucks Co.; ✿. [syn: *P. amarulum* A.Hitchc. & Chase]

Panicum anceps, spikelet (×4)

Panicum anceps Michx.   Panic grass
Perennial; culms 5–10 dm tall, few together from short scaly rhizomes; blades 6–12 mm wide by 2–4 cm long; sheaths and blades glabrous to pilose; panicle 1–4 dm long, branches spreading or ascending; spikelets 2.2–3.9 mm long; first glume ⅓–½ as long as the spikelet; moist, open, sandy soil; mostly SE; Jul–Sep; FAC.

Panicum annulum Ashe   Annulus panic grass

Perennial; vernal culms 3–6 dm tall, solitary or in small tufts, usually purplish, nodes densely bearded; vernal panicle 6–8 cm long; spikelets about 2 mm long, pubescent; autumnal culms declined, with a few short, erect branches from the upper nodes; dry soil of serpentine barrens; SE; ❧.

Panicum bicknellii Nash   Bicknell's panic grass

Perennial; vernal culms 3–5 dm tall, loosely cespitose, glabrous or the nodes slightly bearded; blades 3–8 mm wide by 8–15 cm long, usually ciliate toward

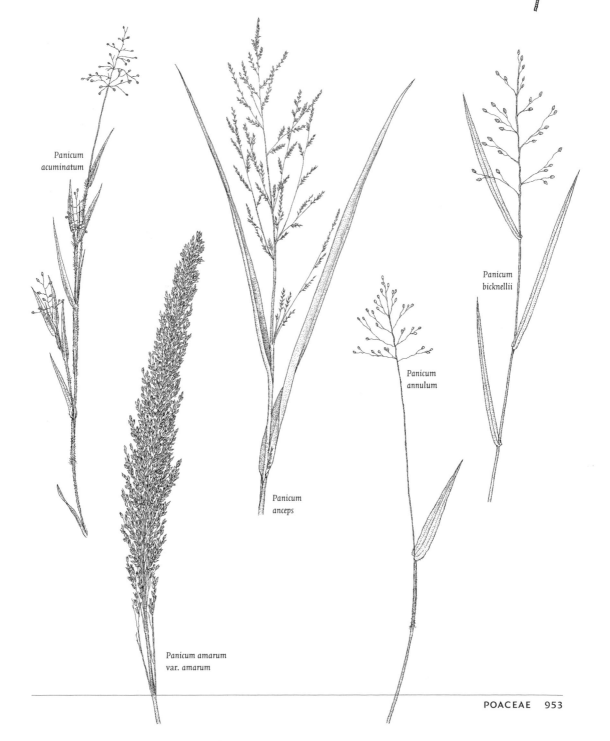

Panicum annulum,
spikelet (×5)

Panicum bicknellii,
spikelet (×5)

Panicum acuminatum

Panicum amarum var. amarum

Panicum anceps

Panicum annulum

Panicum bicknellii

base; vernal panicle 5–8 cm long; spikelets 2.3–2.8 mm long; autumnal culms tufted; blades scarcely reduced, concealing the few-flowered panicles; dry woods; mostly SE; 🌿.

Panicum boreale Nash   Northern panic grass
Perennial; vernal culms 3–6 dm tall, mostly glabrous at the nodes; blades ciliate at the base; vernal panicle 5–10 cm long, few-flowered; spikelets 2–2.2 mm long; bogs and fens; mostly NE, and scattered elsewhere in the mountains; FACU; 🌿.

Panicum boscii,
spikelet (×5)

Panicum boscii Poir.   Panic grass
Perennial; vernal culms 4–7 dm tall, mostly glabrous, bearded at the nodes; blades glabrous or nearly so, sparsely ciliate at the base; vernal panicle 6–12 cm long; spikelets papillose-pubescent; moist woodlands, grassy slopes, and stream banks; mostly S.

Panicum boscii x *commutatum*   Panic grass
Perennial, hybrid with characters intermediate between the parent species; one collection from Berks Co. [syn: *P. commutatum* Schult. var. *ashei* (Pearson) Fernald; *P. ashei* Pearson]

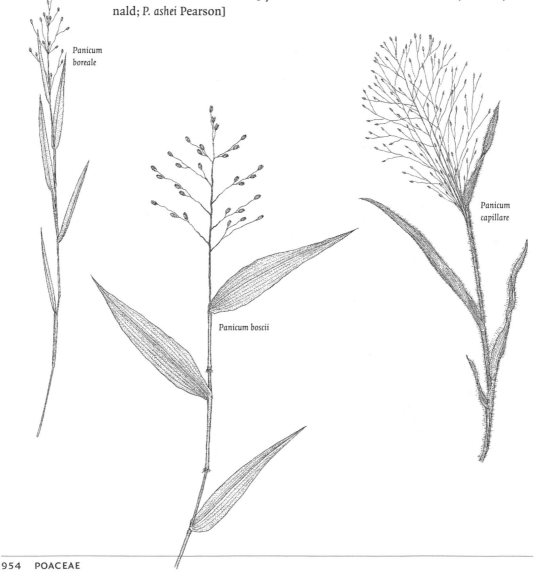

*Panicum
boreale*

Panicum boscii

*Panicum
capillare*

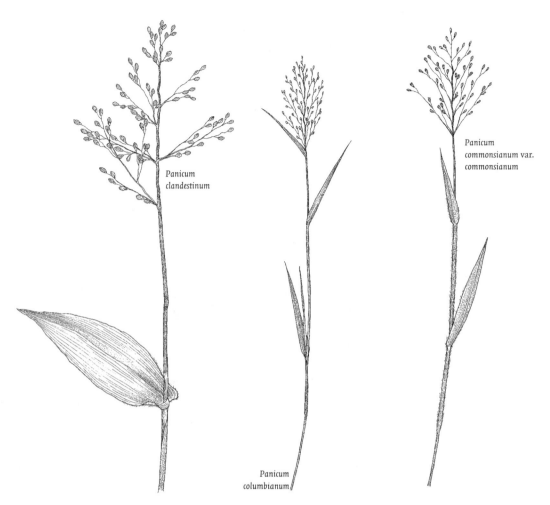

Panicum
clandestinum

Panicum
columbianum

Panicum
commonsianum var.
commonsianum

Panicum capillare L. — Witchgrass

Annual; blades 5–15 mm wide by 10–25 cm long, hispid on both sides; panicle often separating whole and rolling with the wind; spikelets 2–2.5 mm long; cultivated fields, shores, and roadsides; throughout; Aug–Oct; FAC–.

Panicum capillare,
spikelet (×5)

Panicum clandestinum L. — Deer-tongue grass

Perennial; vernal culms 7–15 dm tall, in a dense clump from rhizomes 5–10 cm long; blades 10–20 cm long, usually scabrous on both surfaces; vernal panicle 8–15 cm long; autumnal culms erect or leaning; sheaths mostly enclosing the autumnal panicles; moist, often sandy soil of woods edges and clearings; throughout; FAC+.

Panicum
clandestinum,
spikelet (×5)

Panicum columbianum Scribn. — Panic grass

Perennial; vernal culms 2–5 dm tall, densely puberulent; blades glabrous above, puberulent or glabrous below; vernal panicle 2–4 cm long; autumnal culms decumbent at the base, branching from the middle and upper nodes; roadsides, waste ground, and open woods, in dry sandy soil; in the mountains and SE.

Panicum commonsianum Ashe

Perennial; vernal culms 2–5 dm tall, appressed pilose; blades 4–7 mm wide, broadest near the rounded base; vernal panicle 4–8 cm long, branches stiffly

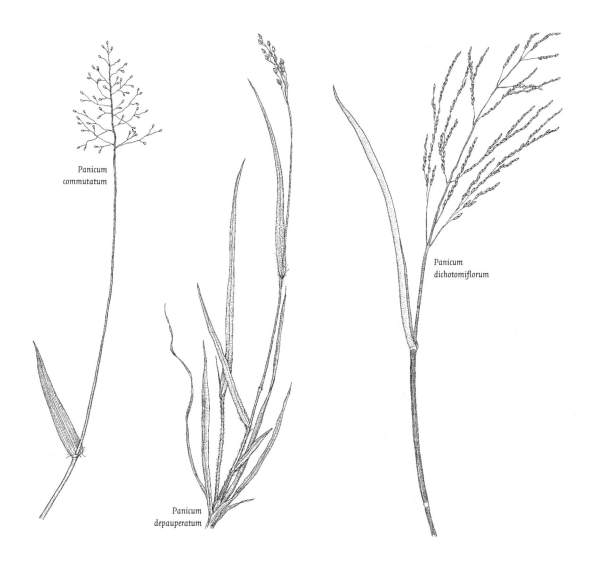

Panicum commutatum

Panicum dichotomiflorum

Panicum depauperatum

Panicum commonsianum var. commonsianum, spikelet (×5)

Panicum commutatum, spikelet (×5)

Panicum depauperatum, spikelet (×5)

spreading; autumnal culms branching from the middle and upper nodes, spreading or decumbent and forming mats; plants of sandy soils; 🌿; 2 varieties:

A. spikelets 2–2.1 mm long .... var. *commonsianum*—cloaked panic grass Philadelphia Co. [syn: *P. commonsianum* var. *addisonii* (Nash) Pohl]

A. spikelets 2.2–2.8 mm long .... var. *euchlamydeum* (Shinners) Pohl—panic grass known in Pennsylvania only from the Lake Erie shore; FACU.

Panicum commutatum Schult.   Panic grass

Perennial; vernal culms 4–8 cm, erect; blades usually glabrous on both sides; sheaths glabrous or nearly so; vernal panicle 6–12 cm long; spikelets about 2.5 mm long; autumnal culms erect or leaning, branching from the middle and upper nodes; dry woods, rocky slopes, and barrens; mostly S; FACU+.

Panicum depauperatum Muhl.   Poverty panic grass

Perennial; culms 1–4 dm tall, cespitose, very slender; blades 2–5 mm wide by 8–15 cm long; sheaths and blades glabrous to pilose; vernal panicles 3–6 cm long, only slightly surpassing the blades; first glume about ⅓ as long as the spikelet;

autumnal panicles much reduced and usually concealed among the leaves; dry open woods, roadside banks, and serpentine barrens; mostly SE.

Panicum dichotomiflorum Michx.   Smooth panic grass
Perennial; culms mostly 5–10 dm tall; blades 3–20 mm wide by 10–50 cm long, with a prominent white mid-rib; panicles mostly 10–40 cm long, partly enclosed by the upper sheath; spikelets about 2.5 mm long, acute; dry to moist, open woods, and meadows; throughout; FACW–.

Panicum dichotomiflorum, spikelet (×5)

Panicum dichotomum L.   Panic grass
Perennial; vernal culms 3–6 dm tall, erect from a knotty base, usually purplish; blades 4–8 mm wide, glabrous; vernal panicle 4–9 cm long, flexuous; autumnal culms much-branched from the middle nodes, mostly bladeless below; moist, sandy woods; throughout; FAC.

Panicum dichotomum, spikelet (×5)

Panicum flexile (Gatt.) Scribn.   Old witchgrass
Annual; culms much-branched from the base, hispid below, pubescent on the nodes; blades 2–6 mm wide, up to 30 cm long, glabrous to sparsely hispid; spikelets 3.1–3.5 mm long; dry fields, moist meadows, banks, and swales; mostly SE; Aug–Sep; FACU; 🌿.

Panicum gattingeri, spikelet (×5)

Panicum gattingeri Nash   Witchgrass
Annual; culms often decumbent and rooting at the lower nodes; terminal panicles 10–15 cm long, lateral ones shorter; spikelets about 2 mm long; culti-

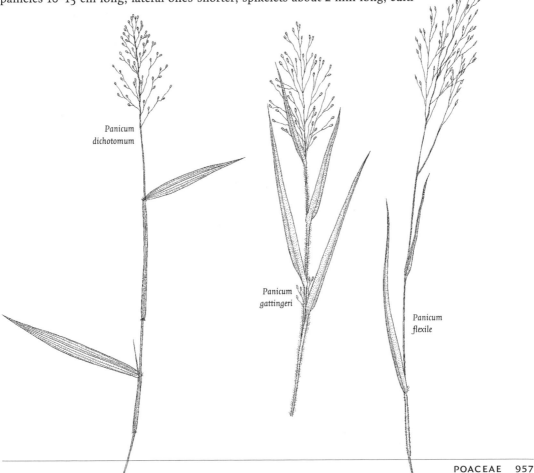

*Panicum
dichotomum*

*Panicum
gattingeri*

*Panicum
flexile*

vated fields, roadsides, and waste ground in sandy soils; throughout; Aug–Oct; FAC.

Panicum latifolium, spikelet (×5)

Panicum latifolium L.   Panic grass
Perennial; vernal culms 4–10 dm tall, glabrous or slightly pubescent below; vernal panicle 7–15 cm long; autumnal culms spreading, branching from the middle nodes; roadsides, shores, and thickets; throughout; FACU–.

Panicum laxiflorum Lam.   Panic grass
Perennial; culms 2–6 dm tall, often bent below; blades 7–12 mm wide by 10–20 cm long; vernal panicle 8–12 cm long, lax, few-flowered; spikelets 2.2–2.3 mm long, papillose-pilose; autumnal blades not much reduced, much exceeding the autumnal panicles; sand barrens; Erie Co.; native south of Pennsylvania, probably adventive; FACU.

Panicum leibergii (Vasey) Scribn.   Leiberg's panic grass
Perennial; vernal culms 3–7 dm tall, pilose or scabrous; blades 7–15 mm wide by 6–15 cm long, rather thin; sheaths papillose-hispid; vernal panicle 8–15 cm long, <½ as wide; spikelets 3.7–4 mm long, papillose-hispid; autumnal culms few-branched from the lower and middle nodes; limestone outcrops; Centre Co.; believed to be extirpated; FACU; 🌿.

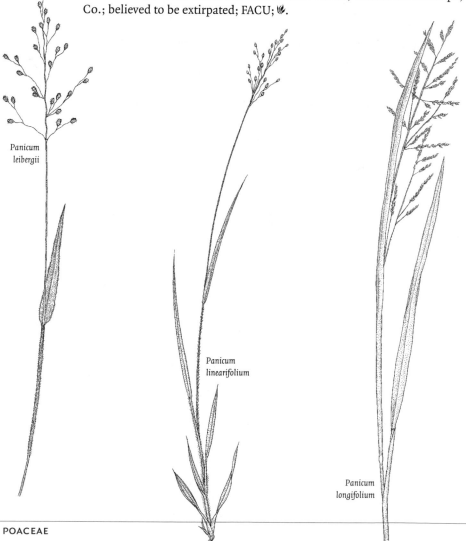

Panicum leibergii

Panicum linearifolium

Panicum longifolium

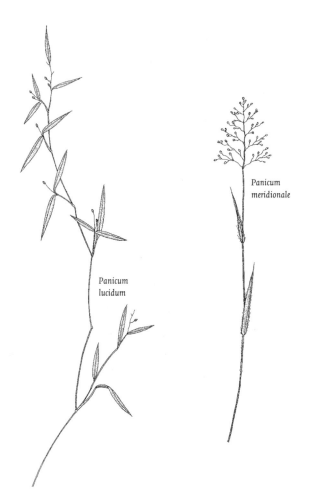

Panicum meridionale

Panicum lucidum

Panicum linearifolium Scribn.   Panic grass
Perennial; culms 2–6 dm tall, densely cespitose, mostly glabrous; blades 2–5 mm wide by 10–20 cm long, glabrous to sparsely pilose; vernal panicle 3–8 cm long, usually surpassing the blades; first glume about 1/3 as long as spikelet; autumnal panicles much reduced, usually concealed among the blades; dry banks, open woods and fields; throughout.

Panicum linearifolium, spikelet (×5)

Panicum longifolium Torr.   Long-leaved panic grass
Perennial; culms to 1 m tall, densely tufted, compressed; blades 2–4 mm wide, up to 3 dm long; sheaths glabrous to villous, compressed; panicle 1–2 dm long; spikelets 2.4–2.7 mm long; peaty, sandy bogs and shores; mostly SE; Jul–Oct; OBL; 🌿.

Panicum longifolium, spikelet (×5)

Panicum lucidum Ashe   Shining panic grass
Perennial; blades 4–6 mm wide, thin, glabrous and shining; vernal panicle 4–10 cm long, flexuous, few-flowered; spikelets about 2 mm long, glabrous; autumnal culms weak, much-branched, forming clumps or mats; bogs, swamps, and boggy swales; extreme SE; FAC; 🌿.

Panicum lucidum, spikelet (×5)

Panicum meridionale Ashe   Panic-grass
Perennial; vernal culms 2–4 dm tall; blades 2–4 mm wide by 1.5–3 cm long, pi-

Panicum
microcarpon

Panicum
oligosanthes

Panicum
miliaceum

lose above; vernal panicle 1.5–4 cm long; dry, sandy soil; mostly SE and scattered in the mountains.

Panicum microcarpon,
spikelet (×5)

Panicum
miliaceum,
spikelet (×5)

Panicum oligosanthes,
spikelet (×5)

Panicum microcarpon Muhl. — Panic grass

Perennial; vernal culms 5–10 dm tall, tufted; blades 8–15 mm wide by 10–12 cm long; vernal panicle 8–12 cm long, many-flowered; spikelets mostly glabrous; autumnal culms much-branched from all nodes; well-drained woods; mostly S, scattered elsewhere; FACU.

Panicum miliaceum L. — Broomcorn millet

Perennial; culms 2–10 dm tall, stout, often decumbent at the base; blades rounded at the base, as much as 2 cm wide by 30 cm long; panicle 10–30 cm long, usually at least partially enclosed in the upper sheath, nodding, very scabrous; spikelets many-nerved; cultivated and occasionally escaped; throughout; Aug–Sep; native to Eurasia.

Panicum oligosanthes Schult. — Panic grass, Heller's witchgrass

Perennial; vernal culms 2–8 dm tall; blades 5–12 mm wide by 6–14 cm long, glabrous above, glabrous to roughly pubescent below; vernal panicle 6–12 cm long;

spikelets 3.2–4 mm long; autumnal culms branching from the middle and upper nodes; thickets, in loamy or clayey soils; mostly E, especially along the Delaware River; FACU; 🌿. [includes *P. scribnerianum* Nash]

Panicum philadelphicum Bernh. ex Trin. Panic grass
Annual; culms slender, light yellowish-green; spikelets 1.7–2 mm long, usually in 2s at the ends of the branchlets; dry, open woods, fields, and roadsides; mostly S; Aug–Oct; FAC–.

Panicum philadelphicum, spikelet (×5)

Panicum polyanthes Schult. Panic grass
Perennial; blades not much reduced upward; spikelets about 1.5 mm long, minutely pubescent; roadsides and open woods; mostly SE; FACU.

Panicum recognitum Fernald Fernald's panic grass
Perennial; vernal culms 6–15 dm tall, glabrous; blades 8–15 mm wide by 6–13 cm long, glabrous or pilose below, ciliate on the margins toward the base; vernal panicle 8–13 cm long, branches ascending and few-flowered; glumes nearly equal, 0.8–1 mm long; woods or wooded slopes; extreme SE; possibly a hybrid of *P. clandestinum* and *P. microcarpon*; OBL; 🌿.

Panicum recognitum, spikelet (×5)

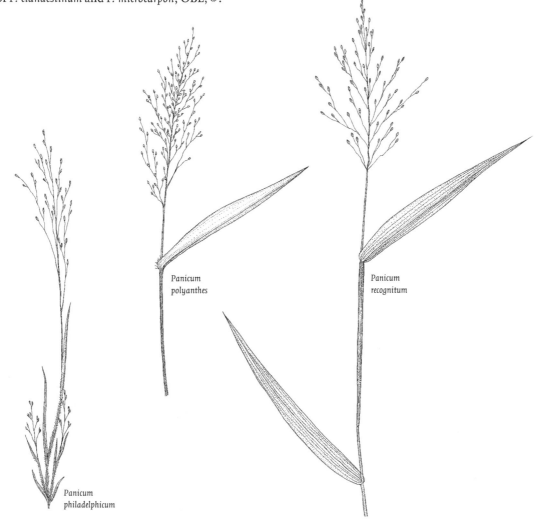

Panicum philadelphicum

Panicum polyanthes

Panicum recognitum

Panicum rigidulum Nees

Panic grass

Perennial; similar to P. stipitatum; panicles 1–3 dm long; spikelets about 2 mm long; moist, open, sandy or peaty ground; mostly S; Jul–Sep; FACW+.

Panicum scoparium, spikelet (×5)

Panicum scoparium Lam.

Velvety panic grass

Perennial; vernal culms 8–13 dm tall, usually bent at the base; blades 10–18 mm wide by 12–20 cm long, rather thick; vernal panicle 8–15 cm long; spikelets about 2.5 mm long, papillose-pubescent; moist meadows and swales; extreme SE; FACW; 🌿.

Panicum sphaerocarpon, spikelet (×5)

Panicum sphaerocarpon Elliot

Panic grass

Perennial; vernal culms light green, spreading, usually pubescent on the nodes; spikelets about 1.7 mm long; dry woods, thickets and old fields; mostly SE, scattered elsewhere; FACU.

Panicum spretum, spikelet (×5)

Panicum spretum Schult.

Panic grass

Perennial; vernal culms 3–9 dm tall; blades 4–8 mm wide, often ciliate near base; sheaths glabrous; vernal panicle 8–12 cm long; branches ascending to appressed; spikelets about 1.5 mm long; autumnal culms mostly reclined; low wet ground on the coastal plain; Bucks Co.; 🌿.

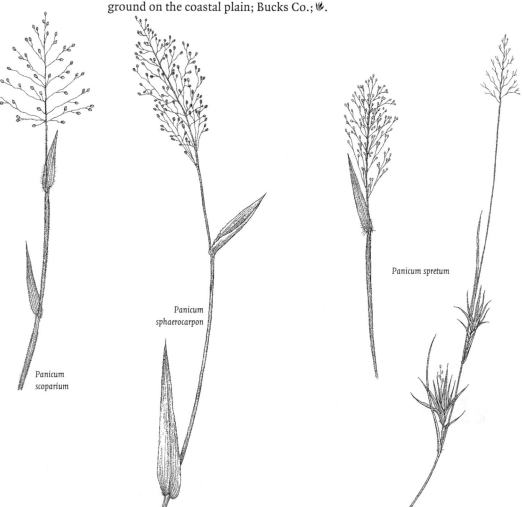

Panicum scoparium

Panicum sphaerocarpon

Panicum spretum

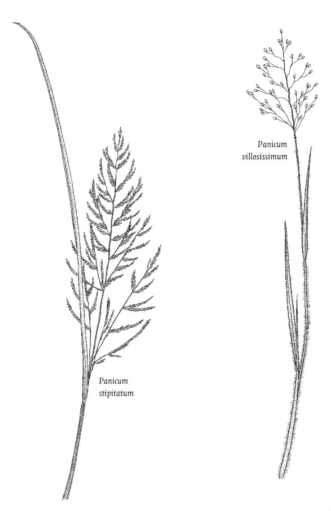

Panicum
villosissimum

Panicum
stipitatum

Panicum stipitatum Nash Panic grass

Perennial; culms 5–10 dm tall, densely cespitose, often purplish; blades folded at the base, flat above, 5–12 mm wide by 20–50 cm long; panicles 10–20 cm long, usually several per culm, narrow, densely flowered; spikelets 2.5–2.8 mm long; wet meadows and sandy shores; scattered throughout except NE; Jul–Sep; FACW+.

Panicum tuckermanii Fernald Tuckerman's panic grass

Annual; similar to *P. philadelphicum*; spikelets usually in 3s or more at the ends of the branchlets; sandy flats; Erie Co.; FAC–; 🍂.

Panicum verrucosum Muhl. Panic grass

Annual; culms 2–15 dm long, widely spreading; blades 4–10 mm wide by 5–20 cm long; panicles 5–30 cm long, about as wide; spikelets 1.8–2.1 mm long; moist, sandy or peaty soil; SE and SC; late Aug–early Oct; FACW.

Panicum
verrucosum,
spikelet (×5)

Panicum villosissimum Nash Long-haired panic grass

Perennial; vernal culms 2.5–5 dm tall, olive-green; blades 5–10 mm wide by 6–10 cm long; ligules 4–5 mm long; vernal panicle 4–8 cm long, branches ascending to spreading; autumnal culms eventually prostrate; dry woods and serpentine barrens; SE and Centre Co.; 🍂.

Panicum
villosissimum,
spikelet (×5)

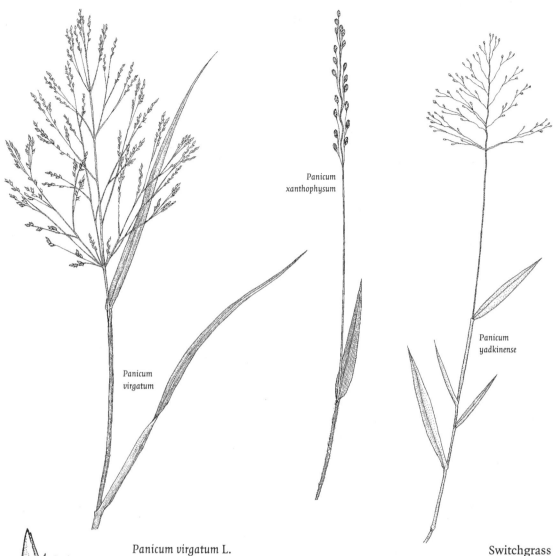

Panicum xanthophysum

Panicum yadkinense

Panicum virgatum

Panicum virgatum, spikelet (×3)

Panicum yadkinense, spikelet (×5)

Panicum virgatum L.   Switchgrass
Perennial; culms to 2 m or more, green or glaucous, usually bunched; blades 3–15 mm wide by 10–60 cm long; panicle 15–50 cm long; spikelets 3.5–5 mm long; first glume ⅔–¾ as long as the spikelet; sandy shores, alluvium, fields, and banks; throughout; late Jun–Sep; FAC.

Panicum xanthophysum A.Gray   Slender panic grass
Perennial; vernal culms 2–6 dm tall, yellowish-green, usually scabrous; blades 10–20 mm wide by 10–15 cm long, rather thin, erect or nearly so; sheaths sparsely papillose-pilose; vernal panicle 5–12 cm long; spikelets 3.7–4 mm long, pubescent; autumnal culms erect or ascending, branching from the second and third nodes; dry, rocky slopes or sandy, open woods; mostly in the mountains; NE; ❦.

Panicum yadkinense Ashe   Yadkin River panic grass
Perennial; vernal phase similar to *P. dichotomum*, somewhat taller; blades and vernal panicle a little larger; autumnal culms loosely branching from the middle nodes; dry woods; mostly SE; ❦.

Paspalum L.

Perennials; spikelet solitary or in pairs; first glume usually absent; second glume and sterile lemma about equal in length.

A. racemes 2, <1 cm apart at the apex of the peduncle .... *P. paspalodes*
A. racemes 1 and terminal or 2–several and >1 cm apart on the peduncle
    B. spikelets 3.9–4.5 mm long; plants usually 1–2 m .... *P. floridanum*
    B. spikelets <3.5 mm long; plants usually <1 m
        C. spikelets 2.5–3.2 mm long .... *P. laeve*
        C. spikelets mostly <2.5 mm long .... *P. setaceum*

Paspalum floridanum Michx.   Florida beadgrass
Culms 1–2 m tall, stout; blades 5–10 mm wide, pubescent at least at the base; racemes 2–5, 6–10 cm long, sparsely silky-hairy at the base; spikelets 3.8–4.3 mm long, $\frac{2}{3}$ to nearly as wide; moist, sandy soils; Lancaster Co.; a coastal species, believed extirpated; FACW; ✹. Ours is var. *glabratum* Engelm. ex Vasey.

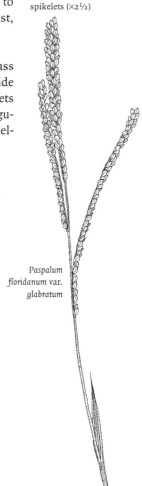

Paspalum laeve,
spikelets (×2½)

Paspalum laeve Michx.   Field beadgrass
Culms 3–8 dm tall, usually several from the base; blades 3–10 mm wide by 5–25 cm long, glabrous to pilose; racemes 2–6, 4–12 cm long; spikelets $\frac{2}{3}$ as wide to fully as wide as long, $\frac{1}{3}$ to $\frac{1}{2}$ as thick; glume and sterile lemma 5-nerved; moist, sandy fields; mostly SE; Aug–Sep; FAC+.

Paspalum paspalodes Michx.   Beadgrass
Culms 2–5 dm tall, ascending from a creeping base; blades mostly 2–5 mm wide by 2–8 cm long, ciliate at the base; racemes often curved, 2–5 cm long; spikelets solitary, 2.6–3 mm long, about ½ as wide; first glume about 1 mm long, triangular, second glume 5-nerved; sterile lemma 5-nerved, glabrous; ballast; Philadelphia Co.; Jul–Sep; native to Eurasia; FACW+. [syn: *P. distichum* L.]

*Paspalum
floridanum* var.
glabratum

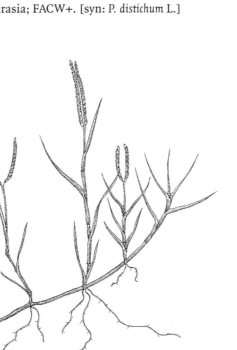

Paspalum paspalodes

Paspalum setaceum Michx.   Slender beadgrass

Culms 2–6 dm tall, erect, tufted; blades and sheaths with stiff pubescence 1–3 mm long; racemes usually solitary (rarely 2), usually curved, 4–7 cm long; spikelets usually paired; glume usually glabrous; sterile lemma glabrous; dry, open ground; mostly SE; Jul–Oct; FACU+; 4 varieties:

A. sterile lemma lacking a midnerve ....
.... var. *psammophilum* (Nash) D.Banks [syn: *P. psammophilum* Nash]
A. sterile lemma with an apparent midnerve
    B. spikelets averaging about 1.5 mm long .... var. *setaceum*
    B. spikelets 2–2.5 mm long
        C. largest leaf blades 7 mm or less wide ....
        .... var. *muhlenbergii* (Nash) D.J.Banks
        C. largest leaf blades 8–12 mm wide .... var. *supinum* (Bosc & Poir.) Trin.

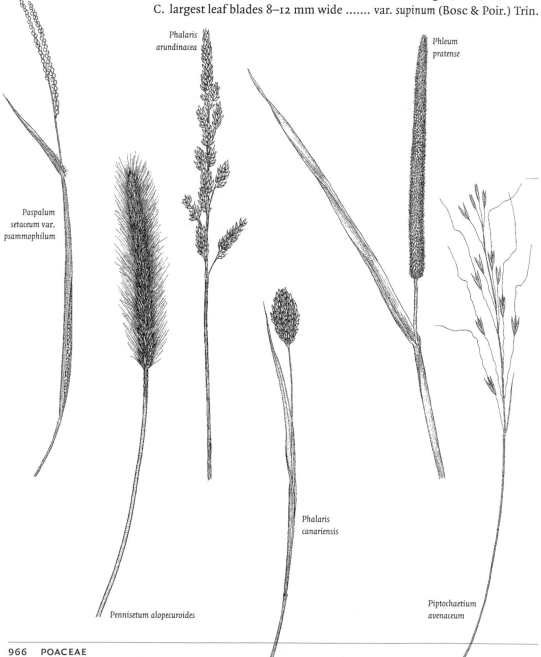

Phalaris arundinacea

Phleum pratense

Paspalum setaceum var. *psammophilum*

Phalaris canariensis

Piptochaetium avenaceum

Pennisetum alopecuroides

Pennisetum Rich.

Pennisetum alopecuroides (L.) Spreng.   Fountaingrass
Perennial; culms to 1 m tall; blades scabrous; panicle 8–15 cm long, softly bristly; bristles to 2 cm long; cultivated and rarely escaped to roadsides and abandoned fields; Berks and Chester Cos., probably elsewhere; native to China.

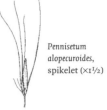

Pennisetum alopecuroides, spikelet (×1½)

Phalaris L.

Spikelets disarticulating above the glumes; glumes about equal, keeled to
slightly winged along the midnerve.

A. plants rhizomatous; inflorescence cylindrical, often lobed, 7–25 cm or more
    long .... *P. arundinacea*
A. plants not rhizomatous; inflorescence ovoid, 1.5–4 cm long .... *P. canariensis*

Phalaris arundinacea L.   Reed canary-grass
Perennial; culms 10–15 dm tall, from rhizomes; principal blades 1–3 dm long by
10–15 mm wide; panicle 5–12 cm long, branched or lobed; glumes 4–6.5 mm
long; lemmas only about 1 mm long, linear; marshes, alluvial meadows, shores,
and ditches; throughout; Jun–early Jul; FACW.

Phalaris arundinacea, spikelet (×2)

Phalaris canariensis L.   Canary-grass
Annual; culms 3–7 dm tall; panicle about 3 cm long; glumes 7–8 mm long,
broadly winged on the keel; fertile lemma about 5 mm long, sterile lemmas
about 2.5 mm long; cultivated and occasionally escaped to waste ground; mostly
SE, probably elsewhere; Jul–Aug; native to Europe; FACU.

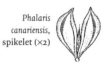

Phalaris canariensis, spikelet (×2)

Phleum L.

Phleum pratense L.   Timothy
Annual; culms 5–10 dm tall, cespitose; blades scabrous on the margins; inflorescence cylindric, 5–10 cm long by 6–8 mm wide; glumes 2.6–3.2 mm long, with
an awn 0.7–1.5 mm long; cultivated and frequently escaped to fields, meadows,
and roadsides; throughout; Jun–Sep; native to Europe; FACU.

Phleum pratense, spikelet (×3)

Phragmites Adans.

Phragmites australis (Cav.) Trin. ex Steud.   Common reed
Perennial; culms 2–4 m tall, usually forming extensive colonies; panicles tawny,
2–4 dm long; spikelets 3–7-flowered; second glume nearly twice as long as the
first; lemmas 8–12 mm long; rachilla hairs as long as the lemma; marshes,
ditches, and moist disturbed ground; throughout; late Jul–Sep; FACW. [syn: P.
communis Trin.]

Piptochaetium J.Presl

Piptochaetium avenaceum (L.) Parodi   Black oatgrass
Perennial; culms 5–10 dm tall, loosely cespitose; blades 2–3 dm long by 1–2 mm
wide, usually glabrous; sheaths glabrous, ligules 2–3 mm long; panicle 1–2 dm
long, loose and open, few-flowered; lemma 7.5–11 mm long in fruit, villous toward the base; awn scabrous, 4.5–7 cm long, twice bent near the middle; thin,

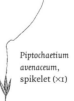

Piptochaetium avenaceum, spikelet (×1)

rocky woods and sandy, open ground; mostly SE; May–early Jun; UPL. [syn: *Stipa avenacea* L.]

Poa L.

Spikelets 2–several-flowered, disarticulating above the glumes and between the lemmas; first glume 1–3-nerved, second glume 3-nerved; lemmas usually 5-nerved, awnless; uppermost florets usually greatly reduced.

A. lemmas with cottony webbing at the base
    B. lemmas pubescent at least on the keel
        C. marginal nerves of lemmas pubescent
            D. plants with prominent creeping rhizomes
                E. culms strongly flattened and keeled; lemmas usually rounded at the apex; culms without basal tufts of long leaves, forming a very loose turf .... *P. compressa*
                E. culms terete; lemmas acute; culms with long, tufted basal leaves, forming a rather dense turf
                    F. lower panicle branches mostly in 5s; lemmas 2.4–3.6 mm long; anthers 1–1.8 mm long .... *P. pratensis*
                    F. lower panicle branches mostly in 2s; lemmas 3.5–5.4 mm long; anthers 2–3.5 mm long .... *P. cuspidata*
            D. plants tufted, lacking rhizomes (although sometimes with decumbent culms)
                G. florets, or many of them, modified into bulblets with an elongate tip .... *P. bulbosa*
                G. florets not modified into bulblets
                    H. upper ligules 2.5–7 mm long
                        I. lemmas greenish-yellow, the intermediate nerves prominent; sheaths mostly scaberulous .... *P. trivialis*
                        I. lemmas green, usually purple or bronzy at the tip, the intermediate nerves obscure; sheaths smooth ...... *P. palustris*
                    H. upper ligules 2 mm or less long
                      J. panicle branches in whorls of 3 or more
                          K. lemmas pubescent between the nerves; lemmas and glumes rounded to an acute apex; lower panicle branches reflexed at maturity .... *P. sylvestris*
                          K. lemmas glabrous between the nerves; glumes acuminate; lower panicle branches not reflexed .... *P. nemoralis*
                      J. panicle branches in 2s .... *P. paludigena*
        C. marginal nerves of lemmas glabrous
            L. upper ligules 3–7 mm long; sheaths mostly scaberulous; some of the panicle branches usually short and floriferous nearly to their bases; intermediate nerves of the lemmas conspicuous .... *P. trivialis*
            L. upper ligules 0.5–2 mm long; sheaths nearly smooth; branches of the panicle elongate, spreading, naked near the base; intermediate nerves of the lemmas obscure .... *P. alsodes*
    B. lemmas glabrous except for the webbing
        M. lemmas acute, rather thin, with prominent nerves .... *P. saltuensis*
        M. lemmas obtuse, rather coriaceous, with obscure nerves .... *P. languida*

A. lemmas not webbed at the base
  N. lemmas pubescent only on the nerves; plants mostly 5–20 cm tall; upper-most leaf blade 1–5 cm long .... *P. annua*
  N. lemmas pubescent on the nerves and on the lower internerves; plants mostly 3–6 dm tall; uppermost leaf blade 5–15 cm long .... *P. autumnalis*

Poa alsodes A.Gray   Woodland bluegrass

Perennial; culms 3–10 dm tall; blades 2–4 mm wide; panicle 10–20 cm long, very open, its branches in whorls of 4–5; spikelets 2–3-flowered, about 5 mm long; cool, moist woods and thickets; throughout, but less frequent SE and SC; May–Jun.

Poa alsodes,
spikelet (×2½)

Poa annua L.   Annual bluegrass

Perennial; culms usually prostrate, or ascending from a decumbent base; panicle 2–8 cm long; spikelets 3–6-flowered, 3–5 mm long; lowest lemma 2.2–3.4 mm long; roadsides, lawns, open woods, and moist, alluvial soils; throughout; Apr–Sep; native to Eurasia.

Poa annua,
spikelet
(×2½)

Poa autumnalis Muhl. ex Elliot   Autumn bluegrass

Annual or biennial; culms in large tufts; leaves crowded near the base; blades 2–3 mm long; panicle 8–15 cm long; spikelets 4–6-flowered; lemmas 3.2–4.4 mm long; moist woods; SE; late May–Jun; FAC; 🍂.

Poa autumnalis,
spikelet (×2½)

Poa alsodes

Poa annua

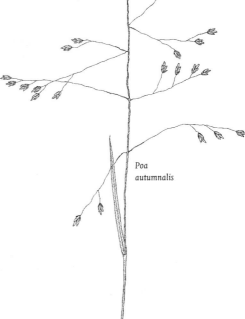

Poa autumnalis

Poa bulbosa,
spikelet (×1)

Poa bulbosa L. Bulbous bluegrass
Perennial; culms 2–5 dm tall, purplish below, from thickened bases; blades 1–2 mm wide, flat or involute; panicle 4–8 cm long; bulblets dark purple at the base; waste ground; Dauphin Co., probably elsewhere; native to Europe.

Poa compressa,
spikelet (×2½)

Poa compressa L. Canada bluegrass
Perennial; culms 2–7 dm tall, not at all cespitose; blades 2–4 mm wide; panicle 2–8 cm long, the branches usually paired, bearing spikelets nearly to the base; lemmas 2–2.8 mm long; dry woods, fields and rock outcrops; throughout; May–Sep; native to Europe; FACU.

Poa cuspidata,
spikelet (×2½)

Poa cuspidata Nutt. Bluegrass
Perennial; culms 3–6 dm tall, loosely tufted; blades 2–4 mm wide; sheaths pubescent at least at the summit; panicle very open, branches bearing a few spikelets mostly above the middle; spikelets 3–4-flowered; dry, wooded hillsides, and banks; mostly S; Apr–early May.

Poa languida,
spikelet (×2½)

Poa languida A.Hitchc. Woodland bluegrass, drooping bluegrass
Perennial; culms 3–10 dm tall, weak; blades 2–5 mm wide; sheaths mostly glabrous; panicle loose, usually nodding, branches bearing a few spikelets beyond the middle; spikelet 2–4-flowered; lemmas 2.4–3.2 mm long; moist woods and fens; scattered, mostly at higher elevations; late May–early Jun; 🍂.

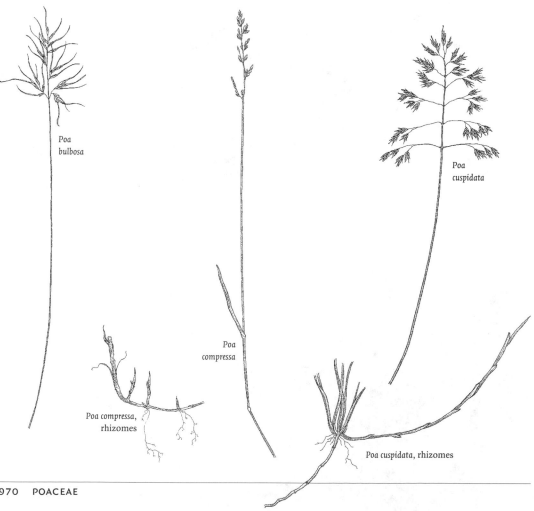

*Poa
bulbosa*

*Poa compressa,
rhizomes*

*Poa
compressa*

*Poa
cuspidata*

Poa cuspidata, rhizomes

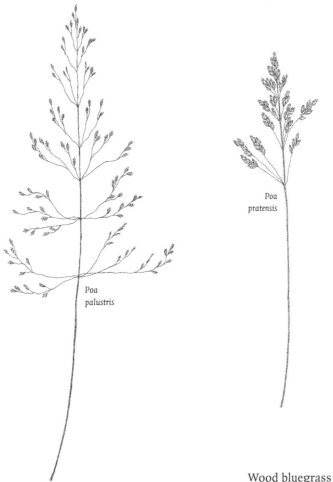

Poa palustris

Poa pratensis

Poa nemoralis L.   Wood bluegrass
Perennial; culms 4–8 dm tall; blades 1–2 mm wide, lax, the upper only slightly
shorter than the lower; ligule about 0.5 mm long; panicle 1–2 dm long, branches
in whorls of about 5; spikelets 2–4-flowered; ligules 2.1–3.1 mm long, 3-nerved;
dry woods and edges; scattered; Jun; native to Europe; FAC.

Poa paludigena Fernald & Wiegand   Bog bluegrass
Perennial; culms 2–6 dm tall; blades 1–2 mm wide; ligule about 1 mm long;
panicle 5–15 cm long, lax, branches in pairs, bearing a few spikelets above the
middle; spikelets 2–5-flowered; lemmas 2.5–3.3 mm long; boggy woods, and
swamps; Berks, Chester and Lancaster Cos. and in the mountains NE, NW, and
SC; late May–Jun; FACW; ❦.

Poa paludigena,
spikelet (×2¹⁄₂)

Poa palustris L.   Fowl bluegrass
Perennial; culms 5–15 dm tall, cespitose, often somewhat decumbent; blades 1–
2 mm wide; ligules 2.5–5 mm long; panicle 1–3 dm long, often nodding,
branches in whorls of about 5; spikelets 2–4-flowered; lemmas 3–5-nerved, 2.1–
3 mm long; wet meadows, shores, and thickets; mostly N; Jun–early Aug; FACW.

Poa palustris,
spikelet
(×2¹⁄₂)

Poa pratensis L.   Kentucky bluegrass
Perennial; culms 3–10 dm tall, cespitose and sod-forming; blades 2–5 mm wide;
sheaths glabrous; spikelets 3–5-flowered; cultivated and widely naturalized in

Poa pratensis,
spikelet (×2¹⁄₂)

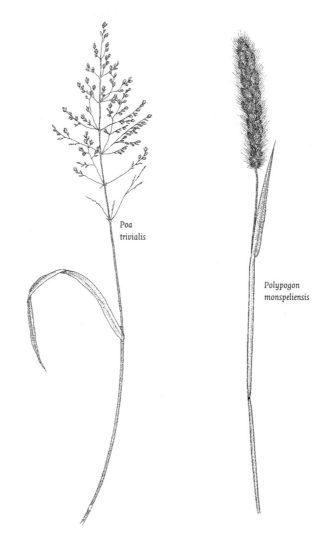

Poa
trivialis

Polypogon
monspeliensis

meadows, roadsides, open woods, and waste ground; throughout; late Apr–Jul;
native to Europe; FACU.

Poa saltuensis,
spikelet (×2½)

Poa saltuensis Fernald & Wiegand — Old-pasture bluegrass

Perennial; similar to P. *languida* in habit; spikelets 3–5-flowered; lemmas 2.4–3.9
mm long; woods; mostly N and at higher elevations; late May–Jun.

Poa sylvestris,
spikelet (×2½)

Poa sylvestris A.Gray — Woodland bluegrass

Perennial; culms 4–8 dm tall; principal blades 3–5 mm wide, mostly cauline;
ligules about 1 mm long; panicle 1–2 dm long, rather narrow; spikelets 2–5-
flowered; lemmas 2.1–3.5 mm long, distinctly 5-nerved; rich woods; mostly S
and W; May–Jul; FACW.

Poa trivialis,
spikelet (×2½)

Poa trivialis L. — Rough bluegrass

Perennial; culms 5–10 dm tall, from a decumbent base, scabrous below the
panicle; blades 2–6 mm wide, soft; sheaths scabrous; panicle branches ascending, in whorls of 5–8; spikelets 2-flowered; lemmas 2.3–3.2 mm long, 5-nerved;
cultivated and frequently established in wet meadows, swamps, and alluvial
woods; throughout; May–Jun; native to Europe; FACW.

Polypogon Desf.

Polypogon monspeliensis (L.) Desf.   Beardgrass, rabbitfoot grass
Annual; culms 1–5 dm tall, sometimes decumbent at the base; blades rather
short, 3–8 mm wide, scabrous; inflorescence spike-like, 2–5 cm long by about 1
cm wide excluding awns; glumes about 2 mm long, silky-hairy; lemma about 1
mm long, with a short, delicate awn; ballast ground and rubbish dumps; Berks
and Philadelphia Cos., probably elsewhere; Jun–Jul; native to Europe; FACW+.

Pseudosasa Makino

Pseudosasa japonica (Sieb. & Zucc. ex Steud.) Makino   Bamboo
Perennial; culms woody, 1–3 m tall, forming dense colonies; cultivated and occasionally spreading to railroad embankments and waste ground; mostly SE, probably elsewhere; native to Japan.

Puccinellia Parl.

Puccinellia distans (L.) Parl.   Alkali grass, goosegrass
Perennial; culms 2–7 dm tall, decumbent at the base; blades flat to involute, 2–4
mm wide; panicle 6–20 cm long, loosely branched, the lower branches bearing

Puccinellia distans,
spikelet (×2½)

Puccinellia
distans

Pseudosasa
japonica

spikelets only beyond the middle; spikelets 4–6-flowered, about 5 mm long; first glume 1-nerved, about 1 mm long, second glume 3-nerved, a little longer; lemmas 1.5–2 mm long; roadsides, waste ground, and ballast; along the Pennsylvania Turnpike and I-80, probably elsewhere; late May–early Jul; native to Europe; OBL.

Schizachne Hack.

Schizachne purpurascens (Torr.) Swallen   Grass

Perennial; culms 4–10 dm from a decumbent base; blades 1–5 mm long, mostly erect; panicle few-branched, each branch bearing 1–3 spikelets; spikelets 2–2.5 cm long; glumes <½ as long as the spikelet, purple at the base; lemmas about 1 cm long; awn as long or longer than the lemma; rich, dry woods; mostly N and southward in the mountains; late May–Jun; FACU–.

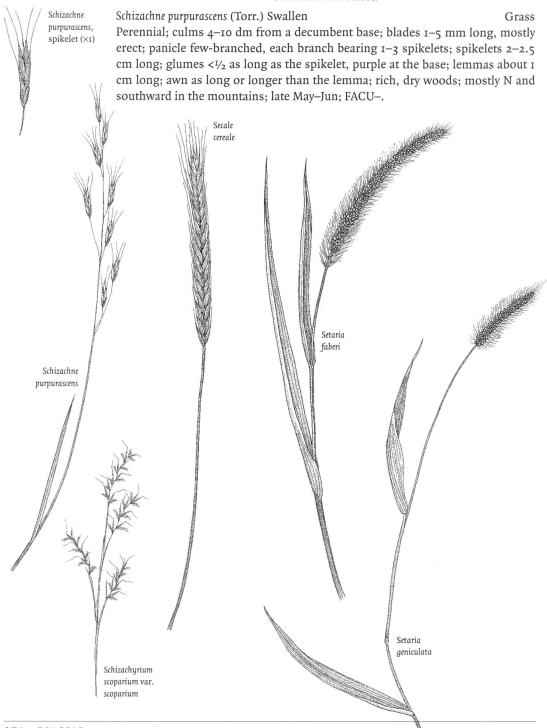

Schizachne purpurascens, spikelet (×1)

Schizachne purpurascens

Secale cereale

Setaria faberi

Setaria geniculata

Schizachyrium scoparium var. scoparium

Schizachyrium Nees.

Schizachyrium scoparium (Michx.) Nash
Perennial; culms 5–12 dm tall, tufted, often branched above, often glaucous or purplish; blades 3–7 mm wide; sheaths glabrous to villous; racemes bearing 5–20 pairs of spikelets; sessile spikelet 6–8 mm long; awn 8–15 mm long; pedicellate spikelet reduced, short-awned; Aug–early Oct; FACU; 2 varieties: [syn: *Andropogon scoparius* Michx.]

Schizachyrium
scoparium var.
scoparium,
spikelet (×1)

A. leaf sheaths not keeled or only weakly so; cilia of the rachis 1–4 mm long .... var. *scoparium*—little bluestem
  old fields, roadsides, and open woods; mostly S.
A. leaf sheaths prominently keeled; cilia of the rachis 5 mm or more long .... var. *littorale* (Nash) Gould—seaside bluestem
  sand dunes along Lake Erie; 🍂.

Secale L.

Secale cereale L.   Rye
Annual; culms usually 1–1.5 m; spikes 8–15 cm long; spikelets usually 2-flowered, disarticulating above the glumes; lemmas 5-nerved; awn 3–8 cm long; cultivated and occasionally occurring in fallow land or field margins; throughout; late May–early Aug; native to Eurasia.

Secale cereale,
spikelet (×1)

Setaria P.Beauv.

Annuals or perennials; inflorescence a spike-like panicle; spikelets awnless, subtended by 1–several bristles, disarticulating above the bristles; first glume 3–5-nerved, about ½ or less as long as the spikelet; second glume longer; sterile lemma about equaling the spikelet, several-nerved.

A. bristles of the inflorescence downward-barbed, rough-feeling when brushed
  upward .... S. *verticillata*
A. bristles of the inflorescence upward-barbed, rough-feeling when brushed
  downward
    B. sheath margins entire (not ciliate); fertile lemmas strongly rugose
      C. culms tufted, arising from a fibrous root system .... S. *pumila*
      C. culms arising separately from a knotted, much-branched rhizome .... S. *geniculata*
    B. sheath margins ciliate; fertile lemmas not rugose
      D. leaf blades pubescent above .... S. *faberi*
      D. leaf blades glabrous above
        E. glumes and sterile lemma falling with fruit .... S. *viridis*
        E. fruit falling without glumes and sterile lemma .... S. *italica*

Setaria faberi,
spikelet
(×2½)

Setaria faberi Herrm.   Giant foxtail
Annual; similar to S. *viridis* in habit; panicle nodding; spikelets about 3 mm long; cultivated fields, roadsides and waste ground; mostly S; Jul–Oct; native to eastern Asia; UPL.

Setaria geniculata (Lam.) P.Beauv.   Perennial foxtail
Annual; similar to S. *pumila* in habit; perennial from short knotty rhizome; blades usually not twisted; dry to moist, open soil; mostly SE; Jul–Sep; FAC.

Setaria geniculata,
spikelet (×2½)

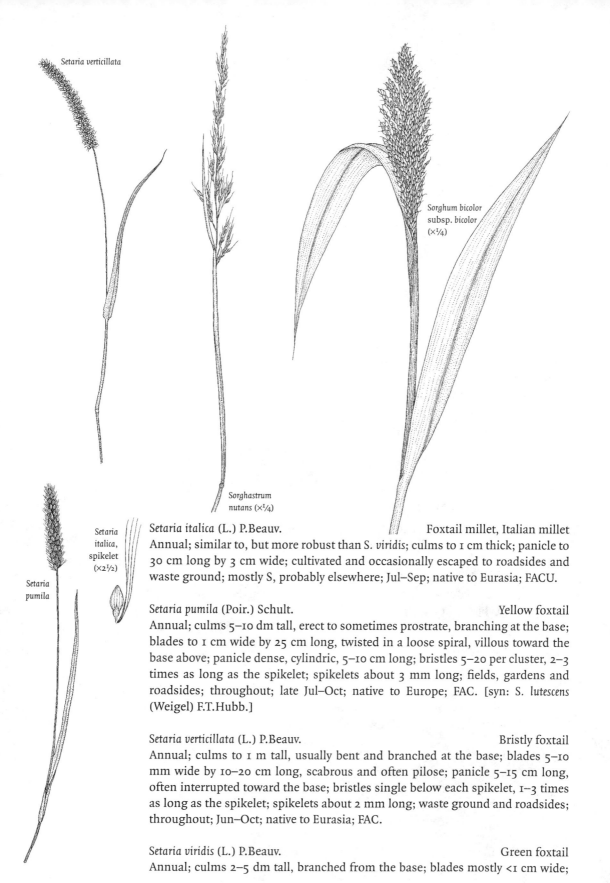

Setaria verticillata

Sorghum bicolor
subsp. bicolor
(×¼)

Sorghastrum
nutans (×¼)

Setaria
italica,
spikelet
(×2½)

Setaria
pumila

Setaria italica (L.) P.Beauv. Foxtail millet, Italian millet
Annual; similar to, but more robust than *S. viridis*; culms to 1 cm thick; panicle to
30 cm long by 3 cm wide; cultivated and occasionally escaped to roadsides and
waste ground; mostly S, probably elsewhere; Jul–Sep; native to Eurasia; FACU.

Setaria pumila (Poir.) Schult. Yellow foxtail
Annual; culms 5–10 dm tall, erect to sometimes prostrate, branching at the base;
blades to 1 cm wide by 25 cm long, twisted in a loose spiral, villous toward the
base above; panicle dense, cylindric, 5–10 cm long; bristles 5–20 per cluster, 2–3
times as long as the spikelet; spikelets about 3 mm long; fields, gardens and
roadsides; throughout; late Jul–Oct; native to Europe; FAC. [syn: *S. lutescens*
(Weigel) F.T.Hubb.]

Setaria verticillata (L.) P.Beauv. Bristly foxtail
Annual; culms to 1 m tall, usually bent and branched at the base; blades 5–10
mm wide by 10–20 cm long, scabrous and often pilose; panicle 5–15 cm long,
often interrupted toward the base; bristles single below each spikelet, 1–3 times
as long as the spikelet; spikelets about 2 mm long; waste ground and roadsides;
throughout; Jun–Oct; native to Eurasia; FAC.

Setaria viridis (L.) P.Beauv. Green foxtail
Annual; culms 2–5 dm tall, branched from the base; blades mostly <1 cm wide;

panicle 1–7 cm long; bristles 1–3 below each spikelet, usually 2–10 mm long; spikelets 2–2.5 mm long; cultivated fields, gardens, and waste ground; throughout; Jul–Sep; native to Eurasia; 2 varieties:

A. inflorescence 10–20 cm long .... var. *major* (Gaudin) Posp.
A. inflorescence 4–8 cm long .... var. *viridis*

Setaria viridis
var. *viridis*,
spikelet
(×2½)

Sorghastrum Nash

Sorghastrum nutans (L.) Nash   Indian-grass
Perennial; culms mostly 1–2 m tall, in loose tufts from short rhizomes; panicle narrow, 15–30 cm long; spikelets 6–8 mm long; first glume villous, second glume glabrous or ciliate; awn 9–15 mm long, twisted below, bent at about ⅓ of its length; moist or dry fields, roadsides, and serpentine barrens; throughout; Aug–Sep; UPL.

Sorghum Moench

Tall annuals or perennials; leaves often splotched with purple; panicles of several jointed racemes.

A. plants rhizomatous .... *S. halepense*
A. plants not rhizomatous .... *S. bicolor*

Sorghum bicolor (L.) Moench
Annual; culms to 2 m or more tall from creeping, scaly rhizomes; panicle open or narrow, 15–50 cm long; sessile spikelet 4.5–5.5 mm long, pediceled spikelet 5–7 mm long; cultivated and occasionally spreading to waste ground; mostly E, probably elsewhere; late Jul–early Nov; native to Africa; UPL; designated as a noxious weed in Pennsylvania; 2 varieties:

A. inflorescence very compact, the branches not spreading ....
.... var. *bicolor*—broom-corn
[syn: *S. vulgare* L.]
A. inflorescence open, the branches spreading ....
.... var. *drummondii* (Steud.) De Wet & Harlan—shattercane
[syn: *S. sudanense* (Piper) Stapf]

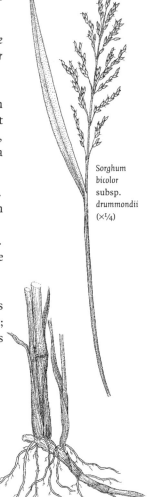

Sorghum
bicolor
subsp.
drummondii
(×¼)

Sorghum halepense (L.) Pers.   Johnsongrass
Similar to *S. bicolor* but perennial and not as robust; cultivated fields; throughout; native to the Mediterranean region; Aug–Oct; FACU; designated as a noxious weed in Pennsylvania.

Sorghum bicolor
subsp. *drummondii*,
base of plant

Sorghum halepense, rhizome

Spartina Schreb.

Spartina pectinata Link   Freshwater cordgrass
Perennial; culms 1–2 m tall, stout; principal blades 5–10 mm wide, very long, scabrous on the margins; panicle 2–4 dm long, with 10–20 spikes; spikelets 1-flowered, crowded, disarticulating below the glumes; glumes 1-nerved, the first about 3/4 as long as the lemma, the second 10–13 mm long, with an awn 3–10 mm long; along waterways throughout; late Jul–Oct; OBL.

Sphenopholis Scribn.

Spikelets 2–3-flowered, disarticulating below the glumes; rachilla somewhat prolonged as a bristle behind the upper palea; first glume 1–3-nerved, second glume 3–5-nerved.

A. second lemma awnless
    B. second lemma scabrous .... *S. nitida*
    B. second lemma not scabrous .... *S. obtusata*
A. second lemma awned
    C. spikelets 5–9.5 mm long ....*S. pensylvanica*
    C. spikelets 1.5–5 mm long .... *S. obtusata* x *pensylvanica*

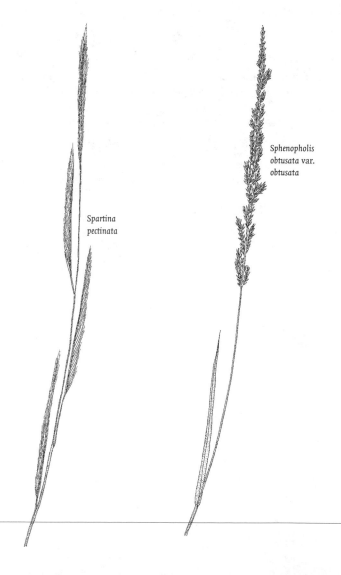

Sphenopholis
obtusata var.
obtusata

Spartina
pectinata

Sphenopholis nitida (Biehler) Scribn.   Wedgegrass
Perennial; culms 3–8 dm tall, cespitose; principal blades 2–5 mm wide, upper
(and sometimes lower) ones pubescent; panicle slender, but not spike-like, to 10
cm long; glumes 2.2–2.9 mm long; lower lemma 2.3–2.9 mm long, upper lemma
scabrous toward the apex; dry, rocky woods and roadside banks; throughout;
May–early Jul.

Sphenopholis nitida,
spikelet (×2½)

Sphenopholis obtusata (Michx.) Scribn.
Annual or short-lived perennial; culms 2–10 dm tall, cespitose; blades 2–5 mm
wide, scabrous to pubescent; panicle dense, spike-like, 5–15 cm long; first
glume subulate, second glume broadly obovate; lower lemma 1.7–2.5 mm long;
FAC−; 2 varieties:

A. second lemma acute at apex ....
.... var. *major* (Torr.) Erdman—Slender wedgegrass
serpentine barrens and other dry, open sites; mostly SE; late May–Jun. [syn: S.
intermedia (Rydb.) Rydb.]

Sphenopholis obtusata
var. *major*, spikelet
(×2½)

A. second lemma rounded at apex .... var. *obtusata*—Prairie wedgegrass
stream banks, open woods, and alluvial thickets; throughout; Jun–early Aug.

Sphenopholis obtusata
var. *obtusata*,
spikelet (×2½)

Sphenopholis obtusata x pensylvanica   Wedgegrass
Perennial; culms 2–6 dm tall; blades 1–2 mm wide, glabrous; upper sheaths minutely pubescent, lower ones glabrous; panicle narrow, compact, often nodding,
15–25 cm long; second lemma awned just below the apex; awn 1–2 mm long;
wet woods, moist slopes, and springheads; scattered; mostly S; FAC. [syn: S.
pallens (Biehler) Scribn.]

Sphenopholis obtusata
x *pensylvanica,*
spikelet (×2½)

Sphenopholis pensylvanica (L.) A.Hitchc.   Swamp-oats
Perennial; culms 4–8 dm tall, slender and weak; blades 2–5 mm wide, somewhat
scabrous; sheaths glabrous; panicle lax, slender, 1–2 dm long; glumes scabrous
on the keel; upper (and sometimes lower) lemma awned from just below the
apex; awn 5–7 mm long; swamps, wet woods, or springy meadows; mostly SE
and SC; May–Jun; OBL. [syn: *Trisetum pennsylvanicum* (L.) P.Beauv.]

*Sphenopholis
pensylvanica,*
spikelet (×2½)

Sporobolus R.Br.

Spikelets 1-flowered, disarticulating above the glumes; glumes lanceolate to
ovate, 1-nerved; lemma obtuse to acuminate; palea as long as or longer than the
lemma; fruit readily separating from the lemma and palea.

A. lemmas pubescent
   B. panicles 1–3 cm long, often wholly or partially covered by uppermost leaf
    .... S. *vaginiflorus*
   B. panicles 3–10 cm long, not covered by uppermost leaf .... S. *clandestinus*
A. lemmas glabrous
   C. panicles contracted and more or less spike-like
      D. glumes and lemma of 3 distinctly different lengths .... S. *asper*
      D. glumes and lemma of nearly equal lengths .... S. *neglectus*
   C. panicles open, with spreading or ascending branches
      E. first glume expanded at the base, long subulate to tip .... S. *heterolepis*
      E. first glume acuminate, not expanded at the base .... S. *cryptandrus*

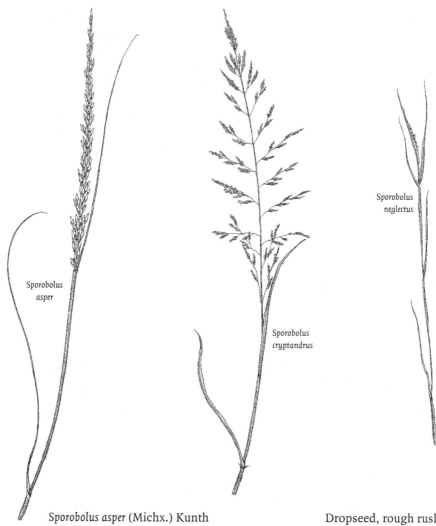

Sporobolus
asper

Sporobolus
cryptandrus

Sporobolus
neglectus

Sporobolus asper,
spikelet (×2½)

Sporobolus asper (Michx.) Kunth   Dropseed, rough rushgrass
Perennial; culms to 1.2 m tall, stout; principal blades very long, 2–4 mm wide,
involute when dry; inflorescence 5–15 cm long by about 1 cm wide, enclosed in
the upper sheath; first glume 2–3.5 mm long, second 2.5–4.6 mm long; lemma
3.5–6 mm long, about equaling the palea; dry, sandy banks; scattered; mostly E;
Aug–early Sep; UPL.

Sporobolus
clandestinus,
spikelet
(×2½)

Sporobolus clandestinus (Biehler) A.Hitchc.   Rough dropseed, rough rushgrass
Perennial; culms 5–10 dm tall from a knotty base, cespitose; principal blades
long, 2–5 mm wide; first glume 2–4 mm long, second a little longer; lemma 4–
6.5 mm long, sparsely villous; palea acuminate to a subulate tip, up to 8 mm
long, usually exceeding the lemma; dry, sandy or rocky soil; Dauphin and Lancaster Cos.; late Aug–Oct; 🍂.

Sporobolus
cryptandrus,
spikelet (×5)

Sporobolus cryptandrus (Torr.) A.Gray   Sand dropseed
Perennial; culms 3–10 dm tall, usually in small tufts, seldom solitary; principal
blades 2–6 mm wide, involute when dry; panicles 1–2 dm long, base partly enclosed by the upper sheath; first glume ½ or less as long as the second; lemma
about equaling the second glume; dry waste ground and sandy shores; E and
NW; Aug–Sep; UPL; 🍂.

Sporobolus heterolepis (A.Gray) A.Gray   Prairie dropseed
Perennial; culms 4–10 dm tall; principal blades usually involute; panicle 1–2 dm
long; first glume about ½ as long as the second; lemma usually slightly exceeded by the second glume; palea usually slightly exceeding the lemma; serpentine barrens; Chester and Lancaster Cos; Aug–Sep; UPL; ✤.

Sporobolus
heterolepis,
spikelet (×2½)

Sporobolus neglectus Nash   Small rushgrass
Annual; culms 2–6 dm tall, branching from the base, cespitose; principal blades
narrow, 1–2 mm wide, often involute; panicle partially or wholly enclosed by the
upper sheath; spikelets 2–3 mm long; dry, sterile, or sandy soil; mostly S; late
Aug–Oct; FACU–.

Sporobolus
neglectus,
spikelet (×2½)

Sporobolus vaginiflorus (Torr. ex A.Gray) A.W.Wood   Poverty grass
Annual; similar to S. neglectus; panicle eventually exsert from the upper leaf
sheath; spikelets somewhat longer than in S. neglectus; dry, sandy, or sterile soil
including serpentine barrens; mostly S, scattered elsewhere; late Aug–Oct; UPL.

Sporobolus
vaginiflorus,
spikelet (×2½)

Stipa L.

Stipa spartea Trin.   Needlegrass, porcupine grass
Perennial; culms tufted, 6–12 dm tall; blades 2–5 mm wide, scabrous and pubescent above, glabrous below; ligules of the upper leaves 4–6 mm long; panicle
narrow, often nodding, 1–2 dm long, few-flowered; mature lemma 1.8–2.1 cm

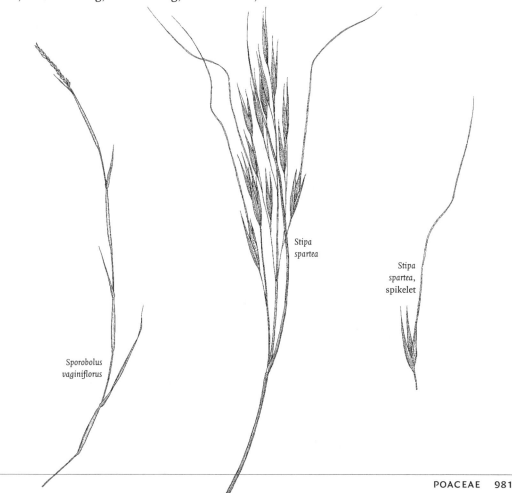

Stipa
spartea

Stipa
spartea,
spikelet

Sporobolus
vaginiflorus

long, pubescent at the base; awn very stiff, 12–20 cm long, twice bent near the middle; roadsides; Lackawanna Co.; a midwestern and western North American species, possibly adventive in Pennsylvania; 🍃.

Torreyochloa Church

Torreyochloa pallida var. pallida, spikelet (×2½)

Torreyochloa pallida (Torr.) Church   Pale meadowgrass
Perennial; culms 3–10 dm tall; blades 2–8 mm wide; panicle open, 5–15 cm long, few-branched; spikelets 4–7-flowered, 6–7 mm long; glumes 1.5–2.5 mm long; lemmas 2.5–3 mm long; OBL; 2 varieties: [syn: *Glyceria pallida* (Torr.) Trin., *Puccinellia pallida* (Torr.) R.T.Clausen]

A. leaf blades to 8 mm wide; flr. Jul–Aug .... var. *fernaldii* (A.Hitchc.) Dore
   lake and stream edges; scattered throughout.
A. leaf blades 2–3 mm wide; flr. May–Jun .... var. *pallida*
   swamps, marshes and ditches; scattered throughout.

Tragus Haller

Tragus racemosus, spikelet (×2½)

Tragus racemosus (L.) All.   Texas bur
Annual; culms 1–3 dm tall, spreading or ascending, densely branched from the

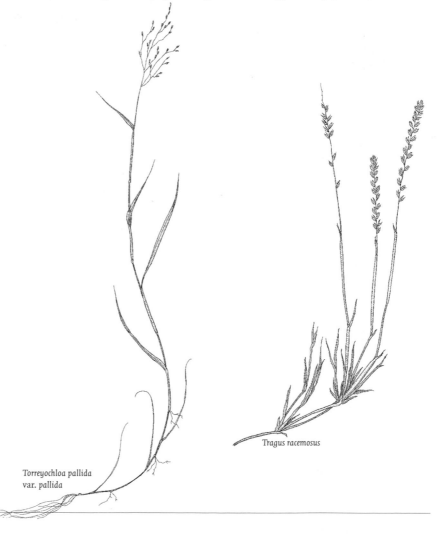

Tragus racemosus

Torreyochloa pallida var. *pallida*

base; blades mostly <5 cm long, stiffly ciliate; second glumes spiny, 3.6–4.5 mm long; ballast ground; Lehigh and Philadelphia Cos.; Jul–early Oct; native to Europe.

Tridens Roem. & Schult.

Tridens flavus (L.) A.Hitchc.   Purpletop

Perennial; culms 10–15 dm tall, solitary or in small tufts; blades 3–8 mm wide; sheaths densely pubescent at summit; panicle 2–4 dm long with spreading or drooping branches, purplish; spikelets 5–10 mm long, 4–9-flowered, only slightly flattened, disarticulating above the glumes and between the lemmas; glumes 2.5–3.5 mm long; lemmas densely villous on the lower ½; meadows, old fields and roadsides; mostly S, especially SE; late Jul–Sep; FACU. [syn: *Triodia flava* (L.) Smyth]

Tridens flavus,
spikelet (×1½)

Triplasis P.Beauv.

Triplasis purpurea (Walter) Chapm.   Purple sandgrass

Annual; culms 2–8 dm tall, cespitose; blades 1–2 mm wide, the upper ones greatly reduced; panicle 2–8 cm long, few-branched; spikelets 2–5-flowered,

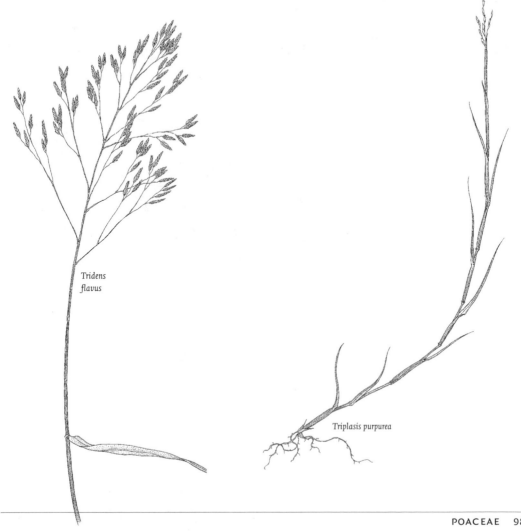

Triplasis purpurea,
spikelet (×2½)

*Tridens
flavus*

Triplasis purpurea

purplish, disarticulating above the glumes and between the lemmas; glumes 2–4 mm long; lemmas 3–4 mm long with a short awn; dry, open, sandy soil; Coastal Plain and Lake Erie shore; Aug–Sep; ❧.

Tripsacum L.

Tripsacum dactyloides (L.) L.   Gammagrass
Perennial; culms to 3 m or more tall, glabrous throughout; spikes 1–4, 10–25 cm long, staminate spikelets above the pistillate; staminate spikelets 2-flowered; pistillate spikelets with 1 functional floret and 1 sterile lemma, sunken in hollows in the thickened rachis; swamps and wet shores; SE; Jul–Oct; FACW; ❧.

Trisetum Pers.

Trisetum spicatum (L.) Richt.   Oatgrass
Perennial; culms 1–5 dm tall, cespitose, pubescent below the panicle; blades 1–3 mm wide, flat or involute; sheaths pubescent; panicle 3–10 cm long, usually interrupted at the base; spikelets usually 2-flowered; first glume 1-nerved, second glume 3-nerved; open, shaly outcrops; Lehigh and Mercer Cos.; circumboreal and possibly adventive in Pennsylvania; Jun–Jul; FACU; ❧.

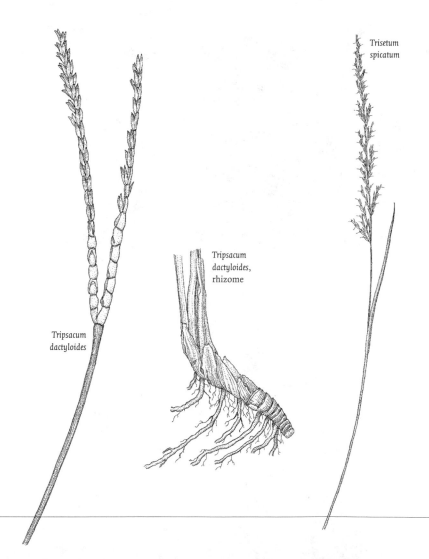

Trisetum
spicatum,
spikelet
(×2½)

Trisetum
spicatum

Tripsacum
dactyloides,
rhizome

Tripsacum
dactyloides

Triticum L.

Triticum aestivum L.   Wheat

Perennial; culms 5–12 dm tall; spikelets 2–5-flowered; lemmas awnless or with an awn up to 8 cm long; cultivated and frequently occurring on roadsides and edges of fields; throughout; late May–early Aug; native to Eurasia.

Vulpia C.C.Gmel.

Annuals; culms slender, erect, usually bent at the base; blades usually involute; spikelets 4–13-flowered, disarticulating above the glumes and between the lemmas.

A. spikelets 5–13-flowered; first glume >½ as long as the second ...... *F. octoflora*
A. spikelets 4–5-flowered; first glume <½ as long as the second .... *F. myuros*

Vulpia myuros (L.) C.C.Gmel.   Foxtail fescue

Culms 1–7 dm tall; panicle 5–20 cm long, often nodding; second glume 3.7–6 mm long; open fields and banks in sandy soil; mostly SE, probably elsewhere; May–Jun; native to Europe; UPL. Ours is var. *myuros*. [syn: *Festuca myuros* (L.)]

Vulpia myuros
var. *myuros*,
spikelet (×1)

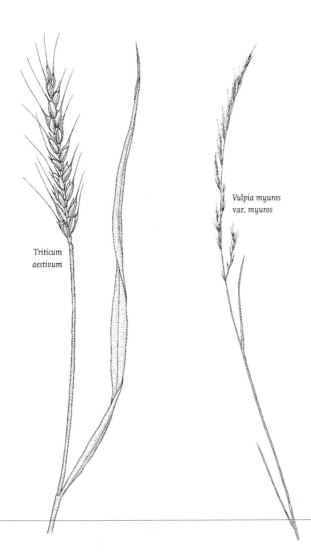

Vulpia myuros
var. *myuros*

Triticum
aestivum

Vulpia octoflora (Walter) Rydb.   Six-weeks fescue
Culms 1–4 dm tall; panicle 3–10 cm long, branches ascending; second glume
2.5–5 mm long; dry, sterile soil; mostly S, especially SE; late May–Jun; UPL. Ours
is var. *glauca* (Nutt.) Fernald. [syn: *Festuca octoflora* (Walter) Rydb. var. *tenella*
(Willd.) Fernald]

Zizania L.

Zizania aquatica L.   Wild-rice
Annual; culms to 3 m or more, often decumbent at the base, occasionally
branched; ligules 10–15 mm long; panicles open, 1–6 dm long, pistillate spikelets above the staminate; pistillate spikelets linear, about 2 cm long, awn 1–6 cm
long; staminate spikelets 6–11 mm long, pendulous, sometimes with a short
awn; late May–early Sep; OBL; ❧; 2 varieties:

A. pistillate lemmas dull, thin and papery, more or less scaberulous over the entire surface; blades mostly 1–5 cm wide .... var. *aquatica*
tidal and nontidal marshes; mostly SE and Erie Co., scattered elsewhere.
A. pistillate lemmas shining, firm and leathery, glabrous except along the margins and on the awn and at its base; blades mostly 1 cm or less wide ....
.... var. *angustifolia* A.Hitchc.
shallow water; Lake Erie shore.

Zizania aquatica
var. *aquatica*,
staminate
spikelet (×1)

Zizania aquatica
var. *aquatica*,
pistillate
spikelet (×1)

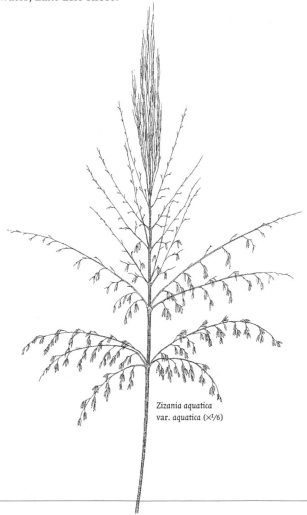

Zizania aquatica
var. *aquatica* (×1/6)

PONTEDERIACEAE Water-hyacinth Family

Emergent or submersed aquatics with varying habit; leaves linear or with an expanded blade with parallel, curved-convergent veins, with or without a sheathing base; flowers perfect, regular or irregular; perianth of 6 colored tepals that are fused at the base to form a short tube; stamens 3–6, filaments fused to the perianth tube; ovary superior, with 2–3 locules; fruit a capsule or achene.

A. leaves narrowly linear, plants totally submersed .... *Zosterella*
A. leaves wider, oblong to kidney-shaped, with a cordate base
    B. plants erect, leaves and inflorescence emergent .... *Pontederia*
    B. plants decumbent, creeping on the mud, submersed, or with floating leaves .... *Heteranthera*

Heteranthera Ruiz & Pavon

Low-growing plants often creeping on the mud, submersed, or with some leaves floating; leaves broadly kidney-shaped, with parallel, curving veins; inflorescence a spike of 2–16 flowers enclosed in a folded spathe; flowers perfect, regular; perianth with a slender tube and 6 narrow, spreading lobes; stamens 3, the 2 lower shorter than the upper 1; filaments hairy; ovary 3-locular; fruit a capsule; flr. Jul–early Oct.

A. hairs on the filaments white; spike barely elongating beyond the spathe .... *H. reniformis*
A. hairs on the filaments purple; spike elongating well beyond the spathe .... *H. multiflora*

Heteranthera multiflora (Griseb.) Horn   Mud-plantain
Very similar to *H. reniformis*; leaf blades slightly longer than wide; inflorescence of 3–16 flowers, all opening within a day or 2 and extending well beyond the

Heteranthera multiflora

Pontederia cordata, leaf (×¼)

Pontederia cordata, inflorescence

spathe; perianth pale purple or white; hairs of the filaments purple; rare on tidal shores and mudflats; SE; OBL; 🌿.

Heteranthera reniformis Ruiz. & Pavon   Mud-plantain
Leaf blades 1–4 cm long by 1–5 cm wide; flowers 2–8, all opening the same day and only the terminal one extending past the tip of the spathe; perianth and filament hairs white; occasional in shallow water and muddy shores; E; OBL.

Pontederia L.

Pontederia cordata L.   Pickerel-weed
Erect emergent aquatic to 1 m tall; leaves long-petioled, broadly cordate to lanceolate with curved-parallel venation; inflorescence a terminal spike 5–15 cm long, subtended by a loosely sheathing spathe; flowers purple, slightly bilabiate; stamens 5, inserted on the perianth tube; fruit an achene with the perianth tube persisting; frequent in the swampy edges of lakes, streams, and tidal shores; E and NW; flr. Jun–Oct; OBL.

Zosterella Small

Zosterella dubia (Jacq.) Small   Water star-grass
Submersed, highly branched, aquatic perennial with sessile, linear leaves lacking a midrib; flowers solitary, enclosed in a spathe, regular; perianth yellow, 1–2 cm wide, tubular with 6 narrow, spreading lobes, the outer 3 somewhat narrower than the inner; stamens 3, with inflated filaments; fruit an indehiscent capsule 1 cm wide; occasional in lakes, rivers, and streams; throughout; flr. Jun–Sep; OBL.

POTAMOGETONACEAE Pondweed Family

Rooted, perennial (except *P. vaseyi*), aquatic herbs; leaves often of 2 types, submersed and floating, alternate or the upper nearly opposite, parallel-veined; stipules fused to form a sheath that clasps the stem and is sometimes fused for part of its length to the base of the leaf; blades of the submersed leaves usually narrower than those of the floating leaves and often with 1 or more rows of translucent cells forming lacunar strips on both sides of the midrib; many species also form winter buds consisting of branch tips with shortened internodes and crowded leaves and stipules; inflorescence of axillary or terminal spikes that project above the water surface in species with floating leaves; flowers small, regular, perfect; perianth a whorl of 4 tepals; stamens 4; ovaries 4, distinct and superior; fruit a cluster of achenes.

Potamogeton L.

Characteristics of the family. We have chosen not to recognize the segregate genus, *Coleogeton*, which would include *P. pectinatus* and *P. filiformis* in our flora. In addition to the species described below, a presumed hybrid, which was previously identified as *P. alpinus*, has been found at several locations in the Delaware River drainage.

A. stipular sheaths of submersed leaves fused to the base of the leaf blade for ⅔ or more of their length
    B. lower stipular sheath inflated; achenes 2–3 mm long, with inconspicuous beaks ....*P. filiformis*
    B. lower stipular sheath not inflated; achenes >3.5 mm long, beaked .... *P. pectinatus*
A. stipular sheaths of submersed leaves fused to the base of the leaf for <½ their length or not at all
    C. stipular sheaths of submersed leaves fused to the base of the leaf with the tips extending as a ligule
        D. leaves conspicuously 2-ranked; blades lobed (auriculate) at the base; margins minutely serrulate .... *P. robbinsii*
        D. leaves not conspicuously 2-ranked, blades not lobed at the base, margins entire
            E. submersed leaves obtuse to acute; floating leaves with rounded tips
                F. submersed leaves 0.5–2 mm wide, obtuse; achene 1.3–2.2 mm wide, not beaked .... *P. spirillus*
                F. submersed leaves <0.5 mm wide, acute; achene about 1 mm wide with a minute beak .... *P. diversifolius*
            E. submersed leaves acute to acuminate; floating leaves with pointed tips
                G. submersed leaves with obvious lacunae; floating leaves 9–23-veined .... *P. tennesseensis*
                G. submersed leaves without obvious lacunae; floating leaves 3–7-veined .... *P. bicupulatus*
    C. stipular sheaths of the submersed leaves mostly free to the base
        H. leaves distinctly wavy-margined and serrulate ....*P. crispus*
        H. leaves not wavy-margined or serrulate
            I. submersed leaves linear-oblong, lanceolate or elliptic
                J. submersed leaves with a weakly to strongly cordate-clasping base; floating leaves not produced
                    K. stipules persistent; leaf tips hooded .... *P. praelongus*
                    K. stipules disintegrating into fibers; leaf tips flat .... *P. perfoliatus*
                J. submersed leaves sessile or petiolate, not cordate-clasping; floating leaves present or absent
                    L. submersed leaves folded and curved along the midrib, 23–37-veined .... *P. amplifolius*
                    L. submersed leaves not folded or curved along the midrib, with <23 veins
                        M. stems black-spotted .... *P. pulcher*
                        M. stems not black-spotted
                            N. submersed leaves sessile
                                O. stipular sheath 4–10 cm long; submersed leaves with 7–19 veins .... *P. illinoensis*
                                O. stipular sheaths 1–3 cm long; submersed leaves with 3–9 veins .... *P. gramineus*

N. submersed leaves distinctly petiolate
 P. larger submersed leaves with a sharp awn-like tip ....
 .... *P. illinoensis*
 P. larger submersed leaves acute but not awn-tipped................
 .... *P. nodosus*
I. submersed leaves ribbon- or thread-like, 0.1–10 mm wide
 Q. floating leaves usually present
  R. floating leaves to 1.5 cm long; submersed leaves 1(3)-veined ... *P. vaseyi*
  R. floating leaves >2 cm long; submersed leaves with 3 or more veins
   S. submersed leaves flat, 2–10 mm wide, with 5 or more veins and distinct lacunar bands .... *P. epihydrus*
   S. submersed leaves to 0.2–2.5 mm wide with 3–5 obscure veins; lacunae present or absent
    T. floating leaves cordate at the base; achenes 3.5–5 mm long........
    .... *P. natans*
    T. floating leaves rounded or cuneate at the base; achenes 2.5–3.5 mm long
     U. floating leaves to 3.5 cm long; achenes with a strongly winged keel .... *P. tennesseensis*
     U. floating leaves to 6 cm long; achenes not winged ....
     .... *P. oakesianus*
 Q. floating leaves not produced
  V. stems strongly flattened .... *P. zosteriformis*
  V. stems round or only slightly flattened
   W. submersed leaves very slender, to 0.3 mm wide
    X. leaves with 1 vein; stems lacking glands at the nodes ....
    .... *P. confervoides*
    X. leaves with 3 veins; stems with small glands at the nodes ....
    .... *P. pusillus*
   W. submersed leaves >0.5 mm wide, with >1 vein
    Y. spikes few-flowered, with at most 3 whorls of flowers; achenes keeled; glands rarely present at the nodes
     Z. leaves bristle-tipped................................... *P. hillii*
     Z. leaves acute but not bristle-tipped.......................... *P. foliosus*
    Y. spikes with >2 whorls of flowers; achenes not keeled; glands often present at the nodes
     AA. stipular sheaths strongly fibrous, becoming shredded and whitish with age
      BB. leaves thin, 1.2–3.2 mm wide, obtuse or rounded with a minute apiculate tip .... *P. friesii*
      BB. leaves firm, 0.6–2 mm wide, acute and usually bristle-tipped .... *P. strictifolius*
     AA. stipular sheaths membranous or delicate, greenish, brownish, or white, not becoming shredded and fibrous
      CC. leaf apex bristle-tipped; spikes few-flowered, subcapitate .... *P. hillii*
      CC. leaf apex blunt or acute; spikes thick-cylindric...........
      .... *P. obtusifolius*

Potamogeton amplifolius Tuck.   Bigleaf pondweed

Stems elongate, simple; submersed leaves 8–20 cm long, folded and curved along the midrib, with 19–37 veins; floating leaves elliptic, 5–10 cm long; stipules free; spikes dense, 2–5 cm; achenes 4–5.5 mm; occasional in ponds and streams; scattered; OBL.

Potamogeton bicupulatus Fernald   Pondweed

Submersed leaves 0.1–0.4 mm wide, without lacunar strips; stipules fused for <½ their length, not becoming fibrous at the tips; floating leaves 0.6–2.3 cm long; achenes 1.1–2.1 mm long, keeled; rare in shallow, quiet water; E; OBL; 🌿. [syn: *P. diversifolius* var. *trichophyllus* Morong]

Potamogeton confervoides Reichenb.   Tuckerman's pondweed

Stems slender, 1–8 dm, much-branched and appearing fan-shaped in the water; leaves all submersed, very thin and delicate, 2–5 cm long and 0.25 mm wide, 1-veined; stipular sheath not fused to the leaf base; spike few-flowered; achenes 2–3 mm long; rare in glacial lakes and boggy ponds; NE; OBL; 🌿.

Potamogeton crispus L.   Curly pondweed

Stems 3–8 dm, lightly branched; leaves all submersed, the blades strongly wavy-margined and serrulate; stipules free; spike dense; achenes approximately 6 mm long, including a 2–2.5-mm beak; common in lakes, ponds, rivers, and streams; rarely flowering; throughout; native to Europe.

Potamogeton diversifolius Raf.   Snailseed pondweed

Stems slender, to 15 dm, clustered and sparsely branched; submersed leaves linear, 0.1–0.5 mm wide, 1(3)-veined, with narrow lacunar bands; floating leaves, when present, 3–8.5 mm wide with 3–7 veins; stipules fused to the leaf base for about ½ their length, the free tips not shredding or fibrous; spikes axillary to

Potamogeton
diversifolius,
achene (×5)

Potamogeton
diversifolius,
stipules (×1½)

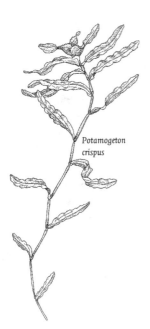

Potamogeton
crispus

Potamogeton
diversifolius

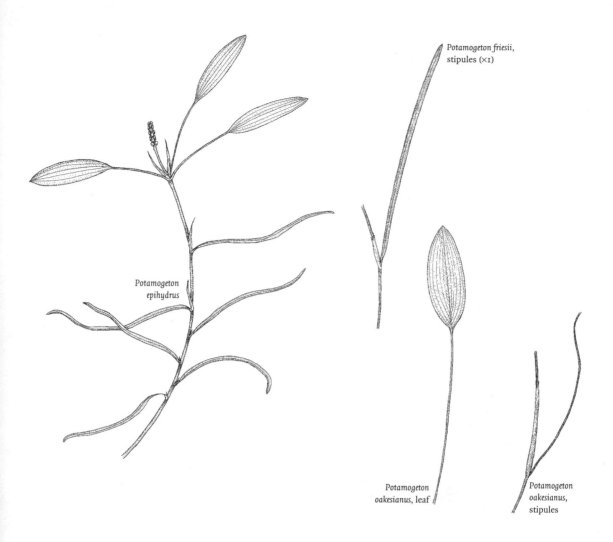

Potamogeton friesii, stipules (×1)

Potamogeton epihydrus

Potamogeton oakesianus, leaf

Potamogeton oakesianus, stipules

submersed leaves shorter and denser than those in the axils of floating leaves; achenes compressed, winged, about 1 mm; frequent in shallow, quiet water; throughout except in the northernmost counties; OBL.

Potamogeton epihydrus Raf.   Ribbonleaf pondweed
Stems to 2 m, slender, slightly flattened, branched; submersed leaves sessile, linear, to 2 dm long and 0.2–1 cm wide, 3–13-nerved with conspicuous lacunar bands; floating leaves elliptic, 2–8 cm long and up to 2 cm wide, 11–41-veined, narrowed to a slender petiole; stipules free, 1–3 cm long; spikes cylindric, dense, 1–3 cm; achenes 2.5–4 mm, keeled; common in lakes, ponds, and streams; throughout; OBL.

Potamogeton filiformis Pers.   Threadleaf pondweed
Stems erect, 1–4 dm, branched from the base; leaves all submersed, narrowly linear, turgid, to 1 mm wide, 1-nerved with an obtuse tip; stipules fused to the leaf base; spikes submersed, with 2–5 whorls of flowers; achenes 2–3 mm; rare in shallow, calcareous water of streams and ponds; scattered; OBL; ✿. Ours is var. *borealis* (Raf.) St. John.

Potamogeton foliosus Raf.   Leafy pondweed

Stems to 7 dm, freely branched and slightly compressed; leaves all submersed, linear, 0.3–2.3 mm wide, acute or apiculate, 1–5-veined with the lacunar strip narrow or lacking; stipular sheaths free, 0.5–2 cm, greenish or brownish, membranous; spikes with 2 or 3 whorls of flowers; achenes 1.5–2.7 mm with a wing-like keel; frequent in lakes and streams; throughout; OBL.

Potamogeton friesii Rupr.   Fries' pondweed

Stems to 1 m, branched, slightly compressed, with glands at the nodes; leaves all submersed, linear, 1.5–3 mm wide, minutely cuspidate at the rounded tip, 5–9-veined with very narrow lacunar strips; stipules free, fibrous; spike slender with 2–5 well-spaced whorls of flowers; achenes 2–3 mm, not keeled; a rare native of calcareous streams; scattered; OBL; ❧.

Potamogeton friesii, leaf venation (×5)

Potamogeton gramineus L.   Grassy or variable pondweed

Stems slender, 3–7 dm, much-branched, somewhat compressed; submersed leaves sessile, 3–10 mm wide, acute to cuspidate, 3–7-veined; floating leaves broadly elliptic, 3–8 cm long with 11–19 veins, rounded or subcordate at the base; stipules free, 1–3 cm long; spikes dense, 1.5–3 cm; achenes 2–2.5 mm; a rare native of lakes and streams; scattered; OBL; ❧.

Potamogeton hillii Morong   Hill's pondweed

Stems very slender, to 1 m, freely branched; leaves all submersed, narrowly linear with 3 veins, the midvein flanked by narrow lacunar strips, bristle-tipped; stipules free, 1–1.5 cm long, rigid and fibrous; spike subcapitate, few-flowered; achenes flattened, 3.5–4 mm; rare in lakes and streams; scattered; OBL; ❧.

Potamogeton illinoensis Morong   Illinois pondweed

Stems stout, branched, to 2 m; submersed leaves 8–20 cm long and 2–5 cm wide, with 9–19 veins, acute; floating leaves, when present, 7–13 cm long and 2–6 cm wide; stipules free, 4–10 cm long; spikes dense, 2.5–6 cm long; achenes 3–4 mm with 3 conspicuous keels; occasional in lakes and streams; throughout; OBL; ❧.

Potamogeton natans L.   Floating pondweed

Stems 1–2 m, simple or lightly branched; submersed leaves 1–2 mm wide with 3–5 faint veins and often gone by flowering time; floating leaves 5–10 cm long and 2–4.5 cm wide with rounded or subcordate bases and short cuspidate tips; stipules free, 4–10 cm long; spike dense, 2–5 cm; achenes 3.5–5 mm; infrequent in lakes and streams; E and W; OBL.

Potamogeton nodosus Poir.   Longleaf pondweed

Stems to 2 m, branched; submersed leaves linear, to 3 dm long and 1–2.5 cm wide with 7–15 veins and conspicuous lacunar bands; floating leaves elliptic, 5–13 cm long; stipules free, 4–10 cm long; spikes dense, 3–5 cm; achenes 3–4 mm; frequent in lakes, ponds, and streams; throughout; OBL.

Potamogeton oakesianus J.W.Robbins   Oakes' pondweed

Stems to 1 m, usually freely branched; submersed leaves 0.3–1 mm wide with 3 veins, lacunar bands lacking, often gone by flowering time; floating leaves elliptic, 2.5–6 cm long and 1–2.5 cm wide; stipules free, 2.5–4 cm long; spike 1.5–3

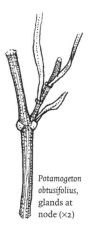

Potamogeton obtusifolius, glands at node (×2)

cm; achenes 2–3.5 mm, rounded on the back; rare in ponds and lakes; mostly E; OBL; 🌿.

Potamogeton obtusifolius Mert. & Koch   Blunt-leaved pondweed
Stems very slender, to 1 m, branched, with a pair of small glands at the nodes; leaves all submersed, linear, 1–3.5 mm wide with 3 veins, lacunar strips present, tip obtuse or rounded; stipules free, 1–2 cm long, thin and membranous; spikes 1.5–2 cm with 3–5 whorls of flowers; achenes 3.8–4 mm long; rare in boggy ponds and lakes; NE; OBL; 🌿.

Potamogeton pectinatus L.   Sago pondweed
Stems 3–8 dm, dichotomously branched; leaves all submersed, narrowly linear, 1-nerved, 0.2–1 mm wide, and tapering to an acute tip; stipules fused to the leaf bases and clasping the stem for 1–3 cm; spikes submersed, 1–4 cm; fruits 3–4.5 mm with 1–3 keels; occasional in shallow brackish or calcareous water; throughout; OBL.

Potamogeton perfoliatus L.   Perfoliate pondweed, redhead-grass
Stems slender, much branched; leaves all submersed, ovate to suborbicular, 1–5(9) cm long, with cordate-clasping to perfoliate bases; stipules free; spikes dense, 1–4.5 cm; achenes 1.6–3 mm; occasional in ponds, lakes and streams; throughout; OBL. Includes Pennsylvania material identified as *P. richardsonii*, which tends to have narrower leaves and fibrous stipules.

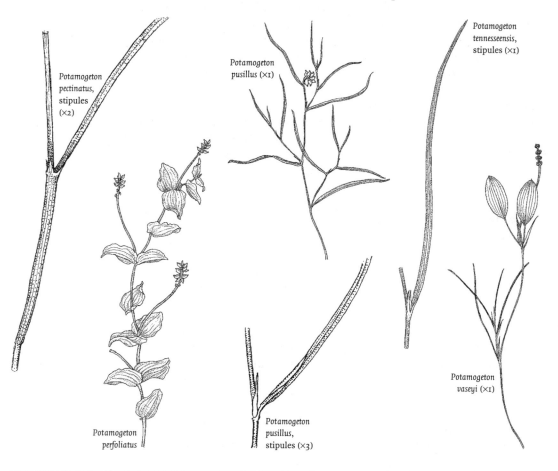

Potamogeton pectinatus, stipules (×2)

Potamogeton pusillus (×1)

Potamogeton tennesseensis, stipules (×1)

Potamogeton perfoliatus

Potamogeton pusillus, stipules (×3)

Potamogeton vaseyi (×1)

Potamogeton praelongus Wulfen White-stem pondweed
Stems to 3 m, freely branched and often somewhat zigzag, especially when the internodes are short; leaves all submersed, sessile, lanceolate to lance-linear, 8–30 cm long and 1–4.5 cm wide, hooded at the tip; stipules free, white, shredding at the tip; spikes 3–7.5 cm, dense or interrupted; achenes 4–5.7 mm; rare in lakes in deep water; NW; believed to be extirpated; OBL; ⚘.

Potamogeton pulcher Tuck. Heartleaf or spotted pondweed
Stems <5 dm, simple, black-spotted; submersed leaves broadly lanceolate, 1–3 cm wide; floating leaves broadly elliptic, obtuse or subcordate at the base; petioles black-spotted; stipules free, to 6 cm long; spikes dense, 2–4 cm; achenes 3.5–4.5 mm with a well-developed keel; rare in shallow acidic streams, swamps, or muddy shores; SE; OBL; ⚘.

Potamogeton pusillus L. Pondweed
Stems slender, to 1.5 m, much branched; leaves all submersed, linear, 0.2–2.5 mm, 1–3-veined with lacunar strips; stipules free, delicate, to 1 cm long; spikes 2–10 mm with 3–5 well-spaced whorls of flowers; achenes 1.9–2.8 mm, rounded; frequent in lakes and ponds; throughout; OBL.

Potamogeton robbinsii Oakes Flat-leaved or fern pondweed
Stems to 1 m, repeatedly branched at flowering; leaves all submersed, strongly 2-ranked, very crowded, linear, 3–8 mm wide with rounded lobes or auricles at the base of the blade and minutely serrulate margins, at least toward the tips; stipules fused to the base of the leaf for 5–15 mm, the free part as long, but soon disintegrating into fibers; spikes numerous, slender, 7–15 mm long with 3–5 whorls of flowers separated by distinct internodes; achenes 3.5–4.5 mm, but rarely produced; rare in quiet water of lakes and ponds; E and NW; OBL; ⚘.

Potamogeton spirillus Tuck. Snailseed pondweed
Very similar to *P. diversifolius*; submersed leaves 0.5–2 mm wide and 1–3-veined with broad lacunar bands; fused portion of the stipular sheath longer than the free part; submersed spikes globular with 1–8 flowers, emersed spikes to 1.3 dm long with up to 35 flowers; achenes 1.3–2.4 mm, with a jagged, winged keel; frequent in shallow water of lakes, ponds, and rivers; scattered; OBL.

Potamogeton strictifolius A.Benn. Narrow-leaved or straight-leaved pondweed
Stems slender, to 1 m, branched; leaves all submersed, linear, 0.5–2 mm wide, acute or bristle-tipped, 3–7-veined; stipules free, white, 7–15 mm, becoming fibrous in age; spikes mostly terminal, slender, 10–15 mm, with 3–4 well-spaced whorls of flowers; achenes 2–3 mm; rare in calcareous ponds or streams; scattered; OBL; ⚘.

Potamogeton tennesseensis Fernald Tennessee pondweed
Floating leaves 2–4 cm long and 5–13 mm wide; submersed leaves linear, 0.2–2 mm wide, 1–3-veined with prominent lacunar bands; stipules fused to the leaf bases for ¼ or less of their length, or free; spikes 1.2–2 cm; achenes orbicular, 2.5–3 mm with a winged keel; rare in ponds and streams; SW; OBL; ⚘.

Potamogeton vaseyi J.W.Robbins Vasey's pondweed
Annual, stems very slender, highly branched, to 5 dm, often with glands at the

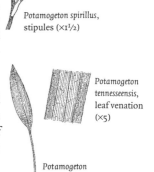

Potamogeton pusillus, achene (×5)

Potamogeton robbinsii, stipules

Potamogeton spirillus

Potamogeton spirillus, stipules (×1½)

Potamogeton tennesseensis, leaf venation (×5)

Potamogeton tennesseensis, leaf

nodes; submersed leaves 0.2–1 mm wide, 1-nerved, without lacunar strips; floating leaves, when present, 0.7–3.5 cm long and 2–13 mm wide; stipules free, 4–10 mm long, becoming weakly fibrous; spikes 3–8 mm with 1–4 whorls of flowers; achenes 1.5–2.5 mm; rare in ponds, lagoons, and other slow-moving water; scattered; OBL; ⚘.

Potamogeton zosteriformis Fernald — Flat-stemmed pondweed
Stems elongate, branched, strongly compressed; leaves all submersed, linear, 2–5 mm wide with an acute or cuspidate tip and 15 or more veins; stipules free, firm, 1–2 cm; spikes 1.5–2.5 cm with 7–11 whorls of flowers; achenes 4–4.5 mm with a sharp, dentate keel; rare in slow-moving streams; scattered; OBL; ⚘.

SCHEUCHZERIACEAE Scheuchzeria Family

Rhizomatous, perennial herbs with narrow, alternate leaves with dilated, open basal sheaths; flowers few, in a terminal raceme, perfect, regular; perianth of 6 yellow-green tepals; stamens 6; carpels 3, fused only at the base, each with a sessile stigma; fruit a recurved, spreading follicle.

Scheuchzeria L.

Scheuchzeria palustris L. — Pod-grass
Stems 2–4 dm tall; leaves basal and cauline; blades erect, flat with a short, terete tip; raceme 3–10 cm with a leaf-like bract at the base, upper bracts reduced to a small sheath; fruit 6–8 mm; very rare in sphagnum bogs; NE and NW; flr./frt. late Jun–early Aug; OBL; ⚘.

SMILACACEAE Catbrier Family

Prickly, deciduous, woody vines, unarmed herbaceous vines, or erect herbs; leaves simple, alternate, with a pair of tendrils arising from the base of the petiole; leaf blades with curved-convergent veins interconnected by a network of smaller veins; margins entire or minutely serrulate; flowers in umbels, 3-merous; perianth parts distinct; ovary superior; fruit a berry with 1–6 seeds.

Smilax L.

Plants dioecious; umbels in the leaf axils; flowers greenish-yellow, malodorous; fruits dark blue to black, sometimes glaucous.

A. stems herbaceous, not prickly
    B. sides of the leaves straight or concave .... *S. pseudochina*
    B. sides of the leaves convex
        C. leaves glabrous, pale and somewhat glaucous beneath; fruit dark blue, glaucous .... *S. herbacea*
        C. leaves bright green and shiny beneath, finely pubescent on the veins; fruit black, not glaucous .... *S. pulverulenta*

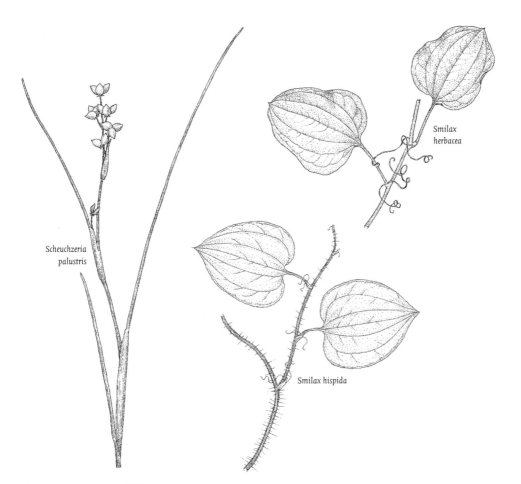

Scheuchzeria palustris

Smilax herbacea

Smilax hispida

A. stems woody, prickly

    D. leaves strongly glaucous beneath .... *S. glauca*

    D. leaves green beneath, not glaucous

        E. stems with broad-based prickles but not densely bristly; leaf margins entire .... *S. rotundifolia*

        E. stems densely bristly, at least below; leaf margins finely serrulate-ciliate .... *S. hispida*

Smilax glauca Walter   Catbrier, greenbrier

Similar to *S. rotundifolia* except the stems round and often densely bristly below; leaves mottled with lighter green above, glaucous beneath; frequent in dry to moist, sandy soil of fields, woods, thickets, swamps, and roadsides; S; flr. late May–early Jun, frt. late Aug–winter; FACU.

Smilax herbacea L.   Carrion-flower

Herbaceous, climbing vine, 1–3 m tall; leaves ovate to lance-ovate with truncate to rounded bases, glabrous and glaucous beneath; umbels from the axils of leaves or bracts; fruits blue-black, glaucous; common in damp thickets, moist woods, and floodplains; throughout; flr. late May–early Jun, frt. Jul–Nov; FAC.

Smilax hispida Muhl. ex Torr.   Bristly greenbrier

Similar to *S. rotundifolia* except stems densely bristly, at least on the basal portions, and only slightly angled; leaves lanceolate with finely serrulate-ciliate mar-

Smilax hispida, leaf margin (×1¹⁄₂)

gins; fruits black, not glaucous; occasional in swamps, moist woods, thickets, and roadsides; throughout; flr. late May, frt. Aug–winter.

Smilax pseudochina L. False chinaroot, long-stalked greenbrier
Climbing herbaceous vine, 1–2 m tall; leaves glabrous, triangular-ovate to almost hastate, the sides concave; umbels on long peduncles from the upper leaf axils; very rare in dry woods; SE; believed to be extirpated; flr. late Jun, frt. Sep; FAC+.

Smilax pulverulenta Michx. Carrion-flower
Very similar to S. *herbacea* but the undersides of the leaves shiny green and pubescent at least on the veins, not glaucous; fruit black, not glaucous; infrequent in moist woods and thickets; SE and SC; flr. May–early Jun, frt. Jul–fall; FACU.

Smilax rotundifolia L. Catbrier, greenbrier
Climbing, thicket-forming, glabrous vine; stems green, more or less 4-angled, with broad-based prickles; leaves semi-evergreen, ovate to ovate-lanceolate with a rounded, cordate, or truncate base and entire margins; fruit bluish-black, glaucous; common in moist to dry woods, thickets, old fields, and serpentine barrens; throughout, except in the northernmost counties; flr. late May–early Jun, frt. Sep–winter; FAC.

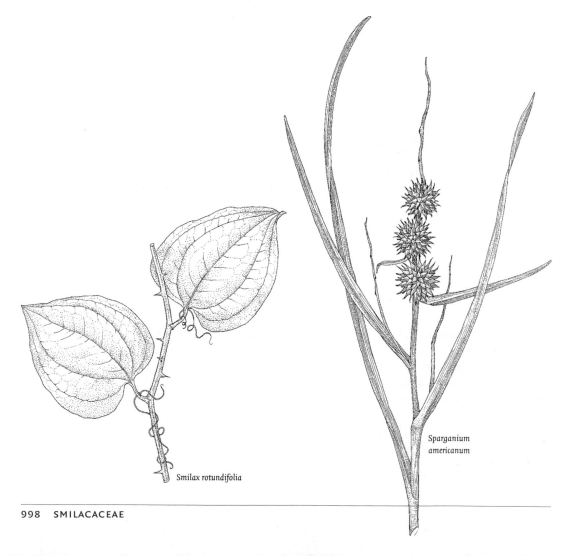

Smilax rotundifolia

Sparganium americanum

SPARGANIACEAE Bur-reed Family

Perennial, emergent or submergent, rhizomatous, aquatic herbs; leaves linear, alternate, 2-ranked, with sheathing bases, stiffly erect or flexuous with floating ends; flowers unisexual, in globose heads that are axillary to leaf-like bracts or inserted above the nodes of the inflorescence (supra-axillary); plants monoecious, staminate heads above, flowers with 3 stamens; pistillate heads below, flowers with 3–6 linear scales comprising the calyx; corolla absent; ovary superior, 1- or 2-celled with 1 or 2 styles; fruits elongate, pointed achenes borne in dense spherical heads; frt. Jul–Sep.

Sparganium L.

A. stigmas 2; mature achenes broader at the top .... S. eurycarpum
A. stigmas 1; achenes tapering toward both ends
  B. beak 0.5–1.2 mm long; fruiting heads 1–1.2 cm thick .... S. minimum
  B. beak 1.5–7 mm long; fruiting heads 1.2–3 cm thick
    C. mature fruiting heads 1.5–2 cm thick; leaves 2–9 mm wide, with floating ends
      D. inflorescence branched; pistillate heads axillary; leaves 5–9 mm wide
      .... S. fluctuans
      D. inflorescence not branched; pistillate heads supra-axillary; leaves 2–3(5) mm wide .... S. angustifolium
    C. mature fruiting heads 2–3 cm thick; leaves erect
      E. pistillate heads supra-axillary; beak of the achene as long as the body
      .... S. chlorocarpum
      E. pistillate heads axillary; beak of the achene shorter than the body
        F. achene body 4–5 mm long, beak 2.5–4 mm .... S. americanum
        F. achene body 6 mm long, beak 5–7 mm .... S. androcladum

Sparganium americanum Nutt.   Bur-reed
Emergent, 3–10 dm tall, with stiffly erect leaves; inflorescence often branched; pistillate heads 2 cm thick, sessile, axillary; achene body 4–5 mm, beak 3–5 mm; common in muddy shores and shallow water of rivers, streams, swamps, and ponds; throughout; OBL.

Sparganium americanum, fruit (×3)

Sparganium androcladum (Engelm.) Morong   Branching bur-reed
Emergent, 4–10 dm tall, with erect, keeled leaves; inflorescence simple or branched; pistillate heads sessile, axillary; achene body 6 mm long, beak 5–7 mm; rare in wet meadows, swales, and stream banks; scattered; OBL; 🌿.

Sparganium angustifolium Michx.   Bur-reed
Stems and leaves elongate and floating; leaves 2–6 mm wide, rounded on the back; inflorescence unbranched, the lowest pistillate heads supra-axillary and usually peduncled, the upper axillary and sessile; body of the achene 3–5 mm, narrowed to a 1-mm beak; rare in lakes and bogs, NE; OBL; 🌿.

Sparganium chlorocarpum Rydb.   Bur-reed
Emergent, 0.5–6 dm tall, with erect, flat, or somewhat keeled leaves 2–7 mm wide; inflorescence unbranched, at least the lowest pistillate heads supra-

Sparganium chlorocarpum, fruit (×3)

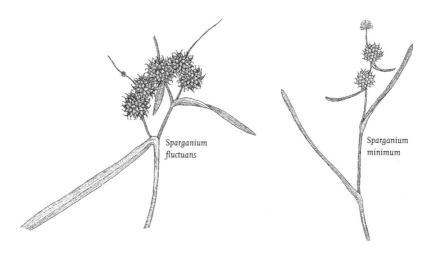

Sparganium
fluctuans

Sparganium
minimum

Sparganium
eurycarpum,
fruit (×3)

axillary, sessile, or short-peduncled; achene with beak about the same length as the body; occasional in slow-moving streams, lakes, bogs, and wet meadows; scattered; throughout, except in the SE and SC; OBL.

Sparganium eurycarpum Engelm. Bur-reed
Emergent, 5–12 dm tall, with leaves 6–12 mm wide; inflorescence branched; pistillate flowers with 2 stigmas; achene broadest at the top; occasional in bogs, swamps, lake margins, ditches, and swampy meadows; throughout; OBL.

Sparganium fluctuans (Morong) B. L. Robins. Bur-reed
Stem slender and flexuous, to 15 dm tall; leaves flat, thin, 3–9 mm wide and floating; inflorescence branching; heads sessile or short-pedunculate, axillary; achene body 3–4 mm, beak 2–3 mm; occasional in lake margins, slow-moving streams, and muddy shores; NE; OBL.

Sparganium
minimum,
fruit (×3)

Sparganium minimum (Hartm.) Fries Small bur-reed
Stems erect or floating; leaves thin, flat, 2–7 mm wide, and elongate when floating; pistillate heads 1–3, axillary, sessile, or the lowest on short peduncles; staminate head 1; achenes 3–4 mm long with a 0.5–1.5-mm beak; very rare in shallow water; NC and NW; believed to be extirpated; OBL; 🌿.

TYPHACEAE Cat-tail Family

Erect plants of wet soils or growing as emergents in shallow water; flowering stems simple; leaves stiff, elongate, with a sheathing base; flowers in spikes terminating the stems; the pistillate flowers forming a dense cylindrical zone below, each flower consisting of a single stalked carpel surrounded by numerous capillary bristles representing the perianth; the staminate flowers above, less compact and withering after the pollen is shed, each with 3 stamens and 3 bristles; fruit small, dry, 1-seeded, wind dispersed; flr. late May–early Jul.

Typha L.

A. pistillate and staminate portions of the inflorescence contiguous; leaves 10–23 mm wide, flat .... *T. latifolia*
A. pistillate and staminate portions of the inflorescence separated by 2–12 cm of bare stem; leaves 5–11 mm wide, somewhat rounded on the back .... *T. angustifolia*

Typha angustifolia L.   Narrow-leaved cat-tail
Stems 1–1.5 m tall; leaves 5–11 mm wide, with auricles at the top of the sheath; staminate portion of the inflorescence separated from the pistillate zone by 2–12 cm; occasional in wet meadows, shores, marshes, and ditches, often in calcareous or brackish habitats, increasingly found along highways where salt is used; throughout; OBL. *Typha* x *glauca* Godr., a hybrid with *T. latifolia*, which is intermediate in characteristics, is present in scattered locations.

Typha latifolia L.   Common cat-tail
Stems 1–3 m tall; leaves flat, 10–23 mm wide; the staminate portion of the inflorescence immediately adjacent to the brown pistillate section, or separated by a gap of <4 mm; common in swamps, marshes, wet shores, and ditches; throughout; OBL.

XYRIDACEAE Yellow-eyed-grass Family

Tufted perennial herbs with narrow basal leaves and erect, leafless flowering stems; flowers in a dense terminal head or spike, subtended by overlapping scales, perfect; sepals 3, the 2 lateral ones keeled, persistent, scarious and clasping the fruit, the upper one thin, membranous, falling when the flower opens; corolla with a narrow tube and 3 lobes; stamens 3, fused to the corolla tube; ovary superior, with 3-branched terminal stigma; fruit a many-seeded capsule.

Xyris L.

Flowers yellow, 3-merous, protruding from behind the subtending scales of the inflorescence when open; flr. Jul–Sep.

A. leaves distinctly spirally twisted; keels of the lateral sepals ciliate most of their length .... *X. torta*
A. leaves only slightly or not at all twisted; keels lacerate or slightly ciliate in the upper half only
  B. leaves 5–15 mm wide .... *X. difformis*
  B. leaves 1–2.5 mm wide .... *X. montana*

Xyris difformis Chapm.   Yellow-eyed-grass
Leaves broadly linear, 1–5 dm long and 5–15 mm wide, spreading; flowering stems 3.5–8 dm, twisted below, straighter and compressed above; spikes 0.5–1.5 cm long; the keel of the lateral sepals lacerate in the upper half; rare in peaty or sandy soil of swamps or bogs; E; OBL.

Xyris torta (×¼)

Xyris montana Ries   Yellow-eyed-grass
Densely tufted with linear leaves 4–15 cm long and 1–2.5 mm wide, straight or
slightly twisted; flowering stems 0.5–3 dm, slightly twisted below; spikes <1 cm
long, few-flowered; keel of the lateral sepals entire or slightly ciliate near the
apex; rare in exposed peat of floating bog mats; NE; OBL; 🌿.

Xyris torta Small   Yellow-eyed-grass
Leaves ascending, 2–5 dm long and 2–5 mm wide, twisted; flowering stems
longer than the leaves, also more or less twisted; spikes 1–2.5 cm long, many-
flowered; keel of the lateral sepals ciliate-scabrous from base to tip; flowers
opening in the morning; rare in sphagnum bogs and peaty meadows; S; OBL; 🌿.

ZANNICHELLIACEAE Horned Pondweed Family

Submergent aquatics with slender branching stems and opposite, linear-filiform
leaves with sheathing stipules; plants monoecious; flowers axillary, unisexual,
lacking a perianth; staminate flowers with a single anther; pistillate flowers with
4 or 5 distinct ovaries; fruit a cluster of stalked achenes.

Zannichellia L.

Zannichellia palustris L.   Horned pondweed
Flowers in pairs (1 of each sex) on a common peduncle; fruits 2–3 mm long,
curved, with a beak-like style; occasional in streams, ponds, lakes, springs, and
tidal mudflats, especially in alkaline or brackish water; mostly S; flr. May–Oct;
OBL.

Zannichellia palustris, fruit (×5)

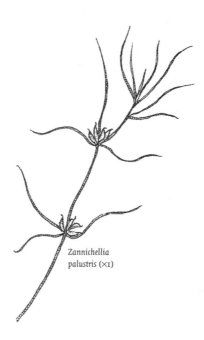

Zannichellia palustris (×1)

GLOSSARY

abaxial: side away from the main axis (e.g., the underside of a leaf)

achene: dry, indehiscent, one-seeded fruit with the seed not adnate to the fruit wall

actinomorphic: radially symmetrical (see also zygomorphic)

acuminate: forming an angle <45° (see illustration on plate 2)

acute: forming an angle >45° and <90° (see illustration on plate 2)

adaxial: side toward the main axis (e.g., the upper side of a leaf)

adnate: condition resulting from the fusion of dissimilar parts (usually stamens to petals)

adventive: locally established outside of a plant's native range

alluvial soils: soils deposited by running water

alternate leaved: having only 1 leaf attached at each node of a stem (see illustration on plate 2)

anastomosing: curving and branching repeatedly

anemophilous: wind pollinated

annual: a plant that completes its life cycle in 1 year

annular: ring-like

anther: sac-like part of a stamen that produces pollen (see illustration on plate 1)

anthesis: period of flowering

antrorse: directed upward or frontward

apetalous: without petals

apex: the top or distal end of a plant organ

apical: of the apex

apiculate: ending in a short, slender point (see illustration on plate 2)

apiculum: a short, slender point

appressed: flat against a plant organ

approximate: close to

areolae: the areas between the veinlets of a leaf

aril: a fleshy covering present on some seeds

arilate: having an aril

aristate: having a bristle at the tip (see illustration on plate 2)

armature: armament (thorns, spines, or prickles)

attenuate: tapering very gradually to a very slender tip or base (see illustration on plate 2)

auricle: ear-shaped appendage

auriculate: having auricles (see illustration on plate 2)

autumnal: occurring in the autumn

awn: long, bristle-like appendage

axillary: occurring in the axil of a leaf

axillary bud: bud occurring in the junction formed by the leaf and stem (see illustration on plate 2)

axil: the junction of a leaf and a stem

achene

actinomorphic

anastomosing

aril

PLATE 1

Flower Parts

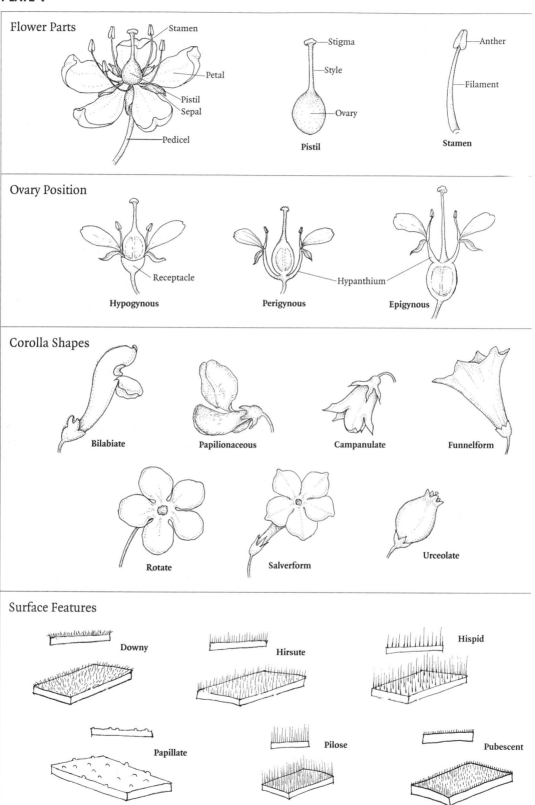

Stamen

Petal

Pistil

Sepal

Pedicel

Stigma

Style

Ovary

Pistil

Anther

Filament

Stamen

Ovary Position

Receptacle

Hypogynous

Hypanthium

Perigynous

Epigynous

Corolla Shapes

Bilabiate

Papilionaceous

Campanulate

Funnelform

Rotate

Salverform

Urceolate

Surface Features

Downy

Hirsute

Hispid

Papillate

Pilose

Pubescent

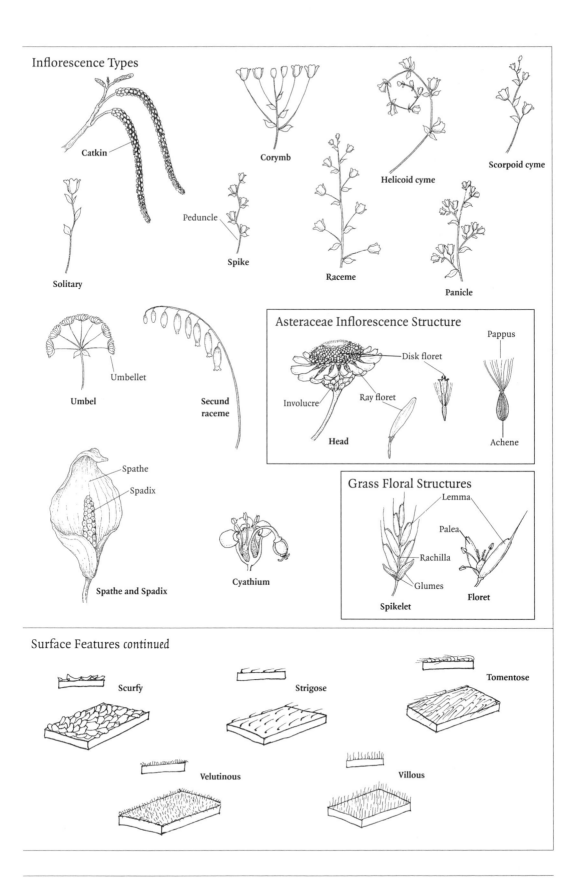

Inflorescence Types

Catkin

Solitary

Peduncle

Spike

Corymb

Helicoid cyme

Scorpoid cyme

Raceme

Panicle

Umbellet

Umbel

Secund raceme

Asteraceae Inflorescence Structure

Disk floret

Pappus

Involucre

Ray floret

Head

Achene

Spathe

Spadix

Grass Floral Structures

Lemma

Palea

Rachilla

Glumes

Spikelet

Floret

Spathe and Spadix

Cyathium

Surface Features *continued*

Scurfy

Strigose

Tomentose

Velutinous

Villous

PLATE 2 Leaf Terminology

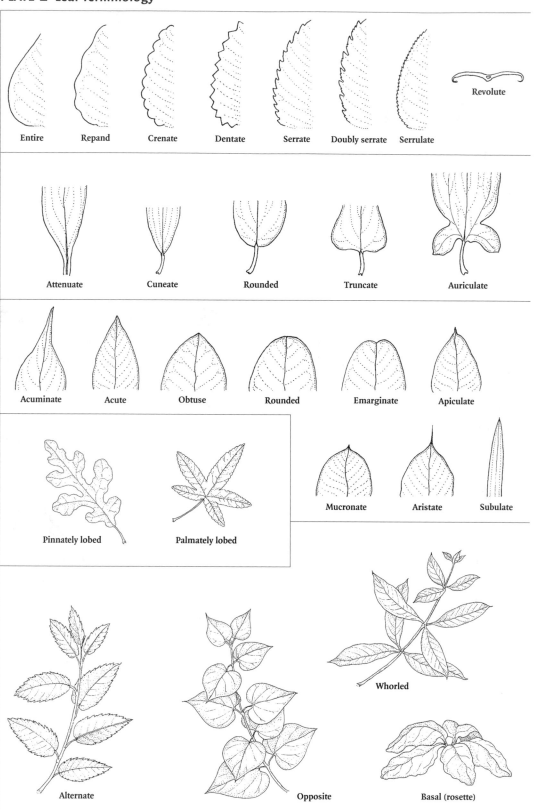

Entire Repand Crenate Dentate Serrate Doubly serrate Serrulate Revolute

Attenuate Cuneate Rounded Truncate Auriculate

Acuminate Acute Obtuse Rounded Emarginate Apiculate

Mucronate Aristate Subulate

Pinnately lobed Palmately lobed

Whorled

Alternate Opposite Basal (rosette)

Cordate

Deltate

Elliptic

Hastate

Lanceolate

Linear

Obcordate

Lyrate

Oblanceolate

Obovate

Ovate

Peltate

Perfoliate

Reniform

Rhombic

Saggitate

Spatulate

Palmately compound

Pinnately compound

Bipinnately compound

Tripinnately compound

Parallel venation

Reticulate venation

Blade

Petiole

Stipule

Axillary bud

Terminal bud

Node

Lenticels

Internode

beak

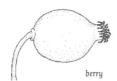 berry

 canaliculate

 capsule

 ciliate

barrens: areas of natural, often shrubby vegetation where tree growth is sparse due to droughty conditions, low nutrient soils, and/or frequent fire

basal: at or near the base

basifixed: attached by the base

basilaminar: occurring at the base of a leaf blade

beak: a narrow, prolonged tip

berry: a fleshy, several-seeded fruit developing from a single pistil

bicarpellate: having 2 carpels

bidentate: having 2 teeth

biennial: a plant that completes its life cycle over a 2-year period

bilabiate: 2-lipped (see illustration on plate 1)

bipinnate: twice pinnately compound (see illustration on plate 2)

bipinnatifid: twice pinnately cleft

blade: the flattened portion of a leaf (see illustration on plate 2)

bog: low nutrient (usually acidic) peatland

bracetole: see bractlet

bract: a leaf-like structure often subtending an inflorescence

bractlet: a small bract subtending a branch of an inflorescence or an individual flower within an inflorescence

calcareous: soils with high calcium content, usually of limestone origin

calyx: collective term for the set of sepals

campanulate: bell-shaped (see illustration on plate 1)

canaliculate: with a groove or channel (U-shaped if viewed in cross section)

capitate: head-like

capsule: dry, dehiscent fruit composed of more than one carpel

carpel: organ that bears the ovules and seeds in angiosperms, basic unit of the pistil

caryopsis: dry, indehiscent, one-seeded fruit with the seed adnate to the fruit wall

catkin: dense, spike-like, often drooping inflorescence of reduced unisexual flowers (see illustration on plate 1)

caudex: short, thick, persistent base of a herbaceous perennial

cauline: of the stem

cespitose: densely tufted or bunched

chasmogamous: flowers that open normally

ciliolate: with minute hairs along the margin

ciliate: bearing long, stiff hairs along an edge or keel

cinereous: grayish due to the presence of hairs

circumscissile: dehiscing along an equatorial line

clasping: partly surrounding the stem at the base of the leaf

clavate: club-shaped

claw: the narrowed base of some petals or sepals

cleistogamous: flowers that never open

collateral bud: a bud occurring beside an axillary bud

colonial: forming spreading patches of plants or stems derived from a single individual

composite: a plant or inflorescence of Asteraceae

compound leaf: a leaf divided into leaflets (see illustration on plate 2)

concave: curved inward

conduplicate: folded and flattened lengthwise

connate: condition resulting from the fusion of similar parts convex: curved outward

cordate: heart-shaped and attached at the lobed end (see illustration on plate 2)

corm: upright, enlarged, fleshy base of a stem

cormose: having a corm

corolla: collective term for the set of petals

corona: an outgrowth from the base of the petals or stamens in some flowers

corymb: a flat- or round-topped, indeterminate inflorescence, usually with the lower pedicels longer than the upper (see illustration on plate 1)

corymbiform: shaped like a corymb

corymbose: corymb-like

crenate: margin bearing rounded teeth (see illustration on plate 2)

crenation: rounded projection along the margin of a leaf or stipule

crown: persistent base of an herbaceous perennial

culm: a stem or stalk in the grasses (Poaceae), sedges (Cyperaceae), or rushes (Juncaceae)

cultivar: form of a plant originating in cultivation

cuneate: wedge-shaped (see illustration on plate 2)

cuspidate: tipped with a short, sharp, abrupt point

cyathium: inflorescence type in Euphorbiaceae consisting of a cup-like involucre that contains a single pistillate flower, reduced to the pistil only and several staminate flowers, each reduced to 1 stamen only (see illustration on plate 1)

cylindroid: like a cylinder, but elliptic in cross section

cyme: a flat- or round-topped determinate inflorescence in which the terminal or central flower blooms first (see illustration on plate 1)

deciduous: shedding leaves seasonally every year

decumbent: reclining on the ground but with the tip ascending

decurrent: extending downward along the stem

decurved: see deflexed

deflexed: bent downward

dehiscent: a fruit opening at maturity to release the seeds

deltate: triangular (see illustration on plate 2)

deltoid: see deltate

dentate: margin with sharp teeth that point directly outward from the midvein (see illustration on plate 2)

depauperate: growing in a reduced form (usually a result of marginal habitat conditions)

diabase: an igneous rock that is high in magnesium and iron

diadelphous: having stamens united into 2 clusters

dichotomously: branching in 2s

digitate: finger-like

dioecious: species with unisexual flowers borne on separate plants

discoid: in Asteraceae, an inflorescence composed entirely of disk florets

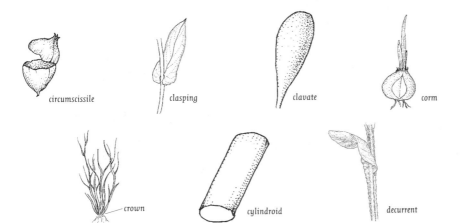

circumscissile   clasping   clavate   corm

crown   cylindroid   decurrent   diadelphous

disk: in Asteraceae, the central portion of an inflorescence composed of disk florets

disk floret: in Asteraceae, those flowers that are regular and tubular (see illustration on plate 1)

distal: toward the tip or end

distichous: in 2 rows on opposite sides

distylous: having 2 styles

divaricate: widely spreading

doubly serrate: margin with sharp, forward-directed teeth, these bearing smaller teeth (see illustration on plate 2)

downy: covered with soft, fine hairs

dredge spoil: sandy or gravelly substrate removed from a stream bed and deposited on land

droughty: subject to periodic drying

drupe: fleshy fruit in which the inner layer of the fruit becomes stony and encloses 1–several seeds

drupelet: a small drupe

echinate: having prickles or spines

ellipsoid: elliptic in 3 dimensions

elliptic: broadest at the middle and narrowing to equal ends (see illustration on plate 2)

emarginate: slightly indented at the apex (see illustration on plate 2)

emersed: said of plants or plant parts occurring beneath the surface of water

entire: without teeth or lobes (see illustration on plate 2)

epicalyx: a whorl of bracts resembling a calyx; an additional whorl of bracts beneath the calyx

epigynous: having an inferior ovary(s) (see illustration on plate 1)

epipetalous: having another floral part (usually stamens) fused to the petals

epiphyte: a plant that uses a host plant for physical support but derives no nutrient from the host

erose: irregularly toothed

even pinnate: pinnately compound leaf terminated by a pair of leaflets

excurrent: extending beyond

exfoliating: peeling in layers

exserted: extending from

extirpated: no longer growing in an area

falcate: sickle-shaped

false indusium: seen in the ferns of the genus *Adiantum*, where the edge of the leaf curls under to form an indusial flap (see also indusium)

farinose: mealy or powdery

fascicled: united into a bundle

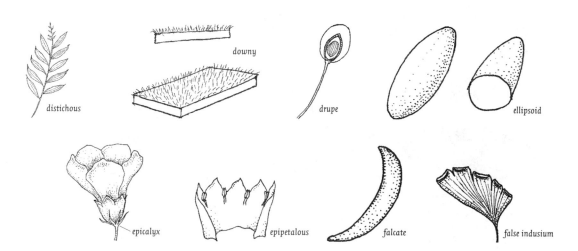

distichous   downy   drupe   ellipsoid

epicalyx   epipetalous   falcate   false indusium

follicle

globose

glomerule

gynostegium

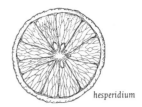
hesperidium

hip

fen: high nutrient peatland

fern allies: spore-bearing plants often grouped with the ferns but not closely related to them

fertile frond: a spore-producing leaf in the ferns, differing in size and/or shape from the sterile fronds (see also sterile frond)

fibrillose: bearing delicate fibers or hairs

fiddlehead: a juvenile fern leaf, curled like the end of a violin

filament: slender stalk of a stamen that supports the anther (see illustration on plate 1)

filamentous: finely thread-like

filiform: thread-like

flexuose: curving or bending

flexuous: see flexuose

floret: a small flower within a dense cluster, as in the Poaceae or Asteraceae

floricane: second-year flowering and fruiting stem of *Rubus* (Rosaceae)

floriferous: bearing flowers

foliaceous: bearing leaves

follicle: dry, dehiscent fruit derived from 1 carpel and splitting along 1 suture

friable: easily torn

frond: a fern leaf

funnelform: funnel-shaped (see illustration on plate 1)

fuscous: dark gray-brown

genera: plural of genus

genus: taxonomic level above species and below family

gibbous: swollen on one side

glabrate: mostly glabrous

glabrescent: see glabrate

glabrous: not bearing hairs

gland: structure that secretes a sticky or oily substance

glandular: bearing glands

glaucous: having a waxy coating, often resulting in a bluish color

globose: globe-shaped

glomerule: dense cluster

glume: one of a pair of bracts that subtend a spikelet in Poaceae (see also lemma and palea) (see illustration on plate 1)

glutinous: sticky or gummy

gynostegium: structure formed by the fusion of the anthers and stigma in Asclepiadaceae

hastate: arrowhead-shaped with the basal lobes turned outward (see also saggitate) (see illustration on plate 2)

head: rounded, densely crowded inflorescence of sessile or subsessile flowers (see illustration on plate 1)

helicoid cyme: spirally coiled cyme (see illustration on plate 1)

hemiparasite: photosynthetic plant that derives part of its nutrient requirements from another plant

hesperidium: thick-skinned berry containing juice filled sacs characteristic of citrus fruits

heterostylous: having 2 or more different sizes or shapes of styles

hip: fruit that is an aggregation of achenes surrounded by an urn-shaped receptacle and hypanthium characteristic of *Rosa* (Rosaceae)

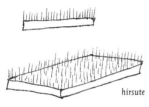
hirsute

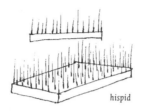
hispid

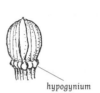
hypogynium

imbricate

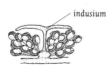
indusium

keel

legume

hirsute: bearing stiff hairs

hirsutulous: bearing very short stiff hairs

hirtellous: see hirsutulous

hispid: bearing very stiff straight hairs

hispidulous: bearing short very stiff hairs

hyaline: thin, colorless, and nearly transparent

hypanthium: cup-like floral structure resulting from the fusion of the calyx, corolla, and stamens, often surrounding or partly fused to the ovary(s) (see illustration on plate 1)

hypogynium: disk at the base of the achene in some Cyperaceae

hypogynous: having a superior ovary(s) (see illustration on plate 1)

imbricate: made up of scales whose edges overlap (usually refers to bud scales) (see also valvate)

indehiscent: a fruit not opening at maturity

indigenous: native to a given area

indusium: a protective flap of tissue associated with a sorus in the ferns

inferior ovary: condition in which the stamens and perianth parts are attached above the ovary (= epigynous)

inflexed: bent downward or inward

inflorescence: a grouping or cluster of flowers

infrastipular: below the stipules

infructescence: an inflorescence in fruit

inrolled: rolled inward (usually leaf margins)

internerves: the area between the nerves or veins (usually of a leaf)

internodal: between the nodes

internode: portion of the stem to which leaves are not attached (see also node) (see illustration on plate 2)

intervenal: see internerve

intranodal: within the node

introduced: a non-native plant brought in intentionally

involucral: of an involucre

involucre: a set of usually small, overlapping bracts subtending an inflorescence or flower

irregular: bilaterally symmetrical, zygomorphic

keel: a longitudinal ridge; the lower fused petals of some Fabaceae

laciniate: deeply divided into long flat segments

lacunar: having an empty, air-filled space

lanceolate: much longer than wide with the widest part below the middle, lance-shaped (see illustration on plate 2)

leaflet: a division of a compound leaf

legume: dry, dehiscent fruit formed from 1 carpel and splitting along 2 sutures, characteristic of Caesalpiniaceae, Fabaceae, and Mimosaceae

lemma: the lower of the 2 bractlets that subtend a floret in a grass spikelet (see also palea and glume) (see illustration on plate 1)

lenticel: corky pore in the bark of a stem (see illustration on plate 2)

lenticular: lens-shaped

ligulate: having ligules

ligule: tongue- or strap-shaped organ (sometimes reduced to a fringe of hairs) at the junction of the blade and sheath in grasses and sedges

linear: very narrow with parallel sides (see illustration on plate 2)

lobulate: with small lobes

locule: chamber within an ovary containing the developing ovules or seeds

loculocidal: dehiscing through the locules

lyrate: lyre-shaped (see illustration on plate 2)

megasporangium: sporangium that produces large, female spores

membranaceous: thin and membrane-like

membraneous: see membranaceous

merous: parts of a set

mesic: moist

microsporangium: sporangium that produces small, male spores

midnerve: the usually prominent central nerve or vein of a leaf or other structure

midvein: see midnerve

monadelphous: having stamens united into 1 cluster or fascicle

monoecious: species with unisexual flowers borne on the same plant

monospecific: a genus or family containing only 1 species

monotypic: any taxonomic level containing only 1 taxon of lower rank

mucronate: abruptly tapered to a short, sharp point (see illustration on plate 2)

mucronulate: very shortly mucronate

mycorrhiza: a symbiotic relationship between a fungus and the roots of a plant

mycotrophic: having a mycorrhizal relationship

native: a plant occurring within its pre-European settlement range

naturalized: non-native plants that have become established and reproduce without human assistance

nectary: a tissue that produces nectar

node: specific region of the stem to which the leaf or leaves are attached (see also internode) (see illustration on plate 2)

nut: dry, indehiscent fruit with a single seed enclosed in a hardened carpel wall

nutlet: a small nut

obconic: cone-shaped and attached at the narrow end

obcordate: heart-shaped and attached at the narrow end (see illustration on plate 2)

oblanceolate: much longer than wide with the widest part above the middle (see illustration on plate 2)

obligate: restricted to certain conditions

obovate: egg-shaped and attached by the narrow end (see illustration on plate 2)

obovoid: egg-like and attached at the narrow end

obpyramidal: pyramid-like and attached at the narrow end

obtuse: forming an angle >90° (see illustration on plate 2)

ocrea: a stipular sheath at the nodes in Polygonaceae

ocreae: plural of ocrea

ocrea

ligule

monadelphous

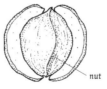

nut

obconic

obovoid

ocreolae: a small stipular sheath at the branches of the inflorescence in some Polygonaceae

odd pinnate: pinnately compound leaf terminated by a single leaflet

old field: an abandoned agricultural field, usually in the early stages of succession

oogonium: an egg bearing cell in some algae

opposite leaved: having 2 leaves attached at each node on opposite sides of the stem (see illustration on plate 2)

outcurved: curved outward

ovary: that portion of the pistil that contains the ovules and seeds (see illustration on plate 1)

ovate: egg shaped and attached at the wider end (see illustration on plate 2)

ovoid: egg-like

ovule: an immature seed

palea: the upper of the 2 bracts that subtend a floret in a grass spikelet (see also lemma and glume) (see illustration on plate 1)

palmate: like the fingers of a hand (see illustration on plate 2)

palustrine: of wet meadows or marshes

panicle: a branched inflorescence with pedicelled flowers (see illustration on plate 1)

papilionaceous: butterfly-shaped, as in pea or bean flowers (see illustration on plate 1)

papilla: a short, rounded bump

papillate: having papillae

pappus: the bristles or scales at the apex of the achenes in Asteraceae (see illustration on plate 1)

parallel venation: having the major veins of the leaf parallel to the axis or margins (see illustration on plate 2)

parasite: a plant (nonphotosynthetic) which derives all of its nutrient requirements from another plant

pedicel: the stalk of an individual flower (see illustration on plate 1)

pedicellate: having a pedicel

peduncle: the stalk of an entire inflorescence, or of a single flower if solitary (see illustration on plate 1)

pellucid dots: translucent dots in the leaves most visible when held up to the light

peltate: leaf with the petiole attached near the middle of the blade (see illustration on plate 2)

perennial: a plant that survives for more than 2 years

perfect flower: a flower having both functional male and female reproductive organs

perfoliate: leaf with the blade surrounding the stem so that the stem appears to pass through the leaf (see illustration on plate 2)

perianth: collective term for the set of petals and sepals

perigynium: a sac-like structure that encloses the pistil or achene in *Carex* (Cyperaceae)

perigynous: a flower in which the ovary sits in a cup formed by the hypanthium (see illustration on plate 1)

petal: an individual segment of the corolla (see illustration on plate 1)

petaliferous: having petals

petaloid: petal-like

petiolate: having a petiole(s)

petiole: the stem-like stalk of a leaf (see also blade) (see illustration on plate 2)

petiolulate: having a petiolule(s)

petiolule: the stem-like stalk of a leaflet

phyllary: an individual bract of an involucre (usually in Asteraceae)

pilose: bearing long, soft, straight hairs

pinnae: the first order divisions of a pinnately compound leaf, especially in the ferns

pinnate: arranged on opposite sides of an elongate axis (see illustration on plate 2)

pinnatifid: pinnately cleft

pinnules: divisions of a pinna

pistil: the female reproductive organ of a flower (see illustration on plate 1)

pistillate: a flower lacking male reproductive organs

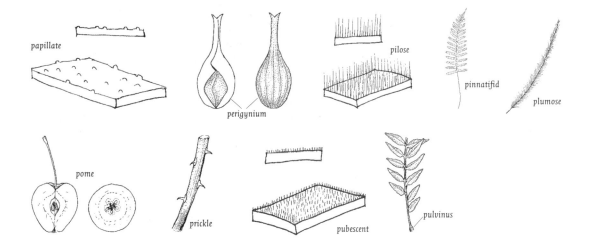

papillate

perigynium

pilose

pinnatifid

plumose

pome

prickle

pubescent

pulvinus

pith: the central soft portion of a stem

placentation: the arrangement of ovules or seeds within a fruit

planoconvex: flat on one side and convex on the other

plumose: feather-like

pollinium: an often sticky packet or cluster of pollen (usually in Orchidaceae or Asclepiadaceae)

polygamodioecious: mostly dioecious, but with some perfect flowers

polygamous: bearing both unisexual and perfect flowers on the same plant

polyploid: having 3 or more complete sets of chromosomes

polyploidy: the condition of being polyploid

pome: a fleshy, berry-like fruit from a multi-carpellate inferior ovary adnate to a modified hypanthium as in *Malus* or *Pyrus* (Rosaceae)

prickle: a sharp spine-like outgrowth of the plant epidermis, occurring anywhere on the plant

primocane: the first year non-flowering stems of *Rubus* (Rosaceae)

procumbent: lying flat on the ground, but not rooting at the nodes

prophyllate: subtended by small bracts

prostrate: see procumbent

protuberence: a rounded bulge or projection

proximal: toward the base

pseudobulb: bulb-like base of the leaves often found in the Orchidaceae

puberulence: minute pubescence

puberulent: minutely pubescent

pubescent: bearing short, soft hairs

pulvinus: a swelling at the base of the petiole or petiolule of some leaves

pyramidal: pyramid-like and attached at the wide end

raceme: unbranched inflorescence with pedicelled flowers (see illustration on plate 1)

racemiform: shaped like a raceme

rachilla: the axis of a grass or sedge spikelet (see illustration on plate 1)

rachis: the main axis of a compound leaf or inflorescence

radiate: spreading from a central point

ray floret: in Asteraceae, those flowers that are irregular and bear a single strap-shaped petal (see illustration on plate 1)

receptacle: the portion of the pedicel to which the flower parts are attached (see illustration on plate 1)

receptacular: of the receptacle

recurved: bent backward or downward

reflexed: see recurved

rachis

regular: radially symmetrical, actinomorphic

reniform: kidney-shaped (see illustration on plate 2)

repand: slightly wavy (see illustration on plate 2)

reticulate venation: with the veins of the leaf branching repeatedly, net-veined (see illustration on plate 2)

retrorse: directed downward or backward

revolute: with the margin rolled under (see illustration on plate 2)

rhizoid: a small root-like structure without vascular tissue

rhizome: horizontal usually below ground stem

rhombic: diamond-shaped (see illustration on plate 2)

rootlet: a small root

rosette: a radiating cluster of leaves at ground level (see illustration on plate 2)

rotate: a flat, circular corolla with widely spreading lobes and little or no tube (see illustration on plate 1)

rounded: without an indentation or point (see illustration on plate 2)

rugose: wrinkled

rugulose: slightly wrinkled

saccate: sac-like

saggitate: arrowhead-shaped with the basal lobes directed downward (see also hastate) (see illustration on plate 2)

salverform: trumpet-shaped (see illustration on plate 1)

samara: a dry, indehiscent winged fruit

scaberulous: slightly rough to the touch

scabrous: rough to the touch

scape: a naked peduncle

scarious: a thin, non-green, membranous tissue

schizocarp: dry, indehiscent fruit splitting into 2 one-seeded segments at maturity

scorpioid cyme: cyme with a zigzag rachis (see illustration on plate 1)

scurfy: bearing small scales

secund: bearing all of the flowers or other structures along 1 side of an elongate rachis (see illustration on plate 1)

sepal: an individual segment of the calyx (see illustration on plate 1)

septate: bearing a partition(s)

septum: a partition

serpentinite: a metamorphic rock high in magnesium and heavy metals

serrate: margin with sharp, forward directed teeth (see illustration on plate 2)

serrulate: margin with very small, sharp, forward directed teeth (see illustration on plate 2)

sessile: attached directly without a stalk

setose: bearing bristles

setulose: bearing short bristles

sheath: the basal portion of a leaf which surrounds the stem

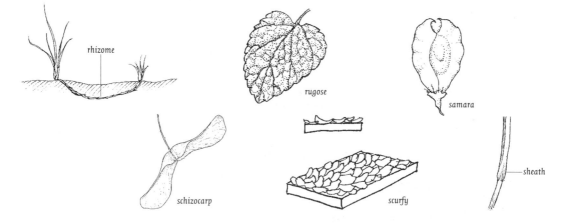

rhizome

rugose

samara

schizocarp

scurfy

sheath

silique

—sorus

spine

stellate

sterigma

sheathing: having a sheath

silique: dry, dehiscent fruit derived from 2 carpels and dehiscing along 2 sutures, leaving a persistent septum, characteristic of Brassicaceae

simple leaf: a leaf not divided into leaflets

sinuous: wavy or snake-like

sinus: the cleft between 2 lobes of a leaf

solitary: an inflorescence consisting of a single flower (see illustration on plate 1)

sorus: aggregation of sporangia in the ferns, commonly appearing as spots on the underside of a frond

spadix: a spike-like inflorescence subtended or enclosed by a sheathing bract (see illustration on plate 1)

spathe: a large, leaf-like bract subtending or enclosing a spadix or an inflorescence (see illustration on plate 1)

spatulate: rounded toward the top and very gradually tapered to the base (see illustration on plate 2)

spike: an inflorescence with sessile flowers along an elongate axis (see illustration on plate 1)

spikelet: small, spike-like inflorescence or section of an inflorescence (see illustration on plate 1)

spine: sharp pointed projection of leaf origin

spinulose: bearing very small spines

sporangiaster: hairs with numerous apical branches mixed with the sporangia in the sori of some ferns of the genus *Polypodium*

sporangium: sac-like structure where spores are produced

spores: dust-sized propagules of non-seed plants

spur: slender sac-like appendage

squarrose: abruptly recurved or spreading

stamen: the male reproductive organ of a flower (see illustration on plate 1)

staminal: of or having to do with stamens

staminate: a flower lacking female reproductive organs

staminode: a modified and usually greatly reduced stamen, not functional as a sexual organ

staminodia: plural of staminode

staminoids: stamen-like structures

stellate: star-shaped, as the pith of *Quercus* (Fagaceae)

sterigma: persistent peg-like base of a leaf remaining on the twig (as seen in the genus *Picea*)

sterigmata: plural of sterigma

sterile frond: in the ferns, a leaf not producing spores

stigma: the portion of the pistil that receives the pollen (see illustration on plate 1)

stipe: the "petiole" of a fern frond

stipel: stipule-like structure at the base of a leaflet

stipellule: see stipel

stipitate: on a stalk

stipule: leaf-like appendage at the base of the petiole (see illustration on plate 2)

stipule scar: linear mark on the twig at the base of the petiole or leaf scar left after the falling of a stipule

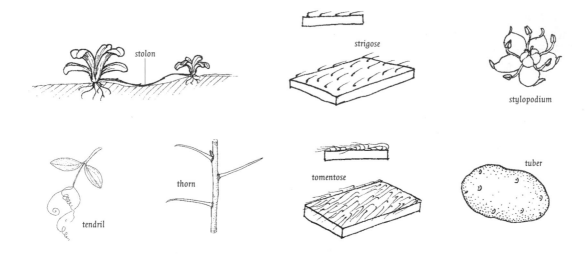

stolon: horizontal stem at ground level, rooting at the nodes

stoloniferous: having stolons

stramineous: straw-colored

striate: with parallel lines

strigose: bearing stiff, straight, appressed hairs

stylal: of the style

style: portion of the pistil that connects the stigma to the ovary (see illustration on plate 1)

stylopodium: disk-like expansion at the base of the style in Apiaceae

sub- (prefix): used to indicate nearly, almost, somewhat, etc.

submergent: remaining under water

subtend: to occur closely below

subulate: tipped by a fine, sharp point (see illustration on plate 2)

succulent: juicy and fleshy, thickened

superior ovary: condition in which the stamens and perianth parts are attached below the ovary (= hypogynous)

superposed bud: a bud occurring directly above an axillary bud

suture: a line of fusion or dehiscence

swale: low, wet area

sympetalous: having fused petals

syncarp: a fruit with united carpels

taproot: large, dominant central root

taxa: plural of taxon

taxon: any level of classification

tendril: a coiled organ used for support

tepal: sepals and petals when the two are indestinguishable

terete: rounded when viewed in cross section

terminal bud: a bud occurring at the apex of a stem (see illustration on plate 2)

ternate: occurring in 3s

thallus: a flattened plant body

thorn: sharp pointed projection of branch origin

tomentose: bearing soft, tangled hairs

trichome: a plant hair

trichotomously: splitting in 3s

trilocular: having 3 locules

tripinnate: 3 times pinnately divided (see illustration on plate 2)

truncate: abruptly flattened as if cut off (see illustration on plate 2)

tuber: an underground storage organ of stem origin

tubercule: tuber-like swelling

tuberosity: small swelling at the base of the lip in the flowers of *Spiranthes* (Orchidaceae)

turion: overwintering short, vegetative shoot in some aquatic plants

twice pinnate: see bipinnate (see illustration on plate 2)

umbel: a more or less flat topped or rounded inflorescence in which the pedicels arise from a common point (see illustration on plate 1)

umbellet: a subdivision of a compound umbel (see illustration on plate 1)

umbelliform: shaped like an umbel

understory: tree layer below the canopy in a forest

unilocular: having 1 locule

urceolate: urn-shaped (see illustration on plate 1)

utricle: small, one-seeded bladdery fruit

valvate: made up of scales whose edges do not overlap (usually refers to bud scales) (see also imbricate)

valve: a segment of a dehiscent fruit

veinlet: a small vein

vein scar: dot-like marks within the leaf scar indicating the number of vascular traces to the leaf

velutinous: bearing short, soft, spreading hairs

vernal: occurring in the spring

vesicle: small bladder-like structure

villous: bearing long, soft hairs

viscid: sticky or gummy

whorled: having 3 or more leaves or other structures attached at a node or point on a stem (see illustration on plate 2)

zygomorphic: bilaterally symmetrical (see also actinomorphic)

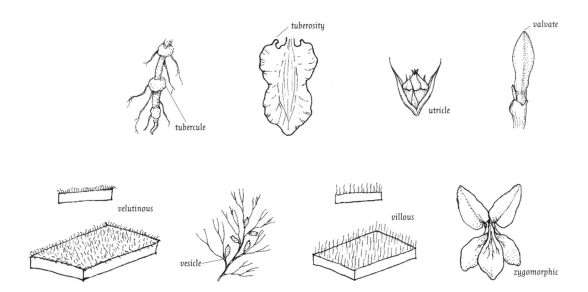

BIBLIOGRAPHY

Barnes, B. V. and W. H. Wagner, Jr. 1981. Michigan trees. University of Michigan Press, Ann Arbor, Michigan.

Blackwell, W. H. 1976. Guide to the woody plants of the tri-state area: southwestern Ohio, southern Indiana, and northern Kentucky. Kendall/Hunt Publishing Company, Dubuque, Iowa.

Blakeslee, A. F. and C. D. Jarvis. 1911. New England trees in winter. Storrs Agricultural Experiment Station bulletin no. 69, Storrs, Connecticut.

Braun, E. L. 1950. Deciduous forests of Eastern North America. Hafner Press, New York, New York.

Braun, E. L. 1989. The woody plants of Ohio. Ohio State University Press, Columbus, Ohio.

Catling, P. M., A. A. Reznicek, and W. J. Crins. 1993. *Carex juniperorum* (Cyperaceae), a new species from northeastern North America, with a key to *Carex* sect. *Phyllostachys*. *Systematic Botany* 18 (3): 496–501.

Clayton, W. D. and S. A. Renvoize. 1986. Genera Graminum: grasses of the world. Kew Botanic Gardens, London.

Coulter, J. M. and J. N. Rose. 1900. Monograph of the North American Umbelliferae. *Contributions from the U.S. National Herbarium* 7 (1): 1–256.

Cronquist, A. 1981. An integrated system of classification of flowering plants. Columbia University Press, New York, New York.

Department of Conservation and Natural Resources. 1987. Pennsylvania Code, Title 17, Chapter 45. Conservation of Pennsylvania native wild plants. Pennsylvania Bulletin 17(49). December 5, 1987. Harrisburg, Pennsylvania.

Farrar, J. L. 1995. Trees in Canada. Fitzhenry and Whiteside, Ltd., Markham, Ontario.

Fernald, M. L. 1970. Gray's manual of botany, 8th ed. D. Van Nostrand Company, New York, New York.

Fike, J. 1999. Terrestrial and palustrine plant communities of Pennsylvania. Pennsylvania Department of Conservation and Natural Resources, Harrisburg, Pennsylvania.

Flora of North America Editorial Committee [eds.]. 1993. Flora of North America North of Mexico, vol. 2. Oxford University Press, New York, New York.

Flora of North America Editorial Committee [eds.]. 1997. Flora of North America north of Mexico, vol. 3. Oxford University Press, New York, New York.

Furlow, J. J. and R. S. Mitchell. 1990. Betulaceae through Cactaceae of New York State: contributions to a flora of New York State VIII. New York State Museum, Albany, New York.

Gleason, H. A. 1952. The new Britton and Brown illustrated flora of the northeastern United States and adjacent Canada. Lancaster Press, Inc., Lancaster, Pennsylvania.

Gleason, H. A. and A. Cronquist. 1991. Manual of the vascular plants of northeastern United States and adjacent Canada, 2nd ed. New York Botanical Garden, New York, New York.

Harlow, W. M. 1946. Fruit key and twig key to trees and shrubs. Dover Publications, Inc., New York, New York.

Harris, J. G. and M. W. Harris. 1994. Plant identification terminology: an illustrated glossary. Spring Lake Publishing, Spring Lake, Utah.

Hellquist, C. B. and G. E. Crow. 1984. Aquatic vascular plants of New England: part 7. Cabombaceae, Nymphaeaceae, Nelumbonaceae, and Ceratophyllaceae. New Hampshire Agricultural Experiment Station bulletin 527. Durham, New Hampshire.

Hickman, J. C. [ed.]. 1993. The Jepson manual: higher plants of California. University of California Press, Berkeley, California.

Hitchcock, A. S. and A. Chase. 1971. Manual of the grasses of the United States, 2nd ed. Dover Publications, New York.

Illick, J. S. 1915. Pennsylvania trees. Department of Forestry, Commonwealth of Pennsylvania, Harrisburg, Pennsylvania.

Li, H. 1972. Trees of Pennsylvania, the Atlantic states, and the lake states. University of Pennsylvania Press, Philadelphia.

Mabberly, D. J. 1993. The plant book: a portable dictionary of the higher plants. Cambridge University Press, Cambridge.

Miller, N. G. 1970. The genera of the Cannabaceae in the southeastern United States. *Journal of the Arnold Arboretum* 51 (2): 185–203.

Mitchell, R. S. 1983. Berberidaceae through Fumariaceae of New York State: contributions to a flora of New York State V. New York State Museum, Albany, New York.

Mitchell, R. S. 1988. Platanaceae through Myricaceae of New York State: contributions to the flora of New York State VI. New York State Museum, Albany, New York.

Muenscher, W. C. 1950. Keys to woody plants. Comstock Publishing Company, Inc., Ithaca, New York.

Naczi, R. F. C. 1999. *Carex planispicata*, a widespread and frequent new species of *Carex* section *Griseae* (Cyperaceae) from the Eastern United States of America. *Journal of the Kentucky Academy of Science* 60 (1): 37–44.

Pohl, R. W. 1947. A taxonomic study on the grasses of Pennsylvania. *The American Midland Naturalist* 38 (3): 513–604.

Pryer, K. M. and L. R. Phillippe. 1989. A synopsis of the genus *Sanicula* in eastern Canada. *Canadian Journal of Botany* 67 (3): 694–707.

Reznicek, A. A. and P. W. Ball. 1974. The taxonomy of *Carex* series *Lupulinae* in Canada. *Canadian Journal of Botany* 52: 2387–2399.

Reznicek, A. A. and P. W. Ball. 1980. The taxonomy of *Carex* section *Stellulatae* in North America north of Mexico. *Contributions from the University of Michigan Herbarium* 14: 153–203.

Rhoads, A. F. and W. M. Klein, Jr. 1993. The vascular flora of Pennsylvania: annotated checklist and atlas. American Philosophical Society, Philadelphia, Pennsylvania.

Rollins, R. C. 1993. The Cruciferae of Continental North America. Stanford University Press, Stanford, California.

Sargent, C. S. 1965. Manual of the trees of North America. Dover Publications, New York.

Small, E. 1978. A numerical and nomenclatural analysis of morpho-geographic taxa of *Humulus*. *Systematic Botany* 3: 37–76.

Standley, L. A. 1985. Systematics of the *Acutae* group of *Carex* (Cyperaceae) in the Pacific Northwest. *Systematic Botany Monographs* 7: 1–106.

Standley, L. A. 1989. Taxonomic revision of the *Carex stricta* (Cyperaceae) complex in eastern North America. *Canadian Journal of Botany* 67: 1–14.

Terrell, E. E. 1996. Revision of *Houstonia* (Rubiaceae—Hedyotideae). *Systematic Botany Monographs* 48: 1–119.

Tutin, T. G., N. A. Burges, A. O. Chater, J. R. Edmondson, V. H. Heywood, D. M. Moore, D. H. Valentine, S. M. Walters, and D. A. Webb [eds.]. 1964. Flora Europaea, vol. 1. Cambridge University Press, Cambridge.

Tutin, T. G., V. H. Heywood, N. A. Burges, D. M. Moore, D. H. Valentine, S. M. Walters, and D. A. Webb [eds.]. 1968. Flora Euopaea, vol. 2. Cambridge University Press, Cambridge.

Tutin, T. G., V. H. Heywood, N. A. Burges, D. M. Moore, D. H. Valentine, S. M. Walters, and D. A. Webb [eds.]. 1972. Flora Europaea, vol. 3. Cambridge University Press, Cambridge.

Tutin, T. G., V. H. Heywood, N. A. Burges, D. M. Moore, D. H. Valentine, S. M. Walters, and D. A. Webb [eds.]. 1976. Flora Europaea, vol. 4. Cambridge University Press, Cambridge.

Weishaupt, C. G. 1971. Vascular plants of Ohio, 3rd ed. Kendall/Hunt Publishing Company, Dubuque, Iowa.

Wilbur, R. L. 1994. The Myricaceae of the Untied States and Canada: genera, subgenera, and series. *Sida* 16 (1): 93–107.

INDEX

empress-tree, 262
enchanter's-nightshade, 518
English ivy, 154
Epifagus virginiana, 527
Epigaea repens, 378
Epilobium
  angustifolium, 519
  ciliatum, 519
  coloratum, 519
  hirsutum, 519
  leptophyllum, 520
  palustre, 520
  parviflorum, 520
  strictum, 520
Epipactis helleborine, 880
Equisetaceae, 59
Equisetum, 59
  arvense, 60
  arvense x fluviatile, 60
  x ferrissii, 60
  fluviatile, 60
  hyemale, 60
  hyemale var. affine x laevigatum,
    60
  x littorale, 60
  sylvaticum, 61
  variegatum, 61
Eragrostis, 928
  capillaris, 929
  cilianensis, 929
  frankii, 929
  hypnoides, 929
  minor, 930
  pectinacea, 930
  pilosa, 930
  poaeoides, 930
  spectabilis, 930
Eranthis hyemalis, 575
Erechtites hieraciifolia, 204
Erianthus
  giganteus, 931
  saccharoides, 931
Ericaceae, 375
Erigenia bulbosa, 140
Erigeron
  annuus, 205
  philadelphicus, 205
  pulchellus, 205
  strigosus, 205
Eriocaulaceae, 837
Eriocaulon
  aquaticum, 837
  decangulare, 837
  parkeri, 837
  septangulare, 837
Eriochloa villosa, 931
Eriophorum
  gracile, 822
  tenellum, 823
  vaginatum, 823

virginicum, 823
viridicarinatum, 823
Erodium
  cicutarium, 442
  moschatum, 442
Erophila verna, 288
Eruca vesicaria, 288
Erucastrum gallicum, 289
Eryngium aquaticum, 141
eryngo, marsh, 141
Erysimum
  cheiranthoides, 289
  hieracifolium, 290
  inconspicuum, 290
  repandum, 290
Erythronium
  albidum, 861
  americanum, 861
Eschscholzia californica, 530
Eubotrys racemosa, 382
Eulalia viminea, 942
eulalia, 943
Euodia
  daniellii, 638
  huphensis, 638
Euonymus
  alatus, 341
  americanus, 341
  atropurpureus, 341
  europaeus, 342
  fortunei, 342
  hamiltonianus, 342
  obovatus, 343
  yedoensis, 343
euonymus, winged, 341
Eupatorium, 206
  album, 207
  altissimum, 207
  aromaticum, 207
  coelestinum, 207
  dubium, 207
  fistulosum, 208
  godfreyanum, 208
  hyssopifolium, 208
  leucolepis, 208
  maculatum, 209
  perfoliatum, 209
  pilosum, 209
  purpureum, 209
  rotundifolium, 210
  rotundifolium, 209
  rugosum, 210
  serotinum, 210
  sessilifolium, 210
  vaseyi, 208
eupatorium, 208
  hyssop-leaved, 208
  late, 210
  ragged, 209
  round-leaved, 210

tall, 207
upland, 210
white-bracted, 207
Euphorbia, 392
  commutata, 393
  corollata, 393
  cyparissias, 394
  dentata, 397
  esula, 394
  falcata, 395
  helioscopia, 395
  ipecacuanhae, 395
  lathyris, 395
  lucida, 395
  marginata, 395
  obtusata, 395
  peplus, 396
  platyphyllos, 396
  purpurea, 396
Euphorbiaceae, 388
Euphrasia stricta, 667
Euthamia
  graminifolia, 211
  tenuifolia, 211
evening-primrose, 524, 525, 526
  cut-leaved, 525
  shale-barren, 524
  white, 526
evening-primrose family, 517
eyebane, 390
eyebright, 667

Fabaceae, 397
Fagaceae, 424
Fagopyrum
  esculentum, 543
  sagittatum, 544
  tataricum, 544
Fagus grandifolia, 425
Falcaria vulgaris, 141
fall-dandelion, 228
false boneset, 192
false buckwheat, climbing, 553
false china-root, 998
false dragonhead, 477
false flax, 281
  small-fruited, 281
false foxglove, 664, 665
  Blue Ridge, 664
  cut-leaf, 665
  downy, 665
  eared, 663
  slender, 664
  small-flowered, 664
  yellow, 664
false gromwell, 270
  Virginia, 270
false indigo, 399
  blue, 401
false loosestrife, 521, 522